Nuclear Medicine Radiation Dosimetry

Brian J. McParland

Nuclear Medicine Radiation Dosimetry

Advanced Theoretical Principles

Springer

Brian J. McParland, BASc MSc PhD
Amersham, Buckinghamshire
United Kingdom

ISBN 978-1-4419-9655-8 (PB)
ISBN 978-1-84882-125-5 (HB) e-ISBN 978-1-84882-126-2
DOI 10.1007/978-1-84882-126-2
Springer Dordrecht Heidelberg London New York

Library of Congress Control Number: 2010923178

Cover design: eStudio Calamar, Figueres/Berlin

Printed on acid-free paper

Springer is part of Springer Science+Business Media (www.springer.com)

For Sharon

"Doubt thou the stars are fire,
Doubt that the sun doth move,
Doubt truth to be a liar,
But never doubt I love"

- Hamlet Act II, Scene II

Preface

This book addresses the applications of radiation dosimetry theory to diagnostic and therapeutic nuclear medicine, medical disciplines which have both contrasting and parallel requirements.

To begin with, modern medical diagnostic imaging technologies have enabled the visualization of both anatomical structure and function to unprecedented degrees of resolution and complexity. However, with the major exceptions of magnetic resonance, ultrasound, and optical imaging, these technologies require the patient to be exposed to ionizing radiation. This presents the medical practitioner with the challenge of optimizing the diagnostic benefit obtained through the imaging procedure against the theoretical risk posed to the patient as a result of this exposure. The diagnostic benefit of the imaging procedure will be known from prior clinical experience or else have been determined empirically in clinical studies which yield measures of diagnostic efficacy such as sensitivity, specificity, and positive/negative predictive values. Evaluation of the theoretical radiation risk, on the other hand, is much more complex and is based upon the knowledge of the radiation fields and their interactions with tissues in combination with an understanding of the biological consequences of these interactions. The impact of the magnitudes of the risks presented by medical imaging has been the subject of debate for decades as their estimation requires extrapolation of radiation dose responses from epidemiological data obtained at high levels of radiation dose to the lower radiation doses associated with radiological or nuclear medicine procedures.[1] Despite the resulting uncertainty over the magnitudes of these extrapolated and theoretical risks, there is an expectation within modern society to minimize the radiation doses associated with diagnostic medicine whenever practicable. This is not an entirely unfounded or unreasonable expectation. For example, consider the hypothetical case of the informed patient willing to consider trivial an excess cancer mortality risk of 0.5% as a consequence of a particular imaging procedure if the immediate diagnostic benefit to him is overwhelming. On the other hand, society may, indeed should, question the application of

[1]Although, it is now common to see radiation absorbed doses received in modern multislice computed tomography comparable to those received by the survivors of the nuclear bombings at Hiroshima and Nagasaki who still provide the bulk of the epidemiological data of the somatic effects of exposure to ionizing radiation.

an imaging procedure with the same level of risk to the screening of an asymptomatic population of 100,000 individuals, which has the theoretical possibility of inducing 500 extra cancer deaths in that population, and expect that the mortality reduction due to screening exceed the estimated number of extra resulting deaths. Thus, to enable both the clinician deciding upon those imaging tests to be included within this diagnostic process and society to judge the suitability of a broadly-based diagnostic imaging program, means of accurately calculating the radiation doses resulting from imaging procedures are necessary.

Radiological imaging studies involve external beams of radiation and the associated dosimetry evaluations are relatively straightforward as the associated radiation fields can be measured directly and the anatomy irradiated identified simply by the physical alignment of the patient and the radiation beam. As a result, many modern diagnostic radiological devices can provide real-time measures of patient dosimetry through metrics such as the estimated entrance skin dose, dose-area product or Computed Tomography Dose Index. In contrast, the estimation of the patient dosimetry resulting from nuclear medicine procedures is far more complex as the irradiation is internal and is coupled with the combination of the biokinetics of the radiopharmaceutical, the nuclear decay scheme of the radioisotope employed, and the types of radiations emitted during the nuclear decay. The dependence of the internal radiation dosimetry upon the biokinetics makes this dosimetry highly personalized and difficult to predict *a priori*.

The application of radiation dosimetry theory to therapeutic nuclear medicine is markedly different to that applied to diagnosis. In the latter case, the ability to calculate the internal radiation dosimetry is fundamental to the requirements of optimization through estimating and then minimizing the risk presented to the diagnostic nuclear medicine patient. The intent of radiation dosimetry calculation in radionuclide therapy is to improve the chances of cure or palliation by maximizing the therapeutic ratio through maximizing the radiation absorbed dose to the neoplasm of concern and minimizing the absorbed dose to normal uninvolved tissues and the risk of any treatment-related sequelae. In addition to intent, applications of radiobiological theory differ between diagnostic and therapeutic nuclear medicine. In diagnosis (involving low administered activities of a radioisotope), one is concerned with stochastic risks such as radiocarcinogenesis and genetic effects resulting from chromosomal or chromatid aberrations. On the other hand, in therapy (which uses high levels of radioisotope administered activities), one wishes to quantify and understand determinisitic effects such as tumor control and normal tissue radiotoxicity minimization resulting from cell death. Yet, perhaps somewhat surprisingly, quantitative dosimetry applied to nuclear medicine therapy is still relatively infrequent and tends to be based upon empirical and clinical experience. However, the field appears to be entering a renaissance where accurate dosimetry, approaching that required in external beam radiotherapy, may become the norm. Therapeutic applications are, by nature, patient-specific and bespoke calculations are required.

Beyond these two clinical applications of nuclear medicine dosimetry is the need of accurate dosimetry in epidemiological studies of the consequences of radiation exposure. As noted earlier, the bulk of society's understanding of the effects of ionizing radiation upon the human has evolved from the monitoring of the survivors of the nuclear bombings of Hiroshima and Nagasaki. As time progresses and the number of these survivors diminishes, one of the remaining dominant populations of humans providing epidemiological data on the effects of exposure to ionizing radiation will be those patients having been exposed during the course of their diagnosis or therapy. In recent years, the numbers of diagnostic nuclear medicine procedures and the associated absorbed doses have grown immensely. As the epidemiological data are provided in the form of the rate of excess cancer incidence or mortality per absorbed dose, the accurate evaluation of nuclear medicine absorbed doses is essential to the accurate assessment of the risks associated with ionizing radiation.

There are a wide variety of textbooks, monographs, and software available for the practicing nuclear medicine physicist or physician to use in order to estimate the internal radiation dosimetry associated with a given radiopharmaceutical. In fact, one can apply tabular data to estimate the internal radiation dosimetry while being unaware of the assumptions and limitations inherent to the underlying physics and risk estimates. This is quite undesirable. The intent of this book is not to supplant the aforementioned literature; indeed, most (if not all) of this literature is cited here as references. Nor is an intent of this book to provide copious amounts of numerical and physical data: these are already available in the literature, with more recent collections accessible through a large number of Internet sites associated with various national nuclear and physics data centers (and which are also referenced to in this book).

Rather, this book provides the underlying theoretical bases for understanding the many interlocking components of nuclear medicine dosimetry and it is to be considered as an adjunct to these references. To demonstrate this, consider the example of the detailed description of the weak interaction provided in Chap. 4 which includes the development of the Fermi theory to calculate β decay rates and electron and positron energy spectra, allowed and forbidden transitions, the V–A interaction and the nonconservation of parity by weak processes. One could argue that the practicing nuclear medicine physicist does not "need" to know the physics of the weak interaction to this level of detail in order to be able to evaluate internal radiation dosimetry. The counterargument to this view, and that which is taken here, is that β decay, the most commonly observed manifestation of the weak interaction, is fundamental to diagnostic and therapeutic nuclear medicine and the nuclear medicine physicist should have an understanding of the underlying physics of the weak interaction affecting β decay and how this physics is manifested in the dosimetry of a β-emitting radionuclide. Similarly, as the interactions between electromagnetic and corpuscular radiations and matter are at the heart of radiation dosimetry, an understanding of quantum scattering theory is a necessary foundation of any advanced understanding of radiation dosimetry theory. It is somewhat dissatisfying to find that many dosimetry textbooks present, for examples, formulae for the Klein–Nishina cross sections for photon–electron scattering or the Bethe–Heitler cross sections for electron-nuclear *bremsstrahlung* but fail to describe the theoretical development or the limitations inherent to the presented results. It is attempted to address these limitations here by providing, wherever possible, full derivations of those interactions that are at the heart of nuclear medicine dosimetry. However, at the same time, this book attempts to remain as pragmatic wherever possible in describing these derivations by avoiding unnecessary detailed discussions of derivational mechanics (e.g., Dirac trace algebra) when these are both provided adequately elsewhere and exposition would not add value to the dosimetric evaluations provided here.

Necessarily, such results are provided without proof but with cited references for the interested reader to pursue if desired.

As the previous paragraph would suggest, the intended audience of this book is the practicing or research nuclear medicine physicist or graduate student with an interest both in internal radiation dosimetry and, in general, the interactions of radiations with matter at the energies typical of nuclear medicine. This latter audience would include those desiring an in-depth understanding of the underlying physics used, for example, in Monte Carlo simulations of radiation transport at photon and electron/positron energies of about 2 MeV and below and of α particles with kinetic energies of about 10 MeV and below. Hence, the reader is presumed to have had considerable exposure to advanced mathematics, including complex variable theory, and to have a significant understanding of nonrelativistic and relativistic quantum theory in order to fully appreciate the development of the radiation physics that is presented here.

It has been stated that nuclear medicine should be categorized as a "mature" discipline in which all of the underlying scientific principles are known and understood and that only the engineering evolution of the relevant technology is of interest. This assumption is disputed here, at least with regards to the radiation dosimetry. There remains much potential for the applications of fundamental science in this field. As a result, this aspect of the discipline is to be considered far from mature and much fruitful research awaits.

Finally, it is necessary to provide two points – one of clarification and one of defense:

- This book considers only the radiation dosimetry of the patient having received the radionuclide and not that of those individuals exposed to that patient.
- To the possible chagrin of many PET colleagues, this book consolidates PET and single-photon imaging within the single term "diagnostic nuclear medicine" – *mea maxima culpa*.

Brian J. McParland, BASc MSc PhD
Amersham, UK

Acknowledgments

During three decades of practice as a medical physicist, I have profited from the intellectual, scholarly, and clinical guidance provided by many teachers and mentors in the physical sciences and the medical arts. It is my hope that this book conveys the combination of scientific rigor, humanity, societal worth, and pure intellectual pleasure of physics applied to medicine that each of them has imparted to me.

The weaknesses that the reader will inevitably find evident in this book would have been far greater had I not been fortunate enough to have benefited from the reviews provided during the course of its writing by colleagues whose expertise span the disciplines of physics, medicine, biology, and pharmacology. I thank (in alphabetical order):

Jonathan Allis, DPhil
Manuel Bardiès, PhD
Mark M. Boyd, MSc
David Hamilton, PhD FIPEM
Mandeep K. Khela, PhD MInstP MIPEM
David B. Lewall, MD FRCR(C)
Orjän Lindhe, PhD
Matthew P. Miller, PhD MInstP
Matthew Morrison, PhD
Rikard Owenius, PhD
Roger D. Pickett, PhD
Simon Roitt PhD, MInstB
Andrea Sheppick, DCR SRR
Jeffry Siegel, PhD
Terry Spinks, PhD
David Tattam, PhD
Jeffrey Winick, PhD
Jan Wolber, PhD
José Zubeldia, MD FABR

Each agreed, without hesitation, to my request to review drafts in preparation – responses for which I am most grateful – as, indeed, should be the reader! However, while the corrections and suggestions they provided have, without question, improved beyond measure my original attempts at writing, those faults and errors that remain are of my responsibility alone.

Finally, it is with the greatest gratitude and humility that I thank my family. I could not have contemplated the genesis and completion of this book without my wife Sharon's understanding, encouragement, patience, and support. For her, there are no bounds to my respect, devotion, and love. I am also beholden beyond measure to our daughters, Siobhán and Áine, who, for 5 years, have tolerated their distracted father in the family library writing this book. Thank you – I hope that this result will be worthwhile for all.

Contents

Glossary and Abbreviations

α	Fine structure constant, $\alpha = e^2/4\pi\varepsilon_0\hbar c \approx 1/37$
	The ^{4}He nucleus
	Incident photon energy in Compton scatter normalized to the electron mass, $\alpha = {}^{k}/{m_e}$
	Recombination coefficient
	Townsend coefficient
	Coefficient of linear-quadratic dose response model
α_e	Electronic polarizability
$\alpha\hbar c$	Conversion factor, $\alpha\hbar c = 1.44$ MeV fm
$\boldsymbol{\alpha}_i$	Dirac matrix, $\boldsymbol{\alpha}_i = \begin{pmatrix} 0 & \boldsymbol{\sigma}_i \\ \boldsymbol{\sigma}_i & 0 \end{pmatrix}$
β	Speed normalized to the speed of light, c
	Beta particle (electron or positron) produced through beta decay
	Coefficient of linear-quadratic dose response model
$\boldsymbol{\beta}_i$	Dirac matrix, $\boldsymbol{\beta} = \begin{pmatrix} \mathbf{1} & 0 \\ 0 & -\mathbf{1} \end{pmatrix}$
χ_0	Screening angle for charged particle-atom interactions
χ_a	Molière characteristic screening angle
χ_c	Molière angle
χ_e	Electric susceptibility
δ	Density effect of a medium to the collision stopping power
	Dirac delta function
	δ Particle (recoil electron)
	Multiplication factor per dynode of a photomultiplier tube
Δ^2	$\equiv \hbar^2\partial^2/\partial t^2 - (\hbar c)^2\nabla^2$
ε	Energy imparted
ε_i	Energy deposit for ith single interaction

ε_0	Dielectric permittivity of free space
ε_P	Plating efficiency
ε_R	Relative dielectric constant
ε_{tr}	Energy transferred
$\hat{\varepsilon}$	Photon polarization unit-vector
$\phi(r_T \leftarrow r_S)$	Absorbed fraction for combination of source region r_S and target region r_T
γ	Relativistic contraction factor, $\gamma = 1/\sqrt{1-\beta^2}$
	Photon
	Gyromagnetic ratio, $\gamma = \frac{\lvert\boldsymbol{\mu}\rvert}{\lvert\mathbf{L}\rvert}$
$\boldsymbol{\gamma}^0$	Dirac matrix, $\gamma^0 = \boldsymbol{\beta}$
$\gamma^i, i = 1,2,3$	Dirac matrix, $\gamma^i \equiv \begin{pmatrix} 0 & \boldsymbol{\sigma}_i \\ -\boldsymbol{\sigma}_i & 0 \end{pmatrix}$ $i = 1,2,3$
γ_5	Dirac matrix, $\gamma_5 = i\gamma^0\gamma^1\gamma^2\gamma^3 = \begin{pmatrix} 0 & \mathbf{1} \\ \mathbf{1} & 0 \end{pmatrix}$
γ_{EM}	Euler-Mascheroni constant
$\Gamma(z)$	Gamma function
Γ_δ	Air kerma rate constant
Φ	Scalar electric potential
	Particle fluence
$\Phi(r_T \leftarrow r_S)$	Specific absorbed fraction for source region r_S and target region r_T
$\boldsymbol{\Phi}$	Vector particle fluence
$d\Phi/dt$	Particle flux density
$d\boldsymbol{\Phi}/dt$	Vector particle flux density
$\hbar$	Reduced Planck constant, $\hbar = h/2\pi$
$\hbar c$	Conversion factor 197.33 MeV fm
φ	Particle flux (particles/cm^2 s)
λ_{Biol}	Biological rate constant
λ_C	Compton wavelength of the electron, $\lambda_C = \frac{h}{m_e c}$
λ_{Eff}	Effective decay constant
λ_{fi}	Transition rate (per unit time) from initial state i to final state f
λ_L	Landau parameter
λ_{LLI}	Clearance constant for lower large intestine contents
λ_{SI}	Clearance constant for small intestine contents
λ_{St}	Clearance constant for stomach contents
λ_{ULI}	Clearance constant for upper large intestine contents
λ_{Phys}	Physical decay constant
λ_{Ruth}	Mean free path between Rutherford scatters
λ_V	Vavilov parameter
$\lambda\!\!\!^{-}$	Reduced de Broglie wavelength, $\lambda\!\!\!^{-} = \hbar c/p$
μ	Linear attenuation coefficient
	Mobility
μ/ρ	Mass attenuation coefficient

μ_{en}/ρ	Mass energy-absorption attenuation coefficient
μ_{tr}/ρ	Mass energy-transfer attenuation coefficient
$\mu^{\pm}$	Muon
μ_o	Permeability of free space
μ_{eff}	Effective linear attenuation coefficient
μ_B	Bohr magneton, $\mu_B = e\hbar/2m_e$
μ_N	Nuclear magneton, $\mu_N = e\hbar/2m_p$
μ_d	Deuteron magnetic dipole moment
μ_n	Neutron magnetic dipole moment
μ_p	Proton magnetic dipole moment
$\boldsymbol{\mu}$	Magnetic dipole moment
$\nu_{e,\mu,\tau}$	Neutrino associated with the electron, muon, and tau lepton, respectively
$\aleph$	Dose point kernel
$\pi^{\pm}$, π^0	π-mesons (pions)
$\boldsymbol{\Pi}$	Parity operator
$\Re$	Rate of recombination of positively and negatively charged moieties following ionization in a gas
$\Re(\mathbf{r})$	Radiant energy at position **r**
$\Re_{CSDA}$	Charged particle range in the continuous slowing-down approximation
ρ	Physical density (mass per unit volume)
ρ_e	Electron density (number of electrons per unit volume)
ρ_f	Phase space factor (density of final states per unit energy level)
σ	Total cross section
σ_{KN}	Klein–Nishina total cross section
$\sigma_{KN,S}$	Klein–Nishina scatter cross section
$\sigma_{KN,Tr}$	Klein–Nishina energy-transfer cross section
σ_{PE}	Total cross section for photoelectric absorption
$\sigma_{PE,Tr}$	Energy-transfer total cross section for photoelectric absorption
σ_{Ray}	Total cross section for Rayleigh scatter
σ_{Tho}	Total cross section for Thomson scatter
σ_i; i = 1,2,3	Pauli matrices
$d\sigma/d\Omega$	Differential cross section in solid angle
$d^2\sigma/d\Omega dk'$	Differential cross section in solid angle and scattered photon energy (Compton scatter)
$d\sigma_{Inc}/d\Omega$	Incoherent scatter differential cross section in solid angle
$d\sigma_{KN}/d\Omega$	Klein–Nishina differential cross section in solid angle
$d\sigma_{KN}/dT'_e$	Klein–Nishina differential cross section in electron kinetic energy
$d\sigma_{KN,Tr}/d\Omega$	Klein–Nishina energy-transfer differential across section in solid angle

$d\sigma_{PE}/d\Omega$	Photoelectric absorption differential cross section in solid angle
$d\sigma_{Ray}/d\Omega$	Differential cross section in solid angle for Rayleigh scattering
$d\sigma_{Ruth}/d\Omega$	Differential cross section in solid angle for Rutherford scattering
$d\sigma_{Ruth}/dq$	Differential cross section in three-vector momentum transfer for Rutherford scattering
$d\sigma_{Tho}/d\Omega$	Differential cross section in solid angle for Thomson scattering
$\Im$	Linear scattering power
$\Im/\rho$	Mass scattering power
τ	τ lepton
	Mean lifetime
$\boldsymbol{\tau}_{\pm}$	Isotopic spin (isospin) ladder operators
ω	Angular frequency
ω_c	Cyclotron frequency
ω_i	Fluorescent yield for orbital i
Ω	Solid angle
Ψ	Logarithmic derivative of the gamma function, Γ
	Energy fluence
$\boldsymbol{\Psi}$	Vector energy fluence
$d\Psi/dt$	Energy flux density
$d\boldsymbol{\Psi}/dt$	Vector energy flux density
a_i	Auger electron yield for orbital i
A	Atomic mass number
	Activity
AAPM	American Association of Physicists in Medicine
$A_F(t)$	Activity in voided feces at time t
$A_{Lung,CF}(t)$	Activity in lung space at time t due to activity introduced by a continuous flow generator
$A_{Lung,S}(t)$	Activity in lung space at time t due to activity introduced through rebreathing apparatus
$A_{Lung,Eq}$	Equilibrium lung activity
$A_{LLI}(t)$	Activity in lower large intestine contents at time t
$A_{SI}(t)$	Activity in small intestine contents at time t
$A_{St}(t)$	Activity in stomach contents at time t
$A_{r_S,Norm}(t)$	Activity in source region r_S at time t normalized to that administered
$A_{r_S,Norm,Corr}(t)$	Activity in source region r_S at time t normalized to that administered and corrected for physical decay
$A_{UB,Norm}(t)$	Activity in urinary bladder contents normalized to that administered, at time t

$A_{UBVU,Corr,Norm}(t)$	Activity in urinary bladder contents and voided urine combined, corrected for physical decay and normalized to that at administered, at time t
$A_{ULI}(t)$	Activity in upper large intestine contents at time t
$\mathbf{A}$	Three-vector electromagnetic potential
A_0	Scalar electromagnetic potential
$\tilde{A}_{Lung,Aerosol,Admin}$	Lung administered cumulated activity due to administration via an aerosol
$\tilde{A}_{Lung,CF,Admin}$	Lung administered cumulated activity due to administration via a continuous flow apparatus
$\tilde{A}_{Lung,S,Admin}$	Lung administered cumulated activity due to administration via a rebreathing apparatus
$\tilde{A}_{r_S}$	Cumulated activity in source region r_S
$\tilde{A}_{r_S,\ Norm}$	Cumulated activity in source region r_S normalized to administered activity (also known as the residence time)
$\tilde{A}_F$	Cumulated activity of feces
$\tilde{A}_{LLI}$	Cumulated activity of lower large intestine contents
$\tilde{A}_{RM,Norm}$	Normalized cumulated activity of red bone marrow
$\tilde{A}_{SI}$	Cumulated activity of small intestine contents
$\tilde{A}_{UB,Norm}$	Normalized cumulated activity of the urinary bladder contents
$\tilde{A}_{ULI}$	Cumulated activity of upper large intestine contents
$[\tilde{A}]_{RM,Norm}$	Normalized cumulated activity concentration of red bone marrow
$[A_{BL}(t)]$	Activity concentration in whole blood at time t
$[A(t)]_{BL,Norm,Corr}$	Activity concentration in whole blood, normalized to the administered activity and corrected for physical decay at time t
$[A(t)]_{P,Norm,Corr}$	Activity concentration in plasma, normalized to the administered activity and corrected for physical decay at time t
$[A(t)]_{RM,Norm,Corr}$	Activity concentration in red bone marrow, normalized to the administered activity and corrected for physical decay at time t
$A^{\mu}(\boldsymbol{X}), \boldsymbol{A}_{\mu}(\boldsymbol{X})$	Four-vector electromagnetic potential
ADME	Administration, distribution, metabolism, and excretion
Ai(t)	Airy function, $\mathrm{Ai}(t) = \frac{1}{\pi}\int_0^{\infty} dy\, \cos\left(yt + \frac{y^3}{3}\right)$
ALARA	As low as reasonably achievable
b	Barn ($1b = 10^{-24}\ cm^2$)
	Impact parameter

becquerel	Unit of activity (transitions per second)
B	Baryon number
	Buildup factor
B(A,Z)	Nuclear binding energy for atomic mass number A and atomic number Z
B_{2k}	Bernoulli number of order 2k
$B_F^{\pm}$	Reduced Fermi transition probability
$B_{GT}^{\pm}$	Reduced Gamow–Teller transition probability
BD	Absorbed dose buildup factor
$B\phi$	Number fluence buildup factor
$B\Psi$	Energy fluence buildup factor
BED	Biologically equivalent dose
BEIR	Biological Effects of Ionising Radiation Panel
B	Magnetic field
Bq	Becquerel
c	Speed of light
cdf	Cumulative density function
cGy	centigray
cpm	Counts per minute
$C_e(\beta)$	Term describing effect of atomic electron shells on collision stopping power
Ci(x)	Cosine integral, $Ci(x) = \gamma_{EM} + \ln x + \int_0^x dt \frac{\cos t - 1}{t}$
CF	Cellularity factor of bone marrow
CFSA	Carrier-free specific activity
CPE	Charged particle equilibrium
CRE	Complete radiation equilibrium
CSDA	Continuous slowing-down approximation
CT	Computed tomography
CZT	Cadmium zinc telluride
dpm	Disintegrations per minute
$d^3\mathbf{r}$	Differential volume element $d^3\mathbf{r} \equiv r^2\, dr\, d\varphi\, d(\cos\theta)$
D	Absorbed dose
	Diffusion coefficient
$D_{r_T}(t)$	Absorbed dose at time t in target region, r_T
$D_{r_T,Norm}(t)$	Absorbed dose at time t in target region, r_T, normalized to the administered activity
D_Q	Quasithreshold absorbed dose
DDREF	Dose and dose-rate effectiveness factor
DSB	Double strand break
DTPA	Diathethylenetriaminepentaacetic acid
eV	Electron volt
E	Total energy (sum of kinetic and rest mass energies)
	Effective dose
E_B	Atomic electron binding energy

$E_{F,p}$	Fermi energy for protons
$E_{F,n}$	Fermi energy for neutrons
EAR	Excess absolute risk
ECMR	Excess cancer mortality rate
ERR	Excess relative risk
$\mathbf{E}_{rad}$	Electric radiation field
$(dE/\rho\, dx)_{Col}$	Collision mass stopping power
$(dE/\rho dx)_{Col,\Delta}$	Restricted collision mass stopping power for electrons or positrons
$(dE/\rho\, dx)_{Col,H}$	Hard-collision mass stopping power
$(dE/\rho dx)_{Col,H,\Delta}$	Restricted hard-collision mass stopping power for electrons or positrons
$(dE/\rho\, dx)_{Col,S}$	Soft-collision mass stopping power
$(dE/\rho\, dx)_{Rad}$	Radiative mass collision stopping power
$\mathbf{D}$	Electric flux density
$\mathbf{E}$	Electric field
$\mathbf{E}_{lm}^{(E)}$	Electric field of electric multipole of order *l*, m
$\mathbf{E}_{lm}^{(M)}$	Electric field of magnetic multipole of order *l*, m
$Ei(x)$	Exponential integral, $Ei(x) = \int_{-x}^{\infty} dt \frac{e^{-t}}{t}$
e	Fundamental unit of electric charge
$f_n(q, Z)$	Generalized oscillator strength (GOS)
$f_{Self\text{-}atten}$	Self-attenuation correction factor
F	Fano factor
FBP	Filtered backprojection
F(L)	Linear energy transfer (L) frequency distribution
F(q)	Scattering form factor for three-vector momentum transfer, q
F(q,Z)	Atomic form factor
$F(Z_Y,E_e)$	Fermi β-decay nuclear Coulomb correction factor for daughter nucleus Z_Y and β particle energy, E_e
FSU	Functional subunit
f(**q**)	Scattering amplitude as a function of three-vector momentum transfer, **q**
f(A,Z)	Nuclear binding energy per nucleon
ft	Comparative half-life of β decay
g	Gram
	Fraction of liberated charged particles' initial kinetic energies that is irradiated as *bremsstrahlung*
	Marinelli geometric factor
G	Lea–Catcheside dose protraction factor
$G(\mathbf{r}, \mathbf{r}')$	Green's function
GCP	Good clinical practice

Gy	Gray
$h_l^{\binom{1}{2}}(x) = \sqrt{\frac{\pi}{2x}}\left(J_{l+\frac{1}{2}}(x) \pm iN_{l+\frac{1}{2}}(x)\right)$	Hankel functions of first and second type of order l
hcR_∞	Rydberg energy
H	Total Hamiltonian
	Magnitude of the magnetic field strength
$H_i(x)$	ith Hermite polynomial
H_0	Steady-state Hamiltonian
H_{50,r_T}	Total equivalent dose received by target region r_T over 50 years postexposure
H(t)	Heaviside function: $H(t) = 0\ t < 0$ $=1\ t > 0$
H_T	Equivalent dose to tissue or organ T
$H_{EM}(\boldsymbol{X})$	Electromagnetic Hamiltonian
$\mathbf{H}$	Magnetic field strength
$\mathbf{H}_{rad}$	Magnetic radiation field
$\mathbf{H}_{lm}^{(E)}$	Magnetic field of the electric multipole of order l and m
$\mathbf{H}_{lm}^{(M)}$	Magnetic field of the magnetic multipole of order l and m
I	Moment of inertia
IAEA	International Atomic Energy Agency
ICH	International Conference on Harmonisation
ICR	International Congress of Radiology
ICRP	International Commission on Radiological Protection
ICRU	International Commission on Radiation Units and Measurement
IMP	Investigational Medicinal Product
IPEM	Institute of Physics and Engineering in Medicine
I_{ds}	Source-to-drain current in MOSFET
$\bar{I}$	Mean ionization energy/mean excitation potential
$\mathbf{J}$	Total angular momentum
	Electric current density
$J_l(x)$	Bessel function of the first kind of order l
$j_l(x)$	Spherical Bessel function of the first kind of order l
$J_\mu^{EM}(\boldsymbol{X})$	Four-vector electromagnetic current
$J(p_{e,z})$	Compton profile for electron moment p_e and atom with atomic number Z
k	Photon energy
	Boltzmann's constant
kVp	Kilovoltage potential
$\mathbf{k}$	Photon three-vector momentum
K	Kerma
K_c	Collision kerma

$K_{c,air}$	In-air collision kerma
$K_{\nu}(x)$	Modified Besself function of the second kind of order ν
l	Quantum angular momentum number
$l_{\mu}^{C}(\boldsymbol{X})$	Weak leptonic current density
L	Total lepton number
L_{∞}	Unrestricted linear energy transfer
L(β)	Stopping number for charged particle with speed normalized to c, β
LET	Linear energy transfer
LNT	Linear no-threshold (dose response model)
LoR	Line of response
LQ	Linear-quadratic (dose response model)
LT	Linear threshold (dose response model)
L	Total angular momentum
L_e	Electronic lepton number
L_{μ}	Muonic lepton number
L_{τ}	Tau lepton number
$Li_n(x)$	Polylogarithm of order n
m_e	Rest mass of the electron
m_{μ}	Rest mass of the muon
m_n	Rest mass of the neutron
m_N	Reset mass of a nucleon
m_{π}	Rest mass of the pion (π-meson)
m_p	Rest mass of the proton
m_{τ}	Rest mass of the tau lepton
mGy	Milligray
M	Multiplication factor (ionization in gases)
M(A,Z)	Atomic mass for atomic mass number A and atomic number Z
M_{fi}	Matrix element, $M_{fi} = \int d^3\mathbf{r}\, \psi_f^* U \psi$, for transition from initial state I to final state f
M_H	Mass of the hydrogen atom
M_{lm}	Magnetic multipole moment of order l,m
MIRD	Medical internal radiation dose
MLEM	Maximum-likelihood expectation maximization
MOSFET	Metal-oxide semiconductor field effect transistor
MR(I)	Magnetic resonance (imaging)
MWPC	Multiwire proportional chamber
n	Neutron Principal quantum number Index of refraction
N	Number of neutrons in a nucleus, N = A – Z
N_A	Avogadro's number
$N_l(x)$	Neumann function of order l
NaI(Tl)	Sodium iodide (doped with thallium)

NECR	Noise-equivalent count rate
NTP	Normal temperature and pressure
NURBS	Nonuniform rational B-spline
$\left\langle \frac{dN}{dt} \right\rangle_{geom}$	Geometric mean count rate
OER	Oxygen enhancement ratio
OLINDA/EXM	Organ level internal dose assessment/ exponential modeling
OSEM	Ordered subsets expectation maximization
p	Proton Magnitude of three-vector momentum vector, **p**
$p_{T,\ j}$	Spectral distribution of particle radiance for particle species j
pdf	Probability distribution function
$p_{F,p}$	Fermi momentum of protons
$P_{F,n}$	Fermi momentum of neutrons
p	Three-vector momentum Electric dipole moment
$\mathbf{p}_T$	Vector particle radiance
p	Four-vector momentum, $\boldsymbol{p} = (E, \mathbf{p})$
P	Pressure
PE	Pulmonary embolism
PET	Positron emission tomography
PI	Product insert
PMT	Photomultiplier tube
P	Poynting vector Polarization (number of electric dipole moments per unit volume)
$P_l m(\mu)$	Associated Legendre polynomial of the first kind
$P_l(\mu)$	Legendre polynomial
P_{rad}	Magnitude of the power radiated by an accelerated charged particle
q	Electric charge Magnitude of three-vector momentum transfer
Q	Electric charge Energy released (>0; exoergic) or absorbed (<0; endoergic) in a transition Kinetic energy transfer Nuclear electric quadrupole moment
Q_C	Energy transfer demarcating the difference between soft and hard charged-particle collisions
$Q/_e$	Reduced nuclear electric quadrupole moment
Q_{lm}	Electric quadrupole moment of order *l*, m
QE	Quantum efficiency (of a photomultiplier tube)
RE	Relative effectiveness
RR	Relative risk
q	Three-vector momentum transfer

r_S	Source region
r_T	Target region
RBE	Relative biological effectiveness
RIDIC	Radiation Internal Dose Information Center
RM	Red bone marrow
RNG	Random number generator
RoI	Region of interest
$R_{10,50,90}$	Nuclear radius at which the nuclear electric charge density has decreased to 10, 50, and 90%, respectively, of the value at the nuclear center
R_N	Nuclear radius
R_{TF}	Atomic radius in Thomas–Fermi model
R_∞	Rydberg constant
r_0	Classical electron radius, $r_0 = e^2/4\pi\varepsilon_0 m_e c^2$
r_∞	Bohr radius, $r_\infty = \hbar c/\alpha m_e$
s	Intrinsic spin
SCD	Source-collimator distance
$SEE(r_T \leftarrow r_S)$	Specific effective energy from source region r_S to target region r_T (ICRP nomenclature)
SF	Scatter fraction (PET) Surviving fraction
SNM	Society of Nuclear Medicine
SPC	Summary of Product Characteristics
SSB	Single strand break (of DNA)
$S(\mathbf{q}, Z)$	Incoherent scattering function
$S(\theta, x)$	Sievert integral, $S(\theta, x) = \int_0^\theta d\theta\, e^{-\frac{x}{\cos\theta}}$
$S(r_T \leftarrow r_S; t)$	S-factor for source region r_S and target region r_T
S_{fi}	S-matrix element for initial state i to final state f
$Si(x)$	Sine integral, $Si(x) = \int_0^x dt \frac{\sin t}{t}$
S_n	Neutron separation energy
S_p	Proton separation energy
$S_F(\boldsymbol{X} - \boldsymbol{Y})$	Feynman propagator, $S_F(\boldsymbol{X} - \boldsymbol{Y}) = \frac{1}{(2\pi)^4} \int d^4\boldsymbol{p} \frac{e^{-\boldsymbol{p}\bullet(\boldsymbol{X}-\boldsymbol{Y})}}{\not{p} - m}$
SI	Système International
SPECT	Single-photon emission computed tomography
t	Isotopic spin (isospin) vector
t_3	Component of isospin vector in isospin-space
T	Kinetic energy of a particle
T_V	Urinary bladder voiding interval
TCPE	Transient charged particle equilibrium
$TD_{5/5}$, $TD_{50/5}$	Tolerance doses for 5 and 50% complication rates, respectively, in 5 years
TLD	Thermoluminescent dosimetry (dosimeter)
TM	Total trabecular marrow space

$T_{\frac{1}{2}}$	Half life
$T_{\frac{1}{2},\mathrm{Biol}}$	Biological half-life
$T_{\frac{1}{2},\mathrm{Eff}}$	Effective half-life
$T_{\frac{1}{2},\mathrm{Phys}}$	Physical half-life
U_{r_S}	Cumulated activity in source region r_S (ICRP nomenclature)
UV	Ultraviolet
$\mathbf{v}$	Velocity
v	Speed (magnitude of velocity vector)
$\overline{v_d}$	Mean drift speed
V	Electric potential
VoI	Volume of interest
$V_\mu^{C\dagger}(\boldsymbol{X})$	Weak hadronic current density
w_R	Radiation weighting factor
w_T	Tissue weighting factor
W	Mean energy to create an ion pair
$W^\pm$	Charged intermediate vector boson, W
WHO	World Health Organization
WMA	World Medical Association
x	Three-vector position
$\boldsymbol{X}$	Four-vector position
X	Exposure
$\mathbf{X}_{lm}(r, \phi)$	Vector spherical harmonic or order *l*,m
y	Lineal energy
$y_l(x)$	Spherical Bessel function of the second kind of order *l* (also known as a Neumann function)
$Y_l(x)$	Bessel function of the second kind of order *l*
$Y_{lm}(\theta, \phi)$	Spherical harmonic
z	Specific energy (imparted)
Z	Atomic number
Z_{eff}	Effective atomic number
Z_0	Neutral intermediate vector boson

1 The Role of Radiation Dosimetry in Nuclear Medicine

Abstract This chapter summarizes the importance and relevance of ionising radiation dosimetry to nuclear medicine.

Contents

1.1 Introduction

Medical internal radiation dosimetry is the discipline of determining the absorbed radiation doses received by an individual as a consequence of the deliberate or accidental intake of radioactive substances. As this book's topic is that of the physical theory of dosimetry applied to diagnostic and therapeutic nuclear medicine, it will be assumed implicitly that the intake is deliberate and that the amount of radioactive substance administered is known.

Diagnostic nuclear medicine provides functional and physiological information of the patient through the in vivo imaging of photons emitted by a radionuclide or following the $e^-e^+ \rightarrow 2\gamma$ annihilation in the case of a positron-emitting radionuclide. Within the context of optimization, it is intended that the amount of radioactive substance administered to the patient be limited to that required for obtaining an image of the necessary diagnostic quality whilst minimizing the radiation dose burden borne by the patient (Early 1995). Although diagnostic nuclear medicine is frequently discussed in terms of imaging, it is possible for a clinical diagnosis to be achieved from measurements not based upon imaging, such as the ^{14}C-urea breath test for *Helicobacter pylori* infection (Balon et al. 1998) or the Schilling test with ^{57,58}Co-labeled cyanocobalamin (vitamin B_{12}) for vitamin B_{12} absorption impairment (Hamilton 2004). On the other hand, the intents of therapeutic nuclear medicine are vastly different from those of diagnosis. Whereas, by definition, the latter seeks to avoid all physiological effect, the former deliberately achieves it in order to treat or palliate neoplastic disease. Examples of therapeutic nuclear medicine include oral administration of Na^{131}I for the treatment of thyroid cancer and ^{90}Y-labeled *ibritumomab tiuxetan* in the treatment of non-Hodgkin's lymphoma (Wiseman et al. 2000).

In both diagnostic and therapeutic nuclear medicine, accurate knowledge of the internal radiation dosimetry resulting from the administration of a radionuclide is essential. For example, it is a fundamental component of the required safety profile in the development of a radiopharmaceutical. Occasionally, in the practical setting of the clinic, it is necessary in the retrospective estimation of patient radiation dose due to, for example, a misadministration. While still not

B.J. McParland, *Nuclear Medicine Radiation Dosimetry*,
DOI 10.1007/978-1-84882-126-2_1,

yet frequently done in therapeutic nuclear medicine, individual patient dosimetry is desirable for maximizing the therapeutic ratio (the ratio of the probability of tumor control to that of normal tissue complications). The intellectual challenge of radiation dosimetry applied to nuclear medicine resides within its being, at its highest and most obvious level, a bond between physics, chemistry, and biology (Stelson et al. 1995; Sgouros 2005; Stabin 2006). As one scrutinizes this junction with greater resolution, it is seen that accurate nuclear medicine radiation dosimetry is dependent upon not only the detailed knowledge of radiation-matter interactions (Simpkin 1999) but also of the highly variable and far less-predictable biokinetics of a radioactive substance in the human body and the broadly-estimated intrinsic radiosensitivities of tissues and organs. This combination of precise physics and imprecise biological variability presents one of the great challenges, and indeed interests, of radiation dosimetry as applied to medicine.

1.2 Diagnostic Nuclear Medicine

Consider diagnostic nuclear medicine and the statement that the focus of developments in modern-day diagnostic nuclear medicine has been the improvement of diagnostic efficacy. This has been attempted through engineering in the development of new imaging technologies, the improvement and evolution of preexisting ones (such as the introduction over the past decade of PET–CT, PET–MR, and SPECT–CT hybrid devices), and through chemistry and biology in the introduction of new radiopharmaceuticals with improved specificities for pathologies of interest. But what cannot be ignored is that, parallel to this drive to yield greater diagnostic utility, there is a growing awareness of the radiation absorbed doses and consequent potential risk to the patient as a result of exposure to ionizing radiation. This awareness has become manifest in government legislation, guidelines and established frameworks of medical practice. Examples include various European Union directives, the United States Code of Federal Regulations and guidance notes provided by the UK's Administration of Radioactive Substances Advisory Committee (ARSAC 2006). The process of maximizing diagnostic imaging efficacy, whilst minimizing radiation absorbed dose and patient risk, has long been followed in all aspects of diagnostic radiology and nuclear medicine. In particular, the reduction of patient radiation dose in diagnostic nuclear medicine can be achieved through a variety of means, including the following.

1.2.1 *Radiation Detector Efficiency*

In diagnostic nuclear medicine, the radiation detector converts a physical process (e.g., scintillation photons resulting from the photoelectric absorption of a γ ray in a crystal) to an analog electronic signal which is ultimately processed to generate a digital image to be analyzed. Detectors of high detective quantum efficiency enable the acquisition of a given signal for a reduced incident photon fluence and, thus, provide an opportunity for a reduced amount of administered activity to be used for a given signal level (Lewellen 2008).

1.2.2 *Radionuclide*

The selection of the radionuclide to label the chemical moiety will profoundly alter not only the radiation dose burden but also imaging efficacy. Perhaps the most obvious example of this is the selection of either ^{123}I or ^{131}I for diagnostic imaging. A ligand capable of iodination can be complexed with both isotopes. However, the combinations of the lower major photon energy (159 keV) and reduced physical half-life (13.2 h) of ^{123}I compared to those of ^{131}I (364 keV and 8.02 days) and its associated low-energy particulate radiations make the former's absorbed dose burden lesser and imaging profile greater than those of the latter. Hence, ^{123}I is preferable for diagnostic applications whereas ^{131}I is preferable in therapy. However, both isotopes can be paired in therapy planning and treatment. For example, a diagnostic amount of Na^{123}I can be administered in order to enable quantification of uptake at active sites of disease so as to allow a calculation of the amount of subsequently administered Na^{131}I required to achieve a given therapeutic absorbed dose.

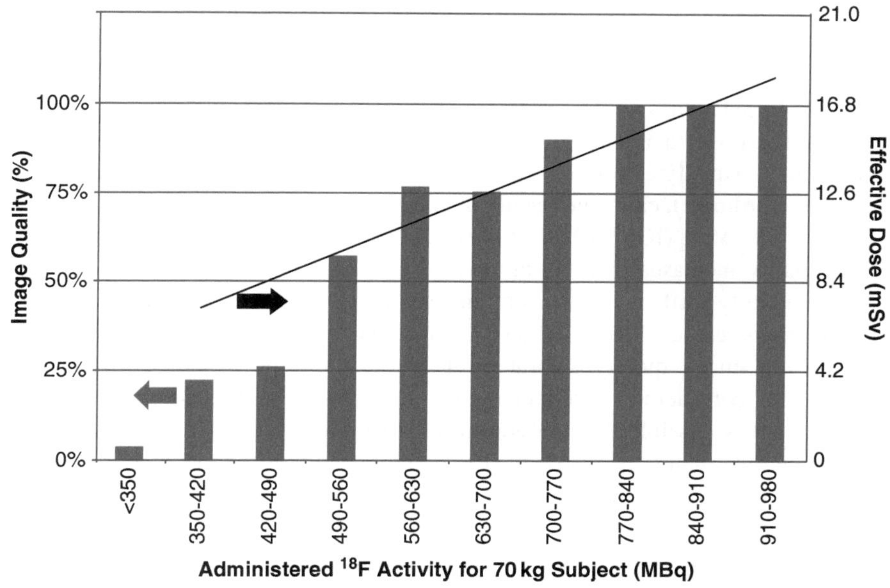

Fig. 1.1 Nonlinear response of image quality and linear response of effective dose as functions of administered ^{18}F activity. Figure is derived using data published by Everaert et al. (2003) and ICRP (1998)

1.2.3 Radiopharmaceutical Design

A radiopharmaceutical is designed so as to increase the differential of the signal (i.e., the number of detected photons) between the pathology of interest relative to the unspecific signal of the surrounding normal pathology (this differential can be an increase in uptake by the pathology relative to background, as in tumor imaging, or a decrease as in myocardial perfusion imaging). Appropriate design leading to increased specific binding will elevate the detected target-to-background ratio and improve the diagnostic efficacy which can enable the amount of administered activity, and radiation dose burden, to be reduced.

1.2.4 Clinical Imaging Practice

Relatively simple modifications of imaging protocols and clinical practice and the development of imaging strategies have been demonstrated to markedly reduce patient radiation exposure in diagnostic radiological procedures such as fluoroscopy (McParland and Lewall 1998) and multidetector CT (Smith et al. 2007). This result can also be true when applied to diagnostic nuclear medicine. A potent example of this can be seen by recognizing that the acquisition of a nuclear medicine image is an integral process: signals resulting from detected photons are accumulated over a period of time to create an image. Whereas the information[1] contained within the image is a nonlinear function of the number of detected photons used to generate the image, the number of detected photons, as is the radiation dose to the patient, is proportional to the amount of administered activity. Eventually, after some number of detected photons has been attained, all of the information that is potentially available from the image has been obtained and any further increase in detected photons provides no further diagnostic utility. This is shown conceptually in Fig. 1.1 which was calculated using data presented by Everaert et al. (2003) who determined the optimal dose of ^{18}F-fluorodeoxyglucose for a fixed acquisition time in a PET scanner using lutetium oxyorthosilicate scintillation detectors. Whole-body ^{18}FDG images were acquired of 186 subjects and image quality was scored on a 5-point scale (poor, reasonable, good, very good, and excellent) by two independent observers. The plotted histogram is the image quality, defined as the

[1] What is meant by "information" in this statement is defined by the context of how the image is assessed. This could be through quantitative assessments such as sensitivity and specificity or signal-to-noise ratio. It can also be determined through qualitative and subjective measures as the degree of the image's diagnostic quality as assessed by a reader.

percentage of images scored good or better averaged over the two observers, as a function of the ^{18}F administered activity per unit body weight (left-hand ordinate). The line shown is a measure of risk, described by the effective dose (a measure of radiation risk and defined in Chap. 10), normalized to body weight (right-hand ordinate), calculated using the proportionality of 19 μSv/MBq (ICRP 1998). It is evident that the image quality increases with the amount of administered ^{18}F activity until the ratio of administered activity per body mass reaches 11–12 MBq/kg. At this level of activity, the image quality saturates: at 100%, the images are considered excellent and no further improvement is possible through administering more ^{18}F. On the other hand, the metric of risk, given by the product of the administered activity normalized to body weight and the proportionality above, increases linearly with administered activity. For administered activities per unit body weight in the range of 12–14 MBq/kg, the risk continues to increase even though image quality has reached a plateau. Thus, the risk-to-benefit ratio begins to increase for ^{18}F activities exceeding about 12 MBq/kg and there would be no justification in administering greater amounts of ^{18}F.

There are practical limits to the minimization of administered activity which are dictated by, for example, patient tolerance to long acquisition times or a busy hospital's need for a high patient throughput rate. It is unlikely that the above argument has been applied explicitly within the clinic, but there are studies described in the literature examining this reciprocal role of acquisition time and administered activity for PET (Halpern et al. 2004) and SPECT (Robinson et al. 2008). A means of examining empirically this effect is described by Bailey and Kalemis (2005).

1.2.5 Dose Reference Levels

Dose reference levels are recommended levels of patient radiation exposure, for both diagnostic radiology and nuclear medicine, which provide clinical guidance for the imaging practitioner. For example, in the United Kingdom, the Administration of Radioactive Substances Advisory Committee provides guidance for the amount of administered activity of given radionuclide for a given diagnostic nuclear medicine study (ARSAC 2006).

Qualitatively, the ways in which these processes can reduce the patient radiation absorbed dose are clearly understood. However, this is perhaps not the case in terms of a quantitative understanding in that, ultimately, the mechanisms of reducing the absorbed doses to the diagnostic nuclear medicine patient rely upon the ability to accurately predict these absorbed doses and to estimate their associated somatic risks. To underline the requirement for accurate radiation dosimetry in diagnostic nuclear medicine, compare it with diagnostic radiology. The small levels of radiation absorbed dose associated with all diagnostic imaging procedures using ionizing radiation limit the risks presented to the patient to *stochastic* effects, so-called as the risk is probabilistic and assumed proportional to absorbed dose. These effects include radiocarcinogenesis and hereditary damage. The UNSCEAR 2000 report on the sources and effects of ionizing radiation included a survey of diagnostic radiological and nuclear medicine procedures performed throughout the world during the period of 1991–1996 (UNSCEAR 2000). Figure 1.2 presents the number of diagnostic nuclear medicine procedures over that period as a percentage of the combined number of diagnostic nuclear medicine and radiology procedures (including dental X-ray examinations). The corresponding percentage *per caput* effective dose is also shown. The figure is derived from data published in the UNSCEAR 2000 report and categorizes these percentages over four levels of health care defined by the ratio of physician number to population. A comparison of the ratios is rather revealing. Averaged over the world, between 1991 and 1996 diagnostic nuclear medicine procedures accounted for only about 1.3% of the total number of diagnostic medical imaging procedures using ionizing radiation. However, the percentage contribution that diagnostic nuclear medicine made to the *per caput* effective dose is nearly five times greater at about 6.1%. In other words, one can crudely (and provocatively) state that, on average, diagnostic nuclear medicine presents a five times greater risk per procedure than diagnostic radiology.

Such a statement and any other interpretation of this observation must, of course, be made whilst being cognizant of a number of *caveats*:

- The risks of radiation carcinogenesis resulting from diagnostic radiology and nuclear medicine procedures are low (Wall et al. 2006).

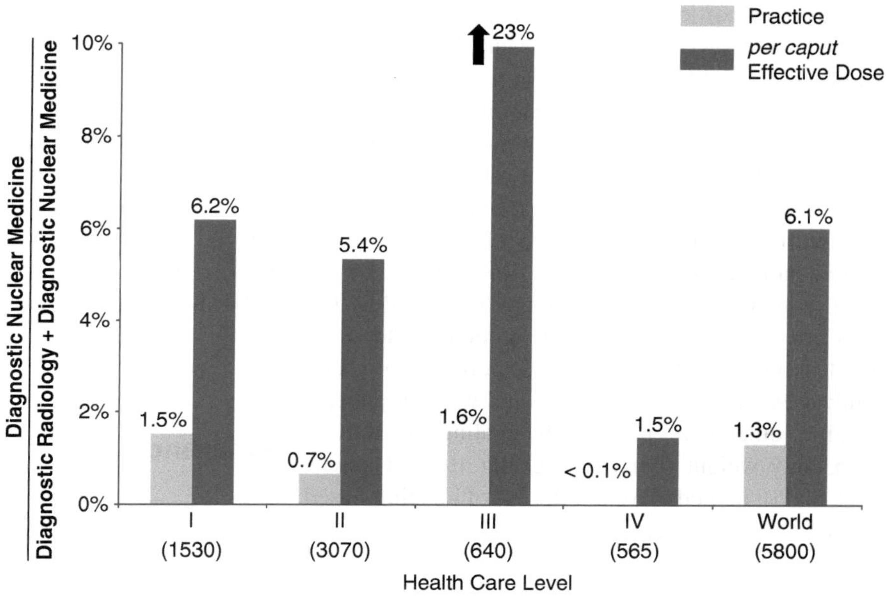

Fig. 1.2 Diagnostic nuclear medicine procedures as a percentage of all diagnostic imaging procedures using ionizing radiation and the corresponding percentage *per caput* effective dose. Stratification is by Health Care Level (HCL) as defined by the number of physicians per 10^6 population: I >1,000, II 300–1,000, III 100–300, and IV <100. World data are all HCLs combined. Figure is derived from data provided in UNSCEAR (2000)

- Diagnostic radiology procedures are dominated by the posterior–anterior and lateral chest X-ray examination which has a markedly low effective dose and which will elevate the nuclear medicine percentage effective dose.
- The effective dose is not expected to be distributed uniformly over the population as a whole as many of these imaging examinations will be received by patients nearing the ends of their lives and who will have a negligible risk of radiation carcinogenesis due to the extended latency period of radiation-induced malignancies.
- The data were obtained in the early 1990s and, as such, new imaging modalities with intrinsically higher dose burdens, such as multidetector CT and PET–CT and which would confuse any interpretation, have become available since.

The position of the last point has been recently accentuated by the survey presented by Fazel et al. (2009) of diagnostic imaging procedures performed in the United States of America during the 3-year period between 2005 and 2007. Of 655,613 patients who underwent at least one imaging procedure associated with ionizing radiation, they determined that myocardial perfusion imaging (using ^{99m}Tc or ^{201}Tl) was the single procedure providing the largest effective dose (22.1% of the total) at 15.6 mSv. As myocardial perfusion imaging is associated with patients with a not unreasonable life expectancy (Mettler et al. 2008), during which a radiogenic malignancy can be expressed, this observation emphasizes the need for accurate nuclear medicine dosimetry with which risk can be quantified and from which optimization can be obtained.

1.3 Therapeutic Nuclear Medicine

On the other hand, the dosimetry interests in therapeutic nuclear medicine are very different in that biological effects are deliberately sought (Zanzonico 2000). These biological effects are deterministic, made manifest through targeted cell-killing, and exhibit an absorbed dose threshold below which no effect is apparent and, above this threshold, a pathological effect proportional to this dose. The levels of activity administered in therapeutic nuclear medicine are intended to be high enough to cause such effects.[2] In addition, the isotopes selected to be used in therapy

[2]Both intracavitary and interstitial placement of radioactive sources (e.g., remote-afterloading ^{137}Cs applicators for the treatment of gynecological cancers or the permanent insertion of ^{125}I 'seeds' into the prostate gland) will be considered as sources of external exposure as the radioactive sources are not integral to the corpus nor are subject to distribution, metabolism or excretion.

emit charged particulate radiations (α and β particles) which have limited range in tissue and, hence, deposit all of their initial kinetic energy in tissue. This makes the design of the targeting capability of the vector carrying the source of radiation especially crucial. The lower limits to the therapeutic levels of activity are dictated by the need to exceed the threshold required for a beneficial biological effect (either curative or palliative) and the upper limits are set by the need to avoid radiotoxicity to uninvolved healthy tissues or organs. Historically, the amount of administered activity used for a therapeutic purpose was set by clinical experience and protocol and was not patient-specific, other than, perhaps, by scaling the activity with body weight (Early and Landa 1995; Thierens et al. 2005). In modern times, the use of imaging-based predictive and patient-specific dosimetry has been slowly growing in therapeutic nuclear medicine, mainly within the research arena (Zanzonico 2002; Flux et al. 2006) This can be achieved, as noted earlier, by the pretherapy administration of a photon-emitting diagnostic radiopharmaceutical expected to have the same biodistribution of the particulate-emitting therapeutic radiopharmaceutical (e.g., by using a different isotope of the intended therapeutic moiety) followed by imaging in order to extract the exact in vivo biodistribution a priori. Should the therapeutic radiopharmaceutical also emit photons in addition to particulate radiation, a diagnostic activity level of the radiopharmaceutical itself can be administered and subsequently imaged using the photon component, under the expectation that the radiation absorbed dose associated with the therapeutic component is negligible in this instance. Quantitative measurement of the in vivo biodistribution can then be used to guide the amount of administered therapeutic levels of activity. Although such an approach may appear obvious, it should be considered with some degree of caution as, for example, in the use of ^{131}I in the treatment of differentiated thyroid cancer. The therapeutic absorbed dose is frequently less than that predicted from the kinetics of the pretherapy tracer amount of ^{131}I due to leakage of the iodine from damaged thyroid cells (Zanzonico et al. 1995) or through thyroid stunning (Coakley 1998). In an even more sophisticated and individualized approach, the measured biodistribution data and a whole-body CT image set demonstrating the patient's anatomy are used as input to an individualized patient-specific Monte Carlo calculation of the organ absorbed doses per unit administered activity of the therapeutic radiopharmaceutical. Knowing the desired minimum and maximum absorbed doses to the target and tissues, one can then use these values of the absorbed dose per unit administered activity to readily calculate the required activity to administer for the patient in question. However, the role of radiobiology in the responses of tumor and normal tissue to internal irradiation cannot be neglected (Goldsmith 2004; Kassis and Adelstein 2005).

References

ARSAC (2006) Notes for guidance on the clinical administration of radiopharmaceuticals and sealed radioactive sources. Administration of Radioactive Substances Advisory Committee, Health Protection Agency, London

Bailey DL, Kalemis A (2005) Externally triggered gating of nuclear medicine acquisitions: a useful method of partitioning data. Phys Med Biol 50:N55–N62

Balon HR, Gold CA, Dworkin HJm McCormick VA, Freitas JE (1998) Procedure guideline for carbon-14 urea breath test. J Nucl Med 39:2012–2014

Coakley AJ (1998) Thyroid stunning. Eur J Nucl Med 25:203–204

Early PJ (1995) Use of diagnostic radionuclides in medicine. Health Phys 69:649–661

Early PJ, Landa ER (1995) Use of therapeutic radionuclides in medicine. Health Phys 69:677–694

Everaert H, Vanhove C, Lahoutte T, Muylie K, Caveliers V, Bossuyt A, Franken PR (2003) Optimal dose of ^{18}F-FDG required for whole-body PET using an LSO PET camera. Eur J Nucl Med 30:1615–1619

Fazel R, Krumholz HM, Wang Y, Ross JS, Chen J, Ting HH et al (2009) Exposure to low-dose ionizing radiation from medical imaging procedures. N Engl J Med 361:849–857

Flux G, Bardies M, Monsieurs M, Savolainen S, Strand S-E, Lassman M (2006) The impact of PET and SPECT on dosimetry for targeted radionuclide therapy. Z Med Phys 16:47–59

Goldsmith SJ (2004) Improving insight into radiobiology and radionuclide therapy. J Nucl Med 45:1104–1105

Halpern BS, Dahlborn M, Quon A, Schiepers C, Waldherr C, Silverman DH, Ratib O, Czernin J (2004) Impact of patient weight and emission scan duration on PET/CT image quality and lesion detectability. J Nucl Med 45:797–801

Hamilton D (2004) Diagnostic nuclear medicine – a physics perspective. Springer, Berlin

ICRP (1998) Radiation dose to patients from radiopharmaceuticals, Addendum to publication 53 – publication 80. International Commission on Radiological Protection. Pergamon, Oxford

Kassis AI, Adelstein SJ (2005) Radiobiologic principles in radionuclide therapy. J Nucl Med 46:3S–12S

Lewellen TK (2008) Recent developments in PET detector technology. Phys Med Biol 53:R287–R317

McParland BJ, Lewall DB (1998) Reductions in fluoroscopy screening times resulting from credentialling and practice surveillance. Br J Radiol 71:461

Mettler FA, Bhargavan M, Thomadsen BR, Gilley DB, Lipoti JA, Mahesh M, McCrohan J, Yoshizumi TT (2008) Nuclear medicine exposure in the United States, 2005–2007: preliminary results. Semin Nucl Med 38:384–391

Robinson CN, van Aswegen J, SA NTO, Thomson WH, Tindale WB, tout DA, Underwood SR (2008) The relationship between administered radiopharmaceutical activity in myocardial perfusion scintigraphy and imaging outcome. Eur J Nucl Med Mol Imaging 35:329–335

Sgouros G (2005) Dosimetry of internal emitters. J Nucl Med 46:18S–27S

Simpkin DJ (1999) Radiation interactions and internal dosimetry in nuclear medicine. RadioGraphics 19:155–167

Smith AB, Dillon WP, Gould R, Wintermark M (2007) Radiation-dose reduction strategies for neuroradiology CT protocols. Am J Neuroradiol 28:1628–1632

Stabin MG (2006) Nuclear medicine dosimetry. Phys Med Biol 51:R187–R202

Stelson ATS, Watson EE, Cloutier RJ (1995) A history of medical internal dosimetry. Health Phys 69:766–782

Thierens HM, Monsieurs MA, Bacher K (2005) Patient dosimetry in radionuclide therapy: the whys and the wherefores. Nucl Med Comm 26:593–599

UNSCEAR (2000) Sources and effects of ionizing radiation – volume I, United Nations Scientific Committee on the Effects of Atomic Radiation 2000 Report to the General Assembly, United Nations, New York

Wall BF, Kendall GM, Edwards AA, Bouffler S, Muirhead CR, Meara JR (2006) What are the risks from medical x-rays and other low dose radiation? Br J Radiol 79:285–294

Wiseman A, White CA, Stabin M, Dunn WL, Erwin W, Dahlbom M, Raubitschek A, Karvelis K, Schultheiss T, Witzig TE, Belanger R, Spies S, Silverman DHS, Berlfein JR, Ding E, Grillo-Lopez AJ (2000) Phase I/II ^{90}Y-Zevalin (yttrium-90 ibritumomab tiuxetan, IDEC-Y2B8) radioimmunotherapy dosimetry results in relapsed or refractory non-Hodgkin's lymphoma. Eur J Nucl Med 27:766–777

Zanzonico PB, Brill AB, Becker DV (1995) Radiation dosimetry. In: Wagner HN, Szabo Z, Buchanan JW (eds) Principles of nuclear medicine, 2nd edn. Saunders, Philadelphia

Zanzonico PB (2000) Internal radionuclide radiation dosimetry: a review of basic concepts and recent developments. J Nucl Med 41:297–308

Zanzonico PB (2002) Model-based versus patient-specific dosimetry: blurring the lines. J Nucl Med 43:1665–1666

Theoretical Tools 2

Abstract In this chapter, the tools to be used in the theoretical derivations provided later in this book are summarized. These include relativistic kinematics, perturbation, and quantum scattering theories and Dirac's equation.

Contents

2.1 Introduction

The absorbed doses received by the tissues and organs of the nuclear medicine patient are the culmination of a series of physical phenomena[1] as shown diagrammatically in Fig. 2.1. The three theoretical tools which to be used in the description of such processes are categorized as:

- Kinematics
- Theory of quantum transitions (perturbation and scattering)
- Dirac relativistic theory

These phenomena are taken to be the simplified culmination of four steps:

Step 1: This is the production of the radioactive substance itself through, for example, the fission of uranium within a reactor following the absorption of a neutron (e.g., to produce ^{99}Mo which subsequently undergoes decays to produce the isomer ^{99m}Tc) or the charged-particle bombardment of a nucleus to produce a radioactive daughter (such as in the charge exchange reaction of ^{18}O(p,n)^{18}F to form the ^{18}F used in PET imaging). Clearly, a complete understanding of radioisotope production requires knowledge of both the kinematics of the nuclear reaction and the quantum-mechanically derived probability of the reaction occurring.

[1]Following the absorbed dose, biological effects occur which lead to the risk associated with the magnitude and type of radiation dose and the intrinsic radiosensitivity of the tissue irradiated.

B.J. McParland, *Nuclear Medicine Radiation Dosimetry*,
DOI 10.1007/978-1-84882-126-2_2,

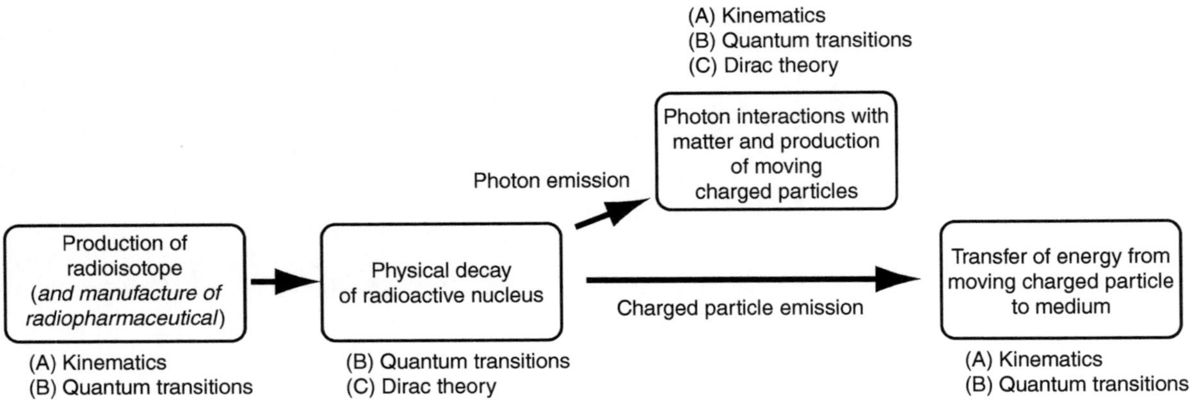

Fig. 2.1 Schematic chain of the physical phenomena from the production of the radioisotope to culminate in the deposition of energy in a medium (i.e., absorbed dose to tissue)

Step 2: This series is the in vivo physical decay of the radioactive nucleus. For diagnostic nuclear medicine, the expectation is that the decay produces either a single photon (as with ^{99m}Tc) or a positron, the antiparticle of the electron, which travels a short distance in tissue before annihilating with an atomic electron to produce two (sometimes, one or three) photons. The photons resulting from radioactive decay are detected externally and then used to form the diagnostic nuclear medicine image. On the other hand, therapeutic nuclear medicine seeks to deliver a high radiation dose to neoplasia and this is most easily achieved by the decay producing a short-ranged charged particle such as an α particle or an Auger/Coster–Krönig electron, either as a direct result of the decay or following atomic relaxation processes induced by the decay. Here, one needs to understand the requirements of an unsteady balance between nuclear mass and charge or the mechanisms through which internal degrees of freedom, such as angular momentum in a spinning nucleus, can couple to external emissions of energy, such as photons, and the relativistic theory of Dirac in explaining the production and existence of the positron. Again, the probability of a quantum event must be evaluated.

Step 3: The third step is a consequence of the interaction of a produced photon with matter and can be divided into two categories in terms of whether or not the photon transfers energy to the medium within which it is traveling. In classical Thomson scatter, for example, the photon does not transfer energy to an atomic electron, but its direction of travel is altered which, while not leading to energy deposition in tissue, can be important to calculating the transport of radiation in the body. A category of interactions in which energy is transferred to the medium can be further subdivided into subcategories of whether or not the photon is conserved. The process of Compton scatter, in which a fraction of the incident photon's energy is transferred to an electron which recoils and the photon is scattered with a reduced energy, retains the photon. On the other hand, a photon can be absorbed by an atomic electron, thus transferring its energy to the electron which is then ejected if the photon energy exceeds the binding energy. Should the photon have an energy exceeding twice the electron rest-mass (plus a little more to allow for nuclear recoil – refer to the Appendices), it can interact with the nuclear electromagnetic field to produce an electron–positron pair. In both of these cases, the abilities to calculate the probabilities of the interaction occurring, their physical observables (differential cross sections) and the kinematics of the process are necessary in order to solve the dosimetry problem.

Step 4: The final step, is the transfer of the secondary charged particle's kinetic energy to the medium as it slows down to be eventually thermalized. This transferred energy, per unit mass of medium, is the absorbed dose that we seek to eventually calculate. The probability and kinematics of energy transfer as a function of charged particle can, depending upon the particular scenario, be calculated classically or through quantum theory.

2.2 Physical Units

The question of units is inevitably contentious as consensus within the community appears to be perpetually unattainable! However, over the past 30 years, the *Systeme International d' Unités* (SI) system[2] has become dominant in the practice of medical radiation dosimetry. Correspondingly, this book will use the SI system in order to maintain as much consistency as possible with the definitions of physical constants provided in the literature.[3] Unfortunately, this exclusivity may have the potential to cause some difficulty and confusion for the reader who is more familiar with Gaussian/"natural" units favored by some (e.g., $\hbar = c = 1$) which provide time and distance with units of length and momentum in units of inverse length. Admittedly, the use of such units can make the derivations of some equations somewhat more transparent, but one should be reluctant to risk the loss of cohesion by switching systems of units within a text linking microscopic and macroscopic dosimetry theory. However, having said that, in order to ensure clarity of the expressions derived here, the rest-mass and three-vector momentum of a particle are implicitly assumed to be given in units of energy. In other words, instead of writing the total energy of a moving particle as $E = \sqrt{p^2c^2 + m^2c^4}$ we will write it as $E = \sqrt{p^2 + m^2}$. For example, the rest-mass of the electron will be given as $m_e = 511$ keV. This reduces the plethora of powers of c which often appear in expressions, but at the slight cost of the frequent use of the conversion factor, $\hbar c = 197.33\,\text{MeV} \bullet \text{fm}$.

Frequent use of the definitions of the fine structure constant $\alpha = e^2/4\pi\varepsilon_0\hbar c$ and the classical radius of the electron $r_0 = e^2/4\pi\varepsilon_0 m_e c^2$ will be made throughout the text in order to simplify expressions and still retain transparency. For example, as a result of these, the magnitude of the electric field due to a point charge e will be written in the succinct form as,

$$E(r) = \frac{\alpha\hbar c}{r}. \tag{2.1}$$

2.3 Mathematical Notations

2.3.1 Vector Notation

Three-component vectors are denoted in bold. If $\mathbf{x}$ is a three-component vector, then the corresponding unit vector is,

$$\hat{\mathbf{x}} = \frac{\mathbf{x}}{|\mathbf{x}|}. \tag{2.2}$$

A four-component vector is denoted in bold italics, e.g., $\boldsymbol{X}$. The contravariant components are denoted by superscripts whereas the covariant components are denoted by subscripts. For example,

$$\boldsymbol{X}^\mu \equiv \left(x^0, x^1, x^2, x^3\right) \equiv \left(x^0, \mathbf{x}\right). \tag{2.3}$$

The covariant and contravariant components are related through,

$$\boldsymbol{X}_\mu = \sum_{\nu=0}^{3} g_{\mu\nu}\boldsymbol{X}^\nu \tag{2.4}$$

where the metric tensor is,

$$g_{\mu\nu} = g^{\mu\nu} = \begin{pmatrix} 1 & 0 & 0 & 0 \\ 0 & -1 & 0 & 0 \\ 0 & 0 & -1 & 0 \\ 0 & 0 & 0 & -1 \end{pmatrix}. \tag{2.5}$$

Hence,

$$\boldsymbol{X}_\mu = \left(x^0, -\mathbf{x}\right). \tag{2.6}$$

The summation convention will be used in which, when an identical superscript and subscript occur, the components are summed over that superscript/subscript. That is,

[2] See, e.g., http://www.bipm.org/en/si/si_brochure/.

[3] An inconsistency will be admitted with respect to interaction coefficients such as attenuation coefficients and stopping powers where this book will use cgs units (e.g., the photon mass attenuation coefficient has units of square centimeter per gram). These are simply more practical and virtually all numerical tabulations of such quantities are provided in these units.

$$p_\mu X^\mu \equiv \sum_{\mu=0}^{3} p_\mu X^\mu. \quad (2.7)$$

In the event that no ambiguity can arise, this summation will be simplified by omitting the indices so as to write $p_\mu X^\mu \equiv pX$. Note also that,

$$p_\mu X^\mu = p^\mu X_\mu = \mathrm{p}^0 \mathrm{x}^0 - \mathbf{p} \bullet \mathbf{x}. \quad (2.8)$$

The four-vector momentum for a particle of rest-mass m is,

$$p^\mu = (\mathrm{E}, \mathbf{p}) \quad (2.9)$$

where E is the total energy (sum of kinetic energy and rest-mass) and **p** is the three-vector momentum. From the above,

$$\begin{aligned} p \bullet p &= \mathrm{E}^2 - \mathbf{p} \bullet \mathbf{p} \\ &= \mathrm{m}^2 \end{aligned} \quad (2.10)$$

and, hence, is an invariant. The product of the four-vector momenta of two distinguishable particles A and B is,

$$p_\mathrm{A} \bullet p_\mathrm{B} = \mathrm{E_A E_B} - \mathbf{p}_\mathrm{A} \bullet \mathbf{p}_\mathrm{B}. \quad (2.11)$$

2.3.2 Complex Conjugation

Let $z = x + iy$ be a complex variable. Its complex conjugate is denoted by,

$$z^* = x - iy. \quad (2.12)$$

2.3.3 Hermitian Conjugation

Let **M** be a matrix with complex elements $(\mathrm{M})_{ij} = m_{ij}$. The Hermitian conjugate of this matrix is denoted by $\mathbf{M}^\dagger$ where $(\mathrm{M}^\dagger)_{ij} = m_{ji}{}^*$. That is, the Hermitian conjugate is the transpose of the original matrix and with the elements replaced by their complex conjugates. If $\mathbf{M} = \mathbf{M}^\dagger$, the matrix is referred to as being Hermitian or self-adjoint.

2.3.4 Adjoint Operator

The adjoint of the spinor $\psi(\mathbf{r}, t)$ is,

$$\bar{\psi}(\mathbf{r}, t) = \psi^\dagger(\mathbf{r}, t)\gamma^0 \quad (2.13)$$

where γ_0 is a Dirac matrix, to be defined later in this chapter.

2.4 Relativistic Kinematics of a Two-Body Elastic Collision

2.4.1 Introduction

As summarized above, the absorbed dose to a medium exposed to radiation is due to the transfer of the kinetic energy of a charged particle to the medium as it slows down. For the charged particles and kinetic energies of interest to nuclear medicine, these energy transfers are predominantly through interactions with atomic electrons. The kinematics of the scattered projectile and recoil target resulting from these interactions are the subject of this subsection.

An elastic scatter between two bodies is that in which the sum of the kinetic energies of the bodies pre- and postcollision is the same. There are no internal degrees of freedom present (such as atomic or nuclear excitation) that can channel away kinetic energy and transfer it to potential energy. Figure 2.2 shows a two-body scatter in which a particle of rest-mass m_1, total energy E_1, three-vector momentum $\mathbf{p}_1$ and kinetic energy T_1 is incident to a particle at rest in the laboratory reference frame with a rest-mass m_2. The pre- and postcollision four-vector momenta of the particles are,

$$p_1 = (\mathrm{E}_1, \mathbf{p}_1) \quad (2.14)$$

$$p_2 = (\mathrm{m}_2, 0) \quad (2.15)$$

$$p_1' = (\mathrm{E}_1', \mathbf{p}_1') \quad (2.16)$$

$$p_2' = (\mathrm{E}_2', \mathbf{p}_2'). \quad (2.17)$$

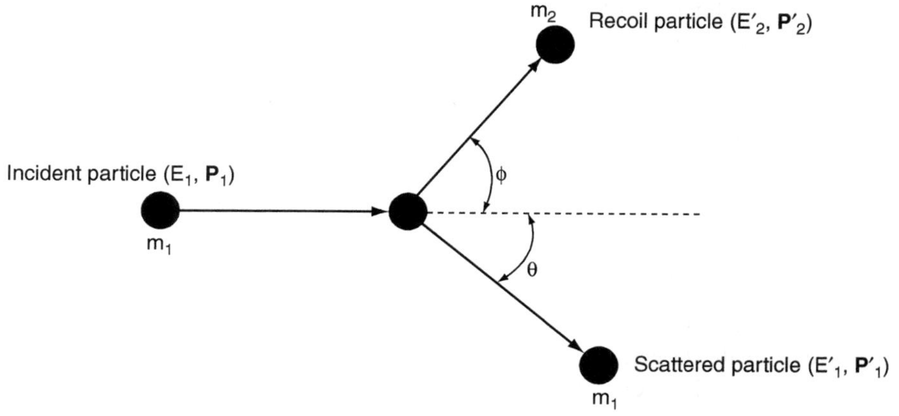

Fig. 2.2 The kinematics of two-body elastic scatter

2.4.2 *Kinetic Energy of the Recoil Particle*

2.4.2.1 Derivation

The conservation of four-vector momenta in the scatter of Fig. 2.2 is,

$$\boldsymbol{p}_1 + \boldsymbol{p}_2 = \boldsymbol{p}'_1 + \boldsymbol{p}'_2. \tag{2.18}$$

By isolating and squaring the four-vector momentum of the scattered particle and recalling its invariance,

$$\begin{aligned} \boldsymbol{p}'^2_1 &= (\boldsymbol{p}_1 + \boldsymbol{p}_2 - \boldsymbol{p}'_2)^2 \\ m_1^2 &= m_1^2 + 2m_2^2 + 2(\boldsymbol{p}_1 \bullet \boldsymbol{p}_2 \\ &\quad - \boldsymbol{p}_1 \bullet \boldsymbol{p}'_2 - \boldsymbol{p}_2 \bullet \boldsymbol{p}'_2). \end{aligned} \tag{2.19}$$

After further algebraic manipulation,

$$p_1 p'_2 \cos\phi = \left(E'_2 - m_2\right)(E_1 + m_2). \tag{2.20}$$

Solving for the total energy of the recoil particle,

$$E'_2 = m_2 \frac{(E_1 + m_2)^2 + p_1^2 \cos^2\phi}{(E_1 + m_2)^2 - p_1^2 \cos^2\phi} \tag{2.21}$$

and the corresponding kinetic energy $T'_2 = E'_2 - m_2$ is,

$$T'_2 = 2m_2 \frac{p_1^2 \cos^2\phi}{(E_1 + m_2)^2 - p_1^2 \cos^2\phi}. \tag{2.22}$$

2.4.2.2 Maximum Recoil Kinetic Energy

Equation (2.22) provides the kinetic energy of the recoil particle as a function of the recoil particle's mass, the incident projectile's total energy and momentum and the recoil angle, ϕ. In many instances, we will be interested in only the maximum recoil kinetic energy (i.e., the maximum energy transferred). This will clearly occur for the condition of $\phi = 0$,

$$\begin{aligned} T'_{2,\max} &= 2m_2 \frac{p_1^2}{(E_1 + m_2)^2 - p_1^2} \\ &= 2m_2 \frac{p_1^2}{m_1^2 + m_2^2 + 2m_2\sqrt{p_1^2 + m_1^2}}. \end{aligned} \tag{2.23}$$

Massive Projectile and Light Target

For the case of a heavy incident particle incident to a light particle, $m_1 \gg m_2$, (e.g., a proton projectile and an electron target), (2.23) reduces to the following expression,

$$T'_{2,\max} \approx \frac{2p_1^2}{\left(\frac{m_1^2}{m_2}\right) + 2\sqrt{p_1^2 + m_1^2}} \tag{2.24}$$

for $m_1 \gg m_2$.

For the combination of a massive projectile and light target, we are able to consider two further conditions of the projectile momentum. For the case of the projectile having a high incident three-vector momentum such that $p_1 \gg (m_1^2/m_2)$, this result reduces considerably to,

$$T'_{2,max} \approx p_1 \\ \approx T_1 \quad \text{for } m_1 \gg m_2, p_1 \gg \left(m_1^2/m_2\right). \tag{2.25}$$

Thus, even though the projectile is much more massive than the target, it is possible for all of the projectile's kinetic energy to be transferred to the target should the incident projectile be relativistic. On the other hand, considering the case of a low incident three-vector momentum $p_1 \ll (m_1^2/m_2)$, for which the above result becomes,

$$T'_{2,max} \approx 2m_2\left(\frac{p_1}{m_1}\right)^2 \\ = 2m_2\gamma_1^2\beta_1^2 \quad \text{for } m_1 \gg m_2, p_1 \ll \left(m_1^2/m_2\right). \tag{2.26}$$

Projectile and Target of Equal Masses

The most frequent case in nuclear medicine dosimetry is that of the incident and target particles having equal masses, such as the combination of an electron or positron projectile and an atomic electron target. Although, in this case, the target will be moving and be bound to the atom, we will neglect these conditions as would be acceptable if the projectile is sufficiently energetic that we can treat the atomic electron as being at rest relative to the projectile. In such a case, the recoil kinetic energy can be rewritten as,

$$T'_{2,max} = 2m\frac{p_1^2}{2m^2 + 2m\sqrt{p_1^2 + m^2}} \tag{2.27}$$

where $m_1 = m_2 \equiv m$. Further manipulation leads to,

$$T'_{2,max} = \frac{p_1^2}{m + \sqrt{p_1^2 + m^2}} \\ = \frac{p_1^2}{m + E_1} \\ = \frac{T_1(T_1 + 2m)}{T_1 + 2m} \\ \approx T_1 \quad \text{for } T_1 \gg m_1, m_2;\ m_1 = m_2. \tag{2.28}$$

Again, in the ultrarelativistic limit for the incident kinetic energy greater than the rest masses of projectile and target, the maximum recoil kinetic energy is equal to that of the incident projectile. However, for the case of an electron projectile and an electron target, special considerations must be taken. In such an extreme collision, the projectile electron is completely stopped and the target electron carries off the total kinetic energy. But a subtle facet to this collision exists and requires one to be cognizant of the indistinguishability between the projectile and target electrons. This inability to differentiate between the two electrons leads to the obvious question of how an observer can tell the difference between when the projectile electron transfers none or all of its energy to the target electron. In both cases, one observes a final state with one electron at rest and another with kinetic energy equal to that incident. This indistinguishability between the electrons leads to the convention that the maximum kinetic energy of the recoil electron in the relativistic region is set to $T_{1/2}$ and the assumption that the electron with the highest kinetic energy in the final state is the projectile electron.

2.4.3 Kinetic Energy of the Scattered Projectile

There will also be interest in the kinetic energy of the scattered projectile which, in terms of radiation transport, provides us with the knowledge of how much energy remains to be transferred to the medium at a distance from the collision. The spatial resolution explored by a scattering reaction will be inversely proportional to the bombarding particle's momentum, as follows from the reduced de Broglie wavelength $\lambda\!\!\!^{-} = \hbar c/p$. For example, an electron with a kinetic energy of 1 GeV has a de Broglie wavelength of about 0.2 fm, which is of the order of 10% of the nuclear dimension making such high-energy electrons useful for probing the nucleus and elucidating understanding of its size and spatial structure, as will be shown in Chap. 3.

To evaluate the scattered projectile's kinetic energy,

$$\begin{aligned} p_1 p_2 &= p'_1 p'_2 \\ &= p'_1 (p_1 + p_2 - p'_1) \\ &= p'_1 p_1 + p'_1 p_2 - m_1^2. \end{aligned} \tag{2.29}$$

Expanding,

$$E_1 m_2 = E'_1 E_1 - p'_1 p_1 \cos\theta + E'_1 m_2 - m_1^2. \tag{2.30}$$

Assume that the projectile is an electron with a de Broglie wavelength smaller than nuclear dimensions thus requiring the electron to be highly relativistic or,

$$E_1 m_2 \approx E'_1 E_1 + E'_1 m_2 - E'_1 E_1 \cos\theta. \tag{2.31}$$

Rearranging and solving for the scattered projectile's total energy,

$$E'_1(\theta) = \frac{E_1}{1 + \left(\frac{E_1}{m_2}\right)(1 - \cos\theta)}. \tag{2.32}$$

Figure 2.3 shows the ratio of the kinetic energy of the scattered projectile to that prior to the interaction for an electron projectile (E_1) incident to light (^1H) and heavy (^{208}Pb) nuclear targets (m_2) as a function of scattering angle for 1 and 10 GeV electrons, the combinations satisfying the above kinematic requirements. For a head-on scatter of $\theta = 0$, the scattered energy equals that of that incident, a result independent of the mass of the target should the projectile be relativistic.

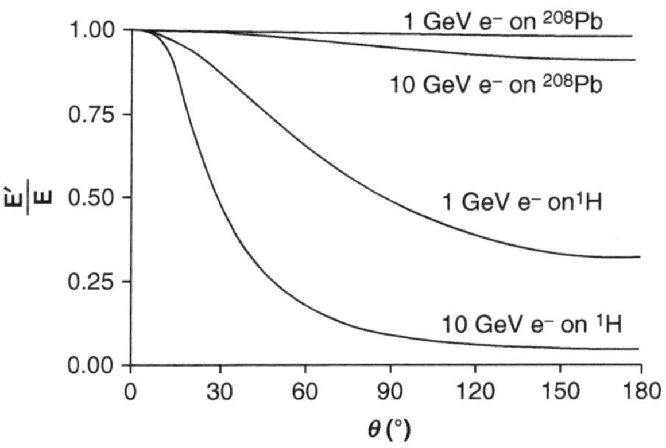

Fig. 2.3 Ratio of the kinetic energy of an elastically-scattered electron to that incident for 1 and 10 GeV electrons incident to hydrogen and lead targets

No energy is transferred to the target as a result, which is of little interest. However, the differences in the angular variations of the scattered energy with scattering angle for a relativistic projectile as functions of projectile energy and target mass are certainly of interest and reflect the role of the target mass. For the two cases of the two electron energies of 1 and 10 GeV incident to the ^{208}Pb target, the target mass is much greater than the total incident energy $m_2 \gg E_1$ ($m_2 \approx 208$ GeV) and the functional dependence upon the scattering angle is suppressed due to the E_1/m_2 factor in the denominator. In other words, for the heavy Pb target, relatively little kinetic energy is taken up by the recoil nucleus, $E'_e(\theta)/E_e \approx 1$ over all scattering angles, and there is little difference in this feature between 1 and 10 GeV electrons. On the other hand, there is a greater dependence of the scattered projectile energy upon scattering angle for the ultrarelativistic electron incident to the relatively light target of a proton. This is indicative of the greater propensity of the light target nucleus to acquire a greater amount of the projectile's kinetic energy.

2.5 Time-Dependent Perturbation Theory

2.5.1 Introduction

In this book, frequent examinations of conditions under which a given quantum state makes a transition to another distinguishable quantum state will be made. The most pertinent examples will be the transition of a radioactive nucleus from one state to another. In such cases, the energy differential between the pre- and posttransition states is manifest in the emission of an energetic particulate radiation (α particles, electrons or positrons), or electromagnetic radiation (photons), which are the essence of nuclear medicine. Fundamental to the study of such transitions is the ability to calculate the probability per unit time of these transitions occurring. For the examples to be considered in this book, this probability (i.e., the transition rate) is calculated using time-dependent perturbation theory.

2.5.2 Transition Rate

In this section, we will derive the transition rate which is the probability per unit time for a given transition to occur. Consider a basic system which is initially in steady-state but then perturbed by a small time-dependent potential. This perturbation induces the quantum system to a new state and we wish to calculate the probability with which this can occur. The calculation begins with the definition of the total Hamiltonian of the system as,

$$H = H_0 + \lambda U(t) \tag{2.33}$$

where H_0 is the steady-state Hamiltonian and U(t) is the time-dependent perturbation noted above. For calculational purposes, a dimensionless weighting parameter λ is explicitly applied to the perturbation as it will be used later in a power-series expansion in this derivation. We now consider Schrödinger's description of this quantum state. Schrödinger's equation for the steady-state component (i.e., without the perturbation) is,

$$i\hbar\frac{\partial\Psi_0}{\partial t} = H_0\Psi_0 \tag{2.34}$$

for which the solution can be written as,

$$\Psi_0 = \sum_k a_{k,0}\psi_k e^{-i\frac{E_k t}{\hbar}} \tag{2.35}$$

where $a_{k,0}$ are constants. As the eigenstates ψ_k form a complete set, the solution for the complete Hamiltonian (i.e., for the combined steady and perturbed state) is,

$$\Psi = \sum_k a_k(t)\psi_k e^{-i\frac{E_k t}{\hbar}} \tag{2.36}$$

where,

$$a_k(t=0) \equiv a_{k,0} \tag{2.37}$$

The coefficients are such that $|a_k(t)|^2$ represents the probability of finding the system in the kth-state at time t. These coefficients can be found by substituting the wavefunction into the Schrödinger equation to give,

$$i\hbar\left[\sum_k\left(\frac{da_k(t)}{dt}\psi_k e^{-i\frac{E_k t}{\hbar}} - i\frac{E_k}{\hbar}a_k(t)\psi_k e^{-i\frac{E_k t}{\hbar}}\right)\right] = (H_0 + \lambda U(t))\sum_k a_k(t)\psi_k e^{-i\frac{E_k t}{\hbar}} \tag{2.38}$$

Isolating the term associated with the free Hamiltonian from that of the perturbation gives

$$\sum_k a_k(t)E_k\psi_k e^{-i\frac{E_k t}{\hbar}} = H_0\sum_k a_k(t)\psi_k e^{-i\frac{E_k t}{\hbar}} \tag{2.39}$$

$$i\hbar\sum_k\frac{da_k(t)}{dt}\psi_k e^{-i\frac{E_k t}{\hbar}} = \lambda\sum_k U(t)a_k(t)\psi_k e^{-i\frac{E_k t}{\hbar}}. \tag{2.40}$$

Ignore the trivial result of (2.39) and instead examine the effect of the perturbation upon the system as described by (2.40). For ease of presentation, the Dirac notation is used so that by multiplying both sides of (2.40) by the bra $\langle\psi_f|$ and by the ket $|\psi_i\rangle$ and integrating over space, gives,

$$i\hbar\sum_k\int d^3\mathbf{r}\frac{da_k(t)}{dt}\langle\psi_f|\psi_k\rangle e^{-i\frac{E_k t}{\hbar}} = \lambda\sum_k\int d^3\mathbf{r}\langle\psi_f|U(t)|\psi_k\rangle a_k(t)e^{-i\frac{E_k t}{\hbar}} \tag{2.41}$$

Using the orthonormality of states, $\langle\psi_n|\psi_m\rangle = \delta_{mn}$, we have

$$i\hbar\frac{da_f}{dt}e^{-i\frac{E_f t}{\hbar}} = \lambda\sum_k U_{fk}(t)a_k(t)e^{-i\frac{E_k t}{\hbar}}. \tag{2.42}$$

where the matrix element is simplified to,

$$U_{fk}(t) \equiv \langle\psi_f|U(t)|\psi_k\rangle \tag{2.43}$$

Rearranging this result,

$$\frac{da_f}{dt} = -i\frac{\lambda}{\hbar}\sum_k U_{fk}(t)a_k(t)e^{-i\omega_{fk}t} \tag{2.44}$$

where,

$$\omega_{fk} \equiv \frac{E_f - E_k}{\hbar}. \tag{2.45}$$

Essential to this derivation is the assumption that the strength of the perturbation is weak so as to allow the coefficients $a_k(t)$ to be expanded in a power-series in λ,

$$a_k(t) = \sum_{m=0}^{\infty} \lambda^m a_{k,m}(t). \tag{2.46}$$

Substituting this expansion into (2.44) gives,

$$\begin{aligned}&\sum_{m=0}^{\infty} \lambda^m \frac{da_{f,m}(t)}{dt}\\ &= -\frac{i}{\hbar}\sum_{m=0}^{\infty}\sum_k U_{fk}(t)\lambda^{m+1}a_{k,m}(t)e^{i\omega_{fk}t}.\end{aligned} \tag{2.47}$$

Then, by equating coefficients of equal powers of λ,

$$\frac{da_{f,0}(t)}{dt} = 0 \tag{2.48}$$

and

$$\frac{da_{f,m}(t)}{dt} = -\frac{i}{\hbar}\sum_k U_{fk}(t)e^{i\omega_{fk}t}a_{k,m-1}; \quad m \geq 1. \tag{2.49}$$

In order to use these results to calculate the coefficients, assume that the perturbing potential is "switched on" at time $t = 0$ and remains constant for $t > 0$,

$$U(t) = U_0 H(t) \tag{2.50}$$

where $H(t)$ is the Heaviside function. Further assume that the system is in a single and well-defined state $|\psi_i\rangle$ before the potential is "switched" on,

$$|a_{i,0}|^2 = 1. \tag{2.51}$$

The remaining coefficients at time t are then found by integrating (2.49) where, as λ has been previously defined as small, the calculation need only be limited to first order $m = 1$,

$$\begin{aligned}a_{f,1}(t) &= -\frac{i}{\hbar}\int_0^t dt' M_{fi} e^{i\omega_{fi}t'}\\ &= -i\frac{M_{fi}}{\hbar}\int_0^t dt' e^{i\omega_{fi}t'}\\ &= \frac{M_{fi}}{\hbar}\left(\frac{1 - e^{i\omega_{fi}t}}{\omega_{fi}}\right)\end{aligned} \tag{2.52}$$

where,

$$M_{fi} = \langle\psi_f|\psi_i\rangle U_0. \tag{2.53}$$

The probability that a transition will be made from state $|\psi_i\rangle$ to state $|\psi_f\rangle$ at time t is the squared modulus of (2.52),

$$\begin{aligned}P_{fi}(t) &= |a_{f,1}(t)|^2\\ &= 2\frac{|M_{fi}|^2}{\hbar^2}\frac{(1 - \cos\omega_{fi}t)}{\omega_{fi}^2}\\ &= 2\frac{|M_{fi}|^2}{\hbar^2}B(\omega_{fi}t).\end{aligned} \tag{2.54}$$

The function $B(\omega,t)$, which is plotted in Fig. 2.4, is sharply peaked at $\omega = 0$ and the sharpness of this peak increases with t. Thus the probability of a transition is greater for a reduced energy difference between the two states.

In many cases of interest, there will in fact be an ensemble of neighboring states around $|\psi_f\rangle$ with energies around E_f. The transition probability is then determined by integrating over this ensemble of energies,

$$P_{fi}(t) = \frac{2}{\hbar^2}\int_{-\infty}^{\infty} dE_{f'}|M_{f'i}|^2 B(\omega_{f'i}, t)\rho_{f'} \tag{2.55}$$

where the number density of states is,

$$\rho_{f'} = \frac{dN}{dE_{f'}}(E_{f'}). \tag{2.56}$$

This can be integrated simply if it is assumed that the matrix element $M_{f'i}$ and the density of states are

Fig. 2.4 The function, $B(\omega, t) = \frac{1-\cos\omega t}{\omega^2}$

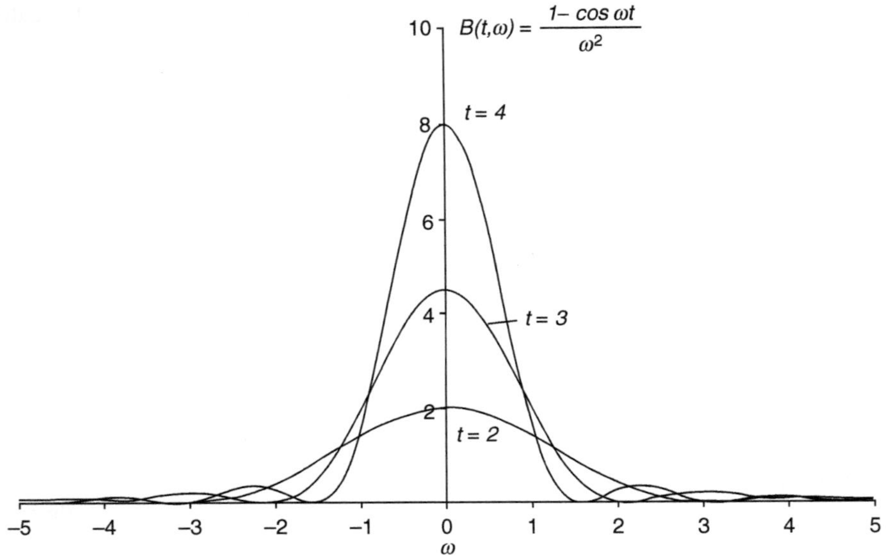

reasonably constant over the region where $B(\omega_{f'i},t)$ is significant, thus allowing them to be extracted from the integral,

$$P_{fi}(t) = \frac{2}{\hbar^2}|M_{fi}|^2\rho_f\int_{-\infty}^{\infty}dE_{f'}B(\omega_{f'i},t) = \frac{2}{\hbar}|M_{fi}|^2\rho_f\int_{-\infty}^{\infty}d\omega_{f'}\frac{(1-\cos\omega_{f'i}t)}{\omega_{f'i}^2}. \quad (2.57)$$

The integral is solved by the substitution of the variable, $y = \omega_{f'} - \omega_i$ to give

$$\int_{-\infty}^{\infty}d\omega_{f'}\frac{(1-\cos\omega_{f'i}t)}{\omega_{f'i}^2} = \int_{-\infty}^{\infty}dy\frac{(1-\cos yt)}{y^2}. \quad (2.58)$$

It is straightforward to solve this integral by parts to yield,

$$\int_{-\infty}^{\infty}dy\frac{(1-\cos yt)}{y^2} = 2t\int_{0}^{\infty}dy\,\frac{\sin yt}{y} \quad (2.59)$$

where this integral is the sine integral, $Si(x) = \int_0^x dy\,\frac{\sin y}{y}$, with the property $Si(\infty) = \pi/2$. Hence,

$$\int_{-\infty}^{\infty}d\omega_{f'}\frac{(1-\cos\omega_{f'i}t)}{\omega_{f'i}^2} = \pi\,t. \quad (2.60)$$

The probability of the transition is now,

$$P_{fi}(t) = \frac{2\pi\,t}{\hbar}|M_{fi}|^2\rho_f. \quad (2.61)$$

The transition rate is the probability of the transition occurring per unit time,

$$\lambda_{fi} = \frac{dP_{fi}}{dt} \quad (2.62)$$

or,

$$\lambda_{fi} = \frac{2\pi}{\hbar}|M_{fi}|^2\rho_f. \quad (2.63)$$

This result is known as Fermi's Golden Rule Number 2.

Equation (2.63) describes a first-order process and, under some conditions, can yield a zero result. One must account for second-order processes as a result. Consider, for example, the elastic scatter of a photon by a free electron at rest (Compton scatter). There are two scenarios is this interaction. In the first, the primary photon is absorbed by the electron and a secondary photon with a reduced energy (corresponding to that scattered) is emitted, with the energy difference appearing as the electron's recoil kinetic energy. In the second scenario, the secondary photon is emitted by the electron and the primary photon is subsequently absorbed. In order to manage such a combination of processes, (2.63) is rearranged to give,

$$\lambda_{fi} = \frac{2\pi}{\hbar}\left|\sum_{n'}\frac{M_{fn'}\,M_{n'i}}{E_i - E_{n'}}\right|^2 \rho_f \qquad (2.64)$$

where the intermediate state of n' differs from the initial and final states by one photon. Equation (2.64), also known as Fermi's Golden Rule Number 1, allows for the two cases of the absorption of a photon by an electron followed by the emission of a photon or that of the emission of a photon followed by the absorption of a photon.

2.6 Quantum Scattering Theory

2.6.1 Introduction

As shown in the diagram of Fig. 2.1, the absorbed dose is the ultimate consequence of the interactions of particles and radiation fields with matter. Photons, in the form of X or γ rays, interact with atomic electrons and nuclei and are absorbed, scattered or, if sufficiently energetic, generate an electron–positron pair through coupling with the nuclear electromagnetic field. On a more fundamental level, the scattering of charged particles sets additional charged particles moving into the medium. In all of these cases, the outcome of the interaction is that an electron or positron is set into motion. As this charged particle moves through the medium, it loses kinetic energy through interactions with matter, the main ones of interest to us being those collisions with atomic electrons individually, with the ensemble of electrons or through violent deflections from its trajectory by the nuclear Coulomb field resulting in the emission of radiation. It is also possible for the charged particle to interact with matter but without the transfer of energy, e.g., elastic Coulomb scattering. Many of the derivations to be faced will be involve understanding the interactions of photons, electrons, and positrons with the medium that they travel through. This subsection develops quantum-mechanical descriptions of how projectiles (photons or particulate) interact with the medium through which they travel.

A quantum-mechanical Hamiltonian can have both discrete and continuous eigenvalues, unlike the classical Hamiltonian which has only continuous eigenvalues. The discrete eigenvalues correspond to bound states of a system which are consequences of an attractive interaction potential confining the system of particles to within a finite volume, the Schrödinger equation in this case being,

$$H\,\Psi_n = E_n\,\Psi_n \qquad (2.65)$$

with energy eigenvalues E_n and eigenfunctions Ψ_n for the Hermitian Hamiltonian. In such cases, the wavefunctions of the bound particles decrease rapidly with growing inter-particle spacing. Continuous eigenvalues, on the other hand, correspond to unbound, or scattering, states in which the wavefunctions can extend asymptotically to infinity: an incident particle, modeled by a plane wave, is incident to a potential (e.g., the static Coulomb field of a nucleus) and is then scattered by the potential so that the final state is the combination of the incident and scattered wavefunctions. Whereas the discrete eigenfunction problem corresponds to a single and confined system, the continuous eigenfunction problem describes the interactions between two subsystems (here, the projectile and the target) which are not mutually bound states. The kinematics of a collision[4] between a projectile and a target have been derived. These results provide only the energy and momentum of the two particles in the final state once a scattering or recoil angle has been defined. They do not shed any light on the likelihood of such a final state occurring. The scattering problem is that of interacting systems representing asymptotically-free states which must exist not just during the

[4]Collision and scatter have been differentiated by some authors so that the former correspond to multi-channel final states and the latter was linked to a single channel final state. Such a distinction is not made here.

time interval about the collision but also for all time ($t \rightarrow \pm\infty$). As with our previous derivation of perturbation theory, we will treat the Hamiltonian affecting the scattered system as the sum of a free-state Hamiltonian and that of a perturbing potential.

Ultimately, in the solution of the scattering problem, one is required to derive an observable, a physical quantity that can be measured in the laboratory, describing the probability of a given final postscattering state. Here, the observable will be the total cross section of the scatter which is the ratio of the relative decrease, due to scatter, in the number of particles per unit area incident to an ensemble of scattering centers with an areal density of ρ scattering centers per unit area normalized to the areal density.[5] As the total cross section is proportional to the relative decrease in fluence due to scatter, it is a direct physical measure of the scattering probability. There are further refinements, such as differential cross sections in solid angle and in energy, which provide more detailed information of the scattering process.

This subsection provides a review of those aspects of quantum scattering theory which we will require in later solutions of dosimetry problems. This begins with the derivation of the Born approximation, which is, in fact, a different manifestation of the first-order approximation used in the perturbation theory leading to Fermi's Golden Rules. The outcome of this derivation is the scattering amplitude which will be shown to be directly linked to the differential cross section in solid angle. As the scattering amplitude is proportional to the Fourier transform of the interaction potential between projectile and target, the Born approximation provides an easy mechanism to solve the many scattering problems we will encounter later. The phase-shift analysis approach to the scattering problem is then reviewed and, inter alia, the Schrödinger equation is solved in spherical coordinates. This will be used in Chap. 3 in solving the bound state problem of the nuclear shell model. Finally, the phase-shift analysis is extended to derive the optical theorem, which provides a simple relationship between the scattering amplitude and the total cross section.

[5]The areal density of a scattering medium with physical density ρ and thickness *t* is the product ρ*t*.

2.6.2 Scattering Amplitude

Consider the geometry of Fig. 2.5 in which a particle/wavefunction is incident to a potential centered at the origin $U(\mathbf{r})$. Assume that the potential is weak. As a result of the interaction between the wavefunction and this potential, it is deflected from its original trajectory and one is interested in the probability of the particle being scattered through the angle θ into the differential solid angle element dΩ at a distance r from the scattering center. In order to calculate this probability, first consider the case of a free particle (i.e., one not subject to a potential) with mass m and the nonrelativistic Hamiltonian,

$$H_0 = \frac{p^2}{2m} \tag{2.66}$$

The eigenvalue equation is, continuing our use of the Dirac notation,

$$H_0|\psi_0\rangle = E_0|\psi_0\rangle \tag{2.67}$$

where $|\psi_0\rangle$ is the energy eigenfunction of H_0. The wavefunction corresponding to the particle is the scalar product,

$$\psi_0(\mathbf{r}) = \langle \mathbf{r}\,|\psi_0\rangle \tag{2.68}$$

The complete Hamiltonian for the particle when it is in the presence of a potential U is $H_0 + U$ and the corresponding Schrödinger equation is,

$$(H_0 + U)|\psi\rangle = E|\psi\rangle \tag{2.69}$$

where $|\psi\rangle$ is an energy eigenstate of the total Hamiltonian. The scattered wavefunction is given by the scalar product,

$$\psi(\mathbf{r}) = \langle \mathbf{r}\,|\psi\rangle. \tag{2.70}$$

The solution to this Schrödinger equation is,

$$|\psi\rangle = |\psi_0\rangle + \frac{1}{E - H_0 + i\varepsilon}U|\psi\rangle \tag{2.71}$$

where ε is a real, positive small number required to eliminate the singularity that occurs when $E = H_0$. This is an exact solution, known as the Lippman–Schwinger equation, and which can be solved via

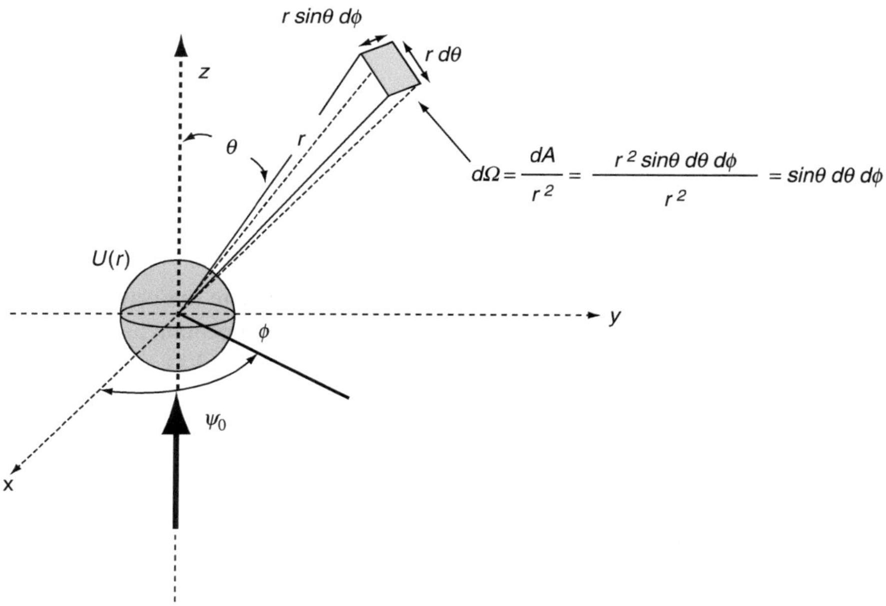

Fig. 2.5 Scattering geometry: the wave function is incident along the *z*-axis to a potential, $U(\mathbf{r})$, centered at the origin. The wave is scattered, producing a spherical outgoing wave, a fraction of which traverses the differential solid angle element, $d\Omega$

perturbation theory. Using a power-series expansion in U leads to,

$$|\psi\rangle = |\psi_0\rangle + \frac{1}{E - H_0 + i\varepsilon} U|\psi_0\rangle + O(U^2). \quad (2.72)$$

The first Born approximation for $|\psi\rangle$ is the truncation at the first order of U, which is possible following our original assumption of a weak interaction potential. It is also possible to derive the Born approximation, and a solution for the scattered wavefunction in coordinate space, by inserting the free Hamiltonian into the Schrödinger equation and using the canonical relationship $\mathbf{p} \equiv -i\hbar c\nabla$, in which the momentum is linked to the gradient. This change results in the Helmholtz equation,

$$\begin{aligned} \left(-\frac{(\hbar c)^2}{2m}\nabla^2 + U\right)\psi(\mathbf{r}) &= E\psi(\mathbf{r}) \\ \left(\nabla^2 + \frac{2mE}{(\hbar c)^2}\right)\psi(\mathbf{r}) &= \frac{2m}{(\hbar c)^2}U\psi(\mathbf{r}) \\ (\nabla^2 + k^2)\psi(\mathbf{r}) &= \Xi\psi(\mathbf{r}) \end{aligned} \quad (2.73)$$

where the wave vector is,

$$k = \frac{\sqrt{2mE}}{\hbar c} \quad (2.74)$$

and,

$$\Xi = \frac{2mU}{(\hbar c)^2}. \quad (2.75)$$

The solution to the Helmholtz equation is,

$$\psi(\mathbf{r}) = \psi_0(\mathbf{r}) + \int d^3\mathbf{r}' G(\mathbf{r}, \mathbf{r}')\langle \mathbf{r}|\Xi|\psi\rangle \quad (2.76)$$

where the free particle wavefunction $\psi_0(\mathbf{r})$ is the solution to the homogeneous equation (i.e., the Schrödinger equation for the free particle),

$$(\nabla^2 + k^2)\psi_0(\mathbf{r}) = 0 \quad (2.77)$$

and $G(\mathbf{r}, \mathbf{r}')$ is the Green's function defined from,

$$(\nabla^2 + k^2)G(\mathbf{r}, \mathbf{r}') = \delta(\mathbf{r} - \mathbf{r}') \quad (2.78)$$

and where $\delta(\mathbf{r})$ is the Dirac delta function. These yield an expression for the Green's function of the form,

$$G(\mathbf{r}, \mathbf{r}') = -\frac{e^{\pm ik|\mathbf{r} - \mathbf{r}'|}}{4\pi|\mathbf{r} - \mathbf{r}'|} \quad (2.79)$$

and the expression for the scattered wavefunction is,

$$\psi^{\pm}(\mathbf{r}) = \psi_0(\mathbf{r}) - \int d^3\mathbf{r}' \frac{e^{\pm ik|\mathbf{r}-\mathbf{r}'|}}{4\pi|\mathbf{r}-\mathbf{r}'|}\langle \mathbf{r}'|\Xi|\psi\rangle \quad (2.80)$$

where the $\pm$ subscript has been added in order to reflect the sign of the exponent in the Green's function, the physical interpretation of which will be discussed shortly. From the definition of Ξ, the matrix element contained within the integrand is written as,

$$\begin{aligned}\psi^{\pm}(\mathbf{r}) &= \psi_0(\mathbf{r}) - \int d^3\mathbf{r}' \frac{e^{\pm ik|\mathbf{r}-\mathbf{r}'|}}{4\pi|\mathbf{r}-\mathbf{r}'|}\langle \mathbf{r}'|\Xi|\psi\rangle \\ &= \psi_0(\mathbf{r}) - \frac{2m}{(\hbar c)^2}\int d^3\vec{r}' \frac{e^{\pm ik|\mathbf{r}-\mathbf{r}'|}}{4\pi|\mathbf{r}-\mathbf{r}'|} \\ &\quad \times \langle \mathbf{r}'|U(\mathbf{r})\delta(\mathbf{r}-\mathbf{r}')|\psi\rangle \\ &= \psi_0(\mathbf{r}) - \frac{m}{2\pi(\hbar c)^2}\int\int d^3\mathbf{r}''d^3\mathbf{r}' \frac{e^{\pm ik|\mathbf{r}-\mathbf{r}'|}}{|\mathbf{r}-\mathbf{r}'|} \\ &\quad \times \langle \mathbf{r}'|U(\mathbf{r})\delta(\mathbf{r}-\mathbf{r}')|\mathbf{r}''\rangle\langle \mathbf{r}''|\psi\rangle \\ &= \psi_0(\mathbf{r}) - \frac{m}{2\pi(\hbar c)^2} \\ &\quad \times \int d^3\mathbf{r}' \frac{e^{\pm ik|\mathbf{r}-\mathbf{r}'|}}{|\mathbf{r}-\mathbf{r}'|}U(\mathbf{r}')\psi(\mathbf{r}'). \end{aligned} \quad (2.81)$$

To further the evaluation of the matrix element, the assumption that the scattering potential is weak can be modified by requiring it to be non-zero only in the vicinity of the origin. Then consider this wavefunction only at large distances from the scattering center (i.e., $r \gg r'$) in which case, the r-dependence of the denominator and exponent can be approximated by,

$$\begin{aligned}|\mathbf{r}-\mathbf{r}'| &= r\sqrt{1+\left(\frac{r'}{r}\right)^2 - 2\left(\frac{r'}{r}\right)\cos\theta} \\ &\approx r\left(1-\frac{r'}{r}\cos\theta\right) \\ &\approx r - \hat{\mathbf{r}}\bullet\mathbf{r}'\end{aligned} \quad (2.82)$$

where $\hat{\mathbf{r}}$ is the unit vector in the direction of $\mathbf{r}$. As a result, the wavefunction expression becomes, for large r,

$$\begin{aligned}\psi^{\pm}(\mathbf{r}) &= \psi_0(\mathbf{r}) - \frac{m}{2\pi(\hbar c)^2}\int d^3\mathbf{r}' \frac{e^{\pm ik(r-\hat{\mathbf{r}}\bullet\mathbf{r}')}}{r-\hat{\mathbf{r}}\bullet\mathbf{r}'}U(\mathbf{r}')\psi(\mathbf{r}') \\ &= \psi_0(\mathbf{r}) - \frac{m}{2\pi(\hbar c)^2}\int d^3\mathbf{r}' \frac{e^{\pm ikr}e^{\mp ik\hat{\mathbf{r}}\bullet\mathbf{r}'}}{r}U(\mathbf{r}')\psi(\mathbf{r}') \\ &= \psi_0(\mathbf{r}) - \frac{m}{2\pi(\hbar c)^2}\frac{e^{\pm ikr}}{r}\int d^3\mathbf{r}' e^{\mp i\mathbf{k}'\bullet\mathbf{r}'}U(\mathbf{r}')\psi(\mathbf{r}'). \end{aligned} \quad (2.83)$$

where the wave vector $\mathbf{k}'$ (for particles with the same momentum as those incident but scattered along $\hat{\mathbf{r}}$) is defined as,

$$\begin{aligned}\mathbf{k}' &= |\mathbf{k}|\hat{\mathbf{r}} \\ &= k\hat{\mathbf{r}}.\end{aligned} \quad (2.84)$$

If the incident particle were to be treated as a plane wave,

$$\psi_0(\mathbf{r}) = e^{i\mathbf{k}\bullet\mathbf{r}} \quad (2.85)$$

then the wavefunction can be rewritten as,

$$\begin{aligned}\psi^{\pm}(\mathbf{r}) &= e^{i\mathbf{k}\bullet\mathbf{r}} - \frac{m}{2\pi(\hbar c)^2}\frac{e^{\pm ikr}}{r} \\ &\quad \times \int d^3\mathbf{r}' e^{\mp i\mathbf{k}'\bullet\mathbf{r}'}U(\mathbf{r}')\psi(\mathbf{r}'). \end{aligned} \quad (2.86)$$

The first term is simply the incident plane wave and the second term corresponds to spherical waves propagating away from the scattering center ($\psi^+(\mathbf{r})$) and towards the scattering center ($\psi^-(\mathbf{r})$). Clearly, the latter solution is nonphysical and will be ignored leaving the physical solution for the wavefunction at large $\mathbf{r}$,

$$\begin{aligned}\psi(\mathbf{r}) &= e^{i\mathbf{k}\bullet\mathbf{r}} - \frac{m}{2\pi(\hbar c)^2}\frac{e^{ikr}}{r} \\ &\quad \times \int d^3\mathbf{r}' e^{-i\mathbf{k}'\bullet\mathbf{r}'}U(\mathbf{r}')\psi(\mathbf{r}') \end{aligned} \quad (2.87)$$

which is the summation of the incident wave and spherical waves emitted from the scattering center, as shown in Fig. 2.6. From the assumption that the potential $U(\mathbf{r})$ is negligible at large $\mathbf{r}$, it is possible to approximate the $\psi(\mathbf{r}')$ in the integrand by the incident wavefunction $\psi_0(\mathbf{r}')$,

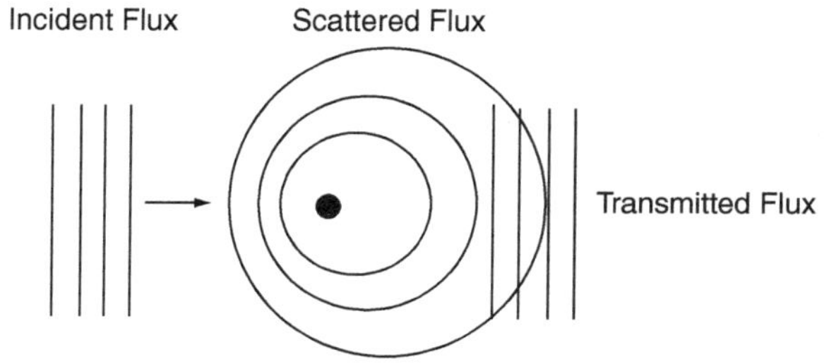

Fig. 2.6 Wave mechanical description of scattering

$$\begin{aligned}\psi(\mathbf{r}) &= e^{i\mathbf{k}\bullet\mathbf{r}} - \frac{m}{2\pi(\hbar c)^2}\frac{e^{ikr}}{r}\int d^3\mathbf{r}' e^{-i\mathbf{k}'\bullet\mathbf{r}'}U(\mathbf{r}')e^{i\mathbf{k}\bullet\mathbf{r}'} \\ &= e^{i\mathbf{k}\bullet\mathbf{r}} - \frac{m}{2\pi(\hbar c)^2}\frac{e^{ikr}}{r}\int d^3\mathbf{r}' e^{i(\mathbf{k}-\mathbf{k}')\bullet\mathbf{r}'}U(\mathbf{r}')\end{aligned} \tag{2.88}$$

which is the first Born approximation. By writing the momentum transfer as,

$$\mathbf{q} = \hbar c(\mathbf{k} - \mathbf{k}') \tag{2.89}$$

the wavefunction can be written in the simpler form,

$$\begin{aligned}\psi(\mathbf{r}) &= e^{i\mathbf{k}\bullet\mathbf{r}} - \frac{m}{2\pi(\hbar c)^2}\frac{e^{ikr}}{r}\int d^3\mathbf{r}' e^{i\frac{\mathbf{q}\bullet\mathbf{r}'}{\hbar c}}U(\mathbf{r}') \\ &= e^{i\mathbf{k}\bullet\mathbf{r}} + f(\mathbf{q})\frac{e^{ikr}}{r}.\end{aligned} \tag{2.90}$$

The first term is, again, the incident plane wave and the second term is the scattered spherical wave with a magnitude specified by $f(\mathbf{q})$, which is known as the scattering amplitude. The scattering amplitude is proportional to the Fourier transform of the scattering potential,

$$f(\mathbf{q}) = -\frac{m}{2\pi(\hbar c)^2}\int d^3\mathbf{r}' e^{i\frac{\mathbf{q}\bullet\mathbf{r}'}{\hbar c}}U(\mathbf{r}'). \tag{2.91}$$

Most of the scattering problems to be encountered in this book reduce to the calculation of the scattering amplitude as it will be shown to be related to the physical observable of the differential cross section in solid angle.

For example, the potentials that will be encountered here are central or, in other words, azimuthally symmetric. In this case, the integral of (2.91) is,

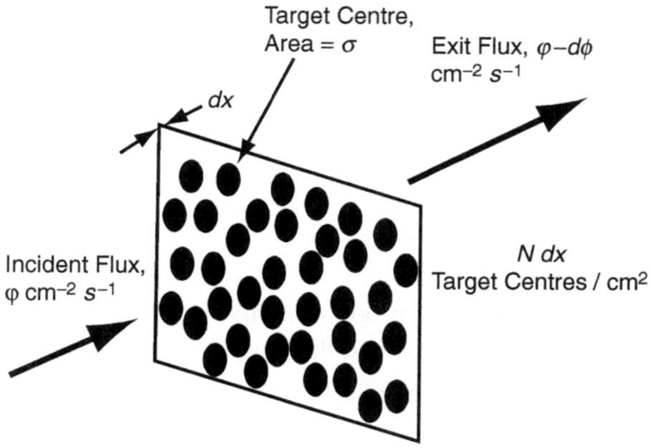

Fig. 2.7 Definition of the reaction total cross section

$$\begin{aligned}\int d^3\mathbf{r}' e^{i\frac{\mathbf{q}\bullet\mathbf{r}'}{\hbar c}}U(\mathbf{r}') &= 2\pi\int_0^\infty dr' r'^2 U(r')\int_{-1}^{1} d(\cos\theta)e^{i\frac{qr\cos\theta}{\hbar c}} \\ &= 4\pi\frac{\hbar c}{q}\int_0^\infty dr' r' \sin\left(\frac{qr'}{\hbar c}\right)U(r')\end{aligned} \tag{2.92}$$

giving the scattering amplitude as,

$$f(\mathbf{q}) = -\frac{2m}{q\hbar c}\int_0^\infty dr' r' \sin\left(\frac{qr'}{\hbar c}\right)U(r'). \tag{2.93}$$

Note that the dimension of the scattering amplitude is that of *length*.

2.6.3 Scattering Cross Sections

Next, consider the physical observables associated with scattering, the differential and total cross sections. The cross section is a quantitative measure of the complete likelihood of a given interaction occurring between a bombarding beam and a target. An initial intuitive description is given by Fig. 2.7 which shows a flux φ of particles per square centimeter per second incident orthogonally to a medium of differential thickness dx containing N "target centers" per cubic centimeter (with an areal density of N dx target centers per square centimeter). Assume that only one type of interaction can occur between a bombarding particle and a target center. The interactions between the incident particles and the target centers will cause a differential flux reduction, $d\varphi$, proportional to the incident flux and the areal density of target centers. The relative flux reduction is,

$$\frac{d\varphi}{\varphi} = -\sigma N dx. \tag{2.94}$$

The constant of proportionality, σ, between the relative flux reduction and the areal density of scattering centers has dimensions of area and is referred to as the reaction total cross section per target center. As each target center has a cross-sectional area σ, the fraction of the total area covered by the target centers is $\sigma N\,dx$ and any of the bombarding particles that hit these areas are removed from the beam. Integrating over the target thickness L (and neglecting any other changes over this thickness, such as energy loss) leads to the exponential attenuation of the beam through the medium,

$$\varphi = \varphi_0\, e^{-\sigma NL} \tag{2.95}$$

where φ_0 is the incident flux.

Reaction cross sections in differential form will be required in most of our calculations. For example, consider a process in which an incident particle is scattered by a target center into the direction (θ,ϕ). The differential flux fraction scattered into the differential solid angle element $d\Omega$ centered on the direction (θ,ϕ) is,

$$\frac{d\varphi}{\varphi} = \frac{d\sigma(\theta,\phi)}{d\Omega} N\, dx\, d\Omega. \tag{2.96}$$

This differential cross section in solid angle is the fractional loss of fluence per areal density per unit solid angle. Similarly, the differential flux fraction within the energy interval between E and E + dE following an interaction is,

$$\frac{d\varphi}{\varphi} = \frac{d\sigma(E)}{dE} N\, dx\, dE. \tag{2.97}$$

The total cross section, σ, is obtained by integrating $d\sigma/d\Omega$ over 4π steradians or by integrating $d\sigma/dE$ over all energies.

The differential cross section with solid angle, $d\sigma/d\Omega$, can be readily shown to be related to the scattering amplitude of the Born approximation due to its definition as the ratio of the number of particles scattered per unit time into the differential solid angle element $d\Omega$ to the number of incident particles per unit time per unit area, or,

$$\begin{aligned}\frac{d\sigma}{d\Omega} &= \frac{r^2 |f(\mathbf{q})|^2 \left|\frac{e^{ikr}}{r}\right|^2}{|e^{i\mathbf{k}\bullet\mathbf{r}}|^2} \\ &= |f(\mathbf{q})|^2.\end{aligned} \tag{2.98}$$

2.6.4 Phase-Shift Analysis

A description of scattering based upon phase shifts of the wavefunctions induced as a result of the interaction with the potential is now provided. Recall from above that, in the Born approximation, the final state is the sum of the incident and scattered wavefunctions. As the spherical coordinate system is more natural for describing the scattering problem, we shall recast this description by expressing the incident plane wave in terms of spherical waves and solve the Schrödinger equation in a region near the scattering center where the potential is non-zero by writing the Laplacian in spherical coordinates

$$\left[\frac{1}{r^2}\frac{\partial}{\partial r}\left(r^2\frac{\partial\psi}{\partial r}\right) + \frac{1}{r^2\sin\theta}\frac{\partial}{\partial\theta}\left(\sin\theta\frac{\partial\psi}{\partial\theta}\right) + \frac{1}{r^2\sin^2\theta}\frac{\partial^2\psi}{\partial\phi^2}\right] + (k^2 - \Xi)\psi = 0 \tag{2.99}$$

where k and Ξ are given above. The wavefunction is solved by first separating it into radial and angular terms,

$$\Psi(r,\theta,\phi) = R(r)Y(\theta,\phi) \tag{2.100}$$

and then substituting the expression into the Schrödinger equation,

$$\frac{Y(\theta,\phi)}{r^2}\frac{d}{dr}\left(r^2\frac{dR}{dr}\right)+\frac{R(r)}{r^2\sin\theta}\frac{\partial}{\partial\theta}\times\left(\sin\theta\frac{\partial Y}{\partial\theta}\right)+\frac{R(r)}{r^2\sin^2\theta}\frac{\partial^2 Y}{\partial\phi^2}+\left(k^2-\Xi\right)R(r)Y(\theta,\phi)=0 \quad (2.101)$$

The radial and angular terms can be isolated by multiplying through by the ratio $r^2/R(r)Y(\theta,\phi)$,

$$\frac{1}{R(r)}\frac{d}{dr}\left(r^2\frac{dR}{dr}\right)+\frac{1}{Y(\theta,\phi)\sin\theta}\frac{\partial}{\partial\theta}\left(\sin\theta\frac{\partial Y}{\partial\theta}\right)+\frac{1}{Y(\theta,\phi)\sin^2\theta}\frac{\partial^2 Y}{\partial\phi^2}+r^2\left(k^2-\Xi\right)=0 \quad (2.102)$$

or,

$$\frac{1}{R(r)}\frac{d}{dr}\left(r^2\frac{dR}{dr}\right)+r^2\left(k^2-\Xi\right)=-\left(\frac{1}{Y(\theta,\phi)\sin\theta}\frac{\partial}{\partial\theta}\left(\sin\theta\frac{\partial Y}{\partial\theta}\right)+\frac{1}{Y(\theta,\phi)\sin^2\theta}\frac{\partial^2 Y}{\partial\phi^2}\right). \quad (2.103)$$

The right- and left-hand sides must both equal the same constant. As the radial equation will be linked to the Bessel equation, this constant will be taken to be $l(l+1)$ where l is an integer equal to 0, 1, 2...,

$$\frac{1}{R(r)}\frac{d}{dr}\left(r^2\frac{dR}{dr}\right)+r^2\left(k^2-\Xi\right)-l(l+1)=0 \quad (2.104)$$

$$\frac{1}{\sin\theta}\frac{\partial}{\partial\theta}\left(\sin\theta\frac{\partial Y}{\partial\theta}\right)+\frac{1}{\sin^2\theta}\frac{\partial^2 Y}{\partial\phi^2}+l(l+1)Y(\theta,\phi)=0. \quad (2.105)$$

The radial equation is first solved for. Following differentiation and rearrangement,

$$r^2\frac{d^2R}{dr^2}+2r\frac{dR}{dr}+\left[\left(k^2-\Xi\right)r^2-l(l+1)\right]R=0$$

$$r^2\frac{d^2R}{dr^2}+2r\frac{dR}{dr}+\left[\lambda^2r^2-l(l+1)\right]R=0 \quad (2.106)$$

where

$$\lambda\equiv\sqrt{k^2-\Xi}=\frac{\sqrt{2m(E-U)}}{\hbar c}. \quad (2.107)$$

By defining $\rho=\lambda r$, (2.106) reduces to the Bessel equation,

$$\rho^2\frac{d^2R}{d\rho^2}+2\rho\frac{dR}{d\rho}+\left[\rho^2-l(l+1)\right]R=0. \quad (2.108)$$

The solution for this equation for a given value of l is the weighted sum of the spherical Bessel functions of the first and second kind,

$$R_l(\rho)=A_lj_l(\rho)+B_ly_l(\rho)$$

or

$$R_l(r)=A_lj_l(\lambda r)+B_ly_l(\lambda r) \quad (2.109)$$

where $j_l(x)$ is the spherical Bessel function of the first kind and $y_l(x)$ is the spherical Bessel function of the second kind (also referred to as a Neumann function) and which are related to the Bessel functions via,

$$j_l(x)=\sqrt{\frac{\pi}{2x}}J_{l+\frac{1}{2}}(x) \quad (2.110)$$

$$y_l(x)=\sqrt{\frac{\pi}{2x}}Y_{l+\frac{1}{2}}(x). \quad (2.111)$$

These functions are plotted in Fig. 2.8 for $l=0$, 1, and 2. From the asymptotic behavior of these spherical Bessel functions, the radial function is,

$$R_l(r)=\frac{1}{\lambda r}\left(A_l\sin\left(\lambda r-\frac{l\pi}{2}\right)-B_l\cos\left(\lambda r-\frac{l\pi}{2}\right)\right)=\frac{\sqrt{A_l^2+B_l^2}}{\lambda r}\left(\frac{A_l}{\sqrt{A_l^2+B_l^2}}\sin\left(\lambda r-\frac{l\pi}{2}\right)-\frac{B_l}{\sqrt{A_l^2+B_l^2}}\cos\left(\lambda r-\frac{l\pi}{2}\right)\right) \quad (2.112)$$

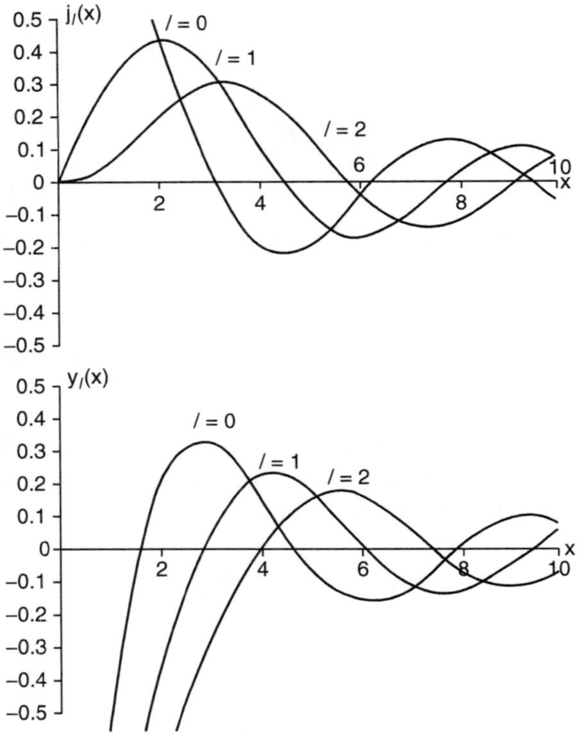

Fig. 2.8 Spherical Bessel functions of the first and second kind for $l = 0$, 1, and 2

The integration constants, A_l and B_l, will be set by the boundary conditions. From the trigonometric identity, $\sin(\alpha - \beta) = \sin\alpha \sin\beta - \cos\alpha \cos\beta$, (2.112) can be written as,

$$R_l(r) = \frac{C_l}{kr} \sin\left(\lambda r - \frac{l\pi}{2} + \delta_l\right). \tag{2.113}$$

The phase shift for the lth partial wave is,

$$\delta_l = -\tan^{-1}\frac{B_l}{A_l} \tag{2.114}$$

and

$$C_l = \sqrt{A_l^2 + B_l^2}. \tag{2.115}$$

For completeness, we solve for the angular component of the wavefunction expression by differentiating and expanding to give,

$$\frac{\partial^2 Y}{\partial\theta^2} + \cot\theta\frac{\partial Y}{\partial\theta} + \frac{1}{\sin^2\theta}\frac{\partial^2 Y}{\partial\phi^2} + l(l+1)Y(\theta,\phi) = 0 \tag{2.116}$$

and, then, by using another separation of variables, $Y(\theta,\phi) = \Theta(\theta)\Phi(\phi)$,

$$\Phi(\phi)\frac{d^2\Theta(\theta)}{d\theta^2} + \Phi(\phi)\cot\theta\frac{d\Theta(\theta)}{d\theta} + \frac{\Theta(\theta)}{\sin^2\theta}\frac{d^2\Phi(\phi)}{d\phi^2} + l(l+1)\Theta(\theta)\Phi(\phi) = 0$$

Finally, by multiplying through by $\sin^2\theta$ and dividing by $\Theta(\theta)\Phi(\phi)$, and isolating the angular terms,

$$\sin^2\theta\left(\frac{1}{\Theta(\theta)}\frac{d^2\Theta(\theta)}{d\theta^2} + \frac{\cot\theta}{\Theta(\theta)}\frac{d\Theta(\theta)}{d\theta} + l(l+1)\right) = -\frac{1}{\Phi(\phi)}\frac{d^2\Phi(\phi)}{d\phi^2} \tag{2.117}$$

As both sides of the equation are each a function of a single angular variable, they must both be equal to the same constant. As the left-hand side of (2.117) can be written as Legendre's differential equation, this constant can be specified to be m^2,

$$\frac{1}{\Phi(\phi)}\frac{\partial^2\Phi(\phi)}{\partial\phi^2} = -m^2 \tag{2.118}$$

$$\frac{d^2\Theta(\theta)}{d\theta^2} + \cot\theta\frac{d\Theta(\theta)}{d\theta} + \left(l(l+1) - m^2\csc^2\theta\right)\Theta(\theta) = 0. \tag{2.119}$$

Equation (2.118) has the general solution,

$$\Phi(\phi) = k_1 e^{im\phi + k_2}. \tag{2.120}$$

As $\Phi(\phi)$ is required to be single valued, m must be an integer or zero. By defining $\mu = \cos\theta$, (2.119) is recast as,

$$(1-\mu^2)\frac{d^2\Theta}{d\mu^2} - 2\mu\frac{d\Theta}{d\mu} + \left(l(l+1) - \frac{m^2}{1-\mu^2}\right)\Theta = 0. \tag{2.121}$$

This is the associated Legendre differential equation, the solutions of which are the associated Legendre polynomials of the first kind, $P_{lm}(\mu)$, with the requirement that $|m| \leq l$. A special case arises for $m = 0$,

$$\left(1-\mu^2\right)\frac{d^2\Theta}{d\mu^2} - 2\mu\frac{d\Theta}{d\mu} + l(l+1)\Theta = 0 \qquad (2.122)$$

which is the Legendre differential equation with solutions given by the Legendre polynomials, $P_l(\mu)$,

$$P_l(\mu) = \frac{(-1)^l}{2^l l!}\frac{d^l}{d\mu^l}\left(\left(1-\mu^2\right)^l\right). \qquad (2.123)$$

The first three Legendre polynomials are,

$$\begin{aligned} P_0(\mu) &= 1 \\ P_1(\mu) &= \mu \\ P_2(\mu) &= \frac{3\mu^2-1}{2}. \end{aligned} \qquad (2.124)$$

The Legendre polynomials are orthogonal and have the recurrence relation,

$$\begin{aligned} (l+1)P_{l+1}(\mu) &= (2l+1)\mu\, P_l(\mu) \\ &\quad - lP_{l-1}(\mu). \end{aligned} \qquad (2.125)$$

The associated Legendre polynomials can be calculated from the Legendre polynomials via,

$$P_{lm}(\mu) = \left(1-\mu^2\right)^{|m|/2}\frac{d^{|m|}}{d\mu^{|m|}}P_l(\mu). \qquad (2.126)$$

Combining this with the angular solutions gives the spherical harmonic,

$$Y_{lm}(\theta,\phi) = k_{lm}P_{lm}(\theta)e^{im\phi} \qquad (2.127)$$

where,

$$k_{lm} = (-1)^m\sqrt{\frac{2l+1}{4\pi}\frac{(l-m)!}{(l+m)!}}. \qquad (2.128)$$

The final form of the wavefunction with the radial and angular terms combined is,

$$\begin{aligned} \psi(r,\theta,\phi) &= R_l(r)Y_{lm}(\theta,\phi) \\ &= (A_l j_l(\lambda r) + B_l y_l(\lambda r))Y_{lm}(\theta,\phi). \end{aligned} \qquad (2.129)$$

2.6.5 Optical Theorem

An important relationship between the scattering amplitude and the total reaction cross section is derived by considering the simple case of azimuthal symmetry (i.e., no ϕ-dependence) so that the wavefunction at a distance can be expanded as a summation of Legendre polynomials,

$$\begin{aligned} \psi(\mathbf{r}) &= \psi(r,\theta) \\ &= \sum_{l=0}^{\infty} R_l(r)P_l(\cos\theta). \end{aligned} \qquad (2.130)$$

By inserting the asymptotic form of the radial function, we achieve,

$$\begin{aligned} \psi(r,\theta) &= \sum_{l=0}^{\infty}\frac{C_l}{kr}\sin\left(kr - \frac{l\pi}{2} + \delta_l\right)P_l(\cos\theta) \\ &= \sum_{l=0}^{\infty} C_l\left[\frac{e^{i\left(kr-\frac{l\pi}{2}+\delta_l\right)} - e^{-i\left(kr-\frac{l\pi}{2}+\delta_l\right)}}{2ikr}\right]P_l(\cos\theta). \end{aligned} \qquad (2.131)$$

As the incident plane wave can be described by the identity,

$$\begin{aligned} \psi_0(\mathbf{r}) &= e^{i\mathbf{k}\bullet\mathbf{r}} \\ &= e^{ikr\cos\theta} \\ &= \sum_{l=0}^{\infty} i^l(2l+1)j_l(kr)P_l(\cos\theta) \\ &= \sum_{l=0}^{\infty} i^l(2l+1)\frac{\sin\left(kr-\frac{l\pi}{2}\right)}{kr}P_l(\cos\theta) \\ &= \sum_{l=0}^{\infty} i^l(2l+1)\left[\frac{e^{i\left(kr-\frac{l\pi}{2}\right)} - e^{-i\left(kr-\frac{l\pi}{2}\right)}}{2ikr}\right]P_l(\cos\theta). \end{aligned} \qquad (2.132)$$

Then the difference between (2.131) and (2.132) must be the scattered wavefunction,

$$\begin{aligned} &\sum_{l=0}^{\infty} C_l\left[\frac{e^{i\left(kr-\frac{l\pi}{2}+\delta_l\right)} - e^{-i\left(kr-\frac{l\pi}{2}+\delta_l\right)}}{2ikr}\right]P_l(\cos\theta) \\ &-\sum_{l=0}^{\infty} i^l(2l+1)\left[\frac{e^{i\left(kr-\frac{l\pi}{2}\right)} - e^{-i\left(kr-\frac{l\pi}{2}\right)}}{2ikr}\right]P_l(\cos\theta) = f(\theta)\frac{e^{ikr}}{r}. \end{aligned} \qquad (2.133)$$

By equating the coefficients of the exponentials,

$$\begin{aligned} C_l &= (2l+1)i^l e^{i\delta_l} \\ &= (2l+1)e^{i\left(\delta_l + \frac{l\pi}{2}\right)}. \end{aligned} \tag{2.134}$$

The scattering amplitude is then given by the summation,

$$f(\theta) = \frac{1}{k}\sum_{l=0}^{\infty}(2l+1)e^{i\delta_l}\sin\delta_l P_l(\cos\theta). \tag{2.135}$$

In order to obtain an expression for the total cross section, one first recalls that the differential cross section in solid angle is the squared-magnitude of the scattering amplitude,

$$\begin{aligned} \frac{d\sigma}{d\Omega} &= |f(\theta)|^2 \\ &= \frac{1}{k^2}\sum_{l=0}^{\infty}\sum_{l'=0}^{\infty}(2l+1)e^{i\delta_l}\sin\delta_l P_l(\cos\theta) \\ &\quad \times (2l'+1)e^{i\delta_{l'}}\sin\delta_{l'} P_{l'}(\cos\theta) \end{aligned} \tag{2.136}$$

The total cross section is obtained by integrating over 4π steradians,

$$\begin{aligned} \sigma &= \int_0^{2\pi} d\phi \int_{-1}^{1} d(\cos\theta)\frac{d\sigma}{d\Omega} \\ &= \frac{2\pi}{k^2}\sum_{l=0}^{\infty}\sum_{l'=0}^{\infty}(2l+1)(2l'+1)e^{i(\delta_l-\delta_{l'})}\sin\delta_l\sin\delta_{l'} \\ &\quad \times \int_{-1}^{1} d(\cos\theta)P_l(\cos\theta)\,P_{l'}(\cos\theta) \\ &= \frac{4\pi}{k^2}\sum_{l=0}^{\infty}(2l+1)\sin^2\delta_l \end{aligned} \tag{2.137}$$

where the orthonormality of the Legendre polynomials has been used. However, note that the imaginary part of the scattering amplitude for $\theta = 0$ is,

$$\begin{aligned} \mathrm{Im}\,f(0) &= \frac{1}{k}\mathrm{Im}\left(\sum_{l=0}^{\infty}(2l+1)e^{i\delta_l}\sin\delta_l P_l(1)\right) \\ &= \frac{1}{k}\sum_{l=0}^{\infty}(2l+1)\sin^2\delta_l. \end{aligned} \tag{2.138}$$

Substituting this into the expression for the total cross section gives,

$$\sigma = \frac{4\pi}{k}\mathrm{Im}\ f(0) \tag{2.139}$$

which is known as the optical theorem. The total cross section is a measure of the reduction in flux which results from the destructive interference between the incident and scattered wavefunctions for $\theta = 0$.

2.7 Dirac's Equation

2.7.1 Introduction

Dirac's equation will be the foundation of subsequent discussions regarding the interactions of electrons with the electromagnet field and of the phenomena of the creation and annihilation of positrons. This section is intended to be only an overview of the derivation and characteristics of the Dirac equation to be called upon later.

This discussion begins by considering Schrödinger's equation which is nonrelativistic and, as a descriptor for free particles, was derived from the quantum-mechanical interpretation of energy and momentum as operators,

$$E \to i\hbar\frac{\partial}{\partial t} \tag{2.140}$$

$$\mathbf{p} \to -i\hbar c\nabla. \tag{2.141}$$

Note that the constants of $\hbar$ and $\hbar c$ have been used in order to maintain our convention that the units of both energy and momentum are those of energy. Substituting these operator expressions into the nonrelativistic relationship between energy and momentum,

$$E = \frac{|\mathbf{p}|^2}{2m}$$

gives the Schrödinger equation,

$$i\hbar\frac{\partial\psi(\mathbf{x},t)}{\partial t} = -\frac{(\hbar c)^2}{2m}\nabla^2\psi(\mathbf{x},t). \tag{2.142}$$

Early attempts at forming a relativistically invariant analog to this Schrödinger equation began with the relativistic relationship between energy and momentum,

$$E^2 = p^2 + m^2. \tag{2.143}$$

If the operator forms of energy and momentum are substituted into this relationship, then the Klein–Gordon equation is obtained,

$$(\Delta^2 + m^2)\psi = 0. \tag{2.144}$$

where the operator $\Delta^2 \equiv \hbar^2\partial^2/\partial t^2 - (\hbar c)^2\nabla^2$ has been defined. This attempt at forming a relativistic result confronted immediate problems. First, the energy-momentum relationship of (2.143) allowed negative-energy solutions, $E = -\sqrt{p^2 + m^2}$. Although this was a difficulty later confronted and surmounted by Dirac in his interpretation of antiparticles, this conundrum was at first assumed to be fatal to this attempt. Second, the Klein–Gordon equation is second-order in that the time derivative leads to a probability density for the wavefunction that was not positive-definite. As a result, the solution represented by the Klein–Gordon equation was rejected, although it was "rehabilitated" in later times as a descriptor of spinless particles (scalar or pseudoscalar wavefunctions) such as the pion.

However, the above result has shown that, in order to derive the required relativistic form of a solution, the use of a first-order time derivative was required.

2.7.2 Derivation of Dirac's Equation

Recognizing the requirement for a first-order time derivative, Dirac began with a linear *ansatz* of the form,

$$E\psi = (\boldsymbol{\alpha}\mathbf{p} + \boldsymbol{\beta}m)\psi. \tag{2.145}$$

The operators $\boldsymbol{\alpha}$ and $\boldsymbol{\beta}$ are required in order to be consistent with the relativistic relationship between energy and momentum given by (2.143),

$$E^2 = (\boldsymbol{\alpha}\mathbf{p})^2 + (\boldsymbol{\beta}m)^2 + (\boldsymbol{\beta}m\boldsymbol{\alpha}\mathbf{p} + \boldsymbol{\alpha}\mathbf{p}\boldsymbol{\beta}m). \tag{2.146}$$

Equating terms to those of (2.143), one finds the requirements, following Dirac's proposal that (2.146) be regarded as a *matrix* equation with the four matrices $\boldsymbol{\alpha}_i$ and $\boldsymbol{\beta}$ obeying the algebra, of,

$$\boldsymbol{\alpha}_i^2 = \boldsymbol{\beta}^2 = \mathbf{1} \tag{2.147}$$

$$\boldsymbol{\alpha}_i\boldsymbol{\alpha}_j + \boldsymbol{\alpha}_j\boldsymbol{\alpha}_i = 2\delta_{ij}\mathbf{1} \tag{2.148}$$

$$\boldsymbol{\alpha}_i\boldsymbol{\beta} + \boldsymbol{\beta}\boldsymbol{\alpha}_i = \mathbf{0}. \tag{2.149}$$

In order to construct these four matrices, a number of conditions must be met. First, they must be Hermitian in order for the Hamiltonian to be so. Second, the eigenvalues of the $\boldsymbol{\alpha}$ and $\boldsymbol{\beta}$ matrices are ± 1. Finally, from their anti-commutation property, the traces of the matrices must be zero. These last two conditions allow the dimensions of the matrices to be specified. As the trace of a matrix is also the sum of its eigenvalues, the number of positive and negative eigenvalues must be equal, requiring that the dimensions of the $\boldsymbol{\alpha}$ and $\boldsymbol{\beta}$ matrices be even. A dimension of 2 is insufficient and the smallest even-numbered dimension for which these matrices can be realized is 4. Hence,

$$\boldsymbol{\alpha}_i = \begin{pmatrix} 0 & \boldsymbol{\sigma}_i \\ \boldsymbol{\sigma}_i & 0 \end{pmatrix} \quad i = 1, 2, 3 \tag{2.150}$$

$$\boldsymbol{\beta} = \begin{pmatrix} \mathbf{1} & 0 \\ 0 & -\mathbf{1} \end{pmatrix} \tag{2.151}$$

where $\boldsymbol{\sigma}_i$ are the 2×2 Pauli matrices and $\mathbf{1}$ is the 2×2 identity matrix. In order to write the Dirac result in a covariant form, we introduce the notation of,

$$\gamma^0 = \boldsymbol{\beta} \tag{2.152}$$

$$\gamma^i \equiv \boldsymbol{\beta}\boldsymbol{\alpha}_i = \begin{pmatrix} 0 & \boldsymbol{\sigma}_i \\ -\boldsymbol{\sigma}_i & 0 \end{pmatrix} \quad i = 1, 2, 3. \tag{2.153}$$

An additional matrix is defined as,

$$\gamma_5 = i\gamma^0\gamma^1\gamma^2\gamma^3 = \begin{pmatrix} 0 & \mathbf{1} \\ \mathbf{1} & 0 \end{pmatrix}. \tag{2.154}$$

One can then write the Dirac equation as,

$$(\gamma^0 E - \gamma \bullet \mathbf{p} - m\mathbf{1})\psi = 0 \tag{2.155}$$

where $\boldsymbol{p}^{\mu} = (\mathrm{E}, \mathbf{p})$ and the covariant is,

$$\begin{aligned} \boldsymbol{p}_{\mu} &\equiv g_{\mu\nu}\boldsymbol{p}^{\nu} \\ &= (\mathrm{E}, -\mathbf{p}). \end{aligned} \tag{2.156}$$

Equation (2.155) can be rewritten in the form,

$$(\gamma^{\mu}\boldsymbol{p}_{\mu} - \mathrm{m}\mathbf{1})\psi = 0. \tag{2.157}$$

One now has the $\psi(\boldsymbol{X})$ as a wavefunction with space–time and spinor degrees of freedom,

$$\psi(\boldsymbol{X}) = \begin{pmatrix} \psi_1(\boldsymbol{X}) \\ \psi_2(\boldsymbol{X}) \\ \psi_3(\boldsymbol{X}) \\ \psi_4(\boldsymbol{X}) \end{pmatrix}. \tag{2.158}$$

There are two solutions to the Dirac equation,

$$\begin{aligned} \psi_{+}(\mathbf{x}, \mathrm{t}) &= \mathrm{u}e^{-i\boldsymbol{p}_{\mu}\boldsymbol{X}^{\mu}} \\ &= \mathrm{u}e^{-i\left(\frac{\mathrm{Et}}{\hbar} - \frac{\mathbf{p}\bullet\mathbf{x}}{\hbar c}\right)} \end{aligned} \tag{2.159}$$

where u is defined as a Dirac spinor and $\hbar$ and $\hbar c$ have been inserted in order to make the exponent dimensionally correct. This solution corresponds to a positive energy. The negative-energy solution is,

$$\begin{aligned} \psi_{-}(\mathbf{x}, \mathrm{t}) &= \mathrm{v}e^{i\boldsymbol{p}_{\mu}\boldsymbol{X}^{\mu}} \\ &= \mathrm{v}e^{+i\left(\frac{\mathrm{Et}}{\hbar} - \frac{\mathbf{p}\bullet\mathbf{x}}{\hbar c}\right)}. \end{aligned} \tag{2.160}$$

$\psi_{-}(\mathbf{x}, \mathrm{t})$ is the wavefunction of a positron or, in the Feynman picture, an electron moving backward in time with momentum $-\mathbf{p}$. The Dirac spinors are,

$$\mathrm{u}(\boldsymbol{p}, \mathrm{s}) = \sqrt{\mathrm{E}+\mathrm{m}} \begin{pmatrix} \chi_s \\ \frac{\boldsymbol{\sigma}\bullet\mathbf{p}}{\sqrt{\mathrm{E}+\mathrm{m}}}\chi_s \end{pmatrix} \tag{2.161}$$

$$\mathrm{v}(\boldsymbol{p}, \mathrm{s}) = \sqrt{\mathrm{E}+\mathrm{m}} \begin{pmatrix} \frac{\boldsymbol{\sigma}\bullet\mathbf{p}}{\sqrt{\mathrm{E}+\mathrm{m}}}\hat{\boldsymbol{\varepsilon}}\chi_s^* \\ \hat{\boldsymbol{\varepsilon}}\chi_s \end{pmatrix} \tag{2.162}$$

where

$$\hat{\boldsymbol{\varepsilon}} = \begin{pmatrix} 0 & -1 \\ 1 & 0 \end{pmatrix} \tag{2.163}$$

and where χ_s is the (nonrelativistic) Pauli two-component spinor for a spin-½ particle. Note that another variable, s, has been introduced and which describes the spin orientation of the particle. Equation (2.161) describes the particle (electron) states and (2.162) describes the antiparticle (positron) states.

Nuclear Properties, Structure, and Stability 3

Abstract Nuclear medicine exists because of the instability of some atomic nuclei which result in the emission of photons or particulate radiations. Later chapters consider the microscopic theories of these radioactive decays and how, in practice, they can be quantified. This chapter reviews the fundamental properties of atomic nuclei, through phenomenology and consideration of the Fermi nuclear gas, the nuclear liquid drop and the nuclear shell models, and how changes in these properties result in the radioactive instability necessary for nuclear medicine.

Contents

3.1 Introduction

The subject of this book is the radiation absorbed dose resulting from medical applications of radionuclides placed within the body. This absorbed dose is the consequence of an unstable, or radioactive, atomic nucleus which is one in a physical state from which it is possible to transit, or decay, to another state with lower energy. If further transitions are not possible, this latter state is referred to as being stable. These changes in nuclear state and energy are accompanied by the emission of radiation or particles which, along with the recoil kinetic energy of the daughter nucleus, carry away the energy difference between the initial and final nuclear states. External detection of long-range radiation (X and γ-rays) which exits the body is used to generate the diagnostic image. On the other hand, particulate radiations (e.g., α particles, electrons, and positrons) are contained within the body and

B.J. McParland, *Nuclear Medicine Radiation Dosimetry*,
DOI 10.1007/978-1-84882-126-2_3,

deposit energy as they slow down and stop. This exchange of particle kinetic energy to the medium is among the subjects of Chap. 7.

As this book provides the theoretical basis of evaluating the absorbed dose in nuclear medicine, then it is clearly evident that an understanding of the properties of radioactive nuclei is integral to achieving this basis. At the most fundamental level, this understanding leads to knowledge of those combinations of nuclear characteristics which lead to a given nucleus being radioactive. These properties are, in particular, the nuclear mass, size, and electric charge (i.e., the relative proportion of protons and neutrons) and its intrinsic energy (e.g., the rotational energy associated with a high-spin state). In addition to appreciating the requirements leading to a radioactive nucleus, the accurate evaluation of the radiation dosimetry resulting from its decay products requires understanding of how, during the process of radioactive decay, the nucleus emits radiations and particles of specific energies or energy spectra. The discussion of this will be covered both in this chapter and that following.

The selection of a radionuclide for a given nuclear medicine application is dependent upon a large number of characteristics. The three most important would be:

- It must have chemical properties that allow it to be labeled to a suitable vector that can transport it to the biological target of interest or that allows metabolism to draw it to the target. These are requirements common to both diagnostic or therapeutic uses of the radionuclide.[1]
- The radioactive properties of the nucleus are critical to the indicated application. Should it be therapeutic, one would preferentially select those radionuclides that emit short-range charged particulate radiations, such as electrons or α particles, which deposit their energies within very small volumes in order to deliver a high therapeutic radiation dose to the cellular target and minimize the absorbed dose to surrounding normal tissue. Associated emissions of X or γ-rays would allow confirmation of the anatomical localization through imaging of the therapeutic isotope. On the other hand, a diagnostic indication requires a nucleus with a decay process that emits either X or γ-rays, which are detected externally, or positrons which annihilate with electrons present in tissue at a short range from the decay site to produce annihilation γ-rays which are subsequently detected. Even though these radiations are detected outside the body, the decay processes will still result in the generation of short-range moving charged particles that deposit energy within the body. Nuclear decay culminating in the emission of a photon frequently includes the ejection of low-energy atomic electrons[2] that deposit energy within the volume immediately surrounding the nucleus. The photon itself can be either absorbed by an atomic electron subsequently ejected or elastically scattered by an atomic electron in tissue through Compton scatter thus setting the electron into motion (with both processes to be discussed in Chap. 6). Hence, photon emission is associated with the generation of short-range charged particles which transfer their kinetic energy to the surrounding tissue as they slow down. Also, positrons transfer energy to tissue as they slow down following emission and approach annihilation with an electron.
- As radioactive decay is a stochastic process, the probability of a decay occurring per unit time (inversely proportional to its half-life) is of extreme importance to the clinical application. In general, a short half-life, such as the 6.02 h of ^{99m}Tc, is frequently of great benefit to diagnostic nuclear medicine as it will tend to limit the absorbed dose. On the other hand, the longer 67 h half-life of ^{111}In is of benefit to the use of autologous ^{111}In-labeled leukocytes in the imaging of unknown sites of sepsis in which it may take several hours for the labeled leukocyte to reach the region of interest. The role of the decay lifetime in therapeutic applications is still somewhat contentious. Short half-lives imply high dose rates and conversely for long half-lives. As the absorbed dose rate affects the biological response of a tissue to irradiation, careful consideration to the appropriate half-life to use in nuclear medicine therapy should be applied.

[1]The vector must have high specificity for its intended target and, either directly or through metabolism, results in a clearance of the radioactive isotope from nontarget tissues in order to increase the target-to-background ratio of the image and to reduce that absorbed dose to uninvolved tissues.

[2]Such as Auger/Coster–Kronig and conversion electrons and which are discussed in Chap. 6.

In general, a proper appreciation of the suitability of a radioactive nucleus for a given nuclear medicine application requires an understanding of the underlying nuclear structure, of what makes a nucleus unstable and of what radiations and particulates are emitted. The foundation for this appreciation is developed in this chapter and further expanded upon in Chaps. 4 and 5.

3.2 Characteristics of Atomic Nuclei: Part I

3.2.1 Introduction

It is probable that the Greek philosopher Democtritus of Abdera (460–370 BC) and the Roman poet Titus Lucretius Carus (ca 99 BC to ca 55 BC) were the first in the West to propose atomic theories of matter that could be considered distant foundations for modern theory. However, the concept of the atom as a fundamental constituent of matter lay dormant for nearly two millennia until when Dalton resurrected the concept in 1808. On the basis of gravimetric measurements of substances involved in chemical reactions, Dalton proposed that the quantitative descriptions of chemical composition could be interpreted in terms of integral numbers of atoms, thus anticipating the concept of the isotope. It was not until the late nineteenth century when the first of the atomic constituents was discovered through Thomson's analysis of cathode rays which demonstrated the existence of negatively-charged electrons. The existence of the atomic nucleus itself became evident following Rutherford's interpretation of the results of the α particle scattering experiments by Geiger and Marsden. The constituents of the nucleus were identified through Rutherford's demonstration of the existence of the proton via α particle-induced fragmentation of the nitrogen nucleus in 1919 and Chadwick's detection of the neutron in 1932. The underlying quark structures of the proton and neutron themselves were not fully elucidated until the 1950s and 1960s through the theoretical work of Gell-Mann, Feynman, and others and the confirmatory experimental work of Hofstadter. It shall be important to note that it was through the Coulomb scattering of projectiles that the substructures of both the nucleus and the proton and neutron were revealed.

While the combination of Thomson's and Rutherford's discovery clearly indicated that the atom was composed of a massive positively-charged nucleus and an ensemble of electrons of equal but negative charge, the size, and makeup of the nucleus itself remained unclear. Up until the 1920s, Thomson's plum pudding model which envisaged the nucleus as being made up of a conglomeration of positive charge (protons), within which electrons were embedded, held sway. The magnitude of the net positive electric charge of this amalgam would, as required, exactly balance the negative charge of atomic electrons. However, Rutherford's observation that the atomic mass (expressed as an integer) was almost double the nuclear electric charge (i.e., the atomic number) began to bring the plum pudding model into doubt. This approximate ratio of 2 between nuclear integral mass and electric charge suggested the existence of a nuclear constituent with a mass very nearly equal to that of the proton but with no charge. It was Ehrenfest's and Oppenheimer's (1931) study of the diatomic homonuclear $^{14}N_2$ molecule which heralded the end of the proton–electron plum pudding nuclear model. Had that model been valid, the nitrogen nucleus would be made up of 14 protons and 7 electrons. Although the net nuclear electric charge exactly balances that of the ensemble of atomic electrons, this combination of nuclear protons and electrons requires the nitrogen nucleus to have half-integral spin. However, Raman spectroscopy data showed that the nitrogen nucleus had indeed an integral spin which was a result that could be readily explained by the combination of seven protons and the seven hypothesized particles similar to protons in mass and intrinsic spin but with no electric charge. The existence of the neutron was confirmed experimentally by Chadwick.

As protons are positively charged and neutrons have no charge, it is immediately evident that an attractive nuclear force is required to compensate for the repulsive Coulomb force between the protons. Our first examination of nuclear properties will culminate in the density of the nucleus and the demonstration of it being very nearly constant. This simple result leads to the conclusion that this attractive strong nuclear force is saturable and has finite range. This culmination will, inter alia, result in the derivations of nuclear mass and size. Our examination of the latter will expend considerable effort on understanding elastic Coulomb scattering and how it can be used to derive not only nuclear size, but nuclear structure.

3.2.2 Fundamental Particles and Interactions

Before progressing into a review of the details of nuclear structure relevant to internal radiation dosimetry, the fundamentals of matter and how its different components interact are considered. The fundamental components of matter are the fermions[3] which are divided amongst the three generations shown in the mass spectrum of Fig. 3.1. Within each generation of fundamental fermions are two flavors of quarks to give a total of six flavors. The quarks have the highest masses within each generation: up and down in generation 1; charm and strange in generation 2; and top and bottom in generation 3. The quarks have electric charge and interact with the electromagnetic force (mediated through photons), the color (or strong force mediated through gluons) and the weak force (mediated through intermediate vector bosons).[4] Quarks have an additional quantum characteristic, that of color. As only the quarks within each generation are subject to the color (or strong) force, those particles composed of quarks (e.g., mesons, protons, and neutrons) are referred to as hadrons (derived from the Greek, *hadros*, or "thick"). Baryons (from, *barys*, or "heavy") are those hadrons composed of three quarks and, hence, are fermions. Obvious examples of baryons are the proton and neutron. Mesons (from, *mesos*, or "middle"),[5] on the other hand, are quark pairs and are bosons.[6] All mesons are unstable.

The next massive particle in each generation is a lepton (from the Greek, *leptos*, or "thin") so named as they are charged point-like fermions subject to only the electromagnetic and weak forces: the electron in generation 1; the muon (μ) in generation 2; and the tau lepton (τ) in generation 3. As an aside, there is an intriguing empirical, and unexplained, relationship between the leptonic masses given by Koide's formula (Koide 1983; Rivero and Gjsponer 2005),

$$\frac{m_e + m_\mu + m_\tau}{\left(\sqrt{m_e} + \sqrt{m_\mu} + \sqrt{m_\tau}\right)^2} = \frac{2}{3}. \tag{3.1}$$

Of the three electrically-charged leptons, only the electron is stable. Finally, each generation contains another lepton, the neutrino (ν), which has no electric charge and is associated with the more massive lepton within each of the generations. Generation 1 represents "permanent" matter in so far as there is no evidence for the decays of the proton (the proton being composed of 2 u quarks and 1 d quark) and the electron. As a result, the fermions of generation 1 (and their intermediary force-exchange particles) and the particles that they compose (protons, neutrons and, eventually, nuclei) will be the focus of this book. The constituents of generations 2 and 3 have ephemeral lifetimes and will be of no consequence to practical nuclear medicine, with the possible exception of the muon which, as a product of cosmic ray interactions with the atmosphere, can contribute to the

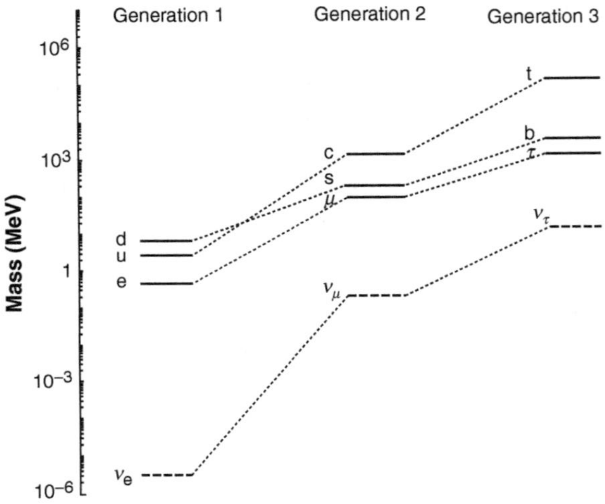

Fig. 3.1 Mass spectra of the elementary fermions. The *heavy-dashed lines* for the neutrinos indicate the upper mass limits and the *light-dashed lines* connect corresponding fermions of different generations (for the quarks, this connection is that of electric charge)

[3]Fermions are particles with half-integral spin and subject to Fermi–Dirac statistics.

[4]Throughout this book, gravity is ignored and reference will be made to the remaining three forces only.

[5]The descriptive *mesos* was selected as a result of Yukawa's prediction of the mass of the boson exchanged between protons and neutrons to be 130 MeV, intermediate between those of the electron and the neutron and proton.

[6]Bosons are those particles with zero or integral spin and subject to Bose–Einstein statistics.

background signal of radiation detectors. For each fundamental fermion (quark, electron, muon, τ lepton, and neutrino), there exists an antiparticle, the existence of which is predicted by relativistic field theory (as demonstrated by solutions to the Dirac equation). The particle and its antiparticle have the same mass, lifetime, intrinsic spin, parity, and isospin but differ in the signs of both electric charge and additive quantum numbers. In this book, particles and antiparticles will in general be differentiated by a horizontal line (e.g., u and ū for the up and anti-up quarks) with the exception to allow for the common practice of referring to the antielectron as the positron which is denoted by e^+.

It is recognized empirically that, throughout nature, there are four fundamental interactions or "forces" as summarized in Table 3.1. Obviously, gravity will not be considered here and all subsequent discussions will be limited to the strong, electromagnetic and weak interactions. The distinction between the "strong color" and "strong nuclear" forces will be expanded upon later. Quarks are unique in that they interact with other particles through the electromagnetic, weak and strong (color) forces. Consequently, those systems made up of quark multiples, the baryons and mesons, will interact via the electromagnetic, weak and strong (nuclear) forces. Charged leptons interact via the electromagnetic and weak forces only whereas the uncharged leptons (neutrinos) interact via the weak force alone. To understand these three interactions, focus on the exchange processes of interest provided in Fig. 3.2.

Table 3.1 Characteristics of the fundamental interactions

Interaction		Interaction boson	Range
Strong	Color	Gluon	Quark confinement
	Nuclear	Pion	$\hbar c/m_\pi \approx 1.4$ fm
Weak		$W^\pm$, Z^0	$\hbar c/m_W \approx 2 \times 10^{-3}$ fm
Electromagnetic		Photon	Infinite
Gravity		Graviton	Infinite

Propagator theory (refer to, e.g., Bjorken and Drell (1964)) describes the probability amplitude of an interaction in terms of the two vertices describing the emission and absorption of the intermediary boson between two interacting fermions and of the field propagator proportional to $(\boldsymbol{q} - \mathrm{M})^{-2}$, where $\boldsymbol{q}$ is the four-vector momentum transfer and M is the mass of the interaction boson. In particular, the range of the interaction will be indicated by the Fourier transform of the propagator from momentum space into coordinate space. For example, as the photon is massless, the electromagnetic potential has infinite range and thus varies as r^{-1}. On the other hand, the masses of the intermediate vector $W^\pm$ and Z^0 bosons are 80.4 and 91.2 GeV, respectively, which results in the weak interaction being very short-ranged (of the order of 10^{-3} fm). As the gluon, like the photon, is massless one would expect, from the above argument, that the strong (color) interaction between quarks will also be of infinite range. But, on the contrary, such an infinite-range interaction is not observed as a consequence of the quarks having an additional quantum number, referred to as color,

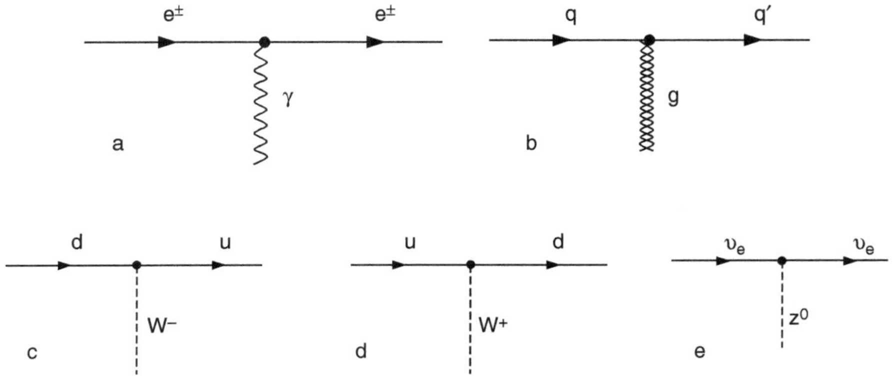

Fig. 3.2 Fermion field quantum vertices: (**a**) electromagnetic interaction with an electron/positron and interacting photon boson; (**b**) color (strong) interaction with a quark and gluon interaction boson; (**c**) weak interaction exchanging quark flavor from *down* to *up* through an interacting W^- boson; (**d**) weak interaction exchanging quark flavor from *up* to *down* through an interacting W^+ boson; and (**e**) weak interaction for neutrino scattering through an interacting Z^0 boson

which has the three degrees of freedom of red, green, and blue.[7] Free quarks have not been detected and groupings of quarks are currently known only to be in three-quark (baryons) and two-quark (mesons) states. The three quarks in a baryon can have any combination of flavor (up, down, strange, charm, top, or bottom) but the combination must be that of colors of red, green, and blue so that, to the outside world, the baryon appears "colorless" or "white" and there is no evidence of the strong (color) interaction. Mesons contain quark–antiquark pairs which, again, can be in any combination of flavor but the colors of the quark–antiquark cancel (e.g., red/antired) so that meson also appears colorless to the outside world. Quantum chromodynamics (QCD) describes the color interactions between quarks. In QCD the gluon, unlike the photon which does not have an electric charge, carries color and, hence, can interact with another gluon. This gluon–gluon interaction due to their carrying color has two important consequences – confinement and asymptotic freedom (Close 1979). Confinement refers to the fact just described that quarks exist only in bound colorless (color singlet) systems. To demonstrate that this should be so, consider, for example, a quark–antiquark pair (meson). Attempting to remove a single quark from this pair will require an increasing amount of potential energy. The gluon–gluon interaction leads to the gluons becoming the source of new gluons to eventually form a new color singlet pair. Asymptotic freedom refers to the coupling between quarks decreasing with decreasing separation and asymptotically reaching zero for zero separation. The strong (color) interaction describes processes within the proton and neutron and leads to the description of the interactions between hadronic multiple-quark states (baryons and mesons) via the strong (nuclear) interaction. This strong (nuclear) interaction does not directly involve gluons and is sometimes referred to as the residual strong interaction.[8] It exists between color-neutral systems and is analogous to the Coulomb-based van der Waals interaction between electrically-neutral atoms caused by momentary polarization of electrically-neutral systems. Yukawa modeled the interaction between nucleons (a generic term describing both protons and neutrons) as the exchange of an intermediate boson (Segrè 1977). It was possible to predict at least the order-of-magnitude of the boson mass from the Heisenberg uncertainty principle by knowing that the strong (nuclear) interaction had a range of about 1.5 fm. As a result, the mass of the boson was of the order of 130 MeV which, because it was intermediate between those of the electron and nucleon, resulted in the boson being referred to as a meson. In particular, Yukawa's boson interacting between the nucleons was the π meson or, as more commonly known, the pion. The pion comes in three charge states of the $\pi^{\pm}$, each with a mass of 139.6 MeV, and the π^0, with a mass of 135 MeV.

The weak interaction vertices are those of (c), (d), and (e) in Fig. 3.2. The neutral current of (e) is not relevant to dosimetry and will not be considered further. However, the flavor-changing interactions of (c) and (d) form the basis of the beta decay described in Chap. 4. As noted earlier, the weak interaction has a range of the order of 10^{-3} fm, or about 0.02% of the diameter of a nucleus, as described in the modern Glashow–Salam–Weinberg electroweak model with the exchange of the $W^{\pm}$ and Z^0 intermediate vector bosons. In addition to the limited interaction distance, a rationale for the use of weak in describing these interactions is obtained by comparing the lifetimes of particles decaying through electromagnetic, strong and weak interactions. For example, the π^0 has a lifetime of 8.4×10^{-17}s, predominantly through the $\pi^0 \rightarrow 2\gamma$ electromagnetic decay. The charged pions

[7] Experimental evidence for the existence of a quantum number of color is provided by the existence of the Δ_{33} resonance which has a mass of about 1,232 Mev, an intrinsic spin of 3/2, isospin of 3/2, and an electric charge of +2e (Close 1979). This combination of spin, isospin and electric charge indicates that the Δ_{33} is made up of three up quarks which have coupled to provide spin 3/2 and isospin of 3/2. In the absence of an additional quantum number, the Pauli exclusion principle would not allow for the existence of such a particle. Hence, for this principle for fermions to remain valid, an additional quantum number distinguishing the three quarks is required. This additional quantum number is color.

[8] Historically, the term "strong interaction" referred to the residual strong force between hadrons as mediated by mesons. With the development of QCD in describing color interactions between hadrons, "strong interactions" became synonymous with these color interactions, thus requiring a differentiation between "strong nuclear" and "strong color" interactions. Throughout the remainder of this book, unless otherwise noted, we will use "strong interaction" to mean "strong nuclear interaction" exclusively.

decay through the weak interaction, e.g., $\pi^+ \rightarrow \mu^+ + \nu_\mu$, with a lifetime of 26 ns, some 30 million times longer.

3.2.3 Quantum Numbers

3.2.3.1 Introduction

Quantum states and the interactions between them are conveniently described by quantum numbers. In most cases, the quantum number of a system following a process will not change, i.e., the quantum number is conserved. When a quantum number is not conserved, for example, the lack of parity conservation in weak interactions, a profound indication of underlying physics is provided.

3.2.3.2 Electric Charge

With the exception of the quarks, electric charge Q is quantized in integer units of the fundamental charge, e. It is an additive quantum number and is conserved in strong, weak and electromagnetic interactions. Two examples demonstrating this are those of the single charge exchange reaction between a proton and negative pion and of free neutron decay,

$$\begin{array}{lcccc} & \pi^- + & p & \rightarrow n + & \pi^0 \\ Q: & -1 & +1 & 0 & 0 \end{array} \quad \text{Strong interaction} \tag{3.2}$$

$$\begin{array}{lcccc} & n \rightarrow & p & + e^- & + \bar{\nu}_e \\ Q: & 0 & +1 & -1 & 0 \end{array} \quad \text{Weak interaction} \tag{3.3}$$

Quarks have electric charges of either $+{}^2\!/_3 e$ or $-{}^1\!/_3 e$.

3.2.3.3 Baryon Number

Baryons are quark triplets and the baryon number is an additive quantum number conserved in strong, weak and electromagnetic interactions. Identifying the baryon number as ± 1, the quark is assigned a baryon number of $1/3$ and the antiquark has a baryon number of $-1/3$. Hence, the proton and neutron each have a baryon number of 1 whereas all mesons, being quark–antiquark doublets, have a baryon number of zero. If the baryon number were not conserved, then proton decay would be observable:

$$\begin{array}{lccc} & p & \rightarrow e^+ + & \pi^0 \\ B: & +1 & 0 & 0 \end{array} \quad \text{Not observed} \tag{3.4}$$

The minimum proton lifetime determined experimentally is in excess of 10^{33} years, indicating the strength of baryon number conservation. As expected, antibaryons, such as the antiproton, have the negative of the baryon number of the baryon.

3.2.3.4 Lepton Number

The lepton number is an additive quantum number with leptons assigned a lepton number of 1, antileptons assigned a lepton number of -1 and particles which are not leptons assigned a lepton number of 0. For our previous example of free neutron decay,

$$\begin{array}{lcccc} & n \rightarrow & p + & e^- + & \bar{\nu}_e \\ L: & 0 & 0 & +1 & -1 \end{array} \tag{3.5}$$

Lepton number can be further subdivided according to the generations of Fig. 3.1 to provide an electronic lepton number for the leptons of generation 1, a muonic lepton number for the leptons of generation 2 and a τ-lepton number for the leptons of generation 3. Evidence for this subdivision comes from the pionic and muonic decays,

$$\begin{array}{lccc} & \pi^- \rightarrow & \mu^- + & \bar{\nu}_\mu \\ L & 0 & +1 & -1 \\ L_e & 0 & 0 & 0 \\ L_\mu & 0 & +1 & -1 \end{array} \tag{3.6}$$

$$\begin{array}{lcccc} & \mu^- \rightarrow & e^- + & \bar{\nu}_e + & \nu_\mu \\ L & +1 & +1 & -1 & +1 \\ L_e & 0 & +1 & -1 & 0 \\ L_\mu & +1 & 0 & 0 & +1 \end{array} \tag{3.7}$$

3.2.3.5 Spin

Spin is a quantum mechanical analog to angular momentum and is quantized as integral or half-integral

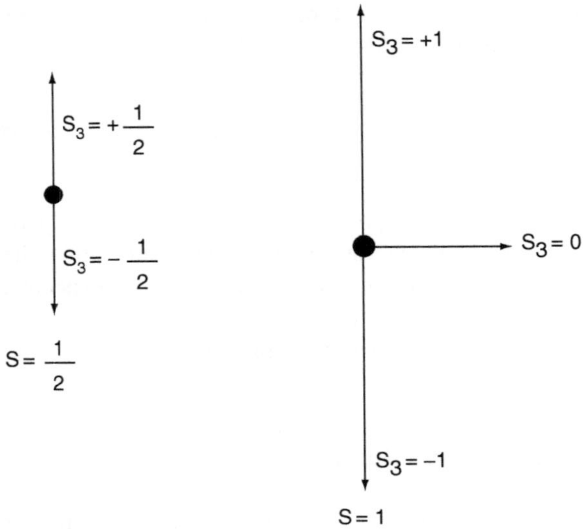

Fig. 3.3 Spin projections for spin-1/2 and spin 1 (in units of $\hbar$)

multiples of $\hbar$. Particles with integral spin are subject to Bose–Einstein statistics and consequently referred to as bosons. Those particles with half-integral spin are subject to Fermi–Dirac statistics and referred to as fermions. As with angular momentum, for spin s there are (2s + 1) different states of the projection s_3 as shown in Fig. 3.3.

3.2.3.6 Parity

Parity is a quantum mechanical concept and describes the effect upon a wavefunction caused by a reversal of spatial coordinates from r to −r. Let Π be the parity operator that, acting upon a wavefunction, performs the result,

$$\Pi\Psi(\mathbf{r}) = \lambda\Psi(-\mathbf{r}). \tag{3.8}$$

The eigenvalue of the operator, λ (also referred to as the eigenparity), is determined by successive applications of the parity operator,

$$\Pi(\Pi\,\Psi(\mathbf{r})) = \lambda^2\,\Psi(\mathbf{r}). \tag{3.9}$$

Hence, the value of the eigenvalue is $\lambda = \pm 1$. A quantum system with $\lambda = +1$ is referred to as having positive parity whereas that with $\lambda = -1$ is said to have negative parity. As parity is a multiplicative

Table 3.2 Meson wavefunction parity characteristics

J^P	Characteristic	Physical examples
0^-	Pseudoscalar	Pressure
0^+	Scalar	Mass, time
1^-	Vector or polar vector	Position **r**, Momentum **p**
1^+	Pseudovector or axial vector	Spin, Angular momentum $\mathbf{L} = \mathbf{r} \times \mathbf{p}$

quantum number, the parity of a system $\Psi(\Psi_1, \Psi_2 \ldots \Psi_N)$ made up of N subsystems is the product of the parities of the N subsystems,

$$P_\Psi = \prod_{i=1}^{N} P_{\psi_i}. \tag{3.10}$$

The parity of a state with an orbital angular momentum quantum number, l, is $(-1)^l$.

Particles and nuclear states can be identified by J^π where J is the angular momentum and $\pi = \pm 1$is the parity. Table 3.2 summarizes the behaviors of meson (integral spin) wavefunctions for different combinations of J and P.

3.2.3.7 Isotopic Spin

Shortly after the discovery of the neutron and its observed similarity of its rest-masses with that of the proton (939.565 and 938.272 MeV, respectively), Heisenberg proposed in 1932 the concept that they be treated as two different quantum states of the same particle known as the nucleon (Blatt and Weisskopf 1979). The proton and neutron thus reflect a dichotomous quantum degree of freedom of the nucleon through the quantum number isotopic spin (or, more commonly, isospin). In order to develop this concept, an isospin space with the isospin vector **t** associated with the nucleon is introduced. Mathematically, **t** is analogous to the nucleon's spin vector **s** in that there are two possible components of **t** in the direction of the third axis in isospin space. That is, the variable t_3 has two values of ±1/2 corresponding to the charge state of the nucleon. For the proton, $t_3 = +1/2$, and for the neutron, $t_3 = -1/2$. In units of the fundamental electric charge, e, the charge of the nucleon is related to t_3 and the baryon number by,

$$Q = t_3 + \frac{B}{2} \tag{3.11}$$

For a composite particle made up of N_u up and N_d down quarks, the third component of the isospin vector is,

$$t_3 = \frac{N_u - N_d}{2}. \tag{3.12}$$

For pions, which contain antiquarks, (3.12) is modified,

$$t_3 = \frac{N_u - N_{\bar{u}} - (N_d - N_{\bar{d}})}{2}. \tag{3.13}$$

For example, the π^+ contains an up quark and an anti-down quark and, hence, has $t_3 = +1$. Isospin is an additive quantum number and, as the strong nuclear force is charge independent, then $\mathbf{t}$ and t_3 are conserved in strong interactions. However, they are not conserved in the weak interaction. For example, the beta decay of a neutron $n \to p + e^- + \bar{\nu}_e$, involves the change of a down quark to an up quark (a quark isospin flip) and, hence, $\mathbf{t}$ and t_3 are not conserved. In an electromagnetic interaction, there is no isospin flip but because of the different electrical charges of the quarks, they can be distinguished. Thus, $\mathbf{t}$ is not conserved, but t_3 is.

Consider a two-nucleon system. The total isospin of this system is made up of the two isospins of the nucleons as it is for ordinary spin,

$$\mathbf{T} = \mathbf{t}(1) \oplus \mathbf{t}(2) \tag{3.14}$$

and the eigenvalues for $\mathbf{T}^2$ are T(T+1) where $t(1) - t(2) \le T \le t(1) + t(2)$. The two-nucleon system can be an isospin triplet (i.e., $T = 1$ with $t_3 = -1$, 0, 1) or an isospin singlet (i.e., $T = 0$ with $t_3 = 0$). The only stable bound two-nucleon system is the deuteron (proton–neutron) which is isospin singlet. As with spin, a nucleon isospin operator is defined through the identity,

$$\mathbf{t} = \frac{1}{2}\boldsymbol{\tau} \tag{3.15}$$

where the three components of $\boldsymbol{\tau}$ are analogous to the Pauli spin matrices (Pauli 1973),

$$\boldsymbol{\tau}_1 = \begin{pmatrix} 0 & 1 \\ 1 & 0 \end{pmatrix} \quad \boldsymbol{\tau}_2 = \begin{pmatrix} 0 & -i \\ i & 0 \end{pmatrix} \quad \boldsymbol{\tau}_3 = \begin{pmatrix} 1 & 0 \\ 0 & -1 \end{pmatrix} \tag{3.16}$$

One defines an isospin space, in which the proton and neutron are doublets of states described by the eigenvalue of the Hermitian operator, $\boldsymbol{\tau}_3$,

$$\boldsymbol{\tau}_3|p\rangle = |p\rangle \tag{3.17}$$

$$\boldsymbol{\tau}_3|n\rangle = -|n\rangle \tag{3.18}$$

where the above kets are,

$$|p\rangle = \begin{pmatrix} 1 \\ 0 \end{pmatrix} \quad |n\rangle = \begin{pmatrix} 0 \\ 1 \end{pmatrix}. \tag{3.19}$$

Ladder operators are defined as those operators which convert one nucleon state to another and are made up from weighted combinations of the isospin operators,

$$\boldsymbol{\tau}_\pm = \frac{1}{2}(\boldsymbol{\tau}_1 \pm i\boldsymbol{\tau}_2) \tag{3.20}$$

These operators lead to descriptions of the transformations,

$$\boldsymbol{\tau}_+|n\rangle = |p\rangle \tag{3.21}$$

$$\boldsymbol{\tau}_+|p\rangle = 0 \tag{3.22}$$

$$\boldsymbol{\tau}_-|p\rangle = |n\rangle \tag{3.23}$$

$$\boldsymbol{\tau}_-|n\rangle = 0. \tag{3.24}$$

The ladder operators perform isospin flips on the nucleon: $\boldsymbol{\tau}_+$ annihilates a proton and transforms a neutron to a proton and $\boldsymbol{\tau}_-$ transforms a proton to a neutron and annihilates a neutron. Hence, these operators will prove useful in the description of beta decay in Chap. 4 (Grotz and Klapdor 1990).

The analogue of (3.12) for a nucleus consisting of A nucleons (Z protons and N = A – Z neutrons) is,

$$\begin{aligned} T_3 &= \frac{Z - N}{2} \\ &= \frac{A}{2} - N. \end{aligned} \tag{3.25}$$

3.2.4 Nuclear Constituents: Nucleons

The atom is known to be composed of a positively-charged nucleus surrounded by electrons. Even though the diameter of the atom is of the order of 10^{-8} cm and that of the nucleus is only of the order of 10^{-12} cm, more than 99.9% of the atomic mass is concentrated in the nucleus. The nucleus itself is made up of positively-charged protons and electrically-neutral neutrons. Like all hadrons, the neutron and proton have an internal quark substructure. The proton consists of two up quarks, each with an electric charge of $+{}^{2}/_{3}e$ and isospin projection $t_3 = +1/2$, and one down quark with an electric charge of $-{}^{1}/_{3}e$ and isospin projection $t_3 = -1/2$, resulting in the proton's electric charge of $+1e$ and isospin projection $t_3 = +1/2$. On the other hand, the neutron is composed of one up quark and two down quarks leading to a net electric charge of zero and an isospin projection of $t_3 = -1/2$. Despite their difference in electric charge, the neutron and proton are very nearly equal in mass (the neutron being only 0.1% more massive than the proton) and have the same intrinsic spin, $\hbar/2$. A consequence of this latter feature is that both particles are fermions subject to Fermi–Dirac statistics and the Pauli exclusion principle. The slight mass difference between the proton and the neutron allows the instability of the free neutron which undergoes beta-decay to yield a proton.[9]

3.2.5 Categorizations of Nuclei

The total number of nucleons contained in a nucleus is characterized by the atomic mass number, A, and the total number of protons which is described by the atomic number, Z. Nuclei with differing combinations of A and Z are categorized as follows.

[9]The converse is not possible for a free proton. However, it is possible for a nuclear proton to be transformed to a neutron, as in β^+ decay, where the energy required is gained from the Fermi momentum of the proton.

3.2.5.1 Isotopes

Isotopes[10] are those nuclei with the same atomic number, Z, but different atomic mass number, A. In other words, isotopes are of the same element but with different masses (i.e., same number of protons, but differing numbers of neutrons). One example is the three isotopes of hydrogen: ^{1}H, ^{2}H, and ^{3}H, each with one proton but containing zero, one, and two neutrons, respectively. Although isotopes have the same atomic number and, hence, the same number of orbital electrons, they can have slightly different physical or chemical properties due to their differing masses. For example, the boiling points of ordinary water ($^{1}H_2O$) and heavy water ($^{2}H_2O$) are 373.2 and 374.6 K. Another difference is manifested by the dependence of molecular vibrational modes upon mass. As a result, the same molecules but with differing isotopes can exhibit differing optical absorption properties.

3.2.5.2 Isobars

Isobars are those nuclei with the same atomic mass number, A (i.e., the same number of nucleons), but differing atomic numbers, Z. An example of a pair of isobars is ^{3}H and ^{3}He, with one proton and two neutrons and two protons and one neutron, respectively. Isobaric nuclei play a prominent role in $\beta^{\pm}$ decay and electron capture.

3.2.5.3 Isotones

Isotones are those nuclei with the same number of neutrons, $N = A - Z$, but with different numbers of protons, Z. An example of an isotone pair is ^{2}H and ^{3}He, each with one neutron but with one and two protons, respectively.

[10]The name comes from the Greek *iso* = same*; topos* = place as the variants would all be at the same position in the periodic table of elements. Anecdotally, the term was suggested by Margaret Todd, a Scottish physician, to Frederick Soddy who had discovered atoms with identical chemical properties but with different atomic masses.

3.2.5.4 Isomers

Isomers are excited states of a nucleus with the same A and Z. These are studied in Chap. 4.

3.2.6 Nuclear Mass

3.2.6.1 Atomic Mass Unit

Each elemental species has been found experimentally to have a mass very nearly equal to an integer multiple of the proton mass (i.e., the mass of the ionized hydrogen atom), where this integer is equal to the total number of nucleons or the atomic mass number,

$$A = Z + N \tag{3.26}$$

where Z is the number of protons and N is the number of neutrons. The fact that the mass of an elemental species is only approximately equal to an integer multiple of the proton mass reflects the underlying presence of isotopes, as discussed below.

The fundamental unit of mass (known as the atomic mass unit, or amu) is defined as 1/12th the mass of the ^{12}C atom (by convention, the atomic mass is used, rather than the nuclear mass, in the tabulation of masses). The mass defect is the difference between the measured atomic mass of an atom (of atomic number Z and atomic mass number A) and that mass if the atom were considered as an ensemble of Z hydrogen atoms and N neutrons,

$$\text{Mass Defect} = M(A, Z) - (ZM_H + Nm_n) \tag{3.27}$$

where M_H is the mass of the hydrogen atom and m_n is the neutron mass.[11] The term mass defect is used as this difference between the mass of the ensemble and the sum of the constituent masses is negative due to binding energy. For example, the atomic mass of ^{12}C is 11,178 MeV whereas the combined mass of 6 hydrogen atoms and 6 neutrons is (neglecting the negligible hydrogen electron binding energy) is 11,270 MeV giving a mass defect of 92 MeV.

The concept of atomic or nuclear masses being integral multiples of a fundamental mass was first proposed by Prout in the early nineteenth century in a hypothesis that elements were made up of combinations of hydrogen atoms. However, refined measurements by others, such as Berzelius, demonstrated that the atomic weights of elements were not always an integer multiple of that of hydrogen. For example, the atomic masses of naturally-occurring magnesium and copper are 24.3 and 63.5, respectively. These deviations from Prout's hypothesis were the first evidence of the existence of isotopes which was demonstrated by Thomson in 1912 through his discovery of two isotopes of neon, using the principle of electromagnetic mass separation. These isotopes have atomic masses of 20 and 22 and a corresponding 9:1 abundance ratio; neglecting the small contribution of ^{21}Ne, the relative atomic mass of naturally-occurring neon is $0.9 \times 20 + 0.1 \times 22 = 20.3$.

3.2.6.2 Determination of Nuclear Mass

The mass of a nucleus or any charged particle can be determined using the method of electromagnetic mass separation, the principle of which is based upon the fact that a charged particle with a nonzero initial speed in a magnetic field orthogonal to the particle's velocity vector will orbit with a radius proportional to the magnitude of its linear momentum. Hence, a measurement of this radius will yield the momentum and, if the particle speed is known, the particle's mass can be obtained. Particles with a range of speeds can be "filtered" through an orthogonal combination of static electric and magnetic fields so that only those particles with a known speed are introduced into this orthogonal static magnetic field. The basic principle behind this method is developed here.

The velocity of a charged particle can be determined by passing the particle through a crossed set of magneto- and electrostatic fields, or a Wein filter, which selects only those particles with a velocity equal to the ratio of the magnitudes of the electric to magnetic field strengths. This can be shown by first calculating the Lorentz force upon a charged particle of

[11] The rest-masses of the neutron, proton, and electron are (to three decimal places) 939.565, 938.272, and 0.511 MeV, respectively. For simplicity in some of the following discussions, both nucleon rest-masses will be approximated by the value of 940 MeV.

mass m, electric charge q and velocity $\mathbf{v}$ in a magnetic field, $\mathbf{B}$,

$$\mathbf{F_B} = q\mathbf{v} \times \mathbf{B} \tag{3.28}$$

with the direction of the force being perpendicular to both the directions of the velocity and that of the magnetic field. In the absence of any other force, the particle will thus orbit with a radius derived by equating the magnitudes of the centripetal and Lorentz forces,

$$\frac{m|\mathbf{v}|^2}{r} = q|\mathbf{v}||\mathbf{B}| \tag{3.29}$$

to yield,

$$\begin{aligned} r &= \frac{m|\mathbf{v}|}{q|\mathbf{B}|} \\ &= \frac{|\mathbf{p}|}{q|\mathbf{B}|}. \end{aligned} \tag{3.30}$$

The orbital frequency of the particle is known as the cyclotron frequency,

$$\begin{aligned} \omega_c &= \frac{|\mathbf{v}|}{r} \\ &= q\frac{|\mathbf{B}|}{m}. \end{aligned} \tag{3.31}$$

A static field $\mathbf{E}$ is next applied in the direction opposing the Lorentz force, $\mathbf{F_B}$. The particle will then experience an additional force,

$$\mathbf{F_E} = q\mathbf{E} \tag{3.32}$$

and the resulting trajectory of the particle will now depend upon its initial velocity and the relative magnitudes of the crossed electro- and magnetostatic fields. For example, consider the initial velocity of the particle to be in the x direction with magnitude v_0, $\mathbf{E}$ directed along the y-axis and $\mathbf{B}$ directed along the z direction. The equation of motion of the particle is,

$$\begin{aligned} m\frac{d\mathbf{v}}{dt} &= q\mathbf{E} + q\mathbf{v} \times \mathbf{B} \\ &= qv_yB\hat{\mathbf{x}} + q(E - v_xB)\hat{\mathbf{y}} \end{aligned} \tag{3.33}$$

the vector components of the acceleration are,

$$\begin{aligned} \frac{dv_x}{dt} &= \frac{q}{m}v_yB \\ &= \omega_c v_y \end{aligned} \tag{3.34}$$

$$\begin{aligned} \frac{dv_y}{dt} &= \frac{q}{m}(E - v_xB) \\ &= \frac{q}{m}E - \omega_c v_x \end{aligned} \tag{3.35}$$

$$\frac{dv_z}{dt} = 0 \tag{3.36}$$

where the definition of the cyclotron frequency has been used. These coupled differential equations are solved by first differentiating dv_y/dt and inserting (3.34) into the result,

$$\begin{aligned} \frac{d^2v_y}{dt^2} &= -\omega_c\frac{dv_x}{dt} \\ &= -\omega_c^2 v_y. \end{aligned} \tag{3.37}$$

As the initial condition is $v_y(t = 0) = 0$, the solution to (3.37) is,

$$v_y(t) = k \sin \omega_c t. \tag{3.38}$$

Differentiating (3.38) and inserting the result into (3.35) yields an expression for v_x,

$$v_x = \frac{1}{\omega_c}\frac{q}{m}E + k \cos \omega_c t. \tag{3.39}$$

The integration constant, k, is determined from the initial condition of the velocity, $v_x(t = 0) = v_0$,

$$\begin{aligned} k &= v_0 - \frac{1}{\omega_c}\frac{q}{m}E \\ &= v_0 - \frac{E}{B}. \end{aligned} \tag{3.40}$$

Thus, expressions for the two components of the particle velocity are obtained as functions of time,

$$v_x(t) = \frac{E}{B} + \left(v_0 - \frac{E}{B}\right) \cos \omega_c t \tag{3.41}$$

and

$$v_y(t) = \left(v_0 - \frac{E}{B}\right) \sin \omega_c t. \tag{3.42}$$

Parametric equations of the spatial position of the particle are then derived by integrating (3.41) and (3.42),

$$x(t) = \left(\frac{E}{B}\right)t + \frac{\left(v_0 - \frac{E}{B}\right)}{\omega_c} \sin \omega_c t \tag{3.43}$$

$$y(t) = -\frac{\left(v_0 - \frac{E}{B}\right)}{\omega_c}(1 - \cos \omega_c t) \tag{3.44}$$

for the initial condition $y(t = 0) = 0$. If the particle is initially at rest ($v_0 = 0$), then the resulting trajectory is a cycloid,

$$x(t) = \left(\frac{E}{B}\right)\left(t - \frac{\sin \omega_c t}{\omega_c}\right) \tag{3.45}$$

$$y(t) = \left(\frac{E}{B}\right)\frac{(1 - \cos \omega_c t)}{\omega_c}. \tag{3.46}$$

For a nonzero initial velocity, (3.43) and (3.44) describe a trochoid trajectory. Examples of trochoid trajectories for $v_0 = \pm 0.5\frac{E}{B}$, $v_0 = \pm 2\frac{E}{B}$ and $v_0 = \frac{E}{B}$ are shown in Fig. 3.4. Nodes in the trajectories ($y = 0$) occur for values of x separated by $\frac{2\pi}{\omega_c}\frac{E}{B}$. Most importantly, a straight line trajectory ($y(t) = 0$) is obtained when the ratio of the electrostatic and magnetostatic fields is matched to the particle speed,

$$v_0 = \frac{E}{B}. \tag{3.47}$$

If a beam of charged particles with a spectrum of velocities enters the crossed static fields, the resulting trajectories will be a combination of trochoids. In the Wein filter, the addition of an arrangement of slits along the x-axis of Fig. 3.4 will select only those particles moving in a straight line, i.e., those with $v_0 = E/B$. If these iso-velocity particles are then passed into a region containing a static magnetic field, B_1, orthogonal to the exit velocity, they will then curve with the radius, $r = p/B_1$. Hence, a measurement of this orbital radius, r, with a spatially-sensitive detector yields the magnitude of the particle momentum, p, as shown in Fig. 3.5. This momentum, when combined with the known velocity from the Wein filter, yields the particle's rest mass. For the nonrelativistic case, this is,

$$\begin{aligned} m &= \frac{p}{v} \\ &= r\frac{B_1 B}{E}. \end{aligned} \tag{3.48}$$

3.2.7 Nuclear Size

3.2.7.1 Introduction

Defining what is exactly meant by the term nuclear size is difficult. If the nucleus were assumed to be a

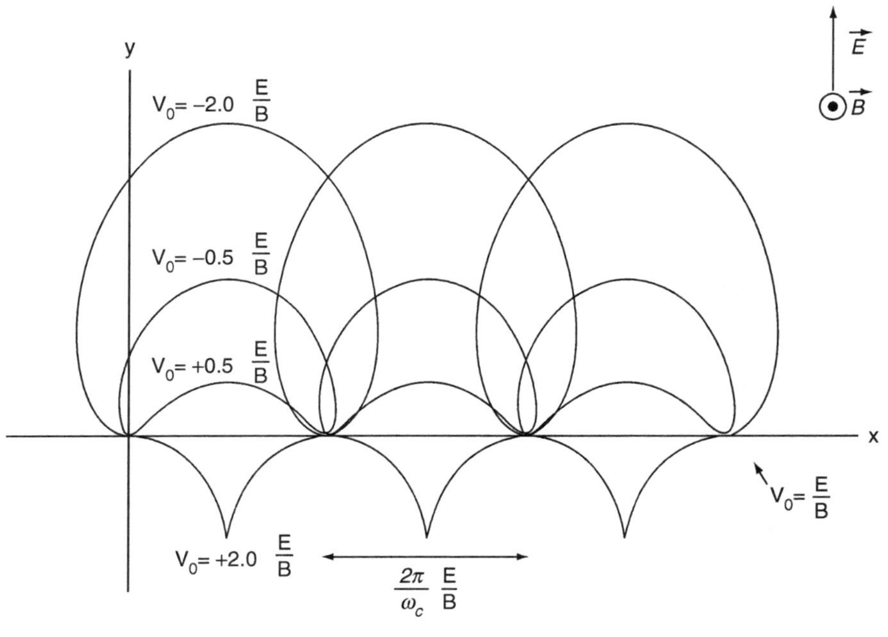

Fig. 3.4 Trajectories of a charged particle in crossed electro-and magnetostatic fields with five different magnitudes of initial velocity (the particle enters the fields with an initial velocity v_0 in the + x-direction)

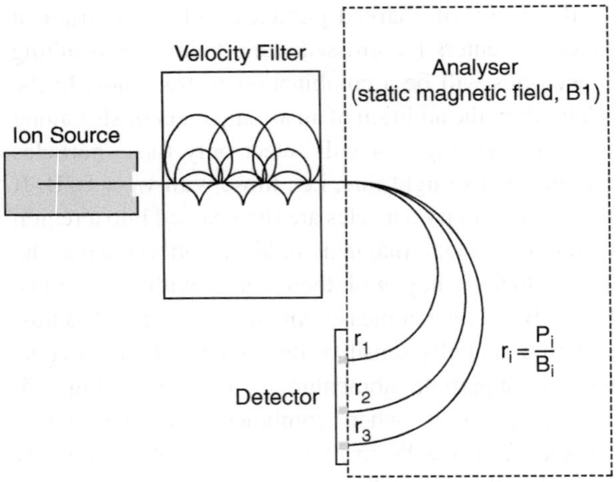

Fig. 3.5 Principle of electromagnetic mass separation

simple sphere with a well-defined radius (and one which will be made frequent use of) then the nuclear size is simply defined by the radius. But it should be recognized that this is obviously a simplistic and unphysical picture and not at all representative of reality. The occasion of when a nucleus is spherical (but not necessarily with a sharply-defined radius) is representative of special properties of stability; nuclei can be nonspherical (oblate or prolate). In this subsection, spherical models of the nucleus and how their size, in particular the root-mean square (RMS) radius, can be determined are reviewed. Discussion of nonspherical nuclei is deferred to the discussion of electric quadrupole moments following derivations of nuclear models.

The theoretical bases for three means of determining nuclear dimensions are reviewed here. The first is a simple means of determining the sizes of mirror nuclei (a subset of isobars consisting of pairs of nuclei for which the numbers of neutrons and protons are interchanged, e.g., ^3H and ^3He) by comparing the difference in nuclear binding energies and assuming that this difference is due solely to the Coulomb interaction between protons. The second method of deducing nuclear size is by measuring the energies of the X-rays that are emitted when a negatively-charged muon is captured by an atom and cascades through Bohr atomic orbitals before it eventually decays. The third method of determining the size and, importantly, the spatial distribution of nuclear matter is through scattering probes from nuclei. As an understanding of the underlying theory of Coulomb scatter of charged particles will be of significant importance later, considerable detail will be paid to this third method of determining nuclear size.

3.2.7.2 Nuclear Size Derived from Nuclear Binding Energies

The very existence of nuclei is clear evidence that the strong nuclear force binding the nucleons counterbalances the Coulomb repulsion between protons. As the nuclear attractive force is evidently the same between the three categories of nucleon pairs (proton–proton, proton–neutron, and neutron–neutron), any differences between the measured binding energies of mirror nuclei should be due solely to the difference due to the Coulomb energy.[12] For example, the measured binding energy of ^3He is 8.482 MeV whereas that of ^3H is 7.718 MeV, the ^3He necessarily having a higher binding energy in order to compensate for the Coulomb repulsion between the protons. This 0.764 MeV difference in binding energy thus reflects the difference between the Coulomb potentials of the two nuclei. The Coulomb potential of a spherical nucleus of radius R_N and electric charge Ze uniformly distributed within it is,[13]

$$E_{Coul} = \frac{3}{5}(\alpha\hbar c)\frac{Z^2}{R_N}. \tag{3.49}$$

Assuming that the nuclear volume scales linearly with the number of nucleons, A, the nuclear radius will be proportional to $A^{1/3}$. Thus, R_N should be the same for ^3H and ^3He and can be calculated from the difference in Coulomb potentials,

$$\Delta E_{Coul} = \frac{3}{5}(\alpha\hbar c)\left(\frac{1}{R_N}\right)(4-1)$$
$$= 0.764\,\text{MeV} \tag{3.50}$$

[12]There would also be a small effect due to the difference between proton and neutron masses, but we will neglect this.

[13]Consider a spherical nucleus of charge Ze and radius R. The charge density is a constant, $\rho_e = 3Ze/4\pi R^3$ and the total work required to assemble the nucleus from differential shells containing charge $dq = 4\pi r^2 \rho_e\, dr$ from infinity to the position of the nucleus is $\frac{\alpha\hbar c}{e^2}\int_0^{Ze} dq\frac{q}{r}$ which, noting that $q(r) = {}^{4\pi}/_{3}\rho_e r^3$, gives the Coulomb repulsion energy as $\left({}^3/_5\right)\alpha\hbar c\frac{Z^2}{R}$.

which yields the nuclear radius of $R_N = 3.4$ fm for ^3H and ^3H nuclei (Sect. 3.2.7.2).

Muons are produced in accelerators through the decay of the charged pions,[14] $\pi^- \rightarrow \mu^- + \bar{\nu}_\mu$ and $\pi^+ \rightarrow \mu^+ + \nu_\mu$, and slow down in matter through energy exchange with atomic electrons (through processes described in Chap. 7). In particular, the μ^-, once having been thermalized, can be captured by an atomic nucleus in a time span of the order of 0.1 ps. As this amount of time is greater than the muon's mean life of 2.2 μs (through the weak decay $\mu^- \rightarrow e^- + \bar{\nu}_e + \nu_\mu$), a muonic atom can be formed as, once captured, the μ^- will cascade to the 1s muonic orbital.[15] Assume, for the time being, that the nucleus is a point-like electric charge, the muonic atom is hydrogen-like but with a Bohr orbital radius of,

$$r_\mu = \frac{n^2}{Z} \frac{r_0}{\alpha^2} \left(\frac{m_e}{m_\mu} \right) \tag{3.51}$$

where r_0 is the classical electron radius and n is the principal quantum number. Because of its mass, the muon 1s orbital radius will be much smaller than that of the atomic electrons so that its interactions with the atomic electrons can be ignored. As the energy of the muon orbital will be m_μ/m_e times that of the corresponding electron orbital, the energies of the X-rays emitted by the muon as it cascades through orbitals will be 207 times greater than those of electrons.

The above description is based upon the assumption that the nucleus is a point-like electric charge. Experimental differences between the muon orbital energies calculated for a point-like nucleus and those measured are a consequence of the finite size of the nucleus, an effect which will be clearly more noticeable for heavy nuclei due to the Z^{-1} dependence of (3.51). For example, consider a lead muonic atom for which, from (3.51), the 1s orbital radius for the muon is 3.1 fm whereas (as shown later) the lead nuclear half-density radius is about 6.5 fm. Hence, the muon wavefunction will overlap the nuclear wavefunction and the point-like nucleus assumption is no longer valid. As the muon in a 1s orbital spends much of its time within the nucleus, the effective Coulomb potential felt by the muon is less than that due to the total nuclear charge of Ze. Hence, the X-ray energies between transitions in a muonic atom will be less than those calculated for a point-like nucleus. Calculating an effective potential due to a reduced effective Ze and fitting to the measured muonic X-ray energy provides a measure of the RMS radius of the electric charge distribution of the nucleus.

This approach of measuring the difference between the X-ray energy calculated for a point nucleus and that measured can also be applied to atomic electron transitions to measure the relative nuclear radii for a set of isotopes (i.e., a fixed Ze).

3.2.7.3 Nuclear Size Derived from Charged Particle Scattering

Introduction

The dimensions of a nucleus and the spatial distribution of its charged constituents (protons) can be determined by scattering charged particle probes off of nuclei and detecting the scattered particle. Measurements of the scattered particle's kinematics and the probability with which the particle is scattered at a given angle can then be used to derive the nuclear size and distribution.[16] The "type" of nuclear size measured will obviously depend upon the nature of the interaction between the nucleus and the probe. For example, electrons interact with a nucleus only through the electrostatic potential (neglecting the weak force and magnetic interactions) and will probe the charge distribution of the nucleus which is proportional to the spatial distribution of protons.[17] Because of the difficulty in measuring the spatial distribution of mass (i.e., the nucleon density) within the nucleus, it is frequently assumed that the nuclear mass and electric charge densities are proportional,

[14]The pion can be produced by, for example, proton bombardment of targets such as beryllium, carbon, and copper.

[15]The formation of muonic atoms was predicted by Fermi and Teller (1947).

[16]While neutron scattering can also be used to infer the nucleon spatial distribution, we will restrict ourselves to the problem of elastic Coulomb scatter from nuclei, which will be more relevant to nuclear medicine.

[17]This statement implicitly assumes that the de Broglie wavelength of the incident electron is of the order of nucleon dimensions. At higher electron momenta, this wavelength becomes of the order of subnucleon dimensions allowing mapping of the electrical distributions of quarks within the nucleons.

$\rho_m(\mathbf{r}) = (^A\!/_Z)\rho_e(\mathbf{r})$. However, it cannot be forgotten that the proton has a finite size with a charge RMS radius of 0.87 fm so that the nuclear charge density, $\rho_e(r)$, will be given by the convolution of the point proton density within the nucleus and the electric charge density of an individual proton. We will return to this later in our discussion of nuclear RMS radii. On the other hand, a neutron probe interacts with the nucleus via the "strong" nuclear force and so will measure the mass distribution (i.e., both protons and neutrons). Charged hadrons, such as protons or α particles (^{4}He nuclei), interact with the nucleus via both electromagnetic and nuclear forces, although the former dominates unless the charged particle is sufficiently energetic to overcome the repulsive Coulomb barrier of the nucleus. Hence, analysis of the scattering data of such probes requires the separation between nuclear and electromagnetic interactions. In the subsequent discussion, only the Coulomb interaction between the probe and the nucleus is considered (Goldberger and Watson 1964; Belkić 2004).

Kinematics

The spatial resolution explored by a scattering reaction is inversely proportional to the bombarding particle's momentum, as follows from the reduced de Broglie wavelength $\lambda\!\!\!^{-} = \hbar c/p$. For example, an electron with a kinetic energy of 1 GeV has a momentum of $p = \sqrt{(T + m_e)^2 - m_e^2} \approx 1$ GeV/c and an associated de Broglie wavelength of the order of 0.2 fm. Moreover, as noted earlier, the electron has the advantage of not being subject to the strong nuclear force. The kinematics of an electron scattered from light and massive nuclei were previously derived in Chap. 2.

Elastic Coulomb Scatter

Spin-0 Projectiles

The first empirical assessment of nuclear size was performed by Rutherford in his analysis of the measurements by Marsden and Geiger of the scattering of α particles (^{4}He nuclei produced by radioactive decays) from thin ($\approx$0.3 μm) gold foils. In this section, perturbation theory is used to derive the differential cross-section of elastic Coulomb scatter of a probe of mass m and charge +ze from an infinitely-massive scattering center of charge +Ze (e.g., a nucleus much more massive than the projectile).[18] There are two assumptions used in this derivation. The first is that we can neglect the spin of the projectile and the angular momentum of the nucleus providing the scattering Coulomb field. This reflects an α particle projectile incident to a spin-0 nucleus. Despite this simplification, the results are valid for electron projectiles with low β and/or low scattering angle (equivalent to low momentum transfer). The second assumption is that the scatterer is a point electric charge of sufficiently large mass that we can ignore its recoil. In this scatter, the projectile is scattered from an initial momentum $\mathbf{p}$ and kinetic energy T through an angle θ to the momentum $\mathbf{p}'$ and, as the collision is deemed to be elastic, the postscatter kinetic energy of the projectile, T', is obviously unchanged from the initial value. Although the magnitudes of the pre- and postscatter momenta are the same, there will be a three-vector momentum transfer due to the directional change resulting from the scatter,

$$\begin{aligned} q &= \left|\mathbf{p} - \mathbf{p}'\right| \\ &= \sqrt{p^2 + p'^2 - 2pp' \cos\theta} \\ &= p\sqrt{2(1 - \cos\theta)} \\ &= 2p \sin\frac{\theta}{2}. \end{aligned} \tag{3.52}$$

Recall from Chap. 2 that the differential cross-section in solid angle is equal to the squared modulus of the scattering amplitude,

$$\frac{d\sigma_{Ruth}}{d\Omega} = |f(\mathbf{q})|^2. \tag{3.53}$$

[18]A collision with a nucleus is strictly elastic if no energy is transferred to any internal energy absorption channels. Even ignoring these energy absorption channels, the simultaneous conservation of energy and momentum requires that the projectile lose a small amount of energy which is transferred to the recoil kinetic energy of the atom. This energy transfer will be of the order of the product of the incident kinetic energy and the ratio of the projectile to target atom rest masses. Clearly, for electron and positron projectiles, the target recoil kinetic energy can be neglected (the ratio of projectile to target electron mass being no smaller than 5×10^{-4}). This may not be the case if the projectile was an energetic alpha particle and the target atom had a comparable atomic mass which would be typical of elements of biological interest (e.g., carbon and oxygen).

In the Born approximation, the scattering amplitude is proportional to the Fourier transform of the potential between the scattering center and the projectile,

$$f(\mathbf{q}) = -\frac{m}{2\pi(\hbar c)^2}\int d^3\mathbf{r}\, e^{i\left(\frac{\mathbf{q}\bullet\mathbf{r}}{\hbar c}\right)} U(\mathbf{r}). \tag{3.54}$$

The Ruth subscript applied to the differential cross-section denotes that it is of elastic Coulomb scatter from an infinitely-massive point-charge scattering center (i.e., Rutherford scatter). For the Coulomb potential between point charges ze and Ze,

$$U(r) = \alpha\hbar c\frac{zZ}{r} \tag{3.55}$$

the scattering amplitude is,

$$f(\mathbf{q}) = -\left(\frac{\alpha}{2\pi\hbar c}\right) mzZ\int d^3\mathbf{r}\frac{e^{i\frac{\mathbf{q}\bullet\mathbf{r}}{\hbar c}}}{|\mathbf{r}|}. \tag{3.56}$$

The integral in (3.56) is frequently referred to as the Bethe integral and can be readily solved by rewriting it as,

$$\int d^3\mathbf{r}\frac{e^{i\frac{\mathbf{q}\bullet\mathbf{r}}{\hbar c}}}{|\mathbf{r}|} = \int d^3\mathbf{r}\frac{e^{-\kappa r}}{|\mathbf{r}|}e^{i\frac{\mathbf{q}\bullet\mathbf{r}}{\hbar c}} \quad \text{as } \kappa \to 0. \tag{3.57}$$

Solving,

$$\begin{aligned}\int d^3\mathbf{r}\frac{e^{-\kappa r}}{|\mathbf{r}|}e^{i\frac{\mathbf{q}\bullet\mathbf{r}}{\hbar c}} &= 2\pi\int_0^\infty dr\, r\, e^{-\kappa r}\int_{-1}^{1} d(\cos\theta)e^{i\frac{qr\cos\theta}{\hbar c}}\\ &= 4\pi\left(\frac{\hbar c}{q}\right)\int_0^\infty dr\, e^{-\kappa r}\sin\frac{qr}{\hbar c}\\ &= \frac{4\pi}{\kappa^2 + \left(q/\hbar c\right)^2}.\end{aligned} \tag{3.58}$$

Thus, the solution to the Bethe integral is,

$$\int d^3\mathbf{r}\frac{e^{i\frac{\mathbf{q}\bullet\mathbf{r}}{\hbar c}}}{|\mathbf{r}|} = 4\pi\left(\frac{\hbar c}{q}\right)^2 \tag{3.59}$$

and the scattering amplitude for the Coulomb potential of a point-like source is,

$$f(q) = -2\,(\alpha\hbar c)\,\frac{mzZ}{q^2}. \tag{3.60}$$

Note that the directional dependence of the momentum transfer has been removed. The differential cross-section, following the substitution of the expression linking scattering angle and momentum transfer, is the familiar Rutherford form,

$$\frac{d\sigma_{Ruth}}{d\Omega} = \frac{\left(\frac{zZ\alpha\hbar c}{p\beta}\right)^2}{4\sin^4\frac{\theta}{2}} \tag{3.61}$$

which, for small scattering angles, further reduces to the form,

$$\frac{d\sigma_{Ruth}}{d\Omega} = 4\left(\frac{zZ\,\alpha\hbar c}{p\beta\,\theta^2}\right)^2 \text{(Small - angle approximation).} \tag{3.62}$$

It will be noted that the Rutherford cross-section diverges as the scattering angle $\theta \to 0$ and a correction for this is discussed later in the context of screening by the atomic electrons. Importantly, it can also be seen that the cross-section is independent of the signs of the electric charges of the probe and the target nucleus and, hence, the attractive or repulsive nature of the Coulomb potential is irrelevant. The Rutherford differential cross-section in solid angle is plotted in Fig. 3.6 for the example of a 7 MeV α particle incident to a gold nucleus, a combination typical of the experiments performed by Geiger and Marsden.

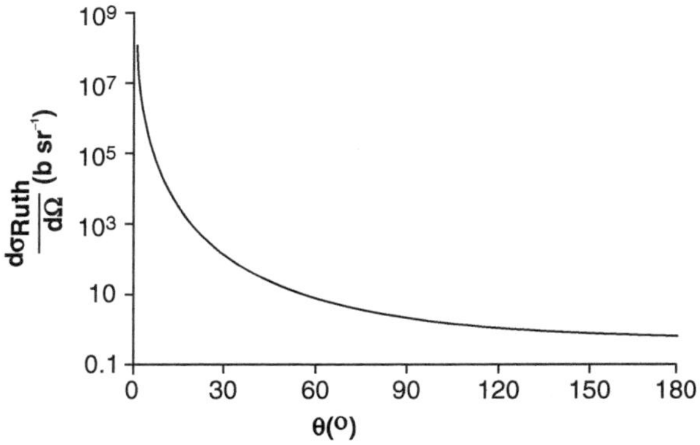

Fig. 3.6 Rutherford differential cross-section in solid angle for 7 MeV α particles incident on a gold target as a function of scattering angle; units are in barns per steradian where 1 b = 10^{-24} cm^2

The elastic Coulomb differential cross-section in momentum transfer is given by the transformation,

$$\frac{d\sigma_{Ruth}}{dq} = \frac{d\sigma_{Ruth}}{d\Omega}\left|\frac{dq}{d\Omega}\right|^{-1} = 2\pi\frac{d\sigma_{Ruth}}{d\Omega}\left|\frac{dq}{d(\cos\theta)}\right|^{-1}. \tag{3.63}$$

From the momentum transfer,

$$\left|\frac{dq}{d(\cos\theta)}\right| = \frac{p}{2\sin\left(\frac{\theta}{2}\right)} \tag{3.64}$$

and the differential cross-section in momentum transfer is,

$$\frac{d\sigma_{Ruth}}{dq} = \frac{8\pi}{q^3}\left(\frac{zZ\alpha\hbar c}{\beta}\right)^2 \tag{3.65}$$

This calculation has modeled the nucleus by a point-like electric charge. If the dimensional scale of the scattering should now be reduced (e.g., by increasing the momentum of the projectile and reducing its de Broglie wavelength) so that the nucleus is to be "seen" by the projectile as a spatial distribution of electric charge and the probe is subject to only the electromagnetic force, then the electric potential felt by the probe will no longer be that of $(\alpha\hbar c)Z/r$ as it approaches the nucleus. Thus, when the scattering cross-section deviates from that derived for a point-like scattering center, a measure of the nuclear radius can be estimated, as will be demonstrated shortly. Similarly, if the probe is subject to both the electromagnetic and strong nuclear forces, as is the α particle, then the deviation from the pure Rutherford result is indicative of the distance from the center of the nucleus at which the probe begins to experience the strong nuclear force.

Spin-1/2 Projectiles

A high-energy electron probe is the most practical means of achieving a reduced de Broglie wavelength comparable to nuclear dimensions. Moreover, unlike an α particle, the electron is not subject to the strong interaction and only a pure electromagnetic interaction is measured. However, the above derivation of elastic Coulomb scatter neglected the role of the projectile's spin. While this is acceptable for the scattering of α particles, the electron has intrinsic spin-1/2 and there are conditions for under which this spin must be accounted for.[19] The differential cross-sections of an electron scattered from a static Coulomb field of electric charge + Ze (referred to as Mott scatter (Mott 1929)) are next derived. To invoke spin-1/2, the Dirac theory is employed to guide the development along the lines of propagator formalism as described by, for example, Bjorken and Drell (1964) and Dyson (2007). The differential cross-section obtained here for the electron (to the lowest order of the fine-structure constant, α) is equal to that of the positron, which can be deduced by charge conjugation invariance or else shown explicitly (Bjorken and Drell 1964). Hence, we need not repeat an explicit calculation for the positron cross-section. The transition matrix element for the elastic Coulomb scatter of an electron describes the interaction of the electron with the electromagnetic field,

$$S_{fi} = -ie\int d^3\mathbf{r}\,dt\,\bar{\psi}_f(\mathbf{r},t)\not{A}(\mathbf{r},t)\psi_i(\mathbf{r},t). \tag{3.66}$$

The Feynman "dagger" notation is,

$$\not{A} = \gamma^0 A_0 - \boldsymbol{\gamma}\bullet\mathbf{A} \tag{3.67}$$

where $(\gamma^o, \boldsymbol{\gamma})$ are the 4×4 Dirac matrices described in Chap. 2 and A^0 and $\mathbf{A}$ are the scalar and vector electromagnetic potentials, respectively. The Coulomb gauge in which $\gamma \bullet \mathbf{A} = 0$ is used,

$$\gamma^0 A_0 = -\frac{(Z\alpha\hbar c)}{e}\frac{\gamma^0}{|\mathbf{r}|}. \tag{3.68}$$

Plane-wave approximations are used for the incident and scattered electron,

$$\psi_i(\mathbf{r},t) = \sqrt{\frac{m_e}{EL^3}}u(\mathbf{p},s)e^{\left(i\frac{\mathbf{p}\bullet\mathbf{r}}{\hbar c}-\frac{Et}{\hbar}\right)} \tag{3.69}$$

$$\psi_f(\mathbf{r},t) = \sqrt{\frac{m_e}{E'L^3}}u(\mathbf{p}',s')\,e^{\left(i\frac{\mathbf{p}'\bullet\mathbf{r}}{\hbar c}-\frac{E't}{\hbar}\right)} \tag{3.70}$$

[19]As the electron and positron are the particles of greatest interest to nuclear medicine dosimetery, derivations of their Coulomb scatter cross sections will be necessary for later descriptions of radiation transport in tissue.

where s and s′ are the pre- and postscattering electron spin orientations and E and E′ are the corresponding total energies. The m_e/E factor is required in the normalization in order to compensate for the Lorentz contraction of the volume element along the projectile's motion and to ensure that the spinor probability density remains invariant. The bar over $\psi_f(\mathbf{r},t)$ denotes it as an adjoint spinor and L^3 is the usual volume of normalization. The transition matrix is then,

$$S_{fi} = iZ(\alpha\hbar c)\frac{m_e}{L^3\sqrt{E'E}}u(\mathbf{p}', s)\gamma^0 u(\mathbf{p}, s) \times \int d^3\mathbf{r}\frac{e^{i\frac{\mathbf{q}\cdot\mathbf{r}}{\hbar c}}}{|\mathbf{r}|}\int_0^\infty dt\, e^{-i\frac{(E-E')t}{\hbar}}. \quad (3.71)$$

The space integral is the Bethe integral solved previously and, recalling the definition of the δ-function, $\delta(k) = \frac{1}{2\pi}\int_{-\infty}^{\infty} d\omega\, e^{ik\omega}$, the time integral is,

$$\int_0^\infty dt\, e^{-i\frac{(E-E')t}{\hbar}} = 2\pi\,\delta\left(\frac{E'-E}{\hbar}\right). \quad (3.72)$$

Substituting these two results into the expression for the S-matrix element gives,

$$S_{fi} = i\,8\pi^2 Z(\alpha\hbar c)\frac{m_e}{L^3\sqrt{E'E}} \times \left(\frac{\hbar c}{q}\right)^3 u(\mathbf{p}', s)\gamma^0 u\,(\mathbf{p}, s)\delta\left(\frac{E'-E}{\hbar}\right). \quad (3.73)$$

Having obtained this S-matrix element, the transition probability per incident electron is,

$$p_{fi} = |S_{fi}|^2 \rho_f \quad (3.74)$$

where the phase–space factor ρ_f is the density of final states per unit energy level dT′,

$$\rho_f\, dT' = \left(\frac{L}{2\pi\hbar c}\right)^3 d^3\mathbf{p}'. \quad (3.75)$$

Expanding the differential, the phase–space factor is,

$$\rho_f = \frac{1}{4\pi^2}\left(\frac{L}{\hbar c}\right)^3 p'^2\frac{dp'}{dT'}d(\cos\theta) = \frac{1}{4\pi^2\beta'}\left(\frac{L}{\hbar c}\right)^3 p'^2 d(\cos\theta). \quad (3.76)$$

As

$$|d(\cos\theta)| = \frac{q}{p^2}dq$$

the phase–space factor is

$$\rho_f = \frac{1}{4\pi^2\beta}\left(\frac{L}{\hbar c}\right)^3 q\,dq \quad (3.77)$$

where $\beta = \beta'$ and $p = p'$ due to the elasticity of the scatter and the negligible recoil of the scatterer. The squared magnitude of the S-matrix element to be used in (3.74) is,

$$|S_{fi}|^2 = 64\pi^4 Z^2(\alpha\hbar c)^2\frac{m_e^2\hbar^2}{L^3E'E}\left(\frac{\hbar c}{q}\right)^4 \times \left|\bar{u}(\mathbf{p}', s)\gamma^0 u(\mathbf{p}, s)\right|^2\left|\delta\left(\frac{E'-E}{\hbar}\right)\right|^2. \quad (3.78)$$

A derivation of the square of the δ-function has been provided by Bjorken and Drell (1964) and by Pauli (1973). In the nomenclature used here, this is,

$$\left|\delta\left(\frac{E'-E}{\hbar}\right)\right|^2 = \frac{T}{2\pi}\delta\left(\frac{E'-E}{\hbar}\right). \quad (3.79)$$

Hence,

$$|S_{fi}|^2 = 32\pi^3 Z^2(\alpha\hbar c)^2\frac{Tm_e^2}{L^3E'E}\left(\frac{\hbar c}{q}\right)^4 \times \left|\bar{u}(\mathbf{p}', s)\gamma^0 u(\mathbf{p}, s)\right|^2\delta\left(\frac{E'-E}{\hbar}\right). \quad (3.80)$$

where T is the time duration of the scattering process. The transition rate corresponding to the transition probability per incident electron is,

$$\lambda_{fi} = \frac{p_{fi}}{T} \quad (3.81)$$

Combining (3.74), (3.77), and (3.80), the transition rate is obtained,

$$\lambda_{fi} = 8\pi Z^2(\alpha\hbar c)^2\frac{m_e^2(\hbar c)}{\beta q^3 L^3 E'E} \times \left|\bar{u}(\mathbf{p}', s)\gamma^0 u(\mathbf{p}, s)\right|^2 dq\,\delta\left(\frac{E'-E}{\hbar}\right). \quad (3.82)$$

Attention is now focused in evaluating $\left|\bar{u}(\mathbf{p}', s)\gamma^0 u(\mathbf{p}, s)\right|^2$. Before doing so, it should be recognized that there is no knowledge of the spin orientation of the incident electron nor is it of interest following the scatter. In other words, one averages over the initial spin states and sums over the final spin states, a process denoted by an overline, $\overline{|\bar{u}(\mathbf{p}', s)\gamma^0 u(\mathbf{p}, s)|^2}$, or,

$$\lambda_{fi} = 8\pi Z^2(\alpha\hbar c)^2 \frac{m_e^2(\hbar c)}{\beta q^3 L^3 E' E} \times \overline{|\bar{u}(\mathbf{p}', s)\gamma^0 u(\mathbf{p}, s)|^2} dq\, \delta\left(\frac{E' - E}{\hbar}\right) \quad (3.83)$$

where,

$$\overline{|\bar{u}(\mathbf{p}', s)\gamma^0 u(\mathbf{p}, s)|^2} = \frac{1}{2}\sum_{\pm s, s'} \left|\bar{u}(\mathbf{p}', s)\gamma^0 u(\mathbf{p}, s)\right|^2. \quad (3.84)$$

This spin sum is next expanded by first reducing it to traces and then using the trace properties of the Dirac matrices. The trace algebra of Dirac matrices are not reviewed here. Instead, the result is presented without proof,

$$\sum_{\pm s, s'} \bar{u}(\mathbf{p}', s')\gamma^0 u(\mathbf{p}, s) = \mathrm{Tr}\left(\left(\frac{\not{p}' + m_e}{2m_e}\right)\gamma^0 \left(\frac{\not{p}' + m_e}{2m_e}\right)\gamma^0\right). \quad (3.85)$$

This trace is,

$$\mathrm{Tr}\left(\left(\frac{\not{p}' + m_e}{2m_e}\right)\gamma^0 \left(\frac{\not{p}' + m_e}{2m_e}\right)\gamma^0\right) = \frac{1}{m_e^2}(2EE' - \mathbf{p} \bullet \mathbf{p}' + m_e^2) = \frac{2EE'}{m_e^2}\left(1 - \beta^2 \sin^2\frac{\theta}{2}\right) \quad (3.86)$$

leading to,

$$\overline{|\bar{u}(\mathbf{p}', s)\gamma^0 u(\mathbf{p}, s)|^2} = \frac{EE'}{m_e^2}\left(1 - \beta^2 \sin^2\frac{\theta}{2}\right) \quad (3.87)$$

Applying this result to the calculation of the transition rate gives,

$$\lambda_{fi} = \frac{8\pi}{q^3} Z^2 (\alpha\hbar c)^2 \frac{(\hbar c)}{\beta L^3} \times \left(1 - \beta^2 \sin^2\frac{\theta}{2}\right) dq\, \delta\left(\frac{E' - E}{\hbar}\right) \quad (3.88)$$

and the differential cross-section in momentum transfer is,

$$\frac{d\sigma}{dq} = \frac{L^3}{\beta c}\frac{\lambda_{fi}}{dq} = \frac{8\pi}{q^3}\left(\frac{Z\alpha\hbar c}{\beta}\right)^2 \left(1 - \beta^2 \sin^2\frac{\theta}{2}\right) \quad (3.89)$$

Comparing this to the unscreened elastic Coulomb scatter result for a spin-0 projectile, accounting for the electron's spin-1/2 has introduced a multiplicative $(1 - \beta^2 \sin^2\theta/2)$ term. Extending this calculation to the Mott differential cross-section in solid angle using (3.63) and (3.64),

$$\frac{d\sigma_{Mott}}{d\Omega} = \frac{d\sigma_{Ruth}}{d\Omega}\left(1 - \beta^2 \sin^2\frac{\theta}{2}\right). \quad (3.90)$$

The electron's intrinsic spin thus reduces the (Rutherford) spin-0 cross-section by the factor $(1 - \beta^2 \sin^2\theta)$ which is shown as a function of electron kinetic energy for scattering angles of 15, 45, and 90° in Fig. 3.7. As expected, this effect becomes significant for energetic electrons and large scattering angles. The electron kinetic energies of most interest to nuclear medicine are typically below about 200 keV and, as the differential cross-section is highly-forward scattered, the effect of the electron spin can often be neglected in practical internal radiation dosimetric and transport calculations. However, electron scattering as a means of elucidating nuclear structure requires high-energy electrons and the $(1 - \beta^2 \sin^2\theta/2)$ factor

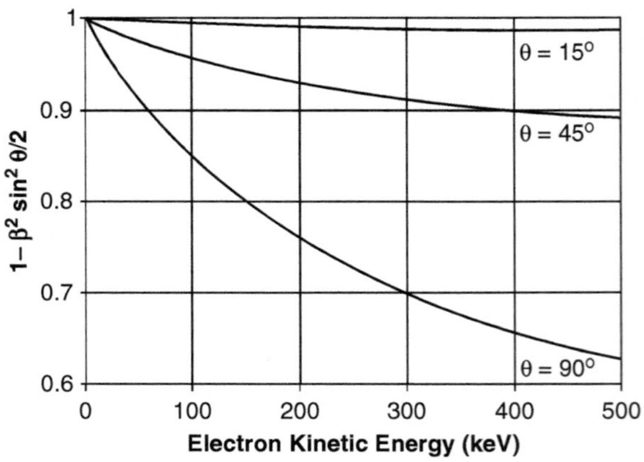

Fig. 3.7 The Mott $\left(1 - \beta^2 \sin^2\theta/2\right)$ modifying factor of the elastic Coulomb scatter differential cross-section for an electron as a function of electron kinetic energy up to 500 keV

cannot be ignored. At relativistic energies ($\beta \longrightarrow 1$), and the cross-section in the relativistic limit becomes,

$$\frac{d\sigma}{d\Omega} \longrightarrow \frac{d\sigma_{Ruth}}{d\Omega}\cos^2\frac{\theta}{2} \tag{3.91}$$

3.2.7.4 Nuclear Scattering Form Factors

It has been assumed in the previous derivations of the elastic Coulomb scatter differential cross-sections that the scattering center was a point electric charge of infinite mass. While the latter assumption is valid for low-energy α particles elastically scattered from heavy nuclei and for electrons/positrons elastically scattered from all nuclei, the former assumption can no longer hold at ultrarelativistic electron probe energies, where the reduced de Broglie wavelength becomes comparable to nuclear dimensions. Here, the nucleus will not appear as a point electric charge but rather as a distribution of electric charge (i.e., the spatial distribution of protons). Hence, measuring the elastic Coulomb scatter of high-energy electrons from nuclei provides information of this spatial distribution. In this subsection, the nuclear form factor which contains this spatial distribution is derived.

The differential cross-section of a charged particle from a spatially-distributed electric charge is calculated, as before, from the Born approximation's scattering amplitude. As this is proportional to the Fourier transform of the scattering potential, the electron-nucleus Coulomb potential is derived first. The spatial distribution of the protons within the nucleus is assumed to be continuous and the nuclear electric charge distribution is $\rho_e(\mathbf{r})$ with the normalization,[20]

$$\int d^3\mathbf{r}\rho_e(\mathbf{r}) = 1. \tag{3.92}$$

The resulting electron-nuclear Coloumb potential is,

$$U(\mathbf{r}) = -Z(\alpha\hbar c)\int d^3\mathbf{r}'\frac{\rho_e\mathbf{r}'}{|\mathbf{r}-\mathbf{r}'|} \tag{3.93}$$

where the integration is over the nuclear volume. Neglecting spin, the scattering amplitude is,

$$\begin{aligned} f(\mathbf{q}) &= -\frac{m_e}{2\pi(\hbar c)^2}\int d^3\mathbf{r}e^{i\left(\frac{\mathbf{q}\bullet\mathbf{r}}{\hbar c}\right)}U(\mathbf{r}) \\ &= \frac{Z m_e\alpha}{2\pi(\hbar c)}\int d^3\mathbf{r}\,e^{i\left(\frac{\mathbf{q}\bullet\mathbf{r}}{\hbar c}\right)}\int d^3\mathbf{r}'\frac{\rho_e(\mathbf{r}')}{|\mathbf{r}-\mathbf{r}'|}. \end{aligned} \tag{3.94}$$

Using the substitution $\mathbf{R} = \mathbf{r} - \mathbf{r}'$,

$$\begin{aligned} f(\mathbf{q}) &= \frac{Z m_e\alpha}{2\pi(\hbar c)}\int d^3\mathbf{R}\,e^{i\left(\frac{\mathbf{q}\bullet(\mathbf{R}+\mathbf{r})}{\hbar c}\right)}\int d^3\mathbf{r}\frac{\rho_e(\mathbf{r})}{|\mathbf{R}|} \\ &= \frac{Z m_e\alpha}{2\pi(\hbar c)}\int d^3\mathbf{R}\frac{e^{i\left(\frac{\mathbf{q}\bullet\mathbf{R}}{\hbar c}\right)}}{|\mathbf{R}|}\int d^3\mathbf{r}\,e^{i\left(\frac{\mathbf{q}\bullet\mathbf{r}}{\hbar c}\right)}\rho_e(\mathbf{r}). \end{aligned} \tag{3.95}$$

By comparing this scattering amplitude with that derived for a point scattering center, it can be deduced that the differential cross-section for elastic Coulomb scatter from a distributed charge scattering center is related to that for a point charge scattering center,

$$\frac{d\sigma_{Dist}}{d\Omega} = \frac{d\sigma_{Point}}{d\Omega}|F(q)|^2 \tag{3.96}$$

where the form factor is the Fourier transform of the electric charge distribution

$$F(q) = \int d^3\mathbf{r}\,e^{i\left(\frac{\mathbf{q}\bullet\mathbf{r}}{\hbar c}\right)}\rho_e(\mathbf{r}). \tag{3.97}$$

The form factor reduces the elastic scattering differential cross-section for high momentum transfer: $|F(q)|^2 \longrightarrow 0$ as $q \longrightarrow \infty$ and $|F(q)|^2 \longrightarrow 1$ as $q \longrightarrow 0$. Measuring the deviation of the measured differential cross-section, $d\sigma_{Dist}/d\Omega$, from that expected for a point scattering center, $d\sigma_{Point}/d\Omega$, provides the $|F(q)|^2$ which can be used to derive the nuclear charge distribution.[21] For a spherically-symmetric charge distribution, the nuclear form factor is,

$$\begin{aligned} F(q) &= \int d^3\mathbf{r}\,e^{i\left(\frac{\mathbf{q}\bullet\mathbf{r}}{\hbar c}\right)}\rho_e(r) \\ &= 2\pi\int dr\,r^2\,\rho_e(r)\int_{-1}^{1} d(\cos\theta)\,e^{i\left(\frac{qr\cos\theta}{\hbar c}\right)} \\ &= 4\pi\left(\frac{\hbar c}{q}\right)\int dr\,r\,\sin\left(\frac{qr}{\hbar c}\right)\rho_e(r). \end{aligned} \tag{3.98}$$

---

[20]Some authors use a normalisation in which the volume integral of the charge density is equal to Ze, which would require the removal of Z from (3.93).

[21]As the derivation of the Mott cross section neglected both the recoil and the angular momentum of the target nucleus, this formula is applicable strictly to only J = 0 nuclei with sufficient mass that the recoil kinetic energy can be neglected.

Form factors for representative models of the spatial distribution of nuclear electric charge are now derived. Consider the simplest model of the point-like scattering center with an electric charge distribution is $k\delta(r)$. The normalization constant is determined from (3.92),

$$4\pi k \int_0^\infty dr\, r^2\, \delta(r) = 1 \tag{3.99}$$

giving $\rho_e(r) = \delta(r)/4\pi$. The form factor is $F(q) = 1$ as expected.

Another model of the electric charge spatial distribution is that in which the charge density decreases exponentially with radial distance from the nuclear center. This is of the form $ke^{-(r/R_N)}$ where R_N is the distance from the center of the nucleus that the charge density is $e^{-1} \cong 0.37$ that of the density at the center and k is the normalization constant found from,

$$4\pi R_N^3 k \int_0^\infty dx\, e^{-x} = 1 \tag{3.100}$$

where the substitution of variable $x = r/R_N$ has been used. The integral is solved through two applications of integration by parts to lead to $k = 1/8\pi R_N^3$ which gives the normalized exponential charge distribution,

$$\rho_e(r) = \frac{e^{-(r/R_N)}}{8\pi R_N^3}. \tag{3.101}$$

The nuclear size for this model is described by the RMS charge radius which, for this distribution, is,

$$\begin{aligned}\sqrt{\overline{r^2}} &= \left(\int d^3\mathbf{r}\, r^2\, \rho_e(r)\right)^{1/2} \\ &= \left(\frac{1}{2R_N^3}\int_0^\infty dr\, r^4\, e^{-(r/R_N)}\right)^{1/2} \\ &= \sqrt{12}\, R_N \\ &\approx 3.464\, R_N.\end{aligned} \tag{3.102}$$

The corresponding form factor is,

$$F(q) = \frac{\hbar c}{2qR_N^3}\int_0^\infty dr\, r \sin\left(\frac{qr}{\hbar c}\right) e^{-(r/R_N)} \tag{3.103}$$

To solve for this, consider the integral, which is of the form,

$$\begin{aligned}\int_0^\infty dr\, r \sin(ar)\, e^{-br} &= \mathrm{Im}\int_0^\infty dr\, r\, e^{-(b-ia)r} \\ &= \mathrm{Im}\int_0^\infty dr\, r\, e^{-cr}\end{aligned} \tag{3.104}$$

where $c = b\text{–}ia$. Using the substitution of variable $x = cr$, the integral is rewritten as,

$$\begin{aligned}\int_0^\infty dr\, r\, e^{-cr} &= \frac{1}{c^2}\int_0^\infty dx\, x\, e^{-x} \\ &= \frac{1}{c^2}\end{aligned} \tag{3.105}$$

where the integral has been solved by integration by parts. This result leads to,

$$\begin{aligned}\int_0^\infty dr\, r \sin(ar) e^{-br} &= \mathrm{Im}\frac{1}{(b-ia)^2} \\ &= \mathrm{Im}\frac{e^{i2\tan^{-1}(a/b)}}{a^2+b^2} \\ &= \frac{\sin 2\tan^{-1} a/b}{a^2+b^2} \\ &= \frac{2ab}{(a^2+b^2)^2}.\end{aligned} \tag{3.106}$$

Then,

$$\int_0^\infty dr\, r \sin\left(\frac{qr}{\hbar c}\right) e^{-(r/R_N)} = 2R_N^2 \frac{(qR_N/\hbar c)}{\left(1+\left(^{qR_N}/_{\hbar c}\right)^2\right)^2} \tag{3.107}$$

and giving the form factor for a charge distribution decreasing exponentially with radius,

$$F(q) = \frac{1}{\left(1+\left(^{qR_N}/_{\hbar c}\right)^2\right)^2}. \tag{3.108}$$

Another simple model of the nucleus is that of a homogeneous, well-defined sphere. In this case, the nuclear charge distribution is constant up a fixed radius R_N and zero beyond. This normalized distribution is,

$$\rho_e(r) = \frac{3}{4\pi R_N^3} \quad r \le R_N \\ = 0 \quad r > R_N. \tag{3.109}$$

The RMS charge radius for this distribution is given by,

$$\sqrt{\overline{r^2}} = \left(\frac{3}{R_N^3}\int_0^{R_N} dr\, r^4\right)^{1/2} = \sqrt{\frac{3}{5}} R_N. \tag{3.110}$$

The form factor for a homogeneous spherical nucleus is,

$$F(q) = 3\frac{\hbar c}{qR_N^3}\int_0^{R_N} dr\, r \sin\frac{qr}{\hbar c} = 3\left(\frac{\hbar c}{qR_N}\right)^2\left(\left(\frac{\hbar c}{qR_N}\right)\sin\left(\frac{qR_N}{\hbar c}\right) - \cos\left(\frac{qR_N}{\hbar c}\right)\right) = 3\frac{j_1(qR_N/\hbar c)}{\left(qR_N/\hbar c\right)} \tag{3.111}$$

where $j_1(x)$ is the spherical Bessel function of the first kind shown in Fig. 3.8. As expected, the sharp cut off of the hard-edge charge distribution will be associated with an oscillatory form factor; in other words, the differential cross-section can be thought of as a diffraction pattern for the incident electron wavefunction. It is noted that the zero of the form factor

$$q_0 R_N/\hbar c \approx 4.493$$

can lead to an estimate of the nuclear radius. Hence, by measuring the momentum transfer q_0 at which the form factor goes to zero provides the nuclear radius for the homogeneous spherical model,

$$R_N \cong 4.493\frac{\hbar c}{q_0} \quad \text{homogeneous sphere model.} \tag{3.112}$$

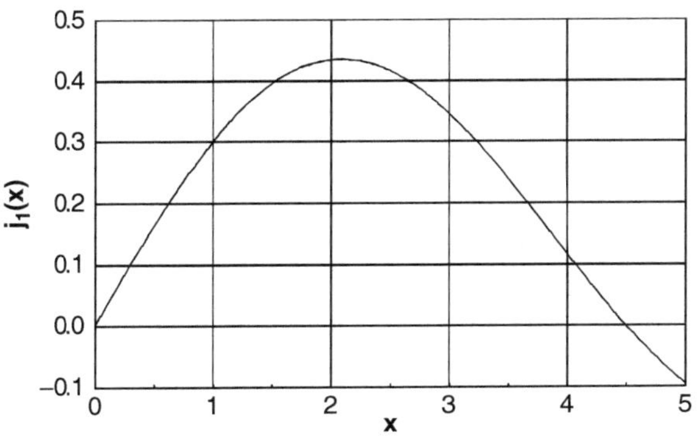

Fig. 3.8 Spherical Bessel function of the first kind

Experimentally, it is found that the experimental form factor does not go to zero as predicted by the hard-edge nucleus model. This is indicative of the nuclear edge being more diffuse than assumed by this model.

Instead of the unphysical sharp edge of the previous nuclear model, one can introduce a less-sharp truncated charge distribution, the simplest of which is Gaussian. The derivations of the normalization, mean-square radius, and form factor for a Gaussian charge distribution are lengthy and are thus relegated to Appendix A. The normalized Gaussian charge distribution is,

$$\rho_e(r) = \frac{e^{-(r/R_N)^2}}{\pi^{3/2} R_N^3} \tag{3.113}$$

for a nuclear radius R_N, defined here as the radius at which the charge distribution density is e^{-1}that of the central value. The RMS radius of the nuclear charge distribution is,

$$\sqrt{\overline{r^2}} = \sqrt{\frac{3}{2}} R_N \tag{3.114}$$

and the form factor is Gaussian in the momentum transfer,

$$F(q) = e^{-(qR_N/2\hbar c)^2}. \tag{3.115}$$

A nuclear charge spatial distribution model which has proven to provide good fits to experimental scattering data for nuclei with atomic mass numbers between 40 and 238 is the two-parameter Woods–Saxon

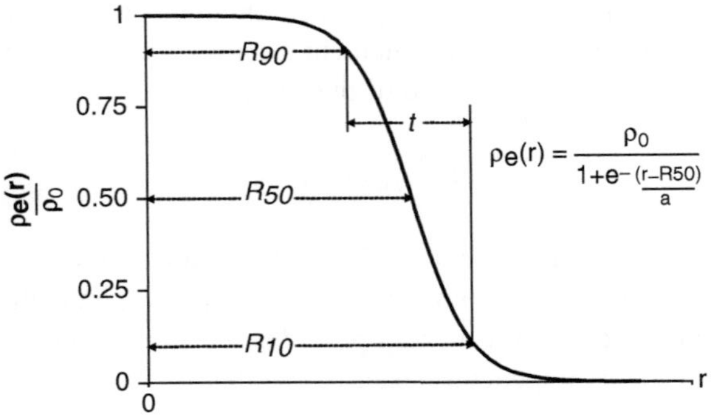

Fig. 3.9 Woods–Saxon distribution of the nuclear electric charge

distribution[22] which represents the nucleus as a sphere with a diffuse surface (or skin),

$$\rho_e(r) = \frac{\rho_0}{1 + e^{\left(\frac{r-R_{50}}{a}\right)}} \quad (3.116)$$

and is shown in Fig. 3.9. R_{50} is the radius at which the nuclear electric charge density has fallen to half of its central value and ρ_0 is determined from the normalization requirement. R_{10} and R_{90} are the radii at which the density has fallen to 10 and 90% of its central value, respectively: the "thickness" between R_{90} and R_{10} is frequently referred to as the "skin" of the nucleus. The parameter a describes the diffuseness of the edge. The normalization and mean-square radius of the Woods–Saxon distribution are derived in Appendix B. The normalized Woods–Saxon charge distribution is,

$$\rho_e(r) = -\frac{1}{8\pi a^3 Li_3\left(-e^{-R_{50}/a}\right)e^{\left(\frac{r-R_{50}}{a}\right)}} \quad (3.117)$$

where $Li_n(x)$ is the polylogarithm of order n and is defined in Appendix B. The mean-square radius for the Woods–Saxon distribution is,

$$\sqrt{\overline{r^2}} = 2a\sqrt{\frac{3Li_5(-e^{-(-R_{50}/a)})}{Li_3(-e^{-(-R_{50}/a)})}}. \quad (3.118)$$

[22]Also known as the Fermi–Dirac distribution in statistical mechanics and solid-state physics.

The form factor for the Woods–Saxon charge distribution does not have a closed analytical form and must be solved for either by numerical integration or by, for example, approximating the Woods–Saxon charge distribution as a summation of Gaussians (each reflecting a proton position) and solving the form factor for each term of the summation (Sick 1974). The resulting form factor is similar to that derived for the homogeneous sphere as it is also oscillatory. However, because of the lack of the sharp-cut off at R_N, the minima are finite and do not go to zero. Empirically, from electron scattering measurements, the nuclear "skin" thickness is quite constant over a large number of atomic mass numbers at about 2.3 fm. From experiment, it has been known that the nuclear volume increases linearly with A, and, hence, the Woods–Saxon nuclear charge radius can be parameterized by,

$$R_{50} = bA^{1/3}; \quad (3.119)$$

where b = 1.2 fm. The other Woods–Saxon constant is a = 0.545 fm. Similarly, the RMS radius of the Fermi distribution as a function of A is,

$$\sqrt{\overline{r^2}} = r_{RMS}A^{1/3} \quad (3.120)$$

where $r_{RMS} = 0.94$ fm. By crudely modeling the nucleus as a homogenous sphere of radius R_N, then this radius can be described in terms of the RMS radius and which provides a linear relationship between the nuclear radius and the cube-root of the atomic mass number,

$$R_N = r_n A^{\frac{1}{3}} \qquad r_n = 1.2\,\text{fm}. \quad (3.121)$$

Examples of calculated nuclear charge distributions for ^{32}S, ^{40}Ca, ^{56}Fe, ^{127}I, and ^{208}Pb are shown in Fig. 3.10. Recall that the parameter t describes the distance over which the nuclear density decreases from 90 to 10% of its maximum, or t = a ln81 ≈ 4.4 a = 3.4 fm.

Figure 3.11 provides a pictorial summary of the nuclear charge distributions and the resulting form factors reviewed in this section (excluding the Woods–Saxon). The point-charge form factor is that of an electron projectile (i.e., it does not demonstrate an internal structure) whereas the exponential form

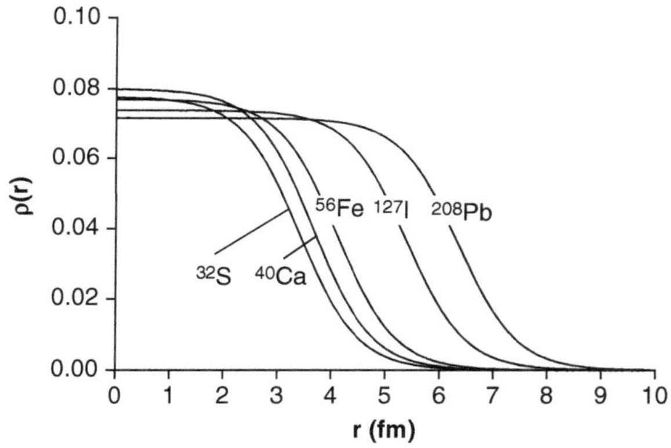

Fig. 3.10 Nuclear charge distributions calculated for the Wood–Saxon distribution. The units of the charge density is Z fm^{-3}

factor well represents that measured for a proton target (hydrogen nucleus). Very light nuclei tend to demonstrate a Gaussian form factor with most others displaying a Woods–Saxon type. In principle, the nuclear electric charge distribution can be obtained through the inverse-Fourier transform of a complete set of form factor data,

$$\rho_e(r) = \frac{1}{(2\pi)^3}\int_0^\infty d^3q\, e^{-i\left(\frac{\mathbf{q}\bullet\mathbf{r}}{\hbar c}\right)}F(q) \tag{3.122}$$

However, this is not practical as the differential cross-section measurement can only be obtained over a limited range of the momentum transfer (or, equivalently, the elastic scattering angle) as the Mott-modified Rutherford scattering cross-sections drops rapidly with increasing q. Figure 3.12 shows the ratio of the elastic Coulomb scattering cross-section to that of the pure Rutherford type for 150 MeV electrons for when the Mott factor is applicable (i.e., the projectile is spin-1/2) and when the target has a Gaussian charge distribution with an RMS charge radius of 3 fm. For clarity, the variable is the cosine of the scattering angle. The Mott factor decreases the cross-section rather slowly for small scattering angles to eventually reduce it to 10% of the Rutherford value for a scattering angle of about 145°. On the other hand, when the Mott factor and the nuclear form factor are combined (as would be the case for an electron projectile incident to a nucleus), the differential cross-section drops off far more rapidly and is only about 1% that of the pure Rutherford cross-section at a 90° scattering angle.

This demonstrates the difficulty of obtaining a direct calculation of the nuclear charge distribution from the inverse-Fourier transform of the measured form factor. In practice, the nuclear charge density is obtained by initially assuming a parametric model of the density and, through optimizing the parameters of the form factor, obtaining agreement with the measured data.

In addition to estimating the nuclear RMS charge radius from the minimum of the form factor, it is possible to calculate this value from a limited set of form factor data, in particular that for when the momentum transfer is low and the scattering angle approaches zero (i.e., the scattering cross-section is maximal). Begin with the definition of the form factor,

$$F(q) = \int d^3\mathbf{r}\, e^{i\left(\frac{\mathbf{q}\bullet\mathbf{r}}{\hbar c}\right)}\rho_e(\mathbf{r})$$

and expand the exponential in the form factor to second order,

$$\begin{aligned}
F(q) &\approx \int d^3\mathbf{r}\left(1 + i\frac{\mathbf{q}\bullet\mathbf{r}}{\hbar c} - \frac{1}{2}\frac{(\mathbf{q}\bullet\mathbf{r})^2}{(\hbar c)^2}\right)\rho_e(r)\\
&= \int d^3\mathbf{r}\rho_e(r) + \frac{i}{\hbar c}\int d^3\mathbf{r}(\mathbf{q}\bullet\mathbf{r})\rho_e(r)\\
&\quad - \frac{1}{2(\hbar c)^2}\int d^3\mathbf{r}(\mathbf{q}\bullet\mathbf{r})^2\rho_e(r)\\
&= 1 + i\,2\pi\left(\frac{q}{\hbar c}\right)\int_0^\infty dr\, r^2\rho_e(r)\int_{-1}^{1} d(\cos\theta)\,\cos\theta\\
&\quad - \pi\left(\frac{q}{\hbar c}\right)^2\int_0^\infty dr\, r^4\rho_e(r)\int_{-1}^{1} d(\cos\theta)\cos^2\theta\\
&= 1 - \frac{2\pi}{3}\left(\frac{q}{\hbar c}\right)^2\int_0^\infty dr\, r^4\rho_e(r)\\
&= 1 - \frac{1}{6}\left(\frac{q}{\hbar c}\right)^2\int d^3\mathbf{r}\, r^2\,\rho_e(r)\\
&= 1 - \frac{1}{6}\left(\frac{q}{\hbar c}\right)^2\overline{r^2}.
\end{aligned} \tag{3.123}$$

Thus the mean-square radius of the nucleus can be approximated by the form factor at a given value of momentum transfer in inelastic Coulomb scatter,

$$\overline{r^2} = 6\left(\frac{\hbar c}{q}\right)^2(1 - F(q)). \tag{3.124}$$

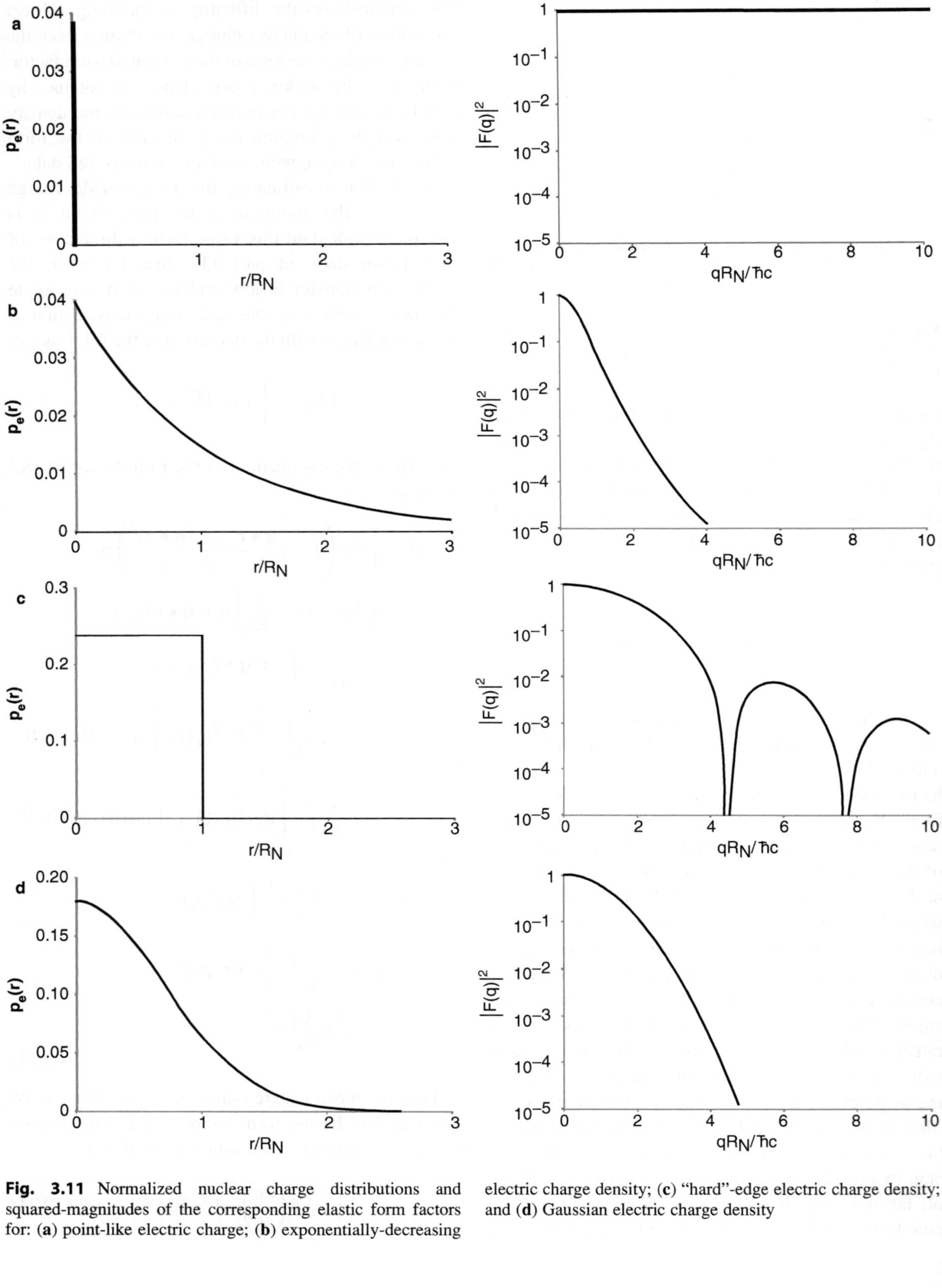

Fig. 3.11 Normalized nuclear charge distributions and squared-magnitudes of the corresponding elastic form factors for: (**a**) point-like electric charge; (**b**) exponentially-decreasing electric charge density; (**c**) "hard"-edge electric charge density; and (**d**) Gaussian electric charge density

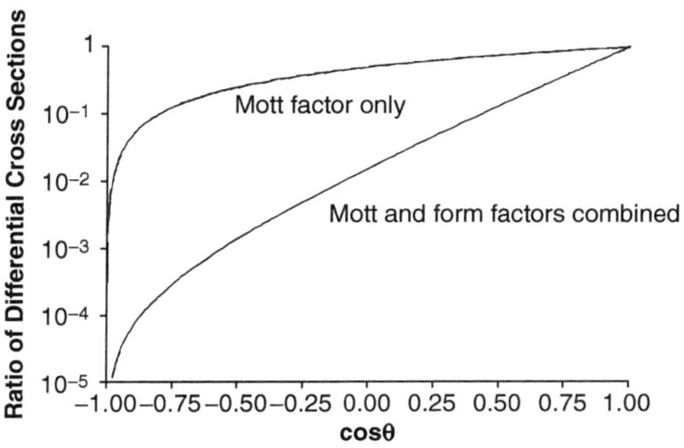

Fig. 3.12 The ratio of the elastic Coulomb scatter differential cross-section to that for the Rutherford cross-section for the Mott factor only and the Mott and form factors combined applied as a function of the cosine of the scattering angle for 150 MeV electrons and a Gaussian form factor with an RMS nuclear charge radius of 3 fm

Equivalently, from the definition of the form factor as the Fourier transform of the nuclear electric charge distribution,

$$F(q) = \int d^3\mathbf{r}\, e^{i\left(\frac{\mathbf{q}\bullet\mathbf{r}}{\hbar c}\right)} \rho_e(r)$$

and

$$\frac{d^2F(q)}{dq^2} = -\frac{1}{(\hbar c)^2}\int d^3\mathbf{r}\, r^2 \cos^2\theta\, e^{i\left(\frac{\mathbf{q}\bullet\mathbf{r}}{\hbar c}\right)} \rho_e(r). \quad (3.125)$$

As $\cos\theta = 1$ is equivalent to $q = 0$, there is a direct relationship between the mean-square radius of the nuclear electric charge distribution and the second-derivative of the nuclear form factor at small scattering angles,

$$\sqrt{\overline{r^2}} = \hbar c \sqrt{-\left.\frac{d^2F(q)}{dq^2}\right|_{q=0}}. \quad (3.126)$$

It is important to recall that the above derivations were based upon the Born approximation and, hence, are subject to the limitations of that theory. While use of the Born approximation is acceptable for calculations of light nuclei, its validity diminishes with increasing Z. Dimensions of heavy nuclei are more accurately determined using phase-shift analysis.

3.2.8 Nuclear Density

Methods of determining nuclear size and mass from which the nuclear density can be derived have been reviewed. Accurate measurements of nuclear mass are essential for determining nuclear binding energies, the importance of which is shown later. For a nucleus composed of A nucleons, the nuclear radius is $R_N = 1.2$ A fm. Assuming a spherical nucleus, the volume is,

$$\frac{4\pi}{3}R_N^3 = \frac{4\pi}{3}(1.2)^3 A\,\text{fm}^3 \approx 6.2\,A\,\text{fm}^3 \quad (3.127)$$

or about 0.16 nucleons fm^{-3}, which implies an average internucleon separation of about 2 fm. This constancy of density is indicative of a short-range, saturable nuclear force and can be used to model of the nucleus as an incompressible nuclear fluid.

3.3 Nucleon Dynamics: The Fermi Gas Model

Some of the simplest aspects of the phenomenology of the nucleus have now been examined: its constituents, electric charge, mass, size, and density. With this information, it is now possible to expand the understanding of nuclear structure, in particular those characteristics that lead to nuclear stability. As an unstable nucleus is, by definition, radioactive, an appreciation of nuclear stability is a foundation to the development of internal radiation dosimetry. In the first nuclear model to be considered, we evaluate the dynamics of the nucleons as a result of their confinement within the nuclear volume $\frac{4\pi}{3}r_n^3A$ with uniform density $\frac{3}{4\pi r_n^3}$. Should the number of nucleons be sufficiently large, statistical mechanics can be used by treating the nucleons as weakly-interacting fermions (subject, of course, to the Pauli exclusion principle) interacting via an attractive potential. In its ground state, the nucleus is a zero-temperature Fermi gas and an appropriate nucleon kinetic energy distribution for medium to heavy nuclei obtained. Consider the nucleus as an ensemble of A nucleons confined within a volume L^3 with the phase space divided into unit cells each of

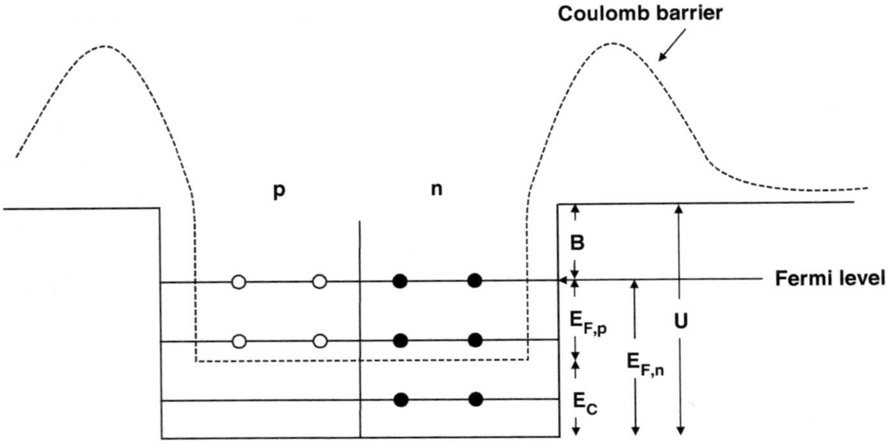

Fig. 3.13 Proton and neutron potential wells in the Fermi gas model. B is the mean binding energy of a nucleon (~8 MeV), U is the depth of the potential well (~41 MeV), $E_{F,n}$ is the neutron Fermi energy (~33 MeV) and $E_{F,p}$ is the proton Fermi energy, shifted up relative to that of the neutron by E_C due to the Coulomb repulsion between the protons

volume $(2\pi\hbar c)^3$. Each unit cell will contain four nucleons due to the spin-isospin degeneracy from the combinations of nucleon spin-1/2 and the isospin-1/2. Because of isospin, the neutrons and protons are distinguishable and will thus occupy separate energy wells. In the ground state (i.e., zero temperature), all available states up to the Fermi momenta, $p_{F,p}$ and $p_{F,n}$ for protons and neutrons, respectively, will be filled. The number of protons in this Fermi nuclear gas can then be related to the Fermi momentum through,

$$\begin{aligned} Z &= 2\left(\frac{L}{2\pi\hbar c}\right)^3 \int d^3\mathbf{p} \\ &= 8\pi\left(\frac{L}{2\pi\hbar c}\right)^3 \int_0^{p_{F,p}} dp\, p^2 \\ &= \frac{1}{3\pi^2}\left(\frac{L\, p_{F,p}}{\hbar c}\right)^3 \end{aligned} \tag{3.128}$$

where the multiplicative factor of 2 reflects the spin degeneracy among the protons. Similarly, the number of neutrons is related to the neutron Fermi momentum by,

$$N = \frac{1}{3\pi^2}\left(\frac{L\, p_{F,n}}{\hbar c}\right)^3. \tag{3.129}$$

As the volume of a spherical nucleus containing A nucleons is,

$$L^3 = \frac{4\pi}{3} r_n^3 A \tag{3.130}$$

the proton and neutron Fermi momenta are,

$$p_{F,p} = \left(\frac{9\pi}{4}\right)^{\frac{1}{3}}\left(\frac{Z}{A}\right)^{\frac{1}{3}}\left(\frac{\hbar c}{r_n}\right) \tag{3.131}$$

$$p_{F,n} = \left(\frac{9\pi}{4}\right)^{\frac{1}{3}}\left(\frac{N}{A}\right)^{\frac{1}{3}}\left(\frac{\hbar c}{r_n}\right) \tag{3.132}$$

For $r_n = 1.2$ fm and a nucleus with equal numbers of neutrons and protons, $N = Z = A/2$, the Fermi momentum is 250 MeV/c. The corresponding Fermi energy,[23] $E_F = p_F{}^2/m \approx 33$ MeV where m is the nucleon mass ($m_p \approx m_n \approx 940$ MeV), represents the energy of the highest occupied nucleon level. The resulting energy levels are shown in Fig. 3.13 for which the nucleons are bound within the nucleus by an energy B (later shown to be about 8 MeV for most nuclei) and the depth of the potential well, U, which is the sum of the Fermi and binding energies and is about 41 MeV. The proton and neutron potential wells as the protons are subject to Coulomb repulsion which neutrons are not. As a consequence, the neutron Fermi energy, $E_{F,n}$, is equal to E_F and the proton Fermi energy, $E_{F,p}$, is less than that of the neutron by the net Coulomb repulsion energy,

[23]The nomenclature used in this book takes a slight diversion here (in order to maintain consistency with that used elsewhere) as the energy level, denoted by E, is equal to the kinetic energy only which would normally be denoted by T. Here, E does not denote the total energy (i.e., the sum of kinetic and rest mass energies).

$$E_{F,p} = E_{F,n} - E_C. \qquad (3.133)$$

Hence, the depth of the proton potential well is less than that of the neutrons. Despite this difference in the depths of the neutron and proton potential wells, the Fermi energy levels for both protons and neutrons must remain the same. Were this not so, the nucleus could not remain stable and would decay to a more energetically favorable state by converting all or the neutrons into protons or vice versa via beta decay. The combination of a deeper neutron potential well and the requirement for the proton and neutron energy levels to be equal leads to the state of there being more occupied neutron levels than proton levels, a feature reflected by the fact that $N > Z$ for stable medium to heavy nuclei.

The mean energy of the occupied proton states is,

$$\begin{aligned}\bar{E}_P &= 2\left(\frac{L}{2\pi\hbar c}\right)^3 \int_0^{p_{F,p}} dp\left(4\pi p^2\right)\frac{p^2}{2m} \\ &= \frac{3}{5} Z E_{F,p}\end{aligned} \qquad (3.134)$$

where $E_{F,p} = p_{F,p}^2/m_p$. Similarly for the lower occupied neutron states,

$$\bar{E}_n = \frac{3}{5} N E_{F,n}. \qquad (3.135)$$

The sum of the proton and neutron mean kinetic energies is,

$$\begin{aligned}\bar{E}_{Tot} &= \bar{E}_p + \bar{E}_n \\ &= \frac{3}{10m}\left(Z p_{F,p}^2 + N p_{F,n}^2\right) \\ &= \frac{3}{10m}\left(\frac{9\pi}{4}\right)^{2/3}\left(\frac{\hbar c}{r_n}\right)^2\left(\frac{Z^{5/3}+N^{5/3}}{A^{2/3}}\right).\end{aligned} \qquad (3.136)$$

The differences in the total mean nucleon kinetic energies between nuclei with equal numbers of protons and neutrons and heavy nuclei with more neutrons than protons are exemplified by the values of 20.03 MeV for ^{12}C ($Z = N = 6$) and the slightly higher value of 20.5 MeV for ^{208}Pb ($Z = 82$; $N = 126$). That $\bar{E}_{Tot}$ should be lower for nuclei with equal numbers of protons and neutrons, provided that Coulomb effects are neglected, can be demonstrated by manipulating (3.136) through first writing $Z = \left(\frac{A}{2}\right) - \left(\frac{(N-Z)}{2}\right)$ and $N = \left(\frac{A}{2}\right) + \left(\frac{(N-Z)}{2}\right)$ so that the term dependent upon atomic number and atomic mass number is,

$$\begin{aligned}\left(\frac{Z^{5/3}+N^{5/3}}{A^{\frac{2}{3}}}\right) &= \frac{\left(\frac{A}{2}-\frac{N-Z}{2}\right)^{5/3}+\left(\frac{A}{2}+\frac{N-Z}{2}\right)^{5/3}}{A^{\frac{2}{3}}} \\ &= \frac{A}{2}\left(\left(1-\frac{N-Z}{A}\right)^{5/3}+\left(1+\frac{N-Z}{A}\right)^{5/3}\right).\end{aligned} \qquad (3.137)$$

As $(N - Z)^2 \ll A$, binomial expansions of the two terms to second order can be used,

$$\left(1 \pm \frac{N-Z}{A}\right)^{5/3} \cong 1 \pm \frac{5}{3}\left(\frac{N-Z}{A}\right) + \frac{5}{9}\left(\frac{N-Z}{A}\right)^2 \qquad (3.138)$$

The $\pm$ on the first-order term of the expansion explains why a second-order expansion was required. This expansion gives,

$$\left(\frac{Z^{5/3}+N^{5/3}}{A^{2/3}}\right) = A\left(1+\frac{5}{9}\left(\frac{N-Z}{A}\right)^2\right). \qquad (3.139)$$

Placing this in the expression for the total Fermi gas energy and rearranging to normalize per nucleon results in,

$$\begin{aligned}\frac{\bar{E}_{Tot}}{A} &= \frac{3}{10m}\left(\frac{9\pi}{4}\right)^{2/3}\left(\frac{\hbar c}{r_n}\right)^2\left(1+\frac{5}{9}\left(\frac{N-Z}{A}\right)^2\right) \\ &= \frac{\bar{E}_{Vol}}{A} + \frac{\bar{E}_{Sym}}{A}.\end{aligned} \qquad (3.140)$$

The volume and symmetry energy terms have been identified,

$$\frac{E_{Vol}}{A} = \frac{3}{10m}\left(\frac{9\pi}{4}\right)^{2/3}\left(\frac{\hbar c}{r_n}\right)^2 = 31.8\,\text{MeV} \qquad (3.141)$$

$$\begin{aligned}\frac{E_{Sym}}{A} &= \frac{1}{6m}\left(\frac{9\pi}{4}\right)^{2/3}\left(\frac{\hbar c}{r_n}\right)^2\left(\frac{N-Z}{A}\right)^2 \\ &= 17.7\left(\frac{N-Z}{A}\right)^2 \text{MeV}.\end{aligned} \qquad (3.142)$$

The volume energy term per nucleon is a constant referred to as the "condensation energy" term as it represents the energy released when a grouping of nucleons combine to form a nucleus. As it must be proportional to the number of nucleons, the E_{Vol}/A term is constant. The symmetry energy term provides the $\left(\frac{(N-Z)}{A}\right)^2$ factor indicating that the nucleons' energy will be minimized for fixed A when the numbers of neutrons and protons are equal (reflecting the observed tendency for zero nuclear isospin in stable nuclei). However, the magnitude of the symmetry term is lower than that determined from measurement, a result reflecting the neglect of proton–neutron interactions when the numbers of neutrons and protons differ. This difference, as we shall see, increases with nuclear mass as more neutrons are required to compensate for the Coulomb repulsion between the protons in order to bind the nucleus.

3.4 Phenomenology of Nuclear Stability

3.4.1 Introduction

Before progressing into further development of nuclear theory, the empirically-determined characteristics of nuclear stability are reviewed.

3.4.2 Average Binding Energy per Nucleon

The mass of an atom of atomic mass number, A, and atomic number, Z, can be written as,

$$M(A,Z) = ZM_H + (A-Z)m_n - B(A,Z) \quad (3.143)$$

where M_H and m_n are the masses of the hydrogen atom and the neutron, respectively, and B(A,Z) is the nuclear binding energy. Rearranging this equation gives the nuclear binding energy in terms of measurable quantities,

$$B(A,Z) = ZM_H + (A-Z)m_n - M(A,Z). \quad (3.144)$$

The nuclear binding energy per nucleon is,

$$\begin{aligned} f(A) &= \frac{B(A,Z)}{A} \\ &= \left(\frac{Z}{A}\right)(M_H - m_n) + m_n - \frac{M(A,Z)}{A}. \end{aligned} \quad (3.145)$$

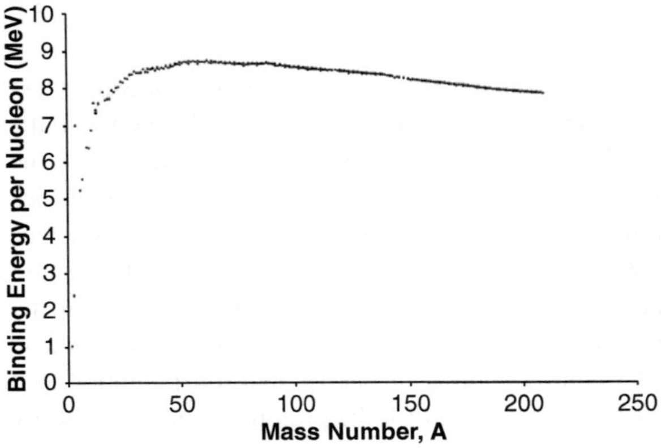

Fig. 3.14 Measured binding energies per nucleon, f(A, Z), of stable nuclei

This result is plotted in Fig. 3.14 for nuclei as a function of A. For nuclei with atomic mass numbers exceeding 10, the binding energy per nucleon is very nearly constant at between 7.5 and 8.8 MeV per nucleon. The lowest binding energy is that of the deuteron (f(A) = 1.11 MeV) which reflects the relatively large separation between the neutron and proton such that the attractive nuclear potential just overcomes the kinetic energies of the nucleons. It is striking that the ^4He nucleus has such a large binding energy (f(A) = 7.08 MeV) which is a consequence of the pairing off of the two neutrons and two protons with opposing spins leading to the results that the ^4He nucleus is spin-0 and isopsin-0.

The nucleus with the highest f(A) value is ^{62}Ni with f(A) = 8.79 MeV per nucleon.

This near constancy of f(A) for $A > 10$ reflects the presence of a short-range and saturable attractive internucleon force. Otherwise, should the internucleon force have infinite range, the binding energy would be proportional to $A(A-1) \approx A^2$ for heavy nuclei and leading to f(A) increasing with A. In practical terms, using the analogy of a liquid drop, a constant binding energy per nucleon is similar to the independence of the latent heat from the size of the drop. Features of this figure which are indicative of underlying nuclear structure include:

- Peaks in f(A) for certain nuclei, typically with atomic mass numbers of integral multiples of 4 (e.g., ^4He, ^{12}C, ^{16}O, ^{20}Ne, and ^{24}Mg), demonstrating nuclei with particularly stable numbers of nucleons.

- The slow decrease in f(A) with A for large A which is a consequence of the growing repulsive Coulomb force between protons (roughly proportional to Z^2).
- The rapid increase in f(A) with A for small A reflecting an effect analogous to the surface tension of a liquid drop: there is a greater proportion of nucleons at the nuclear surface for small A, thus resulting in a smaller binding energy per nucleon. As the number of nucleons increases, the proportion of nucleons at the nuclear surface decreases and the net binding energy per nucleon increases.
- Nuclei with even numbers of both protons and neutrons consistently have higher f(A) values than neighboring nuclei with odd numbers of nucleons.

3.4.3 Nucleon Separation Energy

The nucleon separation energy is that required to separate one or a group of nucleons from the nucleus in its ground state and to move them to infinity, whilst leaving the residual nucleus in its ground state. Unlike beta decay, in which a neutron can be transformed to a proton and vice versa, this process conserves the total individual numbers of neutrons and protons. Consider the general case of the decay of a nucleus with atomic number Z and neutron number N to two daughter nuclei with $(Z_{f,1}, N_{f,1})$ and $(Z_{f,2}, N_{f,2})$ where $Z = Z_{f,1} + Z_{f,2}$ and $N = N_{f,1} + N_{f,2}$. An example of such a decay is α decay: $(Z_{f,1}, N_{f,1}) = (2,2)$ and $(Z_{f,2}, N_{f,2}) = (Z-2, N-2)$.

The condition of one of the produced nuclei to be a single neutron is allowed:

$(Z_{f,1}, N_{f,1}) = (0, 1)$ and $(Z_{f,2}, N_{f,2}) = (Z, N-1)$. The Q of this decay is the energy released by it,

$$Q = M(Z,N) - m_n - M(Z,N-1) \qquad (3.146)$$

where M(Z, N) is the rest-mass energy of a nucleus with Z protons and N neutrons. As the decay can only occur if Q is positive (exoergic), the separation energy for a single neutron is,

$$\begin{aligned} S_n &= -Q_n \\ &= M(Z,N-1) + m_n - M(Z,N) \\ &= B(A,Z) - B(A-1,Z) \\ &= A\,f(A) - (A-1)f(A-1) \\ &= f(A-1) + A(f(A) - f(A-1)). \end{aligned} \qquad (3.147)$$

Consider heavy nuclei ($A > 62$) for which, in general, the change in binding energy per nucleon is negative (i.e., $\Delta f(A) \equiv f(A-1) - f(A) > 0$). Then, the neutron separation energy is,

$$S_n = f(A) - (A-1)\Delta f(A). \qquad (3.148)$$

As only a single neutron is being removed, $\Delta A = -1$ and, taking advantage of the small change in binding energy with A,

$$\begin{aligned} \frac{df(A)}{dA} &\approx \frac{\Delta f(A)}{\Delta A} \\ &= -\Delta f(A) < 0. \end{aligned} \qquad (3.149)$$

Hence, the expression for the neutron separation energy is,

$$S_n = f(A) + (A-1)\frac{df(A)}{dA}. \qquad (3.150)$$

As, in general, $df(A)/dA$ is negative for nuclei heavier than ^{62}Ni, the neutron separation energy will be less than the nuclear binding energy for such nuclei. In particular, should $(A-1)\left|{}^{df(A)}\!/_{dA}\right| > f(A)$, then S_n will be negative or, correspondingly, Q is positive and it is energetically possible for the nucleus to decay through the spontaneous emission of a single neutron.

In the case of a proton being removed from the nucleus, the Q is,

$$Q = M(Z,N) - m_p - M(Z-1,N). \qquad (3.151)$$

But, the separation energy is slightly more complicated as we now must account for the Coulomb repulsion between the released proton and the daughter nucleus,

$$S_p = -Q - (Z-1)\frac{\alpha\hbar c}{R_N}. \qquad (3.152)$$

Repeating the above derivation,

$$S_p \approx S_n - (Z-1)\frac{\alpha\hbar c}{R_N} \qquad (3.153)$$

A negative value of the separation energies means that the nucleus can spontaneously decay through the emission of a single neutron or proton and the boundary between negative and positive values of the separation energies is known as the “drip line.” The significance of this line for single nucleon emission lies when beta decay is considered. A nucleus just inside the neutron drip line (S_n being slightly positive) can undergo β^- decay, (Z, N) →(Z + 1, N − 1). As this is a weak interaction, the decay rate is small (as shown in Chap. 4) so that the half life of the decay is relatively long (of the order of 10^{-3} s or above). Should the nucleus be just outside the neutron drip line (S_n being slightly negative) it can under spontaneous neutron emission, (Z, N) →(Z, N − 1) with, as the decay is due to the strong nuclear interaction, a short transition rate (of the order of 10^{-20}s).

Of particular importance to the understanding of nuclear stability is the recognition that a high positive nucleon separation energy is indicative of a particularly stable nucleus. As an example, Fig. 3.15 shows the neutron separation energy S_n vs. the number of neutrons for the isotopes of Pb. It is evident that the neutron separation energy is higher for an even number of neutrons. As the atomic number of Pb is even, these peaks denoting particularly stable Pb nuclei suggest that the combination of an even number of protons and an even number of neutrons leads to a nucleus with increased stability.

3.4.4 Characteristics of Stable Nuclei

Figure 3.16 is a graph of the number of protons, Z, vs. the number of neutrons, N, for stable nuclei and is known as the Segrè plot. Also shown on the plot is the N = Z line and a variety of isobar lines; the latter will be of interest in later discussion of mass parabolas. Recall the symmetry energy term from the Fermi gas model which predicts a minimum in the nucleons' kinetic energy if the numbers of neutrons and protons in the nucleus are equal. This tendency to the minimum is reflected in the locus of stable nuclei settling to the N = Z line for light nuclei. However, the locus curves upwards (i.e., N > Z) for increasing Z due to the greater number of neutrons required to provide the attractive strong nuclear force to counter the repulsive Coulomb force between protons. Nuclei created through fission or nuclear reactions can appear on

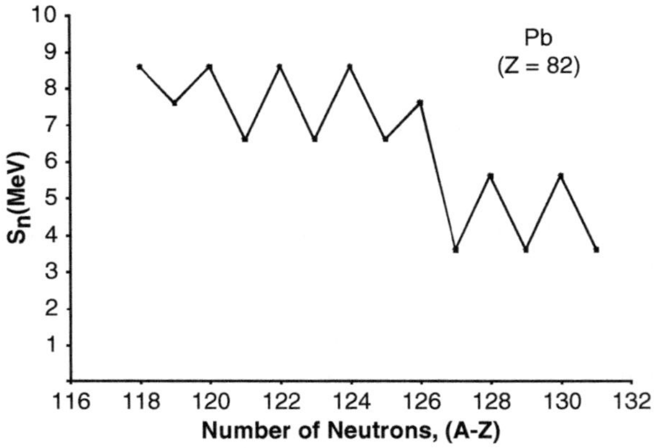

Fig. 3.15 Neutron separation energy vs. number of neutrons for Pb isotopes

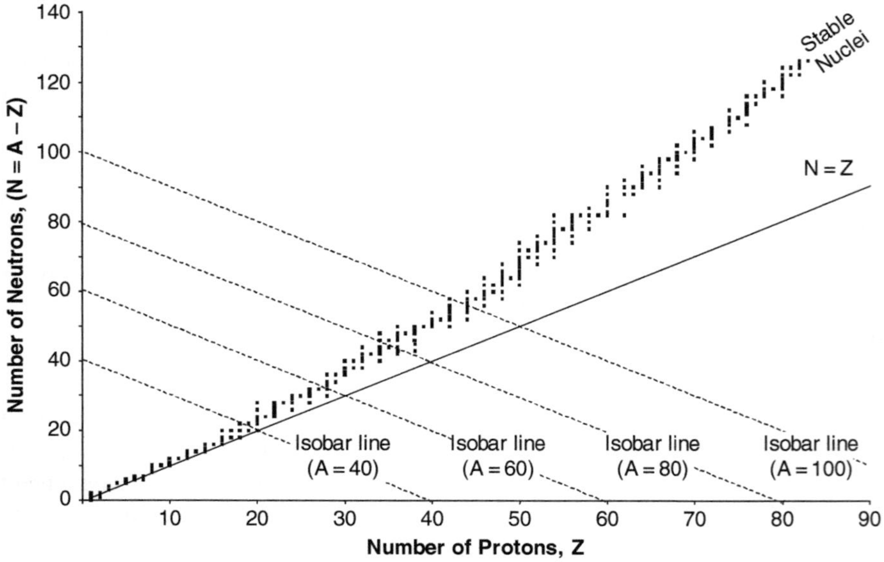

Fig. 3.16 The Segrè plot: the number of protons vs. the number of neutrons for stable nuclei. Also shown is the N = Z line and the *isobar lines* (constant A) for A = 40, 60, 80, and 100

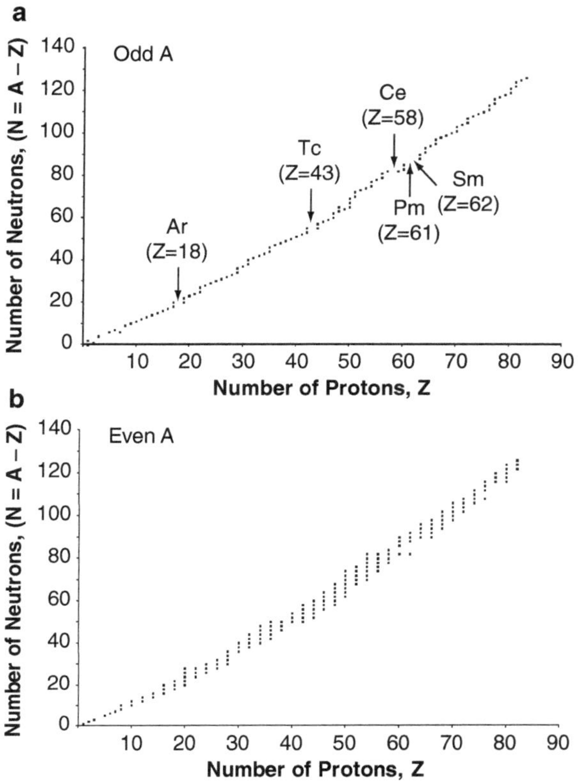

Fig. 3.17 Number of protons vs. the number of neutrons for stable nuclei partitioned into (**a**) odd and (**b**) even values of A

either side of the locus and will be unstable as there are decay routes through which they can maximize their binding energy. As to be discussed in Chap. 4, such nuclei can reach stability through decay processes which alter the number of protons and neutrons. It is these nuclei that are outside this locus and which undergo α decay, β decay, or electron capture to reach the locus of stability which are used in nuclear medicine. As β decay and electron capture occur between isobars, these transitions occur along the diagonal isobar lines in the figure. By partitioning Fig. 3.16 into those nuclei with odd and even A, as shown in Fig. 3.17, further information regarding nuclear stability characteristics can be obtained.

A number of striking features are obvious when the Segrè plot partitioned into plots for even-A and odd-A nuclei:

- There are more even-A stable nuclei than odd-A stable nuclei (which was suggested in the earlier examination of the neutron separation energy for lead isotopes).
- With only four exceptions (^{2}H, ^{6}Li, ^{10}B, and ^{14}N),[24] there are no even-A odd-Z (i.e., odd-N odd-Z) stable nuclei.
- Odd-Z nuclei have only one or two stable isotopes (i.e., there are only one or two stable nuclei in any group of odd-A odd-Z isobars).
- There are no stable odd-A isotopes of argon, cerium, and samarium.
- There are no stable isotopes of technetium and praseodymium.

Table 3.3 Statistics of stable nuclei

	Number of stable nuclei	
	Odd Z	Even Z
Odd N	4	53
	Even A	Odd A
Even N	50	156
	Odd A	Even A

Table 3.3 summarizes the statistics of stable nuclei shown in Fig. 3.17.

It is also evident from Table 3.3 that the stability of an odd-A nucleus is essentially independent of whether or not the combination of nucleons is odd-Z/even-N or even-Z/odd-N. That is, the unpaired nucleon in a stable nucleus can be a proton or a neutron. On the other hand, even-A stable nuclei have an overwhelming preference for the even-N/even-Z combination (as noted earlier, there are only four stable odd-N odd-Z nuclei). This indicates the presence of a nucleon pairing energy.

3.5 Liquid-Drop Model and the Semi-Empirical Nuclear Mass Formula

3.5.1 Introduction

The nuclear mass formula includes a binding energy term, B(A, Z), which is maximized for stable nuclei. Any expression for this binding energy must be able to

[24]Some authors have included the 9^{-} excited state of ^{180}Ta as stable odd-*Z* even-*A* nucleus; however it is unstable with a half life of 1.8×10^{13} years.

reproduce the five main nuclear phenomena demonstrated so far:

- The increase in binding energy with nuclear size and nucleon number
- The decrease in net binding energy due to an increase in the number of nucleons at the nuclear surface
- The decrease in net binding energy with nucleon number due to the Coulomb repulsion between protons
- The decrease in net binding energy reflecting the tendency of the locus of neutron number vs. proton number for stable nuclei to deviate from the $N = Z$ line with increasing A
- A term to account for the observed rarity of odd–odd nuclei and the preference in nature for even–even nuclei

These requirements are consistent with the model of a nucleus as a liquid drop.

3.5.2 Nuclear Binding Energy

Weizsäcker proposed a parameterization of B(A, Z) with terms which accounted for each of the five features above and each weighted by a factor determined by fitting the expression to experimental data. A current expression for this parameterization is of the form,

$$B(A,Z) = a_{Vol}A - a_{Surf}A^{2/3} - a_{Coul}\frac{Z^2}{A^{1/3}} - a_{Sym}\frac{(A-2Z)^2}{A} - \delta \qquad (3.154)$$

where the terms correspond to volume (i.e., constancy of f(A) with A), surface, Coulomb, symmetry, and paired nucleon effects, respectively. The volume term is the previously-derived "condensation" energy and represents the nearest-neighbor interaction between nucleons. The remaining terms can be thought of as "correction" terms which account for the differences between the observed binding energy and the simple condensation energy. The surface term describes the reduction in the binding energy due to the fact that the nucleons at the surface interact only with those in the interior. The Coulomb term describes the reduction in binding energy due to the Coulomb repulsion between protons and the symmetry term represents the minimum in nucleon kinetic energy for equal numbers of protons and neutrons as has been derived in the Fermi gas model. Finally, the paired nucleon term is an empirical reflection of the observed prevalence of stable even–even nuclei shown in Table 3.3. The description of the nuclear mass by the sum of Weizsäcker's formula for the binding energy and the rest masses of the neutrons and protons is also referred to as the semi-empirical formula for nuclear mass as, although some are determined by empirical fits to measured nuclear mass data, the coefficients of (3.154) have a theoretical basis. Even so, it is important to bear in mind that the formula is only an approximation capable of predicting the general trends in variations in nuclear binding energy.

Each of the terms in the formula is discussed below.

3.5.3 Binding Energy Terms

3.5.3.1 Volume Term

As noted earlier, the total binding energy per nucleon is very nearly constant (or slowly decreasing) for $A > 10$ which implies a saturable and short-range force as the total number of pairs of nucleons available from a nucleus of atomic mass number A is equal to $A(A-1)/2$ which would imply a force proportional to A^2. As the volume term is proportional to A, the attractive strong nuclear force must have a short range thus restricting the nucleons to interact with only their nearest partners. Empirically, the coefficient for this term is $a_{Vol} = 15.56$ MeV.

As the volume term exaggerates the net binding energy, the remaining four terms describe the diminution of the attractive effect of the volume term.

3.5.3.2 Surface Term

The average binding energy per nucleon grows rapidly with nucleon number for small A reflecting an incomplete binding of nucleons at the nuclear surface for light nuclei. This incomplete binding is analogous to the surface tension of a liquid drop and hence proportional to the surface area of the nucleus, $4\pi R_N^2$ where R_N is the nuclear radius. As R_N is proportional to $A^{1/3}$, the binding energy surface term is thus proportional to $A^{2/3}$. The coefficient for the surface energy term $a_{Surf} = 17.23$ MeV

3.5.3.3 Coulomb Term

The repulsive Coulomb potential between the protons reduces the net binding energy. This Coulomb potential was derived earlier for a spherical nucleus with a uniformly-distributed electric charge,

$$E_{Coul} = \frac{3}{5}(\alpha\hbar c)\frac{Z^2}{r_N A^{1/3}}. \qquad (3.155)$$

The specification of a uniform distribution of charge would thus lead to the coefficient of the Coulomb term in the Weizsäcker formula to be,

$$\frac{3}{5}\frac{\alpha\hbar c}{r_N} \approx 0.72\,\text{MeV}.$$

However, the actual factor is $a_{Coul} = 0.697$ MeV, the difference being due to the invalidity of the assumption of a uniform distribution of charge within the nucleus which led to the 3/5 multiplicative factor. The fact that the magnitude of the Coulomb term coefficient is an order-of-magnitude less than the others reflects the relative strength of the strong nuclear force compared to the electromagnetic force.

3.5.3.4 Symmetry Term

A result of the derivation of the Fermi gas model was that the total nucleon kinetic energy is minimized for equal numbers of protons and neutrons. This gives the symmetry energy as being proportional to $(N - Z)^2 = (A - 2Z)^2$. This term appears in the Weizsäcker formula with the coefficient, $a_{Sym} = 23.29$ MeV.

3.5.3.5 Paired Nucleons Term

This term in the Weizsäcker semi-empirical formula arises from the observations summarized in Table 3.3. Even-N/even-Z nuclei are the most prevalent making up about 59% of all stable nuclei and even-N/odd-Z and odd-N/even-Z nuclei are the next prevalent making up about 39% of all stable nuclei. The odd-N/odd-Z combination is rare for stable nuclei, equivalent to less than 2% of all stable nuclei. Thus, even-N/even-Z nuclei will have the greatest binding energy, followed by even-N/odd-Z and odd-N/even-Z nuclei and the odd-N/odd-Z nuclei will have the lowest binding energy. To replicate this observation, a representation of the paired nucleon term in the Weizsäcker semi-empirical formula is,

$$\delta = 0 \quad \text{even-N/odd-Z and odd-N/even-Z} \qquad (3.156)$$

$$\delta = +\frac{12}{\sqrt{A}}\,\text{MeV} \quad \text{odd-N/odd-Z} \qquad (3.157)$$

$$\delta = -\frac{12}{\sqrt{A}}\,\text{MeV} \quad \text{even-N/even-Z.} \qquad (3.158)$$

This term cannot be reproduced from the liquid-drop model but can be suggested from the shell model to be discussed.

3.5.4 Contributions of Binding Energy Terms

Figure 3.18 presents the effects of the contributions of the different terms to the binding energy in the semi-empirical mass equation. This plot shows the volume binding energy per nucleon (which is a constant) and the subsequent curves resulting from the subsequent corrections of the surface, Coulomb and symmetry energy terms which diminish the net binding energy. It is clear that the effect of the surface term is more significant for light nuclei where the fraction of nucleons at the surface of the nucleus is greatest, the Coulomb term is greater for heavy nuclei with larger numbers of protons and the symmetry term becomes significant for heavy nuclei where the A/Z ratio is at its greatest.

The final curve of Fig. 3.18 provides general representation of the binding energy per nucleon of the Weizsäcker formula. Comparing this result to the measured binding energies per nucleon of Fig. 3.14, one can see that the Weizsäcker result can replicate the

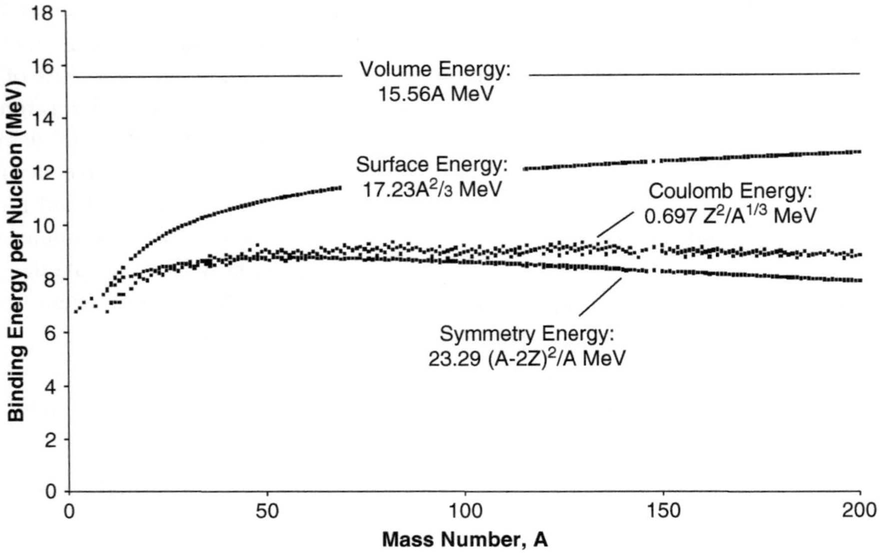

Fig. 3.18 The surface, Coulomb, and asymmetry terms successively subtracted from the volume term to yield the final Weizsäcker semi-empirical approximation to the binding energy per nucleon as a function of the atomic mass number, A

measured binding energy to within about 2 MeV, but that there are particularly large disagreements between these values for certain nuclei. In these cases, the Weizsäcker formula can underestimate significantly the binding energy for these nuclei by values of up to 10 MeV or so. Such discrepancies reflect a quantum mechanical behavior of nucleons which differs from the semiclassical approaches considered so far and which becomes most evident for these certain nuclei, which are referred to as "magic" nuclei. These exceptional nuclei are discussed later.

3.5.5 Mass Parabolae

By combining the expression for the atomic mass and the Weizsäcker semi-empirical formula for the nuclear binding energy, the mass of an atom of atomic mass number A and atomic number Z can be written with the atomic number and atomic mass number as variables,

$$M(A,Z) = ZM_H + (A-Z)m_n - \left[a_{Vol}A - a_{Surf}A^{2/3} - a_{Coul}\frac{Z^2}{A^{1/3}} - a_{Sym}\frac{(A-2Z)^2}{A} - \delta\right] \quad (3.159)$$

It is now shown that, for isobars, the nuclear binding energy can be maximized or, equivalently, the atomic mass minimized, through nuclear decay processes (including beta decay and electron capture) that conserve the total number of neutrons and protons between parent and daughter nuclei. Consider first the cases of odd-A isobars for which the nucleon pairing term δ is equal to zero. The atomic mass expression of can be written as a quadratic function in Z,

$$M(A,Z) = k_2(A)Z^2 + k_1Z + k_0(A) \quad (3.160)$$

with the coefficients,

$$k_2(A) = \left[\frac{a_{Coul}}{A^{1/3}} + \frac{4a_{Asym}}{A}\right] \quad (3.161)$$

$$k_1 = \left[M_H - m_n - 4a_{Asyn}\right] \quad (3.162)$$

$$k_0(A) = A\left[m_n - a_{Vol} + \frac{a_{Surf}}{A^{1/3}}\right] \quad (3.163)$$

where A is a constant for isobars. Equation (3.160) describes the atomic mass as a parabola for constant A in the variable of the atomic number, Z. As the

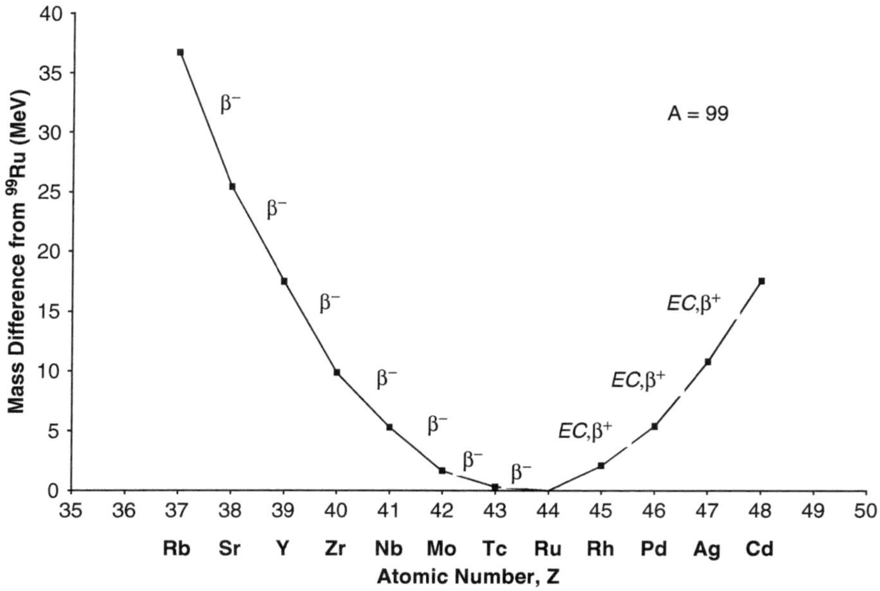

Fig. 3.19 Mass parabola for A = 99

coefficient $k_2(A)$ is always positive, the parabola is inverted with a minimum for the atomic number,

$$Z_{min} = -\left(\frac{k_1}{2k_2(A)}\right) = \left(\frac{A}{2}\right)\left(\frac{4a_{Sym} + (m_n - M_H)}{4a_{Sym} + a_{Coul}A^{2/3}}\right). \quad (3.164)$$

This result gives the atomic number of the stable isobar. Substituting the numerical values of the constants yields the approximation,

$$Z_{min} \approx \left(\frac{A}{2}\right)\left(\frac{94}{93 + 0.7A^{2/3}}\right). \quad (3.165)$$

Hence, for light nuclei where A is small, $Z_{min} \approx A/2$, indicating the tendency for equal numbers of neutrons and protons when the Coulomb repulsion between nucleons can be neglected. Such parabolas lie along the isobar lines in the Segrè plot. Being isobars, the volume and surface terms in the Weizsäcker formula remain constant and only the Coulomb and symmetry terms, being functions of atomic number, vary. An example of the parabolic feature is shown by the measured nuclear masses in Fig. 3.19 for those isobars with $A = 99$. As this value of A is odd, the paired nucleon term δ is zero and the only terms of the Weizsäcker formula that will change with Z for constant odd A are the sum of the Coulomb and symmetry terms,

$$a_{Coul}\frac{Z^2}{A^{1/3}} + a_{Sym}\frac{(A - 2Z)^2}{A}. \quad (3.166)$$

For $A = 99$, the isobar with the smallest mass (i.e., the greatest binding energy) is ^{99}Ru and the differences between the other isobar masses and that of ^{99}Ru is plotted against atomic number, Z. Isobars on either side of ^{99}Ru (i.e., nuclei with an atomic number greater or less than that of ^{99}Ru) are unstable and the binding energies of those to the left of ^{99}Ru can be increased (and, hence, their masses decreased) by the β^- decay of a neutron to a proton.[25] For $A = 99$ and Z increasing from 37 to 43, the sum of the Coulomb and symmetry terms in the Weizsäcker formula decreases which, as these are subtracted from the constant volume and surface terms, result in a corresponding increase in binding energy. For Z greater than 43, the sum of the Coulomb and symmetry terms increase leading to a decrease in binding energy. Hence, isobars

[25]While the excited states of ^{99}Sr, ^{99}Y, and ^{99}Zr can decay via single neutron emission and excited states of ^{99}Ag can decay via proton emission, the probabilities of such exotic decays are extremely small and are neglected here.

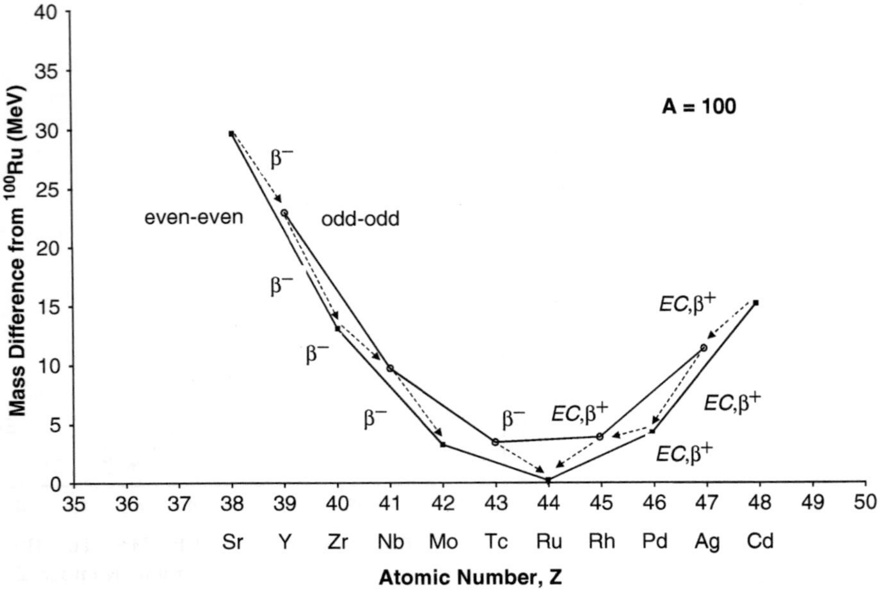

Fig. 3.20 Mass parabolae for odd-Z/odd-N and even-Z/even-N combinations of A = 100 (note that both ^{100}Mo and ^{100}Ru on the even–even parabola are stable)

to the right of ^{99}Ru (i.e., with an atomic number greater than that of ^{99}Ru) will increase their binding energies by decreasing their net electric charge through the conversion of a proton to a neutron through either β^+ decay (which emits a positron) or by the capture of an atomic orbital electron.

Next consider the case of those mass parabolas for even-A nuclei. In such cases, the isobars are other odd-Z/odd-N or even-Z/even-N and the nucleon pairing term is not zero and will differ between the two combinations of isobars with the result of two mass parabolas in Z. Figure 3.20 presents the mass parabolae for A = 100. Because of the pairing term, the even–even parabola contains more nuclei of lower mass than the odd–odd parabola and nuclei in the odd–odd parabola can always decay to those in the even–even parabola. As before, nuclei to the left of the even–even parabola's minimum undergo β^- decay (increase Z) in order to achieve stability (minimum mass) and those to the right undergo β^+decay or electron capture to do the same. However, the even–even parabola for these isobars contains two stable nuclei: ^{100}Mo and ^{100}Ru. While one could expect ^{100}Ru to go through sequential β^- decays through ^{100}Tc to ^{100}Ru, ^{100}Tc is slightly more massive by about 170 keV than ^{100}Mo. Because, as shown in the following chapter, β^- decay cannot occur unless the product nucleus is lighter than the original radioactive nucleus, ^{100}Mo cannot undergo β^- decay to ^{100}Tc, meaning that the ^{100}Mo nucleus is stable. However, ^{100}Tc itself is an unstable nucleus as it is energetically preferable for it to undergo β^- decay to ^{100}Ru. ^{100}Tc is not produced via radioactive decay, but is created by reactions such as the radiative (n, γ) neutron capture on ^{99}Tc or the (p, n) charge exchange reaction on ^{100}Mo.

3.5.6 Prediction of Stable Isobars

Although it is only capable of providing only an approximation to the general tendencies of nuclear binding energy with nuclear mass, the Weizsäcker parameterization of nuclear mass does allow a prediction of what will be the most stable isobar for a given atomic number, *Z*. This can be shown by finding its minimum through differentiating the Weizsäcker formula with respect to *Z* and setting the result to zero to yield the locus of *Z* for minimal isobar mass,

$$(M_H - m_n) + 2a_{Coul}\frac{Z}{A^{1/3}} + \frac{a_{Sym}}{A}(8Z - 4A) = 0. \tag{3.167}$$

Rearranging and using the substitution $y = A^{-1/3}$ yields the cubic equation,

$$y^3 + \left(\frac{a_{Coul}}{4a_{Sym}}\right)y + \left(\frac{M_H - m_n - 4a_{Sym}}{8Za_{Sym}}\right) = 0 \quad (3.168)$$

which is of the form,

$$y^3 + ay + b = 0 \quad (3.169)$$

where the two coefficients are,

$$a = \frac{a_{Coul}}{4a_{Sym}} \quad (3.170)$$

and

$$b = \frac{M_H - m_n - 4a_{Sym}}{8Za_{Sym}}. \quad (3.171)$$

As all of the coefficients are real and the discriminant is positive, there can only be one real root and two conjugate imaginary roots, the latter two being ignored as they are obviously nonphysical. From the real root of the cubic equation, the number of neutrons N = (A – Z) as a function of Z for isobars with minimum mass (i.e., the greatest binding energy), is,

$$N = \frac{1}{\sqrt[3]{-\frac{b}{2} + \sqrt{\frac{b^2}{4} + \frac{a^3}{27}}} + \sqrt[3]{-\frac{b}{2} - \sqrt{\frac{b^2}{4} + \frac{a^3}{27}}}} - Z. \quad (3.172)$$

The calculated neutron number for the most stable isobar as a function of proton number is shown in Fig. 3.21 along with the empirical locus of all stable nuclei. The calculated N(Z) line reproduces quite well the locus stability of the Segrè plot.

The expressions of nuclear mass/binding energy derived from the Weizsäcker formula provide only approximate tendencies of, for example, the relationship between the numbers of neutrons and protons in stable nuclei. As can be seen in Fig. 3.21, whilst able to reproduce the overall functional dependence of neutron number with increasing atomic number, the Weizsäcker-derived result does not predict the unusual cases of relatively large number of stable isotopes (i.e., nuclei with constant Z but variable N) that occur for Z = 20, 50, and 82. The liquid drop model of the nucleus has been shown to be capable of reproducing quantitatively many physical features of the nucleus and, through the contributions of the Coulomb and surface terms, to also allow predictions of the instability of isobaric nuclei. While the model can predict such general and macroscopic trends, it is unable to predict a variety of detailed features of nuclei. These features are addressed by the independent particle/nuclear shell models.

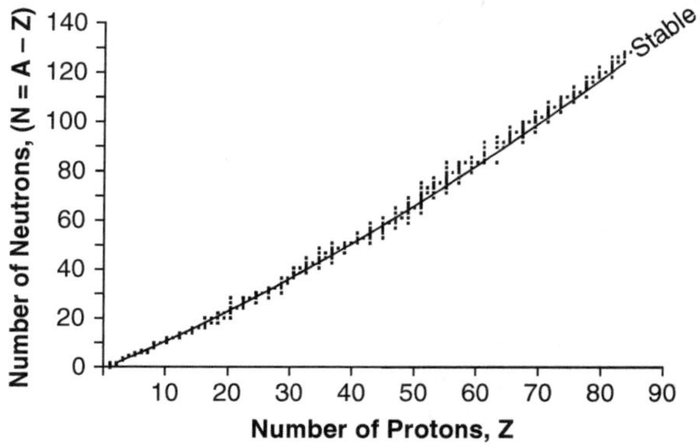

Fig. 3.21 The number of neutrons (A – Z) vs. the number of protons for stable nuclei Z; the curve is that calculated from (3.172) of the number of neutrons for the most stable isobar for a given value of Z

3.6 Nuclear Shell Model

3.6.1 Introduction

Whereas the Fermi gas and liquid drop models can predict the general features of nuclei, they cannot reproduce the properties of nuclei which exhibit characteristics of extraordinary stability and prevalence in nature. Such nuclei being referred to as being "magic," with the Weizsäcker semi-empirical formula underestimating their binding energy by as much as 10 MeV. In order to correct for these differences between theory and measurement, the quantum mechanical properties of intrinsic spin and orbital angular momentum of the nucleons within the nucleus must be invoked.

3.6.2 Magic Nuclei

For nuclei with numbers of neutrons or protons equal to 2, 8, 20, 28, 50, 82, and 126, the Weizsäcker representation underestimates the nuclear binding energy by about 10 MeV, indicating high degrees of stability not predicted by either the Fermi gas and liquid drop models. Such neutron and proton numbers yielding these properties are labeled as being "magic" and nuclei with proton and neutron numbers that are both "magic," e.g., ^{16}O and ^{208}Pb, are exceptionally stable and are referred to as being "doubly magic." Examples of magic numbers include:

- The naturally-occurring radioactive series beginning with ^{238}U, ^{235}U, and ^{232}Th and ending with the stable isotopes ^{206}Pb, ^{207}Pb, and ^{208}Pb, respectively, all of which have Z = 82.
- The element with the greatest number of stable isotopes (10) being Sn (Z = 50).
- The greatest number of stable isotones occurring for N = 82.
- ^{136}Xe (with N = 82) having a low cross-section for neutron absorption whereas ^{135}Xe (with N = 81) has a high cross-section for neutron absorption so as to achieve the magic number of N = 82.

Additional evidence for these magic numbers can be seen in Fig. 3.22 which is a plot of the relative abundance of elements in the solar system. Relative peaks appear for Z = 8, 28, 50, and 82. Moreover, it can be seen that (excluding the obvious case of ^{1}H) even-Z nuclei tend to be more abundant than odd-Z nuclei, a feature already noted.

3.6.3 Calculation of Nucleon Orbitals

One can compare the stability of magic nuclei with the noble gas atoms with closed electron shells. This analogy leads to the nuclear shell model and its refinements which have been successful in reproducing nuclear properties such as the magic numbers, the nuclear magnetic dipole moment and the modeling of excited nuclear states. But, a priori, it is difficult to compare nuclear and atomic electron shells for two reasons. The most obvious is that the atomic electrons are subject to a central Coulomb force whereas nucleons are not subject to such a central force. In addition, as the nuclear density is so great, one struggles with the idea of a nucleon having a well-defined orbit at all as the mean free path between nucleon–nucleon scatters would be much less than the nuclear dimensions. The Pauli exclusion principle presents a solution to these dilemmas as it can be argued that, despite the overall noncentral nuclear force and the

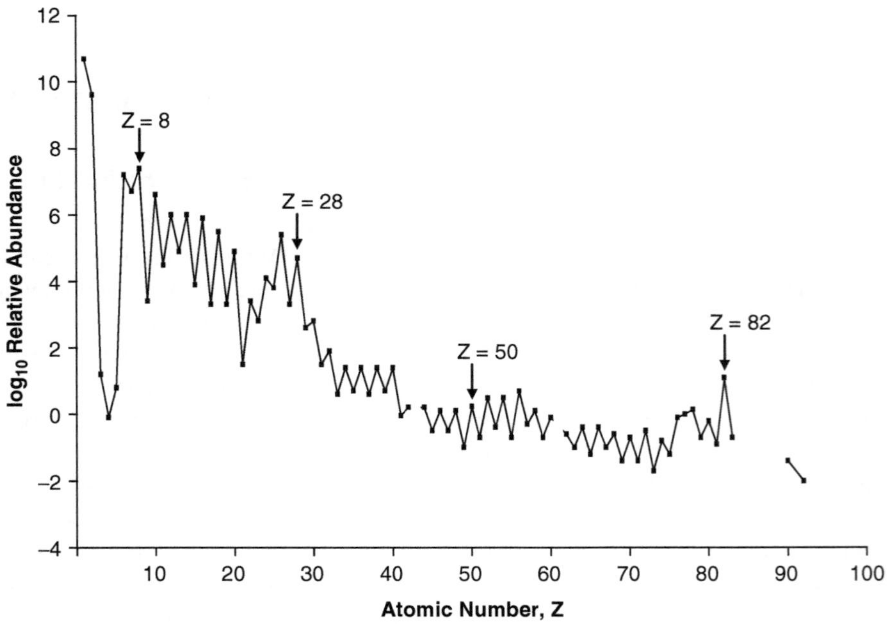

Fig. 3.22 Logarithm of the relative elemental abundance in the solar system as a function of atomic number, Z

high nuclear density, an available quantum state following a nucleon–nucleon interaction may not be present thus extending the intranuclear mean free path with the result that a nucleon would behave as if it were subject to a constant static potential. A simple nonrelativistic calculation for the nucleon wavefunction is now performed under the assumption that the nucleon is subject to such a potential.

Although the measured nuclear form factor would suggest that the nuclear potential varies smoothly with distance from the nuclear center (as with the Woods–Saxon potential), for the simplicity of calculation we will evaluate the nucleon wave function for a spherically-symmetric rectangular potential well,

$$\begin{aligned} U(r) &= -U_0 \quad r < R_N \\ &= 0 \qquad r \geq R_N. \end{aligned} \tag{3.173}$$

There is indeed some justification for invoking such an extreme model of the nuclear potential on the basis of the arguments that the net force on a nucleon at the center of the nucleus is zero (leading to a constant potential) and that the nuclear density is relatively constant. From Chap. 2, the wavefunction for a central potential is of the general form $\psi(r, \theta, \phi) = (A_1 j_1(\lambda r) + B_1 y_1(\lambda r))Y_l^m(\theta, \phi)$ where $j_l(x)$ is the spherical Bessel function of the first kind, $y_l(x)$ is the spherical Bessel function of the second kind (also called a Neumann function) and λ was derived in Chap. 2. The A_l and B_l are integration constants for the radial component of the wavefunction which, as it must remain finite, require that as $y_l(x) \to -\infty$ as $x \to 0$ then $B_l = 0$. The resulting nuclear wavefunction is,

$$\psi_{lm}(r, \theta, \phi) = A_l j_l(\lambda r) Y_{lm}(\theta, \phi) \tag{3.174}$$

where the spherical harmonics are,

$$Y_{lm}(\theta, \phi) = \sqrt{\frac{2l+1}{4\pi}\frac{(l-m)!}{(l+m)!}} P_{lm}(\cos\theta) e^{im\phi}. \tag{3.175}$$

The $P_{lm}(\cos\theta)$ are the associated Legendre polynomials, l is the quantized orbital angular momentum number, $l = 0, 1, 2\ldots$ and $m = -l, -(l-1)\ldots 0\ldots(l-1), l$. Combining these results and allowing for the boundary condition, the nuclear wavefunction is,

$$\psi(r, \theta, \phi) = A_l \sqrt{\frac{2l+1}{4\pi}\frac{(l-m)!}{(l+m)!}} j_l(\lambda r) P_{lm}(\cos\theta) e^{im\phi}. \tag{3.176}$$

From the definition of the radial coefficient λ,

$$\frac{(\hbar c)^2}{2m}\lambda^2 = E - U(r) \tag{3.177}$$

it can be seen that $(\hbar c)^2\lambda^2/2m$ is the energy of the system for a constant potential. Consider the nuclear wavefunction for $l = 0$ and which has zeroes for kr = π, 2π, 3π .... The boundary condition that $\psi(R_N, \theta, \phi) = 0$ (i.e., the nucleon wavefunction is zero at the nuclear surface) leads to the first zero of $\lambda_0 R_N = \pi$, or,

$$E - U_0 = \frac{1}{2m}\left(\frac{\pi\hbar c}{R_N}\right)^2. \tag{3.178}$$

The zeroes for the first six spherical Bessel functions are shown in Fig. 3.23. First consider the zeroth-order Bessel function with the corresponding radial coefficient, λ_0. For $\lambda_0 R_N = \pi$, the radial function has no nodes within the nucleus and for $\lambda_0 R_N = 2\pi$, the radial function has one node, etc. The radial quantum number n is defined as the number of radial function nodes plus one. The usual spectroscopic notation of s, p, d, f, g ... is used for the orbital angular momentum number $l = 0, 1, 2, 3, 4 \ldots$, respectively. For any given Bessel function (3.177), shows that the energy of the quantum state is proportional to λ^2, where λR_N is a zero of the radial function necessary for the boundary condition. Table 3.4 shows the first 8 quantum states thus calculated.

The occupancy of each state, $2(2l + 1)$, will be determined by the Pauli exclusion principle. Four quantum numbers describe each nucleon: the radial quantum number n, the orbital angular quantum number l, the magnetic quantum number m and the intrinsic spin of the nucleon m_s (as the proton and neutron ensembles are being considered separately, isospin can be ignored). For a given (n, l) state there are $(2l + 1)$ values of m and, as the nucleon is spin-1/2, the occupancy is $2(2l + 1)$. Hence, for example, only two neutrons can fill the 1s state ($n = 1$, $l = 0$, $m_s = \pm 1/2$). Figure 3.24 provides the results of the nucleon wavefunction calculation and shows the quantum

Fig. 3.23 Spherical Bessel functions of the first kind. Zeroes of the functions are identified. (**a**) shows the functions of orders 0, 1 and 2; (**b**) shows the functions of orders 3, 4 and 5

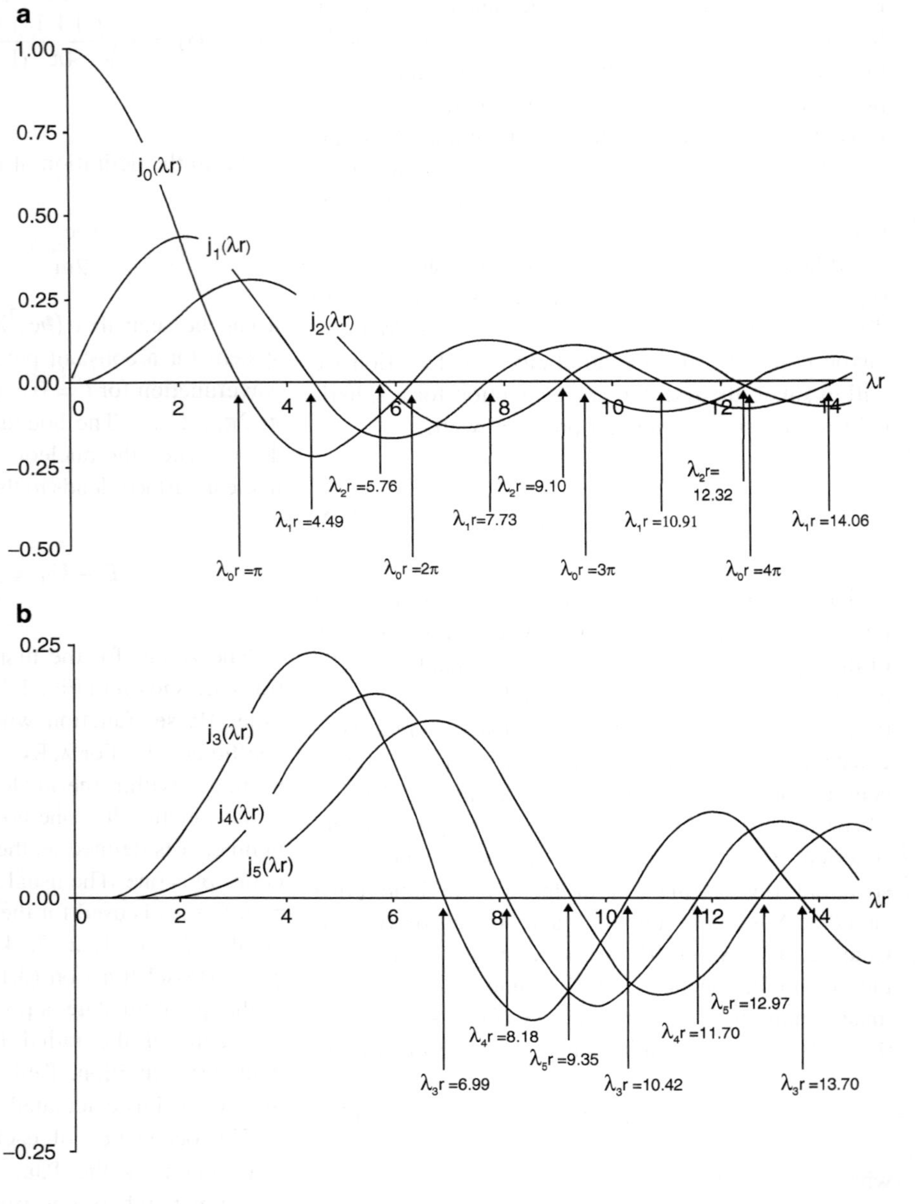

levels available to the nucleons. These states are labeled by the principal quantum number n and the orbital angular momentum quantum number *l*. The occupancy, or the number of available states at each level, is also shown for each level.[26] This simple model can reproduce the observed magic numbers of 2, 8, and 20, reflecting the complete fillings of the 1s, 1p, and 2s levels, but it both fails to reproduce higher values of magic numbers and predicts a nonexistent magic number of 18. These errors can be corrected for by invoking the coupling between the intrinsic spin of the nucleon and its orbital angular momentum.

It was recognized by Mayer (1949, 1950) and by Haxel et al. (1949) that the coupling of the orbital angular momentum ***l*** of a nucleon with its intrinsic angular momentum, or spin **s**, can reproduce the

[26]Strictly speaking, the calculated levels are for the neutron as the Coulomb potential was not accounted for. In practice, this would mean that the proton levels would be slightly higher due to the Coulomb repulsion, but the relative spacing would be the same.

Table 3.4 Shell model quantum states

λR_N	Energy $(\lambda R_N)^2$	n	l	Quantum state	$2\,(2l+1)$	$\Sigma\,2l(l+1)$
π	9.87	1	0	1s	2	2
4.49	20.16	1	1	1p	6	8
5.76	33.18	1	2	1d	10	18
2π	39.48	2	0	2s	2	20
6.99	48.86	1	3	1f	14	34
7.73	59.75	2	1	2p	6	40
8.18	66.91	1	4	1g	18	58
9.10	83.81	2	2	2d	10	68

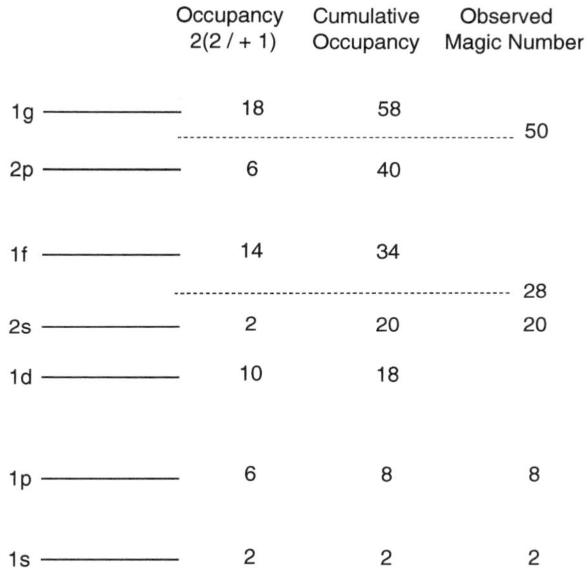

Fig. 3.24 Quantum states for a central attractive nuclear potential. The first column provides the occupancy or the number of states available to the nucleon and the second column provides the cumulative occupancy or the total number of nucleons. The third column provides the experimentally observed magic numbers. Conventional spectroscopic notation is used where each state is identified by the radial quantum number and the orbital angular quantum number of 0, 1, 2 . . . labeled by s, p, d . . .

observed magic numbers. This spin-orbit coupling is a potential proportional to $\boldsymbol{l} \bullet \mathbf{s}$ which each nucleon experiences in addition to the constant radial potential to provide a net potential of the form, $U(r) - f(r)(\boldsymbol{l} \bullet \mathbf{s})$, and which removes the degeneracy in states by generating an energy splitting between the $(l + 1/2)$ and $(l - 1/2)$ states. For example, the 1p state splits into the $1p_{3/2}$ and $1p_{1/2}$ levels, where the subscripts indicate $(l \pm 1/2)$. If the $(l + 1/2)$ state is assumed to have a lower energy level than the $(l - 1/2)$ state (which is opposite to the case of atomic electrons), the nuclear

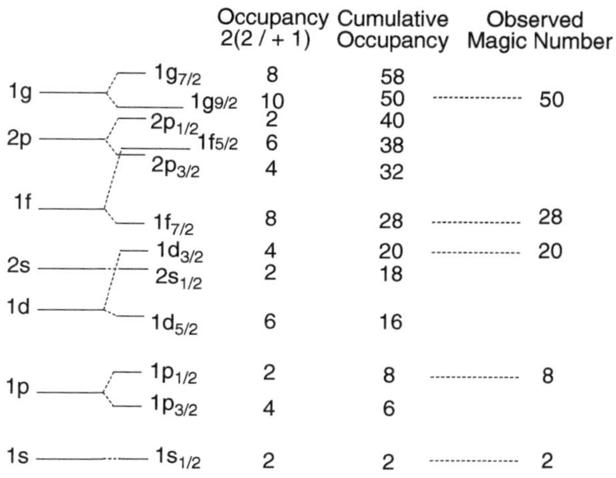

Fig. 3.25 Nucleon energy levels with the inclusion of spin-orbit coupling. Spectroscopic notation is used to identify the levels, with the subscript denoting the value of $(l \pm 1/2)$

magic numbers are reproduced, as shown in Fig. 3.25. The energy differences between the $(l + 1/2)$ and $(l - 1/2)$ states will increase with l such that the splitting can overlap another higher-energy state. This can be seen by comparing the 1d and 2s states in which the 2s state is at a higher level than the 1d state if spin-orbit coupling is neglected. The inclusion of spin-orbit coupling and the splitting between the $1d_{3/2}$ and $1d_{5/2}$ levels leaves the $1d_{3/2}$ level higher than the $2s_{1/2}$ level, thus removing the anomalous magic number of 18. Hence, nuclear stability is dependent upon not only the mass and charge of the nucleus (i.e., the mass parabolas) but also the angular momenta of the nucleons.

The result of Fig. 3.25 can also be used to predict the net angular momentum of a given nucleus. The Pauli exclusion principle requires that the levels of the ground state be sequentially filled by nucleons beginning with the $1s_{1/2}$ level. As will be shown, even-Z/even-N nuclei have zero angular momentum (i.e., zero magnetic moments) indicating that when a level is completely filled or if there is an even number of nucleons in a level, the nucleons "pair off" such that the sum of their angular momenta is zero. This suggests that the nuclear angular momentum of an odd-Z/even-N or even-Z/odd-N nucleus will be that of the single unpaired nucleon. For example, the angular momentum of ^{3}He would be expected to be that of the unpaired neutron in the $1s_{1/2}$ level and, indeed, the measured value is 1/2. Similarly, the angular momentum of ^{17}O is set by that of the unpaired neutron in the

Table 3.5 Angular momenta of ground states of odd-Z/odd-N nuclei

Nucleus	Quantum level of unpaired nucleon		Measured angular momentum
	Proton	Neutron	
^{2}H	$1p_{1/2}$	$1p_{1/2}$	1
^{6}Li	$1p_{3/2}$	$1p_{3/2}$	1
^{10}B	$1p_{3/2}$	$1p_{3/2}$	3
^{14}N	$1p_{1/2}$	$1p_{1/2}$	1

$1d_{5/2}$ state. However, this independent particle model of nuclear angular momentum for odd–even nuclei does not hold universally. For example, whereas the model predicts that the angular momentum of the ^{23}Na nucleus is $5/2$ due to the unpaired proton in the $1d_{5/2}$ state, experiment shows it to be in fact equal to $3/2$. In general, failure of the model to reproduce measured nuclear angular momenta occurs at values of very high angular momentum. For example, the model predicts an angular momentum of $11/2$ for ^{131}Xe due to an unpaired neutron in the $1h_{11/2}$ level whereas measurement assigns a value of $3/2$. This discrepancy could be explained if it is assumed that the nucleon–nucleon pairing energy increases with angular momentum so as to allow the unpaired neutron in the $11/2$ level to split up a neutron pair in the $2d_{3/2}$ level so as to couple with one of these and leave the remaining $2d_{3/2}$ neutron unpaired.

The quantum levels of the unpaired nucleons of the four stable odd-Z/odd-N nuclei are shown in Table 3.5 along with the measured nuclear angular momentum. While the angular momentum of such nuclei cannot be predicted by the shell model, it can be seen that the total angular momentum of the nucleus is equal to or less than the sum of the unpaired nucleons' angular momenta. The deuteron is a particularly special case as it provides insight into the various nuclear forces and is discussed later.

By using this shell model, a number of predictions of nuclear phenomenology may be made:

- The pairing of nucleons in even-Z/even-N nuclei lead to the ground states of these nuclei having zero net angular momentum and positive parity (i.e., 0^{+}), which is corroborated by the strong nucleon pairing term.
- The spin and parity of an odd-A nucleus in its ground state will be that of the unpaired nucleon (recall that a vacancy or hole can be treated as a particle in terms of angular momentum: a single hole in a subshell will have the same angular momentum and parity as would a nucleon in that subshell).[27]
- The spin and parity of odd-Z/odd-N nuclei can be predicted by coupling the spins and parities of the two unpaired nucleons and noting the Brennan and Bernstein rules which are

Angular momenta of the two nucleons are $j_1 = l_1 \pm 1/2$ and $j_2 = l_2$ $1/2$: the net angular momentum is $J = |j_1 - j_2|$. For example, ^{38}Cl, which undergoes β^- decay to ^{38}Ar, has an unpaired proton in the $1d_{3/2}$ orbit and an unpaired neutron in the $1f_{7/2}$ orbit with corresponding l_1 and l_2 values of 2 and 3 respectively. The net angular momentum is $|3/2 - 7/2| = 2$ with negative parity.

Angular momenta of the two nucleons are $j_1 = l_1 \pm 1/2$ and $j_2 = l_2 \pm 1/2$: The net angular momentum is $J = |j_1 + j_2|$. For example, ^{26}Al, which undergoes β^+ decay to ^{26}Mg, has an unpaired proton and an unpaired neutron in the $1d_{5/2}$ shell. The orbital angular momentum is $l = 2$, so $j_1 = 2 + 1/2 = 5/2$ and $j_2 = 2 + 1/2 = 5/2$ and the net angular momentum is $|5/2 + 5/2| = 5$ with positive parity.

3.7 Characteristics of Atomic Nuclei: Part II

3.7.1 Nuclear Moments

Having now gained an understanding of the binding between nucleons and their motion within the nucleus, we now look at the resulting magnetic and electric properties of nuclei.

3.7.2 Nuclear Magnetic Dipole Moments

3.7.2.1 Introduction

"Magic" nuclei exhibit both stability and prevalence in nature and, in the shell model, reflect nuclei with full

[27] Consider a full shell from which a nucleon is removed; the resulting angular momentum is that of the unpaired nucleon remaining in the shell, which is also that of the remnant hole.

nuclear shells. When a shell is filled, the nucleons within that shell pair off to yield zero angular momentum. On the other hand, nuclei with an unpaired nucleon have a nonzero angular momentum and, as a result, will have a magnetic dipole moment.

3.7.2.2 Spin-Orbit Angular Momenta Coupling

Consider a nucleon with orbital angular momentum **L** which has an absolute magnitude of $\sqrt{l(l+1)}\hbar$. As shown in Fig. 3.26, the orbital angular momentum can be considered as a vector that precesses about a specified axis (e.g., the direction of an external magnetic field) and that the projections of this vector onto the axis are quantized in integral multiples of $\hbar$, called the magnetic quantum number, m_l which has $(2l + 1)$ values in the range from $-l$ to $+l$, including zero. But the nucleon also has an intrinsic angular momentum (spin), **s**, which can also precess (although it will have only two projections, $\pm 1/2$). The orbital angular momentum and the intrinsic spin can thus couple to yield a total angular momentum, **J**, as shown in Fig. 3.27.

3.7.2.3 Magnetic Dipole Moment

Classically, the magnetic dipole moment can be defined from consideration of a square current loop of length L carrying a current I in a magnetic field **B**, as shown in Fig. 3.28. The forces on the two arms of the loop which are orthogonal to the direction of the magnetic field as given by $\mathbf{F} = \mathrm{IL}\,\hat{\mathbf{l}} \times \mathbf{B}$, where $\hat{\mathbf{l}}$ is the unit-vector along which the arm is directed. This loop will then experience a net torque of magnitude,

$$\Gamma = \mathrm{IL}^2\mathrm{B}\sin\theta \tag{3.179}$$

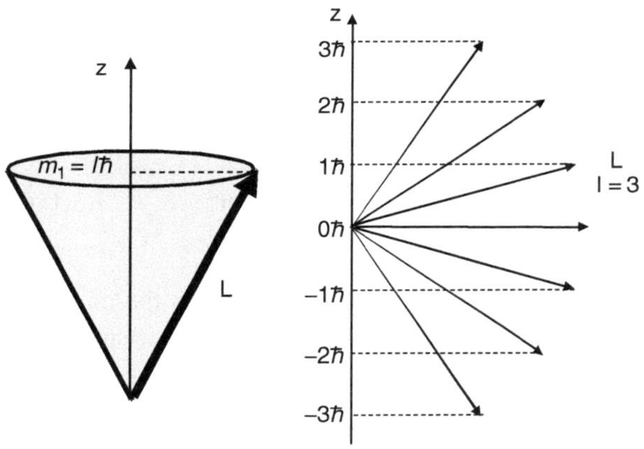

Fig. 3.26 Precession of the orbital angular momentum vector and projections for the example of $l = 3$

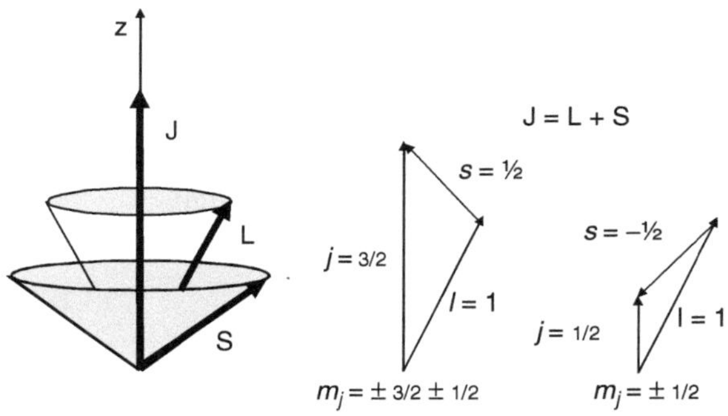

Fig. 3.27 Coupling of orbital angular momentum and intrinsic spin: $l = 1$ and m = 1/2

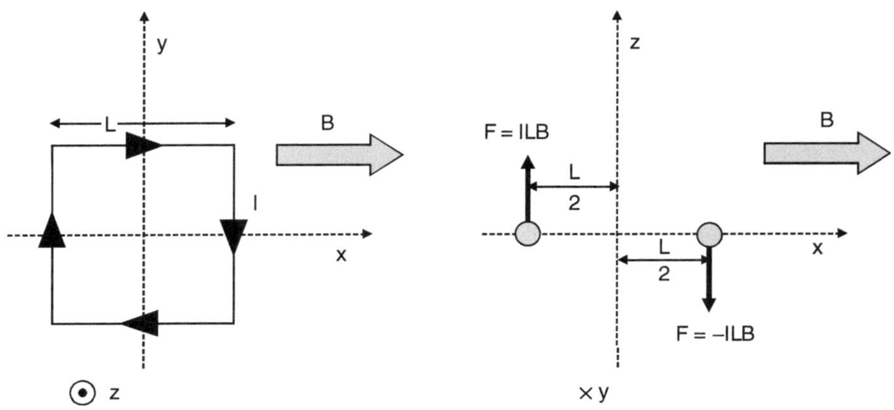

Fig. 3.28 A square current loop in a magnetic field will experience two off-set and opposing forces that create a torque about the y-axis

where θ is the angle between the direction of the magnetic field and the normal unit-vector, $\hat{\mathbf{n}}$, to the loop. The magnetic dipole moment of the loop is,

$$\begin{aligned}\boldsymbol{\mu} &= IL^2\hat{\mathbf{n}} \\ &= IA\hat{\mathbf{n}}\end{aligned} \tag{3.180}$$

where $A = L^2$ is the area of the loop. In vector form, the torque is written as,

$$\boldsymbol{\Gamma} = \boldsymbol{\mu} \times \mathbf{B}. \tag{3.181}$$

The current in the wire loop can be replaced by an electron of mass m_e in a circular orbit of radius r orbiting with an angular frequency of ω radians $s^{-1.}$ In this case, the magnetic dipole moment is,

$$\begin{aligned}\boldsymbol{\mu} &= \left(\frac{e\omega}{2\pi}\right)\pi r^2\hat{\mathbf{n}} \\ &= \frac{e\omega r^2}{2} \times \frac{m_e}{m_e}\hat{\mathbf{n}} \\ &= \left(\frac{e}{2m_e}\right)\mathbf{L}\end{aligned} \tag{3.182}$$

where the angular momentum of the orbiting electron is $\mathbf{L} = m_e\omega r^2\hat{\mathbf{n}}$. The gyromagnetic ratio is defined as the ratio of the magnitude of the magnetic dipole moment to that of the angular momentum,

$$\gamma = \frac{|\boldsymbol{\mu}|}{|\mathbf{L}|} = \frac{e}{2m_e}. \tag{3.183}$$

Note that a positive magnetic dipole moment is in the same direction as that of the angular momentum.

This model can now be extended to understanding the nuclear magnetic dipole moment. Consider the magnetic dipole moment associated with charged sphere of radius R, mass M, and electric charge Q rotating at an angular frequency ω as shown in Fig. 3.29. The sphere's electrical charge and mass can be written in terms of the volume integrals $Q = \int \rho_e dV$ and $M = \int \rho dV$, respectively, where ρ_e and ρ are the charge and mass densities. The magnitude of the magnetic dipole moment of a differential volume dV of the sphere at a perpendicular distance $r_\perp$ from the axis of rotation is,

$$d\mu = \left(\frac{\rho_e dV}{2\rho dV}\right)(\rho dV)\omega r_\perp^2. \tag{3.184}$$

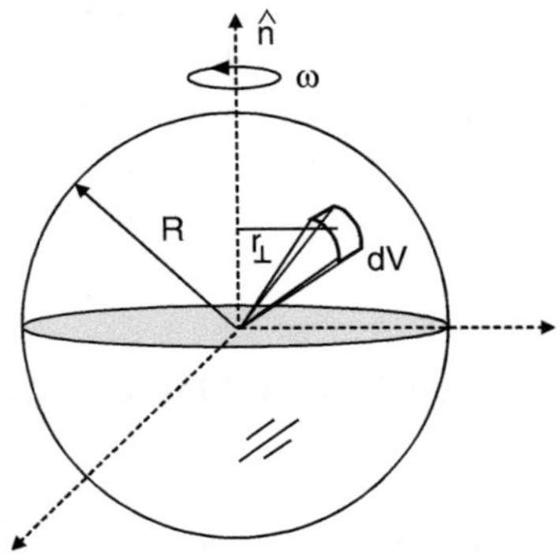

Fig. 3.29 Rotating charged sphere

As the electric charge and density are both assumed to be uniformly distributed throughout the sphere, then $\rho_e/\rho_e = Q/M$ and the differential magnetic moment is,

$$d\mu = \left(\frac{\omega Q}{2M}\right)\rho r_\perp^2 dV \tag{3.185}$$

and the total magnetic moment is then found by integrating over the volume of the sphere,

$$\mu = \left(\frac{Q}{2M}\right)\omega \int dV\rho r_\perp^2. \tag{3.186}$$

It will be recognized that the integral is the moment-of-inertia,[28] I (equal to $2/5MR^2$ for a sphere of uniform mass density). Thus,

$$\begin{aligned}\mu &= \left(\frac{Q}{2M}\right)\omega I \;\square \\ &= \left(\frac{Q}{2M}\right)L\end{aligned} \tag{3.187}$$

where $L = \omega I$ is the angular momentum of the sphere and γ is the gyromagnetic ratio which, for a sphere with uniform mass and charge densities, is $\gamma = \left(\frac{Q}{2M}\right)$.

Now consider the case of a nonuniform charge density (e.g., a model of nucleon with a quark

[28]Consider a particle of mass m in an orbit of radius r with a constant angular velocity ω. The tangential velocity is equal to ωr and the kinetic (rotational) energy is $\frac{m\omega^2 r^2}{2}$. This energy can also be written as $\frac{I\omega^2}{2}$ where I is defined as the moment-of-inertia, mr^2, for the orbiting particle.

substructure). The differential magnetic moment can be written as $d\mu = \frac{\omega}{2}\rho_e r_\perp^2 dV$. As $L = \omega I$, this can be integrated to yield the total magnetic dipole moment,

$$\mu = \frac{L}{2I}\int dV\rho_e r_\perp^2. \tag{3.188}$$

This can be used to demonstrate that a sphere containing two equal amounts of electric charge, but of opposite sign, can have a nonzero magnetic dipole moment even though the net electric charge is zero. Consider the case of a sphere of radius R made up from a sphere of radius r_1 surrounded by a shell as shown in Fig. 3.30, such that the volumes of the core sphere and the shell are the same, or,

$$\frac{4\pi}{3}r_1^3 = \frac{4\pi}{3}(R^3 - r_1^3)$$

yielding,

$$r_1 = \frac{R}{2^{1/3}} \approx 0.629\,R. \tag{3.189}$$

Further, assume that the core sphere is of a positive charge and that the surrounding shell is of an equal charge, but of negative sign. From the isovolume specification, the magnitudes of the charge densities in both components equal $|\rho_e|$. As expected from the isovolume requirement, the net electric charge for the entire sphere is,

$$\begin{aligned}\int dV\rho_e &= 4\pi\rho_e\left(\int_0^{r_1} drr^2 - \int_{r_1}^{R} drr^2\right)\\ &= 4\pi\rho_e\left(\frac{2r_1^3 - R^3}{3}\right) = 0.\end{aligned} \tag{3.190}$$

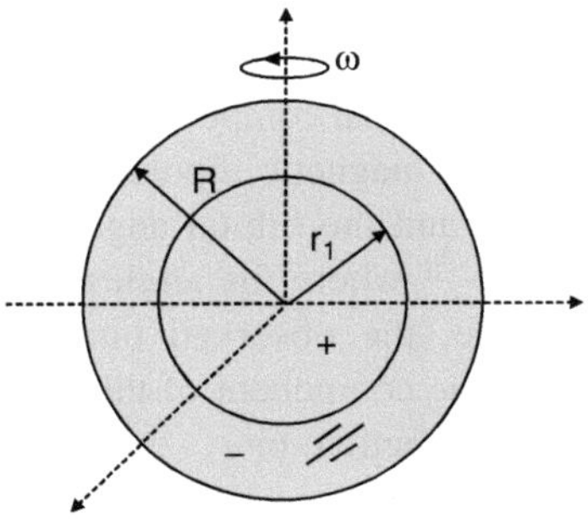

Fig. 3.30 A core sphere of radius r_1 with a positive charge surrounded by a shell with the same volume and the same magnitude of charge, but with negative sign

As the integral in the expression for the magnetic dipole moment is,

$$\begin{aligned}\int dV\rho_e r_\perp^2 &= 4\pi\rho_e\left(\int_0^{r_1} drr^4 - \int_{r_1}^{R} drr^4\right)\int_0^{\frac{\pi}{2}} d\theta\sin^3\theta\\ &= 4\pi\rho_e\left(\frac{2r_1^5 - R^5}{5}\right)\left(\frac{2}{3}\right)\\ &= \left(\frac{8\pi}{15}\right)\left(2^{-\frac{2}{3}} - 1\right)\rho_e R^5\end{aligned}$$

then the ratio of the magnetic dipole moment to the angular momentum is,

$$\frac{\mu}{L} = \frac{1}{2I}\left(2^{-2/3} - 1\right)\left(\frac{2Q}{5}\right)R^2 \tag{3.191}$$

where the definition of the charge density, $\rho_e = 3Q/4\pi r_1^3 = 3Q/2\pi R^3$, has been used. Since $I = 2/5MR^2$ for a sphere of uniform mass density, this is now,

$$\begin{aligned}\frac{\mu}{L} &= \left(\frac{5}{6MR^2}\right)\left(2^{-\frac{2}{3}} - 1\right)\left(\frac{6Q}{5}\right)R^2\\ &= \left(2^{-\frac{2}{3}} - 1\right)\left(\frac{Q}{M}\right)\\ &\approx -0.370\left(\frac{Q}{2M}\right) = g\gamma\end{aligned} \tag{3.192}$$

Hence, the ratio of the magnetic dipole moment to the angular momentum of a rotating sphere with a nonuniform distribution of electric charge is equal to the ratio if the electric charge were uniformly distributed, $Q/2M$, but scaled by a numerical g-factor (here, $g = -0.37$). Equation (3.192) demonstrates that a rotating sphere with electrically-charged substructures but zero net electric charge can have a nonzero magnetic dipole moment. This will have particular relevance in the subsequent discussion of the magnetic dipole moments of the proton and neutron.

3.7.2.4 Nucleon Magnetic Dipole Moments

Returning to the classical picture of an orbiting electron with a magnetic dipole moment magnitude, $\mu_e = \left(\frac{e}{2m_e}\right)L$, in the quantum mechanical counterpart, the angular momentum L is replaced by the projection $m_l\hbar$. By regrouping, we have,

$$\mu_e = \left(\frac{e\hbar}{2m_e}\right)m_l = m_l\mu_B \quad (3.193)$$

where $\mu_B = e\hbar/2m_e$ is defined as the Bohr magneton, equal to 5.788×10^{-11} MeV $T^{-1}$and, similarly, the nuclear magneton is defined as $\mu_N = e\hbar/2m_p$, and is equal to 3.152×10^{-14} MeV/T.

As noted earlier, a magnetic dipole moment will arise from a rotating structure with zero net charge, but with an internal charge structure. Thus, it will be useful to separate the magnetic moment of a nucleon due to its orbital angular momentum,

$$\mu_l = g_l m_l \mu_N \quad (3.194)$$

from that due to its intrinsic spin,

$$\mu_s = g_s m_s \mu_N \quad (3.195)$$

where $m_s = \pm 1/2$ and the g's are dimensionless numerical values. The g_l values for the nucleons are,

$$\begin{aligned} \text{Proton}: \quad & g_{l,p} = 1 \\ \text{Neutron}: \quad & g_{l,n} = 0 \end{aligned} \quad (3.196)$$

where the zero value for the neutron reflects its lack of electric charge. If the proton and neutron were point particles we would expect that $g_{s,p} = 2$ and $g_{s,n} = 0$. Experimentally, however,

$$\begin{aligned} \text{Proton}: \quad & g_{s,p} = +5.58569 \\ \text{Neutron}: \quad & g_{s,n} = -3.82609 \end{aligned} \quad (3.197)$$

where the negative sign for the neutron indicates that the direction of its magnetic dipole moment opposes that of the spin. The difference between these expected values and those measured are evidence of the nucleons' quark substructure, where each quark carries an electric charge and intrinsic spin of its own.

3.7.2.5 Nuclear Magnetic Dipole Moments

Recall the coupling of the orbital and intrinsic spin angular momenta and let a reference axis be defined by an external magnetic field as shown in Fig. 3.31. These vector additions of the angular momentum and intrinsic spin would also represent those of the nucleon magnetic dipole moments if the nucleon orbital and spin *g*-factors were to be equal. However, they do not. The magnetic dipole moment associated with the intrinsic spin will have a component along the direction of the total angular momentum, **J**, and another component orthogonal to **J**. Due to precession, the average value of this orthogonal component will be zero. Similarly, the magnetic dipole moment associated with the orbital angular momentum will have a component along the direction of the total angular momentum and a zero time-averaged orthogonal component.

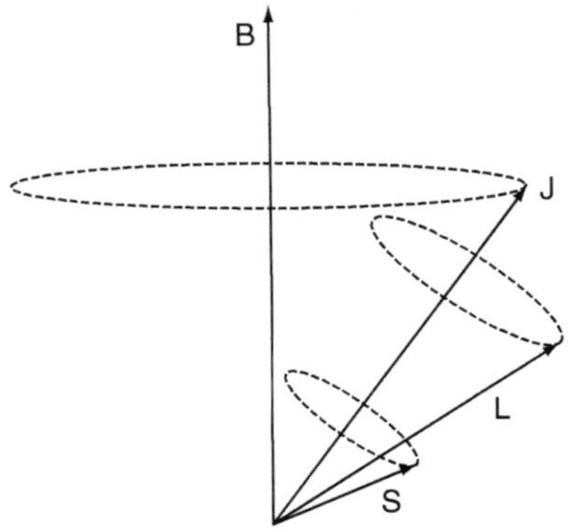

Fig. 3.31 Precessions of the intrinsic spin, orbital angular momentum, and total angular momentum vectors relative to an applied magnetic field

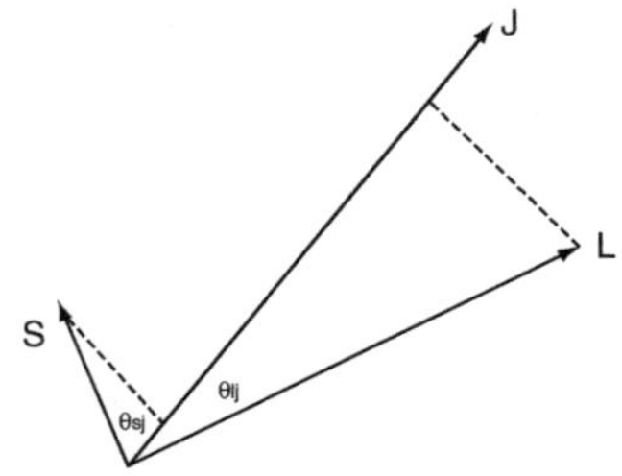

Fig. 3.32 Intrinsic spin and orbital angular momentum vector components along the direction of the total angular momentum **J**

The sum of the magnetic dipole moments due to the intrinsic spin and the orbital angular momentum is $\frac{g_s|\mathbf{s}|\cos\theta_{sj}+g_l|\mathbf{L}|\cos\theta_{lj}}{\hbar}$ where the angles are defined in Fig. 3.32. Hence, the observed nuclear magnetic moment will be the component of this sum along the magnetic field **B** direction, or

$$\mu = \left(\frac{g_s|\mathbf{s}|\cos\theta_{sj} + g_l|\mathbf{L}|\cos\theta_{lj}}{\hbar}\right)\cos\theta_{jB} \quad (3.198)$$

where θ_{jB} is the angle between the total angular momentum vector and the magnetic field direction, or,

$$\mu = \frac{g_s[j(j+1)+s(s+1)-l(l+1)]+g_l[j(j+1)+l(l+1)-s(s+1)]}{2(j+1)}. \quad (3.199)$$

For a nucleon, s = 1/2 and j = $l \pm 1/2$. For the two cases of j,

$$\mu_{l+\frac{1}{2}} = \frac{1}{2}g_s + l\,g_l \quad (3.200)$$

$$\mu_{l-\frac{1}{2}} = \frac{j}{j+1}\left[-\frac{1}{2}g_s + (l+1)g_l\right]. \quad (3.201)$$

The corresponding values for the proton and neutron are,

$$\text{Proton}: \quad \mu_{l+\frac{1}{2}} = j + 2.2928 \quad (3.202)$$

$$\mu_{l-\frac{1}{2}} = j - 2.2928\frac{j}{j+1} \quad (3.203)$$

$$\text{Neutron}: \quad \mu_{l+\frac{1}{2}} = -1.9130 \quad (3.204)$$

$$\mu_{l-\frac{1}{2}} = 1.9130\frac{j}{j+1} \quad (3.205)$$

In the nuclear shell model, where the nuclear magnetic dipole moment would be defined by the unpaired nucleon for odd-A nuclei, these *Schmidt lines* given by the above equations define the magnetic dipole moment of the entire nucleus and are shown as a function of the nuclear angular momentum in Fig. 3.33 (the nuclear angular momentum is assumed to be continuous in order to show the theoretical trend).

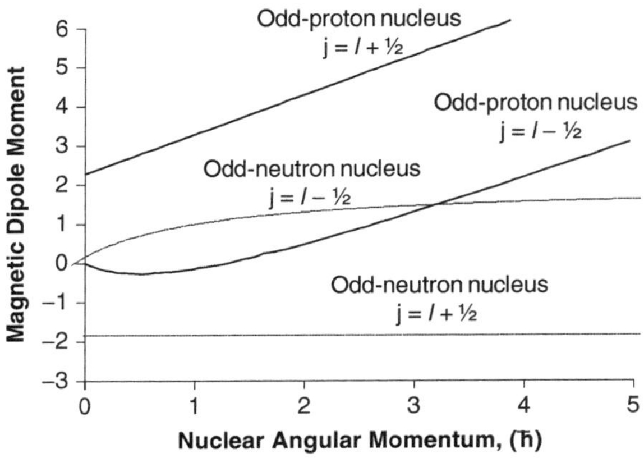

Fig. 3.33 Continuous Schmidt lines of nuclear magnetic dipole moment vs. nuclear angular momentum for nuclei with an odd number of nucleons

Figure 3.34 presents these theoretical Schmidt lines along with the measured magnetic dipole moments for odd-A nuclei. It is apparent that, in reality, the Schmidt lines only set limits to the nuclear magnetic dipole moment for stable nuclei: only ^{3}He, ^{13}C, and ^{15}N have moments exceeding those predicted by the Schmidt lines. Excluding these two exceptions, and the trivial case of the proton, the measured magnetic dipole moments are all less than those predicted from the shell model. The measured moments form two groups with moments with values of about 50–60% of that predicted, demonstrating that more than one nucleon contributes to the nuclear wavefunction of an odd-A nucleus. Hence, while the paired nucleons are indeed compensating each others' angular momentum, they do not compensate each others' magnetic dipole moment.

3.7.3 Nuclear Electric Quadrupole Moments

3.7.3.1 Introduction

The nuclear magnetic dipole moment confirms that the nuclear properties of an odd-A nucleus are dominated by the unpaired nucleon, thus substantiating the fundamentals of the shell model. When a shell is filled, the nucleus becomes spherically symmetric. Because the nucleus is a spatial distribution of electric charge, it can be described as a superposition of electric multipoles with the weights of these multipoles being the static electric moments which provide a measure of the spatial distribution of protons. Hence, analysis of these moments, in particular the quadrupole moment, can determine the sphericity of a given nucleus.

3.7.3.2 Multipole Expansion of the Electric Potential

Consider a nonspherical charged body, such as that in Fig. 3.35, but which is symmetric about the *z*-axis and

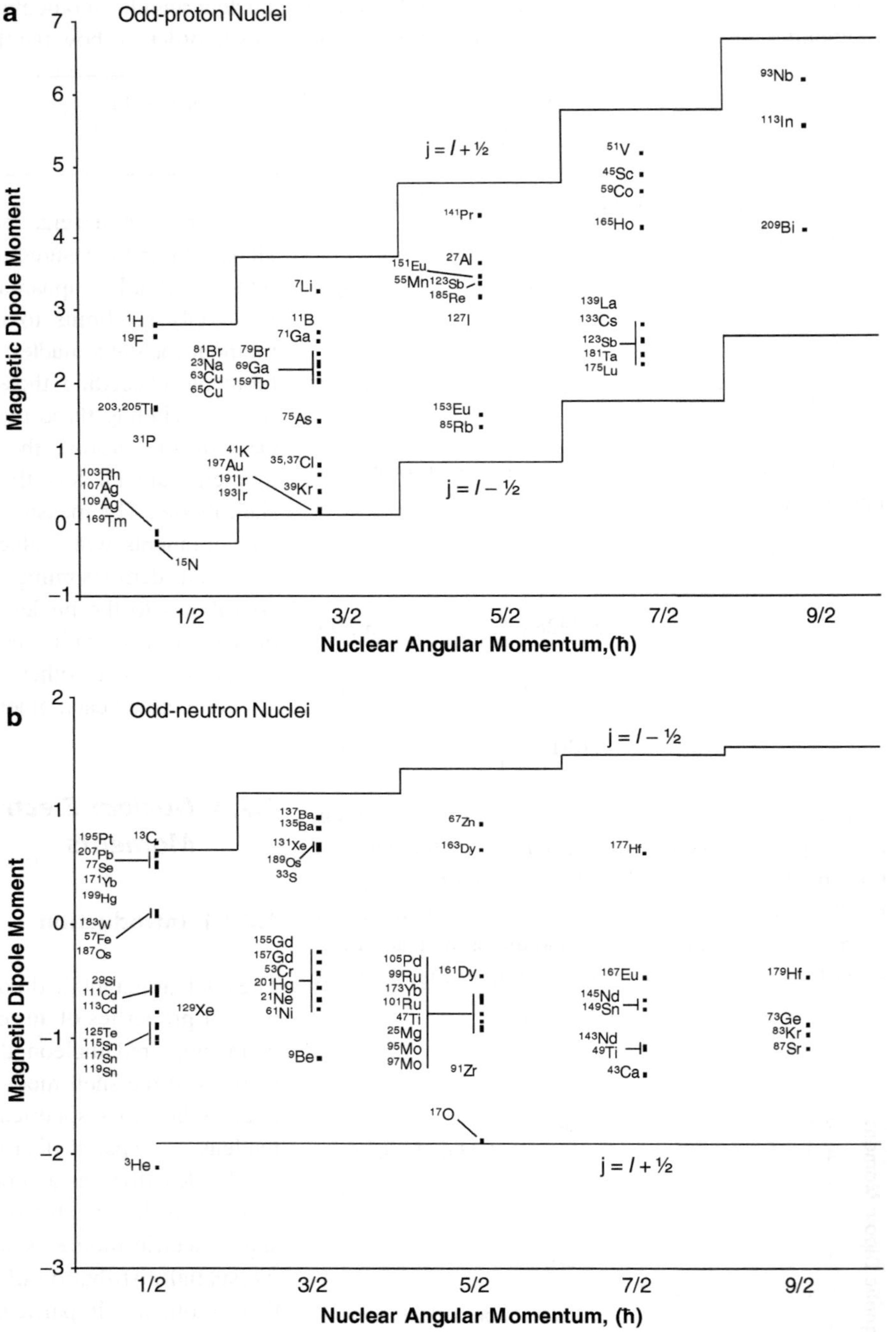

Fig. 3.34 Measured magnetic dipole moment vs. nuclear angular momentum for odd-A nuclei. *Lines* are the Schmidt lines. (**a**) shows nuclei with an odd number of protons and (**b**) shows nuclei with an odd number of neutrons

has total electric charge, q. The electric potential resulting from this body will satisfy Laplace's equation,

$$\nabla^2 V = 0 \tag{3.206}$$

which, in spherical coordinates, is,

$$\left[\frac{1}{r^2}\frac{\partial}{\partial r}\left(r^2\frac{\partial V}{\partial r}\right)+\frac{1}{r^2\sin\theta}\frac{\partial}{\partial\theta}\left(\sin\theta\frac{\partial V}{\partial\theta}\right)+\frac{1}{r^2\sin^2\theta}\frac{\partial^2 V}{\partial\phi^2}\right]=0. \tag{3.207}$$

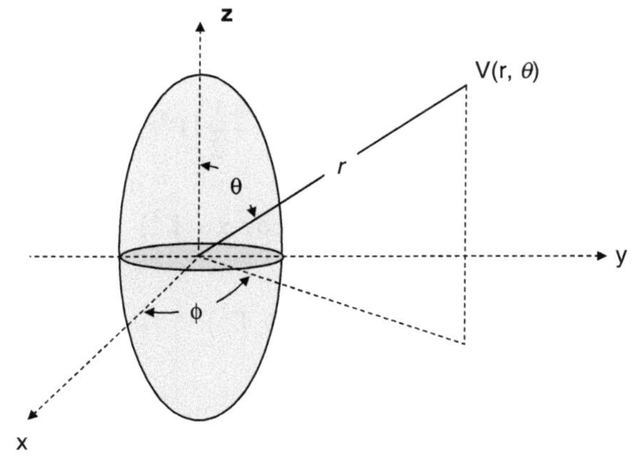

Fig. 3.35 Electric potential of a charged object. Note that, due to symmetry about the *z*-axis, the electric potential V has no Φ-dependence

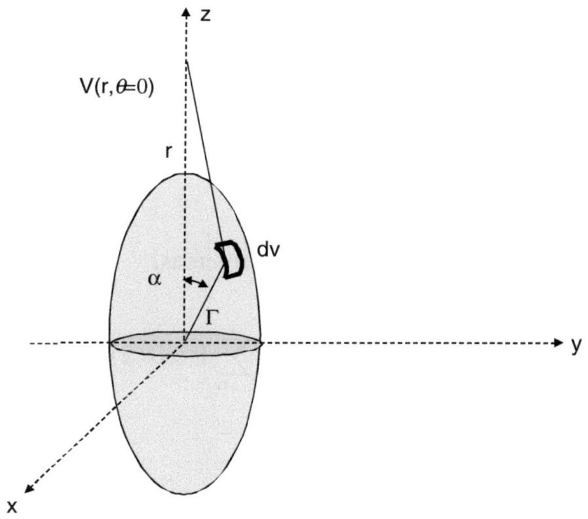

Fig. 3.36 Calculation of the multipole expansion coefficients

Due to symmetry about the *z*-axis, $\partial^2 V/\partial\phi^2 = 0$. Thus, separating V(r, θ) into radial and angular components F(r)G(θ), we have, firstly, the angular component given by Legendre polynomials,

$$G(\theta) = P_l(\cos\theta) \tag{3.208}$$

(for convenience, normalization is ignored as this will be performed later) and, secondly, the radial differential equation,

$$r^2\frac{d^2F}{dr^2} + 2r\frac{dF}{dr} - l(l+1)F = 0 \tag{3.209}$$

which evolves from Chap. 2 and where l is equal to zero or a positive integer. If we write the radial component F(r) as the series,

$$F(r) = \sum_{n=1}^{\infty}\frac{k_n}{r^n} \tag{3.210}$$

where n is a positive integer so as to ensure that the potential goes to zero as $r\rightarrow\infty$, so that the radial differential equation becomes,

$$\begin{aligned}\sum_{n=1}^{\infty} n^2\frac{k_n}{r^n} + \sum_{n=1}^{\infty} n\frac{k_n}{r^n} - 2\sum_{n=1}^{\infty} n\frac{k_n}{r^n}\\ -l(l+1)\sum_{n=1}^{\infty}\frac{k_n}{r^n} = 0\end{aligned} \tag{3.211}$$

It can be seen that $n(n-1) = l(l+1)$ from which $n = l + 1$. The radial function can then be rewritten as,

$$F(r) = \frac{1}{r}\sum_{l=0}^{\infty}\frac{k_l}{r^l}. \tag{3.212}$$

Combining the radial and angular components, the solution for the potential is,

$$V(r,\theta) = \frac{1}{r}\sum_{l=0}^{\infty} k_l\frac{P_l(\cos\theta)}{r^l}. \tag{3.213}$$

The coefficients of (3.213) are determined by taking advantage of the fact that the Legendre polynomials are equal to unity for θ = 0,

$$V(r,0) = \frac{1}{r}\sum_{l=0}^{\infty} k_l\frac{1}{r^l}. \tag{3.214}$$

Next, consider the differential volume dv at (Γ, α) containing an electric charge $\rho_e(\Gamma, \alpha)$ dv, as shown in Fig. 3.36. The differential electric potential at (r, 0) due to this element is,

$$\begin{aligned}dV(r,0) &= \frac{\rho_e(\Gamma,\alpha)}{\sqrt{r^2+\Gamma^2-2r\Gamma\cos\alpha}}dv\\ &= \frac{1}{r}\frac{\rho_e(\Gamma,\alpha)}{\sqrt{1+\left(\frac{\Gamma}{r}\right)^2-2\left(\frac{\Gamma}{r}\right)\cos\alpha}}dv\end{aligned} \tag{3.215}$$

As the generating function for the Legendre polynomial is,

$$\frac{1}{\sqrt{1-2xz+z^2}} = \sum_{l=0}^{\infty} P_l(x)z^l; \quad -1<x<1, |z|<1 \tag{3.216}$$

part of the equation can be rewritten as,

$$\frac{1}{\sqrt{1+\left(\frac{\Gamma}{r}\right)^2 - 2\left(\frac{\Gamma}{r}\right)\cos\alpha}} = \sum_{l=0}^{\infty} P_l(\cos\alpha)\left(\frac{\Gamma}{r}\right)^l. \tag{3.217}$$

Substituting this result into (3.216),

$$dV(r,0) = \frac{1}{r}\rho_e(\Gamma,\alpha)\sum_{l=0}^{\infty} P_l(\cos\alpha)\left(\frac{\Gamma}{r}\right)^l dv \tag{3.218}$$

and integrating over the volume yields the total electric potential at (r, 0),

$$V(r,0) = \frac{1}{r}\sum_{l=0}^{\infty}\int_{Vol} dv\frac{\rho_e(\Gamma,\alpha)P_l(\cos\alpha)\Gamma^l}{r^l}. \tag{3.219}$$

The coefficients can then be extracted by equating (3.214) and (3.219),

$$k_l = \int_{Vol} dv\rho_e(\Gamma,\alpha)P_l(\cos\alpha)\Gamma^l. \tag{3.220}$$

The first three coefficients are,

$$k_0 = \int_{Vol} dv\rho_e(\Gamma,\alpha) = q \tag{3.221}$$

which is the total electric charge;

$$\begin{aligned} k_1 &= \int_{Vol} dv\rho_e(\Gamma,\alpha)\Gamma\cos\alpha \\ &= \int_{Vol} dv\rho_e(\Gamma,\alpha)z \end{aligned} \tag{3.222}$$

which is electric dipole moment; and

$$\begin{aligned} k_2 &= \int_{Vol} dv\rho_e(\Gamma,\alpha)P_2(\cos\alpha)\Gamma^2 \\ &= \int_{Vol} dv\rho_e(\Gamma,\alpha)\left(\frac{3}{2}\cos^2\alpha - \frac{1}{2}\right)\Gamma^2 \\ &= \frac{1}{2}\int_{Vol} dv\rho_e(\Gamma,\alpha)\left(3\Gamma^2\cos^2\alpha - \Gamma^2\right) \\ &= \frac{1}{2}\int_{Vol} dv\rho_e(\Gamma,\alpha)\left(3z^2 - \Gamma^2\right). \end{aligned} \tag{3.223}$$

This latter coefficient is proportional to the quadrupole moment,

$$Q = \int_{Vol} dv\rho_e(\Gamma,\alpha)\left(3z^2 - \Gamma^2\right) \tag{3.224}$$

or,

$$k_2 = \frac{1}{2}Q \tag{3.225}$$

As the nucleus has a definite parity, the odd electric moments (l = 1, 3, 5...) are zero. For example, consider the electric dipole moment

$$\int_{Vol} dv\rho_e(\Gamma,\alpha)P_1(\cos\alpha)\Gamma$$

where $dv = \Gamma^2\, d\Gamma\, \sin\alpha\, d\alpha\, d\phi$ and the integration is over the nuclear volume. As the charge density is proportional to the squared-modulus of the nuclear wavefunction $\rho_e \propto |\psi|^2$, it remains unaltered by a change in parity (x,y,z) → (−x, −y, −z). On the other hand, the first-order Legendre polynomial $P_1(\cos\alpha) = \cos\alpha$ is an odd function and its sign is changed by a parity change. Integrating over the nuclear volume, $P_1(\cos\alpha)$ averages to zero.

3.7.3.3 Electric Quadrupole Moment

It can be shown that the electric quadrupole moment is a measure of the nuclear shape. Consider a charged sphere of radius R for which the electric charge is uniformly distributed as shown in Fig. 3.37. From (3.224), the electric quadrupole moment is,

$$Q = \rho_e\int_{Vol} dv\left(3z^2 - r^2\right) \tag{3.226}$$

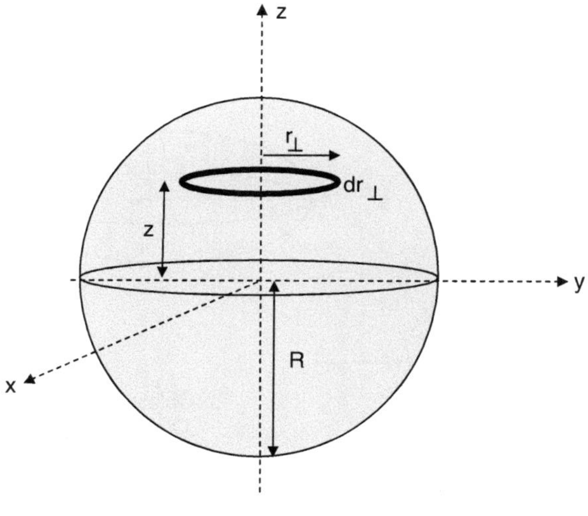

Fig. 3.37 Charged sphere

which is the integral of the volume elements described by annuli of radius $r_\perp$ and thickness $dr_\perp$ at a height z. Thus,

$$\begin{aligned} Q &= 2\pi\rho_e \int_{Vol} dr_\perp dz\left(3z^2 - z^2 - r_\perp^2\right) r_\perp \\ &= 4\pi\rho_e \int_0^R dr_\perp \int_0^{r_{\perp,max}(z)} dz\left(2z^2 r_\perp - r_\perp^3\right) \end{aligned} \tag{3.227}$$

where

$$r_{\perp,max}(z) = \sqrt{R^2 - z^2}. \tag{3.228}$$

Integrating,

$$\begin{aligned} Q &= 4\pi\rho_e \int_0^R dz\left(z^2 r_{\perp,max}^2 - \frac{1}{4} r_{\perp,max}^4\right) \\ &= 4\pi\rho_e \int_0^R dz\left(z^2\left(R^2 - z^2\right) - \frac{1}{4}\left(R^2 - z^2\right)^2\right) \\ &= 4\pi\rho_e \left(\frac{3}{2}R^2 \int_0^R dz\, z^2 - \frac{5}{4}\int_0^R dz\, z^4 - \frac{1}{4}R^4 \int_0^R dz\right) \\ &= 4\pi\rho_e \left(\frac{1}{2}R^5 - \frac{1}{4}R^5 - \frac{1}{4}R^5\right) = 0 \end{aligned}$$

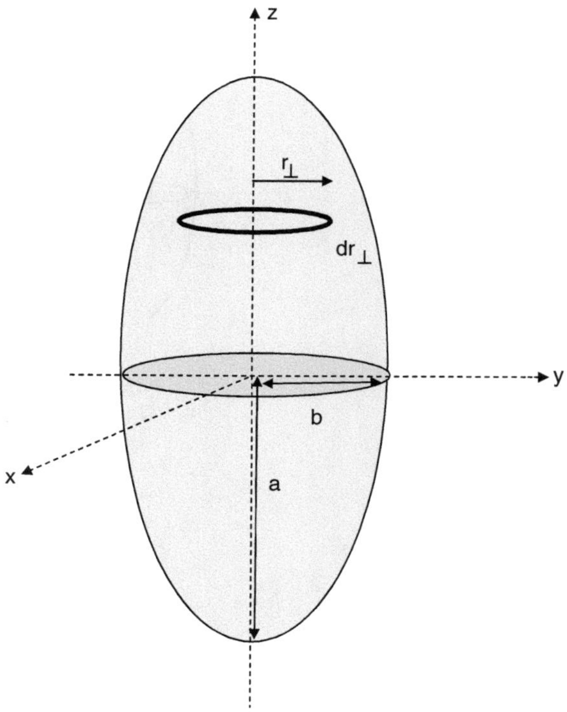

Fig. 3.38 Charged prolate spheroid

Thus a spherical, uniform charge distribution has a zero electric quadrupole moment. Next, consider the charged prolate spheroid of Fig. 3.38, which is generated by the rotation about the *z*-axis of an ellipse with major and minor axes 2a and 2b, respectively,

$$\left(\frac{y}{b}\right)^2 + \left(\frac{z}{a}\right)^2 = 1. \tag{3.229}$$

The electric quadrupole moment is given by,

$$Q = 4\pi\rho_e \int_0^a dr_\perp \int_0^{r_{\perp,max}(z)} dz\left(2z^2 r_\perp - r_\perp^3\right) \tag{3.230}$$

and $r_{\perp,max}(z)$ is determined from (3.229),

$$\begin{aligned} r_{\perp,max}(z) &= b\sqrt{1 - \left(\frac{z}{a}\right)^2} \\ &= \left(\frac{b}{a}\right)\sqrt{a^2 - z^2}. \end{aligned} \tag{3.231}$$

Fig. 3.39 Spheroids

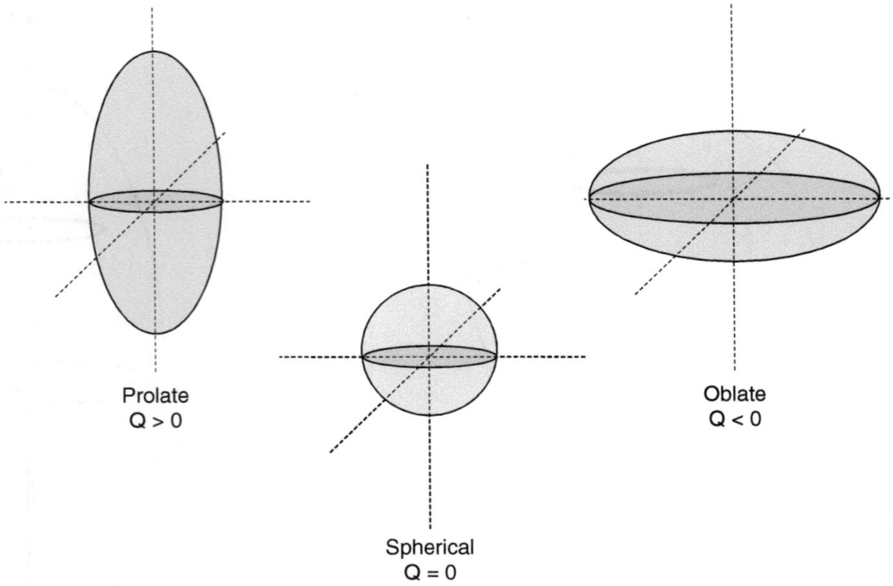

Integrating (3.230),

$$
\begin{aligned}
Q &= 4\pi\rho_e \int_0^a dz\left(z^2 r_{\perp,\max}^2 - \frac{1}{4} r_{\perp,\max}^4\right) \\
&= 4\pi\rho_e \int_0^a dz\left(\left(\frac{b}{a}\right)^2 z^2\left(a^2 - z^2\right) - \frac{1}{4}\left(\frac{b}{a}\right)^4\left(a^2 - z^2\right)^2\right) \\
&= \frac{8\pi}{15}\rho_e ab^2(a+b)(a-b) \\
&= \frac{8\pi}{15}\rho_e ab^2\left(a^2 - b^2\right)
\end{aligned}
\tag{3.232}
$$

As the volume of the spheroid is $(4\pi/3)ab^2$ and the total charge in the spheroid is q, this expression for the quadrupole moment reduces to,

$$
Q = \frac{2}{5} q\left(a^2 - b^2\right). \tag{3.233}
$$

If the spheroid were to represent a nucleus with a total charge Ze, then the ratio Q/e is,

$$
\frac{Q}{e} = \frac{2}{5} Z\left(a^2 - b^2\right). \tag{3.234}
$$

This result shows that the nuclear quadrupole moment Q is,

- Positive, if the nucleus is a prolate spheroid ($a > b$).
- Zero, if the nucleus is a sphere ($a = b$).
- Negative, if the nucleus is a oblate spheroid ($a < b$).

These conclusions are summarized in Fig. 3.39.

The electric quadrupole moment Q has units of the charge-area product and nuclear quadrupole moments are usually tabulated in terms of Q/e which has units of area.[29]

Figure 3.40 shows measured electric quadrupole moments vs. atomic mass number; those nuclei with magic numbers are highlighted. The electric quadrupole moment changes sign at the magic numbers, indicating the sphericity of such nuclei. Between the magic numbers, values of the electric quadrupole moments values can be high, especially for the lanthanides and actinides (values of A between 150 and 200), these nuclei are prolate spheroids (i.e., $Q > 0$) with high quadrupole moment values due to their lying between the major shells. The nuclei with the highest positive and negative values for the quadrupole moment are ^{176}Lu and ^{127}I with 8 b and −0.79 b, respectively. Nuclei which have one nucleon outside a filled subshell have negative electric quadrupole moments whereas those which have a hole (i.e., require another nucleon to fill a shell) have positive

[29]In the literature, this ratio is often simply (and confusingly) written as Q.

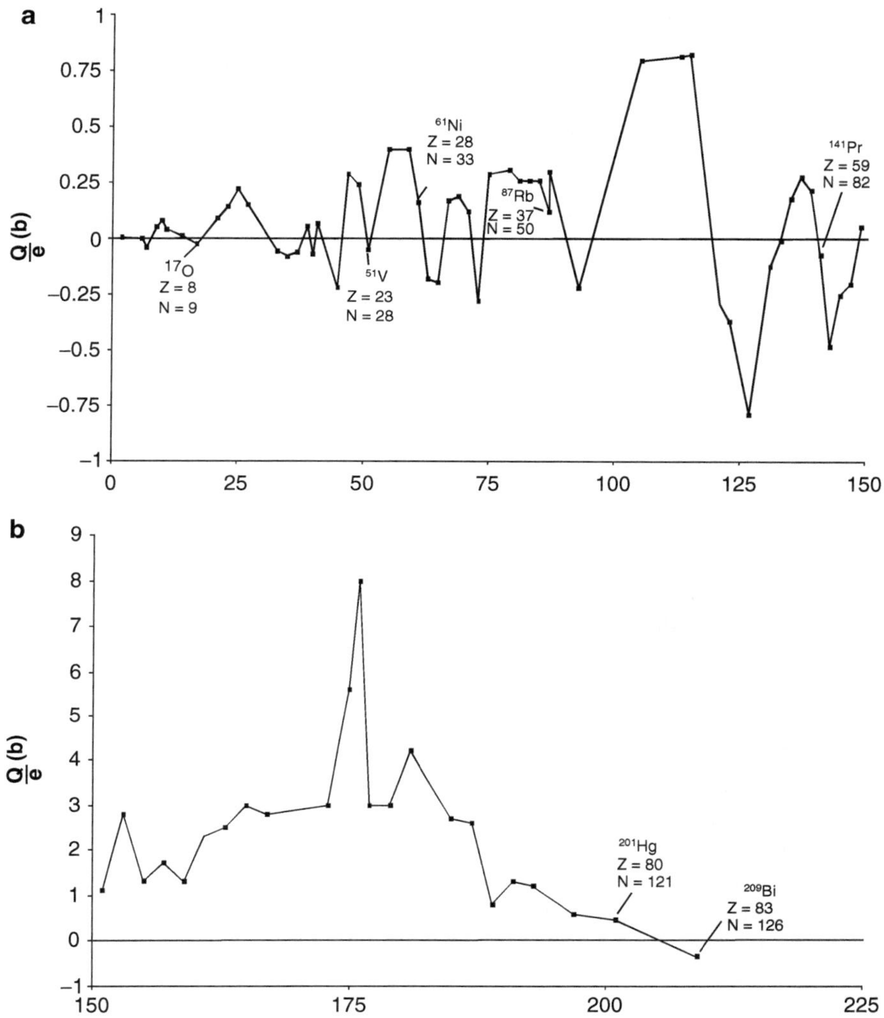

Fig. 3.40 Measured electric quadrupole moments vs. atomic mass number; note the difference in ordinate scales between the two graphs. (**a**) shows the electric quadrupole moment for nuclei with A < 150; (**b**) shows the electric quadrupole moment for nuclei with A > 150

quadrupole moments. An example is of the sulfur isotopes, ^{33}S and ^{35}S with values of Q/e equal to −6.4 and +4.5 mb, respectively. The negative value for ^{33}S is a result of the $1d_{3/2}$ shell which contains a single neutron whereas the positive value for ^{35}S is the result of the these nuclei are prolate spheroids (i.e., $Q > 0$) with high quadrupole moment values due to their lying between the major shells. The nuclei with the highest positive and negative values for the quadrupole moment are ^{176}Lu and ^{127}I with 8 b and −0.79 b, respectively. Nuclei which have one nucleon outside a filled subshell have negative electric quadrupole moments whereas those which have a hole (i.e., require another nucleon to fill a shell) have positive quadrupole moments. An example is of the sulfur isotopes, ^{33}S and ^{35}S with values of Q/e equal to −6.4 and +4.5 mb, respectively. The negative value for ^{33}S is a result of the single neutron whereas the positive value for ^{35}S is the result of the $1d_{3/2}$ shell, which can take four neutrons (i.e., $2(3/2) + 1 = 4$), having only three neutrons.

A more succinct representation is the reduced electric quadrupole moment in which the quadrupole moment is normalized to the atomic number and the square of the nuclear radius, Q/eZR_N^2 which is,

$$\frac{Q}{eZR_N^2} = \frac{2}{5}\left(\frac{a^2 - b^2}{R_N^2}\right) \tag{3.235}$$

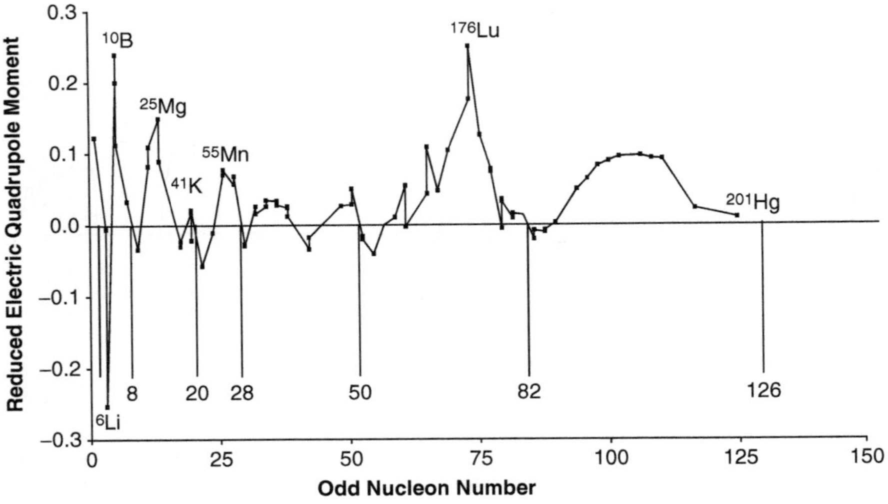

Fig. 3.41 Reduced electric quadrupole moment as a function of odd nucleon number for stable nuclei

In this form, the influences of the atomic number and size are removed and the effects of nonsphericity are isolated. The reduced electric quadrupole moment is shown in Fig. 3.41 as a function of A for odd-A nuclei.

It is clear that the electric quadrupole moment for stable changes sign at the magic numbers of 8, 20, 28, 50, and 83. As there are no stable nuclei heavier than ^{209}Bi with a quadrupole moment, a sign change for the magic number of 126 does not appear.

The semimajor and semiminor axes of an ellipsoid nucleus can be determined from the measured quadrupole moment by first rearranging (3.234),

$$a^2 - b^2 = \left(\frac{5}{2Z}\right)\frac{Q}{e} \tag{3.236}$$

and then by recognizing that the mean radius of a spheroid is $\bar{R} = \sqrt{\frac{a^2+b^2}{2}}$ leading to,

$$a^2 + b^2 = 2r_n^2 A^{\frac{2}{3}}. \tag{3.237}$$

Solving these two equation for semimajor and semiminor axes, a and b,

$$a = \sqrt{\left(\frac{5}{4Z}\right)\frac{Q}{e} + r_n^2 A^{\frac{2}{3}}} \tag{3.238}$$

$$b = \sqrt{r_n^2 A^{\frac{2}{3}} - \left(\frac{5}{4Z}\right)\frac{Q}{e}}. \tag{3.239}$$

For example, consider the nucleus ^{176}Lu where $Q/e = 8 \times 10^{-24}\,cm^2$ for which,

$$a = 7.70\,fm$$

$$b = 5.58\,fm$$

and, for ^{127}I, which has $Q/e = -0.79 \times 10^{-24}\,cm^2$,

$$a = 5.88\,fm$$

$$b = 6.18\,fm$$

3.8 Isomers

Isomers have been defined previously as excited nuclear states with the same atomic mass number and atomic number. An isomer decays to the nuclear ground state via the emission of electromagnetic radiation or through internal conversion (IC).[30] While "prompt" γ-ray emissions are associated with nuclear state mean lives of the order of 10^{-16} s, the probability of a γ ray/IC transition occurring per unit time in an isomeric transition is highly suppressed if the angular momentum differential between the initial and final

[30]Internal conversion is a radiation-less transition for an excited nucleus to a lower-energy state via the transfer of energy directly to an atomic electron which is ejected.

states is high or if the energy difference between is low. This will be demonstrated explicitly in Chap. 4. In such a case, the mean life of the state can range from 100 ms to days or even years. Such relatively long-lived states are referred to as metastable or isomeric states, the most common example in nuclear medicine being that of ^{99m}Tc.

For nuclei with $A < 40$, there are no adjacent states with high angular momentum differentials that can lead to the existence of an isomeric state. As a result, there are no isomers of light nuclei. This feature changes, however, with increasing A. In particular, groupings (or "islands") of isomers occur at values of A just under the magic numbers of 50, 82, or 126 where there are neighboring levels with high angular momentum differentials. For example, just below the $A = 50$ level, the $1g_{9/2}$ state is adjacent to the $2p_{1/2}$ state ($\Delta j = 4$) and just at the $A = 82$ and $A = 126$ levels, there are adjacent orbits with $\Delta j = 5$ and $\Delta j = 6$, respectively. Such excited states often result following β decay, by inelastic nuclear collisions or the absorption of γ-rays.

The most important isomer for diagnostic nuclear medicine is ^{99m}Tc (Seaborg and Segrè 1939).

The electromagnetic transitions associated with isomers are discussed in detail in Chap. 4.

3.9 The Deuteron

The chapter is concluded by briefly considering the simplest stable nucleus containing two or more nucleons. In order to explain nuclear binding and "magic" nuclei, we have had to describe the forces acting upon nucleons contained within the nucleus as being a combination of an attractive term and a term proportional to the spin-orbit coupling of the nucleons. The deuteron is the only bound state of a nucleon pair: the bound states of a diproton and dineutron do not exist. Referring back to the curve of binding energies, we see that the binding energy of the deuteron is quite low. The magnitude of this binding energy was first determined from the ^{2}H(γ, n)p photodisintegration reaction performed by Chadwick and Goldhaber (1934) in which the target was heavy hydrogen gas in a cloud chamber. The binding energy is equal to 2.225 MeV. Now consider the other properties of the deuteron. First, it has an angular momentum of 1, which is consistent with the assumption that the deuteron is in an *S* state (i.e., zero angular momentum) and that the intrinsic spins of the neutron and proton couple in parallel to form the triplet state, $^{3}S_{1}$. But there are two other empirical observations that question this assumption. First, the deuteron has a nonzero, but small, electric quadrupole moment of $Q/e = 0.286\,\text{fm}^2$. The spherically-symmetric s-state would require that this electric quadrupole moment be equal to zero. Second, there is a difference between the sum of the proton and neutron magnetic dipole moments and that of the deuteron: the latter is less than the former, $\mu_p + \mu_n = \mu_d + 0.022\mu_n$.

These experimental results can be explained by the hypothesis that the deuteron ground state wavefunction is an admixture of states with zero and finite angular momenta (Messiah 1958). In order for two spin-1/2 particles to couple to yield a total angular momentum of 1, they must be in the states of $^{3}S_{1}, {}^{1}P_{1}, {}^{3}P_{1}$ or $^{3}D_{1}$. In order to keep consistent parity, the simplest case is an admixture of the two triplet states, $^{3}S_{1}$ and $^{3}D_{1}$, which both have even parity. This admixture of states with different orbital angular momenta but the same total angular momentum indicates the existence of noncentral forces within the nucleus. The derivations of the degree of mixing of the two states are not reproduced here (see, e.g., Preston and Bhaduri (1975)), but the deuteron magnetic dipole moment suggests that the deuteron has a D-state admixture of about 4%. Conflicting experimental data from, for example, nucleon–nucleon scattering and the deuteron form factor imply that this D-state admixture is slightly higher at about 7%.

As the deuteron's total angular momentum is equal to 1 and it has a small electric quadrupole moment, it clearly spends most of its time in the S-state allowing one to temporarily ignore the $^{3}D_{1}$ state of the admixture and consider the spherically symmetric $^{3}S_{1}$ state. In such a state, the forces upon the nucleon must be central and the simplest case would be a potential dependent only upon the inter-nucleon distance, *r*. From Chap. 2, the radially-dependent Schrödinger equation is in the form,

$$\frac{1}{r^2}\frac{d}{dr}\left(r^2\frac{d\psi}{dr}\right) + \frac{2\mu}{(\hbar c)^2}(E - U(r))\psi = 0 \quad (3.240)$$

where μ is the reduced mass of the system,

$$\mu = {}^{m_N}/_2 \tag{3.241}$$

where m_N is the nucleon mass. This equation is solved by using the substitution of variable, $\psi(r) = \frac{u(r)}{r}$ to give,

$$\frac{d^2u}{dr^2} + \frac{m_N}{(\hbar c)^2}(E - U(r))u = 0. \tag{3.242}$$

Using (3.173) to describe the nuclear potential and recognizing that $E = -E_B$, where E_B is the deuteron binding energy, (3.242) is,

$$\frac{d^2u}{dr^2} + \frac{m_N}{(\hbar c)^2}(U_0 - E_B)u = 0 \quad r < b \tag{3.243}$$

$$\frac{d^2u}{dr^2} - \frac{m_N E_B}{(\hbar c)^2}u = 0 \qquad r \geq b. \tag{3.244}$$

This potential is shown in Fig. 3.42. The solutions to these equations, subject to the obvious boundary condition of $u(r) \to 0$ as $r \to 0$, are,

$$u(r) = A \sin\left(\frac{\sqrt{m_N(U_0 - E_B)}}{\hbar c} r\right) \quad r < b \tag{3.245}$$

$$u(r) = B\, e^{-\left(\frac{\sqrt{m_N E_B}}{\hbar c}\right) r} \qquad r \geq b. \tag{3.246}$$

The shape of the deuteron wavefunction is shown in Fig. 3.43. As this function, its derivative and those of its logarithm must be continuous, we have,

$$\frac{d(\ln u(r))}{dr} = \frac{\sqrt{m_N(U_0 - E_B)}}{\hbar c} \times \cot\left(\frac{\sqrt{m_N(U_0 - E_B)}}{\hbar c} r\right) \quad r < b \tag{3.247}$$

$$\frac{d(\ln u(r))}{dr} = -\frac{\sqrt{m_N E_B}}{\hbar c} \qquad r \geq b \tag{3.248}$$

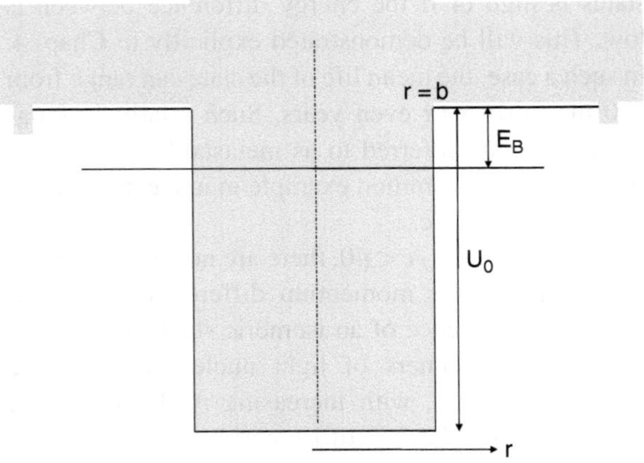

Fig. 3.42 Nuclear rectangular potential of depth U_0, which is the total energy of the system, and E_B, which is the binding energy, for the deuteron

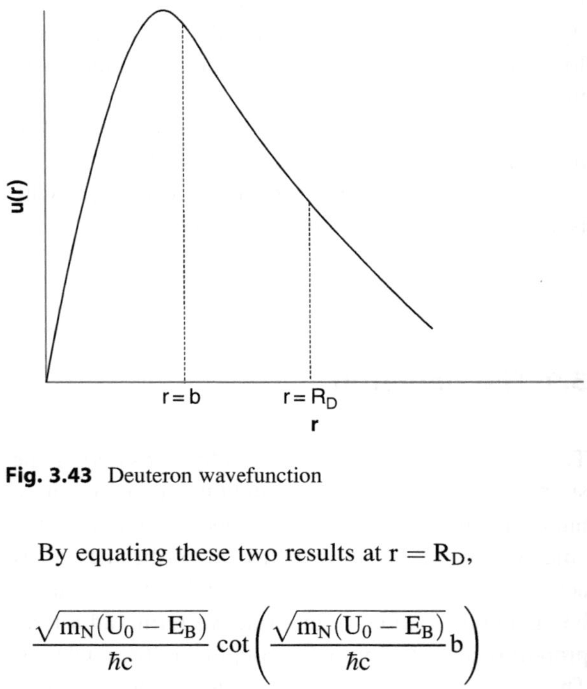

Fig. 3.43 Deuteron wavefunction

By equating these two results at $r = R_D$,

$$\frac{\sqrt{m_N(U_0 - E_B)}}{\hbar c} \cot\left(\frac{\sqrt{m_N(U_0 - E_B)}}{\hbar c} b\right) = -\frac{\sqrt{m_N E_B}}{\hbar c} \tag{3.249}$$

and recalling that deuteron binding energy is small $U_0 \gg E_B$ one obtains, first,

$$\cot\left(\frac{\sqrt{m_N U_0}}{\hbar c} b\right) = -\sqrt{\frac{E_B}{U_0}} \tag{3.250}$$

and, then

$$\frac{\sqrt{m_N U_0}}{\hbar c} b \approx \frac{\pi}{2} \quad (3.251)$$

which leads to,

$$U_0 b^2 \approx \frac{\left(\frac{\pi \hbar c}{2}\right)^2}{m_N} \approx 102\,\text{MeV fm}^2. \quad (3.252)$$

It is known from our previous studies that the range of the nuclear force is small, between about 1.5 and 2 fm. If one were to use an intermediate value of $b \approx 1.7$ fm, then $U_0 \approx 35$ MeV, a value which is consistent both with the $U_0 \gg E_B$ requirement and the Fermi gas model.

From (3.246) it can be seen that the deuteron wave function extends beyond the range of the nuclear force. If the deuteron radius were to be defined as that where the amplitude of the wavefunction is $e^{-1} \approx 0.37$ that of its maximum, then,

$$R_D = \frac{\hbar c}{\sqrt{m_N E_B}} \approx 4.3\,\text{fm} \quad (3.253)$$

Thus, the radius of the deuteron, which is the only stable nucleon pair, is about double the range of the nuclear force, b. This clearly indicates the reason for the low binding energy of the deuteron: about 40% of the time, the proton and neutron are separated beyond the range of the nuclear force.

In order to explain the small nonzero electric quadrupole moment, it is necessary to allow for the deuteron wavefunction to be an admixture of the spherically-symmetric s-wavefunction just calculated for (assuming a central potential) and a small $^3D_1$ component. This requires that, in the $^3D_1$ state, the nuclear force is no longer purely central but depends upon the separation between the neutron and proton and the orientation of their spins with respect to the line joining the two spins. This is a tensor force of the form, $U_T(r)S_{12}$, where,

$$S_{12} = \frac{3(\boldsymbol{\sigma}_1 \bullet \mathbf{r})(\boldsymbol{\sigma}_2 \bullet \mathbf{r})}{r^2} - \boldsymbol{\sigma}_1 \bullet \boldsymbol{\sigma}_2 \quad (3.254)$$

where σ_1 and σ_2 are the neutron and proton Pauli spin operators and $\mathbf{r}$ is the separation vector between the neutron and proton.

References

Belkić DŽ (2004) Principles of quantum scattering theory. Institute of Physics, Philadelphia

Bjorken JD, Drell SD (1964) Relativistic quantum mechanics. McGraw-Hill, New York

Blatt JM, Weisskopf VF (1979) Theoretical nuclear physics. Dover, New York

Chadwick J, Goldhaber M (1934) A 'nuclear photo-effect' – disintegration of the diplon by gamma-rays. Nature 134:237–238

Close FE (1979) An introduction to quarks and partons. Academic, London

Dyson F (2007) Advanced quantum mechanics. World Scientific, Singapore

Ehrenfest P, Oppenheimer JR (1931) Note on the statistics of nuclei. Phys Rev 37:333–338

Fermi E, Teller E (1947) The capture of negative mesotrons in matter. Phys Rev 72:399–408

Goldberger ML, Watson KM (1964) Collision theory. Wiley, New York

Grotz K, Klapdor HV (1990) The weak interaction in nuclear, particle and astrophysics. Institute of Physics, Bristol

Haxel O, Jensen JHD, Suess HE (1949) On the "magic numbers" in nuclear structure. Phys Rev 75:1766

Koide Y (1983) New view of quark and lepton mass hierarchy. Phys Rev D 28:252–254

Mayer MG (1949) On closed shells in nuclei II. Phys Rev 75:1969–1970

Mayer MG (1950) Nuclear configurations in the spin-orbit coupling modelI. Empirical evidence. Phys Rev 78:16–21

Messiah A (1958) Quantum mechanics. Wiley, New York

Mott NF (1929) Scattering of fast electrons by atomic nuclei. Proc Roy Soc A124:425–442

Pauli W (1973) Pauli lectures on physics: selected topics in field quantization. Massachusetts Institute of Technology Press, Cambridge

Preston MA, Bhaduri RK (1975) Structure of the nucleus. Addison-Wesley, Reading

Rivero A, Gjsponer A (2005) The strange formula of Dr Koide. arXiv:hep-ph/0505220v1

Seaborg GT, Segrè E (1939) Nuclear isomerism in element 43. Phys Rev 55:808–814

Segrè E (1977) Nuclei and particles, 2nd edn. Benjamin/Cummings, Reading

Sick I (1974) Model-independent nuclear charge densities from elastic electron scattering. Nucl Phys A 218:509–541

Radioactive Decay: Microscopic Theory 4

Abstract An unstable nucleus has excess energy and can make a transition to a more stable state through a variety of means, each subject to quantum-mechanical and energy restrictions. For diagnostic and therapeutic nuclear medicine, the transitions of major interest are those in which an α particle, β particle (electron or positron), or γ ray are emitted. Subsidiary atomic radiations or processes (electron capture, characteristic X-rays, internal conversion (IC) electrons, and Auger/Coster–Krönig electrons) are consequences of nuclear transitions. This chapter looks at the three nuclear decay schemes highlighted above. Quantum tunneling is derived and used to explain the empirical characteristics of α decay. The Fermi theory of β decay will be fundamental to understanding this type of transition and will be derived in order to predict energy spectra and allowable/forbidden transitions. Within that development, the weak interaction, which is fundamental to the existence of nuclear medicine, is introduced. Electromagnetic transitions, resulting in the emission of γ rays or IC electrons, are introduced and the theory behind them developed.

Contents

4.1 Introduction and History

The previous chapter described the characteristics that make a given nucleus stable or unstable. Clearly, for nuclear medicine purposes, one is interested in unstable or radioactive nuclei. The instability of a nucleus can be manifest in a number of ways. For example, a nucleus may be unstable to the emission of positrons or electrons (β decay) due to its having excessive numbers of protons or neutrons, respectively, for its mass. This particular instability reflects an imbalance between the Coulomb and symmetry terms of the Weizsäcker formula for the binding energy. Should the sum of the binding energies of two protons and two neutrons in a nucleus be less than the binding energy of the ^{4}He nucleus (α particle), then that nucleus is unstable to α decay, in which the α particle

B.J. McParland, *Nuclear Medicine Radiation Dosimetry*,
DOI 10.1007/978-1-84882-126-2_4,

can "quantum tunnel" through the nuclear Coulomb barrier and escape. Excited nuclei, which can be thought of as having high angular momentum or with nucleons excited to higher orbitals, can de-excite through the emission of quanta (γ decay) or by the direct transfer of nuclear energy to an atomic electron, a process known as internal conversion (IC). While there are other exotic, but rarer, cases of decay such as the spontaneous emission of a single proton or neutron or, for particularly heavy nuclei, spontaneous fission, the only radioactive decays of practical interest to diagnostic and therapeutic nuclear medicine are those of α, β and γ decay, and IC.[1]

The study of radioactivity began with Becquerel's original serendipitous discovery in 1896 of the phenomenon, using photographic emulsion as the radiation detector (Martins 1997). The stimulus for Becquerel's experimental work had been Röntgen's observation of the phosphorescence of some materials exposed to X-rays and Becquerel sought to determine if phosphorescent substances emitted similar radiations. He selected uranium as a suitable element for study due to its atomic absorption and emission spectra and noted that a uranium salt contained within an opaque covering and accidentally placed upon a photographic emulsion caused darkening of the emulsion due to radiations able to penetrate the optical cover. Even around this time, it had been noted that these radiations did not vary with time (which we would now recognize as reflecting a long half-life) and could ionize air. Shortly afterwards, the Curies began investigating the radioactivity associated with various minerals containing uranium and thorium. In particular, they determined that pitchblende was more radioactive than the chemical content of uranium or thorium would have allowed. Hypothesizing that this was due to the presence of other, but unknown, radioactive substances, the Curies chemically isolated two radioactive elements from pitchblende which they called polonium and radium. Although present in pitchblende in just trace amounts, their specific activities were some six orders-of-magnitude greater than that of uranium.

At about the same time, Rutherford, whilst at McGill University in Montréal, was also studying the radiations emitted by uranium (Wilson 1983). He had recognized that the photographic emulsion used by Becquerel was an inefficient detector and noted, as did the Curies, that the radiation-induced ionization of air had a faster response than emulsion, could be measured immediately and, moreover, could be easily quantified. This change of detector type aided Rutherford's discovery of the emanation of two types of radiation from samples of uranium compounds (uranium metal, uranium nitrate, uranium oxide, and uranium potassium sulfate) by measuring the changes in the detected ionization caused by them following their transmission through increasing thicknesses of aluminium foil. The decrease in ionization with increasing foil thickness was, at first, monotonic and indicated simple absorption of the radiation. However, the reduction in ionization with foil thickness eventually became limited until, when further foils were added, the radiation intensity began to decrease more rapidly again. Rutherford's original transmission data are shown in Fig. 4.1 in which the transmission of the ionizing radiations from uranium oxide through aluminium is plotted. The transmission of radiation dropped rapidly with absorber thickness to about 20 μm of aluminium and then remained nearly constant up to a thickness of 60 μm. This feature indicated a readily absorbed radiation, termed α radiation by Rutherford, and a more penetrating radiation, that he termed β radiation.[2] As will be seen in Chap. 7, the shorter range of the α rays indicates that they are strongly ionizing particles and are, in fact, absorbed in only a few cm of air or a few μm of soft tissue.

The nature of the α radiation was next studied through its magnetic deflection, which was first achieved by Rutherford using a 0.6 T magnetic field. This demonstrated that α radiation was massive, positively-charged particles. Further measurements of electric charge and mass suggested that these were doubly-ionized helium atoms (He^{++}). This nature of the α particle was finally confirmed by Rutherford and Royds (Rutherford and Royds 1909) in an elegant and simple experiment using an apparatus consisting of a

[1]There are other emissions often associated with radioactive decay, such as characteristic X-rays and Auger electrons. These, however, are due to relaxation processes following an excitation of the atom and are more properly considered in Chap. 6.

[2]It is interesting to note that when he repeated the experiment using thorium nitrate rather than uranium, Rutherford detected a third, more penetrating type of radiation. Thus, it would seem that this observation, now recognized to be of γ radiation, preceded the usually-credited discovery by Villard (Gerward 1999).

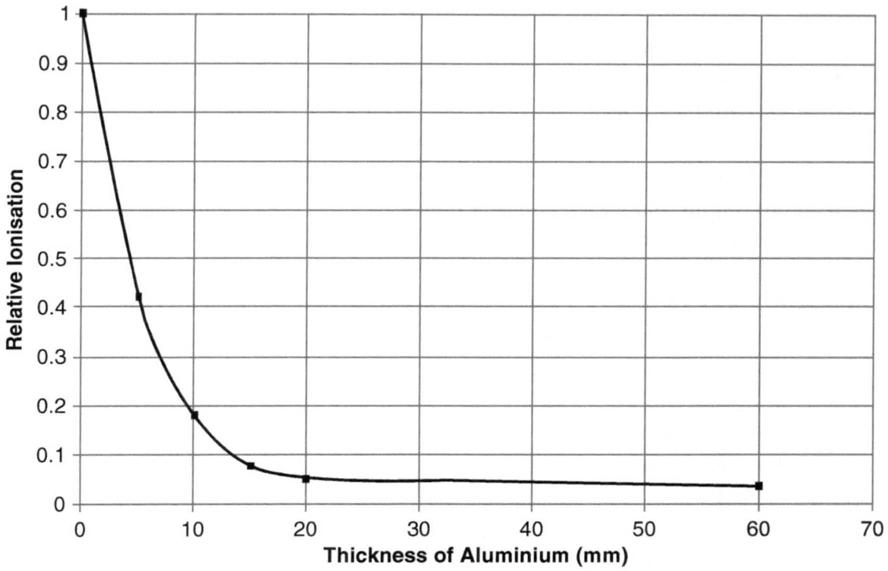

Fig. 4.1 Plot of Rutherford's original measurements of the ionization from uranium oxide following absorption by aluminium. Data points are the original data; the line is provided for guidance

volume of radon gas contained within a capillary tube with thin walls and encapsulated within an evacuated chamber. Radon is the radioactive daughter product of the α decay of radium and the α particles emitted by it came to rest within the thin wall. The He^{++} captured electrons from within the wall to form neutral atoms of helium gas which diffused through the wall into the surrounding chamber. After a sufficient amount of helium gas was formed, it was compressed and, via an electric discharge, demonstrated the characteristic emission spectrum that confirmed its nature.

Measurements by the Curies of the trajectories of the β component through a magnetic field demonstrated that it was made up of light, negatively-charged particles and studies of the β particle trajectories in combined magnetic and electric fields demonstrated that it had a charge-to-mass ratio e/m equal to that of the electron. The most elegant proof that the β particle is indeed an electron was provided by the measurements by Goldhaber and Scharff-Goldhaber (1948) of slow β particles stopped within a lead target. A low β particle energy was required to ensure that it did not eject the K-shell electron and the Pauli exclusion principle would, had β particles and electrons be the same, forbid the capture of the β particle into the full L and K shells. Had β particles and electrons been different, the β particle would cascade through orbitals, emitting X-rays, until it was eventually captured. The fact that no characteristic X-rays were detected gave unambiguous evidence of the identity of the β particle as an electron. The nuclear, rather than atomic, origin of the β particle is evident from its having a kinetic energy of a couple of MeV or less and which, from the uncertainty principle, $\Delta p\,\Delta x \approx \hbar c/2 \approx 100\,\text{MeV} \bullet \text{fm}$ and the fact that the nuclear diameter is of the order of a few fm. Nuclear β decay is the emission of a fast electron (e^-) or positron (e^+) resulting in a daughter nucleus with an atomic number of one greater or one less than the parent, respectively.[3] Associated with these is the process of electron capture in which the overlap of the wavefunctions of an atomic electron and a nuclear proton result in the capture of the electron by the proton leading to a daughter nucleus with an atomic number of one less than the parent. As such, these transitions occur between isobars and a chain of β decays will follow the mass parabola. Electron capture is a process related to β decay in which an atomic electron is "captured" by a nuclear proton to decrease the atomic number of the nucleus by one.

As noted earlier, the emission of γ-rays in nuclear decay was probably first detected from thorium by Rutherford in about 1898 but has not been greatly recognized historically as such. Villard is customarily

[3] While the term "β decay" is also sometimes used to refer to the capture of an atomic orbital electron by a proton-rich nucleus, it does not include internal conversion, a radioactive process described in the context of γ decay.

assigned recognition as the discoverer of γ rays through his measurements of the deflection by a magnetic field of the radiations emitted from radium and using photographic plates as detectors (Gerward 1999). For example, in one experiment, a collimated beam from a radium source was incident to two collinear photographic plates for which a magnetic field was placed between the collimator and the first plate to deflect the beam. The first plate demonstrated two distinctive traces of a sharp line, denoting a beam unaffected by the magnetic field, superimposed upon a second, more diffuse, trace caused by the magnetic deflect. On the second plate, which was further away, there was only the sharp trace as the charged particles creating the diffuse pattern in the first plate had been completely deflected. The undeflected beam had traversed 10 mm of glass without noticeable attenuation and even inserting a 300 μm-thick lead foil had only a slight affect upon this penetrating beam. Villard associated this penetrating and electrically-neutral radiation with Röntgen's X-rays but he did not call them γ rays. This assignation does not seem to have happened until about 1903 and was apparently due to Rutherford. Experimental study of γ rays in the early twentieth century was not as intense as were those of α and β particles and it was not until 1914 that Rutherford and Andrade established the electromagnetic nature of γ rays. Following von Laue's use in 1912 of a crystal as a space diffraction grating for X-rays, Rutherford and Andrade repeated the same approach to measure the wavelength of γ rays. Compton's later work on the elastic scatter of photons from electrons extended the wave-point duality of light photons to γ rays.

Radioactive decay is, obviously, the reason that the discipline of nuclear medicine exists. Hence, we will look at the theoretical principles behind the three major radioactive modes of interest to nuclear medicine: α decay, β decay (including electron capture) and γ emission (including IC).

4.2 α Decay

4.2.1 Introduction

α emission is a two-body process:

$$^{A}_{Z}X \rightarrow {}^{A-4}_{Z-2}Y + {}^{4}_{2}He \tag{4.1}$$

The final state is that of two heavy charged particles in motion, which has significant dosimetric consequences in that, first, both rapidly lose kinetic energy to the medium in a small volume as they slow down leading to a high absorbed dose. The second consequence is that the daughter nucleus is frequently left in an excited state resulting in the subsequent emission of γ rays or conversion electrons. α decay is intrinsically linked to the high binding energy (28.3 MeV) of the α particle. Hence, a nucleus will be unstable to α decay if the sum of the binding energies of a pair or protons and a pair of neutrons within it are less than the α particle binding energy. As a result, the α emitting parent nucleus is necessarily heavy as shown in Fig. 4.2.[4] Early understanding of α decay was confounded by the failure of classical theory to provide an explanation of it even occurring – a failure shown with considerable power by consideration of the magnitude of the nuclear Coulomb potential, as shown in Fig. 4.3. Within the nucleus, the α particle is subject to the conflicting attractive strong nuclear force and the repulsive Coulomb force from the nuclear protons. Beyond the nuclear radius R_N, where the strong nuclear force is negligible, the α particle will be subject solely to the nuclear Coulomb potential. For a nucleus with atomic number Z, the Coulomb potential is,

$$U(r) = 2\alpha\hbar c \frac{Z_Y}{r} \quad r > R_N \tag{4.2}$$

where the factor of 2 is the atomic number of the α particle. It should be noted that that the atomic number is that of the nucleus following α emission.

The Coulomb potential experienced by an α particle at this nuclear radius for a nucleus with $Z \approx 80$ and $A \approx 200$ is about 33 MeV and a classical dilemma arises from the fact that experimental measurements of the kinetic energies of the emitted α particles are typically of the order of only a few MeV, as shown

[4]Note that Fig. 4.2 shows the kinetic energy of the emitted α particle as a function of the parent nucleus' mass; the probability of the α particle occurring at all is not considered in that plot but is the focus of a later subsection.

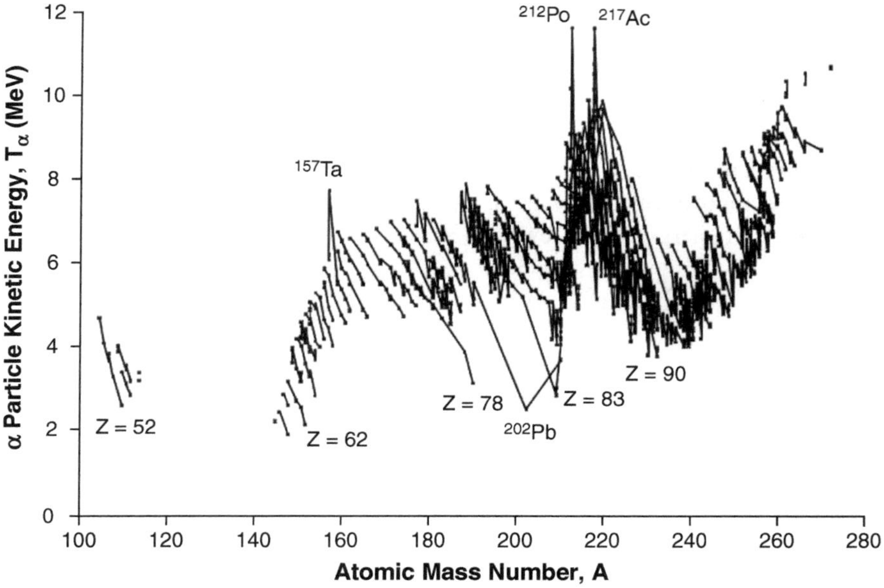

Fig. 4.2 Plot of α particle kinetic energy as a function of atomic mass number for elements between Z = 52 (tellurium) and Z = 110 (darmstadtium). Lines join isotopes. Plot is derived from data from a National Nuclear Data Center (2008) database

in Fig. 4.2.[5] Over the 2,076 α emissions represented in this figure, the mean α particle kinetic energy is 6.1 MeV, with a minimum of 2.55 MeV for ^{202}Pb and a maximum of 11.66 MeV for ^{212}Po.

4.2.2 *Kinematics of α Emission*

4.2.2.1 Kinetic Energy of the α Particle

The energy released in the α-emission of (4.1) is,

$$Q_\alpha = m_X - (m_Y + m_\alpha) \tag{4.3}$$

where the lower-case m's are the nuclear masses and where Q_α must be positive (i.e., (4.1) is exoergic) in order for the α decay to proceed. Both nuclei are assumed to be in their ground states (the fine structure in α particle energy spectra as a result of excited states is discussed in the following subsection) so that the Q_α is distributed amongst the kinetic energies of the α particle and the daughter nucleus. Because of the much larger mass of the daughter nucleus and the simultaneous conservation of linear momentum and

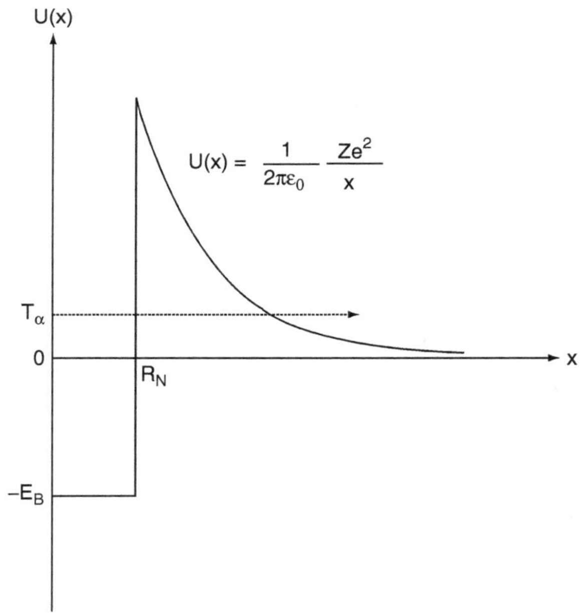

Fig. 4.3 Potential energy of a heavy nucleus for an α particle and a typical kinetic energy of an emitted α particle. Abscissa is given as x rather than r as this is a one-dimensional potential

[5]Gamow (1928) reported that Rutherford had used the Thomson "plum pudding" model of the nucleus in an attempt to avoid this dilemma: the α particle within the nucleus was electrically-neutral as it carried two electrons. This allowed it to travel through the barrier unheeded; once out of the nucleus, the α particle somehow shed these two electrons. This hypothesis was invalidated with the demise of the Thomson model.

energy, virtually all of Q_α is taken up by the kinetic energy of the α particle. The conservation of energy is,

$$m_X = m_Y + m_\alpha + T_Y + T_\alpha. \tag{4.4}$$

As the kinetic energy of the α particle is only a few MeV, nonrelativistic kinematics are applicable and the conservation of energy can be rewritten as,

$$\frac{p_Y^2}{2m_Y} + \frac{p_\alpha^2}{2m_\alpha} = Q_\alpha. \tag{4.5}$$

Using the conservation of momentum, the recoil kinetic energy of the daughter nucleus is then obtained,

$$T_Y = \frac{Q_\alpha}{\left(1 + \frac{m_Y}{m_\alpha}\right)}. \tag{4.6}$$

As the atomic mass of the daughter nucleus is of the order of 200 or more, it is evident that only about 2% of the available energy will appear as the recoil kinetic energy of the daughter nucleus. Hence, for practical purposes, we can ignore the recoil kinetic energy whilst calculating for the α-particle kinetic energy.[6] Equation (4.6) can then be approximated by,

$$T_\alpha \approx Q_\alpha. \tag{4.7}$$

Assuming that the parent and daughter nuclei are both in their ground states, then using results from Chap. 3 and neglecting the atomic electron binding energies, the α particle kinetic energy corresponding to a parent nucleus of atomic number Z and atomic mass A can be written as,

$$\begin{aligned} T_\alpha(A,Z) &\approx Q_\alpha \\ &= M(A,Z) - M(A-4,Z-2) - M(4,2). \end{aligned} \tag{4.8}$$

[6]While this is acceptable for analyzing the kinematics of α decay, we cannot neglect the energy of the recoil nucleus in a dosimetry evaluation. Although its kinetic energy is small, the range of the recoil nucleus is also small, meaning that the recoil energy is deposited within a very small volume leading to a high absorbed dose. Hence, α-emitting radionuclides are important in therapeutic nuclear medicine should the α particle-emitting radiopharmaceutical be covalently-bound to the nuclear deoxyribonucleic acid.

where the atomic mass is given by $M(A, Z) = ZM_H + (A - Z)m_n - B(A, Z)$. The α particle kinetic energy can thus be given in terms of the nuclear binding energies of the parent, daughter, and the α particle,

$$\begin{aligned} T_\alpha(A,Z) &= B(4,2) + B(A-4,Z-2) - B(A,Z) \\ &= B(A-4,Z-2) - B(A,Z) + 28.3\,\text{MeV}. \end{aligned} \tag{4.9}$$

where the numerical value of the α particle binding energy has been inserted. The binding energies of the parent and daughter nuclei can be estimated using the Weizsäcker semi-empirical formula,

$$\begin{aligned} B(A,Z) = a_{Vol}A - a_{Surf}A^{2/3} - a_{Coul}\frac{Z^2}{A^{1/3}} \\ - a_{Sym}\frac{(A-2Z)^2}{A} - \delta(A,Z). \end{aligned}$$

Substituting this into (4.9) gives,

$$\begin{aligned} T_\alpha(A,Z) =& -4a_{Vol} - a_{Surf}\left((A-4)^{2/3} - A^{2/3}\right) \\ & - a_{Coul}\left(\frac{(Z-2)^2 - Z^2}{(A-4)^{1/3} - A^{1/3}}\right) \\ & - a_{Sym}\left(\frac{(A-4-2(Z-2))^2}{A-4} - \frac{(A-2Z)^2}{A}\right) \\ & - \delta(A-4,Z-2) + \delta(A,Z) + 28.3\,\text{MeV} \\ =& -4a_{Vol} - a_{Surf}\left((A-4)^{2/3} - A^{2/3}\right) \\ & - a_{Coul}\left(\frac{(Z-2)^2}{(A-4)^{1/3}} - \frac{Z^2}{A^{1/3}}\right) \\ & - a_{Sym} - \left(\frac{(A-2Z)^2}{A-4} - \frac{(A-2Z)^2}{A}\right) \\ & - \delta(A-4,Z-2) + \delta(A,Z) + 28.3\,\text{MeV} \end{aligned} \tag{4.10}$$

As α decay occurs only for heavy nuclei for which $A \gg 1$ and $Z \gg 1$, the expression can be simplified by approximating the surface, Coulomb and symmetry terms with,

$$\left((A-4)^{2/3} - A^{2/3}\right) \approx -\frac{8}{3}A^{-1/3}$$

$$\left(\frac{(Z-2)^2}{(A-4)^{1/3}}-\frac{Z^2}{A^{1/3}}\right)\approx 4Z^2A^{-1/3}\left(\frac{\frac{1}{Z}-\frac{1}{3A}}{1-\frac{4}{3A}}\right)$$

$$\left(\frac{(A-2Z)^2}{A-4}-\frac{(A-2Z)^2}{A}\right)\approx 4\left(1-\frac{2Z}{A}\right)^2.$$

Recalling that the paired nucleon term is non-zero only for odd–odd and even–even nuclei, it is retained as an explicit item. These results give,

$$\begin{aligned}T_\alpha(A,Z) = {} & -4a_{Vol}+\frac{8}{3}a_{Surf}A^{-1/3}\\ & +4a_{Coul}Z^2A^{-1/3}\left(\frac{\frac{1}{Z}-\frac{1}{3A}}{1-\frac{4}{3A}}\right)\\ & -4a_{Sym}\left(1-{}^{2Z}/_{A}\right)^2-\delta(A,Z)\\ & +\delta(A-4,Z-2)+28.3\,\text{MeV}.\end{aligned}\tag{4.11}$$

The fact that this expression is an approximation will not lead to an exact result of T_α but, even so, the accuracy is more than sufficient to predict the α particle kinetic energies for dosimetry purposes. For example, using (4.11) (and the values of a_{Vol}, a_{Surf}, a_{Coul}, a_{Sym}, and δ) to predict the kinetic energy of α particles emitted by ^{210}Po will yield a value of 4.26 MeV whereas the experimentally-determined value is about 4.52 MeV. This result can also be used to predict the systematics of α particle kinetic energy of Fig. 4.2. For example, for a given isotope, the α particle kinetic energy decreases with increasing atomic mass[7] which can be predicted from,

$$\begin{aligned}\Delta T_\alpha(A,A+1,Z) &= T_\alpha(A,Z)-T_\alpha(A+1,Z)\\ &= \frac{8}{3}a_{Surf}\left(A^{1/3}-(A+1)^{1/3}\right)\\ &\quad +4a_{Coul}Z^2\left(A^{1/3}\left(\frac{\frac{1}{Z}-\frac{1}{3A}}{1-\frac{4}{3A}}\right)\right.\\ &\quad\left.-(A+1)^{1/3}\left(\frac{\frac{1}{Z}-\frac{1}{3(A+1)}}{1-\frac{4}{3(A+1)}}\right)\right)\\ &\quad -8a_{Sym}Z\left(\frac{1}{A+1}-\frac{1}{A}\right)\\ &\quad -\delta(A,Z)+\delta(A+1,Z)\\ &\quad +\delta(A-4,Z-2)-\delta(A-3,Z)\end{aligned}\tag{4.12}$$

which is a negative quantity.

4.2.2.2 Energy Spectrum: Fine Structure

The discussion of the kinematics of α emission has, so far, neglected any additional energy channels which can alter Q_α. Experimentally, however, it has been observed that the α particles emitted from a single nuclide can have multiple energies occurring with different probabilities. An example is shown in Fig. 4.4 for ^{215}Po. The presence of multiple α particle kinetic energies means, from the definition of Q_α and the approximation of T_α by Q_α, that the nuclear masses of the parent and/or daughter must be multivalued, reflecting the excited states of the daughter ^{211}Pb nucleus. However, the relative intensity of each α emission (i.e., the relative probability of an α particle with that particular kinetic energy being emitted) indicates a variable "transparency" of the Coulomb barrier which impedes α emission.[8] For this example of ^{215}Po, this transparency varies by over five orders-of-magnitude.

4.2.3 Barrier Penetration

4.2.3.1 Introduction

The half-lives of α emissions vary from about 0.3 μs for ^{212}Po to 4.5 × 10^9 y for ^{238}U, an astonishing variation of some 23 orders-of-magnitude. This fact, combined with simple consideration of the kinematics of α emission, leads quickly to a conclusion that there

[7] This trend is broken for the conditions 209 < A < 213 which is related to the magic proton and neutron numbers of 82 and 126, respectively.

[8] Note that, despite the different intensities among the α particles emitted, the decay probability per unit time for each is the same.

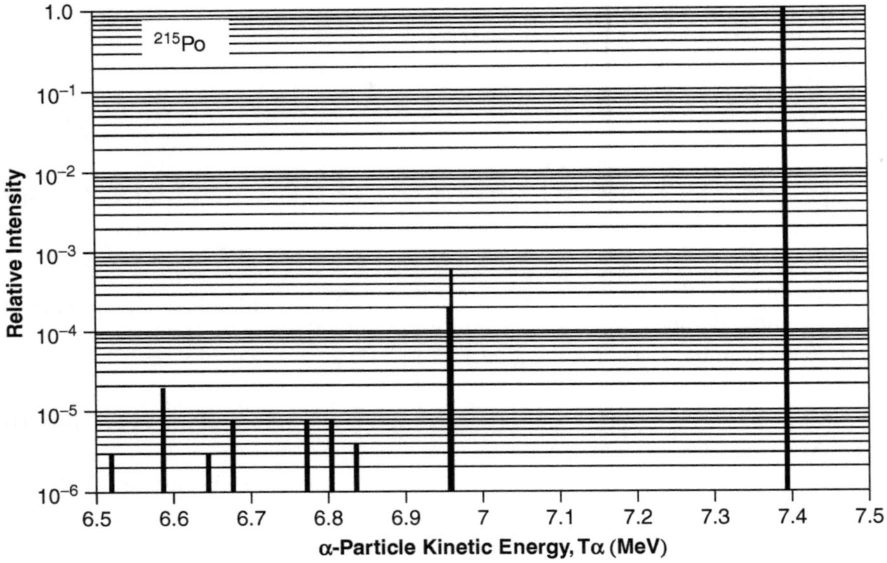

Fig. 4.4 α particle kinetic energy spectrum for ^{215}Po. Plot derived from data from National Nuclear Data Center (2008) data

exists a barrier impeding the emission of the α particle. As the kinetic energy of the α particle is a few MeV, corresponding to a speed of the order of 10^7 m/s, and the radius of a heavy nucleus is of the order of about 10 fm, the α particle should need only 10^{-21}s to traverse the nucleus. As the measured half-lives are much larger than this, there clearly must be some nature of a barrier to α emission, as was shown Fig. 4.3. However, in classical mechanics, the barrier would be completely impervious to the α particle. Gamow (1928) and, independently, Condon and Gurney (1928) showed that α emission could be explained by the quantum-mechanical tunneling of the α particle through this barrier. The solution to this problem is determined by first studying the simple case of a one-dimensional barrier.

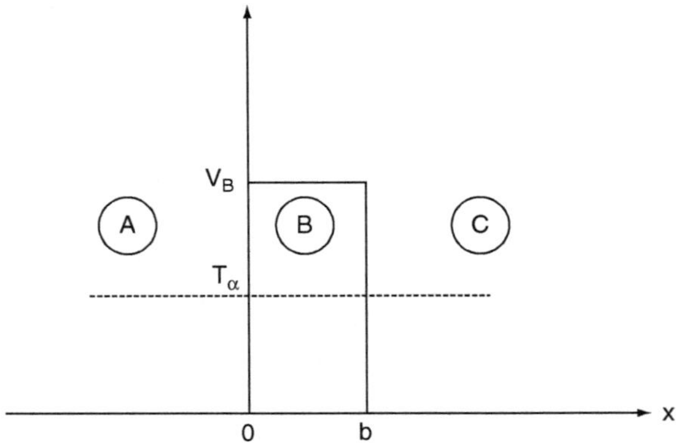

Fig. 4.5 One-dimensional potential barrier of height V_B and width b. Region A corresponds to the interior of the nucleus, B to the barrier and C to the exterior of the nucleus. The α particle has kinetic energy T_α

4.2.3.2 One-Dimensional Rectangular Barrier

The simplest potential barrier to model is the one-dimensional rectangular one of Fig. 4.5. As it is one-dimensional, angular momentum is necessarily excluded. The potential barrier is of height V_B and width b and the α particle is treated as a wave within the nucleus (Zone A) and incident to the barrier from the left. The time-independent one-dimensional Schrödinger equation is,

$$\frac{d^2\Psi}{dx^2} + \frac{2m_\alpha}{(\hbar c)^2}(T_\alpha - V(x))\Psi = 0 \tag{4.13}$$

where

$$\begin{aligned} V(x) &= V_B \quad 0 < x < b \\ &= 0 \quad \text{elsewhere.} \end{aligned} \tag{4.14}$$

Recall that the units of the α particle rest mass and its three-vector momentum are that of energy and the $\hbar c$ factor is required in the denominator to make the equation dimensionally correct. Outside the barrier, in Zones A and C, (4.13) is,

$$\frac{d^2\Psi}{dx^2} + \frac{2m_\alpha T_\alpha}{(\hbar c)^2}\Psi = 0 \tag{4.15}$$

This is rewritten using the reduced de Broglie wavelength of the α particle,

$$\lambda\!\!\!^{-} = \frac{\hbar c}{\sqrt{2m_\alpha T_\alpha}} \tag{4.16}$$

$$\frac{d^2\Psi}{dx^2} + \frac{1}{\lambda\!\!\!^{-2}}\Psi = 0. \tag{4.17}$$

In Zone A, the solution to (4.17) is,

$$\Psi_A(x) = w_{A1}e^{\frac{ix}{\lambda\!\!\!^{-}}} + w_{A2}e^{-\frac{ix}{\lambda\!\!\!^{-}}} \quad x < 0 \tag{4.18}$$

and, similarly for Zone C,

$$\Psi_C(x) = w_{C1}e^{\frac{ix}{\lambda\!\!\!^{-}}} \quad b < x \tag{4.19}$$

where there is no reflected wave from Zone C, i.e., $w_{C2} = 0$. Within the rectangular barrier (Zone B), the Schrödinger equation is,

$$\frac{d^2\Psi}{dx^2} + \frac{2m_\alpha(T_\alpha - V_B)}{(\hbar c)^2}\Psi = 0 \quad 0 < x < b. \tag{4.20}$$

Since $T_\alpha < V_B$,

$$\frac{d^2\Psi}{dx^2} - \frac{2m_\alpha|T_\alpha - V_B|}{(\hbar c)^2}\Psi = 0 \tag{4.21}$$

with the solution,

$$\Psi_B(x) = w_{B1}e^{k_\alpha x} + w_{B2}e^{-k_\alpha x} \tag{4.22}$$

where,

$$k_\alpha = \frac{\sqrt{2m_\alpha|T_\alpha - V_B|}}{\hbar c}. \tag{4.23}$$

The probability of the wave tunneling through the barrier is defined as the transmission given by the ratio of the squared amplitudes of the wavefunction incident to and exiting the barrier,

$$\Im = \frac{|w_{C1}|^2}{|w_{A1}|^2}. \tag{4.24}$$

To solve for $\Im$, the amplitudes w_{A1} and w_{C1} are calculated in terms of the amplitudes w_{B1} and w_{B2}. The remainder of the amplitudes are determined by the boundary conditions which require that the functions and their first-derivatives be continuous at the boundaries of the potential:

$$\Psi_A(0) = \Psi_B(0) \tag{4.25}$$

$$\frac{d\Psi_A(0)}{dx} = \frac{d\Psi_B(0)}{dx} \tag{4.26}$$

$$\Psi_B(b) = \Psi_C(b) \tag{4.27}$$

$$\frac{d\Psi_B(b)}{dx} = \frac{d\Psi_C(b)}{dx}. \tag{4.28}$$

Equations (4.25) and (4.26) give,

$$w_{A1} + w_{A2} = w_{B1} + w_{B2} \tag{4.29}$$

$$\frac{i}{\lambda\!\!\!^{-}}(w_{A1} - w_{A2}) = k_\alpha(w_{B1} - w_{B2}). \tag{4.30}$$

From which, w_{A1} and w_{A2} can be solved,

$$w_{A1} = \frac{w_{B1}(1 - ik_\alpha\lambda\!\!\!^{-}) + w_{B2}(1 + ik_\alpha\lambda\!\!\!^{-})}{2} \tag{4.31}$$

$$w_{A2} = \frac{w_{B1}(1 + ik_\alpha\lambda\!\!\!^{-}) + w_{B2}(1 - ik_\alpha\lambda\!\!\!^{-})}{2}. \tag{4.32}$$

Equations (4.27) and (4.28) give,

$$w_{B1}e^{k_\alpha b} + w_{B2}e^{-k_\alpha b} = w_{C1}e^{\frac{ib}{\lambda\!\!\!^{-}}} \tag{4.33}$$

and

$$k_\alpha\left(w_{B1}e^{k_\alpha b} - w_{B2}e^{-k_\alpha b}\right) = \frac{i}{\lambda\!\!\!^{-}}w_{C1}e^{\frac{ib}{\lambda\!\!\!^{-}}}. \tag{4.34}$$

Solving for w_{B1} and w_{B2},

$$w_{B1} = \frac{w_{C1}}{2} e^{-k_\alpha b(1 - i/k_\alpha \bar{\lambda})}\left(1 + \frac{i}{k_\alpha \bar{\lambda}}\right) \quad (4.35)$$

$$w_{B2} = \frac{w_{C1}}{2} e^{k_\alpha b(1 + i/k_\alpha \bar{\lambda})}\left(1 - \frac{i}{k_\alpha \bar{\lambda}}\right). \quad (4.36)$$

For convenience, the reciprocal of $\Im$ is calculated,

$$\frac{1}{\Im} = \left(\frac{w_{A1}}{w_{C1}}\right)\left(\frac{w_{A1}}{w_{C1}}\right)^*. \quad (4.37)$$

Substituting (4.35) and (4.36) into (4.31) gives,

$$\begin{aligned}
\frac{w_{A1}}{w_{C1}} &= \frac{e^{-k_\alpha b\left(1 - \frac{i}{k_\alpha \bar{\lambda}}\right)}\left(1 + \frac{i}{k_\alpha \bar{\lambda}}\right)(1 - ik_\alpha \bar{\lambda}) + e^{k_\alpha b\left(1 + i/k_\alpha \bar{\lambda}\right)}\left(1 - \frac{i}{k_\alpha \bar{\lambda}}\right)(1 + ik_\alpha \bar{\lambda})}{4} \\
&= \frac{e^{ib/\bar{\lambda}}}{4} e^{-k_\alpha b}\left(1 + \frac{i}{k_\alpha \bar{\lambda}}\right)(1 - ik_\alpha \bar{\lambda}) + e^{k_\alpha b}\left(1 - \frac{i}{k_\alpha \bar{\lambda}}\right)(1 + ik_\alpha \bar{\lambda}) \\
&= e^{ib/\bar{\lambda}}\left(\frac{1 - (k_\alpha \bar{\lambda})^2}{k_\alpha \bar{\lambda}}\right)\left(\cosh k_\alpha b - \frac{i}{2}\sinh k_\alpha b\right) \\
&= \frac{e^{ib/\bar{\lambda}}}{2}\left(\frac{1 - (k_\alpha \bar{\lambda})^2}{k_\alpha \bar{\lambda}}\right)\sinh k_\alpha b(2\coth(k_\alpha b) - i). \quad (4.38)
\end{aligned}$$

It follows that,

$$\begin{aligned}
\frac{1}{\Im} &= \left(\frac{w_{A1}}{w_{C1}}\right)\left(\frac{w_{A1}}{w_{C1}}\right)^* \\
&= \frac{1}{4}\left(\frac{1 - (k_\alpha \bar{\lambda})^2}{k_\alpha \bar{\lambda}}\right)^2 (1 + 4\coth^2 k_\alpha b)\sinh^2 k_\alpha b \\
&= \frac{(V_B/2T_\alpha)^2}{\left(V_B/T_\alpha\right) - 1}(1 + 4\coth^2 k_\alpha b)\sinh^2 k_\alpha b \\
&= \frac{(V_B/2T_\alpha)^2}{\left(V_B/T_\alpha\right) - 1}(1 + 4\coth^2 k_\alpha b)\sinh^2 k_\alpha b \\
&\approx 1.94(1 + 4\coth^2 k_\alpha b)\sinh^2 k_\alpha b. \quad (4.39)
\end{aligned}$$

where a barrier height of 33 MeV and a kinetic energy of 5 MeV have been assumed in order to obtain the numerical result. Determining the product $k_\alpha b$ requires an estimation of the width of the barrier, b. To obtain this, replace the rectangular barrier with the Coulomb barrier as shown in Fig. 4.6 (with the attractive nuclear potential treated as a well potential) and use,

$$2Z\frac{\alpha\hbar c}{R_N + b} = T_\alpha. \quad (4.40)$$

Solving for b,

$$b = 2Z\left(\frac{\alpha\hbar c}{T_\alpha}\right) - R_N. \quad (4.41)$$

Using the previous example of $Z = 80$ and $R_N = 7$ fm, we calculate that for a 5 MeV α particle $b \approx 40$ fm. Hence, for $V_B = 33$ MeV, we have $k_\alpha \approx 2.3\ \text{fm}^{-1}$ leading to $bk_\alpha \approx 92$. For this value, approximations to the hyperbolic trigonometric functions can be used,

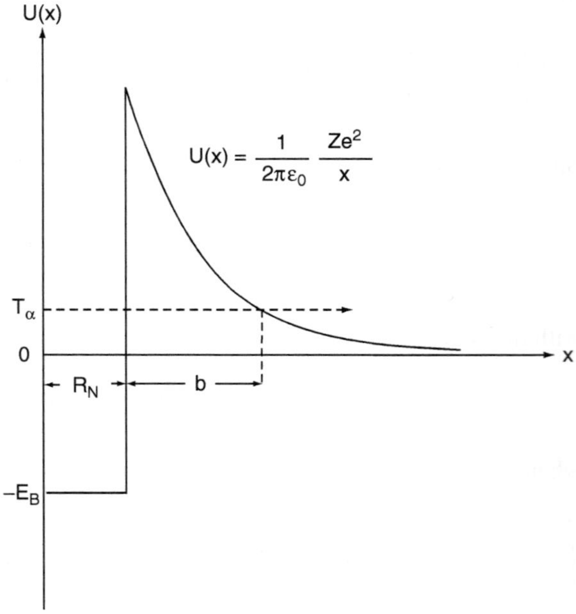

Fig. 4.6 Estimation of the barrier width for simple one-dimensional rectangular barrier calculation

$$\left(1+4\coth^2 k_\alpha b\right) = 1+\left(\frac{2\cosh k_\alpha b}{\sinh k_\alpha b}\right)^2$$
$$= 1+\left(2\frac{\left(e^{k_\alpha b}+e^{-k_\alpha b}\right)}{\left(e^{k_\alpha b}-e^{-k_\alpha b}\right)}\right)^2 \approx 5$$

and

$$\sinh^2 k_\alpha b = \frac{\left(e^{k_\alpha b}-e^{-k_\alpha b}\right)^2}{4}$$
$$\approx \frac{e^{2k_\alpha b}}{4}.$$

As the magnitude of the exponent is $\frac{1}{\Im} \approx 2.4\, e^{2k_\alpha b}$, the transmission is

$$\Im \approx 0.41\, e^{-2k_\alpha b}. \qquad (4.42)$$

And, as the magnitude of the exponent is $e^{-2k_\alpha b} \approx 1.2 \times 10^{-80}$, the transmission is simply written as,

$$\Im \approx e^{-2k_\alpha b}. \qquad (4.43)$$

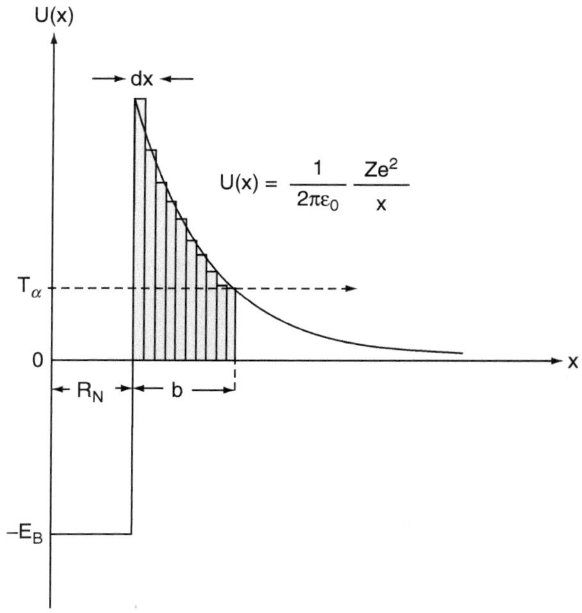

Fig. 4.7 Calculation of penetration through a one-dimensional Coulomb barrier

4.2.3.3 Three-Dimensional Barrier

Gamow Factor

The simple barrier model of the previous section is now extended to calculate the more realistic case of the penetration through the 1/r Coulomb barrier which is treated as a series of infinitesimal rectangular barriers with decreasing heights, as shown in Fig. 4.7. This now requires k_α to have a spatial dependence,

$$k_\alpha(x) = \frac{\sqrt{2m_\alpha|T_\alpha - V_B(x)|}}{\hbar c}. \qquad (4.44)$$

Then, as $\Im \approx e^{-\frac{2x}{\hbar c}\sqrt{2m_\alpha|T_\alpha - V(x)|}}$, the differential transmission through a barrier of thickness *dx* is,

$$d\Im = 4e^{-\frac{2}{\hbar c}dx\sqrt{2m|T_\alpha - V(x)|}}. \qquad (4.45)$$

Integrating this result gives,

$$\Im = 8e^{-2G} \qquad (4.46)$$

where G is defined as the Gamow factor,

$$G = \frac{1}{\hbar c}\int_{R_N}^{b+R_N} dx\sqrt{2m_\alpha|T_\alpha - V_B(x)|}. \qquad (4.47)$$

Hence, in order to determine the transmission, which by extension is the calculation of the decay rate constant and decay half-life, requires an explicit calculation of the Gamow factor. This is now performed for the three-dimensional case.

Changing the spatial variable from the one-dimensional *x* to the three-dimensional *r*, the Gamow factor is,

$$\begin{aligned} G &= \frac{1}{\hbar c}\int_{R_N}^{b+R_N} dr\sqrt{2m_\alpha|V_B(r) - T_\alpha|} \\ &= \frac{\sqrt{2m_\alpha}}{\hbar c}\int_{R_N}^{b+R_N} dr\sqrt{\left|2Z\left(\frac{\alpha\hbar c}{r}\right) - T_\alpha\right|} \\ &= \frac{\sqrt{2m_\alpha T_\alpha}}{\hbar c}\int_{R_N}^{b+R_N} dr\sqrt{\frac{R_N+b}{r} - 1} \\ &= \frac{\sqrt{R_N+b}}{\lambda}\int_{R_N}^{b+R_N} dr\sqrt{\frac{1}{r} - \frac{1}{R_N+b}}. \end{aligned} \qquad (4.48)$$

The integral is solved via the change of variable,

$$\cos\theta = \sqrt{\frac{r}{R_N + b}}$$

which gives,

$$\sqrt{\frac{1}{r} - \frac{1}{R_N + b}} = \frac{\sin\theta}{\sqrt{r}} \\ = \frac{\tan\theta}{\sqrt{R_N + b}}$$

and

$$dr = -2(R_N + b)\sin\theta\,\cos\theta\,d\theta.$$

Substituting these into (4.48) gives,

$$\begin{aligned} G &= \frac{\sqrt{R_N + b}}{\lambda\!\!\!^{-}} \int_{R_N}^{b+R_N} dr \sqrt{\frac{1}{r} - \frac{1}{R_N + b}} \\ &= 2\frac{R_N + b}{\lambda\!\!\!^{-}} \int_0^{\arccos\sqrt{\frac{R_N}{R_N+b}}} d\theta\,\sin^2\theta \\ &= \frac{R_N + b}{\lambda\!\!\!^{-}} \left(\arccos\left(\sqrt{\frac{R_N}{R_N + b}}\right) - \frac{1}{2}\sin\left(2\arccos\left(\sqrt{\frac{R_N}{R_N + b}}\right)\right)\right) \\ &= \frac{R_N + b}{\lambda\!\!\!^{-}} \left(\arccos\left(\sqrt{\frac{R_N}{R_N + b}}\right) - \frac{\sqrt{R_N b}}{R_N + b}\right). \end{aligned} \quad (4.49)$$

This expression is next simplified through the use of the dimensionless quantity,

$$\xi = \frac{T_\alpha}{V_B} = \frac{R_N}{R_N + b}$$

to give,

$$\begin{aligned} G &= \frac{R_N}{\xi\lambda\!\!\!^{-}} \left(\cos^{-1}\left(\sqrt{\xi}\right) - \sqrt{\xi(1-\xi)}\right) \\ &\equiv \frac{R_N}{\xi\lambda\!\!\!^{-}} \gamma(\xi) \end{aligned} \quad (4.50)$$

By noting that $R_N \ll b$, the arccos function is expanded to first-order to give,

$$\gamma(\xi) \approx \frac{\pi}{2} - 2\sqrt{\xi} \quad (4.51)$$

allowing the Gamow factor to be written as,

$$\begin{aligned} G &= \frac{R_N}{\xi\lambda\!\!\!^{-}} \left(\frac{\pi}{2} - 2\sqrt{\xi}\right) \\ &= \frac{\pi R_N}{2\xi\lambda\!\!\!^{-}} - \frac{2R_N}{\sqrt{\xi}\lambda\!\!\!^{-}} \\ &= \frac{\sqrt{R_N + b}}{\lambda\!\!\!^{-}} \left(\frac{\pi}{2}\sqrt{R_N + b} - 2\sqrt{R_N}\right). \end{aligned} \quad (4.52)$$

Numerically, for the example of a heavy nucleus with $R_N = 7$ fm and $b \approx 40$ fm and a 5 MeV α particle, the reduced de Broglie wavelength is $\lambda\!\!\!^{-} = \hbar c/\sqrt{2m_\alpha T_\alpha} \approx 1\text{fm}$ and the Gamow factor is $G \approx 37$.

Angular Momentum

Having now extended the calculation to a three-dimensional barrier, we consider the effect induced by angular momentum by using the three-dimensional Schrödinger equation as shown in Chap. 3.[9] Clearly, as this will increase the Coulomb barrier through the addition of a centrifugal term, $l(l+1)(\hbar c)^2/2m_\alpha r^2$, the angular momentum of the emitted α particle will influence the transmission of the particle through the barrier. This would arise in a step of the derivation of (4.48) in which the Gamow factor would, as a result of the centrifugal term, appear as,

$$G = \frac{\sqrt{2m_\alpha}}{\hbar c} \times \int_{R_N}^{b+R_N} dr \sqrt{\left| 2Z\left(\frac{\alpha\hbar c}{r}\right) + \frac{l(l+1)(\hbar c)^2}{2m_\alpha r^2} - T_\alpha \right|} \quad (4.53)$$

[9] As the α particle has zero spin, the angular momentum carried away is solely orbital in character.

Due to the α particle mass and the square of the radius present in the denominator of the centrifugal term, the effect of the centrifugal term to the barrier transmission will be quite limited. For an example of $Z = 80$, $r = b = 40$ fm and the high angular momentum case of $l = 4$, we find that,

$$\text{Coulomb term } 2Z\frac{\alpha\hbar c}{b} = 5.75\,\text{MeV}$$
$$\text{Centrifugal term } \frac{l(l+1)(\hbar c)^2}{2m_\alpha b^2} = 0.13\,\text{MeV}$$

Clearly for smaller l, the role of the centrifugal term within the Gamow factor can be neglected. But where the importance of the centrifugal barrier does arise is through the fact that the l values available to the emerging α particle are subject to the simultaneous conservation of angular momentum and parity.

4.2.4 Estimation of α Decay Half-Life

Consider an α particle within a spherical well potential well of radius R_N and moving with speed v_0. The rate at which the α particle will hit the "wall" is of the order of v_0/R_N times per second. The decay constant (the transition probability per unit time) for α emission is the product of this "impact rate" and the transmission factor,

$$\begin{aligned}\lambda &= \frac{v_0}{R_N}\Im \\ &= 8\frac{v_0}{R_N}e^{-2G}\end{aligned} \tag{4.54}$$

Taking logarithms of both sides, noting that $|\ln 8| \ll 2G$ and using our derived expression for the Gamow factor, we have,

$$\begin{aligned}\ln\lambda &\cong \ln\frac{v_0}{R_N} - 2G \\ &= \ln\frac{v_0}{R_N} - 2\frac{\sqrt{R_N+b}}{\lambda}\left(\frac{\pi}{2}\sqrt{R_N+b} - 2\sqrt{R_N}\right) \\ &= \ln\frac{v_0}{R_N} - \frac{1}{\lambda}\left(\pi(R_N+b) - 4\sqrt{R_N(R_N+b)}\right)\end{aligned} \tag{4.55}$$

Recalling the definition of the de Broglie wavelength, we can write,

$$\begin{aligned}\ln\lambda &= \ln\frac{v_0}{R_N} - \frac{\sqrt{2m_\alpha T_\alpha}}{\hbar c}\left(2\pi Z\frac{\alpha\hbar c}{T_\alpha} - 4\sqrt{R_N\frac{2Z\alpha\hbar c}{T_\alpha}}\right) \\ &= \ln\frac{v_0}{R_N} - \left(2\pi\alpha\sqrt{2m_\alpha}\right)\frac{Z}{\sqrt{T_\alpha}} + 8\sqrt{\frac{\alpha m_\alpha R_N}{\hbar c}}\sqrt{Z}\end{aligned} \tag{4.56}$$

As shown in the following chapter, the half-life is related to the decay constant via,

$$T_{1/2} = \frac{\ln 2}{\lambda} \tag{4.57}$$

Hence, one can write the α emission half-life as,

$$T_{1/2} = \frac{R_N \ln 2}{v_0}e^{\left(2\pi\alpha\sqrt{2m_\alpha}\right)\frac{Z}{\sqrt{T_\alpha}}}e^{-8\sqrt{\frac{\alpha m_\alpha R_N}{\hbar c}}\sqrt{Z}} \tag{4.58}$$

Because of the $T_\alpha^{-1/2}$ dependence the most energetic α particles will be associated with α decays with short half-lives. This can be seen in Fig. 4.8 which shows the very obvious tendency of the half-lives of α emitting isotopes of polonium to decrease with increasing α particle energy.

In 1911, Geiger and Nuttall reported an empirical relationship between the range of an α particle in air and the decay constant of the nuclide emitting the α particle,

$$\ln\lambda = k_1 + k_2\Re_\alpha \tag{4.59}$$

where $\Re_\alpha$ is the range of the α particle in air. For α particles with kinetic energies between 4 and 15 MeV (a range which covers most α emissions), an empirical relationship between α particle kinetic energy and its range in air is (Tsoulfanidis 1995),

$$\Re_{air} \approx (2.85 + 0.05T_\alpha)T_\alpha^{3/2} \tag{4.60}$$

where the range is in mm and the kinetic energy is in MeV. As $0.05T_\alpha > 2.85$ for the kinetic energies of interest, (4.59) can be rewritten in the form $\ln\lambda = k_1 + k_2T_\alpha^{3/2}$ leading to the Geiger–Nuttall rule relating α emission half-life to the α particle kinetic energy,

$$T_{1/2} = k_1 e^{-k_2T_\alpha^{3/2}} \tag{4.61}$$

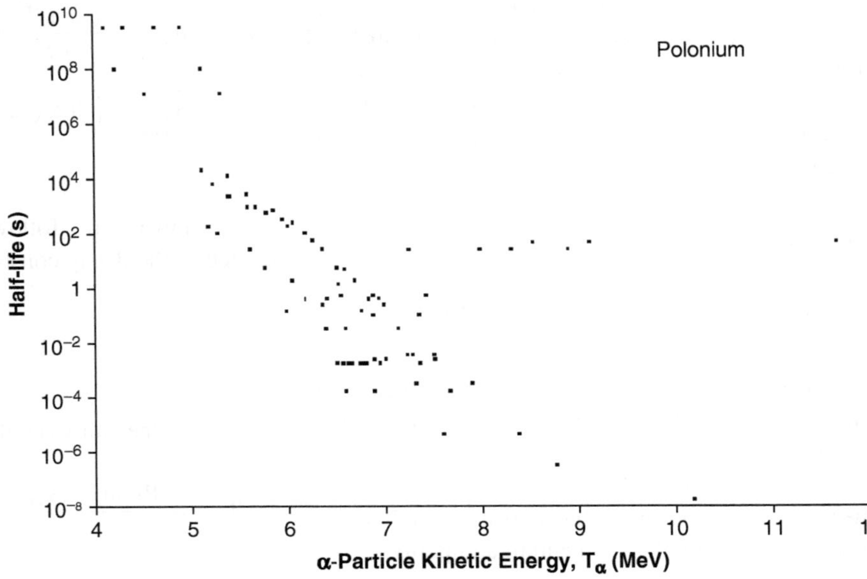

Fig. 4.8 Plot of polonium α emission half-life as a function of α particle kinetic energy. Points indicating multiple α emissions of differing kinetic energies but same half-life reflect multiple α emissions per given decay, as discussed in the text. Plot derived from data from the National Nuclear Data Center database (2008)

(note that the constants k_1 and k_2 are not the same throughout these equations). Although the empirical Geiger–Nuttall rule has $\ln T_{1/2} \propto T_\alpha^{-3/2}$ whereas the result derived from Gamow theory has $\ln T_{1/2} \propto T_\alpha^{-1/2}$, both demonstrate an association between α particles with high kinetic energies and short half-lives of the α emitters that gave rise to them.

4.3 The Weak Interaction: β Decay and Electron Capture

4.3.1 Introduction

β decay is both the most apparent manifestation of the weak force and is greatly important in nuclear medicine. Hence, it, and the weak force itself, are well-deserving of the detail provided in this section.

Early experimental studies of β decay showed that the detected final state contained only the electron and the recoiling daughter nucleus. Such a two-body decay would have required the electron to have a single well-defined kinetic energy equal to the difference between the mass of the parent nucleus and the sum of the masses of the daughter nucleus and electron (excluding any excitation energy taken up by the daughter). Using photographic emulsion as a detector, Hahn and Meitner demonstrated around 1906 that electron energy spectra associated with β decay from isotopes of thorium and actinium showed discrete peaks superimposed upon a weak, continuous background. While a well-defined electron energy was expected from the presumed two-body decay, the peaks observed were, in fact, the result of IC (a two-body final state) and the weak background was due to the electrons produced in β decay. Using a Geiger counter as a more sensitive ionization detector, Chadwick demonstrated in 1914 that the electron energy spectrum was continuous, contrary to the expectation of a two-body decay. This contradiction between the observed final state and the expected energy spectrum was problematic as the missing energy implied a failure of the law of conservation of energy. Experimental evidence demonstrated that the measured electron kinetic energy spectrum ranged in value from 0 to a value equal, to within experimental error, to the maximum dictated by a two-body final state. For example, this could be found by comparing the masses of ^3H and ^3He in the β^- decay of ^3H for which the mass-energy balance of $M_{^3H} = M_{^3He} + m_e + Q$ is achieved exactly by setting Q to the observed maximum electron energy in the spectrum. Another means of demonstrating this was shown by comparing the energies released in the decay of a nucleus through alternative routes (e.g., β decay followed by α decay and vice versa) to a final nucleus. As the initial and final nuclei in the decay schemes are the same, the total energies released though all of the routes must also be the same. It was found that the sum of the α particle kinetic energies and the upper limits of the β particle energy spectra were equal in both decay branches.

Experimental data also implied the nonconservation of both angular momentum and the spin statistics of a composite system. Again, consider the β decay of ^{3}H in which there are only two observed products in the final state, the ^{3}He nucleus and the electron. Initially excluding the possibility of a non-zero orbital angular momentum, the ^{3}He angular momentum (due to the single proton in the $1p_{1/2}$ level) and the electron spin couple to form a system with a total angular momentum of either 0 or 1. In either case, the total angular momentum is an integer. Even if we were to now allow the possibility of an orbital angular momentum contribution to the final state, the total angular momentum of the final state remains an integer. The problem lies in that the angular momentum of ^{3}H in the initial state is 1/2. Thus, the ^{3}H initial state is subject to Fermi-Dirac statistics whereas the ^{3}He-e composite in the final state is subject to Bose–Einstein statistics.

These dilemmas were famously resolved by Pauli. Unable to an invitation to a meeting at Tübingen, on 4 December 1930 he wrote a letter to those attending describing his proposition:

"Dear Radioactive Ladies and Gentlemen,

As the bearer of these lines, to whom I graciously ask you to listen, will explain to you in more detail, how because of the "wrong" statistics of the N and Li^6 nuclei and the continuous beta spectrum, I have hit upon a desperate remedy to save the "exchange theorem" of statistics and the law of conservation of energy. Namely, the possibility that there could exist in the nuclei electrically-neutral particles, that I wish to call neutrons, which have spin 1/2 and obey the exclusion principle and which further differ from light quanta in that they do not travel with the velocity of light. The mass of the neutrons should be of the same order of magnitude as the electron mass and in any event not larger than 0.01 proton masses. The continuous beta spectrum would then become understandable by the assumption that in beta decay a neutron is emitted in addition to the electron such that the sum of the energies of the neutron and the electron is constant...

I agree that my remedy could seem incredible because one should have seen these neutrons much earlier if they really exist. But only the one who dare can win and the difficult situation, due to the continuous structure of the beta spectrum, is lighted by a remark of my honored predecessor, Mr Debye, who told me recently in Bruxelles: "Oh, It's well better not to think about this at all, like new taxes." From now on, every solution to the issue must be discussed. Thus, dear radioactive people, look and judge.

Unfortunately, I cannot appear in Tubingen personally since I am indispensable here in Zurich because of a ball on the night of 6/7 December. With my best regards to you, and also to Mr Back.

Your humble servant,

W. Pauli"

Following Chadwick's discovery of what is now known as the neutron (as a nucleon), Fermi later coined the word *neutrino* (ν_e)[10] to describe Pauli's proposed particle which carried away the "missing" kinetic energy.

While a succinct solution to the empirical problem, it was immediately obvious from the lack of direct evidence for its existence that the neutrino interacted very weakly with matter. Despite this, further consideration of the experimental data at these early times led to the exposition of other properties of the neutrino:

- Mass: As the maximum of the electron/positron energy spectrum was, to within experimental error, equal to the maximum set by two-body decay, the β decay kinematics demonstrated that the neutrino mass was very small or zero.[11]
- Electric charge: The observed conservation of electric charge in β decay demonstrated that the neutrino had no electric charge.
- Spin: Conservation of angular momentum requires the neutrino to be a fermion as indicated in our earlier consideration of the decay of ^{3}He.

It was not until the early 1950s that the existence of the neutrino or, to be exact, the antineutrino was confirmed experimentally. Using the antineutrino flux

[10]The e subscript signifies that the neutrino produced by *β* decay is associated with an electron/positron.

[11]Only in recent years has conclusive evidence for a non-zero neutrino mass been obtained from measurements of neutrinos of cosmic origin (see, e.g., Fukuda et al. 1998).

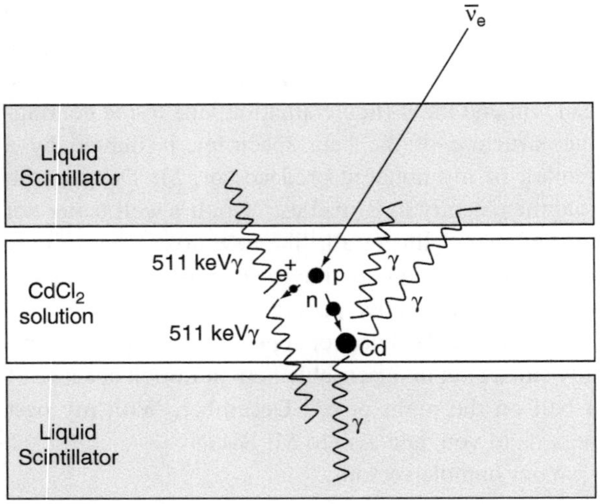

Fig. 4.9 Schematic diagram of the experimental arrangement used by Reines and Cowan to detect the existence of the electron-antineutrino. A description of the detection process is provided in the text

produced by the β decay of fission products in a nuclear reactor, Reines and Cowan (1953) (Reines 1997) detected the positron and neutron resulting from the "inverse β^+ decay" reaction of[12] $\bar{\nu}_e + p \rightarrow n + e^+$. The target consisted of $CdCl_2$ dissolved in water and sandwiched between two tanks containing liquid scintillator each viewed by photomultiplier tubes, as shown schematically in Fig. 4.9. An observed event was of an antineutrino captured by a proton of the hydrogen atom in a target water molecule to generate a positron and a neutron. The positron had a small mean free pathlength in water before it annihilated on a free electron ($e^-e^+ \rightarrow 2\gamma$) within about 1 ns of creation to produce two 511 keV γ rays which were subsequently detected in coincidence by the surrounding liquid scintillator. The neutron product of the inverse β^+decay thermalized through elastic collisions with the protons in the water molecules so as to be eventually captured by a cadmium nucleus some 10 μs later. This "radiative capture" by the cadmium resulted in an excited cadmium nucleus which, during de-excitation, emitted γ rays also detected by the surrounding liquid scintillator. The time signature of the two separate γ ray detections signaled an antineutrino detection.

The history of β decay, and the weak interaction it demonstrates is concluded by considering briefly the complexity of the lepton family beyond the electron and neutrino that resulted from the discovery of the muon in cosmic rays in 1937 by Neddermeyer and Anderson (1937)[13] and the determination that it was not subject to the strong nuclear force and was, hence, a lepton by Conversi et al. (1947). This was the first indication of the existence of the Generations of matter shown in Fig. 3.1 in Chap. 3 and of lepton flavor. Due to the leptonic similarities of the muon with an electron, it was of interest to determine if it also had an associate neutrino. That the muon neutrino existed was demonstrated by Lederman, Schwartz and Steinberger (Danby et al. 1962) in an experiment at Brookhaven National Laboratory in which protons bombarded a beryllium target to produce pions following which each decayed in flight to a muon and a neutrino (e.g., $\pi^+ \rightarrow \mu^+ + \nu_\mu$; the μ subscript indicates that the neutrino is associated with a muon and, in analogy to the electron and positron, the positively-charged muon is regarded as the antiparticle). The combined beam of muons and neutrinos then passed through a 13.5-m thick steel absorber so that only the neutrinos were able to pass through to be detected by a spark chamber on the exit side. Whereas the electron is a stable particle,[14] the muon is not. It decays, with a half-life of 2.197 μs, via $\mu^- \rightarrow e^- + \nu_e + \bar{\nu}_e$. However, the simpler decay process of $\mu^- \rightarrow e^- + \gamma$ has not been observed which empirically indicates the existence of the lepton quantum number. The existence of a charged lepton of the third flavor[15] was shown by Perl et al. (1975) through the topology of the trajectories of the products resulting from collisions between high-energy electron and positron beams leading to $e^-e^+ \rightarrow \tau^-\tau^+ \rightarrow e^\pm + e^\mp + 4\,\text{neutrinos}$. The mean τ-lepton life is 290.6 fs and the existence of a neutrino associated with the τ-lepton was finally

[12]As the total cross-section for this reaction had been estimated from the measured half-life of free neutron decay to be about 6×10^{-44} cm^{-2}, an intense antineutrino flux such as that from a fission reactor was required in order for a detectable number of events to be obtained.

[13]The muon was initially called the mu -mesotron and incorrectly thought to have been the intermediary of the strong force postulated by Yukawa. That particle is the pion which is subject to the strong nuclear force which the muon is not.

[14]The electron has a measured lifetime in excess of 4.6×10^{26} yrs (90% confidence level) (Particle Data Group 2004).

[15]This heavy lepton is referred to as the τ-lepton referring it to being the third lepton discovered.

Table 4.1 Lepton flavors

Electron flavor	
Lepton	Mass (MeV)[a]
e	0.510999
ν_e	<4.6 × 10^{-4} (68% c.l.)
Muon flavor	
Lepton	Lepton
μ	105.658369
ν_μ	<0.19 (90% c.l.)
Tau-lepton flavor	
Lepton	Mass (MeV)[a]
τ	1776.99
ν_τ	<18.2 (95% c.l.)

[a]Masses are averages taken from the Review of Particle Physics (Particle Data Group 2004); neutrino masses are provided with specified confidence levels

confirmed experimentally in 2000 (Kodama et al. 2001). The family of leptons is summarized in Table 4.1.

Within a modern viewpoint, β decay and electron capture can be considered at three different scales of dimension. At the lowest spatial resolution, they can be regarded as transitions between nuclear isobars which change the atomic number of the parent nucleus by unity:

$$(Z, A) \rightarrow (Z+1, A) + e^- + \bar{\nu}_e$$

$$(Z, A) \rightarrow (Z-1, A) + e^+ + \nu_e$$

$$(Z, A) + e^- \rightarrow (Z-1, A) + \nu_e$$

Going beyond the nuclear picture by increasing the spatial resolution to nucleon dimensions, these processes are seen as isospin-projection changing transitions between nucleons:

$$n \rightarrow p + e^- + \bar{\nu}_e$$

$$p \rightarrow n + e^+ + \nu_e$$

$$p + e^- \rightarrow n + \nu_e$$

Finally, at the spatial resolution of sub-nucleon dimensions, the transitions are now seen as a quark flavor change with the emission of a $W^{\pm}$ intermediate vector boson (so called as it has spin 1) which subsequently decays into a lepton pair,

$$d \rightarrow u + W^- \rightarrow u + e^- + \bar{\nu}_e$$

$$u \rightarrow d + W^+ \rightarrow d + e^+ + \nu_e$$

$$u + e^- \rightarrow d + W^+ + e^- \rightarrow d + \nu_e$$

as shown in Fig. 4.10.

As described in the introduction, the small interaction cross-section of the neutrino, the long physical half-lives of nuclei subject to β decay compared to γ decay and the change in nuclear isospin (through the isospin flips of n → p and p → n and the underlying quark flavor changes) led to the postulate that β decay was indeed a manifestation of a new force, termed the weak force. The electron and neutrino are leptons with the electron interacting through both electromagnetic and weak forces and the electrically-neutral neutrino interacting through the weak force alone. Both particles have antiparticles, the positron and antineutrino, the positron having been predicted by Dirac and experimentally detected in cosmic rays by Anderson (1933). Isospin and flavor, which are manifestations of the strong nuclear force, are not conserved by this weak force.

β decay is the most commonly observed weak interaction and is of significance to internal radiation dosimetry for several reasons:

- The daughter nucleus resulting from β decay or EC is often in an excited state which subsequently de-excites by γ ray emission or other radioactive processes. The γ ray emission can be directly useful for nuclear medicine imaging (e.g., ^{99}Mo undergoing β decay to ^{99m}Tc which makes an isomeric transition (IT) to ^{99}Tc through the emission of a 140.5 keV γ ray). For heavy nuclei, these other processes include fission and delayed nucleon emission as shown in Fig. 4.11.
- While the electron resulting from β^- decay does not contribute to the nuclear medicine imaging process,[16] it does deposit energy in tissue while decelerating, creating an absorbed dose.
- The positron created in β^+ decay will also deposit energy as it traverses tissue to annihilate with an electron to produce two collinear 511 keV γ rays

[16]Excluding imaging of the *bremsstrahlung* radiation produced by the decelerating electron.

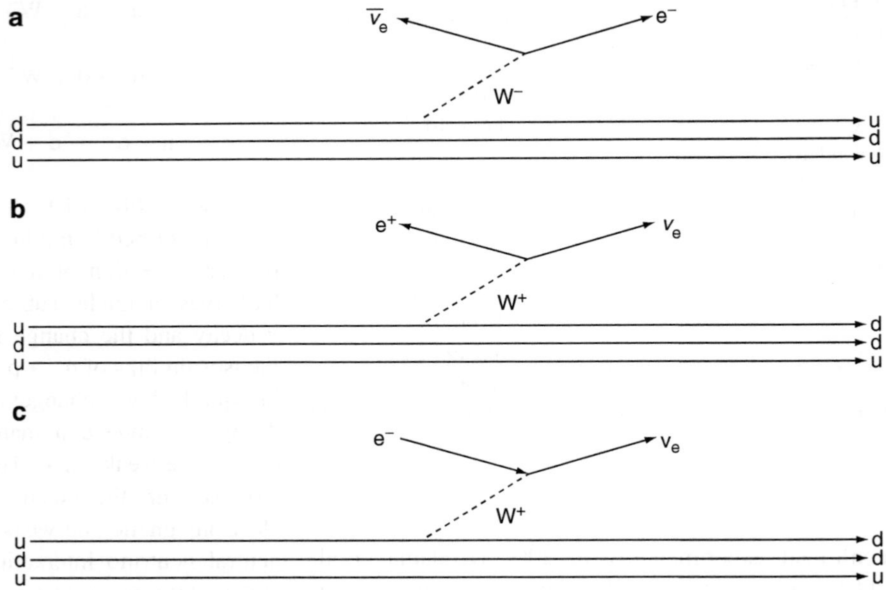

Fig. 4.10 Quark-boson-lepton diagrams for (**a**) β^- decay of a neutron; (**b**) β^+ decay of a proton; and (**c**) electron capture by a proton

which can be detected externally to form a tomographic image.

- Electron capture leaves an atomic electron vacancy inducing fluorescence X-ray production (which can also be detected to form an image) or Auger/Coster–Kronig electron cascades.

4.3.2 Kinematics of β Decay and Electron Capture

4.3.2.1 Neutron β Decay

The study here of β decay kinematics begins by considering free neutron decay, $n \rightarrow p + e^- + \bar{\nu}_e$. The mass-energy balance of a free neutron undergoing β^- decay is,

$$m_n = m_p + m_e + Q_- \tag{4.62}$$

Where the "–" subscript on Q signifies this as the energy released in β^- decay. Throughout this discussion, the masses of the (anti)neutrinos are neglected and, upon substituting the appropriate rest masses, it is found that free neutron decay is exoergic with $Q_- = 0.782$ MeV. Thus, a free neutron is unstable to β^- decay to a proton with a half-life of 885.7 s and the

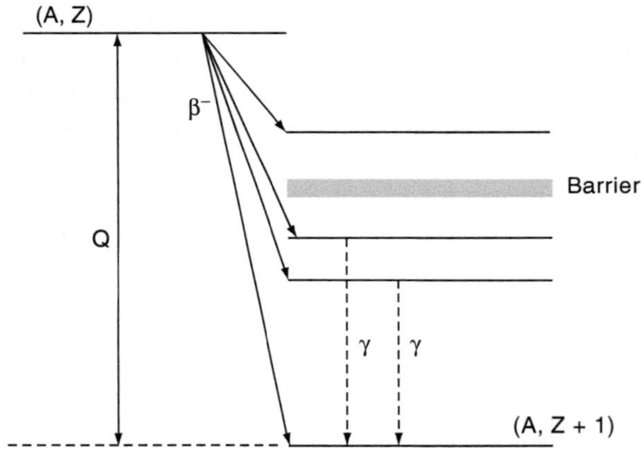

Fig. 4.11 Representative decay scheme for β^- decay with transitions to the ground state and various excited states of the daughter nucleus. The barrier signifies the minimum energy required in order for processes such as delayed nucleon emission and nuclear fission to occur

0.782 MeV of energy available will be distributed amongst the kinetic energies of the three decay products. By contrast, the mass-energy balance for a proposed β^+ decay of a free proton is,

$$m_p = m_n + m_e + Q_+ \tag{4.63}$$

Again, substituting the rest-mass values will show that this reaction is endoergic since $Q_+ = -1.765$ MeV and, hence, $p \rightarrow n + e^+ + \nu_e$ cannot occur for a free proton. It can occur only if the proton is bound within the nucleus such that the additional energy required for the transition arises from the nuclear binding energy.

4.3.2.2 Nuclear β Decay

Introduction

Now consider nuclear β decay between isobars (Grotz and Klapdor 1990). Such β decays can only occur if there is an adjacent isobar with a smaller mass (i.e., greater binding energy). Figure 4.12 presents a summary of the kinematics of nuclear β decay and electron capture. Recall from the discussion of the Segrè plot that, with increasing atomic mass number A in stable nuclei, the number of neutrons N increases relative to the number of protons, Z. An unstable nucleus (i.e., one which is displaced from the locus of stable nuclei in the Segrè plot) can be created either through the bombardment of a stable nucleus with charged particles or neutrons or through the fission of a heavy nucleus. Such an unstable nucleus will lie on either side of the stable nuclei locus. The type of β decay available to the displaced nucleus will be dictated by the magnitudes of the Coulomb and symmetry terms of the Weizsäcker formula and these will follow the mass parabolae.

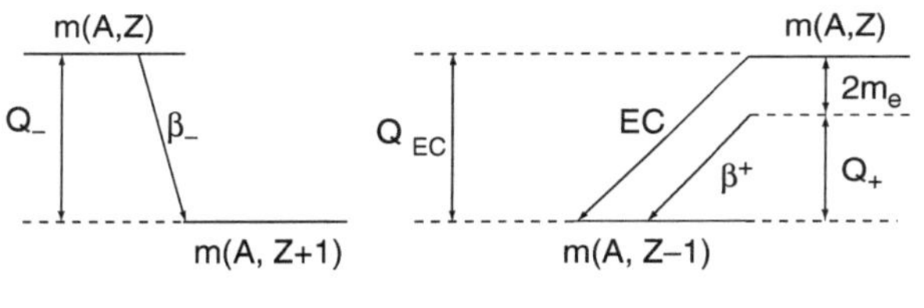

Fig. 4.12 Kinematics of β decay and electron capture

The Coulomb term reduces the net binding energy with increasing Z whereas the symmetry term increases the binding energy with increasing Z. As seen, the result is a maximum in the binding energy indicating the stable nucleus in the mass parabola (which would lie along the isobar line of Fig. 4.13). If the artificially produced nucleus were to be to the left of the locus of stable nuclei (i.e., it has an excess of neutrons or is "neutron-rich"), it has a deficit of electric charge for its mass. Hence, the Coulomb term is reduced and the symmetry term increased. A maximum in the binding energy is then achieved by an increase in the former and a decrease in the latter which results from β^- decay. Should the nucleus instead be to the right of the locus of stable nuclei (i.e., it has an excess of protons or is "proton-rich"), the relative magnitudes of the Coulomb and symmetry terms are reversed and a maximum in the binding energy is achieved by decreasing the magnitude of the Coulomb term and increasing that of the symmetry term through β^+ decay or electron capture.

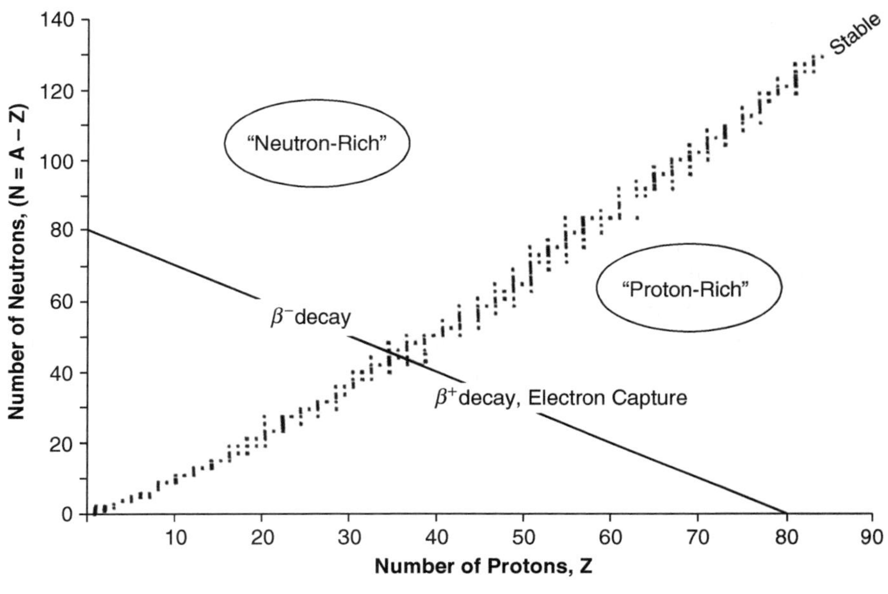

Fig. 4.13 Segrè plot showing the "neutron-rich" region, where β^- decay can occur, and the "proton-rich" region, where β^+ decay or electron capture can occur, are shown. The isobar line of A = 80 is provided as an example to show the directions of these transitions towards the locus of stable nuclei

The mass difference required between parent and daughter nuclei is next calculated in order to determine whether or not nuclear β decay or electron capture is energetically possible.

Nuclear β^- Decay

Nuclear β^- decay between the isobar pair ${}^{A}_{Z}X$ and ${}^{A}_{Z+1}Y$ is,

$$ {}^{A}_{Z}X \rightarrow {}^{A}_{Z+1}Y + e^- + \bar{\nu}_e \quad (4.64) $$

and which has the nuclear mass-energy balance,[17]

$$ m(A, Z) = m(A, Z+1) + m_e + Q_-. \quad (4.65) $$

As the nuclear mass is $m(A,Z) = Zm_p + (A-Z)m_n - B(A,Z)$, (4.65) gives the Q_- value for nuclear β^- decay as,

$$ \begin{aligned} Q_- = {} & (m_n - m_p - m_e) \\ & + (B(A, Z+1) - B(A, Z)) \end{aligned} \quad (4.66) $$

where the first term in parentheses in (4.66) is the Q_- for free neutron decay and the second term is the difference in nuclear binding energies. A simpler expression for Q_- can be extracted from (4.66) by recalling that the nuclear binding energy can be written in terms of atomic masses, $B(A,Z) = ZM_H + (A-Z)m_n - M(A,Z)$, or,

$$ Q_- = M(A, Z) - M(A, Z+1). \quad (4.67) $$

As Q_- must be positive, nuclear β^- decay is energetically permissible should the parent's atomic mass exceeds that of the daughter. An example of β^- decay shown in Fig. 4.14 is that of the radionuclide ^{32}P, which is produced through neutron bombardment of stable ^{31}P, and undergoes β^- decay to ^{32}S with a half-life of 14.263 days and a Q_- of 1.71 MeV.

[17] Note that any excitation energy conferred to the daughter nucleus is ignored.

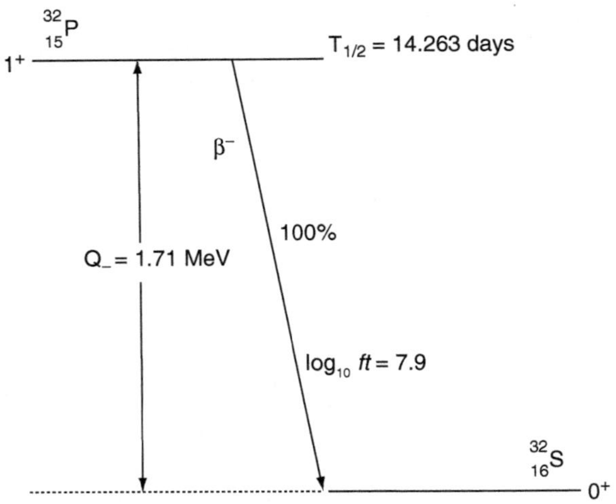

Fig. 4.14 $^{32}P \rightarrow {}^{32}S + e^- + \bar{\nu}_e$. The $\log_{10} ft$ value is a measure of the transition rate and is discussed in the text

Nuclear β^+ Decay

β^+ decay is inherently nuclear as the β^+ decay of a free proton is energetically impossible. A nucleus in the proton-rich region of Fig. 4.13 will have excessive electric charge for its mass and will seek to reduce this charge by either positron emission or electron capture. The β^+ decay between the isobar pair ${}^{A}_{Z}X$ and ${}^{Y}_{Z+1}Y$ is,

$$ {}^{A}_{Z}X \rightarrow {}^{A}_{Z-1}Y + e^+ + \nu_e \quad (4.68) $$

with a nuclear mass-energy balance,

$$ m(A, Z) = m(A, Z-1) + m_e + Q_+. \quad (4.69) $$

Substituting the expression for nuclear mass gives,

$$ \begin{aligned} Q_+ = {} & -(m_n - m_p - m_e) - 2m_e \\ & - (B(A, Z-1) - B(A, Z)) \end{aligned} \quad (4.70) $$

which, in terms of parent and daughter nuclear masses, is,

$$ Q_+ = (M(A, Z) - M(A, Z-1)) - 2m_e. \quad (4.71) $$

Thus, in order for β^+ decay to occur, the parent atom must be heavier than the daughter nucleus by an amount of at least double the electron/positron rest-mass of $2m_e$ (1.022 MeV). An example is shown in

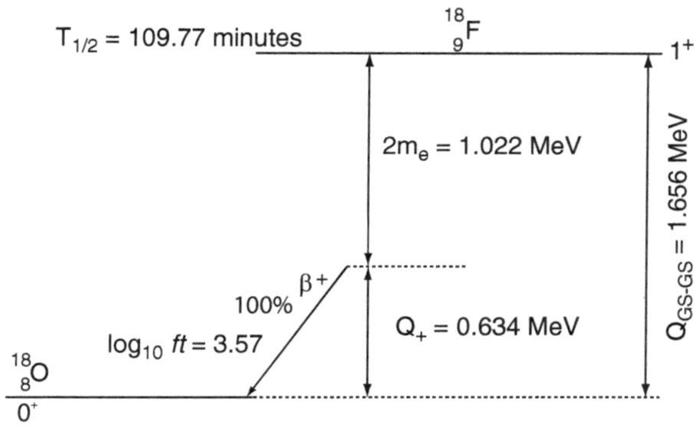

Fig. 4.15 $^{18}F \rightarrow {}^{18}O + e^{+} + \nu_e$

Fig. 4.15 of ^{18}F which undergoes β^+-decay with a 109.77 min half-life to produce ^{18}O.

Electron Capture

The process of electron capture can compete with β^+ decay as the wavefunction of an orbital electron, especially that of the K-orbital, will have a finite extension into the nuclear volume.[18] As a consequence, it is possible for a nucleus to capture an orbital electron through $e^- + Z \rightarrow (Z-1) + \nu_e$, a reaction which occurs preferentially for high-A nuclei which, with their large nuclear radii, will have greater overlaps with the orbital electrons' wavefunctions. The EC process is,

$$ {}^{A}_{Z}X + e^- \rightarrow {}^{A}_{Z-1}Y + \nu_e \quad (4.72)$$

with the nuclear mass-energy balance,

$$m(A,Z) + m_e - B_K = m(A,Z-1) + Q_{EC}. \quad (4.73)$$

The capture of a K-shell electron with B_K binding energy has been assumed. Solving for Q_{EC} in terms of nuclear binding energies,

$$\begin{aligned} Q_{EC} = &-(m_n - m_p - m_e) - B_K \\ &+ (B(A,Z-1) - B(A,Z)) \end{aligned} \quad (4.74)$$

and of atomic masses,

$$Q_{EC} = (M(A,Z) - M(A,Z-1)) - B_K \quad (4.75)$$

Hence, EC is energetically feasible if the parent's atomic mass exceeds the sum of that of the daughter and the electron binding energy (which, in practice, can be neglected). Comparing Q_{EC} and Q_+,

$$Q_+ = Q_{EC} - 2m_e \quad (4.76)$$

Because EC does not produce a positron and the neutrino cannot be readily detected, experimental evidence for EC is obtained indirectly via the emission of a characteristic X-ray daughter nucleus (Alvarez 1937). EC creates a hole in the atomic energy orbital which is subsequently filled by an electron from a higher-order orbital to produce an X-ray with energy equal to the difference between the binding energies of the two orbitals. Although β^+ decay of a proton-rich nucleus yields the same daughter nucleus as that of electron capture, the former cannot occur unless there is a mass-energy differential of 1.022 MeV between the parent and daughter atoms, whereas the threshold for the latter simply requires that the parent atom be heavier than the daughter (if the magnitude of the orbital electron binding energy is neglected). Hence, if the mass-energy difference between the parent and daughter atoms for which EC occurs exceeds 1.022 MeV, β^+ decay can become possible, in which the total transition probability is the sum of the two individual probabilities for each transition type.

Summary of β Decay and Electron Capture Kinematics

The thresholds for β decay and electron capture in terms of the parent (X) and daughter (Y) atomic masses are,

$$M_X > M_Y \qquad \beta^-, \text{EC}$$

$$M_X > M_Y + 2m_e \qquad \beta^+$$

Table 4.2 presents Q values for a variety of β decays of particular interest to diagnostic nuclear medicine and PET imaging.

[18]The probabilities of L and higher orbital wavefunctions extending into the nucleus are much smaller and are not considered here.

Table 4.2 Q values for various β decays of nuclear medicine interest

Parent	Daughter	Decay mode (% of all decays)	Q (MeV)	Applications
^{11}C	^{11}B	β^+ (99.7%)	0.961	PET imaging
		EC (0.3%)	1.983	
^{14}C	^{14}N	β^-	0.156	^{14}C-labelled urea for *Helicobacter pyloris* diagnosis
^{15}O	^{15}N	β^+ (99.9%)	1.732	PET imaging
		EC (0.1%)	2.754	
^{18}F	^{18}O	β^+ (96.9%)	0.634	PET imaging
		EC (3.1%)	1.656	
^{51}Cr	^{51}V	EC	0.752	$Na_2{}^{51}CrO_4$ used to label red blood cells (RBC) for measurement of RBC volume or RBC sequestration
^{59}Fe	^{59}Co	β^-	1.565	^{59}Fe ferrous citrate used to assess gastrointestinal iron absorption by oral administration or of iron metabolism by intravenous injection
^{64}Cu	^{64}Ni	β^+ (18%)	0.653	PET imaging
	^{64}Ni	EC (44.9%)	1.675	
	^{64}Zn	β^- (37.1%)	0.578	
^{67}Ga	^{67}Zn	EC (100%)	1.000	^{67}Ga citrate used for imaging tumors and inflammation with γ-ray scintigraphy
^{68}Ga	^{68}Zn	β^+ (89.1%)	1.899	PET Imaging
		EC (10.9%)	2.921	
^{90}Y	^{90}Zr	β^- (100%)	2.280	^{90}Y-labelled *ibritumomab tiuxetan* for treatment of non-Hodgkin'slymphoma

4.3.3 Fermi Theory of β Decay: Part I

4.3.3.1 Introduction

Having established the kinematic requirements to be fulfilled in order for β decay and electron capture to occur, we now progress to calculating the transition probabilities and decay rates of these processes. Although dating from the 1930s, Fermi's theory (1934a, b) remains an inspired and pragmatic approach to understanding β decay and electron capture, evaluating their transition probabilities and calculating the electron and positron energy spectra. The insight of Fermi's theory was its being an analogy to that describing the interaction between charges in quantum electrodynamics.[19] Even though it has been supplanted by the modern Glashow–Salam–Weinberg electroweak theory, the Fermi theory provides the low-energy limit of that theory which is appropriate and useful for our study of β decay.

4.3.3.2 Nuclear β Decay

Introduction

In this subsection, the transition rates for β decay and electron capture are derived from perturbation theory using Fermi's Golden Rule No 2,[20]

$$\lambda_{\mathrm{fi}} = \frac{2\pi}{\hbar}\left|\int d^3\mathbf{r}\,\psi_{\mathrm{f}}^{*}U\psi_{\mathrm{i}}\right|^2 \rho_{\mathrm{f}} \tag{4.77}$$

Matrix Element

The matrix element, $\int d^3\mathbf{r}\,\psi_{\mathrm{f}}^{*}U\psi_{\mathrm{i}}$, is calculated by first noting that the initial wavefunction is that of the parent

[19] Fermi's paper on his β decay theory was, famously, rejected by *Nature* in 1933 but was subsequently published in Italian in *Nuovo Cimento* and in German in *Zeitschrift für Physik*. Fermi never again published on this subject (Segrè 1970).

[20] For simplicity in this discussion, "electron" will refer to either an electron or a positron and "neutrino" will refer to either a neutrino or antineutrino. An exact assignment will be apparent from the context of the discussion.

nucleus' wavefunction $\psi_i \equiv \psi_{nuc,i}$, and the final wavefunction ψ_f is the product of the three wavefunctions corresponding to the daughter nucleus, electron and neutrino, $\psi_f \equiv \psi_{nuc,f}\psi_e\psi_{\nu_e}$. The lepton wavefunctions are represented by plane waves,

$$\psi_e^*\psi_{\nu_e} = \frac{1}{L^3}e^{i\frac{\mathbf{q}\bullet\mathbf{r}}{\hbar c}} \tag{4.78}$$

where L^3 is the volume containing the system and $\mathbf{q} = \mathbf{p}_{\nu_e} - \mathbf{p}_e$. Because of the distortion induced by the nuclear Coulomb field, a plane wave expansion for the electron is not entirely valid, but its use will corrected for later. Expanding the exponential,

$$e^{i\frac{\mathbf{q}\bullet\mathbf{r}}{\hbar c}} = \sqrt{4\pi}\sum_{l=0}^{\infty}\sqrt{2l+1}\; j_l\left(\frac{qr}{\hbar c}\right)Y_{l0}(\theta,0) \tag{4.79}$$

where θ is the angle between $\mathbf{q}$ and $\mathbf{r}$. As the lepton momenta are of the order of 1–2 MeV/c and the nuclear dimension is of the order of 5 fm, then the argument of the spherical Bessel function is $^{qr}/_{\hbar c} \approx 0.05$ or less. In this case, the small-argument form of the spherical Bessel function, $j_l(x) \approx x^l/(2l+1)!!$, can be used where the double factorial is defined by $(2l+1)!! = 1\times 3\times 5\ldots(2l+1)$. Hence, the low-argument approximation to the spherical Bessel function diminishes rapidly with l for $x \ll 1$ and the expansion of (4.79) can be restricted to $l = 0$ to give $\psi_e^*\,\psi_{\nu_e} \approx 1/L^3$. Hence, a simplified result for the matrix element is obtained,

$$\int d^3\mathbf{r}\,\psi_f^*U\psi_i \approx \frac{1}{L^3}\int d^3\mathbf{r}\,\psi_{nuc,f}^*U\psi_{nuc,i}. \tag{4.80}$$

The perturbation potential inducing the β decay is taken to be weak and to have an infinitely small range[21] allowing the perturbation to be approximated by a scalar constant and removed from the integral,

$$\int d^3\mathbf{r}\,\psi_f^*U\psi_i \approx \frac{g}{L^3}\int d^3\mathbf{r}\,\psi_{nuc,f}^*\psi_{nuc,i} = \frac{g}{L^3}M_{fi} \tag{4.81}$$

[21]In the modern-day picture of the W intermediate vector boson-mediated β decay, the range of the force is of the order of 10^{-3} fm. Hence, Fermi's approximation of a zero-range force was insightful.

where M_{fi} is the nuclear matrix element. The β decay transition rate from initial state i to final state f can now be written as,

$$\lambda_{fi} = \frac{2\pi}{\hbar}\frac{g^2}{L^6}|M_{fi}|^2\rho_f \tag{4.82}$$

Phase Space Factor

From the previous derivation of β decay kinematics, the total kinetic energy available to the electron and neutrino is equal to the Q of the appropriate decay. Neglecting a neutrino rest mass, the total energy made available is the sum of the kinetic and rest-mass energies, $E_0 = Q_\pm + m_e$, which is also equal to the electron and neutrino total energies,

$$E_0 = E_e + E_{\nu_e} \tag{4.83}$$

Equation (4.82) can be generalized into the differential form,

$$d\lambda_{fi}(E_e) = \frac{2\pi}{\hbar}\frac{g^2}{L^6}|M_{fi}|^2\frac{d\rho_f(E_0,E_e)}{dE_e}dE_e \tag{4.84}$$

which is the differential probability that the electron will be emitted with a total energy between E_e and $E_e + dE_e$. As the volume of a unit cell in phase space is $(2\pi\hbar c)^3$ and the electron-antineutrino final state will have a 12-dimensional phase space (three-dimensions each for the electron and antineutrino momenta and position), the phase space differential volume element in electron and neutrino momentum is,

$$\begin{aligned} d^2\rho_f &= \left(\frac{L}{2\pi\hbar c}\right)^3 4\pi p_e^2\, dp_e \bullet \left(\frac{L}{2\pi\hbar c}\right)^3 \\ &\quad\times 4\pi p_{\nu_e}^2\, dp_{\nu_e}\,\delta(Q - T_e - T_{\nu_e}) \\ &= \frac{1}{4\pi^4}\left(\frac{L}{\hbar c}\right)^6 p_e^2\, dp_e\, p_{\nu_e}^2\, dp_{\nu_e}\,\delta(Q - T_e - T_{\nu_e}) \end{aligned} \tag{4.85}$$

where the δ-function specifies the required energy conservation. It is convenient to solve for this expression by using the particles' total energies as the variables instead. As the total energy is $E = \sqrt{p^2 + m^2}$, then $dE/dp = p/E$ and $p^2dp = E\sqrt{E^2 - m^2}dE$, then (4.85) can be written in the form,

$$d^2\rho_f = \frac{1}{4\pi^4}\left(\frac{L}{\hbar c}\right)^6 \times E_e\sqrt{E_e^2 - m_e^2}\, dE_e\, E_{\nu_e}^2 dE_{\nu_e} \delta(E_0 - E_e - E_{\nu_e}). \tag{4.86}$$

The arguments of the δ functions of (4.85) and (4.86) are equivalent. As the neutrino is not detected, this result is subsequently integrated over the neutrino energy,

$$\begin{aligned} d\rho_f &= \frac{1}{4\pi^4}\left(\frac{L}{\hbar c}\right)^6 E_e\sqrt{E_e^2 - m_e^2}\, E_e\, dE_e \\ &\times \int dE_{\nu_e} E_{\nu_e}^2 \delta(E_0 - E_e - E_{\nu_e}) \\ &= \frac{1}{4\pi^4}\left(\frac{L}{\hbar c}\right)^6 E_e\sqrt{E_e^2 - m_e^2}(E_0 - E_e)^2 dE_e. \end{aligned} \tag{4.87}$$

So far, this calculation has neglected the interaction between the product electron and the daughter nucleus. As noted earlier, this will distort the electron's wavefunction, thus invalidating the original assumption of describing its wavefunction as a plane. For β^- emission, the attractive Coulomb interaction slows down the electron and increases the low-energy portion of the spectrum. Thus, the electron is emitted with a higher energy than that detected or, in other words, the phase space is larger than $4\pi p_e^2 dp_e$. On the other hand, for β^+ emission, the Coulomb force between the positron and the nucleus is repulsive, thus suppressing the low-energy portion of the spectrum and correspondingly decreasing the available phase space. In order to account for these Coulomb effects, but still retain the overall form of the result obtained from a plane-wave approximation, a correction to the phase space is inserted into the phase space factor. This correction term is the Fermi factor, $F(Z_Y, E_e)$,

$$\frac{d\rho_f}{dE_e} = \frac{1}{4\pi^4}\left(\frac{L}{\hbar c}\right)^6 E_e\sqrt{E_e^2 - m_e^2}\,(E_0 - E_e)^2 \times F(Z_Y, E_e) \tag{4.88}$$

where Z_Y is the atomic number of the recoil daughter nucleus. The Fermi factor is the ratio of the electron/positron wavefunctions at the centre of the nucleus with the Coulomb interaction to that without,

$$F(Z_Y, E_e) = \frac{|\psi_{e,\text{with}}(0)|^2}{|\psi_{e,\text{without}}(0)|^2}. \tag{4.89}$$

A suitable nonrelativistic approximation of the Fermi factor is,

$$F(Z_Y, E_e) = \frac{2\pi\eta}{1 - e^{-2\pi\eta}} \tag{4.90}$$

where

$$\eta = \pm\alpha Z_Y \frac{E_0}{\sqrt{E_e^2 - m_e^2}} \tag{4.91}$$

and where the ± sign corresponds to electron/positron emission. The nonrelativistic Fermi factor is plotted in Fig. 4.16 as a function of the ratio $\frac{E_e}{E_0} = \frac{T_e + m_e}{Q + m_e}$ for four different β decays. By expanding the exponential term in (4.90) to second order, the features of Fig. 4.16 become more apparent,

$$\begin{aligned} F(Z_Y, E_e) &= \frac{1}{1 - \pi\eta} \\ &= \frac{p_e}{p_e \mp \mu\pi\alpha Z E_0} \\ &= \frac{\sqrt{\left(\frac{E_e}{E_0}\right)^2 - \left(\frac{m_e}{E_0}\right)^2}}{\sqrt{\left(\frac{E_e}{E_0}\right)^2 - \left(\frac{m_e}{E_0}\right)^2} \mp \pi\alpha Z} \end{aligned} \tag{4.92}$$

This result shows that $F(Z_Y, E_e) > 1$ for electrons and $F(Z_Y, E_e) < 1$ for positrons and that both factors approach unity with increasing energy. However, the nonrelativistic approximation fails for high-*Z* nuclei and the Fermi factor must then be calculated from the Dirac equation and an extended Coulomb potential. Tabulated values for $F(Z_Y, E_e)$ can be found in Behrens and Jähnecke (1969).

Energy Spectra

The transition rate can now be written as,

$$\frac{d\lambda_{fi}(E_e)}{dE_e}dE_e = \left(\frac{g^2}{2\pi^3\hbar(\hbar c)^6}\right)|M_{fi}|^2 E_e\sqrt{E_e^2 - m_e^2} \times (E_0 - E_e)^2 \times F(Z_Y, E_e)dE_e \tag{4.93}$$

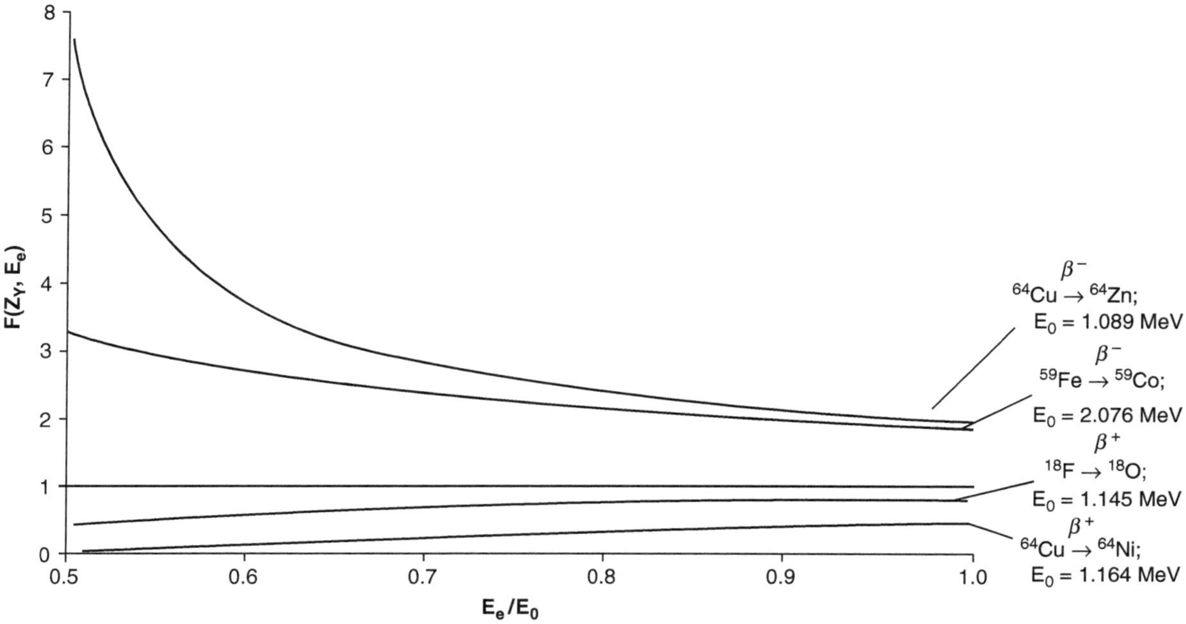

Fig. 4.16 Nuclear Coulomb correction factor used in the phase space factor for Fermi β decay theory to correct for electron/positron plane-wave assumption for two different types each of β^- and β^+ decays as a function of the total electron/positron energy (sum of kinetic and rest mass energies) normalized to the maximum total energy available from the decay kinematics. $F(Z_Y, E_e) = 1$ is shown as a *horizontal line*

As a result of the zeroth-order truncation of the lepton wavefunction expansions, the nuclear matrix element M_{fi} will be independent of the electron and neutrino momenta (those cases of when the expansion of the wavefunctions going beyond first-order and the nuclear matrix element no longer independent of the electron and neutrino momenta are considered later). The shape of the electron/positron energy spectrum is thus defined by the combination of the phase space of the emitted electron/positron and the Fermi factor. Now consider the example of ^{64}Cu, an interesting radionuclide used in a number of nuclear medicine applications and which undergoes all three types of β decay: β^- decay to ^{64}Zn and both β^+ decay and electron capture to ^{64}Ni, as shown in Fig. 4.17.

The ^{64}Cu electron and positron energy spectra are shown in Fig. 4.18 and the differences in the shapes of the spectra at low energies due to the differing electron-nuclear and positron-nuclear Coulomb interactions are readily apparent. The proportion of low-energy electrons exceeds that of low-energy positrons due to the nuclear Coulomb attraction and repulsion, respectively.

Kurie Plot

Additional information about β decay can be had by analyzing the algebraic rearrangement of (4.93),

$$\left(\frac{g^2}{2\pi^3\hbar(\hbar c)^6}\right)|M_{fi}|^2(E_0 - E_e)^2 = \left[\frac{\left(\frac{d\lambda_{fi}}{dE_e}\right)}{F(Z_Y, E_e)E_e\sqrt{E_e^2 - m_e^2}}\right]. \tag{4.94}$$

By taking the square-root of both sides, one can form the linearized equation,

$$K(E_e) \equiv C(E_0 - E_e) \tag{4.95}$$

where,

$$K(E_e) \equiv \sqrt{\frac{\left(\frac{d\lambda_{fi}}{dE_e}\right)}{F(Z_Y, E_e)E_e\sqrt{E_e^2 - m_e^2}}}. \tag{4.96}$$

Fig. 4.17 $\beta^{\pm}$ and EC decay schema for ^{64}Cu

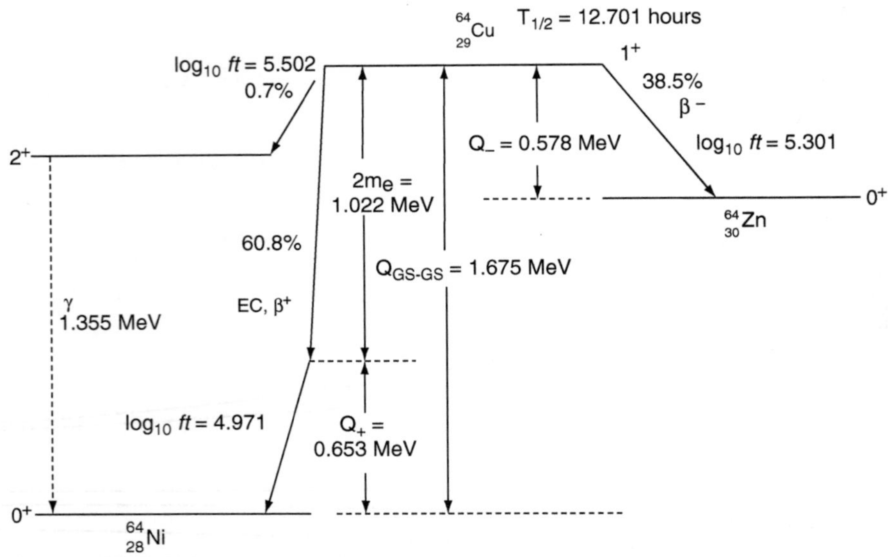

The constant is,

$$C = \sqrt{\left(\frac{g^2}{2\pi^3\hbar(\hbar c)^6}\right)|M_{fi}|^2}. \tag{4.97}$$

A plot of (4.96) vs. electron energy, known as the Kurie plot, can yield considerable information concerning a β decay, as shown schematically in Fig. 4.19. Linear extrapolation to $K(E_e) = 0$ yields $E_e = E_0$ or the Q of the decay for a zero neutrino mass. For a non-zero neutrino mass, the Kurie curve would become nonlinear at high electron energies. This is, in fact, one method used for determining the neutrino mass. Another cause for deviation from linearity arises from C being a constant as a result of, for example, the nuclear matrix element being dependent upon the electron energy. Recall that the independence of M_{fi} from E_e was due to the truncation of the electron wavefunction at zeroth-order due to the assumption of a point nucleus. When nuclear size is not neglected, higher orders of the plane wave expansion are required and the matrix element becomes dependent upon the electron momentum/energy. Such nonlinearities occur for forbidden transitions, so-called for their reduced transition rates.

Decay Constant

In order to calculate the decay constant, λ, it should be recognized that it will be for all final states and is thus obtained by integrating (4.93) over all possible β particle total energies,

$$\begin{aligned}
\lambda &= \int_{m_e}^{E_0} dE_e \frac{d\lambda_{fi}(E_e)}{dE_e} \\
&= \left(\frac{g^2}{2\pi^3\hbar(\hbar c)^6}\right)|M_{fi}|^2 \int_{m_e}^{E_0} dE_e E_e \sqrt{E_e^2 - m_e^2} \\
&\quad \times (E_0 - E_e)^2 F(Z_Y, E_e) \\
&= \left(\frac{g^2}{2\pi^3\hbar(\hbar c)^6}\right)|M_{fi}|^2 \int_{m_e}^{E_0} dE_e E_e \sqrt{E_e^2 - m_e^2} \\
&\quad \times (E_0 - E_e)^2 F(Z_Y, E_e).
\end{aligned} \tag{4.98}$$

Using the change of variable $x = E_e/m_e$, the transition rate can be rewritten as,

$$\begin{aligned}
\lambda &= \left(\frac{g^2 m_e^5}{2\pi^3\hbar(\hbar c)^6}\right)|M_{fi}|^2 \\
&\quad \times \int_1^{x_0} dx\, x\sqrt{x^2 - 1}(x_0 - x)^2 F(Z_Y, x) \\
&= \left(\frac{g^2 m_e^5}{2\pi^3\hbar(\hbar c)^6}\right)|M_{fi}|^2 f(Z_Y, x_0)
\end{aligned} \tag{4.99}$$

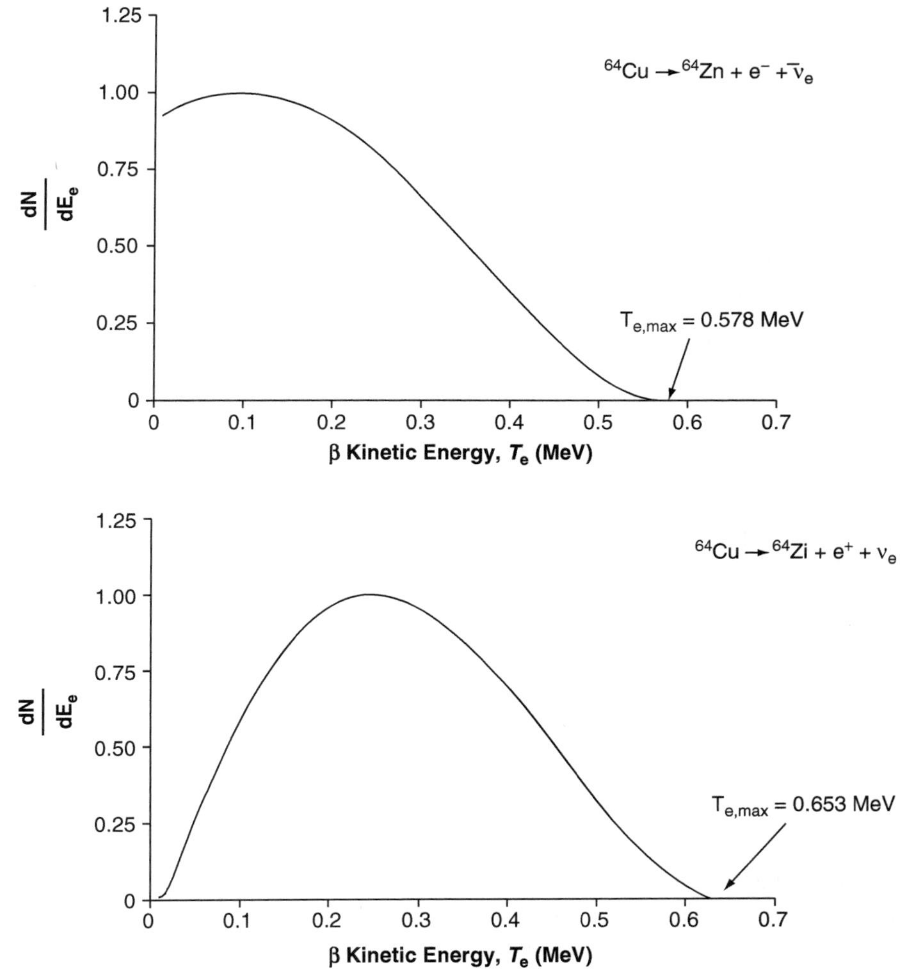

Fig. 4.18 Calculated electron and positron energy spectra for the β^- and β^+ decays of ^{64}Cu. Curves are not normalized

where

$$x_0 = \frac{E_0}{m_e} = 1 + \frac{Q}{m_e} \tag{4.100}$$

and where the Fermi integral is,

$$f(Z_Y, x_0) = \int_1^{x_0} dx\, x\sqrt{x^2 - 1}(x_0 - x)^2 F(Z_Y, x). \tag{4.101}$$

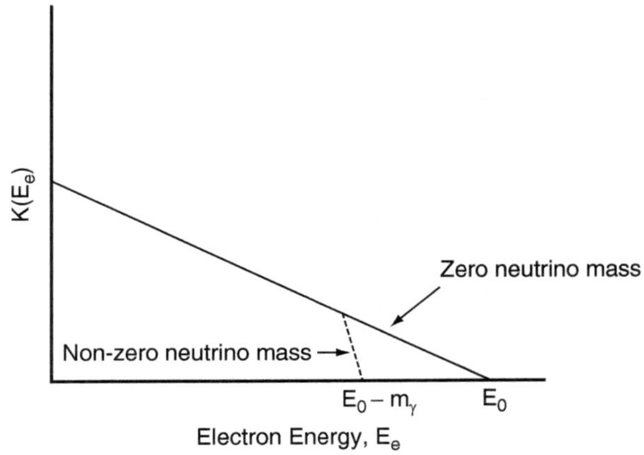

Fig. 4.19 Plot of Kurie function vs. electron energy for zero and non-zero neutrino masses

While this integral itself does not have an analytical solution, an approximation to it can be readily solved should the daughter nucleus have a sufficiently small Z_Y such that the Fermi factor is $F(Z_Y, x) \approx 1$,

$$
\begin{aligned}
f(Z_Y, x_0) &\approx \int_1^{x_0} dx\, x\sqrt{x^2-1}(x_0-x)^2 \\
&\approx x_0^2 \int_1^{x_0} dx\, x\sqrt{x^2-1} + \int_1^{x_0} dx\, x^3\sqrt{x^2-1} \\
&\quad - 2x_0 \int_1^{x_0} dx\, x^2\sqrt{x^2-1} \\
&\approx \frac{x_0^2}{3}(x_0^2-1)^{\frac{3}{2}} + \frac{1}{15}(x_0^2-1)^{\frac{3}{2}}(3x_0^2+2) \\
&\quad - \frac{x_0^2}{4}\sqrt{x_0^2-1}2(x_0^2-1) \\
&\approx \frac{\sqrt{x_0^2-1}}{60}(2x_0^4-9x_0^2-8).
\end{aligned}
\tag{4.102}
$$

From this result, it can be see that an approximate proportionality arises for the transition rate,

$$
\lambda \propto \frac{E_0^5}{30} \tag{4.103}
$$

which predicts a strong dependence of the transition rate upon the magnitude of the energy released by the decay and that the transition rate will increase markedly with released energy. Equation (4.103) is Sargent's rule (1933). This strong dependence of the transition rate upon the decay kinematics must be addressed if one wishes to compare different β-decays with varying kinematics. This dependence upon Q can be removed by defining the comparative half-life as the product of the physical half-life and the integral of (4.101),

$$
\begin{aligned}
ft &\equiv f(Z_Y, x_0)T_{1/2} \\
&= \frac{\ln 2}{\left(\frac{g^2 m_e^5}{2\pi^3\hbar(\hbar c)^6}\right)|M_{fi}|^2} \propto \frac{1}{|M_{fi}|^2}.
\end{aligned}
\tag{4.104}
$$

As values of ft range considerably in magnitude, it is conventional to work with the logarithm of the comparative half-life, $\log_{10}$ ft (where t is in units of seconds). This large variation of the comparative half-life is indicative of the dependence upon the nuclear wavefunctions, as shown by (4.104). The simplest β decays (generally involving low-Z nuclei) involve the greatest overlap of the initial- and final-state nuclear wavefunctions and, from (4.102), the smallest comparative half-life values. $\log_{10}$ ft values are provided in the β decay schema and in Table 4.3.

4.3.3.3 Electron Capture

While the decay constant for EC will be calculated in the same manner as that for β decay, two distinct differences between EC and β decay should be noted in the calculation of the matrix element and the phase space factor. First, the neutrino is the only lepton in the final state. Second, the electron is present in the initial state but, because it is an orbital atomic electron, it cannot be approximated by a plane-wave wavefunction. The EC transition rate is,

$$
\lambda_{fi,EC} \approx \frac{2\pi}{\hbar}\frac{1}{L^3}\left|\int d^3\mathbf{r}\psi^*_{nuc,f}\, U\psi_{nuc,i}\psi_e\right|^2 \rho_f \tag{4.105}
$$

where the plane-wave approximation has been applied to the neutrino only with, again, truncation to zeroth-order. As before, the potential is set equal to a coupling constant g within the nucleus and zero beyond, giving,

$$
\lambda_{fi,EC} \approx \frac{2\pi}{\hbar}\frac{g^2}{L^3}\left|\int d^3\mathbf{r}\psi^*_{nuc,f}\, \psi_{nuc,i}\, \psi_e\right|^2 \rho_f. \tag{4.106}
$$

As lower orbitals predominate in electron capture due to their closer proximity to the nucleus and the resultant greater degree of overlap between the nuclear and electronic wavefunctions, we calculate $\lambda_{fi,EC,K}$ for a K-shell electron. The wavefunction for a 1s-orbital electron in a hydrogen-like atom with atomic number, Z_X, is,

$$
\psi_{e,K}(r) = \sqrt{\frac{Z_X^3}{\pi r_\infty^3}}e^{-\left(\frac{Z_X r}{r_\infty}\right)} \tag{4.107}
$$

where r_∞ is the Bohr radius and Z_x is the atomic number for the parent nucleus. As r is of the order of 5 fm, the ratio $r/r_\infty \approx 10^{-4}$ allowing the exponential term to be neglected and the electron wavefunction treated as being spatially invariant within the nucleus,

$$\psi_{e,K} \approx \sqrt{\frac{Z_X^3}{\pi r_\infty^3}}. \tag{4.108}$$

This simplifies the expression of the EC transition rate to,

$$\lambda_{fi,EC,K} \approx \left(\frac{2g^2}{\hbar r_\infty^3}\right) Z_X^3 \frac{1}{L^3} |M_{fi}|^2 \rho_f \tag{4.109}$$

where the subscript K for the decay constant indicates that the calculation is explicitly for the capture of a K-orbital electron. The differential phase space term for this 2-body decay is,

$$d\rho_f = \frac{1}{(2\pi\hbar c)^3}(4\pi L^3) p_{\nu_e}^2 dp_{\nu_e} \delta(Q - T_{\nu_e})$$

which, for the massless neutrino, reduces to,

$$\frac{d\rho_f}{dE_{\nu_e}} = \frac{L^3}{2\pi^2(\hbar c)^3} E_{\nu_e}^2 \delta(E_0 - E_{\nu_e}).$$

Integrating over the neutrino energy gives the complete phase space factor (as this is a 2-body decay with a negligible recoil kinetic energy, the neutrino will take all of the available energy),

$$\rho_f(E_0) = \frac{L^3}{2\pi^2(\hbar c)^3} E_0^2 \tag{4.110}$$

which yields the electron capture transition rate,

$$\lambda_{fi,EC,K} = \left(\frac{g^2}{\pi^2 \hbar (\hbar c)^3 r_\infty^3}\right) Z_X^3 |M_{fi}|^2 E_0^2.$$

As $r_\infty = \hbar c/\alpha m_e$,

$$\lambda_{fi,EC,K} = \left(\frac{g^2 m_e^3 \alpha^3}{\pi^2 \hbar (\hbar c)^6}\right) Z_X^3 |M_{fi}|^2 E_0^2. \tag{4.111}$$

By defining the dimensionless quantity

$$f_{EC,K}(Z_X, E_0) = (\alpha Z_X)^3 \left(\frac{E_0}{m_e}\right)^2 \tag{4.112}$$

the electron capture decay constant can be written in the form,

$$\lambda_{fi,EC,K} \equiv \left(\frac{g^2 m_e^5}{\pi^2 \hbar (\hbar c)^6}\right) |M_{fi}|^2 f_{EC,K}(Z_X, E_0). \tag{4.113}$$

Note that electron capture decay rate has a Z^3-dependence contained within the $f_{EC,K}(Z_X, E_0)$ term.

4.3.4 Selection Rules for β Decay

4.3.4.1 Introduction

Having established the kinematic requirements for β decay to occur and the probability with which it will occur, it is now necessary to include the two so far neglected two important and linked features: first, the constituents of a β decay have angular momentum and, second, the decaying nucleus has a finite size. These have significant effects upon the probability of a given β decay (Lipkin 1962).

Conservation of the total angular momentum in β decay requires that the difference between the angular momenta of the parent and daughter states equals the total angular momentum carried away by the lepton pair. This investigation of the deep role angular momentum plays in β decay is begun by considering the simplest case of β^- decay, free neutron decay, $n \rightarrow p + e^- + \bar{\nu}_e$, as shown in Fig. 4.20. As the neutron is considered to be a point entity, no orbital angular momentum is involved.

There are two possible final states resulting from free neutron decay which correspond to the emitted lepton pair having either antiparallel or parallel spins. The transition leading to the final state in which the lepton pair forms a spin singlet (i.e., coupled spin of 0) is referred to as a Fermi transition, for a reason to be shown later. As a consequence of the lepton spin coupling, the spin state of the proton must be the same as that of the neutron. The other final state, that in which the lepton pair form a triplet (coupled spin of 1) and the spin states of the neutron and proton are consequently opposed, is the result of a Gamow–Teller transition (Gamow and Teller 1936). In a

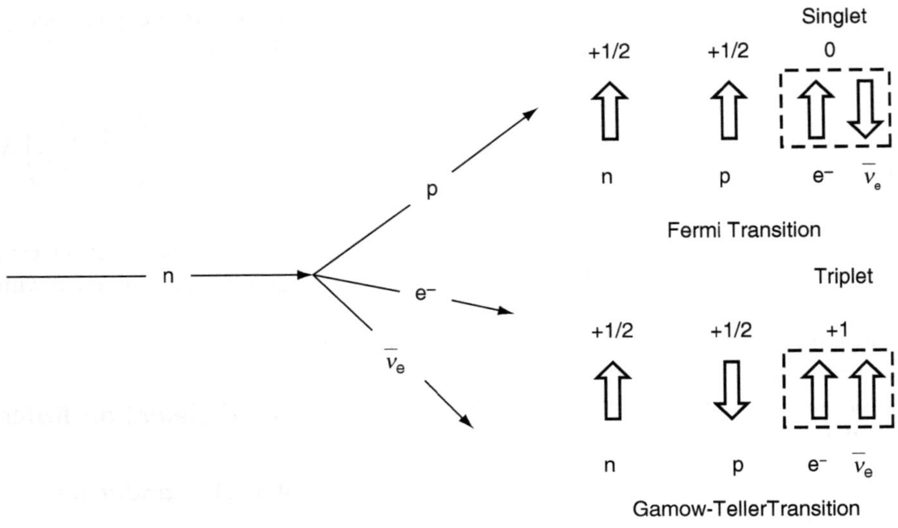

Fig. 4.20 β decay in the four-point vertex approximation of a free neutron and the combinations of final spin states for Fermi and Gamow–Teller transitions

Fermi transition for neutron decay, the operator transforming the neutron spin state to that of a proton is the unit operator. On the other hand, in the Gamow–Teller transition the transformation of the neutron spin to the proton spin is linked through the Pauli spin matrices. Purely on the basis of this result, one would expect the probability of a Gamow–Teller transition for neutron decay to be three times greater than that of the Fermi transition. This is not quite the case and the ratio of the Gamow–Teller to Fermi transitions is actually equal to $3c_A^2$, where $c_A = 1.26$, a consequence of the quark structure of the nucleon.

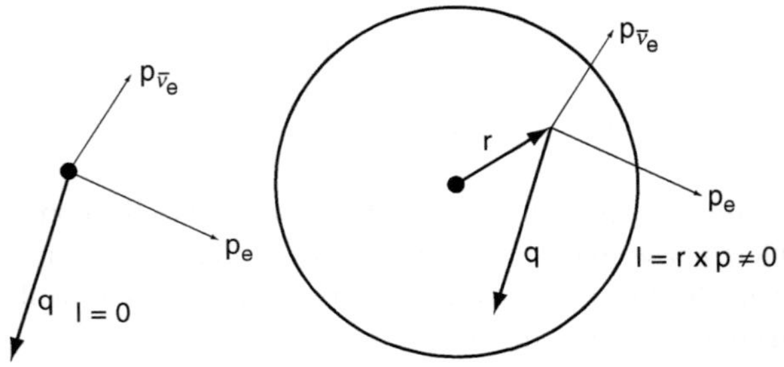

Fig. 4.21 Effects of angular momentum in β decay. In the first case, the nucleus is considered as a point and the matrix element as being independent of electron and antineutrino momenta. The orbital angular momentum of the exiting leptons is zero. In the second case, the nucleus has a finite size and the lepton pair leaves with a non-zero angular momentum

4.3.4.2 Selection Rules in Nuclear β decay

Introduction

Next consider the role of angular momentum in nuclear β decay. Throughout this study of nuclear β decay so far, it has been noted several times that the finite size of the nucleus has been ignored. This allows the lepton wavefunction expansions to be truncated to zeroth-order and, implicitly, for the lepton pair to be taken to have been emitted from the nucleus with zero angular momentum. As, classically, $\mathbf{l} = \mathbf{r} \times \mathbf{p} = \mathbf{0}$, this latter condition is equivalent to saying that the leptons are emitted radially (i.e., s wave) from the nucleus as shown in Fig. 4.21. Thus, as in free neutron decay, the change in nuclear angular momentum will equal the coupling of the leptons into either a singlet or triplet state.

When nuclear size is not neglected, there are two significant consequences to the description of nuclear β decay. First, non-zero orbital angular momentum is now made available to the lepton pair and the change in nuclear angular momentum can be greater than one. Second, these non-zero orbital angular momentum β decays contribute to the transition rate through the higher-order terms of the lepton wavefunctions' expansions. As these additional contributions are of the order of $(qR_N)^l$, the non-zero orbital angular momentum transition will be highly suppressed relative to the s-wave transitions. Consequently, β decays associated with $l = 0$ are referred to as being "allowed" whereas those with $l > 0$ are labeled as being "forbidden."

Of the allowed transitions, the Fermi requirement of a singlet lepton pair obviously dictates that there be no change in the nuclear angular momentum between parent and daughter nuclei, $\Delta J = 0$. As the lepton pair is in the triplet state following a Gamow–Teller decay, the vector difference between the parent's and daughter's nuclear angular momenta must equal unity and the change in nuclear angular momentum will be $\Delta J = 0,\ 1$ but with no $0 \rightarrow 0$ transition. In either allowed transition, there will be no change in nuclear parity due to $l = 0$.

Table 4.3 Selection rules for Fermi and Gamow–Teller allowed β decays

Transition	Change in Angular momentum, ΔJ	Parity ΔΠ	Isospin ΔT
Fermi	0	0	0
Gamow–Teller	$\lvert\Delta J\rvert = 0,1$ No $0 \rightarrow 0$	0	$\lvert\Delta T\rvert = 0,1$ No $0 \rightarrow 0$

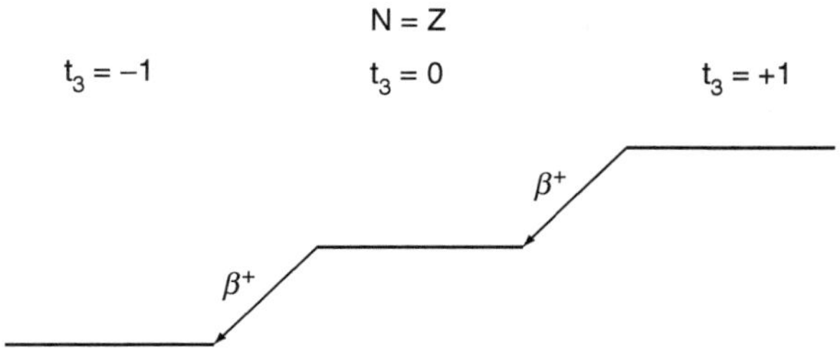

Fig. 4.22 Fermi transitions for β^+ decay between isobaric analogue states. Due to the Coulomb interaction between protons, corresponding energy levels in an isobar multiplet are higher for an isobar with a greater number of protons

Allowed Nuclear β decays

As β decay results in the change of isospin-projection, isospin operators are required in describing these transitions. Recalling the isospin ladder operators of Chap. 3, $\tau_\pm = \frac{\tau_1 \pm i\tau_2}{2}$, one can write operators for Fermi and Gamow–Teller transitions for the nucleus as summations of the isospin ladder operators over all nucleons (i.e., the operators act over the entire nuclear wavefunction),

$$\mathbf{T}_\pm = \sum_{i=1}^{A} \boldsymbol{\tau}_\pm(i) \quad \text{Fermi transition} \tag{4.114}$$

$$\mathbf{Y}_\pm = \sum_{i=1}^{A} \boldsymbol{\sigma}(i)\boldsymbol{\tau}_\pm(i) \quad \text{Gamow – Teller transition.} \tag{4.115}$$

As the Gamow–Teller transition results in a change in angular momentum, the Pauli spin matrix operator, $\boldsymbol{\sigma}(i)$, must be included. Table 4.3 summarizes the selection rules for Fermi and Gamow–Teller transitions.

Fermi's β decay theory, to which this discussion shall return, predicted only $\Delta J = 0$ transitions and the characteristics of the first row of Table 4.3. However, there was experimental evidence of the β^- decay $^6He \rightarrow {}^6Li + e^- + \bar{\nu}_e$, in which the J^π of the parent and daughter nuclei are 0^+ and 1^+ and the change in nuclear isospin was $|\Delta T| = 1$. Such a transition would be forbidden by the Fermi selection rules but obey the Gamow–Teller selection rule. As there are β decays that satisfy the Fermi selection rule (e.g., the $^{14}O \rightarrow {}^{14}N^* + e^+ + \nu_e$ for which the J^π of the parent and daughter nuclei are both 0^+), both transitions existed in nature. Moreover, the summary of the transition rules of Table 4.3 shows that some β decays are permitted by both Fermi and Gamow–Teller selection rules, an obvious example being that of free neutron decay. In other words, every Fermi transition contains an admixture of a Gamow–Teller transition, except for a $0 \rightarrow 0$ transition.

From (4.114), one sees that the Fermi transition occurs through the isospin ladder operator and, hence, can only occur between isospin multiplets. This sets a severe restriction upon this transition type. As these transitions can only occur between isobaric analogue states, it is necessary to reflect upon the energy differences between the nuclear levels of the parent and daughter nuclei. The energies of matching levels will increase with atomic number due to the Coulomb repulsion between protons. Hence, Fermi transitions cannot occur in β^- decays, as shown by Fig. 4.22. But even within the permitted β^+ decays, Fermi transitions can occur only for nuclei with more protons than neutrons since, in the case of $N > Z$, the isospin of the daughter nucleus exceeds that of the parent. In practice, Fermi transitions are relegated to β^+ decays in light nuclei.

The β-decay transition matrix element can be split up into Fermi and Gamow–Teller components,

$$|M_{fi}|^2 = |C_F|^2|M_F|^2 + |C_{GT}|^2|M_{GT}|^2 \tag{4.116}$$

where one averages over the initial spin states and sums over the final spin states,

$$|M_F|^2 = \frac{1}{2J_i+1}\sum_{i,f}|\langle\psi_f|\mathbf{T}_\pm|\psi_i\rangle|^2 \tag{4.117}$$

$$|M_{GT}|^2 = \frac{1}{2J_i+1}\sum_{i,f}|\langle\psi_f|\mathbf{Y}_\pm|\psi_i\rangle|^2. \tag{4.118}$$

The reduced transition probabilities are defined as,

$$B_F^\pm = \frac{|\langle\psi_f|\mathbf{T}_\pm|\psi_i\rangle|^2}{2J_i+1} \tag{4.119}$$

$$B_{GT}^\pm = \frac{c_A^2|\langle\psi_f|\mathbf{Y}_\pm|\psi_i\rangle|^2}{2J_i+1}. \tag{4.120}$$

The effects of the nuclear structure upon the allowed transition are contained within $B_F^\pm$ and $B_{GT}^\pm$. The transition rate of (4.99) can thus be rewritten as,

$$\lambda = \left(\frac{g^2 m_e^5}{2\pi^3\hbar(\hbar c)^6}\right)(B_F + B_{GT})f(Z_Y, x_0) \tag{4.121}$$

Forbidden Nuclear β decays

Forbidden transitions are those with large log ft values, involve parity change and an angular momentum change of greater than one (Marshak 1942). The total angular momentum taken by the lepton pair is $\mathbf{J} = \mathbf{L} \oplus \mathbf{S}$, which, for a lepton pair, assumes values of $(l-1), l, (l+1)$. A first-forbidden transition is one for $l = 1$, a second-forbidden transition is one for $l = 2$, etc. Parity can change in forbidden transitions as $\Delta\Pi = \Pi_f\Pi_i = (-1)^l$ and $l > 0$. Table 4.4 presents the categories of allowed and forbidden β-decay transitions and the corresponding observed ranges of $\log_{10} ft$ values, and Table 4.5 gives transition data for a variety of β decays.

Table 4.4 Categories of β decay transitions

Decay type	ΔJ	ΔΠ	Typical range of $\log_{10} ft$
Superallowed (Fermi and GT)	0	No	3
Allowed	0, 1; not $0 \to 0$	No	4–6
First forbidden	0, 1, 2	Yes	5–10
Second forbidden	2, 3	No	11–16
Third forbidden	3, 4	Yes	17–22
Fourth forbidden	4, 5	No	22–24

4.3.5 Fermi Theory of β Decay: Part II

4.3.5.1 Four-Fermion Interaction Vertex

Modern weak interaction theory underlying β decay is based upon the exchange of the massive intermediate vector bosons, $W^\pm$ and Z^0, between fermions. Fermi's theory did not include these exchanges but instead described the weak interaction in analogy to the electromagnetic interaction (Fig. 4.23) but as occurring at a vertex at which four fermions meet. Because the intermediate vector bosons are massive (the $W^\pm$ has a rest mass of over 80 GeV), the corresponding interaction distance is small (of the order of 10^{-3} fm) and, hence, Fermi's use of a point interaction is valid for β decay. In fact, the Fermi theory represents the low-energy limit of the modern Glashow–Weinberg–Salam theory. Here, we look at Fermi's relativistic theory applied to the simple case of neutron decay and the weak interaction.

To work through the Fermi theory, one can begin by considering it, as did Fermi, as an analogue to the electromagnetic interaction. Electrons produce an electromagnetic four-current density of the form, $J_\mu^{EM}(\boldsymbol{X}) = -\overline{\psi}_e(\boldsymbol{X})\gamma_\mu\psi_e(\boldsymbol{X})$ which couples to the electromagnetic field $A^\mu(\boldsymbol{X})$ to form the electromagnetic Hamiltonian density, $H_{EM}(\boldsymbol{X}) = -e\overline{\psi}_e(\boldsymbol{X})\gamma_\mu\psi_e(\boldsymbol{X})A^\mu(\boldsymbol{X})$. Before deriving the Fermi theory, recall that the matrix

Table 4.5 Transition data for β decays (in order of increasing log *ft*)

Parent	Decay mode	Daughter	$J_i^\pi \to J_f^\pi$	Transition	Half-life	$\log_{10} ft$
^6He	β^-	^6Li	$0^+ \to 1^+$	GT	797 ms	2.77
^3H	β^-	^3He	$\frac{1}{2}^+ \to \frac{1}{2}^+$	F/GT	12.33 years	3.05
^{14}O	β^-	$^{14}N^a$	$0^+ \to 0^+$	F	71.36 s	3.495
^{18}F	96.86% β^+, 3.14% EC	^{18}O	$1^+ \to 0^+$	GT	1.8295 h	3.57
^{11}C	99.7% β^+, 0.3% EC	^{11}B	$\frac{3}{2}^- \to \frac{3}{2}^-$	F/GT	20.39 min	3.592
^{15}O	99.9% β^+, 0.1% EC	^{15}N	$\frac{1}{2}^- \to \frac{1}{2}^-$	F/GT	2.037 min	3.6
^{13}N	99.8% β^+, 0.2% EC	^{13}C	$\frac{1}{2}^- \to \frac{1}{2}^-$	F/GT	9.965 min	3.654
^{64}Cu	18% β^+, 44.9% EC	^{64}Ni	$1^+ \to 0^+$	GT	12.701 h	4.97
	37.1% β^-	^{64}Zn				5.29
^{14}C	β^-	^{14}N	$0^+ \to 1^+$	GT	5,730 years	9.04[a]
^{39}Ar	β^-	^{39}K	$\frac{7}{2}^- \to \frac{3}{2}^+$	Forbidden	269 years	9.03
^{10}Be	β^-	^{10}B	$0^+ \to 3^+$		1.6×10^6 years	12.08
^{40}K	β^-	^{40}Ca	$4^- \to 0^+$		1.26×10^9 years	15.6
^{115}In	β^-	^{115}Sn	$\frac{9}{2}^+ \to \frac{1}{2}^+$		6×10^{14} years	23.0

[a]Although $^{14}C \xrightarrow{\beta^-} {}^{14}N$ is an allowed transition, the high $\log_{10}$ ft value is due to a small matrix element

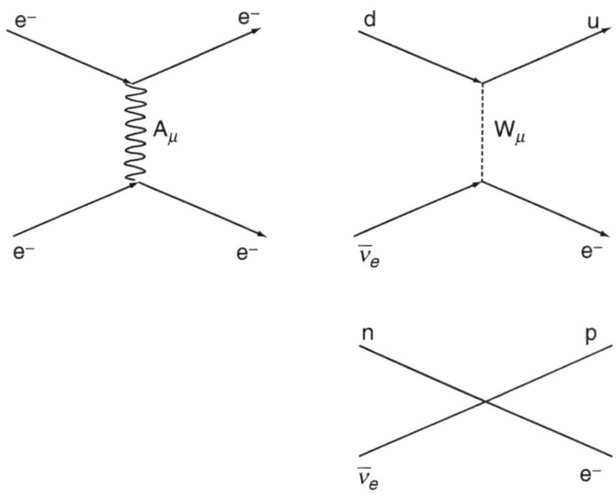

Fig. 4.23 The electromagnetic and weak interactions shown as the exchange of bosons; the Fermi four-fermion vertex approximates the short distance flavor-changing exchange of the intermediate vector $W^\pm$ boson as a single interaction point

element for β decay contains the wavefunctions of four spin-½ particles corresponding to,

$$n \to p + e^- + \bar{\nu}_e$$

$$p \to n + e^+ + \nu_e$$

$$p + e^- \to n + \nu_e.$$

In each case, there are hadronic and leptonic components[22] and, hence, weak hadronic and weak leptonic current densities must be constructed. The weak hadronic current density applied to free neutron decay is,

$$V_\mu^{C\dagger}(\boldsymbol{X}) = \overline{\psi}_p(\boldsymbol{X})\gamma_\mu\,\psi_n(\boldsymbol{X}) \qquad (4.122)$$

and a leptonic current density is,

$$l_\mu^C(\boldsymbol{X}) = \overline{\psi}_e(\boldsymbol{X})\gamma_\mu\,\psi_{\nu_e}(\boldsymbol{X}). \qquad (4.123)$$

Recall that the Dirac 4 × 4 matrices are,

$$\gamma_0 = \begin{pmatrix} \mathbf{1} & 0 \\ 0 & -\mathbf{1} \end{pmatrix}$$

$$\gamma_i = \begin{pmatrix} 0 & \boldsymbol{\sigma}_i \\ -\boldsymbol{\sigma}_i & 0 \end{pmatrix} \quad i = 1, 2, 3$$

and where the C superscript indicates that the electric charge of the hadron is being changed. For a point-like

[22]A full examination of the weak interaction would consider the pure leptonic currents of, say, $\mu^- \to e^- + \nu_e + \bar{\nu}_e$ or the pure hadronic currents of $K^+ \to \pi^+ + \pi^+ + \pi^-$.

interaction between these two currents, a Hermitian Hamiltonian density can be constructed as,

$$\begin{aligned} H(\boldsymbol{X}) &= g\left(V^{C\mu}(\boldsymbol{X})l_{\mu}^{C\dagger}(\boldsymbol{X}) + l^{C\mu}(\boldsymbol{X})V_{\mu}^{C\dagger}(\boldsymbol{X})\right) \\ &= g(\overline{\psi}_n(\boldsymbol{X})\gamma^{\mu}\psi_p(\boldsymbol{X})\overline{\psi}_{\nu_e}(\boldsymbol{X})\gamma_{\mu}\psi_e(\boldsymbol{X}) \\ &\quad + \overline{\psi}_e(\boldsymbol{X})\gamma^{\mu}\psi_{\nu_e}(\boldsymbol{X})\overline{\psi}_p(\boldsymbol{X})\gamma_{\mu}\psi_n(\boldsymbol{X})) \\ &= gV. \end{aligned} \quad (4.124)$$

The V denotes this to be a vector interaction as $\bar{\psi}\,\gamma_{\mu}\,\psi$ transforms like a polar vector when undergoing a Lorentz transformation. The first term of this Hamiltonian describes β^{+} decay and electron capture and the second term describes β^{-} decay. Importantly, the vector Hamiltonian describes a Fermi transition. To demonstrate this, consider the nonrelativistic case for which the nucleon wavefunction is $\psi = \begin{pmatrix}\phi \\ 0\end{pmatrix}$. Then, the $\mu = 0$ component of the nucleon current is,

$$\begin{aligned} \overline{\psi}_p\gamma_0\psi_n &= \psi_p^{+}\gamma_0\gamma_0\psi_n \\ &= \begin{pmatrix}\phi_p^{+} & 0\end{pmatrix}\begin{pmatrix}\mathbf{1} & 0 \\ 0 & \mathbf{1}\end{pmatrix}\begin{pmatrix}\phi_n \\ 0\end{pmatrix} \\ &= \phi_p^{+}\phi_n \end{aligned} \quad (4.125)$$

and the $\mu = 1, 2, 3$ space component is,

$$\begin{aligned} \overline{\psi}_p\gamma_{\mu}\psi_n &= \psi_p^{+}\gamma_0\gamma_{\mu}\psi_n \\ &= \begin{pmatrix}\phi_p^{+} & 0\end{pmatrix}\begin{pmatrix}\mathbf{1} & 0 \\ 0 & -\mathbf{1}\end{pmatrix}\begin{pmatrix}0 & \boldsymbol{\sigma}_{\mu} \\ -\boldsymbol{\sigma}_{\mu} & 0\end{pmatrix}\begin{pmatrix}\phi_n \\ 0\end{pmatrix} \\ &= \begin{pmatrix}\phi_p^{+} & 0\end{pmatrix}\begin{pmatrix}\mathbf{1} & 0 \\ 0 & -\mathbf{1}\end{pmatrix}\begin{pmatrix}0 \\ -\boldsymbol{\sigma}_{\mu}\phi_n\end{pmatrix} \\ &= \begin{pmatrix}\phi_p^{+} & 0\end{pmatrix}\begin{pmatrix}0 \\ \boldsymbol{\sigma}_{\mu}\phi_n\end{pmatrix} \\ &= 0. \end{aligned} \quad (4.126)$$

Hence, $\Delta J = 0$. However, as Gamow–Teller transitions exist in nature, it is clear that this vector–vector (VV) coupling cannot be the only interaction involved in weak decays. Gamow and Teller noted that other Lorentz-invariant current densities beyond the VV coupling of $\gamma^{\mu}\gamma_{\mu}$ can be made. From these constructs, there are five types of current density structure,

Scalar (S)	$\overline{\psi}_p\psi_n$
Pseudoscalar (P)	$\overline{\psi}_p\gamma_5\psi_n$
(Polar) vector (V)	$\overline{\psi}_p\gamma_{\mu}\psi_n$
(Axial) vector (A)	$\overline{\psi}_p\gamma_5\gamma_{\mu}\psi_n$
Tensor (T)	$\overline{\psi}_p\gamma_{\mu}\gamma^{\nu}\psi_n$

As has been done for the VV current coupling, Hamiltonians can be constructed from these structures. However, a real Hamiltonian must reduce to a single component and, hence, can only be a scalar or pseudoscalar quantity. The only resulting possible couplings yielding such quantities are SS, VV, TT, AA, PS and VA. A pseudoscalar weak Hamiltonian was not considered acceptable as it violated parity conservation (recall Table 3.2 in Chap. 3) and, as the PS and VA couplings yielded such a parity-nonconserving Hamiltonian, they were rejected until the 1950s. It was not considered possible that parity would not be conserved in weak interactions as it was in the strong and electromagnetic interactions. Up until that time, the weak interaction was assumed to be made up of SS and TT components which permitted a Gamow–Teller transition with a scalar Hamiltonian. But before this is demonstrated to not be the case, the experimental evidence of parity violation by the weak interaction during the 1950s and which led to the appropriate current coupling is reviewed.

4.3.5.2 Evidence for Parity Nonconservation in Weak Interactions

The θ-τ Dilemma

θ and τ mesons were discovered in cosmic rays (they are now known as K mesons, or kaons). Experiment demonstrated that while they had the same mass, spin, and half-life, they differed in their decay processes,

$$\tau^{+} \rightarrow \pi^{+} + \pi^{0}$$

$$\theta^{+} \rightarrow \pi^{+} + \pi^{0} + \pi^{0} \text{ or } \theta^{+} \rightarrow \pi^{+} + \pi^{+} + \pi^{-}.$$

These are weak decays (with mean lifetimes of the order of 10^{-8} s) and it was within this difference between decay products that a fundamental dilemma arose. The pion has zero spin and a negative intrinsic parity as shown experimentally by pion absorption in deuterium. Hence the dipion final state of τ meson decay has a parity of +1 whereas the tripion final

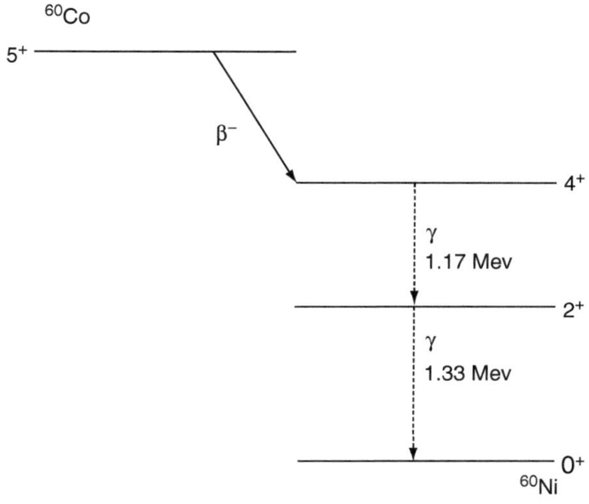

Fig. 4.24 ^{60}Co β^- decay to ^{60}Ni

state of θ meson decay has a parity of −1. The only evident difference between the θ and τ mesons was their parity, and this indicated the nonconservation of parity. This result was problematic as parity selection rules had worked well in nuclear and atomic physics (i.e., parity was conserved in strong and electromagnetic interactions). Triggered by this dilemma, Lee and Yang (1956) postulated the possibility that parity was in fact not conserved in weak interactions. They reviewed the results of all known experiments of the weak interaction performed to that time to detect evidence for parity nonconservation. Their review concluded that all prior experiments had in fact measured scalar quantities and, thus, unable to detect parity violation.[23] They recommended a number of tests to determine if parity is conserved in weak interactions. The critical one was performed by Wu et al. (1957) and is discussed next.

Parity Nonconservation in β Decay

On the basis of Lee's and Yang's suggestions, Wu et al. (1957) designed and performed a fundamental

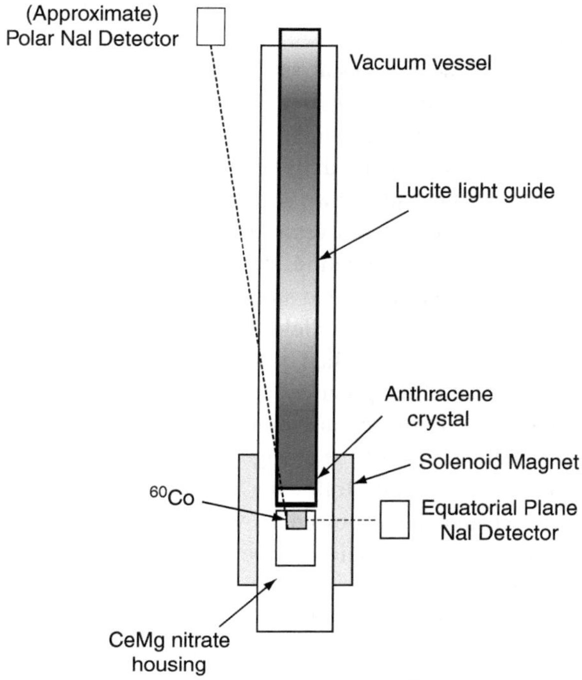

Fig. 4.25 Schematic diagram of experimental apparatus used by Wu et al. (1957) to detect parity nonconservation in β decay

experiment to prove the nonconservation of parity by measuring a pseudoscalar quantity in the Gamow–Teller β^- decay of ^{60}Co to ^{60}Ni of Fig. 4.24. It will be noted that photon emissions are associated with this decay.[24] The experiment was designed to measure the expectation value of the scalar product of the nuclear spin and the electron momentum, which is a pseudoscalar variable.

The experiment is shown schematically in Fig. 4.25. The apparatus was designed to measure the mean value of the pseudoscalar quantity given by the scalar product of the velocity of the electron emitted in the β^- decay of ^{60}Co and the orientation of the nuclear spin, $\overline{\mathbf{v} \bullet \mathbf{J}}$. Should parity be conserved, this mean value would be zero, indicating that β decay is symmetric in space. In order to orient the spin of the ^{60}Co nucleus, a ^{60}Co sample in the form of a 50 μm thick layer coated on the top of a paramagnetic crystal cerium magnesium nitrate was cooled to 0.003°K using adiabatic demagnetization. The entire structure was encased within a vacuum vessel and surrounded

[23] However, Cox et al. (1928) had unknowingly provided evidence for parity conservation in their measurements of the longitudinal polarization of electrons from β decay. This result was disbelieved at the time due to, for example, the failure of attempts to reproduce the measurements but by using unpolarized electrons arising from thermionic emission.

[24] The 1.17 and 1.33 MeV γ rays are used in radiotherapy.

by a solenoid coil that oriented the spin of the ^{60}Co nuclei.

An anthracene crystal set immediately above the source detected the β particles and a lucite light guide carried the scintillation light from the crystal resulting from electrons bombarding the crystal to an external photomultiplier tube. Two sodium iodide scintillators external to the vacuum vessel were used to detect the degree of ^{60}Co polarization. Regardless of the orientation of the polarized ^{60}Co spin being parallel or antiparallel to the magnetic field, the count-rate from the equatorial sodium iodide (NaI) scintillator counter exceeds that of the polar counter and, hence, provides a measure of the polarization of the nucleus. The measurement was of the change in β particle counting rate during the time following when the nuclei were oriented and then allowed to warm. As the sample warmed, the polarization of the ^{60}Co nuclei was lost, as can be seen in the schematic results of Fig. 4.26. This loss of photon anisotropy was a direct measure of nuclear polarization. As also shown in the figure, the detected β particle rate was greater in one orientation of the ^{60}Co nuclei than in the other: i.e., the emission of β particles is more favored in the direction opposite to that of the nuclear spin. Hence, $\overline{\mathbf{v} \bullet \mathbf{J}} \neq 0$ and the nonconservation of parity in β decay was demonstrated.

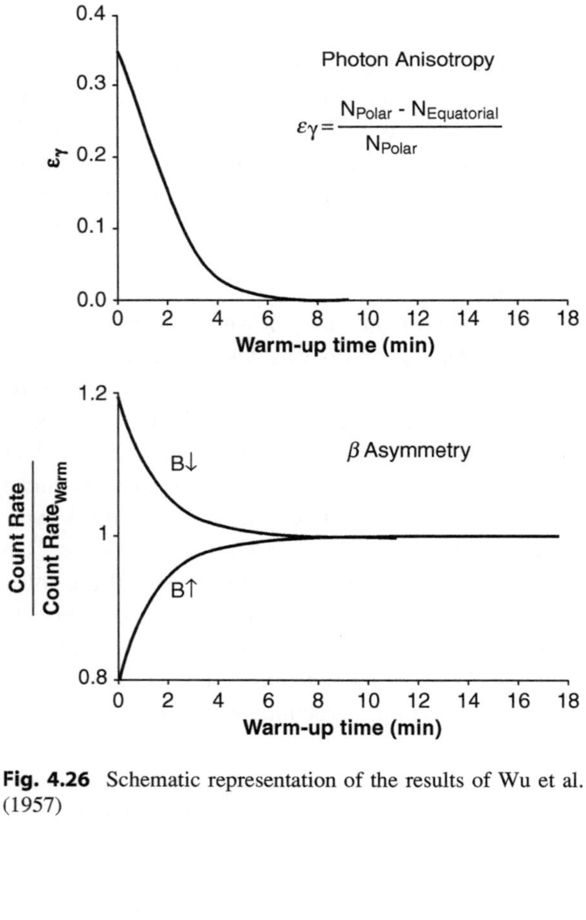

Fig. 4.26 Schematic representation of the results of Wu et al. (1957)

Neutrino Helicity

Here, what is properly regarded as one of the most elegant and cleverest physics experiments ever performed is described. This experiment demonstrated that the orientation of the spin of a neutrino is antiparallel to the neutrino's direction of motion. As the spin is an axial vector and the momentum is a polar vector, a non-zero longitudinal polarization of the neutrino is indicative of parity violation. The helicity operator for a fermion is defined as,

$$\aleph = \frac{\mathbf{p} \bullet \boldsymbol{\sigma}}{|\mathbf{p} \bullet \boldsymbol{\sigma}|} = 2\hat{\mathbf{p}} \bullet \boldsymbol{\sigma} \tag{4.127}$$

where $\boldsymbol{\sigma}$ is the spin operator and $\mathbf{p}$ is the momentum operator. $\aleph$ has the eigenvectors $|\psi_{\pm}\rangle$, which are states in which the spin is parallel or antiparallel to the particle's direction of motion with eigenvalues ± 1,

$$\aleph|\psi_{\pm}\rangle = \pm|\psi_{\pm}\rangle. \tag{4.128}$$

Parity invariance would require that neutrinos with positive or negative helicities be equally probable and any type of measured neutrino helicity asymmetry would thus reflect parity violation. The Goldhaber–Grodzins–Sunyar experiment (Goldhaber et al. 1958) demonstrated this helicity asymmetry for ν_e. Their method was to begin with a nucleus with angular momentum 0 which decays through allowed EC (i.e., $l = 0$) to an excited state of a daughter nucleus with angular momentum 1. As the decay is through electron capture, only the daughter nucleus and neutrino are present in the final state and, hence, the neutrino will have a fixed energy due to the at-rest initial state and the two-body final state. The combined angular momenta of the daughter nucleus (angular momentum of 1) and neutrino must equal that of the K-shell electron prior to capture, so one need only measure the polarization of the daughter nucleus in order to determine that of the neutrino. The initial nucleus was the metastable isomer ^{152m}Eu, produced through neutron bombardment of europium, which undergoes $\beta^{\pm}$ decay and electron capture and Fig. 4.27 shows that part of the decay chain from ^{152m}Eu to ^{152}Sm.

The EC from the 0^- state of ^{152}Eu to the excited 1^- state of ^{152}Sm (which is denoted as ^{152}Sm*) requires the following angular momentum balance. The initial state angular momentum of ½ (that of the K-shell electron) must equal the final state angular momentum made up of the angular momentum of 1 from the ^{152}Sm nucleus and the ½ – spin of the neutrino. This is shown schematically in Fig. 4.28. As the final state of EC is a two-body state with the initial state being at rest, the conservation of momentum requires that the momenta of the ^{152}Sm* nucleus and of the neutrino oppose each other. Further, as the conservation of angular momentum requires that the angular momentum of ^{152}Sm* and the spin of the neutrino oppose each other, the polarizations of the neutrino and the ^{152}Sm* nucleus must be the same. Hence, it is only needed to measure the polarization of the ^{152}Sm* nucleus in order to determine the neutrino polarization. The measurement of the ^{152}Sm* polarization is enabled by the fact that ^{152}Sm* decays electromagnetically through the emission of a γ ray to the ground state of ^{152}Sm, which has zero angular momentum. As the photon is spin-1, the photon spin must be parallel to the angular momentum of ^{152}Sm*, as shown in Fig. 4.29. Hence, photons emitted in the same direction as the ^{152}Sm* nucleus will have the same polarization, to within a factor of ½) as the neutrino. As a result, the measurement of the neutrino polarization/helicity reduces to measurements of the polarization of the photon and its direction relative to that of the ^{152}Sm* nucleus. How this was achieved is shown schematically in Fig. 4.30.

The basic structure of the apparatus was of a source of ^{152m}Eu placed within an iron analyzing magnet and a detector consisting of a sodium iodide scintillator coupled to a photomultiplier tube. Surrounding the scintillator was the annular conical frustum made of Sm_2O_3 and the scintillator was shielded from the direct γ rays from the ^{152m}Eu source by lead. The γ rays from the ^{152}Sm* de-excitation that were transmitted through the analyzing magnet were scattered by the Sm_2O_3 annulus through nuclear resonant scattering to be detected by the scintillator. Recall that the

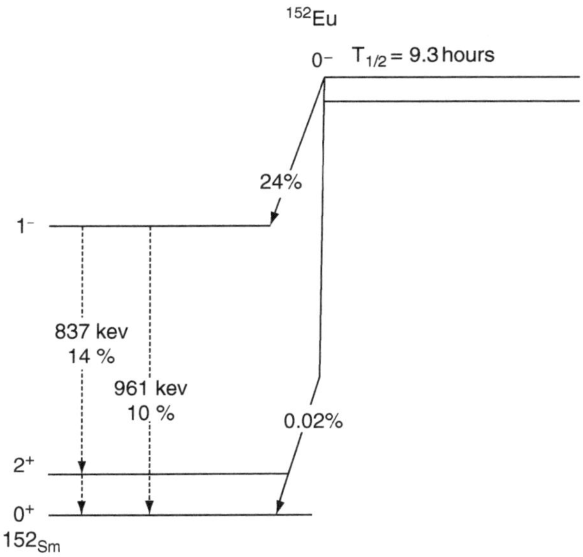

Fig. 4.27 Partial decay schema of ^{152}Eu to ^{152}Sm

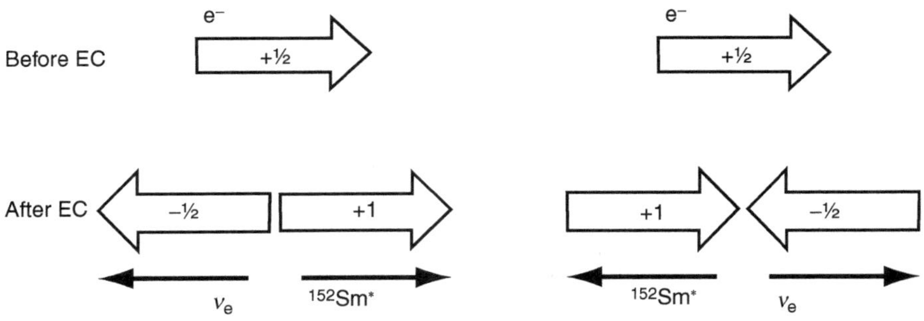

Fig. 4.28 This shows schematically the possible arrangement of the z-components of the spins/angular momenta (*full arrows*) and the momenta (*thin arrows*) of the excited 1^- state of the ^{152}Sm* nucleus and neutrino following electron capture by the 0^- state of ^{152m}Eu. Those on the left correspond to a right-handed neutrino in which the neutrino spin orientation is parallel to the neutrino's direction of motion. Those on the right correspond to a left-handed neutrino in which the neutrino spin orientation is antiparallel to the neutrino's direction of motion

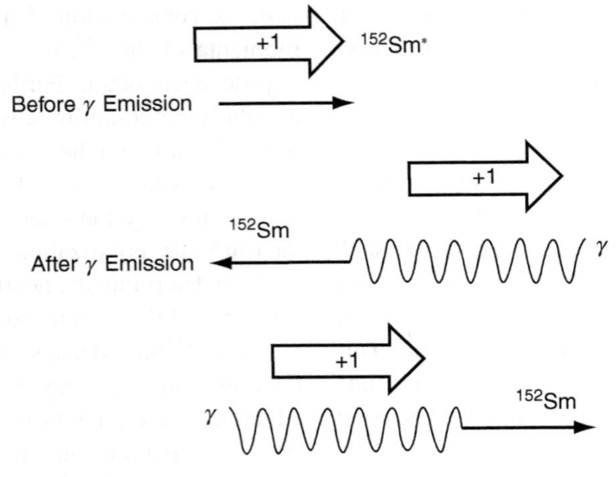

Fig. 4.29 This shows schematically the possible arrangement of the z-components of the spins/angular momenta (*full arrows*) and the momenta (*thin arrows*) of the excited 1^- state of the $^{152}Sm^*$ nucleus and neutrino following electron capture by the 0^- state of ^{152m}Eu. The photon spin is always in the same direction as the angular momentum of $^{152}Sm^*$

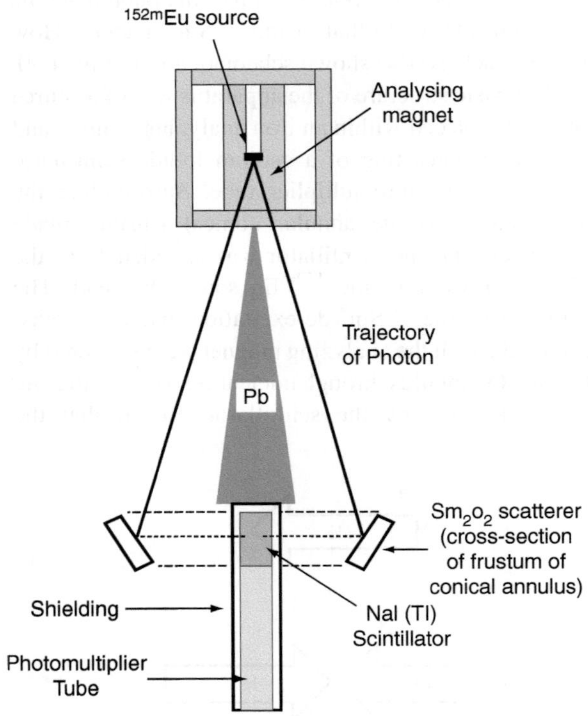

Fig. 4.30 Schematic representation of the Goldhaber–Grodzins–Sunyar experiment to measure the neutrino helicity. See text for description

experiment is to measure the polarization of the photons and their direction relative to the recoil ^{152}Sm nucleus. Depending upon the orientations of the directions of the photon polarization and the spin of the electrons in the magnetized iron, the γ rays resulting from the $^{152}Sm^*$ de-excitation interact with the electrons in the magnetic material differently. If the photon and electron spins are parallel, the incident photon cannot affect the electron spin and does not interact with the electron. If the electron spin is anti-parallel to that of the γ ray, then the photon induces a spin-flip in the electron and is consequently absorbed. Hence, those γ rays with spins parallel to those of the atomic electrons in the magnetic will penetrate the magnet and exit to impinge on the Sm_2O_3 scatterer. By selecting the direction of the field of the analyzing magnet, it was then possible to select the polarization of the transmitted γ rays.

Next, in order to measure the direction of the γ ray relative to the $^{152}Sm^*$ nucleus recoiling from the emission of the neutrino as a result of the EC, advantage was taken over the phenomenon of nuclear resonant scattering. Consider the case in which a nucleus in an excited state of energy E_i decays to a state of energy E_f with the emission of a photon with an energy equal to $\Delta E = E_i - E_f$ in the reference frame of the nucleus. Should another nucleus in the state of energy E_f and of the same species be present in that same reference frame, it can absorb this photon (to within a natural transition width, equal to 23 meV for $^{153}Sm^*$) to yield an excited nucleus with a state of energy E_i . This is the phenomenon of resonant absorption fundamental to the Mössbauer effect. However, an absorbing nucleus in the ground state will not be found in the same reference frame as the emitting nucleus and, in the laboratory reference frame, the emitted photon will not have an energy equal toΔE but, rather an energy reduced by $(\Delta E)^2/2m_f$, where m_f is the rest mass of the daughter nucleus (this result is derived in the following section). This energy reduction is that due to the recoil energy of the daughter nucleus which, in the case of ^{152}Sm, is equal to about 3.2 eV. This recoil energy is more than enough to shift the photon energy to ensure that resonant scattering cannot occur. However, if the $^{152}Sm^*$ nucleus is already in motion prior to the γ ray de-excitation, it can compensate for this reduction in photon energy by transferring some of the recoil energy resulting from the EC and neutrino emission to the photon (i.e., a Doppler shift). Such a transfer will be a maximum if the photon is emitted in the direction of the $^{152}Sm^*$ motion, thus enabling a reabsorption of the photon by ^{152}Sm in the scatterer and a reemission (nuclear resonant scatter). Thus, those photons detected by the NaI scintillator were generated by those directed in the direction of the $^{152}Sm^*$ motion, the polarization of which were given by the orientation of the magnetic field in the analyzer. The analyzing power, equal to the difference in photon detection rate with the magnetic field directed upwards and that with the field directed downwards normalized to the average, was

$$P = 2\frac{N_+ - N_-}{N_+ + N_-} = 0.017 \pm 0.003$$

This was a non-zero result; in fact, Goldhaber, Grodzins and Sunyar determined, from the estimated three mean free paths for photons through the magnetic material, that 68% of the photons were polarized with negative helicity. Hence, it was demonstrated that neutrinos had negative helicity (i.e., were left-handed). This asymmetry, as noted at the beginning of this discussion, demonstrated the lack of parity conservation in weak interactions.

4.3.5.3 V-A Interaction

The experimental results clearly demonstrated the nonconservation of parity in β decay. The helicity operator $\aleph$ of (4.127) is not relativistically invariant for massive fermions; for such fermions, it is always possible to have a Lorentz boost from one reference frame in which the helicity is positive/negative to another reference frame in which the helicity is, correspondingly, negative/positive. However, this discussion concerns a massless fermion, the neutrino. For massless fermions, the operator γ_5 is equivalent to the helicity operator $\aleph$, as shall now be demonstrated Recall that the Dirac spinor for a particle of mass m is,

$$u(\boldsymbol{p},s) = \sqrt{E+m}\begin{pmatrix} \chi_s \\ \frac{\boldsymbol{\sigma}\bullet\mathbf{p}}{\sqrt{E+m}}\chi_s \end{pmatrix}.$$

Thus, the Dirac spinor of a massless particle is,

$$u = \sqrt{p}\begin{pmatrix} \chi \\ \left(\frac{\boldsymbol{\sigma}\bullet\mathbf{p}}{p}\right)\chi \end{pmatrix}. \tag{4.129}$$

Applying the helicity operator to this spinor,

$$\begin{aligned} \aleph u &= -\frac{\boldsymbol{\sigma}\bullet\mathbf{p}}{|\mathbf{p}|}\sqrt{p}\begin{pmatrix} \chi \\ \left(\frac{\boldsymbol{\sigma}\bullet\mathbf{p}}{p}\right)\chi \end{pmatrix} \\ &= \sqrt{p}\begin{pmatrix} \left(\frac{\boldsymbol{\sigma}\bullet\mathbf{p}}{|\mathbf{p}|}\right)\chi \\ \left(\frac{\boldsymbol{\sigma}\bullet\mathbf{p}}{p}\right)^2\chi \end{pmatrix} \\ &= \sqrt{p}\begin{pmatrix} \left(\frac{\boldsymbol{\sigma}\bullet\mathbf{p}}{|\mathbf{p}|}\right)\chi \\ \chi \end{pmatrix}. \end{aligned} \tag{4.130}$$

It is also noted that

$$\begin{aligned} \gamma_5 u &= \sqrt{p}\begin{pmatrix} 0 & \mathbf{1} \\ \mathbf{1} & 0 \end{pmatrix}\begin{pmatrix} \chi \\ \left(\frac{\boldsymbol{\sigma}\bullet\mathbf{p}}{p}\right)\chi \end{pmatrix} \\ &= \sqrt{p}\begin{pmatrix} \left(\frac{\boldsymbol{\sigma}\bullet\mathbf{p}}{p}\right)\chi \\ \chi \end{pmatrix}. \end{aligned} \tag{4.131}$$

The chirality operator $(1-\gamma_5)/2$ projects out the left-handed components of any spinor for massless particles with negative helicity, i.e., neutrinos. This can be shown using the Pauli two-component spinors $\chi_\pm$ which are described by,

$$\frac{\boldsymbol{\sigma}\bullet\mathbf{p}}{p}\chi_\pm = \pm\chi_\pm \tag{4.132}$$

where ± describes the spin orientation of a spin-½ particle. Thus, from (4.129) the corresponding Dirac spinors are,

$$u_\pm = \sqrt{p}\begin{pmatrix} \chi_\pm \\ \pm\chi \end{pmatrix} \qquad (4.133)$$

from which is obtained,

$$\left(\frac{1-\gamma_5}{2}\right)u_- = -u_- \qquad (4.134)$$

and

$$\left(\frac{1-\gamma_5}{2}\right)u_+ = 0. \qquad (4.135)$$

Returning to the hadronic and leptonic current densities of β decay given by (4.122) and (4.123) and using the chirality operator, one can now construct a leptonic current density which produces left-handed neutrinos,

$$l^C_\mu(\boldsymbol{X}) = \bar{\psi}_e(\boldsymbol{X})\gamma_\mu(1-\gamma_5)\psi_{\bar{\nu}_e}(\boldsymbol{X}) \qquad (4.136)$$

and a similar form of the hadronic current density,

$$V^{C\dagger}_\mu(\boldsymbol{X}) = \bar{\psi}_p(\boldsymbol{X})\gamma_\mu(1-c_A\gamma_5)\psi_n(\boldsymbol{X}). \qquad (4.137)$$

Both weak currents contain the Fermi polar vector component and an axial vector component which is the V-A (pronounced V minus A) structure of the weak interaction.

4.4 γ Transitions and Internal Conversion

4.4.1 Introduction

The daughter nucleus resulting from α or β decay will be left in one of three states. It may be stable to further radioactive decay: the maximum nuclear binding energy for the isobar series (i.e., the mass parabola's minimum) will have been attained. The nucleus may still yet undergo further α or β decay. Finally, the nucleus may be such that while further α or β decay is not energetically feasible, it is in an excited state in which one or more nucleons are placed in higher orbitals above the ground state. The nucleus can de-excite through the release of electromagnetic energy as the nucleon cascades to the ground state. There are two energy-transfer channels available. In the first, energy is released in the form of γ rays with energies, in the reference frame of the nucleus, equal to the energy differences between the orbitals that the nucleon transits between. This γ transition is a nuclear process. On the other hand, the second channel is an atomic process and is one in which the excess nuclear energy is transferred directly to an atomic electron (through the exchange of a virtual photon) but without the emission of electromagnetic energy. Should this transferred energy exceed the electron binding energy, the electron is ejected from the atom, inducing a variety of atomic relaxation processes as discussed in Chap. 6. This second channel is known as IC.

While a full description of nuclear electromagnetic processes is inherently quantum-mechanical, considerable understanding can be achieved through the use of classical and semiclassical theories. As a result, these will be the foundation of the derivations of the transition probabilities of γ rays and the selection rules that arise through the conservation of angular momentum and parity. Links to quantum theory will be forged where appropriate and necessary. This section begins with a review of the kinematics of γ decay and, using classical multi-pole expansions of the electromagnetic field, derives transition selection rules (similar to those of β decay) based upon the angular momentum carried away by the emitted photon (equal to the difference between those of the initial and final nuclear states) and parity. These expansions also allow a derivation of the transition rate (the reciprocal of the mean lifetime) of a given nuclear state and this is examined as a function of photon energy and photon multi-pole type. Following this study of γ decay, IC is reviewed.

4.4.2 γ Decay

4.4.2.1 Kinematics

Consider an excited nucleus with energy E_i making a transition to a state with energy E_f through the emission of a γ ray. While it is frequently stated that the γ ray is emitted with an energy equal to the energy difference between the two states, $k = \Delta E = E_i - E_f$, this is only true in the reference frame of the nucleus. In the laboratory reference frame, however, consideration of the simultaneous conservations of energy and linear momentum shows the nucleus takes some of this energy difference as a recoil kinetic energy. The conservation of energy in the laboratory is,

$$m_i = m_f + k + T_f. \tag{4.138}$$

where $m_{i,f}$ are the rest masses of the nucleus before and after the transition such that $\Delta E = m_i - m_f$ and T_f is the nuclear recoil kinetic energy. The conservation of momentum is,

$$p_f = k. \tag{4.139}$$

As the nucleus is sufficiently massive to assume that its recoil is nonrelativistic, then,

$$\Delta E = k + \frac{k^2}{2m_f}. \tag{4.140}$$

Solving the resulting quadratic equation $k^2 + 2m_f k - 2m_f \Delta E = 0$ for the photon energy gives the full result for the photon energy,

$$k = \frac{m_f}{2}\left(\sqrt{1 + 2\frac{\Delta E}{m_f}} - 1\right). \tag{4.141}$$

Expanding the square-root to second order,

$$k = \Delta E\left(1 - \frac{\Delta E}{2m_f}\right) \tag{4.142}$$

Hence, the photon energy will be reduced by, to second order, an amount of $\Delta E^2/2m_f$ due to the recoil of the nucleus. But, as $\Delta E \ll m_f$, $k = \Delta E$ can be assumed in practice.

4.4.2.2 Multi-Pole Radiation

Introduction

The nucleus with excitation energy ΔE prior to γ decay can be thought of as distributions of electric charge, current and (as the nucleons have intrinsic spin) magnetization moving with periodic motion at a frequency $\omega = \Delta E/\hbar$ and confined to a region of the order of nuclear dimensions. As there is a marked range in measured γ decay lifetimes, selection rules based upon the conservations of angular momentum and parity are suggested. Hence, nuclear radiation emission can be investigated in terms of classical theory of electric and magnetic multi-poles.

The multi-pole expansion in free space is initially applied to nuclear γ emission in order to first establish these selection rules and then to calculate energy, angular momentum and spatial distribution of multi-pole radiation. This derivation shows, inter alia, that $0 \to 0$ γ transitions are impossible in γ emission, although they are permissible in IC. These results are extended to the case of where the source is present and used to estimate the rate at which γ emission occurs as a function of photon energy, atomic mass number, and multi-pole order.

Multi-Pole Expansion in Free Space

Modelling the excited nucleus as a conglomeration of periodically moving electric charges with frequency ω leads to induction and radiation zones based upon the inequalities of $r \ll {}^{c}/_{\omega}$ and $r \gg {}^{c}/_{\omega}$, respectively. One can use analogous inequalities of $r \ll \hbar c/k$ and $r \gg \hbar c/k$ where $\hbar c/k \approx 200\,\text{fm}$ and $\hbar c/k \approx 20\,\text{fm}$ for 1 and 10 MeV photons, respectively. Within the induction zone, the electric and magnetic fields are calculated directly from Maxwell's equations knowing the positions and velocities of the moving charges. Extending into the radiation zone, retardation (i.e., the time delay between when an electromagnetic wave is "emitted" and when it is "detected" at a distance) must be accounted for. The multi-pole expansion in the radiation zone in vacuo and with no source present in the region is evaluated first. In this case, Maxwell's equations are,

$$\nabla \bullet \mathbf{E} = 0 \tag{4.143}$$

$$\nabla \bullet \mathbf{H} = 0 \tag{4.144}$$

$$\nabla \times \mathbf{E} = -\mu_0 \frac{\partial \mathbf{H}}{\partial t} \tag{4.145}$$

$$\nabla \times \mathbf{H} = \varepsilon_0 \frac{\partial \mathbf{E}}{\partial t} \tag{4.146}$$

The fields harmonically vary as $e^{-i\omega t}$ (where it is understood that the real part of the complex quantity is taken) so that the time derivatives are,

$$\frac{\partial \mathbf{E}}{\partial t} = -i\omega \mathbf{E} \tag{4.147}$$

$$\frac{\partial \mathbf{H}}{\partial t} = -i\omega \mathbf{H}. \tag{4.148}$$

The curls of the fields can be written as,

$$\nabla \times \mathbf{E} = i\kappa \sqrt{\frac{\mu_0}{\varepsilon_0}} \mathbf{H} \tag{4.149}$$

$$\nabla \times \mathbf{H} = -i\kappa \sqrt{\frac{\varepsilon_0}{\mu_0}} \mathbf{E}. \tag{4.150}$$

Note that the wave-number $\kappa = \omega/c$ and the relationship $c = 1/\sqrt{\varepsilon_0 \mu_0}$ have been used. In order to develop the multi-pole expansion of the electromagnetic field, the Helmholtz equation is derived for the individual vector components of the fields by taking the curl of (4.149) and substituting (4.150)

$$\begin{aligned} \nabla \times (\nabla \times \mathbf{E}) &= i\kappa \sqrt{\frac{\mu_0}{\varepsilon_0}} \nabla \times \mathbf{H} \\ &= \kappa^2 \mathbf{E}. \end{aligned} \tag{4.151}$$

As the vector triple product is,

$$\begin{aligned} \nabla \times (\nabla \times \mathbf{E}) &= (\nabla \bullet \mathrm{E}) \nabla - \nabla^2 \mathbf{E} \\ &= -\nabla^2 \mathbf{E} \end{aligned} \tag{4.152}$$

then,

$$(\nabla^2 + \kappa^2)\mathbf{E} = \mathbf{0} \tag{4.153}$$

and, similarly,

$$(\nabla^2 + \kappa^2)\mathbf{H} = \mathbf{0}. \tag{4.154}$$

As these are vector equations, it is the Cartesian coordinates of the electric and magnetic fields that will satisfy the Helmholtz equation,

$$(\nabla^2 + \kappa^2) f = 0 \tag{4.155}$$

where f is a scalar function. The Helmholtz equation is solved via a separation of variables and by writing the scalar function as an expansion in spherical harmonics,

$$f(\mathbf{r}) = \sum_{l=1}^{\infty} \sum_{m=-l}^{l} f_l(r) Y_{lm}(\theta, \phi). \tag{4.156}$$

The radial term is solved through a form of (2.104),

$$\left(r^2 \frac{d^2}{dr^2} + 2r \frac{d}{dr} + r^2 \kappa^2 - l(l+1) \right) f_l(r) = 0 \tag{4.157}$$

which is rearranged and, following the substitution $u_l(r) = \sqrt{r} f_l(r)$, arrives at Bessel's equation for half-integer order $(l+1/2)$,

$$\left(\frac{d^2}{dr^2} + \frac{1}{r}\frac{d}{dr} + \kappa^2 - \frac{(l+1/2)^2}{r^2} \right) u_l(r) = 0. \tag{4.158}$$

From (2.109) to (2.111), one can then write a general solution to the Helmholtz equation of the form,

$$\begin{aligned} f(\mathbf{r}; k) = \frac{1}{\sqrt{r}} \sum_{l=1}^{\infty} \sum_{m=-l}^{l} \Big(& A_l J_{l+1/2}(\kappa r) \\ & + B_l N_{l+1/2}(\kappa r) \Big) Y_{lm}(\theta, \phi) \end{aligned} \tag{4.159}$$

where the coefficients A_l and B_l are defined by the boundary conditions. This solution can be simplified by using the Hankel functions which are defined by,

$$h_l^{(1)}(kr) = \sqrt{\frac{\pi}{2\kappa r}} (J_{l+1/2}(\kappa r) + i N_{l+1/2}(\kappa r)) \tag{4.160}$$

$$h_l^{(2)}(\kappa r) = \sqrt{\frac{\pi}{2\kappa r}}(J_{l+1/2}(\kappa r) - iN_{l+1/2}(\kappa r)). \quad (4.161)$$

The general solution to the Helmholtz equation is,

$$f(\mathbf{r};\kappa) = \sum_{l=1}^{\infty}\sum_{m=-l}^{l}\left(A_l h_l^{(1)}(\kappa r) + B_l h_l^{(2)}(\kappa r)\right)Y_{lm}(\theta,\phi). \quad (4.162)$$

Rather than calculate for the field vector Cartesian components from the above, a simpler approach originally devised by Bowkamp and Casimir (1954) is followed in which the scalar products $\mathbf{r} \bullet \mathbf{E}$ and $\mathbf{r} \bullet \mathbf{H}$ are instead solved for and the electric and magnetic fields of the electric and magnetic multi-poles subsequently extracted.[25] To show the suitability of these substitutions, consider the Laplacian acting upon the scalar product $\mathbf{r} \bullet \mathbf{H}$,

$$\nabla^2(\mathbf{r} \bullet \mathbf{E}) = \mathbf{r} \bullet \nabla^2\mathbf{E} + 2\nabla \bullet \mathbf{E} \quad (4.163)$$

so that,

$$(\nabla^2 + \kappa^2)(\mathbf{r} \bullet \mathbf{E}) = \mathbf{r} \bullet \nabla^2\mathbf{E} + 2\nabla \bullet \mathbf{E} + \kappa^2(\mathbf{r} \bullet \mathbf{E}). \quad (4.164)$$

As the second term is equal to zero for the sourceless case and the sum of the first and third terms is equal to zero, then $\mathbf{r} \bullet \mathbf{E}$ satisfies the Helmhotz equation as does $\mathbf{r} \bullet \mathbf{H}$.

The multi-pole expansions of the electromagnetic fields are determined by first calculating for the magnetic multi-pole field of order (l, m). The conditions,

$$\mathbf{r} \bullet \mathbf{H}_{lm}^{(M)} = \frac{l(l+1)}{\kappa} \times \left(A_l h_l^{(1)}(\kappa r) + B_l h_l^{(2)}(\kappa r)\right)Y_{lm}(\theta,\phi) \quad (4.165)$$

$$\mathbf{r} \bullet \mathbf{E}_{lm}^{(M)} = 0 \quad (4.166)$$

are specified.

[25]Jackson (1999) provides a detailed derivation.

The rationale for the inclusion of the $l(l+1)/\kappa$ factor in (4.165) will soon be demonstrated.[26] One can then write a relationship between the scalar product $\mathbf{r} \bullet \mathbf{H}$ and the electric field from (4.149),

$$i\kappa\sqrt{\frac{\mu_0}{\varepsilon_0}}\mathbf{r} \bullet \mathbf{H} = \mathbf{r} \bullet (\nabla \times \mathbf{E})$$

or

$$\kappa\sqrt{\frac{\mu_0}{\varepsilon_0}}\mathbf{r} \bullet \mathbf{H} = \mathbf{L} \bullet \mathbf{E} \quad (4.167)$$

where the operator $\mathbf{L} = -i(\mathbf{r} \times \nabla)$ has been defined. It will be noted that this operator is $\hbar^{-1}$ times the quantum-mechanical angular momentum operator and its properties are briefly reviewed here. From Chap. 2, one has, for the spherical harmonic,

$$-\left(\frac{\partial^2}{\partial\theta^2} + \cot\theta\frac{\partial}{\partial\theta} + \frac{1}{\sin^2\theta}\frac{\partial^2}{\partial\phi^2}\right) \times Y_{lm}(\theta,\phi) = l(l+1)Y_{lm}(\theta,\phi) = L^2\, Y_{lm}(\theta,\phi). \quad (4.168)$$

It is apparent that the $\mathbf{L}$ operator acts only upon the angular variables. Raising and lowering operators can be created from the components of $\mathbf{L}$,

$$L_{\pm} = L_x \pm iL_y = e^{\pm i\phi}\left(\pm\frac{\partial}{\partial\theta} + i\cot\theta\frac{\partial}{\partial\phi}\right) \quad (4.169)$$

$$L_z = -i\frac{\partial}{\partial\phi}. \quad (4.170)$$

Applying these to the spherical harmonics,

$$L_{\pm}Y_{lm}(\theta,\phi) = \sqrt{(l \mp m)(l \pm m + 1)}Y_{l m\pm1}(\theta,\phi) \quad (4.171)$$

[26]As the electric field is transverse to the radius vector, the magnetic multipole field is sometimes referred to, especially in engineering textbooks, as a transverse electric (TE) field. Similarly, the electric multi-pole field is also referred to as a transverse magnetic (TM) field.

$$L_z Y_{lm}(\theta,\phi) = m\, Y_{lm}(\theta,\phi). \tag{4.172}$$

L acts upon the spherical harmonic to transform the m value leaving the l value unaffected. Consequently, the electric field of the magnetic multi-pole must satisfy,

$$\mathbf{L}\bullet\mathbf{E}^{(M)}_{lm} = l(l+1)\sqrt{\frac{\mu_0}{\varepsilon_0}} \times \left(A_l h_l^{(1)}(\kappa r) + B_l h_l^{(2)}(\kappa r)\right) Y_{lm}(\theta,\phi) \tag{4.173}$$

which demonstrates the convenience of the $l(l+1)/\kappa$ factor introduced in (4.165). Recalling the properties of the **L** operator, the electric field of the magnetic multi-pole can be extracted from this result. First, as the **L** operator acts only upon the angular variables,

$$\mathbf{L}\left(\mathbf{L}\bullet\mathbf{E}^{(M)}_{lm}\right) = l(l+1)\sqrt{\frac{\mu_0}{\varepsilon_0}} \times \left(A_l h_l^{(1)}(\kappa r) + B_l h_l^{(2)}(\kappa r)\right)\mathbf{L}Y_{lm}(\theta,\phi). \tag{4.174}$$

As $\mathbf{L}\left(\mathbf{L}\bullet\mathbf{E}^{(M)}_{lm}\right) = L^2\mathbf{E}^{(M)}_{lm} = l(l+1)\mathbf{E}^{(M)}_{lm}$, this result then gives the electric field of the magnetic multi-pole as,

$$\mathbf{E}^{(M)}_{lm} = \sqrt{\frac{\mu_0}{\varepsilon_0}}\left(A_l h_l^{(1)}(\kappa r) + B_l h_l^{(2)}(\kappa r)\right) \times \mathbf{L}Y_{lm}(\theta,\phi) = \sqrt{\frac{\mu_0}{\varepsilon_0}}\sqrt{l(l+1)} \times \left(A_l h_l^{(1)}(\kappa r) + B_l h_l^{(2)}(\kappa r)\right)\mathbf{X}_{lm}(r,\phi) \tag{4.175}$$

where the normalized vector spherical harmonic is defined as,

$$\mathbf{X}_{lm}(r,\phi) = \frac{1}{\sqrt{l(l+1)}}\mathbf{L}Y_{lm}(r,\phi). \tag{4.176}$$

The normalized vector spherical harmonic has the orthonormality,

$$\int d\Omega\, \mathbf{X}^*_{l'm'}(\theta,\phi)\bullet\mathbf{X}_{lm}(\theta,\phi) = \delta_{ll'}\delta_{mm'} \tag{4.177}$$

and

$$\int d\Omega\, \mathbf{X}^*_{l'm'}(\theta,\phi)\bullet(\mathbf{r}\times\mathbf{X}_{lm}(\theta,\phi)) = 0. \tag{4.178}$$

The magnetic field of the magnetic multi-pole is then defined from (4.149),

$$\mathbf{H}^{(M)}_{lm} = -\frac{i}{\kappa}\sqrt{\frac{\varepsilon_0}{\mu_0}}\nabla\times\mathbf{E}^{(M)}_{lm}. \tag{4.179}$$

Repeating the above derivation for the electric and magnetic fields of the electric multi-pole gives,

$$\mathbf{H}^{(E)}_{lm} = \sqrt{l(l+1)} \times \left(C_l h_l^{(1)}(\kappa r) + D_l h_l^{(2)}(\kappa r)\right)\mathbf{X}_{lm}(\theta,\phi) \tag{4.180}$$

$$\mathbf{E}^{(M)}_{lm} = \frac{i}{\kappa}\sqrt{\frac{\mu_0}{\varepsilon_0}}\nabla\times\mathbf{H}^{(E)}_{lm}. \tag{4.181}$$

These allow the electric and magnetic fields to be written as the sums of the multi-pole fields,

$$\mathbf{E}(\mathbf{r}) = \sum_{l=1}^{\infty}\sum_{m=-l}^{l}\left[\frac{i}{\kappa}a^{(E)}_{lm}(\nabla\times f_l(\kappa r)\mathbf{X}_{lm}(\theta,\phi)) + a^{(M)}_{lm}g_l(\kappa r)\mathbf{X}_{lm}(\theta,\phi)\right] \tag{4.182}$$

$$\mathbf{H}(\mathbf{r}) = \sum_{l=1}^{\infty}\sum_{m=-l}^{l}\left[a^{(E)}_{lm}f_l(\kappa r)\mathbf{X}_{lm}(\theta,\phi) - \frac{i}{\kappa}a^{(M)}_{lm}(\nabla\times g_l(\kappa r)\mathbf{X}_{lm}(\theta,\phi))\right] \tag{4.183}$$

where

$$f_l(kr) = \left(A_l h_l^{(1)}(\kappa r) + B_l h_l^{(2)}(\kappa r)\right) \tag{4.184}$$

and

$$g_l(kr) = \left(C_l h_l^{(1)}(\kappa r) + D_l h_l^{(2)}(\kappa r)\right). \quad (4.185)$$

The coefficients $a_{lm}^{(E)}$ and $a_{lm}^{(M)}$ indicate the contributing amounts of electric and magnetic multi-poles, respectively, to the fields and are specified by the boundary conditions and sources as determined through,

$$a_{lm}^{(E)} f_l(r) = -\frac{\kappa}{\sqrt{l(l+1)}}\sqrt{\frac{\varepsilon_0}{\mu_0}} \times \int d\Omega\, Y_{lm}^*(\theta,\phi)\mathbf{r} \bullet \mathbf{E}(\mathbf{r}) \quad (4.186)$$

$$a_{lm}^{(M)} g_l(r) = \frac{\kappa}{\sqrt{l(l+1)}} \times \int d\Omega\, Y_{lm}^*(\theta,\phi)\mathbf{r} \bullet \mathbf{H}(\mathbf{r}). \quad (4.187)$$

Equations (4.182) and (4.183) demonstrate the dual transformation relationship between the fields of an electric multi-pole and those of a magnetic multi-pole through which one can transfer from one multi-pole to another by interchanging the electric and magnetic fields and changing the sign of the electric field.

Before concluding this introduction to multi-pole radiation, consider one characteristic of the vector spherical harmonic which will be of importance. This is, that for $l = 0$,

$$-i(\mathbf{r} \times \nabla)Y_{00}(\theta,\phi) = 0 \quad (4.188)$$

with the result that there is no multi-pole radiation for $l = 0$. In other words, there is no transition between nuclear states both with $l = 0$ that can result in γ-ray emission

Energy and Angular Momentum of Multi-Pole Radiation

The energy and angular momentum of multi-pole radiation are derived through calculating the electric and magnetic fields of the electric and magnetic multi-poles in the radiation zone. This result will then be used to demonstrate that the radiation emitted by a multi-pole of order (l, m) carries away $m\hbar$ of the z-component of angular momentum for each emitted photon of energy $\hbar\omega$ (in the reference frame of the nucleus). Begin by calculating the multi-pole fields in the radiation zone, within which the asymptotic forms of the spherical Bessel functions are, from Chap. 2,

$$j_l(\kappa r) \rightarrow \frac{\sin(\kappa r - l\pi/2)}{\kappa r} \quad \kappa r \rightarrow \infty \quad (4.189)$$

$$y_l(\kappa r) \rightarrow -\frac{\cos(\kappa r - l\pi/2)}{\kappa r} \quad \kappa r \rightarrow \infty \quad (4.190)$$

and the Hankel functions become,

$$h_l^{(1)}(\kappa r) \approx \frac{\sin(\kappa r - l\pi/2)}{\kappa r} - i\frac{\cos(\kappa r - l\pi/2)}{\kappa r} \approx -ie^{i\kappa r} \approx -i^{(l+1)}\frac{e^{i\kappa r}}{\kappa r} \quad \kappa r \gg 1 \quad (4.191)$$

$$h_l^{(2)}(\kappa r) \approx \frac{\sin(\kappa r - l\pi/2)}{\kappa r} + i\frac{\cos(\kappa r - l\pi/2)}{\kappa r} \approx -i^l\frac{e^{-i\kappa r}}{\kappa r} \quad \kappa r \gg 1. \quad (4.192)$$

These far-field approximations are then applied to the previously-derived expressions for the electric and magnetic fields of multi-pole radiation. From (4.175), the electric field of the magnetic multi-pole in the radiation zone is,

$$\mathbf{E}_{lm}^{(M)} = -i^l\sqrt{\frac{\mu_0}{\varepsilon_0}} \times \frac{\left(B_1 e^{-i\kappa r} + iA_1 e^{i\kappa r}\right)}{\kappa r}\mathbf{X}_{lm}(r,\phi) \quad (4.193)$$
$$\kappa r \gg 1$$

and, for an outgoing wave,

$$\mathbf{E}_{lm}^{(M)} = -i^{(l+1)}\sqrt{\frac{\mu_0}{\varepsilon_0}}\sqrt{l(l+1)}\frac{e^{i\kappa r}}{\kappa r}\mathbf{X}_{lm}(\theta,\phi) \quad (4.194)$$
$$\kappa r \gg 1$$

and the magnetic field of the magnetic multi-pole will be given by,

$$\mathbf{H}_{lm}^{(M)} = \sqrt{\frac{\varepsilon_0}{\mu_0}}\mathbf{E}_{lm}^{(M)} \times \hat{\mathbf{r}}. \tag{4.195}$$

From (4.180), the magnetic field of the electric multi-pole in the radiation zone is,

$$\mathbf{H}_{lm}^{(E)} = -\mathrm{i}^l\sqrt{l(l+1)}\frac{(\mathrm{D}_l\mathrm{e}^{-\mathrm{i}\kappa r} + \mathrm{i}\mathrm{C}_l\mathrm{e}^{\mathrm{i}\kappa r})}{\kappa r} \times \mathbf{X}_{lm}(\theta,\phi) \quad \kappa r \gg 1 \tag{4.196}$$

and, for an outgoing wave,

$$\mathbf{H}_{lm}^{(E)} = -\mathrm{i}^{(l+1)}\sqrt{l(l+1)}\frac{\mathrm{e}^{\mathrm{i}\kappa r}}{\kappa r}\mathbf{X}_{lm}(\theta,\phi) \quad kr \gg 1 \tag{4.197}$$

and the electric field of the electric multi-pole is,

$$\mathbf{E}_{lm}^{(E)} = \frac{\mathrm{i}^l}{\kappa^2}\sqrt{\frac{\mu_0}{\varepsilon_0}}\sqrt{l(l+1)}\nabla \times \left(\frac{\mathrm{e}^{\mathrm{i}\kappa r}}{r}\mathbf{X}_{lm}(\theta,\phi)\right) \tag{4.198}$$

As these are in the far-field region, the curl is calculated to powers no greater than $1/r$ and the identity $\nabla \times \mathbf{L} = -\mathrm{i}\mathbf{r}\nabla^2 - \nabla\left(1 + r\frac{\partial}{\partial r}\right)$ used to give, for the electric field of the electric multi-pole,

$$\mathbf{E}_{lm}^{(E)} = -\mathrm{i}^{(l+1)}\sqrt{\frac{\mu_0}{\varepsilon_0}}\frac{\mathrm{e}^{\mathrm{i}\kappa r}}{\kappa r}\left(\sqrt{l(l+1)}\hat{\mathbf{r}} \times \mathbf{X}_{lm}(\theta,\phi) - \frac{1}{\kappa}\left(\mathbf{r}\nabla^2 - \nabla\right)\mathrm{Y}_{lm}\left(\theta,\phi\right)\right) \tag{4.199}$$

where $\hat{\mathbf{r}}$ is the unit radial vector. The second term can be neglected,

$$\mathbf{E}_{lm}^{(E)} = -\mathrm{i}^{(l+1)}\sqrt{\frac{\mu_0}{\varepsilon_0}}\frac{\mathrm{e}^{\mathrm{i}\kappa r}}{\kappa r} \times \left(\sqrt{l(l+1)}\hat{\mathbf{r}} \times \mathbf{X}_{lm}(\theta,\phi)\right). \tag{4.200}$$

Using (4.198),

$$\mathbf{E}_{lm}^{(E)} = \sqrt{\frac{\mu_0}{\varepsilon_0}}\mathbf{H}_{lm}^{(E)} \times \hat{\mathbf{r}}. \tag{4.201}$$

It is now demonstrated that the radiation emitted by a multi-pole of order (l,m) carries away $m\hbar$ of the z-component of angular momentum for each photon of energy $\hbar\omega$, doing so for a pure multi-pole field (in this case, the electric). Following the example of Jackson (1999), consider the electromagnetic field to be a linear superposition of electric multi-poles of order (l,m) but with the requirement that l be the same for the multi-poles but each have a different value of m. The magnetic field of a pure electric multi-pole (i.e., $a_{lm}^{(M)} = 0$) is, assuming a harmonic time dependence,

$$\mathbf{H}(\mathbf{r}) = \sum_{m=-l}^{l} a_{lm}^{(E)}h_{lm}^{(1)}(\kappa r)\mathbf{X}_{lm}(\theta,\phi)\mathrm{e}^{-\mathrm{i}\omega t} \tag{4.202}$$

which, in the radiation zone is,

$$\mathbf{H}(\mathbf{r}) = \sum_{m=-l}^{l} a_{lm}^{(E)}h_{lm}^{(1)}(\kappa r)\mathbf{X}_{lm}(\theta,\phi)\mathrm{e}^{-\mathrm{i}\omega t} \approx -\mathrm{i}^{(l+1)}\frac{\mathrm{e}^{\mathrm{i}\kappa r}}{\kappa r} \quad \kappa r \gg 1 \tag{4.203}$$

and the corresponding electric field is,

$$\mathbf{E}(\mathbf{r}) = \frac{\mathrm{i}}{\kappa}\sqrt{\frac{\mu_0}{\varepsilon_0}}\nabla \times \mathbf{H}(\mathbf{r}). \tag{4.204}$$

The time-averaged energy density is,

$$u = \frac{1}{4}\left[\varepsilon_0\mathbf{E}(\mathbf{r}) \bullet \mathbf{E}^*(\mathbf{r}) + \mu_0\mathbf{H}(\mathbf{r}) \bullet \mathbf{H}^*(\mathbf{r})\right] \tag{4.205}$$

and the change in energy with radius is,

$$\frac{dE}{dr} = \frac{\mu_0}{2\kappa^2}\sum_{m=-l}^{l}\left|a_{lm}^{(E)}\right|^2. \tag{4.206}$$

In the case of the field being made up of electric and magnetic multi-poles, this would be,

$$\frac{dE}{dr} = \frac{\mu_0}{2\kappa^2} \sum_{m=-l}^{l} \left(\left| a_{lm}^{(E)} \right|^2 + \left| a_{lm}^{(M)} \right|^2 \right). \qquad (4.207)$$

Next consider the angular momentum density which one obtains from the electromagnetic field linear momentum. This follows from the Lorentz equation and Newton's second law for a spatially-distributed charge,

$$\frac{d\mathbf{P}}{dt} = \int d^3\mathbf{r}(\rho_e \mathbf{E} + \mathbf{J} \times \mathbf{B}) \qquad (4.208)$$

where ρ_e and $\mathbf{J}$ are the electric charge and current densities. Using Maxwell's equations for the case of the fields in free space and with a source present,

$$\nabla \bullet \mathbf{E} = \frac{\rho_e}{\varepsilon_0} \qquad (4.209)$$

$$\mathbf{J} = \nabla \times \mathbf{H} - \varepsilon_0 \frac{\partial \mathbf{E}}{\partial t} \qquad (4.210)$$

the integrand can be rewritten, using $\varepsilon_0 \mu_0 = 1/c^2$, as,

$$\begin{aligned} &\rho_e \mathbf{E} + \mathbf{J} \times \mathbf{B} \\ &= \varepsilon_0 \left[\mathbf{E}(\nabla \bullet \mathbf{E}) + \mathbf{B} \times \frac{\partial \mathbf{E}}{\partial t} - c^2 \mathbf{B} \times (\nabla \times \mathbf{B}) \right]. \end{aligned} \qquad (4.211)$$

With some additional manipulations, the momentum of the electromagnetic field arises from the volume integral,

$$\begin{aligned} \mathbf{p} &= \varepsilon_0 \int d^3\mathbf{r}(\mathbf{E} \times \mathbf{B}) \\ &= \frac{1}{c^2} \int d^3\mathbf{r}(\mathbf{E} \times \mathbf{H}). \end{aligned} \qquad (4.212)$$

The angular momentum then follows as,

$$\mathbf{m} = \frac{1}{c^2} \int d^3\mathbf{r}\, \mathbf{r} \times (\mathbf{E} \times \mathbf{H}) \qquad (4.213)$$

from which one can obtain the time-averaged angular momentum density,

$$\mathbf{m} = \frac{1}{2c^2} \mathrm{Re}\left[\mathbf{r} \times (\mathbf{E} \times \mathbf{H}^*)\right]. \qquad (4.214)$$

Expanding the vector triple cross product, $\mathbf{r} \times (\mathbf{E} \times \mathbf{H}^*) = (\mathbf{r} \bullet \mathbf{H}^*)\mathbf{E} - (\mathbf{r} \bullet \mathbf{E})\mathbf{H}^*$, leads to,

$$\mathbf{m} = \frac{\mu_0}{2\omega} \mathrm{Re}\left[\mathbf{H}^*(\mathbf{L} \bullet \mathbf{H})\right]. \qquad (4.215)$$

The change in the z-component of the angular momentum with radius is,

$$\frac{dm_z}{dr} = \frac{\mu_0}{2\omega\,\kappa^2} \sum_{m=-l}^{l} m \left| a_{lm}^{(E)} \right|^2. \qquad (4.216)$$

Comparing this result with the derivative of (4.207) leads to,

$$\begin{aligned} \frac{dm_z}{dr} &= \frac{m}{\omega} \frac{dE}{dr} \\ &= \frac{\hbar m}{\hbar \omega} \frac{dE}{dr}. \end{aligned} \qquad (4.217)$$

This result is interpreted as representing that, for multi-pole radiation of order (l, m), a photon emitted with an energy $\hbar\omega$ carries away $m\hbar$ units of the z-component of angular momentum.

Selection Rules for Multi-Pole Radiation

In a radiative transition, the state changes from ψ_i to ψ_f and the overall angular momentum is conserved. If the quantum carries away angular momentum l with z-component m (all in units of $\hbar$), then the conservation of angular momentum stipulates that,

$$\mathbf{J}_i = l \oplus \mathbf{J}_f \qquad (4.218)$$

where $\mathbf{J}_i$ and $\mathbf{J}_f$ are the total angular momenta of the initial and final nuclear states leading to multi-pole radiation of order (l,m) being emitted only if,

$$|\mathbf{J}_i - \mathbf{J}_f| \le l \le |\mathbf{J}_i + \mathbf{J}_f| \qquad (4.219)$$

and

$$m_{z,i} - m_{z,f} = m \qquad (4.220)$$

As, from (4.188), there is no $l = 0$ multi-pole radiation, a radiative transition between nuclear states with $J_i = J_f = 0$ will be forbidden. In addition to this angular momentum selection rule, an additional selection rule arises as a result of the conservation of parity. The parity of the final state is equal to the product of the parity of the final nuclear state and that of the multi-pole radiation. That is, the matrix element of the transition will be non-zero when,

$$\Pi_i = \Pi_f \quad \text{for even} - \text{parity radiation} \tag{4.221}$$

$$\Pi_i = -\Pi_f \quad \text{for odd} - \text{parity radiation.} \tag{4.222}$$

The magnetic field can be used to specify the parity, the rationale for which becomes apparent by noting that the matrix element of the transition is proportional to $\int d^3\mathbf{r}\, \psi_f(\mathbf{J} \bullet \mathbf{A})\psi_i$ (where the $\mathbf{J}$ is the current operator and is the electromagnetic vector potential). $\mathbf{A}$ has the same parity as the electric field, $\mathbf{E}$. As $\mathbf{J}$ is a polar vector, it will have negative parity and the scalar product $\mathbf{J} \bullet \mathbf{A}$ will thus have the opposite parity of the electric field. Because the curl operation changes parity, we then have the result that $\mathbf{J} \bullet \mathbf{A}$ has the same parity as $\mathbf{H}$. Consider, first, the electric multi-pole field where the electric and magnetic fields of the multi-pole are

$$\mathbf{H}^{(E)}_{lm} = \sqrt{l(l+1)}\Big(C_l h^{(1)}_l(\kappa r) + D_l h^{(2)}_l(\kappa r)\Big)\mathbf{X}_{lm}(\theta, \phi)$$

and

$$\mathbf{E}^{(E)}_{lm} = \frac{i}{\kappa}\sqrt{\frac{\mu_0}{\varepsilon_0}}\nabla \times \mathbf{H}^{(E)}_{lm}.$$

As the spherical harmonic is $\mathbf{X}_{lm}(\theta, \phi) = \frac{i}{\sqrt{l(l+1)}}(\mathbf{r} \times \nabla)Y_{lm}(\theta, \phi)$ and $\mathbf{r} \times \nabla$ is invariant under inversion, the parity properties of $\mathbf{H}^{(E)}_{lm}$ are given by those of the spherical harmonic, $Y_{lm}(r, \phi)$, which has parity $(-1)^l$ and the parity of electric multi-pole radiation will be given by $(-1)^{l+1}$. Table 4.6 summarizes the selection rules for multi-pole electromagnetic radiation. Let us use a pragmatic example to describe these selection rules. Consider an excited nuclear state with $J_i = 1/2$ which de-excites to a state with $J_f = 3/2$ through the emission of a γ ray. From (4.219), $l = 1, 2$ so that the only allowed transitions are dipole ($l = 1$) and quadrupole ($l = 2$). If the parities of the initial and final nuclear states are the same ($\Pi_i = \Pi_f$), then only the magnetic dipole (M1) and electric quadrupole (E2) transitions can occur. Should, the parities of the initial and final nuclear states differ ($\Pi_i = -\Pi_f$), then only electric dipole (E1) and magnetic quadrupole (M2) transitions are allowed.

Table 4.6 Selection rules for electromagnetic multipole radiation

Rule	Dipole E1	Quadrupole E2	Sextupole E3	Octupole E4
	E1	E2	E3	E4
$\Delta\Pi$	1	0	1	0
$\lvert\Delta J\rvert$	1	2	3	4
	M1	M2	M3	M4
$\Delta\Pi$	0	1	0	1
$\lvert\Delta J\rvert$	1	2	3	4

Angular Distributions of Multi-Pole Radiation

From the above, it can be seen that the time-averaged power radiated will be proportional to $|\mathbf{X}_{lm}(\theta, \phi)|^2$. Table 4.7 presents the angular distributions of some dipole and quadrupole radiations and which are shown in Fig. 4.31. The sum of the squared magnitudes of the vector spherical harmonics for a set of multi-poles of order l is,

Table 4.7 Angular distributions of dipole and quadrupole radiations

Type	l	m 0	±1	±2
Dipole	1	$\left(\frac{3}{8\pi}\right)\sin^2\theta$	$\left(\frac{3}{16\pi}\right)(1+\cos^2\theta)$	–
Quadrupole	2	$\left(\frac{15}{8\pi}\right)\sin^2\theta\cos^2\theta$	$\left(\frac{5}{16\pi}\right)(1-3\cos^2\theta+4\cos^4\theta)$	$\left(\frac{5}{16\pi}\right)(1-\cos^4\theta)$

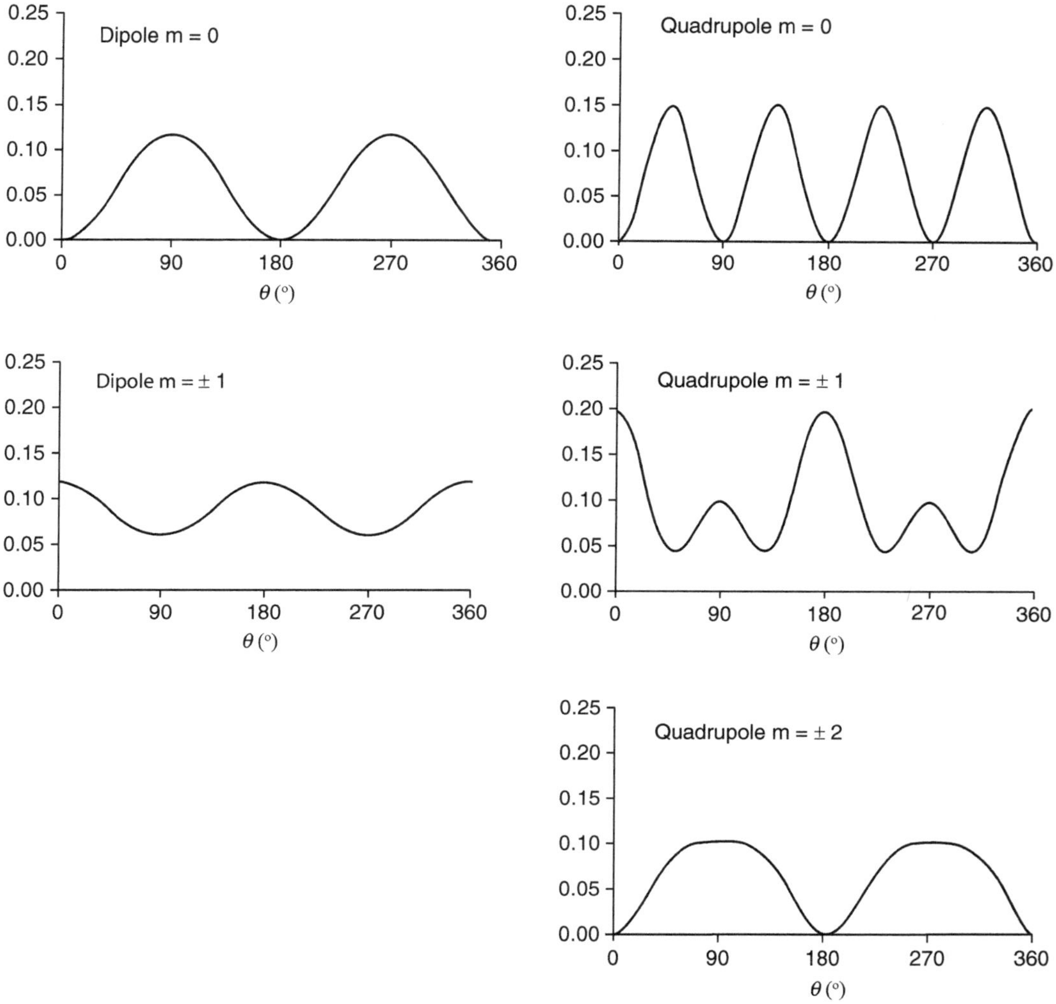

Fig. 4.31 Angular distributions of dipole and quadrupole radiations

$$\sum_{m=-l}^{l} |\mathbf{X}_{lm}(\theta, \phi)|^2 = \frac{2l+1}{4\pi} \tag{4.223}$$

which demonstrates that the radiation is isotropic from a radiating source consisting of this set of multi-poles with multi-poles added incoherently.

Multi-Pole Expansion With Source Present

It is necessary to connect the multi-pole fields with the sources that have generated them; i.e., the coefficients $a_{lm}^{(E)}$ and $a_{lm}^{(M)}$ are related to the source that has produced the multi-pole radiation. The source, which is the nucleus, is considered to be a harmonically-varying distribution of charge, current and magnetization,

$$\rho(\mathbf{r}, t) = \rho(\mathbf{r})e^{-i\omega t} \tag{4.224}$$

$$\mathbf{J}(\mathbf{r}, t) = \mathbf{J}(\mathbf{r})e^{-i\omega t} \tag{4.225}$$

$$\mathbf{M}(\mathbf{r}, t) = \mathbf{M}(\mathbf{r})e^{-i\omega t} \tag{4.226}$$

where it is understood that the real part of the complex quantities are to be taken.[27] Maxwell's equations are now,

[27] It is also possible to simply write, for example, $\rho(\mathbf{r}, t) = 1/2(\rho(\mathbf{r})e^{-i\omega t} + \rho^*(\mathbf{r})e^{+i\omega t})$ to achieve the same result.

$$\nabla \bullet \mathbf{E} = \frac{\rho}{\varepsilon_0} \tag{4.227}$$

$$\nabla \bullet \mathbf{B} = 0 \tag{4.228}$$

$$\nabla \times \mathbf{E} = i\omega \mathbf{B} \tag{4.229}$$

$$\nabla \times \mathbf{B} = \mu_0(\mathbf{J} + \nabla \times \mathbf{M}) - i\frac{\omega}{c^2}\mathbf{E}. \tag{4.230}$$

Some simplification is possible by defining $\mathbf{H}' = \mathbf{B}/\mu_0$ and $\mathbf{E}' = \mathbf{E} + \frac{i}{\omega\varepsilon_0}\mathbf{J}$ to yield the set of equations,

$$\nabla \bullet \mathbf{E}' = 0 \tag{4.231}$$

$$\nabla \bullet \mathbf{H}' = 0 \tag{4.232}$$

$$\nabla \times \mathbf{E} = \frac{i}{\omega\varepsilon_0}\nabla \times \mathbf{J} + \omega\mu_0\mathbf{H}' \tag{4.233}$$

$$\nabla \times \mathbf{H}' = \nabla \times \mathbf{M} - i\kappa\sqrt{\frac{\varepsilon_0}{\mu_0}}\mathbf{E}'. \tag{4.234}$$

From the two curl equations,

$$(\nabla^2 + \kappa^2)\mathbf{E}' = -i\kappa\sqrt{\frac{\mu_0}{\varepsilon_0}}\nabla \times \left(\mathbf{M} + \frac{\nabla \times \mathbf{J}}{\kappa^2}\right) \tag{4.235}$$

$$(\nabla^2 + \kappa^2)\mathbf{H}' = -\nabla \times (\mathbf{J} + \nabla \times \mathbf{M}). \tag{4.236}$$

As before, the scalars $\mathbf{r} \bullet \mathbf{E}'$ and $\mathbf{r} \bullet \mathbf{H}'$ are solved for. Using (4.163), (4.235) and (4.236), and the relationship for an arbitrary vector field $\mathbf{F}$,

$$\begin{aligned} \mathbf{r} \bullet (\nabla \times \mathbf{F}) &= (r \times \nabla) \bullet \mathbf{F} \\ &= i\mathbf{L} \bullet \mathbf{F} \end{aligned} \tag{4.237}$$

one obtains the inhomogeneous equations,

$$(\nabla^2 + \kappa^2)\mathbf{r} \bullet \mathbf{E}' = \omega\mu_0\mathbf{L} \bullet \left(\mathbf{M} + \frac{\nabla \times \mathbf{J}}{\kappa^2}\right) \tag{4.238}$$

$$(\nabla^2 + \kappa^2)\mathbf{r} \bullet \mathbf{H}' = -i\mathbf{L} \bullet (\mathbf{J} + \nabla \times \mathbf{M}). \tag{4.239}$$

These can be solved using Green's functions (cf Chap. 2). However, as the solution is involved, it will not be repeated here for clarity; the interested reader is referred to Chap. 9 of Jackson (1999) or Chap. XII of Blatt and Weisskopf (1979). The result of the calculation is that exact results of the multipole coefficients are given by the expressions,

$$a_{lm}^{(E)} = -i\frac{\kappa^2}{\sqrt{l(l+1)}}\int d^3\mathbf{r}\, Y_{lm}(\theta,\phi) \times \left\{ \begin{array}{c} c\rho(\mathbf{r})\frac{\partial}{\partial r}(rj_l(\kappa r)) + i\kappa(\mathbf{r} \bullet \mathrm{J}(\mathbf{r}))j_l(\kappa r) \\ -i\kappa\nabla \bullet (\mathbf{r} \times \mathbf{M})j_l(\kappa r) \end{array} \right\} \tag{4.240}$$

$$a_{lm}^{(M)} = -i\frac{\kappa^2}{\sqrt{l(l+1)}} \times \int d^3\mathbf{r}\, Y_{lm}(\theta,\phi) \left\{ \begin{array}{c} \nabla \bullet (\mathbf{r} \times \mathbf{J}(\mathbf{r}))j_l(\kappa r) + \\ \nabla \bullet \mathbf{M}(\mathbf{r})\frac{\partial}{\partial r}(rj_l(\kappa r)) \\ -k^2(\mathbf{r} \bullet \mathrm{M}(\mathbf{r}))j_l(\kappa r) \end{array} \right\}. \tag{4.241}$$

Significant simplifications of these exact results are possible when it is recognized that the source dimensions (i.e., nuclear size) are much smaller than the photon wavelength. In this case, the small-argument limit of the spherical Bessel function,

$$j_l(x) \approx \frac{x^l}{(2l+1)!!} \quad x \ll 1 \tag{4.242}$$

can be used. Keeping the smallest powers in κr in the integrals, simplified expressions for the multi-pole coefficients of the form are obtained,

$$a_{lm}^{(E)} \cong -i\frac{c\kappa^{l+2}}{(2l+1)!!}\sqrt{\frac{l+1}{l}}\left(Q_{lm} + Q'_{lm}\right) \tag{4.243}$$

$$a_{lm}^{(M)} \cong i\frac{c\kappa^{l+2}}{(2l+1)!!}\sqrt{\frac{l+1}{l}}\left(M_{lm} + M'_{lm}\right). \tag{4.244}$$

Q_{lm} and M_{lm} are the electric and magnetic multipole moments, respectively, given by,

$$Q_{lm} = \int d^3\mathbf{r}\, r^l\, Y_{lm}^*(\theta,\phi)\rho_e(r) \tag{4.245}$$

$$M_{lm} = \int d^3\mathbf{r}\, r^l\, Y^*_{lm}(\theta, \phi) \frac{\nabla \bullet (\mathbf{r} \times \mathbf{J}(\mathbf{r}))}{l+1}. \quad (4.246)$$

The integrations are over the nuclear volume. In addition to these radiation contributions, it is also necessary to consider those linked to the nucleons' intrinsic spins which, classically, produce a spatial distribution of magnetization **M** changing with time. The corresponding electric and magnetic multi-pole moments are,

$$Q'_{lm} = -i\left(\frac{\kappa}{l+1}\right) \int d^3\mathbf{r}\, r^l\, Y^*_{lm}(\theta, \phi) \times \nabla \bullet (\mathbf{r} \times \mathbf{M}) \quad (4.247)$$

$$M'_{lm} = -i \int d^3\mathbf{r}\, r^l\, Y^*_{lm}(\theta, \phi) \nabla \bullet \mathbf{M} \quad (4.248)$$

The multi-pole expansions of the electric and magnetic fields are,

$$\mathbf{E}(\mathbf{r}) = \sum_{l=1}^{\infty} \sum_{m=-l}^{l} \left(a^{(E)}_{lm} \mathbf{E}^{(E)}_{lm}(\mathbf{r}) + a^{(M)}_{lm} \mathbf{E}^{(M)}_{lm}(\mathbf{r}) \right) \quad (4.249)$$

$$\mathbf{H}(\mathbf{r}) = \sum_{l=1}^{\infty} \sum_{m=-l}^{l} \left(a^{(E)}_{lm} \mathbf{H}^{(E)}_{lm}(\mathbf{r}) + a^{(M)}_{lm} \mathbf{H}^{(M)}_{lm}(\mathbf{r}) \right) \quad (4.250)$$

Transition Rates for Multi-Pole Radiation

The above derivations of the multi-pole transitions can be used to estimate the rate at which a nucleus can decay through γ emission and what nuclear characteristics dictate this rate. The power radiated in multi-pole radiation is,

$$P^{(E)}(l, m) = \frac{1}{2\kappa^2} \sqrt{\frac{\mu_0}{\varepsilon_0}} \left| a^{(E)}_{l,m} \right|^2 \quad (4.251)$$

Electric radiation

$$P^{(M)}(l, m) = \frac{1}{2\kappa^2} \sqrt{\frac{\mu_0}{\varepsilon_0}} \left| a^{(M)}_{l,m} \right|^2 \quad (4.252)$$

Magnetic radiation.

As the transition rate will be this radiated power divided by the photon energy $\hbar\omega$, the transition rates for El and Ml radiation are, respectively,

$$\lambda^{(E)}(l, m) = \frac{8\pi(l+1)}{\hbar l((2l+1)!!)^2} \left(\frac{\omega}{c}\right)^{2l+1} \left| Q_{lm} + Q'_{lm} \right|^2 \quad (4.253)$$

$$\lambda^{(M)}(l, m) = \frac{8\pi(l+1)}{\hbar l((2l+1)!!)^2} \left(\frac{\omega}{c}\right)^{2l+1} \left| M_{lm} + M'_{lm} \right|^2 \quad (4.254)$$

The moments are calculated as follows (see, e.g., Segrè (1977)). From (4.245) to (4.248), for a nucleus of mass m with A nucleons and Z protons,

$$Q_{lm} = e \sum_{n=1}^{Z} \int d^3\mathbf{r}\, r^l_n\, Y^*_{lm}(\theta_n, \phi_n) \psi^*_f(\mathbf{r}) \psi_i(\mathbf{r}) \quad (4.255)$$

$$M_{lm} = -\left(\frac{l}{l+1}\right)\left(\frac{e\hbar}{m_n c}\right) \times \sum_{n=1}^{Z} \int d^3\mathbf{r}\, r^l_n\, Y^*_{lm}(\theta_n, \phi_n) \nabla \bullet \left(\psi^*_f(\mathbf{r}) \mathbf{L}_n \psi_i(\mathbf{r})\right) \quad (4.256)$$

$$Q'_{lm} = -i\frac{(\omega/c)}{l+1}\left(\frac{e\hbar}{2m_n c}\right) \sum_{n=1}^{A} \int d^3\mathbf{r}\, r^l_n\, Y^*_{lm}(\theta_n, \phi_n) \times \nabla \bullet \left(\psi^*_f(\mathbf{r})(\mathbf{r}_n \times \boldsymbol{\sigma})_n \psi_i(\mathbf{r})\right) \quad (4.257)$$

$$M'_{lm} = -\left(\frac{e\hbar}{2m_n c}\right) \times \sum_{n=1}^{A} \int d^3\mathbf{r}\, \mu_n\, r^l_n Y^*_{lm}(\theta_n, \phi_n) \nabla \bullet \left(\psi^*_f(\mathbf{r}) \boldsymbol{\sigma}_n \psi_i(\mathbf{r})\right) \quad (4.258)$$

where μ_n is the magnetic moment of the nth nucleon in units of $e\hbar/2m_n c$. For the moments due to charge, the summations are over the protons only; for those due to spin, the summations are over all nucleons. The transition probabilities are obtained by averaging $\lambda^{(E)}(l, m)$ and $\lambda^{(M)}(l, m)$ over the initial m_i states and summed over the final m_f states,

$$\lambda^{(E)}(l) = \frac{1}{2J_i + 1} \sum_{m_i} \sum_{m_f} \lambda^{(E)}(l, m) \tag{4.259}$$

and

$$\lambda^{(M)}(l) = \frac{1}{2J_i + 1} \sum_{m_i} \sum_{m_f} \lambda^{(M)}(l, m). \tag{4.260}$$

A transition rate calculation based upon the above expressions for the multi-pole moments is clearly quite involved and requires detailed knowledge of the nucleus. However, calculations for simple scenarios are possible, e.g., that of a single nucleon outside closed shells making an electromagnetic transition. Following the approach of Weisskopf (1951), a number of simplifications can be employed in order to solve for the moments so as to obtain expressions for electric and magnetic transition rates. The results are, while not accurate, elucidating and useful. One begins by invoking the independent-particle model of Chap. 3 in which there is a single nucleon outside a closed shell. Consider the case of the single proton being in an initial state with orbital angular momentum l making a transition to a final state with zero orbital angular momentum through the emission of electric multi-pole radiation of order l. Further, assume that the nucleon's spin is parallel to the orbital angular momentum (i.e., the transition is parity favored). Then the nucleon wavefunctions in the initial and final states are,

$$\psi_i = g_i(r) Y_{lm}(\theta, \phi) s \tag{4.261}$$

$$\psi_f = \frac{g_f(r)}{\sqrt{4\pi}} s. \tag{4.262}$$

where s is a spin function. If these are substituted into the expression for Q_{lm} given by (4.255) (only the first term in the summation contributes), then,

$$Q_{lm} = \frac{e}{\sqrt{4\pi}} \int_0^\infty dr\, r^{l+2} g_i(r) g_f(r). \tag{4.263}$$

A rough order-of-magnitude estimate of this integral is obtained by assuming that radial components of the nucleon wavefunctions are,

$$\begin{aligned} g_i(r) = g_f(r) &= C \quad r < R_N \\ &= 0 \quad\;\; r > R_N \end{aligned} \tag{4.264}$$

where R_N is the nuclear radius. From the normalization of the wavefunctions, $C = \sqrt{3/R_N^3}$ and we obtain an approximation to the electric multi-pole moment,

$$Q_{lm} \approx \frac{3e}{\sqrt{4\pi}} \frac{R_N^l}{l + 3}. \tag{4.265}$$

The other moments are determined from ratios with this expression. Note that, in (4.256), the divergence operator can be approximated by $1/R_N$ and that the operator $\mathbf{L}_n$ will approximately cancel the $(l + 1)$ factor in the denominator. These then provides the approximate ratio,

$$\frac{M_{lm}}{Q_{lm}} \approx \frac{\hbar c}{m_n R_n}. \tag{4.266}$$

The magnetic multi-pole moment M'_{lm}, which is due to the nucleons' spins, is roughly 2–3 times greater than that due to the nucleon orbits, M_{lm}, as can be surmised by comparing the ratio $\mathbf{L}/l + 1$ with $\mu_n \sigma$. Hence, it would not be unreasonable to approximate the ratio,

$$\frac{\left|M_{lm} + M'_{lm}\right|^2}{\left|Q_{lm}\right|^2} \approx 10 \frac{\hbar c}{m_n R_n}. \tag{4.267}$$

The spin electric multi-pole moment Q'_{lm} can be neglected in this approximation as can be shown by replacing the divergence operator in (4.257) with l/R_N to find that,

$$\frac{Q'_{lm}}{Q_{lm}} \approx \frac{\hbar\omega}{m_n} \approx 10^{-3} \tag{4.268}$$

for photon energies typical in γ transitions. Using these approximations and (3.121), the transition probabilities for the Weisskopf single-proton model are, for photon energy k (replacing $\hbar\omega$) in MeV,

$$\begin{aligned} \lambda^{(E)}(l) = {} & \frac{4.4(l+1)}{l((2l+1)!!)^2} (1.2)^{2l} \left(\frac{k}{197}\right)^{2l+1} \\ & \times A^{2l/3} 10^{2l} s^{-1} \end{aligned} \tag{4.269}$$

$$\begin{aligned} \lambda^{(M)}(l) &= \frac{1.9(l+1)}{l((2l+1)!!)^2} (1.2)^{2l-2} \left(\frac{k}{197}\right)^{2l+1} \\ & \quad \times A^{(2l-2)/3} 10^{2l} s^{-1} \\ &= \frac{0.3}{A^{2/3}} \lambda^{(E)}(l) \end{aligned} \tag{4.270}$$

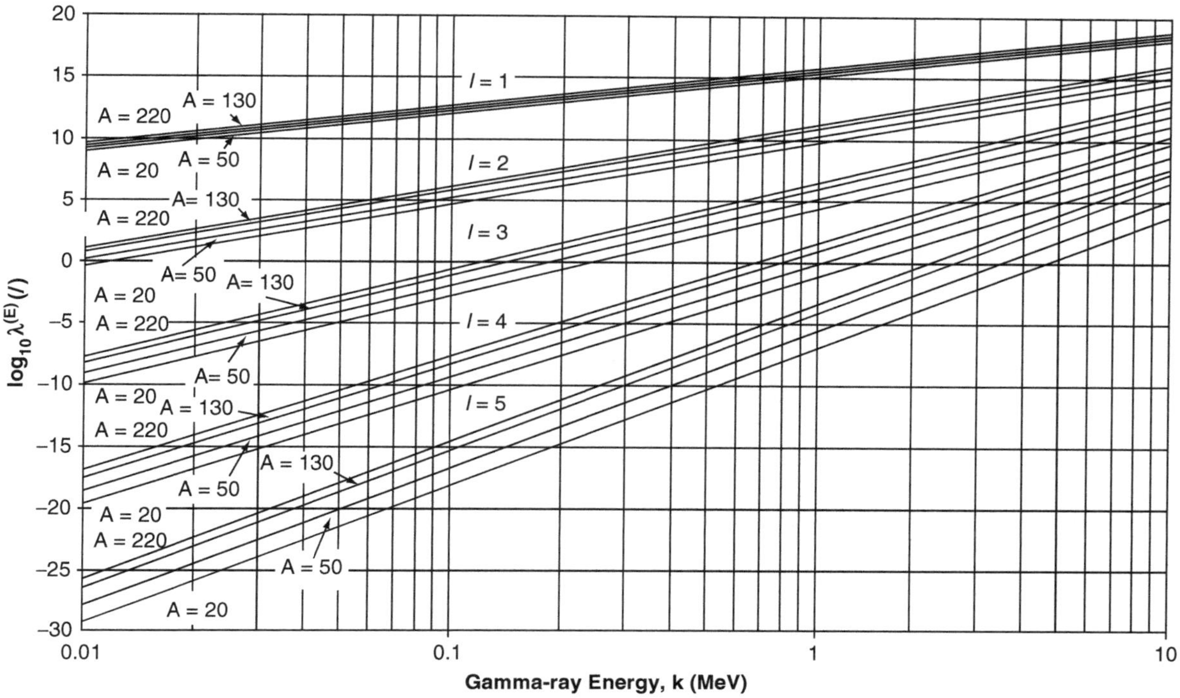

Fig. 4.32 Logarithm of the electric multi-pole transition probability (in s^{-1}) as a function of γ-ray energy for multi-pole orders of 1 through 5 and nuclei with atomic mass numbers of 20, 50, 130, and 220 calculated from the Weisskopf single-proton model. Internal conversion contributions are excluded

These transition probabilities are plotted in Figs. 4.32 and 4.33.

The general characteristics of these transition probabilities are as follows:

- The transition probabilities increase with γ-ray energy, with the rate of increase increasing with multi-pole order.
- For a given γ-ray energy, both transition probabilities decrease by up to about six orders-of-magnitude for each unit increase in l – hence, the dipole and quadrupole transitions will dominate.
- With the exception of the magnetic dipole radiation, the transition probability increases with atomic mass number.
- The magnetic multi-pole transition probability is at least an order-of-magnitude less than that of the electric multi-pole (for a given multi-pole order and photon energy) and decreases with increasing atomic number.

It is important to recall the approximations used to calculate the multi-pole moments, including the assumption of a single transiting nucleon, in the derivation of these multi-pole transition rates. Hence, one should not expect these results to be accurate in predicting γ-transition rates. In fact, Blatt and Weisskopf suggest that the above expressions overestimate the actual transition rates by factors of up to three orders-of-magnitude. Even despite the size of this error, the calculations are of some benefit, if not just for understanding the qualitative aspects of γ-transition rates, but when one recognizes that for γ-ray energies of less than about 1 MeV, there is a six orders-of-magnitude decrease in the multi-pole transition rates per unit increase in multi-pole order.

4.4.3 Internal Conversion

4.4.3.1 Introduction

In the discussion of γ decay, the excited nucleus was treated in isolation. One of the consequences of this

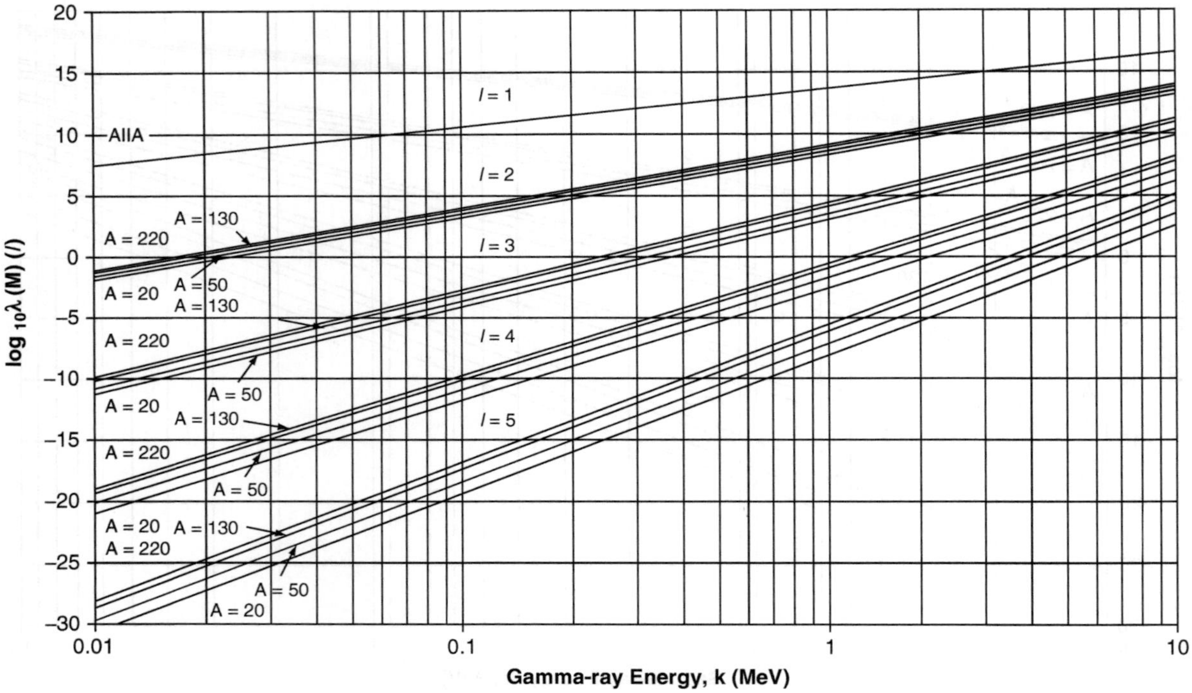

Fig. 4.33 Logarithm of the magnetic multi-pole transition probability (in s^{-1}) as a function of γ-ray energy for multipole orders of 1 through 5 and nuclei with atomic mass numbers of 20, 50, 130, and 220 calculated from the Weisskopf single-proton model. Contributions from internal conversion are excluded

consideration was that it was impossible for an excited nucleus to make a transition between states each with zero angular momentum through the emission of electromagnetic radiation. In reality, however, the nucleus is not isolated but is contained within the atom and interacts with the orbiting atomic electrons. This interaction provides two additional channels for the excited nucleus to lose energy: IC, in which there is direct energy-transfer to an atomic electron through a virtual photon (i.e., it is not an internal photoelectric process in which a photon is emitted from the nucleus and absorbed by the electron), and, should the energy of the transition exceed twice the electron rest mass, internal pair production. The latter process is infrequent and is not considered here.

As noted earlier, if the energy of the transition exceeds the atomic electron binding energy (which is usually the case), the electron is ejected. The IC coefficient is the ratio of the mean number of IC electrons ejected to the mean number of γ rays emitted,

$$\alpha^{(IC)} = \frac{\overline{N}_{ICe}}{\overline{N}_{\gamma}}. \tag{4.271}$$

The coefficient is decomposed into those contributions from electrons in different orbitals,

$$\alpha^{(IC)} = \alpha_K^{(IC)} + \alpha_L^{(IC)} + \alpha_M^{(IC)} + \cdots. \tag{4.272}$$

A full and proper evaluation of the transition probabilities of the previous section would include a multiplicative factor of $1 + \alpha^{(IC)}$ to include the contributions of IC.

The contributions of IC to internal radiation dosimetry are significant (Smith et al. 1965). The ejected IC electrons travel short distances in tissue transferring energy to tissue as they slow down and, hence, produce an absorbed dose. The immediate atomic consequence of the ejection of the IC electron is that a vacancy is produced amongst the atomic electron

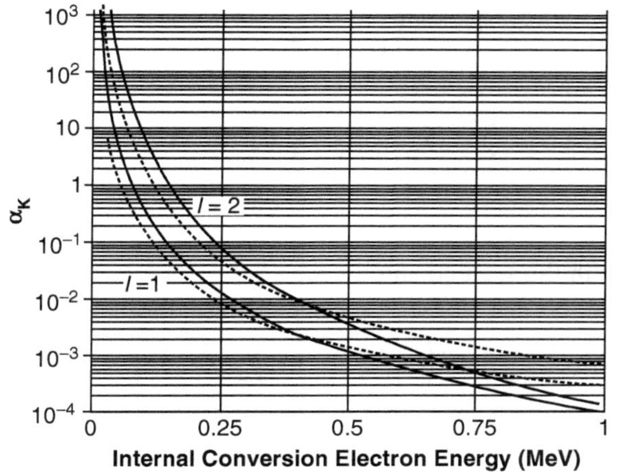

Fig. 4.34 K-orbital internal conversion coefficients for electric multi-poles $l = 1{,}2$ calculated from (4.289) (*solid lines*) and from the tabulated values of Widman and Powsner (1970) (*dashed line*) for Z = 40 (zirconium)

orbitals. This vacancy is filled by the transition of an electron from another orbital and the consequent emission of a characteristic (or fluorescent) X-ray or an Auger electron cascade. These also contribute to the absorbed dose to tissue and are the subjects of Chap. 6.

4.4.3.2 Calculation of the Internal Conversion Coefficient

Introduction

A full calculation of the IC coefficient will be quite involved, requiring detailed knowledge of the pre- and posttransition nuclear wavefunctions. Here, a simplified calculation of the conversion coefficient will be presented. However, it is one which is sufficient to demonstrate the gross dependence of the coefficient upon multi-pole order, excitation energy and atomic number. Tables of IC coefficients have been provided, for example, by Hager and Seltzer (1968) and current values are provided by the National Nuclear Data Center website http://www.nndc.bnl.gov/hsicc/; Widman and Powsner (1970) have provided tables of these coefficients for internal absorbed dose calculations. A critical evaluation of published IC coefficients has been provided recently by Kibédi et al. (2007).

The main contributor to the IC process is the static Coulomb interaction between the protons and the atomic electron. As IC will be more probable for K-shell electrons due to the greater overlap of nuclear and electronic wavefunctions, a calculation of $\alpha_K^{(IC)}$ is developed here using perturbation theory. The calculation is simplified by assuming that the ejected electron is nonrelativistic and, thus, can be modeled by a plane wave. The initial wavefunction is the product of the nuclear initial wavefunction and that of the K-shell electron and the final wavefunction is the product of the nuclear final wavefunction and the electron plane wave. Hence,

$$\psi_i = \psi_{nuc,i}\sqrt{\frac{Z^3}{\pi r_\infty^3}}\, e^{-(ZR/r_\infty)} \tag{4.273}$$

where r_∞ is the Bohr radius and,

$$\psi_f = \psi_{nuc,f}\left(\frac{1}{L^{3/2}}\right)e^{i\left(\frac{\mathbf{p}_e \bullet \mathbf{R}}{\hbar c}\right)} \tag{4.274}$$

where $\mathbf{R}$ is the position vector of the electron. The Coulomb interaction potential between the protons and the atomic electron is,

$$U = \alpha\hbar c\sum_{i=1}^{Z}\frac{1}{|\mathbf{R}-\mathbf{r}_i|} \tag{4.275}$$

where $\mathbf{r}_i$ is the position vector of the *l*th proton.

Calculation of the Matrix Element

The matrix element is,

$$M_{if} = \alpha\hbar c\sqrt{\frac{Z^3}{\pi L^3 r_\infty^3}}\sum_{n=1}^{Z}\int d^3\mathbf{R}\int d^3\mathbf{r}_n\psi^*_{nuc,f}(\mathbf{r}_n) \times e^{-\left(\frac{\mathbf{p}_e \bullet \mathbf{R}}{\hbar c}\right)}\frac{1}{|\mathbf{R}-\mathbf{r}_n|}\times e^{-(ZR/r_\infty)}\psi_{nuc,i}(\mathbf{r}_n) \tag{4.276}$$

where the integration over $\mathbf{R}$ is over the electron position. The double integral is solved by first using the substitution of variable $\mathbf{R}' = \mathbf{R} - \mathbf{r}_n$ and noting that $Zr \gg r_\infty$ which allows the approximation $e^{-(ZR/r_\infty)} \approx 1$,

$$\int d^3\mathbf{R}\int d^3\mathbf{r}_n\psi^*_{nuc,f}(\mathbf{r}_n)e^{-\left(\frac{\mathbf{p}_e\bullet\mathbf{R}}{\hbar c}\right)}\frac{1}{|\mathbf{R}-\mathbf{r}_n|}$$
$$\times\, e^{-(ZR/a_\infty)}\psi_{nuc,i}(\mathbf{r}_n)$$
$$\approx\int d^3\mathbf{r}_n\psi^*_{nuc,f}(\mathbf{r}_n)e^{-\frac{\mathbf{p}_e\bullet\mathbf{r}_n}{\hbar c}}\psi_{nuc,i}(\mathbf{r}_n)\int d^3\mathbf{R}'\frac{e^{-\frac{\mathbf{p}_e\bullet\mathbf{R}''}{\hbar c}}}{R'}$$
$$=4\pi\left(\frac{\hbar c}{p_e}\right)^2\int d^3\mathbf{r}_n\psi^*_{nuc,f}(\mathbf{r}_n)e^{-\frac{\mathbf{p}_e\bullet\mathbf{r}_n}{\hbar c}}\psi_{nuc,i}(\mathbf{r}_n) \tag{4.277}$$

to give,

$$M_{if}=4\pi\alpha\hbar c\sqrt{\frac{Z^3}{\pi L^3 r^3_\infty}}\left(\frac{\hbar c}{p_e}\right)^2$$
$$\times\sum_{n=1}^{Z}\int d^3\mathbf{r}_n\psi^*_{nuc,f}(\mathbf{r}_n)e^{-\left(\frac{\mathbf{p}_e\bullet\mathbf{r}_n}{\hbar c}\right)}\psi_{nuc,i}(\mathbf{r}_n). \tag{4.278}$$

The remaining integral is solved by, first expanding the plane wave into spherical waves,

$$e^{i\mathbf{b}\bullet\mathbf{r}}=4\pi\sum_{l=0}^{\infty}\sum_{m=-l}^{l}i^l j_l(br)Y^*_{lm}(\theta_b,\phi_b)Y_{lm}(\theta_r,\phi_r). \tag{4.279}$$

As $(p_e r/\hbar c)\ll 1$, the small-argument approximation to the spherical Bessel function can be used,

$$e^{-\frac{\mathbf{p}_e\bullet\mathbf{r}_n}{\hbar c}}=4\pi\sum_{l=0}^{\infty}\sum_{m=-l}^{l}\frac{(-i)^l}{(2l+1)!!}\left(\frac{p_e}{\hbar c}\right)^l r^l_n$$
$$\times\, Y_{lm}(\theta_e,\phi_e)Y^*_{lm}(\theta_n,\phi_n) \tag{4.280}$$

where the e and n subscripts of the angular variables refer to the electron and nucleon, respectively. Substituting (4.280) into (4.278) gives,

$$M_{if}=16\pi^2\alpha\hbar c\sqrt{\frac{Z^3}{\pi L^3 r^3_\infty}}\sum_{n=1}^{Z}\sum_{l=0}^{\infty}\sum_{m=-l}^{l}\frac{(-i)^l}{(2l+1)!!}\left(\frac{p_e}{\hbar c}\right)^{l-2}$$
$$\times\int d^3\mathbf{r}_n\psi^*_{nuc,f}(\mathbf{r}_n)r^l_n\psi_{nuc,i}(\mathbf{r}_n)$$
$$\times\, Y_{lm}(\theta_e,\phi_e)Y^*_{lm}(\theta_n,\phi_n). \tag{4.281}$$

Recalling the definition of the electric multi-pole moment for electric charge given by (4.255), (4.281) can be reduced to,

$$M_{if}=16\pi^2\left(\frac{\alpha\hbar c}{e}\right)\sqrt{\frac{Z^3}{\pi L^3 r^3_\infty}}$$
$$\times\sum_{l=0}^{\infty}\sum_{m=-l}^{l}\frac{(-i)^l}{(2l+1)!!}\left(\frac{p_e}{\hbar c}\right)^{l-2}Y_{lm}(\theta_e,\phi_e)Q_{lm} \tag{4.282}$$

with the squared amplitude,

$$|M_{if}|^2=256\pi^3\left(\frac{\alpha\hbar c}{e}\right)^2\frac{Z^3}{L^3 r^3_\infty}\sum_{l=0}^{\infty}\sum_{m=-l}^{l}\sum_{l'=0}^{\infty}$$
$$\times\sum_{m=-l'}^{l'}\frac{(-i)^l i^{l'}}{((2l+1)!!)((2l+1)!!)}$$
$$\times\left(\frac{p_e}{\hbar c}\right)^{l+l'-4}Y_{lm}(\theta_e,\phi_e)Y^*_{l'm'}(\theta_e,\phi_e)Q_{lm}\,Q^*_{l'm'}. \tag{4.283}$$

Calculation of the Phase Space Factor

This is a straightforward calculation as only a single particle is ejected The number of states available in the differential element of momentum dp_e and differential solid angle $d\Omega_e$ is,

$$d^2N=\left(\frac{L}{2\pi\hbar c}\right)^3 p_e^2\,dp_e\,d\Omega_e. \tag{4.284}$$

As $p_e dp_e = m_e dT_e$ (the electron is considered non-relativistic), then,

$$d\rho_f=\frac{d^2N}{dT_e}$$
$$=\left(\frac{L}{2\pi\hbar c}\right)^3 m_e\,p_e\,d\Omega_e. \tag{4.285}$$

Internal Conversion Transition Rate

The IC transition rate for a K-orbital electron is,

$$\lambda_K^{(IC)}=2\frac{2\pi}{\hbar}|M_{if}|^2\rho_f \tag{4.286}$$

where the additional factor of 2 is due to the two electrons in the K orbit. Inserting (4.283) and (4.285) into (4.286), integrating $d\rho_f$ over Ω_e and noting the orthonormalization of the spherical harmonics, $\int d\Omega_e \, Y_{lm}(\theta_e, \phi_e) Y^*_{l'm'}(\theta_e, \phi_e) = \delta_{ll'}\delta_{mm'}$, an expression for the K-orbital transition rate is obtained,

$$\lambda_K^{(IC)} = \frac{128\,\pi\, m_e\, \alpha^2 Z^3}{\hbar\, e^2\, r_\infty^3} \times \sum_{lm} \frac{1}{((2l+1)!!)^2} \left(\frac{p_e}{\hbar c}\right)^{2l-3} |Q_{lm}|^2. \quad (4.287)$$

As only the static Coulomb interaction has been considered and, from (4.268), one can neglect the contribution Q'_{lm}, comparing the K-orbital transition rate directly with the electric multi-pole transition rate of (4.253), a crude approximation to the IC coefficient for K-orbital electrons is had,

$$\alpha_K \approx 16\left(\frac{Z^3}{r_\infty^4}\right)\left(\frac{l}{l+1}\right)\left(\frac{p_e}{\hbar c}\right)^{2l-3}\left(\frac{\hbar c}{k}\right)^{2l+1} \quad (4.288)$$

replacing $\hbar\omega$ with k. It is reasonable to assume that the nuclear excitation energy exceeds the binding energy of the K-orbital electrons, in which case, $p_e = \sqrt{2m_e k}$, and,

$$\alpha_K \approx \left(\frac{\hbar c}{m_e r_\infty}\right)^4 Z^3 \left(\frac{l}{l+1}\right)\left(\frac{2m_e}{k}\right)^{l+5/2} \quad (4.289)$$

This result is a reasonable approximation when the electron can be considered as nonrelativistic and the transition energy greatly exceeds the atomic electron binding energy. The main considerations of this formula are that IC increases strongly with atomic number and multi-pole order, but decreases with increasing transition energy. Hence, IC is important for high-Z nuclei in low-energy transitions of high multi-pole order. Figure 4.33 shows the IC coefficient for K-orbital electrons for zirconium as a function of IC electron kinetic energy as calculated from (4.289) for the electric multi-poles $l = 1, 2$. Full calculation of the IC coefficients would consider the total contributions of all electromagnetic interactions and, hence, would include the effects of the magnetic multi-pole moments. While the K-orbital electrons will dominate in IC, the L- and M-orbital electrons will make contributions, albeit with lower probability.

0 → 0 Transitions

As discussed earlier, electromagnetic transitions between nuclear states with zero angular momentum are not possible as there are no $l = 0$ multi-poles in the radiation field. However, closer examination of the above derivation of the IC coefficient will demonstrate that such a transition is feasible in which the K-orbital electron takes away the energy. This statement, at first sight, contradicts the above result of (4.286) in which the multi-pole moment would go to zero for a 0 → 0 transition. However, the above derivation excluded the contribution to the matrix element when $R > R_n$, i.e., when the electron is within the nucleus. Inclusion of this contribution leads to the small probability of a 0 → 0 transition. The fact that this is possible demonstrates that IC cannot be interpreted as a photoelectric effect in which the photon emitted from the nucleus is absorbed by an atomic electron which is subsequently ejected.

4.4.4 Nuclear Isomerism

4.4.4.1 Introduction

The γ-transition selection rules described earlier can delay a γ transition significantly such that the excited nucleus can have a long half-life, up to years. Such nuclei are described as being in an *isobaric* state or as being *metastable*. Clearly, such isomeric transitions (ITs) are associated with large changes in nuclear angular momentum ΔJ and small transition energies. But we have also seen that these conditions are preferential for IC. ITs are frequently associated with β decay which leaves the nucleus with a high angular momentum, as shown schematically in Fig. 4.35.

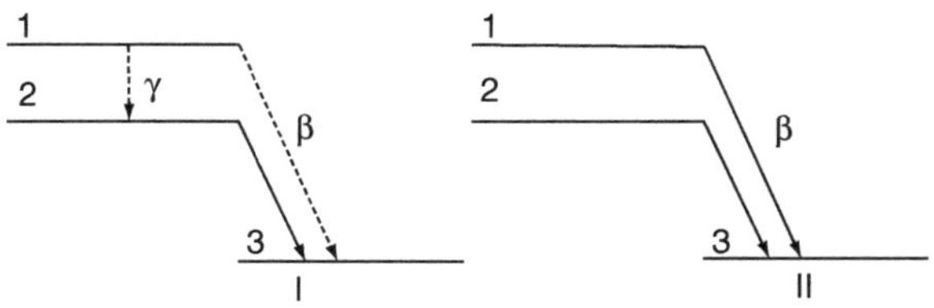

Fig. 4.35 Schematic representation of nuclear isomerism. In example I, level 1 decays predominantly through γ emission to level 2 which subsequently decays to level 3 through β decay. In example II, levels 1 and 2 undergo β decay independently to level 3

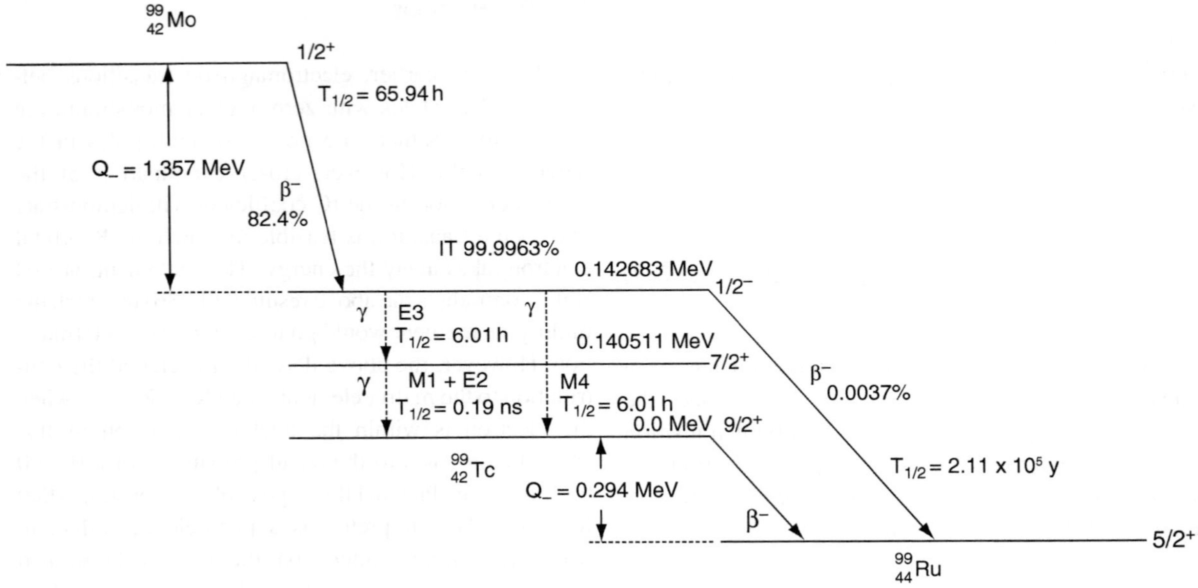

Fig. 4.36 Decay scheme ^{99}Mo→^{99,99m}Tc→^{99}Ru. (IT ≡ isomeric transition)

In Example I, the probability of level 1 decaying to level 2 through an IT exceeds the probability of it decaying to level 3 through β decay; in Example II, levels 1 and 2 transit to level 3 independently through β decay. If, in Example I, the transition rate for the γ transition from level 1 to level 2 is much greater or much less than that for the β-decay to level 3, then the β spectrum of level 2 with a transition rate equal to the IT of 1 → 2 is demonstrated. In the case of Example II, the levels 1 and 2 decay independently to level 3.

It was Weizsäcker who proposed the theory of nuclear isomers to explain the observation of long-lived γ-ray emitting isotopes. Equations (4.269) and (4.270) show that γ transitions with half-lives of a second or more for photon energies of about 0.1–0.5 MeV (typical for nuclear medicine) are between nuclear states with ΔJ = 3. A common example of an IT used in diagnostic nuclear medicine is the de-excitation of the ^{99m}Tc nucleus (the m superscript refers to its metastable state) to the ground state ^{99}Tc with a 6.01 h half-life (while ^{99m}Tc can undergo β^- decay to ^{99}Ru, the associated branching ratio is a negligible 3.7×10^{-5}); this is shown in Fig. 4.36. ^{99}Tc and ^{99m}Tc both result from the β^- decay of ^{99}Mo. The spin/parity (J^π) of the ^{99}Tc ground state is $9/2^+$ due to an unpaired proton in the $1g_{9/2}$ shell. The excited ^{99m}Tc nucleus is formed in the $J^\pi = 1/2$ level and can de-excite directly to the $9/2^+$ ground state via the emission of a 142.683 keV photon, with a branching ratio of about 3×10^{-4}. However, the difference in the angular momenta of the two states is significant: $\Delta J = 4$ and the transition is M4 (as there is also a change in parity) with a 6.01 h half-life. It can also de-excite with the same half-life to the intermediate $7/2^+$ excited state at the 140.511 keV level with $\Delta J = 3$ via an E3 transition (as there is no change in parity). This is followed by the de-excitation from this $7/2^+$ state to the $9/2^+$ ground state through a combination of M1 and E2 transitions ($\Delta J = 1$) with the emission of a 140.511 keV γ ray which is used for imaging.

References

Alvarez LW (1937) Nuclear K electron capture. Phys Rev 52:134–135

Anderson CD (1933) The positive electron. Phys Rev 43:491–494

Behrens H, Jähnecke J (1969) Numerische Tabellen für Beta-Zerfall und Elektronen-Einfang. In: Schopper H (ed) Landolt-Börnstein Gruppe I, Bd 4. Springer, Berlin

Blatt JM, Weisskopf VF (1979) Theoretical nuclear physics. Dover, New York

Bowkamp JJ, Casimir HBG (1954) On multipole expansions in the theory of electromagnetic radiation. Physica XX: 539–554

Condon EU, Gurney RW (1928) Wave mechanics and radioactive disintegration. Nature 122:439

Conversi M, Pancini E, Piccioni O (1947) On the disintegration of negative mesons. Phys Rev 71:209–210

Cox RT, McIlwraith CG, Kurrelmeyer B (1928) Apparent evidence of polarization in a beam of β-rays. Proc Nat Acad Sci 14:544–549

Danby G, Gaillard JM, Goulianos K, Lederman LM, Mistry NB, Schwartz M, Steinberger J (1962) Observations of high-energy neutrino reactions and the existence of two kinds of neutrinos. Phys Rev Lett 9:36–44

Fermi E (1934a) Tentativo di una teoria dei raggi β. Nuovo Cimento 11:1–19

Fermi E (1934b) Versuch einer theorie der β-strahlen I. Z Physik 1934:161–171

Gamow G (1928) Zur Quantentheorie des Atomkernes. Z Phys 51:204–212

Gamow G, Teller E (1936) Selection rules for the β-disintegration. Phys Rev 49:895–899

Gerward L (1999) Paul Villard and his discovery of gamma rays. Phys Perspect 1:367–383

Goldhaber M, Grodzins L, Scharff-Goldhaber G (1948) Identification of beta-rays with atomic electrons. Phys Rev 73:1472–1473

Goldhaber GL, Grodzins L, Sunyar AW (1958) Helicity of neutrinos. Phys Rev 109:1015–1017

Grotz K, Klapdor HV (1990) The weak interaction in nuclear, particle and astrophysics. Adam Hilger, Bristol

Hager RS, Seltzer EC (1968) Nucl Data A4:1–235

Jackson JD (1999) Classical electrodynamics. Wiley, New York

Kibédi T, Burrows TM, Trzhaskovskay MB et al (2007) Internal conversion coefficients – how good are they now? In: International Conference on Nuclear Data for Science and Technology 2007. http://nd2007.edpsciences.org/index.php?option=article&access=standard&Itemid=129&url=/articles/ndata/abs/2007/01/ndata07771/ndata07771.html. Accessed 2008

Kodama K, Ushida N, Anderopoulos C et al (2001) Observation of tau neutrino interactions. Phys Lett B 504:218–224

Lee TD, Yang CN (1956) Question of parity conservation in weak interactions. Phys Rev 104:254–258

Lipkin HJ (1962) Beta decay for pedestrians. North-Holland, Amsterdam

Marshak RE (1942) Forbidden transitions in β-decay and orbital electron capture and spins of nuclei. Phys Rev 61: 431–449

Martins RA (1997) Becquerel and the choice of uranium compounds. Arch Hist Exact Sci 51:67–81

National Nuclear Data Center (2008) Chart of Nuclides. http://www.nndc.bnl.gov. Accessed 2005–2008

Neddermeyer SH, Anderson CD (1937) Note on the nature of cosmic-ray particles. Phys Rev 51:884–886

Particle Data Group (2004) Review of particle physics. Phys Lett B 592:1–1109

Perl ML, Abrams GS, Boyarski AM et al (1975) Evidence for anomalous lepton production in e^+–e^- annihilation. Phys Rev Lett 35:1489–1492

Reines F (1997) The neutrino: from poltergeist to particle. In: Ekspong G (ed) Nobel lectures, physics 1991–1995. World Scientific, Singapore

Reines F, Cowan CL Jr (1953) Detection of the free neutrino. Phys Rev Let 92:830–831

Rutherford E, Royds T (1909) The nature of the α-particle from radioactive substances. Phil Mag 17:281–286

Sargent BW (1933) The maximum energy of the β-rays from uranium X and other bodies. Proc Roy Soc (London) A139:659–673

Segrè E (1970) Enrico Fermi – physicist. University of Chicago, Chicago

Segrè E (1977) Nuclei and particles. Benjamin/Cummings, Reading

Smith EM, Harris CC, Rohrer RH (1965) Calculation of local energy deposition due to electron capture and internal conversion. J Nucl Med 7:23–31

Tsoulfanidis N (1995) Measurement and detection of radiation, 2nd edn. Taylor and Francis, Washington

Weisskopf VF (1951) Radiative transition probabilities in nuclei. Phys Rev 83:1073

Widman JC, Powsner ER (1970) Internal conversion coefficients for absorbed dose calculations. Phys Med Biol 15:99–106

Wilson D (1983) Rutherford – simple genius. Hodder and Stoughton, London

Wu CS, Ambler E, Hayward RW, Hoppes DD, Hudson RP (1957) Experimental test of parity conservation in beta decay. Phys Rev 105:1413–1415

Radioactive Decay: Macroscopic Theory

5

Abstract This chapter presents the macroscopic theory of radioactivity in which ensembles of radioactive nuclei are considered rather than single nuclei. It begins with a review of the fundamentals of radioactivity and radioactive decay chains. The determination of the activity biodistribution requires measurements of the activity through either in vivo imaging or through in vitro assays of activity content in blood, plasma, and excreta. The theory behind such measurements is presented along two paths. The first is the derivation of corrections factors required in practical cases: background correction, decay compensation, and reference standard normalization. The second is through decision theory in which, on the basis of measurements, one decides whether or not activity is present and then quantifies the result if the decision is positive. Electronic radiation detectors (well counters, gamma cameras, and PET scanners) are all subject to dead time and this concept and the compensatory factors for it are derived. Paralyzable and nonparalyzable detectors are both considered.

Contents

5.1 Introduction

The previous chapter was concerned with the microscopic theories describing α decay, β decay, γ emission, electron capture, and internal conversion. The theories are referred to as being microscopic as they are applicable to single nuclei or atoms. In nuclear medicine dosimetry, one must contend with macroscopic ensembles of radioactive nuclei,[1] hence a "macroscopic" theoretical viewpoint is required. In most clinical cases, the number of nuclei is so large that radioactive decay can be considered to be a continuous (or non-stochastic) phenomenon. This is traceable to the original work on radioactivity by

[1]For example, a typical clinically administered activity of 740 MBq of ^{99m}Tc consists of 2×10^{13} ^{99m}Tc nuclei.

B.J. McParland, *Nuclear Medicine Radiation Dosimetry*, DOI 10.1007/978-1-84882-126-2_5,

Rutherford and Soddy (1902). However, as radioactive transitions are inherently random and described by a probability per unit time of a transition occurring, λ, a stochastic approach is required when dealing with either small numbers of radioactive nuclei, small numbers of detected radiations or small time intervals. Applicable to such approaches are various probability distribution functions (pdfs): in the discrete case, the binomial and Poisson pdfs, and in the continuous case, the normal and lognormal pdfs. As the derivations and descriptions of these are provided amply elsewhere (see, e.g., James (2006)), these will not be repeated here but will be called upon where necessary throughout this book. The stochastic approach is applicable to "counting statistics" and the practical means of detecting and measuring activity.

5.2 Physical Decay Constant and Activity

Consider a species of atomic nucleus in a particular state for which it is energetically favorable to make a transition to another nuclear state or nucleus. The probability of such a transition (or "decay") occurring during the differential time interval dt is proportional to the time interval, λdt, where λ is the "physical decay constant" with units of reciprocal time and given by perturbation theory in the form of Fermi's Golden rule No. 2,

$$\lambda_{fi} = \frac{2\pi}{\hbar}\left|\int d^3\mathbf{r}\psi_f^* U\psi_i\right|^2 \rho_f$$

where, recall, U is the perturbative potential causing the transition, the ψ_i and the ψ_f are the wavefunctions of the initial and final states and the phase-space term, ρ_f, is the density of states available to the transition. The decay constant is singular to the nuclear species and to the decay mode (although it is possible to observe very small changes in the decay due to the surrounding chemical environment if any of the wavefunctions include that of an orbital electron). The proportion of radioactive decays in an ensemble of N identical radioactive nuclei making this transition is equal to the probability of the transition occurring,

$$\frac{dN}{N} = -\lambda dt \tag{5.1}$$

where a negative sign is inserted to indicate that is the decay results in a net decrease of the number of nuclei. The number of remaining radioactive nuclei at time t is determined by integrating (5.1) with the initial condition of N_0 being the number of nuclei at time $t = 0$, to yield,

$$N(t) = N_0 e^{-\lambda t}. \tag{5.2}$$

Equation (5.1) can also be rearranged to give the transition rate of the ensemble,

$$\frac{dN}{dt} = -\lambda N \tag{5.3}$$

where it can be seen that the transition rate is proportional to the number of radioactive nuclei. The absolute value of this rate is the "activity," A(t), or the number of decays per unit time,

$$A(t) = \lambda N(t) \tag{5.4}$$

and substituting (5.2) into (5.4) yields,

$$A(t) = A_0 e^{-\lambda t} \tag{5.5}$$

where $A_0 \equiv \lambda N_0$ is the activity at time $t = 0$. The exponential nature of the radioactive decay is shown in Fig. 5.1. The unit for activity is the number of

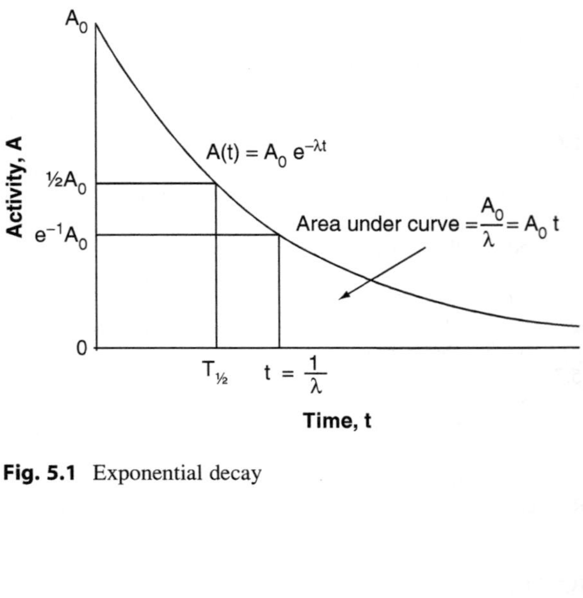

Fig. 5.1 Exponential decay

transitions (decays) per unit time or the Becquerel (Bq), where[2] Bq = 1 transition (disintegration) per second.

Equations (5.2) and (5.5) show that both the number of remaining radioactive nuclei and the rate at which they decay decrease exponentially with time.

5.3 Physical Half-Life, Effective Half-Life, and Mean Lifetime

5.3.1 Physical Half-Life

The time at which the number of remaining radioactive nuclei is reduced to 1/2 of the original number is the half-life, $T_{1/2}$,

$$T_{1/2} = \frac{\ln 2}{\lambda} \approx \frac{0.693}{\lambda} \tag{5.6}$$

Because the activity is proportional to the number of radioactive nuclei, the activity at time $T_{1/2}$ is 1/2 the activity at time t = 0, as shown in Fig. 5.1.

5.3.2 Effective Half-Life

In nuclear medicine dosimetry, one is interested in the temporal variation of activity associated with a radiopharmaceutical incorporated within an organ or tissue. This will be a combination of the physical decay (as given by (5.5)) and the washout or clearance by biological processes. A common time-activity behavior is a rapid uptake of activity followed by washout, with the latter usually well-described by a multiple exponential function in time. The activity in the organ during washout is,

$$A(t) = e^{-\lambda_{Phys} t} \sum_{i=1}^{N} k_i e^{-\lambda_{Biol,i} t} \tag{5.7}$$

where λ_{Phys} is the physical decay constant and the $\lambda_{Biol,i}$ are the biological rate constants. Hence,

$$A(t) = \sum_{i=1}^{N} k_i e^{-\lambda_{eff,i} t} \tag{5.8}$$

where the effective decay constants $\lambda_{eff,i}$ are,

$$\lambda_{eff,i} = \lambda_{Phys} + \lambda_{Biol,i}. \tag{5.9}$$

As an example, consider the simplest case of a monoexponential washout where the physical and biological half-lives are,

$$T_{1/2,Phys} = \frac{\ln 2}{\lambda_{Phys}}$$

$$T_{1/2,Biol} = \frac{\ln 2}{\lambda_{Biol}}$$

The effective half-life is thus,

$$T_{1/2,Eff} = \frac{\ln 2}{\lambda_{eff}} = \frac{\ln 2}{\lambda_{Phys} + \lambda_{Biol}} = \frac{T_{1/2,Phys} T_{1/2,Biol}}{T_{1/2,Phys} + T_{1/2,Biol}}. \tag{5.10}$$

5.3.3 Mean Lifetime

As noted before, the mean lifetime, τ, is the reciprocal of the decay constant, λ,

$$\tau = \frac{1}{\lambda} \tag{5.11}$$

which is related to the half-life by,

$$\tau = \frac{1}{\lambda} = \frac{T_{1/2}}{\ln 2} \approx 1.443\, T_{1/2} \tag{5.12}$$

[2]The curie (Ci) is still in common use as a unit of activity. It was originally defined as the number of disintegrations per second of a 1 gram sample of radium measured at 3.7×10^{10} disintegrations per second, although later measurements demonstrated that this was 3.61×10^{10} disintegrations per second. Jennings (2007) noted that, in 1930, the unit of the Rutherford (rd) was defined as the amount of any radioactive isotope that disintegrated at the rate of 10^6 disintegrations per second. Currently, this would give 1 rd = 1 MBq.

At time τ, both the number of remaining nuclei and the activity are $e^{-1} \approx 0.368$ of their values at time $t = 0$, as shown in Fig. 5.1. An alternative interpretation of the mean lifetime is shown in Fig. 5.2 where, again, activity is plotted against time. The area under the exponential curve is simply equal to the original number of nuclei (as all of the nuclei will have disintegrated as time $t \to \infty$),

$$A_0 \int_0^\infty e^{-\lambda t} = \frac{A_0}{\lambda} = N_0 \tag{5.13}$$

If one were to imagine a scenario in which the transition rate was constant at A_0 for some finite length of time and were to then go to zero, as shown in Fig. 5.2, the length of time that yields the same area under the curve as the physical exponential decrease is the mean lifetime, τ. In addition, the initial slope of the exponential decay curve of the activity is,

$$\left.\frac{dA(t)}{dt}\right|_{t=0} = -\lambda A_0 = -\lambda^2 N_0$$

and the time required for a linear decrease in activity to zero at this slope is taken from the linear equation set to zero,

$$A_0 - \lambda A_0 t = \lambda N_0 - \lambda^2 N_0 t = 0$$

$$\therefore t = \frac{\lambda N_0}{\lambda^2 N_0} = \frac{1}{\lambda} = \tau. \tag{5.14}$$

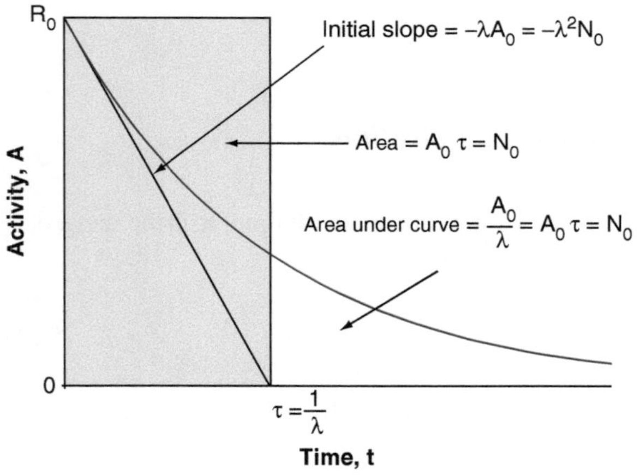

Fig. 5.2 Definition of mean lifetime

The time taken for all of the nuclei to decay if the rate of decay stays constant at its initial value is τ.

5.4 Variability of the Physical Decay Constant

From Fermi's Golden Rule No. 2, the radioactive decay constant is $\lambda_{fi} = \frac{2\pi}{\hbar}\left|\int d^3\mathbf{r}\psi_f^* U \psi_i\right|^2 \rho_f$. Any variation in it due to the external environment can enter only through the initial and final wavefunctions or through the potential. This is possible as the EC and IC processes result from the interaction of the nuclear and atomic electrons. Segrè (1947) suggested that, since the molecular environment surrounding the atom will alter the orbital electron wavefunction, it was possible for the electron density at the nucleus to be varied. Hence, the physical decay constant could be very slightly modified by the chemical state in which the radioactive atom was in. Early measurements of the ^{99m}Tc half-life revealed very small differences (of the order of 10^{-3} to 10^{-4}) in the decay constant between the compounds $KTcO_4$ and Tc_2S_7 (Bainbridge et al. 1953) and between technetium metal at temperatures of 293 and 4.2°K (Byers and Stump 1958). This was due to changes in the internal conversion coefficients for L- and M-shell electrons between the $1/2^-$ and $7/2^+$ nuclear states induced by the surrounding electronic environment (Tuli et al. 2001). More recent high precision measurements of the EC decay constant of ^{7}Be determined $T_{1/2}$ values of 53.69, 53.42, and 54.23 d for ^{7}Be in $Be^{2+}(OH_2)_4$, $Be(OH)_2$, and BeO, respectively (Huh 1999).

5.5 Specific Activity

The specific activity is that per unit mass of a radioactive substance. If this is a pure sample of only the radioactive isotope (i.e., stable isotopes or any other substances are not present), the sample is referred to as being *carrier free* and the specific activity is the *carrier-free specific activity (CFSA)*. On the other hand, the sample could also contain other nonradioactive substances, which are *carriers*, in addition to the radioactive isotope. The value of the specific activity for this case is less than that of the CFSA.

The CFSA is the activity per unit mass of the radioisotope in isolation and may be derived as follows. As the activity A is proportional to the number of nuclei, the activity per atom in an ensemble of N radioactive atoms is,

$$\frac{A}{N} = \lambda \tag{5.15}$$

and the mass of these N atoms is,

$$M = \frac{N}{N_A} \times \text{Atomic Mass}. \tag{5.16}$$

Substituting the expression for N from (5.15) into (5.16) and solving for the activity per unit mass, or CFSA, A/M, yields

$$\text{CFSA} = \frac{\lambda N_A}{\text{Atomic Mass}} \tag{5.17}$$

For example, ^{99m}Tc has an atomic mass of 98.91 g/mole and a half-life of 6.015 h, (physical decay constant of $3.2 \times 10^{-5}\ s^{-1}$). The CFSA for ^{99m}Tc is 1.95×10^{17} Bq/g.

5.6 Radioactive Parents and Progeny

5.6.1 General Case

As seen in Chap. 4, it is common for the product of a radioactive decay (referred to as the "daughter") to be itself radioactive and for a given radioactive nucleus to have more than one possible transition mode, with the probability of a given transition occurring described as the branching ratio. This section considers the relationships between radioactive parent and daughter nuclei.

Figure 5.3 presents a model of a simple decay chain: the sequential decays of radioactive isotope X to radioactive daughter Y to radioactive granddaughter Z. The decay constants for X, Y, and Z are λ_X, λ_Y, and λ_Z, respectively, and the rates of decay of each nuclear species are described by the set of coupled first-order differential equations,

$$\frac{dN_X}{dt} = -\lambda_X N_X(t) \tag{5.18}$$

$$\frac{dN_Y}{dt} = \lambda_X N_X(t) - \lambda_Y N_Y(t) \tag{5.19}$$

$$\frac{dN_Z}{dt} = \lambda_Y N_Y(t) - \lambda_Z N_Z(t) \tag{5.20}$$

These equations may be solved most simply by substitution with the initial conditions of there being no nuclei of Y or Z present at time $t = 0$ and that $N_X(t = 0) \equiv N_{X,0}$. The solution for $N_X(t)$ has already been derived,

$$N_X(t) = N_{X,0} e^{-\lambda_X t}. \tag{5.21}$$

$N_Y(t)$ can be solved by writing it in the form,

$$N_Y(t) = k\left(e^{-\lambda_X t} - e^{-\lambda_Y t}\right) \tag{5.22}$$

which satisfies the initial condition, $N_Y(t = 0) = 0$. The first derivative is,

$$\frac{dN_Y}{dt} = k(-\lambda_X e^{-\lambda_X t} + \lambda_Y e^{-\lambda_Y t}). \tag{5.23}$$

Solving for k,

$$k = N_{X,0}\left(\frac{\lambda_X}{\lambda_X - \lambda_Y}\right)$$

and the solution for $N_Y(t)$ is,

$$N_Y(t) = N_{X,0}\left(\frac{\lambda_X}{\lambda_Y - \lambda_X}\right)\left(e^{-\lambda_X t} - e^{-\lambda_Y t}\right). \tag{5.24}$$

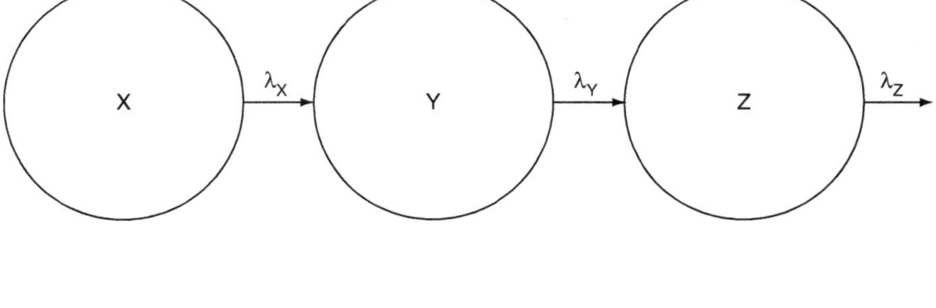

Fig. 5.3 Sequential radioactive decay without branching

In a similar fashion,

$$N_Z(t) = N_{X,0}\left(\frac{\lambda_X\lambda_Y}{(\lambda_Y-\lambda_X)(\lambda_Z-\lambda_X)(\lambda_Z-\lambda_Y)}\right)\times\Big((\lambda_Z-\lambda_Y)e^{-\lambda_X t}-(\lambda_Z-\lambda_X)e^{-\lambda_Y t}+(\lambda_Y-\lambda_X)e^{-\lambda_Z t}\Big). \quad (5.25)$$

Whereas these solutions are for the number of nuclei of the three species, what are usually of practical interest are the activities of the species. As activity is proportional to the number of nuclei with the constant of proportionality being the decay constant for that nuclear species, the activities are

$$A_X(t) = A_{X,0}e^{-\lambda_X t} \quad (5.26)$$

$$A_Y(t) = A_{X,0}\left(\frac{\lambda_Y}{\lambda_Y-\lambda_X}\right)\left(e^{-\lambda_X t}-e^{-\lambda_Y t}\right) = A_X(t)\left(\frac{\lambda_Y}{\lambda_Y-\lambda_X}\right)\left(1-e^{-(\lambda_Y-\lambda_X)t}\right) \quad (5.27)$$

$$A_Z(t) = A_{X,0}\left(\frac{\lambda_Y\lambda_Z}{(\lambda_Y-\lambda_X)(\lambda_Z-\lambda_X)(\lambda_Z-\lambda_Y)}\right)\times\Big((\lambda_Z-\lambda_Y)e^{-\lambda_X t}-(\lambda_Z-\lambda_X)e^{-\lambda_Y t}+(\lambda_Y-\lambda_X)e^{-\lambda_Z t}\Big). \quad (5.28)$$

It has been previously shown that the integral of (5.26) is equal to the original number of X nuclei, $N_{X,0}$. Similarly, it can be shown that the integrals of (5.27) and (5.28) (i.e., the areas under the activity-time curves of the progeny Y and Z, respectively) are also equal to $N_{X,0}$, due to the decay of one X nucleus producing one Y nucleus which, following its own decay, produces one Z nucleus.

Equations (5.26)–(5.28) describe the time dependencies of the activities of the three nuclear species, and it is of special interest to consider in isolation the relationship between the radioactive parent, X, and the radioactive daughter, Y and, in particular, how this relationship varies with their relative half-lives of each.

5.6.2 Parent Half-Life Much Greater than that of Daughter

Here, $T_{X,1/2} \gg T_{Y,1/2}$ or, equivalently, $\lambda_X \ll \lambda_Y$. Hence, $\lambda_Y/(\lambda_Y-\lambda_X)\approx 1$ and $1-e^{-(\lambda_Y-\lambda_X)t}\approx 1-e^{-\lambda_Y t}$, which simplify the expression for the Y activity to,

$$A_Y(t)\approx A_X(t)\left(1-e^{-\lambda_Y t}\right) \quad \lambda_X \ll \lambda_Y. \quad (5.29)$$

This shows that the activity of the daughter, Y, grows asymptotically with a time constant λ_Y equal to that of the parent, X, and subsequently decays with that of the parent at the same decay constant λ_X. This equality between parent and daughter activities is referred to as *secular equilibrium*. Figure 5.4 shows the decay of a fictitious parent radionuclide, with a half-life of 10^3 h, and the growth and subsequent decay of its daughter, with a half-life of 2 h. The daughter activity increases asymptotically to approach that of the parent and, by about 15 h, the activities of both species are, for practical purposes, the same.

5.6.3 Parent Half-Life Greater than that of Daughter

Here, $T_{X,1/2} > T_{Y,1/2}$ or, equivalently, $\lambda_X < \lambda_Y$. Hence, $\lambda_Y/(\lambda_Y-\lambda_X)\approx 1$ and $\left(1-e^{-(\lambda_Y-\lambda_X)t}\right)$ approaches unity with increasing time. The activity of the daughter,

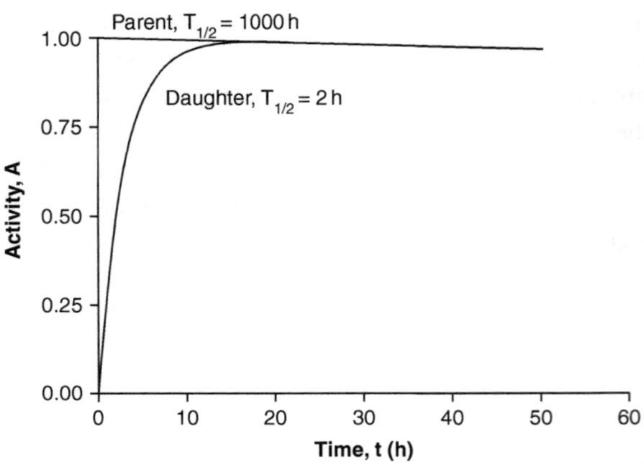

Fig. 5.4 Secular equilibrium for a fictitious parent isotope with a half-life of 10^3 h and a daughter with a half-life of 2 h

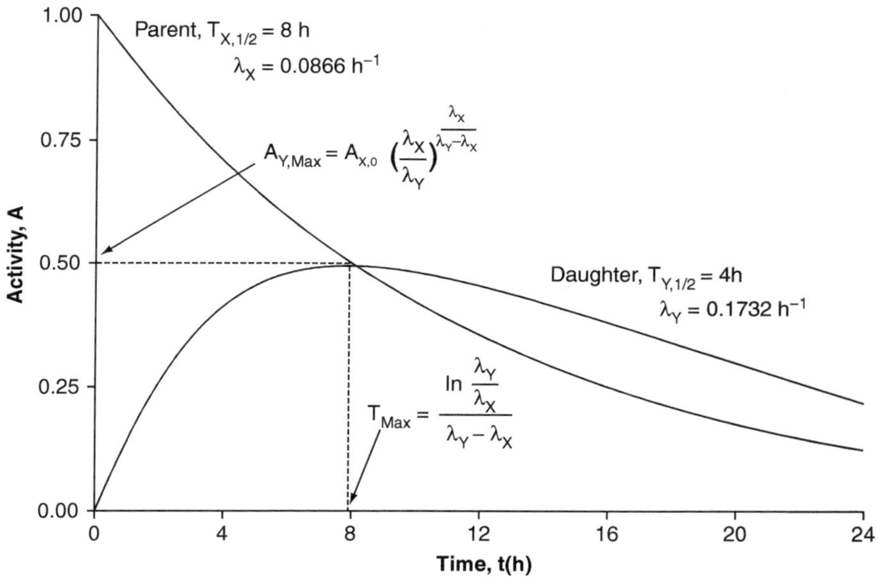

Fig. 5.5 Transient equilibrium for a fictitious parent with a half-life of 8 h and a daughter with a half-life of 4 h

$A_Y(t)$, will increase to a value greater than that of the parent (the ratio being equal to $\lambda_Y/(\lambda_Y - \lambda_X)$ and then subsequently decay at the same rate as that of the parent. This is referred to as transient equilibrium. Figure 5.5 shows an example of the activity-time curves for a fictitious parent with an 8-h half-life with a radioactive daughter with a 4-h half-life. The ratio of the daughter-to-parent activities approaches $\lambda_Y/\lambda_Y - \lambda_X = 2$ with increasing time. The time at which the activity of the daughter reaches its maximum, T_{Max}, can be calculated by differentiating the expression for $A_Y(t)$, setting the result equal to zero and solving for t to give,

$$T_{Max} = \frac{\ln\left(\frac{\lambda_Y}{\lambda_X}\right)}{\lambda_Y - \lambda_X} \tag{5.30}$$

and the maximum activity of the daughter Y is,

$$A_{Y,Max} = A_{X,0}\left(\frac{\lambda_X}{\lambda_Y}\right)^{\left(\frac{\lambda_X}{\lambda_Y - \lambda_X}\right)} \tag{5.31}$$

5.6.4 Daughter Half-Life Greater than that of Parent

Here, $T_{X,1/2} < T_{Y,1/2}$ or, equivalently, $\lambda_X > \lambda_Y$. Rewriting (5.27) by changing signs, the expression for $A_Y(t)$ is,

$$\begin{aligned} A_Y(t) &= A_X(t)\left(\frac{\lambda_Y}{\lambda_X - \lambda_Y}\right)\left(e^{(\lambda_X - \lambda_Y)t} - 1\right) \\ &= A_{X,0}\left(\frac{\lambda_Y}{\lambda_X - \lambda_Y}\right)\left(e^{-\lambda_Y t} - e^{-\lambda_X t}\right). \end{aligned} \tag{5.32}$$

An example is shown in Fig. 5.6 for the fictitious case of a parent with a 6-h half-life and a daughter with a 10-h half-life. Equilibrium cannot be achieved for $\lambda_X > \lambda_Y$.

5.6.5 Decay Branching

Another transition series is shown in Fig. 5.7 in which a parent nucleus decays to a product nucleus directly and through an intermediate daughter state. Here, X decays with a decay constant λ_X into Y and Z, with corresponding branching ratios, w and (1 − w). Y itself decays to Z with a decay constant λ_Y and Z decays with a decay constant λ_Z. The rates of decay of each nuclear species are described by the set of coupled first-order differential equations,

$$\frac{dN_X}{dt} = -\lambda_X N_X(t) \tag{5.33}$$

$$\frac{dN_Y}{dt} = w\lambda_X N_X(t) - \lambda_Y N_Y(t) \tag{5.34}$$

Fig. 5.6 Activity decay for a fictitious parent with a 6-h half-life and a daughter with a 10-h half-life. Equilibrium cannot be achieved

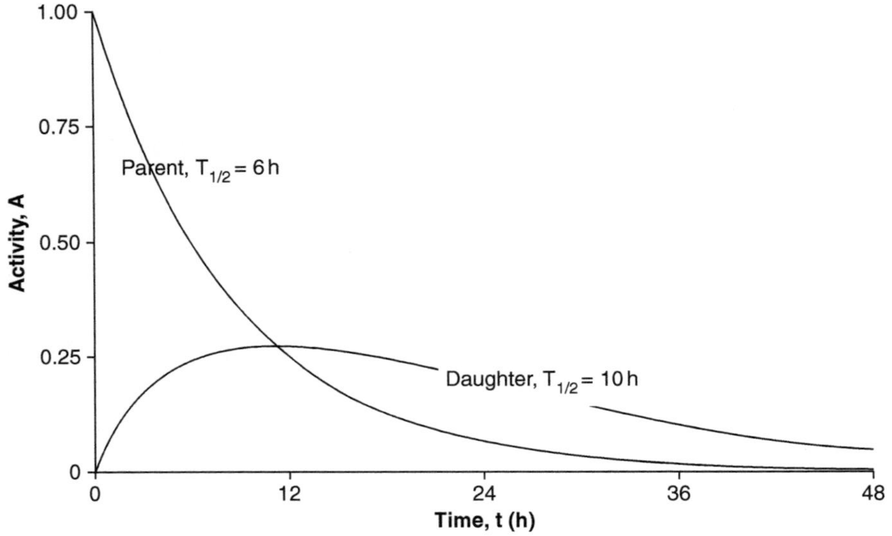

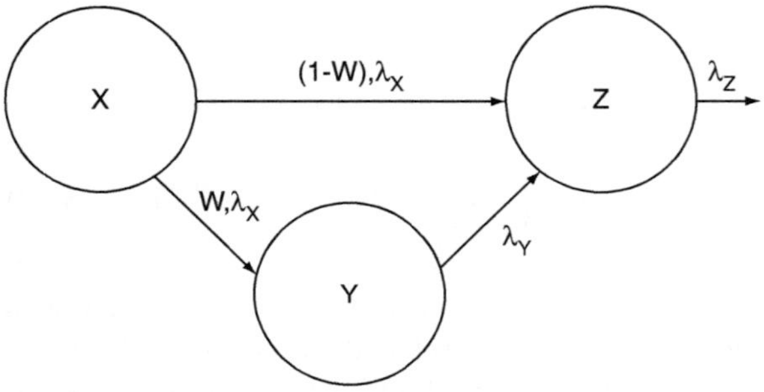

Fig. 5.7 Decay chain with branching to two daughters

$$\frac{dN_Z}{dt} = (1-w)\lambda_X N_X(t) + \lambda_Y N_Y(t) - \lambda_Z N_Z(t). \tag{5.35}$$

Again using the method of substitution, the derived solutions (for the initial conditions of $N_X(t=0) = N_{X,0}$ and $N_Y(t=0) = N_Z(t=0) = 0$) are,

$$N_X(t) = N_{X,0}e^{-\lambda_X t} \tag{5.36}$$

$$N_Y(t) = wN_{X,0}\left(\frac{\lambda_X}{\lambda_Y - \lambda_X}\right)(e^{-\lambda_X t} - e^{-\lambda_Y t}) \tag{5.37}$$

$$\begin{aligned} N_Z(t) = N_{X,0}\Bigg(&\left(\frac{\lambda_X}{\lambda_Z - \lambda_X}\right)\left(1 + w\frac{\lambda_X}{\lambda_Y - \lambda_X}\right)e^{-\lambda_X t} \\ &- w\left(\frac{\lambda_X}{\lambda_Z - \lambda_Y}\right)\left(\frac{\lambda_Y}{\lambda_Y - \lambda_X}\right)e^{-\lambda_Y t} \\ &- \left(\frac{\lambda_X}{\lambda_Z - \lambda_X}\right)\left(1 - w\frac{\lambda_Z}{\lambda_Z - \lambda_Y}\right)e^{-\lambda_X t}\Bigg) \end{aligned} \tag{5.38}$$

and the corresponding activities are,

$$A_X(t) = A_{X,0}e^{-\lambda_X t} \tag{5.39}$$

$$A_Y(t) = wA_{X,0}\left(\frac{\lambda_Y}{\lambda_Y - \lambda_X}\right)(e^{-\lambda_X t} - e^{-\lambda_Y t}) \tag{5.40}$$

$$\begin{aligned} A_Z(t) = A_{X,0}\Bigg(&\left(\frac{\lambda_Z}{\lambda_Z - \lambda_X}\right)\left(1 + w\frac{\lambda_X}{\lambda_Y - \lambda_X}\right)e^{-\lambda_X t} \\ &- w\left(\frac{\lambda_Y}{\lambda_Z - \lambda_Y}\right)\left(\frac{\lambda_Z}{\lambda_Y - \lambda_X}\right)e^{-\lambda_Y t} \\ &- \left(\frac{\lambda_Z}{\lambda_Z - \lambda_X}\right)\left(1 - w\frac{\lambda_Z}{\lambda_Z - \lambda_Y}\right)e^{-\lambda_X t}\Bigg). \end{aligned} \tag{5.41}$$

While the time at which the activity of the daughter Y is at a maximum is that given by (5.30), the value of this maximum activity is given by (5.31) but now weighted by the branching ratio for the X – Y transition,

$$A_{Y,Max} = wA_{X,0}\left(\frac{\lambda_X}{\lambda_Y}\right)^{\left(\frac{\lambda_X}{\lambda_Y - \lambda_X}\right)} \tag{5.42}$$

A practical example of a multiple transition is that of ^{99m}Tc and the relevant decay chain is shown in Fig. 5.8 (refer also to the discussion of nuclear isomerism given in Chap. 4). ^{99}Mo undergoes a series of forbidden β decays to excited states of ^{99}Tc with a half-life of 65.9 h.

Figure 5.9 shows the relative number of ^{99}Mo, ^{99}Tc, and ^{99m}Tc nuclei as a function of time. The number of ^{99}Mo parent nuclei decays exponentially and the number of ^{99m}Tc daughter nuclei initially increases to a maximum and then decays exponentially with the same decay constant as ^{99}Mo once transient equilibrium is attained. However, as ^{99}Tc is the product of both the β^- decay of ^{99}Mo and the isomeric transition of ^{99m}Tc, the number of ^{99}Tc nuclei grows asymptotically to the original number of ^{99}Mo nuclei. Figure 5.9 also shows the relative activities of ^{99}Mo and ^{99m}Tc as functions of time (due its small decay constant, the ^{99}Tc activity is negligible and is not shown). From (5.40), the ratio of the ^{99m}Tc to ^{99}Mo activities approaches the asymptotic value of,

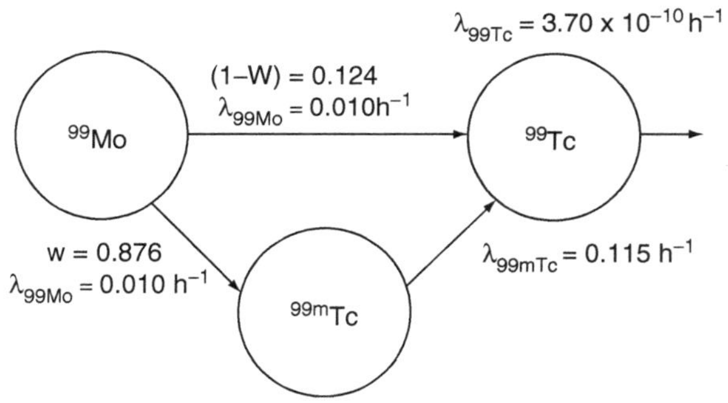

Fig. 5.8 Decay chain from ^{99}Mo to ^{99}Tc directly and via the ^{99m}Tc isomer

$$w\left(\frac{\lambda_Y}{\lambda_Y - \lambda_X}\right) = 0.876\left(\frac{0.115\,h^{-1}}{0.115\,h^{-1} - 0.010\,h^{-1}}\right) = 0.959$$

Because not all of the ^{99}Mo decays result in the production of ^{99m}Tc, the ^{99m}Tc activity can never equal that of ^{99}Mo. The maximum activity of ^{99m}Tc

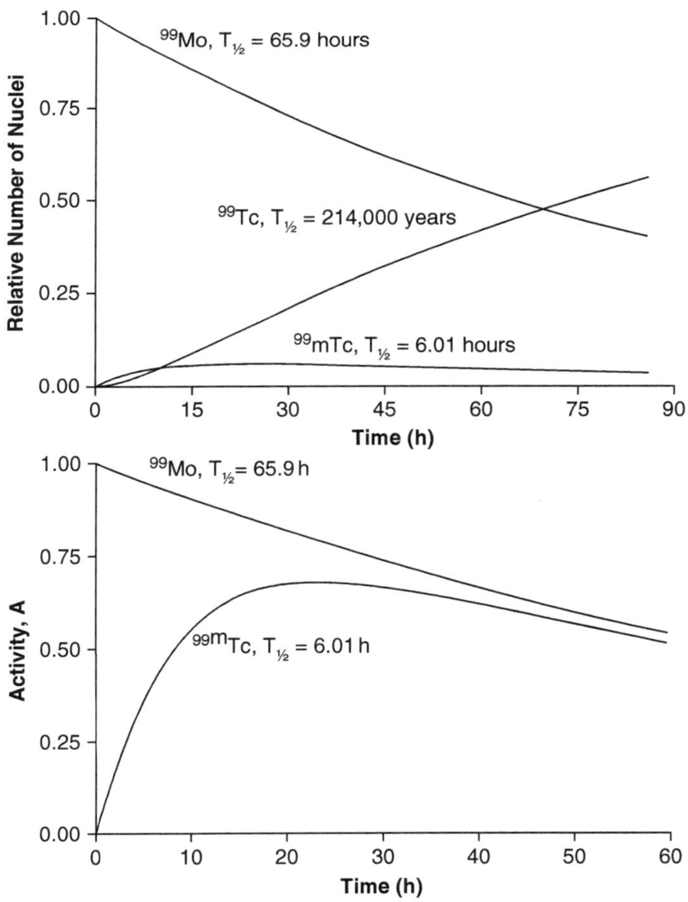

Fig. 5.9 The number of nuclei and activities in the ^{99}Mo decay chain to ^{99m}Tc and ^{99}Tc. (**a**) shows the relative number of nuclei whereas (**b**) shows the activity

is 69% of the initial ^{99}Mo activity and this maximum is reached at 22.9 h.

5.7 Applications

5.7.1 Introduction

Evaluating the radiation absorbed dose received by the nuclear medicine patient begins with measuring the temporal variation of the radioactive isotope in vivo (almost exclusively through imaging) and measuring in vitro the amounts of activity excreted. In this section, the applications of radioactive decay measurement theory to such measurements are considered.

5.7.2 Measurement of Radioactivity

5.7.2.1 Correction for Radioactive Decay During Measurement

At an elementary level, the activity of a radioactive sample is given by the ratio of the number of detected emitted radiations,[3] N, over a period of time, Δt. However, the ratio, $N/\Delta t$ will underestimate the actual activity should the rate of decay over the time period be significantly long compared to the lifetime of the radioactive nuclei. This can be demonstrated using Fig. 5.10, where the number of radioactive decays, N, between time T and $T + \Delta t$ is measured. The activity could be calculated as $N/\Delta t$ at the start of the measurement, T. However, the sample will have decayed during the time of measurement and this must be accounted for. The area under the curve between T and $T + \Delta t$ is equal to the number of radioactive decays that occur during Δt,

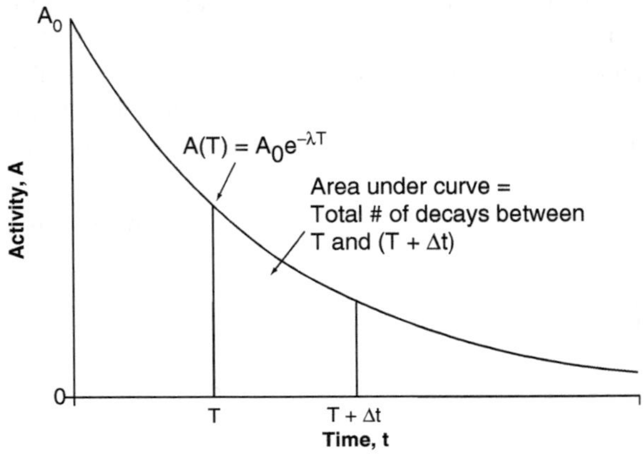

Fig. 5.10 Correction for radioactive decay during time of measurement

$$N = A_0 \int_T^{T+\Delta t} dt\, e^{-\lambda t} = \left(\frac{A_0}{\lambda}\right) e^{-\lambda T}\left(1 - e^{-\lambda \Delta t}\right). \quad (5.43)$$

and the "measured" activity given by the ratio, $N/\Delta t$, is,

$$A_{Meas} = \frac{N}{\Delta t} = A_0 e^{-\lambda T}\left(\frac{1 - e^{-\lambda \Delta t}}{\lambda \Delta t}\right) \quad (5.44)$$

$N/\Delta t$ is an underestimate of the actual activity at the time of the beginning of the measurement, $A_0 e^{-\lambda T}$, as can be seen by rearranging (5.44) and expanding the exponential to second-order,

$$\frac{\left(\frac{N}{\Delta t}\right)}{A_0 e^{-\lambda T}} = \left(\frac{1 - e^{-\lambda \Delta t}}{\lambda \Delta t}\right) \cong \frac{\lambda \Delta t - \frac{1}{2!}(\lambda \Delta t)^2}{\lambda \Delta t} = 1 - \frac{\lambda \Delta t}{2} = 1 - \frac{\ln 2}{2}\frac{\Delta t}{T_{1/2}} \quad (5.45)$$

For example, the measured activity underestimates the actual activity at the time of the beginning of the measurement by about 10% for a measurement duration of 1.7 h for ^{99m}Tc ($T_{1/2} = 6.02$ h) and 0.5 h for ^{18}F ($T_{1/2} = 109.8$ min). Hence, the measured activity must be corrected for this decay if the measurement duration is significant compared to the mean life time. The actual activity at the time of the

[3] Accounting for the inefficiencies of the detector. In the case of a well-detector commonly used for nuclear medicine assays, inefficiencies arise due to absorption within the sample, "dead time" (discussed later), differences in sample volume from that of the calibration sample, etc.

beginning of the measurement, $A_{Actual}(T)$, is given by correcting the activity measured for the time Δt beginning at time T,

$$A_{Actual}(T) = \left(\frac{\lambda \Delta t}{1 - e^{-\lambda \Delta t}}\right) A_{Meas}. \tag{5.46}$$

5.7.2.2 Background Correction

Most cases of in vivo radioactive sample measurement are done in a well-counter. The measured "count rate" (i.e., the rate of event[4] detection) is used to estimate the activity in the sample. The first consideration is that the measurement will have contributions from both the activity in the sample and from background sources. The latter include background radiation due to surrounding natural sources, such as cosmic rays or ambient radon, and electronic noise of the detector and must be subtracted from the measured signal in order to isolate those counts[5] due to the sample activity alone. This is done by measuring the count rate with the sample present, the count rate being due to both the sample and the background noise and the count rate due to the background alone.

The background contribution to the measurement is assumed to be random and statistically independent of that due to the activity contained within the sample being measured. The measured count rate is the ratio of counts measured over a time interval,

$$\dot{N}_{S+B} = \frac{N_{S+B}}{T_{S+B}} \tag{5.47}$$

where the S + B subscript notes that the measured counts are due to both the sample and background. The background count rate is that measured by the apparatus without the sample present,

$$\dot{N}_B = \frac{N_B}{T_B}. \tag{5.48}$$

The count rate due to the sample only is given by background subtraction,

$$\dot{N}_S = \dot{N}_{S+B} - \dot{N}_B = \frac{N_{S+B}}{T_{S+B}} - \frac{N_B}{T_B}. \tag{5.49}$$

From the propagation of errors, the uncertainty in the background-corrected sample count rate is, assuming Poisson statistics and assuming that the measurement times are known exactly,

$$\begin{aligned}
\sigma_{\dot{N}_S} &= \sqrt{\sigma^2_{\dot{N}_{S+B}}\left(\frac{\partial \dot{N}_S}{\partial \dot{N}_{S+B}}\right)^2 + \sigma^2_{\dot{N}_B}\left(\frac{\partial \dot{N}_S}{\partial \dot{N}_B}\right)^2} \\
&= \sqrt{\sigma^2_{\dot{N}_{S+B}} + \sigma^2_{\dot{N}_B}} \\
&= \sqrt{\frac{N_{S+B}}{T^2_{S+B}} + \frac{N_B}{T^2_B}} \\
&= \sqrt{\frac{\dot{N}_{S+B}}{T_{S+B}} + \frac{\dot{N}_B}{T_B}}.
\end{aligned} \tag{5.50}$$

A practical problem to solve in the laboratory is the decision of the time allocated to the sample measurement and that to measuring the background. The time allocations can be optimized by calculating the time allocated to measuring the background counts, T_B, as a proportion of the total measurement time, $(T_{S+B} + T_B)$, using the above results. Obviously, if the sample count rate is much greater than that of the background, then the background measurement time, T_B, would be reduced as a fraction of the total measurement time. Conversely, if the sample and background count rates are comparable, then T_{S+B} and T_B should also be comparable (and both being longer in absolute terms in order to detect the count rate differential). Rewrite the background measurement time as,

$$T_B = T - T_{S+B} \tag{5.51}$$

where T is the fixed total measurement time (sum of the sample plus background and background). Then, the standard deviation of the background-corrected count rate is,

$$\sigma_{\dot{N}_S} = \sqrt{\frac{\dot{N}_{S+B}}{T_{S+B}} + \frac{\dot{N}_B}{(T - T_{S+B})}} \tag{5.52}$$

[4] In the context of this discussion, an "event" is the decay of a radioactive nucleus.

[5] A "count" is a detected event.

and then differentiating with respect to T_{S+B} and setting the result to zero finds the minimum of $\sigma_{\dot{N}_S}$,

$$\frac{d\sigma_{\dot{N}_S}}{dT_{S+B}} = \frac{d}{dT_{S+B}}\sqrt{\frac{\dot{N}_{S+B}}{T_{S+B}} + \frac{\dot{N}_B}{(T - T_{S+B})}}$$
$$= \frac{1}{2\sqrt{\frac{\dot{N}_{S+B}}{T_{S+B}} + \frac{\dot{N}_B}{T_B}}}\left(\frac{\dot{N}_B}{T_B^2} - \frac{\dot{N}_{S+B}}{T_{S+B}^2}\right) = 0$$

yielding the optimum ratio of the background measurement time to that of the sample,

$$\frac{T_B}{T_{S+B}} = \sqrt{\frac{\dot{N}_B}{\dot{N}_{S+B}}}. \quad (5.53)$$

5.7.2.3 Reference Standard

In practice, the physical quantity of activity is not directly measurable but is linked to an observable (e.g., the electronic display of the number of counts in a given time interval). This linkage is through a calibration factor which is obtained from the measurement of a reference standard of known activity. The ratio of the absolute activity of the sample to that of the reference standard is given by the ratio,

$$\Re = \frac{\dot{N}_S}{\dot{N}_R} \quad (5.54)$$

where $\dot{N}_S$ is the sample count rate and $\dot{N}_R$ is that of the reference standard. An application of this is when the reference source has a known activity and the background-corrected count rate of the sample is scaled by $\Re$ to yield the activity of the sample. The uncertainty in $\Re$ is,

$$\sigma_{\Re} = \sqrt{\sigma_{\dot{N}_S}^2\left(\frac{\partial\Re}{\partial\dot{N}_S}\right)^2 + \sigma_{\dot{N}_R}^2\left(\frac{\partial\Re}{\partial\dot{N}_R}\right)^2}$$
$$= \sqrt{\frac{\dot{N}_S}{\dot{N}_R^2}\frac{1}{T_S} - \frac{\dot{N}_S^2}{\dot{N}_R^3}\frac{1}{T_R}}. \quad (5.55)$$

Again, the optimum proportions of measurement time allocated to the sample and reference standard counting can be determined by setting $T_R = (T - T_S)$, where T is the total measurement time, differentiating (5.55) with respect to T_S and setting the result to zero. This yields,

$$\frac{T_R}{T_S} = \sqrt{\frac{\dot{N}_S}{\dot{N}_R}}. \quad (5.56)$$

This result differs from that of (5.53) in that it is reciprocal: i.e., a small reference standard count rate requires a greater reference standard measurement time.

5.7.2.4 Decision Theory

Introduction

Measurements over time of activity in the blood, plasma, and excreta are essential to the derivation of the biodistribution of the radionuclide and are necessary to the evaluation of the internal radiation dosimetry. Practical problems are associated with these measurements. Ideally, a diagnostic radiopharmaceutical should be rapidly excreted in order to reduce both the confounding background signal and the absorbed dose to the patient. Urinary excretion is the preferred route due to its rapidity and measuring the activity in the urinary bladder contents (through in vivo imaging) and the voided urine (through in vivo assays) is of prime importance in evaluating the biodistribution. The potentially high amount of activity present in the in vivo sample can impact the measurement device's performance through its "dead time" response as reviewed later in this chapter. At the opposite extreme, the amount of activity in a blood and plasma samples is relatively small – the amount of activity per ml of blood/plasma will be 0.02% or less of that administered.[6] In this section, we focus on the detection and quantification of small amounts of activity.

[6]Consider, for example, a typical activity of 370 MBq of ^{18}F administered intravenously. The concentration of ^{18}F activity in whole blood will be less than about 70 kBq/mL, further reduced by clearance, any specific binding and physical decay.

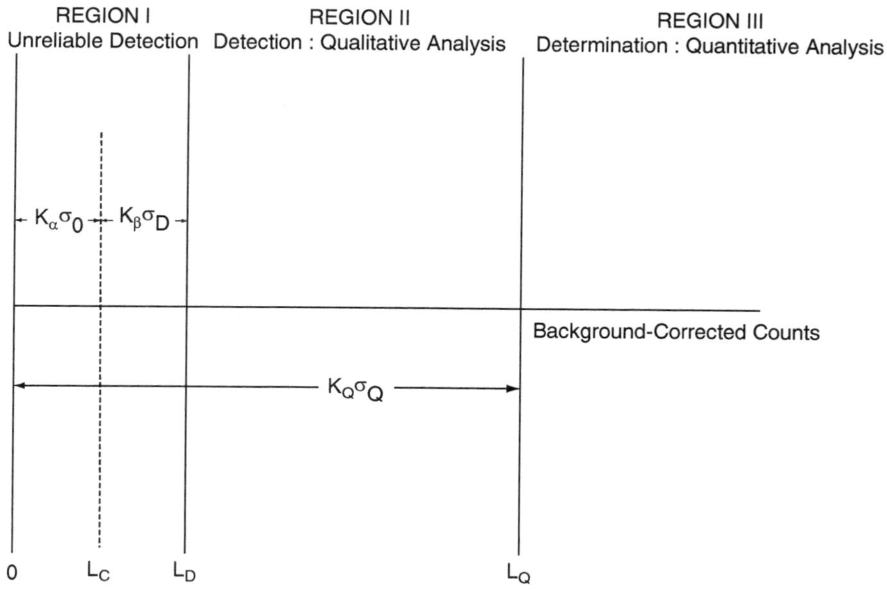

Fig. 5.11 Currie's (1968) definitions of the three principal analytical regions. Quantities are defined in the text

Qualitative Detection and Quantitative Determination

In his seminal paper, Currie (1968) applied statistical decision theory directly to qualitative detection and quantitative determination. This theory has its basis in the statistical detection of a signal in background noise. He proposed that a complete description of the lower limits for a measurement process requires three quantities (reworded in the context of detecting radioactivity):

- Decision level: at which a decision is made as to whether radioactivity is present or not.
- Detection level: at which a measurement may be considered reliable so to lead to detection.
- Determination level: at which a measurement is decided upon to be sufficiently precise so as to yield an acceptable quantitative result.

The three analytical regions proposed by Currie on the basis of these limits are shown in Fig. 5.11. Region I, labeled as that of unreliable detection, is where one applies decision theory to decide whether a measured count (or count rate) is indicative of radioactivity or is background noise. For a measured count within Region II (i.e., exceeding an a priori detection level, L_D), a decision is made that radioactivity is present, but with an established Type II error (discussed below). Finally, in Region III (exceeding a specified quantification level, L_Q), the measured net count rate is considered sufficient to not only allow a statistically-justified decision that radioactivity is present but to also be quantifiable.

While Currie's theory has been revisited and refined since, especially in relation to the transition from a Poisson pdf to a Gaussian pdf at small levels of activity (the interested reader is recommended to refer to, in particular, Zorn et al. (1997); Currie (2004); De Geer (2004); Weise et al. (2006); and Alvarez (2007)[7]), the discussion here will be limited to the original formulation of the theory which provides the basis for specifying these detection limits. The discussion is begun by defining the following quantities:

Signal and background	
Measured number of counts	N_{S+B}
"True" mean number of counts	μ_{S+B}
Variance	σ^2_{S+B}
Background	
Measured number of counts	N_B
"True" mean number of counts	μ_B
Variance	σ^2_B
Background-subtracted signal	
Derived number of counts	$N_S = N_{S+B} - N_B$
"True" mean number of counts	$\mu_S = \mu_{S+B} - \mu_S$
Variance	$\sigma^2_S = \sigma^2_{S+B} + \sigma^2_B$

[7]With the following associated correspondence by Bramlitt and Shonka (2008) and Alvarez (2008).

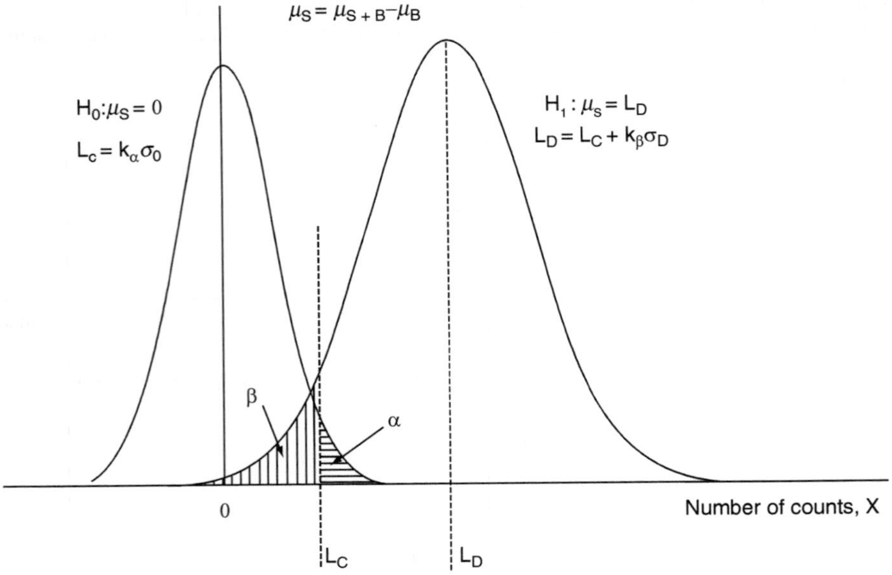

Fig. 5.12 Decision thresholds for the detection of activity. Gaussian *pdfs* are assumed as approximations to the expected Poisson pdfs

It is recognized that both sets of measured counts are Poisson distributed but it is assumed that the counts are sufficiently large that the difference between the two Poisson distributions can be approximated by the difference between two Gaussian distributions (hence, the variance of the background-subtracted signal being given by the quadrature sum of the variances of the total signal and the background signal). Figure 5.12 presents the problem in the form of hypothesis testing (for clarity, the figure shows continuous pdfs, consistent with the approximation of the Poisson distribution by a Gaussian). The null hypothesis H_0 is that the mean of the background-subtracted signal is zero (indicating the absence of activity) and the counterhypothesis H_1 is that the mean is nonzero, i.e., the sample is radioactive. It is important to note that in this detection problem one must, first, a priori estimate a minimum background-corrected count μ_S that is sufficiently large to signify the detection of activity in the sample and, second, to then make an a posteriori decision after recording a number of counts from the apparatus and subtracting the background of whether or not activity is present in the sample. This a posteriori decision is made once one defines an acceptable probability of a Type I error (i.e., rejecting the null hypothesis when it is valid or, in other words, deciding that the number of recorded counts is indicative of activity in the sample when there is in fact none present). The probability of the Type I error is given by the area α shown in Fig. 5.12 for a critical level, L_C,

$$\frac{1}{\sqrt{2\pi}\,\sigma_0}\int_{L_C}^{\infty} dx\, e^{-(x^2/2\sigma_0^2)} = \alpha \tag{5.57}$$

where x is the number of counts and σ_0 is the standard deviation of the background-subtracted pdf when no sample is present.[8] When the number of background-subtracted counts exceeds L_C it is decided that radioactivity has been detected. The critical level is defined by,

$$L_C = k_\alpha\, \sigma_0 \tag{5.58}$$

where k_α is the abscissa for which the area under the normal curve[9] is equal to $1 - \alpha$, and which is sometimes referred to as the coverage parameter (Hurtgen et al. 2000); for example, $k_\alpha = 1.645$ for $\alpha = 0.05$.

[8]It is assumed that the background counts are due to background radiation only and that any random, non-Poisson contributions due to, for example, electrical interference are not present.

[9]The normal curve is the Gaussian distribution for zero mean and unity standard deviation; it is also sometimes referred to as the standardized normal distribution.

Having established the critical level, the a priori detection limit, L_D, is defined by L_C, the standard deviation σ_D of the background-corrected count pdf when its "true" mean is $\mu_S = L_D$ and an acceptable probability of a Type II error, β, (i.e., accepting the null hypothesis when it is invalid or, in other words, deciding that there is no radioactivity present when in fact there is). From Fig. 5.12,

$$\frac{1}{\sqrt{2\pi}\,\sigma_D}\int_{-\infty}^{L_C} dx\, e^{-\left((x-L_D)^2/2\sigma_D^2\right)} = \beta. \qquad (5.59)$$

From (5.58) and (5.59), the detection limit can be written as,

$$L_D = L_C + k_\beta\,\sigma_D \qquad (5.60)$$

where k_β is the abscissa for which the area under the normal curve is equal to $1 - \beta$, as shown in Fig. 5.12. To derive expressions for L_C and L_D, assume that the mean number of background counts, $\bar{N}_B$, is obtained from n independent measurements without the sample present. With the sample in place, the variance of the number of counts is,

$$\begin{aligned}\sigma^2 &= \sigma_{S+B}^2 + \sigma_{\bar{B}}^2 = \sigma_{S+B}^2 + \sigma_B^2 + \sigma_{\bar{B}}^2\\ &= \mu_S + \mu_B + \frac{\mu_B}{n}.\end{aligned} \qquad (5.61)$$

From these,

$$L_C = k_\alpha\,\sigma_0 = k_\alpha\sqrt{\mu_B + \sigma_{\bar{B}}^2} \qquad (5.62)$$

and

$$L_D = L_C + k_\beta\,\sigma_D = L_C + k_\beta\sqrt{L_D + \sigma_0^2}. \qquad (5.63)$$

An expression for the detection limit can be then be derived from these results,

$$\begin{aligned}L_D &= L_C + \frac{k_\beta^2}{2}\left(1 + \sqrt{1 + \frac{4L_C}{k_\beta^2} + \frac{4\sigma_0^2}{k_\beta^2}}\right)\\ &= L_C + \frac{k_\beta^2}{2}\left(1 + \sqrt{1 + \frac{4L_C}{k_\beta^2} + \frac{4L_C^2}{k_\alpha^2 k_\beta^2}}\right).\end{aligned} \qquad (5.64)$$

If, as is usually done, the acceptable levels of the Type I and II errors are equated, $k_\alpha = k_\beta \equiv k$, then this expression for the detection limit simplifies to,

$$L_D = 2L_C + k^2. \qquad (5.65)$$

As the above derivations are only for the detection of radiation (Region II of Fig. 5.11), Currie extended them to define a determination limit, L_Q, satisfactory for quantitative analysis. He defined this as,

$$L_Q = k_Q\sigma_Q \qquad (5.66)$$

where L_Q is the true value of the background-subtracted counts, μ_S, corresponding to a standard deviation, σ_Q. The factor k_Q is the relative standard deviation for which Currie suggests a value of 10 (corresponding to a 10% relative standard deviation). In this case,

$$L_Q = k_Q\sqrt{L_Q + \sigma_0^2} \qquad (5.67)$$

from which one can solve for the determination limit,

$$L_Q = \frac{k_Q^2}{2}\left(1 + \sqrt{1 + \frac{4\sigma_0^2}{k_Q^2}}\right). \qquad (5.68)$$

Currie recognized that difficulties arose from this analysis when the number of counts was small with the validity of the Gaussian approximation to the Poisson distribution becoming questionable. This problem was addressed in the later papers previously cited.

As a brief numerical example of Currie's formulae, consider the measurement of a ^{99m}Tc sample using a well-counter with an efficiency of $\varepsilon = 70\%$; it is specified that $k_\alpha = k_\beta = 1.645$ for $\alpha = \beta = 0.05$ and that $k_Q = 10$. The background count rate is 100 cpm and the counting duration T is 2 min (hence, corrections for physical decay during the measurement time can be neglected). The minimum detectable activity is given by $A_D = L_D/\varepsilon T$ (and, similarly, for A_C and A_Q) and the limiting mean of the background is $\mu_B = 100\times\,2 = 200$ counts. Then, from (5.62), $L_C = \sqrt{2}k_\alpha\sqrt{\mu_B} = 33$ counts, and (5.63), $L_D = (2\times 32.9) + 2.71 = 69$ counts; from (5.68), $L_Q = 50\left(1 + \sqrt{1 + {}^{\mu_B}/_{12.5}}\right) = 256$ counts. These results lead to the corresponding activities,

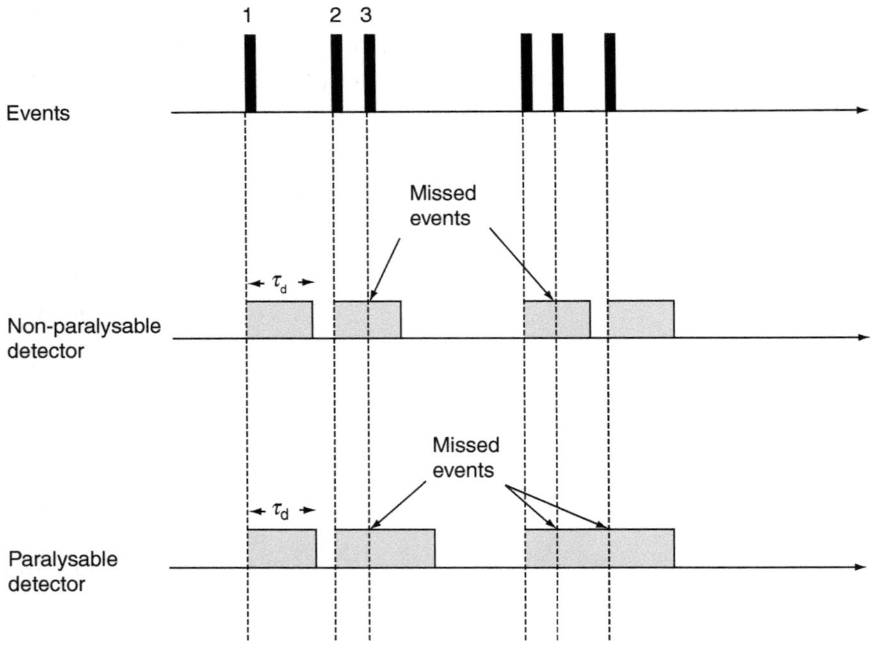

Fig. 5.13 Event detection failures due to dead times of nonparalyzable and paralyzable detectors

$A_C = 24$ dpm[10]
$A_D = 49$ dpm
$A_C = 183$ dpm

Suppose that a measurement of a sample plus the background resulted in a total of 250 counts; as the background is 200 counts, the net background-corrected signal would be $250 - 200 = 50$ counts with an estimated standard deviation of $\sqrt{250 + 50} = 17.3$ counts. As $L_C = 33$ counts, the decision would be that activity has been detected in the sample.

5.7.2.5 Detector Dead Time

A radiation detector converts a physical phenomenon associated with a radiation event (e.g., ionization) into a signal that can be recorded or interpreted. After such an event, some categories of electronic radiation detectors are temporarily incapable of detecting any events: the period of time during which the detector is insensitive to radiation events is known as the dead time (Driver 2002). As further events occurring during the dead time are not detected, this is an inefficiency that increases with event rate. If, during the dead time, an additional event occurs which extends the dead time further, the detector response is referred to as being paralyzable. If the dead time is not extended, the response is termed nonparalyzable. These two modes of event detection failure are shown in Fig. 5.13.

Paralyzable Response

In this type of response, the total dead time is of variable length, depending upon the number of events which occur during the dead time following the original event. Let $\dot{\mu}$ be the mean actual event rate and τ_d be the detector dead time. As the probability that x events will occur during τ_d is given by the Poisson distribution $e^{-\dot{\mu}\tau_d}(\dot{\mu}\tau_d)^x/x!$, the probability that no events occurring during the dead time is $e^{-\dot{\mu}\tau_d}$. This probability is also equal to the ratio of the measured event rate, $\dot{N}$, to the actual event rate, $\dot{\mu}$, or,

$$\dot{N} = \dot{\mu} e^{-\dot{\mu}\tau_d}. \tag{5.69}$$

While $\dot{\mu}$ cannot be calculated in an analytical solution of this result, if the dead time loss is small due to a small $\dot{\mu}$ and/or small τ_d, the exponential can be expanded to give the approximation, $\dot{N} \cong \dot{\mu}(1 - \dot{\mu}\tau_d)$. Solving for the true event rate,

[10] dpm = disintegrations per minute.

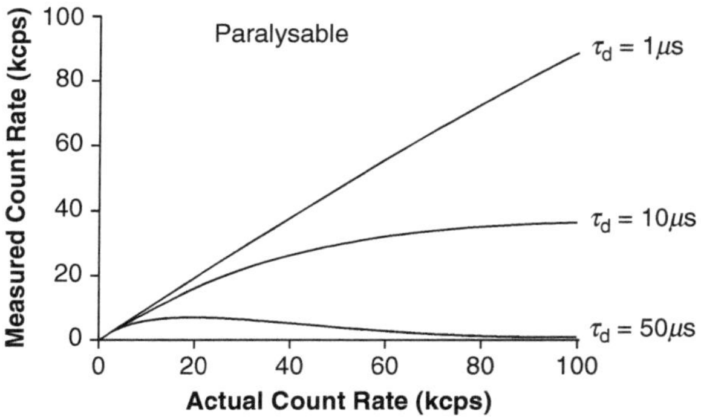

Fig. 5.14 Measured count rate for a paralyzable detector as a function of count rate for different dead times. (kcps = thousands of counts per second)

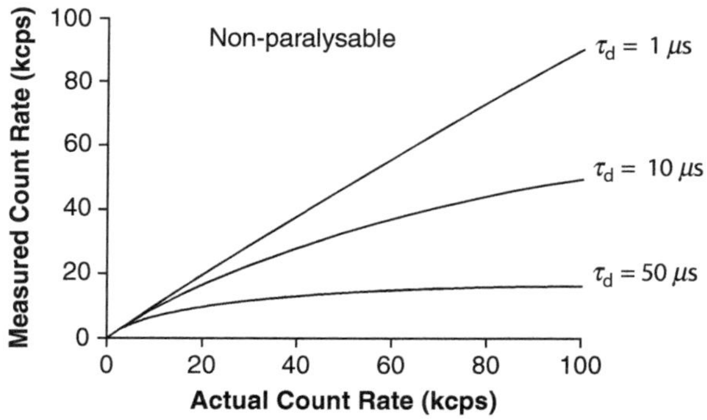

Fig. 5.15 Detector response for a nonparalyzable detector as a function of actual count rate for different dead times

$$\dot{\mu} \cong \frac{1}{2\tau_d} + \sqrt{\frac{1}{4\tau_d^2} - \frac{\dot{N}}{\tau_d}}. \tag{5.70}$$

Figure 5.14 shows the measured count rate $\dot{N}$ for a paralyzable detector as a function of the actual count rate $\dot{\mu}$ for dead times of 1, 10, and 50 μs. It is evident that in the limit of zero dead time, $\dot{N} = \dot{\mu}$, as expected. With increasing τ_D, $\dot{N} \approx \dot{\mu}$ only at small $\dot{\mu}$ and, in fact, for a sufficiently large dead time (the 50 μs value in this example), the observed detector response can correspond to two values of the actual count rate: a true response for a low count rate (to the left of the maximum) and a false response for a high count rate (to the right of the maximum).

Nonparalyzable Response

For this mode, the total dead time following an event is fixed at τ_d, so a probabilistic calculation is not necessary. For each detected event, the number of events which fail to be detected is $\dot{\mu}\tau_d$, so the measured event rate will be,

$$\dot{N} = \left(\frac{\dot{\mu}}{1 + \dot{\mu}\tau_d}\right) \tag{5.71}$$

and solving for $\dot{\mu}$,

$$\dot{\mu} = \frac{\dot{N}}{1 - \dot{N}\tau_d} \tag{5.72}$$

For small dead time losses, this result can be approximated by $\dot{N} \cong \dot{\mu}(1 - \dot{\mu}\tau_d)$. Hence, the responses of paralyzable and nonparalyzable detectors are the same for small dead time losses. Figure 5.15 shows the measured count rate $\dot{N}$for a nonparalyzable detector as a function of the actual count rate $\dot{\mu}$ for dead times of 1, 10, and 50 μs.

In order to correct for dead time losses using the above expressions, τ_d, must be known. It can be determined from knowledge and analysis of the detector and its associated electronics or estimated by empirical measurement. The simplest method is the "2-source method" in which count rates are measured for two individual sources with different activities separately and then in combination. Let $\dot{N}_1$, $\dot{N}_2$ and $\dot{N}_{12}$ be the measured count rates for source 1, source 2, and source 1 and 2 in combination, respectively, and $\dot{N}_{Bkgd}$ is the measured background count rate. Let $\dot{\mu}_1$, $\dot{\mu}_2$, $\dot{\mu}_{12}$, and $\dot{\mu}_{Bkgd}$ be the corresponding mean actual count rates. Clearly,

$$\dot{\mu}_1 + \dot{\mu}_2 - 2\dot{\mu}_{Bkgd} = \dot{\mu}_{12} - \dot{\mu}_{Bkgd}$$

from which,

$$\dot{\mu}_1 + \dot{\mu}_2 = \dot{\mu}_{12} + \dot{\mu}_{Bkgd}. \tag{5.73}$$

For the nonparalyzable model, (5.73) can be rewritten in terms of the measured count rates,

$$\frac{\dot{N}_1}{1 - \dot{N}_1\tau_d} + \frac{\dot{N}_2}{1 - \dot{N}_2\tau_d} = \frac{\dot{N}_{12}}{1 - \dot{N}_{12}\tau_d} + \frac{\dot{N}_{Bkgd}}{1 - \dot{N}_{Bkgd}\tau_d}. \tag{5.74}$$

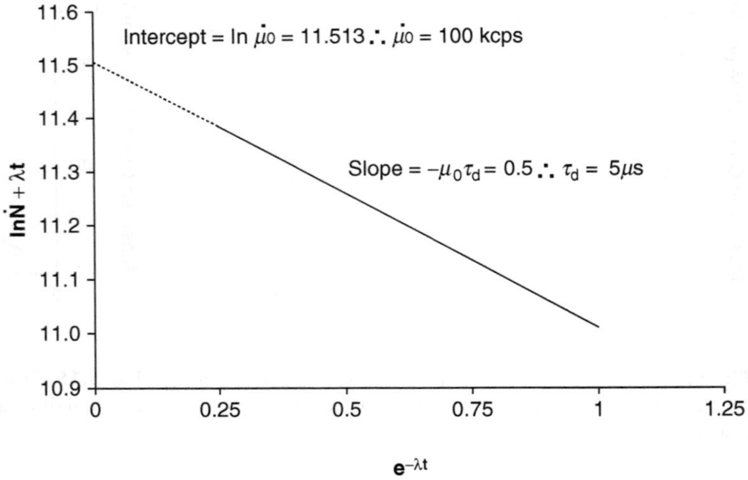

Fig. 5.16 Example of determining the dead time of a paralyzable detector with a 5 μs dead time using a ^{99m}Tc source with an initial actual count rate of 100 kcps

Solving for τ_d,

$$\tau_d = \frac{\beta\left[1 - \sqrt{1 - \frac{4\alpha\gamma}{\beta^2}}\right]}{2\alpha} \quad (5.75)$$

where

$$\alpha = \dot{N}_1\dot{N}_2\left(\dot{N}_{12} + \dot{N}_{Bkgd}\right) - \dot{N}_{12}\dot{N}_{Bkgd}\left(\dot{N}_1 + \dot{N}_2\right)$$

$$\beta = 2\left[\dot{N}_{12}\dot{N}_{Bkgd} - \dot{N}_1\dot{N}_2\right]$$

$$\gamma = \left(\dot{N}_1 + \dot{N}_2\right) - \left(\dot{N}_{12} + \dot{N}_B\right).$$

There is no such solution for the paralyzable model using two sources, but a second method in which a radioactive source is counted as it decays (with decay constant λ) can evaluate the dead time for the paralyzable model. Neglecting any background contribution, the actual mean count rate is,

$$\dot{\mu} = \dot{\mu}_0 e^{-\lambda t}. \quad (5.76)$$

Then the measured count rate for the paralyzable model is,

$$\dot{N} = \dot{\mu}_0 e^{-\lambda t} e^{-\dot{\mu}_0\tau_d e^{-\lambda t}}$$

and taking logarithms,

$$\ln\dot{N} = \ln\dot{\mu}_0 - \lambda t - \dot{\mu}_0\tau_d e^{-\lambda t}. \quad (5.77)$$

Plotting $\ln(\dot{N}) + \lambda t$ on the ordinate and $e^{-\lambda t}$ on the abscissa yields a straight line with the slope equal to $-\dot{\mu}_0\tau_d$ and an intercept of $\ln\dot{\mu}_0$. A numerical example is shown in Fig. 5.16 in which a ^{99m}Tc source, with an initial mean actual count rate of 10^5 cps, decays over a period of two half-lives whilst being counted by a detector with a 5 μs dead time. The plot of $\ln(\dot{N}) + \lambda t$ vs. $e^{-\lambda t}$ is a straight line, from which the values of $\dot{\mu}_0$ and τ_d are extracted.

5.7.3 Verification of Statistical Distribution of Measured Data

A valuable quality control measurement of a counter is to compare the measured data distribution to an expected statistical distribution (Poisson or Gaussian, depending upon the mean number of counts). A comparative parameter is the χ^2. For example, consider an example of source with an activity which is measured M times in succession. It is assumed that the duration of the M measurements is sufficiently short in relation to the mean lifetime of the radionuclide. Each measurement consists of acquiring counts from the source over equal time intervals; hence, $\dot{N}_i$ is the count rate over the ith time interval. The χ^2 parameter is,

$$\chi^2 \equiv \frac{1}{\bar{\dot{N}}}\sum_{i=1}^{M}\left(\dot{N}_i - \bar{\dot{N}}\right)^2 \quad (5.78)$$

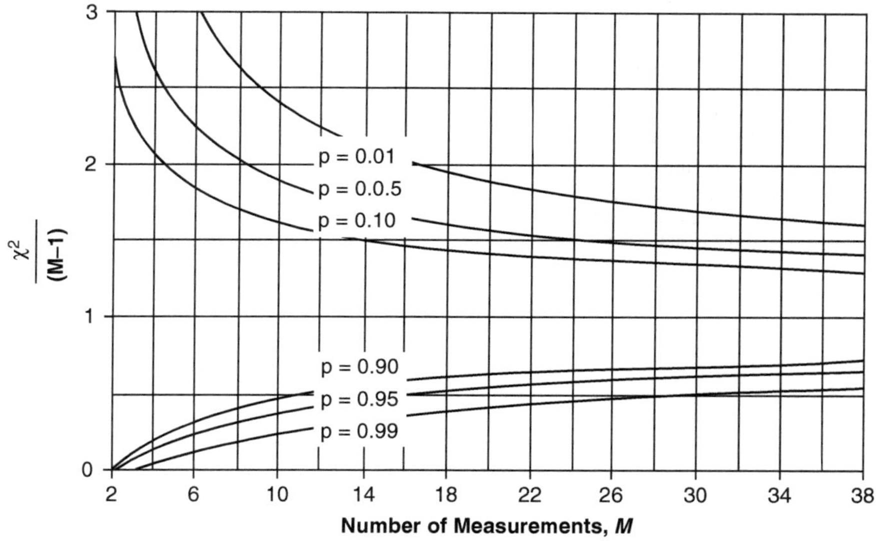

Fig. 5.17 Reduced-χ^2 as a function of the number of measurements, M. The *p*-values shown are the probabilities that a random sample of M measurements would yield a reduced-χ^2 value greater than the corresponding value of the curve

where $\bar{\dot{N}}$ is the expected mean count rate. The χ^2 is also related to the variance of the count rates,

$$\chi^2 = \frac{(M-1)s_{\dot{N}}^2}{\bar{\dot{N}}}. \tag{5.79}$$

If the count rates are (or are nearly) Poisson distributed, then $s_{\dot{N}}^2 \cong \bar{\dot{N}}$, and (Walpole and Myers 1997),

$$\chi^2 \approx (M-1) \tag{5.80}$$

Thus, deviations of the measured χ^2 value from (M − 1) would be indicative of errors in the equipment acquisition (Hamilton 2004).

Figure 5.17 shows the reduced χ^2 (i.e., the χ^2 normalized to the number of degrees-of-freedom, M − 1) as a function of the number of measurements, M. The *p*-values shown are the probabilities that a random sample of M measurements would yield a reduced χ^2 value greater than the corresponding value of the curve. An example of how this figure could be used in a calculation is to consider M = 10 measurements of count rates, from which a χ^2 is calculated. A χ^2 value of less than 3.33 (i.e., $\chi^2/(M-1)0.37$) or a χ^2 value of greater than 16.92 (i.e., $\chi^2/(M-1)$ 1.88) have a 5% probability or less of having occurred randomly if the underlying distribution were Poisson: i.e., they imply greater variation than that due to a Poisson distribution and would suggest further investigation of the equipment and/or sample.

References

Alvarez JL (2007) Poisson-based detection limit and signal confidence intervals for few counts. Health Phys 93:120–126

Alvarez JL (2008) Reply to Bramlitt and Shonka. Health Phys 94:87–88

Bainbridge KT, Goldhaber M, Wilson E (1953) Influence of the chemical state on the lifetime of a nuclear isomer, Tc^{99m}. Phys Rev 90:430–439

Bramlitt ET, Shonka JT (2008) Comment on Poisson-based detection limit and signal confidence intervals for few total counts. Health Phys 94:86–87

Byers DH, Stump R (1958) Effects of low temperature on the technetium-99m lifetime. Phys Rev 112:77–79

Currie LA (1968) Limits for qualitative detection and quantitative determination. Anal Chem 40:586–592

Currie LA (2004) Detection and quantification limits: basic concepts, international harmonization, and outstanding ("low-level") issues. Appl Rad Isotopes 61:145–149

De Geer L-E (2004) Currie detection limits in gamma-ray spectroscopy. App Rad Isotopes 61:151–160

Driver I (ed) (2002) Radioactive sample counting – principles and practice. Institute of Physics and Engineering in Medicine, York

Hamilton D (2004) Diagnostic nuclear medicine. Springer, Berlin

Huh C-A (1999) Dependence of the decay rate of ^{7}Be on chemical forms. Earth Planet Sci Lett 171:325–328

Hurtgen C, Jerome S, Woods M (2000) Revisiting Currie – how low can you go? Appl Rad Isotopes 53:45–50

James F (2006) Statistical methods in experimental physics, 2nd edn. World Scientific Publishing, Singapore

Jennings WA (2007) Evolution over the past century of quantities and units in radiation dosimetry. Phys Med Biol 27:5–16

Rutherford E, Soddy F (1902) The cause and nature of radioactivity. Phil Mag 4:370–396

Segrè E (1947) Possibility of altering the decay rate of a radioactive substance. Phys Rev 71:274–275

Tuli JK, Reed G, Singh B (2001) Nuclear data sheets for 99Tc. Nucl Data Sheets 93:1–32

Walpole RE, Myers RH (1997) Probability and statistics for engineers and scientists, 5th edn. Prentice-Hall, Englewood Cliffs

Weise K, Hübel K, Rose E et al (2006) Bayesian decision threshold, detection limit and confidence limits in ionising-radiation measurement. Rad Prot Dos 121:52–63

Zorn ME, Gibbons RD, Sonzogni WC (1997) Weighted least-squares approach to calculating limits of detection and quantification by modeling variability as a function of concentration. Anal Chem 69:3096–3075

Photon Interactions with Matter

6

Abstract The radiation-matter interactions of interest to nuclear medicine are those of photons (X or γ rays) and of charged particles (α particles and electrons). This chapter reviews the photon–matter interaction classes of interest to nuclear medicine dosimetry and classifies them in terms of whether or not the incident photon is preserved through the process. In practice, this will be those major interactions that occur at photon energies below 1 MeV. The cross sections for Thomson and Rayleigh scatter, which are classical in nature, are derived; insignificant energy transfer results from such scatters although they do lead to attenuation of a photon beam. Compton scatter is reviewed extensively, including through the derivation of the Klein–Nishina cross sections using the Feynman propagator method. Photoelectric absorption is next examined and the cross sections for photon absorption on the K-shell electrons derived. The excited atom must relax through either radiative or nonradiative means and these are reviewed and characteristic X-rays and Auger/Coster–Kronig electrons introduced. Finally, the interaction coefficients used in dosimetry to describe photon–matter interactions are introduced.

Contents

6.1 Introduction

A photon travelling through a medium will interact with the atoms and nuclei in that medium through various mechanisms and usually transfer energy in the process. The types and likelihoods of these interactions are functions of photon energy, photon polarization, and the atomic and nuclear physical properties of the medium. In the medical context, knowledge of the physics of photon–matter interactions is essential to understanding, for example, the modeling of image production and the associated radiation dosimetry (Hubbell 1999; Zaidi 2000). This chapter looks at those categories of photon–matter interactions relevant to nuclear medicine and which all occur at photon

B.J. McParland, *Nuclear Medicine Radiation Dosimetry*,
DOI 10.1007/978-1-84882-126-2_6,

energies below the threshold energy for $\gamma \rightarrow e^- e^+$ pair production in the nuclear Coulomb field. There are four such interactions, all of which are exclusively photon–electron processes, and which will be categorized in terms of photon number conservation.

Should the photon be conserved during the interaction, it will be scattered from its original trajectory with or without energy loss. Here, the relevant processes are Thomson, Rayleigh, and Compton scatter. In both Thomson and Rayleigh scatter, the photon is elastically scattered from a single or an ensemble of atomic electrons, respectively, with negligible energy transfer to the electrons.[1] The incident and scattered photons are treated classically as electromagnetic waves and, as there is no energy transfer, the phases of the outgoing and incoming waves are the same leading these to modes of scatter being referred to as coherent. This coherence allows interference effects to arise between the scattered waves produced in Rayleigh. While no energy is transferred to the medium and the photon is not absorbed, these elastic processes are still relevant to photon transport calculations as the total photon number in the incident beam will be reduced by these scattering processes due to the photon removal. The assumption of complete elasticity is entirely reasonable due to the scattering atom being part of a lattice which distributes the recoil over a massive entity leading to a negligible recoil kinetic energy. Compton scatter, on the other hand, is a quantum mechanical interaction in which the quantum is scattered from a single electron (assumed to be free) which subsequently recoils and the photon exits the interaction with reduced energy. Thus, unlike coherent scatter, Compton scatter results in both the attenuation of the incident photon beam and the transfer of energy to the medium through the recoiling electron. As a result of the pre- and post-scatter photon energies differing, it is also referred to as incoherent scatter. Special considerations are required if the electron is not free but is bound within an atom.

Some photon–matter interactions can result in the absorption of the photon. As it is a boson, the photon can interact directly with an atomic electron or the Coulomb field of the nucleus[2], being absorbed in both processes. The latter interaction results in the production of an electron–positron pair is known as pair production and has a photon energy threshold, as shown in Appendix of about 1.022 MeV. As this is above virtually all of the photon energies of interest to nuclear medicine, $\gamma \rightarrow e^- e^+$ will not be a significant contributor to internal radiation dosimetry and, hence, will not be considered here. Photoelectric absorption (also known as the photoelectric effect) is the absorption of a photon by an atomic electron. This interaction, on the other hand, is highly relevant at the photon energies typical of nuclear medicine.

Compton scatter or photoelectric absorption transfers all or part of the photon energy to the electron. Should this energy transfer exceed the electron's binding energy, it is ejected leaving a vacancy in the atomic electron orbitals. This vacancy is filled through a variety of radiative and nonradiative relaxation processes resulting in the production of X-rays or ejected low-energy Auger/Coster–Kronig electrons.

In this chapter, the kinematics and cross sections for these four types of photon–matter interactions and the descriptors of the energy transferred and absorbed are derived. In addition, the relaxation mechanisms of an excited atom and the characteristic X-rays and low-energy Auger and Coster–Kronig electrons resulting are reviewed. Reviews of photon–matter interactions and the subsequent atomic relaxation processes can be found in the articles by Bergstrom and Pratt (1997), Harding (1997), Hubbell et al. (1994, 2006a, b), and Roy et al. (1999). Numerical values of photon–matter interaction cross sections and coefficients are not tabulated here but the reader can easily access these at the databases provided by Berger et al. (2005) and by Hubbell and Seltzer (1996). The classic texts by Attix (1986) and by Johns and Cunningham (1983) also provide tabulated photon–matter interaction data.

Throughout this chapter, examples of calculations will be provided for photon interactions with carbon and lead, representing low- and high-Z materials, respectively. The former is a simple model for biological

[1]Nuclear Thomson scatter is the analog to Thomson scatter but with the nucleus as a point charged target. As the cross section is inversely proportional to the square of the mass of the scattering centre, nuclear Thomson scatter is negligible.

[2]The incident photon can also interact with the Coulomb field of an atomic electron to create an electron-positron pair and conferring energy to the recoil electron; such a process is known as triplet production and has an energy threshold slightly greater than 2.044 MeV.

materials (water and other compounds can be calculated for using Bragg's additivity rule).

6.2 Photon-Conserving Interactions

6.2.1 Thomson Scatter

Thomson scatter is the elastic interaction between a photon, considered classically as an electromagnetic wave, and a "free" electron[3] in which the electron is set into oscillatory motion and no energy is transferred. The oscillating electron radiates electromagnetic energy with the same frequency as that of the incident wave and with the angular distribution of an electric dipole.

Consider an electron at the origin, as shown in Fig. 6.1, with a polarized electromagnetic wave incident along the *y*-axis with its polarization such that the electric field is parallel to the *z*-axis,

$$\mathbf{E}_i = E_0 e^{-i\omega t}\hat{\mathbf{z}}. \tag{6.1}$$

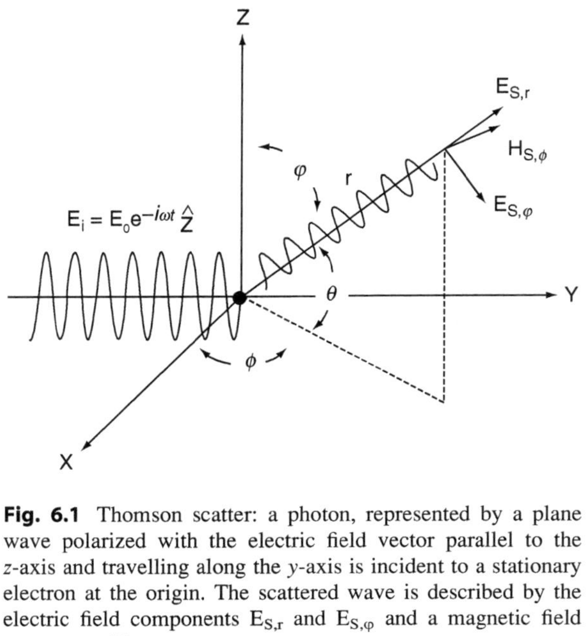

Fig. 6.1 Thomson scatter: a photon, represented by a plane wave polarized with the electric field vector parallel to the *z*-axis and travelling along the *y*-axis is incident to a stationary electron at the origin. The scattered wave is described by the electric field components $E_{S,r}$ and $E_{S,\varphi}$ and a magnetic field component $H_{S,\phi}$

The electron is set into oscillation by the electric field with an equation of motion about the origin,

$$m_e \frac{d^2\mathbf{r}'}{dt^2} = -e\mathbf{E}_i \tag{6.2}$$

with the solution,

$$\mathbf{r}'(t) = \frac{e}{m_e\omega^2} e^{-i\omega t} E_0 \hat{\mathbf{z}}. \tag{6.3}$$

The scattered photon is described by the electromagnetic energy radiated by the oscillating electron (the Poynting vector), the calculation of which requires that the corresponding magnetic and electric fields be known. To obtain these, the retarded vector potential is first derived and then used to calculate the magnetic and electric field strengths from Maxwell's equations. The derivation of the retarded vector potential follows from the current density of the oscillating electron which, assuming that the oscillation amplitude is small and centered at the origin, is,

$$\begin{aligned}\mathbf{J} &= -e\frac{d\mathbf{r}'}{dt}\delta(\mathbf{r}') \\ &= i\frac{e^2E_0}{m_e\omega} e^{-i\omega t}\delta(\mathbf{r}')\hat{\mathbf{z}}\end{aligned} \tag{6.4}$$

where $\delta(\mathbf{r})$ is the Dirac delta function. The retarded vector potential at the point $\mathbf{r}$ is,

$$\begin{aligned}\mathbf{A}_S(\mathbf{r},t) &= \frac{\mu_0}{4\pi}\int d^3\mathbf{r}' \frac{\mathbf{J}\left(\mathbf{r}, t - \frac{|\mathbf{r}-\mathbf{r}'|}{c}\right)}{|\mathbf{r}-\mathbf{r}'|} \\ &= i\frac{\mu_0}{4\pi}\frac{e^2E_0}{m_e\omega}\frac{e^{-i\omega\left(t-\frac{r}{c}\right)}}{r}\hat{\mathbf{z}}\end{aligned} \tag{6.5}$$

where the S subscript denotes correspondence to the scattered photon and μ_0 is the permeability of free space. Equation (6.5) is simplified using the definition of the classical radius of the electron, r_o, and $c = 1/\sqrt{\varepsilon_0\mu_0}$, where ε_0 is the permittivity of free space,

$$\mathbf{A}_S(\mathbf{r},t) = i\left(\frac{r_0}{r}\right)\left(\frac{E_0}{\omega}\right)e^{-i\omega\left(t-\frac{r}{c}\right)}\hat{\mathbf{z}}. \tag{6.6}$$

In spherical coordinates, the components of the retarded vector potential are,

$$\begin{aligned}A_{S,r} &= |\mathbf{A}_S|\cos\varphi \quad A_{S,\varphi} = -|\mathbf{A}_S|\sin\varphi \\ A_{S,\phi} &= 0.\end{aligned} \tag{6.7}$$

[3]In reality, the electron is not "free" as its recoil is ignored. The electron is loosely bound to the atom which, in turn, may be bound to within a crystal lattice, for example. As a result, the recoil is shared with the atom and can be considered negligible.

There is no ϕ-dependence as a result of the symmetry about the dipole (z) axis. As,

$$\nabla \times \mathbf{A_S} = \mu_0 \mathbf{H_S} \tag{6.8}$$

then,

$$\mu_0 H_{S,r} = 0 \quad \mu_0 H_{S,\varphi} = 0$$
$$\mu_0 H_{S,\phi} = \frac{1}{r}\left[\frac{\partial}{\partial r}(rA_{S,\varphi}) - \frac{\partial A_{S,r}}{\partial \varphi}\right]. \tag{6.9}$$

Substituting the non-zero component into (6.8) and solving for $H_{S,\phi}$,

$$H_{S,\phi}(r,t) = \left(\frac{1}{\mu_0 c}\right) \times \left(\frac{r_0}{r}\right) E_0 \sin\varphi\, e^{-i\omega\left(t-\frac{r}{c}\right)}\left(1 - i\frac{c}{\omega r}\right). \tag{6.10}$$

The electric field of the irradiated wave is calculated from,

$$\varepsilon_0 \frac{d\mathbf{E_S}}{dt} = \nabla \times \mathbf{H_S} \tag{6.11}$$

where the two non-zero components are,

$$\frac{dE_{S,r}}{dt} = \frac{1}{\varepsilon_0 r \sin\varphi}\frac{\partial}{\partial \varphi}(\sin\varphi H_{S,\phi}) = -2c\left(\frac{r_0}{r}\right)E_0 \cos\varphi e^{-i\omega\left(t-\frac{r}{c}\right)}\left(1 - i\frac{c}{\omega r}\right) \tag{6.12}$$

and

$$\frac{dE_{S,\varphi}}{dt} = -\frac{1}{\varepsilon_0 r}\frac{\partial}{\partial r}(rH_{S,\phi}) = -c\left(\frac{r_0}{r}\right)E_0 \sin\varphi\, e^{-i\omega\left(t-\frac{r}{c}\right)}\left(\frac{1}{r} + i\left(\frac{\omega}{c} + \frac{c}{\omega r^2}\right)\right). \tag{6.13}$$

Integrating these yields the electric field components of the scattered wave,

$$E_{S,r}(r,t) = -2c\left(\frac{r_0}{r}\right)E_0 \cos\varphi\, e^{-i\omega\left(t-\frac{r}{c}\right)} \times \left(\frac{c}{\omega^2 r^2} + i\frac{1}{\omega r}\right) \tag{6.14}$$

$$E_{S,\varphi}(r,t) = -\left(\frac{ic}{\omega}\right)\left(\frac{r_0}{r}\right)E_0 \sin\varphi\, e^{-i\omega\left(t-\frac{r}{c}\right)} \times \left(\frac{1}{r} + i\left(\frac{\omega}{c} + \frac{c}{\omega r^2}\right)\right). \tag{6.15}$$

The Poynting vector is the average intensity, or the average rate of energy flow per unit area, of the scattered radiation,

$$\mathbf{P} = \frac{1}{2}\mathbf{E} \times \mathbf{H}^*. \tag{6.16}$$

The two components of this vector are,

$$P_{S,r}(r) = \frac{1}{2}E_{S,\varphi}(r,t)H^*_{S,\phi}(r,t) \tag{6.17}$$

$$P_{S,\varphi}(r) = \frac{1}{2}E_{S,r}(r,t)H^*_{S,\phi}(r,t) \tag{6.18}$$

Substituting the previously-derived expressions gives,

$$P_{S,r}(r) = \frac{1}{2\mu_0}\left(\frac{r_0}{r}\right)^2 E_0^2 \sin^2\varphi \left[\frac{1}{c} + \frac{2c}{\omega^2 r^2} + i\frac{c^2}{\omega^3 r^3}\right] \tag{6.19}$$

$$P_{r,\varphi}(r) = -\frac{i}{\mu_0}\left(\frac{r_0}{r}\right)^2 E_0^2 \cos\varphi \sin\varphi \left[\frac{1}{\omega r} + \frac{c^2}{\omega^3 r^3}\right]. \tag{6.20}$$

Those terms in the square brackets which are inversely dependent upon r represent near-field components of the dipole radiation and diminish rapidly with distance. Hence, $P_{r,\varphi}(r)$ can be ignored for large r and the radial component of the Poynting vector $P_{S,r}(r)$ simplified to,

$$P_{S,r}(r) \approx \frac{1}{2c\mu_0}\left(\frac{r_0}{r}\right)^2 E_0^2 \sin^2\varphi = \frac{\varepsilon_0 c}{2}\left(\frac{r_0}{r}\right)^2 E_0^2 \sin^2\varphi. \tag{6.21}$$

As $r \gg r_0$ for atomic dimensions, the intensity of the scattered wave will be much less than of that incident.

The above derivation for a polarized incident wave can be extended to the case of an unpolarized electromagnetic wave, as shown in Fig. 6.2. The

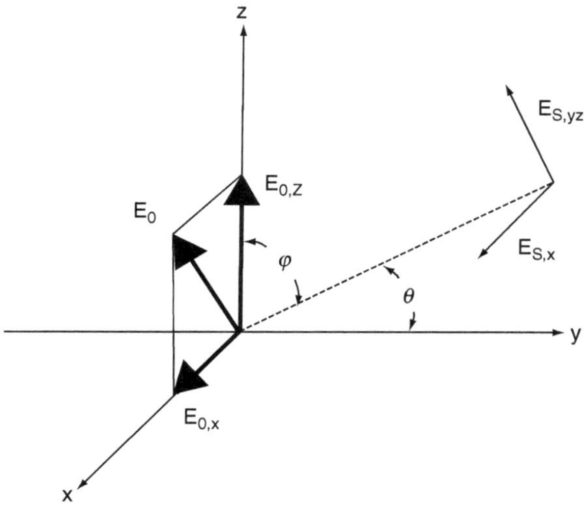

Fig. 6.2 Thomson scatter for an unpolarized electromagnetic wave

electric field of the incident wave is first decomposed into two orthogonal components, one parallel to the *z*-axis and parallel to the *x*-axis, as shown. The scattering problem for the former component has already been solved to yield the radial component of the Poynting vector. The calculation is repeated for the electric field polarized parallel to the *x*-axis but scattered in the *yz*-plane. The magnitude of the scattered wave's retarded vector potential is the same as that of the polarized case, but in the *x*-direction. As the electron oscillation is also now along the *x*-axis, there is no φ-dependence due to symmetry about the dipole axis. In spherical coordinates, the components of this retarded vector potential are,

$$\begin{aligned} A_{S,r} &= |\mathbf{A}_S|\cos\phi \quad A_{S,\varphi} = 0 \\ A_{S,\phi} &= |\mathbf{A}_S|\sin\phi. \end{aligned} \tag{6.22}$$

Due to the above described φ-symmetry, there is only one component of the magnetic field strength which, from $\mu_0\mathbf{H}_S = \nabla \times \mathbf{A}_S$, is,

$$\begin{aligned} H_{S,\varphi}(r,t) &= \left(\frac{1}{\mu_0 c}\right) \\ &\times \left(\frac{r_0}{r}\right) E_0 \sin\phi e^{-i\omega\left(t-\frac{r}{c}\right)}\left(1 - i\frac{c}{\omega r\sin\varphi}\right). \end{aligned} \tag{6.23}$$

The electric field components of the scattered wave are calculated by integrating $\varepsilon_0\frac{d\mathbf{E}_S}{dt} = \nabla \times \mathbf{H}_S$,

$$\begin{aligned} E_{S,r}(r,t) &= -\frac{c}{\sin\varphi}\left(\frac{r_0}{r}\right) E_0 \cos\phi\, e^{-i\omega\left(t-\frac{r}{c}\right)} \\ &\times \left(\frac{c}{\omega^2 r^2 \sin\varphi} + i\frac{1}{\omega r}\right) \end{aligned} \tag{6.24}$$

$$\begin{aligned} E_{S,\phi}(r,t) &= \left(\frac{ic}{\omega}\right)\left(\frac{r_0}{r}\right) E_0 \sin\varphi\, e^{-i\omega\left(t-\frac{r}{c}\right)} \\ &\times \left(\frac{1}{r\sin\varphi} + i\left(\frac{\omega}{c} + \frac{c}{\omega r^2\sin\varphi}\right)\right). \end{aligned} \tag{6.25}$$

The Poynting vector of the scattered wave thus has two components,

$$P_{S,r} = -\frac{1}{2}E_{S,\phi}H^*_{S,\varphi} \quad P_{S,\phi} = -\frac{1}{2}E_{S,r}H^*_{S,\varphi}.$$

If terms of r^{-n} with $n \geq 3$ are ignored, only the radial component need be considered,

$$P_{S,r}(r) = \frac{\varepsilon_0 c}{2}\left(\frac{r_0}{r}\right)^2 E_0^2. \tag{6.26}$$

If the incident wave is completely unpolarized, then,

$$E_{0,x}^2 = E_{0,z}^2 = \frac{E_0^2}{2} \tag{6.27}$$

and the intensity of the scattered wave will be the sum of the radial components of the Poynting vector due to both electric fields,

$$P_{S,r,unpol}(r) = \frac{\varepsilon_0 c}{2}\left(\frac{r_0}{r}\right)^2 E_0^2\left(\frac{1+\sin^2\varphi}{2}\right). \tag{6.28}$$

As the Poynting vector of the incident wave is

$$P_{0,r,unpol}(r) = \frac{\varepsilon_0 c}{2}E_0^2 \tag{6.29}$$

the ratio of the intensities is,

$$\begin{aligned} \frac{P_{S,r,unpol}}{P_{0,r,unpol}} &= \left(\frac{r_0}{r}\right)^2 \frac{1+\sin^2\varphi}{2} \\ &= \left(\frac{r_0}{r}\right)^2 \left(\frac{1+\cos^2\theta}{2}\right) \end{aligned} \tag{6.30}$$

where the complement of the angle φ, the angle θ, has been substituted for convenience as θ is also the scattering angle (i.e., the angle between the direction of the incident photon and that of the scattered photon, as shown in Fig. 6.1). This is the ratio of the incident photon energy that is scattered into the differential solid angle given by $d\Omega = 1/r^2$, allowing (6.30) to be written as,

$$r^2\, d\sigma_{Tho} = r_0^2 \left(\frac{1 + \cos^2\theta}{2} \right) \tag{6.31}$$

to yield the differential cross section in scattering angle for Thomson scatter,

$$\frac{d\sigma_{Tho}}{d\Omega} = \frac{r_0^2}{2}\left(1 + \cos^2\theta\right). \tag{6.32}$$

This differential cross section is independent of the incident photon energy and exhibits a symmetry about the 90° scattering angle, as shown in Fig. 6.3. A measure of the total probability that a Thomson scatter will occur is given by integrating the differential cross section over the complete solid angle of 4π steradians,

$$\begin{aligned} \sigma_{Tho} &= \frac{r_0^2}{2} \int_0^{2\pi} \int_{-1}^{1} d\phi\, d(\cos\theta)\left(1 + \cos^2\theta\right) \\ &= \frac{8\pi}{3} r_0^2 \\ &= 665\,\text{mb}. \end{aligned} \tag{6.33}$$

The Thomson total cross section is a universal constant independent of the incident photon energy for the elastic scattering of a photon by an electron.

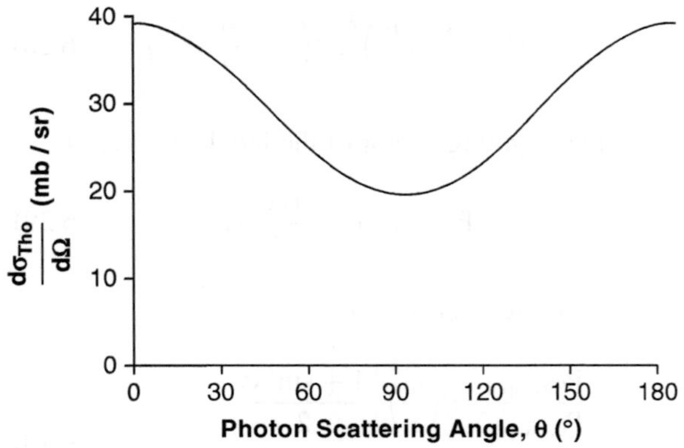

Fig. 6.3 Angular dependence of the Thomson scatter differential cross section

6.2.2 Rayleigh (Coherent) Scatter

While the Thomson formulae describe the elastic scatter of a photon by a single electron, of more practical interest is the likelihood that the photon will be incident to a spatially-distributed ensemble of electrons such as, for example, the atomic electrons orbiting the nucleus. As a result, now allow for the possibility of the scattered waves interacting leading to constructive and destructive interference. This scattering process is referred to as Rayleigh scatter.

Consider a plane wave incident to a pair of electrons, as shown in Fig. 6.4. The electrons oscillate as a result of the interaction with the wave and emit a pair of electromagnetic waves at an angle θ from the direction of the initial electromagnetic wave. The path-length difference between the two waves is equal to $\mathbf{q} \bullet \mathbf{r}/\hbar c$, where the momentum transfer is $\mathbf{q} = (\mathbf{k} - \mathbf{k}')$ and $\mathbf{r}$ is the radial vector between the two scattering centers. The phase difference between the two scattered waves is $e^{i(\mathbf{q}\bullet\mathbf{r}/\hbar c)}$. As the scatter is elastic, the momentum transfer is,

$$q = 2k \sin\frac{\theta}{2} \tag{6.34}$$

Extending the picture from a pair of electrons to an electron number distribution of density $\rho_e(\mathbf{r})$, the amplitude of the scattered wave normalized to that incident is given by integrating the phase differences over the electron spatial distribution, or,

$$F(\mathbf{q}, Z) = \int d^3\mathbf{r}\, e^{i\frac{\mathbf{q}\bullet\mathbf{r}}{\hbar c}} \rho_e(\mathbf{r}) \tag{6.35}$$

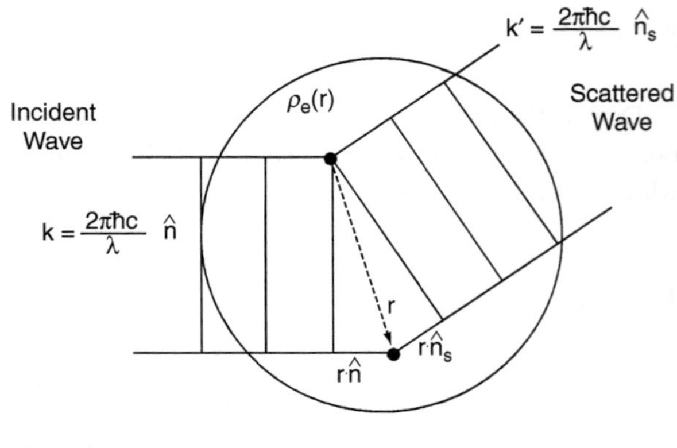

Fig. 6.4 Rayleigh scattering

which will be immediately recognized as the Fourier transform of the electron spatial distribution to momentum space. This expression is referred to as the atomic form factor. The Z of the argument refers to the number of electrons in the ensemble. As the intensity of the scattered wave is given by the square of the amplitude, the differential cross section in photon scattering angle for the coherent scatter from this ensemble of electrons is the coherent superposition of the amplitudes of these scattered waves, or,

$$\frac{d\sigma_{Ray}}{d\Omega}(\theta, Z) = \frac{d\sigma_{Tho}}{d\Omega}(\theta)|F(q, Z)|^2 \tag{6.36}$$

where the Ray subscript refers to Rayleigh scattering. The atomic form factor can also be written as the matrix elements summed over all of the electrons,

$$F(\mathbf{q}, Z) = \sum_{n=1}^{Z} \left\langle \psi_{gs} \middle| e^{i\frac{\mathbf{q}\bullet\mathbf{r}_n}{\hbar c}} \middle| \psi_{gs} \right\rangle \tag{6.37}$$

where ψ_{gs} is the atomic ground-state wavefunction and $\mathbf{r}_n$ is the vector from the origin (i.e., the center of the nucleus) to the nth electron. Only the ground-state wavefunction is used as this is an elastic scatter without excitation or ionization of the atom. If one writes the electron density as,

$$\rho_e(\mathbf{r}) = \left\langle \psi_{gs} \middle| \sum_{n=1}^{Z} \delta(\mathbf{r} - \mathbf{r}_n) \middle| \psi_{gs} \right\rangle \tag{6.38}$$

then the matrix element leads immediately to the Fourier transform.

As the coherent scattering of electromagnetic waves will be a diffraction phenomenon, consider a central electron distribution (i.e., one which is only a function of radial distance). The corresponding atomic form factor is the same representation as that derived for the nuclear form factor in Chap. 3 except that $\rho_e(r)$ represents the charge density of the atomic electrons rather than that of the protons. Rewriting that result slightly,

$$F(\mathbf{q}, Z) = 4\pi \int dr\, r^2 \frac{\sin\left(\frac{qr}{\hbar c}\right)}{\left(\frac{qr}{\hbar c}\right)} \rho_e(r)$$

and noting that

$$\lim_{x \to 0} \frac{\sin x}{x} = 1$$

the atomic form factor for the magnitude of the momentum transfer $q = 0$ (which corresponds to $\theta = 0$) is,

$$F(0, Z) = 4\pi \int_0^\infty dr\, r^2 \rho_e(r) = Z. \tag{6.39}$$

This can also be seen from Fig. 6.4 in which the phase differences go to zero for a zero scattering angle such that there is constructive interference between the waves so that F(0, Z) = Z.

General features of the atomic form factor as a function of momentum transfer can be observed by calculating that for atomic hydrogen. The atomic wavefunction of the hydrogen atom in its ground state is,

$$\psi_{gs} = \frac{1}{\sqrt{\pi r_\infty^3}} e^{-\frac{r}{r_\infty}} \tag{6.40}$$

where r_∞ is the Bohr radius. The corresponding electron density will equal its squared modulus,

$$\begin{aligned} \rho_e(r) &= |\psi_{gs}|^2 \\ &= \frac{1}{\pi r_\infty^3} e^{-2\frac{r}{r_\infty}}. \end{aligned} \tag{6.41}$$

The atomic form factor for hydrogen is the Fourier transform of this result,

$$F(q, Z = 1) = \frac{1}{\left(1 + \left(\frac{r_\infty q}{2\hbar c}\right)^2\right)^2} \tag{6.42}$$

and is plotted in Fig. 6.5 as a function of q.

As $F(q, Z) \to Z$ as the scattering angle $\theta \to 0$, then,

$$\frac{d\sigma_{Ray}}{d\Omega}(0, Z) = Z^2 \frac{d\sigma_{Tho}}{d\Omega}(0) \tag{6.43}$$

Thus the forward-angle coherent scatter from an atom of atomic number Z is Z^2 times that for a single electron. Figure 6.6 shows the coherent differential cross section for carbon (Z = 6) as a function of photon scattering angle for incident photon energies of 10, 20, and 30 keV. For $\theta = 0$, the coherent differential cross section is equal to 2.86 b/sr, which is $Z^2 = 36$ times greater than the Thomson value of 79.5 mb/sr. As the Thomson differential cross section

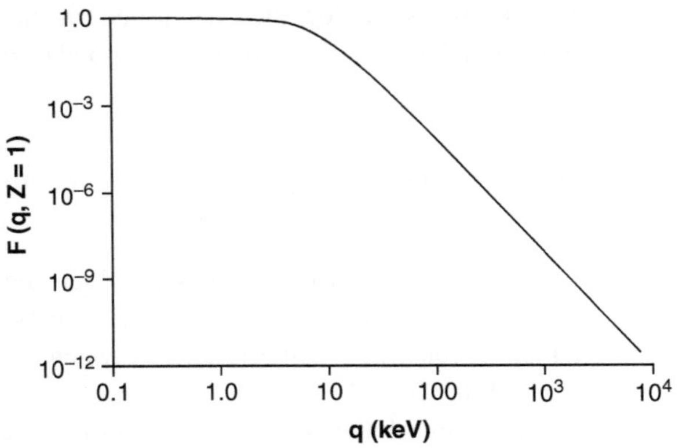

Fig. 6.5 Atomic form factor for hydrogen as a function of momentum transfer

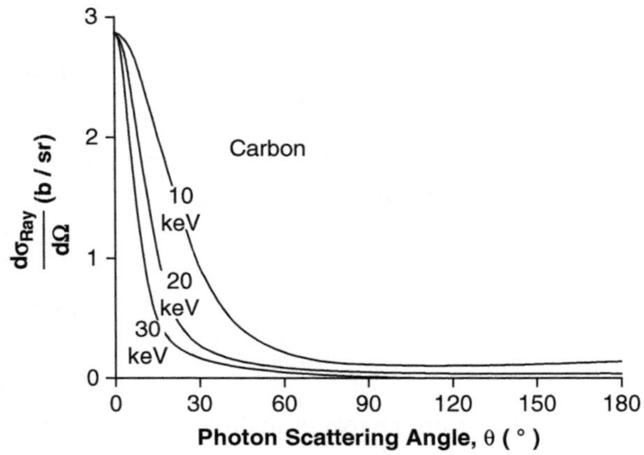

Fig. 6.6 Rayleigh (coherent) scatter differential cross section as a function of photon scattering angle for photon energies of 10, 20, and 30 keV. Ordinate units are in barns (1 b $= 10^{-24}$ cm^2) per steradian

is independent of photon energy and the atomic form factor has no kinematic dependencies at a 0° scattering angle, the 0° coherent scatter cross section will have no energy dependence. Figure 6.6 also shows that, for a fixed non-zero scattering angle, the Rayleigh differential cross section will decrease with increasing photon energy or, equivalently, decreasing photon wavelength. This is due to the phase differences between the scattered waves increasing with decreasing wavelength so that the interference becomes more destructive overall and suppressing the differential cross section. The decrease in the Rayleigh differential cross section with increasing photon energy for a fixed scattering angle can also be interpreted as being due to the corresponding decrease in photon wavelength leading to an increase in the phase differences between the scattered waves.

While the total coherent cross section can be obtained by integrating the differential cross section with solid angle over 4π steradians, it is more conveniently performed by integrating over momentum transfer rather than solid angle,

$$\begin{aligned}\sigma_{Ray} &= \int_0^{2k} \frac{dq}{\left(\frac{dq}{d\Omega}\right)} \frac{d\sigma_{Ray}}{d\Omega} \\ &= \int_0^{2k} dq \frac{d\sigma_{Tho}}{d\Omega} \frac{|F(q,Z)|^2}{\left(\frac{dq}{d\Omega}\right)} \\ &= \frac{r_0^2}{2} \int_0^{2k} dq\left(1+\cos^2\theta(q)\right) \frac{|F(q,Z)|^2}{\left(\frac{dq}{d\Omega}\right)}\end{aligned} \tag{6.44}$$

The upper limit of the integral follows for $\theta = \pi$ and the derivative is calculated,

$$\begin{aligned}\frac{dq}{d\Omega} &= \frac{1}{2\pi} \frac{dq}{d(\cos\theta)} \\ &= \frac{k^2}{2\pi q}\end{aligned} \tag{6.45}$$

which leads to,

$$\sigma_{Ray} = \frac{\pi r_0^2}{k^2} \int_0^{2k} dq\left(1+\cos^2\theta(q)\right) q|F(q,Z)|^2 \tag{6.46}$$

This integral can be simplified by noting that $|F(q,Z)|^2$ is appreciable only for low *q* or low scattering angles. In this case, $(1+\cos^2\theta) \approx 2$,

$$\sigma_{Ray} = \frac{2\pi r_0^2}{k^2} \int_0^{2k_\gamma} dq \bullet q|F(q,Z)|^2 \tag{6.47}$$

Hence, a representation of the atomic form factor is required to calculate the Rayleigh total cross section and this requires detailed knowledge of the ground-state atomic wavefunction.[4] As a result, it is typically

[4]This evaluation is summarized by Hubbell et al. (1975).

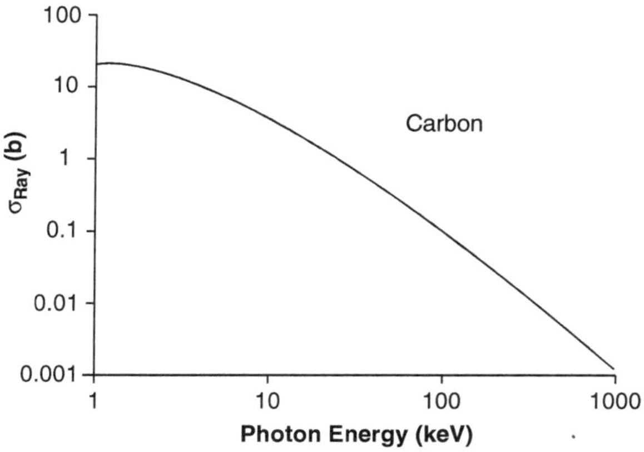

Fig. 6.7 Total Rayleigh (coherent) cross section for carbon

calculated using numerical integration. An example of the total coherent cross section for carbon is shown in Fig. 6.7 as a function of photon energy ranging from 1 keV to 1 MeV. At high photon energies, the total cross section varies approximately as $1/k^2$.

6.2.3 Compton (Incoherent) Scatter

6.2.3.1 Introduction

Thomson and Rayleigh scatter are classical electrodynamic processes in which no energy is transferred to the medium. Compton scatter, on the other hand, is a quantum electrodynamic interaction and results in the direct transfer of energy from the photon to a recoil electron. Experimental suggestions of an inelasticity in the scattering of electromagnetic radiation[5] could not be explained using classical electrodynamic theory but is possible through quantum electrodynamics (QED) in which the photon is considered as a quantum. This was demonstrated independently by Compton (1923a, b, 1925) and Debye (1923) who both derived a relationship between the scattered photon wavelength (inversely proportional to its energy) and the scattering angle from a free electron at rest which subsequently recoils with a fraction of the projectile photon's energy. These results, however, are purely of the kinematics of the photon–electron interaction and

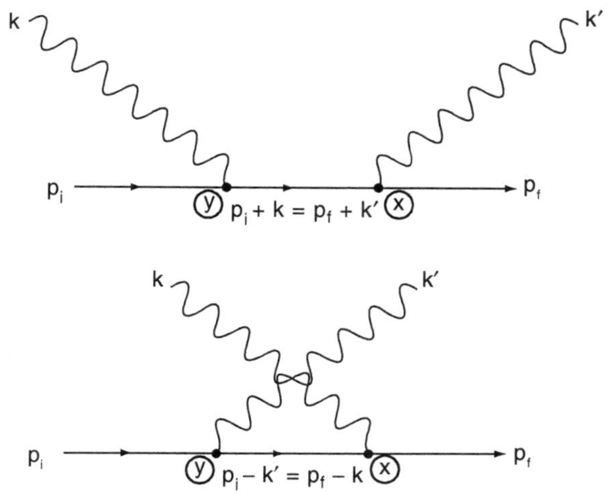

Fig. 6.8 Feynman diagrams for Compton (incoherent) scatter

do not shed light on the probability of the scattering occurring. Initial derivations of the Compton scatter differential cross section were made by Dirac (1926) and Gordon (1927). However, these results were only in agreement with experimental data for the combination of small angle scattering and light elements. It was the collaboration between Klein and Nishina (1929) who, using Dirac's electron theory (Chap. 2), provided the complete analytical expressions for the Compton scatter cross section. The Feynman diagrams for Compton scatter are shown in Fig. 6.8.

These diagrams show that Compton scatter is a second-order process with an intermediate state differing from the initial and final states by one quantum. Compton scatter proceeds through either the absorption of the incident photon by an electron at one vertex (the upper diagram with no photon present in the intermediate state) and the emission of the final photon at the second or through the emission of the final photon by the electron at one vertex (the lower diagram with two photons present in the intermediate state) followed by the absorption of the incident photon at the second vertex. Before Fig. 6.8 is used to derive the Klein–Nishina cross section, the kinematics of Compton scatter must first be derived.

6.2.3.2 Compton Kinematics

Figure 6.9 shows the kinematics of a photon of energy k and momentum **k** incident to a free electron of mass

[5]The scatter was considered inelastic in that the pre- and postscatter photon energies differed.

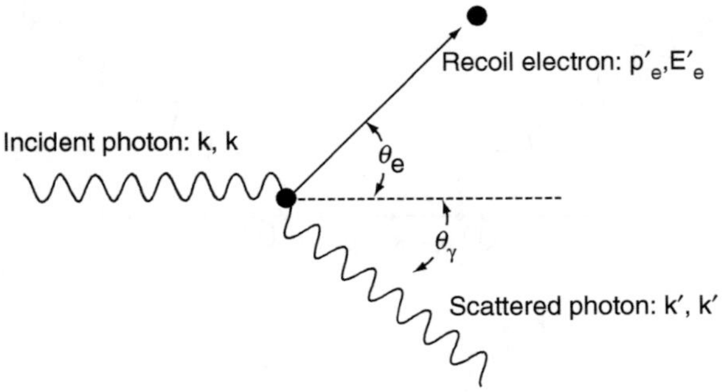

Fig. 6.9 Compton scatter kinematics

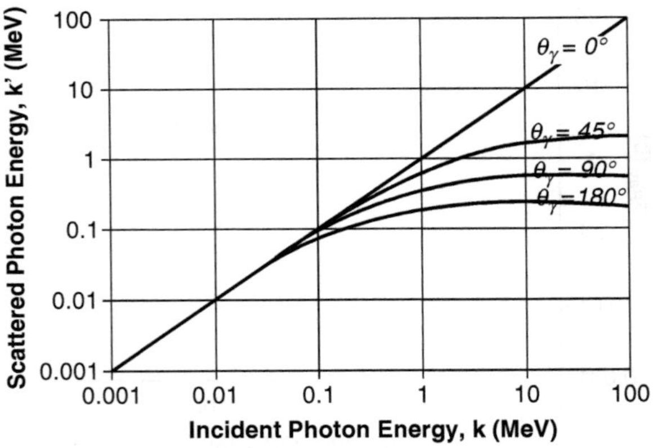

Fig. 6.10 Compton (incoherent) scatter: scattered photon energy as a function of incident photon energy for different scattering angles

m_e at rest. The photon is scattered through an angle θ_γ with a reduced energy k' and momentum $\mathbf{k}'$ and the electron recoils at an angle θ_e with a total energy E'_e and momentum $\mathbf{p}'_e$. The conservations of momentum and energy are,

$$\mathbf{k} = \mathbf{k}' + \mathbf{p}'_e \tag{6.48}$$

$$k + m_e = k' + E'_e. \tag{6.49}$$

The squared momentum transfer to the electron is,

$$|\mathbf{p}'_e|^2 = (\mathbf{k} - \mathbf{k}')^2 \tag{6.50}$$

and the square of the total energy of the recoil electron is,

$$\begin{aligned} E'^2_e &= |\mathbf{p}'_e|^2 + m_e^2 \\ &= k^2 + k'^2 - 2kk' \cos\theta_\gamma + m_e^2. \end{aligned} \tag{6.51}$$

The scattered electron energy is also obtained from the conservation of energy,

$$E'^2_e = (k - k' + m_e)^2. \tag{6.52}$$

Equating these two expressions gives, with some further algebraic manipulation, the scattered photon energy,

$$k' = \frac{k}{1 + \alpha(1 - \cos\theta_\gamma)} \tag{6.53}$$

where the ratio of the incident photon energy to the electron rest mass is conventionally[6] described by $a = k/m_e$. Figure 6.10 presents the scattered photon energy k as a function of the incident photon energy k for a variety of scattering angles. For $\theta_\gamma = 0$, the Thomson scatter condition of $k = k'$ is met. However, for non-zero photon scattering angles, the scattered photon energy is less than that incident with the amount of decrease growing with photon energies. For practical considerations, at incident photon energies below about 200 keV, the scattered photon energy is very nearly equal to that incident. This limited difference has practical implications when energy discrimination is used to exclude scattered photons in emission image acquisition. Above this energy, a greater fraction of the photon energy is transferred to the recoil electron and $k' < k$. With increasing incident photon energy, the amount of energy carried away by the scattered photon approaches a constant value which is a function only of scattering angle. This asymptotic feature can be more clearly understood by rewriting (6.53),

$$k' = m_e \frac{k}{m_e + k(1 - \cos\theta_\gamma)} \tag{6.54}$$

[6]A nomenclature which, unfortunately, can be confused with the fine-structure constant. However, which of the two quantities is being referred to should be clear due to the context of the discussion.

which, for high photon energies ($k \gg m_e$), simplifies to,

$$k' \approx \left(\frac{m_e}{1-\cos\theta_\gamma}\right) \quad \text{for } k \gg m_e. \tag{6.55}$$

This gives the maximum scattered photon energies for scattering angles of 45, 90, and 180° of,

$$k'(\theta_\gamma = 45°) \approx 1.74 \text{ MeV}$$
$$k'(\theta_\gamma = 90°) \approx m_e = 0.511 \text{ MeV}$$
$$k'(\theta_\gamma = 180°) \approx 0.5\, m_e = 0.256 \text{ MeV}.$$

The vertical differences in Fig. 6.10 between the diagonal line ($k = k'$) and the curves are equal to the energies transferred to the electron. The maximum energy transfer will obviously occur for a backscattered photon ($\theta_\gamma = 180°$) and approaches a value of $k - m_e/2$ for very large incident photon energies. While most of the photon energy is transferred, not all of it is given up to the electron.

It is of interest to look at Compton scatter in terms of the wavelengths of the incident and scattered photons. The expression for the scattered photon energy can be rearranged to give,

$$\frac{1}{k'} - \frac{1}{k} = \frac{1}{m_e}\left(1-\cos\theta_\gamma\right) \tag{6.56}$$

which can be then used to yield the increase in photon wavelength resulting from the incoherent scatter,

$$\begin{aligned}\Delta\lambda = \lambda' - \lambda &= \frac{h}{m_e c}\left(1-\cos\theta_\gamma\right)\\ &= \lambda_C\left(1-\cos\theta_\gamma\right)\\ &= 2\lambda_C \sin^2\theta_\gamma\end{aligned} \tag{6.57}$$

where λ_C is the Compton wavelength of the electron with a value equal to 2.46 pm. Note that the change in photon wavelength is independent of the incident photon wavelength and is determined only by the photon scattering angle. Hence, by measuring the scattered photon wavelength at a given scattering angle, the incident photon energy can be determined by,

$$k = \frac{hc}{\lambda' - \lambda_C\left(1-\cos\theta_\gamma\right)}. \tag{6.58}$$

Having looked at the scattered photon energy, the recoil electron's kinetic energy, which is of consequence to the radiation dosimetry, is considered next. The recoil kinetic energy of the electron is equal to the difference between the incident and scattered photon energies,

$$T'_e = k - k' = k\frac{\alpha\left(1-\cos\theta_\gamma\right)}{1+\alpha\left(1-\cos\theta_\gamma\right)} \tag{6.59}$$

and the maximum recoil kinetic energy will occur for a backscattered photon, $\theta_\gamma = \pi$,

$$T'_{e,max} = k\frac{2\alpha}{1+2\alpha}. \tag{6.60}$$

The photon scattering and electron recoil angles are related by,

$$\cot\theta_e = (1+\alpha)\tan\frac{\theta_\gamma}{2}. \tag{6.61}$$

This angular relationship is plotted in Fig. 6.11 for various values of α (the recoil angle is obviously less than 90°). For low photon energies (i.e., $\alpha \ll 1$),

$$\cot\theta_e \approx \tan\frac{\theta_\gamma}{2} \quad \alpha \ll 1. \tag{6.62}$$

Manipulation of this result gives the electron recoil angle for low photon energies,

$$\theta_e = \frac{\pi - \theta_\gamma}{2} \quad \alpha \ll 1 \tag{6.63}$$

which is the linear relationship between the scattered and recoil angles for $\alpha = 0$ seen in the figure. Equation (6.61) gives the scattered and recoil angle relationship at high photon energies,

$$\cot\theta_e \approx \alpha\tan\frac{\theta_\gamma}{2} \quad \text{Large } \alpha. \tag{6.64}$$

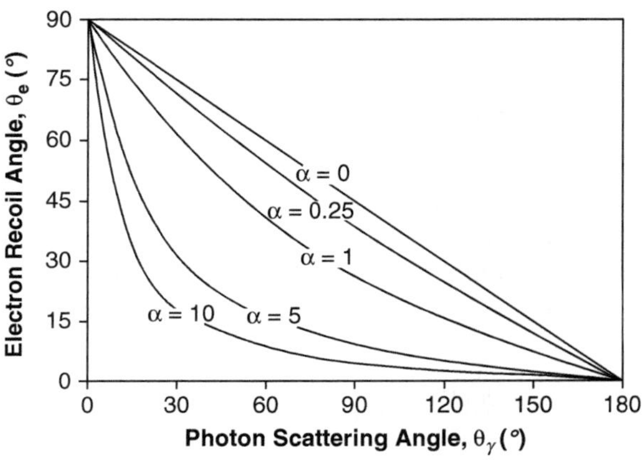

Fig. 6.11 Compton (incoherent) scatter electron recoil angle as a function of photon scattering angle for various values of $\alpha = k/m_e$

Hence, if a forward scatter occurs at high photon energies, the recoil angle of the angle approaches 90°. At low incident photon energies, the electron recoil angle smoothly decreases from 90° to 0° as the photon scattering angle increases from 0° to 180°. But, for high photon energies, the electron recoil angle stays relatively low and constant over a wide range of photon scattering angles only to rapidly grow to 90° as the photon scattering angle approaches 0°.

6.2.3.3 Klein–Nishina Cross Section

The derivation of the Klein–Nishina differential cross section for Compton scatter is complicated and tedious. Such derivations, with varying degrees of detail and complexity, can be found in a variety of quantum electrodynamic texts such as those by Heitler (1984), Bjorken and Drell (1964), Gribov and Nyiri (2001), Gould (2006), and Dyson (2007). Although the derivations of the Klein–Nishina cross section in these texts have provided the genesis of the derivation provide here, the propagator theory as described by Bjorken and Drell is the foundation of this derivation (but with the omission of certain details such as trace algebra).

The incident photon is described as a four-vector potential of a photon with momentum $\boldsymbol{K}_\mu$ and polarization $\boldsymbol{\varepsilon}_\mu$,

$$A_\mu(\boldsymbol{X};\boldsymbol{K}) = \frac{\boldsymbol{\varepsilon}_\mu}{\sqrt{2\mathrm{kL}^3}}\left(\mathrm{e}^{\mathrm{i}\boldsymbol{K}\bullet\mathbf{x}} + \mathrm{e}^{-\mathrm{i}\boldsymbol{K}\bullet\mathbf{x}}\right). \quad (6.65)$$

Similarly, that for the scattered photon is,

$$A_\mu(\boldsymbol{X}';\boldsymbol{K}') = \frac{\boldsymbol{\varepsilon}'_\mu}{\sqrt{2\mathrm{kL}^3}}\left(\mathrm{e}^{\mathrm{i}\boldsymbol{K}'\bullet\mathbf{x}'} + \mathrm{e}^{-\mathrm{i}\boldsymbol{K}'\bullet\mathbf{x}'}\right). \quad (6.66)$$

L^3 is the usual volume of normalization and the Feynman dagger notation from Chap. 3 has been used. The initial and final electron Dirac wavefunctions are,

$$\psi_\mathrm{i}(\boldsymbol{X}) = \sqrt{\frac{\mathrm{m_e}}{\mathrm{E_i L^3}}}\,\mathrm{u}(\boldsymbol{p}_\mathrm{i},\mathrm{s_i})\mathrm{e}^{-\boldsymbol{p}\bullet\boldsymbol{Y}} \quad (6.67)$$

$$\psi_\mathrm{f}(\boldsymbol{X}) = \sqrt{\frac{\mathrm{m_e}}{\mathrm{E_f L^3}}}\,\mathrm{u}(\boldsymbol{p}_\mathrm{f},\mathrm{s_f})\mathrm{e}^{-\boldsymbol{p}\bullet\boldsymbol{X}}. \quad (6.68)$$

For clarity,

$$\boldsymbol{p}\bullet\boldsymbol{X} \equiv \frac{\mathrm{Et}}{\hbar} - \frac{\mathbf{p}\bullet\boldsymbol{X}}{\hbar\mathrm{c}}. \quad (6.69)$$

The S-matrix for the two Feynman diagrams describing Compton scatter shown in Fig. 6.8 is,

$$\begin{aligned}\mathrm{S_{fi}} = \mathrm{e}^2\int \mathrm{d}^4\boldsymbol{X}\,\mathrm{d}^4\boldsymbol{Y}\,\bar{\psi}_\mathrm{f}(\boldsymbol{X})\Big\{&-\mathrm{i}\gamma^\mu \boldsymbol{A}_\mu(\boldsymbol{X};\boldsymbol{K})\mathrm{iS_F}\\ &\times(\boldsymbol{X}-\boldsymbol{Y})(-\mathrm{i})\gamma_\mu\boldsymbol{A}^\mu(\boldsymbol{Y};\boldsymbol{K}') + (-\mathrm{i})\gamma^\mu\boldsymbol{A}_\mu(\boldsymbol{X};\boldsymbol{K}')\\ &\times \mathrm{iS_F}(\boldsymbol{X}-\boldsymbol{Y})\big(-\mathrm{i}\gamma^\mu\boldsymbol{A}_\mu(\boldsymbol{Y};\boldsymbol{K})\big)\Big\}\psi_\mathrm{i}(\boldsymbol{Y}).\end{aligned} \quad (6.70)$$

$\mathrm{S_F}(\boldsymbol{X}-\boldsymbol{Y})$ is the Feynman propagator,

$$\mathrm{S_F}(\boldsymbol{X}-\boldsymbol{Y}) = \frac{1}{(2\pi)^4}\int \mathrm{d}^4\boldsymbol{p}\,\frac{\mathrm{e}^{-\boldsymbol{p}\bullet(\boldsymbol{X}-\boldsymbol{Y})}}{\not{\boldsymbol{p}}-\mathrm{m}}. \quad (6.71)$$

The result of the integration is,

$$\begin{aligned}\mathrm{S_{fi}} = &\frac{\mathrm{e}^2}{\mathrm{L}^6}\frac{\mathrm{m_e}(2\pi)^4}{\sqrt{2\mathrm{E_iE_f k k'}}}\delta^4\left(\boldsymbol{p}_\mathrm{e,f}+\boldsymbol{K}'-\boldsymbol{p}_\mathrm{e,i}-\boldsymbol{K}\right)\\ &\times\bar{\mathrm{u}}(\boldsymbol{p}_\mathrm{f},\mathrm{s_f})\left\{(-\mathrm{i}\not{\boldsymbol{\varepsilon}}')\frac{\mathrm{i}}{\not{\boldsymbol{p}}_\mathrm{i}+\boldsymbol{K}-\mathrm{m_e}}(-\mathrm{i}\not{\boldsymbol{\varepsilon}})\right.\\ &\left.+(-\mathrm{i}\not{\boldsymbol{\varepsilon}})\frac{\mathrm{i}}{\not{\boldsymbol{p}}_\mathrm{i}-\boldsymbol{K}'-\mathrm{m_e}}(-\mathrm{i}\not{\boldsymbol{\varepsilon}}')\right\}\mathrm{u}(\boldsymbol{p}_\mathrm{i},\mathrm{s_i}).\end{aligned} \quad (6.72)$$

The Klein–Nishina scatter differential cross section follows by first forming a rate by squaring this result and dividing by $(2\pi)^4\delta^4(0)$ and then dividing by the incident electron flux and electron number density, $|\mathbf{v}|/\mathrm{L}^3$ and L^{-3}, respectively. This is then integrated over the phase–space of the final state of the recoil electron and scattered photon, $(\mathrm{L}^6/(2\pi)^6)\mathrm{d}^3\mathbf{p}_\mathrm{e,f}\,\mathrm{d}\mathbf{k}'$. This lengthy procedure yields,

$$\begin{aligned}\mathrm{d}\sigma_\mathrm{KN} = &\frac{\mathrm{e}^4\mathrm{m_e^2}}{16\pi^2\mathrm{kE_i}|\mathbf{v}|}\int \mathrm{d}^3\mathbf{p}_\mathrm{e,f}\,\mathrm{d}\mathbf{k}'\\ &\times\left|\bar{\mathrm{u}}(\boldsymbol{p}_\mathrm{f},\mathrm{s_f})\left\{\frac{\not{\boldsymbol{\varepsilon}}'\not{\boldsymbol{\varepsilon}}}{\not{\boldsymbol{p}}_\mathrm{i}+\boldsymbol{K}-\mathrm{m_e}}+\frac{\not{\boldsymbol{\varepsilon}}\not{\boldsymbol{\varepsilon}}'}{\not{\boldsymbol{p}}_\mathrm{i}-\boldsymbol{K}'-\mathrm{m_e}}\right\}\mathrm{u}(\boldsymbol{p}_\mathrm{i},\mathrm{s_i})\right|^2\\ &\times\frac{\delta^4\left(\boldsymbol{p}_\mathrm{e,f}+\boldsymbol{K}'-\boldsymbol{p}_\mathrm{e,i}-\boldsymbol{K}\right)}{\mathrm{E_f}\boldsymbol{K}'}.\end{aligned} \quad (6.73)$$

Calculating for the laboratory reference frame (where the electron is initially at rest, $\mathbf{p}_i = 0$), this reduces to,

$$\frac{d\sigma_{KN}}{d\Omega} = \alpha^2 \left(\frac{k'}{k}\right)^2 \left| \bar{u}(\boldsymbol{p}_f, s_f) \right.$$
$$\left. \times \left\{ \frac{\not{\varepsilon}' \not{\varepsilon}}{\not{p}_i + \boldsymbol{K} - m_e} + \frac{\not{\varepsilon} \not{\varepsilon}'}{\not{p}_i - \boldsymbol{K}' - m_e} \right\} u(\boldsymbol{p}_i, s_i) \right|^2 . \qquad (6.74)$$

Further simplification is achieved through the use of a gauge in which the incident and scattered photons are transversely polarized in the laboratory reference frame. By averaging over the initial electron spins s_i and summing over the final electron spins s_f, and doing the same for the photon polarizations, the result is (after considerable and tedious trace algebra) the Klein–Nishina differential cross section in solid angle is,

$$\frac{d\sigma_{KN}}{d\Omega} = \frac{r_0^2}{2} \left(\frac{k'}{k}\right)^2 \left(\frac{k'}{k} + \frac{k}{k'} - \sin^2\theta_\gamma\right). \qquad (6.75)$$

This result can be seen to reduce to the Thomson classical differential cross section in the low-energy limit, $k' \approx k$,

$$\begin{aligned} \frac{d\sigma_{KN}}{d\Omega} &\approx \frac{r_0^2}{2} \left(2 - \sin^2\theta_\gamma\right) \\ &\approx \frac{r_0^2}{2} \left(1 + \cos^2\theta_\gamma\right) \quad k' \approx k. \end{aligned} \qquad (6.76)$$

The Klein–Nishina differential cross section per electron is plotted in Fig. 6.12 as a function of photon scattering angle for various values of $\alpha = k/m_e$. Note that, unlike the Thomson cross section, which has no energy dependence and is symmetric about $\theta_\gamma = 90°$, the Klein–Nishina cross section becomes more forward directed with increasing photon energy.

The Klein–Nishina total cross section represents the probability that a photon will undergo a Compton scatter with a single free electron and, as usual, is calculated by integrating the differential cross section over 4π steradians. Avoiding, again, a tedious calculation, the result is provided without proof,

$$\sigma_{KN} = \frac{3}{4}\sigma_{Tho} \left[\frac{1+\alpha}{\alpha^2} \left(\frac{2(1+\alpha)}{(1+2\alpha)} - \frac{\ln(1+2\alpha)}{\alpha} \right) + \frac{\ln(1+2\alpha)}{2\alpha} - \frac{1+3\alpha}{(1+2\alpha)^2} \right]. \qquad (6.77)$$

where σ_{Tho} is the total Thomson cross section. σ_{KN} is plotted in Fig. 6.13 as a function of photon energy. At low photon energies ($\alpha \ll 1$), $\sigma_{KN} \approx \sigma_{Tho}$ and σ_{KN} decreases with photon energy, eventually exhibiting a 1/k dependence. As the total Klein–Nishina cross section is per electron and was calculated for a free electron target, it is independent of the atomic number, Z, of the medium in which the scattering occurs. However, Compton scatter will obviously have, in practice, a dependence upon Z and this is evaluated later.

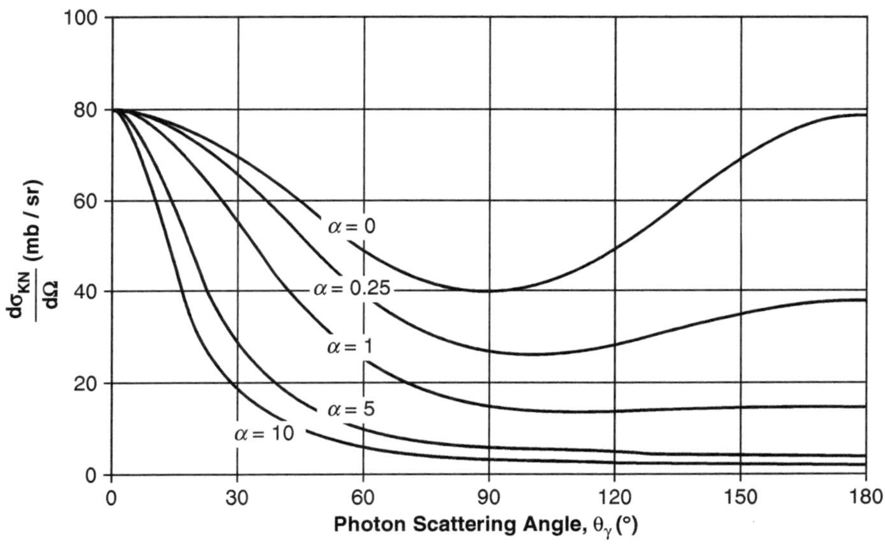

Fig. 6.12 Klein–Nishina differential cross section as a function of photon scattering angle

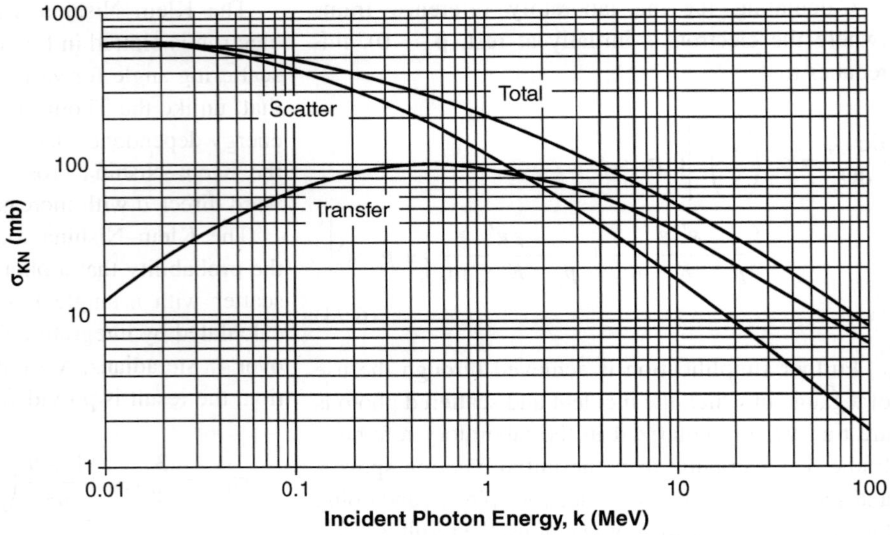

Fig. 6.13 Klein–Nishina total, energy-transfer and scatter cross sections

Of particular interest to dosimetry is the overall fraction of the photon energy k_γ that is transferred to the recoil electron's kinetic energy, T'_e. This is determined by first weighting the Klein–Nishina differential cross section by this fraction,

$$\frac{d\sigma_{KN,Tr}}{d\Omega}(\theta_\gamma) = \frac{d\sigma_{KN}}{d\Omega}(\theta_\gamma)\left(1 - \frac{k'(\theta_\gamma)}{k}\right). \quad (6.78)$$

Integrating this differential cross section over a solid angle of 4π steradians yields the energy-transfer cross section,

$$\sigma_{KN,Tr} = \frac{3}{4}\sigma_{Tho}\left[\frac{2(1+\alpha)^2}{\alpha^2(1+2\alpha)} - \frac{1+3\alpha}{(1+2\alpha)^2} + \frac{(1+\alpha)(1+2\alpha-2\alpha^2)}{\alpha^2(1+2\alpha)^2} - \frac{4\alpha^2}{3(1+2\alpha)^3} - \left(\frac{1+\alpha}{\alpha^3} - \frac{1}{2\alpha} + \frac{1}{2\alpha^3}\right)\ln(1+2\alpha)\right] \quad (6.79)$$

$\sigma_{KN,Tr}$ is plotted in Fig. 6.13 as a function of incident photon energy. $\sigma_{KN,Tr}$ increases with photon energy up to a broad maximum at about 500 keV, reflecting the increase in the recoil kinetic energy with photon energy which is eventually overwhelmed by the drop off in the Klein–Nishina total cross section with photon energy. The scatter cross section is that for the energy carried by the scattered photon and is simply the difference between the total and energy-transfer cross sections,

$$\sigma_{KN,S} = \sigma_{KN} - \sigma_{KN,Tr} \quad (6.80)$$

and which is also plotted in Fig. 6.13. For low incident photon energies, $\sigma_{KN,S} \approx \sigma_{KN}$, reflecting the fact that there is little energy transferred to the recoil electron at these energies. With increasing photon energy, $\sigma_{KN,S}$ decreases as the fraction of energy taken up by the recoil electron increases. The ratio of the Klein–Nishina transfer to total cross sections is equal to the mean fraction of photon energy transferred to the electron,

$$\frac{\sigma_{KN,Tr}}{\sigma_{KN}} = \frac{\bar{T}'_e}{k} \quad (6.81)$$

and is shown in Fig. 6.14 as a function of photon energy. This mean fraction increases relatively slowly with photon energy (recall from Fig. 6.10 that the fraction of incident photon energy transferred to the electron only becomes significant for high photon energies and non-zero photon scattering angles) and is only equal to about 50% of the incident photon energy at about 1.5 MeV. Also plotted in Fig. 6.14 is the maximum electron energy as a fraction of the

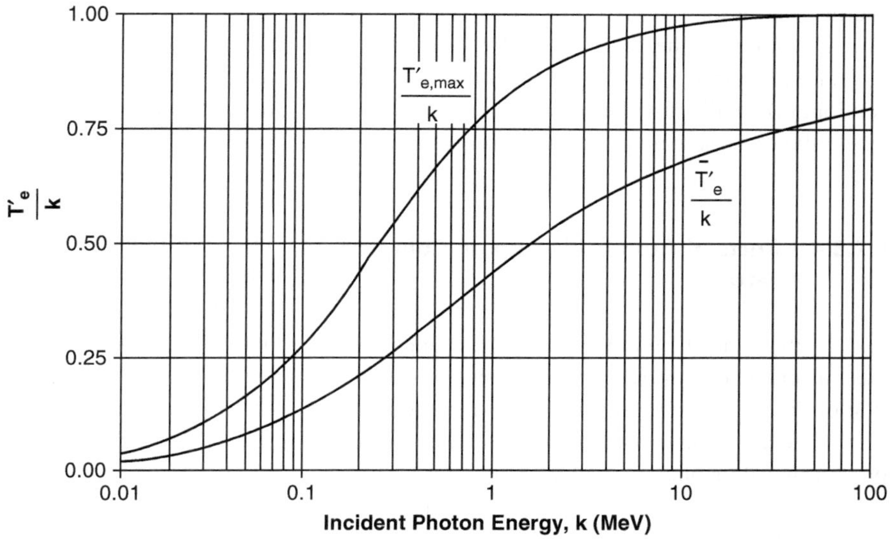

Fig. 6.14 The ratio of the mean and maximum recoil electron energies to the incident photon energy in Compton scatter

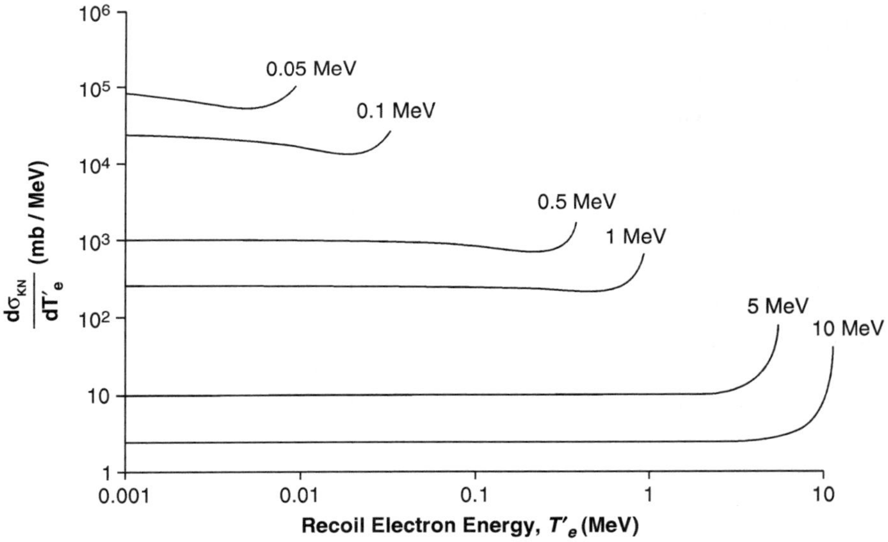

Fig. 6.15 The Klein–Nishina differential cross section in recoil electron kinetic energy as a function of the recoil kinetic energy for incident photon energies between 0.05 and 10 MeV

incident photon energy for a backscattered photon ($\theta_\gamma = \pi$),

$$T'_{e,max} = k\frac{2\alpha}{1+2\alpha}. \quad (6.82)$$

The Klein–Nishina differential cross section in electron energy will also be of some interest when we derive a variety of dosimetric quantities and can be calculated from,

$$\frac{d\sigma_{KN}}{dT'_e} = \frac{d\sigma_{KN}}{d\Omega}\left[\frac{dT'_e}{d\Omega}\right]^{-1} = \frac{3}{8}\left(\frac{\sigma_0}{\alpha k}\right) \times \left[2 - \frac{2T'_e}{\alpha(k-T'_e)} + \frac{T'^2_e}{\alpha^2(k-T'_e)^2} + \frac{T'^2_e}{k(k-T'_e)}\right]. \quad (6.83)$$

This expression also represents the energy distribution of the scattered electrons and is plotted in Fig. 6.15 as a function of the recoil electron kinetic energy for various values of k. The curves exhibit similar features as the variation of $d\sigma_{KN}/dT'_e$ with T_e is relatively small (this is known as the Compton continuum) until just before the value of $T_{e,max}$ where the cross section peaks (the Compton peak).

6.2.3.4 Effects of Atomic Binding on Compton Scatter

Recall that the Compton kinematics and Klein–Nishina cross sections were derived under the assumption of the target electron being both free and initially at rest. While such an assumption may yield accurate predictions of incoherent scatter for photons with energies greater than about 1 MeV and for low-atomic number scatters, electrons are in reality both bound to the atom and have momentum distributions prior to interacting with the photon. Kinematic and electron binding considerations must then be accounted for (Bergstrom and Pratt 1997; Cooper 1997). The electron momentum will either add to or subtract from the scattered photon momentum with the result that, for a monoenergetic incident photon beam and a fixed scattering angle, polyenergetic scattered photons will be detected rather than monoenergetic photons as calculated from the Compton kinematics. Such Doppler broadening of scattered photons was first demonstrated experimentally by Dumond (1929) and Fig. 6.16 provides a schematic representation of the effects of an initial electron momentum distribution in which the double differential cross section in photon scatter solid angle and scattered photon energy $d^2\sigma/d\Omega dk'$ is

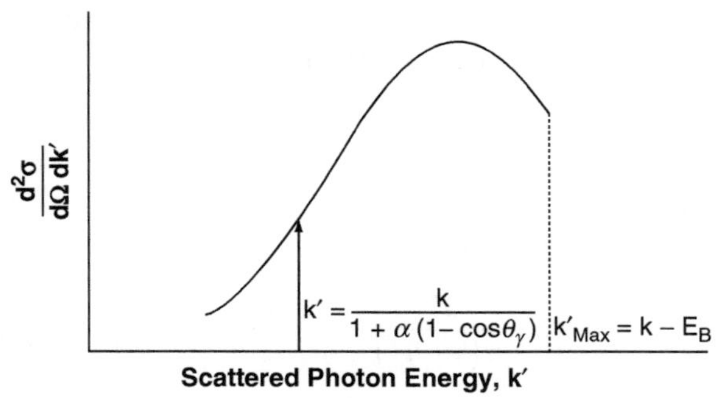

Fig. 6.16 Schematic of the double differential Klein–Nishina cross section in solid angle and scattered photon energy as a function of scattered photon energy for a photon incident to an atomic electron

shown as a function of k′. For a free electron initially at rest, this double differential cross section would be a delta-function at the scattered photon energy k for the electron at rest. This spectrum is broadened as the target atomic electron has a momentum spectrum. The degree of broadening will be an inverse function of the degree of binding. For a tightly bound electron (i.e., constrained to a small spatial distribution), the $d^2\sigma/d\Omega dk'$ spectrum will be broad. This narrows with decreasing binding (approaching the state of a free electron). Binding also leads to a shift in the peak of the spectrum away from the free electron value, a difference known as the Compton defect. Binding of the target electron also dictates photon kinematic limits for this Compton scatter. The incident photon energy must be sufficient to excite the electron to a higher orbital or to the continuum. This specifies the lower limit of the required photon energy and, conversely, the upper limit of the observed scattered photon energy to $k_{Max} = k - E_B$, where E_B is the electron binding energy.

As with the atomic form factor which extends the calculation of elastic scatter from a single electron to that from multiple atomic electrons, an incoherent scattering factor for Compton scatter from atomic electrons can also be defined (Hubbell et al. 1975; Hubbell 1997). This, unlike Compton scatter from a free electron, is inherently an inelastic process as atomic excitation and ionization energy channels are now present and not all of the energy lost by the photon will appear in the kinetic energy of the recoil electron. An appropriate form factor is,

$$F_\varepsilon(\mathbf{q}, Z) = \left\langle \psi_\varepsilon \middle| \sum_{j=1}^{Z} e^{i\frac{\mathbf{q}\bullet \mathbf{r}_j}{\hbar c}} \middle| \psi_{gs} \right\rangle \tag{6.84}$$

where ψ_ε is the wavefunction of the atom (which we now allow to be excited or ionized) and ε is its energy. It will be assumed that the scatter necessarily results in the transfer of sufficient energy to lead to the electron's ejection, so those final atomic states which are merely excited (corresponding to Raman scattering) are ignored. The incoherent scattering function is the sum of the squared moduli of the above form factor for non-zero ε, or,

$$S(\mathbf{q}, Z) = \sum_{\varepsilon > 0} |F_\varepsilon(\mathbf{q}, Z)|^2. \tag{6.85}$$

This summation is over all possible excitation energies. By using the closure property,

$$\sum_{\varepsilon} |\psi_\varepsilon\rangle\langle\psi_\varepsilon| = 1 \tag{6.86}$$

and including the $\varepsilon = 0$ state in the summation and then explicitly subtracting that term,

$$S(\mathbf{q}, Z) = \sum_{m=1}^{Z}\sum_{n=1}^{Z}\left\langle \psi_{gs}\left| e^{-i\frac{\mathbf{q}\bullet(\mathbf{r}_m-\mathbf{r}_n)}{\hbar c}}\right|\psi_{gs}\right\rangle - |F(\mathbf{q}, Z)|^2 \tag{6.87}$$

where the summations are over the number of electrons and (6.37) has been used. The differential cross section for incoherent scatter from an atom with Z electrons is the Klein–Nishina differential cross section for a single electron scaled by the incoherent scattering function,

$$\frac{d\sigma_{Inc}}{d\Omega} = \frac{d\sigma_{KN}}{d\Omega} S(\mathbf{q}, Z) \tag{6.88}$$

Further understanding of the incoherent scattering function can be obtained using the nonrelativistic impulse approximation in which the bound, nonstationary electron is treated as a free electron but with the same momentum distribution as that calculated from the wavefunction of a bound atomic electron. The kinematics are shown in Fig. 6.17.

In the impulse approximation, the electron feels the same binding potential before and after the scatter. Another way of interpreting this is that the photon interacts with a single atomic electron with the remainder acting as spectators. As a result, only the change in kinetic energy need be considered. The conservation of momentum is,

$$\mathbf{p}_e + \mathbf{k} = \mathbf{p}'_e + \mathbf{k}' \tag{6.89}$$

from which the momentum transfer, $\mathbf{q}$, is,

$$\begin{aligned} \mathbf{q} &= \mathbf{p}'_e - \mathbf{p}_e \\ &= \mathbf{k} - \mathbf{k}'. \end{aligned} \tag{6.90}$$

Assuming nonrelativistic kinematics, the conservation of energy is,

$$\frac{p_e^2}{2m_e} + k = \frac{p_e'^2}{2m_e} + k'^2 - E_B \tag{6.91}$$

where E_B is the electron binding energy. This leads to the change in photon energy,

$$k - k' = \frac{1}{2m_e}\left(p_e'^2 - p_e^2\right) - E_B. \tag{6.92}$$

From (6.90),

$$|\mathbf{q}|^2 = p_e'^2 - p_e^2 - 2\mathbf{p}_e \bullet \mathbf{q}. \tag{6.93}$$

Using this result in (6.92), the change in photon energy following an inelastic scatter from an electron initially in motion and bound to an atom is found,

$$k - k' = \frac{q^2}{2m_e} + \frac{\mathbf{p}_e \bullet \mathbf{q}}{m_e} - E_B. \tag{6.94}$$

The first term is that of a nonrelativistic electron initially at rest and unbound; the second and third terms account for changes due to electron motion and binding, respectively. The term due to electron motion can either increase or decrease the scattered photon energy depending upon the sign of the scalar product, $\mathbf{p}_e \bullet \mathbf{q}$, or the direction between the initial electron momentum and the net momentum transfer. If the z-axis is defined to be that of the direction of the scattered photon, then,

$$p_{e,z} = \frac{\mathbf{p}_e \bullet \mathbf{q}}{q} \tag{6.95}$$

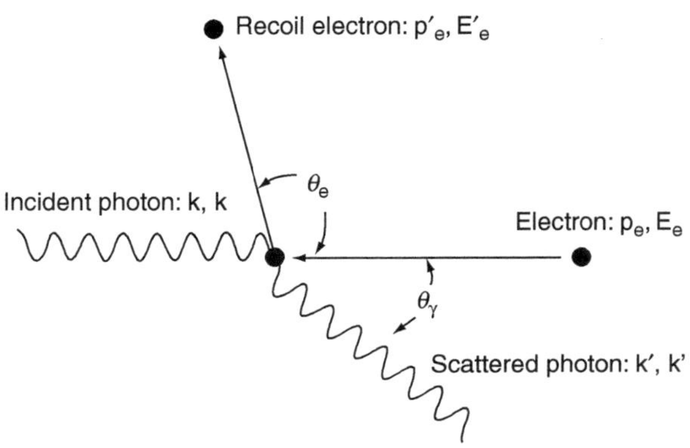

Fig. 6.17 Kinematics of the impulse approximation in which a photon is incident to a moving electron

is the projection of the electron's initial momentum vector onto the scattering vector. The Compton profile is the integral,

$$J(p_{e,z}) = \int dp_{e,x} dp_{e,y} |\psi(\mathbf{p})|^2 \quad (6.96)$$

where $\psi(\mathbf{p})$ is the wavefunction of the bound electron in momentum space and $|\psi(\mathbf{p})|^2$ is the electron's momentum distribution. The contributions from different orbitals to the Compton profile are calculated separately and then summed over the orbitals,

$$J(p_{e,z}) = \sum_i J_i(p_{e,z}) Z_i H(k - k' - E_{B,i}) \quad (6.97)$$

where $J_i(p_{e,z})$ is the Compton profile for the ith orbital, Z_i is the number of electrons in that orbital, $E_{B,I}$ is the corresponding binding energy and H(x) is the Heaviside function. The inclusion of the Heaviside function is necessary to account for electron binding, i.e., incoherent scatter cannot occur unless $k - k' > E_B$. The individual Compton profiles for a given orbit have the normalization,

$$\int_{-\infty}^{\infty} dp_{e,z} J_i(p_{e,z}) = 1 \quad (6.98)$$

and the incoherent scattering factor is the integral of the Compton profile,

$$S(\mathbf{q}, Z) = \int_{-\infty}^{p_{e,z,Max}} dp_{e,z} J(p_{e,z}). \quad (6.99)$$

The limiting values of the incoherent scattering factor are next considered. As $q \rightarrow 0$, $F(q,Z) \rightarrow Z$ and $\sum_{m=1}^{Z} \sum_{n=1}^{Z} \langle \psi_{gs} | e^{i\frac{\vec{q}\bullet(\vec{r}_m - \vec{r}_n)}{\hbar c}} | \psi_{gs} \rangle \rightarrow Z^2$, then $S(q,Z) \rightarrow 0$. On the other hand, for large q, the rapid oscillatory behavior of $e^{i\frac{q\bullet(r_m - r_n)}{\hbar c}}$ will average to zero except when the indices are equal, in which case $S(q,Z) \rightarrow Z$ as $q \rightarrow \infty$.

Values of incoherent scattering factors and incoherent scatter cross sections are provided by Hubbel et al. (1975). Figure 6.18 shows total cross sections per atom for carbon (Z = 6) for incoherent scatter and the Klein–Nishina total cross section for a single free electron scaled by the atomic number of carbon, $Z\sigma_{KN}$. The two calculations agree for photon energies greater than about 0.05 MeV; below this energy, the Klein–Nishina free electron approximation markedly overestimates the cross section. Note that the free electron cross section approaches the Thomson free electron value of $Z\sigma_{Tho} = 3990$ mb.

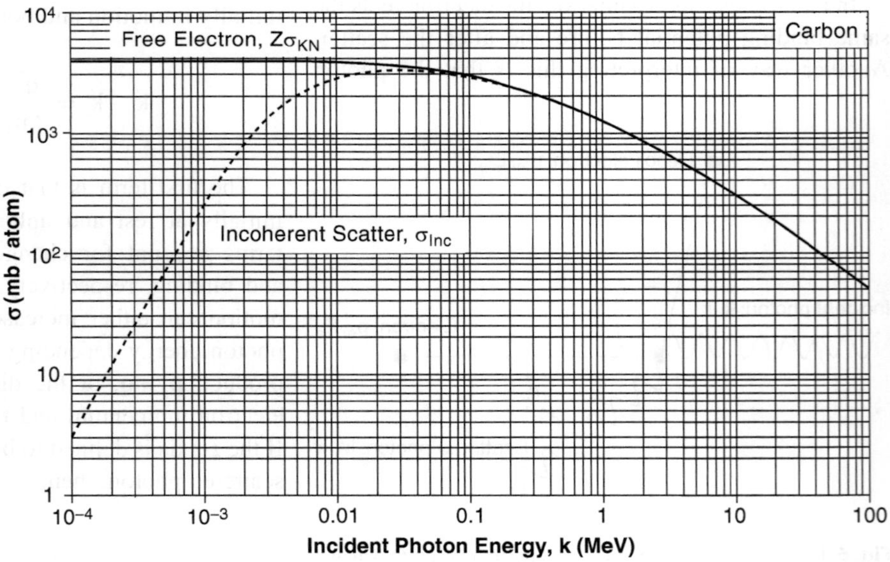

Fig. 6.18 Total incoherent cross section for carbon as a function of photon energy. The free electron curve is the Klein–Nishina total cross section scaled by the atomic number of carbon

6.3 Photon Nonconserving Interaction

6.3.1 Photoelectric Absorption

6.3.1.1 Introduction

As the photon is a boson and the electron is a fermion, it is possible for a photon to be absorbed by an atomic electron and the electron gain a kinetic energy equal to the difference between the incident photon energy and the sum of the electron binding energy and the atomic recoil kinetic energy.

The photon is also not conserved in the pair production $\gamma \to e^+e^-$ process in the nuclear Coulomb field and in the triplet production $\gamma \to e^+ + 2e^-$ process in the atomic electron Coulomb field. However, as shown in the Appendix, the minimum photon energy required for the pair production interaction is equal to just above $2m_e$, or about 1.022 MeV. This photon energy is above that of practical nuclear medicine interest and, as a result, pair and triplet production are not considered here.

6.3.1.2 Kinematics

In the photoelectric process (Fig. 6.19), the incident photon is absorbed by a bound atomic electron followed by its ejection, with an exchange of photon energy to the emitted electron's kinetic energy (minus the atomic binding energy and the kinetic energy carried away by the recoil nucleus). The total energy of the ejected electron is,

$$E'_e = k - E_B - T'_{Rec} + m_e. \qquad (6.100)$$

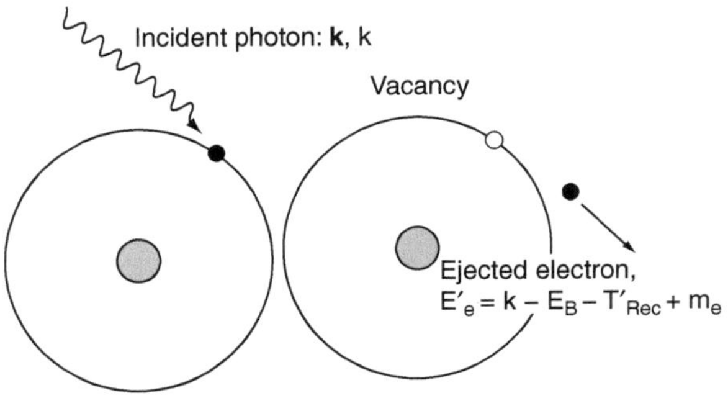

Fig. 6.19 Photoelectric absorption

Hence, the kinematic specification for the minimum photon energy required for photoelectric absorption to occur is,

$$k > E_B + T'_{Rec} \qquad (6.101)$$

It should be recognized that photoelectric absorption is necessarily a 3-body reaction (incident photon, atomic electron, and atom) as a free electron cannot absorb a photon due to the requirement of simultaneous conservations of linear momentum and energy. In such a case, the conservations of linear momentum and energy are $k = p'_e$ and $k + m_e = E'_e$. Substituting these two expressions into the relativistic invariant, $E^2 = p^2 + m^2$, yields $k^2 + {m_e}^2 + 2km_e = k^2 + {m_e}^2$ which leads immediately to the contradictory conclusion, $2\ km_e = 0$: i.e., linear momentum is not conserved. Hence, a third body (the nucleus) is required to conserve linear momentum. However, while the recoil nucleus absorbs the momentum, it gains negligible kinetic energy as result of its mass. Hence, in practical terms, the kinematic requirement for photoelectric absorption to occur is that the photon energy exceed the atomic binding energy,

$$k > E_B. \qquad (6.102)$$

6.3.1.3 Cross Section

From this kinematic threshold, the photoelectric absorption cross section for a given orbital will be zero if the incident photon energy does not exceed the binding energy. In order to provide an understanding of the nature of the photoelectric absorption cross section, it is advantageous to consider photon energies far from these thresholds. As the nucleus is necessary for the simultaneous conservation of momentum and energy, the photoelectric absorption cross section should increase with increasing electron proximity to the nucleus which will correspond to increased binding energy. Superimposed upon this energy dependence is an expectation that the absorption probability will decrease with increasing photon energy. An intuitive argument for the decrease in the photoelectric absorption cross section with energy is based upon a model of a bound electron absorbing a photon being subject to an oscillating force proportional to $m_e\omega^2$, where $\omega = k/\hbar$. Equating this force to the Coulomb force

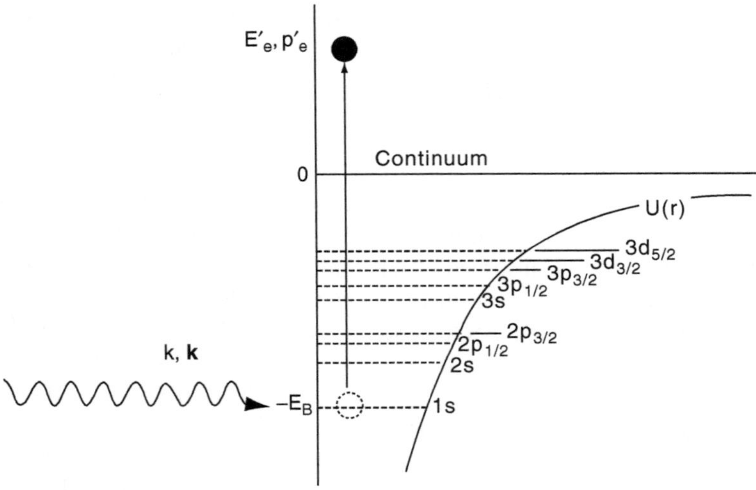

Fig. 6.20 Photoelectric absorption by a K-shell electron

between the electron and the nucleus, $Z\alpha\hbar c/r^2$, yields the reciprocal relationship $r \propto k^{-1}$. As the corresponding volume is r^3, this implies that the photoelectric absorption cross section should vary approximately as k^{-3}.

Here, a derivation of the photoelectric absorption cross section for K-shell absorption, which is the elevation of one of the 1s-electrons into the continuum in an atom of atomic number, Z, as shown in Fig. 6.20, is provided. The 1s-electron is bound with a binding energy E_B given by (neglecting relativistic and spin-orbit coupling effects),

$$E_B = \frac{(\alpha Z)^2}{2} m_e. \tag{6.103}$$

The electron absorbs the photon and, if the photon energy satisfies the threshold requirement, is elevated into the continuum with a total energy E_e and momentum $\mathbf{p}_e$. The cross section can be calculated, as usual, from,

$$\lambda_{fi} = \frac{2\pi}{\hbar} |M_{fi}|^2 \rho_f$$

Two assumptions will be used in order to simplify the cross section calculation. The first is that the photon energy is much greater than the binding energy of the electron. This permits the ejected electron to be treated as a plane wave unperturbed by the atomic potential. The second assumption is that the photon energy is small compared with the electron rest mass so that relativistic effects can be ignored. When combined, these two assumptions set boundaries on the photon energy,

$$E_B = \frac{(\alpha Z)^2}{2} m_e \ll k \ll m_e. \tag{6.104}$$

Phase–Space Calculation

The phase–space factor is,

$$\begin{aligned} \rho_f dE_e &= \left(\frac{L}{2\pi\hbar}\right)^3 d^3\mathbf{p}_e \\ &= \left(\frac{L}{2\pi\hbar}\right)^3 p_e^2 dp_e d\Omega \end{aligned} \tag{6.105}$$

where $d\Omega$ is the differential solid angle element that the electron is ejected into. The differential electron momentum dp_e is determined from,

$$\begin{aligned} dp_e &= \frac{dp_e}{dE_e} dE_e \\ &= \frac{d}{dE_e}\sqrt{E_e^2 - m_e^2}\, dE_e \\ &= \frac{E_e}{p_e} dE_e \end{aligned} \tag{6.106}$$

leading to,

$$\rho_f dE_e = \left(\frac{L}{2\pi\hbar}\right)^3 m_e p_e dE_e d\Omega \tag{6.107}$$

where, due to the nonrelativistic assumption of electron energy, the approximation $E_e \approx m_e$ has been used.

Matrix Element Calculation

To solve for the transition matrix element, the three components of the integrand will be considered separately, beginning with the initial and final wavefunctions. For a 1s-orbital electron in a hydrogen-like atom with atomic number, Z, the initial state wavefunction can be approximated by,

$$\psi_i(r) = \sqrt{\frac{Z^3}{\pi r_\infty^3}}\, e^{-\left(\frac{Zr}{r_\infty}\right)}. \tag{6.108}$$

The final state wavefunction is that of the free electron which is taken to be a plane wave, as it has been assumed that it is sufficiently energetic to be unperturbed by the atomic potential,

$$\psi_f = \frac{1}{L^{3/2}} e^{i\frac{\mathbf{p_e}\bullet\mathbf{r_e}}{\hbar c}}. \tag{6.109}$$

The perturbing potential, U, is determined from the Hamiltonian of an electron/photon system,

$$\begin{aligned} H &= H_e + H_{rad} \\ &= \frac{|\mathbf{p}|^2}{2m_e} + H_{rad}. \end{aligned} \tag{6.110}$$

In the presence of an electromagnetic field, the momentum becomes $\mathbf{p} - e\mathbf{A}$ where $\mathbf{A}$ is the vector potential and the Hamiltonian becomes,

$$H = \frac{|\mathbf{p}|^2}{2m_e} + H_{rad} + \frac{e}{m_e}\mathbf{A}\bullet\mathbf{p} + \frac{e^2}{2m_e}|\mathbf{A}|^2 \tag{6.111}$$

where the last two terms make up the interaction Hamiltonian. Neglecting the second-order term in $\mathbf{A}$, the perturbative potential (interaction Hamiltonian) reduces to,

$$U = \frac{e}{m_e}\mathbf{A}\bullet\mathbf{p}_e. \tag{6.112}$$

Writing the vector potential as a plane wave,

$$\mathbf{A} = A_0 e^{i\left(\frac{\mathbf{k}\bullet\mathbf{r}}{\hbar c} - \frac{kt}{\hbar}\right)}\hat{\boldsymbol{\varepsilon}}. \tag{6.113}$$

where $\hat{\boldsymbol{\varepsilon}}$ is the polarization of the electric field, thus yielding,

$$U = \frac{e}{m_e} A_0 e^{-i\omega t} e^{i\frac{\mathbf{k}\bullet\mathbf{r}}{\hbar c}}\hat{\boldsymbol{\varepsilon}}\bullet\mathbf{p}_e. \tag{6.114}$$

The matrix element for photoelectric absorption on a K-shell electron is,

$$\begin{aligned} M_{fi} &= \frac{1}{L^{3/2}}\left(\frac{e}{m_e}\right)\sqrt{\frac{Z^3}{\pi r_\infty^3}} A_0 e^{-i\omega t}(\hat{\boldsymbol{\varepsilon}}\bullet\hat{\mathbf{p}}_e) \\ &\times \int d^3\mathbf{r}\, e^{i\frac{\mathbf{q}\bullet\mathbf{r}}{\hbar c}} e^{-\left(\frac{Z}{r_\infty}\right)r} \end{aligned} \tag{6.115}$$

where $\hat{\boldsymbol{\varepsilon}}\bullet\mathbf{p}_e$ is the projection of the electron momentum onto the direction of the incident electric field polarization. The integral of (6.115) has been seen before in many guises,

$$\int d^3\mathbf{r}\, e^{i\frac{\mathbf{q}\bullet\mathbf{r}}{\hbar c}} e^{-\left(\frac{Z}{r_\infty}\right)r} = \frac{8\pi\left(\frac{Z}{r_\infty}\right)}{\left(\left(\frac{Z}{r_\infty}\right)^2 + \left(\frac{q}{\hbar c}\right)^2\right)^2}. \tag{6.116}$$

The matrix element thus reduces to,

$$\begin{aligned} M_{fi} &= 8\sqrt{\frac{\pi}{L^3}}\left(\frac{e}{m_e}\right)\left(\frac{Z}{r_\infty}\right)^{\frac{5}{2}} \\ &\times A_0 e^{-i\omega t}\frac{(\hat{n}\bullet\vec{p}_e)}{\left(\left(\frac{Z}{r_\infty}\right)^2 + \left(\frac{q}{\hbar c}\right)^2\right)^2} \end{aligned} \tag{6.117}$$

The denominator can be simplified beginning with the nonrelativistic form, $k \approx p_e^2/2m_e$, which gives $k/p_e \approx p_e/2m_e = \beta_e/2$ where β_e is the electron speed normalized to that of light. Then, the denominator can be rewritten as,

$$\begin{aligned} \left(\frac{Z}{r_\infty}\right)^2 + \left(\frac{q}{\hbar c}\right)^2 &= \left(\frac{Z}{r_\infty}\right)^2 + \frac{k^2 + p_e^2 - 2kp_e\cos\theta}{(\hbar c)^2} \\ &= \left(\frac{Z}{r_\infty}\right)^2 + \left(\frac{p_e}{\hbar c}\right)^2 \\ &\times \left[\left(\frac{k}{p_e}\right)^2 + 1 - 2\left(\frac{k}{p_e}\right)\cos\theta\right] \\ &= \left(\frac{Z}{r_\infty}\right)^2 + \left(\frac{p_e}{\hbar c}\right)^2\left[\frac{\beta_e^2}{4} + 1 - \beta_e\cos\theta\right]. \end{aligned} \tag{6.118}$$

Following from the assumption of a nonrelativistic electron, the β_e^2 term can be neglected, which leads to,

$$\left(\frac{Z}{r_\infty}\right)^2+\left(\frac{q}{\hbar c}\right)^2=\left(\frac{Z}{a_0}\right)^2+2\frac{m_e k}{(\hbar c)^2}\times(1-\beta_e\cos\theta). \quad (6.119)$$

It is trivial to show that, in the energies of interest,

$$\left(\frac{Z}{r_\infty}\right)^2<2\frac{m_e k}{\hbar c} \quad (6.120)$$

resulting in,

$$\left(\frac{Z}{r_\infty}\right)^2+\left(\frac{q}{\hbar c}\right)^2=2\frac{m_e k_\gamma}{(\hbar c)^2}(1-\beta_e\cos\theta). \quad (6.121)$$

The final expression for the matrix element is,

$$M_{fi}=8\sqrt{\frac{\pi}{L^3}}\left(\frac{e}{m_e}\right)\left(\frac{Z}{r_\infty}\right)^{\frac{5}{2}}\times A_0e^{-i\omega t}\frac{(\hat{\boldsymbol{\varepsilon}}\bullet\mathbf{p}_e)}{\left(2\frac{m_e k_\gamma}{(\hbar c)^2}(1-\beta_e\cos\theta)\right)^2}. \quad (6.122)$$

ϕ is the angle between the plane carrying the photon and electron momenta and that plane formed by the photon momentum and the direction of polarization. Then the projected component of $\mathbf{p}_e$ is,

$$\hat{\boldsymbol{\varepsilon}}\bullet\mathbf{p}_e=p_e\sin\theta\cos\phi \quad (6.123)$$

giving the matrix element as,

$$M_{fi}=2\sqrt{\frac{\pi}{L^3}}\left(\frac{e}{m_e}\right)\left(\frac{Z}{r_\infty}\right)^{\frac{5}{2}}\frac{p_e(\hbar c)^4}{(m_e k)^2}A_0e^{-i\omega t}\times\frac{\sin\theta\cos\phi}{((1-\beta_e\cos\theta))^2} \quad (6.124)$$

Transition Probability and Cross Section

The transition probability per unit time is,

$$\lambda_{fi}=\frac{1}{\hbar^4}\left(\frac{e}{m_e}\right)^2\left(\frac{Z}{r_\infty}\right)^5\left(\frac{p_e^3}{m_e^3k^4}\right)\times A_0^2\frac{\sin^2\theta\cos^2\phi}{(1-\beta_e\cos\theta)^4}d\Omega. \quad (6.125)$$

The quantity A_0^2 is determined from the magnitude of the time-averaged Poynting vector which represents the energy flow rate per unit area,

$$P=2\varepsilon_0\omega^2cA_0^2. \quad (6.126)$$

Thus, the transition probability per unit time becomes,

$$\lambda_{fi}=\frac{1}{\hbar^4}\left(\frac{e}{m_e}\right)^2\left(\frac{Z}{r_\infty}\right)^5\left(\frac{p_e^3}{m_e^3k^4}\right)\left(\frac{P}{2\varepsilon_0\omega^2c}\right)\times\frac{\sin^2\theta\cos^2\phi}{(1-\beta_e\cos\theta)^4}d\Omega. \quad (6.127)$$

The differential cross section in solid angle is then calculated by noting that the cross section is the transition rate per unit flux (the number of incident particles per unit area per unit time),

$$\sigma=\frac{\lambda_{fi}}{\left(\frac{n}{L^3}\right)|\mathbf{v}|} \quad (6.128)$$

where n/L^3 is the number of particles per unit volume and $|\mathbf{v}|$ is the speed. Thus, the differential cross section for the photoelectric effect on a K-shell electron is,

$$\frac{d\sigma_{PE}}{d\Omega}=r_0^2\left(4\sqrt{2}\right)\alpha^4Z^5\left(\frac{m_e}{k}\right)^{\frac{7}{2}}\frac{\sin^2\theta\cos^2\phi}{(1-\beta_e\cos\theta)^4} \quad (6.129)$$

As this differential cross section is maximized for the combination of $\theta=90°$ and $\theta=0°$, which is also the direction of the photon polarization, the electron will tend to be ejected in this direction (although for relativistic electrons, $\beta_e\to 0$, and the electrons will be ejected in a more forward direction). The total cross section is determined by integrating $d\sigma_{PE}/d\Omega$ over a solid angle of 4π steradians, which is simplified by using $(1-\beta_e\cos\theta)^4\approx 1$. A factor of 2 is inserted so as to account for the two electrons in the K-shell in order to yield the total photoelectric cross section per atom,

$$\sigma_{PE}=2r_0^2\left(4\sqrt{2}\right)\alpha^4Z^5\left(\frac{m_e}{k}\right)^{\frac{7}{2}}\int_0^\pi d\theta\sin^3\theta\int_0^{2\pi}d\phi\cos^2\phi$$
$$=\sigma_{Tho}\left(4\sqrt{2}\right)\alpha^4Z^5\left(\frac{m_e}{k}\right)^{\frac{7}{2}}. \quad (6.130)$$

This result yields a $k^{-(7/2)}$ energy dependence for the probability of a photoelectric absorption, which is closely approximated by the k^{-3} dependence given in the earlier intuitive argument. This $k^{-(7/2)}$ dependence is valid so long as the assumptions of a photon energy being greater than the electron binding energy and a nonrelativistic electron are met. Equation (6.130) also predicts a marked dependence of the photoelectric absorption total cross section upon the atomic number, $\propto Z^5$. While this derivation highlights the strong atomic number and photon energy dependencies of the photoelectric effect, experimental data have shown that there are some modifications:

- The total cross section per atom varies approximately with $Z^{4.8}$ for low-atomic number materials, decreasing to a Z^4 dependence for high-Z values.
- The $k^{-(7/2)}$ dependence predicted becomes less rapid at higher photon energies.

Figure 6.21 shows the total cross sections per atom for photoelectric absorption for carbon (Z = 6) and

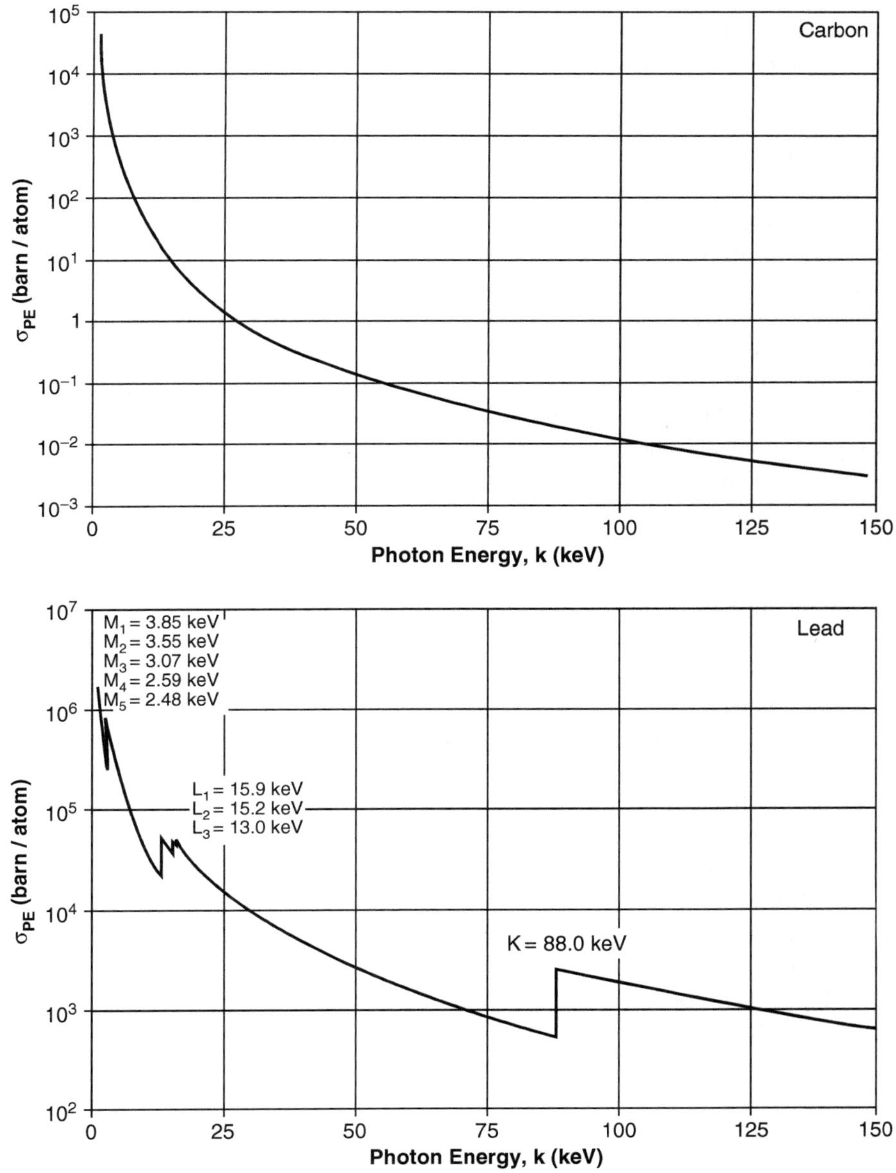

Fig. 6.21 Photoelectric total cross sections as functions of photon energy for carbon and lead. Note difference in ordinate scales, reflecting the $\sim Z^5$ dependence

lead (Z = 82) as a function of incident photon energy. The $k^{-(7/2)}$ dependence appears in both curves, but is perhaps more clearly evident for carbon due to the lack of discontinuities. These discontinuities correspond to the various binding energies of the K, L, and M orbitals and are more apparent when superimposed upon the $k^{-(7/2)}$ dependence of lead and show the energy thresholds for electron ejection. For example, an incident photon with an energy of just under 88 keV has insufficient energy to eject a K-shell electron (i.e., has a cross section of 532 b). However, a photon with an energy just above the K-shell binding energy of 88 keV has sufficient energy to eject the bound electron and thus providing the sudden increase by a factor of 5 in the photoelectric absorption probability (i.e., a cross section of 2,520 b).

As with Compton scatter, one is also interested in knowing the fraction of the incident energy which is transferred to the medium through the kinetic energies of the photoelectron and the recoil nucleus following a photoelectric absorption. This is described by the photoelectric energy-transfer cross section which is the total photoelectric cross section weighted by the fraction of energy received by the recoil atom and ejected electrons. It excludes the energy resulting from the relaxation processes which yield characteristic, or fluorescent, X-rays. Atomic relaxation processes (both radiative and nonradiative) are to be discussed: these resolve vacancies created in atomic electron orbitals following the ejection of an electron. It will suffice to present here the fluorescence yield, ω_i, where the subscript i refers to the orbital filled by an electron making a transition from a higher orbital with the emission of an X-ray in the process. For example, consider a vacancy created in the K orbital. As there are many orbitals that can feed this vacancy, it is useful to define a mean X-ray energy resulting from an electron transiting to this orbital, $\bar{k}_K$. If P_K is the fraction of all vacancies created in the K-orbital, then the mean energy carried away from the atom per photoelectric absorption by K-orbital characteristic X-rays is given by the product, $P_K\omega_K\bar{k}_K$. As the probability that energy will be carried away by L- or M-orbital characteristic X-rays is negligible for $k > (E_B)_K$, then, for $k > (E_B)_K$ the mean fraction of energy transferred to the photoelectron is,

$$\bar{f}_K = 1 - \frac{P_K\omega_K\bar{k}_K}{k}. \tag{6.131}$$

For a photon energy between the K- and L-orbital edges (i.e., $(E_B)_K > k > (E_B)_L$), K-orbital photoelectron absorption cannot occur and the mean fraction of energy transferred to the photoelectron is now,

$$\bar{f}_L = 1 - \frac{P_L\omega_L\bar{k}_L}{k} \tag{6.132}$$

which is essentially negligible except for high-*Z* materials.

The energy-transfer cross sections per atom are, then,

$$\sigma_{PE,Tr} = \sigma_{PE}\left(1 - \frac{P_K\omega_K\bar{k}_K}{k}\right); \quad k > (E_B)_K \tag{6.133}$$

$$\sigma_{PE,Tr} = \sigma_{PE}\left(1 - \frac{P_L\omega_L\bar{k}_L}{k}\right); \quad (E_B)_K > k > (E_B)_L \tag{6.134}$$

6.4 Atomic Relaxation

6.4.1 Introduction

An atomic electron vacancy is created through photoelectric absorption, electron capture, internal conversion and, as to be described in the following chapter, can result from an ionizing collision by a moving charged particle. The creation of an orbital electron vacancy produces an excess of energy above the atom's ground state and a number of subsequent processes can occur in order to remove this energy. These relaxation processes will be categorized in terms of whether or not electromagnetic radiation is emitted by the atom as it de-excites.

6.4.2 Radiative Transitions

In the simple picture of Fig. 6.22, an electron vacancy in the K-orbit is filled by an L-orbital electron. The emitted photon has a definite energy given by the difference between the two electron binding energies (neglecting, as usual, the recoil of the atom). This radiative transition creates a new vacancy in the L

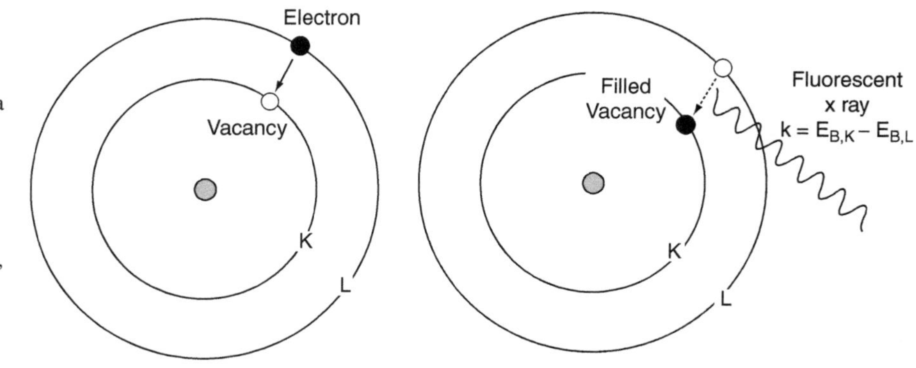

Fig. 6.22 The emission of a fluorescent (characteristic) X-ray due to the filling of a K-orbital electron vacancy via the cascade of an L-orbital electron. The energy of the emitted X-ray is equal to the difference between the two atomic binding energies (neglecting recoil) and, hence, is characteristic of the atom

orbit which can be filled by an electron from a higher orbit with a subsequent emission of a second photon, etc. Following the discussion of nuclear radiative transitions of Chap. 4, quantum selection rules are also applicable to atomic radiative transitions.

By systematically measuring the K X-ray energy as a function of atomic number, Z, Moseley determined in 1913 that it scaled with Z^2. Historically, this empirical relationship was given in terms of the frequency of the X-ray,[7] or,

$$\sqrt{\nu} = A(Z - \sigma) \tag{6.135}$$

where A and σ are constants. An interpretation of Moseley's law can be had using Bohr's model of the atom, in which the electrons are grouped into orbitals (and suborbitals) of decreasing binding energy identified as K, L, M orbitals, etc, with corresponding principal quantum numbers of n = 1, 2, 3. Ignoring relativistic and spin effects, the binding energy of electrons in a hydrogen-like atom for principal quantum number n is,

$$E_n = -hcR_\infty \frac{Z_{eff}^2}{n^2} \tag{6.136}$$

where the Rydberg energy is $hcR_\infty = 13.61$ eV and an infinitely-massive atom is assumed. Z_{eff} is an effective atomic number modified so as to account for the screening of the nucleus by the other orbital electrons and which can be approximated by reducing the atomic number by a constant,

$$Z_{eff} = Z - \sigma_1 \tag{6.137}$$

[7]Recall the Einstein relationship $k = h\nu$.

The energy of the characteristic X-ray formed by the filling of a K-orbital vacancy (n = 1) by an L-orbital electron (n = 2) is,

$$\begin{aligned} k &= E_2 - E_1 \\ &= \frac{3hcR_\infty}{4}(Z - \sigma_1)^2 \end{aligned} \tag{6.138}$$

which replicates Moseley's empirical law. However, this expression does not account for electron degeneracy due to the electrons' orbital angular momenta and spins. Momentarily neglecting screening, the complete energy for a state of principal quantum number n, orbital angular momentum number l (where $l = 0, 1, 2 \ldots n - 1$) and total angular momentum, $j = l \pm 1/2$, is given by,

$$\begin{aligned} E_{nlj} = hcR_\infty \Bigg[& \frac{Z^2}{n^2} - \alpha^2 \frac{Z^4}{n^3}\left(\frac{3}{4n} - \frac{1}{l + \frac{1}{2}}\right) \\ & - \alpha^2 \frac{Z^4}{n^3}\left(\frac{\boldsymbol{l} \bullet \mathbf{s}}{l(l+1)\left(l + \frac{1}{2}\right)}\right)\Bigg] \end{aligned} \tag{6.139}$$

where s is the intrinsic spin (± 1/2) of the electron. The first term, or Balmer term, is dominant and describes the binding as a function of principal quantum number only. The second term is the correction term accounting for orbital angular momentum and the third term accounts for the coupling between the orbital angular momentum and the intrinsic spin of the electron. From Chap. 3, this spin-orbit coupling term can be written as,

$$\boldsymbol{l} \bullet \mathbf{s} = \frac{1}{2}[j(j+1) - l(l+1) - s(s+1)]. \tag{6.140}$$

Noting that $j = l \pm 1/2$ and allowing for the screening of the nucleus by other electrons so as to reduce the effective atomic number, (6.139) becomes,

$$E_{nl j=l\pm\frac{1}{2}} = hcR_{\infty} \times \left[\frac{(Z-\sigma_1)^2}{n^2} - \alpha^2 \frac{(Z-\sigma_2)^4}{n^4}\left(\frac{3}{4} - \frac{n}{j+\frac{1}{2}}\right)\right] \quad (6.141)$$

The screening constants for the Balmer and the spin-relativity terms, σ_1 and σ_2, respectively, are not equal. σ_1 increases with atomic number and orbit reflecting increasing screening with more atomic electrons present or for the given orbit being at a greater distance from the nucleus. On the other hand, σ_2 is almost independent of atomic number but increases with orbit. This difference is due to σ_1 being due to the total screening effects of electrons internal and external to the orbit whereas σ_2 is due to those electrons between the orbit and the nucleus. The quantum numbers for the K, L, and M orbitals are given in Table 6.1.

Table 6.1 Atomic electron orbits and quantum numbers

Quantum numbers			Orbital/suborbital	X-ray notation
n	l	j (=$l \pm 1/2$)		
1	0	1/2	1s	K
2	0	1/2	$2s_{1/2}$	L_I
2	1	1/2	$2p_{1/2}$	L_{II}
2		3/2	$2p_{3/2}$	L_{III}
3	0	1/2	$3s_{1/2}$	M_I
3	1	1/2	$3p_{1/2}$	M_{II}
3		3/2	$3p_{3/2}$	M_{III}
3	2	3/2	$3d_{3/2}$	M_{IV}
3		5/2	$3d_{5/2}$	M_V

As with nuclear radiative transitions, selection rules restrict the transitions between orbitals. The most likely processes are the allowed electric dipole transitions with $l = \pm 1$ and $j = 0, \pm 1$. Any changes in the principal quantum number are allowed. Electric quadrupole ($\Delta l = \pm 2$) and magnetic dipole transitions are second-order processes and are "forbidden" as they occur much less frequently than the dipole transition. Electron dipole transitions for the orbitals described in Table 6.1 are shown in Fig. 6.23.

In addition to the selection rules, which are based upon changes in orbital angular momentum and total angular momentum, the probability of characteristic X-ray production is also dictated by the competition between different mechanisms of dissipating the binding energy following the creation of an atomic electron vacancy. These other transitions do not involve

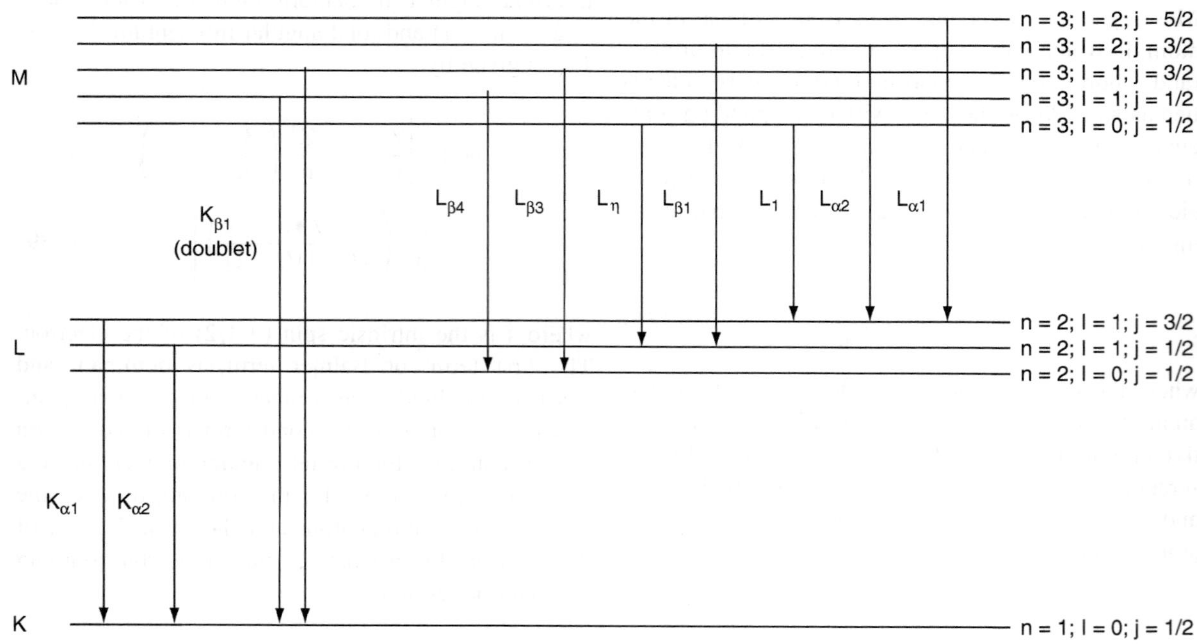

Fig. 6.23 Allowed electric dipole transitions ($\Delta l = \pm 1$; $\Delta j = 0, \pm 1$) between the K, L, and M orbitals. Note that the vertical scale is not proportional to energy. Multiplets are groups of lines with either the same initial (n, l) or final (n, l) values

the emission of radiation and are the subjects of the following subsection. Recall that the fluorescence yield represents the probability that the filling of the orbital vacancy is accompanied by the emission of radiation and is defined for the ith suborbital of the Xth orbit as,

$$\omega_{X,i} = \frac{f_{X,i}(Z)}{v_{X,i}(Z)} \quad (6.142)$$

where $f_{X,i}(Z)$ is the total number of characteristic photons emitted as a result of the filling of $v_{X,i}(Z)$ vacancies in the ith suborbital of the Xth orbit. A further definition is the mean fluorescence yield for the ith orbital which is the weighted sum over the suborbitals,

$$\bar{\omega}_X = \sum_{i=1}^{k} \eta_{X,i}\omega_{X,i} \quad (6.143)$$

where $\eta_{X,i}$ is the relative number of vacancies in the ith suborbital of the Xth orbital,

$$\eta_{X,i} = \frac{v_{X,i}}{\sum_{i=1}^{k} v_{X,i}} \quad (6.144)$$

and the summations are over all k suborbitals of the Xth orbital.

It is possible to calculate a result which demonstrates how the fluorescence yield varies with atomic number by treating the radiating atom as a damped harmonic oscillator. The K-shell fluorescence yield can be described by the ratio,

$$\omega_K = \frac{N_{rad}}{N_{non-rad} + N_{rad}} \quad (6.145)$$

in terms of the probabilities of possible radiative and nonradiative transitions, where the nonradiative transitions, as to be shown, decrease relatively slowly with Z. The relaxation time for the harmonic oscillator is,

$$T_{relax} = \frac{E}{\left(d\bar{E}/dt\right)} \quad (6.146)$$

where E is the energy of the oscillator and $d\bar{E}/dt$ is the mean rate at which energy is dissipated. For an accelerating charge emitting radiation $d\bar{E}/dt \propto a_{max}^2$ where the maximum acceleration is $a_{max} \approx \omega v_0$, where v_0 is the maximum velocity and, from Moseley's law, $\omega \propto Z^2$. As a result, the relaxation time is proportional to Z^{-4}. As the transition probability is proportional to the reciprocal of the relaxation time, this leads to the approximation of the fluorescence yield as,

$$\omega_K \approx \frac{Z^4}{a + Z^4} \quad (6.147)$$

where a is a constant. This result predicts that the fluorescence yield will increase with atomic number to approach unity if $Z^4 \gg a$. Hubbell et al. (1994) have summarized various parameterizations of the K-orbital fluorescence yield. For example, one parameterization follows directly,

$$\left(\frac{\omega_K}{1 - \omega_K}\right)^{\frac{1}{4}} = \xi(Z) \quad (6.148)$$

where ξ is a polynomial in the atomic number Z,

$$\xi(Z) = \sum_{n=0}^{3} c_n Z^n \quad (6.149)$$

with coefficient values,

$$c_0 = 0.0370 \pm 0.0052$$
$$c_1 = 0.03112 \pm 0.00044$$
$$c_2 = (5.44 \pm 0.11) \times 10^{-5}$$
$$c_3 = -(1.25 \pm 0.07) \times 10^{-6}$$

Equation (6.148) can be rewritten in a more convenient form,

$$\omega_K = \frac{\xi^4(Z)}{1 + \xi^4(Z)} \quad (6.150)$$

This form can also represent the L-orbital mean fluorescence yield with corresponding parameter values provided by Hubbell et al.,

$$c_0 = 0.177650$$
$$c_1 = 0.00298932$$
$$c_2 = 8.91297 \times 10^{-5}$$
$$c_3 = -2,67184 \times 10^{-7}$$

The K- and L-orbital fluorescence yields given by these parameterizations are plotted in Fig. 6.24. It can

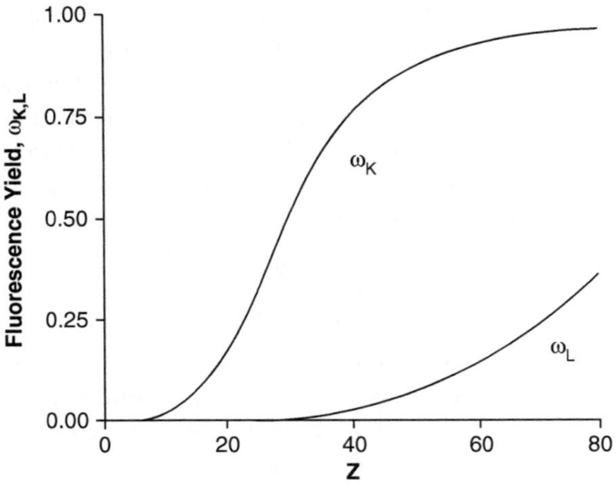

Fig. 6.24 Fluorescence yields for K and L orbitals as functions of atomic number calculated from the parameterizations of Hubbell et al. (1994)

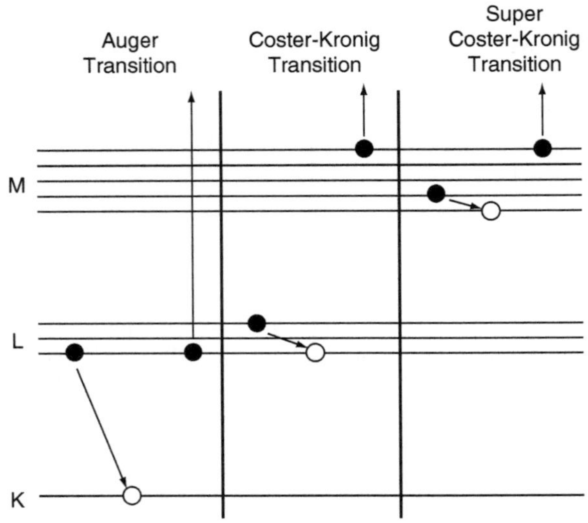

Fig. 6.25 Types of nonradiative atomic transitions resulting in atomic electron emission

be seen that the K-orbital fluorescence yield increases rapidly for $Z > 10$, reaching 90% for barium ($Z = 56$). On the other hand, the L-orbital fluorescence yield is much smaller and is negligible for $Z < 30$ (zinc), only reaching a value of 34% for mercury ($Z = 80$). The mean fluorescence yield for the M-orbital is essentially negligible. Where M-orbital transitions are not negligible, Hubbell et al. recommend a parameterization of the form,

$$\bar{\omega}_M = 1.29 \times 10^{-9}(Z - 13)^4 \tag{6.151}$$

For the example of lead ($Z = 82$), $\bar{\omega}_M = 2.7\%$. Thus, an L- or M-orbital vacancy would be most likely filled through a nonradiative transition. This has important dosimetry implications as these nonradiative transitions release low-energy atomic electrons with very limited ranges which subsequently deposit a large localized absorbed dose.

6.4.3 Nonradiative Transitions

In addition to the radiative processes through which the excess energy of the excited atom is manifested as an X-ray, this energy can also be directed to eject one or more atomic electrons from the outer orbits. This is known, generally, as the Auger effect and the ejected electron(s) are referred to as Auger electron(s). There are further subclassifications of these transitions as shown in Fig. 6.25. Strictly speaking, the Auger effect is that of a vacancy in a lower major shell being filled by an electron from that higher major shell with an electron ejected from that higher major shell. The Coster–Kronig transition is that in which a vacancy is filled by an electron from the same major shell but with an electron ejected from a higher major shell. Finally, the super Coster–Kronig transition is that in which the vacancy and the transiting and ejected electrons are all in the same major shell. Unless when differentiation is necessary, all three transition types will be referred to generically as Auger transitions.

The consequence of an Auger transition is two or more atomic orbital vacancies as can be shown by considering a vacancy present in the K-orbital which is filled by an electron from the pth suborbital of the Xth orbital[8] followed by the ejection of an Auger electron from the qth suborbital of the Yth orbital.[9] Assume that the K-orbital vacancy is filled by an electron from the L-orbital; an energy excess of

[8]In analogy to electron conversion, this electron could come from the $2s_{1/2}$ orbital because, while forbidden for a radiative transition, it is allowed for a nonradiative transition. This would be a Coster–Kronig type of transition.

[9]A large number of Auger-type transitions are possible: if the Xth and Yth orbital are both the L-orbital, then a possible nine transitions exist, three of which are indistinguishable.

$|(E_B)_K| - |(E_B)_L|$ is available. If the Auger electron is from the M-orbital, it will have a kinetic energy equal to,

$$T_{e,M} = |(E_B)_K| - |(E_B)_L| - |(E_B)_M| \quad (6.152)$$

Now consider the case of two electrons from the N-orbital filling the vacancies in the L- and M-orbitals, following which two more Auger electrons are ejected with kinetic energies equal to (assuming that they are emitted from the N-orbital),

$$T_{e,N_1} = |(E_B)_L| - 2|(E_B)_N| \quad (6.153)$$

$$T_{e,N2} = |(E_B)_M| - 2|(E_B)_N| \quad (6.154)$$

Thus, four electron vacancies now exist as do three Auger electrons with a total kinetic energy equal to $|(E_b)_K| - 4|(E_b)_N|$. In other words, the combined kinetic energies of the Auger electrons is equal to the difference between the binding energy of the K-orbital and the sum of the binding energies of the remaining vacancies. As with internal conversion, one could erroneously consider the Auger process as a result of the energy transfer to the Auger electron directly via a photon with no intermediate states. However, the existence of Auger transitions, such as a $2s_{1/2} \rightarrow 1s_{1/2}$ transition, which are forbidden for radiative transitions, demonstrates that the interaction cannot involve an intermediate real photon.

Figure 6.24 showed that the fluorescence yield increases with atomic number. This is due both to the previously-provided argument that the probability of X-ray emission will increase with Z^4 and to the probability of an Auger transition will decrease with atomic number due to greater binding. The Auger yield is,

$$a_{X,i} = \frac{f_{X,i}(Z)}{v_{X,i}(Z)} \quad (6.155)$$

where $f_{X,i}(Z)$ is the total number of Auger electrons emitted as a result of the filling of $v_{X,i}(Z)$ vacancies in the ith suborbital of the Xth orbit and the mean Auger yield is,

$$\bar{a}_X = \sum_{i=1}^{k} \eta_{X,i}\omega_{X,i} \quad (6.156)$$

Table 6.2 Total yield of Auger-type electrons per decay for commonly-used isotopes in nuclear medicine (data from Howell (1992))

Isotope	Total yield of Auger electrons per decay
^{99m}Tc	4.0
^{67}Ga	4.7
^{111}In	14.7
^{123}I	14.9
^{201}Tl	36.9

Auger transitions are of particular interest to internal radiation dosimetry as many radionuclides used in radiopharmaceuticals emit Auger electrons (in addition to internal conversion electrons). The total Auger electron yield per decay for common isotopes used in diagnostic nuclear medicine are summarized in Table 6.2. The kinetic energies of the Auger electrons are small and, thus, they have limited range and consequently deposit their energy within small, localized volumes in tissue. This has two consequences: the dose distribution becomes nonuniform and the localized energy deposition can have significant biological effects, especially if the Auger electron-emitting radionuclide is incorporated within the radiosensitive nucleus of the cell (Boswell and Brechbiel 2005).

As shown in Fig. 6.25, the Coster–Kronig and super Coster–Kronig transitions take place between suborbitals and cause a redistribution of electrons in a given orbital by transferring energy to a suborbital of a lower energy. This releases energy which causes the ejection of an electron from a higher-order orbital with a binding energy less than that made available. Because the Coster–Kronig transitions alter the vacancy distribution, the mean fluorescence yield must be altered as a result. Rewriting this as,

$$\bar{\omega}'_X = \sum_{i=1}^{k} \eta'_{X,i}\omega_{X,i} \quad (6.157)$$

where the primes indicate quantities modified as a result of a Coster–Kronig transition. Recall that $\eta_{X,i}$ is the relative number of vacancies in the ith suborbital of the Xth orbital; as the Coster–Kronig transition will create more orbital vacancies,

$$\sum_{i=1}^{k} \eta'_{X,i} > 1 \quad (6.158)$$

If $\rho_{X,ij}$ is defined as the probability of a Coster–Kronig transition of a vacancy from the ith suborbital to the jth suborbital (all in the Xth orbital), then the modified relative numbers of vacancies are,

$$\begin{aligned}\eta'_{X,1} &= \eta_{X,1}\\ \eta'_{X,2} &= \eta_{X,2} + \rho_{X,12}\eta_{X,1}\\ \eta'_{X,3} &= \eta_{X,3} + \rho_{X,23}\eta_{X,2} + \left(\rho_{X,13} + \rho_{X,12}\rho_{X,23}\right)\eta_{X,1}.\end{aligned} \tag{6.159}$$

The product of a cascade of Auger transitions is that of a remnant highly-charged ion. In nuclear medicine, the Auger electron-emitting radionuclide is typically incorporated within a molecular carrier and the transfer of neighboring electrons may lead to partial neutralization of the ion. This may lead to a Coulomb fragmentation of the moiety due to the proximity of multiple positive charges.

6.5 Photon Interaction Coefficients

6.5.1 Introduction

The significant mechanisms through which a photon can interact with matter at the photon energies relevant to nuclear medicine – coherent (Thomson and Rayleigh) scatter, incoherent (Compton) scatter, and photoelectric absorption – have been discussed. Here, the probabilities of these interactions occurring are summarized in terms of the total cross section (i.e., the total probability of the interaction) and the energy-transfer cross section (i.e., the probability of transferring energy to the medium):

Thomson scatter: $\sigma_{Tho} = \frac{8\pi}{3} r_0^2$ (per electron)

Rayleigh scatter: $\sigma_{Ray} = \frac{2\pi r_0^2}{k^2} \int_0^{2k_\gamma} dq \bullet q|F(q, Z)|^2$ (per atom)

Compton scatter:

$$\sigma_{KN} = \frac{3}{4}\sigma_{Tho}\left[\frac{1+\alpha}{\alpha^2}\left(\frac{2(1+\alpha)}{(1+2\alpha)} - \frac{\ln(1+2\alpha)}{\alpha}\right) + \frac{\ln(1+2\alpha)}{2\alpha} - \frac{1+3\alpha}{(1+2\alpha)^2}\right] \quad \text{(per electron)}$$

$$\begin{aligned}\sigma_{KN,Tr} = \frac{3}{4}\sigma_{Tho}\Bigg[&\frac{2(1+\alpha)^2}{\alpha^2(1+2\alpha)} - \frac{1+3\alpha}{(1+2\alpha)^2} + \frac{(1+\alpha)(1+2\alpha-2\alpha^2)}{\alpha^2(1+2\alpha)^2} - \frac{4\alpha^2}{3(1+2\alpha)^3}\\ &- \left(\frac{1+\alpha}{\alpha^3} - \frac{1}{2\alpha} + \frac{1}{2\alpha^3}\right)\ln(1+2\alpha)\Bigg]\end{aligned}$$

(per electron)

Incoherent scatter (Compton scatter from bound atomic electrons):

$$\sigma_{Inc} = \int d\Omega \frac{d\sigma_{KN}}{d\Omega} S(\mathbf{q}, Z) \quad \text{(per atom)}$$

Photoelectric absorption:

$$\sigma_{PE} = \sigma_{Tho}\left(4\sqrt{2}\right)\alpha^4 Z^5 \left(\frac{m_e}{k}\right)^{\frac{7}{2}} \quad \text{(per atom)}$$

$$\sigma_{PE,Tr} = \sigma_{PE,T}\left(1 - \frac{P_K\omega_K\bar{k}_K}{k}\right) \quad k > (E_b)_K \quad \text{(per atom)}$$

$$\sigma_{PE,Tr} = \sigma_{PE,T}\left(1 - \frac{P_L\omega_L\bar{k}_{\gamma,L}}{k_\gamma}\right) \times (E_b)_K > k_\gamma > (E_b)_L \quad \text{(per atom)}$$

Photon interaction coefficients are non-stochastic quantities describing the macroscopic behavior of the interactions of photons with matter. Formal definitions can be found in ICRU Report 33 (1980); note that there is a deviation from the general use of SI units to the centimeter–gram–second system in this discussion of interaction coefficients.

6.5.2 Mass Attenuation Coefficient

The mass attenuation coefficient is a measure of the reduction of the number of photons in a beam as it traverses a medium which is due to the combined effects of photons being scattered out of the beam (i.e., through the photon-number conserving processes of coherent and incoherent scatter) and by the removal of photons through photoelectric absorption and pair production. Recall from Chap. 2 that the differential reduction in the flux φ of particles per cm^2 per s passing through a scattering/absorbing medium of differential thickness dx containing N scattering or absorption centers per cubic centimeter is $d\varphi = -\sigma\varphi N\ dx$, where the constant of proportionality σ is the total cross section (in units of unit area per scattering/absorption center). Integrating this over a thickness L of the medium leads to the exponential decrease as a function of thickness,

$$\begin{aligned}\varphi &= \varphi_0 e^{-\sigma NL}\\ &= \varphi_0 e^{-\mu L}.\end{aligned} \tag{6.160}$$

where φ_0 is the incident flux. The product of the total cross section and the interaction center density, σN, has units of inverse length and is defined as the linear attenuation coefficient, μ, and represents the mean number of interactions (scatters or absorptions) per unit length of medium traversed by the photon. For example, consider the combination of a photon and medium for which $\mu = 0.1/\text{cm}$ which means that the photon has a 10% probability per cm of interacting with a center in the medium. Initially, let the medium have a density of 1 g/cm^3 and be 1 cm thick; the ratio of the flux of photons exiting the 1 cm thick medium to that incident will be,

$$\frac{\varphi}{\varphi_0} = e^{-0.1\,\text{cm}^{-1}\times 1\,\text{cm}} = e^{-0.1} = 0.904$$

Now, imagine that the medium is compressed from a thickness from 1 to 0.5 cm resulting in a density increase from 1 to 2 g/cm^3. The exiting flux of photons will be unaffected with the net transmission being $\varphi/\varphi_0 = 0.904$, but the attenuation coefficient μ will now have doubled to 0.2/cm. To remove this dependence upon the medium's density, we normalise the linear attenuation coefficient to the density of the medium, μ/ρ, to yield the mass attenuation coefficient. For both densities in the above example,

$$\frac{\mu}{\rho} = \frac{0.1/\text{cm}}{1\,\text{g/cm}} = 0.1\ \text{cm}^2/\text{g}$$

$$\frac{\mu}{\rho} = \frac{0.2/\text{cm}}{2\,\text{g/cm}} = 0.1\ \text{cm}^2/\text{g}$$

The total mass attenuation coefficient is the sum of the contributions due to coherent (Rayleigh) scatter, incoherent (Compton) scatter, and photoelectric absorption; in commonly-used nomenclature in radiation dosimetry literature these correspond to the terms in the following equation,

$$\frac{\mu}{\rho} = \frac{\sigma_{Coh}}{\rho} + \frac{\sigma_{Inc}}{\rho} + \frac{\tau}{\rho} \tag{6.161}$$

In order to match the units of each of these terms, we will address each contribution individually.

Coherent scatter: To convert the coherent scatter cross section per atom to that per gram, one must multiply σ_{Coh} by $N_A Z/A$ atoms per g, where N_A is Avogadro's number,

$$\frac{\sigma_{Coh}}{\rho} = \sigma_{Coh}\frac{N_A}{A} \tag{6.162}$$

The contribution of coherent scatter to the total mass attenuation coefficient is appropriate only where the photon beam geometry is such that the scattering of a photon out of the beam results in it not being detected.

Incoherent scatter: At the low photon energies where the Z-scaled Klein–Nishina cross section overestimates the incoherent scatter cross section from the atom, coherent scatter, and photoelectric absorption dominate over the Compton processes. As a result, it is often practical and easier to use the Klein–Nishina cross sections by the atomic number of the medium. In order to convert this to per gram, we must multiply σ_{KN} by $N_A Z/A$ electrons per gram,

$$\frac{\sigma_{Inc}}{\rho} = \sigma_{KN}\frac{N_A Z}{A} \tag{6.163}$$

Recall from Chap. 3 that $N_A Z/A \approx 1/2$ for the low-Z materials which are of primary dosimetric interest. As a result, the multiplicative factor $N_A Z/A$ is very nearly constant for such materials.

Photoelectric absorption: The total cross section of photoelectric absorption per atom is given per atom. The mass attenuation coefficient due to photoelectric absorption is,

$$\frac{\tau}{\rho} = \sigma_{PE}\frac{N}{A} \tag{6.164}$$

In radiation dosimetry, one does not always deal with a medium that is of a single element. For compounds, the Bragg additivity rule is used to calculate the attenuation coefficient,

$$\frac{\mu}{\rho} = \sum_{i}^{N} f_i \left(\frac{\mu}{\rho}\right)_i \qquad (6.165)$$

where $(\mu/\rho)_i$ is the mass attenuation coefficient of the ith element of the compound and f_i is the corresponding weight fraction. The summation is over the N elements making up the compound.[10]

Now that the probabilities of the different photon–matter interactions have been normalized, they can be compared directly. Figure 6.26 shows the relative contributions of coherent scatter, incoherent scatter, photoelectric absorption, and pair production to the mass attenuation coefficient as a function of photon energy for carbon and lead. For carbon ($Z = 6$), photoelectric absorption dominates up to about 20 keV, upon where incoherent scatter becomes the dominant process. The mass attenuation coefficient for coherent scatter exceeds that for incoherent scatter up to about 10 keV in this low-Z material. The overall total mass attenuation coefficient for carbon decreases rapidly with photon energy (due to photoelectric absorption) up to about 0.1 MeV, where the decrease is very limited. The total mass attenuation coefficient for lead ($Z = 82$) is higher than that for carbon due, primarily, to the photoelectric effect and its $\sim Z^5$ dependence. Photoelectric absorption dominates over all other process up to about 600 keV; coherent scatter dominates over incoherent scatter up to about 150 keV. It will be noted that the mass attenuation coefficient due to incoherent scatter for carbon is about 20–30% greater than that for lead. The reason for this follows by recalling that the incoherent scatter cross section is per electron and must be multiplied by the number of electrons per gram of material in order to yield its contribution to the mass attenuation coefficient. However, as shown in Chap. 3, the ratio of the number of protons to the number of neutrons in stable nuclei, $Z/A - Z$, decreases with increasing atomic mass. Hence, the number of protons (and, hence, the number of electrons) per gram of lead is less than that of carbon, resulting in a lower mass attenuation coefficient due to incoherent scatter.

For completeness, the relative contributions of the various photon–matter interaction types to the attenuation coefficient are plotted in Fig. 6.27.

For both elements, photoelectric absorption is the most dominant process at low energies, but is superseded by incoherent scatter at photon energies of about 0.025 MeV for carbon and about 0.55 MeV for lead. For carbon, the relative contribution of coherent scatter to the attenuation coefficient increases smoothly to a maximum of about 15% at 0.02 MeV. On the other hand, that for lead exhibits a more complex dependence due to the increase in photoelectric absorption cross section when the photon energy reaches the energy threshold of a given atomic electron orbital. Even so, its maximum contribution to the attenuation coefficient is similar to that for carbon at about 14% at the K-absorption edge of 0.088 MeV. What is of particular interest is the difference between the ranges of photon energies over which incoherent scatter is dominant for carbon and lead. For carbon, incoherent scatter is the most significant contributor to the total attenuation coefficient between photon energies of about 0.025 MeV and beyond (up to about 28 MeV, not shown here, where pair production begins to dominate); on the other hand, incoherent scatter is dominant for lead only over the limited photon range of 0.55 MeV and above (to about 5 MeV where pair production again dominates).

6.5.3 Mass Energy-Transfer Coefficients

The attenuation coefficients just discussed provide information only on the probability of a coherent or incoherent scatter or a photoelectric absorption occurring. In dosimetry, one is interested in the amount of energy transferred from the photon to the medium. The mass energy-transfer coefficient is a measure of the amount of energy transferred to a medium by a beam of photons traversing the medium. As Thomson and Rayleigh scatter do not transfer energy to the medium, they are excluded from this discussion. The mass energy-transfer coefficient is,

$$\frac{\mu_{tr}}{\rho} = \frac{\tau_{tr}}{\rho} + \frac{\sigma_{tr}}{\rho}. \qquad (6.166)$$

[10]The Bragg additivity rule assumes that atoms in a compound act independently of each other. This, in general, is a sufficiently reasonable assumption in most dosimetry applications.

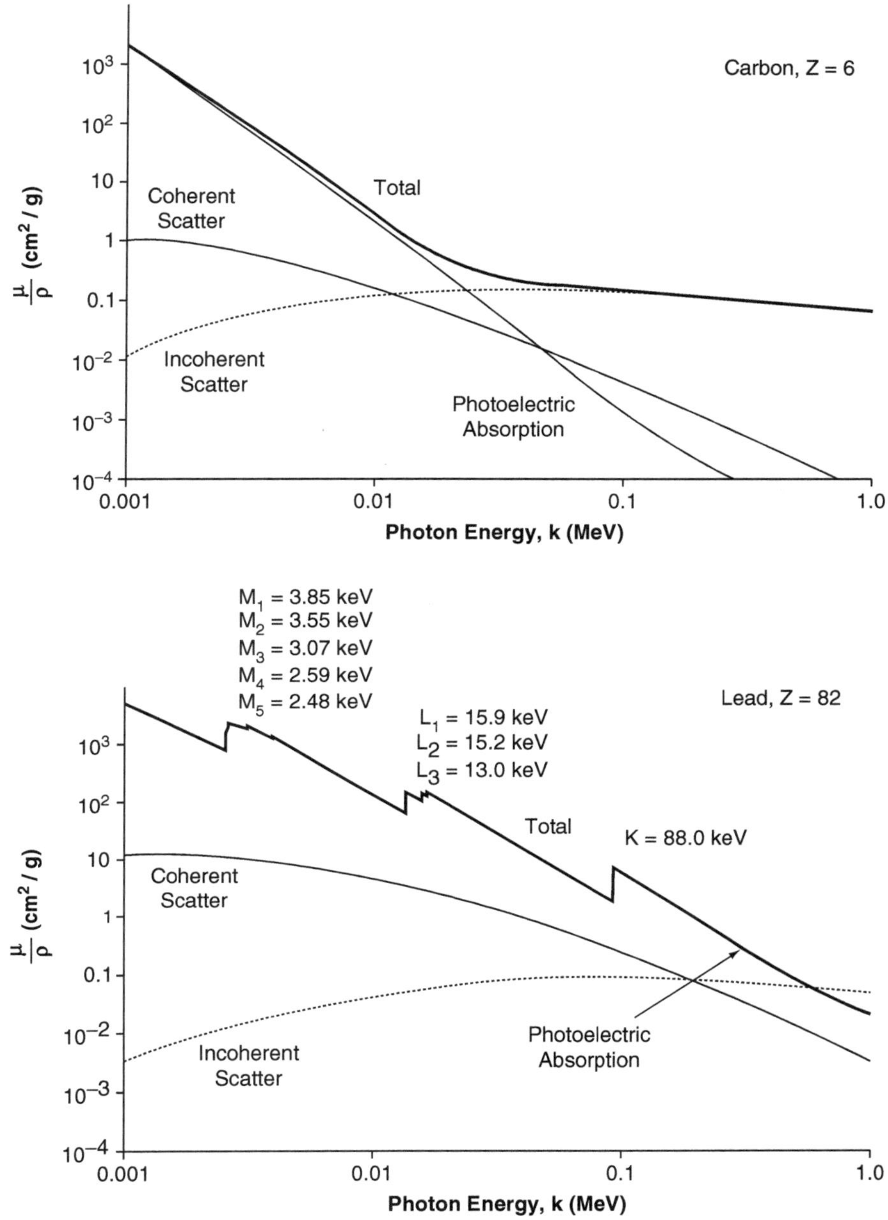

Fig. 6.26 Mass attenuation coefficients for carbon and lead and contributions from photon–matter interactions at photon energies of interest to nuclear medicine

Again, following the derivations of the previous section, the mass energy-transfer coefficient is, for when the photon energy is above the K-absorption edge,

$$\frac{\mu_{tr}}{\rho} = \frac{\tau}{\rho}\left(1 - \frac{P_K \omega_K \bar{k}_K}{k}\right) + \frac{\sigma}{\rho}\left(\frac{\bar{T}_e}{k}\right); \quad k > (E_b)_K. \tag{6.167}$$

For lower photon energies between the L- and K-absorption edges,

$$\frac{\mu_{tr}}{\rho} = \frac{\tau}{\rho}\left(1 - \frac{P_L \omega_L \bar{k}_L}{k}\right) + \frac{\sigma}{\rho}\left(\frac{\bar{T}_e}{k}\right); \quad \times (E_B)_K > k > (E_B)_L. \tag{6.168}$$

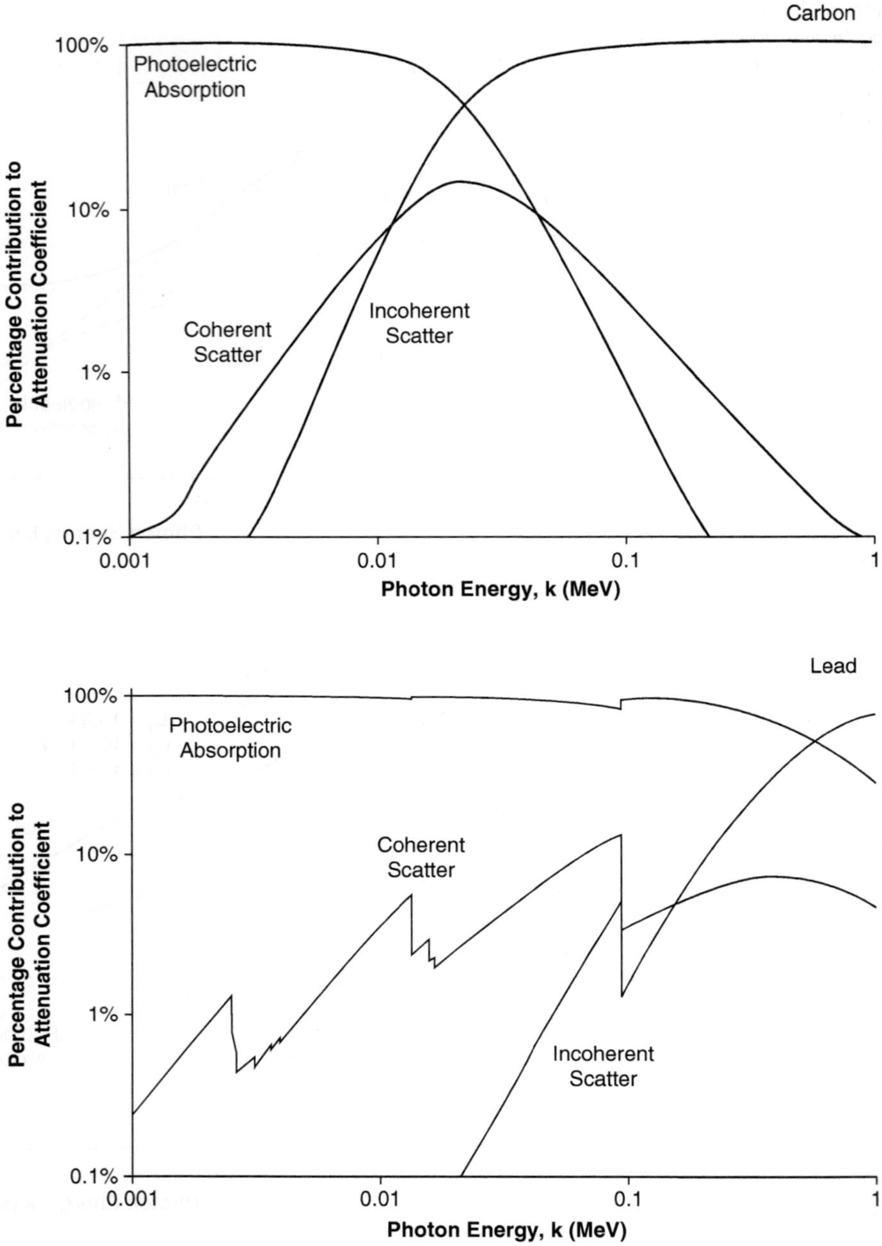

Fig. 6.27 The percentage contributions to the total attenuation coefficient of photoelectric absorption and coherent and incoherent scatter for carbon and lead

where energy emitted as fluorescence has been accounted for.

6.5.4 Mass Energy-Absorption Coefficients

The absorbed dose as a result of the exposure to ionizing radiation is the amount of energy absorbed by the medium per unit mass (see Chap. 8). Photons transfer energy to electrons or positrons (through Compton scatter, photoelectric absorption, and pair production) which subsequently deposit energy in the medium as they slow down (Chap. 7). Not all of the energy transferred to the electron or positron will be deposited in the medium: because they carry electric charge and are in motion, they will radiate electromagnetic energy (*bremsstrahlung*) as they decelerate and are scattered (Haug and Nakel 2004). Thus only a fraction of the energy transferred from the photon to the electron is deposited locally in the medium.

The mass energy-absorption coefficient is related to the mass energy-transfer coefficient by,

$$\frac{\mu_{en}}{\rho} = \frac{\mu_{tr}}{\rho}(1 - g) \qquad (6.169)$$

where g is the mean fraction of electron or positron energy that is lost through radiative processes described above and where it is assumed that the medium in which the original photon–matter interaction occurred is sufficiently large that all of the secondary electrons and positrons stop completely within it. Although g is discussed in Chap. 7, note that it is a function of both photon energy k and atomic number of the medium through which the charged particle is moving through. For low-Z and low k, g is very small and there is little difference between the mass energy-transfer and energy–absorption coefficients. g increases

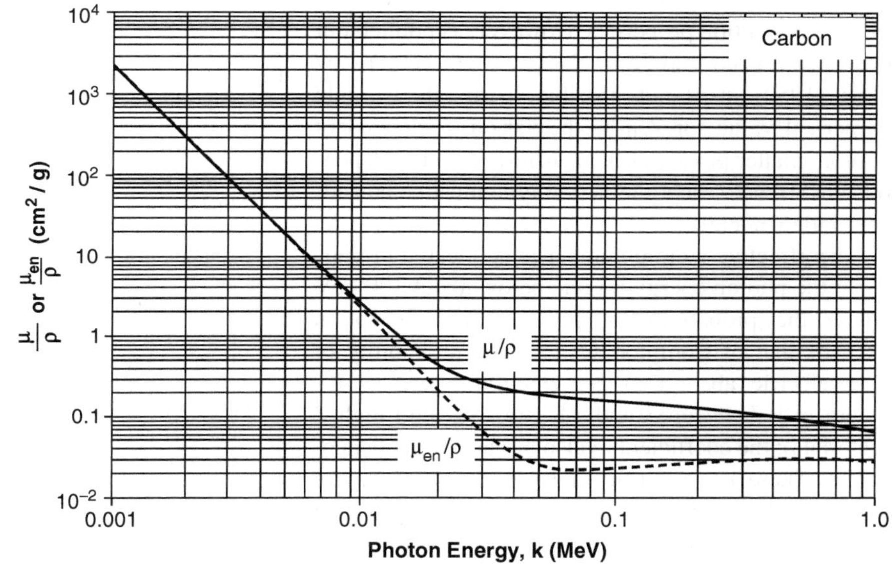

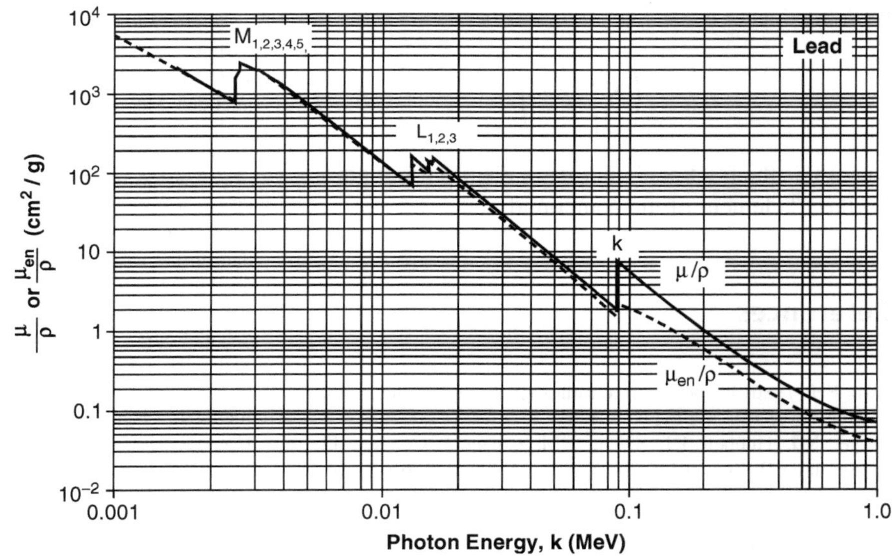

Fig. 6.28 Mass attenuation and mass energy-absorption coefficients for carbon and lead. Curves were calculated using tabulated data from Hubbell and Seltzer (1996), available at http://physics.nist.gov/PhysRefData/XrayMassCoef/cover.html

with photon energy and atomic number. Figure 6.28 shows the mass attenuation coefficients and mass energy-absorption coefficients for carbon and lead.

6.5.5 Effective Atomic Number

It is frequently convenient to characterize a compound material by its effective atomic number, Z_{eff}. One example of the usefulness of a Z_{eff} is in matching the material used in the wall of an ionization chamber with the contained gas (such as air which is 78% N_2, 21% O_2, and 1% Ar) so that the wall and gas have the same photon absorption properties. However, as pointed out originally in the middle of the twentieth century by Hine (1952), it is not possible to assign a single numerical value for Z_{eff} as the various Z-dependent photon–matter processes have differing energy dependencies. Hence, each interaction type must be considered independently.

The effective atomic number is applicable primarily to the photon energy regime below about 100 keV where the probability of photoelectric absorption exceeds that of Compton scatter. The probability of photoelectric absorption per electron varies as Z^m where m is between 3 and 3.8 (for low- and high-Z materials); the coefficient m also has an energy dependence. There is no applicability of Z_{eff} to Compton scatter, which is independent of Z and only dependent upon the electron density. The effective atomic number for a compound consisting of N elements can thus be defined as,

$$Z_{eff} = \left(\sum_{i=1}^{N} f_i Z_i^m \right)^{1/m} \tag{6.170}$$

where f_i is the fractional number of electrons per gram for the ith element of the compound.

References

Attix FH (1986) Introduction to radiological physics and radiation dosimetry. Wiley, New York

Berger MJ, Hubbell JH, Seltzer SM, Chang J, Coursey JS, Sukumar R, Zucker DS (2005) XCOM: photon cross section database (version 1.3). http://physics.nist.gov/xcom. National Institute of Standards and Technology, Gaithersburg, MD. Accessed 24 October 2008

Bergstrom PM, Pratt RH (1997) An overview of the theories used in Compton scattering calculations. Radiat Phys Chem 50:3–29

Bjorken JD, Drell SD (1964) Relativistic quantum mechanics. McGraw-Hill, New York

Boswell CA, Brechbiel MW (2005) Auger electrons: lethal, low energy, and coming soon to a tumor cell nucleus near you. J Nucl Med 46:1946–1947

Compton AH (1923a) A quantum theory of the scattering of X-rays by light elements. Phys Rev 21:483–502

Compton AH (1923b) The spectrum of scattered X-rays. Phys Rev 22:409–413

Compton AH (1925) On the mechanism of X-ray scattering. Proc Nat Acad Sci 11:303–306

Cooper MJ (1997) Compton scattering and the study of electron momentum density distributions. Radiat Phys Chem 50: 63–76

Debye P (1923) Zerstreuung von Röntgenstrahlen und quantentheorie. Physik Z 24:161–166

Dirac PAM (1926) Relativity quantum mechanics with an application to Compton scattering. Proc Royal Soc 111: 405–423

Dumond JWM (1929) Compton modified line structure and its relation to the electron theory of solid bodies. Phys Rev 33:643–658

Dyson F (2007) Advanced quantum mechanics. World Scientific Publishing, Singapore

Gordon W (1927) Der Comptoneffektnach der Schrödingerschen theorie. Z Physik 40:117–133

Gould RJ (2006) Electromagnetic processes. Princeton University Press, Princeton

Gribov VN, Nyiri J (2001) Quantum electrodynamics. Cambridge University Press, Cambridge

Harding G (1997) Inelastic photon scattering: effects and applications in biomedical science and industry. Radiat Phys Chem 50:91–111

Haug E, Nakel W (2004) The elementary process of bremsstrahlung. World Scientific Publishing, Singapore

Heitler W (1984) The quantum theory of radiation. Dover, New York

Hine GJ (1952) The effective atomic numbers of materials for various γ-ray interactions. Phys Rev 85:725 (abstract)

Howell RW (1992) Radiation spectra for Auger-emitting radionuclides. Med Phys 19:1371–1383

Hubbell JH (1997) Summary of existing information on the incoherent scattering of photons, particularly on the validity of the use of the incoherent scattering function. Radiat Phys Chem 50:113–124

Hubbell JH (1999) Review of photon interaction cross section data in the medical and biological context. Phys Med Biol 44:R1–R22

Hubbell JH (2006) Review and history of photon cross section calculations. Phys Med Biol 51:R245–R262

Hubbell JH (2006) Electron-positron pair production by photons: a historical overview. Rad Phys Chem 75:614–623

Hubbell JH, Veigele WJ, Briggs EA, Brown RT, Cromer DT, Howerton RJ (1975) Atomic form factors, incoherent scattering functions, and photon scattering cross sections. J Phys Chem Ref Data 4:471–538

Hubbell JH, Trehan PN, Singh N, Chand B, Mehta D, Garg ML, Garg RR, Singh S, Puri S (1994) A review, bibliography and tabulation of K, L and higher atomic shell X-ray fluorescence yields. J Phys Chem Ref Data 23:339–364

Hubbell JH, Seltzer SM (1996) Tables of X-ray mass attenuation coefficients and mass energy-absorption coefficients (version 1.4). http://physics.nist.gov/xaamdi. National Institute of Standards and Technology, Gaithersburg, MD. Accessed 24 October, 2008

ICRU Report 33 (1980) Radiation quantities and units. International Commission on Radiation Units and Measurements, Washington, DC

Johns HE, Cunningham JR (1983) The physics of radiology. Charles C Thomas, Springfield

Klein O, Nishina T (1929) Über die streuung von strahlung durch freie electronen nach der neuen relativistischen quantendynamik von Dirac. Z Phys 52:853–868

Roy SC, Kissel L, Pratt RH (1999) Elastic scattering of photons. Rad Phys Chem 56:3–26

Zaidi H (2000) Comparative evaluation of photon cross-section libraries for materials of interest in PET Monte Carlo simulations. IEEE Trans Nucl Sci 47:2722–2735

Charged Particle Interactions with Matter

7

Abstract In nuclear medicine, moving charged particles are released in tissue through either the radioactive decay processes of Chap. 4 or as a result of the photon–matter interactions of Chap. 6. Being charged, these particles interact significantly with the medium, transferring their kinetic energy resulting in an absorbed dose to the medium as they slow down to thermal energies. Hence, the study of charged particle interactions with matter is the fundamental core of ionizing radiation dosimetry. In this chapter, the two mechanisms of energy loss are presented. Collision energy losses between the particle and atomic electrons are derived through the Bohr classical and the Bethe quantum-mechanical means; hard collisions losses are derived independently from various quantum-mechanical results. Radiative energy losses resulting from *bremsstrahlung* are initially derived from classical theory which then progresses to the Bethe–Heitler quantum-mechanical theory. The polarization effects of a charged particle upon the medium will limit the collision energy losses and are derived. As energy loss is inherently stochastic, energy straggling models are also presented. In particular, the Vavilov theory of energy straggling is derived as are the Gaussian and Landau results which are treated limiting conditions to that theory. Multiple scatter strongly affects electrons and positrons and the Fermi–Eyges theory is derived as a means of justifying the Gaussian model. The Goudsmit–Saunderson and Moliére theories of multiple scattering are derived. Finally, the mechanisms through which a positron can annihilate on an electron are derived.

Contents

B.J. McParland, *Nuclear Medicine Radiation Dosimetry*,
DOI 10.1007/978-1-84882-126-2_7,

7.1 Introduction

It has been shown in Chaps. 4 and 6 that moving charged particles are products of radioactive decay and photon scattering. ^{4}He nuclei are emitted in α decay, moving electrons or positrons are created in β-decay and atomic electrons are ejected into the medium through internal conversion or atomic relaxation processes. Compton scatter sets electrons into motion and atomic electrons are ejected following photoelectron absorption. These moving charged particles slow down by transferring energy to the medium and, as this is the cause of the absorbed dose to tissue, charged particle–matter interactions with matter will be extensively reviewed.

Charged particles interact with the atoms in the medium they are traversing through the Coulomb component of the Lorentz force, $q(\mathbf{E} + \mathbf{v} \times \mathbf{B})$, by either transferring energy to (predominantly) the atomic electrons or else by being scattered by the nuclear and atomic electron Coulomb fields. These processes have markedly different consequences. In the former, a significant fraction of kinetic energy can be transferred to the atomic electron with, should the incident particle have a mass m much greater than that of the electron, negligible effect upon the projectile's trajectory. If the projectile is an electron or positron, the large mass differential between it and the nucleus can result in a significant deflection from its trajectory by the nuclear Coulomb field resulting in the emission of electromagnetic energy known as *bremsstrahlung* (braking radiation).

There are two fundamental differences between how photons and charged particles interact with matter that should be kept in mind. Being a boson, the photon interacts with matter through either direct absorption by a charged particle or through second-order effects such as Compton scatter. On the other hand, a charged particle will interact with other charged particles at extended distances through its surrounding Coulomb field (or, in another picture, the exchange of photons) resulting in the gradual transfer of energy to the medium and the eventual stopping of the particle. Thus, the cross section of the electromagnetic interaction between a charged particle and matter will be much greater than that for photons. This difference is made clearly evident by comparing the 10 cm mean free path of a 1 MeV photon in water with the few millimeter range of a 1 MeV electron in water.[1] The second fundamental difference between how photons and charged particles interact with matter is that, as a photon loses energy, its wavelength will increase and its frequency decrease but its speed remains constant.[2] Because a charged particle has mass, its speed is diminished through each instance of energy exchange to the medium until it reaches energies below the minimum ionization level and thus attains thermal equilibrium with the medium (although if the charged particle is a positron, it can annihilate with an electron either in-flight or once having been thermalized). Thus, unlike a photon, a charged particle will have a finite range.

[1]The photon mean free path is the reciprocal of the linear attenuation coefficient. The exact range of an electron in a medium is more difficult to define as, due to multiple scatter, its path will be tortuous.

[2]Assuming a constancy of photon speed ignores the frequency dependence of the medium's index of refraction. Such an effect, however, is negligible for photons with sufficient energy to ionize.

Should the charged particle be hadronic, additional energy loss channels become available as a result of the strong nuclear interaction between the projectile and the atomic nucleus. These include absorption, nuclear excitation, and even fragmentation of either the projectile or target nucleus. As previously discussed, the only hadron used in nuclear medicine is the α particle in therapy. α particles emitted in radioactive decay have kinetic energies of only a few MeV with the result that the α particle's penetration into tissue is small and its kinetic energy will be transferred to a very small volume leading to a high absorbed dose. The low kinetic energy will also mean that it is highly unlikely for the α particle to penetrate the nuclear Coulomb barrier in order to reach the ~1 fm separation from the nucleus in order to interact via the strong nuclear force and open up any nuclear energy loss channels.[3] Because there is no interest in nuclear interactions in clinical nuclear medicine, these will be neglected.

The mechanisms through which a moving charged particle loses energy in a medium through collisions can be broadly characterized in terms of the impact parameter, b, which is the perpendicular distance between the projectile's trajectory and a scattering center, such as an atom, as shown in Fig. 7.1 (in Bethe's theory, this categorization is in terms of the momentum transfer q which is approximately related to the impact parameter via $b \approx \hbar c/q$). For large impact parameters (i.e., b much greater than the atomic radius), the incident particle will interact with the atom as whole. The atom can be temporarily polarized (i.e., the electron cloud displaced from its equilibrium position) or atomic electrons excited into empty quantum states or, infrequently, into the continuum. As only small amounts of energy are transferred to the atom in such an encounter, it is described as a soft collision.

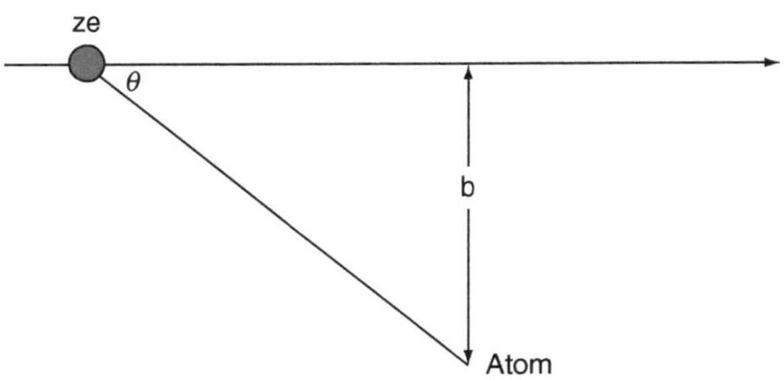

Fig. 7.1 Definition of the impact parameter b during an interaction between a passing charged particle and an atom

[3]Although it is possible for an α particle to penetrate the Coulomb barrier through quantum tunneling, but the likelihood of this occurring at kinetic energies of a few MeV typical of therapeutic nuclear medicine is extremely small.

If b is comparable to atomic dimensions, the particle will interact with a single atomic electron rather than the entire ensemble. This knock on or hard collision can result in the transfer of substantial energy to the atomic electron which can be ejected, leading to the atom's ionization and subsequent relaxation processes. The ejected electron, also referred to as a δ ray, is capable of carrying energy a considerable distance away from the event site, a feature of considerable practical importance in dosimetry calculations. As the energy transfer in this collision type is high, the interaction can be well-approximated using the assumption that the atomic electron is unbound. Because it is more likely that a given charged particle will be at an extended distance from a given atom rather than in close proximity to it, the probability of a soft collision occurring is high and that of a hard collision will be much lower. However, the net energy transfers due to soft and hard collisions are roughly equal in that the cumulative energy transferred through high-probability low energy transfers is approximately equal to that for low-probability high-energy transfers.

If the impact parameter b is much smaller than atomic dimensions, the particle interacts with the nucleus rather than the atomic electrons. A light incident particle, such as an electron or positron, can be deflected violently from its trajectory by the nuclear Coulomb field. An accelerated electric charge will emit electromagnetic radiation. This radiation is known as *bremsstrahlung* with the kinetic energy lost by the projectile carried away by the photon (assuming negligible nuclear recoil). The *bremsstrahlung* energy spectrum is continuous and decreases with increasing photon energy, thus reflecting the higher probability of small deflections and a greater production yield of low energy (or "soft") photons. The maximum *bremsstrahlung* photon energy in the spectrum equals the kinetic energy of the incident charged particle and is the result of the stopping of the particle and the complete conversion of its kinetic energy to radiation (assuming, again, zero nuclear recoil).

An obvious metric of interest in describing the energy loss of a moving charged particle is the rate, averaged over many particles, at which energy is transferred to the medium per unit path length. This ratio is known as the linear stopping power.[4] So as to remove the influence of the medium's physical density, it is useful to define the mass stopping power as the linear stopping power normalized to the physical density of the medium. Stopping powers due to soft and hard collisions with atomic electrons are summed to yield the collision stopping power.[5] The radiative stopping power is a measure of the rate of energy loss due to *bremsstrahlung* alone and the total collision stopping power is given by the sum of the collision and radiative stopping powers. As alluded to above, a refinement to the collision stopping power of significant practical interest to radiation dosimetry considers only that energy deposited locally (i.e., neglecting the energy carried away by a δ ray). The restricted stopping power, or linear energy transfer, is that fraction of the collision stopping power in which the kinetic energy of the δ ray is less than a specified cut-off value Δ.

Because a charged particle loses energy as it penetrates a medium, it slows down and (excluding positron annihilation in-flight or the nuclear absorption of hadronic projectiles) is eventually stopped. The range of the charged particle can be calculated as the integral of the reciprocal of the linear stopping power between the limits of zero and its initial kinetic energy. This is known as the continuous slowing-down approximation (CSDA) range. There are other refinements of the particle range which reflect the stochastic nature of a large number of interactions and multiple scattering events and these will be discussed. Both the stopping power and CSDA range are mean quantities resulting from large numbers of individual interactions which involve the transfer of small amounts of energy and small angle.

As with the discussion of photon interactions with matter in the previous chapter, the quantitative examples of charged particle interactions provided will be those of carbon (Z = 6) and lead (Z = 82) media. Extension of these elemental results to compound media such as soft tissue and bone is provided through Bragg's additivity rule. As the charged particles of interest to nuclear medicine are electrons, positrons, and α particles, these will be emphasized in these discussions.[6]

7.2 Coulomb Scattering With no Energy Transfer to the Medium

7.2.1 Introduction

Interactions between a charged particle and matter are frequently categorized in terms of elasticity. Whether or not a given collision can be defined as being elastic or inelastic depends upon the number of degrees-of-freedom available to the system. For example, the Coulomb scatter of an electron by a free electron (Møller scatter) is elastic as the kinetic energy lost by the projectile electron is made manifest as the target electron's postcollision kinetic energy. Similarly, a charged particle, having been scattered from an infinitely-massive scattering center is considered to have been elastically scattered as it retains its kinetic energy. In both of these cases, the pre- and postcollision (or sum of postcollision) kinetic energies are equal. Despite both being elastic, Møller scatter results in the transfer of energy to the medium whereas Coulomb scatter from an infinitely-massive scattering center does not. On the other hand, if the target electron in Møller scatter is an atomic electron bound to a nucleus, additional energy channels of ionization or excitation will arise and not all of the lost incident energy will appear in the exiting electron's kinetic energy[7] and, as such, the scatter from an atomic electron is referred to as inelastic scatter. While inelastic scatter will always result in the transfer of energy to the medium, only some elastic scatters can (e.g., the projectile and target masses are comparable). Hence, for dosimetry purposes, it is more reasonable to characterize charged particle scattering processes in terms of whether or not they result in the transfer of energy to the medium rather than elasticity.

[4]It has been argued that this nomenclature incorrect as the ratio has the units of force (i.e., 1 N = 1 J/m) rather than power (i.e., 1 W = 1 J/s). While this proposal is dimensionally correct, it does not seem realistic to accept it given the decades-long use of the term stopping power in the context of a charged particle slowing down.

[5]Also called the electronic stopping power.

[6]Protons will also figure in these derivations due to their historical significance.

[7]While, strictly speaking, such an interaction would be considered inelastic, a hard collision can be modeled as being elastic if the projectile's incident kinetic energy sufficiently exceeds the electron binding energy such that the latter can be ignored.

7.2.2 Elastic Coulomb Scatter

7.2.2.1 Spin-0 Projectiles

Unscreened Potential (Rutherford Scatter)

This was previously studied to yield the differential cross sections of the Coulomb scatter of a spin-0 charged projectile from an infinitely-massive charged scattering center given by (3.61), (3.62), and (3.65).

Screened Potential

The elastic Coulomb scatter differential cross sections in solid angle of (3.61) and (3.62) for an unscreened Coulomb potential diverge as $\theta \to 0$. This problem can be managed by recalling that small scattering angles are associated with large impact parameters through the relationship $b \propto \cot\theta/2$. At large impact parameters, the projectile will find the nuclear Coulomb potential screened by the atomic electrons: the screening parameter κ used in the derivation of the scattering amplitude and appearing in the integral of (3.58) is non-zero. Repeating the derivation of (3.61) with $\kappa \neq 0$ results in,

$$\frac{d\sigma_{Ruth}}{d\Omega} = \frac{1}{4}\frac{\left(\frac{zZ\,\alpha\hbar c}{p\beta}\right)^2}{\left[\left(\frac{\hbar c\kappa}{2p}\right)^2 + \sin^2\frac{\theta}{2}\right]^2} = \frac{1}{4}\frac{\left(\frac{zZ\,\alpha\hbar c}{p\beta}\right)^2}{\left[\left(\frac{\lambda\!\!\!^{-}\kappa}{2}\right)^2 + \sin^2\frac{\theta}{2}\right]^2} \quad \text{Screened potential} \tag{7.1}$$

where $\lambda\!\!\!^{-} = \hbar c/p$ is the reduced de Broglie wavelength of the projectile. Equation (7.1) can, for later convenience in the discussion of multiple elastic Coulomb scatter, be rewritten in the form using the half-angle identity, $\sin\theta/2 = \sqrt{(1-\cos\theta)/2}$,

$$\frac{d\sigma_{Ruth}}{d\Omega} = \frac{\left(\frac{zZ\,\alpha\hbar c}{p\beta}\right)^2}{\left[\frac{\chi_o^2}{2} + 1 - \cos\theta\right]^2} \tag{7.2}$$

where the effect of screening is described by the dimensionless parameter, $\chi_0 = \lambda\!\!\!^{-}\kappa$, the nature of which can be identified by rewriting the differential cross section in the small-angle approximation,

$$\frac{d\sigma_{Ruth}}{d\Omega} = 4\frac{\left(\frac{zZ\alpha\hbar c}{p\beta}\right)^2}{\left(\chi_0^2 + \theta^2\right)^2} \quad \text{(Small - angle approximation, screened potential)}. \tag{7.3}$$

χ_0 can be interpreted as a screening angle and represents a minimum scattering angle so that the differential cross section remains finite as the scattering angle $\theta \to 0$,

$$\frac{d\sigma_{Ruth}}{d\Omega} \to 4\frac{\left(\frac{zZ\,\alpha\hbar c}{p\beta}\right)^2}{\chi_0^4} \quad \text{as} \quad \theta \to 0 \tag{7.4}$$

as shown schematically in Fig. 7.2.

An expression for the screening angle χ_0 can be derived using the statistical Thomas–Fermi model of the atom by first equating κ to the reciprocal of that model's atomic radius which is,

$$R_{TF} = \frac{1}{2}\left(\frac{3\pi}{4}\right)^{2/3}\frac{r_\infty}{Z^{1/3}} \approx 0.885\frac{r_\infty}{Z^{1/3}}. \tag{7.5}$$

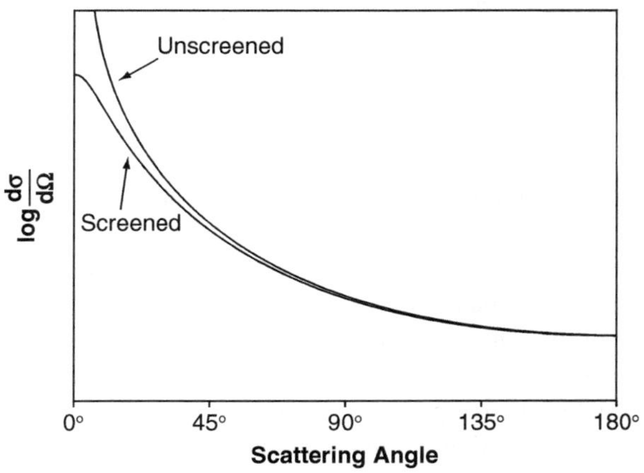

Fig. 7.2 Representative plot of screened and unscreened elastic Coulomb scatter differential cross section for spin-0 charged particles

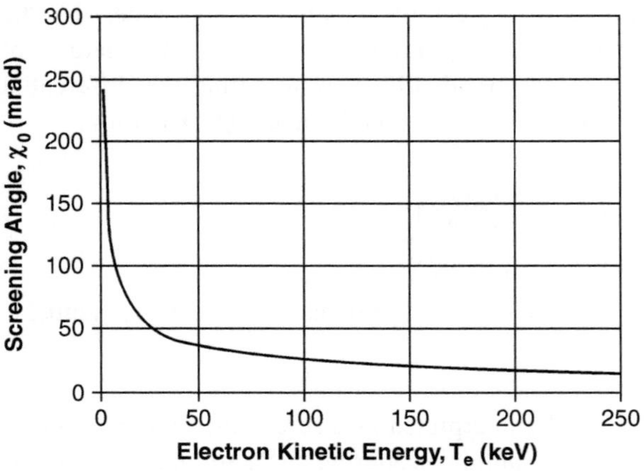

Fig. 7.3 Screening angle for electrons in carbon (neglecting spin)

From the definition of χ_0,

$$\begin{aligned}\chi_0 &= \lambda\!\!\!^{-}\kappa \\ &= \frac{\lambda\!\!\!^{-}}{R_{TF}} \\ &\approx 1.130\left(\frac{\lambda\!\!\!^{-}}{a_\infty}\right)Z^{1/3}\end{aligned} \tag{7.6}$$

This expression for the screening angle χ_0 is plotted in Fig. 7.3 for electrons in carbon as a function of electron kinetic energy. The range of electron kinetic energies shown in this graph is typical of that following Compton scatter, photoelectric absorption, or β-decay in nuclear medicine applications. It can be seen that for electrons with a kinetic energy of 50 keV or higher, the screening angle is less than about 30 mrad.

Mean Free Path Between Elastic Scatters

Using the above results, it is possible to evaluate the mean free path between each elastic scatter. This quantity, which is also referred to as the macroscopic cross section, is of particular importance in that it is equal to the reciprocal of the probability of an elastic scatter occurring per unit pathlength, or,

$$\lambda_{Ruth} = \left(\frac{A}{\rho N_A \sigma_{Ruth}}\right) \tag{7.7}$$

where N_A is Avogadro's number, A is the atomic number of the medium and ρ is its physical density. The total Rutherford cross section is found by integrating the Rutherford cross section over a solid angle of 4π steradians,

$$\begin{aligned}\sigma_{Ruth} &= \int d\Omega \frac{d\sigma_{Ruth}}{d\Omega} \\ &= 2\pi\left(\frac{zZ\alpha\hbar c}{p\beta}\right)^2 \int_{-1}^{1} \frac{d(\cos\theta)}{\left[\frac{\chi_0^2}{2}+1-\cos\theta\right]^2}\end{aligned} \tag{7.8}$$

to give,

$$\sigma_{Ruth} = \left(\frac{zZ\alpha\hbar c}{p\beta}\right)^2 \frac{16\pi}{\chi_0^2(\chi_0^2+4)}. \tag{7.9}$$

Figure 7.4 shows the elastic scatter mean free path as a function of electron kinetic energy in a medium representative of carbon (i.e., Z = 6, A = 12 and $\rho = 2\ g/cm^3$).

The combination of this small mean free path length and the θ^{-4} dependence of the Rutherford cross section leads to the dominance of forward-directed multiple elastic scattering of charged particles traversing a medium.

Elastic Scatter from an Atom

Instead of modeling the atom as a nucleus of infinite mass with the surrounding atomic electrons treated as a continuous screening function in radial distance through the use of a Yukawa-type potential, it is possible to explicitly calculate for the discrete contributions of the atomic electrons. Spin is still neglected. The combined interaction potential is the sum of the Coulomb potentials due to the nucleus and the individual atomic electrons,

$$U = z\alpha\hbar c\left[\frac{Z}{R} - \sum_{j=1}^{Z}\frac{1}{|\mathbf{R}-\mathbf{r}_j|}\right] \tag{7.10}$$

where **R** is the position vector of the projectile and $\mathbf{r}_j$ is that of the jth electron. The origin of the system is fixed at the center of the atomic nucleus and the overall system vector is $\mathbf{r} = (\mathbf{R}, \mathbf{r}_1, \mathbf{r}_2 \ldots \mathbf{r}_Z)$. The pre- and

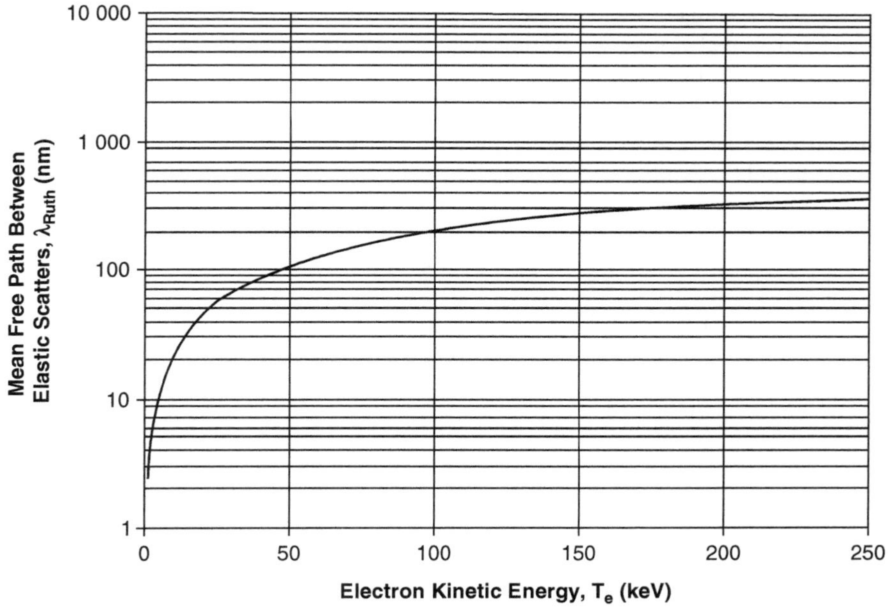

Fig. 7.4 Elastic Coulomb scatter mean free path for electrons in a medium representative of carbon (neglecting spin)

postscattering states of the projectile-atom system are, in bra-ket notation,

$$\begin{aligned}\langle \mathbf{r}|\mathbf{p},0\rangle &= \frac{1}{\sqrt{L^3}} e^{i\frac{\mathbf{p}\bullet\mathbf{R}}{\hbar c}}|0\rangle \\ &= \frac{1}{\sqrt{L^3}} e^{i\frac{\mathbf{p}\bullet\mathbf{R}}{\hbar c}}\psi_0(\mathbf{r}_1,\mathbf{r}_2,\ldots\mathbf{r}_Z)\end{aligned} \quad (7.11)$$

and

$$\langle \mathbf{r}|\mathbf{p}',0\rangle = \frac{1}{\sqrt{L^3}} e^{i\frac{\mathbf{p}'\bullet\mathbf{R}}{\hbar c}}\psi_0(\mathbf{r}_1,\mathbf{r}_2,\ldots\mathbf{r}_Z) \quad (7.12)$$

where $\psi_0(\mathbf{r}_1,\mathbf{r}_2,\ldots\mathbf{r}_Z)$ is the ground state atomic wavefunction and L^3 is the usual volume containing the system and which is to be used for normalization. As the scatter is elastic, the atomic ground state wavefunction appears in both initial and final system states. The differential cross section is obtained in the usual way from the transition rate given by Fermi's Golden Rule No. 2,

$$\lambda_{fi} = \frac{2\pi}{\hbar}|\langle \mathbf{p}',0|U|\mathbf{p},0\rangle|^2\rho_f$$

where the phase space factor ρ_f is the density of final states per energy interval dT,

$$\rho_f\, dT' = \left(\frac{L}{2\pi\hbar c}\right)^3 d^3\mathbf{p}'. \quad (7.13)$$

By expanding the vector differential $d^3\mathbf{p}'$,

$$\begin{aligned}\rho_f\, dT' &= \left(\frac{L}{2\pi\hbar c}\right)^3 2\pi p'^2\, dp'\, d(\cos\theta) \\ &= \frac{1}{4\pi^2}\left(\frac{L}{\hbar c}\right)^3 p'^2\, dp'\, d(\cos\theta).\end{aligned} \quad (7.14)$$

The phase–space factor is,

$$\begin{aligned}\rho_f &= \frac{1}{4\pi^2}\left(\frac{L}{\hbar c}\right)^3 p'^2\left(\frac{dp'}{dT'}\right) d(\cos\theta) \\ &= \frac{1}{4\pi^2\beta'}\left(\frac{L}{\hbar c}\right)^3 p'^2 d(\cos\theta)\end{aligned} \quad (7.15)$$

where $(dp'/dT') = 1/\beta'$ has been used. From Chap. 3, $|d(\cos\theta)| = q/p^2\, dq$, where q is the momentum transfer and the phase–space factor then becomes,

$$\begin{aligned}\rho_f &= \frac{1}{4\pi^2\beta'}\left(\frac{L}{\hbar c}\right)^3\left(\frac{p'}{p}\right)^2 q\, dq \\ &= \frac{1}{4\pi^2\beta}\left(\frac{L}{\hbar c}\right)^3 q\, dq\end{aligned} \quad (7.16)$$

where $\beta = \beta'$ and $p = p'$ due to the elasticity of the scatter and the negligible recoil of the scattering center.

The matrix element for both Coulomb interactions is next calculated for by separating the interactions into the projectile electron and projectile-nucleus components. The matrix element due to the Coulomb interaction between the projectile and the atomic electrons is found by expanding it in position space,[8]

$$
\begin{aligned}
\langle \mathbf{p}',0|U|\mathbf{p},0\rangle &= \int d^3\mathbf{r}\, d^3\mathbf{r}' \langle \mathbf{p}',0|\mathbf{r}\rangle\langle \mathbf{r}|U|\mathbf{r}'\rangle\langle \mathbf{r}'|\mathbf{p},0\rangle \\
&= \int d^3\mathbf{r}\, d^3\mathbf{r}' \langle \mathbf{p}',0|\mathbf{r}\rangle U(\mathbf{r})\langle \mathbf{r}'|\mathbf{p},0\rangle \\
&= \frac{1}{L^3}\int d^3\mathbf{R}\, e^{i\frac{\mathbf{q}\bullet\mathbf{R}}{\hbar}} \prod_{j=1}^{Z} \int d^3\mathbf{r}_j\, \psi_0^*(\mathbf{r}_1,\mathbf{r}_2\ldots\mathbf{r}_Z)U(\mathbf{r}) \\
&\quad \times \psi_0(\mathbf{r}_1,\mathbf{r}_2\ldots\mathbf{r}_Z) \\
&= -\frac{z\alpha\hbar c}{L^3}\prod_{j=1}^{Z} \int d^3\mathbf{r}_j\, \psi_0^*(\mathbf{r}_1,\mathbf{r}_2\ldots\mathbf{r}_Z)\psi_0(\mathbf{r}_1,\mathbf{r}_2\ldots\mathbf{r}_Z) \\
&\quad \times \int d^3\mathbf{R}\frac{e^{i\frac{\mathbf{q}\bullet\mathbf{R}}{\hbar c}}}{|\mathbf{R}-\mathbf{r}_j|} \\
&= -\frac{z\alpha\hbar c}{L^3}\prod_{j=1}^{Z} \int d^3\mathbf{r}_j\, \psi_0^*(\mathbf{r}_1,\mathbf{r}_2\ldots\mathbf{r}_Z)\psi_0(\mathbf{r}_1,\mathbf{r}_2\ldots\mathbf{r}_Z) \\
&\quad \times \int d^3\mathbf{R}'\frac{e^{i\frac{\mathbf{q}\bullet(\mathbf{R}'+\mathbf{r}_j)}{\hbar c}}}{|\mathbf{R}'|} \\
&= -\frac{z\alpha\hbar c}{L^3}\prod_{j=1}^{Z} \int d^3\mathbf{r}_j\, \psi_0^*(\mathbf{r}_1,\mathbf{r}_2\ldots\mathbf{r}_Z)\psi_0(\mathbf{r}_1,\mathbf{r}_2\ldots\mathbf{r}_Z) \\
&\quad \times e^{i\frac{\mathbf{q}\bullet\mathbf{r}_j}{\hbar}} \int d^3\mathbf{R}'\frac{e^{i\frac{\mathbf{q}\bullet\mathbf{R}'}{\hbar c}}}{|\mathbf{R}'|}
\end{aligned} \tag{7.17}
$$

The solution to the integral was given by (3.59) and the matrix element is,

$$
\begin{aligned}
\langle \mathbf{p}',0|U|\mathbf{p},0\rangle &= -4\pi\frac{z\,\alpha\hbar c}{L^3}\left(\frac{\hbar c}{q}\right)^2\prod_{j=1}^{Z} \int d^3\mathbf{r}_j\, \psi_0^*(\mathbf{r}_1,\mathbf{r}_2\ldots\mathbf{r}_Z) \\
&\quad \times e^{i\frac{\mathbf{q}\bullet\mathbf{r}_j}{\hbar c}}\psi_0(\mathbf{r}_1,\mathbf{r}_2\ldots\mathbf{r}_Z) \\
&= -4\pi\frac{z\,\alpha\hbar c}{L^3}\left(\frac{\hbar c}{q}\right)^2\left\langle 0\left|\sum_{j=1}^{Z} e^{i\frac{\mathbf{q}\bullet\mathbf{r}_j}{\hbar c}}\right|0\right\rangle \\
&= -4\pi\frac{z\,\alpha\hbar c}{L^3}\left(\frac{\hbar c}{q}\right)^2 F_0(\mathbf{q},Z)
\end{aligned} \tag{7.18}
$$

where $F_0(\mathbf{q},Z)$ is the elastic scattering form factor, which was seen before in the derivation of the photon coherent scatter cross section (and in a different guise in elastic scatter from the nucleus). Assuming a continuous electron spatial distribution rather than the discrete set of above, the form factor is the Fourier transform of the electron density. Here, the normalization of,

$$
\begin{aligned}
F(0,Z) &= \int d^3\mathbf{r}\, \rho_e(r) \\
&= Z
\end{aligned} \tag{7.19}
$$

will be imposed. As an example, following from the Yukawa approximation of the screened nuclear Coulomb potential, the atomic electron density can be modeled by,

$$\rho_e(r) = \Theta\, e^{-\kappa r} \tag{7.20}$$

where the normalization constant Θ is found via,

$$
\begin{aligned}
\int d^3\mathbf{r}\, \rho_e(r) &= \Theta\int d^3\mathbf{r}\, e^{-\kappa r} \\
&= 2\pi\,\Theta\int_0^\infty dr\, r^2\, e^{-\kappa r} \\
&= 4\pi\frac{\Theta}{\kappa^3}.
\end{aligned} \tag{7.21}
$$

From these, $\Theta = \kappa^3 Z/4\pi$ and the electron density is now written in the form,

$$\rho_e(r) = \frac{\kappa^3 Z}{4\pi}e^{-\kappa r}. \tag{7.22}$$

[8]This implicitly neglects any electrostatic correlations between the atomic electrons.

The expression for the elastic scattering form factor is,

$$\begin{aligned}F(\mathbf{q},Z) &= \frac{\kappa^3 Z}{4\pi}\int d^3\mathbf{r}\, e^{-\left(\kappa r + i\frac{\mathbf{q}\bullet\mathbf{r}}{\hbar c}\right)} \\ &= \frac{\kappa^3 Z}{2}\int dr\, r^2 e^{-\kappa r}\int d(\cos\theta) e^{-i\frac{qr\cos\theta}{\hbar c}} \\ &= \frac{\kappa^3 Z}{2}\frac{\hbar c}{q}\int dr\, r\, e^{-\kappa r}\sin\frac{qr}{\hbar c} \\ &= Z\frac{\kappa^4}{\left(\kappa^2 + \left(\frac{q}{\hbar c}\right)^2\right)^2}.\end{aligned} \tag{7.23}$$

Using the momentum transfer expression of (3.52), the elastic form factor can be written in terms of the scattering angle as,

$$\begin{aligned}F_0(\theta,Z) &= Z\frac{\kappa^4}{\left(\kappa^2 + \frac{4p^2}{(\hbar c)^2}\sin^2\frac{\theta}{2}\right)^2} \\ &= \frac{Z}{\left(1 + \left(\frac{2p}{\hbar c\kappa}\sin\frac{\theta}{2}\right)^2\right)^2} \\ &= \frac{Z}{\left(1 + \left(\frac{2}{\chi_0}\sin\frac{\theta}{2}\right)^2\right)^2}.\end{aligned} \tag{7.24}$$

Writing the form factor in this manner demonstrates that $F(\theta = 0, Z) = Z$ at its maximum and decreases to $Z/\left(1 + 4/\chi_0^2\right)^2$ for $\theta = \pi$ (or to zero for the unscreened potential which corresponds to $\chi_0 = 0$). Now return to the evaluation the component of the matrix element due to the projectile-nuclear Coulomb potential,

$$\begin{aligned}\langle \mathbf{p}',0|U|\mathbf{p},0\rangle &= \int d^3\mathbf{r}\, d^3\mathbf{r}' \langle \mathbf{p}',0|\mathbf{r}\rangle\langle \mathbf{r}|U|\mathbf{r}'\rangle\langle \mathbf{r}'|\mathbf{p},0\rangle \\ &= \frac{zZ\alpha\hbar c}{L^3}\int d^3\mathbf{R}'\frac{e^{i\left(\frac{\mathbf{q}\bullet\mathbf{R}'}{\hbar c}\right)}}{R'}\prod_{j=1}^{Z} \\ &\quad \int d^3\mathbf{r}_j\, \psi_0^*(\mathbf{r}_1,\mathbf{r}_2\ldots\mathbf{r}_Z)\,\psi_0(\mathbf{r}_1,\mathbf{r}_2\ldots\mathbf{r}_Z) \\ &= \frac{zZ\alpha\hbar c}{L^3}\int d^3\mathbf{R}'\frac{e^{i\left(\frac{\mathbf{q}\bullet\mathbf{R}'}{\hbar c}\right)}}{R'} \\ &= 4\pi\frac{zZ\alpha\hbar c}{L^3}\left(\frac{\hbar c}{q}\right)^2\end{aligned} \tag{7.25}$$

The complete matrix element is the sum of the projectile electron and projectile-nucleus components,

$$\langle \mathbf{p}',0|U|\mathbf{p},0\rangle = 4\pi\frac{z\,\alpha\hbar c}{L^3}\left(\frac{\hbar c}{q}\right)^2(Z - F_0(\mathbf{q},Z)). \tag{7.26}$$

The transition rate is,

$$\begin{aligned}\lambda_{fi} &= \frac{2\pi}{\hbar}\left|\langle \mathbf{p}',0|U|\mathbf{p},0\rangle\right|^2\rho_f \\ &= \frac{8\pi c}{\beta}\frac{(z\,\alpha\hbar c)^2}{q^3L^3}\left|Z - F_0(\mathbf{q},Z)\right|^2 dq.\end{aligned} \tag{7.27}$$

The differential cross section in momentum transfer and scattered kinetic energy is obtained from the transition rate in the usual way by normalizing it to the incident particle flux,

$$d\sigma = \frac{L^3}{\beta c}\lambda_{fi} \tag{7.28}$$

leading to,

$$\frac{d\sigma}{dq} = \frac{8\pi}{q^3}\left(\frac{z\,\alpha\hbar c}{\beta}\right)^2\left|Z - F_0(\mathbf{q},Z)\right|^2 \tag{7.29}$$

A comparison of this result with the differential cross section in momentum transfer obtained previously for a single scattering center demonstrates that the Z^2 term for the unscreened potential has been replaced by $|Z - F_0(\mathbf{q},Z)|^2$. That is, the contributions of the atomic electrons, through interference as described by the elastic scattering form factor, lead to a reduction of the cross section and, in particular, the elastic scatter cross section is subject to the destructive interference between the nuclear and electronic amplitudes.

It has been seen that for zero momentum transfer (i.e., zero scattering angle), the elastic form factor is equal to Z. In the general case of small momentum transfer, the elastic form factor can be expanded to second order,

$$\begin{aligned}F(\mathbf{q},Z) &= \left\langle 0\left|\sum_{j=1}^{Z} e^{i\frac{\mathbf{q}\bullet\mathbf{r}_j}{\hbar c}}\right|0\right\rangle \\ &\approx \left\langle 0\left|\sum_{j=1}^{Z}\left(1 + i\left(\frac{\mathbf{q}\bullet\mathbf{r}_j}{\hbar c}\right) - \frac{1}{2}\left(\frac{\mathbf{q}\bullet\mathbf{r}_j}{\hbar c}\right)^2\right)\right|0\right\rangle.\end{aligned} \tag{7.30}$$

As the momentum transfer is small, then $\cos\theta \approx 1$ so $\mathbf{q} \bullet \mathbf{r}_j \approx q r_j$. In addition, the matrix element of the first term of the expansion is equal to zero due to symmetry. Hence, for small q,

$$F(\mathbf{q}, Z) \approx Z - \frac{1}{2}\left(\frac{q}{\hbar c}\right)^2 \sum_{j=1}^{Z} \langle 0|r_j^2|0\rangle \approx Z - \frac{1}{2}\left(\frac{q}{\hbar c}\right)^2 \overline{r^2} \tag{7.31}$$

and the differential cross section in momentum transfer is,

$$\frac{d\sigma}{dq} = \frac{8\pi}{q^3}\left(\frac{z\,\alpha\hbar c}{\beta}\right)^2 |Z - F_0(\mathbf{q}, Z)|^2 \approx 2\pi\left(\frac{z\,\alpha\hbar c}{\beta}\right)^2 \frac{q}{(\hbar c)^4}\left(\overline{r^2}\right)^2 \quad \text{(small momentum transfer).} \tag{7.32}$$

The differential cross section with solid angle can next be derived in the usual way and by replacing $F_0(\mathbf{q}, Z)$ with the $F_0(\theta, Z)$ of,

$$\frac{d\sigma}{d\Omega} = \frac{8\pi}{q^3}\frac{(z\,\alpha\hbar c)^2}{\beta}|Z - F_0(\theta, Z)|^2\left(\frac{p^2}{2\pi q}\right) = \frac{4p^2}{q^4}\frac{(z\alpha\hbar c)^2}{\beta}|Z - F_0(\theta, Z)|^2 \tag{7.33}$$

As $q^4 = 16p^4 \sin^4\theta/2$ the differential cross section can be written in the form,

$$\frac{d\sigma}{d\Omega} = \frac{1}{4}\left(\frac{z\,\alpha\hbar c}{p\beta}\right)^2 \frac{|Z - F_0(\theta, Z)|^2}{\sin^4 \theta/2} \tag{7.34}$$

which is the analog to the point scattering center result.

Comparison of Atomic Scattering Results

In later consideration of charged particle transport, in particular multiple scattering, the elastic Coulomb scatter process will be significant. Hence, a review of the elastic Coulomb scatter differential cross sections calculated so far for an atomic target is now provided. Figure 7.5 shows the differential cross section in solid angle as a function of scattering angle for a 100-keV electron in carbon (the electron spin is, of course, neglected here but its effects will be considered explicitly in the next subsection). The two calculations are for the screened potential and the elastic form factor with the parameter κ defined as the reciprocal of the Thomas–Fermi atomic radius. For scattering angles exceeding about 3°, the two methods yield the same differential cross section, but diverge at smaller angles, both approaching approach a finite value at $\theta = 0$ due to screening. The differential cross section

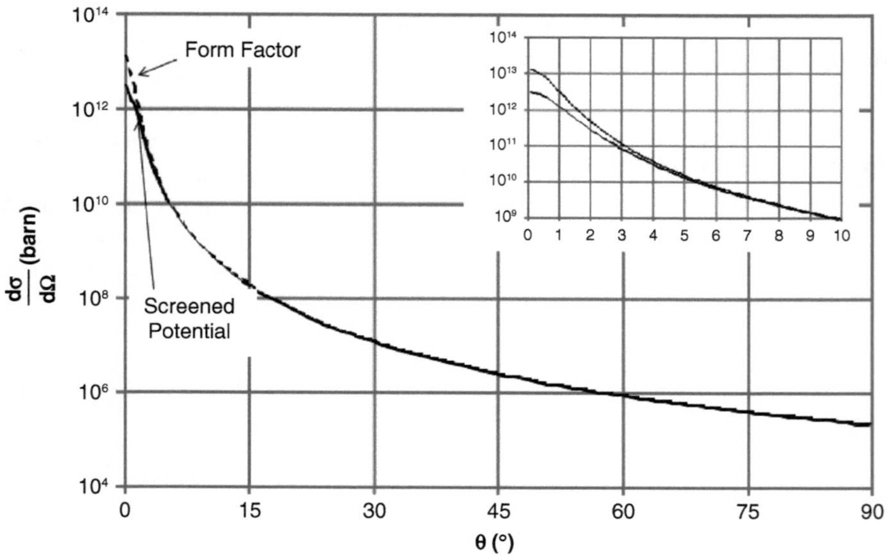

Fig. 7.5 Elastic Coulomb scatter differential cross sections for the screened nuclear Coulomb potential (*solid line*) and the discrete contributions of the atomic electrons (form factor; *dotted line*) as functions of the scattering angle for a 100 keV electron scattered from a carbon atom. Spin is neglected. The inset shows the differential cross sections for scattering angles below 10°

calculated by explicitly including the contributions of the individual atomic electrons only becomes significantly greater than that assuming a continuous screening of the potential for scattering angles of less than 3°. However, it is clear that elastic Coulomb scatter is highly forward directed with a very high differential cross section (exceeding 10^8 b for scattering angles of less than 1°). This feature is highly significant in describing the transport of charged particles in a medium. Charged particles will, as a result, undergo multiple scatters and result in a probability distribution describing both their spatial distribution and their angular direction. This will be discussed in detail later.

7.2.2.2 Spin-1/2 Projectiles

The Mott differential cross sections (Mott 1929, 1932) derived in Sect. 3.2.7 describes the interactions of a spin-1/2 projectile, such as an electron or positron, with a Coulomb field. The relevant expressions are given by (3.89) and (3.90).

7.3 Coulomb Scattering With Energy Transfer to the Medium

7.3.1 Introduction

While the results of the previous section are important input to charged particle transport calculations (through, e.g., evaluation of multiple scatter or range straggling), they do not lead to energy transfer. On the other hand, for example, following Møller scatter with an atomic electron, the atom is left in an excited or even ionized state. Energy is thus transferred to the medium. This section focuses on the inelastic collisions between a projectile and an atom (i.e., impact parameters comparable to or greater than atomic dimensions).

7.3.2 Rutherford Collision Formula

The initial derivation is of the classical formula describing the energy transfer to an electron via a Coulomb interaction with a moving heavy charged particle using the following assumptions:

- The energy loss is local (i.e., emission of electromagnetic energy is neglected).
- The particle is not deflected from its straight-line trajectory (i.e., the impulse approximation is used and multiple scatter is neglected).
- The ion is not "dressed" (i.e., it is completely stripped of electrons).[9]
- The speed of the particle is much higher than the orbital speed of any atomic electron (allowing the electron to be treated as being at rest) but is sufficiently low that nonrelativistic kinematics can be assumed.

Consider the passage of a particle with charge ze, speed v, and a mass $m \gg m_e$ in a medium with physical density ρ, atomic number Z, and atomic mass number A. It interacts with an electron through the Coulomb potential at an impact parameter b, as shown in Fig. 7.1. The force felt by the target electron is decomposed into two orthogonal components, one parallel, and the other perpendicular, to the particle's trajectory,

$$\mathbf{F} = -e\left(\mathbf{E}_{||} + \mathbf{E}_{\perp}\right) \tag{7.35}$$

where $\mathbf{E}_{||}$ and $\mathbf{E}_{\perp}$ are the two orthogonal electric field components at the position of the electron,[10]

$$\mathbf{E}_{||} = -\left(\frac{z\,(\alpha\hbar c)}{e}\right)\frac{\gamma v t}{\left[b^2 + (\gamma v t)^2\right]^{\frac{3}{2}}}\hat{\mathbf{v}}_{||} \tag{7.36}$$

$$\mathbf{E}_{\perp} = \left(\frac{z\,(\alpha\hbar c)}{e}\right)\frac{\gamma b}{\left[b^2 + (\gamma v t)^2\right]^{\frac{3}{2}}}\hat{\mathbf{v}}_{\perp} \tag{7.37}$$

---

[9]The presence of electrons in an ion projectile will have two effects upon the rate of energy loss. The first is that the effective charge will be reduced to the screening by these electrons. The second is that the excitation or ionization of the projectile itself will provide an additional energy loss channel.

[10]These transformed values of the electric field result from the Lorentz transformation corresponding to the boost along an axis with a speed βc for the particle in one reference frame to that containing the electron at rest. These are provided here without derivation, but one may refer to, for example, that provided by Jackson (1999).

$\hat{\mathbf{v}}_{||}$ and $\hat{\mathbf{v}}_{\perp}$ are the unit-vectors in the directions parallel and perpendicular to the projectile's trajectory. The magnitudes of these two electric field components are plotted in Fig. 7.6 as a function of time (weighted by $\gamma v/b$, which is a constant in this impulse approximation since the projectile speed is considered to be unaffected). The momentum transferred to the electron is given by the integral over all time of the force that the electron is subject to,

$$q_{||} = -e \int_{-\infty}^{\infty} dt\, E_{||} \quad (7.38)$$

$$q_{\perp} = -e \int_{-\infty}^{\infty} dt\, E_{\perp}. \quad (7.39)$$

It is clear from both the figure and the γvt multiplicative term in the expression for $\mathbf{E}_{||}$ that the net

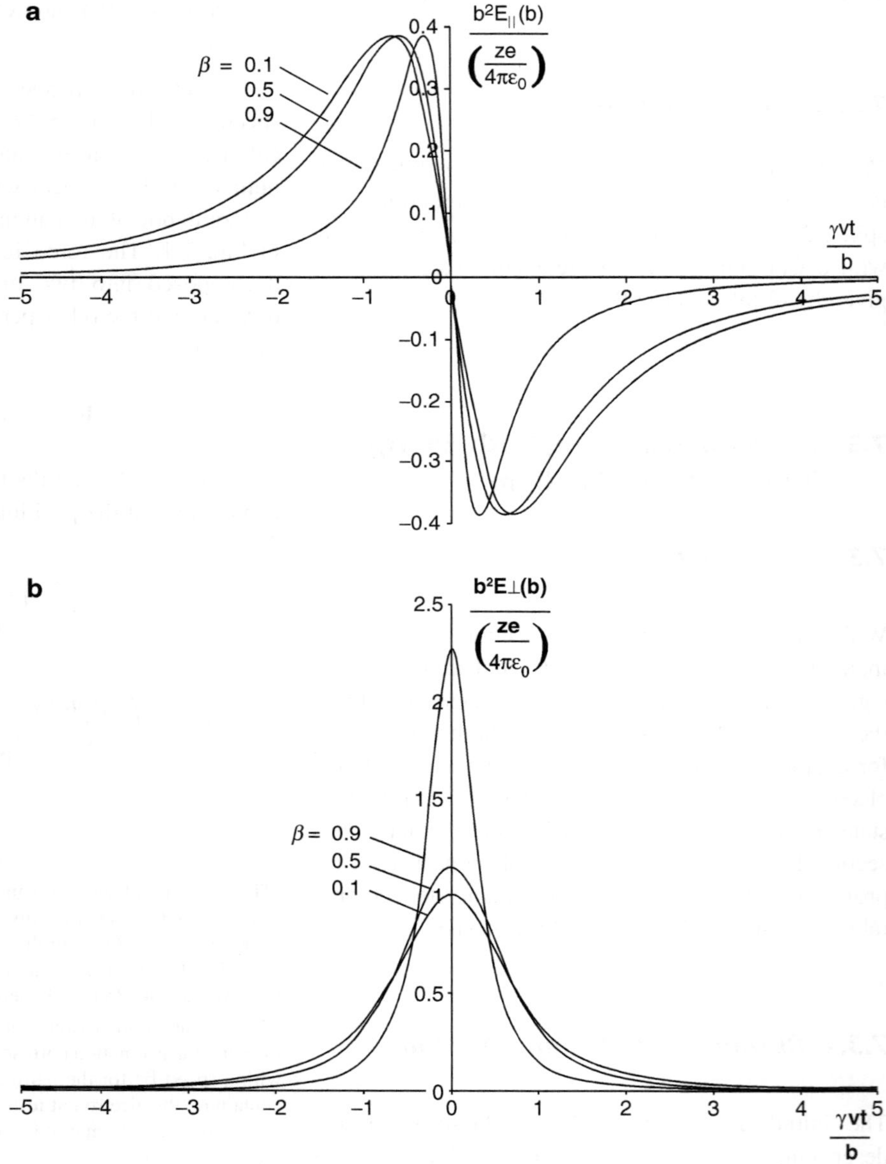

Fig. 7.6 Electric field components parallel and perpendicular to the charged-particle trajectory for different particle speeds

momentum component parallel to the particle trajectory is equal to zero and that the net non-zero momentum transfer will be that perpendicular to the particle's trajectory,

$$
\begin{aligned}
q &= q_{\perp} \\
&= -e \int_{-\infty}^{\infty} dt\, E_{\perp} \\
&= -(z\alpha\hbar c)\gamma b \int_{-\infty}^{\infty} \frac{dt}{\left[b^2 + (\gamma v t)^2\right]^{\frac{3}{2}}} \\
&= -(z\alpha\hbar c)\frac{\gamma}{b^2} \int_{-\infty}^{\infty} \frac{dt}{\left[1 + \left(\frac{\gamma v t}{b}\right)^2\right]^{\frac{3}{2}}} \\
&= -(z\alpha\hbar c)\frac{1}{bv} \int_{-\infty}^{\infty} \frac{dx}{[1 + x^2]^{\frac{3}{2}}} \\
&= -\frac{(2z\alpha\hbar c)}{bv}
\end{aligned}
\tag{7.40}
$$

where the substitution of variables $x = \gamma vt/b$ has been used to solve the integral. As the recoil electron is treated as being nonrelativistic, the energy transferred to it is,

$$
\begin{aligned}
Q &= \frac{q^2}{2m_e} \\
&= \frac{(2z\alpha\hbar c)^2}{2m_e v^2 b^2} \\
&= 2m_e\left(\frac{zr_0}{\beta b}\right)^2.
\end{aligned}
\tag{7.41}
$$

It can be immediately seen that the energy transfer decreases with $1/\beta^2$ (i.e., slower particles lose energy more rapidly) and with $1/b^2$ (i.e., energy loss decreases with increasing distance from the atom) and increases with z^2. This result can be used to demonstrate that collision energy losses will be dominated by the interactions of the projectile with atomic electrons rather than with the nucleus (whereas it is the converse with elastic scatter which is dominated by the interaction with the nucleus). The energy transfer to a nucleus of charge Ze and mass Am_N, where m_N is the nucleon mass, will scale from that to an electron by the ratio,

$$
\frac{z^2Z^2}{Am_N}\frac{m_e}{z^2} \approx \frac{Z}{2}\frac{m_e}{m_N} \approx \frac{Z}{3760}
$$

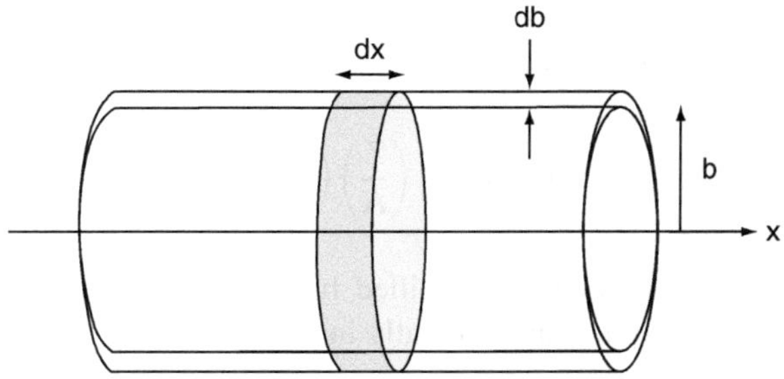

Fig. 7.7 Geometry for calculation of the Rutherford collision formula

As an example, the energy transferred to the nucleus of a carbon atom will be less than 0.2% of that transferred to an atomic electron. Nonradiative energy transfer is the result of an electromagnetic interaction with the nucleus will thus be neglected.

It is now necessary to move beyond consideration of the energy transfer to a single electron to the more realistic case of multiple energy transfers to an ensemble of electrons. Assume that the particle is moving through a sea of electrons and then isolate those electrons contained within a cylinder with its axis coincident with the particle's trajectory, as shown in Fig. 7.7. Because of the correspondence between the energy transfer and the impact parameter, the probability of an energy transfer between Q and Q + dQ to an electron occurring in a given differential pathlength dx of the traveling particle is equal to the probability of a collision with an impact parameter between b and b + db. This latter probability is equal to the number of electrons contained within the differential volume formed by the cylindrical shell of thickness db and length dx,

$$
\Pr(b)\, db\, dx = \rho N_A\left(\frac{Z}{A}\right) 2\pi b\, db\, dx \tag{7.42}
$$

where $\rho N_A(Z/A)$ is the electron number density of the medium. The product $b\,db$ is obtained by differentiating (7.41),

$$
|b\, db| = m_e\left(\frac{zr_0}{\beta}\right)^2 \frac{dQ}{Q^2} \tag{7.43}
$$

From these two expressions, the probability of an energy transfer between Q and Q + dQ to an electron occurring in a given differential pathlength dx of the traveling particle is,

$$\Pr(b) = \left(2\pi\, m_e\, N_A\, r_0^2\right)\left(\frac{Z}{A}\right)\left(\frac{z}{\beta}\right)^2 \frac{dQ}{Q^2}\,\rho\, dx \tag{7.44}$$

This expression is simplified by defining a constant which will appear frequently in this chapter,

$$\begin{aligned} C &= 2\pi\, m_e N_A r_0^2 \\ &= 0.154\,\text{MeV atom cm}^2/\text{mole} \end{aligned} \tag{7.45}$$

to give,

$$\Pr(b) = C\left(\frac{Z}{A}\right)\left(\frac{z}{\beta}\right)^2 \frac{dQ}{Q^2}\,\rho\, dx \tag{7.46}$$

This is Rutherford's formula for energy loss and demonstrates the major features of collision energy loss through electromagnetic interactions with atomic electrons:

- The probability of a collision with energy transfer Q is proportional to Q^{-2}, demonstrating that soft collisions (small Q) are more likely than hard collisions.
- The probability is proportional to β^{-2} or, crudely, is greater for short collision times (i.e., the probability of an energy transfer decreases if the electrons are allowed to react adiabatically).
- The probability is proportional to the electron density of the medium.
- The probability is independent of projectile mass
- The probability increases with the square of the incident particle charge, z^2.

7.3.3 Soft Collision Stopping Power

7.3.3.1 Introduction

The soft collision stopping power is the energy loss due to soft collisions per unit distance traveled by a projectile in a medium. In this subsection, the two most prominent theories as developed by Bohr and Bethe are derived. In Bohr's model, energy is transferred to the atomic electrons which are treated as charged harmonic oscillators. The energy loss is calculated using classical electrodynamics for a heavy projectile interacting with a single electron of a single atom. On the other hand, the Bethe theory is the quantum-mechanical description of the inelastic projectile-atom collision.

7.3.3.2 Bohr Theory

Introduction

The main assumptions of the Bohr theory of soft collisions are that, firstly, the nucleus has infinite mass and, secondly, the projectile transfers energy to harmonically-bound atomic electrons. The Bohr mass soft collision stopping power is derived in three steps:

- Deriving the role that the impact parameter plays in separating the regions of soft- and hard collisions.
- Deriving the energy transfer to a harmonically-bound electron using the assumption that only the projectile's electric field acts upon the electron and that the field is spatially uniform at the position of the electron.
- Using these results to derive the energy transferred per unit pathlength traveled.

The Bohr theory uses classical mechanics to calculate the energy transfer due to soft collisions with the atomic electron orbital frequencies dictating the energy transfer to the atomic electrons in the adiabatic limit.

Impact Parameter

The impact parameter, b, is fundamental to the Bohr theory. For b greater than some maximum value, b_{max}, the projectile will be unable to transfer sufficient energy to the atom in order to excite or ionize it. On the other hand, for b less than some minimum value, b_{min}, the particle will interact with an individual electron (i.e., undergo a hard collision). As a result, b_{min} and b_{max} set the boundaries for a soft collision during which the projectile interacts with the entire ensemble of atomic electrons.

One method of estimating b_{min} is to use the impulse approximation implicit to the Rutherford formula in which the target electron is assumed to be stationary or, in effect, does not recoil a significant distance compared to the impact parameter. One calculates the recoil distance and, by fixing it to be much smaller than the impact parameter, obtain b_{min}. As the time domain over which an electron with speed $v = \beta c$ experiences the electrostatic force of the projectile is of the order of $b/\gamma v$, the recoil distance of the electron will be of the order of $qb/m_e\gamma v$, where q is the momentum transfer. The impulse approximation requires this to be much less than the impact parameter,

$$\frac{q}{m_e}\frac{b}{\gamma v} \ll b$$

to give

$$\frac{q}{m_e \gamma v} \ll 1. \tag{7.47}$$

Inserting the expression for the momentum transfer gives,

$$\frac{2z\,\alpha\hbar c}{b m_e \gamma v^2} \ll 1. \tag{7.48}$$

Rearrangement gives the inequality in terms of the ratio of the electrostatic potential and kinetic energies,

$$\frac{1}{\gamma}\frac{\left(\frac{z\,\alpha\hbar c}{b}\right)}{\left(\frac{1}{2}m_e v^2\right)} \ll 1 \tag{7.49}$$

or,

$$\left(\frac{1}{\gamma}\frac{2z}{\beta^2}\right)\frac{r_0}{b} \ll 1 \tag{7.50}$$

from which an expression for b_{min} can be defined,

$$b_{min} = 2\left(\frac{z}{\gamma\beta^2}\right)r_0 \tag{7.51}$$

As expected, b_{min} decreases with increasing projectile speed (or, equivalently, decreasing de Broglie wavelength).

Another approach to calculating b_{min} recognizes that only hard collisions will occur at impact parameters below this value. From Chap. 2, the maximum energy transferred to an electron as a result of a head-on collision with a massive projectile is $2m_e\gamma^2\beta^2$. Using (7.41),

$$2m_e\gamma^2\beta^2 = 2m_e\left(\frac{zr_0}{\beta b_{min}}\right)^2$$

and solving for b_{min} gives,

$$b_{min} = \left(\frac{z}{\gamma\beta^2}\right)r_0 \tag{7.52}$$

Two classical mechanical proposals for b_{min} have been derived, each differing by a factor of 2. As both show that b_{min} is associated with increasing particle momentum (and, hence, reduced de Broglie wavelength), the quantum-mechanical nature of the interaction cannot be ignored. Moreover, the impulse approximation assumes that the momentum transfer to the electron is negligible and that the projectile trajectory is unaffected. A negligible momentum transfer is clearly unrealistic and, from the Heisenberg uncertainty principle, the uncertainty of the impact parameter will be of the order of $\hbar c/m_e\gamma\beta$. One can thus specify a quantum-mechanical minimum of the impact parameter based upon the magnitude of this uncertainty,

$$b_{min}^{QM} = \frac{\hbar c}{m_e\gamma\beta} \tag{7.53}$$

Now review the three values of b_{min} that have been derived recognizing that, for any given situation, one must select the largest of the three values for b_{min}. Hence, that expression derived from two-body elastic scattering is excluded. This leaves the two expressions derived from the impulse approximation and from the uncertainty principle. The ratio of these two expressions can be used as the metric for determining which of the two to use,

$$\begin{aligned}\frac{b_{min}^{Q}}{b_{min}} &= \left(\frac{\hbar c}{m_e\gamma\beta}\right)\left(\frac{\gamma\beta^2}{2zr_0}\right)\\ &= \frac{\beta}{2\alpha z}.\end{aligned} \tag{7.54}$$

The appropriate minimum impact parameter expression is,

$$b_{min} = \frac{2z}{\gamma\beta^2}r_0 \quad \text{if } \beta < 2\,\alpha z \tag{7.55}$$

$$b_{min} = \frac{\hbar c}{m_e \gamma \beta} \quad \text{if } \beta > 2\alpha z \tag{7.56}$$

As an example, for the case of a proton projectile, the classically-derived impact parameter is used for $\beta < 2/137 = 0.0118$, equivalent to a kinetic energy of less than 100 keV.

An expression for the maximum impact parameter b_{max} which specifies the impact parameter beyond which the projectile cannot ionize or excite the atom is now derived. A simple approach would note that the energy transferred must exceed some (as yet-to-be specified) mean ionization energy, $\bar{I}$, in order for an electron to be elevated into the continuum. Equating this to the energy transfer gives,

$$\bar{I} = 2m_e \left(\frac{z r_0}{\beta b_{max}} \right)^2 \tag{7.57}$$

and solving for b_{max},

$$b_{max} = \frac{z r_0}{\beta} \sqrt{\frac{2m_e}{\bar{I}}} \tag{7.58}$$

A related approach to defining b_{max} was originally proposed by Bohr and sets the upper limit to the impact parameter such that the atomic electrons respond adiabatically. The duration of the interaction is of the order of $b/\gamma\beta c$ and, should it be sufficiently long, the natural motion of the electron can be ignored. The natural frequency of motion of a bound electron can be written as,

$$\omega_0 = \frac{E_B}{\hbar} \tag{7.59}$$

where E_B is an effective binding energy. An expression for b_{max} can be obtained by relating the duration of the interaction to the reciprocal of this frequency,

$$\frac{b_{max}}{\gamma \beta c} \approx \frac{1}{\omega_0} \tag{7.60}$$

to give,

$$\begin{aligned} b_{max} &= \frac{\gamma \beta c}{\omega_0} \\ &= \gamma \beta \frac{\hbar c}{E_B}. \end{aligned} \tag{7.61}$$

There are fundamental differences between these two expressions of b_{max}. Equation (7.58) predicts that the maximum impact parameter will be dependent upon the charge of the projectile and will decrease with increasing projectile speed. On the other hand, the result of (7.61) has no projectile charge dependence and predicts an increase in b_{max} with projectile speed.

Energy Transfer to a Harmonically-Bound Electron

The next steps of Bohr result of the soft collision energy loss are:

- Solve for the equation of motion for a harmonically-bound electron perturbed by the projectile's electric field.
- Use this result to relate the energy loss of the projectile to the electric field.
- Calculate the rate of energy loss with pathlength.

Equation of Motion of Target Electron

Consider a single atomic electron target harmonically bound to the atom with an oscillator natural frequency, ω_0. As the charged particle passes by, the electron is subject to a spatially- and time-dependent electric field, $\mathbf{E}(\mathbf{x},t)$ and the resulting equation of motion is,

$$m_e \frac{d^2\mathbf{x}}{dt^2} = -e\mathbf{E}(\mathbf{x}, t) - m_e \omega_0^2 \mathbf{x} - m_e \Gamma \frac{d\mathbf{x}}{dt}. \tag{7.62}$$

The second and third terms on the right-hand side are the restorative and damping forces upon the electron, respectively, where the latter is assumed small (i.e., $\omega_0 \gg \Gamma$) in order to simplify later derivations. As, by definition, soft collisions occur at large impact parameters, the spatial variation of the electric field at the position of the electron is neglected allowing the removal of the $\mathbf{x}$ dependence of the electric field. The electric field can thus be treated as being spatially uniform at the position of the electron making it possible to replace $\mathbf{E}(\mathbf{x},t)$ with $\mathbf{E}(t)$. This equation of motion is solved using the method of Fourier transform pairs,

$$\mathbf{E}(\omega') = \frac{1}{\sqrt{2\pi}} \int_{-\infty}^{\infty} dt\, e^{i\omega' t}\, \mathbf{E}(t) \tag{7.63}$$

and

$$\mathbf{E}(t) = \frac{1}{\sqrt{2\pi}} \int_{-\infty}^{\infty} d\omega' \, e^{-i\omega' t} \, \mathbf{E}(\omega'). \tag{7.64}$$

The differential equation of motion is first rearranged,

$$\frac{d^2\mathbf{x}}{dt^2} + \Gamma \frac{d\mathbf{x}}{dt} + \omega_0^2 \mathbf{x} = -\left(\frac{e}{m_e}\right)\mathbf{E} \tag{7.65}$$

where functional dependencies are omitted for clarity. Writing both sides of the equation of motion in terms of the inverse- transforms and differentiating with respect to time gives,

$$\int_{-\infty}^{\infty} d\omega' (-\omega'^2 - i\omega'\Gamma + \omega_0^2) e^{-i\omega' t} \, \mathbf{x}(\omega') = -\left(\frac{e}{m_e}\right) \int_{-\infty}^{\infty} d\omega' \, e^{-i\omega' t} \, \mathbf{E}(\omega'). \tag{7.66}$$

Multiplying both sides by $e^{-i\omega t}$ and integrating over time,

$$\int_{-\infty}^{\infty} dt \int_{-\infty}^{\infty} d\omega' \, e^{i(\omega-\omega')t} \left(-\omega'^2 - i\omega'\Gamma + \omega_0^2\right) \mathbf{x}(\omega') = -\left(\frac{e}{m_e}\right) \int_{-\infty}^{\infty} dt \int_{-\infty}^{\infty} d\omega' \, e^{i(\omega-\omega')t} \, \mathbf{E}(\omega')$$

results in the integral equation,

$$\int_{-\infty}^{\infty} d\omega' \, \delta(\omega - \omega')(-\omega'^2 - i\omega'\Gamma + \omega_0^2)\mathbf{x}(\omega') = -\left(\frac{e}{m_e}\right) \int_{-\infty}^{\infty} d\omega' \, \delta(\omega - \omega')\mathbf{E}(\omega') \tag{7.67}$$

where the definition of the δ-function has been used. The integration over ω' is trivial and the resulting frequency-space solution to the equation of motion is,

$$\mathbf{x}(\omega) = -\left(\frac{e}{m_e}\right)\left(\frac{\mathbf{E}(\omega)}{(\omega_0^2 - \omega^2) - i\Gamma\omega}\right). \tag{7.68}$$

Energy Transfer as a Function of the Electric Field

The projectile energy loss is equal to the energy transferred to the electron,

$$Q = -e \int_{-\infty}^{\infty} dt \frac{d\mathbf{x}(t)}{dt} \mathbf{E}(t). \tag{7.69}$$

Writing the integrand in terms of the inverse Fourier transforms,

$$\begin{aligned} Q &= -e \int_{-\infty}^{\infty} dt \left(\frac{1}{\sqrt{2\pi}} \frac{d}{dt} \int_{-\infty}^{\infty} d\omega \, e^{-i\omega t} \mathbf{x}(\omega) \right) \\ &\quad \times \left(\frac{1}{\sqrt{2\pi}} \int_{-\infty}^{\infty} d\omega' \, e^{-i\omega' t} \, \mathbf{E}(\omega') \right) \\ &= i\left(\frac{e}{2\pi}\right) \int_{-\infty}^{\infty} d\omega' \int_{-\infty}^{\infty} d\omega \left(\int_{-\infty}^{\infty} dt \, e^{-i(\omega+\omega')t} \right) \\ &\quad \times \omega \mathbf{x}(\omega)\mathbf{E}(\omega') \\ &= ie \int_{-\infty}^{\infty} d\omega' \int_{-\infty}^{\infty} d\omega \, \delta(\omega + \omega')\omega \, \mathbf{x}(\omega)\mathbf{E}(\omega') \\ &= ie \int_{-\infty}^{\infty} d\omega \, \omega \, \mathbf{x}(\omega)\mathbf{E}(-\omega). \end{aligned} \tag{7.70}$$

The electric field is a real quantity (i.e., $\mathbf{E}(-\omega) = \mathbf{E}^*(\omega)$), so,

$$\begin{aligned} Q &= ie \int_{-\infty}^{\infty} d\omega \, \omega \, \mathbf{x}(\omega)\mathbf{E}^*(\omega) \\ &= 2e \, \mathrm{Re}\left(i \int_{0}^{\infty} d\omega \, \omega \, \mathbf{x}(\omega)\mathbf{E}^*(\omega) \right). \end{aligned} \tag{7.71}$$

The energy transfer as a function of the electric field can now be obtained by substituting the expression for $\mathbf{x}(\omega)$ into the integrand,

$$\begin{aligned} Q &= \left(-\frac{2e^2}{m_e}\right)\mathrm{Re}\left(i\int_0^\infty d\omega \frac{\omega|\mathbf{E}(\omega)|^2}{(\omega_0^2-\omega^2)-i\Gamma\omega}\right) \\ &= \left(-\frac{2e^2}{m_e}\right)\mathrm{Re}\left(i\int_0^\infty d\omega \frac{\omega|\mathbf{E}(\omega)|^2\left((\omega_0^2-\omega^2)+i\Gamma\omega\right)}{(\omega_0^2-\omega^2)^2+\Gamma^2\omega^2}\right) \\ &= \left(-\frac{2e^2}{m_e}\right)\mathrm{Re}\left(\int_0^\infty d\omega \frac{\omega|\mathbf{E}(\omega)|^2\left(-\Gamma\omega+i(\omega_0^2-\omega^2)\right)}{(\omega_0^2-\omega^2)^2+\Gamma^2\omega^2}\right) \\ &= \left(\frac{2e^2}{m_e}\right)\left(\int_0^\infty d\omega \frac{\Gamma\omega^2|\mathbf{E}(\omega)|^2}{(\omega_0^2-\omega^2)^2+\Gamma^2\omega^2}\right). \end{aligned} \tag{7.72}$$

Solving this integral is simplified as the damping of the electron motion is small (i.e., $\Gamma \ll \omega$). As a result of this, $\mathbf{E}(\omega_0) \approx \mathbf{E}(\omega)$ and $|\mathbf{E}(\omega)|^2$ is extracted from the integrand to give,

$$Q = \left(\frac{2e^2}{m_e}\right)|\mathbf{E}(\omega_0)|^2 \int_0^\infty d\omega \frac{\Gamma\omega^2}{(\omega_0^2-\omega^2)^2+\Gamma^2\omega^2}. \tag{7.73}$$

To complete the derivation, the integral is solved using the substitution of variable, $u = \omega/\Gamma$

$$\begin{aligned} &\int_0^\infty d\omega \frac{\Gamma\omega^2}{(\omega_0^2-\omega^2)^2+\Gamma^2\omega^2} \\ &= \int_0^\infty du \frac{u^2}{(k^2-u^2)^2+u^2} \end{aligned} \tag{7.74}$$

where $k = \omega_0/\Gamma$. The integrand is rearranged,

$$\begin{aligned} \int_0^\infty du \frac{u^2}{(k^2-u^2)^2+u^2} &= \int_0^\infty \frac{du}{\left(\frac{(u-k)(u+k)}{u}\right)^2+1} \\ &= \int_{-k}^\infty \frac{dx}{\left(\frac{x(x+2k)}{x+k}\right)^2+1} \end{aligned} \tag{7.75}$$

where the substitution $x = u - k$ has been used. The small damping force requirement is equivalent to $x \ll k$ which enables the approximation,

$$\begin{aligned} \int_{-k}^\infty \frac{dx}{\left(\frac{x(x+2k)}{x+k}\right)^2+1} &\approx \int_{-k}^\infty \frac{dx}{\left(\frac{1}{k^2}\right)(x^2+2kx)^2+1} \\ &\approx \int_{-k}^\infty \frac{dx}{1+4x^2} \end{aligned} \tag{7.76}$$

and to set the lower integration limit to $-\infty$. Hence,

$$\int_0^\infty d\omega \frac{\Gamma\omega^2}{(\omega_0^2-\omega^2)^2+\Gamma^2\omega^2} \approx \int_{-\infty}^\infty \frac{dx}{1+4x^2}. \tag{7.77}$$

Residue theory is used to solve this integral, which is of the form $\int_{-\infty}^{\infty} \frac{dz}{\left(\frac{1}{4}\right)+z^2}$ where $z = x + iy$. The integrand has poles at $z = \pm\, i/2$ and is holomorphic everywhere else. Consider the semicircular contour of radius R in the upper half-plane as shown in Fig. 7.8. For $R > 1/2$, the singularity of the integrand lies within the interior of the contour bounded by the segment of $-R \le x \le R$ with $y = 0$ and the upper half C_R of the circle $|z| = R$. Integrating counter-clockwise over this contour,

$$\begin{aligned} \int_{-R}^{R} \frac{dx}{\left(\frac{1}{4}\right)+x^2} + \int_{C_R} \frac{dz}{\left(\frac{1}{4}\right)+z^2} &= \int_{-\infty}^{\infty} \frac{dz}{\left(\frac{1}{4}\right)+z^2} \\ &= 2\pi i\, B \end{aligned} \tag{7.78}$$

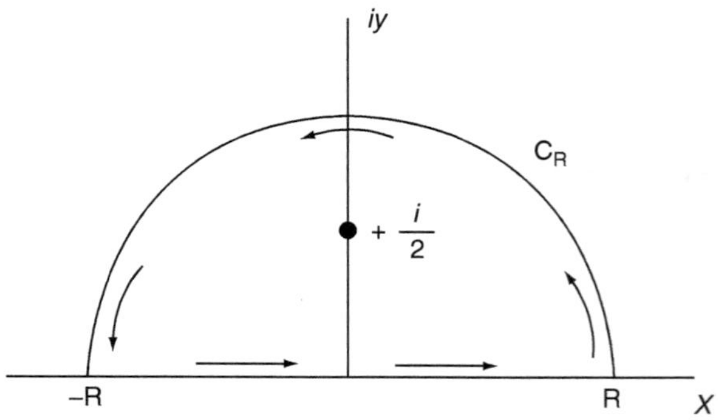

Fig. 7.8 Integration contour in the complex plane $z = x + iy$ for calculating the integral of (7.77)

where B is the residue of the integrand at the point $z = \pm\, i/2$,

$$B = \lim_{z \to \frac{i}{2}} \left(z - \frac{i}{2}\right)\left(\frac{1}{\left(\frac{1}{4}\right) + z^2}\right) = -i. \qquad (7.79)$$

Hence,

$$\int_{-\infty}^{\infty} \frac{dx}{\left(\frac{1}{4}\right) + x^2} = 2\pi - \int_{C_R} \frac{dz}{\left(\frac{1}{4}\right) + z^2} = 2\pi \qquad (7.80)$$

which is valid for all $R > 1/2$ as the integral along the contour on the right-hand side is equal to zero as can be shown by considering the point z on the contour C_R for which,

$$\int_{C_R} \frac{dz}{1/4 + z^2} \le \frac{\pi R}{1/4 + R^2} \qquad (7.81)$$

where πR is the length of the contour. Thus, as $\frac{\pi R}{1/4+R^2} \to 0$ as $R \to \infty$, then $\int_{C_R} \frac{dz}{1/4+z^2} \to 0$ as $R \to \infty$ and the energy transfer to the electron is,

$$Q = \left(\pi \frac{e^2}{m_e}\right) |\mathbf{E}(\omega_0)|^2. \qquad (7.82)$$

Calculation of the Electric Field

An expression for the squared modulus of the electric field at the position of the atomic electrons is found by first calculating the Fourier transforms of the two electric field components for impact parameter b, parallel and orthogonal to the particle trajectory, by rewriting (7.36) and (7.37),

$$\mathbf{E}_\perp(t) = \left(\frac{z\,\alpha\hbar c}{eb^2}\right) \frac{\gamma}{\left[1 + \left(\frac{\gamma v t}{b}\right)^2\right]^{\frac{3}{2}}} \hat{\mathbf{v}}_\perp \qquad (7.83)$$

$$\mathbf{E}_{||}(t) = -\left(\frac{z\,\alpha\hbar c}{eb^3}\right) \frac{\gamma v t}{\left[1 + \left(\frac{\gamma v t}{b}\right)^2\right]^{\frac{3}{2}}} \hat{\mathbf{v}}_{||}. \qquad (7.84)$$

The Fourier transform of the perpendicular component of the electric field is,

$$\mathbf{E}_\perp(\omega) = \frac{1}{\sqrt{2\pi}} \int_{-\infty}^{\infty} dt\, e^{i\omega t}\, \mathbf{E}_\perp(t) = \frac{\gamma}{\sqrt{2\pi}} \left(\frac{z\,\alpha\hbar c}{eb^2}\right) \int_{-\infty}^{\infty} dt \frac{e^{i\omega t}}{\left[1 + \left(\frac{\gamma v t}{b}\right)^2\right]^{\frac{3}{2}}} \hat{\mathbf{v}}_\perp. \qquad (7.85)$$

The integral is solved by using the substitution of variable, $x = \gamma v t/b$,

$$\mathbf{E}_\perp(\omega) = \frac{z\,\alpha\hbar c}{\sqrt{2\pi}\, ebv} \int_{-\infty}^{\infty} dx \frac{e^{i\left(\frac{\omega b}{\gamma v}\right)x}}{[1 + x^2]^{\frac{3}{2}}} \hat{\mathbf{v}}_\perp = \sqrt{\frac{2}{\pi}} \frac{z\,\alpha\hbar c}{ebv} \int_0^{\infty} dx \frac{\cos\left(\frac{\omega b}{\gamma v} x\right)}{[1 + x^2]^{\frac{3}{2}}} \hat{\mathbf{v}}_\perp \qquad (7.86)$$

with the last step following from $\cos\left(\frac{\omega b}{\gamma v} x\right)$ and $\sin\left(\frac{\omega b}{\gamma v} x\right)$ being even and odd functions, respectively. The integral is a form of the modified Bessel function of the second kind (Abramowitz and Stegun 1972),

$$K_\nu(y) = \frac{2^\nu \Gamma\left(\nu + \frac{1}{2}\right)}{\sqrt{\pi} y^\nu} \int_0^{\infty} dx \frac{\cos(xy)}{[1 + x^2]^{\nu + \frac{1}{2}}} \qquad (7.87)$$

where ν is 0 or an integer. Use of $K_\nu(y)$ for $\nu = 0, 1$ will be required,

$$K_0(y) = \int_0^{\infty} dx \frac{\cos(xy)}{\sqrt{1 + x^2}} \qquad (7.88)$$

$$K_1(y) = \frac{1}{y} \int_0^{\infty} dx \frac{\cos(xy)}{[1 + x^2]^{\frac{3}{2}}}. \qquad (7.89)$$

The expression of the electric field orthogonal to the projectile trajectory is now,

$$\mathbf{E}_\perp(\omega) = \sqrt{\frac{2}{\pi}} \frac{z\,\alpha\hbar c}{ebv} \int_0^{\infty} dx \frac{\cos\left(\frac{\omega b}{\gamma v} x\right)}{[1 + x^2]^{\frac{3}{2}}} \hat{\mathbf{v}}_\perp = \sqrt{\frac{2}{\pi}} \frac{z\,\alpha\hbar c}{ebv} \left(\frac{\omega b}{\gamma v}\right) K_1\left(\frac{\omega b}{\gamma v}\right) \hat{\mathbf{v}}_\perp = \frac{e}{(2\pi)^{3/2} \varepsilon_0} \left(\frac{z\omega}{\gamma v^2}\right) K_1\left(\frac{\omega b}{\gamma v}\right) \hat{\mathbf{v}}_\perp \qquad (7.90)$$

and the Fourier transform of the electric field component parallel to the particle's trajectory is,

$$\mathbf{E}_{\|}(\omega) = -\frac{\gamma v}{\sqrt{2\pi}}\frac{z\alpha\hbar c}{eb^3}\int_{-\infty}^{\infty} dt \frac{te^{i\omega t}}{\left[1+\left(\frac{\gamma vt}{b}\right)^2\right]^{\frac{3}{2}}}\hat{\mathbf{v}}_{\|}. \quad (7.91)$$

Using, again, the substitution of variable, $x = \gamma vt/b$,

$$\mathbf{E}_{\|}(\omega) = -\frac{z\alpha\hbar c}{\sqrt{2\pi}\, eb\gamma v}\int_{-\infty}^{\infty} dx \frac{xe^{i\left(\frac{\omega b}{\gamma v}\right)x}}{[1+x^2]^{\frac{3}{2}}}\hat{\mathbf{v}}_{\|}. \quad (7.92)$$

This integral is solved by parts (for clarity, $y = \omega b/\gamma v$),

$$\begin{aligned}\int_{-\infty}^{\infty} dx \frac{xe^{ixy}}{[1+x^2]^{\frac{3}{2}}} &= \int_{-\infty}^{\infty} dr\, s \\ &= s\,r\big|_{-\infty}^{\infty} - \int_{-\infty}^{\infty} ds\, r\end{aligned} \quad (7.93)$$

where

$$s \equiv e^{ixy} \qquad ds = dx\, i\, y\, e^{ixy}$$
$$dr \equiv dx \frac{x}{[1+x^2]^{\frac{3}{2}}} \qquad r = -\frac{1}{\sqrt{1+x^2}}$$

to give,

$$\begin{aligned}\int_{-\infty}^{\infty} dx \frac{xe^{i\left(\frac{\omega b}{\gamma v}\right)x}}{[1+x^2]^{\frac{3}{2}}} &= 2i\left(\frac{\omega b}{\gamma v}\right)\int_{0}^{\infty} dx \frac{\cos\left(\frac{\omega b}{\gamma v}x\right)}{\sqrt{1+x^2}} \\ &= 2i\left(\frac{\omega b}{\gamma v}\right)K_0\left(\frac{\omega b}{\gamma v}\right).\end{aligned} \quad (7.94)$$

The Fourier transform of the electric field component parallel to the particle's trajectory is now had,

$$\mathbf{E}_{\|}(\omega) = -\frac{i}{(2\pi)^{3/2}\varepsilon_0}\frac{ze\omega}{(\gamma v)^2}K_0\left(\frac{\omega b}{\gamma v}\right)\hat{\mathbf{v}}_{\|}. \quad (7.95)$$

The squared modulus of the electric field in frequency space is,

$$\begin{aligned}|\mathbf{E}(\omega_0)|^2 &= |\mathbf{E}_{\perp}(\omega_0)|^2 + |\mathbf{E}_{\|}(\omega_0)|^2 \\ &= \frac{1}{(2\pi)^3\varepsilon_0^2}\left(\frac{ze\omega_0}{v}\right)^2\left[K_1^2\left(\frac{\omega_0 b}{\gamma v}\right) + \frac{K_0^2\left(\frac{\omega_0 b}{\gamma v}\right)}{\gamma^2}\right].\end{aligned} \quad (7.96)$$

The frequency ω_0 is written in terms of the maximum impact parameter, b_{max}, beyond which no energy is transferred to the atom, as determined from the adiabatic response result, $\omega_0 = \gamma v/b_{max}$. Hence, the electric field can be written as a function of the impact parameter,

$$|\mathbf{E}(b)|^2 = \frac{1}{(2\pi)^3\varepsilon_0^2}\left(\frac{ze\gamma}{b_{max}}\right)^2\left[K_1^2\left(\frac{b}{b_{max}}\right) + \frac{K_0^2\left(\frac{b}{b_{max}}\right)}{\gamma^2}\right]. \quad (7.97)$$

The energy transfer to a harmonically-bound electron is,

$$\begin{aligned}Q &= \left(\pi\frac{e^2}{m_e}\right)|\mathbf{E}(b)|^2 \\ &= 2m_e\left(\frac{r_0}{b_{max}}\right)^2\left(\frac{z}{\beta}\right)^2\left[K_1^2\left(\frac{b}{b_{max}}\right) + \frac{K_0^2\left(\frac{b}{b_{max}}\right)}{\gamma^2}\right].\end{aligned} \quad (7.98)$$

Consider the dependence of the energy transfer as a function of impact parameter. For small impact parameters, the low-argument limits of the modified Bessel functions are required,

$$yK_0(y) \to 0 \quad \text{as} \quad y \to 0 \quad (7.99)$$

$$yK_1(y) \to 1 \quad \text{as} \quad y \to 0. \quad (7.100)$$

In this case,

$$Q \to 2m_e\left(\frac{zr_0}{\beta b}\right)^2 \quad \text{as} \quad b \to 0. \quad (7.101)$$

Note that this is the same result of the energy transfer calculated from the impulse approximation. Invoking a minimum impact parameter, an expression for the maximum energy transfer in a soft collision corresponding to the minimum impact parameter can be written,

$$Q_{max} = 2m_e\left(\frac{zr_0}{\beta b_{min}}\right)^2. \quad (7.102)$$

For example, the magnitude of the energy transfer from a soft collision from an α particle with a kinetic energy of 5 MeV is $Q_{max} \approx 200$ eV. For large impact parameters, we can use the large-argument approximations of the modified Bessel functions $K_{0,1}(y) \approx \sqrt{\frac{\pi}{2y}}e^{-y}$ for $y \gg 1$ to give,

$$Q \approx \pi m_e \left(\frac{r_0}{b_{max}}\right)^2 \left(\frac{z}{\beta}\right)\left[1 + \frac{1}{\gamma^2}\right] \times \left(\frac{b_{max}}{b}\right) e^{-2b/b_{max}} \tag{7.103}$$

The exponential term in this expression introduces the desired rapid cut-off for very large impact parameters at which the energy transfer becomes inefficient beyond the adiabatically-limited impact parameter, b_{max}.

Bohr Soft Collision Mass Stopping Power

Using the Bohr soft energy transfer expression, the soft collision stopping power, which is the energy loss per unit pathlength due solely to soft collisions, is calculated. In a medium of electron density ρ_e, the number of electrons in a differential cylinder section of length dx with impact parameters between b and b + db is $2\pi\ \rho_e$ b dx db and the double-differential energy loss of a particle traversing this section is,

$$\begin{aligned} d^2E &= -2\pi \rho_e b Q\, db\, dx \\ &= -4\pi \rho_e m_e \left(\frac{r_0}{b_{max}}\right)^2 \left(\frac{z}{\beta}\right)^2 \times \left[K_1^2\left(\frac{b}{b_{max}}\right) + \frac{K_0^2\left(\frac{b}{b_{max}}\right)}{\gamma^2}\right] b\, db\, dx \\ &= -4\pi \rho_e m_e \left(\frac{zr_0}{\beta}\right)^2 y \left[K_1^2(y) + \frac{K_0^2(y)}{\gamma^2}\right] dy\, dx \end{aligned} \tag{7.104}$$

where the substitution of variable, $y = b/b_{max}$, has been used. Integrating over y, the linear soft collision stopping power[11] is

$$\begin{aligned} \left(\frac{dE}{dx}\right)_{Col,S} &= -4\pi \rho_e m_e \left(\frac{zr_0}{\beta}\right)^2 \times \int_{\left(\frac{b_{min}}{b_{max}}\right)}^{\infty} dy\, y \left[K_1^2(y) + \frac{K_0^2(y)}{\gamma^2}\right] \\ &= -4\pi \rho_e m_e \left(\frac{zr_0}{\beta}\right)^2 \times \int_{\left(\frac{b_{min}}{b_{max}}\right)}^{\infty} dy\, y \left[K_1^2(y) + K_0^2(y) - \beta^2 K_0^2(y)\right]. \end{aligned} \tag{7.105}$$

The mass soft collision stopping power will be the linear collision stopping power normalized to the physical density of the medium the charged particle is moving through,

$$\begin{aligned} \left(\frac{dE}{\rho\, dx}\right)_{Col} &= -4\pi N_A \left(\frac{Z}{A}\right) m_e \left(\frac{zr_0}{\beta}\right)^2 \int_{\left(\frac{b_{min}}{b_{max}}\right)}^{\infty} dy\, y \left[K_1^2(y) + K_0^2(y) - \beta^2 K_0^2(y)\right] \\ &= -2C \left(\frac{Z}{A}\right)\left(\frac{z}{\beta}\right)^2 \int_{\left(\frac{b_{min}}{b_{max}}\right)}^{\infty} dy\, y \left[K_1^2(y) + K_0^2(y) - \beta^2 K_0^2(y)\right]. \end{aligned} \tag{7.106}$$

The integral is solved by simplifying the integrand using the properties of the derivatives of $K_0(y)$ and $K_1(y)$,

$$\frac{dK_0(y)}{dx} = -K_1(y) \tag{7.107}$$

and

$$\frac{dK_1(y)}{dy} = -K_0(y) - \frac{K_1(y)}{y}. \tag{7.108}$$

Then,

$$\begin{aligned} \frac{d}{dy}[y K_0(y) K_1(y)] &= K_0(y)K_1(y) + y\frac{dK_0(y)}{dy}K_1(y) + yK_0(y)\frac{dK_1(y)}{dy} \\ &= -y\left(K_0^2(y) + K_1^2(y)\right) \end{aligned} \tag{7.109}$$

[11] Although the stopping power is also written as S and the mass collision stopping power as S/ρ, it will be written here as a differential.

and

$$\frac{d}{dy}\left[y^2\left(K_1^2(y)-K_0^2(y)\right)\right]=-2yK_0^2(y). \quad (7.110)$$

Incorporating these results into the integral of (7.106) gives,

$$\begin{aligned}
&\int_{\left(\frac{b_{min}}{b_{max}}\right)}^{\infty} dy\, y\left[K_1^2(y)+K_0^2(y)-\beta^2K_0^2(y)\right] \\
&= \int_{\left(\frac{b_{min}}{b_{max}}\right)}^{\infty} dy\left[y\left(K_1^2(y)+K_0^2(y)\right)-\beta^2 yK_0^2(y)\right] \\
&= \int_{\left(\frac{b_{min}}{b_{max}}\right)}^{\infty} dy\left[-\frac{d}{dy}\left[yK_0(y)K_1(y)\right]+\frac{\beta^2}{2}\right. \\
&\quad \left.\times\frac{d}{dy}\left[y^2(K_1^2(y)-K_0^2(y))\right]\right]
\end{aligned} \quad (7.111)$$

from which the Bohr mass soft collision stopping power is obtained,

$$\begin{aligned}
\left(\frac{dE}{\rho\, dx}\right)_{Col,S} =& -2C\left(\frac{Z}{A}\right)\left(\frac{z}{\beta}\right)^2\left[\frac{b_{min}}{b_{max}}K_0\left(\frac{b_{min}}{b_{max}}\right)K_1\left(\frac{b_{min}}{b_{max}}\right)\right. \\
&\left.-\frac{\beta^2}{2}\frac{b_{min}^2}{b_{max}^2}\left(K_1^2\left(\frac{b_{min}}{b_{max}}\right)-K_0^2\left(\frac{b_{min}}{b_{max}}\right)\right)\right]
\end{aligned} \quad (7.112)$$

The variable of interest is the ratio of the minimum to maximum impact parameters,

$$\left(\frac{b_{min}}{b_{max}}\right)=\frac{b_{min}}{\left(\frac{\gamma v}{\omega_0}\right)}=\frac{b_{min}\omega_0}{\gamma v}. \quad (7.113)$$

A form of the minimum impact parameter has not been explicitly provided as this will be dependent upon the projectile speed and electric charge. The magnitude of this ratio of impact parameters can be estimated by recognizing that the resonant frequency of a harmonically-bound electron is approximated by $\omega_0 = E_B/\hbar$ which leads to,

$$\begin{aligned}
\left(\frac{b_{min}}{b_{max}}\right) &= \frac{b_{min}E_B}{\hbar c\,\gamma\beta} \\
&= \frac{2zr_0E_B}{\hbar c\,\gamma^2\beta^3} \quad \beta < 2\alpha Z && (7.114) \\
&= \frac{E_B}{m_e(\gamma\beta)^2} \quad \beta > 2\alpha Z. && (7.115)
\end{aligned}$$

As the binding energy can be written in approximate form, $E_B = -\hbar cR_\infty Z$, then $b_{min}/b_{max} \ll 1$and it is then possible to use the properties of the modified Bessel functions of the second kind for small arguments,

$$\begin{aligned}
K_0(y) &\approx -\ln\frac{y}{2}-\gamma_{EM} \\
&\approx \ln\left(\frac{2e^{-\gamma_{EM}}}{y}\right) \quad \text{for} \quad 0<y\ll 1
\end{aligned} \quad (7.116)$$

where $\gamma_{EM} \approx 0.5772\ldots$ is the Euler–Mascheroni constant, and

$$K_1(y)\approx\frac{1}{y} \quad \text{for} \quad 0<y \ll \sqrt{2}. \quad (7.117)$$

Using these expressions and recalling that $b_{min}/b_{max} \ll 1$,

$$\begin{aligned}
\left(\frac{dE}{\rho\, dx}\right)_{Col,S} &\approx -2C\left(\frac{Z}{A}\right)\left(\frac{z}{\beta}\right)^2\left[\ln\left(2e^{-\gamma_{EM}}\frac{b_{max}}{b_{min}}\right)-\frac{\beta^2}{2}\right] \\
&\approx -2C\left(\frac{Z}{A}\right)\left(\frac{z}{\beta}\right)^2\left[\ln\left(1.123\frac{b_{max}}{b_{min}}\right)-\frac{\beta^2}{2}\right].
\end{aligned} \quad (7.118)$$

Briefly return to the ratio of impact parameters,

$$\begin{aligned}
\left(\frac{b_{max}}{b_{min}}\right) &= \frac{(\gamma\beta c/\omega_0)}{\left(2zr_0/\gamma\beta^2\right)} \\
&= \frac{\gamma^2\beta^3 c}{2zr_0\omega_0}
\end{aligned} \quad (7.119)$$

where a single harmonic oscillator with a resonance frequency, ω_0, which corresponds to a single atomic electron has been calculated for. For atoms other than hydrogen, this should be replaced by the geometric average resonance frequency for the Z atomic electrons,

$$\ln \bar{\omega} = \frac{1}{Z} \sum_j f_j \ln \omega_j \tag{7.120}$$

where the Z electrons have been partitioned into groups each having the same resonance frequency, ω_j. Detailed discussion of the oscillator strengths is deferred until the derivation of the Bethe theory, but it will be noted here that the oscillator strengths must satisfy the requirement,

$$\sum_{j=1}^{Z} \omega_j \frac{df_j}{d\omega} = Z. \tag{7.121}$$

The final expression for the classical Bohr mass soft collision stopping power is now,

$$\left(\frac{dE}{\rho\, dx}\right)_{Col,S} = -2C\left(\frac{Z}{A}\right)\left(\frac{z}{\beta}\right)^2\left[\ln\left(\frac{1.123\,\gamma^2\beta^3 c}{z r_0 \bar{\omega}}\right) - \frac{\beta^2}{2}\right] \tag{7.122}$$

7.3.3.3 Bethe Theory

Introduction

During the 1920s, various attempts were made to provide a quantum-mechanical description of the energy loss in inelastic charged-particle collisions with atoms. Bethe was the first to develop a successful quantum-mechanical theory[12] and which is based upon the first Born approximation.

This derivation of the Bethe soft collision stopping power will, for calculational ease, be limited to the nonrelativistic case.

[12]Reviews of the derivation of the Bethe theory can be found in Fano (1964), Inokuti (1971) and Ahlen (1980).

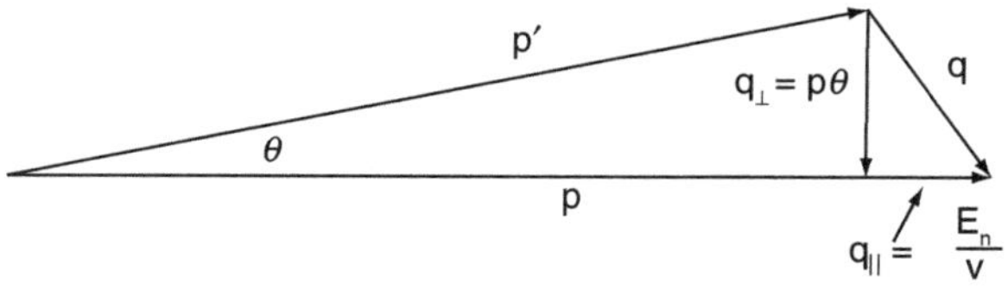

Fig. 7.9 Momentum transfer in the small-angle approximation

Collision Kinematics

It is necessary to first define the kinematics of the collision between the projectile and an atomic electron. A projectile with momentum **p** and kinetic energy T collides with an atomic electron and scatters through the angle θ with momentum **p** and kinetic energy T′. The momentum transfer is given through the derivation of (3.52) for elastic scatter but now allowing for energy loss, $p \neq p'$,

$$q = \sqrt{p^2 + p'^2 - 2pp' \cos\theta} \tag{7.123}$$

for the interaction kinematics of Fig. 7.9 for a small scattering angle. In this derivation, the target atomic electron is treated as being unbound and initially at rest, which is a reasonable assumption for high projectile incident kinetic energies. However, if the projectile speed is comparable to the orbital speeds of the atomic electrons, this assumption becomes untenable and corrections must be made as will be discussed later in this chapter. Following the collision, the kinetic energy of the recoil electron is,

$$Q = T - T' \tag{7.124}$$

which is related to its three-vector momentum through the relativistic relationship,

$$(Q + m_e)^2 = q^2 + m_e^2. \tag{7.125}$$

Upon rearrangement,

$$Q\left(1 + \frac{Q}{2m_e}\right) = \frac{q^2}{2m_e} \tag{7.126}$$

which, for the condition of $Q \ll 2m_e$, reduces to the familiar nonrelativistic form,

$$Q = \frac{q^2}{2m_e}. \tag{7.127}$$

Following the collision, the atom will be excited from its ground state 0 with energy eigenvalue E_0, which will be taken as being equal to zero, to a final state n with the energy eigenvalue, E_n. The energy transfer Q need not equal E_n as the recoil electron cannot be considered in isolation as a consequence of it being part of an ensemble of atomic electrons as energy transferred to it can be shared amongst others.

It is straightforward to calculate the component of the momentum transfer **q** that is parallel to the incident momentum **p** from the excitation energy,

$$\begin{aligned} E_n &\approx \frac{p^2}{2m} - \frac{|\mathbf{p}-\mathbf{q}|^2}{2m} \\ &\approx \frac{2\mathbf{p}\bullet\mathbf{q} - q^2}{2m} \\ &\approx \frac{\mathbf{q}\bullet\mathbf{p}}{m} \end{aligned} \tag{7.128}$$

where m is the projectile mass and, having assumed soft collisions, terms of the order of q^2 are neglected. This leads to,

$$q_{||} = \frac{E_n}{v} \equiv \frac{E_n}{\beta} \tag{7.129}$$

where both q and E_n are in units of energy. As this is a soft collision and q is small, the small-angle approximation can be used and the component of **q** perpendicular to **p** is,

$$q_{\perp} = p\theta. \tag{7.130}$$

In this small-angle approximation, the squared magnitude of the momentum transfer can be written as,

$$q^2 = q_{||}^2 + q_{\perp}^2 = \left(\frac{E_n}{\beta}\right)^2 + (p\theta)^2 \tag{7.131}$$

and the energy transfer to the electron is,

$$Q = \frac{q^2}{2m_e} = \frac{E_n^2}{2m_e\beta^2} + \frac{(p\theta)^2}{2m_e}. \tag{7.132}$$

Bethe Soft Collision Cross Section

The calculation of the Bethe mass soft collision stopping power follows that of the elastic atomic Coulomb scattering cross section, but using both of Fermi's Golden Rules as there is no direct coupling between the initial and final atomic states. Consider the projectile to have an electric charge ze and the atom to have atomic number Z and to be in its ground state, $|0\rangle$. The projectile is treated as a plane wave and the atom is excited to the state $|n\rangle$ as a result of the collision. Hence, the pre- and postcollision system states are,

$$\begin{aligned} \langle \mathbf{r}|\mathbf{p},0\rangle &= \frac{1}{\sqrt{L^3}} e^{i\frac{\mathbf{p}\bullet\mathbf{R}}{\hbar c}}|0\rangle \\ &= \frac{1}{\sqrt{L^3}} e^{i\frac{\mathbf{p}\bullet\mathbf{R}}{\hbar c}}\psi_0(\mathbf{r}_1,\mathbf{r}_2,\ldots\mathbf{r}_Z) \end{aligned} \tag{7.133}$$

and

$$\begin{aligned} \langle \mathbf{r}|\mathbf{p}',n\rangle &= \frac{1}{\sqrt{L^3}} e^{i\frac{\mathbf{p}'\bullet\mathbf{R}}{\hbar c}}|n\rangle \\ &= \frac{1}{\sqrt{L^3}} e^{i\frac{\mathbf{p}'\bullet\mathbf{R}}{\hbar c}}\psi_n(\mathbf{r}_1,\mathbf{r}_2,\ldots\mathbf{r}_Z). \end{aligned} \tag{7.133}$$

The system position vector, with the origin specified at the center of the atom, is given by $\mathbf{r} = (\mathbf{R}, \mathbf{r}_1 \ldots \mathbf{r}_Z)$ where **R** is the position vector of the projectile and the $\mathbf{r}_j$ is the position vector of the jth electron. The overall interaction between the projectile and the atomic electrons is handled in the Coulomb gauge in which there are two types of electromagnetic interactions. The first is the static unretarded potential between the projectile and electrons with a direct coupling between the initial and final states allowing the transition rate to be given by Fermi's Golden Rule No. 2. This is also referred to as the longitudinal excitation as it is directed parallel to the momentum transfer. The second type of interaction is through the emission and absorption of virtual photons between the projectile and atomic electrons which become significant at relativistic speeds. Because this is the result of the interaction between the particle currents with the quantized transverse vector potential, this is often referred to as transverse excitation. As there is no direct coupling between the initial $|0\rangle$ and final $|n\rangle$ states, the transition rate through an intermediate state $|i\rangle$ will be given by Fermi's Golden Rule No. 1. Combining these two categories, the transition rate is,

$$\lambda_{fi} = \frac{2\pi}{\hbar}\left|\langle \mathbf{p}', n|U|\mathbf{p}, 0\rangle + \sum_j \frac{\langle \mathbf{p}', n|U|\mathbf{k}, j\rangle\langle \mathbf{k}, j|U|\mathbf{p}, 0\rangle}{E_0 - E_j}\right|^2 \rho_f \tag{7.135}$$

where E_0 and E_j are the energy eigenvalues for states $|0\rangle$ and $|j\rangle$, respectively, and the summation is over all available intermediate states.

While the projectile will interact with both the ensemble of atomic electrons and the nucleus, it can be proven that the projectile-nucleus interaction does not lead to atomic excitation. The Coulomb potential between the projectile and the nucleus (both taken to be point-like charges) is $U(R) = -zZ\,\alpha\hbar c/R$ and the matrix element of the corresponding perturbation in position space is,

$$\begin{aligned}
&\langle \mathbf{p}', n|U|\mathbf{p}, 0\rangle \\
&= \int d^3\mathbf{r}\, d^3\mathbf{r}' \langle \mathbf{p}', n|\mathbf{r}\rangle\langle \mathbf{r}|U|\mathbf{r}'\rangle\langle \mathbf{r}'|\mathbf{p}, 0\rangle \\
&= \frac{1}{L^3}\int d^3\mathbf{R}\, e^{i\frac{\mathbf{q}\bullet\mathbf{R}}{\hbar c}} \prod_{j=1}^{Z} d^3\mathbf{r}_j\, \psi_n^*(\mathbf{r}_1, \ldots \mathbf{r}_Z) U(\mathbf{R}) \psi_0(\mathbf{r}_1, \ldots \mathbf{r}_Z) \\
&= -zZ\frac{\alpha\hbar c}{L^3}\int d^3\mathbf{R} \frac{e^{i\frac{\mathbf{q}\bullet\mathbf{R}}{\hbar c}}}{R} \prod_{j=1}^{Z} d^3\mathbf{r}_j\, \psi_n^*(\mathbf{r}_1, \ldots \mathbf{r}_Z) \psi_0(\mathbf{r}_1, \ldots \mathbf{r}_Z) \\
&= 0
\end{aligned} \tag{7.136}$$

due to the orthonormality of the two states.

The Coulomb potential between the projectile and the Z atomic electrons is given by,

$$U(\mathbf{r}) = -(z\,\alpha\hbar c)\sum_{j=1}^{Z} \frac{1}{|\mathbf{R} - \mathbf{r}_j|}. \tag{7.137}$$

The exchange process of transverse photons between the projectile and an electron has two intermediate states (assume that the electron can be treated as at rest). In the first, the projectile emits a photon of momentum $\mathbf{q} = \mathbf{p} - \mathbf{p}'$ which is absorbed by the electron to give it a momentum $\mathbf{q}$. In the second, the electron emits a photon with momentum $-\mathbf{q}$, to give it a momentum $\mathbf{q}$. The photon is absorbed by the projectile to give it a momentum $\mathbf{p}' = \mathbf{p} - \mathbf{q}$. From the derivation of the Klein–Nishina cross section in Chap. 6, the photon emission by the projectile is proportional to the matrix element of $z(\alpha\hbar c/e)(\boldsymbol{\alpha} \bullet \hat{\boldsymbol{\varepsilon}}_m)e^{-i\frac{\mathbf{q}\bullet\mathbf{r}}{\hbar c}}$ where $\boldsymbol{\alpha}$ is the Dirac velocity operator of the projectile and $\hat{\boldsymbol{\varepsilon}}_m$ is the photon's unit polarization vector and where $m = 1,2$ (see Sect. 24 of Heitler (1984) for a more detailed description). The absorption of this photon by the jth atomic electron is proportional to a matrix element of $\frac{z\alpha\hbar c}{e}(\boldsymbol{\alpha}_j \bullet \hat{\varepsilon}_m)e^{i\frac{\mathbf{q}\bullet\mathbf{r}_j}{\hbar c}}$. For clarity, the Bethe collision stopping power due to the static Coulomb potential (longitudinal excitation) only will be derived and the final form, which includes the relativistic term, provided. Full derivations of the latter can be found in the review articles by Fano (1964) and Ahlen (1980).

The transition rate from the ground state $|0\rangle$, with an energy eigenvalue considered here to be zero, to the state $|n\rangle$, with energy eigenvalue E_n, through longitudinal excitation is given conveniently by Fermi's Golden Rule No. 2,

$$\lambda_{fi,long} = \frac{2\pi}{\hbar}|\langle \mathbf{p}', n|U|\mathbf{p}, 0\rangle|^2 \rho_f.$$

The phase–space term is common to both longitudinal and transverse excitations and is of the usual form $\rho_f\, dT' = \left(L/2\pi\hbar c\right)^3 d^3\mathbf{p}'$. This derivation of the phase–space term will parallel that of the elastic Coulomb scatter calculation except that the energy transfer rather than the momentum transfer will be used as the kinematic variable. The density of final sates is, as calculated before,

$$\rho_f = \frac{1}{4\pi^2\beta'}\left(\frac{L}{\hbar c}\right)^3 p'^2\, dp'\, d(\cos\theta). \tag{7.138}$$

From the inelastic momentum transfer, $d(\cos\theta) = (q/pp')$ and the momentum transfer is $q = \sqrt{Q(Q + 2m_e)}$ from which one obtains $q\,dq = m_e(1 + Q/m_e)dQ$. This gives the density of final states as,

$$\rho_f = \frac{m_e}{4\pi^2\beta'}\left(\frac{L}{\hbar c}\right)^3\left(\frac{p'}{p}\right)\left(1 + \frac{Q}{m_e}\right)dQ'. \tag{7.139}$$

For soft collisions, the energy transfer is much less than the projectile's kinetic energy (i.e., $Q \ll T$) and we can approximate $p'/p \approx \beta'/\beta \approx 1$ to give,

$$\rho_f \approx \frac{m_e}{4\pi^2\beta}\left(\frac{L}{\hbar c}\right)^3\left(1+\frac{Q}{m_e}\right)dQ. \tag{7.140}$$

The matrix element due to the unretarded Coulomb interactions between the projectile and the atomic electrons is calculated in the same fashion as in our calculation of the elastic scatter cross section,

$$\langle \mathbf{p}', n|U|\mathbf{p}, 0\rangle = \int d^3\mathbf{r}\, d^3\mathbf{r}' \langle \mathbf{p}', n|\mathbf{r}\rangle\langle \mathbf{r}|U|\mathbf{r}'\rangle\langle \mathbf{r}'|\mathbf{p}, 0\rangle. \tag{7.141}$$

The method of solving this matrix element is identical to that used in previous derivations in Chaps. 3 and 6,

$$\begin{aligned}\langle \mathbf{p}', n|U|\mathbf{p}, 0\rangle &= -4\pi\frac{z\,\alpha\hbar c}{L^3}\left(\frac{\hbar c}{q}\right)^2\left\langle n\left|\sum_{j=1}^{Z} e^{i\frac{\mathbf{q}\bullet\mathbf{r}_j}{\hbar c}}\right|0\right\rangle \\ &= -4\pi\frac{z\,\alpha\hbar c}{L^3}\left(\frac{\hbar c}{q}\right)^2 F_n(\mathbf{q}, Z).\end{aligned} \tag{7.142}$$

where $F_n(\mathbf{q}, Z)$ is the inelastic scattering form factor. The transition rate can be considered independent of azimuthal angle if the target atom is in the s-state (i.e., spherically symmetric) or if the target atoms in the medium are randomly oriented (as is the case in medical irradiation). This permits one to replace the vector momentum transfer $\mathbf{q}$ in the argument of the inelastic form factor with its scalar value, q, to give $F_n(q, Z)$. The squared amplitude of the matrix element is, in terms of the energy transfer,

$$\begin{aligned}|\langle \mathbf{p}', n|U|\mathbf{p}, 0\rangle|^2 &= 16\pi^2\frac{(z\,\alpha\hbar c)^2}{L^6}\left(\frac{\hbar c}{q}\right)^4|F_n(q, Z)|^2 \\ &= 4\pi^2\frac{(z\,\alpha\hbar c)^2}{L^6}\frac{(\hbar c)^4}{m_e^2Q^2\left(1+\frac{Q}{2m_e}\right)^2} \\ &\quad\times |F_n(q, Z)|^2.\end{aligned} \tag{7.143}$$

The inelastic Coulomb scatter transition rate for longitudinal excitation is,

$$\begin{aligned}\lambda_{fi,long} &= \frac{2\pi}{\hbar}|\langle \mathbf{p}', n|U|\mathbf{p}, 0\rangle|^2\rho_f \\ &= 2\pi\frac{(z\,\alpha\hbar c)^2 c}{\beta L^3 m_e}\frac{|F_n(q, Z)|^2}{Q^2\left(1+\frac{Q}{2m_e}\right)^2}\left(1+\frac{Q}{m_e}\right)dQ\end{aligned} \tag{7.144}$$

The differential cross section is the transition rate normalized to the incident particle flux, v/V, or,

$$\begin{aligned}d\sigma_{long} &= \frac{L^3}{\beta c}\lambda_{fi,long} \\ &= 2\pi\frac{z^2 m_e r_0^2}{\beta^2}\frac{|F_n(q, Z)|^2}{Q^2\left(1+\frac{Q}{2m_e}\right)^2}\left(1+\frac{Q}{m_e}\right)dQ.\end{aligned} \tag{7.145}$$

The inelastic form factor $F_n(q, Z)$ is related to the generalized (dipole) oscillator strength, GOS (which has already been seen in the derivation of the Bohr energy loss), which is a generalization of the optical oscillator strength (see, e.g., Fernández-Varea 1998) given, in the nomenclature used here, by,

$$f_n(q, Z) = \frac{E_n}{Q}|F_n(q, Z)|^2. \tag{7.146}$$

Analytical representations for the GOS are available for atomic hydrogen and the free electron gas for which the initial and final states are analytically calculable. In the more practical cases of heavier atoms and molecules, these wavefunctions are calculated numerically. This discussion will be limited to the simple details of the GOS required for obtaining an expression for the stopping power. The inelastic form factor can be related to the GOS per unit energy transfer,

$$\frac{df_n(q, Z)}{dE_n} = \frac{1}{Q}|F_n(q, Z)|^2. \tag{7.147}$$

The energy-weighted sum of the oscillator strengths equals the total number of electrons in the atom which, in integral form, is,

$$\int dE_n\frac{df_n(q, Z)}{dE_n} = Z. \tag{7.148}$$

The GOS can also be used to describe the mean ionization energy of the atomic system as the first energy moment of the oscillator strength distribution,

$$\ln \bar{I} = \frac{\int dE \frac{df}{dE} \ln E}{\int dE \frac{df}{dE}} = \frac{1}{Z} \int dE \frac{df_n}{dE} \ln E. \tag{7.149}$$

The GOS per unit energy transfer has a useful interpretation at low momentum transfers. Writing the GOS per unit excitation energy in bra-ket form,

$$\begin{aligned} \frac{df_n(q,Z)}{dE_n} &= \frac{1}{Q} |F_n(q,Z)|^2 \\ &= \frac{1}{Q} \left| \left\langle n \left| \sum_{j=1}^{Z} e^{i\frac{q r_j}{\hbar c}} \right| 0 \right\rangle \right|^2 \\ &= \frac{2m_e}{q^2} \left| \left\langle n \left| \sum_{j=1}^{Z} e^{i\frac{q r_j}{\hbar c}} \right| 0 \right\rangle \right|^2 \end{aligned} \tag{7.150}$$

where, for low q, the nonrelativistic relationship between momentum and energy transfer has been used. Expanding the exponential to first order for small q,

$$\begin{aligned} \frac{df_n(q,Z)}{dE_n} &\approx \frac{2m_e}{q^2} \left| Z\langle n|0\rangle + \frac{i}{\hbar c} \left\langle n \left| \sum_{j=1}^{Z} q\, r_j \right| 0 \right\rangle \right|^2 \\ &\approx -\frac{2m_e}{(\hbar c)^2} \left| \left\langle n \left| \sum_{j=1}^{Z} r_j \right| 0 \right\rangle \right|^2 \quad \text{small } q \end{aligned} \tag{7.151}$$

where the orthogonality relationship $\langle n|0\rangle = 0$ has been used. $df_n(q,Z)/dE_n \to df_n(Z)/dE_n$ for small q, where $df_n(Z)/dE_n$ is the optical oscillator strength per unit excitation energy and which is proportional to the square of the dipole-matrix element.

Having introduced the GOS and its properties at low q, the expression of the differential cross section for longitudinal excitation is then simplified,

$$\begin{aligned} d\sigma_{long} &= 2\pi \frac{z^2 m_e r_0^2}{\beta^2} \frac{|F_n(q,Z)|^2}{Q^2\left(1+\frac{Q}{2m_e}\right)^2} \left(1+\frac{Q}{m_e}\right) dQ \\ &= 2\pi \frac{z^2 m_e r_0^2}{\beta^2} \frac{\left(1+\frac{Q}{m_e}\right)}{Q^2\left(1+\frac{Q}{2m_e}\right)^2} \frac{df_n(q,Z)}{dE_n} dQ \\ &= \frac{C}{N_A} \left(\frac{z}{\beta}\right)^2 \frac{\left(1+\frac{Q}{m_e}\right)}{Q^2\left(1+\frac{Q}{2m_e}\right)^2} \frac{df_n(q,Z)}{dE_n} dQ \end{aligned} \tag{7.152}$$

For $Q \ll m_e$, the cross section simplifies to,

$$d\sigma_{long} = \frac{C}{N_A} \left(\frac{z}{\beta}\right)^2 \frac{dQ}{Q} \frac{df_n(q,Z)}{dE_n}. \tag{7.153}$$

For calculational convenience, the kinetic variable is temporarily changed from the energy transfer to the momentum transfer,

$$d\sigma_{long} = \frac{2C}{N_A} \left(\frac{z}{\beta}\right)^2 \frac{dq}{q} \frac{df_n(q,Z)}{dE_n}. \tag{7.154}$$

Integrating over the momentum transfer yields the total cross section,

$$\sigma_{long} = \frac{2C}{N_A} \left(\frac{z}{\beta}\right)^2 \int_{q_{min}}^{q_{max}} dq \frac{1}{q} \frac{df_n(q,Z)}{dE_n}. \tag{7.155}$$

As q is small for soft collisions, $df_n(q,Z)/dE_n$ is replaced with the optical oscillator strength per unit excitation energy, $df_n(Z)/dE_n$. Then it is removed from the integrand to obtain the total cross section,

$$\begin{aligned} \sigma_{n,long} &= \frac{2C}{N_A} \left(\frac{z}{\beta}\right)^2 \frac{df_n(Z)}{dE_n} \int_{q_{min}}^{q_{max}} \frac{dq}{q} \\ &= \frac{2C}{N_A} \left(\frac{z}{\beta}\right)^2 \frac{df_n(Z)}{dE_n} \ln\left(\frac{q_{max}}{q_{min}}\right). \end{aligned} \tag{7.156}$$

This result highlights the fundamental difference between the Bohr and Bethe theories. Bohr's theory uses the impact parameter to distinguish between soft and hard collisions. This is clearly not possible in quantum theory in which the localization of a wave packet of a particle with well-defined momentum is limited by the uncertainty principle. Hence, one would expect the classical theory to break down for small impact parameters. In the Bethe theory, momentum (or energy) transfer is used to separate the soft and hard collision regimes. For later convenience when expressions for soft and hard collision stopping powers are merged to determine the complete collision stopping power, instead of using the momentum transfer, the use the energy transfer will be returned to as a means of defining soft and hard collisions. The collision is said to be "soft" if $Q < Q_C$ and to be "hard" if $Q > Q_C$. The exact specification of the transition energy transfer Q_C is insignificant as this quantity will cancel out when the expressions for the soft and hard collision stopping powers are summed. Even so, limits should be applied to Q_C in order to ensure that the necessary approximations used in the derivations remain valid. Q_C must exceed atomic binding energies but it must also not be sufficiently great that the projectile's de Broglie wavelength becomes comparable to nuclear dimensions. A value of between 10 and 100 keV for Q_C would allow both conditions to be simultaneously met (Uehling 1954).

The limits of the momentum transfers which define a soft collision are now calculated. The lower limit, q_{min}, is given by $q_{\|} = E_n/v$. Clearly, q_{max} will be set by the energy transfer separation between soft and hard collisions,

$$q_{max} = \sqrt{2m_e Q_C} \tag{7.157}$$

Applying these limits to the momentum transfer,

$$\begin{aligned}\sigma_{n,long} &= \frac{2C}{N_A}\left(\frac{z}{\beta}\right)^2 \frac{df_n}{dE_n} \ln\left(\frac{\sqrt{2m_e Q_C}}{E_n} v\right) \\ &= \frac{C}{N_A}\left(\frac{z}{\beta}\right)^2 \frac{df_n}{dE_n} \ln\left(\frac{2m_e Q_C}{E_n^2} \beta^2\right)\end{aligned} \tag{7.158}$$

Bethe Soft Collision Stopping Power

The mean energy transfer per unit fluence is,

$$\begin{aligned}\overline{\Delta E}_{long} &= \int dE_n\, \sigma_{n,long} \\ &= \frac{C}{N_A}\left(\frac{z}{\beta}\right)^2 \int dE_n \frac{df_n}{dE_n} \ln\left(\frac{2m_e Q_C}{E_n^2} \beta^2\right).\end{aligned} \tag{7.159}$$

The logarithm is split,

$$\ln\left(\frac{2m_e Q_C}{E_n^2} \beta^2\right) = -2 \ln E_n + \ln(2m_e Q_C \beta^2) \tag{7.160}$$

where it is implicitly required for E_n, m_e, and Q_C to have the same units of energy. This enables the integral of (7.159) to be written as,

$$\begin{aligned}\int dE_n \frac{df_n}{dE_n} \ln\left(\frac{2m_e Q_c}{E_n^2} \beta^2\right) &= -2 \int dE_n \frac{df_n}{dE_n} \ln E_n \\ &\quad + \ln(2m_e Q_c \beta^2) \int dE_n \frac{df_n}{dE_n} \\ &= -2Z \ln \bar{I} + Z \ln(2m_e Q_c \beta^2) \\ &= Z \ln\left(\frac{2m_e Q_c}{\bar{I}^2} \beta^2\right).\end{aligned} \tag{7.161}$$

Then, the mean energy loss per interaction as,

$$\overline{\Delta E}_{long} = C\left(\frac{Z}{N_A}\right)\left(\frac{z}{\beta}\right)^2 \ln\left(\frac{2m_e Q_C}{\bar{I}^2} \beta^2\right) \tag{7.162}$$

and the mass soft collision stopping power due to longitudinal excitations only is,

$$\left(\frac{dE}{\rho\, dx}\right)_{Col,S,long} = C\left(\frac{Z}{A}\right)\left(\frac{z}{\beta}\right)^2 \ln\left(\frac{2m_e Q_C}{\bar{I}^2} \beta^2\right) \tag{7.163}$$

This is the quantum-mechanical result of the energy transfer between a charged projectile and an atom due to an unretarded Coulomb potential and which is the nonrelativistic result of the Bethe theory. As shown by Fano (1964) and Ahlen (1980), the full form of the Bethe mass soft collision stopping power, accounting for both longitudinal and transverse excitations, is slightly modified from this result,

$$\begin{aligned}\left(\frac{dE}{\rho dx}\right)_{Col,S} &= C\left(\frac{Z}{A}\right)\left(\frac{z}{\beta}\right)^2 \left[\ln\left(\frac{2m_e Q_C}{\bar{I}^2} \gamma^2 \beta^2\right) - \beta^2\right] \\ &= C\left(\frac{Z}{A}\right)\left(\frac{z}{\beta}\right)^2 \left[\ln\left(\frac{2m_e Q_C}{\bar{I}^2} \beta^2\right) \right. \\ &\quad \left. + \ln\left(\frac{1}{1-\beta^2}\right) - \beta^2\right]\end{aligned} \tag{7.164}$$

where the $\ln(1/1-\beta^2)$ and β^2 terms arise from retarded transverse photon interactions. It can be seen that, since these two terms go to zero as $\beta \rightarrow 0$, this relativistic form reduces to the nonrelativistic result.

7.3.3.4 Comparison of Bohr and Bethe Soft Collision Theories

The Bohr classical and Bethe quantum-mechanical results are compared by considering the mass soft

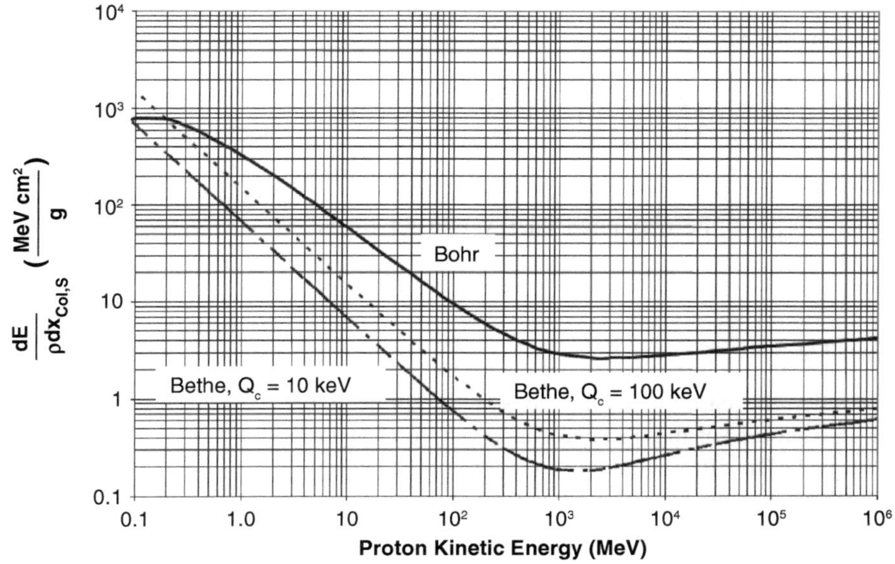

Fig. 7.10 Mass soft collision stopping powers calculated from the Bohr and Bethe theories for protons ranging in kinetic energy from 100 keV to 1 TeV in carbon. The Bethe results are shown for two values of Q_C which separates the energy transfers of soft and hard collisions

collision stopping powers for protons in carbon as given in Fig. 7.10. For the Bethe result, the mean ionization potential has been set to 6hcR, where the Rydberg energy is $hcR_\infty = 13.61$ eV, and stopping powers calculated for the extrema for Q_C equal to 10 and 100 keV. For the Bohr result, the mean oscillator frequency has been set to $\bar{\omega} = 6\,hcR_\infty/\hbar = 1.24 \times 10^{17}\,s^{-1}$. It should be recalled that the derivations of the Bohr and Bethe soft collision theories have been with single ground state atoms or electrons or, in other words, the medium through which the projectile travels has been treated as a cold and dilute monatomic gas rather than as a condensed medium. The results of Fig. 7.10 are those for "carbon" in so far as we have calculated for a homogeneous medium in which $Z = 6$ and $A = 12$. The Bohr result is explicitly truncated for kinetic energies less than 0.1 MeV: the calculated soft collision stopping power changes sign, corresponding to the unphysical condition of the gain of energy by the particle, due to the condition of $1.123\,\gamma^2\beta^3 c/(zr_0\bar{\omega}) \leq \beta^2/2$ at low energies.

Although the two theories do not agree quantitatively (except at low projectile kinetic energies), they exhibit similar behaviors by demonstrating a decreasing stopping power with increasing projectile energy proportional to β^{-2}. In both cases the mass collision stopping powers reach broad minima at proton kinetic energies of about 3 GeV and then exhibit a slow increase. The magnitude of the Bohr mass soft collision stopping power is greater than that obtained from the Bethe theory by roughly a factor of 5. The Bethe result shows a slight dependence upon the selection of Q_C with that calculated for $Q_C = 100$ keV being about a factor of 2 greater than that calculated for $Q_C = 10$ keV.

7.3.4 Hard Collision Stopping Power

7.3.4.1 Introduction

In a hard collision, the projectile interacts with a single atomic electron at a speed much greater than the orbital speed thus allowing the target electron to be assumed to be at rest. In this case, the collision can be treated as being elastic.

7.3.4.2 Differential Cross Sections in Energy Transfer

Massive Projectile Electron Scatter ($m \gg m_e$)

Spin-0

Consider a massive spin-zero projectile (e.g., an α particle) of charge ze with kinetic energy T. The differential cross section for the energy transfer between Q and Q + dQ to an electron at rest is (Bhabha 1938),

$$\frac{d\sigma}{dQ} = 2\pi r_0^2 m_e \left(\frac{z}{\beta}\right)^2 \frac{1}{Q^2}\left(1 - \beta^2 \frac{Q}{Q_{max}}\right) \quad \text{Spin-0} \tag{7.165}$$

where Q_{max} is the maximum energy transfer to the electron

Spin-1/2

Consider the case of a massive spin-1/2 projectile (e.g., a proton) of mass m and charge ze with kinetic energy T. The differential cross section for the energy transfer between Q and Q + dQ to the electron at rest is (Bhabha 1938; Massey and Corben 1939),

$$\frac{d\sigma}{dQ} = 2\pi r_0^2 m_e \left(\frac{z}{\beta}\right)^2 \frac{1}{Q^2} \times \left[1 - \beta^2 \frac{Q}{Q_{max}} + \frac{1}{2}\left(\frac{Q}{T+m}\right)^2\right] \quad \text{Spin-}^1\!/_2. \tag{7.166}$$

Spin-1

Finally, for completeness, consider the case of a massive spin-1 particle with mass m, charge ze, and kinetic energy T. The differential cross section for the energy transfer between Q and Q + dQ to an electron at rest is (Massey and Corben 1939; Oppenheimer et al., 1940) is,

$$\frac{d\sigma}{dQ} = 2\pi r_0^2 m_e \left(\frac{z}{\beta}\right)^2 \frac{1}{Q^2}\left[\left(1 - \beta^2 \frac{Q}{Q_{max}}\right)\left(1 + \frac{1}{3}\frac{Q}{Q_0}\right) + \frac{1}{3}\left(\frac{Q}{T+m}\right)^2\left(1 + \frac{1}{2}\frac{Q}{Q_0}\right)\right] \quad \text{Spin-1} \tag{7.167}$$

where the energy Q_0 is defined as $Q_0 = m^2/m_e$.

It will be noted that, for low projectile energies and low-recoil kinetic energies, the above differential cross sections for spin-0, spin-1/2, and spin-1 massive projectiles reduce to the classical Rutherford result. Hence, spin contributions to the differential cross section become significant only at high projectile energies.

Electron–Electron (Møller) Scatter

Now consider the case of the projectile being an electron with kinetic energy, T. The Feynman diagrams of electron–electron (Møller) scatter are shown in Fig. 7.11. Two graphs necessarily arise as a result of the inability to distinguish between the two exiting electrons of which was the projectile or the target. From Chap. 2, the maximum energy transferred to the target electron is equal to the kinetic energy of the incident. Because of the indistinguishability between the projectile and target electrons, the exiting electron with the highest energy is assumed to be the primary. As the electron is a fermion, the wavefunction of an electron pair system must be antisymmetric in the interchange of the two electrons. Should the electron spins be parallel (i.e., the system is in the triplet state), the system will be symmetric under the exchange of spins thus requiring the spatial wavefunctions to be antisymmetric under the exchange of the electron's relative coordinates. Hence, the triplet state scattering amplitude is,

$$f_t(\theta) = f(\theta) - f(\pi - \theta) \tag{7.168}$$

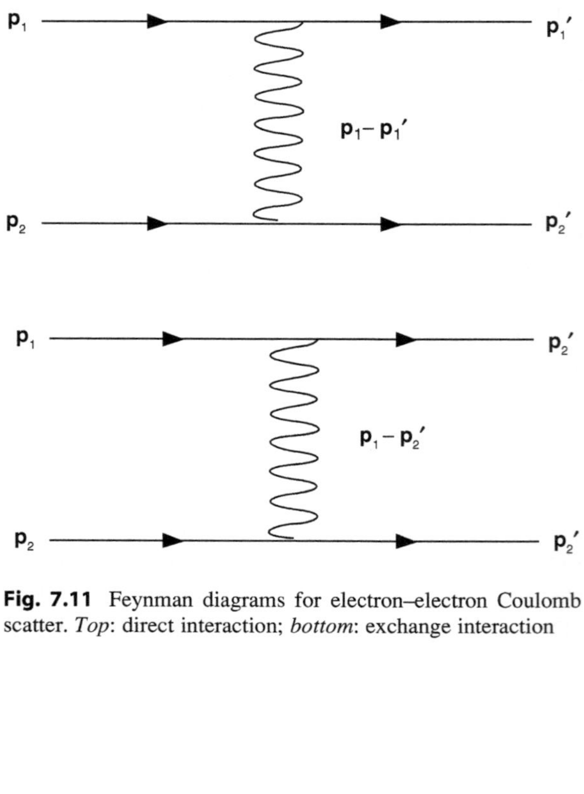

Fig. 7.11 Feynman diagrams for electron–electron Coulomb scatter. *Top*: direct interaction; *bottom*: exchange interaction

where $f(\theta)$ is the scattering amplitude of elastic Coulomb scatter. The transformation from θ to $\pi - \theta$ is the equivalent of the interchange of the two electrons. For antiparallel spins (i.e., the singlet state), the system is antisymmetric under the interchange of spins and, as a result, the spatial wavefunctions must be symmetric,

$$f_s(\theta) = f(\theta) + f(\pi - \theta). \qquad (7.169)$$

In dosimetry calculations, the projectile and target electrons are considered to be unpolarized and e^-e^- scatter has a random distribution of spins. Hence, singlet and triplet states will have a ratio of relative probabilities of 1:3 and the differential cross section is,

$$\begin{aligned}\frac{d\sigma}{d\Omega} &= \frac{1}{4}|f_s(\theta)|^2 + \frac{3}{4}|f_t(\theta)|^2 \\ &= \frac{1}{4}|f(\theta) + f(\pi-\theta)|^2 + \frac{3}{4}|f(\theta) - f(\pi-\theta)|^2 \\ &= |f(\theta)|^2 + |f(\pi-\theta)|^2 - |f(\theta)||f(\pi-\theta)| \\ &= \left(\frac{m_e\,\alpha\hbar c}{2p^2}\right)^2\left(\frac{1}{\sin^4\frac{\theta}{2}} + \frac{1}{\cos^4\frac{\theta}{2}} - \frac{1}{\sin^2\frac{\theta}{2}\cos^2\frac{\theta}{2}}\right) \\ &= \left(\frac{\alpha\hbar c}{4T}\right)^2\left(\frac{1}{\sin^4\frac{\theta}{2}} + \frac{1}{\cos^4\frac{\theta}{2}} - \frac{1}{\sin^2\frac{\theta}{2}\cos^2\frac{\theta}{2}}\right).\end{aligned} \qquad (7.170)$$

The first term is that of the elastic Coulomb scatter cross section, whereas the second reflects the impossibility of distinguishing between the incident and scattered electrons. The third "cross" term is the exchange term. The full relativistic Møller differential cross section in energy transfer is, for an incident electron of kinetic energy T transferring an energy between Q and Q + dQ to another electron (Møller 1932; Rohrlich and Carlson 1954),

$$\begin{aligned}\frac{d\sigma}{dQ} = \frac{\pi}{T}\left(\frac{\alpha\hbar c}{T}\right)^2\Bigg[&\left(\frac{T}{Q}\right)^2 + \left(\frac{Q}{T-Q}\right) + \left(\frac{\gamma-1}{\gamma}\right)^2 \\ &- \left(\frac{2\gamma-1}{\gamma^2}\right)\left(\frac{T}{Q}\right)\left(\frac{Q}{T-Q}\right)\Bigg]\end{aligned} \qquad (7.171)$$

This describes the probability that, following Møller scatter, one electron has a kinetic energy of Q and the other has T − Q. Thus, all possible outcomes of the scatter are obtained for Q ranging in value from 0 to *T*/2. The nonrelativistic form of the differential cross section is had by setting $\gamma = 1$,

$$\frac{d\sigma}{dQ} = \frac{\pi}{T}\left(\frac{\alpha\hbar c}{T}\right)^2\left[\left(\frac{T}{Q}\right)^2 + \left(\frac{Q}{T-Q}\right) - \left(\frac{T}{Q}\right)\left(\frac{Q}{T-Q}\right)\right] \qquad (7.172)$$

Electron–Positron (Bhabha) Scatter

Now consider the case of a positron projectile with kinetic energy, T. The Feynman diagrams for positron–electron (Bhabha) scatter are given in Fig. 7.12. While the first graph is similar to that of Møller scatter, the fact that it is possible for the electron and positron to annihilate requires the provision of an additional graph accounting for the production of a virtual photon from which the exiting electron–positron pair is created. The differential cross section that an incident positron with kinetic energy T will suffer a kinetic energy loss between and Q and Q + dQ which is transferred to the target electron is (Bhabha 1936),

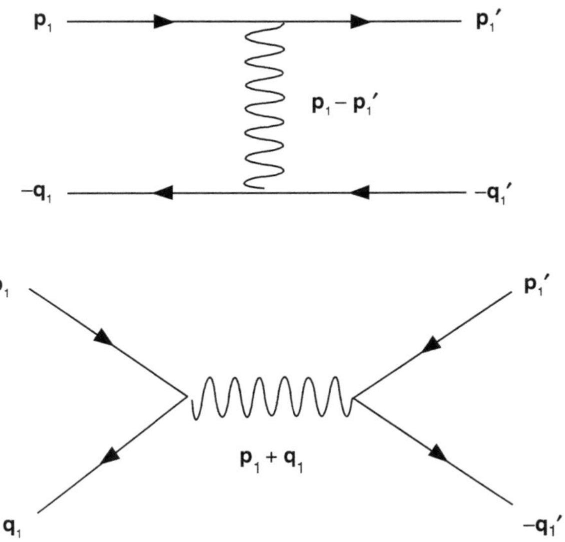

Fig. 7.12 Feynman diagrams for positron–electron Coulomb scatter. *Top*: direct interaction; *bottom*: annihilation intermediate-state interaction

$$\frac{d\sigma}{dQ}=\frac{2\pi r_0^2 m_e}{\beta^2 T^2}\left[\left(\frac{T}{Q}\right)^2-\left(\frac{\gamma^2-1}{\gamma^2}\right)\left(\frac{T}{Q}\right)+\frac{1}{2}\left(\frac{\gamma-1}{\gamma}\right)^2\right.$$
$$-\left(\frac{\gamma-1}{\gamma+1}\right)\left\{\left(\frac{\gamma+2}{\gamma}\right)\left(\frac{T}{Q}\right)-2\left(\frac{\gamma^2-1}{\gamma^2}\right)\right.$$
$$\left.+\left(\frac{\gamma-1}{\gamma}\right)^2\left(\frac{Q}{T}\right)\right\}+\left(\frac{\gamma-1}{\gamma+1}\right)^2$$
$$\left.\times\left\{\frac{1}{2}+\frac{1}{\gamma}+\frac{3}{2\gamma^2}-\left(\frac{\gamma-1}{\gamma}\right)^2\left(\frac{Q}{T}\right)\left(1-\left(\frac{Q}{T}\right)\right)\right\}\right]. \tag{7.173}$$

7.3.4.3 Hard Collision Stopping Powers

Massive Projectiles ($m \gg m_e$)

As the differential cross sections in energy transfer for massive spin-1/2 and spin-1 projectiles reduce to that for a massive spin-0 projectile at low kinetic energies, the spin-0 case is considered first. The general expression for the mass hard collision stopping power is,

$$\left(\frac{dE}{\rho dx}\right)_{col,H}=N_A\left(\frac{Z}{A}\right)\int_{Q_C}^{Q_{max}} dQ\,Q\frac{d\sigma}{dQ} \tag{7.174}$$

where the integral limits are Q_C, which separates soft and hard collisions, and the maximum energy transferred to the target electron, Q_{max}, which is set by the relevant kinematics as shown in Chap. 2. The mass hard collision stopping power for a massive spin-0 particle is,

$$\left(\frac{dE}{\rho dx}\right)_{col,H}=N_A\left(\frac{Z}{A}\right)\int_{Q_C}^{Q_{max}} dQ\,Q\frac{d\sigma}{dQ}$$
$$=C\left(\frac{Z}{A}\right)\left(\frac{z}{\beta}\right)^2\int_{Q_C}^{Q_{max}}\frac{dQ}{Q}\left[1-\beta^2\frac{Q}{Q_{max}}\right]$$
$$=C\left(\frac{Z}{A}\right)\left(\frac{z}{\beta}\right)^2\left[\ln\left(\frac{Q_{max}}{Q_C}\right)\right.$$
$$\left.-\beta^2\left(\frac{Q_{max}-Q_C}{Q_{max}}\right)\right]. \tag{7.175}$$

In a hard collision, the energy transfer is assumed to be sufficiently high ($Q_{max} \gg Q_C$) to allow this to be simplified to the form,

$$\left(\frac{dE}{\rho dx}\right)_{Col,H}=C\left(\frac{Z}{A}\right)\left(\frac{z}{\beta}\right)^2\left[\ln\left(\frac{Q_{max}}{Q_C}\right)-\beta^2\right]\quad\text{Spin-0.} \tag{7.176}$$

Now consider the mass hard collision stopping power for a massive spin-1/2 projectile ,

$$\left(\frac{dE}{\rho dx}\right)_{col,H}=C\left(\frac{Z}{A}\right)\left(\frac{z}{\beta}\right)^2\int_{Q_C}^{Q_{max}}$$
$$\times\frac{dQ}{Q}\left[1-\beta^2\frac{Q}{Q_{max}}+\frac{1}{2}\left(\frac{Q}{T+m}\right)^2\right]$$
$$=C\left(\frac{Z}{A}\right)\left(\frac{z}{\beta}\right)^2\times\left[\ln\left(\frac{Q_{max}}{Q_C}\right)\right.$$
$$\left.-\beta^2\left(\frac{Q_{max}-Q_C}{Q_{max}}\right)+\frac{1}{4}\frac{Q_{max}^2-Q_C^2}{(T+m)^2}\right]$$
$$=C\left(\frac{Z}{A}\right)\left(\frac{z}{\beta}\right)^2\left[\ln\left(\frac{Q_{max}}{Q_C}\right)-\beta^2\right.$$
$$\left.+\left(\frac{Q_{max}}{2(T+m)}\right)^2\right]\quad\text{Spin-}1/2. \tag{7.177}$$

where $Q_{max} > Q_C$. Note that, for a massive spin-1/2 projectile with a rest mass much greater than the maximum energy transfer, the last squared-term in the square brackets can be neglected and this result reduces to the simpler spin-0 expression.

Electron and Positron Projectiles

The restricted mass hard collision stopping power for an electron projectile is,

$$\left(\frac{dE}{\rho\, dx}\right)_{Col,H,\Delta}=N_A\left(\frac{Z}{A}\right)\int_{Q_C}^{\Delta} dQ\,Q\frac{d\sigma}{dQ} \tag{7.178}$$

where the Møller differential cross section is to be used in the integral. Unlike the derivations of the massive particle hard collision stopping powers, an upper limit of integration $\Delta Q_{max} = T/2$ has been specified so as to ignore those energy transfers greater than Δ. Although the restricted mass collision stopping power can be defined for any projectile, the discussion here will be limited to the electron projectile for dosimetric interest.

The mass collision stopping power for the Møller cross section is,

$$\begin{aligned}\left(\frac{dE}{\rho dx}\right)_{Col,H,\Delta} &= C\left(\frac{Z}{A}\right)\frac{1}{\beta^2 T^2}\int_{Q_C}^{\Delta} dQQ \\ &\times\left[\left(\frac{T}{Q}\right)^2+\left(\frac{Q}{T-Q}\right)+\left(\frac{\gamma-1}{\gamma}\right)^2\right. \\ &\left.-\left(\frac{2\gamma-1}{\gamma^2}\right)\left(\frac{T}{Q}\right)\left(\frac{Q}{T-Q}\right)\right] \\ &= C\left(\frac{Z}{A}\right)\frac{1}{\beta^2}\left[\ln\frac{\Delta(T-\Delta)}{Q_C(T-Q)}\right. \\ &+\left(\frac{T}{T-\Delta}-\frac{T}{T-Q}\right)+\left(\frac{\gamma-1}{\gamma}\right)^2 \\ &\left.\times\left(\frac{\Delta^2-Q_C^2}{2T^2}\right)-\left(\frac{2\gamma-1}{\gamma^2}\right)\ln\left(\frac{T-Q_C}{T-\Delta}\right)\right].\end{aligned} \tag{7.179}$$

This expression can be simplified by defining the normalized kinematic variables,

$$\delta \equiv \frac{\Delta}{T} \tag{7.180}$$

$$\tau_C \equiv \frac{Q_C}{T} \tag{7.181}$$

to obtain,

$$\begin{aligned}\left(\frac{dE}{\rho dx}\right)_{Col,H,\Delta} &= C\left(\frac{Z}{A}\right)\frac{1}{\beta^2}\left[\ln\frac{\delta(1-\delta)}{\tau_C(1-\tau_C)}\right. \\ &+\left(\frac{1}{1-\delta}-\frac{1}{1-\tau_C}\right)+\left(\frac{\gamma-1}{\gamma}\right)^2 \\ &\left.\times\left(\frac{\delta^2-\tau_C^2}{2}\right)-\left(\frac{2\gamma-1}{\gamma^2}\right)\ln\left(\frac{1-\tau_C}{1-\delta}\right)\right].\end{aligned} \tag{7.182}$$

As the selection of Q_C is somewhat arbitrary, its value should ensure that $\tau_C \ll 1$ which allows the $\tau_C{}^2$ term to be neglected and to reduce this expression for the electron restricted mass hard collision stopping power to,

$$\begin{aligned}\left(\frac{dE}{\rho\, dx}\right)_{Col,H,\Delta} &= C\left(\frac{Z}{A}\right)\frac{1}{\beta^2}\left[\ln\frac{\delta(1-\delta)}{\tau_C}\right. \\ &+\left(\frac{1}{1-\delta}-1\right)+\left(\frac{\gamma-1}{\gamma}\right)^2\left(\frac{\delta^2}{2}\right) \\ &\left.+\left(\frac{2\gamma-1}{\gamma^2}\right)\ln(1-\delta)\right] \\ &- \text{Electron (restricted).}\end{aligned} \tag{7.183}$$

The unrestricted mass hard collision stopping power is simply that for the parameter $\delta = 1/2$, and,

$$\begin{aligned}\left(\frac{dE}{\rho\, dx}\right)_{Col,H,\Delta=T/2} &= C\left(\frac{Z}{A}\right)\frac{1}{\beta^2} \\ &\times\left[\ln\frac{1}{4\tau_C}+1+\frac{1}{8}\left(\frac{\gamma-1}{\gamma}\right)^2\right. \\ &\left.-\left(\frac{2\gamma-1}{\gamma^2}\right)\ln 2\right] \\ &- \text{Electron(unrestricted).}\end{aligned} \tag{7.184}$$

which corresponds to the result originally given by Rohrlich and Carlson (1954).

The unrestricted mass hard collision stopping power for a positron projectile is calculated as for an electron project but with the Bhabha differential cross section.

7.3.5 Combined Mass Hard and Soft Collision Stopping Powers

7.3.5.1 Introduction

Having now derived the soft and hard collision stopping powers, these can be combined to form the complete collision stopping power. The Bethe quantum-mechanical result will be used for the soft collision stopping power expression.

7.3.5.2 Massive Projectiles ($m \gg m_e$)

The Bethe mass collision stopping power, for when spin is neglected, is given by the sum of the Bethe mass soft collision and the mass hard collision stopping power of a massive spin-0 projectile,

$$\begin{aligned}\left(\frac{dE}{\rho dx}\right)_{Col} &= \left(\frac{dE}{\rho dx}\right)_{Col,S} + \left(\frac{dE}{\rho dx}\right)_{Col,H}\\ &= C\left(\frac{Z}{A}\right)\left(\frac{z}{\beta}\right)^2\left[\ln\left(\frac{2m_e Q_C}{\bar{I}^2}\gamma^2\beta^2\right)\right.\\ &\quad\left. -\beta^2 + \ln\left(\frac{Q_{max}}{Q_C}\right) - \beta^2\right]\\ &= C\left(\frac{Z}{A}\right)\left(\frac{z}{\beta}\right)^2\left[\ln\left(\frac{2m_e Q_{max}}{\bar{I}^2}\gamma^2\beta^2\right) - 2\beta^2\right].\end{aligned} \tag{7.185}$$

From Chap. 2, the maximum energy transfer to an electron for the case of a heavy projectile is $Q_{max} = 2m_e\gamma^2\beta^2$. Inserting this gives the complete mass collision stopping power for a massive spin-0 projectile,

$$\left(\frac{dE}{\rho\, dx}\right)_{Col} = 2C\left(\frac{Z}{A}\right)\left(\frac{z}{\beta}\right)^2\left[\ln\left(\frac{2m_e\gamma^2\beta^2}{\bar{I}}\right) - \beta^2\right] \quad \text{Spin-0.} \tag{7.186}$$

For a massive spin-1/2 projectile (proton),

$$\begin{aligned}\left(\frac{dE}{\rho dx}\right)_{Col} &= \left(\frac{dE}{\rho dx}\right)_{Col,S} + \left(\frac{dE}{\rho dx}\right)_{Col,H}\\ &= C\left(\frac{Z}{A}\right)\left(\frac{z}{\beta}\right)^2\left[\ln\left(\frac{2m_e Q_C}{\bar{I}^2}\gamma^2\beta^2\right) - \beta^2\right.\\ &\quad\left. + \left(\frac{Q_{max}}{2(T+m)}\right)^2 + \ln\left(\frac{Q_{max}}{Q_C T_{e,C}}\right) - \beta^2\right]\\ &= C\left(\frac{Z}{A}\right)\left(\frac{z}{\beta}\right)^2\left[\ln\left(\frac{2m_e Q_{max}}{\bar{I}^2}\gamma^2\beta^2\right)\right.\\ &\quad\left. -2\beta^2 + \left(\frac{Q_{max}}{2(T+m)}\right)^2\right] = 2C\left(\frac{Z}{A}\right)\left(\frac{z}{\beta}\right)^2\\ &\quad\times\left[\ln\left(\frac{2m_e\gamma^2\beta^2}{\bar{I}}\right) - \beta^2\right.\\ &\quad\left. + \left(\frac{1}{2}\frac{m_e}{(T+m)}\gamma^2\beta^2\right)^2\right] \quad \text{Spin-}^1\!/_2.\end{aligned} \tag{7.187}$$

As $m \gg m_e$, it is clear that the mass collision stopping power for a massive spin-1/2 projectile reduces to that for a spin-0 projectile at low kinetic energies (i.e., $\gamma^2\beta^2 \to 0$). Figure 7.13 shows the mass collision stopping power calculated for a proton in carbon and lead. Here, the mean ionization potentials for carbon and lead are taken to be equal to 78 and 823 eV, respectively. Both curves of $(dE/\rho\, dx)_{Col}$ exhibit the same characteristic behavior of a decrease

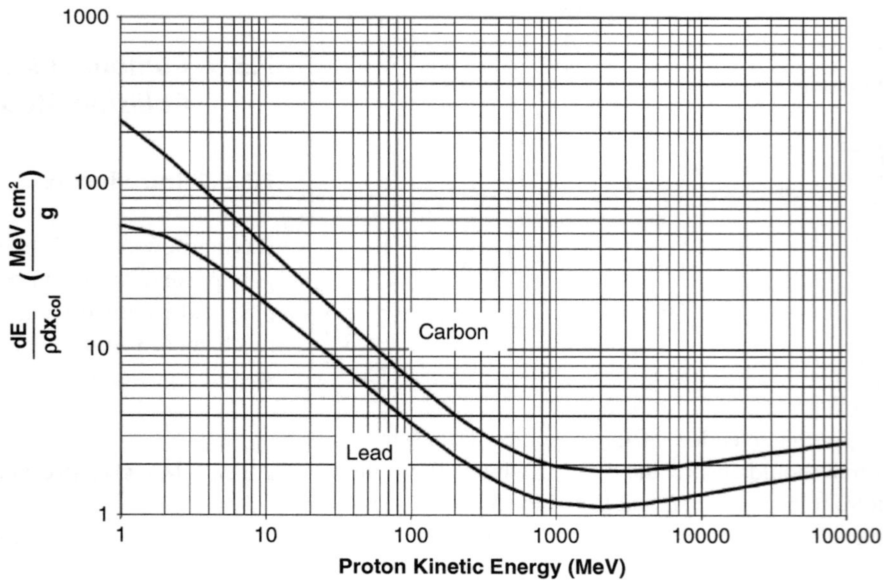

Fig. 7.13 Mass collision stopping powers for protons in carbon and lead

with increasing kinetic energy, due to the β^{-2} factor, to a broad minimum of about 1–2 MeV cm^2/g which occurs at a particle kinetic energy equal to about three times its rest mass. This is referred to as the minimally-ionizing region. With increasing kinetic energy, the mass collision stopping power increases logarithmically due to the $\ln(\gamma^2\beta^2) - \beta^2$ term. This increase is monotonic as the medium is still considered to be a dilute monatomic gas, but could be quenched in condensed media due to polarization of the medium. It should be noted that, for a given projectile kinetic energy, the mass collision stopping power for carbon ($Z = 6$) is greater than that for lead ($Z = 82$). This is a result of the energy loss being dominated by interactions with atomic electrons over those with the nucleus and the electron density being proportional to the ratio Z/A. As a result, the ratio of the carbon to lead mass collision stopping powers is (excluding the effect of the mean ionization potential which is limited due to its placement within the logarithm),

$$\frac{\left(\frac{Z}{A}\right)_C}{\left(\frac{Z}{A}\right)_{Pb}} = \frac{\left(\frac{6}{12}\right)}{\left(\frac{82}{208}\right)} = 1.268$$

In other words, the mass collision stopping power for carbon is greater than that for lead due to the greater number of electrons per unit mass.

7.3.5.3 Electron and Positron Projectiles

To obtain the unrestricted mass collision stopping power of an electron projectile, the Bethe mass soft collision stopping power and the Møller mass hard collision stopping power are summed,

$$\begin{aligned}\left(\frac{dE}{\rho dx}\right)_{Col} &= \left(\frac{dE}{\rho dx}\right)_{Col,S} + \left(\frac{dE}{\rho dx}\right)_{Col,H} \\ &= C\left(\frac{Z}{A}\right)\left(\frac{1}{\beta}\right)^2\left[\ln\left(\frac{2m_eQ_C}{\bar{I}^2}\gamma^2\beta^2\right)\right. \\ &\quad -\beta^2 + \ln\left(\frac{T}{4Q_C}\right) + 1 + \frac{1}{8}\left(\frac{\gamma-1}{\gamma}\right)^2 \\ &\quad \left.-\left(\frac{2\gamma-1}{\gamma^2}\right)\ln 2\right] = C\left(\frac{Z}{A}\right)\left(\frac{1}{\beta}\right)^2 \\ &\quad \times\left[\ln\left(\frac{m_eT}{2\bar{I}^2}\gamma^2\beta^2\right) + f^-(\gamma)\right]\end{aligned} \tag{7.188}$$

where T is the electron kinetic energy and

$$f^-(\gamma) = 1 - \beta^2 + \frac{1}{8}\left(\frac{\gamma-1}{\gamma}\right)^2 - \left(\frac{2\gamma-1}{\gamma^2}\right)\ln 2. \tag{7.189}$$

By using the relativistic relationship,

$$\begin{aligned} m_e\gamma^2\beta^2 &= m_e(\gamma^2 - 1) \\ &= T(\gamma + 1)\end{aligned}$$

the electron mass hard collision stopping power is simplified to,

$$\left(\frac{dE}{\rho dx}\right)_{Col} = C\left(\frac{Z}{A}\right)\left(\frac{1}{\beta}\right)^2\left[\ln\left(\frac{T^2(\gamma+1)}{2\langle I\rangle^2}\right) + f^-(\gamma)\right] \quad - \text{Electron(unrestricted)}. \tag{7.190}$$

The electron restricted mass collision stopping power (i.e., that which excludes energy transfers to the medium greater than Δ) is,

$$\begin{aligned}\left(\frac{dE}{\rho dx}\right)_{Col,\Delta} &= \left(\frac{dE}{\rho dx}\right)_{Col,S} + \left(\frac{dE}{\rho dx}\right)_{Col,H,\Delta} \\ &= C\left(\frac{Z}{A}\right)\left(\frac{1}{\beta}\right)^2\left[\ln\left(\frac{2m_eQ_C}{\bar{I}^2}\gamma^2\beta^2\right)\right. \\ &\quad -\beta^2 + \ln\left(\frac{\Delta\left(1-\frac{\Delta}{T}\right)}{Q_C}\right) + \left(\frac{1}{1-\left(\frac{\Delta}{T}\right)} - 1\right) \\ &\quad \left.+\left(\frac{\gamma-1}{\gamma}\right)^2\frac{\Delta^2}{2T^2} + \left(\frac{2\gamma-1}{\gamma}\right)\ln\left(1-\frac{\Delta}{T}\right)\right] \\ &= C\left(\frac{Z}{A}\right)\left(\frac{1}{\beta}\right)^2\left(\ln\left(\frac{2\Delta(T-\Delta)}{\langle I\rangle^2}(\gamma+1)\right)\right. \\ &\quad \left.+f^-(\gamma,\Delta)\right) \\ &\quad - \text{Electron (restricted)}.\end{aligned} \tag{7.191}$$

where,

$$f^-(\gamma,\Delta) = \left(\frac{\Delta}{T-\Delta}\right) - \beta^2 + \left(\frac{2\gamma-1}{\gamma}\right) \times \ln\left(\frac{T-\Delta}{T}\right) + \left(\frac{\gamma-1}{\gamma}\right)^2 \frac{\Delta^2}{2T^2}. \quad (7.192)$$

Similarly, to obtain the positron complete collision stopping power, the Bethe mass soft and the Bhabha mass hard collision stopping powers are summed,

$$\left(\frac{dE}{\rho dx}\right)_{\substack{Col\\ \text{Positron}}} = C\left(\frac{Z}{A}\right)\frac{1}{\beta^2}\left[\ln\left(\frac{T^2(\gamma+1)}{2\bar{I}^2}\right) + f^+(\gamma)\right] \quad (7.193)$$

where

$$f^+(\gamma) = 2\ln 2 - \frac{\beta^2}{12} \times \left[23 + \frac{14}{(\gamma+1)} + \frac{10}{(\gamma+1)^2} + \frac{4}{(\gamma+1)^3}\right] \quad (7.194)$$

A comparison of the (unrestricted) electron and positron mass collision stopping powers shows that the difference between the electron and positron mass collisions stopping powers resides within the difference between $f^-(\gamma)$ and $f^+(\gamma)$. These two terms are plotted in Fig. 7.14 as functions of the kinetic energy of the incident electron/positron (only the unrestricted version of $\Delta = T/2$ is used for the electron term). At low energies, $f^+(\gamma)$ exceeds $f^-(\gamma)$ by about a factor of 3 and decreases with positron kinetic energy to equal $f^-(\gamma)$ at about 0.32 MeV. Note that both $f^-(\gamma)$ and $f^+(\gamma)$ become negative with increasing kinetic energy. However, when the electron and positron unrestricted mass collision stopping powers for carbon and lead are compared, as shown in Fig. 7.15, it is sees that there is little difference between the two stopping powers which reflects the dominance of the logarithmic term in the collision stopping power expression over the magnitudes of $f^-(\gamma)$ and $f^+(\gamma)$. The mass collision stopping power curves exhibit the same characteristics as those derived earlier for the proton. A clear comparison of the curves shows that while the mass collision stopping power of protons, electrons, and positrons are roughly equal in the minimally-ionizing region, the kinetic energy at which this region occurs is equal to roughly equal to 3 times the particle rest mass (3 GeV for protons and

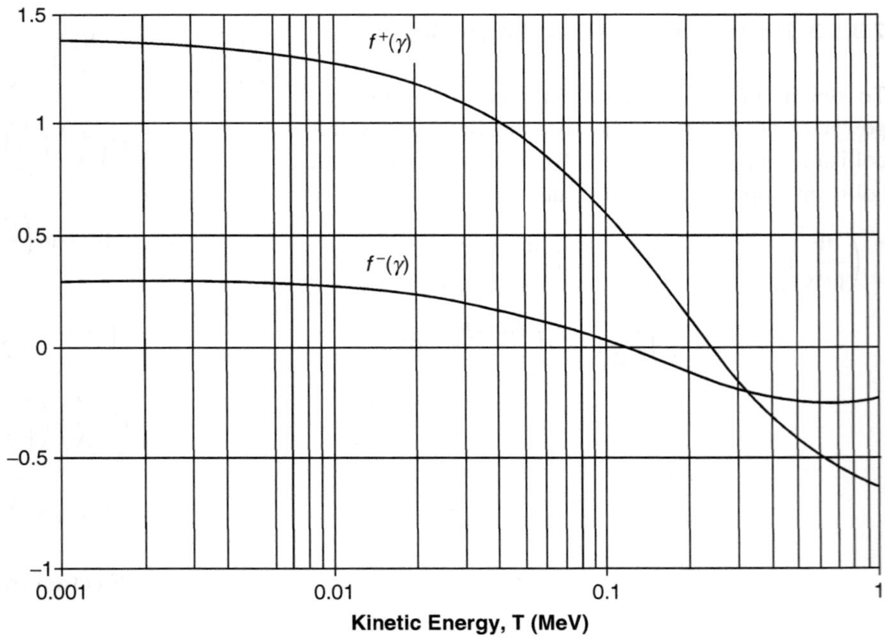

Fig. 7.14 The functions $f^-(\gamma)$ and $f^+(\gamma)$ of the mass collision stopping powers of electrons and positrons, respectively, as a function of the kinetic energy of the electron or positron

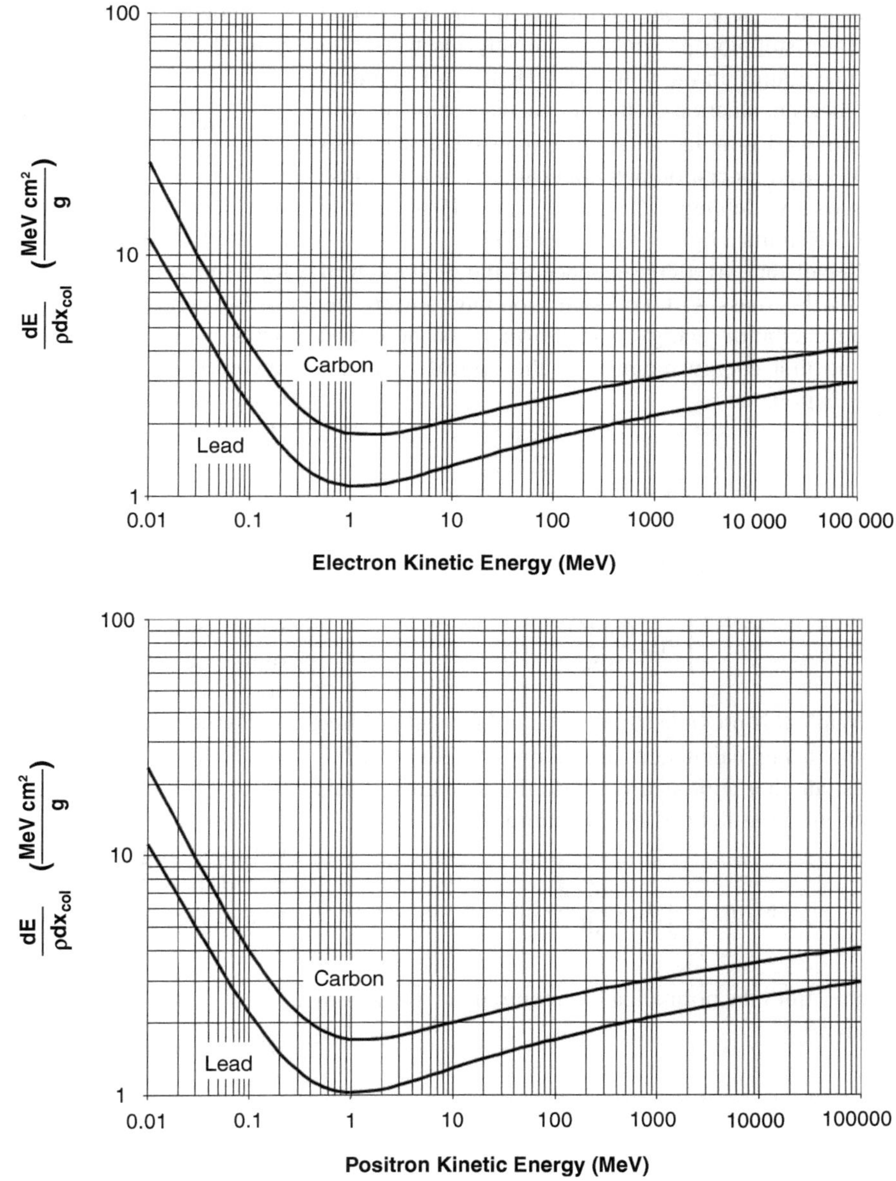

Fig. 7.15 Mass collision stopping powers of electrons and positrons in carbon and lead. Note that no density effect corrections have been applied

1.5 MeV for electrons and positrons) or for a particle speed of the order of $\beta \approx 0.8$.

7.3.6 Mean Excitation Energy

Except for the atomic number and atomic mass number, the only other explicit property of the medium which appears in the expression for the mass collision stopping power is the mean excitation energy (or ionization potential) $\bar{I}$, which is contained within the logarithmic term. The logarithm of the mean ionization potential can be obtained by ab initio calculations for a gas of free atoms or using measured optical oscillator distribution data (Nobel et al. 2005). In a closely related fashion, it is also possible to calculate it from measured dielectric properties of the medium and, as explicitly shown later, to extract it from measured stopping power data. The placement of $\bar{I}$ within the logarithm ensures that variations in the calculated collision stopping power are relatively insensitive to the uncertainty in the value of $\bar{I}$. Hence, for practical dosimetry evaluations, a theoretical discussion of $\bar{I}$ can be provided in reduced detail. Expansive discussions of the mean ionization potential can be found in those reviews by Uehling (1954) and Fano (1964) and in ICRU Reports 37 (1984) and 49 (1993).

Recall that, as a consequence of the derivation of the Bethe mass soft collision stopping power, the mean excitation energy was shown to be described by the first energy moment of the optical oscillator distribution,

$$\ln \bar{I} = \frac{1}{Z} \int dE_n \frac{df_n(Z)}{dE_n} \ln E_n$$

with the normalization,

$$\int dE_n \frac{df_n(Z)}{dE_n} = Z.$$

While it is possible to calculate $\ln \bar{I}$ for a gas of free atoms using calculated oscillator strength distributions, another approach is to use the moments of this distribution as defined by (Dalgarno 1960),

$$M(m) = \int dE_n \frac{df_n(Z)}{dE_n} E_n^m \qquad (7.195)$$

where, from the normalization requirement, the zeroth moment is,

$$M(0) = Z.$$

The logarithm of the mean ionization potential can be written by using $dE_n^m/dm = E_n^m \ln E_n$,

$$\ln \bar{I} = \frac{\left.\frac{dM(m)}{dm}\right|_{m=0}}{M(0)}. \qquad (7.196)$$

ICRU Report 37 (1993) notes that the moments for m equal to -1, 1, and 2 can be determined theoretically and that for m equal to -2 can be extracted from measured data. An analytical fit to these four moments and can be made, from which the ratio in (7.196) can be determined.

Another method was proposed by Lindhard and Scharff (1953) using a free electron gas model,

$$\ln \bar{I} = \frac{1}{Z} \int d^3\mathbf{r}\, \rho_e(r) \ln\left(\sqrt{2}\hbar\omega_p(r)\right) \qquad (7.197)$$

where ω_p is the electron plasma frequency for the electron density $\rho_e(r)$[13] and $\ln \bar{I}$ and $\hbar\omega_p(r)$ must have the same units of energy.

The approaches above are limited to gaseous media within which the positions of the electrons considered uncorrelated. For condensed media, $\ln \bar{I}$ can be calculated from an expression originally derived by Fano (1956),

$$\ln \bar{I} = \frac{2}{\pi\omega_p^2} \int_0^\infty d\omega\, \mathrm{Im}\left[-\frac{1}{\varepsilon_R(\omega)}\right] \omega \ln \hbar\omega \qquad (7.198)$$

where $\varepsilon_R(\omega)$ is the complex relative dielectric permittivity and the imaginary component describes the absorption of electromagnetic energy.

Parametric expressions of $\bar{I}$ are useful for dosimetry calculations. Because $\bar{I}$ is found within the logarithmic term of the collision stopping power, any effects upon the stopping power due to uncertainties in $\bar{I}$ will be correspondingly limited. Some authors have approximated the mean excitation potential as a linear function of the medium's atomic number,

$$\bar{I} = Z\, hcR_\infty \qquad (7.199)$$

where hcR is the Rydberg energy, 13.6 eV. A number of semi-empirical representations of $\bar{I}$ as a function of the medium's atomic number have been provided in the literature. One particularly useful one is presented by Segrè (1977),

$$\bar{I} = 9.1\left(1 + \frac{1.9}{Z^{2/3}}\right) \text{eV} \quad \text{for} \quad Z \geq 4 \qquad (7.200)$$

[13] The plasma frequency describes the oscillatory motion of free electrons in a plasma displaced from a uniform background of ions. The equation of motion for an electron in the simplest case is given by $m_e d^2x/dt^2 = -eE - (\rho_e e^2/\varepsilon_0)x$ where we have taken the restoring electric field to be equal to $-\mathbf{P}/\varepsilon_0$ where $\mathbf{P}$ is the polarization. Solving this equation yields the plasma frequency, $\omega_p = \sqrt{\rho_e e^2/\varepsilon_0 m_e}$.

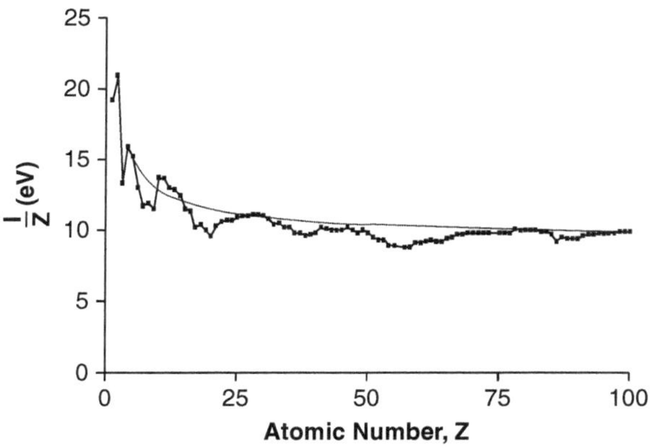

Fig. 7.16 Mean ionization potential normalized to atomic number as a function of atomic number. The curve is the parameterization of Segrè and the data points with connecting lines are from tabulated data provided by ICRU Report 37 (1984)

Segrè's parameterization of $\bar{I}$, normalized to the atomic number of the medium, is shown in Fig. 7.16. Also shown are values from ICRU Report 37 for condensed media. While $\bar{I}/Z$ rapidly decreases with atomic number to an approximately constant value of about 10 eV, it can be seen that there are irregularities attributable to atomic shell structure.

7.3.7 Stopping Number

7.3.7.1 Introduction

The mass collision stopping powers derived to this point are what can be crudely considered to be "zeroth"-order results is that they were obtained on the basis of three simplifying assumptions:

- The projectile speed is much higher than the atomic electron orbital speed allowing the collision kinematics to be derived with the electron assumed to be at rest.
- For a single atom, the atomic electron "cloud" is not displaced by the electric field of the moving charged particle.
- The medium is treated as a cold dilute gas which is not polarized by the projectile's electric field.

In order to extend the expressions of the mass collision stopping powers to enable calculations for more realistic cases, "higher order" correction terms are applied in order to compensate for these simplifying assumptions. A transparent way of doing so is by writing the mass collision stopping power as a series in order to isolate these individual corrections,

$$\left(\frac{dE}{\rho\, dx}\right)_{Col} = 2C\left(\frac{Z}{A}\right)\left(\frac{z}{\beta}\right)^2 L(\beta) \tag{7.201}$$

where $L(\beta)$ is defined as the stopping number given in terms of a summation weighted by powers of the projectile electric charge (normalized to the unit charge, e),

$$L(\beta) = \sum_{j=0}^{2} z^j L_j(\beta). \tag{7.202}$$

For convenience, the zeroth-order term of this expansion is written as a series itself,

$$L_0(\beta) = \sum_{k=0}^{2} L_{0k}(\beta) \tag{7.203}$$

to give,

$$L(\beta) = \sum_{k=0}^{2} L_{0k}(\beta) + \sum_{j=1}^{2} z^j L_j(\beta). \tag{7.204}$$

Note that the stopping number and its terms are explicit functions of the projectile speed through momentum or kinetic energy. From the expression of the mass collision stopping power for a spin-less massive projectile, the zeroth-order term in the $L_0(\beta)$ summation is,

$$L_{00}(\beta) = \ln\left(\frac{2m_e\gamma^2\beta^2}{\bar{I}}\right) - \beta^2 \tag{7.205}$$

The $L_{01}(\beta)$ term accounts for the reduction in stopping power as a result of the projectile slowing down to speeds comparable to those of the target atomic electrons,

$$L_{01}(\beta) = -\frac{C_e(\beta)}{Z} \tag{7.206}$$

and is known as the shell correction term. As it is to be independent of the atomic number of the medium, it contains that quantity in the denominator in order to cancel the corresponding factor in the leading multiplicative term of (7.201). In a dense medium, the projectile polarizes the atoms and reduces the penetration of the electric field into the medium thus diminishing the stopping power. The $L_{02}(\beta)$ term accounts for this reduction in stopping power,

$$L_{02}(\beta) = \delta/2. \tag{7.207}$$

This can be an important effect for fast electrons in soft tissue. The first-order term of the stopping number, $zL_1(\beta)$ is actually proportional to the cube of the projectile's atomic number due to the z^2 in the leading multiplicative term. Hence, this term accounts for the minute difference in stopping powers between a particle and its antiparticle as a result of the differential displacement of the atomic electron cloud by each: a positively-charged projectile will attract the atomic electrons to bring them closer to its trajectory to yield a slightly greater stopping power than its negatively-charged antiparticle which repels the electrons. Finally, the second-order term, $z^2L_2(\beta)$, arises from the reconciliation by Bloch of the Bohr and Bethe theories.

The $L_{01}(\beta)$, $zL_1(\beta)$, and $z^2L_2(\beta)$ terms are now discussed; the polarization effect accounted for by $L_{02}(\beta)$ is discussed separately.

7.3.7.2 Atomic Electron Shell Correction

Up to this point, in the calculation of the collision stopping power the orbital speeds of the atomic electrons have been assumed to be much less than the projectile speed or, equivalently, the electrons are considered to be initially at rest. Clearly, this simplified the kinematics of the hard stopping power calculation. However, for low-energy projectiles with a speed comparable to the orbital speeds, this simplification no longer holds. Moreover, if the slow projectile is an ion (e.g., an α particle), it can capture these electrons, reducing its effective charge and even further diminishing the stopping power.

The first electrons to be affected are those in the K-shell which are the most tightly bound and have the greatest speeds, followed by the L-shell electrons, etc. Thus, as the particle speed decreases, the contributions to the stopping power decrease sequentially. The term accounting for this effect is $L_{01}(\beta)$ given by where the $C_e(\beta)$ term is averaged over the contributions of all atomic electrons. Should the density effect described by $L_{02}(\beta)$ be negligible, $L_{01}(\beta)$ can be taken from the definition of the stopping number,

$$\begin{aligned} L_{01}(\beta) &= \frac{1}{2C}\left(\frac{A}{Z}\right)\left(\frac{\beta}{z}\right)^2\left(\frac{dE}{\rho dx}\right)_{Col} - L_{00}(\beta) \\ &= \frac{1}{2C}\left(\frac{A}{Z}\right)\left(\frac{\beta}{z}\right)^2\left(\frac{dE}{\rho dx}\right)_{Col} \\ &\quad - \ln\left(\frac{2m_e\gamma^2\beta^2}{\bar{I}}\right) + \beta^2. \end{aligned} \tag{7.208}$$

Thus,

$$\begin{aligned} C_e(\beta) &= -Z\,L_{01}(\beta) \\ &= \frac{1}{2C}A\left(\frac{\beta}{z}\right)^2\left(\frac{dE}{\rho\, dx}\right)_{Col} \\ &\quad + Z\left(\ln\left(\frac{2m_e\gamma^2\beta^2}{\bar{I}}\right) - \beta^2\right) \end{aligned} \tag{7.209}$$

Details of how $C_e(\beta)$ can be derived are provided by Ziegler (1999) and ICRU Report 37 (1984). A semi-empirical parameterization of $C_e(\beta)$, useful for calculational purposes, was provided by Barkas (1962),

$$\begin{aligned} C_e = \bar{I}^2 \Bigg\{ &\left(\frac{0.42237}{(\gamma\beta)^2} + \frac{0.0304}{(\gamma\beta)^4} - \frac{0.00038}{(\gamma\beta)^6}\right) \times 10^{-6} \\ &+ \bar{I}\left(\frac{3.858}{(\gamma\beta)^2} - \frac{0.1668}{(\gamma\beta)^4} + \frac{0.00158}{(\gamma\beta)^6}\right) \times 10^{-9} \Bigg\} \end{aligned} \tag{7.210}$$

where $\bar{I}$ is in units of electron volt. The validity of this expression for C_e is limited to $\gamma\beta > 0.13$. Figure 7.17

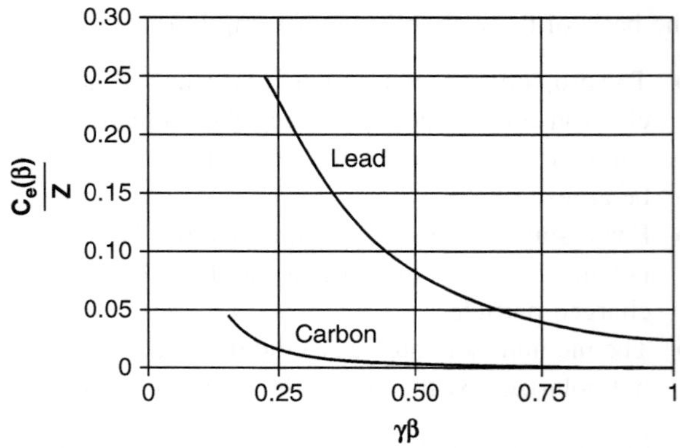

Fig. 7.17 Shell correction terms for carbon and lead as a function of $\gamma\beta$

shows the shell correction term $C_e(\beta)/Z$ for carbon and lead as functions of proton kinetic energy. The $C_e(\beta)/Z$ values decrease with increasing kinetic energy (i.e., increasing particle speed beyond the atomic electron speeds) with $C_e(\beta)/Z$ greater for lead than for carbon due to the higher atomic electron speeds.

7.3.7.3 Barkas Correction Term

As noted earlier, the first-order term of the stopping number is proportional to z^3 due to the z^2 weighting applied to the series. As a result, it will be sensitive to the sign of that charge. The fact that this should be so (even though the Bethe result predicts no such dependence due to the z^2 term) was first apparent in measurements by Barkas et al. (1956) of the kaon decay to three pions which showed that the range of the product π^+ was slightly less (of the order of 0.4%) than that of the π^- for the same initial speed. This indicated that the stopping power for the positively-charged particle was greater than its negatively-charged antiparticle, an effect due to the respective repulsion and attraction of the atomic electrons with a corresponding decrease and increase in energy transfer. Additional measurements (Barkas et al. 1963) of the $K^- + p \rightarrow \Sigma^\pm + \pi^\mp$ reaction repeated this observation. Precise work by Andersen et al. (1969) showed that the α particle stopping power was greater than the factor-of-four multiple over those of protons and deuterons as predicted by the z^2-dependence. Ashley et al. (1972, 1973) provided a thorough theoretical evaluation of the Barkas effect and Lindhard (1976) gave an explicit representation of this effect. The $zL_1(\beta)$ term in the stopping number expansion reduces the stopping power for a negatively-charged projectile relative to its positively-charged antiparticle. From the work by Ashley et al., the Barkas correction term can be written in the form,

$$zL_1(\beta) = \left(\frac{\alpha}{\beta}\right)^3 zZ\,F\left(\frac{\alpha}{\beta}b\sqrt{Z}\right) \qquad (7.211)$$

where F(x) is a numerically-evaluated function and b is related to the minimum impact parameter. Values for this correction term are provided in ICRU Report 49 (1993). For high atomic media, Bichsel (1990) extracted the Barkas correction term from measured stopping power data and found that it could be accurately described by a power-law dependence upon the particle speed,

$$L_1(\beta) = k_1\,\beta^{-k_2} \qquad (7.212)$$

where, for the example of a gold absorber,

$$k_1 = 0.002833$$

$$k_2 = 1.2$$

Figure 7.18 shows $L_1(\beta)$ for protons and antiprotons in gold as a function of proton kinetic energy. Clearly, the Barkas term becomes significant at low projectile speeds only. The only particle–antiparticle pair of interest to nuclear medicine is that of the electron and positron. As the target particle is also an electron, the Barkas effect would be swamped by the differences between the Møller and Bhabha cross sections.

7.3.7.4 Bloch Correction Term

In the early 1930s, Bloch reconciled the Bohr classical and Bethe quantum-mechanical calculations of the soft collision stopping power by demonstrating that the Bohr result was valid quantum-mechanically if the Bohr energy loss were to be interpreted as a mean value over all possible atomic electron transitions (Ahlen 1980). Bloch then looked at close collisions without the assumption that the target electron being considered as a plane wave in the center-of-mass reference frame and allowed them to be perturbed by the projectile's Coulomb field. This Bloch refinement produced a correction term that was overall proportional to z^4 or, in terms of the stopping number expansion,

$$z^2L_2(\beta) = \psi(1) - \mathrm{Re}\,\psi\left(1 + i\alpha\frac{z}{\beta}\right) \qquad (7.213)$$

where $\psi(z)$ is the logarithmic derivative of the gamma function,[14]

$$\psi(z) = \frac{d\ln\Gamma(z)}{dz} = \frac{1}{\Gamma(z)}\frac{d\Gamma(z)}{dz}. \qquad (7.214)$$

[14]This is also referred to as the digamma function.

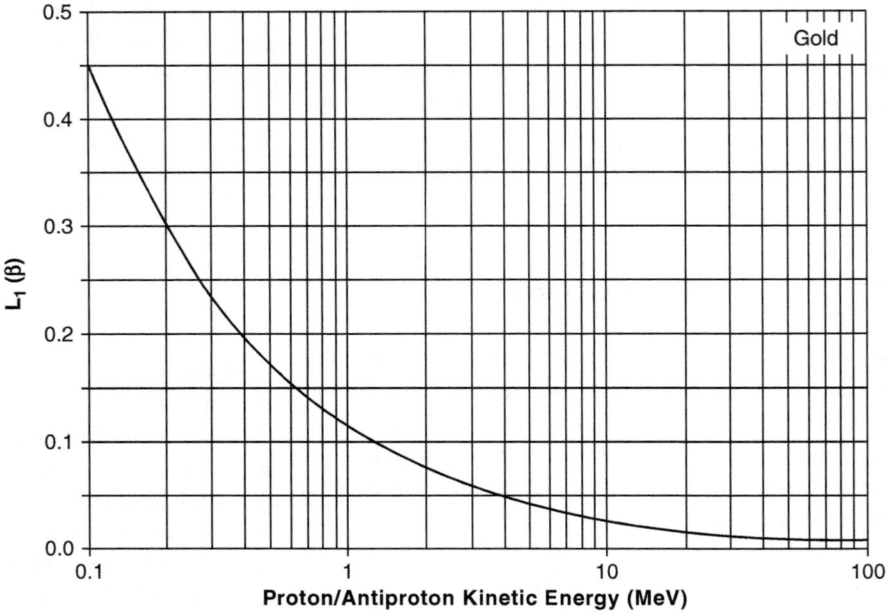

Fig. 7.18 Barkas correction factor in gold for protons and antiprotons calculated from Bichsel's (1990) parameterization

It is possible to rewrite the Bloch correction term as a series by using the identity,

$$\psi(1) = \gamma_{EM}$$

and the series expansion,

$$\mathrm{Re}\psi(1 + iy) = -0.5772\cdots + y^2 \sum_{n=1}^{\infty} \frac{1}{n(n^2 + y^2)},$$

to give,

$$z^2 L_2(\beta) = -\left(\alpha\frac{z}{\beta}\right)^2 \sum_{n=1}^{\infty} \frac{1}{n\left(n^2 + \left(\alpha\frac{z}{\beta}\right)^2\right)}. \quad (7.215)$$

Now consider the asymptotic behaviors of the Bloch correction term due to the projectile charge and speed. To do so, use the asymptotic formula,

$$\mathrm{Re}\psi(1 + iy) = \ln y + \frac{1}{12y^2} + \frac{1}{120y^4} + \frac{1}{252y^6} + \cdots \quad \text{for } y \to \infty.$$

Thus, for a slow heavy charged particle (i.e., $\alpha z/\beta \gg 1$), the Bloch correction term is,

$$z^2 L_2(\beta) \approx -\gamma_{EM} - \ln\left(\alpha\frac{z}{\beta}\right) \quad (7.216)$$

In this limit, the Bloch correction leads to the Bohr classical form of the collision stopping power. On the other hand, one does not achieve the Bethe result in the relativistic case of $\alpha z/\beta \to 0$ due to an error in the original derivation and which is discussed by Ahlen (1980). This has a negligible calculational consequence, however, as the Bloch term becomes insignificant in such a case.

7.3.7.5 Complete Stopping Number (excluding density effect)

If the zeroth-, first-, and second-order terms of the stopping number of the past three sections are summed, one obtains the complete stopping power, excluding the $L_{02}(\beta)$ term,

$$L(\beta) = \ln\left(\frac{2m_e}{\bar{I}}\gamma^2\beta^2\right) - \beta^2 - \frac{C_e(\beta)}{Z} - \frac{\delta}{2} + \left(\frac{\alpha}{\beta}\right)^3 zZ\,F\left(\frac{\alpha}{\beta}b\sqrt{Z}\right) - \left(\alpha\frac{z}{\beta}\right)^2 \times \sum_{n=1}^{\infty} \frac{1}{n\left(n^2 + \left(\alpha\frac{z}{\beta}\right)^2\right)}. \quad (7.217)$$

7.3.7.6 Effect of Medium Polarization Upon the Stopping Power

Introduction

So far, the projectile's electric field has been implicitly assumed to be in vacuo and, using this approximation, the medium has been assumed to be a cold and dilute monatomic gas. Such simplifications ignore any response of a realistic medium to this moving electric field. Should the particle be traveling in a dense dielectric medium (such as tissue), atoms will be polarized, creating an array of electric dipoles which generate a secondary electric field limiting the particle's electric field at a distance and reducing the stopping power. This effect is expected to be significant at high particle speeds, as shown by Fig. 7.2, where, as $\beta \rightarrow 1$, the electric field parallel to the trajectory flattens and the orthogonal component extends such that the dielectric response of the medium limits the relativistic rise in stopping power. Because the magnitude of this dielectric response will be directly related to the number of secondary electric dipoles, it will clearly depend upon the medium's physical density.[15] This reduction in stopping power is referred to as the polarization or density effect and is characterized by a correction term, δ, which is treated as a higher order correction term to the zeroth-order term of the Born series description of the stopping power.

The density effect leads to a reduction in the stopping power of fast electrons in tissue and its consequences are to be investigated in detail. The semiclassical work of Fermi (1940) will guide this derivation.

Electronic Polarization

In order to characterize the properties of a dielectric medium in a time-varying electric field, begin with those of an individual atom. The simplest model of the medium is that of an isotropic monatomic gas with

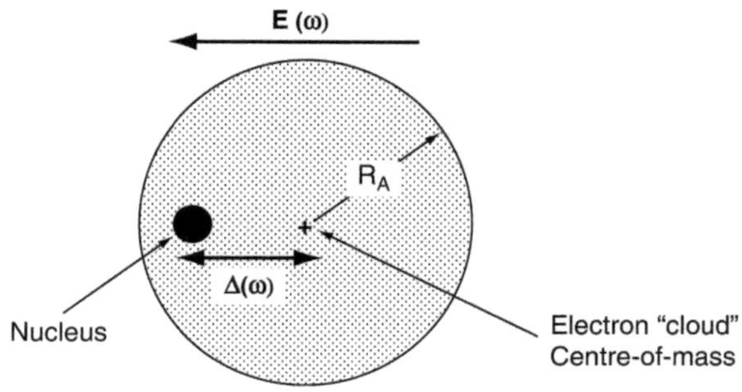

Fig. 7.19 Electronic polarization due to an external electric field

interatomic distances sufficiently large that an atom can be treated in isolation. The atom is taken to be a negatively-charged and mobile spherical electron cloud of charge $-Ze$ and mass Zm_e which, in its unperturbed state, is centered on a fixed nucleus of charge Ze. The electric field of the passing charged particle will perturb the position of this electron cloud and polarize the atom, as shown in Fig. 7.19. At high particle speeds, the shortness of the time duration of the electric field "pulse" experienced by the atom, as shown in Fig. 7.2, will be dominated by high-frequency components and, in frequency space, one can consider the atom as being exposed to a high-frequency electric field. The displacement, Δ, will be limited due to the short duration of the pulse and one can effectively ignore the spatial variation of the field over the atom as a result. The electron cloud will experience a restorative force due to the mutual attraction between it and the nucleus and thus go into oscillatory motion. The attractive force can be calculated by modeling the electron cloud as a uniformly charged sphere of radius R_A with an electric charge density, $(3/4\pi)Ze/R_A^3$. The radial electric field due to this charge at a distance Δ from the center of the atom is found by applying Gauss' law to the sphere of radius Δ,

$$\varepsilon_0 \left(4\pi\Delta^2\right) E = \frac{4\pi}{3}\Delta^3 \left(\frac{3}{4\pi}\frac{Ze}{R_A^3}\right)$$

from which one obtains an expression for the electric field in terms of the displacement of the atomic electron cloud,

$$E = \frac{Ze}{4\pi\varepsilon_0}\frac{\Delta}{R_A^3} \tag{7.218}$$

[15] Because the interatomic spacings in a gas are greater than those in a solid or liquid, the wider dispersion of atoms in the gaseous phase will limit the dielectric response such that the mass collision stopping power for a given medium will be greater for the condensed phases than for the vapor phase and, hence, this interest in this phenomenon in tissue.

The restoring force is F = ZeE, or

$$F = \frac{(Ze)^2}{4\pi\,\varepsilon_0 R_A^3}\Delta = \eta\Delta. \quad (7.219)$$

This gives,

$$ZeE = \frac{(Ze)^2}{4\pi\,\varepsilon_0}\frac{\Delta}{R_A^3}$$

from which the magnitude of the product of the charge and its displacement, electric dipole moment, is obtained,

$$p = Ze\,\Delta = 4\pi\,\varepsilon_0\,R_A^3\,E \quad (7.220)$$

or, in terms of vector quantities,

$$\mathbf{p} = 4\pi\,\varepsilon_0\,R_A^3\,\mathbf{E} \equiv \alpha_e\,\mathbf{E} \quad (7.221)$$

where α_e is defined as the electronic polarizability. It will be noted that the magnitude of the ratio, $\alpha_e/\varepsilon_0 = 4\pi R_A^3$, is the atomic volume.

The resulting equation of motion of the displaced electron cloud is,

$$Zm_e\frac{d^2\Delta(t)}{dt^2} = Ze\mathbf{E}(t) - \eta\Delta(t) - \Gamma Zm_e\frac{d\Delta(t)}{dt} \quad (7.222)$$

where a damping force $\Gamma Zm_e\,d\Delta(t)/dt$ with a positive damping constant Γ has been allowed for. This differential equation is readily solvable, as demonstrated earlier in the derivation of the Bohr soft collision stopping power through the Fourier transform. The result in frequency space is,

$$\Delta(\omega) = \frac{\left(\frac{e}{m_e}\right)\mathbf{E}(\omega)}{(\omega_0^2 - \omega^2) + i\Gamma\omega} \quad (7.223)$$

with a resonant frequency given by,

$$\omega_0 = c\sqrt{Z\frac{r_0}{R_A^3}}. \quad (7.224)$$

As the atomic radius R_A of the order of 10^{-10} m, the resonant frequency will consequently be of the order of 10^{16}–10^{17}/s with a corresponding wavelength in the ultraviolet range of the electromagnetic spectrum. The electric dipole moment in frequency-space is,

$$\begin{aligned}\mathbf{p}(\omega) &= Z\,e\,\Delta(\omega)\\ &= Z\frac{\left(\frac{e^2}{m_e}\right)\mathbf{E}(\omega)}{(\omega_0^2 - \omega^2) + i\Gamma\omega}.\end{aligned} \quad (7.225)$$

Next, the scale is expanded from that of the individual atom to the macroscopic medium which is taken to be linear and isotropic.[16] The polarization is the number of electric dipoles per unit volume, or,

$$\mathbf{P}(\omega) = \rho_e\frac{\left(\frac{e^2}{m_e}\right)\mathbf{E}(\omega)}{(\omega_0^2 - \omega^2) + i\Gamma\omega}. \quad (7.226)$$

The electric-flux density vector is related to the electric field and the polarization in frequency space by,

$$\begin{aligned}\mathbf{D}(\omega) &= \varepsilon_0\mathbf{E}(\omega) + \mathbf{P}(\omega)\\ &\equiv \varepsilon_0\mathbf{E}(\omega) + \chi_e(\omega)\varepsilon_0\mathbf{E}(\omega)\\ &= \varepsilon_0(1 + \chi_e(\omega))\mathbf{E}(\omega)\\ &= \varepsilon_0\,\varepsilon_R(\omega)\mathbf{E}(\omega)\end{aligned} \quad (7.227)$$

where $\chi_e(\omega)$ is the electric susceptibility and $\varepsilon_R(\omega) = 1 + \chi_e(\omega)$ is the relative dielectric constant. This result is algebraically manipulated to relate the polarization to the electric field by,

$$\mathbf{P}(\omega) = \varepsilon_0(\varepsilon_R(\omega) - 1)\mathbf{E}(\omega). \quad (7.228)$$

By equating these two expressions for the polarization, the relative dielectric constant can be written as,

$$\begin{aligned}\varepsilon_R(\omega) &= 1 + \frac{\left(\frac{\rho_e}{\varepsilon_0}\frac{e^2}{m_e}\right)}{(\omega_0^2 - \omega^2) + i\Gamma\omega}\\ &= 1 + \frac{\omega_P^2}{(\omega_0^2 - \omega^2) + i\Gamma\omega}.\end{aligned} \quad (7.229)$$

[16]Should this not be the case, the electric susceptibility and dielectric permittivity scalars would be replaced by tensors.

Separate $\varepsilon_R(\omega)$ into real and imaginary components in order to form the set of dielectric dispersion formulae,

$$\begin{aligned}\varepsilon_R(\omega) &= \left(1+\omega_P^2\left(\frac{\omega_0^2-\omega^2}{\left(\omega_0^2-\omega^2\right)^2+\Gamma^2\omega^2}\right)\right)\\ &\quad -i\,\omega_P^2\left(\frac{\Gamma\omega}{\left(\omega_0^2-\omega^2\right)^2+\Gamma^2\omega^2}\right)\\ &\equiv \varepsilon_R{}'(\omega)-i\varepsilon_R{}''(\omega)\end{aligned} \tag{7.230}$$

As the imaginary component of the relative dielectric permittivity is proportional to Γ, it is also proportional to the power loss within the dielectric.

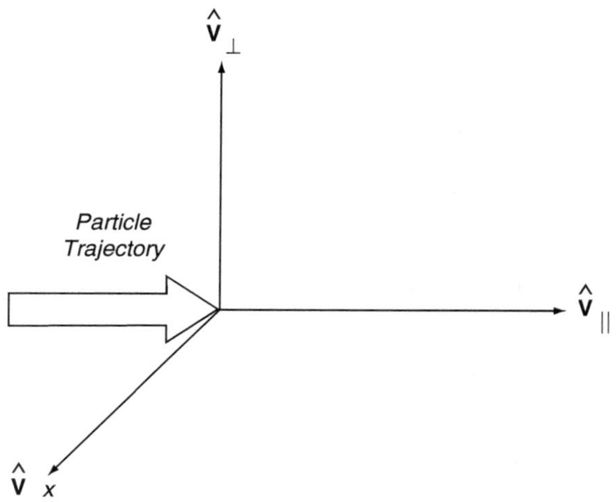

Fig. 7.20 Coordinate system used in the calculation of the electromagnetic fields of a moving charged particle in a dielectric medium

Electromagnetic Fields in a Dielectric Medium

The next step to calculating the response of a dielectric medium to a moving charged particle requires the calculation of the electromagnetic fields of the projectile within the medium. To do this, Maxwell's equations are solved in Fourier space,

$$\nabla \bullet \mathbf{D} = \rho \tag{7.231}$$

$$\nabla \bullet \mathbf{B} = 0 \tag{7.232}$$

$$\nabla \times \mathbf{E} = -\frac{\partial \mathbf{B}}{\partial t} \tag{7.233}$$

$$\nabla \times \mathbf{H} = \mathbf{J} + \frac{\partial \mathbf{D}}{\partial t} \tag{7.234}$$

where

$$\mathbf{D} = \varepsilon_R\, \varepsilon_0\, \mathbf{E} \tag{7.235}$$

and where it is assumed that the magnetic polarizability of the medium is negligible,

$$\mathbf{B} = \mu_0\, \mathbf{H}. \tag{7.236}$$

The electric field and the magnetic flux density are also defined through scalar and vector potentials,

$$\mathbf{E} = -\nabla\Phi - \frac{\partial \mathbf{A}}{\partial t} \tag{7.237}$$

$$\mathbf{B} = \nabla \times \mathbf{A}. \tag{7.238}$$

It is possible to determine these potentials using the four-dimensional Fourier transform,

$$F(\mathbf{k},\omega) = \frac{1}{4\pi^2}\int_{-\infty}^{\infty} d^3\mathbf{x}\int_{-\infty}^{\infty} dt\, F(\mathbf{x},t)e^{-i(\mathbf{k}\bullet\mathbf{x}-\omega t)} \tag{7.239}$$

where k is the wavenumber and ω is the frequency. The coordinate system of Fig. 7.20 where the unit vector $\hat{\mathbf{v}}_\parallel$ lies along the particle's trajectory is used. The Fourier transforms of the divergence of **D** and the curl of **B** are,

$$i\mathbf{k} \bullet \mathbf{E}(\mathbf{k},\omega) = \frac{\rho(\mathbf{k},\omega)}{\varepsilon_R(\omega)\varepsilon_0} \tag{7.240}$$

$$-i\,\mathbf{k} \times \mathbf{H}(\mathbf{k},\omega) = \mathbf{J}(\mathbf{k},\omega) + i\omega\varepsilon_R(\omega)\varepsilon_0\mathbf{E}(\mathbf{k},\omega). \tag{7.241}$$

The Fourier transforms of the fields as defined by the potentials are,

$$\mathbf{E}(\mathbf{k},\omega) = -i\mathbf{k}\Phi(\mathbf{k},\omega) + i\omega\mathbf{A}(\mathbf{k},\omega) \tag{7.242}$$

$$\mathbf{H}(\mathbf{k},\omega) = -\frac{i}{\mu_0}\mathbf{k}\times\mathbf{A}(\mathbf{k},\omega). \quad (7.243)$$

Expressions for $\mathbf{E}(\mathbf{k},\omega)$ and $\mathbf{H}(\mathbf{k},\omega)$ are derived by first determining the scalar and vector potentials. Substituting (7.242) into (7.240),

$$-\,i\mathbf{k}\bullet(i\mathbf{k}\Phi(\mathbf{k},\omega) - i\omega\mathbf{A}(\mathbf{k},\omega)) = \frac{\rho(\mathbf{k},\omega)}{\varepsilon_R(\omega)\varepsilon_0}$$

which leads to,

$$k^2\Phi(\mathbf{k},\omega) - \omega\,\mathbf{k}\bullet\mathbf{A}(\mathbf{k},\omega) = \frac{\rho(\mathbf{k},\omega)}{\varepsilon_R(\omega)\varepsilon_0}. \quad (7.244)$$

As both the scalar and vector potentials appear in this equation, we can decouple them by applying the Lorentz gauge condition. This sets the divergence of the vector potential to being proportional to the time derivative of the scalar potential,

$$\begin{aligned}\nabla\bullet\mathbf{A}(\mathbf{x},t) &= -\mu_0\varepsilon_R\varepsilon_0\frac{\partial\Phi(\mathbf{x},t)}{\partial t}\\ &= -\frac{\varepsilon_R}{c^2}\frac{\partial\Phi(\mathbf{x},t)}{\partial t}\end{aligned} \quad (7.245)$$

where $1/\sqrt{\mu_0\varepsilon_0} = c$ has been used. By transforming this Lorentz gauge condition into wavenumber- and frequency-space,

$$\mathbf{k}\bullet\mathbf{A}(\mathbf{k},\omega) = \frac{\omega\,\varepsilon_R(\omega)}{c^2}\Phi(\mathbf{k},\omega). \quad (7.246)$$

Applying this to (7.244) results in the wave equation for the scalar potential,

$$\left(k^2 - \frac{\omega^2\,\varepsilon_R(\omega)}{c^2}\right)\Phi(\mathbf{k},\omega) = \frac{\rho(\mathbf{k},\omega)}{\varepsilon_R(\omega)\varepsilon_0}. \quad (7.247)$$

The corresponding wave equation in the vector potential is next derived by substituting the Fourier transforms of the electromagnetic fields into the Fourier transform of $\nabla\times\mathbf{B}$,

$$\begin{aligned}\frac{1}{\mu_0}\mathbf{k}\times(\mathbf{k}\times\mathbf{A}(\mathbf{k},\omega)) &= \mathbf{J}(\mathbf{k},\omega) + i\omega\varepsilon_R(\omega)\varepsilon_0\\ &\quad\times(i\mathbf{k}\Phi(\mathbf{k},\omega) - i\omega\mathbf{A}(\mathbf{k},\omega)).\end{aligned}$$

Expanding the vector triple cross-product gives,

$$\begin{aligned}\frac{1}{\mu_0}&\left(\mathbf{k}(\mathbf{k}\bullet\mathbf{A}(\mathbf{k},\omega)) - k^2\mathbf{A}(\mathbf{k},\omega)\right)\\ &= \mathbf{J}(\mathbf{k},\omega) - \omega\varepsilon_R(\omega)\varepsilon_0\mathbf{k}\Phi(\mathbf{k},\omega)\\ &\quad + \omega^2\varepsilon_R(\omega)\varepsilon_0\mathbf{A}(\mathbf{k},\omega).\end{aligned}$$

Applying the Lorentz gauge condition to this result gives,

$$\left(k^2 - \frac{\omega^2\,\varepsilon_R(\omega)}{c^2}\right)\mathbf{A}(\mathbf{k},\omega) = \mu_0\mathbf{J}(\mathbf{k},\omega). \quad (7.248)$$

These wave equations provide the first steps in determining the potentials $\Phi(\mathbf{k},\omega)$ and $\mathbf{A}(\mathbf{k},\omega)$. These are obtained by first calculating the Fourier transforms of the charge and current densities, $\rho(\mathbf{k},\omega)$ and $\mathbf{J}(\mathbf{k},\omega)$. The net charge distribution is that of the projectile,

$$\rho(\mathbf{x},t) = ze\,\delta\left(\mathbf{x} - \beta ct\ \hat{\mathbf{v}}_{||}\right) \quad (7.249)$$

with the Fourier transform,

$$\begin{aligned}\rho(\mathbf{k},\omega) &= \frac{1}{4\pi^2}\int_{-\infty}^{\infty} d^3\mathbf{x}\int_{-\infty}^{\infty} dt\,\rho(\mathbf{x},t)e^{-i(\mathbf{k}\bullet\mathbf{x}-\omega t)}\\ &= \frac{ze}{4\pi^2}\int_{-\infty}^{\infty} d^3\mathbf{x}\int_{-\infty}^{\infty} dt\,e^{-i\left(\beta c\,\mathbf{k}\bullet\hat{\mathbf{v}}_{||}-\omega\right)t}\\ &= \frac{ze}{2\pi}\delta\left(\omega - \beta ck_{||}\right).\end{aligned} \quad (7.250)$$

The current density is,

$$\mathbf{J}(\mathbf{x},t) = \beta c\,\rho(\mathbf{x},t)\hat{\mathbf{v}}_{||}$$

and its Fourier transform is,

$$\begin{aligned}\mathbf{J}(\mathbf{k},\omega) &= \beta c\,\rho(\mathbf{k},\omega)\hat{\mathbf{v}}_{||}\\ &= \frac{ze\,\beta c}{2\pi}\delta\left(\omega - \beta ck_{||}\right)\hat{\mathbf{v}}_{||}.\end{aligned} \quad (7.251)$$

Having obtained these, it is possible to write the Fourier transforms of the scalar and vector potentials as,

$$\begin{aligned}\Phi(\mathbf{k},\omega) &= \frac{\rho(\mathbf{k},\omega)}{\varepsilon_R(\omega)\varepsilon_0\left(k^2 - \frac{\omega^2\,\varepsilon_R(\omega)}{c^2}\right)}\\ &= \frac{ze}{2\pi\,\varepsilon_R(\omega)\varepsilon_0\left(k^2 - \frac{\omega^2\,\varepsilon_R(\omega)}{c^2}\right)}\\ &\quad\times\delta\left(\omega - \beta ck_{||}\right)\end{aligned} \quad (7.252)$$

and

$$\mathbf{A}(\mathbf{k},\omega) = \frac{\mu_0 \mathbf{J}(\mathbf{k},\omega)}{\left(k^2 - \frac{\omega^2 \varepsilon_R(\omega)}{c^2}\right)} = \frac{ze\,\mu_0 \beta c}{2\pi\left(k^2 - \frac{\omega^2 \varepsilon_R(\omega)}{c^2}\right)} \delta\left(\omega - \beta c k_{||}\right) \hat{\mathbf{v}}_{||} \tag{7.253}$$

and the fields in wavenumber- and frequency-space,

$$\begin{aligned}\mathbf{E}(\mathbf{k},\omega) &= i\left(\omega \frac{ze\,\mu_0 \beta c}{2\pi\left(k^2 - \frac{\omega^2 \varepsilon_R(\omega)}{c^2}\right)} \delta\left(\omega - \beta c k_{||}\right) \hat{\mathbf{v}}_{||} - \mathbf{k}\frac{ze}{2\pi \varepsilon_R(\omega)\varepsilon_0\left(k^2 - \frac{\omega^2 \varepsilon_R(\omega)}{c^2}\right)}\delta\left(\omega - \beta c k_{||}\right)\right) \\ &= i\frac{ze\delta\left(\omega - \beta c k_{||}\right)}{2\pi\varepsilon_0\varepsilon_R(\omega)\left(k^2 - \frac{\omega^2 \varepsilon_R(\omega)}{c^2}\right)} \times \left(\omega\varepsilon_R(\omega)\frac{\beta}{c}\hat{\mathbf{v}}_{||} - \mathbf{k}\right)\end{aligned} \tag{7.254}$$

and

$$\mathbf{H}(\mathbf{k},\omega) = i\frac{ze\,\beta c}{2\pi\left(k^2 - \frac{\omega^2 \varepsilon_R(\omega)}{c^2}\right)}\delta\left(\omega - \beta c k_{||}\right)\mathbf{k} \times \hat{\mathbf{v}}_{||}. \tag{7.255}$$

Energy Loss in a Dielectric Medium

Now that the electromagnetic fields $\mathbf{E}(\mathbf{k},\omega)$ and $\mathbf{H}(\mathbf{k},\omega)$ arising from a moving charged particle in a dielectric medium have been evaluated, the energy loss to a single electron at the position associated with the impact parameter, b, is calculated at the vectorial position,

$$\mathbf{x} = b\,\hat{\mathbf{v}}_{\perp}. \tag{7.256}$$

This energy loss is equal to the electromagnetic energy flow which is described by the magnitude of the Poynting vector. In order to calculate this vector, begin by taking the inverse Fourier transform of the previously-derived electric field into the spatial domain,

$$\begin{aligned}\mathbf{E}(b\hat{\mathbf{v}}_{\perp},\omega) &= \frac{1}{(2\pi)^{3/2}}\int_{-\infty}^{\infty} d^3\mathbf{k}\,\mathbf{E}(\mathbf{k},\omega)e^{ib\mathbf{k}\bullet\hat{\mathbf{v}}_{\perp}} \\ &= i\frac{ze}{(2\pi)^{5/2}\varepsilon_0\varepsilon_R(\omega)} \times \int_{-\infty}^{\infty} d^3\mathbf{k}\frac{\left(\omega\varepsilon_R(\omega)\frac{\beta}{c}\hat{\mathbf{v}}_{||} - \mathbf{k}\right)}{\left(k^2 - \frac{\omega^2 \varepsilon_R(\omega)}{c^2}\right)} \times e^{ib\mathbf{k}\bullet\hat{\mathbf{v}}_{\perp}}\delta\left(\omega - \beta\, c k_{||}\right) \\ \mathbf{E}(b\hat{\mathbf{v}}_{\perp},\omega) &= i\frac{ze}{(2\pi)^{5/2}\varepsilon_0\varepsilon_R(\omega)}\int_{-\infty}^{\infty} d^3\mathbf{k} \times \frac{\left(-k_x\hat{\mathbf{v}}_x + \left(\omega\varepsilon_R(\omega)\frac{\beta}{c} - k_{||}\right)\hat{\mathbf{v}}_{||} - k_{\perp}\hat{\mathbf{v}}_{\perp}\right)}{\left(k_x^2 + k_{||}^2 + +k_{\perp}^2 - \frac{\omega^2 \varepsilon_R(\omega)}{c^2}\right)} \times e^{ibk_{\perp}}\delta\left(\omega - \beta c k_{||}\right)\end{aligned} \tag{7.257}$$

In order to evaluate the integral, first use the substitution of variable $\xi = \beta c k_{||}$ to take it to the form,

$$\mathbf{E}(b\hat{\mathbf{v}}_{\perp},\omega) = i\frac{ze}{(2\pi)^{5/2}\varepsilon_0\varepsilon_R(\omega)\beta c}\int_{-\infty}^{\infty} dk_x \int_{-\infty}^{\infty} d\xi \int_{-\infty}^{\infty} dk_{\perp} \times \frac{-k_x\hat{\mathbf{v}}_x + \left(\omega\varepsilon_R(\omega)\frac{\beta}{c} - \frac{\xi}{\beta c}\right)\hat{\mathbf{v}}_{||} - k_{\perp}\hat{\mathbf{v}}_{\perp}}{\left(k_x^2 + \left(\frac{\xi}{\beta c}\right)^2 + +k_{\perp}^2 - \frac{\omega^2\varepsilon_R(\omega)}{c^2}\right)} \times e^{ibk_{\perp}}\delta(\omega - \xi) \tag{7.258}$$

The integration over ξ is trivial due to the δ-function and results in,

$$\mathbf{E}(b\hat{\mathbf{v}}_{\perp},\omega) = i\frac{ze}{(2\pi)^{5/2}\varepsilon_0\varepsilon_R(\omega)\beta c}\int_{-\infty}^{\infty} dk_x \int_{-\infty}^{\infty} dk_{\perp} \times \frac{-k_x\hat{\mathbf{v}}_x + \frac{\omega}{\beta c}\left(\varepsilon_R(\omega)\beta^2 - 1\right)\hat{\mathbf{v}}_{||} - k_{\perp}\hat{\mathbf{v}}_{\perp}}{\left(k_x^2 + k_{\perp}^2 + \lambda(\omega)^2\right)}e^{ibk_{\perp}} \tag{7.259}$$

where the quantity,

$$\lambda^2(\omega) = \left(\frac{\omega}{\beta c}\right)^2 \left(1 - \beta^2 \varepsilon_R(\omega)\right) \tag{7.260}$$

has been defined. In a convenient approach to solving this integral, it is split it up into the integral expression for the Fourier transform of the electric field of three integrals along each orthogonal direction and each are solved separately,

$$\mathbf{E}(b\hat{\mathbf{v}}_\perp, \omega) = i \frac{ze}{(2\pi)^{5/2} \varepsilon_0 \varepsilon_R(\omega) \beta c} \times \left(I_x \hat{\mathbf{v}}_x + I_{||} \hat{\mathbf{v}}_{||} + I_\perp \hat{\mathbf{v}}_\perp\right). \tag{7.261}$$

The integrals are,

$$\begin{aligned} I_x &= -\int_{-\infty}^{\infty} dk_\perp e^{ibk_\perp} \int_{-\infty}^{\infty} dk_x \frac{k_x}{\left(k_x^2 + k_\perp^2 + \lambda(\omega)^2\right)} \\ &= 0 \end{aligned} \tag{7.262}$$

and

$$\begin{aligned} I_{||} &= \frac{\omega}{\beta c}\left(\varepsilon_R(\omega)\beta^2 - 1\right) \int_{-\infty}^{\infty} dk_\perp e^{ibk_\perp} \\ &\quad \times \int_{-\infty}^{\infty} \frac{dk_x}{\left(k_x^2 + k_\perp^2 + \lambda(\omega)^2\right)} \\ &= \frac{\pi\omega}{\beta c}\left(\varepsilon_R(\omega)\beta^2 - 1\right) \int_{-\infty}^{\infty} dk_\perp \frac{e^{ibk_\perp}}{\sqrt{k_\perp^2 + \lambda(\omega)^2}} \\ &= \frac{2\pi\omega}{\beta c}\left(\varepsilon_R(\omega)\beta^2 - 1\right) \int_{-\infty}^{\infty} dk_\perp \frac{\cos bk_\perp}{\sqrt{k_\perp^2 + \lambda(\omega)^2}} \\ &= \frac{2\pi\omega}{\beta c}\left(\varepsilon_R(\omega)\beta^2 - 1\right) \int_{0}^{\infty} \frac{dk_\perp}{\lambda(\omega)} \frac{\cos\left(\lambda(\omega) b\left(\frac{k_\perp}{\lambda(\omega)}\right)\right)}{\sqrt{1 + \left(\frac{k_\perp}{\lambda(\omega)}\right)^2}} \\ &= \frac{2\pi\omega}{\beta c}\left(\varepsilon_R(\omega)\beta^2 - 1\right) K_0(\lambda(\omega) b) \end{aligned} \tag{7.263}$$

and

$$\begin{aligned} I_\perp &= -\int_{-\infty}^{\infty} dk_\perp k_\perp e^{ibk_\perp} \int_{-\infty}^{\infty} \frac{dk_x}{\left(k_x^2 + k_\perp^2 + \lambda(\omega)^2\right)} \\ &= -\pi \int_{-\infty}^{\infty} dk_\perp \frac{k_\perp e^{ibk_\perp}}{\sqrt{k_\perp^2 + \lambda(\omega)^2}} \\ &= -i\pi \frac{d}{db} \int_{-\infty}^{\infty} dk_\perp \frac{e^{ibk_\perp}}{\sqrt{k_\perp^2 + \lambda(\omega)^2}} \\ &= -i\, 2\pi \frac{d}{db} \int_{-\infty}^{\infty} dk_\perp \frac{\cos bk_\perp}{\sqrt{k_\perp^2 + \lambda(\omega)^2}} \\ &= -i\, 2\pi \frac{d}{db} K_0(\lambda(\omega) b) \\ &= i\, 2\pi\, \lambda(\omega) K_1(\lambda(\omega) b). \end{aligned} \tag{7.264}$$

The electric field at the position of the electron $b\hat{\mathbf{v}}_\perp$ is, then,

$$\begin{aligned} \mathbf{E}(b\hat{\mathbf{v}}_\perp, \omega) &= \frac{ze}{(2\pi)^{3/2} \varepsilon_0 \varepsilon_R(\omega) \beta c} \\ &\quad \times \Big(\lambda(\omega) K_1(\lambda(\omega) b) \hat{\mathbf{v}}_\perp - i \frac{\omega}{\beta c} \\ &\quad \times \left(1 - \varepsilon_R(\omega)\beta^2\right) K_0(\lambda(\omega) b) \hat{\mathbf{v}}_{||}\Big). \end{aligned} \tag{7.265}$$

One can see from the definition of λ^2 that, for a real relative dielectric constant,

$$\lambda^2 > 0 \quad \text{for} \quad \beta < \frac{1}{\sqrt{\varepsilon_R(\omega)}} \tag{7.266}$$

$$\lambda^2 < 0 \quad \text{for} \quad \beta > \frac{1}{\sqrt{\varepsilon_R(\omega)}}. \tag{7.267}$$

In other words, λ can only be real if the particle speed is less than the phase velocity of the medium. It should be noted that, if polarization is neglected, i.e.,

$\varepsilon_R(\omega) = 1$, then $\lambda^2(\omega) = (\omega/\gamma\beta c)^2$ and the expression for the electric field reduces to,

$$\begin{aligned}
\mathbf{E}(b\hat{\mathbf{v}}_\perp, \omega) &= \frac{ze}{(2\pi)^{3/2}\varepsilon_0\beta c}\left(\frac{\omega}{\gamma\beta c}K_1\left(\frac{\omega}{\gamma\beta c}b\right)\hat{\mathbf{v}}_\perp\right.\\
&\quad\left. - i\frac{\omega}{\gamma^2\beta c}K_0\left(\frac{\omega}{\gamma\beta c}b\right)\hat{\mathbf{v}}_{||}\right)\\
&= \frac{ze}{(2\pi)^{3/2}\varepsilon_0}\frac{\omega}{\gamma(\beta c)^2}\\
&\quad\times\left(K_1\left(\frac{\omega}{\gamma\beta c}b\right)\hat{\mathbf{v}}_\perp - i\frac{1}{\gamma}K_0\left(\frac{\omega}{\gamma\beta c}b\right)\hat{\mathbf{v}}_{||}\right)\\
&\quad \text{for} \quad \varepsilon_R(\omega) = 1
\end{aligned} \tag{7.268}$$

which is the result which was derived previously for a heavy charged particle interacting with an harmonically-bound electron.

In a fashion similar to the above calculation for the electric field, one can next calculate the magnetic field strength at the position of the target electron,

$$\begin{aligned}
\mathbf{H}(b\hat{\mathbf{v}}, \omega) &= -i\frac{ze\beta c}{(2\pi)^{5/2}}\int_{-\infty}^{\infty} d^3\mathbf{k}\frac{\mathbf{k}\times\hat{\mathbf{v}}_{||}}{\left(k^2 - \frac{\omega^2\varepsilon_R(\omega)}{c^2}\right)}\\
&\quad\times e^{ib\mathbf{k}\bullet\hat{\mathbf{v}}_\perp}\delta\left(\omega - \beta c k_{||}\right)\\
&= -i\frac{ze\beta c}{(2\pi)^{5/2}}\int_{-\infty}^{\infty} dk_x\int_{-\infty}^{\infty} dk_{||}\int_{-\infty}^{\infty} dk_\perp\\
&\quad\times\frac{k_x\hat{\mathbf{v}}_\perp - k_\perp\hat{\mathbf{v}}_x}{\left(k_x^2 + k_{||}^2 + k_\perp^2 - \frac{\omega^2\varepsilon_R(\omega)}{c^2}\right)}\\
&\quad\times e^{ibk_\perp}\delta\left(\omega - \beta c k_{||}\right)\\
&= -i\frac{ze}{(2\pi)^{5/2}}\int_{-\infty}^{\infty} dk_\perp e^{ibk_\perp}\int_{-\infty}^{\infty} dk_x\\
&\quad\times\frac{k_x\hat{\mathbf{v}}_\perp - k_\perp\hat{\mathbf{v}}_x}{\left(k_x^2 + k_\perp^2 + \lambda^2(\omega)\right)}\\
&= -i\frac{ze}{(2\pi)^{5/2}}\left(I_x\hat{\mathbf{v}}_x + I_\perp\hat{\mathbf{v}}_\perp\right).
\end{aligned} \tag{7.269}$$

where

$$\begin{aligned}
I_x &= -\int_{-\infty}^{\infty} dk_\perp\, k_\perp\, e^{ibk_\perp}\\
&\quad\times\int_{-\infty}^{\infty}\frac{dk_x}{\left(k_x^2 + k_\perp^2 + \lambda^2(\omega)\right)}\\
&= i\,2\pi\,\lambda(\omega)K_1(\lambda(\omega)b)
\end{aligned} \tag{7.270}$$

and

$$\begin{aligned}
I_\perp &= \int_{-\infty}^{\infty} dk_\perp\, e^{ibk_\perp}\int_{-\infty}^{\infty} dk_x\frac{k_x}{\left(k_x^2 + k_\perp^2 + \lambda^2(\omega)\right)}\\
&= 0
\end{aligned} \tag{7.271}$$

to give,

$$\mathbf{H}(b\hat{\mathbf{v}}, \omega) = \frac{ze}{(2\pi)^{3/2}}\lambda(\omega)K_1(\lambda(\omega)b)\hat{\mathbf{v}}_x. \tag{7.272}$$

Now that the electric and magnetic fields have been derived in Fourier space at the position of the electron at $b\,\hat{\mathbf{v}}_\perp$, the energy loss due to collisions at an impact parameter of $b \geq b_{min}$ can be readily calculated from the energy flow through a cylinder of radius b_{min} centered on the particle's trajectory through the inverse Fourier transform. The energy flow through this cylinder is equal to the power loss of the particle,

$$\left(\frac{dE}{dx}\right)_{b\geq b_{min}} = \frac{1}{\beta c}\frac{dE}{dt}. \tag{7.273}$$

The power flow is given by the outgoing component of the Poynting vector, $\mathbf{P} = \mathbf{E}\times\mathbf{H}$,

$$\begin{aligned}
\mathbf{P} &= \left(E_{||}\hat{\mathbf{v}}_{||} + E_\perp\hat{\mathbf{v}}_\perp\right)\times H_x\hat{\mathbf{v}}_x\\
&= E_\perp H_x\hat{\mathbf{v}}_{||} - E_{||}H_x\hat{\mathbf{v}}_\perp.
\end{aligned} \tag{7.274}$$

The outgoing component of the vector is $-E_{||}H_x$ and,

$$
\begin{aligned}
\left(\frac{dE}{dx}\right)_{b>b_{min}} &= -\frac{2\pi\, b_{min}}{\beta c}\int_{-\infty}^{\infty} dx\, E_{||} H_x \\
&= -2\pi\, b_{min}\int_{-\infty}^{\infty} dt\, E_{||} H_x \\
&= -2\pi\, b_{min}\left(\frac{1}{2\pi}\int_{-\infty}^{\infty} dt \int_{-\infty}^{\infty} d\omega \int_{-\infty}^{\infty} d\omega' \right.\\
&\quad \left. \times\, E_{||}(\omega) H_x(\omega) e^{-i(\omega+\omega')t}\right) \\
&= -2\pi\, b_{min}\int_{-\infty}^{\infty} d\omega\, E_{||}(\omega)\, H_x^*(\omega) \\
&= -4\pi\, b_{min}\, \mathrm{Re}\left(\int_0^{\infty} d\omega\, E_{||}(\omega)\, H_x^*(\omega)\right) \\
&= \frac{(ze)^2\, b_{min}}{2\pi^2 \varepsilon_0 (\beta c)^2}\mathrm{Re}\left(\int_0^{\infty} d\omega (i\omega\lambda^*(\omega))\right.\\
&\quad \times \left(\frac{1}{\varepsilon_R(\omega)} - \beta^2\right) K_0(\lambda(\omega) b_{min}) \\
&\quad \left. \times\, K_1\left(\lambda^*(\omega) b_{min}\right)\right)
\end{aligned} \tag{7.275}
$$

Although the i can be removed from the integrand through the use of $\mathrm{Re}\, iz = -\mathrm{Im}\, z$, it is retained in anticipation of a future complex integration. In order to have the result of a non-zero stopping power, this integral must have a real component which requires that either λ or $\varepsilon_R(\omega)$ be complex. Even if the relative dielectric permittivity were to be real, it is possible for the stopping power to be non-zero when λ is complex which is a result of the particle speed exceeding the phase velocity of the medium. This energy loss is manifested as the Čerenkov radiation discussed below. On the other hand, if the particle speed were to be less than the phase velocity, λ would be complex only if $\varepsilon_R(\omega)$ was. Because of the $\beta^2\varepsilon_R(\omega)$ term in the expression for λ^2, this effect becomes significant at high projectile speeds. One can simplify the expression for the stopping power in a dielectric medium by taking advantage of this and limiting the derivation to the extreme relativistic case of $\beta \approx 1$ for which,

$$
\lambda^2 \approx \left(\frac{\omega}{c}\right)^2 (1 - \varepsilon_R(\omega)) \quad \text{for} \quad \beta \approx 1. \tag{7.276}
$$

From the earlier discussion of electronic polarization, the integral will be significant for ω in the ultraviolet region (10^{16}–10^{17}/s). As b_{min} is of the order of the atomic radius, R_A, the argument of the modified Bessel function will thus be of the order of,

$$
|\lambda\, b| < \left(\frac{\omega_0 R_A}{c}\right) \approx 0.003.
$$

As a result, one can use the small-argument limits,

$$
K_0(y) \approx \ln\frac{1.123}{y} \text{ and } K_1(y) \approx \frac{1}{y}.
$$

Substituting these limits into the stopping power expression gives,

$$
\begin{aligned}
\frac{dE}{dx} &= \frac{(ze)^2}{2\pi^2\, \varepsilon_0 (\beta c)^2}\mathrm{Re}\left(\int_0^{\infty} d\omega (i\omega)\left(\frac{1}{\varepsilon_R(\omega)} - \beta^2\right)\right.\\
&\quad \left. \times \ln\frac{1.123}{\lambda(\omega) b_{min}}\right) \\
&= \frac{(ze)^2}{2\pi^2 \varepsilon_0 (\beta c)^2}\mathrm{Re}\left(\int_0^{\infty} d\omega (i\omega)\left(\frac{1}{\varepsilon_R(\omega)} - \beta^2\right)\right.\\
&\quad \left. \times \ln\frac{1.123}{\left(\frac{\omega b_{min}}{\beta c}\right)\sqrt{1 - \beta^2 \varepsilon_R(\omega)}}\right) \\
&= \frac{(ze)^2}{2\pi^2\, \varepsilon_0 (\beta c)^2}\mathrm{Re}\left(\int_0^{\infty} d\omega (i\omega)\left(\frac{1}{\varepsilon_R(\omega)} - \beta^2\right)\right.\\
&\quad \left. \times \left(\ln\left(\frac{1.123\, \beta c}{\omega b_{min}}\right) - \frac{1}{2}\ln(1 - \beta^2 \varepsilon_R(\omega))\right)\right)
\end{aligned} \tag{7.277}
$$

(Note that, in the limit of $\varepsilon_R(\omega) \to 1$, $-\frac{1}{2}\ln(1 - \beta^2\varepsilon_R(\omega)) \to \ln\gamma$). In making the extreme relativistic case even more explicit, set $\beta = 1$,

$$
\begin{aligned}
\frac{dE}{dx} &= \frac{(ze)^2}{2\pi^2\, \varepsilon_0\, c^2}\mathrm{Re}\left(\int_0^{\infty} d\omega\, (i\omega)\left(\frac{1}{\varepsilon_R(\omega)} - 1\right)\right.\\
&\quad \left. \times \left(\ln\left(\frac{1.123\, c}{\omega b_{min}}\right) - \frac{1}{2}\ln(1 - \varepsilon_R(\omega))\right)\right)
\end{aligned} \tag{7.278}
$$

For a small damping constant Γ, the imaginary part of the relative dielectric permittivity can be neglected and the expression for the stopping power becomes,

$$\frac{dE}{dx} = \frac{(ze)^2}{2\pi^2 \varepsilon_0 c^2} \mathrm{Re}\left(\int_0^\infty d\omega\,(i\omega) \left(\frac{-\frac{\omega_P^2}{(\omega_0^2-\omega^2)+i\Gamma\omega}}{1+\frac{\omega_P^2}{(\omega_0^2-\omega^2)+i\Gamma\omega}} \right) \times \left(\ln\left(\frac{1.123c}{\omega b_{min}}\right) - \frac{1}{2}\ln\left(-\frac{\omega_P^2}{(\omega_0^2-\omega^2)+i\Gamma\omega} \right) \right) \right)$$

$$= -\frac{(ze)^2}{2\pi^2 \varepsilon_0 c^2} \mathrm{Re}\left(i \int_0^\infty d\omega\,\omega \left(\frac{\omega_P^2}{(\omega_0^2-\omega^2+\omega_P^2)+i\Gamma\omega} \right) \times \left(\ln\left(\frac{1.123c}{\omega b_{min}}\right) + \frac{1}{2}\ln\left(\frac{(\omega^2-\omega_0^2)+i\Gamma\omega}{\omega_P^2} \right) \right) \right) \tag{7.279}$$

The integral of (7.279),

$$i \int_0^\infty d\omega\,\omega \left(\frac{\omega_P^2}{(\omega_0^2-\omega^2+\omega_P^2)+i\Gamma\omega} \right) \times \left(\ln\left(\frac{1.123\,c}{\omega b_{min}}\right) + \frac{1}{2}\ln\left(\frac{(\omega^2-\omega_0^2)+i\Gamma\omega}{\omega_P^2} \right) \right)$$

$$= i \int_0^\infty d\omega\,\omega \left(\frac{\omega_P^2}{(\omega_0^2-\omega^2+\omega_P^2)+i\Gamma\omega} \right) \times \left(\ln\left(\frac{1.123\,c}{\omega_P b_{min}}\right) + \frac{1}{2}\ln\left(\frac{(\omega^2-\omega_0^2)+i\Gamma\omega}{\omega^2} \right) \right) \tag{7.280}$$

is determined using the Cauchy–Goursat theorem by changing the integration over positive real ω (i.e., $0 \le \omega \le \infty$) to that over positive imaginary ω minus the integration over the quarter-circle to infinity, as shown in Fig. 7.21, with the intent of isolating the integration I_X. It can be seen from the integrand that poles occur for,

$$(\omega_0^2 - \omega^2 + \omega_P^2) + i\Gamma\omega = 0$$

and

$$\frac{(\omega^2 - \omega_0^2) + i\Gamma\omega}{\omega^2} = 0.$$

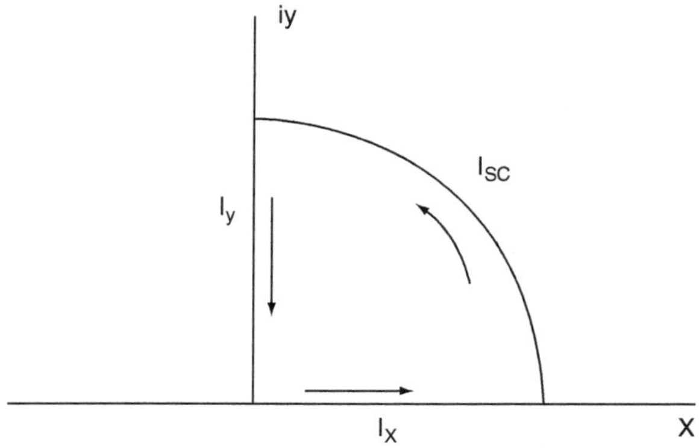

Fig. 7.21 Integration contour in the complex plane z = x + iy used to evaluate the integral of (7.280)

In both cases, ω would be imaginary and the poles of the integrand exist in the lower half-plane. As there are no poles within the quarter-circle of the contour, the total integral is equal to zero, or,

$$\mathrm{Re}(I_X + I_Y + I_{SC}) = 0$$

The integral down the imaginary axis is evaluated first using $\omega = i\Omega$,

$$I_Y = -i \int_0^\infty d\Omega\,\Omega \left(\frac{\omega_P^2}{(\omega_0^2+\Omega^2+\omega_P^2)-i\Gamma\Omega} \right) \times \left(\ln\left(\frac{1.123\,c}{\omega_P b_{min}}\right) + \frac{1}{2}\ln\left(\frac{-(\Omega^2+\omega_0^2+\Gamma\Omega)}{(i\Omega)} \right) \right) \tag{7.281}$$

It is clear that, as this integral is pure imaginary (i.e., Re $I_Y = 0$), its contribution to the stopping power will be zero. As a result,

$$\mathrm{Re}\, I_X = -\mathrm{Re}\, I_{SC}. \tag{7.282}$$

In order to calculate the integral over the quarter-circle, I_{SC}, write ω in terms of complex polar coordinates, or,

$$\omega = re^{i\theta} \tag{7.283}$$

where r is a constant. Thus,

$$\begin{aligned}\mathrm{Re}\, I_X &= -\mathrm{Re}\, I_{SC} \\ &= -\mathrm{Re}\int_0^{\pi/2} d\theta\, r^2 e^{i2\theta}\left(\frac{\omega_P^2}{(\omega_0^2 - r^2e^{i2\theta} + \omega_P^2) + i\Gamma re^{i\theta}}\right) \\ &\quad\times\left(\ln\left(\frac{1.123\,c}{\omega_P b_{min}}\right) + \frac{1}{2}\ln\left(\frac{(r^2e^{i2\theta} - \omega_0^2) + i\Gamma re^{i\theta}}{r^2}e^{-i2\theta}\right)\right).\end{aligned} \tag{7.284}$$

Maintaining the assumption that the damping Γ is small and constant,

$$\begin{aligned}&-\mathrm{Re}\int_0^{\pi/2} d\theta\, r^2 e^{i2\theta}\left(\frac{\omega_P^2}{(\omega_0^2 - r^2e^{i2\theta} + \omega_P^2) + i\Gamma re^{i\theta}}\right) \\ &\times\left(\ln\left(\frac{1.123\,c}{\omega_P b_{min}}\right) + \frac{1}{2}\ln\left(\frac{(r^2e^{i2\theta} - \omega_0^2) + i\Gamma re^{i\theta}}{r^2}e^{-i2\theta}\right)\right) \\ &= \mathrm{Re}\, I_1 + \mathrm{Re}\, I_2\end{aligned}$$

where, by rearranging the variables within the integrand,

$$\begin{aligned}I_1 &= \omega_P^2 \ln\left(\frac{1.123\,c}{\omega_P b_{min}}\right)\int_0^{\pi/2} d\theta \frac{e^{i2\theta}}{\left(\frac{\omega_0^2+\omega_P^2}{r^2}\right) - e^{i2\theta} + i\left(\frac{\Gamma}{r}\right)e^{i\theta}} \\ &\approx -\frac{\pi\omega_P^2}{2}\ln\frac{1.123}{b\omega_P}.\end{aligned} \tag{7.285}$$

The facts that the ratio $\Gamma/|\omega|$ is negligible and that $\omega^2 \gg |\omega_0^2 + \omega_P^2|$ have been used. The second integral is,

$$\begin{aligned}I_2 &= \frac{1}{2}\int_0^{\pi/2} d\theta\, r^2 e^{i2\theta}\left(\frac{\omega_P^2}{(\omega_0^2 - r^2e^{i2\theta} + \omega_P^2) + i\Gamma re^{i\theta}}\right) \\ &\quad\times\left(\ln\left(\frac{(r^2e^{i2\theta} - \omega_0^2) + i\Gamma re^{i\theta}}{r^2}e^{-i2\theta}\right)\right) \\ &= \frac{1}{2}\int_0^{\pi/2} d\theta \frac{e^{i2\theta}}{\left(\frac{\omega_0^2+\omega_P^2}{r^2}\right) - e^{i2\theta} + i\left(\frac{\Gamma}{r}\right)e^{i\theta}} \\ &\quad\times\left(\ln\left(\frac{(r^2e^{i2\theta} - \omega_0^2) + i\Gamma re^{i\theta}}{r^2}e^{-i2\theta}\right)\right) \\ &\approx -\frac{1}{2}\int_0^{\pi/2} d\theta\left(\ln\left(\frac{(r^2e^{i2\theta} - \omega_0^2) + i\Gamma re^{i\theta}}{r^2}e^{-i2\theta}\right)\right) \\ &\approx -\frac{1}{2}\int_0^{\pi/2} d\theta\, \ln\left(\left(e^{i2\theta} - \left(\frac{\omega_0}{r}\right)^2 + i\frac{\Gamma}{r}e^{i\theta r}\right)e^{-i2\theta}\right) \\ &\approx -\frac{1}{2}\int_0^{\pi/2} d\theta\, \ln\left(1 - \left(\frac{\omega_0}{r}\right)^2 e^{-i2\theta} + i\frac{\Gamma}{r}e^{-i\theta r}\right) \\ &\approx 0.\end{aligned}$$

Thus,

$$\begin{aligned}&\mathrm{Re}\left(i\int_0^{\infty} d\omega\,\omega\left(\frac{\omega_P^2}{(\omega_0^2 - \omega^2 + \omega_P^2) + i\Gamma\omega}\right)\right. \\ &\quad\left.\times\left(\ln\left(\frac{1.123\,c}{\omega_P\, b_{min}}\right) + \frac{1}{2}\ln\left(\frac{(\omega^2 - \omega_0^2) + i\Gamma\omega}{\omega^2}\right)\right)\right) \\ &\approx \frac{\pi\omega_P^2}{2}\ln\frac{1.123}{b\omega_P}.\end{aligned} \tag{7.286}$$

One now writes the linear stopping power in a dielectric medium in the extreme relativistic limit as,

$$\left(\frac{dE}{dx}\right)_{b>b_{min}} = \frac{(ze)^2\omega_P^2}{4\pi\varepsilon_0 c^2}\ln\frac{1.123\,c}{b_{min}\,\omega_P}. \tag{7.287}$$

One can recast this as the mass collision stopping power

$$\left(\frac{dE}{\rho\, dx}\right)_{Col} = 2C\left(\frac{Z}{A}\right)z^2\ln\frac{1.123\,c}{b_{min}\,\omega_P} \tag{7.288}$$

where the identifier of $b > b_{min}$ has been removed and the identifier "Col" added to indicate that this is a stopping power due to collisions (i.e., nonradiative interactions). A significant difference between this expression, which accounts for the medium's dielectric response to the charged particle, and the relativistic forms of those that do not is that the former is no longer a function of the medium's atomic structure, which appears in the form of $\bar{I}$ or $\bar{\omega}$ within the logarithmic term, but rather of the medium's electron density which appears through the electron plasma frequency. That is, in the extreme relativistic regime, the mass collision stopping powers of two media are equal provided that the electron densities and the Z/A ratios are the same.

In the absence of any polarization effect, the corresponding relativistic mass collision stopping power is, where for clarity the Bohr result is used,

$$\left(\frac{dE}{\rho\, dx}\right)_{b>b_{min}} = 2C\left(\frac{Z}{A}\right)z^2 \ln\frac{1.123\,\gamma\, c}{b_{min}\,\bar{\omega}}. \quad (7.289)$$

An analytical form of the density correction term given by the difference between these two mass collision stopping powers is,

$$\delta = 2C\left(\frac{Z}{A}\right)z^2 \ln\frac{\gamma\omega_P}{\bar{\omega}} \quad (7.290)$$

Sternheimer–Peierls Parameterization of the Density/Polarization Effect

While the above provides a theoretical expression for the polarization/density effect, for practical purposes, a parameterization of the effect is frequently required in dosimetry calculations. A frequently-used parameterization which can accelerate the calculation of δ is that of Sternheimer and Peierls (1971) who presented an expression for δ applicable to both condensed media and gases

$$\delta = (2\ln 10)x + \Phi \qquad x \geq x_1 \quad (7.291)$$

$$\delta = (2\ln 10)x + \Phi + a(x_1 - x)^n \qquad x_0 \leq x \leq x_1 \quad (7.292)$$

$$\delta = 0 \qquad x < x_0 \quad (7.293)$$

where the kinematic variable is, for a particle of mass m and momentum p,

$$x = \log_{10}\left(\frac{p}{m}\right) \quad (7.294)$$

and where x_0 and x_1 are defined below. This reproduces the logarithmic increase in the density effect with $p/m \approx E/m \approx \gamma$ at high energies. The remaining quantities are,

$$\Phi = -2\ln\left(\frac{\bar{I}}{\hbar\omega_P}\right) - 1 \quad (7.295)$$

$$a = \frac{-\Phi - (2\ln 10)x_0}{(x_1 - x_0)^n} \quad (7.296)$$

The values of x_0 and x_1 also depend upon the values of the mean ionization potential, $\bar{I}$, and the phase of the medium. For liquid and solid media, these are:

$$\begin{aligned} x_0 &= 0.2 && \text{if } \bar{I} < 100\,\text{eV and } -\Phi \leq 3.681 \\ &= -0.326\Phi - 1.0 && \text{if } \bar{I} < 100\,\text{eV and } -\Phi > 3.681 \\ &= 0.2 && \text{if } \bar{I} \geq 100\,\text{eV and } -\Phi \leq 5.215 \\ &= -0.326\Phi - 1.5 && \text{if } \bar{I} \geq 100\,\text{eV and } -\Phi > 5.215 \end{aligned} \quad (7.297)$$

$$\begin{aligned} x_1 &= 2.0 && \text{if } \bar{I} < 100\,\text{eV} \\ &= 3.0 && \text{if } \bar{I} \geq 100\,\text{eV} \end{aligned} \quad (7.298)$$

In all cases, n = 3. The relative magnitude of δ and its growth with particle speed are shown in Fig. 7.22 which plots fitted values for δ for carbon and lead as a function of p/m; for example, the value of δ for carbon reaches 2 for p/m = 9.8, corresponding to a kinetic energy of about 5 MeV for electrons. In the context of examples of radionuclides used in nuclear medicine, the maximum β-particle energy of ^{131}I is 606 keV, for which the value of δ in carbon is equal to 0.124. The maximum recoil electron energy for the 140.5 keV γ ray from ^{99m}Tc is 49.6 keV, for which δ is essentially negligible.

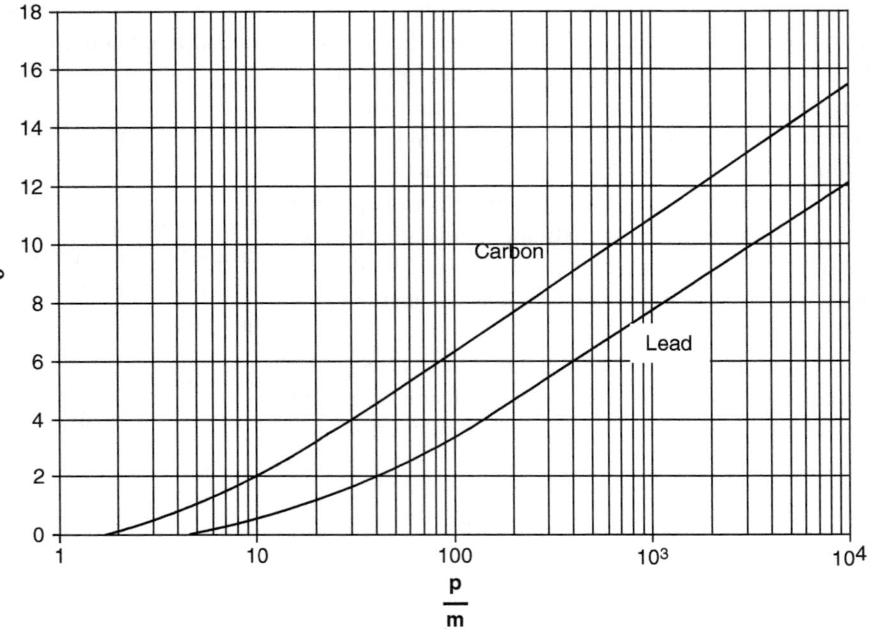

Fig. 7.22 Polarization/density correction terms for carbon and lead as functions of the ratio of the particle momentum normalized to particle mass

Čerenkov Radiation

Čerenkov (1934) reported his observations of the eponymous radiation following the irradiation of various liquids to γ rays. Tamm and Frank (1937) published their interpretation of this observation shortly afterwards. As will be demonstrated, the magnitude of the energy loss associated with Čerenkov radiation is negligible for dosimetry considerations and, as a result, its interest to medical physics applications will be limited. However, as it is straightforward to extend the above derivation of the density effect to explain this phenomenon, Čerenkov radiation is derived here.

Equation (7.287) gives the energy loss per unit distance traveled to regions within the medium with impact parameters greater than a value, b_{min}. It was assumed that b_{min} is of the order of atomic dimensions and that $|\lambda b_{min}| \ll 1$, so that the small-argument limits of the modified Bessel functions could be used. Consequently, the final result represents the rate of local energy deposition with distance. On the other hand, by permitting $|\lambda b_{min}| \ll 1$, one obtains the rate of energy deposited at great distances per distance traveled from the projectile's trajectory. Recalling the large-argument expression for the modified Bessel function,

$$\left(\frac{dE}{dx}\right)_{b>b_{min}} = \frac{(ze)^2}{4\pi\varepsilon_0(\beta c)^2} \times \mathrm{Re}\left(\int_0^\infty d\omega \left(i\omega\sqrt{\frac{\lambda^*(\omega)}{\lambda(\omega)}}\right)\left(\frac{1}{\varepsilon_R(\omega)} - \beta^2\right) \times e^{-(\lambda(\omega)+\lambda^*(\omega))\, b_{min}}\right). \tag{7.299}$$

Consider,

$$\lambda(\omega) = \left(\frac{\omega}{\beta c}\right)\sqrt{1-\beta^2\varepsilon_R(\omega)}. \tag{7.300}$$

In general, Re $\lambda(\omega) > 0$ and the exponential term in the integrand will, as a consequence, rapidly attenuate the energy loss with distance from the trajectory as expected. However, $\lambda(\omega)$ will be purely imaginary if the damping constant Γ is negligible, which allows $\varepsilon_R(\omega)$ to be purely real resulting in $\beta\varepsilon_R{}^2(\omega) > 1$. For purely imaginary $\lambda(\omega)$,

$$\sqrt{\frac{\lambda^*(\omega)}{\lambda(\omega)}} = i$$

and

$$e^{-(\lambda(\omega)+\lambda^*(\omega))b_{min}} = 1.$$

Thus, for the conditions of purely real $\varepsilon_R(\omega)$ and $\beta^2\varepsilon_R(\omega) > 1$, the stopping power expression simplifies to,

$$\left(\frac{dE}{dx}\right) = z^2\hbar \int_{\omega_1}^{\infty} d\omega\, \omega \left(1 - \frac{1}{\beta^2 \varepsilon_R(\omega)}\right) \qquad (7.301)$$

where the lower frequency limit is specified by the requirement of

$$\varepsilon_R(\omega > \omega_1) > \frac{1}{\beta^2} \qquad (7.302)$$

and there is no longer a dependence of the energy transfer upon b_{min}. This result describes the rate per distance traveled by the projectile at which energy is radiated. The projectile's speed must exceed the phase velocity of the medium for the given frequency, ω, for Čerenkov radiation to occur.

In Fig. 7.23 the real component of $\varepsilon_R(\omega)$ is plotted as a function of ω for negligible Γ (i.e., negligible energy absorption). There is a discontinuity at $\omega = \omega_0$ and the "Čerenkov band" is shown for which $\varepsilon_R(\omega) > 1/\beta^2$ and which identifies the lower frequency integration limit.

Before investigating the spectrum of Čerenkov radiation, the frequently-used geometrical description of a charged particle moving through a dielectric medium at a speed βc exceeding the phase velocity is investigated. The index of refraction of the medium is $n(\omega) = \sqrt{\varepsilon_R(\omega)}$; however, for conciseness, the frequency dependence is ignored. The electric field will thus propagate with speed c/n and, for the condition of $\beta c > c/n$, Huygens' constructions are used for the electromagnetic waves emitted by the particle as shown in Fig. 7.24. Let the particle be at point A at time $t = 0$. In the time that it takes the particle to travel the distance AB, which is equal to,

$$T = \frac{AB}{\beta c} \qquad (7.303)$$

(for temporary convenience units where $c \neq 1$ are used) a wavefront emitted at $t = 0$ (i.e., at point A) will have reached point C, where

$$AC = T\frac{c}{n}. \qquad (7.304)$$

Thus,

$$\theta = \cos^{-1}\frac{AC}{AB} = \cos^{-1}\frac{1}{\beta n}. \qquad (7.305)$$

Note that Fig. 7.24 shows a single plane and that, due to axial symmetry about AB, the Čerenkov wavefronts form a cone. As Čerenkov radiation can only occur if the particle speed $\beta > 1/n$, this effect will be limited to high energies in most media of dosimetric interest (e.g., as $n = 1.5$ for Perspex, the threshold value for β is 0.67). The threshold energy for which

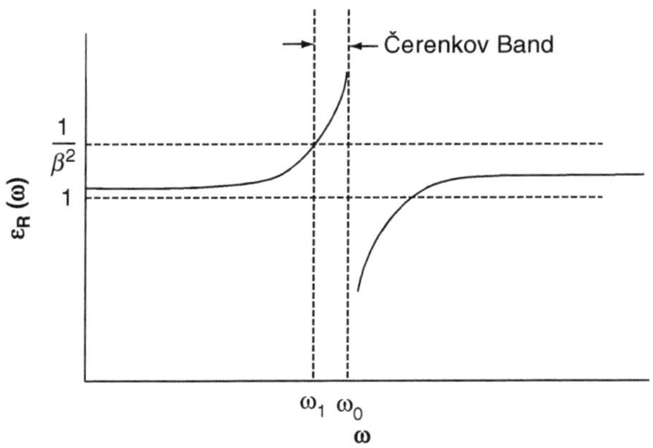

Fig. 7.23 The relative dielectric permittivity (assumed to be purely real) of a medium as a function of frequency. The Čerenkov band, within which Cerenkov radiation can occur, is shown

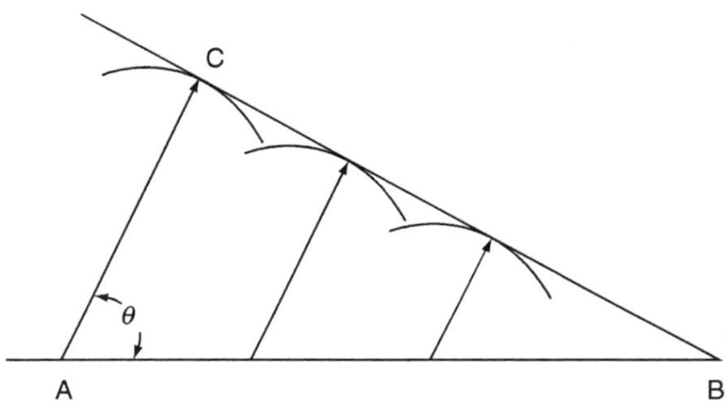

Fig. 7.24 Huygens' reconstruction of the coherent wavefront of Čerenkov radiation

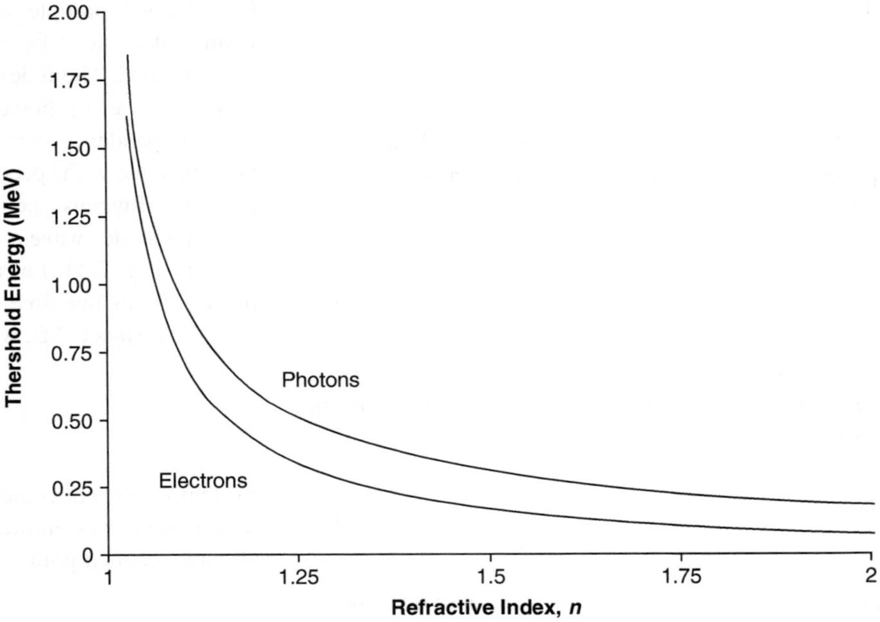

Fig. 7.25 Čerenkov radiation energy threshold for electrons and for Compton-scattered electrons set in motion by backscattered photons (labeled as "photons") as functions of the refractive index

Čerenkov radiation will occur follows as $\beta_{Thr} = 1/n$ where

$$\beta_{Thr} = \sqrt{1 - \left(\frac{m}{T_{Thr} + m}\right)^2} \tag{7.306}$$

where T_{Thr} is the threshold kinetic energy of the charged particle and m is its mass.

Figure 7.25 shows the variation of the threshold energy as a function of refractive index for electrons and for those electrons set in motion by backscattered photons following incoherent scatter. It can be seen that, within the energy range of 0.1–0.3 MeV, Čerenkov radiation presents a means of detecting γ rays in low-Z media where the probability of a photoelectric absorption is much less than that of incoherent scatter.

The differential spectrum in frequency of Čerenkov radiation is provided by the integrand of (7.301). The number of photons emitted per centimeter of projectile path length with a frequency between ω and $\omega + d\omega$ is,

$$dN = \frac{\alpha}{c}\left(1 - \frac{1}{(\beta n)^2}\right)d\omega. \tag{7.307}$$

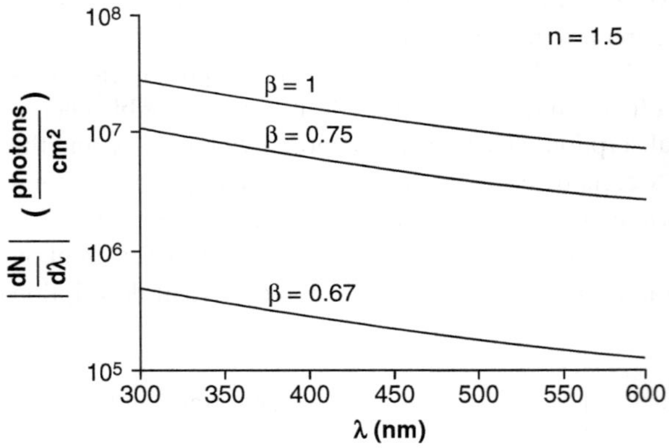

Fig. 7.26 Number of Čerenkov photons emitted per unit path-length per unit wavelength in Perspex (n = 1.5) for three values of β

As $d\omega = -2\pi c\, d\lambda/\lambda^2$, we can rewrite this more conveniently in terms of the photon wavelength,

$$\frac{dN}{d\lambda} = -\frac{2\pi\alpha}{\lambda^2}\left(\beta^2 n^2 - 1\right). \tag{7.308}$$

This spectrum is plotted as a function of photon wavelength for Perspex (n = 1.5) for values of β of 0.67 (threshold), 0.75 and 1 in Fig. 7.26. For protons, this threshold value corresponds to a kinetic energy of about 375 MeV. The λ^{-2} dependence shows that the Čerenkov spectrum is dominated by short wavelength photons. Integrating the Čerenkov spectrum yields the number of Čerenkov photons emitted per centimeter between the spectrum limits of λ_{min} and λ_{max},

$$N = 2\pi\alpha\left(1 - \frac{1}{(\beta n)^2}\right)\left(\frac{1}{\lambda_{min}} - \frac{1}{\lambda_{max}}\right). \quad (7.309)$$

Using $\cos\theta = 1/\beta n$, one can also write the spectrum as,

$$\frac{dN}{d\lambda} = 2\pi\,\alpha\sin^2\theta\left(\frac{1}{\lambda_{max}} - \frac{1}{\lambda_{min}}\right). \quad (7.310)$$

Within the spectrum limits of 300 and 600 nm, a total of 764 $\sin^2\theta$ Čerenkov photons are emitted per centimeter of particle path length, which clearly demonstrates the low level of light output. As for minimally-ionizing particles in a low-Z medium this corresponds to a conversion efficiency of about 0.2%, the contribution of Čerenkov radiation to the overall energy loss of a charged particle traversing a medium is small and can be neglected for dosimetry purposes.

7.3.7.7 Empirical Determination of Mean Excitation Energy and Shell Correction Factor

It has been noted earlier that, instead of calculating $\bar{I}$ and $C_e(\beta)/Z$ for a given medium, it is possible to empirically evaluate their combined effect upon the collision stopping power from detailed experimental measurements of the energy loss of a charged particle (typically a proton) traversing a thin foil of the medium in question (Ammi et al. 2005). At sufficiently low projectile energies, the diminishment of the stopping power due to polarization of the medium (described by $L_{02}(\beta)$) is negligible and the Barkas $L_2(\beta)$ term may also be neglected. Radiative energy transfer is also negligible and the energy loss will be due solely to collisions between the projectile and the atomic electrons. Under these conditions, the mass collision stopping power reduces to,

$$\left(\frac{dE}{\rho\,dx}\right)_{Col} = 2C\left(\frac{Z}{A}\right)\left(\frac{z}{\beta}\right)^2 L_0(\beta). \quad (7.311)$$

Consider a proton (z = 1) with speed βc traversing a thin foil of medium with an areal thickness $\rho\,\Delta x$ sufficiently small that the proton's measured energy loss is,[17]

$$\begin{aligned}\Delta E &= (\rho\,\Delta x)\left(\frac{dE}{\rho\,dx}\right)_{Col} \\ &= 2C\left(\frac{Z}{A}\right)\frac{1}{\beta^2}L_0(\beta).\end{aligned} \quad (7.312)$$

Inverting this to obtain the zeroth-order stopping power number,

$$L_0(\beta) = -\left(\frac{A}{Z}\right)\left(\frac{\beta^2}{2C}\right)\frac{1}{\rho\,\Delta x}\Delta E_{meas} \quad (7.313)$$

where ΔE_{meas} is the measured energy loss of the proton. For the measurement conditions specified,

$$\begin{aligned}L_0(\beta) &= L_{00}(\beta) + L_{01}(\beta) \\ &= \ln\left(\frac{2m_e}{\bar{I}}\gamma^2\beta^2\right) - \beta^2 - \frac{C_e(\beta)}{Z} \\ &= -\left(\ln\bar{I} + \frac{C_e(\beta)}{Z}\right) + \ln(2m_e\gamma^2\beta^2) - \beta^2\end{aligned} \quad (7.314)$$

where $\bar{I}$ and m_e implicitly have the same units of energy. Equating these two expressions for $L_0(\beta)$ and solving for $(\ln\bar{I} + C_e(\beta)/Z)$ gives,

$$\begin{aligned}\left(\ln\bar{I} + \frac{C_e(\beta)}{Z}\right) &= \left(\frac{A}{Z}\right)\left(\frac{\beta^2}{2C}\right)\frac{1}{\rho\,\Delta x}\Delta E_{meas} \\ &\quad + \ln(2m_e\gamma^2\beta^2) - \beta^2.\end{aligned} \quad (7.315)$$

It is clear that the mean excitation potential and the shell correction effect are not separated (although at sufficiently high projectile energies, we can neglect the latter). However, since both quantities appear in this combination in the collision stopping power expression, this is not problematic for calculating the collision stopping power.

[17]ICRU Report 49 (1993) provides a comprehensive historical summary of the various experimental techniques used to measure the stopping power.

7.3.8 Mean Energy Required to Create an Ion Pair

The immediate consequence of ionization in a liquid or gaseous medium is the creation of an electron-ion pair. In some solids, ionization can elevate electrons into the conduction band, thus forming an electron-hole pair (following the convention used by ICRU Report No 31 (1979), the term "ion pair" will be used to describe both electron-ion and electron-hole pairs). Of fundamental interest to experimental radiation dosimetry is the mean energy expended to create an ion pair, W, by electrons. Knowledge of W coupled with the collision stopping power is required in order to estimate the number of ion pairs produced per unit pathlength. Measurement of this number (in, e.g., an ionization chamber) can be used to infer the stopping power or, equivalently, the energy absorption in the medium.

It was seen in the derivation of the collision stopping power that the electromagnetic interaction between an incident charged particle and an atom can lead to both atomic excitation and the elevation of an atomic electron into the continuum. Hence, the competing effects of nonionization energy channels leads to the fact that the value of W will exceed the first ionization potential of the atom. Consider a charged particle with kinetic energy T that has been completely stopped in a gaseous medium. The equation of energy balance is,

$$T = N_{Ion}(\bar{E}_{Ion} + \bar{\varepsilon}) + N_{Exc}\bar{E}_{Exc} \tag{7.316}$$

where N_{Ion} is the total number of electrons generated through ionization and N_{Exc} is the total number of excited atomic states. $\bar{E}_{Ion}$ is the mean energy required to produce an ion, $\bar{\varepsilon}$ is the mean energy of the secondary electrons (δ rays) which are not energetic enough to cause further ionizations and $\bar{E}_{Exc}$ is the mean energy of the discrete excited atomic states. By definition, the mean energy expended to produce an ion pair is given by,

$$\begin{aligned} W &= \frac{T}{N_{Ion}} \\ &= (\bar{E}_{Ion} + \bar{\varepsilon}) + \left(\frac{N_{Exc}}{N_{Ion}}\right)\bar{E}_{Exc} \end{aligned} \tag{7.317}$$

and the ratio of W to the ionization energy I is,

$$\frac{W}{I} = \frac{\bar{E}_{Ion}}{I} + \frac{\bar{\varepsilon}}{I} + \left(\frac{N_{Exc}}{N_{Ion}}\right)\frac{\bar{E}_{Exc}}{I}. \tag{7.318}$$

It is possible to describe qualitatively the magnitude of the three terms on the right-hand side of (7.318) from first principles. First, one would expect,

$$\frac{\bar{E}_{Ion}}{I} > 1 \tag{7.319}$$

due to the fact that excitation and other nonionization channels exist, especially for molecules where there are rotational and vibrational modes available. Similarly, one would also expect that,

$$\frac{\bar{E}_{Exc}}{I} > 1 \tag{7.320}$$

although this ratio is not that much different from unity as the energy levels of most discrete excited states are relatively near I. The ratio $\bar{\varepsilon}/I$ will be small with values typically about 0.3 for noble gases and smaller for molecular gases. Finally, the ratio of the number of discrete excited states to the number of ionizations, N_{Exc}/N_{Ion} is estimated in ICRU Report Number 31 (1979) to be about 0.3 for closed-shell atoms and close to 1 for closed-shell molecules. Overall, the magnitude of W/I is greater than unity with values ranging from about 1.7 for noble gases to up to 3.2 for alkaline earths.

The total number of electrons produced is also expected to be a function of the particle's kinetic energy. Consider a medium in which the atoms have a single ionization energy, I, through which a single electron with kinetic energy T slows down and stops. The number of electrons produced as a result is (Fowler 1923),

$$\begin{aligned} N_e(T) &= \frac{\sigma_{Ion}(T)}{\sigma_{Inel}(T)} + \frac{1}{\sigma_{Inel}(T)}\sum_n \sigma_n(T)\,N(T-\Delta E_n) \\ &+ \frac{1}{\sigma_{Inel}(T)}\int_I^{\frac{1}{2}(T+I)} d(\Delta E_n)\frac{d\sigma_{Ion}(T,\Delta E_n)}{d(\Delta E)_n} \\ &\times (N_{Ion}(T-\Delta E_n) + N_{Ion}(\Delta E_n - I)) \end{aligned} \tag{7.321}$$

where $\sigma_{Inel}(T)$ is the total inelastic cross section, $\sigma_{Ion}(T)$ is the total ionization cross section and

$\sigma_n(T)$ is the total cross section for excitation to the nth atomic state. Obviously, the total inelastic cross section is the sum of the ionization and excitation cross sections,

$$\sigma_{Inel}(T) = \sigma_{Ion}(T) + \sigma_n(T). \tag{7.322}$$

The first term on the right-hand side of (7.321) describes the number of secondary electrons produced as a consequence of the first inelastic collision. If this collision leads to the excitation of the atom to an energy level ΔE_n, the incident electron is scattered with a kinetic energy $T - \Delta E_n$ and the mean of the total number of ion pairs produced by the scattered electron is given by $(\sigma_N(T)/\sigma_{Inel}(T))\,N(T - \Delta E_n)$. The sum of this contribution over all excited states is the second term on the right-hand side of the equation. Finally, one must account for the fact that the first inelastic collision is itself ionizing and results in an electron-ion pair: the scattered electron with kinetic energy $T - \Delta E_n$ and an ejected electron (δ ray) with kinetic energy $\Delta E_n - I$. This event contributes both to the first and third terms on the right-hand side.

Obviously, knowledge of the energy dependence of the inelastic, ionization, and excitation cross sections is required in order to evaluate (7.322). Another approach to calculating for $N_e(T)$ was proposed by Spencer and Fano (1954) and uses the concept of the degradation spectrum. Imagine a monoenergetic beam of electrons with kinetic energy T incident to a gas. Within this medium, these electrons will, through deceleration and the production of secondary electrons, yield a net flux of electrons with an energy spectrum. The pathlength of all electrons with kinetic energies between T' and $T' + dT'$ is $y(T, T')dT'$, which is also a descriptor of the electron spectrum. This spectrum will not be derived here; one can note that in the extreme case of the CSDA for $T \gg 1$ in which the projectile loses only a small amount of energy, the number of produced electrons is,

$$N_e(T) = \rho_{Molec} \int_I^T dT'\, \sigma_{Ion}(T')y(T, T') \tag{7.323}$$

where ρ_{Molec} is the number of molecules per unit volume in the medium (i.e., the molecular number density). A scaling property of $y(T, T')dT'$ was found by Douthat (1975) who showed that the quantity $\rho_{Molec}(T'/T)\ln(T/I)\sigma_{Ion}(T')y(T, T')$ plotted as a function of the variable,

$$\xi = \frac{\ln\left(\frac{T'}{I}\right)}{\ln\left(\frac{T}{I}\right)},$$

was virtually independent of the projectile electron's kinetic energy, Fano and Spencer (1975) subsequently defined the quantity

$$z(\xi) = \rho_{Molec}\, \sigma_{St}(T') \left(\frac{\ln\left(\frac{T'}{I}\right)}{\xi}\right) \left(\frac{T'}{T}\right) y(T', T) \tag{7.324}$$

where $\sigma_{St}(\xi)$ is the stopping cross section. Using this, the expression for $N_e(T)$ can now be rewritten as,

$$N_e(T) = T \int_0^1 d\xi\, z(\xi) \frac{\sigma_{Ion}(\xi)}{\sigma_{St}(\xi)} \tag{7.325}$$

which leads to an expression of the mean energy required to produce an ion pair,

$$W(T) = \frac{1}{\int_0^1 d\xi\, z(\xi) \frac{\sigma_{Ion}(\xi)}{\sigma_{St}(\xi)}}. \tag{7.326}$$

It should be noted that $\Delta T' \frac{\sigma_{Ion}(\xi)}{\sigma_{St}(\xi)}$ is the number of ion pairs produced directly by the projectile electron within the energy interval $T' - \Delta T'$ to T'.

W has a limited sensitivity to the charge, mass and kinetic energy of the projectile, although the dependence upon projectile kinetic energy increases when the projectile speed becomes comparable to those of the valence electrons. This general insensitivity to projectile kinetic energy at high energies is a consequence of the fact that the ratio of the ionization cross section to the sum of excitation cross sections has a limited energy dependence. For dry air as a medium,

$$W_{Dry\,Air} = 33.8_5 \pm 0.15\,\text{eV}$$

Thus a 5 MeV α particle completely stopped in dry air will create about 1.25×10^5 electron-ion pairs.

Values of W for solids will be much less than for gases as a result of the difference between the ~1 eV gap between the valence and electron bands of a solid and the ~10 eV ionization energy of gases. For example, W for electrons in solid silicon is

$$W_{Si} = 3.68 \pm 0.02\,\text{eV}$$

7.3.9 Restricted Mass Collision Stopping Power for Electrons

Here, the discussion of the restricted mass collision stopping power for electrons is returned to,

$$\left(\frac{dE}{\rho dx}\right)_{Col,\Delta} = -C\left(\frac{Z}{A}\right)\left(\frac{1}{\beta}\right)^2 \times \left(\ln\left(\frac{2\Delta(T-\Delta)}{\bar{I}^2}(\gamma+1)\right) + f^-(\gamma,\Delta)\right)$$

where,

$$f^-(\gamma,\Delta) = \left(\frac{\Delta}{T-\Delta}\right) - \beta^2 + \left(\frac{2\gamma-1}{\gamma}\right)\ln\left(\frac{T-\Delta}{T}\right) + \left(\frac{\gamma-1}{\gamma}\right)^2\frac{\Delta^2}{2T^2}$$

and where Δ is the kinetic energy of the knock-on electron, $\Delta < T/2$. The restricted collision stopping power focuses our attention on the local energy deposition along the electron's track. A given value for Δ will denote the extent of the region around the projectile electron's trajectory that we are interested in knowing the energy transferred to or absorbed within the medium. In particular, if one is interested in the energy deposited in a small volume that the electron is passing through, then the use of the unrestricted collision stopping power will overestimate the deposited energy (unless the condition of charged particle equilibrium exists, in which the energy removed from the volume by the δ rays is compensated for by energy brought into the volume by δ rays generated from outside the volume). This will have important consequences in microdosimetry which we will be considering in following chapters. Tables of restricted and unrestricted collision stopping powers for electrons can be found in ICRU Report 37 (1984).

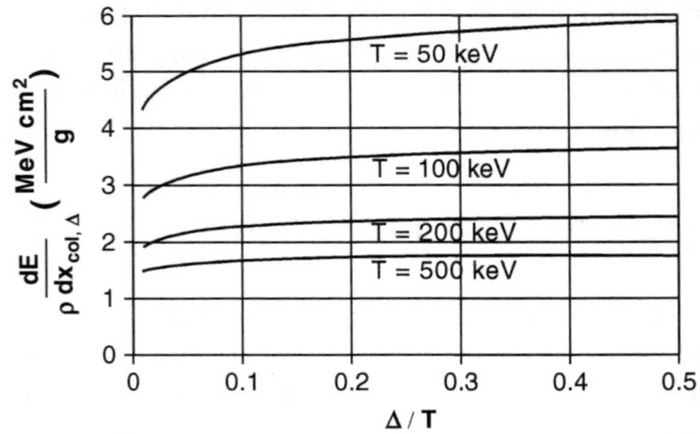

Fig. 7.27 Restricted mass collision stopping power for electrons in carbon as a function of the secondary electron kinetic energy normalized to the incident kinetic energy

Figure 7.27 shows $(dE/\rho dx)_{Col,\Delta}$ calculated as a function of $\Delta/T \leq 1/2$ for 50, 100, 200, and 500 keV electrons in carbon (excluding shell and polarization correction effects in order to show the effect of secondary electron energy restriction more clearly). The unrestricted mass collision stopping power is that for $\Delta = T/2$.[18] One sees that the restricted collision stopping power is always less than the unrestricted and that this difference decreases with increasing T, as to be expected. The increase in $(dE/\rho dx)_{Col,\Delta}$ with increasing Δ/T is the result of including more collisions which result in the transfer of energy to the medium. On the other hand, the decreasing difference between the unrestricted and restricted collision stopping powers with increasing T as shown in the figure is due to the approaching region of minimal ionization.

7.3.10 Summary of the Mass Collision Stopping Power

As the Barkas polarization term is negligible for practical dosimetry purposes, it is neglected in this summary of the collision stopping power. Including the shell-, density-, and Bloch correction factors, the mean

[18] A closely related quantity is the linear energy transfer, or LET, which is simply the restricted linear collision stopping power.

collision stopping power for a massive spin-0 projectile with charge ze is,

$$\left(\frac{dE}{\rho dx}\right)_{Col} = -2C\left(\frac{Z}{A}\right)\left(\frac{z}{\beta}\right)^2\left[\ln\left(\frac{2m_e}{\bar{I}}\gamma\beta^2\right) - \beta^2 - \frac{C_e}{Z} - \frac{\delta}{2} - \gamma_{EM} - \ln\left(\alpha\frac{z}{\beta}\right)\right]. \quad (7.327)$$

The mean collision stopping power for electrons and positrons is,

$$\left(\frac{dE}{\rho dx}\right)_{Col\pm} = -C\left(\frac{Z}{A}\right)\left(\frac{1}{\beta}\right)^2\left[\ln\left(\frac{T^2(\gamma+1)}{2\bar{I}^2}\right) + f_\pm(\gamma) - \frac{C_e}{Z} - \frac{\delta}{2} - \gamma_{EM} - \ln\left(\alpha\frac{z}{\beta}\right)\right] \quad (7.328)$$

The functional dependencies of the mass collision stopping power are:

- At low projectile energies, the mass collision stopping power decreases with increasing kinetic energy as β^{-2} until reaching the minimally-ionizing region which corresponds to $\beta \approx 0.8$. At very low energies where the projectile speed is comparable to those of the atomic electrons, $(dE/\rho dx)_{Col}$ decreases with decreasing projectile energy due to the shell correction factor.
- This general β^{-2} dependence competes with the relativistic increase in the collision stopping power leading to a minimum of about 1.5 Mev • cm^2/g at a kinetic energy equal to about 3 times the projectile mass.
- The relativistic increase in $(dE/\rho dx)_{Col}$ due to the $\ln(\gamma^2\beta^2)$ and $\ln(\gamma + 1)$ terms are quenched by the density correction $\delta/2$ term leading to what is also known as the "Fermi Plateau."

Neglecting the small effect of the Barkas correction, further generalizations can be made:

Projectile rest mass, m: There is no dependence of the mass collision stopping power upon the particle's mass. Thus, for example, the mass collision stopping powers of a proton and a single-ionized helium ion (i.e., equal charges) at the same speed are the same.

Projectile charge, z: The mass collision stopping power increases with the square of the particle charge. Extending the previous example to a proton and an α particle at the same speed, the $(dE/\rho dx)_{Col}$ for the α particle will be four times greater than that of the proton.

Medium atomic number Z and atomic mass number A: For low atomic media $Z/A \approx 1/2$ decreasing with increasing Z (e.g., $Z/A = 0.5$ and approaches 0.4 for carbon and lead, respectively). As a result the mass collision stopping power at a given kinetic energy is greater for a low-Z medium than for one with high-Z. There is an additional dependence upon the atomic number contained within the $\approx -\ln\bar{I}$ term. The mean ionization potential $\bar{I}$ increases with Z thus further contributing to the decrease in $(dE/\rho dx)_{col}$ with Z.

Finally, it is of interest to consider the energy loss resulting from the Coulomb interaction between the heavy charged particle and the nucleus rather than an atomic electron. There will be an immediate increase of a factor of Z^2. However, there is a reciprocal dependence upon the target mass, thus the net change in the collision stopping power will be by a factor of Z^2m_e/M where M is the nuclear mass. As $Z^2m_e/M = 1.6 \times 10^3$ for carbon and 1.7×10^2 for lead, it can be seen that the contribution of the nucleus to the collision stopping power can be neglected.

7.4 Stochastic Collision Energy Loss: Energy Straggling

7.4.1 Introduction

It has been assumed in the prior derivations that energy loss is a continuous function or, for a particle traversing a distance t through a medium,

$$\Delta E = \int_0^t dx\left(\frac{dE}{dx}\right)_{Col}(x) \quad (7.329)$$

However, the energy is lost by the particle through discrete interactions with atomic electrons and, as a result, the energy loss process is not continuous but stochastic. In the simplest approximation, for a beam of monoenergetic particles incident to the medium, (7.329) provides the mean energy lost by the ensemble of particles and the actual energy loss will described by a probability distribution function with the exiting particle beam having an energy spectrum reflecting this pdf. This phenomenon is commonly referred to as energy straggling and the energy loss pdfs describing this phenomenon are the subject of this section.

Practical applications of these pdfs in nuclear medicine dosimetry reside primarily in Monte Carlo codes used to simulate radiation transport and calculate energy deposition in a medium. Such Monte Carlo codes have been categorized in terms of how they calculate for the energy straggling of moving charged particles in a medium (Chibani 2002). Some codes, such as MCNP (Briesmeister 2000) and ETRAN (Seltzer 1991), use a pdf to sample the energy loss corresponding to a given pathlength. Other codes simulate the inelastic collision between the projectile and the secondary electron as a distinct event should the energy transfer Q exceed the energy level Q_c (i.e., the collision is consider hard). A mean energy loss due to soft collisions is calculated through the use of the restricted collision stopping power. An example of such a code widely used in medical physics is EGSnrc (Kawrakow and Rogers 2003).[19] The GEANT code can be considered to straddle both categories as pdfs are sampled to evaluate hard-energy losses, but the user can also treat hard collisions as independent events (GEANT Team 2001).[20]

The pdfs describing these energy losses are derived in this section.

7.4.2 One-Dimensional Continuity Equation

Figure 7.28 shows a slab of material with physical density ρ to which is incident a monoenergetic beam of charged particles with kinetic energy T_0. The model is one-dimensional and multiple scatter is neglected, i.e., the particles travel in straight lines. Let N(x,ΔE) be the number of particles that have penetrated a depth x with a net energy loss ΔE. Now consider the number

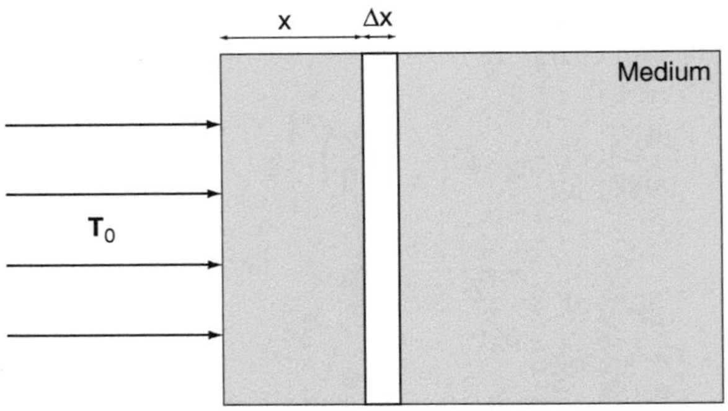

Fig. 7.28 Geometry for derivation of one-dimensional continuity equation. A monoenergetic beam of charged particles with kinetic energy T_0 is incident from the left to a medium with physical density ρ

of particles which, traversing an additional distance Δx, will also have a net energy loss ΔE (the thickness Δx is sufficiently small that a particle can only lose energy within it as the result of a single collision). This number will be the original number of particles with energy loss ΔE at x and which did not lose energy crossing Δx minus the number of particles with energy loss ΔE at x which suffered any additional energy loss as a result of traversing Δx, and plus the number of particles with energy loss $\Delta E' < \Delta E$ at x and which undergo an additional energy loss $\Delta E - \Delta E'$ (where $\Delta E' - \Delta E \leq Q_{max}$) traversing Δx so that they have a net energy loss ΔE at a depth x+Δx.

Define $\Pr(\Delta x, \Delta E \to \Delta E')$ to be the probability that a particle with an energy loss ΔE in penetrating to x will experience an additional energy loss $\Delta E' - \Delta E$ through crossing x to have a total energy loss ΔE′ at a depth of x + Δx. Hence, the number of particles with energy loss ΔE at the depth x + Δx is,

$$\begin{aligned} N(x+\Delta x,\Delta E) &= N(x,\Delta E) \\ &- \int_{\Delta E}^{\infty} d(\Delta E')\Pr(\Delta x,\Delta E \to \Delta E')N(x,\Delta E) \\ &+ \int_{0}^{\Delta E} d(\Delta E')\Pr(\Delta x,\Delta E' \to \Delta E)N(x,\Delta E'). \end{aligned} \tag{7.330}$$

The first integral describes the reduction in particle number with energy loss ΔE at the depth x + Δx due to particles with energy loss ΔE at x experiencing an energy loss while passing through Δx. The second integral yields the increase in the number of particles

[19]Both classes represent conceptual difficulties: a Class I code does not link the sampling of the energy loss pdf with secondary recoil electrons and the Class II code neglects any energy straggling associated with soft collisions.

[20]Further discussion of the practical matters of the Monte Carlo simulation of the transport of charged particles can be found in the articles by Salvat et al. (1999) and Chibani (2002).

with energy loss ΔE at the depth $x + \Delta x$ due to particles with energy loss $\Delta E' < \Delta E$ undergoing that energy loss within the thickness Δx to bring the net energy loss at $x + \Delta x$ to ΔE. The differential change with distance in particle number with energy loss ΔE at depth x is,

$$\begin{aligned}\frac{\partial N(x,\Delta E)}{\partial x} &\equiv \lim_{\Delta x \to 0} \frac{N(x+\Delta x,\Delta E)-N(x,\Delta E)}{\Delta x}\\ &= \lim_{\Delta x \to 0}\left[-\int_{\Delta E}^{\infty} d(\Delta E') \times \frac{Pr(\Delta x,\Delta E \to \Delta E')}{\Delta x} N(x,\Delta E) + \int_{0}^{\Delta E} d(\Delta E')\frac{Pr(\Delta x,\Delta E' \to \Delta E)}{\Delta x} N(x,\Delta E')\right]\\ &= -\int_{\Delta E}^{\infty} d(\Delta E')\omega(\Delta E'-\Delta E)N(x,\Delta E) + \int_{0}^{\Delta E} d(\Delta E')\omega(\Delta E'-\Delta E)N(x,\Delta E')\end{aligned} \tag{7.331}$$

where $\omega(\delta E)$ is the probability per unit pathlength of the particle losing energy δE. One can use the change of variables to rewrite this integrodifferential equation, noting that $\omega(\delta E) = 0$ for $\delta E > Q_{max}$, to obtain

$$\frac{\partial N(x,\Delta E)}{\partial x} = -\int_{0}^{Q_{max}} dQ\,\omega(Q)N(x,\Delta E) + \int_{0}^{\min(Q_{max},\Delta E)} dQ\,\omega(Q)N(x,\Delta E - Q). \tag{7.332}$$

As the probability distribution function describing the collision energy loss is,

$$f(x,\Delta E) = \frac{N(x,\Delta E)}{\int_0^{T_0} d(\Delta E')N(x,\Delta E')} \tag{7.333}$$

it is possible to write the continuity equation for the pdf in the integrodifferential form,

$$\frac{\partial f(x,\Delta E)}{\partial x} = -f(x,\Delta E)\int_{0}^{Q_{max}} dQ\,\omega(Q) + \int_{0}^{\min(Q_{max},\Delta E)} dQ\,\omega(Q)f(x,\Delta E - Q). \tag{7.334}$$

7.4.3 Gaussian Probability Distribution Function for ΔE

The first solution to the integrodifferential continuity equation is provided by following an approach first described by Rossi (1952) and expanded upon by Kase and Nelson (1978) (Segrè (1977) provides a simpler and perhaps more intuitive method based on the central-limit theorem and which is traceable to early work by Bohr). A complete derivation of the Gaussian pdf as a solution to the continuity equation is provided here. By the use of a number of simplifying assumptions, the integrodifferential continuity equation is converted into a differential equation which can be solved via Fourier transform pairs. The assumptions for this solution method are:

- The mean energy lost by the particle penetrating to depth x is small, $\overline{\Delta E} \ll T_0$.
- The collision stopping power can be approximated as a constant over the distance traveled x so that the mean energy loss at the depth of penetration x is given by

$$\overline{\Delta E} = x\left(\frac{dE}{dx}\right)_{Col} \tag{7.335}$$

- The probability distribution function $f(x,\Delta E)$ varies slightly with energy loss enabling $f(x,\Delta E - Q)$ to be expanded into a second-order Taylor's series

$$f(x,\Delta E - Q) \cong f(x,\Delta E) - \frac{\partial f(x,\Delta E)}{\partial(\Delta E)}Q + \frac{1}{2}\frac{\partial^2 f(x,\Delta E)}{\partial(\Delta E)^2}Q^2. \tag{7.336}$$

By noting that $\omega(Q) = 0$ for $Q > Q_{max}$, the continuity equation can be rewritten using the Taylor series expansion to form a differential equation in $f(x, \Delta E)$,

$$\begin{aligned}\frac{\partial f(x,\Delta E)}{\partial x} &= -\int_0^\infty dQ\,\omega(Q) f(x,\Delta E) \\ &\quad + \int_0^\infty dQ\,\omega(Q) f(x,\Delta E - Q) \\ &= \int_0^\infty dQ\,\omega(Q)(f(x,\Delta E - Q) - f(x,\Delta E)) \\ &= \int_0^\infty dQ\,\omega(Q)\left(-\frac{\partial f(x,\Delta E)}{\partial(\Delta E)}Q \right. \\ &\quad \left. + \frac{1}{2}\frac{\partial^2 f(x,\Delta E)}{\partial(\Delta E)^2}Q^2\right) \\ &= -\frac{\partial f(x,\Delta E)}{\partial(\Delta E)}\int_0^\infty dQ\,\omega(Q)Q \\ &\quad + \frac{1}{2}\frac{\partial^2 f(x,\Delta E)}{\partial(\Delta E)^2}\int_0^\infty dQ\,\omega(Q)\,Q^2 \\ &= -k_1\frac{\partial f(x,\Delta E)}{\partial(\Delta E)} + \frac{k_2^2}{2}\frac{\partial^2 f(x,\Delta E)}{\partial(\Delta E)^2}\end{aligned} \tag{7.337}$$

where,

$$\begin{aligned} k_1 &= \int_0^\infty dQ\,\omega(Q)Q \\ &= \frac{\overline{\Delta E}}{x}\end{aligned} \tag{7.338}$$

and

$$k_2^2 = \int_0^\infty dQ\,\omega(Q)Q^2. \tag{7.339}$$

As the expression for k_1 contains the factor Q whereas that for k_2^2 contains Q^2, the role of soft collisions will be less significant in the evaluation of k_2^2 than for k_1, so one consequently calculates k_1 using the total collision stopping power and uses only the hard collision stopping power in the determination of k_2^2. In other words, the width of the pdf is a function of the hard collision energy transfer alone. This observation becomes important when considering asymmetric pdfs for ΔE.

In order to solve this differential equation, one again uses the Fourier transform method,

$$\xi(x,\tau) = \frac{1}{\sqrt{2\pi}}\int_{-\infty}^{\infty} d(\Delta E)e^{-i\tau\Delta E} f(x,\Delta E) \tag{7.340}$$

$$f(x,\Delta E) = \frac{1}{\sqrt{2\pi}}\int_{-\infty}^{\infty} d\tau\, e^{i\tau\Delta E}\,\xi(x,\tau). \tag{7.341}$$

The Fourier transform of (7.337) is,

$$\frac{\partial \xi(x,\tau)}{\partial x} = -\left(ik_1\tau + \frac{k_2^2\tau^2}{2}\right)\xi(x,\tau) \tag{7.342}$$

which is straightforward to solve,

$$\xi(x,\tau) = \xi(0,\tau)e^{-\left(ik_1\tau + \frac{k_2^2\tau^2}{2}\right)x}. \tag{7.343}$$

The initial condition is determined by noting that

$$f(0,\Delta E) = \delta(\Delta E). \tag{7.344}$$

The Fourier transform of this initial condition is,

$$\xi(0,\tau) = \frac{1}{\sqrt{2\pi}} \tag{7.345}$$

Inserting this and the definition of k_1 into (7.343) gives the Fourier transform of the probability distribution function,

$$\xi(x,\tau) = \frac{1}{\sqrt{2\pi}}e^{-\left(i\tau\overline{\Delta E} + \frac{k_2^2\tau^2}{2}\right)x}. \tag{7.346}$$

The pdf is recovered through the inverse Fourier transform,

$$\begin{aligned} f(x,\Delta E) &= \frac{1}{\sqrt{2\pi}}\int_{-\infty}^{\infty} d\tau\, e^{i\tau\Delta E}\,\xi(x,\tau) \\ &= \frac{1}{2\pi}\int_{-\infty}^{\infty} d\tau\, e^{-i\tau(\overline{\Delta E}-\Delta E) - \frac{k_2^2\tau^2}{2}x}.\end{aligned} \tag{7.347}$$

This integral is solved by first algebraically rearranging the exponent,

$$i\tau\left(\overline{\Delta E}-\Delta E\right)+\frac{k_2^2\tau^2}{2}x=\left(k_2\tau\sqrt{\frac{x}{2}}+i\frac{\overline{\Delta E}-\Delta E}{k_2\sqrt{2x}}\right)^2+\frac{\left(\overline{\Delta E}-\Delta E\right)^2}{2k_2^2x} \quad (7.348)$$

which allows the pdf to be rewritten as,

$$f(x,\Delta E)=\frac{1}{2\pi}e^{-\frac{\left(\overline{\Delta E}-\Delta E\right)^2}{2k_2^2x}}\int_{-\infty}^{\infty}d\tau\, e^{-\left(k_2\tau\sqrt{\frac{x}{2}}+i\frac{\overline{\Delta E}-\Delta E}{k_2\sqrt{2x}}\right)^2}. \quad (7.349)$$

Changing variables,

$$z=k_2\tau\sqrt{\frac{x}{2}}+i\frac{\overline{\Delta E}-\Delta E}{k_2\sqrt{2x}} \quad (7.350)$$

$$d\tau=\sqrt{\frac{2}{x}}\frac{dz}{k_2} \quad (7.351)$$

the integral can be simplified to,

$$\int_{-\infty}^{\infty}d\tau\, e^{-\left(k_2\tau\sqrt{\frac{x}{2}}+i\frac{\overline{\Delta E}-\Delta E}{k_2\sqrt{2x}}\right)^2}=\frac{1}{k_2}\sqrt{\frac{2}{x}}\int_{-\infty+iy_0}^{\infty+iy_0}dz\, e^{-z^2} \quad (7.352)$$

where

$$y_0=\frac{\overline{\Delta E}-\Delta E}{k_2\sqrt{2x}} \quad (7.353)$$

The integral is solved by writing it in the complex form $\int_{-K+iy_0}^{K+iy_0}dz\, e^{-z^2}$ where K is a real constant to be later allowed to go to ∞. As e^{-z^2} is holomorphic on and within the contour and there are no singularities within the contour, the Cauchy–Goursat theorem states that,

$$\int_C dz\, e^{-z^2}=0. \quad (7.354)$$

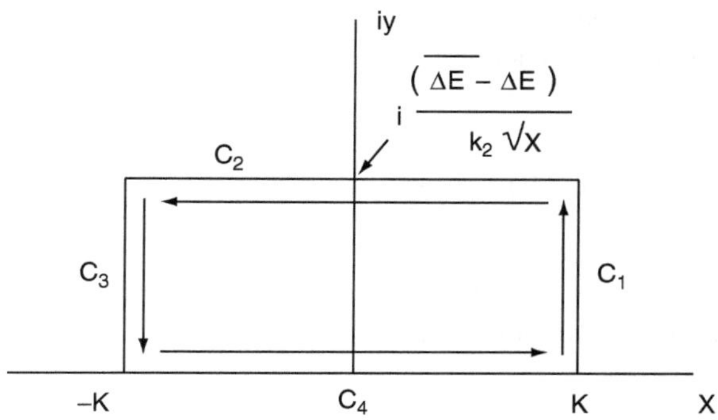

Fig. 7.29 Rectangular contour in the complex $z=x+iy$ plane for solving the integral of (7.352)

The closed contour in the complex plane is shown in Fig. 7.29 and is made up of four individual contours,

$$C=C_1\cup C_2\cup C_3\cup C_4. \quad (7.355)$$

Thus,

$$\int_{C_1}dz\, e^{-z^2}+\int_{C_2}dz\, e^{-z^2}+\int_{C_3}dz\, e^{-z^2}+\int_{C_4}dz\, e^{-z^2}=0. \quad (7.356)$$

This then leads to,

$$\int_{-K+iy_0}^{K+iy_0}dz\, e^{-z^2}=-\int_{C_2}dz\, e^{-z^2}=\int_{C_1}dz\, e^{-z^2}+\int_{C_3}dz\, e^{-z^2}+\int_{C_4}dz\, e^{-z^2}. \quad (7.357)$$

This type of integral is solved in the Appendices,

$$\int_{-\infty+iy_0}^{\infty+iy_0}dz\, e^{-z^2}=\sqrt{\pi} \quad (7.358)$$

leading to,

$$\int_{-\infty}^{\infty}d\tau\, e^{-\left(k_2\tau\sqrt{\frac{x}{2}}+i\frac{\overline{\Delta E}-\Delta E}{k_2\sqrt{2x}}\right)^2}=\frac{1}{k_2}\sqrt{\frac{2\pi}{x}}. \quad (7.359)$$

The above result provides the Gaussian probability distribution function for energy loss ΔE at a depth x,

$$f(x, \Delta E) = \frac{1}{\sqrt{2\pi k_2^2 x}} e^{-\frac{\left(\overline{\Delta E} - \Delta E\right)^2}{2k_2^2 x}}. \tag{7.360}$$

In order to verify that this result is a pdf, it is necessary to require the normalization,

$$\int_0^{\infty} d(\Delta E)\, f(x, \Delta E) = 1$$

or, since the maximum energy loss cannot exceed the incident kinetic energy,

$$\int_0^{T_0} d(\Delta E)\, f(x, \Delta E) = 1.$$

Inserting the expression for the pdf into this integral,

$$\frac{1}{\sqrt{2\pi k_2^2 x}} e^{-\frac{\left(\overline{\Delta E} - \Delta E\right)^2}{2k_2^2 x}} = \frac{1}{\sqrt{\pi}} \int_{y_1}^{y_2} dy\, e^{-y^2} \tag{7.361}$$

where the change of variables,

$$y = \frac{\overline{\Delta E} - \Delta E}{\sqrt{2k_2^2 x}} \tag{7.362}$$

has been used. The limits of integration are,

$$y_1 = \frac{\overline{\Delta E} - T_0}{\sqrt{2k_2^2 x}} \tag{7.363}$$

and

$$y_2 = \frac{\overline{\Delta E}}{\sqrt{2k_2^2 x}}. \tag{7.364}$$

If $T_0 - \overline{\Delta E} \gg \sqrt{2k_2^2 x}$ and $\overline{\Delta E} \gg \sqrt{2k_2^2 x}$, then the result of (7.361) can be written as,

$$\frac{1}{\sqrt{\pi}} \int_{-\infty}^{\infty} dy\, e^{-y^2} = 1$$

thus demonstrating that $f(x, \Delta E)$ is the pdf sought and is a Gaussian distribution with mean and most probable energy loss,

$$\overline{\Delta E} = x \left(\frac{dE}{dx}\right)_{Col} \tag{7.365}$$

and variance,

$$\sigma^2 = k_2^2 x. \tag{7.366}$$

The width of this Gaussian (2σ) must be greater than the maximum energy transferred in a single collision (i.e., $2k_2\sqrt{x} \gg Q_{max}$, which follows from the central-limit theorem) but smaller than the incident kinetic energy ($2k_2\sqrt{x} \ll T_0$) and the mean energy loss, or $2k_2\sqrt{x} \ll \overline{\Delta E}$. As an example, calculate the variance for the limiting case for a massive spin-0 projectile,

$$\begin{aligned} \frac{d\sigma}{dQ} &= 2\pi r_0^2 m_e \left(\frac{z}{\beta}\right)^2 \frac{1}{Q^2}\left(1 - \beta^2 \frac{Q}{Q_{max}}\right) && Q \le Q_{max} \\ &= 0 && Q > Q_{max} \end{aligned}$$

and the probability of energy loss per unit distance traveled is,

$$\begin{aligned} \omega(Q) &= N_A \rho \left(\frac{Z}{A}\right) \frac{d\sigma}{dQ} \\ &= C\rho \left(\frac{Z}{A}\right)\left(\frac{z}{\beta}\right)^2 \frac{1}{Q^2}\left(1 - \beta^2 \frac{Q}{Q_{max}}\right) \\ &\equiv \frac{\zeta}{Q^2}\left(1 - \beta^2 \frac{Q}{Q_{max}}\right) \end{aligned} \tag{7.367}$$

where C is the constant pervading this chapter and the factor,

$$\zeta \equiv C\rho \left(\frac{Z}{A}\right)\left(\frac{z}{\beta}\right)^2 \tag{7.368}$$

has been defined. This constant has units of MeV per centimeter. Hence, the variance of the energy loss pdf is,

$$\begin{aligned} \sigma^2 &= x \int_0^{\infty} dQ\, \omega(Q) Q^2 \\ &= x\zeta \int_0^{Q_{max}} dQ \left(1 - \beta^2 \frac{Q}{Q_{max}}\right) \\ &= x\zeta\, Q_{max}\left(1 - \frac{\beta^2}{2}\right) \end{aligned} \tag{7.369}$$

The energy inequalities resulting from above are,

$$\frac{Q_{max}}{2} \ll k_2\sqrt{x} \ll \frac{T_0}{2} \tag{7.370}$$

and

$$k_2\sqrt{x} \ll \frac{\bar{E}}{2}. \tag{7.371}$$

These can be grouped into the requirement that the ratio

$$\kappa = \frac{x\zeta}{Q_{max}} \tag{7.372}$$

must be large in order for the probability distribution function for energy loss to be approximated by a Gaussian.

7.4.4 *Asymmetric Probability Distribution Functions for ΔE*

7.4.4.1 Introduction

Having demonstrated that a symmetric Gaussian pdf for the energy loss is the consequence of a large value of κ which corresponds to a thick pathlength x and, from the definition of ζ, to nonrelativistic particles ($\beta \ll 1$). For small κ, which corresponds to a thin absorber and/or increasingly relativistic charged particles, the mean energy loss will also decrease. However, because the probability of a hard collision remains constant, its relative contribution to the statistical distribution of energy losses will increase as κ decreases, leading to increasing asymmetry in the pdf for ΔE. In this case, the integrodifferential form of the continuity equation cannot be reduced to a simple differential equation and must, instead, be solved directly.

7.4.4.2 Vavilov Probability Distribution Function

Historically, it was Landau (1944) who first derived a solution to the general integrodifferential continuity equation leading to an asymmetric pdf through the use of Laplace transforms. Landau's result, however, was derived for the Rutherford differential cross section (which is equivalent to neglecting the $\beta^2 Q/Q_{max}$ term in the Bhabha cross section) and was dependent upon a variety of approximations. A more general and exact solution to the one-dimensional continuity equation was provided by Vavilov (1957) which reaches the Gaussian pdf as a limit for $\kappa \to \infty$ and the Landau result for $\kappa \to 0$. Hence, the Landau probability distribution function for energy loss will be treated as a special limiting case of the Vavilov. This derivation follows Vavilov's approach and solves the integrodifferential continuity equation for f(x,ΔE) by using Laplace transform pairs,

$$\xi(x,\tau) = \int_0^\infty d(\Delta E)\, e^{-\tau \Delta E}\, f(x,\Delta E) \tag{7.373}$$

$$f(x,\Delta E) = \frac{1}{2\pi i}\int_{K-i\infty}^{K+i\infty} d\tau\, e^{\tau \Delta E}\xi(x,\tau) \tag{7.374}$$

$$\Omega(\tau) = \int_0^\infty dQ\, e^{-\tau Q}\,\omega(Q) \tag{7.375}$$

$$\omega(Q) = \frac{1}{2\pi i}\int_{K-i\infty}^{K+i\infty} d\tau\, e^{\tau Q}\,\Omega(\tau) \tag{7.376}$$

where K is an arbitrary real constant. For a thin absorber, the energy loss will be small and it is reasonable to assume that the pdf will vary slowly with energy over this range thus allowing us to make the approximation f(x,ΔE) ≈ f(x,ΔE − Q). Applying the convolution theorem, then,

$$\frac{\partial \xi(x,\tau)}{\partial x} = \Omega(\tau)\xi(x,\tau) - \xi(x,\tau) \times \int_0^{Q_{max}} dQ\,\omega(Q) \tag{7.377}$$

which has the solution,

$$\xi(x,\tau) = \xi(0,\tau)\exp\left(\left(\Omega(\tau) - \int_0^{Q_{max}} dQ\,\omega(Q)\right)x\right). \tag{7.378}$$

From the obvious result,

$$\xi(0,\tau) = \int_0^\infty d(\Delta E) e^{-\tau \Delta E}\,\delta(\Delta E) = 1 \tag{7.379}$$

the Laplace transform of the pdf is obtained,

$$\xi(x,\tau) = \exp\left(\left(\Omega(\tau) - \int_0^{Q_{max}} dQ\omega(Q)\right) x\right) = \exp\left(\left(\int_0^{\infty} dQ e^{-\tau Q}\omega(Q) - \int_0^{Q_{max}} dQ\omega(Q)\right) x\right) = \exp\left(x\int_0^{\infty} dQ\omega(Q)\left(e^{-\tau Q} - 1\right)\right) \tag{7.380}$$

as $\omega(Q) = 0$ for $Q > Q_{max}$. The pdf is recovered from this result by calculating its inverse Laplace transform,

$$f(x, \Delta E) = \frac{1}{2\pi i}\int_{K-i\infty}^{K+i\infty} d\tau\, e^{\tau\Delta E}\,\xi(x,\tau) = \frac{1}{2\pi i}\int_{K-i\infty}^{K+i\infty} d\tau \exp \times \left(\tau\Delta E + x\int_0^{\infty} dQ\omega(Q)\left(e^{-\tau Q} - 1\right)\right) \tag{7.381}$$

To solve for this, the exponent is first evaluated,

$$\tau\Delta E + x\int_0^{\infty} dQ\omega(Q)\left(e^{-\tau Q} - 1\right) = \tau\Delta E + x\int_0^{\infty} dQ\omega(Q)e^{-\tau Q} - x\int_0^{\infty} dQ\omega(Q) = \tau\left(\Delta E - \overline{\Delta E}\right) - x\int_0^{Q_{max}} dQ\,\omega(Q) \times \left(1 - e^{-\tau Q} - \tau Q\right) \tag{7.382}$$

where $\overline{\Delta E}$ is the mean energy loss over the distance traversed, x. For the Bhabha cross section of a massive spin-0 projectile, this exponent becomes,

$$\tau\left(\Delta E - \overline{\Delta E}\right) - x\int_0^{Q_{max}} dQ\,\omega(Q)\left(1 - e^{-\tau Q} - \tau Q\right) = \tau\left(\Delta E - \overline{\Delta E}\right) - x\zeta\int_0^{Q_{max}} \frac{dQ}{Q^2}\left(1 - \beta^2\frac{Q}{Q_{max}}\right) \times \left(1 - e^{-\tau Q} - \tau Q\right) = \tau\left(\Delta E - \overline{\Delta E}\right) - x\zeta\int_0^{Q_{max}} dQ\frac{1 - e^{-\tau Q}}{Q^2} + \kappa\beta^2\int_0^{Q_{max}} dQ\frac{1 - e^{-\tau Q}}{Q} + \tau x\zeta\int_0^{Q_{max}}\frac{dQ}{Q} - \tau\beta^2 x\zeta. \tag{7.383}$$

Three separate integrals must now be solved. The first integral is, following a change of variables,

$$\int_0^{Q_{max}} dQ\frac{1 - e^{-\tau Q}}{Q^2} = \tau\int_0^{\tau Q_{max}} du\frac{1 - e^{-u}}{u^2} \tag{7.384}$$

which is readily solved by parts,

$$f(u) = -\frac{1}{u} \qquad \frac{df}{du} = \frac{1}{u^2} \qquad \frac{dg}{du} = e^{-u} \qquad g(u) = 1 - e^{-u} \tag{7.385}$$

to give,

$$\tau\int_0^{\tau Q_{max}} du\frac{1-e^{-u}}{u^2} = \tau\left(-\left(\frac{1-e^{-u}}{u}\right)\Bigg|_0^{\tau Q_{max}} + \int_0^{\tau Q_{max}} du\frac{e^{-u}}{u}\right). \tag{7.386}$$

From l'Hôpital's rule,

$$\lim_{u\to 0}\frac{1 - e^{-u}}{u} = 1 \tag{7.387}$$

then,

$$\int_0^{Q_{max}} dQ\frac{1 - e^{-\tau Q}}{Q^2} = \tau - \frac{1 - e^{-\tau Q_{max}}}{Q_{max}} + \tau\int_0^{\tau Q_{max}} du\frac{e^{-u}}{u}. \tag{7.388}$$

The second integral of (7.383) can be found from tables (Abramowitz and Stegun 1972),

$$\int_0^{Q_{max}} dQ \frac{1-e^{-\tau Q}}{Q} = -Ei(-\tau Q_{max}) + \ln \tau Q_{max} + \gamma_{EM} \tag{7.389}$$

where $Ei(x)$ is the exponential integral, $Ei(x) = \int_{-x}^{\infty} dt \frac{e^{-t}}{t}$.

The third integral is rewritten as,

$$\int_0^{Q_{max}} \frac{dQ}{Q} = \int_0^{\tau Q_{max}} \frac{du}{u}. \tag{7.390}$$

Using these results, the following form of the exponential term in the inverse Laplace transform of the pdf is given,

$$\begin{aligned} &\tau(\Delta E - \overline{\Delta E}) - x\zeta \int_0^{Q_{max}} dQ \frac{1-e^{-\tau Q}}{Q^2} + \kappa\beta^2 \\ &\times \int_0^{Q_{max}} dQ \frac{1-e^{-\tau Q}}{Q} + \tau x\zeta \int_0^{Q_{max}} \frac{dQ}{Q} - \tau\beta^2 x\zeta \\ &= \tau(\Delta E - \overline{\Delta E}) - \tau x\zeta(1+\beta^2) + \kappa(1-e^{-z}) \\ &+ (\kappa\beta^2 + \tau x\zeta)(-Ei(-z) + \ln z + \gamma_{EM}) \end{aligned} \tag{7.391}$$

where $z = \tau Q_{max}$. Inserting this expression for the exponent into the inverse Laplace transform for the energy loss pdf yields the form attributable to Vavilov,

$$\begin{aligned} f(x, \Delta E) &= \frac{1}{2\pi i Q_{max}} e^{\kappa(1+\beta^2\gamma_{EM})} \int_{K-i\infty}^{K+i\infty} dz \\ &\times \exp\left[z\lambda_V + \kappa\left((z+\beta^2)\right.\right. \\ &\times \left.\left.(-Ei(-z) + \ln z) - e^{-z}\right)\right] \end{aligned} \tag{7.392}$$

where the dimensionless Vavilov parameter is,

$$\lambda_V = \frac{\Delta E - \overline{\Delta E}}{Q_{max}} - \kappa(1 + \beta^2 - \gamma_{EM}). \tag{7.393}$$

The solutions of this expression of the Vavilov pdf will be considered in a variety of ways. First, integrate this expression over the imaginary axis (i.e., set $K = 0$ and $z = iy$),

$$\begin{aligned} f(x, \Delta E) &= \frac{1}{2\pi Q_{max}} e^{\kappa(1+\beta^2\gamma_{EM})} \int_{-\infty}^{\infty} dy \\ &\times \exp\left[iy\lambda_V + \kappa\left((iy+\beta^2)\right.\right. \\ &\times \left.\left.(-Ei(-iy) + \ln iy) - e^{-iy}\right)\right] \end{aligned} \tag{7.394}$$

From complex variable theory (Churchill et al. 1974),

$$\begin{aligned} \ln iy &= i\frac{\pi}{2} + \ln|y| \quad & y > 0 \\ &= -i\frac{\pi}{2} + \ln|y| \quad & y < 0 \end{aligned} \tag{7.395}$$

and

$$\begin{aligned} Ei(-iy) &= Ci(|y|) - iSi(y) + i\frac{\pi}{2} \quad & y > 0 \\ &= Ci(|y|) - iSi(y) - i\frac{\pi}{2} \quad & y < 0 \end{aligned} \tag{7.396}$$

where $Ci(x)$ and $Si(x)$ are the cosine and sine integrals, respectively,

$$Ci(x) = \gamma_{EM} + \ln x + \int_0^x dt \frac{\cos t - 1}{t}$$

$$Si(x) = \int_0^x dt \frac{\sin t}{t}.$$

By inserting these into (7.396) and defining the two functions,

$$g_1(y) = \beta^2(\ln|y| - Ci(y)) - \cos y - ySi(y) \tag{7.397}$$

$$g_2(y) = y\ (\ln|y| - Ci(y)) + \sin y + \beta^2 Si(y), \tag{7.398}$$

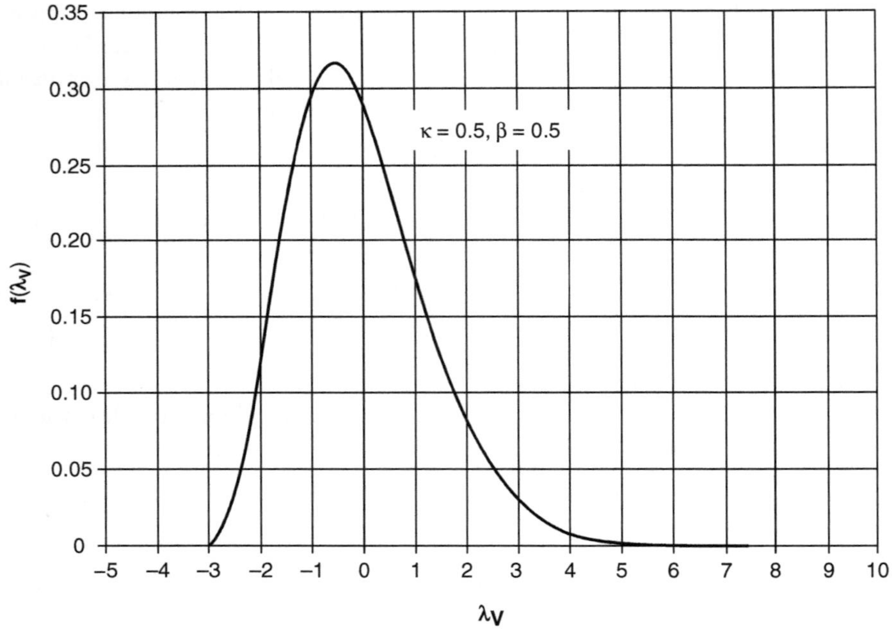

Fig. 7.30 Vavilov energy loss pdf for $\kappa = 0.5$, $\beta = 0.5$ calculated using the Edgeworth polynomial expansion described in Sect. 7.4.4.5

a simpler integral form of the Vavilov pdf is obtained,

$$f(x, \Delta E) = \left(\frac{\kappa}{\pi x \zeta}\right) e^{\kappa\left(1+\beta^2\gamma_{EM}\right)} \int_0^\infty dy\, e^{\kappa g_1(y)} \times \cos(\lambda_V y + \kappa g_2(y)) \tag{7.399}$$

This result can be solved numerically and through the use of approximations for the cosine and sine integrals. A variety of other approximations also exist and these will be discussed shortly. Figure 7.30 shows the Vavilov pdf (calculated using an Edgeworth expansion method) for $\kappa = 0.5$ and $\beta = 0.5$ as a function of the Vavilov parameter λ_V.

7.4.4.3 Gaussian Limit to the Vavilov Probability Distribution Function

It can now be shown that the Vavilov pdf approaches the Gaussian limit for $\kappa \to \infty$. Begin with another approach to solving the pdf given by the inverse Laplace transform by expanding the exponential. Because of the asymmetry of the Vavilov pdf, the expansion is made to third order in $\tau\, Q$,

$$\begin{aligned} f(x, \Delta E) &= \frac{1}{2\pi i} \int_{K-i\infty}^{K+i\infty} d\tau \exp \\ &\quad \times \left(\tau\,\Delta E - x \int_0^\infty dQ\omega(Q)\left(1 - e^{-\tau Q}\right)\right) \\ &= \frac{1}{2\pi i} \int_{K-i\infty}^{K+i\infty} d\tau \exp\left(\tau\,\Delta E - x \int_0^\infty dQ\omega(Q)\right. \\ &\quad \left.\times \left(\tau Q - \frac{(\tau Q)^2}{2} + \frac{(\tau Q)^3}{6}\right)\right) \\ &\approx \frac{1}{2\pi i} \int_{K-i\infty}^{K+i\infty} d\tau \exp\left(\tau\left(\Delta E - \overline{\Delta E}\right)\right. \\ &\quad \left. + \tau^2 \frac{x}{2} \vartheta - \tau^3 \frac{x}{6} \varphi\right) \end{aligned} \tag{7.400}$$

where,

$$\vartheta = \int_0^{Q_{max}} dQ\omega(Q) Q^2 \tag{7.401}$$

and

$$\varphi = \int_0^{Q_{max}} dQ\omega(Q)Q^3 \tag{7.402}$$

Note that ϑ corresponds to $k_2{}^2$ in the derivation of the Gaussian pdf from the pure differential form of the continuity equation. As a Gaussian result for the pdf will be obtained if the integral is evaluated only up to ϑ, the $\exp(-\tau^3 x/6\varphi)$ term provides the asymmetry of the pdf which, as it is proportional to e^{-x}, must decrease with increasing absorber thickness so that the pdf approaches the Gaussian form. This can be shown explicitly for the Bhabha cross section for a massive spin-0 projectile,

$$\vartheta = \zeta Q_{max}\left(1 - \frac{\beta^2}{2}\right) \tag{7.403}$$

and,

$$\varphi = \zeta\, Q_{max}^2\left(\frac{1}{2} - \frac{\beta^2}{3}\right) \tag{7.404}$$

The pdf is solved for by the substitution of variables. First, define,

$$u = \left(\frac{x\varphi}{2}\right)^{1/3}\left(\tau - \frac{\vartheta}{\varphi}\right) \tag{7.405}$$

from which,

$$d\tau = \left(\frac{2}{x\varphi}\right)^{1/3} du \tag{7.406}$$

For simplicity, define,

$$\eta = \left(\frac{x\varphi}{2}\right)^{1/3} \tag{7.407}$$

which leads to a simplified expression of the Vavilov pdf,

$$f(x, \Delta E) = \frac{e^{at-\frac{a^3}{3}}}{2\pi i\,\eta}\int_{K-i\infty}^{K+i\infty} du\, e^{ut-\frac{u^3}{3}} \tag{7.408}$$

where

$$\begin{aligned} a &= \eta\frac{\vartheta}{\varphi} \\ &= \left(1 - \frac{\beta^2}{2}\right)\left(\frac{2\kappa}{\left(1 - \frac{2}{3}\beta^2\right)^2}\right)^{\frac{1}{3}} \end{aligned} \tag{7.409}$$

and

$$t = \frac{\Delta E - \overline{\Delta E}}{\eta} + a^2. \tag{7.410}$$

The integral of (7.408) is integrated over the imaginary axis,

$$\begin{aligned} \frac{1}{2\pi i}\int_{K-i\infty}^{K+i\infty} du\, e^{ut-\frac{u^3}{3}} &= \frac{1}{2\pi}\int_{-\infty}^{\infty} dy\, e^{\frac{y^3}{3}+iyt} \\ &= \frac{1}{\pi}\int_0^{\infty} dy \cos\left(yt + \frac{y^3}{3}\right) \end{aligned} \tag{7.411}$$

which will be recognized as being the Airy function, Ai(t). The pdf is now written as,

$$f(x, \Delta E) = \frac{e^{at-\frac{a^3}{3}}}{\eta} Ai(t). \tag{7.412}$$

Consider the function for large values of κ. Equation (7.409) shows that $a \to (2\kappa)^{1/3}$ for $\kappa \to \infty$ and (7.410) shows that $t \to \infty$ for $a \to \infty$. Thus, for $\kappa \to \infty$, one can use the limiting form of the Airy function,

$$Ai(t) \to \frac{e^{-\frac{2}{3}t^{3/2}}}{2\sqrt{\pi}\, t^{1/4}} \quad \text{as } t \to \infty \tag{7.413}$$

so that the pdf will be,

$$f(x, \Delta E) = \frac{e^{\left(at-\frac{a^3}{3}-\frac{2}{3}t^{3/2}\right)}}{2\eta\sqrt{\pi}^{\,1/4}} \quad \text{as } t \to \infty \tag{7.414}$$

This result is further manipulated by noting that,

$$t^{\frac{1}{4}} \to \sqrt{a} \quad \text{as } \kappa \to \infty \tag{7.415}$$

and that

$$at - \frac{2}{3}t^{\frac{3}{2}} \to \frac{a^2}{3} - \frac{z^2}{4a} \quad \text{as } \kappa \to \infty \tag{7.416}$$

where

$$z = \frac{\Delta E - \overline{\Delta E}}{\eta} \ll a \qquad (7.417)$$

to give,

$$f(x, \Delta E) = \frac{e^{-\frac{z^2}{4a}}}{2\eta\sqrt{\pi a}}. \qquad (7.418)$$

The final result for $\kappa \to \infty$ is,

$$f(x, \Delta E) \approx \frac{1}{\sqrt{2\pi\vartheta\, x}} e^{-\left(\frac{(\Delta E - \overline{\Delta E})^2}{2\vartheta x}\right)} \qquad (7.419)$$

Thus, the Vavilov expression reduces to the Gaussian pdf for large κ.

7.4.4.4 Landau Limit to the Vavilov Probability Distribution Function

Having established that the Vavilov pdf approaches the Gaussian pdf for large κ, now look at the opposite limit of the pdf for $\kappa \to 0$. Recall the original Vavilov pdf,

$$f(x, \Delta E) = \frac{1}{2\pi i Q_{max}} e^{\kappa(1+\beta^2\gamma_{EM})} \times \int_{K-i\infty}^{K+i\infty} dz \exp\left[z\lambda_V + \kappa\left((z+\beta^2) \times (-Ei(-z) + \ln z) - e^{-z}\right)\right].$$

By changing variables, $p = \kappa z$, the pdf is,

$$\begin{aligned} f(x, \Delta E) &= \frac{1}{2\pi i x\zeta} e^{\kappa(1+\beta^2\gamma_{EM})} \times \int_{K-i\infty}^{K+i\infty} dp \\ &\times \exp\left[p\frac{\lambda_V}{\kappa} + \kappa\left(\left(\frac{p}{\kappa} + \beta^2\right)\left(-Ei\left(-\frac{p}{\kappa}\right) + \ln p - \ln\kappa\right) - e^{-\frac{p}{\kappa}}\right)\right] \\ &= \frac{1}{2\pi i x\zeta} e^{\kappa(1+\beta^2\gamma_{EM})} \int_{K-i\infty}^{K+i\infty} dp \\ &\times \exp\left[p\left(\frac{\lambda_V}{\kappa} - \ln\kappa\right) + p\left(-Ei\left(-\frac{p}{\kappa}\right) + \ln p\right) + \kappa\beta^2\left(-Ei\left(-\frac{p}{\kappa}\right) + \ln p - \ln\kappa\right) - \kappa e^{-p/\kappa}\right]. \end{aligned} \qquad (7.420)$$

For $\kappa \to 0$, this reduces to the Landau pdf,

$$f(x, \Delta E) = \frac{1}{2\pi i x\zeta} \int_{K-i\infty}^{K+i\infty} dp\, e^{p(\lambda_L + \ln p)} \qquad (7.421)$$

where the Landau and Vavilov parameters are related to each other by,

$$\begin{aligned} \lambda_L &= \frac{\lambda_V}{\kappa} - \ln\kappa \\ &= \frac{\Delta E - \langle\Delta E\rangle}{x\zeta} - (1+\beta^2) + \gamma_{EM} - \ln\kappa. \end{aligned} \qquad (7.422)$$

By changing variables,

$$p = iy \qquad (7.423)$$

the Landau pdf can be written in a form readily amenable to numerical integration,

$$f(\lambda_L) = \frac{1}{\pi x\zeta} \int_0^\infty dy\, e^{-\left(\frac{\pi}{2}\right)y} \cos(y\ln y + \lambda_L y) \qquad (7.424)$$

Figure 7.31 shows the product $xf(\lambda_L)$ as a function of the Landau parameter λ_L. The maximum of $f(\lambda_L)$ occurs at $\lambda_{L,Max} = -0.22278$ (Kölbig and Schorr 1984) from which the most probable energy loss can be calculated by noting that the Landau parameter can be written as,

$$\lambda_L = \frac{\Delta E - (\Delta E)_{MP}}{x\zeta} + \lambda_{L,Max} \qquad (7.425)$$

where $(\Delta E)_{MP}$ is the most probable energy loss which can be solved for, using (7.422) and (7.425),

$$(\Delta E)_{MP} = x\left[\left(\frac{dE}{dx}\right)_{Col} + \zeta(\lambda_{L.Max} + 1 + \beta^2 + \ln\kappa - \gamma_{EM})\right]. \qquad (7.426)$$

A closed analytic form of a pdf has been presented by Moyal (1955) as a representation of the Landau pdf,

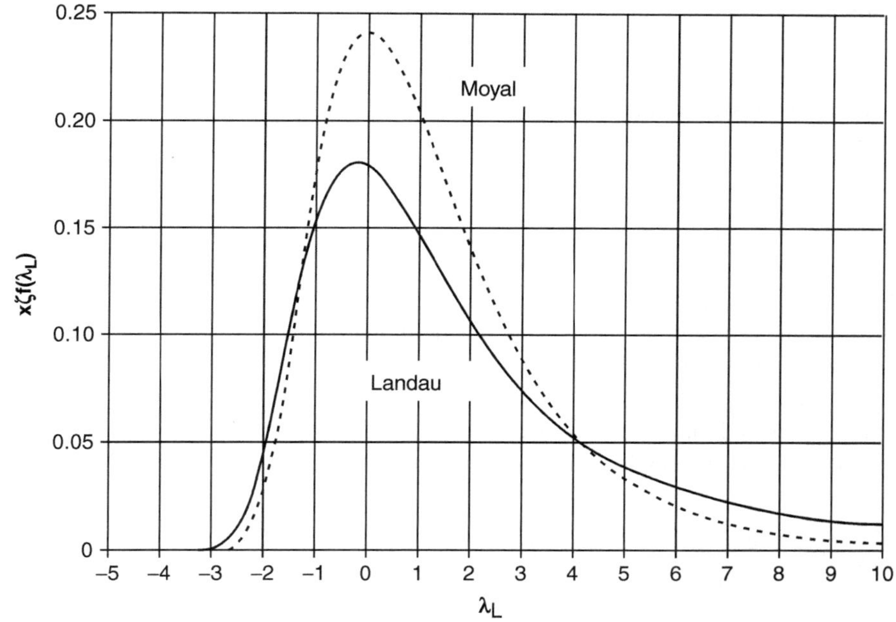

Fig. 7.31 Weighted Landau and Moyal probability distribution functions as functions of the Landau parameter

$$f(\lambda) = \frac{1}{x\zeta\sqrt{2\pi}} e^{-\frac{1}{2}(\lambda + \exp(-\lambda))} \tag{7.427}$$

where

$$\lambda = \frac{\Delta E - (\Delta E)_{MP}}{x\zeta}. \tag{7.428}$$

The Moyal result $x\zeta f(\lambda)$ is also plotted in Fig. 7.31. As can be seen, it is not an entirely accurate reproduction of the Landau result due to both the differing positions of the most probable energy loss (the Moyal curve peaks at $\lambda_L = 0$) and the different magnitudes of the maximum values of the two pdfs. Importantly, the Moyal approximation underestimates the high-energy loss tail. Even so, the Moyal approximation clearly has calculational advantages over the Landau integral result and is used as its approximation.[21] Rotondi and Montagna (1990) have proposed an improvement upon the Moyal approximation,

$$f_{\kappa,\beta}(\lambda) = \left(\frac{a_1}{x\zeta}\right) \exp\left[-a_2(\lambda + a_5\lambda^2) - a_3 \exp(-a_4(\lambda + a_6\lambda^2))\right] \tag{7.429}$$

[21]See, for example, Sauli's description of multiwire proportional chambers (1977).

where the six parameters, a_i, are functions of and β and were determined by fitting (7.427) to the numerical solution for the Vavilov pdf for $\kappa \leq 3$. They are provided as the weighted sum of the products of two Tchebyshev polynomials and the reader is referred to that publication for further details.

7.4.4.5 Practical Methods of Calculating the Vavilov pdf

Introduction

The form of the pdf to be used to describe the energy loss is dictated by the value of κ where, by convention, the appropriate pdf to use for a given range of values of κ is,

$\kappa \leq 0.01$	Landau
$0.01 \leq \kappa \leq 10$	Vavilov
$10 \leq \kappa$	Gaussian

Direct analytical solutions of the Vavilov integral are difficult and one typically resorts to approximations, numerical methods, or parameterizations to solve them. This is especially true in Monte Carlo applications where calculation speed is critical. Three methods of calculating the Vavilov pdf are summarized here; no derivations will be provided, but the interested reader can refer to the original publications.

Edgeworth Series

For values of κ that are not too small, the Vavilov distribution is Gaussian-like where the perturbation from the Gaussian described in terms of an Edgeworth series (Symon 1948; Rotondi and Montagna 1990; Van Ginneken 2000). In the nomenclature of Rotondi and Montagna, the Vavilov pdf is approximated by,

$$\begin{aligned} f(x,\Delta E) = \frac{e^{-\frac{\Delta E^2}{2\sigma^2}}}{\sqrt{2\pi}\sigma}\Bigg[& 1 + \frac{1}{3!}\frac{\mu_3}{\sigma^3}H_3\left(\frac{\Delta E}{\sigma}\right) + \frac{1}{4!}\left(\frac{\mu_4}{\sigma^4}-3\right) \\ & \times H_4\left(\frac{\Delta E}{\sigma}\right) + \frac{1}{5!}\left(\frac{\mu_5}{\sigma^5}-10\frac{\mu_3}{\sigma^3}\right)H_5\left(\frac{\Delta E}{\sigma}\right) \\ & + \frac{10}{6!}\left(\frac{\mu_3}{\sigma^3}\right)^2 H_6\left(\frac{\Delta E}{\sigma}\right) + \frac{35}{7!}\frac{\mu_3}{\sigma^3}\left(\frac{\mu_4}{\sigma^4}-3\right) \\ & \times H_7\left(\frac{\Delta E}{\sigma}\right) + \frac{280}{9!}\left(\frac{\mu_3}{\sigma^3}\right)^3 H_9\left(\frac{\Delta E}{\sigma}\right)\Bigg] \end{aligned} \tag{7.430}$$

where the H_i are the Hermite polynomials and the μ_i and σ are related to the moments of the Vavilov distribution,

$$\mu_n = x\int_0^{Q_{max}} dQ\,\omega(Q)Q^n \quad n = 2,3 \tag{7.431}$$

$$\mu_4 = 3\mu_2^2 + \left(x\int_0^{Q_{max}} dQ\,\omega(Q)Q^4\right)^2 \tag{7.432}$$

$$\mu_5 = 10\mu_2\mu_3 + x\int_0^{Q_{max}} dQ\,Q^5\,\omega(Q) \tag{7.433}$$

$$\sigma^2 = \mu_2 \tag{7.434}$$

This expansion is reported to be valid for $0.29 \le \kappa$ and for $\lambda_L \le \lambda \le \lambda_H$ where the limits λ_L and λ_H define the limits of 0 and 1 in the cumulative distribution function and are determined from empirical fits.

Fourier Series Solution

Schorr (1974, 1975) developed an algorithm for calculating both the Landau and Vavilov pdfs using a Fourier series methodology. The approximation to the Vavilov pdf, written in the form,

$$f(\lambda_V) = \frac{1}{2\pi i}\int_{K-i\infty}^{K+i\infty} ds\,\phi(s)e^{\lambda_V s} \tag{7.435}$$

with

$$\phi(s) = e^{\kappa(1+\beta^2\gamma_G)}e^{\psi(s)}$$

and

$$\psi(s) = s\ln\kappa + (s+\beta^2\kappa) \times \left[\int_0^{Q_{max}} ds\frac{1-e^{-s/\kappa}}{s} - \gamma_{EM}\right] - \kappa e^{-s/\kappa}$$

is

$$g(\lambda_L) = \frac{\omega}{\pi}\left(\frac{1}{2} + \sum_{k=1}^{\infty}[A_k\cos(k\omega\lambda_V) + B_k\sin(k\omega\lambda_V)]\right) \tag{7.436}$$

where

$$\omega = \frac{2\pi}{T_+ - T_-} \tag{7.437}$$

$$A_k = \mathrm{Re}\phi(ik\omega) \tag{7.438}$$

$$B_k = -\mathrm{Im}\phi(ik\omega) \tag{7.439}$$

Schorr provides the methodology for calculating T_- and T_+ so as to minimize the difference $g(\lambda_L) - f(\lambda_L)$. As this method requires a point-by-point calculation of the pdfs, it is not suitable as a sampling method for Monte Carlo simulations.

Distorted Log-Normal Distribution

Chibani (1998, 2002) has described two algorithms for calculating the Vavilov pdf in the interval $\kappa \in [0.01,1.0]$. The first algorithm is valid for $\kappa \in [0.01,0.3]$ and describes the Vavilov pdf as the convolution of log-normal and Poisson distributions. The second method, for $\kappa \in [0.3,1.0]$, takes advantage of the similarity of the shape of the log-normal distribution to that of the Vavilov pdf. The reader is referred to these two papers for further details.

7.4.4.6 Vavilov pdf for Electron Projectiles

Monte Carlo simulations of practical dosimetric interest will concern those cases of electron and positron projectiles. In such cases, energy straggling is accounted for by modifying the Landau or Vavilov distributions. In particular, the Vavilov pdf is calculated using the Møller cross section for electron–electron collisions. Chibani simplifies this calculation by approximating the Møller cross section by a fourth-order polynomial in order to allow the Vavilov pdf to be calculated analytically (this is also done for positron projectiles by doing the same for the Bhabha positron–electron collision cross section).

7.4.4.7 Atomic Electron Binding Effects

In the case of a high-Z thin absorber, energy losses through resonant transfers to atomic electrons become important. Such effects can be managed by convolving the Landau distribution with a Gaussian function. In this application, because it is easy to convolve two Gaussian functions, the Landau *pdf* is represented by a weighted sum of four Gaussian pdfs (Blunck and Leisegang, 1951). Even though there are difficulties with this approach that require addressing (Findlay and Dusautoy 1980), they have relatively little immediate application to nuclear medicine dosimetry.

7.5 Multiple Elastic Scattering

7.5.1 Introduction

In addition to being able to calculate for the transfer of energy to the medium from a moving charged particle, it is necessary to know both the number and phase space of the particles. Transport is dominated by the elastic Coulomb scatter in which, for electrons and positrons, negligible amounts of energy are transferred in these scatters and can be neglected. The three most important results obtained from the derivation of the differential cross sections for a single elastic Coulomb scatter are that:

- The θ^{-4} dependence showing that forward-directed elastic scatter will dominate.
- The mean free pathlength between elastic scatters is small due to the large total cross section.
- The differential cross section has a $(p\beta)^{-2}$ dependence.

The combination of the first two results leads to the dominance of forward-directed multiple scatter. The last result shows that, at energies typical of nuclear medicine, electrons and positrons are more subject to multiple scatter than are heavier α particles. Hence, the interest in this section will be on e^-/e^+ multiple scatter.

By using the small-angle approximation, recalling the θ^{-4} dependence and noting that $d\Omega \cong 2\pi\,\theta\,d\theta$ (which implicitly assumes azimuthal symmetry), one can write the mean-square angle of a single elastic Coulomb scatter (assuming that screening at small scattering angles invokes a cut-off in angle),

$$\overline{\theta^2} = \frac{\int d\Omega\,\theta^2 \frac{d\sigma_{Ruth}}{d\Omega}}{\int d\Omega \frac{d\sigma_{Ruth}}{d\Omega}} \approx \frac{\int_{\chi_0}^{\theta_{max}} \frac{d\theta}{\theta}}{\int_{\chi_0}^{\theta_{max}} \frac{d\theta}{\theta^3}} = \frac{\ln(\theta_{max}/\chi_0)}{\left(1/\chi_0^2\right) - \left(1/\theta_{max}^2\right)}. \tag{7.440}$$

The minimum scattering angle is the screening angle χ_0 of (7.6) for the Thomas–Fermi model. As $\chi_0 \ll 1$ and $\chi_0 \ll \theta_{max}$ (recall Fig. 7.3), the root-mean square (RMS) scattering angle is approximately,

$$\sqrt{\overline{\theta^2}} = -\chi_0 \ln \chi_0 \tag{7.441}$$

Hence, $\sqrt{\overline{\theta^2}}$ will be a relatively small multiple of χ_0 and the net deflection of the electron will be small (e.g., from Fig. 7.3 $\chi_0 \approx 25$ mrad for a 100 keV electron in carbon which leads to $\sqrt{\overline{\theta^2}} = 3.7\,\chi_0$ or about 93 mrad). Because the number of elastic Coulomb scatters in a pathlength of practical interest will be high, the central-limit theorem indicates that the probability distribution function describing the scattering angle will be Gaussian with a small variance, $\overline{\theta^2}$. This Gaussian approximation is not entirely accurate as it neglects the small, but not negligible, probability of large-angle Coulomb scatter that increases the "tail" of the pdf. But the Gaussian assumption will provide the basis upon which to explore the three main multiple scattering theories in common use in modern-day calculations of charged-particle transport.

7.5.2 Multiple Elastic Scattering Theory

7.5.2.1 Introduction

Since the 1940s, a number of theories describing the multiple scatter of charged particles have been developed. Two theories (Goudsmit and Saunderson 1940, Molière 1947, 1948) are predominant in the Monte Carlo codes currently used to model charged-particle transport in medical applications. Prior to deriving these theories, the simpler Fermi–Eyges theory (Eyges 1948) is derived (which, while not used in transport calculations for nuclear medicine applications, has been widely employed in software developed for external electron beam treatment planning software in radiation oncology (Hogstrom et al. 1981; McParland et al. 1988). Importantly, the Fermi–Eyges theory justifies the expectation of a Gaussian pdf for the spatial deflection and angular distribution of multiply-scattered charged particles which appears as the "zeroth-order" case in the Goudsmit–Saunderson and Molière multiple scatter theories.

7.5.2.2 Fermi–Eyges Theory

The genesis of this theory was the derivation by Fermi of the diffusion equation for a calculation of the transport of cosmic rays in the atmosphere (given in the review by Rossi and Greisen (1941)). Because of the high kinetic energies of the cosmic rays, ionizational energy losses were neglected in Fermi's derivation and Eyges (1948) extended Fermi's result by allowing for the energy losses suffered particles through ionization. The derivation below of the Fermi diffusion equation will follow that of Jette (1988) which is itself a slightly more general version of the original. Having obtained the diffusion equation, Eyges' approach of allowing for an energy dependence will be used to solve the diffusion equation using the methods developed earlier in this chapter.

Before deriving Fermi's diffusion equation, some preparatory work is required. Consider Fig. 7.32 in which a beam of monoenergetic charged particles with zero lateral width is incident along the $+z$-axis to a semi-infinite scattering medium with the entrance plane defined by the $x - y$ plane. The particle is scattered through the polar θ and azimuthal ϕ angles as shown. Due to the central Coulomb potentials associated with the scattering centers within the medium, azimuthal symmetry exists and ϕ will be uniformly distributed between 0 and 2π. The angles θ_x and θ_y are the projections of the polar scattering angle onto the xz- and yz-planes, respectively, as shown, where,

$$\theta_x = \tan^{-1}(\tan\theta\,\cos\phi) \qquad (7.442)$$

$$\theta_y = \tan^{-1}(\tan\theta\,\sin\phi) \qquad (7.443)$$

and the mean-square values of these projected scattering angles are,

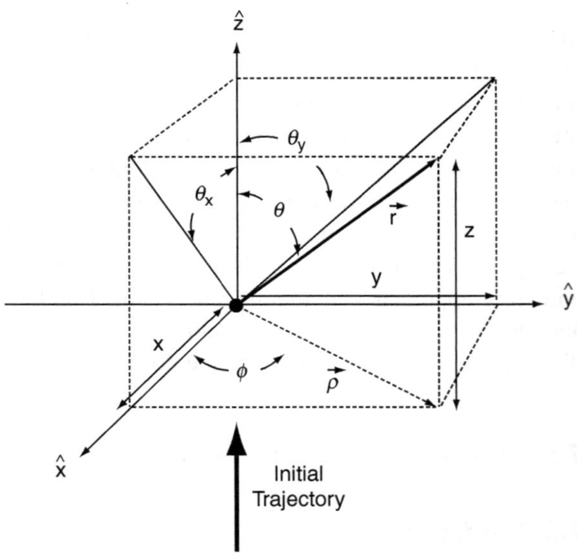

Fig. 7.32 Scattering of a particle with an initial trajectory along the z-axis through (θ,ϕ). θ_x and θ_y are the projections of the polar scattering angle onto the $x - z$ and $y - z$ planes, respectively

$$\begin{aligned}\overline{\theta_x^2} &= \frac{\int d\Omega\, \theta_x^2(\theta,\phi)\frac{d\sigma_{Ruth}}{d\Omega}}{\int d\Omega\, \frac{d\sigma_{Ruth}}{d\Omega}} \\ &\approx \frac{\int d\Omega\, \theta^2 \cos^2\phi\, \frac{d\sigma_{Ruth}}{d\Omega}}{\int d\Omega\, \frac{d\sigma_{Ruth}}{d\Omega}} \\ &= \left(\frac{1}{2\pi}\int_0^{2\pi} d\phi \cos^2\phi\right)\overline{\theta^2} \\ &= \frac{\overline{\theta^2}}{2}\end{aligned} \tag{7.444}$$

and, similarly,

$$\overline{\theta_y^2} = \frac{\overline{\theta^2}}{2}. \tag{7.445}$$

Hence, in the small-angle approximation, the mean-square scatter angle is,

$$\overline{\theta^2} = \overline{\theta_x^2} + \overline{\theta_y^2}. \tag{7.446}$$

Recalling the earlier invocation of the central-limit theorem, one conjectures that the scattering pdf of the two projected scattering angles are Gaussian. For example,

$$f(\theta_x) = \frac{1}{\sqrt{2\pi\overline{\theta_x^2}}}\exp\left(-\frac{1}{2}\frac{\theta_x^2}{\overline{\theta_x^2}}\right) \tag{7.447}$$

with zero mean and variance $\overline{\theta_x^2}$. At this point, the linear scattering power is defined as the derivative of the mean-square scattering angle with respect to the pathlength traveled,

$$\Im = \frac{d\overline{\theta_s^2}}{ds}. \tag{7.448}$$

Of course, this can only be an approximation in that it assumes that the change in the mean-square scattering angle with distance is continuous whereas the processes are stochastic. For calculational purposes, this will be approximated by a continuous derivative with the ratio of the mean-square scattering angle over a pathlength Δs,

$$\Im = \frac{\overline{\theta^2}}{\Delta s} \tag{7.449}$$

where,

$$\begin{aligned}\Delta s &= \frac{\Delta z}{\cos\theta} \\ &\approx \Delta z\end{aligned} \tag{7.450}$$

for forward-directed scatter. The Gaussian pdf for θ_x can now be written as,

$$f(\theta_x) = \frac{1}{\sqrt{\pi\,\Im\,\Delta z}}\exp\left(-\frac{\theta_x^2}{\Im\,\Delta z}\right). \tag{7.451}$$

It is reasonable to presume, for future use, that the pdf for a change in scattering angle from the mean, $\Delta\theta_x = \theta_x - \overline{\theta_x}$, over Δz is also Gaussian,

$$g(\Delta\theta_x;\Delta z) = \frac{1}{\sqrt{\pi\,\Im\,\Delta z}}\exp\left(-\frac{(\Delta\theta_x)^2}{\Im\,\Delta z}\right). \tag{7.452}$$

In the derivation of the Gaussian pdf, use will be needed of the first three moments of $g(\Delta\theta_x;\Delta z)$,

$$\begin{aligned}&\int_{-\infty}^{\infty} d(\Delta\theta_x) g(\Delta\theta_x;\Delta z) \\ &= \frac{1}{\sqrt{\pi\,\Im\,\Delta z}}\int_{-\infty}^{\infty} d(\Delta\theta_x)\exp\left(-\frac{(\Delta\theta_x)^2}{\Im\,\Delta z}\right) \\ &= 1\end{aligned} \tag{7.453}$$

$$\begin{aligned}&\int_{-\infty}^{\infty} d(\Delta\theta_x)\,\Delta\theta_x\, g(\Delta\theta_x;\Delta z) \\ &= \frac{1}{\sqrt{\pi\Im\Delta z}}\int_{-\infty}^{\infty} d(\Delta\theta_x)\Delta\theta_x \exp\left(-\frac{(\Delta\theta_x)^2}{\Im\Delta z}\right) \\ &= 0\end{aligned} \tag{7.454}$$

$$\begin{aligned}&\int_{-\infty}^{\infty} d(\Delta\theta_x)(\Delta\theta_x)^2 g(\Delta\theta_x;\Delta z) \\ &= \frac{1}{\sqrt{\pi\Im\Delta z}}\int_{-\infty}^{\infty} d(\Delta\theta_x)(\Delta\theta_x)^2 \exp\left(-\frac{(\Delta\theta_x)^2}{\Im\Delta z}\right) = \frac{\Im\Delta z}{2}.\end{aligned} \tag{7.455}$$

The limits of integration of $\Delta\theta_x$ can be extended to $\pm\infty$ as $g(\Delta\theta_x;\Delta z)$ is sharply peaked around $\Delta\theta_x = 0$.

The Fermi diffusion equation is next derived. Assume that the initial condition for the particle is $\delta(x)\delta(y)\delta(z)\delta(\theta)$ and define $f(x,\theta_x;y,\theta_y;z)\Delta x\,\Delta\theta_x\,\Delta y\,\Delta\theta_y$ as the probability at a depth z that the particle will be between x and $x+\Delta x$ and between y and $y+\Delta y$ and have projected scattering angles between θ_x and $\theta_x+\Delta\theta_x$ and between θ_y and $\theta_y+\Delta\theta_y$. In order

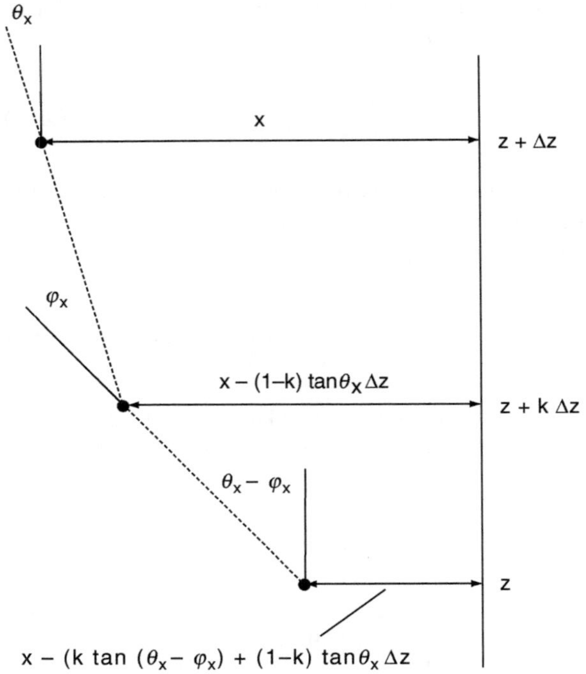

Fig. 7.33 Trajectory of a charged particle twice scattered between depths of z and z + Δz

to simplify the derivation, the calculation is initially restricted to the two-dimensional case and then extended to three dimensions. The intent is to find the variation in $f(x, \theta_x; z)$ with increasing penetration into the medium, z. To do so, consider the trajectory of a particle in two-dimensions as shown in Fig. 7.33. The particle's end-point is at a depth $z + \Delta z$ with projected lateral displacement x and projected polar direction θ_x following an intermediate scatter occurring at a depth $z + k\Delta z$, where $0 < k < 1$. This position is projected back to the particle's original position at a depth z. At the intermediate scattering point, the particle was at a projected lateral displacement,

$$x - (1 - k)\tan\theta_x\,\Delta z \approx x - (1 - k)\theta_x\,\Delta z \quad (7.456)$$

(using the small-angle approximation for θ_x) and was scattered through an angle φ_x. If one further back projects the trajectory to the depth z, the particle had been scattered through an angle $\theta_x - \varphi_x$ and was at a projected lateral displacement,

$$\begin{aligned} &x - (\tan\theta_x + k(\tan(\theta_x - \varphi_x) - \tan\theta_x))\Delta z \\ &\quad \approx x - (\theta_x - k\varphi_x)\Delta z. \end{aligned} \quad (7.457)$$

For any value of k in the range (0,1), one obtains the pdf $f(x,\theta_x;z+\Delta z)$ by convolving over all φ_x,

$$\begin{aligned} f(x, \theta_x; z + \Delta z) = \int_{-\infty}^{\infty} d\varphi_x\, f(x - (\theta_x - k\varphi_x) \\ \times \Delta z, \theta_x - \varphi_x; z) g(\varphi_x; \Delta z) \end{aligned} \quad (7.458)$$

where $g(\varphi_x; \Delta z)$ is given by (7.452).

This integral is solved by first expanding $f(x - (\theta_x - k\varphi_x)\Delta z, \theta_x - \varphi_x; z)g(\varphi_x; \Delta z)$ to second order in θ_x, φ_x and Δz,

$$\begin{aligned} &f(x - (\theta_x - k\varphi_x)\Delta z, \theta_x - \varphi_x; z)g(\varphi_x; \Delta z) \\ &\quad \cong f(x, \theta_x; z) - \frac{\partial f(x, \theta_x; z)}{\partial x}(\theta_x - k\varphi_x) \\ &\quad \times \Delta z - \frac{\partial f(x, \theta_x; z)}{\partial \theta_x}\varphi_x + \frac{1}{2}\frac{\partial^2 f(x, \theta_x; z)}{\partial \theta_x^2}\varphi_x^2. \end{aligned} \quad (7.459)$$

Substituting this into the integrand of (7.458) gives,

$$\begin{aligned} &f(x,\theta_x;z+\Delta z) \\ &= \int_{-\infty}^{\infty} d\varphi_x\, f(x,\theta_x;z)g(\varphi_x;\Delta z) \\ &\quad - \int_{-\infty}^{\infty} d\varphi_x \frac{\partial f(x,\theta_x;z)}{\partial x}(\theta_x - k\varphi_x)\Delta z\, g(\varphi_x;x) \\ &\quad - \int_{-\infty}^{\infty} d\varphi_x \frac{\partial f(x,\theta_x;z)}{\partial \theta_x}\varphi_x\, g(\varphi_x;x) \\ &\quad + \frac{1}{2}\int_{-\infty}^{\infty} d\varphi_x \frac{\partial^2 f(x,\theta_x;z)}{\partial \theta_x^2}\varphi_x^2(\varphi_x;x) \\ &= f(x,\theta_x;z)\int_{-\infty}^{\infty} d\varphi_x\, g(\varphi_x;\Delta z) \\ &\quad - \frac{\partial f(x,\theta_x;z)}{\partial x}\int_{-\infty}^{\infty} d\varphi_x(\theta_x - k\varphi_x)\Delta z\, g(\varphi_x;x) \\ &\quad - \frac{\partial f(x,\theta_x;z)}{\partial \theta_x}\int_{-\infty}^{\infty} d\varphi_x\,\varphi_x\, g(\varphi_x;x) \\ &\quad + \frac{1}{2}\frac{\partial^2 f(x,\theta_x;z)}{\partial \theta_x^2}\int_{-\infty}^{\infty} d\varphi_x\varphi_x^2\, g(\varphi_x;x). \end{aligned} \quad (7.460)$$

This expression is markedly simplified by using the moments of $g(\varphi_x;\Delta z)$ calculated previously,

$$f(x,\theta_x;z+\Delta z)=f(x,\theta_x;z)-\theta_x\frac{\partial f(x,\theta_x;z)}{\partial x}\Delta z+\frac{\Im}{4}\frac{\partial^2 f(x,\theta_x;z)}{\partial\theta_x^2}\Delta z. \tag{7.461}$$

Note that in this final expression for the pdf there is no dependence upon k and, as a result, the selection of the intermediate scatter position is purely arbitrary. From this result, one obtains the differential change in the pdf with penetration,

$$\frac{\partial f(x,\theta_x;z)}{\partial z}=\lim_{\Delta z\to 0}\frac{f(x,\theta_x;z+\Delta z)-f(x,\theta_x;z)}{\Delta z} \tag{7.462}$$

and the Fermi diffusion equation in two-dimensions follows,

$$\frac{\partial f(x,\theta_x;z)}{\partial z}=-\theta_x\frac{\partial f(x,\theta_x;z)}{\partial x}+\frac{\Im}{4}\times\frac{\partial^2 f(x,\theta_x;z)}{\partial\theta_x^2}. \tag{7.463}$$

The extension of the Fermi diffusion equation to 3 dimensions is straight forward,

$$\frac{\partial f}{\partial z}=-\theta_x\frac{\partial f}{\partial x}-\theta_y\frac{\partial f}{\partial y}+\frac{\Im}{4}\left(\frac{\partial^2 f}{\partial\theta_x^2}+\frac{\partial^2 f}{\partial\theta_y^2}\right) \tag{7.464}$$

where, for clarity, the functional dependencies of the pdf have been omitted.

Note that this derivation of the Fermi diffusion equation has ignored the fact that, as a particle penetrates within the medium, it loses energy. As the elastic single scatter cross section has a $(p\beta)^{-2}$ dependence, the scattering power will vary with the projectile kinetic energy and, hence, become a function of the depth of penetration. For the two-dimensional form of the Fermi diffusion equation,

$$\frac{\partial f(x,\theta_x;z)}{\partial z}=-\theta_x\frac{\partial f(x,\theta_x;z)}{\partial x}+\frac{\Im(z)}{4}\frac{\partial^2 f(x,\theta_x;z)}{\partial\theta_x^2} \tag{7.465}$$

where the explicit functional dependency of the scattering power upon z is noted.[22] This expression neglects the fact that once the particle has reached a depth z it will have actually traveled a distance greater than z due to multiple scattering events.

The two-dimensional Fourier pair will be used to solve for the pdf,

$$f(x,\theta_x;z)=\frac{1}{2\pi}\int_{-\infty}^{\infty}\int_{-\infty}^{\infty}d\xi\,d\Lambda\,\omega(\xi,\Lambda;z)e^{i(x\xi+\theta_x\Lambda)} \tag{7.466}$$

$$\omega(\xi,\Lambda;z)=\frac{1}{2\pi}\int_{-\infty}^{\infty}\int_{-\infty}^{\infty}dx\,d\theta_x\,f(x,\theta_x;z)\,e^{-i(x\xi+\theta_x\Lambda)}. \tag{7.467}$$

Applying these to the two-dimensional Fermi diffusion equation gives a differential equation in the Fourier transform of the pdf,

$$\frac{\partial\omega}{\partial z}=\xi\frac{\partial\omega}{\partial\Lambda}-\frac{\Lambda^2\Im(z)}{4}\omega \tag{7.468}$$

which can be simplified by defining the variables,

$$\kappa=z+\frac{\Lambda}{\xi} \tag{7.469}$$

$$z'=z \tag{7.470}$$

to yield,

$$\frac{\partial\omega}{\partial z'}=-\frac{\xi^2(\kappa-z)^2\Im(z)}{4}\omega. \tag{7.471}$$

The solution is,

$$\omega=k(\kappa)\exp\left(-\xi^2\int_{\lambda}^{z'}d\eta\frac{(\kappa-\eta)^2\Im(\eta)}{4}\right). \tag{7.472}$$

---

[22] In an inhomogeneous medium (such as the body), the scattering power would also be a function of x and y. That level of complexity is not required for this discussion.

The lower limit of the integral has been set to some arbitrary valuer, λ, and $k(\kappa)$ is specified from the initial condition of the problem. Assuming a single incident particle, this is, as before,

$$f(x, \theta_x; z = 0) = \delta(x)\delta(\theta_x) \tag{7.473}$$

with the usual Fourier transform,

$$\omega(\xi, \Lambda; z = 0) = \frac{1}{2\pi}. \tag{7.474}$$

Writing out the solution for $\omega(\xi, \Lambda; z)$ in the original variables gives,

$$\omega(\xi, \Lambda; z) = k\left(z + \frac{\Lambda}{\xi}\right) \times \exp\left(-\xi^2 \int_\lambda^z d\eta \frac{\left(z + \frac{\Lambda}{\xi} - \eta\right)^2 \Im(\eta)}{4}\right). \tag{7.475}$$

From the initial conditions,

$$k\left(\frac{\Lambda}{\xi}\right) \exp\left(-\xi^2 \int_\lambda^0 d\eta \frac{\left(\frac{\Lambda}{\xi} - \eta\right)^2 \Im(\eta)}{4}\right) = \frac{1}{2\pi} \tag{7.476}$$

which enables the Fourier transform of the pdf to be written as,

$$\omega(\xi, \Lambda; z) = \frac{1}{2\pi} \exp\left(-\xi^2 \int_\lambda^z d\eta \frac{\left(z + \frac{\Lambda}{\xi} - \eta\right)^2 \Im(\eta)}{4}\right). \tag{7.477}$$

Prior to taking the inverse Fourier transform in order to recover the pdf, this expression is simplified by defining the functions,

$$A_0(z) = \int_0^z d\eta \frac{\Im(\eta)}{4} \tag{7.478}$$

$$A_1(z) = \int_0^z d\eta (z - \eta) \frac{\Im(\eta)}{4} \tag{7.479}$$

$$A_2(z) = \int_0^z d\eta (z - \eta)^2 \frac{\Im(\eta)}{4} \tag{7.480}$$

so as to write a simpler form of the Fourier transform of the pdf,

$$\omega(\xi, \Lambda; z) = \frac{1}{2\pi} \exp(-A_0\Lambda^2 + 2A_1\Lambda\xi + A_2\xi^2) \tag{7.481}$$

where the z-dependencies of the A_i have been omitted for clarity. Taking the inverse Fourier transform of the result will result in the multiple scattering pdf,

$$f(x, \theta_x; z) = \frac{1}{2\pi\sqrt{A_0A_2 - A_1^2}} \times \exp\left(-\frac{\theta_x^2 A_2 - 2A_1 x\theta_x + x^2 A_0}{4A_0(A_0A_2 - A_1^2)}\right) \tag{7.482}$$

which is the Fermi–Eyges result for two dimensions. The three-dimensional result is given by the product of the two-dimensional pdfs in (x,θ_x) and (y,θ_y).

Equation (7.482) provides a Gaussian pdf; as the result of the small-angle approximation, in terms of the functions $A_0(z)$, $A_1(z)$, and $A_2(z)$. The depth of penetration, z, is now explicitly included in the pdf through these functions. As the linear scattering power will have an energy dependence (i.e., a less-energetic particle will be more readily scattered than an energetic one), a direct functional dependence upon particle energy is implicit.

7.5.2.3 Scattering Power

Introduction

The linear scattering power is the change in mean-square scattering angle per unit distance traveled by the particle. In analogy to the linear and mass stopping powers, one removes the dependence of the scattering power upon the physical density of the medium by defining the mass scattering power, $\Im/\rho$, which is the change in mean-square scattering angle per unit mass thickness traveled,

$$\frac{\Im}{\rho} = \frac{d\overline{\theta^2}}{\rho\, ds} \tag{7.483}$$

Or,

$$\frac{\Im}{\rho} = 2\pi \frac{N_A}{A} \int_{\chi_0}^{1} d\theta\, \theta^2 \frac{d\sigma}{d\Omega} \quad (7.484)$$

where, since the differential cross is in units of square centimeter·radian square per atom, the N_A/A factor is required for the mass collision stopping power to have units of square centimeter·radian square per gram of medium. χ_0 is the screening angle of (7.6).

Spin-0 Projectile Scattering

The mass scattering powers for the unscreened and screened Rutherford cross sections for an electron projectile (i.e., $z = 1$), but where spin is neglected, are derived. Beginning with the unscreened cross section the corresponding mass scattering power is,

$$\frac{\Im}{\rho} = 8\pi \left(\frac{N_A Z^2}{A}\right)\left(\frac{\alpha\hbar c}{p\beta}\right)^2 \ln\frac{1}{\chi_0} \quad (7.485)$$

Unscreened cross section.

By incorporating the definition of the screening angle χ_0,

$$\begin{aligned} \chi_0 &\approx 1.130\left(\frac{\hbar c}{p r_B}\right) Z^{1/3} \\ &\approx \left(\frac{4.22 \times 10^{-3}}{p}\right) Z^{1/3} \end{aligned} \quad (7.486)$$

where p is in units of MeV. One can rewrite the electron mass scattering power in the form,

$$\frac{\Im}{\rho} = 8\pi \left(\frac{N_A Z^2}{A}\right)\left(\frac{\alpha\hbar c}{p\beta}\right)^2 \ln\left(237\, p\, Z^{-1/3}\right). \quad (7.487)$$

The screened cross section leads to an expression of the mass scattering power of the form,

$$\begin{aligned} \frac{\Im}{\rho} &= 4\pi \left(\frac{N_A Z^2}{A}\right)\left(\frac{\alpha\hbar c}{p\beta}\right)^2 \\ &\times \left[\ln\left(1 + \frac{1}{\chi_0^2}\right) - \frac{1}{2}\left(\frac{1-\chi_0^2}{1+\chi_0^2}\right) - \ln 2\right] \end{aligned} \quad (7.488)$$

Screened Rutherford cross section.

This expression can be simplified for electrons typical of that encountered in nuclear medicine in low-Z media (i.e., $p \ll 1.6\,\text{MeV}$) for which $(1 + 1/\chi_0^2) \approx 1/\chi_0^2$ and $(1 - \chi_0^2)/(1 + \chi_0^2) \approx 1$ to give,

$$\begin{aligned} \frac{\Im}{\rho} &\approx 4\pi \left(\frac{N_A Z^2}{A}\right)\left(\frac{\alpha\hbar c}{p\beta}\right)^2 \\ &\times \left[\ln\left(5.6 \times 10^4 p^2 Z^{-2/3}\right) - 1.193\right] \end{aligned} \quad (7.489)$$

where p is in units of MeV.

The mass scattering powers of electrons in carbon evaluated from the screened and unscreened cross sections above are shown in Fig. 7.34 as functions of electron kinetic energy.

Mott Cross Section

To include the effect of the electron's intrinsic spin, the Mott elastic scatter cross section is used,

$$\frac{d\sigma_{Mott}}{d\Omega} = \left(\frac{Z\alpha\hbar c}{2p\beta}\right)^2 \frac{F(\theta, Z)}{\sin^4\left(\frac{\theta}{2}\right)}$$

where the multiplicative factor

$$F(\theta, Z) = 1 - \beta^2 \sin^2\left(\frac{\theta}{2}\right)$$

is a consequence of the electron's intrinsic spin. McKinley and Feshbach (1948) have expanded Mott's original result to include a corrective term to F(θ,Z) in order to correct for the Born approximation used in the original derivation,

$$F(\theta, Z) = 1 - \beta^2 \sin^2\left(\frac{\theta}{2}\right) + \pi\alpha\beta Z\left(1 - \sin\frac{\theta}{2}\right)\sin\frac{\theta}{2} \quad (7.490)$$

This expression is valid for $\beta \approx 1$ and $\alpha Z < 0.2$ (i.e., $Z < 27$) which are conditions suitable for most instances of nuclear medicine dosimetry. McParland (1989) derived an analytical expression for the electron mass scattering power using this expanded Mott

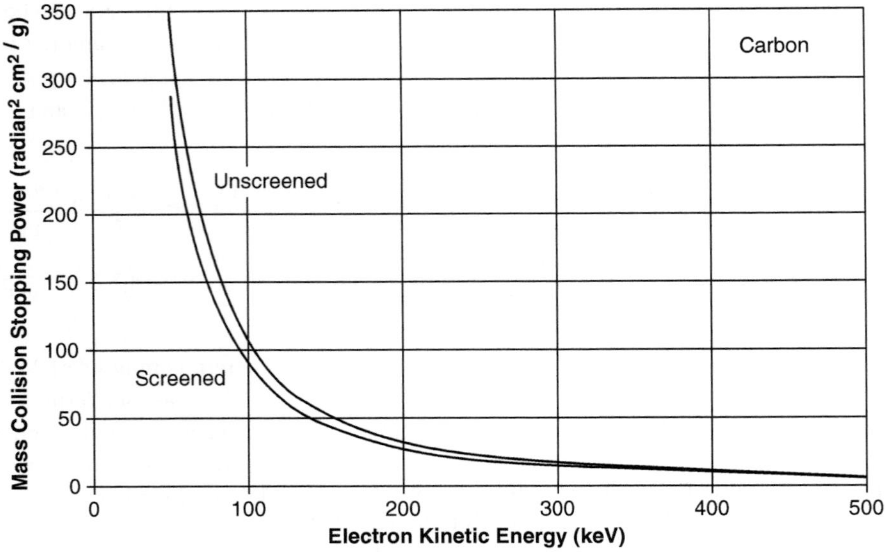

Fig. 7.34 Mass scattering power for electrons in carbon as a function of electron kinetic energy calculated using unscreened and screened potentials and neglecting spin

cross section and excluding any small-angle approximations,

$$\frac{\mathfrak{T}}{\rho}=8\pi\left(\frac{N_AZ^2}{A}\right)\left(\frac{r_0}{\gamma\beta}\right)^2(I_1+\pi\alpha\beta ZI_2-\beta(\beta+\pi\alpha Z)I_3) \tag{7.491}$$

where

$$\begin{aligned} I_1 &= \ln\left(\frac{\sin(1/2)}{\sin(\chi_0/2)}\right) \\ &\quad +\frac{1}{8}\left(\frac{\chi_0^2}{\sin^2(\chi_0/2)}-\frac{1}{\sin^2(1/2)}\right) \\ &\quad +\frac{1}{2}\left(\chi_0\cot\frac{\chi_0}{2}-\cot\frac{1}{2}\right) \end{aligned} \tag{7.492}$$

$$\begin{aligned} I_2 &= \chi_0\left(\frac{(\chi_0/2)}{\sin(\chi_0/2)}-1+2\sum_{k=1}^{\infty}(-1)^k c_{2k}\left(\frac{\chi_0}{2}\right)^{2k}\right) \\ &\quad -\left(\frac{1}{2\sin(\chi_0/2)}-1+2\sum_{k=1}^{\infty}(-1)^k c_{2k}\left(\frac{1}{2}\right)^{2k}\right) \end{aligned} \tag{7.493}$$

$$\begin{aligned} I_3 &= \left(1+\sum_{k=1}^{\infty}(-1)^k d_{2k}\left(\frac{1}{2}\right)^{2k}\right) \\ &\quad -\chi_0^2\left(1+\sum_{k=1}^{\infty}(-1)^k d_{2k}\left(\frac{\chi_0}{2}\right)^{2k}\right). \end{aligned} \tag{7.494}$$

Table 7.1 Values of the coefficients c_{2k} and d_{2k}

k	c_{2k}	d_{2k}
1	2.78×10^{-2}	8.30×10^{-2}
2	-1.94×10^{-3}	-3.71×10^{-3}
3	1.46×10^{-4}	2.65×10^{-4}
4	-1.17×10^{-5}	-2.12×10^{-5}
5	9.70×10^{-7}	1.78×10^{-6}
6	-8.32×10^{-8}	-1.55×10^{-7}
7	7.31×10^{-9}	1.37×10^{-8}

The coefficients in these expressions are,

$$c_{2k}=\left(\frac{2^{2k-1}-1}{(2k+1)!}\right)B_{2k} \tag{7.495}$$

$$d_{2k}=\left(\frac{2^{2k}}{(2+2k)(2k)!}\right)B_{2k} \tag{7.496}$$

B_{2k} is the 2kth Bernoulli number ($B_2 = 1/6$, $B_4 = -1/30$, $B_6 = 1/42$, etc.) The coefficients c_{2k} and d_{2k} rapidly diminish with k, as shown in Table 7.1, and, for practical calculation purposes, the summations for integrals I_2 and I_3 can be truncated at $k = 4$.

Contributions to the Scattering Power from Atomic Electrons

In the previous derivation of electron mass scattering powers for which a term of the form Z^2 appears, the scattering has been calculated from an infinitely-massive point scattering center with charge Ze. This is a reasonable approximation, in most cases, from the nucleus. However, for an atomic scattering center made up from a nucleus with charge Ze and Z electrons, an accounting of the contributions of the atomic electrons to the mass scattering power is required. This inclusion of the Møller scatter is most frequently performed by assuming that it is coherent with the nuclear Rutherford/Mott results derived previously. This assumptions leads to the approximation in which the Z^2 term is replaced by $Z(Z + 1)$, resulting in an increase in our previously-calculated mass scattering powers for carbon and lead by 16.7 and 1.2%, respectively.

7.5.2.4 Specific Electron Multiple Scattering Theories

Introduction

Although a considerable number of multiple scattering theories of varying complexity and utility have been developed, in this subsection, the review of such theories are limited to two – those of Goudsmit and Saunderson (1940) and Molière (1947, 1948), which are two of the more popular theories employed in Monte Carlo simulations of charged particle transport.

Goudsmit–Saunderson Theory

Goudsmit and Saunderson (1940) derived a multiple scatter pdf using multiple independent Coulomb single scatters and the addition theorem of spherical harmonics. The scattering angle is assumed to be small so that the electron's pathlength is equal to the thickness of the scatterer (in other words, the result is strictly valid only for thin foils or short discrete steps in a Monte Carlo simulation) and collisions resulting in energy loss are neglected. The theory does have the advantage in that any underlying single scattering differential cross section can be used. The derivation begins by defining the normalized single scatter angular distribution,

$$f_1(\theta) = \frac{\frac{d\sigma}{d\Omega}(\theta)}{\int d\Omega \frac{d\sigma}{d\Omega}(\theta)}. \tag{7.497}$$

The subscript "1" refers to a single Coulomb scatter and axial symmetry is assumed, which is valid for spherically symmetric atomic scattering centers or randomly-oriented molecules. In order to calculate the distribution for $n > 1$ scatters the distribution is first expanded as a weighted series of Legendre polynomials,

$$f_1(\theta) = \frac{1}{4\pi}\sum_{l=0}^{\infty}(2l+1)F_l\, P_l(\cos\theta). \tag{7.498}$$

The coefficients are given by,

$$\begin{aligned} F_l &= \int d\Omega\, f_1(\theta)P_l(\cos\theta) \\ &= \overline{P_l(\cos\theta)}. \end{aligned} \tag{7.499}$$

The single scatter angular distribution can now be written in the form,

$$f_1(\theta) = \frac{1}{4\pi}\sum_{l=0}^{\infty}(2l+1)\left[\overline{P_l(\cos\theta)}\right]P_l(\cos\theta). \tag{7.500}$$

In order to evaluate the coefficients of the expansion for $n > 1$ scatters, the addition property of spherical harmonics when written in terms of associated Legendre polynomials is used. Let the electron be first scattered through an angle of deflection θ_1 with a corresponding azimuthal angle ϕ_1. It then undergoes a second scatter through the corresponding angles (θ_2,ϕ_2). The total scattering angle is thus $\theta_1 + \theta_2$ and the addition property of spherical harmonics gives,

$$\begin{aligned} P_l(\cos(\theta_1+\theta_2)) &= P_l(\cos\theta_1)P_l(\cos\theta_2) \\ &\quad + \sum_{m=-l}^{l} P_l^m(\cos\theta_1) \\ &\quad \times P_l^m(\cos\theta_2)\sin(m(\phi_2-\phi_1)). \end{aligned} \tag{7.501}$$

Averaging both sides leads to,

$$\overline{P_l(\cos(\theta_1+\theta_2))} = \overline{P_l(\cos\theta_1)P_l(\cos\theta_2)}. \tag{7.502}$$

The generalization of this result for n scatters is,

$$\overline{P_l(\cos(n\theta))} = \left[\overline{P_l(\cos\theta_1)}\right]^n. \tag{7.503}$$

Applying this to (7.500), one obtains the angular distribution following n scatters,

$$f_n(\theta) = \frac{1}{4\pi}\sum_{l=0}^{\infty}(2l+1)\left[\overline{P_l(\cos\theta)}\right]^n P_l(\cos\theta). \quad (7.504)$$

Consider the combination of a pathlength t and a mean free path between elastic collisions λ. The mean number of collisions is given by (t/λ) and the probability that an electron will undergo n collisions while traversing t is Poisson distributed,

$$p(n;t) = \frac{e^{-(t/\lambda)}\left(\frac{t}{\lambda}\right)^n}{n!}. \quad (7.505)$$

The probabilities of electron elastic scatter for $n = 0$, 1, and 2 scatters are provided in Fig. 7.35 as functions of the ratio of the pathlength to the elastic scatter mean free path. In order to calculate the angular distribution of the electrons exiting the foil, one must sum over the probabilities of all possible collisions,

$$\begin{aligned} f_{GS}(\theta;t) &= \sum_{n=0}^{\infty} p(n;t) f_n(\theta) \\ &= \frac{1}{4\pi}\sum_{n=0}^{\infty} e^{-(t/\lambda)}\frac{(t/\lambda)^n}{n!} \\ &\times \sum_{l=0}^{\infty}(2l+1)\left[\overline{P_l(\cos\theta)}\right]^n P_l(\cos\theta) \\ &= \frac{1}{4\pi}\sum_{l=0}^{\infty}(2l+1)P_l(\cos\theta)e^{-(t/\lambda)} \\ &\times \sum_{n=0}^{\infty}\frac{(t/\lambda)^n}{n!}\left[\overline{P_l(\cos\theta)}\right]^n \\ &= \frac{1}{4\pi}\sum_{l=0}^{\infty}(2l+1)e^{-(t/\lambda)}\, e^{-(t/\lambda)\overline{P_l(\cos\theta)}} P_l(\cos\theta) \\ &\equiv \frac{1}{4\pi}\sum_{l=0}^{\infty}(2l+1)e^{-tG_l}P_l(\cos\theta) \end{aligned} \quad (7.506)$$

where the series expansion of e^x has been used and the coefficient,

$$G_l = \frac{1 - \overline{P_l(\cos\theta)}}{\lambda} \quad (7.507)$$

has been defined. This is referred to as the l th-order transport coefficient, as will be made evident shortly.

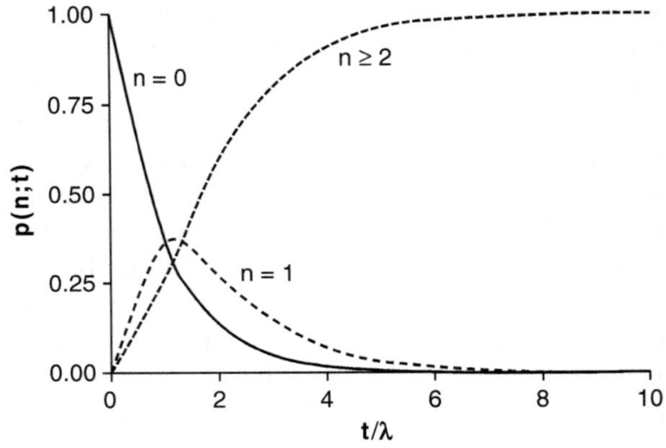

Fig. 7.35 The probability of electron elastic scatter for $n = 0,1$ and 2 scatters as a function of the ratio of the pathlength t to the elastic scatter mean free path

As $\overline{P_0(\cos\theta)} = 1$ then $e^{-tG_0} = 1$ and the zeroth-order term of the expansion $f_{GS}(\theta;t)$ is $1/4\pi$. As $P_l(\cos\theta)$ decreases with increasing l due to the growing oscillatory nature of the Legendre polynomial, the e^{-tG_l} term will tend to $e^{-(t/\lambda)}$ for increasing l. $f_{GS}(\theta;t)$ is forward peaked for small foil thickness t values but, as $t \to \infty$ only the $l = 0$ term contributes to the pdf and $f_{GS}(\theta;t) \to 1/4\pi$. Hence, as expected and because energy loss is ignored, the angular distribution becomes isotropic.

Now look at the first-order transport coefficient,

$$\begin{aligned} G_1 &= \frac{1}{\lambda}\left(1 - \overline{P_1(\cos\theta)}\right) \\ &= \frac{1}{\lambda}\int d\Omega\, f_1(\theta)(1 - \cos\theta). \end{aligned} \quad (7.508)$$

Recalling the definition of the mean free path between collisions given by (7.7), this result is rewritten as,

$$G_1 = \frac{\rho N_A}{A}\int d\Omega \frac{d\sigma}{d\Omega}(\theta)(1 - \cos\theta). \quad (7.509)$$

Invoking the small-angle approximation, $1 - \cos\theta = 2\sin^2(\theta/2) \approx \theta^2/2$ (which is justifiable in this application as the differential cross section is highly forward-peaked),

$$\begin{aligned} G_1 &= \frac{\rho N_A}{2A}\int d\Omega \frac{d\sigma}{d\Omega}(\theta)\theta^2 \\ &= \frac{\mathfrak{I}}{2} \end{aligned} \quad (7.510)$$

which, of course, is proportional to the linear scattering power.

It will be noted that the Goudsmit–Saunderson pdf is exact in that the single scatter differential cross section appears within the transport coefficient. Hence, these coefficients need only be calculated. Goudsmit and Saunderson provide examples of such calculations for analytical forms of the single scatter differential cross section and a series approximation of the Legendre polynomials. The number of terms required in the summation over l to achieve convergence in the calculation of $f_{GS}(\theta;t)$ will increase as the pathlength t decreases due to the e^{-tG_l} term. It is possible to improve the convergence of the calculation for small pathlengths by isolating the contributions from unscattered electrons.

Molière Theory

The Goudsmit–Saunderson result provides the electron multiple scatter pdf as the weighted summation of Legendre polynomials where the underlying single scatter theory is incorporated through the mean free pathlength between elastic collisions and the averaged Legendre polynomial. The Molière multiple scattering theory (Molière 1947, 1948) evolved from consideration of consecutive scatters which, in practice, is the solution of the transport equation. Like the previous theory, the result of Molière's theory is both independent of an individual form of the single scatter cross section and neglects energy loss. On the other hand, the single scatter cross section is input to the Molière theory through only a single parameter, the Molière screening angle, χ_a'. The shape of the multiple scatter pdf is dependent upon a single parameter, b, which is primarily a function of the areal thickness of the medium the particle is traversing and is largely dependent upon the medium's atomic number for most media of dosimetric interest.

Molière's result is derived here using Bethe's (1953) approach, which is mathematically more transparent than the original, and incorporating further improvements suggested by Andreo et al. (1993).[23]

[23]Fernández-Varea et al. (1993) have provided an additional and shorter derivation of the theory beginning with the Goudsmit–Saunderson result.

Again, because of the θ^{-4} dependence of the scattering cross section and its consequently being forward peaked, the derivation is simplified using the small-angle approximation. Consider a monodirectional beam of electrons incident to a medium of physical density ρ, atomic number[24] Z, and atomic mass A. The number of scattering centers per unit volume is given by $\rho N_A/A$. χ is the scattering angle after a single scatter and θ is the cumulative scattering angle after multiple scatters an electron undergoes traversing a finite thickness of medium. $f_M(\theta;t)\theta d\theta$ is the number of scattered electrons in the angular interval dθ following traveling a distance t of the medium. By equating the multiple scattering problem to the diffusion of electrons in the scattering plane, the electron transport equation is, for the scattering pdf,

$$\frac{\partial f(\theta;s)}{\partial s} = \rho\left(\frac{N_A}{A}\right)\left(\int d\chi\, f\left(\hat{\boldsymbol{\theta}}';s\right)\frac{d\sigma(\chi)}{d\chi} - f(\theta;s)\int dX\,\chi\frac{d\sigma(\chi)}{d\chi}\right) \tag{7.511}$$

where $\hat{\boldsymbol{\theta}}' = \hat{\boldsymbol{\theta}} - \hat{\mathbf{X}}$ is the direction vector of the electron prior to the last scatter at t and $dX = \chi\, d\chi\, d\phi/2\pi$ where ϕ is the azimuthal angle of χ in the prescattering plane of the electron. Defining the transforms,

$$f_M(\theta;t) = \int_0^\infty d\eta\,\eta\, J_0(\eta\theta)g(\eta;t) \tag{7.512}$$

and

$$g(\eta;t) = \int_0^\infty d\theta\,\theta\, J_0(\eta\theta)f(\theta;t) \tag{7.513}$$

and applying them to the transport equation gives,

$$\frac{\partial g(\eta;t)}{\partial t} = -g(\eta;t)\left(\rho\frac{N_A}{A}\right) \times \int_0^\infty d\chi\,\chi(1 - J_0(\eta\chi))\frac{d\sigma}{d\chi}. \tag{7.514}$$

[24]Unlike Molière, Bethe included the contributions of the atomic electrons and assumed these to be coherent so that Z is replaced by Z(Z + 1). This is repeated here.

The solution to this equation is,

$$g(\eta;t) = e^{(\Omega(\eta;t)-\Omega_0)} \tag{7.515}$$

where

$$\Omega(\eta;t) = \left(\rho\frac{N_A}{A}\right)t\int_0^\infty d\chi\,\chi\frac{d\sigma}{d\chi}J_0(\eta\chi). \tag{7.516}$$

Note that, as $J_0(0) = 1$, the parameter

$$\Omega_0 = \Omega(0;t) = \left(\rho\frac{N_A}{A}\right)t\int_0^\infty d\chi\,\chi\frac{d\sigma}{d\chi} \tag{7.517}$$

is equal to the total number of collisions occurring along the pathlength t. One then obtains an expression for the Molière multiple scattering pdf,

$$f_M(\theta;t) = \int_0^\infty d\eta\,\eta\,J_0(\eta\theta) \times \exp\left[-\rho\left(\frac{N_A}{A}\right)t\int_0^\infty d\chi\,\chi\frac{d\sigma}{d\chi}(1-J_0(\eta\chi))\right]. \tag{7.518}$$

Because of the explicit inclusion of the scattering differential cross section in this expression, it is apparent that, as with the Goudsmit–Saunderson theory, the multiple scattering pdf will not be restricted to a particular single Coulomb scatter theory.

A simpler form of the Molière pdf can be obtained by taking advantage of the fact that the elastic single scatter differential cross section is proportional to θ^{-4}, becoming complicated only (as shown in Fig. 7.2) when the scattering angle is of the order of, or less than, the screening angle. Using Bethe as a guide, $f_M(\theta;t)$ is calculated using the unscreened Rutherford cross section beginning with evaluating the ratio of the actual to Rutherford cross section,

$$q(\chi) = \frac{(d\sigma_{Act}/d\chi)}{\left(d\sigma_{Ruth}/d\chi\right)} \tag{7.519}$$

where the axial symmetry of the Coulomb interaction is noted. The subscript Act identifies the actual cross section. For the unscreened Rutherford cross section,

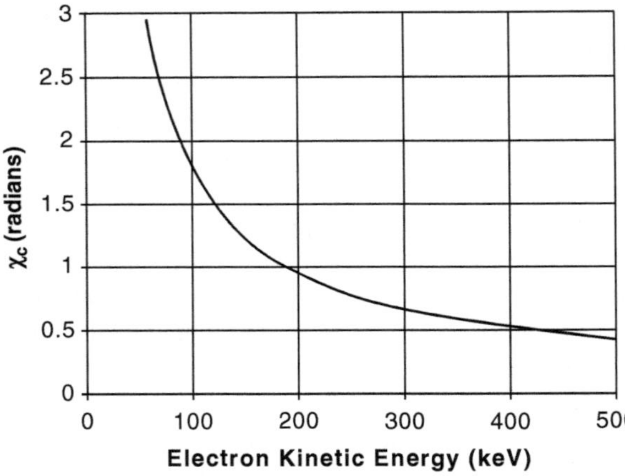

Fig. 7.36 Molière angle χ_c for carbon ($Z = 6$, $A = 12$, $\rho = 2$ g/cm^3, $t = 0.1$ cm) as a function of electron kinetic energy

$$q(\chi) = \left(\rho\frac{N_A}{A}\right)t\left(\frac{\chi^4}{2\chi_c^2(t)}\right)\frac{d\sigma_{Act}}{d\chi} \tag{7.520}$$

where the unit-probability angle χ_c is defined by,

$$\chi_c^2(t) = \left(\rho\frac{N_A}{A}\right)t\left(\frac{\alpha\hbar c}{p\beta}\right)^2 Z(Z+1). \tag{7.521}$$

The Molière angle χ_c has the physical interpretation of being the angle beyond which the probability of a single elastic scatter occurring is equal to unity; it is plotted as a function of electron kinetic energy in 1 mm carbon in Fig. 7.36.

$q(\chi)$ has properties that will prove useful in the derivation. Clearly, $q(\chi) \to 0$ as $\chi \to 0$ due to the χ^{-4} dependence of the unscreened cross section. By extension, as the screened and unscreened cross sections converge for large scattering angles, $q(\chi) \to 1$ as $\chi \to \infty$ (again, for calculational purposes, the maximum scattering angle is set to infinity and advantage is taken of the sharp forward-angle peak of the single scatter cross section). As suggested by Figs. 7.2 and 7.3, the most rapid change in $q(\chi)$ occurs for $\chi \approx 0$. The next step of the derivation is to return to (7.512) and, by taking logarithms of (7.515) and using (7.516), obtain,

$$\begin{aligned} -\ln g(\eta;t) &= \Omega_0 - \Omega(\eta) \\ &= \rho\left(\frac{N_A}{A}\right)t\int_0^\infty d\chi\,\chi(1-J_0(\eta\chi))\frac{d\sigma_{Act}}{d\Omega} \\ &= 2\chi_c^2(t)\int_0^\infty d\chi\,\chi\frac{(1-J_0(\eta\chi))}{\chi^3}q(\chi) \end{aligned} \tag{7.522}$$

As $q(\chi) < 1$ for $\chi \approx \chi_0$ where $\chi \ll \chi_c$, the integral can be split by defining an intermediate angle $\chi_0 \ll \chi_S \ll \chi_c$, for which $q(\chi)$ can be set equal to unity for scattering angles greater than χ_S,

$$\begin{aligned}
&\int_0^\infty d\chi\,\chi\frac{(1-J_0(\eta\chi))}{\chi^3}q(\chi) \\
&= \int_0^{\chi_s} d\chi\,\chi\frac{(1-J_0(\eta\chi))}{\chi^3}q(\chi) + \int_{\chi_s}^\infty d\chi\,\chi\frac{(1-J_0(\eta\chi))}{\chi^3}q(\chi) \\
&\cong \int_0^{\chi_s} d\chi\,\chi\frac{(1-J_0(\eta\chi))}{\chi^3}q(\chi) + \int_{\chi_s}^\infty d\chi\,\chi\frac{(1-J_0(\eta\chi))}{\chi^3} \\
&\equiv I_1(\eta;\chi_s) + I_2(\eta;\chi_s).
\end{aligned} \tag{7.523}$$

The first integral is solved by the change of variable $x = \eta\chi_s$ and using the small-argument approximation of the Bessel function (as $\eta\chi_s$ is small), $J_0(x) \approx 1 - (x/2)^2$,

$$I_1(\eta;\chi_s) = \frac{\eta^2}{4}\int_0^{\chi_s} d\chi\frac{q(\chi)}{\chi}. \tag{7.524}$$

The second integral is,

$$\begin{aligned}
I_2(\eta;\chi_s) &= \int_{\chi_s}^\infty d\chi\frac{(1-J_0(\eta\chi))}{\chi^3} \\
&= \int_{\chi_s}^\infty \frac{d\chi}{\chi^3} - \int_{\chi_s}^\infty d\chi\frac{J_0(\eta\chi)}{\chi^3} \\
&= \frac{1}{2\chi_s^2} - I_3(\eta;\chi_s).
\end{aligned} \tag{7.525}$$

The indefinite form of the integral $I_3(\eta;\chi_s)$ can be expressed in terms of the Meijer G function, but here Bethe's approach of straightforward, but tedious, sequential integrations by parts and ignoring terms of order $(\eta\chi_s)^2$ is followed. Beginning with the change of variable, $x = \eta\chi_s$, the integral is rewritten as,

$$I_3(\eta;\chi_S) = \eta^2\int_{\eta\chi_S}^\infty dx\frac{J_0(x)}{x^3} \tag{7.526}$$

where the integral is,

$$\begin{aligned}
\int_{\eta\chi_s}^\infty dx\frac{J_0(x)}{x^3} &= \int_{\eta\chi_s}^\infty dr\,s \\
&= s\,r\Big|_{\eta\chi_s}^{\infty} - \int_{\eta\chi_s}^\infty ds\,r
\end{aligned} \tag{7.527}$$

where

$$s \equiv J_0(x) \quad ds = -dx\,J_1(x)$$

$$dr \equiv \frac{dx}{x^3} \qquad r \equiv -\frac{1}{2x^2}$$

to give

$$\begin{aligned}
\int_{\eta\chi_s}^\infty dx\frac{J_0(x)}{x^3} &= \frac{J_0(\eta\chi_s)}{2(\eta\chi_s)^2} - \frac{1}{2}\int_{\eta\chi_s}^\infty dx\frac{J_1(x)}{x^2} \\
&= \frac{J_0(\eta\chi_s)}{2(\eta\chi_s)^2} - \frac{1}{2}I_4(\eta;t).
\end{aligned} \tag{7.528}$$

This new integral is also solved by parts,

$$\begin{aligned}
I_4(\eta;t) &= \int_{\eta\chi_s}^\infty dx\frac{J_1(x)}{x^2} \\
&= s\,r\Big|_{\eta\chi_s}^{\infty} - \int_{\eta\chi_s}^\infty ds\,r
\end{aligned} \tag{7.529}$$

where

$$s \equiv J_1(x) \quad ds = dx\left(J_0(x) - \frac{J_1(x)}{x}\right)$$

$$dr \equiv \frac{dx}{x^2} \qquad r \equiv -\frac{1}{x}.$$

Then,

$$\begin{aligned}
I_4(\eta;t) &= \frac{J_1(\eta\chi_s)}{\eta\chi_s} + \int_{\eta\chi_s}^\infty dx\frac{J_0(x)}{x} - I_4(\eta;t) \\
&= \frac{1}{2}\left(\frac{J_1(\eta\chi_s)}{\eta\chi_s} + \int_{\eta\theta\chi}^\infty dx\frac{J_0(x)}{x}\right).
\end{aligned} \tag{7.530}$$

This result is simplified by using the small-argument approximation, $J_1(x) \cong x/2$, and the solution to the definite integral $\int_t^\infty dx \frac{J_0(x)}{x} = \int_0^t dx \frac{1-J_0(x)}{x} - \gamma_{EM} - \ln\frac{t}{2}$,

$$I_4(\eta;t) = \frac{1}{2}\left(\frac{1}{2} - \gamma_{EM} - \ln\frac{\eta\chi_s}{2}\right) + \mathbf{O}\left((\eta\chi_s)^2\right). \tag{7.531}$$

Terms of order $(\eta\chi_s)^2$ are neglected. Working backwards,

$$I_3(\eta;\chi_S) = \frac{\eta^2}{2}\frac{J_0(\eta\chi_s)}{(\eta\chi_s)^2} - \frac{\eta^2}{4}\left(\frac{1}{2} - \gamma_{EM} - \ln\frac{\eta\chi_s}{2}\right). \tag{7.532}$$

Using the small-argument form of $J_0(\eta\chi_s)$, one then arrives at,

$$\begin{aligned} I_2(\eta;\chi_S) &= \frac{1}{2\chi_s^2} - I_3(\eta;\chi_s) \\ &= \frac{\eta^2}{4}\left(1 - \gamma_{EM} - \ln\frac{\eta\chi_s}{2}\right) \end{aligned} \tag{7.533}$$

One can now return to the original integral to give,

$$\begin{aligned} &\int_0^\infty d\chi\,\chi\frac{(1 - J_0(\eta\chi))}{\chi^3}q(\chi) \\ &= \frac{\eta^2}{4}\left(\int_0^{\chi_s} d\chi\frac{q(\chi)}{\chi} + 1 - \gamma_{EM} - \ln\eta - \ln\chi_s + \ln 2\right). \end{aligned} \tag{7.534}$$

Molière next defined a characteristic screening angle,

$$-\ln\chi_a = \lim_{\chi_s\to\infty}\left(\int_0^{\chi_s} d\chi\frac{q(\chi)}{\chi} + \frac{1}{2} - \ln\chi_s\right)$$

allowing (7.534) to be written in the much simpler form,

$$\begin{aligned} &\int_0^\infty d\chi\,\chi\frac{(1 - J_0(\eta\chi))}{\chi^3}q(\chi) \\ &= \frac{\eta^2}{4}\left(\frac{1}{2} - \gamma_{EM} - \ln\eta\chi_a + \ln 2\right) \end{aligned} \tag{7.535}$$

Then, the exponent of (7.515) can be written as,

$$\Omega_0 - \Omega(\eta) = \frac{(\eta\chi_c)^2}{2}\left(\frac{1}{2} - \gamma_{EM} - \ln\eta\chi_a + \ln 2\right). \tag{7.536}$$

In order to simplify the integral $f_M(\theta;t) = \int_0^\infty d\eta\,\eta\,J_0(\eta)e^{-(\Omega_0 - \Omega(\eta;t))}$, Molière further defined the quantity,

$$\begin{aligned} b &= \ln\left(\frac{\chi_c}{\chi_a}\right)^2 + 1 - 2\gamma_{EM} \\ &\equiv \ln\left(\frac{\chi_c}{\chi_a{}'}\right)^2 \end{aligned} \tag{7.537}$$

where the modified characteristic scattering angle is,

$$\begin{aligned} \chi_a{}' &= \chi_a\,e^{\gamma_{EM} - 1/2} \\ &\approx 1.08\,\chi_a \end{aligned} \tag{7.538}$$

leading to a simpler expression,

$$b = \ln\left(\frac{\chi_c^2}{1.167\,\chi_a^2}\right). \tag{7.539}$$

Using the change of variable $y = \eta\chi_c$, (7.536) is recast as,[25]

$$\Omega_0 - \Omega(\eta;t) = \frac{y^2}{4}\left(b - 2\ln\frac{y}{2}\right). \tag{7.540}$$

Finally, by writing $\lambda = \theta/\chi_c$, one at last obtains the transformed form of Molière's pdf,

$$\begin{aligned} f_M(\theta)\theta\,d\theta &= \lambda\,d\lambda\int_0^\infty dy\,y\,J_0(\lambda\,y) \\ &\times \exp\left(\left(\frac{y}{2}\right)^2\left(2\ln\left(\frac{y}{2}\right) - b\right)\right). \end{aligned} \tag{7.541}$$

It is necessary to modify the upper limit of this integration as the exponent goes to infinity for $y \to \infty$, as shown in Fig. 7.37 for $b = 3$. The exponent

[25] Molière achieved this result using an expansion of Hankel functions.

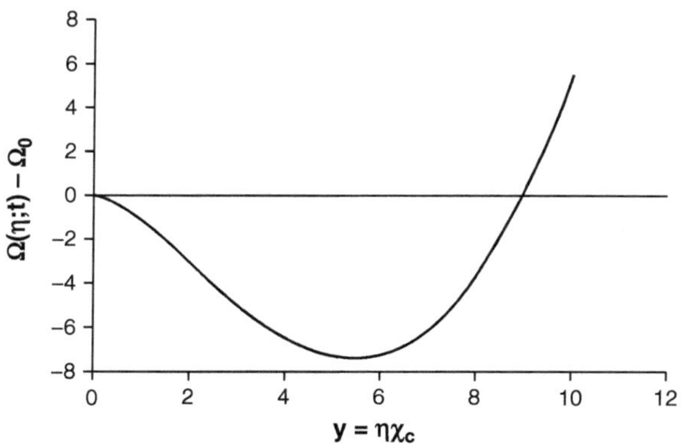

Fig. 7.37 Exponent in the transformed expression of the Molière pdf for b = 3

has a minimum at $y = 2e^{(b-1)/2}$, which is a consequence of the specified inequality $\chi_0 \ll \chi_s \ll \chi_c$. The upper limit of the integral is set to this minimum and the transformed Molière pdf is written as,

$$f_M(\theta)\theta\, d\theta = \lambda\, d\lambda \int_0^{2e^{(b-1)/2}} dy\, y\, J_0(\lambda\, y) \times \exp\left(\left(\frac{y}{2}\right)^2\left(2\ln\left(\frac{y}{2}\right) - b\right)\right). \tag{7.542}$$

This truncation of the integration limit will have a negligible effect as $e^b \approx (\chi_c/\chi_a)^2$ which is of the order of the total number of collisions in the pathlength considered. As the mean free pathlength between elastic collisions is small, this truncation of the integration is allowed.

Molière solved for the angular pdf by first defining a variable B through the transcendental equation,

$$B - \ln B = b. \tag{7.543}$$

Approximations to B have since been given by Scott (1963),

$$B = 1.153 + 2.583\log_{10}\left(\frac{\chi_c}{\chi_a}\right)^2 \tag{7.544}$$

and Tabata and Ito (1976),

$$B = 2.6 + 2.3863\log_{10}\left(\frac{\chi_c}{\chi_a}\right)^2 - \frac{3.234}{\log_{10}\left(\frac{\chi_c}{\chi_a}\right)^2 + 0.994} \tag{7.545}$$

(the value of B typically ranges from between 5 and 20). To complete the derivation, Molière defined the reduced angle,

$$\vartheta = \frac{\lambda}{B} = \frac{\theta}{\chi_c\sqrt{B}}. \tag{7.546}$$

The integration variable of the multiple scatter pdf is then changed to,

$$u = y\sqrt{B} \tag{7.547}$$

and the pdf then expanded in a power series in B^{-1},

$$f_M(\theta)\theta\, d\theta = \vartheta\, d\vartheta \sum_{n=0}^{\infty} f_M^{(n)}(\vartheta) B^{-n} \tag{7.548}$$

where the coefficients of the expansion are,

$$f_M^{(n)}(\vartheta) = \frac{1}{n!}\int^{2e^{(b-1)/2}} du\, u\, J_0(u)\left(\frac{u^2}{4}\ln\frac{u^2}{4}\right)^n \exp\left(-\frac{u^2}{4}\right). \tag{7.549}$$

The zeroth-order coefficient can be calculated from the following property of integer-order Bessel functions,

$$\int_0^\infty dt\, t^{\nu+1} e^{-a^2t^2} J_\nu(bt) = \frac{b^\nu}{(2a^2)^{\nu+1}} e^{-b^2/4a^2}; \quad \mathrm{Re}\ a^2 > 0 \tag{7.550}$$

to give,

$$f_M^{(0)}(\vartheta) = \int^\infty du\, u\, J_0(u)\exp\left(-\frac{u^2}{4}\right) = 2e^{-\vartheta^2} \tag{7.551}$$

where the upper limit of the integral has been allowed to go to infinity due to the $e^{-u^2/4}$ term in the integrand. Clearly, this first term corresponds to a Gaussian multiple scattering pdf. Whereas Molière provided an analytical representation of $f^{(1)}(\vartheta)$ which was proportional to ϑ^{-4} for large ϑ, Bethe reported that he was only able to find numerical solutions to $f^{(n)}(\vartheta)$ for

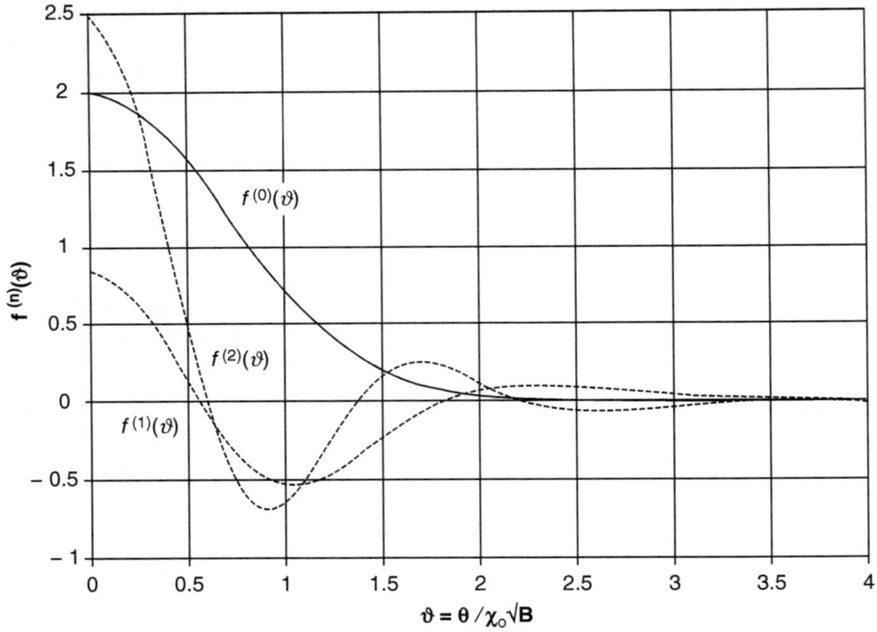

Fig. 7.38 The coefficients $f^{(0)}(\vartheta)$, $f^{(1)}(\vartheta)$, and $f^{(2)}(\vartheta)$ of the Molière expansion of the multiple scattering probability distribution function as functions of the reduced scattering angle ϑ

$n \geq 2$. Values of $f^{(0)}(\vartheta)$, $f^{(1)}(\vartheta)$, and $f^{(2)}(\vartheta)$ are plotted as functions of ϑ in Fig. 7.38. For, $\vartheta < 2$ $f^{(0)}(\vartheta)$ dominates and the multiple scattering pdf is Gaussian for small scattering angles. The $f^{(0)}(\vartheta)$ term decreases exponentially such that at larger values of ϑ, the $f^{(1)}(\vartheta)$ term dominates and, as it is proportional to ϑ^{-4} for large ϑ, it goes over into the Rutherford single scatter angular distribution.

One would expect from the expansion of (7.548) that the accuracy of the pdf would increase with the inclusion of an increasing number of terms. This is not the case due mainly to the omission of electron intrinsic spin and relativity, the effects of which grow at large single scattering angles. Andreo et al. (1993) have exhaustively studied the limitations to the Molière theory.

Now return to the transformed Molière result. In the Molière theory, the scatter is described by the characteristic screening angle and that the final angular distribution is a function of the ratio of this screening angle to the unit-probability scattering angle. The single scatter differential cross section enters the theory through the ratio of differential cross sections, $q(\chi)$. Molière provided a result for the Thomas–Fermi atom and Fernández-Varea et al. (1993) did so for the Yukawa-type screening. Molière's form is,

$$\chi_a^2 = 1.13 + 3.76\left(Z\frac{\alpha}{\beta}\right)^2. \tag{7.552}$$

The second term in the series accounts for deviation from the Born approximation. From the definition of b, for an electron of speed βc in a medium with atomic number Z and atomic mass number A and physical density ρ,

$$\begin{aligned} e^b &= \frac{\chi_c^2}{\chi_a' 2} \\ &= \frac{6680}{\beta^2}\rho t \frac{Z^{1/3}(Z+1)}{A\left(1 + 3.34\left(Z\frac{\alpha}{\beta}\right)^2\right)} \end{aligned} \tag{7.553}$$

where ρt is the pathlength given in units of cm^2/g. Bethe showed that the Z-dependence does not deviate from unity by more than about ±30% for Z ranging from 1 (for deuterium) to 92 (uranium), an observation that indicates that the number of collisions per square centimeter/gram is reasonably constant for all elements.

7.6 Bremsstrahlung

7.6.1 Introduction

Throughout this chapter, only the elastic scatter and the collision energy losses of a charged particle moving through a medium have been considered. Attention is now turned to radiative energy losses through which a charged particle is deflected by the nuclear Coulomb field and emits a photon as a result (*bremsstrahlung*).

In this section, discussions will be limited to those of the radiative energy losses of electron and positron projectiles with kinetic energies of the order of a few MeV or less in low-Z media representative of tissue[26] and omit a derivation of electron–electron *bremsstrahlung* as it is not a significant phenomenon at the energies of interest here.[27] Hence, only the *bremsstrahlung* resulting from electron-atom interactions is considered (i.e., the atom is treated as a nuclear Coulomb field screened by the atomic electrons).

Detailed elucidations of *bremsstrahlung* may be found in Koch and Motz (1959), Heitler (1984), Pratt et al. (1977), and Haug and Nakel (2004). Numerical data are also available from ICRU Publication 37 (1984) and Berger and Seltzer (1983).

[26]This limitation is reasonable for nuclear medicine purposes as the maximum electron energy resulting from the Compton scatter of a 511 keV photon is 340 keV for a backscattered photon and the maximum β^- kinetic energies of isotopes typical of clinical nuclear medicine interest are below a couple of MeV.

[27]There are two classical arguments that will allow electron-electron *bremsstrahlung* to be neglected. On the simplest level, in the dipole approximation, the energy radiated away by an accelerated charged particle is proportional to the dipole moment. As the dipole moment is also proportional to the center-of-mass (which is stationary of particles of identical mass), our first approximation is that electron-electron *bremsstrahlung* will be zero. One can also think of the accelerations of an electron projectile and electron target resulting in electromagnetic radiations of equal magnitude but opposite phase resulting in cancellation.

7.6.2 Classical Electron-Atom Bremsstrahlung Theory

7.6.2.1 Introduction

While the exact understanding of electron-atom *bremsstrahlung* requires a quantum-mechanical treatment, classical theory proves useful despite some fundamental differences. For example, classical theory demonstrates that a charged particle will radiate electromagnetic energy only when accelerated, but also states that this emission will occur at *any* time the particle is accelerated. On the other hand, the quantum-mechanical result shows that there can only be a finite probability that radiation occurs. Classical *bremsstrahlung* theory also fails to reproduce the cutoff of radiation at high frequency (corresponding to the full stopping of the moving charged particle and the complete transfer of its kinetic energy to radiation, neglecting nuclear recoil). In other words, in the classical theory the Fourier transforms of the time-dependent field strengths extend to infinite frequency. Nevertheless, a review of the classical *bremsstrahlung* theory is a useful foundation to the full quantum-mechanical development of the phenomenon.

7.6.2.2 Liénhard–Wiechert Retarded Potentials

In order to demonstrate that electromagnetic energy is radiated by an electron only when accelerated, one begins with the derivation of the Liénhard–Wiechert retarded potentials. Consider an electron moving *in vacuo* along the trajectory $\mathbf{r}(t)$ parametric in time t as shown in Fig. 7.39. It is desired to determine the electromagnetic field at point P (with position vector $\mathbf{x}$) associated with the electron at time t. At this time t, the electron will be at point A with position vector $\mathbf{r}(t)$. However, due to the finite propagation time of the radiation, the field at P at time t will be that due to the radiation emitted at the earlier time t′ when the electron was at point A′ with position vector $\mathbf{r}(t')$. The time taken for the radiation to travel from point A to point P is equal to $|\mathbf{x} - \mathbf{r}(t')|/c$ and the retarded time is the difference between this time and that when the radiation is observed at point P,

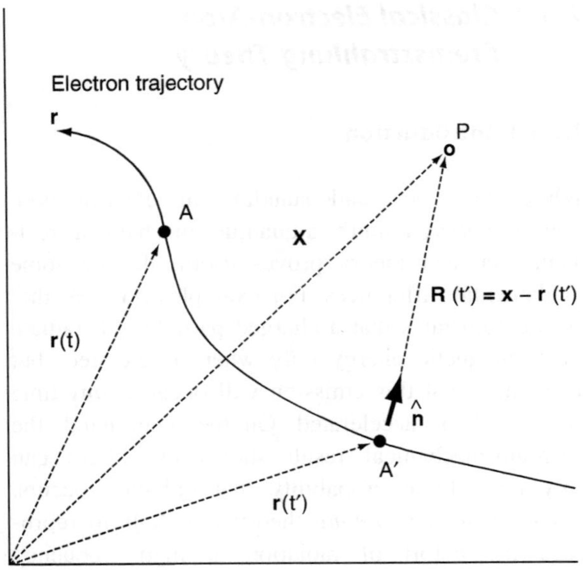

Fig. 7.39 Trajectory of a moving electron for calculation of the Liénhard–Wiechert retarded potentials

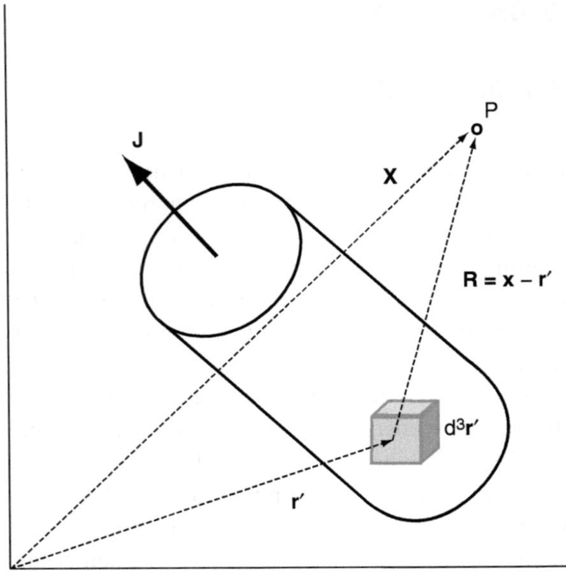

Fig. 7.40 Replacement of the single moving electron with an electric current density

$$\begin{aligned} t' &= t - \frac{|\mathbf{x} - \mathbf{r}(t')|}{c} \\ &\equiv t - \frac{|\mathbf{R}(t')|}{c} \\ &= t - \frac{R(t')}{c} \end{aligned} \tag{7.554}$$

where the vector $\mathbf{R}(t')$ has been defined. The unit vector directed along $\mathbf{R}(t')$ is

$$\hat{\mathbf{n}}(t') = \frac{\mathbf{R}(t')}{R(t')}. \tag{7.555}$$

To enable the calculation of the electromagnetic field at P, return to first principles, neglect retardation and replace the single moving electron with a uniformly moving current density as shown in Fig. 7.40. The resulting vector and scalar potentials at P will be,

$$\mathbf{A}(\mathbf{x}) = \frac{\mu_0}{4\pi}\int d^3\mathbf{r}' \frac{\mathbf{J}(\mathbf{r}')}{R} \tag{7.556}$$

and

$$\Phi(\mathbf{r}) = \frac{1}{4\pi\varepsilon_0}\int d^3\mathbf{r}' \frac{\rho(\mathbf{r}')}{R} \tag{7.557}$$

where $\mathbf{J}(\mathbf{r}')$ and $\boldsymbol{\rho}(\mathbf{r}')$ are the current and charge densities, respectively, and the integration is over a small volume element as shown. If the current and charge are allowed to vary with time and a finite propagation time is accounted for, these potentials can be rewritten in the forms,

$$\mathbf{A}(\mathbf{x}, t) = \frac{\mu_0}{4\pi}\int d^3\mathbf{r} \frac{\mathbf{J}\left(\mathbf{r}, t - \frac{R}{c}\right)}{R} \tag{7.558}$$

$$\Phi(\mathbf{x}, t) = \frac{1}{4\pi\varepsilon_0}\int d^3\mathbf{r} \frac{\rho\left(\mathbf{r}, t - \frac{R}{c}\right)}{R} \tag{7.559}$$

Having derived these potentials, return now to the example of a moving electron for which the potentials are evaluated following the approach of Feynman et al. (1963). First, the charge distribution is replaced with a cube of side dimension L moving towards the point P (i.e., $\hat{\mathbf{n}} = \boldsymbol{\beta}$) and it assumed that $L \ll R$. The integral of (7.559) is replaced by the finite summation,

$$\int d^3\mathbf{r}_0 \frac{\rho\left(\mathbf{r}_0, t - \frac{R}{c}\right)}{R} \cong \rho L^2 \Delta L \sum_{i=1}^{N} \frac{1}{r_i} \tag{7.560}$$

where the cube has been divided into N elements of thickness ΔL and r_i is the distance from the ith element to R. This summation is,

$$\begin{aligned} \rho L^2 \Delta L \sum_{i=1}^{N} \frac{1}{r_i} &= \rho L^2 \Delta L \sum_{i=1}^{N} \frac{1}{r_{eff}} \\ &= \frac{\rho L^3}{r_{eff}} \frac{N\,\Delta L}{L} \\ &= \frac{\rho L^3}{r_{eff}} \frac{L_{eff}}{L} \end{aligned} \tag{7.561}$$

where r_{eff} is the effective distance from the volume elements to P and L_{eff} is the length of the volume traversed by the moving cube over a time Δt, or,

$$L_{eff} = \frac{L}{1-\beta}. \tag{7.562}$$

As ρL^3 is the total charge (taken to be equal to e for an electron),

$$\begin{aligned}\frac{1}{4\pi\,\varepsilon_0}\int d^3\mathbf{r}_0 \frac{\rho(\mathbf{r}_0, t - R/c)}{R} \\ = \frac{e}{4\pi\,\varepsilon_0}\frac{1}{R(t')(1-\beta(t'))}.\end{aligned} \tag{7.563}$$

Hence, the potentials can be written as,

$$\Phi(\mathbf{x}, t) = \frac{e}{4\pi\varepsilon_0}\left\{\frac{1}{R\left(1 - \hat{\mathbf{n}} \bullet \hat{\boldsymbol{\beta}}\right)}\right\}_{t'} \tag{7.564}$$

and

$$\mathbf{A}(\mathbf{x}, t) = \frac{e}{4\pi\varepsilon_0 c}\left\{\frac{\hat{\boldsymbol{\beta}}}{R\left(1 - \hat{\mathbf{n}} \bullet \hat{\boldsymbol{\beta}}\right)}\right\}_{t'}. \tag{7.565}$$

Note that the $\mathbf{x}$ and t dependencies implicitly arise through the definition of the retarded time and that the use of brackets with the subscript t has been introduced in order to signify that the quantities within the brackets are to be evaluated at the retarded time, t'. $\hat{\boldsymbol{\beta}}$ is the electron velocity normalized to the speed of light (as $\beta \leq 1$, the normalized velocity is written as a unit vector for convenience) and the scalar product $\hat{\mathbf{n}} \bullet \hat{\boldsymbol{\beta}}$ is that component of the normalized electron velocity directed towards the detection point, P. Equations (7.564) and (7.565) are the Liénhard–Wiechert retarded potentials which are used to evaluate the radiation field at point P. However, before doing so, that derivation is anticipated by highlighting two features of the results provided of the vector and scalar potentials:

- As both potentials decrease as $1/R$, the resulting fields otherwise fall as $1/R^2$ leading to a net zero electromagnetic energy flow flux for $R \to \infty$. However, recalling that the retarded time has an explicit R-dependence, this leads to a net $1/R$ dependence upon differentiation. Hence, retardation is necessary in order to allow for the radiation of electromagnetic energy at a distance.
- The $\left(1 - \hat{\mathbf{n}} \bullet \hat{\boldsymbol{\beta}}\right)$ term in the denominators of the expressions of the potentials predicts "geometric beaming" as $\beta \to 1$ with the field reaching a maximum along the direction of travel at high electron speeds.

7.6.2.3 Radiation Emission

Electromagnetic Fields at a Distance

The electric field strength is calculated from the potentials through $\mathbf{E} = -\nabla\Phi - \partial\mathbf{A}/\partial t$ (Jackson 1999). The differential with respect to the time t is found by differentiating the expression for $R(t')$,

$$\begin{aligned}R(t') &= \sqrt{(\mathbf{x} - \mathbf{r}(t')) \bullet (\mathbf{x} - \mathbf{r}(t'))} \\ &= c(t - t')\end{aligned} \tag{7.566}$$

to give,

$$\frac{\partial R(t')}{\partial t} = c\left(1 - \frac{\partial t'}{\partial t}\right). \tag{7.567}$$

Applying the chain rule,

$$\begin{aligned}\frac{\partial R(t')}{\partial t} &= \frac{\partial R(t')}{\partial t'}\frac{\partial t'}{\partial t} \\ &= -\frac{\partial r(t')}{\partial t'}\frac{\partial t'}{\partial t} \\ &= -c\left\{\hat{\mathbf{n}} \bullet \hat{\boldsymbol{\beta}}\right\}_{t'}\frac{\partial t'}{\partial t}.\end{aligned} \tag{7.568}$$

Equating these two results gives,

$$c\left(1 - \frac{\partial t'}{\partial t}\right) = -c\left\{\hat{\mathbf{n}} \bullet \hat{\boldsymbol{\beta}}\right\}_{t'}\frac{\partial t'}{\partial t}$$

leading to,

$$\frac{\partial t'}{\partial t} = \left\{\frac{1}{1 - \hat{\mathbf{n}} \bullet \hat{\boldsymbol{\beta}}}\right\}_{t'} \tag{7.569}$$

which is then used to write the differential with respect to t,

$$\begin{aligned}\frac{\partial}{\partial t} &= \frac{\partial t'}{\partial t}\frac{\partial}{\partial t'} \\ &= \left\{\frac{1}{1 - \hat{\mathbf{n}} \bullet \hat{\boldsymbol{\beta}}}\right\}_{t'}\frac{\partial}{\partial t'}.\end{aligned} \tag{7.570}$$

The gradient operator ∇ is evaluated by first separating it into spatial and time components,

$$\nabla = \nabla_x + \nabla_{t'} \tag{7.571}$$

where the first component refers to differentiation with respect to the observation point position vector (disregarding any retardation effects) and the second term accounts for retardation. This latter term is $\nabla_{t'} = \nabla t' \frac{\partial}{\partial t'}$ where,

$$\begin{aligned}\nabla t' &= -\frac{1}{c}\nabla R(t') \\ &= -\frac{1}{c}\left(\hat{\mathbf{n}} + \frac{\partial R(t')}{\partial t'}\nabla t'\right) \\ &= -\frac{1}{c}\left(\hat{\mathbf{n}} - \left(\hat{\mathbf{n}} \bullet \hat{\boldsymbol{\beta}} c\right)\nabla t'\right).\end{aligned} \tag{7.572}$$

Solving for $\nabla t'$ gives,

$$\nabla t' = -\frac{1}{c}\left\{\frac{1}{1 - \hat{\mathbf{n}} \bullet \hat{\boldsymbol{\beta}}}\right\}_{t'} \tag{7.573}$$

which leads to the expression for the gradient operator,

$$\nabla = \nabla_x - \frac{1}{c}\left\{\frac{\hat{\mathbf{n}}}{1 - \hat{\mathbf{n}} \bullet \hat{\boldsymbol{\beta}}}\right\}_{t'}\frac{\partial}{\partial t'}. \tag{7.574}$$

Having established these operators, one can now evaluate the electric and magnetic fields,

$$\begin{aligned}\mathbf{E}(\mathbf{x},t) &= -\nabla\Phi - \partial\mathbf{A}/\partial t \\ &= -\left(\nabla_x - \frac{1}{c}\left\{\frac{\hat{\mathbf{n}}}{1 - \hat{\mathbf{n}} \bullet \hat{\boldsymbol{\beta}}}\right\}_{t'}\frac{\partial}{\partial t'}\right)\Phi(\mathbf{x},t) \\ &\quad - \left\{\frac{1}{1 - \hat{\mathbf{n}} \bullet \hat{\boldsymbol{\beta}}}\right\}_{t'}\frac{\partial \mathbf{A}(\mathbf{x},t)}{\partial t'} \\ &= -\nabla_x\Phi(\mathbf{x},t) + \frac{1}{c}\left\{\frac{\hat{\mathbf{n}}}{1 - \hat{\mathbf{n}} \bullet \hat{\boldsymbol{\beta}}}\right\}_{t'}\frac{\partial\Phi(\mathbf{x},t)}{\partial t'} \\ &\quad - \left\{\frac{1}{1 - \hat{\mathbf{n}} \bullet \hat{\boldsymbol{\beta}}}\right\}_{t'}\frac{\partial \mathbf{A}(\mathbf{x},t)}{\partial t'} \\ &= \frac{e}{4\pi\varepsilon_0}\left(-\nabla_x\left\{\frac{1}{R}\frac{1}{\left(1 - \hat{\mathbf{n}} \bullet \hat{\boldsymbol{\beta}}\right)}\right\}_{t'}\right. \\ &\quad + \frac{1}{c}\left\{\frac{\hat{\mathbf{n}}}{1 - \hat{\mathbf{n}} \bullet \hat{\boldsymbol{\beta}}}\right\}_{t'}\frac{\partial}{\partial t'}\left\{\frac{1}{R}\frac{1}{\left(1 - \hat{\mathbf{n}} \bullet \hat{\boldsymbol{\beta}}\right)}\right\}_{t'} \\ &\quad \left. - \frac{1}{c}\left\{\frac{1}{1 - \hat{\mathbf{n}} \bullet \hat{\boldsymbol{\beta}}}\right\}_{t'}\frac{\partial}{\partial t'}\left\{\frac{1}{R}\frac{\hat{\boldsymbol{\beta}}}{\left(1 - \hat{\mathbf{n}} \bullet \hat{\boldsymbol{\beta}}\right)}\right\}_{t'}\right) \\ &= \frac{e}{4\pi\varepsilon_0}\left\{\frac{\left(1-\beta^2\right)\left(\hat{\mathbf{n}} - \hat{\boldsymbol{\beta}}\right)}{\left(1 - \hat{\mathbf{n}} \bullet \hat{\boldsymbol{\beta}}\right)^3 R^2} + \frac{1}{c}\frac{\hat{\mathbf{n}} \times \left(\left(\hat{\mathbf{n}} - \hat{\boldsymbol{\beta}}\right) \times \dot{\hat{\boldsymbol{\beta}}}\right)}{\left(1 - \hat{\mathbf{n}} \bullet \hat{\boldsymbol{\beta}}\right)^3 R}\right\}_{t'}\end{aligned} \tag{7.575}$$

where, in order to simplify the expression, an overlying dot is used to indicate differentiation with time. The magnetic field strength follows from $\mathbf{B} = \nabla \times \mathbf{A}$.

Consider the above expression for the electric field strength. The first term on the right-hand side is proportional to $1/R^2$ and the particle's velocity and is the Coulomb field for a uniformly moving electric charge. As the resulting radiated power is proportional to $|\mathbf{E}|^2$, the power will drop off as $1/R$ and can be neglected as the energy flow per unit area will simply go to zero at infinity as a result. On the other hand, the second term is proportional to $1/R$ and the particle's acceleration. Because of the latter feature, the energy flow per unit area will thus remain finite as $R \to \infty$. This second term is the radiation field of an accelerating electric charge (which only arises because of retardation). Having recognized this, isolate the electric and magnetic radiation fields accordingly,

$$\mathbf{E}_{rad}(\mathbf{x},t) = \frac{e}{4\pi\varepsilon_0 c}\left\{\frac{\hat{\mathbf{n}} \times \left(\left(\hat{\mathbf{n}} - \hat{\boldsymbol{\beta}}\right) \times \dot{\hat{\boldsymbol{\beta}}}\right)}{\left(1 - \hat{\mathbf{n}} \bullet \hat{\boldsymbol{\beta}}\right)^3 R}\right\}_{t'} \tag{7.576}$$

$$\mathbf{B}_{rad}(\mathbf{x},t) = \frac{\hat{\mathbf{n}} \times \mathbf{E}_{rad}(\mathbf{x},t)}{c} \tag{7.577}$$

Radiated Power: Larmor Formula

The radiated power is calculated using the Poynting vector in the nonrelativistic case (i.e., $\beta \ll 1$) in which

$(\hat{\mathbf{n}}-\hat{\boldsymbol{\beta}}) \to \hat{\mathbf{n}}$ and $(1-\hat{\mathbf{n}}\bullet\hat{\boldsymbol{\beta}})^3 \to 1$. The radiative electric field reduces to the simpler form,

$$\mathbf{E}_{rad}(\mathbf{x},t) = \frac{e}{4\pi\varepsilon_0 c}\left\{\frac{\hat{\mathbf{n}}\times\left(\hat{\mathbf{n}}\times\dot{\boldsymbol{\beta}}\right)}{R}\right\}_{t'}; \quad \beta \ll 1. \tag{7.578}$$

The Poynting vector is,[28]

$$\begin{aligned}\mathbf{P}_{rad} &= \mathbf{E}_{rad}\times\mathbf{H}^*_{rad} \\ &= \frac{1}{\mu_0}\mathbf{E}_{rad}\times\mathbf{B}^*_{rad}\end{aligned} \tag{7.579}$$

From (7.576) and (7.577),

$$|\mathbf{E}_{rad}(\mathbf{x},t)| = \frac{e}{4\pi\varepsilon_0 c}\left\{\frac{\dot{\beta}}{R}\sin\theta\right\}_{t'} \tag{7.580}$$

and

$$|\mathbf{B}_{rad}(\mathbf{x},t)| = \frac{e}{4\pi\varepsilon_0 c^2}\left\{\frac{\dot{\beta}}{R}\sin\theta\right\}_{t'} \tag{7.581}$$

where θ is the angle between the unit vectors $\hat{\mathbf{n}}$ and $\dot{\hat{\boldsymbol{\beta}}}$. Inserting these results into the expression of the Poynting vector provides the magnitude of the radiated power,

$$\begin{aligned}P_{rad} &= \frac{e^2}{16\pi^2\varepsilon_0^2\mu_0 c^3}\left\{\frac{\dot{\beta}^2}{R^2}\sin^2\theta\right\}_{t'} \\ &= \frac{\alpha\hbar}{4\pi}\left\{\frac{\dot{\beta}^2}{R^2}\sin^2\theta\right\}_{t'}\end{aligned} \tag{7.582}$$

where $1/\varepsilon_0\mu_0 = c^2$ has been used. This angular distribution of (7.582) is clearly that of dipole radiation. Energy is radiated with a maximum orthogonal to the direction of travel and with none directed along the acceleration vector. This pattern will, however, alter in the laboratory frame-of-reference due to the Lorentz transformation as the electron becomes relativistic. As the magnitude of the Poynting vector is the electromagnetic energy radiated per unit time and per unit area,

$$|\mathbf{P}_{rad}| = \frac{d^2E}{R^2 dt\, d\Omega} \tag{7.583}$$

one can calculate the total instantaneous radiated power from the accelerated charge,

$$\begin{aligned}\frac{dE}{dt} &= \int d\Omega\, R^2|\mathbf{P}_{rad}| \\ &= \frac{\alpha\hbar}{2}\left\{\dot{\beta}^2\right\}_{t'}\int_0^\pi d\theta\sin^3\theta \\ &= \frac{2}{3}\alpha\hbar\left\{\dot{\beta}^2\right\}_{t'}.\end{aligned} \tag{7.584}$$

This result is the Larmor formula for a nonelativistic accelerated electron (with explicit recognition that the acceleration is that of the electron at the time of emission). Note that the radiated power is proportional to the square of the particle's acceleration at the time of emission.

Classical Radiative Stopping Power

A classical expression of the radiative stopping power (i.e., the energy loss due to *bremsstrahlung* per unit length traveled) can now be derived. Assume that the acceleration is due to a deflection of the electron at an impact parameter b by a nucleus with charge Ze and that the duration of this deflection, τ_{Defl}, is short. The acceleration will be given by the ratio of the Coulomb force experienced by the electron and its mass,

$$\dot{\beta} = \left(\frac{\alpha\hbar c}{m_e}\right)\frac{Zc}{b^2}. \tag{7.585}$$

Inserting this expression for the normalized acceleration into the Larmor formula gives the radiated power,

$$\begin{aligned}\frac{dE}{dt} &= \frac{2}{3}\alpha\hbar\left\{\dot{\beta}^2\right\}_{t'} \\ &= \frac{2}{3}\alpha\hbar\left(\frac{\alpha\hbar c}{m_e}\right)^2\frac{(Zc)^2}{b^4} \\ &= \frac{2}{3}(\alpha\hbar c)^3\frac{Z^2c}{m_e^2b^4}.\end{aligned} \tag{7.586}$$

[28] As only the instantaneous power flow is being considered, the 1/2 multiplicative factor is excluded.

As the duration of the interaction is approximately

$$\tau_{Coll} \approx \frac{b}{\beta c} \tag{7.587}$$

then the total radiated energy as a result of the deflection is,

$$\begin{aligned}\Delta E &\approx \frac{dE}{dt}\tau_{Coll}\\ &\approx \frac{2}{3}(\alpha\hbar c)^3 \frac{Z^2}{\beta m_e^2 b^3}.\end{aligned} \tag{7.588}$$

This is the energy loss for a single interaction with a nucleus. To calculate the radiative stopping power, it is necessary to account for all of the nuclei that the electron may interact with,

$$\begin{aligned}\frac{dE}{\rho\, dx} &= \frac{2\pi}{3}(\alpha\hbar c)^3 \frac{N_A Z^2}{A\,\beta\, m_e^2}\int_{b_{min}}^{b_{max}} \frac{db}{b^2}\\ &= \frac{4\pi}{3}(\alpha\hbar c)^3 \frac{N_A Z^2}{A\,\beta\, m_e^2}\left(\frac{1}{b_{max}} - \frac{1}{b_{min}}\right).\end{aligned} \tag{7.589}$$

Using the expressions for the minimum impact parameter and the constant C of derived previously,

$$\frac{dE}{\rho\, dx} = \frac{2\alpha C}{3}\frac{Z^2}{A\sqrt{1-\beta^2}} \tag{7.590}$$

This predicts that the mass radiative stopping power increases to infinity with increasing electron speed and increases with atomic number.

Angular Distribution of Radiation Emission

In evaluating the angular distribution and the frequency spectrum of the radiation emission, the approach described by Jackson (1999) will be approximately followed, but with the use of SI units retained. Recall that the derivations of the nonrelativistic Larmor formula and subsequent radiative mass stopping power were nonrelativistic and, hence, based upon the assumption that $\beta \ll 1$. As a result, only the acceleration $\dot{\beta}$ appeared in the final expression of the radiated power and the angular distribution was that of a dipole, $\sin^2\theta$. As the electron's speed increases, it can no longer be ignored (in particular that of the effect of "beaming" caused by the $\left(1 - \hat{\mathbf{n}} \bullet \hat{\boldsymbol{\beta}}\right)^3$ term in the denominator). To evaluate the angular distribution, recalculate the magnitude of the radial component of the Poynting vector (i.e., along the direction of $\hat{\mathbf{n}}$) without using the previous nonrelativistic approximation,

$$\begin{aligned}\mathbf{P}_{rad} \bullet \hat{\mathbf{n}} &= \frac{1}{\mu_0}\left(\mathbf{E}_{rad} \times \mathbf{B}^*_{rad}\right) \bullet \hat{\mathbf{n}}\\ &= \frac{1}{\mu_0}\left(\mathbf{E}_{rad} \times \left(\hat{\mathbf{n}} \times \mathbf{E}^*_{rad}\right)\right) \bullet \hat{\mathbf{n}}\\ &= \frac{e^2}{16\pi^2\varepsilon_0 c}\left\{\frac{1}{R^2}\left|\frac{\hat{\mathbf{n}} \times \left(\left(\hat{\mathbf{n}} - \hat{\boldsymbol{\beta}}\right) \times \dot{\hat{\boldsymbol{\beta}}}\right)}{\left(1 - \hat{\mathbf{n}} \bullet \hat{\boldsymbol{\beta}}\right)^3}\right|^2\right\}_{t'}.\end{aligned} \tag{7.591}$$

This scalar product is the detected power per unit area at a distant point at time t of radiation that had been emitted by the electron at the earlier time t′. Assume that the acceleration of the electron at time t′ resulting in the emission of radiation was due to an interaction with a nucleus of atomic number Z over a short finite time interval, τ_{Coll}, and further approximate $\hat{\boldsymbol{\beta}}$ and $\dot{\hat{\boldsymbol{\beta}}}$ as both being constant in magnitude and direction. The energy radiated during a finite time interval from $t' = 0$ to $t' = \tau_{Coll}$ will be,

$$\begin{aligned}E &= \int_{(R(t=0)/c)}^{\tau_{Coll}+(R(\tau_{Coll})/c)} dt\, \mathbf{P}_{rad} \bullet \hat{\mathbf{n}}\\ &= \int_0^{\tau_{Coll}} dt' \frac{dt}{dt'}\mathbf{P}_{rad} \bullet \hat{\mathbf{n}}\end{aligned} \tag{7.592}$$

The quantity $\frac{dt}{dt'}\left(\mathbf{P}_{rad} \bullet \hat{\mathbf{n}}\right)$ is the radiated power per unit area in terms of the electron's time. This can be used to give the power radiated per unit solid angle as,

$$\begin{aligned}\frac{dE}{dt'\, d\Omega} &= R^2\left(\mathbf{P}_{rad} \bullet \hat{\mathbf{n}}\right)\frac{dt}{dt'}\\ &= R^2\left(\mathbf{P}_{rad} \bullet \hat{\mathbf{n}}\right)\left\{1 - \hat{\mathbf{n}} \bullet \hat{\boldsymbol{\beta}}\right\}_{t'}.\end{aligned} \tag{7.593}$$

Substituting the expression for $\mathbf{P}_{rad} \bullet \hat{\mathbf{n}}$,

$$\frac{dE}{dt'd\Omega} = \frac{e^2}{16\pi^2\varepsilon_0 c}\left\{\left|\frac{\hat{\mathbf{n}}\times\left(\left(\hat{\mathbf{n}}-\hat{\boldsymbol{\beta}}\right)\times\dot{\hat{\boldsymbol{\beta}}}\right)}{\left(1-\hat{\mathbf{n}}\bullet\hat{\boldsymbol{\beta}}\right)^3}\right|^2\left(1-\hat{\mathbf{n}}\bullet\hat{\boldsymbol{\beta}}\right)\right\}_{t'}$$
$$= \frac{e^2}{16\pi^2\varepsilon_0 c}\left\{\frac{\left|\hat{\mathbf{n}}\times\left(\left(\hat{\mathbf{n}}-\hat{\boldsymbol{\beta}}\right)\times\dot{\hat{\boldsymbol{\beta}}}\right)\right|^2}{\left(1-\hat{\mathbf{n}}\bullet\hat{\boldsymbol{\beta}}\right)^5}\right\}_{t'}. \tag{7.594}$$

This provides the angular distribution of the radiation emission provided that $\hat{\mathbf{n}}$ and R are reasonably constant (a requirement equivalent to the measurement point being at a sufficiently large distance from the electron). Considering the case of $\hat{\boldsymbol{\beta}}$ and $\dot{\hat{\boldsymbol{\beta}}}$ being collinear and defining θ as the angle between the direction of $\hat{\boldsymbol{\beta}}$ and $\dot{\hat{\boldsymbol{\beta}}}$ and the direction of emission, then (7.594) reduces to,

$$\frac{dE}{dt'\,d\Omega} = \frac{\alpha\hbar c}{4\pi}\dot{\beta}^2\frac{\sin^2\theta}{(1-\beta\cos\theta)^5} \tag{7.595}$$

For $\beta \ll 1$, this result returns the $\sin^2\theta$ dependence of the nonrelativistic Larmor formula but, because of the $(1-\beta\cos\theta)^5$ term, the angular distribution becomes highly forward peaked at high electron speeds. The angle at which the radiation is at a maximum, θ_{max} is,

$$\theta_{max} = \cos^{-1}\frac{\sqrt{1+15\beta^2}-1}{3\beta} \tag{7.596}$$

This maximum emission angle is shown as a function of the electron β in Fig. 7.41. At low electron energies, the maximum emission angle is near to 90° (reflecting the dipole radiation pattern in the laboratory reference frame) but the radiation becomes focused into a cone of decreasing angle with increasing β.

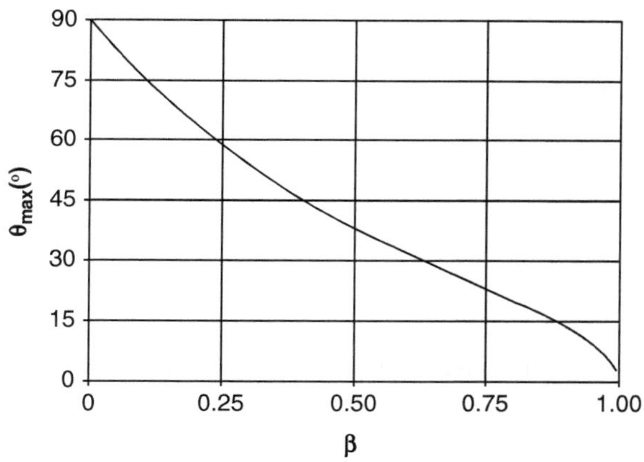

Fig. 7.41 Angle of maximum *bremsstrahlung* for an electron as a function of β

Spectrum of Radiation Emission

Nonrelativistic Case

The frequency (energy) spectrum of the radiated energy of classical *bremsstrahlung* is now calculated in the case of a nonrelativistic incident electron. Here, the shape of the energy spectrum is taken to reflect the shape of the impulse of the electron deflection. If one assumes that the impulse has a duration given by, $\tau_{Coll} = b/\beta c$, then the corresponding frequency spectrum can be approximated as being uniform up to an angular frequency of,

$$\omega \approx \frac{1}{\tau_{Coll}} \approx \frac{\beta c}{b}. \tag{7.597}$$

From this,

$$\frac{dE}{d\omega} \approx \Delta E\,\tau_{Coll}$$
$$\approx \frac{2}{3c}(\alpha\hbar c)^3\frac{Z^2}{\beta^2 m_e^2 b^2}. \tag{7.598}$$

The energy radiated per unit frequency and unit areal density is obtained in the usual manner,

$$\left(\frac{d^2E}{\rho\,dx\,d\omega}\right)_{Rad} = \frac{2}{3c}\left(\frac{N_A}{A}\right)(\alpha\hbar c)^3\frac{Z^2}{m_e^2\beta^2}\left|\int_{b_{min}}^{b_{max}} db\,2\pi b\frac{1}{b^2}\right|$$
$$= \frac{4\pi}{3c}\left(\frac{N_A}{A}\right)(\alpha\hbar c)^3\frac{Z^2}{m_e^2\beta^2}\left|\int_{b_{min}}^{b_{max}}\frac{db}{b}\right|$$
$$= \frac{2}{3}C\left(\frac{Z^2}{A}\right)\frac{r_0}{\beta^2 c}\ln\frac{b_{max}}{b_{min}}. \tag{7.599}$$

To avoid the divergence problem, the lower limit of the impact parameterization is modeled by,

$$b_{min} = \frac{\hbar c}{m_e\beta}$$

and the upper limit can be taken to be the ratio of the incident electron speed to the cutoff frequency (which is necessary as the electron cannot emit infinite energy). The ratio of impact parameters is,

$$\frac{b_{max}}{b_{min}} = \frac{\beta c}{\omega}\frac{m_e \beta}{\hbar c} = \frac{\beta^2 m_e}{\hbar\omega}.$$

For convenience, this will be written as,

$$\frac{b_{max}}{b_{min}} = \frac{m_e v_0^2}{\hbar\omega}$$

where v_0 is the incident electron speed and the electron rest mass is now given in units of mass rather than energy. This gives the expression for the radiated energy spectrum per unit length traversed as,

$$\left(\frac{d^2E}{\rho\, dx\, d\omega}\right)_{Rad} = \frac{2}{3}C\left(\frac{Z^2}{A}\right)\frac{r_0 c}{v_0^2}\ln\frac{m_e v_0^2}{\hbar\omega}. \qquad (7.600)$$

However, as the electron does lose kinetic energy as a result of the radiative collision, it would be more appropriate to replace the v_0^2 term in the logarithm with the square of the mean of the pre- and postdeflection speeds,

$$\frac{1}{2}\left(\sqrt{\frac{2T_0}{m_e}} + \sqrt{\frac{2(T_0 - \hbar\omega)}{m_e}}\right),$$

where T_0 is the incident electron kinetic energy and $\hbar\omega$ is the energy of the bremsstrahlung photon which will be now denoted by the usual symbol, k. To change the differential from frequency to photon energy, divide through the above result with the reduced Planck's constant. Thus, one obtains a classical result for the energy spectrum (the use of $\hbar$ is only for convenience here and does not imply a quantum-mechanical basis to the result),

$$\left(\frac{d^2E}{\rho\, dx\, dk}\right)_{Rad} = \frac{C}{2}\left(\frac{Z^2}{A}\right)\frac{r_0 c}{\hbar v_0^2} \times \ln\left(\frac{\left(\sqrt{T_0} + \sqrt{T_0 - k}\right)^2}{2k}\right) \quad k \ll T_0 \qquad (7.601)$$

It will be seen later that this result is markedly similar to the quantum-mechanical derivation of the Bethe–Heitler theory. Figure (7.42) shows the bremsstrahlung spectrum for 100 keV electrons in lead calculated from (7.601).

Relativistic Case: Weizsäcker–Williams (Virtual Quanta) Method

This is a semiclassical approach to calculating electron-nucleus *bremsstrahlung* performed in the reference frame of the moving electron. The nuclear electromagnetic field is thus experienced by the electron as a pulse, or virtual photon, which is Thomson

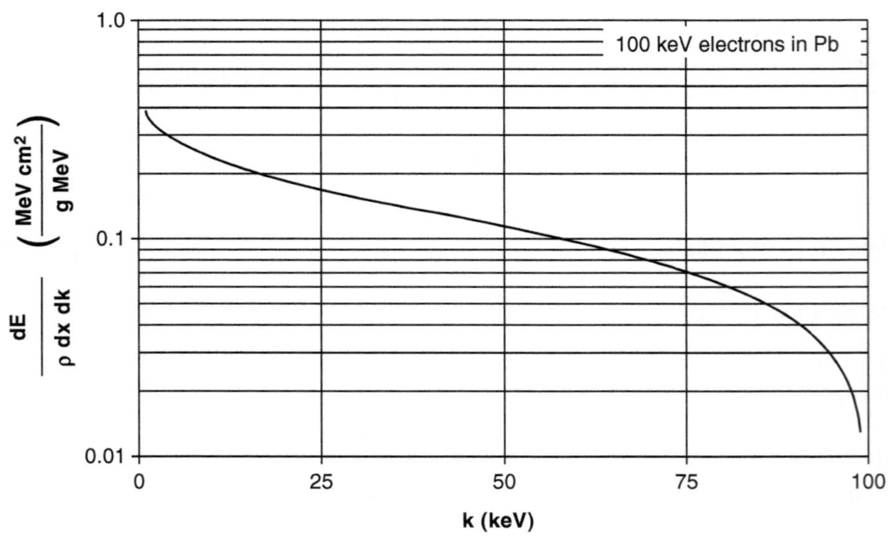

Fig. 7.42 *Bremsstrahlung* spectrum calculated from classical theory for 100 keV electrons in lead

scattered by the electron. This innovative approach is originally attributable to the independent work of von Weizsäcker and Williams, following Fermi.

In the reference frame of the electron, the nucleus with charge Ze is the "projectile." The route of the Bohr soft collision stopping power calculation is followed; recall that only the component of the nuclear electric field at the position of the electron perpendicular to the trajectory need be calculated. This is given by (7.37) except that z is replaced by Z. Thus, in this reference frame, it is the nucleus passing a stationary electron at an impact parameter b, speed v, and relativistic factor γ. The energy transported per unit area and unit frequency is given by the Poynting vector,

$$|\mathbf{P}| = \frac{dI}{dA\,d\omega} = \sqrt{\frac{\varepsilon_0}{\mu_0}}|\mathbf{E}_\perp(t)|^2 \tag{7.602}$$

Parseval's theorem is invoked,

$$\int_{-\infty}^{\infty} dt|\mathbf{E}_\perp(t)|^2 = \int_{-\infty}^{\infty} d\omega|\mathbf{E}_\perp(\omega)|^2 = 2\int_0^{\infty} d\omega|\mathbf{E}_\perp(\omega)|^2 \tag{7.603}$$

where a real electric field, $E_\perp(-w) = E_\perp^*(w)$, has been allowed for. As a result,

$$\frac{dI}{dA\,d\omega} = 2\sqrt{\frac{\varepsilon_0}{\mu_0}}|\mathbf{E}_\perp(\omega)|^2. \tag{7.604}$$

The Fourier transform of the electric field component is given by (7.90),

$$\frac{dI}{dA\,d\omega} = \alpha\hbar c^2\left(\frac{Z\omega}{\pi\gamma v^2}\right)^2 K_1^2\left(\frac{\omega b}{\gamma v}\right). \tag{7.605}$$

This is the electromagnetic energy per unit area and per unit frequency incident to the electron in its reference frame, which represents a virtual photon that can be scattered by the electron to create a *bremsstrahlung* photon. It will be assumed that this scatter is through the classical Thomson elastic process. The energy spectrum in the electron rest frame is,

$$\frac{dI}{d\omega} = \sigma_{Tho}\frac{dI}{dA\,d\omega}. \tag{7.606}$$

As the impact parameter will vary from a minimum value to infinity, note the result of (7.605) in low- and large-argument cases of the modified Bessel function,

$$\frac{dI}{dA\,d\omega} = \alpha\hbar c^2\left(\frac{Z}{\pi\,vb}\right)^2 \quad \text{for } \frac{\omega b}{\gamma v} \ll 1 \tag{7.607}$$

$$\frac{dI}{dA\,d\omega} = \frac{\alpha\hbar c^2}{2\pi}\frac{Z^2\omega}{\gamma b v^3}e^{-2\frac{\omega b}{\gamma v}} \quad \text{for } \frac{\omega b}{\gamma v} \gg 1. \tag{7.608}$$

The exponential cut-off of (7.608) allows a specification of a minimum impact parameter. On the basis of the cut-off, the maximum frequency can be approximated by $\omega_{max} \approx \gamma c/b$. As the scatter (in the electron's reference frame) is nonrelativistic, $\hbar\omega \ll m_e$, then $\gamma c/b \ll m_e/\hbar$ leading to a minimum impact parameter,

$$b_{min} = \frac{\gamma\hbar c}{m_e} \tag{7.609}$$

The power spectrum is then transformed to the laboratory reference frame in which the nucleus is at rest and the electron has a relativistic speed βc. The spectrum is the ratio of the energy to frequency and remains invariant as, for photons, the frequency and energy remain equivalent to within a factor of $\hbar$. Using the Doppler relativistic shift, the frequency in the laboratory frame is $\omega' \approx \gamma\omega$. Averaging over scatter angle (allowing for the isotropy of Thomson scatter) and approximating $\beta \approx 1$, the energy spectrum in the laboratory reference frame is,

$$\frac{dI'}{d\omega'} = \sigma_{Tho}\,\alpha\hbar c^2\left(\frac{Z\omega'}{\pi\gamma^2c^2}\right)^2 K_1^2\left(\frac{\omega' b}{\gamma^2 c}\right) \tag{7.610}$$

Following the mechanics of the derivation of the Bohr soft collision stopping power, the *bremsstrahlung* differential cross section in photon energy is,

$$\frac{d\sigma}{dk} = \frac{2\pi}{\hbar k}\int_{b_{min}}^{\infty} db\,b\frac{dI'}{d\omega'} = \frac{2\alpha\sigma_{Tho}}{\pi}\frac{Z^2}{k}\int_{x_{min}}^{\infty} dx\,x\,K_1^2(x) \tag{7.611}$$

where $x_{min} = \omega' b_{min}/\gamma^2 c$. This integral is solved for using the properties of the derivatives of the modified Bessel functions (as used in the Bohr soft collision stopping power derivation) and recalling the exponential cutoff of (7.608), the Weizsäcker–Williams *bremsstrahlung* spectrum is thus of the form,

$$\frac{d\sigma}{dk} \approx \frac{2\alpha\sigma_{Tho}}{\pi}\frac{Z^2}{k}\left(\ln\left(\frac{1.223\gamma m_e}{k}\right) - \frac{1}{2}\right) \quad (7.612)$$

It is immediately evident that this cross section will diverge as $k \to \infty$. This corresponds to the impact parameter $b \to 0$, but this divergence is not achieved as the screening of the nucleus by atomic electrons, thus reducing its effective charge seen by the electron, has been ignored in this derivation.

7.6.3 Quantum Electron-Nuclear Bremmstrahlung: Bethe–Heitler Theory

7.6.3.1 Introduction

As the Bethe–Heitler theory of *bremsstrahlung* is based upon the Born approximation, the extent of its validity will be defined by limits of this approximation. Refinements and further extensions of the theory are necessary in order to extend beyond these restrictions. The Bethe–Heitler theory was the first relativistic quantum description of *bremsstrahlung* and, whilst being cognizant of its limitations, it is presented here because of its historical importance and the experience gained in the use of the Born approximation throughout this book. Advanced approaches to providing more accurate calculations *bremsstrahlung* cross sections can be found in Haug and Nakel (2004).

7.6.3.2 Derivation of the Triple Differential Cross Section

Interaction

The interaction to be calculated for is shown by the Feynman diagrams of Fig. 7.43.

An electron with momentum **p** and total energy E is incident to an infinitely-massive point charge Ze, which approximates the nucleus. In the first diagram, the electron interacts with the static field to reach a momentum of $\mathbf{k} + \mathbf{p}'$ and then interacts with the radiation field to emit a photon of energy k and momentum

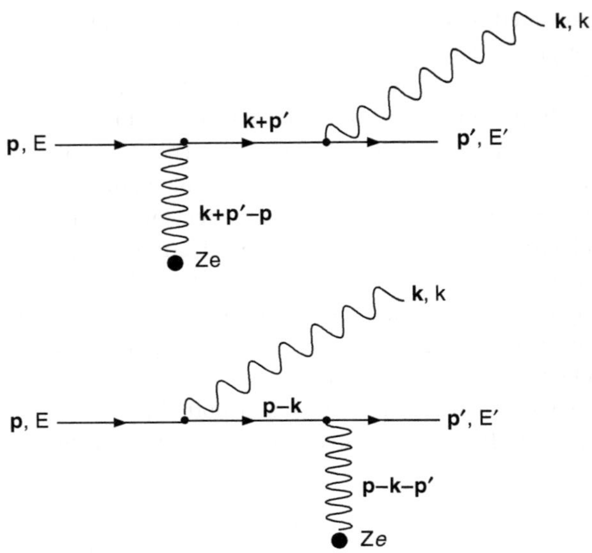

Fig. 7.43 Feynman diagrams for electron-nuclear *bremsstrahlung*

k and exit the interaction with momentum **p**′ and total energy E′. In the second diagram, the electron first interacts with the radiation field to emit the photon of energy k and momentum **k** and reduce its momentum to **p**′– **k**′ before interacting with the static field and exit the interaction with momentum **p** and total energy E. The aim is to calculate the differential cross section in photon energy k, solid angle $\Omega_{p'}$ of the scattered electron and solid angle Ω_k of the emitted photon. Heitler (1984) and Bjorken and Drell (1964) derive this cross section by separately accounting for the interactions with the radiation field and with the Coulomb field of the scattering center, the former using Fermi's Golden Rule No. 1 and the latter using S-matrix theory and the Feynman propagator. Haug and Nakel (2004) instead treat the interaction with the radiation field as the perturbation, to first order, and then correct the wavefunctions of the incident and scattered electrons for the effects of the Coulomb potential of the scattering center. In order to maintain some consistency with the previous derivations of the Klein–Nishina and elastic Coulomb scatter cross sections, the derivation of the Bethe–Heitler theory will follow the Bjorken and Drell S-matrix approach using Feynman propagators. The triple differential cross section is the ratio of the transition rate and the incident electron flux,

$$d^3\sigma = \frac{\lambda_{fi}}{\phi}. \quad (7.613)$$

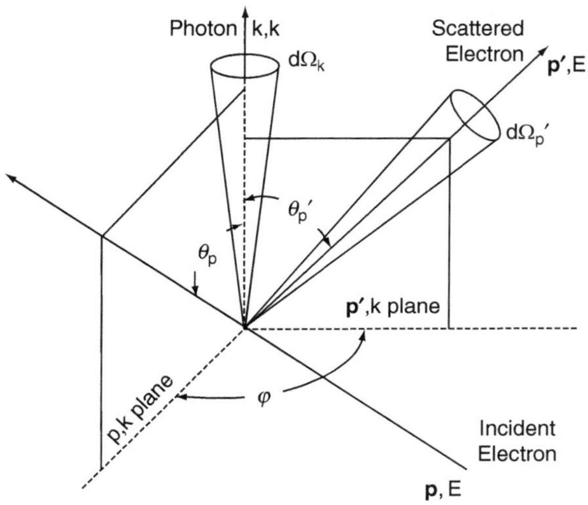

Fig. 7.44 Geometry of the *bremsstrahlung* process. The interaction occurs at the origin and angles are specified by the direction of the emitted photon

The transition rate is,

$$\lambda_{\mathrm{fi}} = \frac{|S_{\mathrm{fi}}|^2}{T}\rho_f \tag{7.614}$$

where T is the time duration of the interaction. The incident electron flux is,

$$\begin{aligned}\phi &= \frac{\beta c}{L^3}\\ &= \frac{pc}{EL^3}\end{aligned} \tag{7.615}$$

where L^3 is the usual volume used for normalization and ρ_f is the phase space factor, which is next calculated for.

Phase–Space Factor

The geometry of the *bremsstrahlung* process is shown in Fig. 7.44. The density of final states,

$$\rho_f\, dT' = \left(\frac{L}{2\pi\hbar c}\right)^6 d^3\mathbf{p}'\, d^3\mathbf{k} \tag{7.616}$$

where T' is the scattered electron's kinetic energy. Expanding,

$$\begin{aligned}\rho_f\, dT' &= \left(\frac{L}{2\pi\hbar c}\right)^6 p'^2\, dp'\, d\Omega_{p'}\, k^2\, dk\, d\Omega_k\\ &= \left(\frac{L}{2\pi\hbar c}\right)^6 p'\, E'\, dT'\, d\Omega_{p'}\, k^2\, dk\, d\Omega_k.\end{aligned} \tag{7.617}$$

The density of final states is,

$$\rho_f = \left(\frac{L}{2\pi\hbar c}\right)^6 p'\, E'\, k^2\, dk\, d\Omega_k d\Omega_{p'}. \tag{7.618}$$

S-Matrix Calculation

For the graphs of Fig. 7.43, the S-matrix element is,

$$\begin{aligned}S_{\mathrm{fi}} = e^2 \int d^4\boldsymbol{r}\, d^4\boldsymbol{\rho}\, \bar{\psi}_f(\boldsymbol{r},\boldsymbol{p}')\{&-i\not{A}(\boldsymbol{r};\boldsymbol{k})\, i\, S_F(\boldsymbol{r}-\boldsymbol{\rho})\\ &\times(-i\gamma^0)A_0^{\mathrm{Coul}}(\boldsymbol{\rho}) + (-i\gamma^0)A_0^{\mathrm{Coul}}(\mathbf{r})\, i\, S_F(\boldsymbol{r}-\boldsymbol{\rho})\\ &- i\not{A}(\boldsymbol{\rho};\boldsymbol{k})\}\psi_i(\boldsymbol{\rho};\boldsymbol{p}).\end{aligned} \tag{7.619}$$

The constituents of the integrand require detailed introduction. The two components within the curly brackets correspond to the two Feynman diagrams of Fig. 7.43. Because of the existence of two graphs and two vertices in each, for clarity, a four-dimensional description is used rather than that for which spatial and temporal variables are explicit. The initial and final electron wavefunctions are,

$$\psi_i(\boldsymbol{\rho};\boldsymbol{p}) = \sqrt{\frac{m_e}{EL^3}}\, u(\mathbf{p},s)\, e^{i\boldsymbol{p}\bullet\boldsymbol{\rho}} \tag{7.620}$$

$$\psi_f(r;\boldsymbol{p}') = \sqrt{\frac{m_e}{E'L^3}}\, u(\mathbf{p}',s')\, e^{i\boldsymbol{r}\bullet\boldsymbol{p}'} \tag{7.621}$$

where

$$\boldsymbol{\rho}\bullet\boldsymbol{p} \equiv \frac{\boldsymbol{\rho}\bullet\mathbf{p}}{\hbar c} - \frac{E\tau}{\hbar} \tag{7.622}$$

and

$$\boldsymbol{r}\bullet\boldsymbol{p}' \equiv \frac{\mathbf{r}\bullet\mathbf{p}'}{\hbar c} - \frac{E't}{\hbar}. \tag{7.623}$$

The four-vector potential of the photon with four-vector momentum k_μ and 4-component polarization ε_μ is,

$$\not{A}(\boldsymbol{r};\boldsymbol{k}) = \frac{\not{\varepsilon}}{\sqrt{2k\varepsilon_0 L^3}}\left(e^{-i\boldsymbol{k}\bullet\boldsymbol{r}} + e^{-i\boldsymbol{k}\bullet\boldsymbol{r}}\right). \tag{7.624}$$

The Coulomb interaction between the projectile electron and the nucleus is,

$$A_0^{\mathrm{Coul}}(\mathbf{r}) = -\frac{Ze}{4\pi\varepsilon_0|\mathbf{r}|}. \tag{7.625}$$

The Feynman relativistic propagator describes the electron between the two vertices and is derived by Bjorken and Drell (1964),

$$S_F(\boldsymbol{r}-\boldsymbol{\rho}) = \lim_{\varepsilon \to 0^+} \frac{1}{(2\pi)^4} \times \int d^4\boldsymbol{p} \frac{e^{-i\boldsymbol{p}\bullet(\boldsymbol{r}-\boldsymbol{\rho})}}{p^2 - m_e^2 + i\varepsilon} \times (\not{p} + m_e\mathbf{1}) \tag{7.626}$$

where $\mathbf{1}$ is the 4×4 unity matrix and, following the integrations, the S-matrix is,

$$\begin{aligned} S_{fi} = &-\frac{m_e Ze^3}{2\varepsilon_0 L^3\sqrt{2\varepsilon_0 kEE'L^3}} \frac{\delta\left(\frac{E'+k-E}{\hbar}\right)}{|\mathbf{q}|^2} \bar{u}(\mathbf{p}', s') \\ &\times \left\{(-i\not{\varepsilon}) \frac{i}{\not{p}' + \not{k} - m_e\mathbf{1}}(-i\gamma_0) + (-i\gamma_0)\frac{i}{\not{p}' - \not{k} - m_e\mathbf{1}}(-i\not{\varepsilon})\right\} u(\mathbf{p}, s) \end{aligned} \tag{7.627}$$

where $\mathbf{q} = \mathbf{p}' + \mathbf{k} - \mathbf{p}$. This result can be simplified in the soft photon limit (i.e., k→0). Consider the quantities in the curly brackets. For example,

$$\frac{1}{\not{p}' + \not{k} - m_e\mathbf{1}} = \frac{\not{p}' + \not{k} - m_e\mathbf{1}}{(\not{p}' + \not{k})^2 - m_e^2\mathbf{1}} \cong \frac{\not{p}' + \not{k} - m_e\mathbf{1}}{2\not{p}'\not{k}}. \tag{7.628}$$

Applying this (and that for the other term in the curly brackets), after considerable algebra,

$$\begin{aligned} &\bar{u}(\mathbf{p}', s')\left\{(-i\not{\varepsilon})\frac{i}{\not{p}' + \not{k} - m_e\mathbf{1}}(-i\gamma_0) + (-i\gamma_0) \times \frac{i}{\not{p}' - \not{k} - m_e\mathbf{1}}(-i\not{\varepsilon})\right\} u(\mathbf{p}, s) \\ &\cong -i\bar{u}(\mathbf{p}', s')\gamma_0 u(\mathbf{p}, s)\left(\frac{\hat{\boldsymbol{\varepsilon}} \bullet \mathbf{p}'}{\mathbf{k} \bullet \mathbf{p}'} + \frac{\hat{\boldsymbol{\varepsilon}} \bullet \mathbf{p}}{\mathbf{k} \bullet \mathbf{p}}\right). \end{aligned} \tag{7.629}$$

Thus, the squared magnitude of the S-matrix is,

$$|S_{fi}|^2 = \frac{m_e^2 Z^2 e^6}{8\pi L^6 kEE'\varepsilon_0^3} \frac{T}{|\mathbf{q}|^4} \delta\left(\frac{E'+k-E}{\hbar}\right) \times |\bar{u}(\mathbf{p}', s')\gamma_0 u(\mathbf{p}, s)|^2 \left(\frac{\hat{\boldsymbol{\varepsilon}} \bullet \mathbf{p}'}{\mathbf{k} \bullet \mathbf{p}'} + \frac{\hat{\boldsymbol{\varepsilon}} \bullet \mathbf{p}}{\mathbf{k} \bullet \mathbf{p}}\right)^2 \tag{7.630}$$

Triple Differential Cross Section in the Soft Photon Limit

Combining the above and performing the usual averaging and summing over electron spins and photon polarizations, the *bremsstrahlung* triple differential cross section is,

$$\frac{d^3\sigma}{dk\, d\Omega_k\, d\Omega_{p'}} = \alpha\left(\frac{Zr_0 m_e}{2\pi}\right)^2 \left(\frac{p'}{kpq^4}\right) \times F(p, p', k; \theta_p, \theta_{p'}, \varphi) \tag{7.631}$$

where

$$\begin{aligned} F(p, p', k; \theta_p, \theta_{p'}, \varphi) = &\frac{\left(4E'^2 - q^2\right)p^2 \sin^2\theta_p}{(E - p\cos\theta_p)^2} + \frac{(4E^2 - q^2)p'^2 \sin^2\theta_{p'}}{(E' - p'\cos\theta_{p'})^2} \\ &- (4EE' - q^2 + 2k^2) \times \frac{2pp' \sin\theta_p \sin\theta_{p'} \cos\varphi}{(E - p\cos\theta_p)(E' - p'\cos\theta_{p'})} \\ &+ \frac{2k^2\left(p^2\sin^2\theta_p + p'^2\sin^2\theta_{p'}\right)}{(E - p\cos\theta_p)(E' - p'\cos\theta_{p'})}. \end{aligned} \tag{7.632}$$

In parallel to the soft photon limit, one applies the nonrelativistic limit in order to simplify $F(p, p', k; \theta_p, \theta_{p'}, \varphi)$ by neglecting those terms with k, p and p relative to m_e and approximating $E \approx m_e$ and $E' \approx m_e$ to give,

$$F(p, p', k; \theta_p, \theta_{p'}, \varphi) \approx 4\left(p^2 \sin^2\theta_p + p'^2 \sin^2\theta_{p'} - 2pp' \sin\theta_p \sin\theta_{p'} \cos\varphi\right). \tag{7.633}$$

The final expression for the Bethe–Heitler *bremsstrahlung* triple differential cross section in this limit is,

$$\frac{d^3\sigma}{dk\, d\Omega_k\, d\Omega_{p'}} = \alpha\left(\frac{Zr_0 m_e}{\pi}\right)^2 \left(\frac{p'}{kpq^4}\right) \times \left(p^2\sin^2\theta_p + p'^2\sin^2\theta_{p'} - 2pp' \sin\theta_p \sin\theta_{p'} \cos\varphi\right) \tag{7.634}.$$

It will be noted that this result predicts a $1/k$ dependence of the cross section – the cross section diverges for low photon energies, a result termed the "infrared catastrophe." Bjorken and Drell note that an experimental device detecting the inelastically-scattered electrons for $k = 0$ will also detect elastically scattered electrons and that additional radiative corrections to the elastic scattering cross section (so that the contributions of *bremsstrahlung* and elastic scatter are both considered to the same order of e) will exactly cancel this $1/k$ factor. Heitler also notes this correction arising from consideration of higher orders of the calculation.

Bethe–Heitler Bremsstrahlung Differential Cross Section in Photon Energy

The differential cross section in photon energy alone is obtained by integrating the triple differential cross section over the two solid angles. Whilst a straightforward procedure, it is tedious and only the result is presented here,

$$\frac{d\sigma}{dk} = \alpha\frac{(Zr_0)^2}{k}\left(\frac{p'}{p}\right)\left\{\frac{4}{3} - 2EE'\left(\frac{p^2+p'^2}{p^2p'^2}\right) + \frac{\kappa m_e^2E'}{p^3} + \frac{\kappa' m_e^2 E}{p'^3} - \frac{\kappa\kappa' m_e^2}{pp'} + K\left[\frac{8EE'}{3pp'} + k^2\left(\frac{E^2E'^2+p^2p'^2}{p^3p'^3}\right) + \frac{k}{2pp'}\left(\kappa\frac{EE'+p^2}{p^3} - \kappa'\frac{EE'+p'^2}{p'^3} + \frac{2kEE'}{p^2p'^2}\right)\right]\right\} \tag{7.635}$$

where

$$\kappa = 2\ln\left(\frac{E+p}{m_e}\right) \tag{7.636}$$

$$\kappa' = 2\ln\left(\frac{E'+p'}{m_e}\right) \tag{7.637}$$

$$K = 2\ln\left(\frac{EE'+pp'-m_e^2}{m_e k}\right). \tag{7.638}$$

In the nonrelativistic limit, this becomes,

$$\frac{d\sigma}{dk} = \alpha\frac{(Zr_0)^2}{k}\frac{8}{3}\ln\left(\frac{\left(\sqrt{T}+\sqrt{T-k}\right)^2}{_ek}\right). \tag{7.639}$$

Screening Effects

The above Bethe–Heitler result for electron-nucleus *bremsstrahlung* neglected the reduction in the nuclear Coulomb potential experienced by the electron due to the screening by atomic electrons. This can be accounted for by changing the Coulomb potential into a Yukawa type,

$$A_0^{Yuk}(\mathbf{r}) = -\left(\frac{Ze}{4\pi\varepsilon_0|\mathbf{r}|}\right)e^{-\lambda r} \tag{7.640}$$

and repeating the calculation. This introduces a $(1-F(q;\, Z))$ multiplicative factor into the expressions for the differential cross section, where $F(q;\, Z)$ is the atomic form factor.

Deviations from the Born Approximation

From Chap. 2, the Born approximation used by the Bethe–Heitler calculation is valid only if the inequalities $(\alpha Z/\beta) \ll 1$ and $(\alpha Z/\beta') \ll 1$ where β and β are the electron speeds (normalized to the speed of light) before and after the interaction, respectively, are met. Hence, the result is valid only for low-Z media or relativistic electrons. At low electron energies, it is invalid to approximate the electron wavefunctions by plane waves and Coulomb distortion must be accounted for. An approximate solution to this dilemma is to multiply the differential cross sections by the Elwert factor,

$$F_{Elw} = \left(\frac{\beta}{\beta'}\right)\frac{1-e^{-2\pi(\alpha Z/\beta)}}{1-e^{-2\pi(\alpha Z/\beta')}} \tag{7.641}$$

7.6.3.3 Further Considerations

The derivations of the Bethe–Heitler *bremsstrahlung* theory are limited by the restrictions just noted, which are particular to the low electron energies of interest to nuclear medicine. At these energies, the assumption of a plane wave for the electron wavefunction is not entirely valid as it neglects the distortion induced by the nuclear Coulomb field. It has been estimated that, in the electron kinetic energy range of 200 keV to

1 MeV (of most interest to nuclear medicine), the Bethe–Heitler calculation agrees with experiment to within ±20% (Morgan 1970).

7.6.4 Electron–Electron Bremsstrahlung

Earlier, in the discussion of classical *bremsstrahlung*, arguments were provided to show that electron–electron *bremsstrahlung* could be neglected at low energies, certainly within the classical framework. However, electron–electron *bremsstrahlung* is not an entirely negligible process. There are two significant differences between the electron-nucleus and electron–electron *bremsstrahlung* interactions. First, the recoil of the target body cannot be neglected and, second, exchange effects must be allowed for. Cross sections for electron–electron *bremsstrahlung* are derived in Haug and Nakel (2004); they show that 8 Feynman diagrams contribute to the calculation of the S-matrix element, rather than 2 for electron-nucleus *bremsstrahlung*. Hence, we will not pursue a derivation of the cross section and the interested reader is referred to that book.

It is convenient (ICRU 1984) to use dimensionless radiative energy-loss cross sections for both electron-nuclear and electron–electron *bremsstrahlung* for an incident electron with total energy, E,

$$\phi_{\mathrm{rad,n}} = \frac{1}{\alpha r_0^2 Z^2} \int_0^{E-m_e} dk \frac{k}{E} \frac{d\sigma_n}{dk} \tag{7.642}$$

$$\phi_{\mathrm{rad,e}} = \frac{1}{\alpha r_0^2} \int_0^{E-m_e} dk \frac{k}{E} \frac{d\sigma_e}{dk} \tag{7.643}$$

The ratio, $\phi_{\mathrm{rad,e}}/\phi_{\mathrm{rad,n}}$, has a value of about 0.5 at electron kinetic energies of 700 keV and vanishes at low kinetic energies.

7.6.5 Positron-Nucleus Bremsstrahlung

Positrons are repelled by the nucleus and attracted to the atomic electrons; hence, the positron *bremsstrahlung* cross section will differ from that of the electron, primarily at low kinetic energies where it is significantly less. ICRU Publication 37 summarizes calculations of the positron *bremsstrahlung* cross section and notes that a universal curve can be derived of the ratio of the dimensionless cross sections for positron to electron radiative losses, $\phi^{+}_{\mathrm{rad,n}}/\phi^{-}_{\mathrm{rad,n}}$, as a function of the variable $\ln T/Z^2$, where T is the electron/positron kinetic energy and Z is the atomic number of the medium, exists. Figure 7.45 presents the ratio of the positron radiative cross section to that for the electron as a function of kinetic energy for carbon.

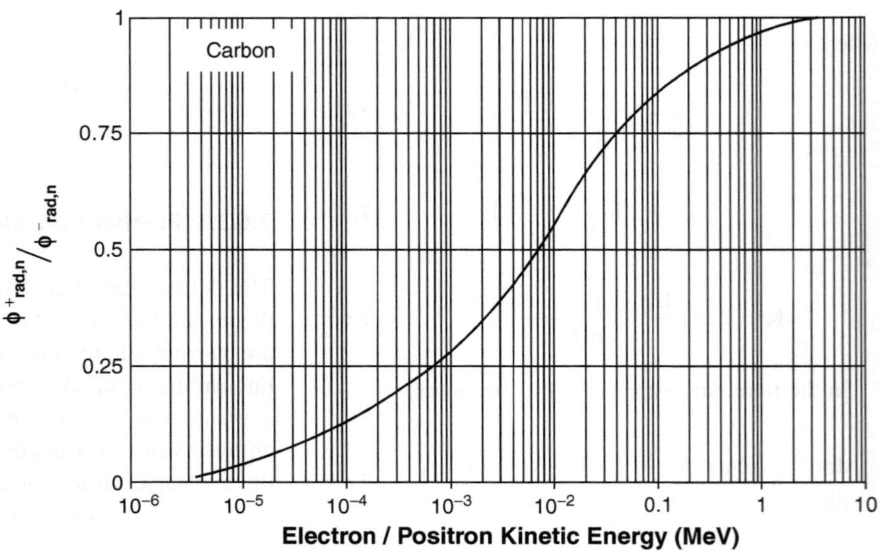

Fig. 7.45 The ratio of the positron to electron-nuclear *bremsstrahlung* cross section in carbon as a function of kinetic energy. Curve drawn from a calculation using tabulated data in ICRU Publication 37 (1984)

7.6.6 Mass Radiative Stopping Power for Electrons

Having calculated differential cross sections for electron-nuclear *bremsstrahlung*, we next evaluate the radiative stopping power. As with energy transfer as a result of collisions with atomic electrons, we can define the probability that an incident electron with kinetic energy T will emit a photon with an energy between k and k + dk is,

$$\Phi_{Rad}(T,k)dk = 4\alpha r_0^2\left(\frac{N_A Z^2}{A}\right)F(T,k;Z)\frac{dk}{k} \quad (7.644)$$

where F(E, k; Z) is a function accounting for the screening of the nucleus by the atomic electrons. The extent of screening is defined by the dimensionless parameter,

$$\zeta = 100\left(\frac{m_e}{T+m_e}\right)\left(\frac{\left(\frac{k}{T+m_e}\right)}{1-\left(\frac{k}{T+m_e}\right)}\right)Z^{-1/3}. \quad (7.645)$$

The degree of screening by the atomic electrons is an inverse function of ζ. That is, $\zeta = 0$ may be described as "complete screening" and that of $\zeta \gg 1$ as "no screening." Note that for a given photon energy k, ζ will decrease (i.e., screening increases) as the incident electron energy increases ($k_{max} = T$). For large T, the function F(T, k; Z) has the following forms for different values of ζ. For total screening, $\zeta = 0$,

$$F(T,k;Z) = \left(1+\left(1-\frac{k}{T+m_e}\right)^2 - \frac{2}{3}\left(1-\frac{k}{T+m_e}\right)\right)^2 \ln 183\, Z^{-1/3} + \frac{1}{9}\left(1-\frac{k}{m_e}\right). \quad (7.646)$$

For no screening (large ζ),

$$F(T,k;Z) = \left(1+\left(1-\frac{k}{T+m_e}\right)^2 - \frac{2}{3}\left(1-\frac{k}{T+m_e}\right)\right)^2 \times \ln\left(\frac{2(T+m_e)}{m_e}\frac{\left(1-\frac{k}{T+m_e}\right)}{\left(\frac{k}{T+m_e}\right)} - \frac{1}{2}\right). \quad (7.647)$$

The mass radiative stopping power for an electron with kinetic energy T is,

$$\frac{dE}{\rho\, dx_{Rad}} = -\frac{1}{\rho}\int_0^{T_e} dk\, k\, \Phi_{Rad}(T,k) \quad (7.648)$$

where, again, a negative sign is used to indicate that energy is lost by the particle. From the above expressions, one can write the mass collision stopping power as,

$$\frac{dE}{\rho\, dx_{Rad}} = -4\alpha r_0^2\left(\frac{N_A Z^2}{A}\right)(T+m_e) \times \ln\left(\frac{2(T+m_e)}{m_e} - \frac{1}{3}\right) \quad \text{for } m_e \ll T+m_e \ll \frac{m_e}{\alpha}Z^{-1/3} \quad (7.649)$$

or as,

$$\frac{dE}{\rho\, dx_{Rad}} = -4\alpha r_0^2\left(\frac{N_A Z^2}{A}\right)T\left(\ln\left(183Z^{-1/3}\right) + \frac{1}{18}\right) \quad \text{for } \frac{m_e}{\alpha}Z^{-1/3} \ll T+m_e. \quad (7.650)$$

The mass radiative stopping powers calculated for electrons in carbon and lead are shown as functions of electron kinetic energy in Fig. 7.46. Comparing Figures 7.15 and 7.46, it can be seen that, for a given electron energy, the mass radiative stopping power of lead exceeds that of carbon (due to the Z^2/A multiplicative factor), whereas the mass collision stopping power of carbon exceeds that of lead (as the multiplicative factor in that case is only Z/A). For the electron energies of interest to nuclear medicine in a low-Z medium such as tissue, the mass radiative stopping power is of the order of about 0.1% of the mass collision stopping power. This indicates the challenge of using *bremsstrahlung* to image the biodistribution of a β-emitting therapeutic radiopharmaceutical, as discussed later in this book. For electron energies below about 1 MeV, the lead and carbon mass radiative stopping powers slowly increase with energy, but after this energy they increase with energy in almost constant proportion.

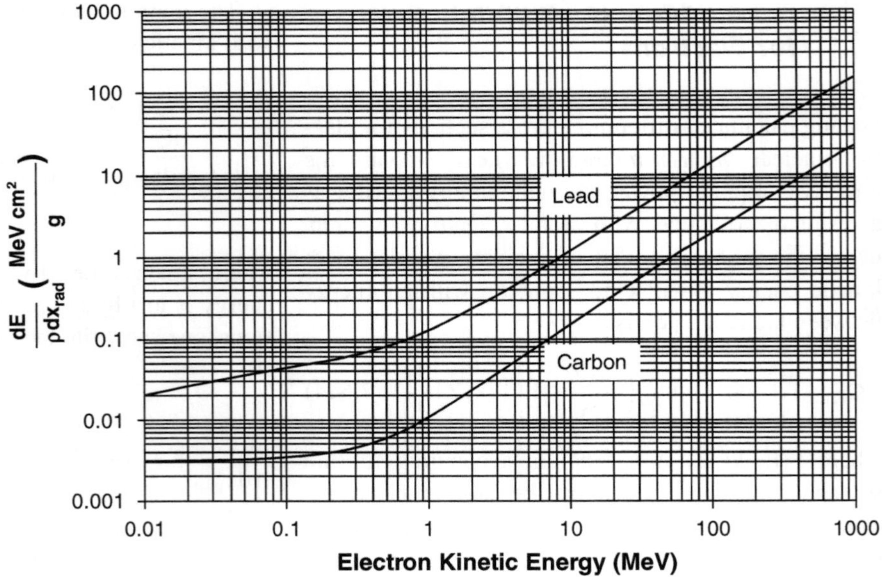

Fig. 7.46 Mass radiative stopping powers for electrons in carbon and lead

7.6.7 Radiation Length

A comparison of Figs. 7.15 and 7.46 demonstrates that the energy loss of an electron slowing down in a medium is predominantly through *bremsstrahlung* at high electron kinetic energies, $\frac{m_e}{\alpha} Z^{-1/3} \ll T + m_e$. Hence, (7.650) would provide a reasonable expression for the total energy loss rate at these electron kinetic energies. If this equation were rewritten as the ratio of the incident electron energy to a length (given in centimeter/gram), the result is,

$$\frac{dE}{\rho\, dx_{Rad}} = -4\alpha r_0^2 \left(\frac{N_A Z^2}{A}\right) T \left(\ln\left(183 Z^{-1/3}\right) + \frac{1}{18}\right) \equiv -\left|\frac{T}{X_0}\right|. \tag{7.651}$$

This length, X_0, is defined as the *radiation length*, with its reciprocal given by,

$$\frac{1}{X_0} = 4\alpha r_0^2 \left(\frac{N_A Z^2}{A}\right) \times \left(\ln\left(183 Z^{-1/3}\right) + \frac{1}{18}\right). \tag{7.652}$$

It is a constant for a given material.

7.7 Collision and Radiative Stopping Powers: A Summary

Figure 7.47 shows the collision and radiative mass stopping powers for electrons in carbon and lead along with their sum (total mass stopping power).

While the overall morphological features of the two graphs are similar, there are distinctive differences. The most important is the difference between the collision and radiative stopping powers. For carbon, an element representative of soft tissue, electron energy losses through *bremsstrahlung* exceed those through collision for kinetic energies above about 100 MeV; the energy threshold is much lower for lead at about 10 MeV. This is characterized by the *bremsstrahlung* efficiency or radiation yield.[29] In theapproximation that the energy loss of the electron in the medium is continuous as it slows down

[29]There is also a radiation yield calculation associated with positrons, although this is not considered here. Customarily, the in-flight $e^- e^+ \rightarrow 2\gamma$ is excluded from the calculation of the positron radiation yield.

Fig. 7.47 Mass collision, radiative, and total stopping powers for electrons in carbon and lead

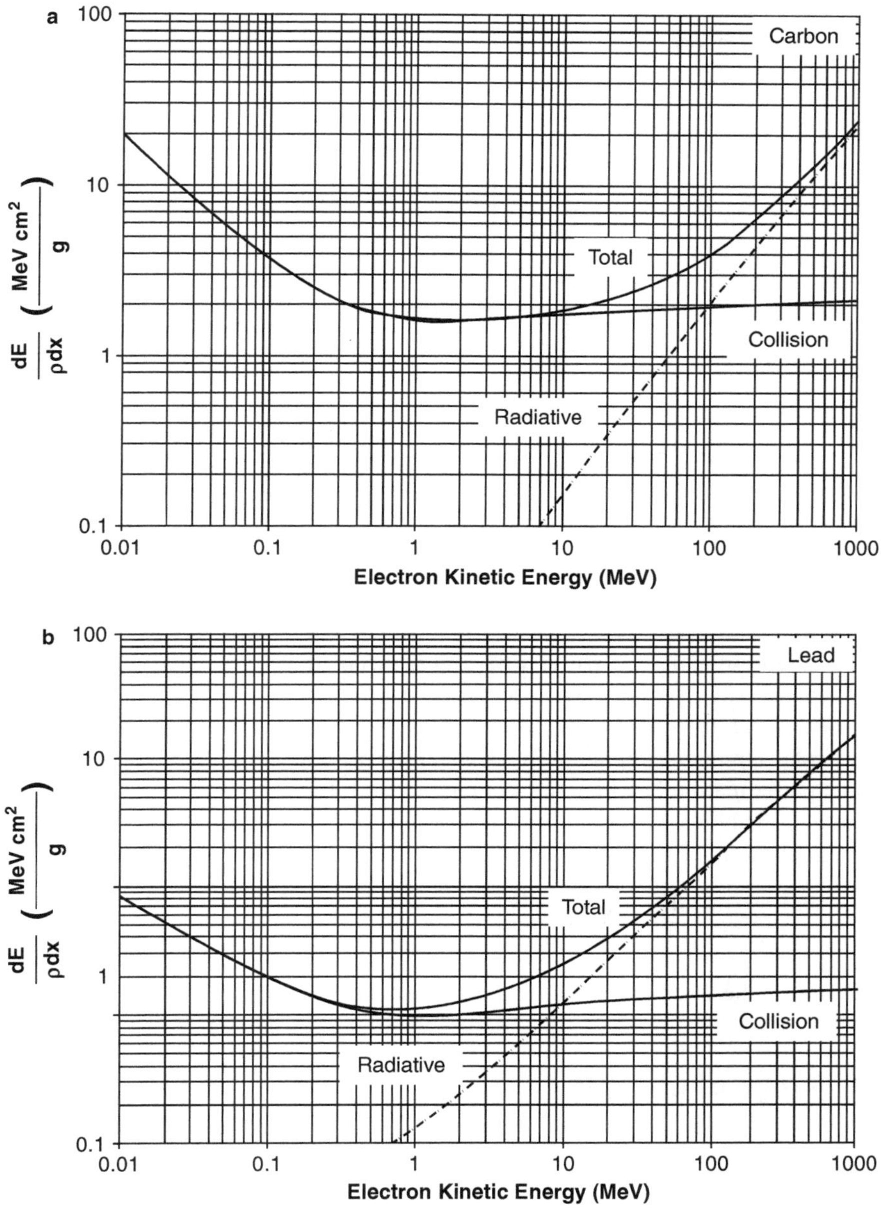

(the CSDA), this is, for an electron with initial kinetic energy,

$$Y(T) = \frac{1}{T}\int_0^T dT' \frac{(dE/dx)_{Rad}}{(dE/dx)_{Col} + (dE/dx)_{Rad}}. \quad (7.653)$$

Recall the fundamental features of the collision and radiative mass stopping powers: the former increases with Z and increases logarithmically with electron energy whereas the latter increases with Z^2 and increases linearly with energy. As a result, the radiation yield Y(T) will increase with electron energy and the atomic number of the medium (Fig. 7.48).

Throughout the above derivations, graphical examples of stopping powers have been provided for carbon and lead elemental media in order to display the two extremes of atomic number dependencies (carbon can

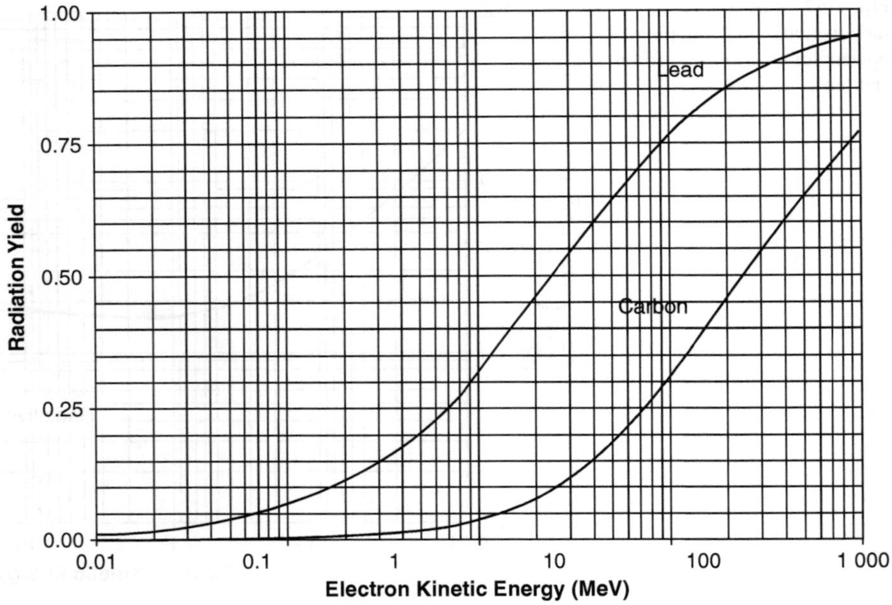

Fig. 7.48 Radiation yield Y (T) for electrons in lead and carbon

also be representative of tissue when normalized to physical density). To calculate stopping powers for compound media (such as soft tissue), Bragg's additivity rule is frequently applied. Bragg's additivity rule is an approximation in which the stopping power of a compound is given by the mass-weighted sum of the stopping powers of the atomic constituents,

$$\frac{dE}{\rho\, dx} = \sum_i w_i \left.\frac{dE}{\rho\, dx}\right|_i \qquad (7.654)$$

where w_i is the fraction by weight of the element.

7.8 Range of Charged Particles

7.8.1 Introduction

The range of a charged particle slowing down in a medium is, at the simplest level, the depth of penetration until its kinetic energy reaches thermal levels. However, we must recall that energy transfer to the medium is a stochastic process and, hence, the range is the expectation value of the pathlength that the particle follows until it is thermalized. The projected range is defined as the effects of multiple scattering must also be considered, especially with electron and positron projectiles. This quantity is the expectation value of the greatest penetration of the particle in the medium. A graphical comparison of the range and the projected range is provided in Fig. 7.49.

7.8.2 Continuous Slowing-Down Approximation (CSDA) Range

The CSDA range (Berger and Seltzer 1983; ICRU 1984) is similar to the concept of the pathlength described above, but neglects the effects of multiple scatter and assumes a straight-line trajectory for a particle with an initial kinetic energy T,

$$\Re_{CSDA} = \int_0^T \frac{dE}{dE/dx|_{Total}} \qquad (7.655)$$

where the total stopping power is the sum of the collision and radiative stopping powers. As $dE/dx|_{Total}$ is the expectation value of the rate of energy loss, $\Re_{CSDA}$ represents an expectation value of the particle's range. For the materials and energy ranges of interest to nuclear medicine, the radiative energy loss contribution can usually be ignored. As the stopping power is relatively constant in the minimally-ionizing region and has a β^{-2} dependence at lower energies, a charged particle penetrating a medium will lose energy at near a constant rate with depth until, as

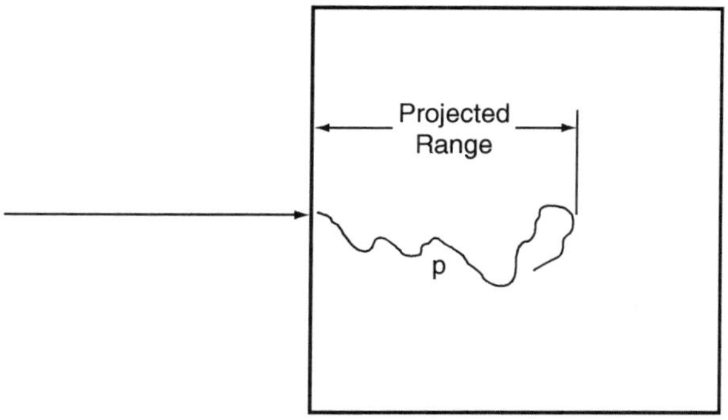

Fig. 7.49 A charged particle enters a medium from the left. The *range* (the expectation value of the pathlength *p* between the point where the particle enters the medium and where it is thermalized) can be tortuous, depending upon multiple scattering of the particle. The *projected range* is the depth of maximum penetration into the medium

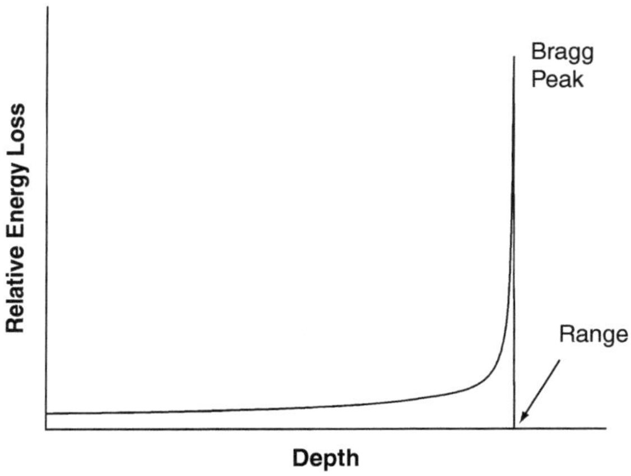

Fig. 7.50 Energy loss (deposition) as a function of depth for a heavy charged particle (e.g., proton or α particle)

it slows down beyond the minimally-ionizing region, it loses its energy at ever-an increasing rate until it has thermalized. Figure 7.50 shows the curve of relative energy deposition as a function of depth for a heavy charged particle, such as an α particle, where multiple scattering can be neglected and the particle travels very nearly along a straight line. Only near the end of its range does the relative energy loss increase from its near continuous value; this rapid increase is known as the "Bragg" peak.

Recall that, for a projectile of charge ze in a medium with atomic number Z and atomic mass number A, the collision stopping power is proportional to $(zZ/\beta A)^2$. Hence, for a given projectile and kinetic energy, the CSDA range, in gram per square centimeter, will be proportional to $(A/Z)^2$. Because of the charge and mass dependencies of the stopping power, it is possible to estimate the CSDA range in a medium of a particle of rest mass m_2 and charge z_2e knowing the CSDA range of a particle of rest mass m_1 and charge z_1e,

$$\Re_{CSDA,2} = \frac{m_1}{m_2}\left(\frac{z_1}{z_2}\right)^2 \Re_{CSDA,1} \tag{7.656}$$

The CSDA ranges of electrons and positrons will differ due to the former's use of the Møller cross section and the latter's use of the Bhabha cross section. As the collision stopping powers differ, the electron range in a medium is greater than that of a positron at lower energies and approximately equalizes at high energies Fig. 7.51 shows the CSDA range of electrons and positrons in carbon and lead as a function of kinetic energy in the range of interest to nuclear medicine applications. As shown, due to the $(A/Z)^2$ dependence, the range (in gram per square centimeter) is greater in lead than in carbon; the positron range is slightly less than that of the electron for both media and equalizes, and slightly exceeds, at higher kinetic energies.

As electrons and positrons are also subject to multiple scatter (which will be significant due to their relatively low rest masses), the $\Re_{CSDA}$ will be an approximation only of the actual range of these particles. As the spatial resolution of PET imaging will be limited by the distance between the emission of the positron and its annihilation, an evaluation of the positron range is of particular practical importance in nuclear medicine. It is most efficiently done within a Monte Carlo calculation which incorporates the stochastic nature of the electromagnetic interactions between the positron and the medium. Examples can be found in the papers by Levin and Hoffman (1997) and Champion and Le Loirec (2007).

7.8.3 Projected Range

The projected range $\bar{t}$ is the maximum perpendicular penetration of the charged aprticle into the medium and is defined as,

$$\bar{t} = \int_0^\infty dt\, t \left|\frac{dN}{dt}(t)\right| \tag{7.657}$$

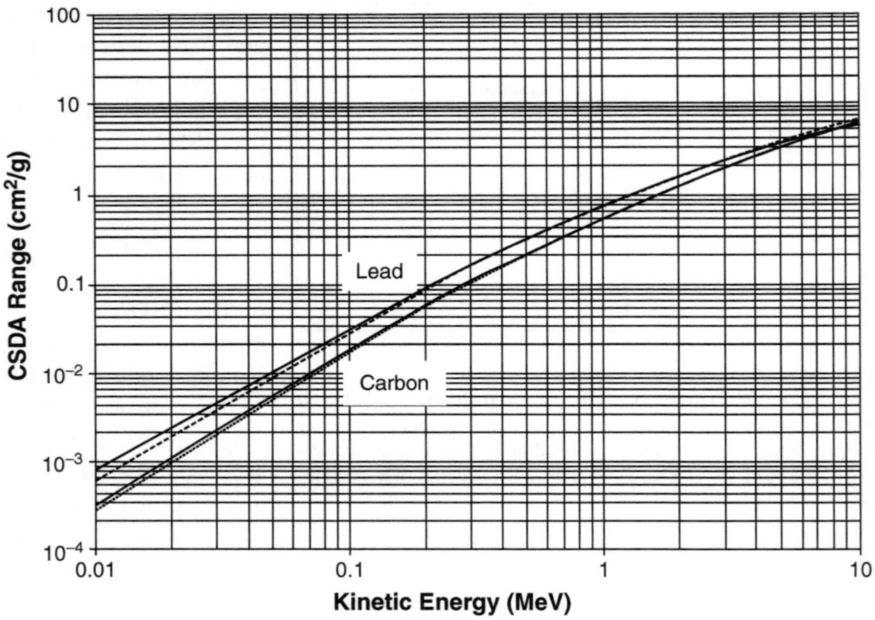

Fig. 7.51 CSDA ranges of electrons (*solid lines*) and positrons (*dashed lines*) in carbon and lead as a function of kinetic energy relevant to nuclear medicine applicatios. Curves are drawn using tabulated numerical data in ICRU Publication 37 (1984)

where $dN/dt(t)$ is the rate at which particles are stopped per unit depth and which is normalized to,

$$\int_0^\infty dt \left| \frac{dN}{dt}(t) \right| = 1. \quad (7.658)$$

The absolute value sign is a formality as the rate is negative due to the loss of particles as they stop.

7.8.4 Range Straggling

Both the CSDA and projected means of calculating the range of a charged particle penetrating a medium are expectation values of an assumed continuous energy losses of the charged particles. However, as demonstrated in Sect. 7.4, energy loss is stochastic and is described by a probability distribution function. As the range is inversely proportional to the stopping power and is also affected by multiple scatter (predominantly in the case of electrons and positrons), the range of a charged particle is also a stochastic quantity. Range straggling is a consequence of the energy loss pdf only.

7.9 Positron–Electron Annihilation

7.9.1 Introduction

The interactions of positrons with atomic electrons resulting in their annihilation and the production of photons are now considered. In terms of internal radiation dosimetry, the annihilation has a limited effect as the result is high-energy (511 keV) γ-ray pairs or triplets. The process is, of course, fundamental to PET imaging. Whereas a moving electron or α-particle will slow down to thermal equilibrium, a positron will eventually annihilate with an electron in the medium, either whilst in-flight or following thermalization. One can consider the annihilation process to be the opposite of electron–positron pair production and, using the hole theory, treat positron annihilation as the transition of an ordinary electron from a positive energy state to a negative energy state with the emission of quanta with a combined energy $\geq 2m_e$ (the inequality accounts for the contributions of any incoming kinetic energy), as shown in Fig. 7.52.

The annihilation processes results in the production of one or more γ rays in order to conserve energy and momentum; in fact up to three photons can be emitted.

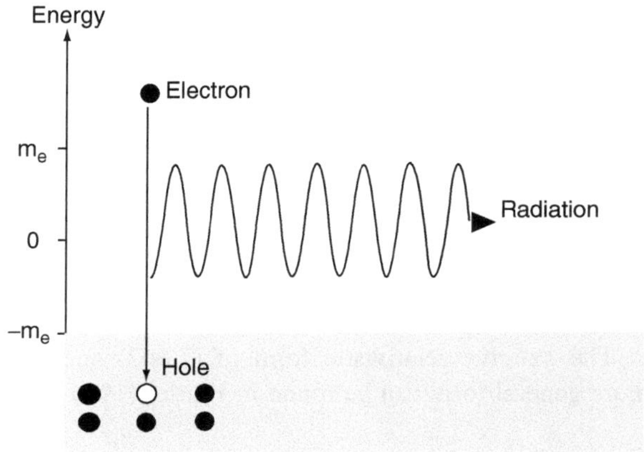

Fig. 7.52 Positron–electron annihilation described by hole theory. A positive energy electron falls into a negative energy hole with the result of radiation being emitted. This radiation will be in the form of two or more photons

For example, a single γ ray can result from the positron annihilation on a bound atomic electron such that the recoil nucleus is available for momentum conservation; γ-ray pairs are created when the positron annihilates with a free electron or is initially bound in the singlet $^1S_0$ state of positronium and γ ray triplets can be produced if the electron and positron are initially bound in the triplet $^3S_1$ state of positronium.

Figure 7.53 shows the Feynman diagrams for positron–electron pair annihilation. A comparison of these graphs with those of incoherent scatter in Chap. 6 show that they are the same if turned on their sides, allowing a similarity in the calculation of the cross section.

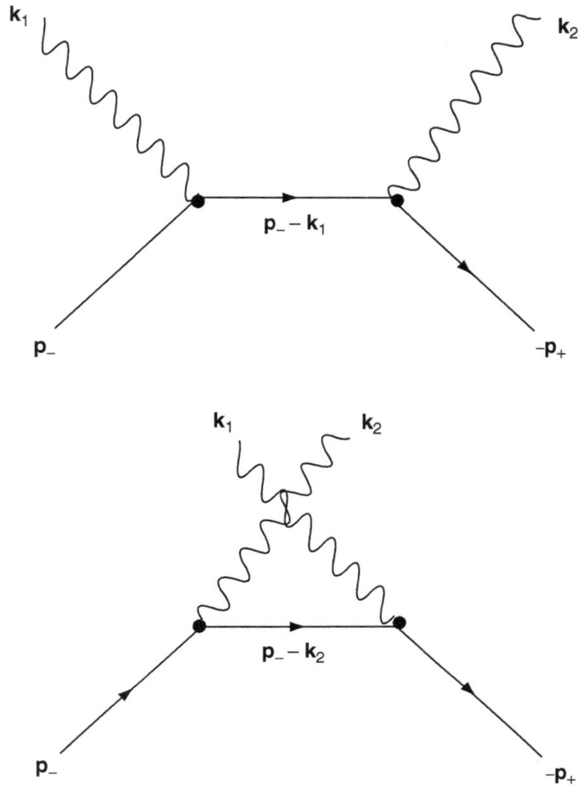

Fig. 7.53 Feynman diagrams for positron–electron annihilation

7.9.2 Annihilation Probabilities and Cross Sections

7.9.2.1 General Features

It is possible to generate estimations of the positron lifetime following emission from β decay and its annihilation in-flight within the medium. The probability of positron annihilation per unit length should obviously be proportional to the electron density of the medium,

$$\frac{d\Phi_{ann}}{dx} = \sigma\left(\frac{N_A Z}{A}\right)\rho \qquad (7.659)$$

where the constant of proportionality, σ, is the annihilation cross section. The probability of annihilation per unit time is,

$$\frac{d\Phi_{ann}}{dt} = \sigma\left(\frac{N_A Z}{A}\right)\rho\,\beta\,c. \qquad (7.660)$$

By crudely approximating the cross section by $\sigma \approx \pi r_0^2$, the probability of positron annihilation per unit length can be estimated to be 0.15 and 0.67/cm in carbon and lead, respectively. For relativistic positrons ($\beta \approx 1$), the estimated probability of in-flight annihilation per unit time is 4.5×10^9/s and 2×10^{10}/s for the same respective elements, corresponding to positron lifetimes of 220 and 50 ps.

These, of course, are relevant only for in-flight annihilation. The positron, like any other charged particle, transfers kinetic energy to the medium as it slows down to eventually thermalize to annihilate or to form a bound system with a free electron known as positronium. This state has a finite lifetime. Due to the intrinsic spin-1/2 of the electron and positron, positronium can exist in either a singlet ($^1S_0$) or triplet ($^3S_1$) state. In order to maintain parity conservation, singlet positronium (parapositronium) will decay into two photons and triplet positronium (orthopositronium) will decay into three photons.

7.9.2.2 Positron Annihilation on a Bound Atomic Electron

Positron annihilation on a bound atomic electron can result in the emission of a single electron as the recoil nucleus is available to take up momentum and kinetic energy. The nonrelativistic calculation of this positron annihilation cross section is easily obtained from the nonrelativistic photoelectric absorption cross section for a K-shell electron,

$$\sigma_{PE} = \sigma_{Tho}\left(4\sqrt{2}\right)\alpha^4 Z^5 \left(\frac{m_e}{k}\right)^{\frac{7}{2}}$$

for a photon of energy k absorbed by K-shell electron in an atomic of atomic number Z. In the positron annihilation case, we consider from Dirac hole theory that the atomic electron following annihilation transits to a state of negative energy and momentum $-\mathbf{p}$, where $+\mathbf{p}$ is the three-vector momentum of the incident positron. Following annihilation, a single photon of energy,

$$k = \sqrt{m_e^2 + p^2} + m_e - E_B \tag{7.661}$$

where E_B is the electron binding energy. The recoil kinetic energy of the atom is neglected. The phase space factor for the photoelectric absorption case, $(L/2\pi\hbar c)^3 m_e p_e\, d\Omega$, is replaced by that for the positron annihilation, $(L/2\pi\hbar c)^3 k^2\, d\Omega$. The energy of the incident photon in the photoelectric absorption case is replaced by the sum of the total energy of the incident positron and the electron/positron rest mass $k \rightarrow \sqrt{p^2 + m_e^2} + m_e \approx 2m_e$, where the nonrelativistic case has been assumed in the last step. Inserting this into the $(m_e/k)^{7/2}$ factor leads to $(m_e/k)^{7/2} \approx 1/8\sqrt{2}$. Combining this and the p/m_e multiplicative factor arising from the change in phase space expressions gives the total cross section (in the nonrelativistic limit) for the annihilation of a positron with momentum p in a medium of atomic number Z on a bound atomic electron leading to a single emitted photon as,

$$\sigma_{1\gamma,K} = \frac{\sigma_{Tho}}{2}\alpha^4 Z^5 \frac{p}{m_e} \tag{7.662}$$

As with photoelectric absorption, this result displays a Z^5 dependence indicating that it will only be of importance with high-Z media and of limited concern to nuclear medicine dosimetry.

The extreme relativistic form of (7.662) and the more general form can be found in Heitler (1984).

7.9.2.3 Positron Annihilation on a Free Electron

The Feynman diagrams of Compton scatter and positron annihilation on a free electron are very similarly in their architecture and, consequently, the calculational procedures of both map closely to each other. Using 4-vectors, the S-matrix element for a positron annihilating with a free electron is, from Bjorken and Drell (1964),

$$\begin{aligned} S_{fi} = &-\frac{(2\pi)^4\; m_e e^2}{\varepsilon_0 L^6 \sqrt{2\varepsilon_0 k_1 k_2 E_+ E_-}}\delta(\boldsymbol{k}_1 + \boldsymbol{k}_2 - \boldsymbol{p}_+ - \boldsymbol{p}_-) \\ &\times \bar{v}(\mathbf{p}_+, s_+) \times \left\{(-i\not{\epsilon}_2)\frac{i}{\not{p}_- - \not{k}_1 - m_e\mathbf{1}}(-i\not{\epsilon}_1)\right. \\ &\left. + (-i\not{\epsilon}_1)\frac{i}{\not{p}_- - \not{k}_2 - m_e\mathbf{1}}(-i\not{\epsilon}_2)\right\}u(\mathbf{p}_-, s_-). \end{aligned} \tag{7.663}$$

This expression satisfies Bose–Einstein statistics by being symmetric under the exchange of the two photons. A long and tedious calculation is avoided in presenting the total cross section for a positron of kinetic energy T_+ annihilating with an electron at rest to produce two photons (refer to Bjorken and Drell (1964) and Heitler (1984) for details),

$$\begin{aligned} \sigma_{2\gamma} = &\left(\frac{\pi r_0^2}{\gamma + 1}\right)\left(\frac{\gamma^2 + 4\gamma + 1}{\gamma^2 - 1}\right. \\ &\left.\times \ln\left(\gamma + \sqrt{\gamma^2 - 1}\right) - \frac{\gamma + 3}{\sqrt{\gamma^2 - 1}}\right) \end{aligned} \tag{7.664}$$

where $\gamma = (T_+ + m_e)m_e$. This result predicts a diverging cross section at low energies ($\gamma \rightarrow 1$)

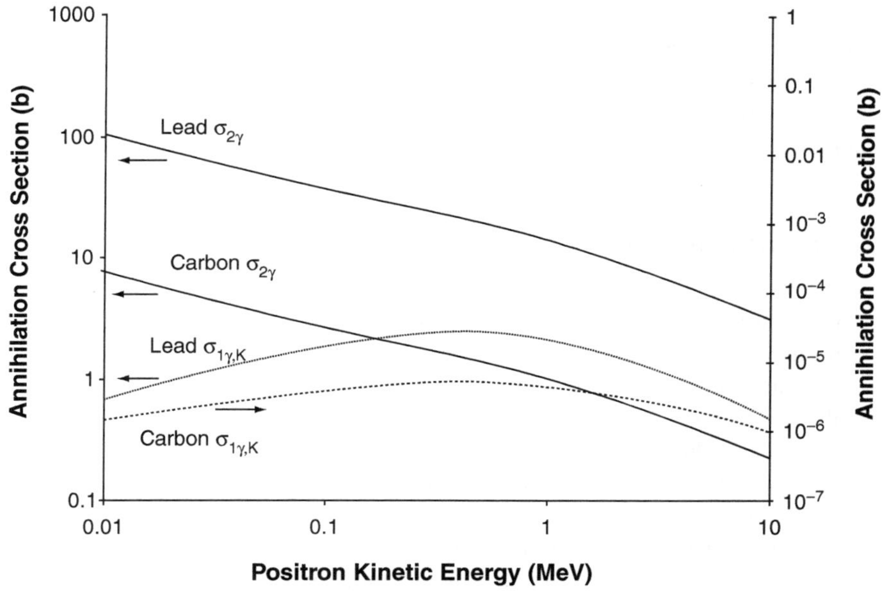

Fig. 7.54 $e^+e^- \rightarrow n\gamma$ cross sections as a function of positron kinetic energy for one- and two-photon annihilation. Note the different scale for the single-photon annihilation in carbon cross section

which is a consequence of the use of plane waves to describe the electron and positron wavefunctions (the calculation is performed in the center-of-mass system and then Lorentz-transformed to the laboratory reference frame); Coulomb wavefunctions should be used at such energies.

Figure 7.54 shows the positron annihilation cross sections per atom for one- and two-photon processes in carbon and lead as functions of the positron kinetic energy. The two-photon annihilation process cross section decreases exponentially with positron energy, thus showing that in-flight annihilation is much less probable than annihilation once the positron has slowed down towards the end of its range. On the other hand, the single-photon annihilation cross sections for both carbon and lead show maxima at positron kinetic energies of about 0.4 MeV. It is of particular interest to compare the relative magnitudes of the cross sections. The dual-photon annihilation cross sections of lead and carbon differ only by the ratio of the number of electrons available in each atom. On the other hand, the Z^5 dependence of the single-photon annihilation cross section leads to the lead cross section being about 5×10^5 times greater than that for carbon. Hence, positron annihilation in low-Z media resulting in a single photon final state is a negligible process, which is not the case for high-Z media. As a means of further comparison, Fig. 7.55 shows the ratio of the single- to dual-photon cross sections for lead expressed as a percentage and as a function of positron kinetic energy. The largest value of the lead single-photon annihilation cross section is about 17% of the dual-photon cross section at a positron kinetic energy of about 3 MeV.

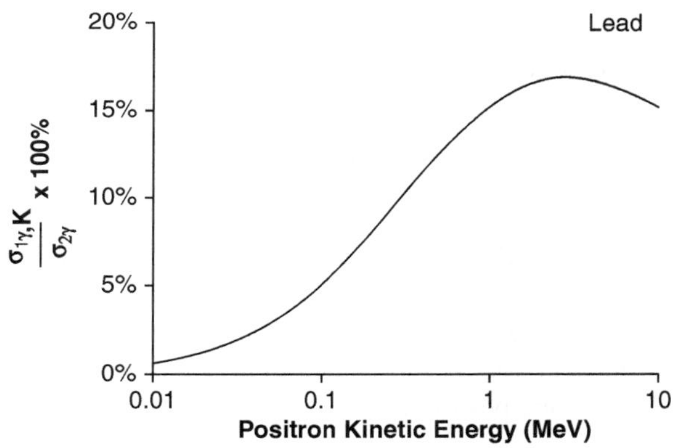

Fig. 7.55 Ratio of the single-photon annihilation cross section to that for two-photon annihilation as a function of positron energy in lead

The above results show that the probability of annihilation in-flight for a fast positron is small with the result that the positron slows down (thermalizes) and is then annihilated nearly at rest. As the kinetic energy of the positron will be non-zero, although small, and the atomic energy with which it annihilates has a finite speed, the combined kinetic energies of the two photons emitted will slightly exceed the value of $2m_e$ and will be emitted at an angle slightly different than 180°.

7.9.3 Positronium

As positron annihilation occurs predominantly at low positron energies, it is possible for the positron and an electron to form positronium, an unstable bound state similar to that of the hydrogen atom. Due to the reduced mass, the positronium Bohr radius is $2r_\infty$. The spins of the two leptons can couple to form a singlet ($^1S_0$) state or a triplet ($^3S_1$) state for antiparallel and parallel spins, respectively. Positronium is unstable. From the conservation of angular momentum, the singlet state of positronium produces two photons whereas the triplet state produces three or more. The density of electrons in (7.659), $(N_A Z/A)\rho$, is replaced by the density of the electron, calculated from its wavefunction,

$$|\psi(r=0)|^2 = \frac{1}{8\pi r_\infty^3}. \qquad (7.665)$$

Equation (7.663) does not include the contribution of the triplet decay. Hence, the probability per unit time of $^1S_0$ annihilation is, including a factor-of-four to account for the combined spin directions of the electron and positron,

$$\begin{aligned}\frac{d\Phi_{ann}}{dt} &\approx 4\pi r_0^2\left(\frac{1}{8\pi r_\infty^3}\right)c \\ &\approx \frac{r_0^2 c}{2r_\infty^3}\end{aligned} \qquad (7.666)$$

which is about 8×10^9/s. The probability per unit time of the triplet state annihilation can be shown to be about 1,110 times less. These estimates of positronium lifetime are, of course, *in vacuo*. In a condensed medium, the positron wavefunction can overlap those of surrounding electrons sufficiently to increase the annihilation rate.

The number of photons that the positronium can decay to will depend upon the total angular momentum of the bound system. This is a consequence of the charge conjugation operator, **C**, which exchanges a particle with its antiparticle. A system containing an equal number of particles and antiparticles (which is, by definition, electrically neutral) with total spin s and orbital angular momentum number l is an eigenstate of **C** with eigenvalue $(-1)^{l+s}$. **C** has eigenvalues of +1 and −1 for the singlet $^1S_0$ and triplet $^3S_1$ states of positronium, respectively. As **C** exchanges the signs of all electric charges, it will also change the direction of the electric field **E**. For a single photon, this is equivalent to $\mathbf{C}|\gamma\rangle = |\gamma\rangle$ or, for an ensemble of n photons, the eigenvalue of **C** is $(-1)^n$. Hence, as the eigenvalue of **C** is +1 for the $^1S_0$ state, the number of photons resulting from the annihilation of the singlet state must be even. The simplest case is n = 2. Similarly, as the eigenvalue of **C** is −1 for the $^3S_1$ state, the number of photons resulting from the annihilation of the triplet state must be odd. As n = 1 is not permissible by the conservations of momentum and energy, then n = 3.

References

Abramowitz M, Stegun IA (eds) (1972) Handbook of mathematical functions. Dover, New York

Ahlen SP (1980) Theoretical and experimental aspects of the energy loss of relativistic heavily ionizing particles. Rev Mod Phys 52:121–173 (erratum Rev Mod Phys 1980; 52:653)

Ammi H, Zemih R, Mammeri S, Allab M (2005) Mean excitation energies extracted from stopping power measurements of protons in polymers by using the modified Bethe-Bloch formula. Nucl Instr Meth B 230:68–72

Andreo P, Medin J, Bielajew AF (1993) Constraints of the multiple-scattering theory of Molière in Monte Carlo simulations of the transport of charged particles. Med Phys 20:1315–1325

Ashley JC, Ritchie RH, Brandt W (1972) Z_1^3 effect in the stopping power of matter for charged particles. Phys Rev B 5:2393–2397

Ashley JC, Ritchie RH, Brandt W (1973) Z_1^3-dependent stopping power and range contributions. Phys Rev B 8:2402–2408

Andersen HH, Simonsen H, Sørensen H (1969) An experimental investigation of charge-dependent deviations from the Bethe stopping power formula. Nucl Phys A 125:171–175

Barkas WH (1962) Technical Report UCRL-10292, University of California Lawrence Radiation Laboratory, August 1962.

Barkas WH, Birnbaum W, Smith FM (1956) Mass ratio method applied to the measurement of K-meson masses and the energy balance in pion decay. Phys Rev 101:778–795

Barkas WH, Dyer JN, Heckman HH (1963) Resolution of the Σ^--mass anomaly. Phys Rev Lett 11:26–28 (erratum Phys Rev Lett 1963; 11:138)

Berger MJ, Seltzer SM (1983) Stopping powers and ranges of electrons and positrons. NBSIR 82-2550-A, National Bureau of Standards, Washington, DC

Bethe HA (1953) Molière's theory of multiple scattering. Phys Rev 89:1256–1266

Bhabha HJ (1936) The scattering of positrons by electrons with exchange on Dirac's theory of the positron. Proc Royal Soc A 154:195–206

Bhabha HJ (1938) On the penetrating component of cosmic radiation. Proc Roy Soc A164:257–294

Bichsel H (1990) Barkas effect and effective charge in the theory of stopping power. Phys Rev A 41:3642–3647

Bjorken JD, Drell SD (1964) Relativistic quantum mechanics. McGraw-Hill, New York

Blunck O, Leisegang S (1951) Zum Energieverlust energiereicher Elektronen in dunnen Schichten. Z Phys 130:641–649

Briesmeister JF (2000) MCNP – A general Monte Carlo N-particle transport code, Version 4C. Report LA-13709-M, Los Alamos National Laboratory, Los Alamos

Champion C, Le Loirec C (2007) Positron follow-up in liquid water: II. Spatial and energetic study for the most important radioisotopes used in PET. Phys Med Biol 52:6605–6625

Čerenkov PA (1934) Visible emission of clean liquids by action of γ radiation. Doklady Akad Nauk SSSR 2:451

Chibani O (1998) New algorithms for the Vavilov distribution calculation and the corresponding energy loss sampling. IEEE Trans Nucl Sci 45:2288–2292

Chibani O (2002) Energy-loss straggling algorithms for Monte Carlo electron transport. Med Phys 29:2374–2383

Churchill RV, Brown JW, Verhey RF (1974) Complex variables and applications, 3rd edn. McGraw-Hill, New York

Dalgarno A (1960) The stopping powers of atoms. Proc Phys Soc (London) 76:422

Douthat DA (1975) Electron degradation spectra in helium. Radiat Res 61:1–20

Eyges L (1948) Multiple scattering with energy loss. Phys Rev 74:1534–1535

Fano U (1956) Atomic theory of electromagnetic interactions in dense materials. Phys Rev 103:1202–1218

Fano U (1964) Studies in penetration of charged particles in matter. Nuclear Science Series Report No. 39, National Academy of Sciences, Washington, DC, 1–338

Fano U, Spencer LV (1975) Quasi-scaling of electron degradation spectra. Int J Radiat Phys Chem 7:63–76

Fermi E (1940) The ionization loss of energy in gases and in condensed materials. Phys Rev 57:485–493

Fernández-Varea JM, Mayol R, Baro J, Salvat F (1993) On the theory and simulation of multiple elastic scattering of electrons. Nucl Instr Meth Phys Res B73:447–473

Fernández-Varea JM (1998) Monte Carlo simulation of the inelastic scattering of electrons and positrons using optical-data models. Rad Phys Chem 53:235–245

Feynman RP, Leighton RB, Sands M (1963) The Feynman lectures on physics. Addison-Wesley, Reading

Findlay DJS, Dusautoy AR (1980) Improvements to the Blunck-Leisegang energy loss straggling distribution. Nucl Instr Meth 174:531–533

Fowler RH (1923) Contributions to the theory of the motion of α-particles through matter. Part II. Ionizations. Proc Cambridge Philos Soc 21:531

GEANT Team (2001) GEANT Version 4.0, physics reference manual. CERN-Information Technology Division, Application Software and Databases, Geneva

Goudsmit S, Saunderson JL (1940) Multiple scattering of electrons. Phys Rev 57:24–29

Haug E, Nakel W (2004) The elementary process of bremsstrahlung. World Scientific Publishing, Singapore

Heitler W (1984) The quantum theory of radiation. Dover, New York

Hogstrom KR, Mills MD, Almond PR (1981) Electron beam dose calculations. Phys Med Biol 26:445–459

ICRU (1979) Average energy required to produce an ion pair, ICRU Report 31, International Commission on Radiation Units and Measurements, Bethesda

ICRU (1984) Stopping powers and ranges for electrons and positrons. ICRU Report 37. International Commission on Radiation Units and Measurements, Bethesda

ICRU (1993) Stopping powers and ranges for protons and alpha particles ICRU Report 49. International Commission on Radiation Units and Measurements, Bethesda

Inokuti M (1971) Inelastic collisions of fast charged particles with atoms and molecules – the Bethe theory revisited. Rev Mod Phys 43:297–347

Jackson JD (1999) Classical electrodynamics, 3rd edn. Wiley, New York

Jette D (1988) Electron dose calculation using multiple-scattering theory. A. Gaussian multiple-scattering theory. Med Phys 15:123–137

Kase KR, Nelson WR (1978) Concepts of radiation dosimetry. Pergamon, New York

Kawrakow I, Rogers DWO (2003) The EGSnrc Code System: Monte Carlo simulation of electron and photon transport. National Research Council of Canada Report PIRS-701, Ottawa

Koch HW, Motz JW (1959) Bremsstrahlung cross-section formulas and related data. Rev Mod Phys 31:920–955

Kölbig KS, Schorr B (1984) Asymptotic expansions for the Landau density and distribution functions. Comput Phys Commun 32:121–131

Landau LD (1944) On the energy loss of fast particles by ionization. J Phys USSR 8:201

Levin CS, Hoffman EJ (1997) Calculation of positron range and its effect on the fundamental limit of positron emission tomography system spatial resolution. Phys Med Biol 44:781–799

Lindhard J, Scharff M (1953) Energy loss in matter by fast particles of low charge. K Dansk Vidensk Selsk, Mat-Fys Medd 34, No. 15

Lindhard J (1976) The Barkas effect – or Z_1^3, Z_1^4-corrections to stopping of swift charged particles. Nucl Instr Meth 132:1–5

McKinley WA, Feshbach H (1948) The Coulomb scattering of relativistic electrons by nuclei. Phys Rev 74:1759–1763

McParland BJ (1989) A derivation of the electron mass scattering power for electron dose calculations. Nucl Instr Meth in Phys Res A274:592–596

McParland BJ, Cunningham JR, Woo MK (1988) Electron beam dose calculations for heterogeneous media. Med Phys 15:489–497

Massey HSW, Corben HS (1939) Elastic collisions of mesons with electrons and protons. Proc Camb Philol Soc 35: 463–473

Molière VG (1947) Theorie der Streuung schneller geladener Teilchen, I. Einzelstreuung am ambgeschirmten Coulomb-field. Z Naturforschg 2a:133–145

Molière VG (1948) Theorie der Streuung schneller geladener Teilchen, II. Mehrfach-und Veilfachstreuung. Z Naturforschg 3a:78–97

Møller C (1932) Zur theorie des durchgangs schneller electronen durch materie. Ann Physik 14:531–585

Morgan SH Jr (1970) Coulomb corrections to the Bethe–Heitler cross sections for electron-nucleus bremsstrahlung. NASA Technical Note D-6038. National Aeronautics and Space Administration, Washington, DC, October 1970

Mott NF (1929) Scattering of fast electrons by atomic nuclei. Proc Roy Soc A124:425–442

Mott NF (1932) The elastic scattering of fast positrons by heavy nuclei. Proc Roy Soc A135:429

Moyal JE (1955) Theory of ionization fluctuations. Phil Mag 46:263–280

Nobel JA, Trickey SB, Sabin JR, Oddershede J (2005) Basis set limitations on the ab initio calculation of stopping cross-sections via generalized oscillator strengths. Chem Phys 209:89–94

Oppenheimer JR, Snyder H, Serber R (1940) The production of soft secondaries by mesotrons. Phys Rev 57:75–81

Pratt RH, Tseng HK, Lee CM, Kissel L, MacCallum C, Riley M (1977) Bremsstrahlung energy spectra from electrons of kinetic energy 1 keV T_1 2000 keV incident on neutral atoms 2 Z 92. At Data Nucl Data Tables 20:175

Rohrlich F, Carlson BC (1954) Positron-electron differences in energy loss and multiple scattering. Phys Rev 93:38–44

Rossi B (1952) High-energy particles. Prentice-Hall, New York

Rossi B, Greisen K (1941) Cosmic-ray theory. Rev Mod Phys 13:240–309

Rotondi A, Montagna P (1990) Fast calculation of Vavilov distribution. Nucl Instr Meth in Phys Res B47:215–223

Salvat F, Fernandéz-Varea JM, Sempau J, Mazurier J (1999) Practical aspects of Monte Carlo simulation of charged particle transport: mixed algorithms and variance reduction techniques. Radiat Environ Biophys 38:15–22

Segrè E (1977) Nuclei and particles, 2nd edn. Benjamin/Cummings, Reading

Sternheimer RM, Peierls RF (1971) General expression for the density effect for the ionization loss of charged particles. Phys Rev B 3:3681–3692

Symon KR (1948) Fluctuations in energy lost by high energy charged particles in passing through matter. Thesis, Harvard University, Cambridge

Tabata T, Ito R (1976) An improved interpolation formula for the parameter B in Molière's theory of multiple scattering. Jpn J Appl Phys 15:1583–1584

Tamm IE, Frank IM (1937) Coherent radiation of fast electrons in a medium. Doklady Akad Nauk SSR, 14:107

Van Ginneken A (2000) Edgeworth series for collision energy loss and multiple scattering. Nucl Instr Meth B160: 460–470

Vavilov PV (1957) Ionization losses of high-energy heavy particles. Sov Phys JETP 5:749–751

Sauli F (1977) Principles of operation of multiwire proportional and drift chambers. CERN Report 77–09, CERN, Geneva

Schorr B (1974) Programs for the Landau and the Vavilov distributions and the corresponding random numbers. Comput Phys Comm 7:215–224

Schorr B (1975) Numerical inversion of a class of characteristic functions. BIT Numer Math 15:94–102

Scott WT (1963) The theory of small-angle multiple scattering of fast charged particles. Rev Mod Phys 35:231–313

Seltzer SM (1991) Electron-photon Monte Carlo calculations: the ETRAN code. Appl Radiat Isot 42:917–935

Spencer LV, Fano U (1954) Energy spectrum resulting from electron slowing down. Phys Rev 93:1172–1181

Uehling EA (1954) Penetration of heavy charged particles in matter. Annu Rev Nucl Sci 4:315–350

Ziegler JF (1999) The stopping of energetic light ions in elemental matter. J Appl Phys 85:1249–1272

Radiation Fields and Radiometrics 8

Abstract This chapter summarizes the descriptors of radiation fields and the definitions of the dosimetric and radiometric quantities to be used in later discussions of ionizing radiation dosimetry.

Contents

8.1 Introduction

The emphasis to this point has been upon the microscopic theories describing individual particles, nuclei and their interactions. This chapter begins focuses upon macroscopic theory in which moving ensembles of photons and charged particles are considered as radiation fields varying continuously in space and time and interacting continuously with matter. Clearly, this description is of the average (non-stochastic) behavior of radiation and the quantities used to describe it will be the expectation values defined in differential form. There are also non-stochastic attributes of radiation that arise at small spatial dimensions (e.g., that of the cell) and these are described by microdosimetry theory.

Before embarking on this discussion, the definitions of indirectly- and directly-ionizing radiation recommended by the International Commission on Radiation Units and Measurements (ICRU 1971) are summarized:

Indirectly-ionizing radiation: These are uncharged particles (photons or neutrons) which transfer their energy through intermediate charged particles (e.g., electrons) in a relatively few large interactions, reflected by the long mean-free path.

Directly-ionizing radiation: These are charged particles which transfer their energy directly to matter through multiple small-energy transfer Coulomb interactions.

However, ionization is not the only process resulting from the interaction between radiation and matter in which energy is transferred. Excitation is a process which can have significant biological, physical or chemical effects. The demarcation between excitation and ionization is set by an energy threshold which will vary between media and will become important to consider when dealing with low-energy radiations or, indeed, when comparing positron and electron interactions with matter.

With the exception of linear energy transfer (LET), the Système International (SI) units will be used in describing dosimetric and radiometric quantities.

B.J. McParland, *Nuclear Medicine Radiation Dosimetry*,
DOI 10.1007/978-1-84882-126-2_8, © Springer-Verlag London Limited 2011

Further reviews of radiation fields and their metrics in non-stochastic and stochastic applications are provided by Alm Carlsson (1985) and Kellerer (1985).

8.2 Radiation Fields

8.2.1 Phase Space

A radiation field is defined by a six-dimensional phase space (position and momentum). The differential element of this phase space is given by $d^3\mathbf{r}\, d^3\mathbf{p}$ where the momentum vector is given by,

$$\mathbf{p} = |\mathbf{p}|\hat{\boldsymbol{\Omega}} \tag{8.1}$$

where $\hat{\boldsymbol{\Omega}}$ is the unit vector in the direction of motion of the field. The differential $d^3\mathbf{p} = p^2\, dp\, d^2\hat{\boldsymbol{\Omega}}$ can be rewritten using the relativistic relationship between momentum, kinetic energy T and rest mass m, $p^2 = \left((T+m)^2 - m^2\right)$,

$$d^3\mathbf{p} = m^2\sqrt{x^4 + 4x^3 + 5x^2 + 2x}\; dT\, d^2\hat{\boldsymbol{\Omega}} \tag{8.2}$$

where $x = (T/m)$ and $d^2\hat{\boldsymbol{\Omega}} = d(\cos\theta)d\varphi$.

8.2.2 Particle Number, Radiant Energy, and Particle Radiance

The number of particles present, emitted, transferred or received in the radiation field at a point $\mathbf{r}$ is N. In nuclear medicine, the radiation field at a point internal to the body is typically a mixed field consisting of a combination of photons, electrons and/or positrons. The particle number will generally exhibit a kinetic energy spectrum,

$$N_T(T;\mathbf{r}) = \frac{dN(T;\mathbf{r})}{dT} \tag{8.3}$$

so that $N_T(T;\mathbf{r})\, dT$ represents the number of particles with kinetic energies between T and T + dT at the point $\mathbf{r}$.

The radiant energy is the energy of particles (excluding rest masses) emitted, transferred, or received at a point $\mathbf{r}$,

$$\Re(\mathbf{r}) = \int dT\, T\, N_T(T;\mathbf{r}). \tag{8.4}$$

In order to describe completely a polyenergetic mixed radiation field at point $\mathbf{r}$ and time t, it is necessary to specify the number N and species j of particles,[1] their kinetic energies T and their direction of motion $\hat{\boldsymbol{\Omega}}$. The expectation value of the number of particles N of species j with a kinetic energy in the interval dT about T and a direction of motion in the interval $d\hat{\boldsymbol{\Omega}}$ about $\hat{\boldsymbol{\Omega}}$ passing through an area element $dA_\perp$ at the point $\mathbf{r}$ orthogonal to the direction of motion about $\hat{\boldsymbol{\Omega}}$ during the time interval dt around t is given by the quantity $p_{T,j}(\mathbf{r})\, dt\, dT\, d\hat{\boldsymbol{\Omega}}\, dA_\perp$ where $p_{T,j}$ is the spectral distribution of the particle radiance, p, with respect to kinetic energy. In other words, for species j,

$$\begin{aligned} p_{T,j}(\mathbf{r}) &\equiv \frac{dp_j(\mathbf{r})}{dT} \\ &\equiv \frac{d^4N_j\left(t, T, \hat{\boldsymbol{\Omega}}; \mathbf{r}\right)}{dt\, dT\, d\hat{\Omega}\, dA_\perp}. \end{aligned} \tag{8.5}$$

The particle radiance is simply this spectrum integrated over the particle kinetic energies,

$$\begin{aligned} p_j(\mathbf{r}) &= \int dT\, p_{T,j}(\mathbf{r}) \\ &= \frac{d^3N_j\left(t, \hat{\boldsymbol{\Omega}}; \mathbf{r}\right)}{dt\, d\hat{\Omega}\, dA_\perp} \end{aligned} \tag{8.6}$$

These are non-stochastic quantities and successive integrations over time, kinetic energy and direction of motion are used to define the most common metrics describing the radiation field.

For the remainder of the discussion concerning the radiance, a single particle species is assumed. For fields consisting of multiple species, calculations are simply summed over the fractional contributions of each species, j.

[1] "Particles" includes photons in this discussion.

8.2.3 Scalar Radiometric Quantities

8.2.3.1 Particle Flux Density (Particle Fluence Rate)

The particle flux density is mean the number of particles moving through a unit area perpendicular to the direction of motion at a point **r** per unit time and is given by the integrals of the particle radiance spectral function over kinetic energy and direction,

$$\begin{aligned}\frac{d\Phi}{dt}(\mathbf{r}) &= \int dT \int d\hat{\boldsymbol{\Omega}}\, p_T(\mathbf{r}) \\ &= \frac{d^2N(t;\mathbf{r})}{dt\, dA}.\end{aligned} \tag{8.7}$$

The particle radiance can also be written in terms of the quotient of the particle flux density to the differential solid angle,

$$p(\mathbf{r}) = \frac{d^2\Phi}{dt\, d\Omega}. \tag{8.8}$$

The SI unit for the particle flux density is $m^{-2}\, s^{-1}$.

8.2.3.2 Particle Fluence

The particle fluence is a non-stochastic quantity and is the mean number of particles moving through a unit area perpendicular to the direction of motion at a point **r** and is given by the time integral of the particle flux density,

$$\Phi(\mathbf{r}) = \int dt \int dT \int d\hat{\boldsymbol{\Omega}}\, p_T(\mathbf{r}) = \frac{dN(\mathbf{r})}{dA} \tag{8.9}$$

which is the number of particles traversing a sphere with cross-sectional area dA centered at **r**. The SI unit for particle fluence is m^{-2}.

8.2.3.3 Energy Flux Density (Energy Fluence Rate)

In addition to characterizing the transport of individual particles, it is necessary to consider the transport of energy by the radiation field. The energy flux density is obtained by weighting the particle radiance spectral distribution by the energy,

$$\begin{aligned}\frac{d\Psi}{dt}(\mathbf{r}) &= \int dT \int d\hat{\boldsymbol{\Omega}}\, T\, p_T(\mathbf{r}) \\ &= \int dT\, T \frac{d^3N_j(t,T;\mathbf{r})}{dt\, dT\, dA}.\end{aligned} \tag{8.10}$$

From (8.4), the energy flux density can also be written in terms of the radiant energy,

$$\frac{d\Psi}{dt}(\mathbf{r}) = \frac{d\,\Re(\mathbf{r})}{dt\, dA}. \tag{8.11}$$

The SI unit for the energy flux density is $W\, m^{-2}$.

8.2.3.4 Energy Fluence

Integrating the energy fluence rate over time yields the energy fluence,

$$\begin{aligned}\Psi(\mathbf{r}) &= \int dt \int dT \int d\hat{\boldsymbol{\Omega}}\, T \frac{d^4N_j\left(t,T,\hat{\boldsymbol{\Omega}};\mathbf{r}\right)}{dt\, dT\, d\hat{\Omega}\, dA_\perp} \\ &= \int dT\, T \frac{d^2N_j(T;\mathbf{r})}{dT\, dA}\end{aligned} \tag{8.12}$$

or, in terms of the radiant energy,

$$\Psi(\mathbf{r}) = \frac{d\,\Re(\mathbf{r})}{dA}. \tag{8.13}$$

The SI unit for the energy fluence is $J\, m^{-2}$.

8.2.4 Vector Radiometric Quantities

8.2.4.1 Vector Radiance

The basic quantity for describing the vector attributes of the radiation field is the vector spectral distribution of particle radiance which is the scalar spectral distribution particle radiance multiplied by the unit vector in the direction of motion,

$$\mathbf{p}_T(\mathbf{r}) = p_T(\mathbf{r})\hat{\boldsymbol{\Omega}}. \tag{8.14}$$

8.2.4.2 Vector Particle Flux Density (Vector Particle Fluence Rate)

As in the scalar case, integrating (8.14) over kinetic energy and direction of motion yields the vector particle flux density,

$$\begin{aligned}\dot{\boldsymbol{\Phi}}(\mathbf{r}) &= \int \mathrm{d}T \int \mathrm{d}\hat{\boldsymbol{\Omega}}\, \mathbf{p}_T(\mathbf{r}) \\ &= \int \mathrm{d}\hat{\Omega}\hat{\boldsymbol{\Omega}} \frac{\mathrm{d}^3 N\left(t, \hat{\boldsymbol{\Omega}}; \mathbf{r}\right)}{\mathrm{d}t\, \mathrm{d}\hat{\Omega}\, \mathrm{d}A_\perp}.\end{aligned} \tag{8.15}$$

8.2.4.3 Vector Particle Fluence

Integrating (8.15) over time yields the vector particle fluence,

$$\boldsymbol{\Phi}(\mathbf{r}) = \int \mathrm{d}\hat{\Omega}\hat{\boldsymbol{\Omega}} \frac{\mathrm{d}^2 N\left(\hat{\boldsymbol{\Omega}}; \mathbf{r}\right)}{\mathrm{d}\hat{\Omega}\, \mathrm{d}A_\perp}. \tag{8.16}$$

The vector particle fluence is the sum of vectors whereas, from (8.9), the scalar particle fluence is the scalar sum of vector lengths. Thus,

$$\begin{aligned}|\boldsymbol{\Phi}| &= \sqrt{\boldsymbol{\Phi} \bullet \boldsymbol{\Phi}} \\ &\le \Phi\end{aligned} \tag{8.17}$$

Equation 8.17 will become an equality for monodirectional radiation only and, indeed, $|\boldsymbol{\Phi}| = 0$ for an isotropic radiation field.

8.2.4.4 Vector Energy Flux Density (Vector Energy Fluence Rate)

Again, by analogy, this is given by,

$$\dot{\boldsymbol{\Psi}} = \int \mathrm{d}T\, T\, \hat{\boldsymbol{\Omega}} \frac{\mathrm{d}^3 N_j(t, T; \mathbf{r})}{\mathrm{d}t\, \mathrm{d}T\, \mathrm{d}A}. \tag{8.18}$$

8.2.4.5 Vector Energy Fluence

Integrating (8.18) over time gives the vector energy fluence,

$$\boldsymbol{\Psi} = \int \mathrm{d}T\, T\, \hat{\boldsymbol{\Omega}} \frac{\mathrm{d}^2 N_j(t, T; \mathbf{r})}{\mathrm{d}T\, \mathrm{d}A}. \tag{8.19}$$

8.3 Energy Exchange

8.3.1 Introduction

In Chaps. 6 and 7, the microscopic theories of the energy transfer and absorption interactions between a medium and individual photons and charged particles were derived. These were described by non-stochastic interaction coefficients. For photons, these are the attenuation, energy transfer and energy absorption coefficients. For charged particles, these were the collision and radiative stopping powers. Most dosimetry calculations deal with macroscopic ensembles of photons and charged particles where these non-stochastic coefficients are applicable. It is frequently necessary to calculate energy transfer and absorption at spatial dimensions (such as the cellular nucleus) that are comparable to the dimensional scales of individual interaction sites. As a result, stochastic descriptors are required where, as the spatial dimensions under consideration increase, the values of the stochastic quantities approach their associated non-stochastic means. In this section, macroscopic stochastic and non-stochastic descriptors of energy exchange from radiation to a medium are described. By definition, a stochastic radiometric is defined for finite domains only, is discontinuous in both space and time and is defined through a probability distribution function. A non-stochastic radiometric is a continuous and differentiable function in both space and time and is considered as the mean of the associated stochastic radiometric.

8.3.2 Stochastic Quantities

8.3.2.1 Energy Deposit

The energy deposit is the fundamental stochastic quantity and is defined as the energy deposited at a point as the result of the *i*th single interaction and is given by the difference between the energy of the incoming ionizing particle (neglecting any rest mass) and the sum of the energies of the outgoing ionizing particles (again neglecting any rest mass) plus any changes in rest mass of all particles and atoms involved in the interaction,

$$\varepsilon_i = T_{in} - T_{out} + Q_{\Delta m} \tag{8.20}$$

where $Q_{\Delta m}$ is negative if there is an increase in rest mass and positive if there is a decrease. The three conditions of $Q_{\Delta m} < 0$, $Q_{\Delta m} = 0$ and $Q_{\Delta m} > 0$ are considered next.

$Q_{\Delta m} < 0$

A negative $Q_{\Delta m}$ means that part of the incoming kinetic (radiant) energy is expended to increase the rest masses of nuclei and particles. An example is electron–positron pair production through a photon interaction with the nuclear electromagnetic field. The energy deposit for a photon with energy k and electron and positron kinetic energies of T_- and T_+, respectively, is,

$$\begin{aligned}\varepsilon_i &= T_{in} - T_{out} + Q_{\Delta m}\\ &= k - T_- - T_+ - 2m_e \\ &= T_{Rec}\end{aligned} \tag{8.21}$$

where T_{Rec} is the recoil kinetic energy of the nucleus and where $T_- + T_+ = k - 2m_e - T_{Rec}$ has been used. As T_{out} is the total kinetic energy of the outgoing ionizing particles, a subtle consideration of the electron and positron products must be considered. The positron is always capable of ionization but if the electron is produced with T_- below the ionization potential of the atom, it can no longer ionize and will thus be part of ε_i.

$Q_{\Delta m} = 0$

A zero $Q_{\Delta m}$ arises when matter is neither created nor annihilated. Consider an incoming charged particle with kinetic energy T_{in} which knocks out an atomic electron (δ ray) with kinetic energy T_δ. The excited atom then de-excites by the emission of a characteristic x ray of energy k_k and an Auger electron with kinetic energy T_A. The original charged particle exits the interaction with kinetic energy $T_{in} - T$, where T is the kinetic energy lost by the incident particle through the interaction Then,

$$\begin{aligned}\varepsilon_i &= T_{in} - T_{out}\\ &= T_{in} - (k_k + T_\delta + T_A + T_{in} - T)\\ &= T - T_\delta - T_A - k_k.\end{aligned} \tag{8.22}$$

The energy deposit is less than the energy lost by the incident charged particle.

$Q_{\Delta m} > 0$

A positive $Q_{\Delta m}$ is the result of rest mass being converted into, for example, kinetic energy of ionizing particles. An obvious example is positron annihilation. As $Q_{\Delta m}$ is positive, the definition of the energy deposit is rewritten as,

$$\varepsilon_i = Q_{\Delta m} - (T_{out} - T_{in}) \tag{8.23}$$

where the quantity in parentheses is that amount of the released rest-mass energy which appears as kinetic or radiant energy of the exiting ionizing particles. For example, consider the annihilation $e^-e^+ \rightarrow 2\gamma$ on an atomic electron. An excited atom remains which, in this example, de-excites through the emission of a characteristic x ray with energy k_k and an Auger electron with kinetic energy T_A. The energy deposit is

$$\begin{aligned}\varepsilon_i &= 2m_e - (T_{out} - T_{in})\\ &= 2m_e - (k_1 + k_2 + k_k + T_A - T_{in})\end{aligned} \tag{8.24}$$

where k_1 and k_2 are the radiant energies of the two annihilation photons. These are,

$$k_1 + k_2 = 2m_e - E_B - T_{in} \tag{8.25}$$

where E_B is the atomic electron binding energy. Substituting this into the expression for the energy deposit gives,

$$\varepsilon_i = E_B - T_A - k_k. \tag{8.26}$$

This is the same energy deposit result for a zero $Q_{\Delta m}$ (8.22) by identifying E_B as $T - T_\delta$. Importantly, T_{in} does not appear in the result of (8.26) so, hence, the energy deposit is the same whether the positron annihilation occurs after the positron has thermalized ($T_{in} \approx 0$) or occurs in flight ($T_{in} > 0$).

Another example of a $Q_{\Delta m} > 0$ process is β decay. In fact, it is possible for the energy deposit to be negative in β^- decay as the resulting increase in the atomic number leads to an increase in the binding energy of the atomic electrons.

8.3.2.2 Energy Imparted

The energy imparted to a volume within the medium is the summation of the individual energy deposits within that volume occurring within a specified time interval,

$$\varepsilon = \sum_{i} \varepsilon_i. \tag{8.27}$$

With the exception of that incident, the ionizing particles interacting within the volume have been produced either by interactions due to the incident particle or through those emitted by radioactive decay should the volume contain radioactive nuclei. A pragmatic consideration within nuclear medicine dosimetry is that of K ionizing particles entering a volume containing radioactive nuclei of which N decay during the time interval of interest (Alm Carlsson 1979). The energy imparted to the matter within this volume is given by,

$$\varepsilon = \sum_{j=1}^{K+N} \sum_{i=1}^{n_j} \varepsilon_{ij} \tag{8.28}$$

where n_j is the number of correlated interactions within the volume due to the *j*th incoming particle or the *j*th decaying nucleus. Inserting (8.20)

$$\begin{aligned} \varepsilon &= \sum_{j=1}^{K+N} \sum_{i=1}^{n_j} \{T - T_{out,k} + Q_{\Delta m}\}_{ij} \\ &= \sum_{j=1}^{K} T_{in,j} - \sum_{m=1}^{L} T_{out,m} + \sum_{n=1}^{M} Q_{\Delta m,n}. \end{aligned} \tag{8.29}$$

The first summation gives the sum of the kinetic (radiant) energies of the incident particles that interact within the volume; the second is the sum of kinetic (radiant) energies of ionizing particles created within the volume and have exited without interactions and the last is the release of rest mass energies.

8.3.2.3 Energy Transferred

The energy transferred in a volume is defined as (ICRU 1980; Attix 1983),

$$\varepsilon_{tr} = \Re_{in}^{unch} - \Re_{out}^{unch} + Q_{\Delta m} \tag{8.30}$$

where $\Re_{in}^{unch}$ is the radiant energy of uncharged particles entering the volume and $\Re_{out}^{unch}$ is the radiant energy of uncharged particles leaving the volume except that which is due to *bremsstrahlung* production or in-flight positron annihilation by charged particles in the volume. Hence, the energy transferred is simply the kinetic energy received by charged particles in the volume[2]

8.3.2.4 Specific Energy

The specific energy (imparted) is the quotient of the energy imparted to the medium of mass m,

$$z = \frac{\varepsilon}{m}. \tag{8.31}$$

z may be due to one or more energy deposition events. The distribution function $F_1(z)$ is the conditional probability that a specific energy $\leq$ z is deposited if a single event has occurred; the generalization F(z) is the conditional probability that a specific energy $\leq$ z has been deposited.

8.3.2.5 Lineal Energy

The lineal energy is the quotient of the energy imparted ε to a volume by a single energy deposition to the mean chord length $\bar{l}$ in that volume,

$$y = \frac{\varepsilon}{\bar{l}}. \tag{8.32}$$

The mean chord length is the mean value of the lengths of randomly-oriented chords. For example, the mean chord length of a convex body of volume V and surface area is $\bar{l} = 4V/A$.

As y is a stochastic quantity, it is described by a probability distribution function. F(y) is the distribution function and is equal to the probability that the lineal energy is less than or equal to y. The probability density function, f(y), is equal to the derivative of F(y) with respect to y, $f(y) = dF(y)/dy$.

[2]Kinetic energy transferred from one charged particle to another is excluded from the calculation of ε_{tr}.

The lineal energy transfer should not be confused with the LET. The latter is a non-stochastic quantity equal to the mean energy loss per unit pathlength travelled by a charged particle and subject to an energy transfer cut-off.

8.3.3 Non-Stochastic Quantities

The expectation values of the stochastic quantities just introduced are non-stochastic metrics. These quantities are defined here.

8.3.3.1 Kerma

Kerma is the acronym for "kinetic energy released per unit mass" and is applicable only to indirectly-ionizing radiations (photons or neutrons) and is the quotient of the sum of the kinetic energies, dE_{tr}, of all charged particles released by uncharged ionizing particles in a volume of medium of mass dm,

$$K = \frac{dE_{tr}}{dm}. \tag{8.33}$$

The kerma includes both the energies of Auger electrons and that radiated by the secondary charged particles as *bremsstrahlung*. Kerma is related to the energy fluence by,

$$K = \Psi\left(\frac{\mu_{tr}}{\rho}\right) \tag{8.34}$$

where (μ_{tr}/ρ) is the mass energy-transfer coefficient. For a polyenergetic photon beam, the kerma can be written as,

$$K = \int dk \frac{d\Phi}{dk} k \frac{\mu_{tr}}{\rho}(k) \tag{8.35}$$

The SI unit for kerma is J kg^{-1} or gray (Gy) where 1 Gy = 1 J kg^{-1}.

8.3.3.2 Air Kerma-Rate Constant

The air kerma-rate constant for a given radioactive photon-emitting isotope is the ratio of $l^2\dot{K}_\delta$ to the activity A of the isotope,

$$\Gamma_\delta = \frac{l^2\dot{K}_\delta}{A} \tag{8.36}$$

where $\dot{K}_\delta$ is the in-air kerma rate due to photons with an energy greater than δ at a point a distance l from the radioactive source. The SI unit for the air kerma-rate constant is m^2 J kg^{-1}. Γ_δ will be specific to the radionuclide and accounts for the contributions due to characteristic x rays and internal *bremsstrahlung*. It supplants the specific γ-ray constant and exposure-rate constant previously defined by the ICRU. Knowing the photon energies emitted by the radionuclide, it is possible to calculate the air kerma-rate constant from first principles using (8.35) and (8.36).

8.3.3.3 Absorbed Dose

The noun "dose" is multi-varied in medicine. Unlike the pharmacological dose of a therapeutic drug, the absorbed dose in radiation physics is the consequence of the interaction between ionizing radiation and the medium. The absorbed dose to a volume of medium of mass dm is the quotient of the mean energy imparted by ionizing radiation, $d\bar{\varepsilon}$, to that volume of the medium of mass dm,

$$D = \frac{d\bar{\varepsilon}}{dm}. \tag{8.37}$$

An alternative definition of the absorbed dose arises from the mean specific energy,

$$\bar{z} = \int_0^\infty dz\, z\, f(z) \tag{8.38}$$

which is a non-stochastic quantity leading to,

$$D = \lim_{m \to o} \bar{z}. \tag{8.39}$$

The SI unit for absorbed dose is Gy.

8.3.3.4 Exposure

Exposure is both an historical and a practical quantity used in radiation field measurements. It is the quotient of the absolute value of the total electric charge of the ions of one sign produced in air by photons, dQ, when all of the resulting electrons and positrons released in a differential mass of air dm are stopped in air,

$$X = \frac{dQ}{dm}. \tag{8.40}$$

This definition of exposure excludes any ionization resulting from the absorption of *bremsstrahlung* emitted by the released electrons, but this is only significant at photon energies much higher than those considered in nuclear medicine and, hence, can be neglected in our discussion.

Exposure can also be defined by,

$$X = \Psi\left(\frac{\mu_{en}}{\rho}\right)\frac{e}{W}. \tag{8.41}$$

The SI unit for exposure is C kg^{-1}; the special unit for exposure is the röntgen (R), where 1R = 2.58×10^{-4} C kg^{-1}.

References

Alm Carlsson G (1979) Definition of energy imparted. A new formulation adapted to exact solutions of the absorbed dose equation under nonequilibrium conditions. Radiat Res 77:209–220

Alm Carlsson G (1985) Theoretical basis for dosimetry. In: Kase KR, Bjärngard BE, Attix FH (eds) The dosimetry of ionizing radiation, vol I. Academic, Orlando

Attix FH (1983) Energy imparted, energy transferred and net energy transferred. Phys Med Biol 28:1385–1390

ICRU (1971) Radiation quantities and units. Report 19, International Commission on Radiological Units and Measurements, Bethesda, MD

ICRU (1980) Radiation quantities and units. Report 33, International Commission on Radiological Units and Measurements, Bethesda, MD

Kellerer AM (1985) Fundamentals of microdosimetry. In: Kase KR, Bjärngard BE, Attix FH (eds) The dosimetry of ionizing radiation, vol I. Academic, Orlando

Kellerer AM (1990) A generalised formulation of microdosimetric quantitites. Rad Prot Dosim 31:9–16

Radiation Dosimetry: Theory, Detection, and Measurement

9

Abstract This chapter provides the bridge from the microscopic theories of radiation-matter interactions to the macroscopic theory of ionising radiation dosimetry and creates the foundation for presentation of nuclear medicine radiation dosimetry. As such, the chapter is divided into two sections. The first provides the theoretical basis of radiation dosimetry. Radiation equilibria and the important role they play in dosimetry evaluations are, in particular, emphasised. The means of calculating the radiation flux, essential to any dosimetry calculation, are reviewed through analytical means, Monte Carlo simulation and, in particular, the semi-empirical build-up factor method. The reciprocity theorem, which plays an implicit role in some nuclear medicine dosimetry calculations, is presented and studied as is the fundamental Bragg–Gray cavity theory. The theory section concludes with an overview of microdosimetry. The second section presents the underlying theories of operation of radiation detectors used in nuclear medicine. These include gaseous detectors based upon the collection of ions generated within an irradiated gas volume (ionisation chambers, proportional counters, and Geiger–Müller counters), scintillation and photoconversion devices, semiconductor detectors, and thermoluminescence.

Contents

9.1 Introduction

Chapters 6 and 7 presented the microscopic theories of photon and charged particle interactions with matter which lead to the absorption of energy in the medium. These theories are categorized as being microscopic in that they describe the interactions of a single quantum or charged particle with an individual atom, electron, or nucleus. In practice, radiation dosimetry must extend these descriptions of the interaction of radiations with matter and the absorbed dose to the macroscopic level. Here, the adjective macroscopic is used as the theory becomes that of one of ensembles of particles or photons that are sufficiently large that they can be described as radiation fields which are both continuous and differentiable with respect to

B.J. McParland, *Nuclear Medicine Radiation Dosimetry*,
DOI 10.1007/978-1-84882-126-2_9,

position and time. Intermediate to these two theories is a mesoscopic theory, referred to as microdosimetry, which is probabilistic. Microdosimetry[1] is the study of the deposition of energy by ionizing particles at dimensions of between about 1 nm (the diameter of the DNA molecule) to about 20–30 μm (the size of the largest mammalian cells).

The discipline of radiation dosimetry can be traced back to the discoveries and classifications of various types of ionizing radiations by Röntgen, Becquerel, the Curies, Rutherford and Villard. Following these discoveries, there was an appreciated need to quantify radiation. Becquerel's serendipitous discovery of radioactivity through the opacification of a photographic emulsion by the radiations emitted by uranium led obviously to the consideration of the degree of photographic film darkening as a means of quantifying exposure to radiation. Similarly, the Curies' observation that radiation could ionize air and that the resulting electric charges collected led to the suggestion by Villard (1908) of ionization as a means of quantifying the radiation causing the ionization. Ionization as a phenomenon for quantifying a radiation field has subsequently dominated the discipline to this day, but it was certainly not the only metric employed. Optical means, other than photographic emulsion, of quantifying radiation were also employed, including irradiation-induced color changes in platino-barium-cyanide capsules (Jennings 2007). A biological response to photon irradiation, the skin erythema dose, was briefly considered as a means of quantifying radiation dose (Failla 1921).

Although ionization became the dominant means of quantifying radiation (and not dose as we currently interpret the quantity), there were considerable differences between the different definitions of units used to describe ionization. For example, Villard defined his ionization unit as the quantity of X-rays which liberated 1 electrostatic unit (equal to 0.334 nC) of "electricity" per 1 cm^3 of air at normal temperature and pressure (NTP). This definition was highly predictive of the röntgen. Villard's unit was eventually to become the "Germany röntgen", R, whereas the corresponding "French röntgen" was defined as the ionization in air at a point 2 cm from a 1 g source of radium. The lack of coordination between the uses of units in quantifying radiation was extremely troublesome in radiation therapy and was first addressed by the International Congress of Radiology (ICR). The ICR formed in 1925 an "International X-ray Units Committee," which was to be the predecessor of the ICRU (International Commission on Radiation Units and Measurement). In 1928, the ICR formed an "International X-ray and Radium Committee," which was to then evolve to become the ICRP (International Commission on Radiological Protection). Between the 1928 meeting of the ICR in Stockholm and its 1937 meeting in Chicago, a final definition of the röntgen as a unit of exposure as a means to quantify X- and γ-irradiation was at last proposed, refined, and agreed upon.

Extending upon the use of ionization as a means of quantifying radiation, interest next focused upon energy transfer, deposition, and absorption. Following the Second World War, the term dose was proposed as the amount of energy absorbed by an irradiated medium. At the 1950 ICR meeting in London, the absorbed dose was defined to describe the amount of energy absorbed per unit mass of an irradiated medium at a point in that medium. The centimeter-gram-second (CGS) unit of erg per gram was applied and, at the following ICR meeting in 1953 in Copenhagen, the rad was defined as the unit of absorbed dose The rad was an acronym for radiation absorbed dose where 1 rad = 100 erg/g. Beginning in the 1970s, the Système International (SI) units were adopted by the radiation dosimetry and metrology communities and led to the SI definitions in ICRU Report 33 (1980) of absorbed dose and kerma as the gray where 1 Gy = 1 J/kg. However, the practice of using the unit rad , however, has been difficult to kill off and one often finds that prescribed doses in radiation oncology departments are persistently (yet conveniently and transparently) given in units of centigrays where 1 cGy = 1 rad.

9.2 Radiation Dosimetry: Theory

9.2.1 Primary and Scattered Radiation Fields

It is frequently necessary in dosimetry to separate ionizing radiation within a medium into two categories.

[1]Microdosimetry is a radiological dosimetry discipline and should not be confused with the pharmacological concept of microdosing.

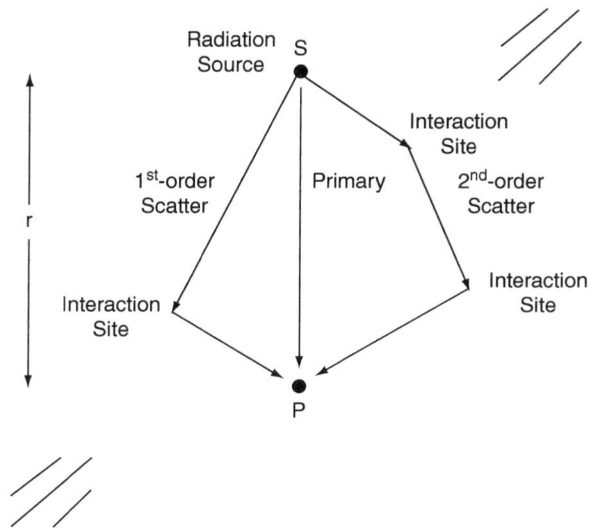

Fig. 9.1 Definition of primary, first-order and second-order scatter components for radiation traveling from the source S to a detector at P a distance r away

In one, the radiation has interacted with the medium it is traveling through and, in the other category, it has not. The latter component is referred to as "primary" radiation and the former is referred to as "scattered" radiation. The differentiation between these radiation components is explained in Fig. 9.1. This shows an isotropic source of photons at S and a radiation detector at P, a distance r away. Should both the source and the detector be in vacuo the radiation detected at P will be that which had traveled along the line SP. Because of the isotropicity of the source, the intensity of this radiation will vary as r^{-2}. However, once S and P are placed within a scattering and absorbing medium, the detector at P will detect not only those primary photons which have traveled along SP without interaction, but also those which left S in directions other than SP and which were subsequently scattered through one or more interactions in order to eventually reach the detector at P. In addition to having reached the detector at the point P from directions different than SP, this scattered radiation will differ from the primary radiation in its having a reduced energy spectrum due to energy losses incurred as a result of scattering. The primary fluence at P will have decreased due to attenuation along the line SP whereas the scattered fluence at P will have increased.

It is possible to isolate experimentally the primary and scattered components of the total radiation reaching the point P if the source of radiation S is external to the medium containing P. This is done by measuring the radiation fluence (or some other metric such as absorbed dose or kerma) at the point P for a variety of field sizes and to then extrapolate the results as a function of field size to zero width which will leave only the primary component (Clarkson 1941; Cunningham 1972). Analytical means of calculating the primary and secondary components through, for example, the Boltzmann equation and other direct approaches, are also possible (Berger and Spencer 1959; Wong et al. 1981), but these are in practice limited to low orders of scatter due to the growing complexity of the calculation with increasing contributions of scatter. The technique of Monte Carlo simulation is the most practical means of calculating these components for medical applications (e.g., McParland 1981; Ljungberg et al. 1998; Zaidi and Andreo 2003).

Knowledge of the scatter contribution to the radiation fluence at P is of significant importance in both imaging and dosimetry calculations. Increased scatter radiation will reduce the spatial resolution of both emission and transmission imaging, thus requiring the use of devices such as the gamma-camera collimator or Bucky filter in radiographic imaging to remove the contribution of scattered photons to the image. Scatter can also increase the absorbed dose at P due to its contribution to the net radiation fluence at P. This is an effect quantified by the buildup factor.

9.2.2 Kerma and Absorbed Dose

9.2.2.1 Introduction

The irradiation of a medium by photons results in the liberation of charged particles. At the photon energies of interest to diagnostic and therapeutic nuclear medicine (which are largely below the electron–positron production threshold), these charged particles are exclusively electrons. This transfer of energy from the photons to the electrons is described by the kerma. These electrons subsequently slow down in the medium by transferring kinetic energy to atomic electrons. The absorption of the electron energies by the medium is described by the absorbed dose.[2] A calculation of this absorbed dose can be performed through analytical, numerical or Monte Carlo means, depending upon the complexity of the medium (in terms of geometry and composition) and the required degree of precision. Such calculations require knowledge of the particle fluences and energies and the coefficients which describe the interaction between the particles and the medium. Direct measurement of the absorbed dose is performed through calorimetry in which the temperature rise due to the absorbed energy is measured and the absorbed dose calculated directly, accounting for any available chemical energy loss. This is a challenging experimental measurement to perform and calorimetry is largely relegated to national physical standards laboratories. In clinical practice, absorbed dose is not measured directly but instead inferred from a variety of physical phenomena, such as ionization.

9.2.2.2 Radiation-Field Based Definitions

The acronym of kerma was defined as the "kinetic energy released per unit mass" by indirectly-ionizing radiation (photons and neutrons). For photon radiation, kerma is the sum of the kinetic energies of all of the charged particles released by photons in photon-matter interactions in a volume of medium per unit mass of that medium. It also includes the energies of any Auger electrons[3] released following atomic relaxation. This definition of kerma can be refined by applying radiation field descriptors to the geometry of a volume V of a medium of density ρ and surface area A which is exposed to a photon radiation field. The direction vector is defined as being positive if exiting from the volume. The net transport of radiant energy (8.4) through a differential surface area element dA into the volume (i.e., in the direction $-\mathbf{dA}$) is equal to the scalar product $-\boldsymbol{\Psi} \bullet \mathrm{d}\mathbf{A}$, where $\boldsymbol{\Psi}$ is the vectorial energy fluence of the radiation. The mean energy imparted to the volume will be the sum of the surface integral of this scalar product and the mean of the sum of any rest-mass energies released,

$$\bar{\varepsilon} = -\oiint_A \mathrm{d}\mathbf{A} \bullet \boldsymbol{\Psi} + \overline{\sum_i Q_i}. \tag{9.1}$$

The surface integral of $\boldsymbol{\Psi}$ in (9.1) is equal to the volume integral of the divergence of $\boldsymbol{\Psi}$ through Gauss' (or divergence) theorem,

$$\oiint_A \mathrm{d}\mathbf{A} \bullet \boldsymbol{\Psi} = \iiint_V \mathrm{dV}\, \nabla \bullet \boldsymbol{\Psi}. \tag{9.2}$$

Applying this theorem to (9.1) and separating the vectorial energy fluence into photon $\boldsymbol{\Psi}_\gamma$ and charged particle $\boldsymbol{\Psi}_c$ components, the mean energy imparted is,

$$\begin{aligned}\bar{\varepsilon} &= -\iiint_V \mathrm{dV}\, \nabla \bullet \boldsymbol{\Psi} + \overline{\sum_i Q_i} \\ &= \iiint_V \mathrm{dV} \left(-\nabla \bullet \boldsymbol{\Psi} + \frac{\mathrm{d}\left(\overline{\sum_i Q_i}\right)}{\mathrm{dV}} \right) \\ &= \iiint_V \mathrm{dV} \left(-\nabla \bullet (\boldsymbol{\Psi}_\gamma + \boldsymbol{\Psi}_c) + \frac{\mathrm{d}\left(\overline{\sum_i Q_i}\right)}{\mathrm{dV}} \right)\end{aligned} \tag{9.3}$$

By incorporating this result into the definition of the absorbed dose in a homogeneous medium provided in (8.39),

$$\begin{aligned}D &= \frac{1}{\rho} \lim_{V \to 0} \left(\frac{\bar{\varepsilon}}{V}\right) \\ &= -\frac{\nabla \bullet (\boldsymbol{\Psi}_\gamma + \boldsymbol{\Psi}_c)}{\rho} + \frac{1}{\rho} \frac{\mathrm{d}\left(\overline{\sum_i Q_i}\right)}{\mathrm{dV}} \\ &= -\frac{\nabla \bullet \boldsymbol{\Psi}_\gamma}{\rho} - \frac{\nabla \bullet \boldsymbol{\Psi}_c}{\rho} + \frac{1}{\rho} \frac{\mathrm{d}\left(\overline{\sum_i Q_i}\right)}{\mathrm{dV}} \\ &= K - \frac{\nabla \bullet \boldsymbol{\Psi}_c}{\rho} + \frac{1}{\rho} \frac{\mathrm{d}\left(\overline{\sum_i Q_i}\right)}{\mathrm{dV}}\end{aligned} \tag{9.4}$$

[2]Rossi and Zaider (1991) have pointedly compared the term "absorbed dose" of radiation physics with the term "dose" in posology, noting that the former describes the actual interaction between tissue and the administered entity (radiation) whereas the latter is simply an administered entity.

[3]Note that, for convenience, Auger, Coster–Krönig and super Coster–Krönig electrons will generally be grouped together and referred to as Auger electrons.

where the kerma has been identified as the divergence of the photon vector energy fluence normalized to the physical density of the medium,

$$K = -\frac{\nabla \bullet \boldsymbol{\Psi}_{\gamma}}{\rho}. \quad (9.5)$$

Equation (9.4) describes the absorbed dose as the kerma minus the divergence of the charged particle vector energy fluence (normalized to the physical density) plus the rest-mass energies released per unit mass of the medium. Excluding this latter contribution, the magnitude of the energy deposited is less than the amount released. There are some conditions, however, where it is reasonable to consider the two quantities as being equal, as to be shown in the next subsection.

9.2.2.3 Kerma, Absorbed Dose, and Charged Particle Equilibrium

Equation (9.4) shows that, in the absence of any rest-mass energy release, the magnitude of the kerma is reduced from that of the absorbed dose if the divergence of the charged particle fluence is nonzero. Such a condition reflects a state of charged particle nonequilibrium. Should a state of charged particle equilibrium (CPE) exist, then the divergence of the charged-particle vectorial energy fluence is equal to zero, $\nabla \bullet \boldsymbol{\Psi}_c/\rho = 0$. Hence, under CPE the absorbed dose and kerma are related by,

$$D = K + \frac{1}{\rho}\frac{\overline{d\left(\sum_i Q_i\right)}}{dV} \quad (\text{under CPE conditions}). \quad (9.6)$$

Consequently, if CPE exists and if there are no rest-mass energies released in the photon-medium interactions, the values of absorbed dose and kerma will be equal.

9.2.2.4 Kerma Per Unit Photon Fluence

The ratio of the kerma of a photon beam to the fluence of photons producing this kerma is important to the understanding of the energy dependence of photon ionization in a medium. Beginning by assuming a monoenergetic photon beam with energy k, the kerma is the product of the photon energy fluence and the mass energy-transfer coefficient,

$$\begin{aligned} K &= \Psi\frac{\mu_{tr}}{\rho}(k) \\ &= \Phi k\frac{\mu_{tr}}{\rho}(k). \end{aligned} \quad (9.7)$$

Recall that the mass energy-transfer coefficient is calculated using the total cross sections of those photon-matter interactions through which the photon transfers energy to the medium. In the photon energy regime used in nuclear medicine, these processes are photoelectric absorption and incoherent (Compton) scatter. The ratio of kerma per unit photon fluence for monoenergetic photons is then given by,

$$\frac{K}{\Phi} = k\frac{\mu_{tr}}{\rho}(k). \quad (9.8)$$

For the general case of polyenergetic photons with an energy spectrum with a maximum energy k_{max}, this ratio of kerma to photon fluence requires integration over the photon spectrum,

$$\frac{K}{\Phi} = \frac{\int_{k_{ion}}^{k_{max}} dk\frac{d\Phi(k)}{dk}k\frac{\mu_{tr}}{\rho}(k)}{\int_{k_{ion}}^{k_{max}} dk\frac{d\Phi(k)}{dk}} \quad (9.9)$$

where k_{ion} is the energy below which the photon is incapable of causing ionization in the medium. In most practical radiation dosimetry calculations, this lower limit of integration can generally be set equal to zero. Equations (9.8) and (9.9) are calculable quantities. To account for radiative losses, one can rewrite (9.8) and (9.9) in terms of the energy-absorption coefficient for a monoenergetic photon beam,

$$\frac{K}{\Phi} = k\frac{\frac{\mu_{en}}{\rho}(k)}{1 - g(k)} \quad (9.10)$$

and for a polyenergetic photon beam,

$$\frac{K}{\Phi} = \frac{\int_{k_{ion}}^{k_{max}} dk\frac{d\Phi(k)}{dk}k\frac{\frac{\mu_{en}}{\rho}(k)}{1-g(k)}}{\int_{k_{ion}}^{k_{max}} dk\frac{d\Phi(k)}{dk}} \quad (9.11)$$

where g(k) is the fraction of energy lost through radiative processes for the photon energy k. As demonstrated

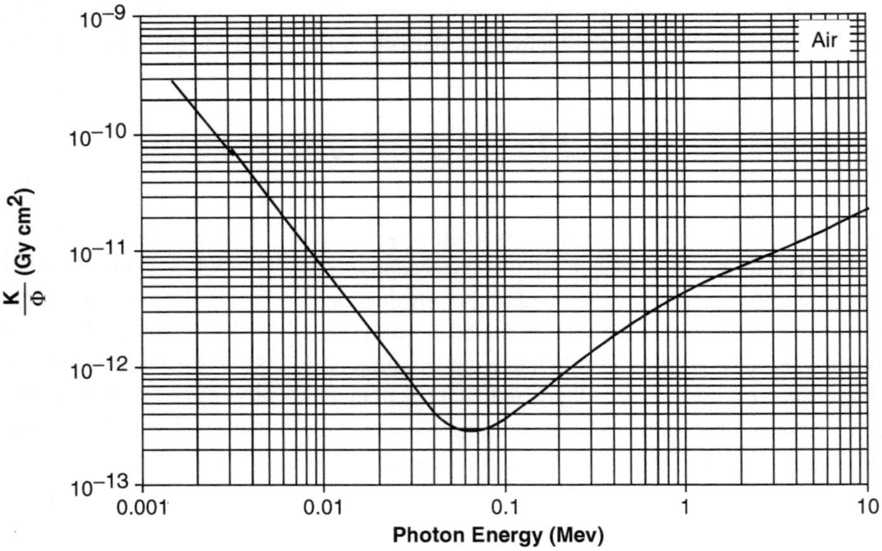

Fig. 9.2 Ratio of kerma to photon fluence for air

for low-Z media (such as soft tissue) and photon energies of interest to nuclear medicine (typically <1 MeV), the magnitude of this quantity is negligible. Figure 9.2 shows the ratio of kerma to photon fluence, K/Φ, calculated for air as a function of photon energy.

The photon energy dependence of K/Φ in air shown in Fig. 9.2 can be readily explained. From (9.8), this energy dependence is given by the product $k\mu_{Tr}(k)/\rho$. At low photon energies, photoelectric absorption dominates the photon-matter interaction processes. As this interaction has a $k^{-7/2}$ energy dependence, K/Φ in air will decrease with photon energy approximately as $k^{-5/2}$. For low-Z media and at photon energies in the region of 100 keV, the contribution of incoherent scatter to the total interaction coefficient exceeds that of photoelectric absorption. As the Klein–Nishina transfer coefficient has a limited energy dependence between 100 keV and 1 MeV, the change in $\mu_{Tr}(k)/\rho$ with photon energy is small. Hence, the ratio K/Φ will then begin to increase roughly linearly with photon energy, k.

Figure 9.2 also represents a calibration curve for converting a measurement of the collected ionization in an ionization chamber exposed to photon radiation to photon fluence. For example, consider an ionization chamber with infinitely-thin walls exposed to monoenergetic photons for which radiative losses through interactions with the gas within the chamber are negligible. These photon-gas interactions generate a measurable ionization current dQ/dt. An electric field across the gas volume will make these charged ionization products drift to corresponding electrodes. If it is assumed that these products are collected with 100% efficiency, the kerma rate in the gas is $dK/dt = dQ/dt(W/e)$, where W is the mean energy expended in the chamber's gas to create an electron-ion pair. The photon fluence rate (flux) is $d\Phi/dt = dQ/dt((W/e)/(K/\Phi))$ where K/Φ is evaluated at the photon energy of interest. Hence, the known W/e ratio for the gas and the energy-dependent calibration factor of K/Φ enables the photon flux to be calculated from the measured current. This is the principle behind the dose calibrator where $d\Phi/dt$ is proportional to the activity of the source being measured. The selection of the radionuclide being calibrated on the dose calibrator's control panel selects the appropriate value of the K/Φ ratio so as to scale the measured dQ/dt current to the appropriate activity of the source being calibrated.

9.2.3 Radiation Equilibria

9.2.3.1 Introduction

Many problems in radiation dosimetry can be solved only when a state of radiation equilibrium exists in the region of interest. For example, formal relationships between absorbed dose and collision kerma can only be established if charged particle exists or else exists transiently. Importantly, for applications in internal radiation dosimetry, when a state of radiation equilibrium exists in a medium containing a radionuclide, the absorbed dose is simply the expectation value of the energy released in radioactive decay per unit mass.

The state of radiation equilibrium can be described, in general terms, as that occurring when a quantity of radiation (energy or particle number) entering a volume of interest is exactly balanced by that quantity exiting the volume. In practice, such a state of perfect equilibrium is infrequently achieved and only an approximate (or transient) state of equilibrium will actually be present. Even so, this condition of approximate radiation equilibria frequently allows many dosimetric problems to be solved. For example, consider the case of a homogeneous sphere containing a uniform distribution of a β-emitting isotope. The sphere's radius is assumed to exceed the maximum range of these β particles. It is intuitive that radiation equilibrium will exist within a spherical volume with a radius equal to the difference between the radius of the sphere and the maximum β-particle range. Should this difference be very small, the absorbed dose to the total spherical volume can be well-approximated by the energy deposited in the total sphere divided by its mass. This implicit assumption of approximate radiation equilibrium is the basis of the Marinelli–Quimby–Hine internal dosimetry method.

9.2.3.2 Complete Radiation Equilibrium

Complete radiation equilibrium (CRE) can be formally defined by first considering the spherical volume V_1 containing a medium which is homogeneous in both density and atomic composition and which is uniformly radioactive (Fig. 9.3). Consider the point of interest, P, within V_1 and a second volume, V_2, which is centered on P. The minimum separation between the boundaries of the two volumes is S. This separation must be greater than the maximum distance of penetration, R_{Max}, of the radiation emitted by the radioactive nuclei within the volume. Thus, R_{Max} will define the radius of a sphere S centered on the tangential point T of V_2. There will be reciprocity (in the non-stochastic limit) of radiations crossing the tangential plane passing through T and bisecting the sphere S. This will indeed be the case for any tangential plane that is oriented around V_2. In other words, the radiant energies of charged and uncharged radiations entering V_2 equal those that exit V_2. This condition defines the state of CRE.

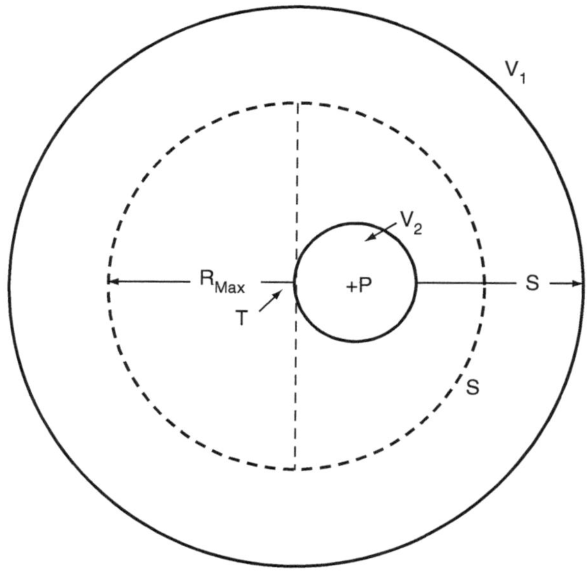

Fig. 9.3 Spherical geometry used to define radiation equilibrium (after Attix (1986))

The balance of ingoing and outgoing radiant energies between the two volumes gives the mean energy imparted as the sum of the mean rest-mass energies released,

$$\bar{\varepsilon} = \sum_{n=1}^{M} \overline{Q_n} \quad \text{(under CRE conditions)}. \tag{9.12}$$

The absorbed dose to volume V_2 is,

$$\begin{aligned} D &= \frac{d\bar{\varepsilon}}{dm} \\ &= \frac{d\overline{Q_{\Delta m}}}{dm}. \end{aligned} \tag{9.13}$$

Thus, under CRE, the absorbed dose to a subvolume of a homogeneous radioactive medium is given by the mean energy released per unit mass of that volume. It is also possible to obtain the result of (9.13) from (9.4) by noting that, from the definition of CRE, $\nabla \bullet \boldsymbol{\Psi}_\gamma = \nabla \bullet \boldsymbol{\Psi}_e = 0$.

9.2.3.3 Charged-Particle Equilibrium

If CRE exists, then the more specific state of CPE, by definition, must also exist. The equilibrium is of the charged particles alone.[4] Practical dosimetric interest in CPE is primarily in those conditions in which an approximate CPE exists. For example, consider a photon beam incident to a medium. The photons have a mean-free path in that medium given by the reciprocal of the linear attenuation coefficient, $1/\mu$. Should this mean-free path exceed the maximum range R_{Max} of the secondary electrons produced by photon-matter interactions within the medium (i.e., the product $\mu R_{Max} \ll 1$), then an approximate CPE can be established within the medium at those depths greater than R_{Max} if the attenuation of the photon beam can be considered negligible. This condition links with the demonstration of (9.6) that, under CPE conditions and assuming that no changes in rest mass occur, the values of the absorbed dose and the kerma are equal.

Reconsider the geometry of Fig. 9.3 and assume that the radionuclide contained in the volume emits only α or β^- particles (i.e., the emitted radiation has a finite range within the volume) and that *bremsstrahlung* losses are negligible. From the argument presented above, CPE will be established at the point P provided that the medium is homogeneous and the radionuclide is uniformly distributed. If the requirement that the radionuclide emits only charged particles is relaxed and it is now allowed to emit photons, there is now a case of more practical interest to diagnostic nuclear medicine. Let the emitted photon mean-free path in the medium be such that $1/\mu \gg R_{Max}$, where R_{Max} is the maximum range of the charged particle emitted in the radioactive decay. The outer volume V_1 is then reduced such that the minimum separation of the two spheres, s, just exceeds R_{Max}. Under such a condition, it can be assumed that there are negligible photon-matter interactions as the photons escape from V_2. Hence, while CPE will exist under this condition, CRE cannot as the outgoing photon radiant energy for V_1 will exceed that of any incoming photons. However, if V_1 is allowed to expand in size to allow $s \gg 1/\mu$, then an approximate CRE will be established.

[4]While CPE can exist in a volume containing a uniformly distributed β-emitting radionuclide, it cannot exist when the volume is irradiated by an external beam of charged particles.

9.2.3.4 Absorbed Fraction

In internal radiation dosimetry, the absorbed fraction $\phi(r_T \leftarrow r_S)$ is the fraction of energy emitted by a radionuclide in a source region r_S that is absorbed by a designated target region, r_T. This discussion of the absorbed fraction is limited to that fraction absorbed by the source region itself, i.e., the "self-dosing" contribution, $\phi(r_S \leftarrow r_S)$. The $\phi(r_S \leftarrow r_S)$ is an implicit measure of radiation equilibrium to the entire source region. Should $\phi(r_S \leftarrow r_S) = 1$ in a volume containing a β-emitting radionuclide, then CPE exists. From the above discussions, $\phi(r_S \leftarrow r_S)$ will never equal 1.0 exactly if the β-emitting radionuclide is uniformly distributed throughout the region and the region is placed in vacuo (i.e., backscattering of the β particles into the volume is not possible). Those radioactive nuclei at the periphery of the region can emit β particles which exit the region but are not compensated by β particles entering the region. However, a unity value of the $\phi(r_S \leftarrow r_S)$ is obtainable in the case of the region containing the radionuclide central to the entire source region and the shortest distance from the radioactive region to the borders is less than the maximum range of the β particle, R_{Max}.

The above is of importance in radionuclide therapy in which a tumor is assumed to be spherical and to have taken up a specified amount of activity. To estimate the amount of absorbed dose received by the tumor, an estimated of the absorbed fraction is required. Values of these have been provided in the literature over the years (e.g., Brownell et al. 1968; Ellett and Humes 1971; Siegel and Stabin 1994; Stabin and Konijnenberg 2000). While it may be reasonable to assume that, under some conditions, $\phi(r_S \leftarrow r_S) = 1$ for all β-emitting radionuclides, the validity of the assumption is dependent upon the combination of R_{Max} and the diameter of the spherical volume. Two examples of this are given in Fig. 9.4 in which the absorbed fractions of energies of β particles in unit-physical density spheres containing uniformly-distributed ^{131}I or ^{90}Y are shown as functions of the diameter of the sphere. The curves are derived from tabular data provided by Siegel and Stabin (1994) which, in turn, were derived from dose point kernel (DPK) calculations performed by Berger (1970). The mean β-particle kinetic energies of ^{131}I and ^{90}Y are 183 and 937 keV, respectively. As can be seen, the $\phi(r_S \leftarrow r_S)$ for ^{131}I approaches unity for spherical

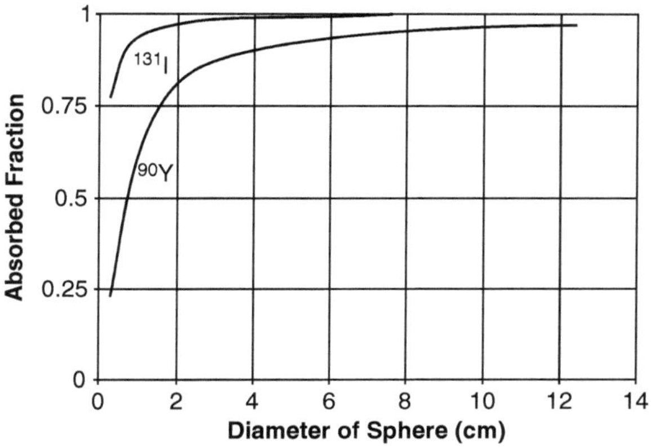

Fig. 9.4 Absorbed fraction for the β particles emitted by ^{131}I and ^{90}Y uniformly distributed within unit-density spheres as functions of the sphere diameter. *Curves* are derived from the data of Siegel and Stabin (1994)

diameters exceeding about 8 cm whereas that for ^{90}Y is still only equal to 0.97 for a 12.4 cm diameter sphere. This reflects the lower β-particle range relative to the sphere diameter for ^{131}I.

9.2.3.5 Collision Kerma and Charged Particle Equilibrium

The role of CPE in the definition of kerma is now considered. In this subsection, the approach taken by Alm Carlsson (1981, 1985) is followed.

Photon interactions with atoms in a medium are considered to be a source of electrons. As these electrons can cause further ionizations in the medium as they slow down and liberate further electrons (δ rays), classification of moving electrons on the basis of how they are produced is useful. Primary electrons are defined as those emitted in radioactive decay or else produced through photon interactions (e.g., Compton scatter). Secondary electrons are those produced by subsequent electron interactions in the medium, i.e., δ rays.[5] The expectation value of the imparted energy resulting from the interactions of a primary electron with kinetic energy T and its secondary electrons which are completely stopped in the medium is,

$$\overline{\varepsilon_{T,c}} = T - \overline{\sum_n k_n} + \overline{\sum_n Q_n} \qquad (9.14)$$

where $\overline{\sum_n k_n}$ is the mean of the sum of the energies of photons created by the charged particle interactions (characteristic X-rays and *bremsstrahlung*) and $\overline{\sum_n Q_n}$ is the mean of the sum of any rest-mass energies released. Consider a homogeneous medium within which primary electrons are liberated to form a uniform fluence. CPE is established by definition and the absorbed dose due to these electrons is,

$$D = n_c \overline{\overline{\varepsilon_{T,c}}} \qquad (9.15)$$

where n_c is the number of charged particles released by radioactive decay per unit mass and the double overbar indicates that the mean imparted energy from the charged particle has been averaged over the energy spectrum of the emitted primary charged particles. Substituting (9.14) into (9.15) results in,

$$\begin{aligned} D &= n_c \overline{\overline{\varepsilon_{T,c}}} \\ &= n_c\bar{T} - n_c\left(\overline{\overline{\sum_n k_n}} - \overline{\overline{\sum_n Q_n}}\right) \\ &= K - n_c\left(\overline{\overline{\sum_n k_n}} - \overline{\overline{\sum_n Q_n}}\right) \end{aligned} \qquad (9.16)$$

where the kerma has been identified as $K = n_c\bar{T}$. Even if CPE should not exist, the right-hand side of this result remains and is referred to as the collision kerma, K_c.

From (9.7), the kerma for a polyenergetic photon beam is,

$$K = \int_{k_{Ion}}^{\infty} dk \frac{d\Psi}{dk} \frac{\mu_{tr}}{\rho}(k). \qquad (9.17)$$

By analogy, the collision kerma can be defined in a similar fashion using the mass energy-absorption coefficient,

$$K_c = \int_{k_{Ion}}^{\infty} dk \frac{d\Psi}{dk} \frac{\mu_{en}}{\rho}(k). \qquad (9.18)$$

[5]This classification is also applicable to neutron interactions with tissue where the primary charged particle would be a proton resulting from neutron-proton collisions and the secondary charged particles would be protons and electrons.

One can now write an expression for the mass–energy absorption coefficient as,

$$\frac{\mu_{en}}{\rho} = \frac{\overline{\overline{\varepsilon_{T,c}}}}{\overline{\overline{T}}}\frac{\mu_{tr}}{\rho} = \left(1 - \frac{\overline{\overline{\sum_n k_n}} - \overline{\overline{\sum_n Q_n}}}{\overline{\overline{T}}}\right)\frac{\mu_{tr}}{\rho}. \tag{9.19}$$

The ICRU's (1980) relationship between the energy-transfer and energy-absorption coefficients,

$$\frac{\mu_{en}}{\rho} = (1-g)\frac{\mu_{tr}}{\rho} \tag{9.20}$$

where g is the fraction of the liberated charged particles' initial kinetic energies that is irradiated as *bremsstrahlung*. Neglecting rest-mass energy releases, combining (9.19) and (9.20) results in an expression for g,

$$g = \frac{\overline{\overline{\sum_n k_n}}}{\overline{\overline{T_c}}}. \tag{9.21}$$

9.2.3.6 Collision Kerma and Transient Charged Particle Equilibrium

Recall that it was stated that a state of an approximate CPE exists for secondary electrons set in motion by photon interactions if $\mu R_{Max} \ll 1$, where μ is the photon's linear attenuation coefficient and R_{Max} is the maximum electron range. For high-energy photons, μR_{Max} is small (≈ 0.1) and the magnitudes of the absorbed dose and the corresponding kerma and collision kerma at a point in a medium irradiated by high-energy photons will differ. To demonstrate the relationship between absorbed dose, kerma, and collision kerma under such conditions, consider a semi-infinite homogeneous medium irradiated by a planar photon beam sufficiently broad that lateral electronic equilibrium exists at the central axis of the beam. This lateral electronic equilibrium is defined as the state in which there is an exact balance between ingoing and outgoing laterally-scattered electrons in volumes along this central axis. Such a geometry is shown in Fig. 9.5. The absorbed dose at the point P, which is at a

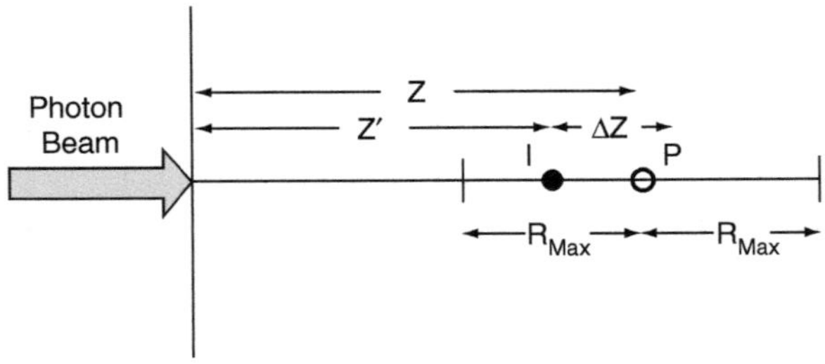

Fig. 9.5 Geometry describing the definition of transient charged particle equilibrium. A photon beam is incident from the left to a semi-infinite medium; it is sufficiently broad that lateral electronic equilibrium exists along the central axis. The absorbed dose at the point P at a depth z is due to electrons liberated at depths between $z - R_{Max}$ and $z + R_{Max}$, such as point I at a depth z′, where R_{Max} is the maximum range of these electrons.

depth z, is due to primary electrons liberated by photon-matter interactions and to secondary electrons set in motion within a distance R_{Max} of P. An electron released from an arbitrary point I at a depth z′ within this interval due to a photon interaction will impart an energy $d\varepsilon(\Delta z)$ in a layer with thickness $d(\Delta z)$, where $\Delta z = z - z'$, at a distance Δz from z′. The average energy imparted at P by a single electron is the integral of the distribution $d\varepsilon(\Delta z)/d(\Delta z)$ over distance,

$$\overline{\overline{\varepsilon_{T,c}}} = \int_{-R_{Max}}^{R_{Max}} d(\Delta z)\frac{d\varepsilon(\Delta z)}{d(\Delta z)}. \tag{9.22}$$

If there are $n_c(z')$ electrons liberated per unit mass of medium at the depth z′, then (9.22) can be rewritten to yield the absorbed dose at a depth z,

$$D(z) = \int_{z-R_{Max}}^{z+R_{Max}} dz'\, n_c(z')\frac{d\varepsilon(\Delta z)}{d(\Delta z)}. \tag{9.23}$$

Assuming that the photons are attenuated purely exponentially as they penetrate into the medium, then, $n_c(z) = n_c(z')\, e^{-\mu\Delta z}$ and the expression for the absorbed dose of (9.23) becomes,

$$D(z) = n_c(z)\int_{-R_{Max}}^{R_{Max}} d(\Delta z)\frac{d\varepsilon(\Delta z)}{d(\Delta z)}e^{\mu\Delta z}. \tag{9.24}$$

This assumption, of course, neglects the fact that the photon energy spectrum will change with depth due to the increase in lower-energy Compton-scattered photons, the resulting depth-dependence of the linear attenuation coefficient and ignores buildup of secondary photons along the central axis due to Compton scatter. The validity of this result thus requires a balance between the need for sufficient lateral extent of the beam to ensure lateral electronic equilibrium along the central axis and a minimum lateral extent so as to ensure that the buildup effects of scattered photons can be ignored. The collision kerma is,

$$K_c(z) = n_c(z)\bar{\bar{\varepsilon}}_{T,c}. \tag{9.25}$$

From these two results, one can now write the absorbed dose of (9.24) as,

$$D(z) = \frac{K_c(z)}{\bar{\bar{\varepsilon}}_{T,c}} \int_{-R_{Max}}^{R_{Max}} d(\Delta z) \frac{d\varepsilon(\Delta z)}{d(\Delta z)} e^{\mu \Delta z}. \tag{9.26}$$

If the inequality $\mu R_{Max} \ll 1$ is met, then the exponential can be expanded to first order, $e^{\mu\Delta z} \approx 1+\mu\Delta z$, and (9.26) simplifies to,

$$\begin{aligned} D(z) &= K_c(z)\left(1+\mu\left(\frac{\int_{-R_{Max}}^{R_{Max}} d(\Delta z)\frac{d\varepsilon(\Delta z)}{d(\Delta z)}\Delta z}{\bar{\bar{\varepsilon}}_{T,c}}\right)\right) \\ &= K_c(z)(1+\mu m) \quad \mu R_{Max} \ll 1; \quad z \geq R_{Max} \end{aligned} \tag{9.27}$$

where m is the first moment of the frequency function, $d\varepsilon(\Delta z)/d(\Delta z)$, normalized to $\bar{\bar{\varepsilon}}_{T,c}$. For high-energy photons, the liberated electrons will be forward-directed (recall the directional dependence of the Klein–Nishina cross section) with the result that $d\varepsilon(\Delta z)/d(\Delta z)$ is large for positive Δz. This results in the magnitude of the absorbed dose at a point being greater than that of the collision kerma, $D(z) > K_c(z)$. On the other hand, for low-energy photons, $d\varepsilon(\Delta z)/d(\Delta z)$ is small which results in $D(z) \approx K_c(z)$. Recall that the elastic Coulomb scatter cross section increases with square of the medium's atomic number. As a consequence, the ratio $d\varepsilon(\Delta z)/d(\Delta z)$ will tend to be more symmetric in high-Z media (as a result of the greater isotropicity of scatter) than in low-Z media for the same photon beam. Thus, the degree of CPE which is achieved will depend upon the combination of the photon mean-free path $1/\mu$, the maximum range of electrons R_{Max}, the isotropy of the production of secondary electrons and the scattering properties of the medium. To summarize these effects, Fig. 9.6 shows the relationships between kerma, collision kerma, and absorbed dose for high- and low-energy photons. The quantities are along the central axis of a photon beam which is incident in vacuo from the left to the vacuum-medium interface and are plotted as a function of depth in the medium. The interface is at depth $z = 0$. The observed decreases in kerma and collision kerma are a result of the exponential attenuation of the photon fluence. For high-energy photons (Fig. 9.6a), there is a clear demarcation between kerma, collision kerma, and absorbed dose. The ratio of the collision kerma to the kerma is a constant equal to the ratio of the mass–energy transfer to mass–energy absorption coefficients which, from (9.20), is equal to,

$$\frac{(\mu_{tr}/\rho)}{(\mu_{en}/\rho)} = \frac{1}{1-g}. \tag{9.28}$$

In Fig. 9.6a, it has been implicitly assumed that there is a fraction of the incident photon energy which eventually appears as *bremsstrahlung* irradiated by primary and secondary electrons Should that fraction be negligible (i.e., $g \approx 0$), then the mass–energy absorption and transfer coefficients would be equal and the kerma and collision kerma curves of Fig. 9.6a would be coincident.

The depth-dependence of the absorbed dose for high-energy photons is a result of energy transfer and absorption from the primary and secondary electrons. As, at high energies the primary photons are forward-directed, the absorbed dose will be zero at the interface between the vacuum and the medium and will increase up to a maximum at a depth equal to R_{Max}. The point at which the absorbed dose and collision kerma are numerically equal occurs at a slightly lower depth. The absorbed dose curve then decreases exponentially for depths $z \geq R_{Max}$ with $D(z) = K_c(z - m)$ or $D(z)/K_c(z) = 1 + \mu m$. One should recognize that, even without the above derivations that culminated in (9.27), $D(z)$ must be greater than $K_c(z)$ at depths $z \geq R_{Max}$ simply from the conservation of energy. The areas under the $D(z)$ and $K_c(z)$ curves in Fig. 9.6a must,

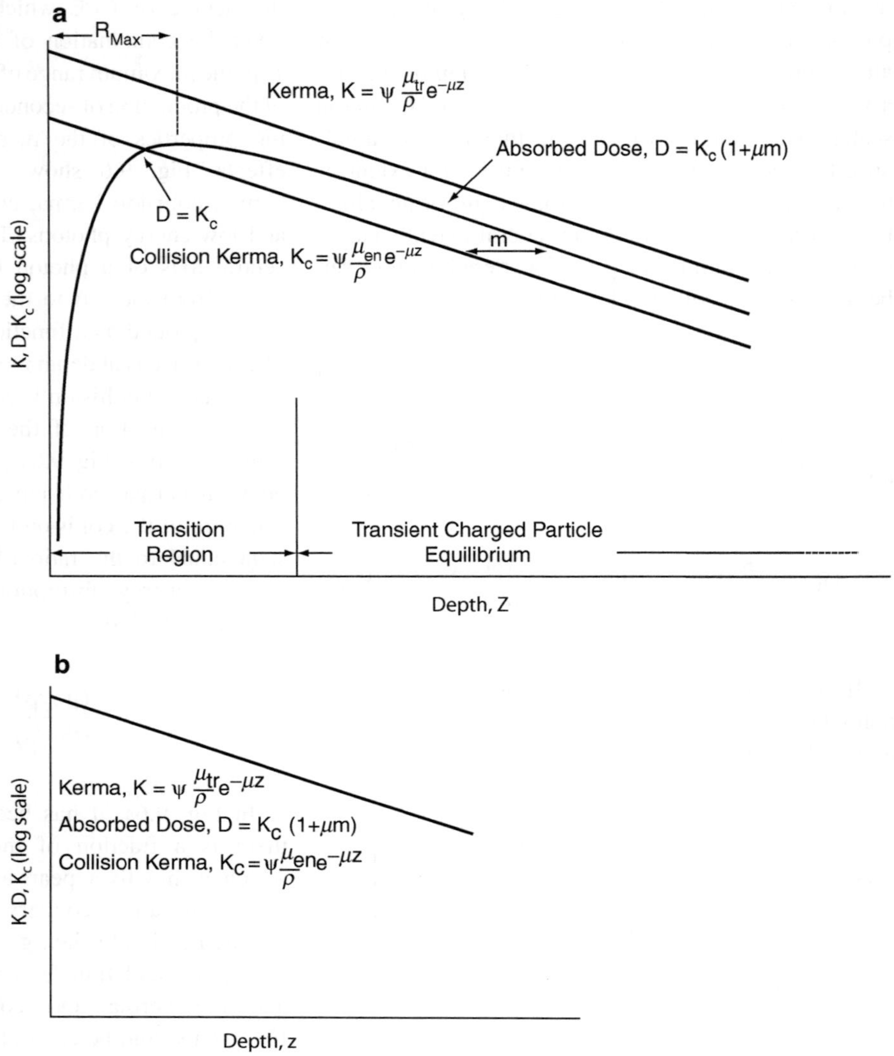

Fig. 9.6 Depth dependencies of kerma, collision kerma, and absorbed dose for photons incident to a medium for (**a**) high-energy photons and (**b**) low-energy photons. Refer to the text for detailed descriptions

as a result, be equal and as $K_c(z) > D(z)$ near the interface, then $D(z) > K_c(z)$ at greater depths. In the transition region shown, CPE does not exist. Beyond the depth of maximum absorbed dose, a state of transient charged particle equilibrium (TCPE) exists.[6] This state is defined as the condition in which the ratio of the absorbed dose to the collision kerma is equal to a constant (assuming that the decrease in both with depth is exponential and the photon energy spectrum is unchanged).

Figure 9.6b shows the kerma, collision kerma, and absorbed dose curves for the case of low-energy photons incident to the medium. Under such conditions, the electron ranges are much smaller and *bremsstrahlung* losses negligible. Thus, the mass–energy absorption and transfer coefficients are equal and the values of the kerma, collision kerma, and absorbed dose are all equal at any given depth. This is the condition in which all of the energy transferred from the photon interactions with the medium and the energy absorbed from the primary and secondary electrons occur locally.

[6] Also referred to as "quasi-equilibrium."

9.2.3.7 In-Air Collision Kerma and Exposure

As described in the introduction to this chapter, the radiometric unit of exposure was defined in 1937 in terms of the amount of ionization in air generated by photons. Recall from Chap. 8,

$$X = \Psi \left(\frac{\mu_{en}}{\rho} \right)_{air} \frac{e}{W}.$$

The unit of exposure is defined for air only whereas, the metric of the collision kerma K_c has broader applications in that it can be defined for any medium. Should that medium be air, the in-air collision kerma and exposure are related by,

$$X = K_{c,air} \frac{e}{W} \quad (9.29)$$

where the in-air collision kerma has been assigned the symbol, $K_{c,air}$.

9.2.3.8 Air Kerma-Rate Constant

The air kerma-rate constant, Γ_δ, is defined (ICRU 1980) for a given photon-emitting radionuclide. It is used to calculate the air kerma rate at a distance r from a point photon-emitting source of activity A,

$$\frac{dK_\delta}{dt} = \Gamma_\delta \frac{A}{r^2} \quad (9.30)$$

where dK_δ/dt is the air kerma rate due to photons with energies exceeding δ. From the definitions of kerma and the flux of a point source of an isotope emitting a single photon per decay, we can also write this air kerma rate as,

$$\frac{dK_\delta}{dt} = k \left(\frac{\mu_{tr}}{\rho}(k) \right)_{air} \frac{A}{4\pi r^2} \quad (k \geq \delta). \quad (9.31)$$

where attenuation of the photons in air has been neglected. For multiple photon emissions per nuclear decay, this result can be generalized,

$$\frac{dK_\delta}{dt} = \frac{A}{4\pi r^2} \sum_{i=1}^{N} Y_i k_i \left(\frac{\mu_{tr}}{\rho}(k_i) \right)_{air} \quad (k \geq \delta) \quad (9.32)$$

Table 9.1 Photon Air Kerma-rate constants for radionuclides in common use in diagnostic nuclear medicine (δ = 20 keV)

Isotope	Γ_δ (Gy cm²/Bq s)
^{11}C, ^{13}N, ^{15}O, ^{18}F	3.89×10^{-13}
^{99m}Tc	3.97×10^{-14}
^{111}In	2.07×10^{-13}

where Y_i is the number of photons of energy k_i emitted per decay. The air kerma-rate constant can then be defined from (9.30) as,

$$\Gamma_\delta = \frac{r^2}{A} \frac{dK_\delta}{dt} \quad (9.33)$$

to yield the final expression,

$$\Gamma_\delta = \frac{1}{4\pi} \sum_{i=1}^{N} Y_i k_i \left(\frac{\mu_{tr}}{\rho}(k_i) \right)_{air} \quad (k \geq \delta). \quad (9.34)$$

Values of Γ_δ for radionuclides in common use in diagnostic nuclear medicine are provided in Table 9.1

Attix (1986) has suggested that a superior definition could have been had if the in-air collision kerma, $K_{c,air}$, had been used in the definition instead. This suggestion amounts to replacing the mass energy-transfer coefficient for air with the mass energy-absorption coefficient for air in the above derivation. Thus, *bremsstrahlung* would escape and the result would be reflective of that which would be obtained by measurement.

9.2.3.9 Fano's Theorem

In the examples of CRE and CPE considered in discussions so far, the medium has been homogeneous. Should this homogeneous medium be uniformly irradiated by photons, the liberated electron fluence will also be uniform. Fano's theorem (Fano 1954) states that this electron fluence will remain uniform even if the requirement for medium homogeneity is relaxed. The theorem states:

If an infinite medium of given atomic composition is exposed to a uniform photon fluence, the resulting electron field will also be uniform and independent of the medium's density and any spatial variations of density within the medium.[7]

The proof of the theorem is as follows. Begin, as Fano did, by considering an infinite homogeneous medium irradiated by a uniform photon fluence which subsequently sets electrons in the medium into motion. Define $\Phi_{e,h}\left(T, \hat{\boldsymbol{\Omega}}\right)$ as the electron number fluence with kinetic energy T moving in the direction of the unit vector $\hat{\boldsymbol{\Omega}}$. The h subscript indicates that this is the electron number fluence in the homogeneous medium. The source term $S_e\left(T, \hat{\boldsymbol{\Omega}}\right)$ represents the number of primary electrons of kinetic energy T created by photon interactions in the medium per unit mass and which move in the direction of the unit vector $\hat{\boldsymbol{\Omega}}$. As the photon irradiation is defined to be uniform, the source term $S_e\left(T, \hat{\boldsymbol{\Omega}}\right)$ must also be uniform throughout the medium as a result. The probability per unit areal density (i.e., mass per unit area) that an electron of energy T moving in direction $\hat{\boldsymbol{\Omega}}$ will be scattered so as to have a postscatter kinetic energy between T′ and T′ + dT′ and moving in the direction between $\hat{\boldsymbol{\Omega}}'$ and $\hat{\boldsymbol{\Omega}}' + d\hat{\boldsymbol{\Omega}}'$ is given by $\Pr\left(T \rightarrow T';\ \hat{\boldsymbol{\Omega}} \rightarrow \hat{\boldsymbol{\Omega}}'\right) dT'\, d\hat{\boldsymbol{\Omega}}'$. From these definitions, the primary electron transport equation is, under the assumption of CPE,

$$S_e\left(T, \hat{\boldsymbol{\Omega}}\right) - \Phi_{e,h}\left(T, \hat{\boldsymbol{\Omega}}\right) \int_0^T dT' \int d\Omega' \Pr\left(T \rightarrow T'; \hat{\boldsymbol{\Omega}} \rightarrow \hat{\boldsymbol{\Omega}}'\right) + \int_T^\infty dT' \int d\Omega' \Pr\left(T \rightarrow T'; \hat{\boldsymbol{\Omega}} \rightarrow \hat{\boldsymbol{\Omega}}'\right) \Phi_{e,h}\left(T', \hat{\boldsymbol{\Omega}}'\right) = 0 \quad (9.35)$$

If the condition of CPE were not to be imposed a priori, then the right-hand side of (9.35) will be nonzero. Further, if the density of the medium is allowed to vary, a spatial dependence is assigned to the density, $\rho(\mathbf{r})$. As a result, the electron number fluence will also have a spatial dependence, $\Phi_{e,i}\left(\mathbf{r}, T, \hat{\boldsymbol{\Omega}}\right)$, where the i subscript indicates that this is the electron fluence in what is now the inhomogeneous medium. The resulting transport equation allowing for a spatially-variant density and without a state CPE being assumed a priori is,

$$\rho(\mathbf{r}) S_e\left(T, \hat{\boldsymbol{\Omega}}\right) - \rho(\mathbf{r}) \Phi_{e,i}\left(\mathbf{r}, T, \hat{\boldsymbol{\Omega}}\right) \int_0^T dT' \int d\Omega' \Pr\left(T \rightarrow T'; \hat{\boldsymbol{\Omega}} \rightarrow \hat{\boldsymbol{\Omega}}'\right) + \rho(\mathbf{r}) \int_T^\infty dT' \int d\Omega' \Pr\left(T \rightarrow T'; \hat{\boldsymbol{\Omega}} \rightarrow \hat{\boldsymbol{\Omega}}'\right) \Phi_{e,i}\left(\mathbf{r}, T', \hat{\boldsymbol{\Omega}}'\right) = \hat{\boldsymbol{\Omega}}' \bullet \nabla \Phi_{e,i}\left(\mathbf{r}, T', \hat{\boldsymbol{\Omega}}'\right) \quad (9.36)$$

Fano's theorem states that the electron number fluences in the homogeneous and inhomogeneous media will be the same, i.e., $\Phi_{e,h}\left(T, \hat{\boldsymbol{\Omega}}\right) = \Phi_{e,i}\left(\mathbf{r}, T, \hat{\boldsymbol{\Omega}}\right)$. This conclusion can be reached by replacing the electron fluence in the inhomogeneous medium, $\Phi_{e,i}\left(\mathbf{r}, T, \hat{\boldsymbol{\Omega}}\right)$, in (9.37) with that in the homogeneous medium, $\Phi_{e,h}\left(T, \hat{\boldsymbol{\Omega}}\right)$. As $\Phi_{e,h}\left(T, \hat{\boldsymbol{\Omega}}\right)$ has no spatial dependence due to the homogeneity of the medium and the uniformity of the incident photon fluence, then its gradient $\nabla \Phi_{e,h}\left(T, \hat{\boldsymbol{\Omega}}\right) = 0$. Hence, the right-hand side of (9.36) will equal zero. The physical density term then divides out across both sides of (9.36), thus leading to the homogeneous medium result of (9.35). Hence, CPE will be established in an inhomogeneous medium under the condition of uniform photon irradiation.

9.2.3.10 Absorbed Doses at Interfaces Between Different Media

Clinical radiation dosimetry calculations are rarely associated with the simple homogeneous media studied in the examples so far. In order to provide accurate results, radiation dosimetry calculations must account for the radiological heterogeneity of the human body. For example, the physical densities of skeletal muscle and bone are 1.04 and 1.65 g/cm^3, respectively, and the corresponding effective atomic numbers are 7.64 and 12.31 (Johns and Cunningham 1983). Clearly, a simplifying assumption of the body as being homogeneous neglects the spatial dependencies of the photon

[7] Attix (1986) and others have highlighted that Fano's derivation of this theorem neglects the polarization effect upon charged-particle energy loss discussed in Chap. 7.

coefficients of attenuation, energy transfer, and energy absorption coefficients and the electron/positron stopping and scattering powers throughout the body. It is evident that the radiation fluences calculated for a given geometry and an assumed homogeneous medium will markedly differ from those that exist in the actual heterogeneous case. For photon irradiation, the absorbed dose at a point in a medium in which CPE exists is readily calculated from $D = \psi\mu_{en}/\rho$. However, at an interface between two radiologically dissimilar media, CPE does not exist and the absorbed dose cannot be calculated in this fashion. In many practical cases, this does not necessarily constitute a problem. For example, consider a 1-cm thick slab of bone embedded within two 1-cm slabs of muscle, as shown in Fig. 9.7. One hundred and forty kiloelectron volt photons (i.e., ^{99m}Tc γ rays) are incident from the left and photon scatter is neglected. As these photons travel through muscle, they are attenuated with a linear attenuation coefficient of 0.15/cm, whereas in bone they are attenuated with a linear attenuation coefficient of 0.26/cm. Moreover, the absorbed dose will be higher in bone than in muscle due to the respective mass energy absorption coefficients of 0.102 and 0.027cm^2/g, reflecting the greater amount of photoelectric absorption in bone due to its higher effective atomic number. As the maximum ranges of Compton electrons produced by 140 keV photons in muscle and bone are about 0.04 and 0.09 mm, respectively, all of the energy liberated through these Compton electrons is absorbed at the site of energy transfer.

The dashed lines indicate the approximately factor of 3 change in absorbed dose that occur at the bone–muscle interfaces. As suggested by Fig. 9.7,

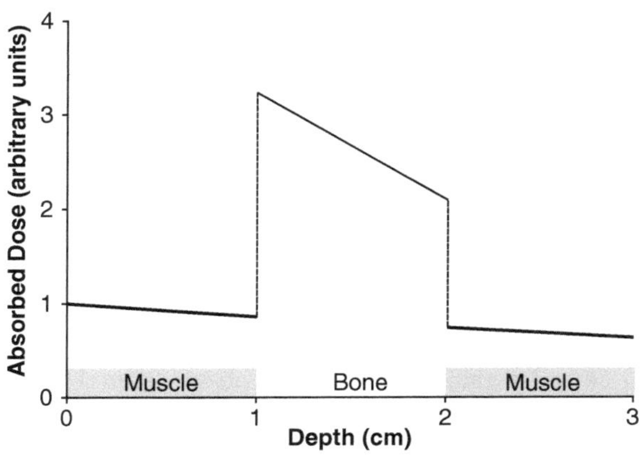

Fig. 9.7 Absorbed doses in 1-cm-thick slabs of muscle and bone irradiated by ^{99m}Tc (140 keV) photons incident from the left. Photon scatter is neglected (after Johns and Cunningham (1983))

with the change in physical density and atomic number, CPE cannot be expected to exist at the interface between two radiologically-dissimilar media. Indeed, there will be a rapidly-changing gradient in absorbed dose at the interfaces.

The general problem of how the fluence is altered at the interface between low- and high-Z media is important in practical dosimetry, as we will see when we consider the absorbed dose to red bone marrow. The influence of an interface between low- and high-Z media upon the electron fluences created by a photon beam is shown conceptually in Fig. 9.8. In Fig. 9.8a, a photon beam is incident from the high-Z medium to

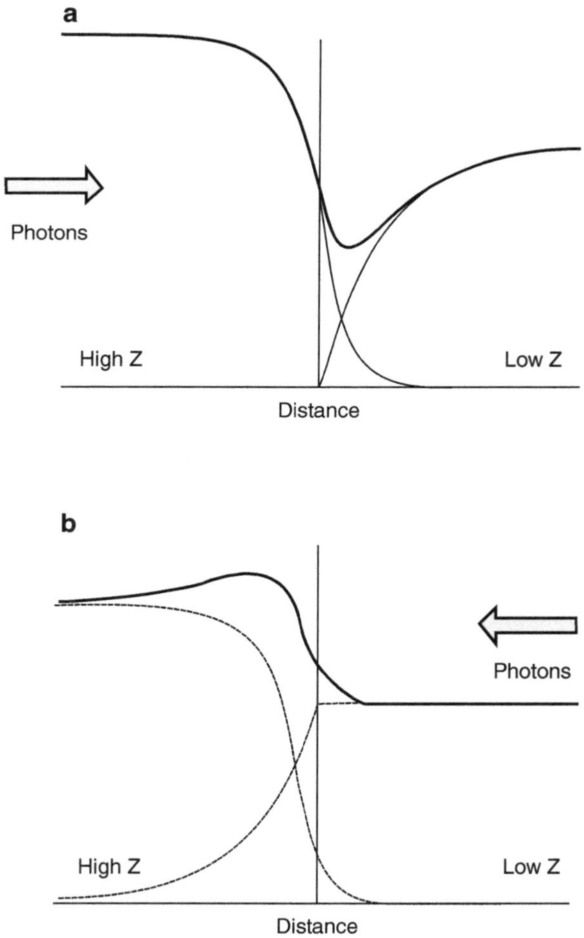

Fig. 9.8 Electron fluences in the vicinity of the interface between high- and low-Z media as a function of distance from the interface. The *solid lines* are the total fluences. In (**a**), photons are incident to the interface from the high-Z medium; in (**b**), photons are incident to the interface from the low-Z medium (after Dutreix and Bernard 1966)

the low-Z medium (e.g., photons traversing from bone to soft tissue). The photons will liberate electrons in both media and the differing properties and the direction of photon irradiation create a nonequilibrium state (although at distances of the order of R_{Max} beyond the interface, TCPE exists). From the study of the Z-dependence of elastic Coulomb scatter in Chap. 6, it is reasonable to neglect any electrons that are backscattered from the low-Z medium across the interface into the high-Z medium. Hence, the fluence of electrons arising in the high-Z medium will decrease as the interface is approached. Once the interface is crossed, this fluence in fact diminishes to zero beyond about R_{Max} (in the low-Z medium) of the forward-directed electrons produced in the high-Z medium. As photons cross the interface, electrons are set in motion due to interactions in the low-Z medium to reach a maximum once CPE is achieved. The combined electron fluences in the vicinity of the interface are shown by the solid line with a minimum on the low-Z side of the interface.

In Fig. 9.8b, the photons are now incident from the low-Z side of the interface. The fluence of electrons set in motion in the low-Z medium is constant up to the medium and then diminishes to zero inside the high-Z medium. The fluence of electrons set in motion by photon interactions in the high-Z medium penetrate into the low-Z medium due to backscatter. The combined fluences show a maximum in the high-Z medium at a distance from the interface which is less than the equilibrium depth (i.e., the maximum range of the electrons in the high-Z medium). These figures were derived from data obtained by Dutreix and Bernard (1966) for a carbon–copper interface irradiated by ^{60}Co γ rays (with a mean energy of 1.25 MeV).

9.2.4 Methods of Calculating the Radiation Flux

9.2.4.1 Introduction

The absorbed dose at a point due to photons or electrons/positrons is proportional to the relevant particle fluence at that point; thus, the calculation of the fluence (or the fluence rate, flux, for a radioactive source) is fundamental to the evaluation of the absorbed dose. In this subsection, analytical and Monte Carlo (stochastic) means of calculating the particle fluence are reviewed. In most cases, the source of radiation is a radionuclide so a time rate must be considered; hence the object of the calculation is the flux. Fluence is obtained by integrating the flux over time. Numerical solutions based upon the solution of the Boltzmann transport equation are not considered here. While these numerical solutions have historically been performed, the Boltzmann equation method is impractical in contemporary nuclear medicine dosimetry calculations, having been surpassed by modern, faster and more precise Monte Carlo methods.[8]

9.2.4.2 Analytical Solutions for Geometric Radiation Sources

Introduction

Analytical calculations of the fluxes due to a radioactive point source, linear source, and planar disc source begin this discussion. The intent of this discussion of analytical methods is to present the general features of the dependence of the flux upon source geometry. Although these results are readily obtainable through Monte Carlo simulation, the analytical derivation provides the background to understanding these dependencies. In addition, as the initial calculations will be for the sources in vacuo, the results are geometric and are applicable to both photons and charged particles. The interposition of absorbers between the sources and the flux calculation points of interest is then allowed. As a result, these extended calculations will be limited to photons. These calculations demonstrate the balance between the contributions of photon attenuation and scattered radiation to the flux at the calculation point.

In addition to the pedagogical attributes of these analytical calculations, they can also provide some practical results. It is frequently necessary in the nuclear medicine department to know at what distance an extended radioactive source can be approximated

[8]Although a Boltzmann-type radiation transport equation was used in the analysis of charged-particle energy straggling and the proof of Fano's theorem.

by a point source, for which the photon flux can be readily calculated. Another application is in nuclear medicine radiation protection, an example being the calculation of the absorbed dose received by others from a radioactive patient.

Point Source

The simplest geometric source to calculate for is that of a point radioactive source. As shall be demonstrated, it is also an approximation of an extended source if the point at which we wish to calculate the flux is at a sufficient distance from the source. Figure 9.9 shows the geometry of a source of activity A sited at the origin and the flux (fluence rate) at the point P a distance r from the source is, in vacuo,

$$\frac{d\Phi}{dt} = \frac{A}{r^2 \int d\Omega} = \frac{A}{4\pi r^2}. \tag{9.37}$$

As this example is in vacuo, scatter and energy transfer cannot occur. Thus, the result is applicable to both photons and charged particles. Now consider the changes in flux at point P that will occur if the source and detection point were no longer in vacuo. As the mean-free paths between interactions for electrons/positrons are significantly smaller than those for photons, this discussion will be limited to a photon-emitting source. Should there be a thin absorber of thickness t placed between the source and point P, then the attenuated photon flux at P is,

$$\frac{d\Phi}{dt} = \frac{A}{4\pi r^2} e^{-\mu t}. \tag{9.38}$$

This result is a simplification based upon the significant assumption that the absorber is sufficiently thin and far enough from P such that negligible scattered radiation from the absorber reaches P. If this were not the case (due to a thicker absorber), then (9.38) must be modified to account for scatter contributions by multiplying it by the buildup factor, B,

$$\frac{d\Phi}{dt} = \frac{A}{4\pi r^2} B e^{-\mu t}. \tag{9.39}$$

As the contributions of scattered radiation will lead to an increase in the flux at the point P, then $B \geq 1$.

The above result is immediately applicable to extended sources which are considered as integrals of differential elements modeled as points.

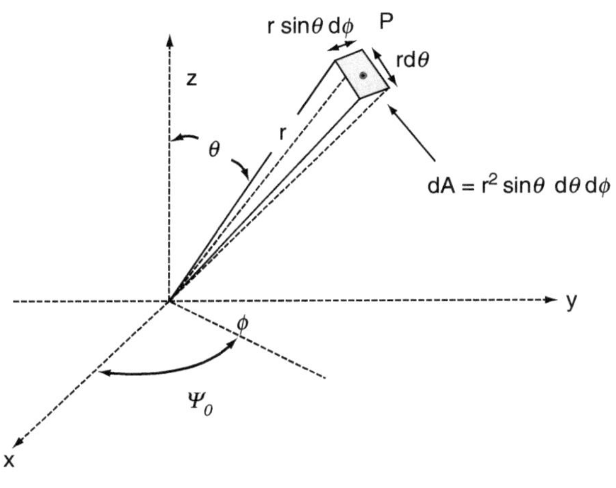

Fig. 9.9 Geometry for calculating the photon flux at the point P from a point source at the origin

Linear Source

A continuous linear source of activity is modeled by a linear array of point sources. Consider an infinitely-thin linear source of length L and activity A uniformly distributed along it (with an activity per unit length $\Gamma_A \equiv A/L$) and evaluate the flux at three points P_1, P_2, and P_3 as shown in Fig. 9.10. Again, as the calculation of the flux is initially in vacuo, the results are equally valid for photons and charged particles.

The differential flux at point P_1 due to the activity contained within the differential element $d\lambda$ is,

$$\left.\frac{d^2\Phi}{dt}\right|_{P_1} = \frac{\Gamma_A}{4\pi r^2} d\lambda. \tag{9.40}$$

From the geometry shown,

$$\lambda = r \tan\theta \tag{9.41}$$

$$r = \frac{h}{\cos\theta} \tag{9.42}$$

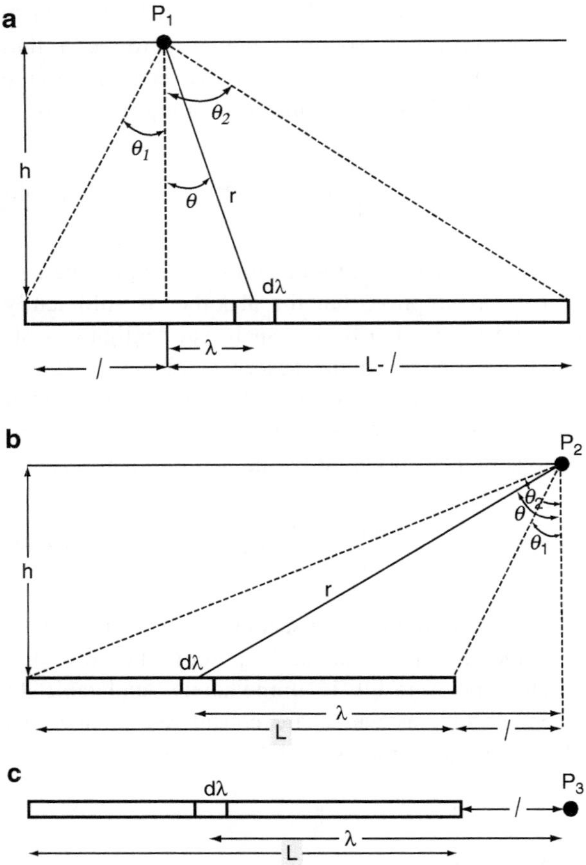

Fig. 9.10 Calculation of photon flux at three points from a linear source of activity

$$\frac{d\lambda}{d\theta} = \frac{h}{\cos^2\theta}. \tag{9.43}$$

Substituting these geometrical relationships into (9.40) results in,

$$\left.\frac{d^2\Phi}{dt}\right|_{P_1} = \frac{\Gamma_A}{4\pi h} d\theta. \tag{9.44}$$

The total radiation flux at P_1 is obtained by integrating over the azimuthal angle,

$$\begin{aligned} \left.\frac{d^2\Phi}{dt}\right|_{P_1} &= \frac{\Gamma_A}{4\pi h} \int_{-|\theta_1|}^{|\theta_2|} d\theta \\ &= \frac{\Gamma_A}{4\pi h}(|\theta_1| + |\theta_2|) \end{aligned} \tag{9.45}$$

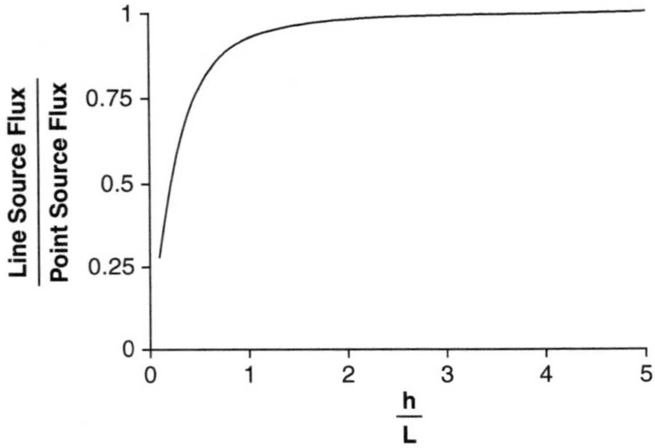

Fig. 9.11 The ratio of the flux at a midline point for a linear source of length L at a distance h from the midline of the linear source to that of a point source at the same distance h as a function of the distance to the linear source length. Both sources are of the same total activity

where the angles are $\theta_1 = \tan^{-1} l/h$ and $\theta_2 = \tan^{-1}\frac{L-l}{h}$. This result can be shown to reduce to the point-source expression when P_1 is at a great distance from the linear source where $h \gg l$ and $h \gg L - l$ allow the arctangents to be expanded to first order,

$$\begin{aligned} |\theta_1| + |\theta_2| &\approx \frac{l}{h} + \frac{L-l}{h} \\ &= \frac{L}{h}. \end{aligned} \tag{9.46}$$

Under such conditions, the flux at P_1 is,

$$\left.\frac{d\Phi}{dt}\right|_{P_1} = \frac{A}{4\pi h^2} \quad (h \gg L) \tag{9.47}$$

which is equivalent to the point source result given by (9.37). Figure 9.11 shows the ratio of the fluxes calculated for the line source at P_1 at midline to the source and the point source as a function of the ratio of h/L, both with the same total activity. The linear source can be approximated by a point source to within 0.5% if the ratio $h/L > 4$.

The differential photon flux at point P_2 (which is at the same perpendicular distance from the source as P_1 but at an axial distance beyond the end of the source) is also given by (9.44) and the integrated photon flux is,

$$\left.\frac{d^2\Phi}{dt}\right|_{P_2} = \frac{\Gamma_A}{4\pi h}(|\theta_2| - |\theta_1|). \tag{9.48}$$

Similarly, the photon flux at the axial point P_3 is,

$$\left.\frac{d\Phi}{dt}\right|_{P_3} = \frac{\Gamma_A}{4\pi}\left(\frac{1}{l} - \frac{1}{L+l}\right) = \frac{\Gamma_A}{4\pi}\frac{L}{l(L+l)} \qquad (9.49)$$

which, for $L \gg l$, simplifies to,

$$\left.\frac{d\Phi}{dt}\right|_{P_3} = \frac{\Gamma_A}{4\pi l} \quad (L \gg l) = \frac{A}{4\pi L l}. \qquad (9.50)$$

These flux calculations were for in vacuo. Consider, now, the case of an absorber of thickness t interposed between the linear source and the calculation point P_1 (clearly, $t \leq h$). Initially, ignore scatter and assume the source to be of photons only. Then, photon attenuation leads to a modification of (9.38),

$$\left.\frac{d^2\Phi}{dt}\right|_{P_1} = \frac{\Gamma_A}{4\pi h} e^{-\frac{\mu t}{\cos\theta}} d\theta. \qquad (9.51)$$

The final form of an expression for the complete photon flux at P_1 accounting for this attenuation is obtained by integrating over the angle θ,

$$\begin{aligned}\left.\frac{d\Phi}{dt}\right|_{P_1} &= \frac{\Gamma_A}{4\pi h}\int_{-|\theta_1|}^{|\theta_2|} d\theta\, e^{-\frac{\mu t}{\cos\theta}} \\ &= \frac{\Gamma_A}{4\pi h}\left(\int_0^{|\theta_2|} d\theta\, e^{-\frac{\mu t}{\cos\theta}} + \int_0^{|\theta_1|} d\theta\, e^{-\frac{\mu t}{\cos\theta}}\right) \\ &= \frac{\Gamma_A}{4\pi h}(S(|\theta_2|, \mu t) + S(|\theta_1|, \mu t))\end{aligned} \qquad (9.52)$$

where $S(\theta, x) = \int_0^\theta d\theta\, e^{-\frac{x}{\cos\theta}}$ is the Sievert integral (Sievert 1930). As there is no analytical solution to this integral, it must be solved either numerically (Worthley and Nicholls 1972) or through Monte Carlo means (Williamson et al. 1983). Tabulated values of the Sievert integral and its relationships to other integrals are provided by Stegun (1972).

If it is next assumed that the source is one of photons and that it and the point of interest, P_1, are both embedded in a homogeneous medium with linear attenuation coefficient, μ, the fluence rate at P_1 is, from (9.51),

$$\left.\frac{d^2\Phi}{dt}\right|_{P_1} = \frac{\Gamma_A}{4\pi h}(S(|\theta_2|, \mu h) + S(|\theta_1|, \mu h)). \qquad (9.53)$$

This is an incomplete solution, however. While it does account for photon attenuation by the medium, it does not include the effects of photon scatter that occurs between the source and the point P_1. As before, this is achieved by the use of the buildup factor through which the contribution of scatter will increase the flux at P_1. The buildup factor can be approximated by,

$$B(\theta) \approx 1 + \alpha(k)\frac{\mu h}{\cos\theta} \qquad (9.54)$$

where $\alpha(k)$ is a coefficient dependent upon the photon energy, k. Including this buildup factor into the integral of (9.52) results in the expression for the photon flux at P_1 as,

$$\left.\frac{d^2\Phi}{dt}\right|_{P_1} = \frac{\Gamma_A}{4\pi h}\left(S(|\theta_2|, \mu h) + S(|\theta_1|, \mu h) + \alpha(k)\mu h \int_{|\theta_1|}^{|\theta_2|} d\theta \frac{e^{-\frac{\mu h}{\cos\theta}}}{\cos\theta}\right) \qquad (9.55)$$

where the last integral is to be solved numerically.

Disc Source

Extending the complexity of the radiation source geometry being calculated for, now consider the flux at an off-axis point P due to activity uniformly distributed across a disc of radius R, as shown in Fig. 9.12. P is at a height h above the disc and is displaced by a distance d from the disc's center. Again, the calculation is performed initially in vacuo and thus remains valid for both photons and charged particles.

The disc source has a total activity A with a uniform areal activity density given by $\Sigma_A = A/\pi R^2$. The differential element of area d^2w at a distance r from the center of the source is $d^2w = r\,dr\,d\theta$ and the

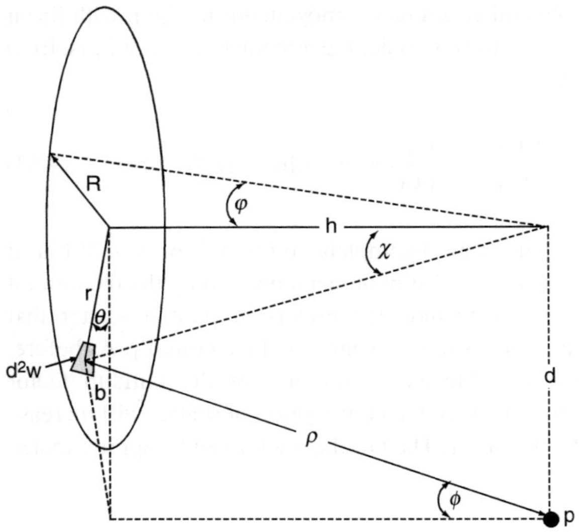

Fig. 9.12 Disc source of activity

differential flux at P from this element is, from the point-source calculation, equal to,

$$\frac{d^3\Phi}{dt}=\frac{\Sigma_A}{4\pi\rho^2}d^2w. \quad (9.56)$$

From the geometry of Fig. 9.12,

$$\rho^2=h^2+b^2 \quad (9.57)$$

where

$$b^2=d^2+r^2-2rd\cos\theta. \quad (9.58)$$

The differential flux at P is thus equal to,

$$\frac{d^3\Phi}{dt}=\frac{\Sigma_A}{4\pi}\frac{r}{h^2+d^2+r^2-2rd\cos\theta}\,dr\,d\theta. \quad (9.59)$$

The total photon flux at P is found by integrating (9.59) over the source area,

$$\begin{aligned}\frac{d\Phi}{dt}&=\frac{\Sigma_A}{4\pi}\int_0^R dr\,r\int_0^{2\pi}\frac{d\theta}{h^2+d^2+r^2-2rd\cos\theta}\\&=\frac{\Sigma_A}{4\pi}\int_0^R dr\frac{r}{h^2+d^2+r^2}\int_0^{2\pi}\frac{d\theta}{1+\left(-\frac{2rd}{h^2+d^2+r^2}\right)\cos\theta}.\end{aligned} \quad (9.60)$$

The angular integral is of the form,

$$\int_0^{2\pi}\frac{d\theta}{1+a\cos\theta}=\frac{2\pi}{\sqrt{1-a^2}} \quad \text{for } a^2<1. \quad (9.61)$$

Employing this identity, the flux is written as a single integral,

$$\frac{d\Phi}{dt}=\frac{\Sigma_A}{2}\int_0^R dr\frac{r}{(h^2+d^2+r^2)\sqrt{1-\frac{4r^2d^2}{h^2+d^2+r^2}}}. \quad (9.62)$$

Using the change of variable within the integral, $x=r^2$, the flux is,

$$\begin{aligned}\frac{d\Phi}{dt}&=\frac{\Sigma_A}{4}\int_0^{R^2}\frac{dx}{\sqrt{x^2+2(h^2-d^2)x+(h^2+d^2)^2}}\\&=\frac{\Sigma_A}{4}\ln\left|\frac{R^2+h^2-d^2+\sqrt{(R^2+h^2-d^2)^2+4d^2h^2}}{2h^2}\right|.\end{aligned} \quad (9.63)$$

In the event that the point P is axial to the disc (i.e., directly over the center of the circular planar source or $d=0$), this expression reduces to,

$$\frac{d\Phi}{dt}=\frac{\Sigma_A}{4}\ln\left(1+\left(\frac{R}{h}\right)^2\right). \quad (9.64)$$

If the point at which the flux is calculated for is sufficiently removed from the source such that $R\ll h$, then this result reduces to, using $\ln(1+x)\approx x$ for small x,

$$\begin{aligned}\frac{d\Phi}{dt}&\approx\frac{\Sigma_A}{4}\frac{R^2}{h^2}\\&\approx\frac{A}{4\pi h^2}.\end{aligned} \quad (9.65)$$

Again, the point-source result is returned. Figure 9.13 presents the ratio of the flux calculated for the disc source to that of the point source for points along the axis as a function of $h/2R$ for the same total activity. Approximating the disc source as a point will introduce an error in the flux calculation of less than 0.5% for $h/2R>5$.

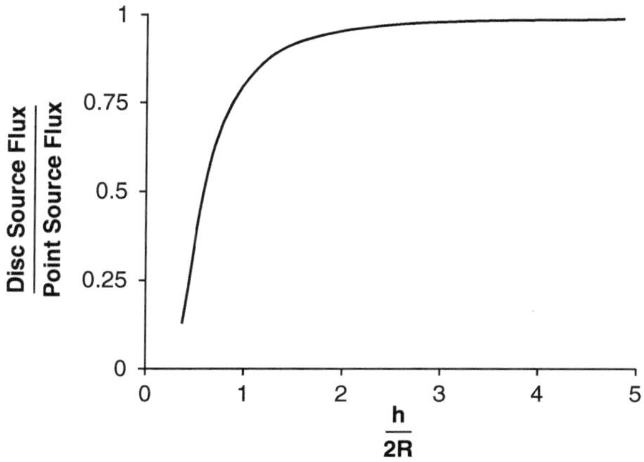

Fig. 9.13 The ratio of the flux at a midline point for a disc source of radius R at a distance h along the central axis of the source to that of a point source at the same distance h as a function of the distance to the linear source length. Both sources are of the same total activity

If an absorber of thickness t is now placed between the disc source (considered to be a source of photons) and the point of calculation, the differential flux is modified

$$\frac{d^3\Phi}{dt} = \frac{\Sigma_A}{4\pi\rho^2} e^{-\mu t \sec\chi} d^2 w. \tag{9.66}$$

The attenuated photon flux for the axial point for which $d = 0$ is now calculated. In this case, $\rho^2 = h^2 + r^2$ and $\cos\chi = h/\rho$ and,

$$\frac{d\Phi}{dt} = \frac{\Sigma_A}{2} \int_0^R dr \frac{r}{h^2 + r^2} e^{-\mu t \sqrt{h^2+r^2}/h} \quad \text{(with absorber, } d = 0) \tag{9.67}$$

Although the integral appears complicated, it is straight-forward to solve by rewriting in terms of ρ and substituting $y = (\mu t/h)\rho$,

$$\begin{aligned} \frac{d\Phi}{dt} &= \frac{\Sigma_A}{2} \int_{\mu t}^{\frac{\mu t}{h}\sqrt{h^2+R^2}} dy \frac{e^{-y}}{y} \\ &= \frac{\Sigma_A}{2} \left(\int_{\mu t}^{\infty} dy \frac{e^{-y}}{y} - \int_{\frac{\mu t}{h}\sqrt{h^2+R^2}}^{\infty} dy \frac{e^{-y}}{y} \right) \\ &= \frac{\Sigma_A}{2} \left(\mathrm{Ei}(\mu t) - \mathrm{Ei}\left(\frac{\mu t}{h} \sqrt{h^2+R^2} \right) \right) \\ &\text{(with absorber, } d=0) \end{aligned} \tag{9.68}$$

where Ei(x) is the exponential integral. For a disc of infinite radius, the second exponential integral clearly goes to zero and one is left with the simple result,

$$\frac{d\Phi}{dt} = \frac{\Sigma_A}{2} \mathrm{Ei}(\mu t) \quad \left(\text{with absorber of thickness } t, \text{ infinite planar source, } d = 0 \right). \tag{9.69}$$

As an endnote, the flux from a spherical source of activity can be calculated for by modeling the sphere as a series of discs with differential thicknesses and integrating over their individual contributions. However, the inclusion of self-absorption within the three-dimensional source magnifies the complexity of the problem. As a consequence, the flux calculation for radionuclides distributed over a three-dimensional geometry is best handled by Monte Carlo simulations.

9.2.4.3 An Overview of Monte Carlo Methods

Introduction

The Monte Carlo simulation method of calculating for radiation transport has a long history, having been originally developed during the 1940s in nuclear weapons design (Metropolis and Ulam 1949). As a full investigation of Monte Carlo methods applied to medical radiation problems is far beyond the intention and scope of this book, this subsection provides an overview only of the principles of the Monte Carlo method and how the method can be applied to radiation transport and dosimetry calculations. Comprehensive and contemporary reviews of the applications of the Monte Carlo method to diagnostic and therapeutic nuclear medicine can be found in the compilations edited by Ljungberg et al. (1998) and Zaidi and Sgouros (2003).

Recall from Chaps. 6 and 7 that the interactions between photons and charged particles with matter are stochastic and that total cross sections describe the probabilities of scatter, energy transfer, irradiation, etc. Thus, chance is key to the descriptions of such interactions. For example, a photon interaction with matter can lead to coherent scatter, photoelectric absorption, Compton scatter, or electron–positron pair production with the likelihoods of these interactions being dictated by the relative magnitudes of the total cross sections. For example, consider a photon-matter interaction at an energy below the pair production threshold. The probability of the interaction being that of incoherent (Compton) scatter is,

$$\text{Probability of incoherent scatter} = \frac{\sigma_{\mathrm{Inc}}}{\sigma_{\mathrm{Coh}} + \sigma_{\mathrm{PE}} + \sigma_{\mathrm{Inc}}} \quad (9.70)$$

where the denominator is the total cross section for a photon interaction.

This gaming aspect led to this method of modeling radiation transport and its interactions with matter as being described as "Monte Carlo" by von Neumann and Ulam during their work in the Manhattan Engineering District project which developed the first nuclear fission weapons. The underlying principle of the Monte Carlo technique is that of random sampling which can be demonstrated in a simple example. The Monte Carlo method can be used to provide an estimate of the value of π as shown in Fig. 9.14. A circle of radius r is centered within a square with side length 2r. The ratio of the area of the circle to that of the enclosing square is equal to π/4. If one were to randomly drop, say, a coin within the square and count the fraction of drops that resulted in the coin falling within the circle, this fraction will approach the ratio of π/4 as the number of drops approaches infinity assuming, of course, that the drops are truly random.

Random Number Generator

From the above discussion, Monte Carlo simulations will require random input in the form of random numbers. These could be derived from a physical phenomenon, e.g., electronic noise or the time intervals between radioactive decays. However, these are not reproducible phenomena and reproducibility is required, for example, in debugging code. Hence, one looks to a mathematical algorithm to provide a random sequence of numbers. The sequence must be reproducible and defined by an initial selection of a "seed" number to start the algorithm. As a result, these random number generators (RNGs) are referred to as being pseudorandom in that they produce random numbers in a reproducible sequence. Three practical additional requirements of an RNG are that the generation of random numbers should be fast (as large numbers of simulations are required to be calculated in order to generate a statistically-valid result), the sequences generated are uncorrelated with each other and that the period of the sequence (i.e., the number of numbers until the sequence is repeated) be long.

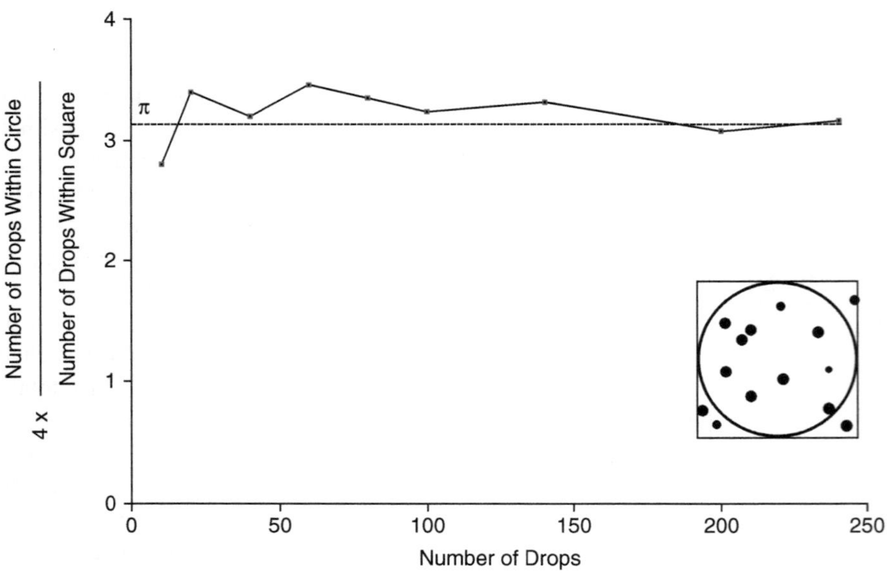

Fig. 9.14 An example of the principle of Monte Carlo method as means of estimating the value of π. A circle of radius r is centered within a square of side dimension 2r. As the area of the circle is πr^2 and the area of the square is $4r^2$, the ratio of areas equals π/4. If one were to randomly drop a coin within the square and tally the number of times it fell within the circle, the ratio of this number to the total number of coin drops will approach π/4 with an increasing number of drops (or "samples") as shown in the graph

Perhaps the most effective RNG is the linear congruence (or residue) method in which the (i + 1)th random number in a sequence is produced is,

$$\xi_{i+1} = (\lambda\xi_i + \eta)(\mathrm{mod}\,P) \tag{9.71}$$

where the result is divided by P−1 to give $\xi_{i+1} \in [0,1]$. For example, the IBM RANDU subroutine in FORTRAN uses values for the parameters, $\lambda = 2^{16} + 3$, $\eta = 0$ and $P = 2^{31}$. Morin et al. (1979) reported that the cycle of the random sequence produced by this algorithm exceeded 10^7.

Analog Sampling

The analog approach used in Monte Carlo simulations of radiation transport models directly the transport and interactions of the radiation. A random physical process is described by probability and cumulative density functions and, by drawing random samples of these functions, one can form a statistical approximation to the process (Emshoff and Sisson 1970; Raeside 1976). For example, the probability distribution function (pdf) for the random variable X is,

$$f(x) = \Pr(X = x) \tag{9.72}$$

and the cumulative density function (cdf) is,

$$\begin{aligned} F(x) &= \int_{-\infty}^{x} dt\, f(t) \\ &= \Pr(x \le X). \end{aligned} \tag{9.73}$$

By this definition, the cdf is such that $F(x) \in [0, 1]$. Thus, the integral of (9.73) can be equated to a randomly-generated number, ξ, uniformly distributed within the interval [0, 1]. Then, the sampled value of the random variable, x, associated with ξ is given by the inverse of the cumulative density function,

$$x = F^{-1}(\xi). \tag{9.74}$$

By repeatedly applying (9.74) to a large group of randomly selected ξs, a statistical approximation of X is obtained. For example, the distance a photon moving in a medium will travel before interacting with the medium can be simulated using this method. The cumulative probability that the photon will traverse a distance x in the medium before interacting is given by,

$$F(x) = 1 - e^{-\mu x} \tag{9.75}$$

where μ is the linear attenuation coefficient in water appropriate for the photon's energy (recall that the linear attenuation coefficient represents the mean number of photon interactions per unit distance traveled and, hence, is the reciprocal of the photon's mean-free pathlength). This cumulative probability of (9.75) can be set equal to a random number ξ where the uniformly-distributed $\xi \in [0,1]$ and the equation inverted to obtain a sample of the distance traveled to the next interaction site,

$$\begin{aligned} x &= -\frac{1}{\mu}\ln(1-\xi) \\ &\equiv -\frac{1}{\mu}\ln\xi \end{aligned} \tag{9.76}$$

where the argument of the logarithm, $1-\xi$, has been replaced by the equally random ξ.

Another example of Monte Carlo analog sampling is the modeling of a Compton scatter. From the Compton relation, the postscatter photon energy for a photon with an initial energy k is,

$$k' = \frac{k}{1 + \alpha(1 - \cos\theta)} \tag{9.77}$$

where $\alpha = k/m_e$ and θ is the photon scattering angle. In the Monte Carlo simulation, k′ is first determined and the scattering angle deduced from (9.77). This is performed by noting that the probability of the photon being scattered from initial energy k to k′ is,

$$\Pr(k' \leftarrow k) = \int_{k'}^{k} d\kappa \frac{d\sigma(\kappa)/d\kappa}{\sigma(\kappa)} \tag{9.78}$$

where $d\sigma(\kappa)/d\kappa$ is the Klein–Nishina differential cross section and $\sigma(\kappa)$ is the Klein–Nishina total cross section. This integral can be set equal to a uniformly-distributed random number $\xi \in [0,1]$ and inverted to yield the scattered photon energy. This inversion was obtained by Carlson (1953),

$$k' = \frac{k}{1 + s\xi + \left(2\frac{k}{m_e} - s\right)\xi^3} \tag{9.79}$$

where

$$s = \frac{(k/m_e)}{1 + 0.5625\left(\frac{k}{m_e}\right)} \tag{9.80}$$

Having determined the scattered photon energy k', the scattering angle θ is determined from (9.77) and the random azimuthal angle ϕ is given by $2\pi\xi$, where again ξ is a random number uniformly distributed on the unit interval (Cashwell and Everett 1959). The postscatter trajectory of the photon is then described by the direction cosines,

$$u' = \frac{\sin\theta(wu\cos\phi - v\sin\phi)}{\sqrt{1 - w^2}} + u\cos\theta \tag{9.81}$$

$$v' = \frac{\sin\theta(wv\cos\phi - v\sin\phi)}{\sqrt{1 - w^2}} + v\cos\theta \tag{9.82}$$

$$w' = -\frac{\sin\theta\cos\phi}{\sqrt{1 - w^2}} + w\cos\theta. \tag{9.83}$$

It may not be possible, in some instances, to easily obtain the inverse of the cdf . The rejection method presents another approach, although it is more computationally intensive. It begins by normalizing the pdf to its maximum value,

$$\tilde{f}(x) = \frac{f(x)}{f_{max}(x)}. \tag{9.84}$$

Two random numbers uniformly distributed between 0 and 1 are then selected, ξ_1 and ξ_2. The random variable x is then calculated from,

$$x = x_{min} + \xi_1(x_{max} - x_{min}) \tag{9.85}$$

where x_{min} and x_{max} are the minimum and maximum possible values for x, respectively. $\tilde{f}(x)$ for this value of x is then compared with ξ_2. If $\tilde{f}(x) \leq \xi_2$, the sampled value of x is accepted. Otherwise, a new set of random numbers are sampled and the process repeated until the condition $\tilde{f}(x) \leq \xi_2$ has been met.

Variance Reduction

While intuitive, the analog sampling method may not always be efficient and can often result in very long computation times in order to yield a result with statistical relevance. Various methods exist to bias the sampling of the Monte Carlo simulation (also known as nonanalog sampling) in order to improve the efficiency of the calculation. The result itself will be biased and a postsimulation correction must be applied. A simple example is the use of inherent geometrical symmetry to reduce the number of regions being calculated for (e.g., Boston et al. 2001).

One means of increasing the efficiency of simulating photon interactions is through interaction forcing (Bielajew and Rogers 1989). Recall from the derivation of (9.76) that the interaction distance for a photon is given by $x = -\ln(1 - \xi)/\mu$. A photon has a maximum range of infinity whereas the number of mean-free paths (per micrometer) of the photons within a typical geometry of interest will be finite. Hence, a Monte Carlo analog simulation of photon transport within a finite geometry will, in most samples, result in large numbers of photons escaping the geometry without interaction. This is clearly an inefficient outcome as the escaping electrons fail to provide information on interactions within the geometry. One avoids this inefficiency by forcing the photons to interact within the geometry by defining a finite distance x_{max}, which could be the maximum spatial dimension of the geometry being simulated, and which limits the maximum distance traveled by a photon. This enforcement can then be used to reduce the distance between interactions,

$$x = -\frac{\ln(1 - \xi(1 - e^{-\mu x_{max}}))}{\mu}. \tag{9.86}$$

The resultant bias must be compensated for through a reduction in the "weight" of the photon's interaction with the medium (i.e., the magnitude of its contribution) by scaling it by the factor $1 - e^{-\mu x_{max}}$. A reduction in the value of x_{max} maps a reduction in the physical contribution of the photon interaction.

Monte Carlo Codes: Examples

A variety of Monte Carlo codes are available for the nuclear medicine physicist to easily and efficiently simulate radiation transport and dosimetry without being required to write their own code. The growth in "off-the-shelf" Monte Carlo power and the ease of its use in medical physics applications have grown remarkably over the past few decades.[9] This subsection summarizes three of the Monte Carlo codes within the public domain that are most frequently used in medical radiation dosimetry calculations. The websites of the institutions that provide and support these codes are given.

MCNP

The MCNP (Monte Carlo N-particle) code was developed at Los Alamos and its genesis is directly traceable to Ulam, Metropolis, and von Neumann. The code is general purpose and can be found at http://mcnp-green.lanl.gov/. It is used for calculating the transport and energy transfer of photons, neutrons, and electrons. For photons, it allows for incoherent and coherent scatter, photoelectric absorption and subsequent characteristic X-ray emission, electron–positron pair product with annihilation and *bremsstrahlung*. The CSDA is used to model electron transport.

GEANT4

The GEANT4 (geometry and tracking) code is the successor to the GEANT software tool kits developed for high-energy physics applications at CERN. Its development and support are provided by an international collaboration and a description of the code can be found at http://www.geant4.org/geant4/. As Monte Carlo methods are inherently parallel in nature, investigations of distributed processors running GEANT4 through CERN's Large Hadron Collider computing grid as a means of speeding up dosimetry calculations have been performed (Chauvie et al. 2007).

EGS

The EGS (electron gamma shower) code was originally developed at the Stanford Linear Accelerator Center but it is no longer supported by SLAC. The code was developed for simulating the coupled transport of photons and electrons for energies between 1 keV and 10 GeV. Current maintenance and development of the code provided by the National Research Council of Canada, where the code is known as EGSnrc and which can be downloaded from http://www.irs.inms.nrc.ca/EGSnrc/EGSnrc.html) and the KEK laboratory in Japan and which can be found at http://rcwww.kek.jp/research/egs/kek/.

9.2.5 Buildup Factor

9.2.5.1 Introduction

Recall that in the previous analytical derivations of the photon fluxes at a point in space arising from radioactive geometrical sources, the problem was first considered in vacuo where the solution was obtained solely on geometrical grounds. Then, the influence upon the photon flux due to the interposition of an absorbing material between the radioactive source and the point of calculation was derived. There were two effects upon the detected flux as a result. First, the absorber attenuates the primary photon flux received at the point. Second, it behaves as a scattering medium which serves to increase the net photon flux. The first effect is described using the linear attenuation coefficient. The second effect is clearly more complicated and the use of the buildup factor as a means of describing the increase in radiation flux as a result of scatter has already been alluded to. In this subsection, the buildup factor is formally introduced as are the methods of calculating for and measuring it.

Consider a point P behind a slab of material of thickness t which is irradiated by a photon beam, as shown in Fig. 9.15. The two cases of "narrow" and

[9] For example, the author recalls writing his own Monte Carlo code in FORTRAN in 1980 to simulate photon transport in a cubic water phantom exposed to ^{60}Co γ rays. Even though the output of the code was simply the kerma due to primary and scattered photons, a simulation of 10^7 photons (without electron transport) on a time-shared Data General ECLIPSE S200 minicomputer took the best part of a week. In 2001, the author was using the MCNP code to simulate the total photon and electron transport and complete energy deposition for a ^{125}I brachytherapy source to even greater statistical precision on a desktop PC in the space of an afternoon.

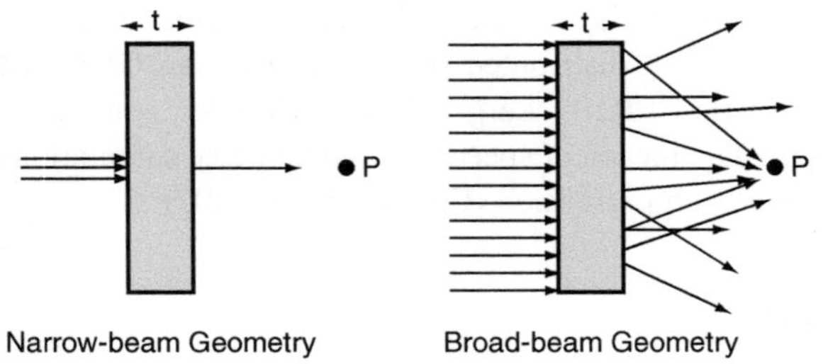

Fig. 9.15 Definitions of narrow- and broad-beam geometries for a photon beam incident to an absorbing and scattering medium. In the narrow-beam geometry, scatter reaching the point P is negligible and thus attenuation need only be considered. In the broad-beam geometry, photons can be scattered to reach P, thus increasing the net fluence at it

"broad" beam geometries are considered, where the narrow-beam geometry is such that there is negligible scattered radiation reaching P, as shown. This increase in the number of photons reaching P due to scatter can be quantified by the buildup factor, B, where $B \geq 1$. As the width of the beam approaches zero, as does the contribution of scatter, then $B \rightarrow 1$. Being interested in energy deposition, we can define the buildup factor in terms of the photon energy fluence. As the buildup factor can be related to other radiation quantities, we denote this definition by adding an appropriate subscript, B_Ψ . For the narrow-beam geometry, the transmitted energy fluence incident to P is $\Psi_0 e^{-\mu t}$, where Ψ_0 is the fluence incident to the slab and μ is the narrow-beam attenuation coefficient. For any geometry, then, the energy fluence incident to P is,

$$\Psi_t = B_\Psi \Psi_0 e^{-\mu t} \tag{9.87}$$

The energy fluence buildup factor $B_\Psi \rightarrow 1$ as the thickness of the slab $t \rightarrow 0$, leading to $\Psi_t \rightarrow \Psi_0$ as expected. Further definitions of the buildup factor relative to specific applications are provided in terms of number fluence, energy fluence, and absorbed dose. For polyenergetic photon beams, these buildup factors are (integrations are over the relevant photon spectrum),

$$B_\phi = \frac{\int dk\, \phi(k)}{\int dk\, \phi_0(k)} \tag{9.88}$$

$$B_\Psi = \frac{\int dk\, \Psi(k)}{\int dk\, \Psi_0(k)} \tag{9.89}$$

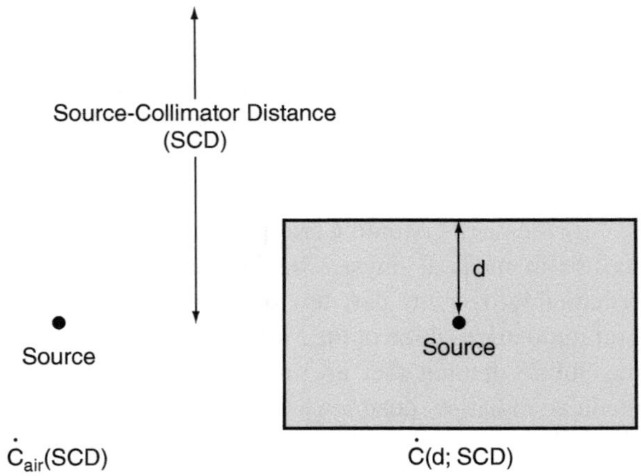

Fig. 9.16 Method for measuring the buildup factor to use in quantitative activity determination in gamma-camera scintigraphy

$$B_D = \frac{\int dk \frac{\mu_{en}}{\rho}(k)\Psi(k)}{\int dk \frac{\mu_{en}}{\rho}(k)\Psi_0(k)}. \tag{9.90}$$

Again, the 0 subscript indicates the uncollided photon flux.

The buildup factor plays two important roles in nuclear medicine dosimetry. The first is in the measurement of the absolute in vivo activity through gamma camera scintigraphy (Wu and Siegel 1984; Siegel et al. 1985; Siegel 1985). The second role is in the use of DPKs in the calculation of the internal radiation dosimetry due to a distributed photon-emitting radionuclide.

9.2.5.2 Methods of Determining the Buildup Factor

Measurement

This is the most direct method in nuclear medicine of determining B for a fixed radionuclide and geometry, including collimator type. Consider Fig. 9.16. Rearranging (9.87) and replacing the energy fluence with the number of counts (photons which are photoelectrically absorbed in the gamma camera's scintillation crystal), the measured buildup factor for a given depth d in tissue and source-collimator distance (SCD) is,

$$B(d; SCD) = \frac{\dot{C}(d; SCD)}{\dot{C}_{Air}(SCD)} e^{\mu d} \tag{9.91}$$

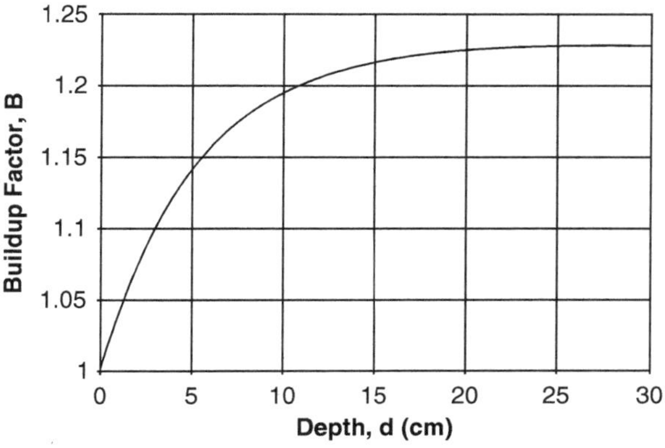

Fig. 9.17 Buildup factor for ^{99m}Tc photons in a tissue-equivalent phantom as a function of depth. *Curve* was derived from the coefficients of a fit presented by Siegel et al. (1985) to measured data obtained for a low-energy parallel hole collimator and a 15% energy-acceptance window

These quantities are shown in Fig. 9.16.$\dot{C}_{Air}(SCD)$ is the measured count rate for a source placed in air (i.e., negligible scatter and attenuation present) at a given SCD and $\dot{C}(d; SCD)$is that measured with the source at a depth d in a phantom radiologically representative of tissue (e.g., water) at the same SCD. μ is the linear attenuation coefficient appropriate for the energy of photons emitted by the source under the broad-geometry measurement conditions. The $e^{\mu d}$ factor corrects for attenuation so that, as required, the derived buildup factor accounts for scattered radiation only.[10] Figure 9.17 shows an example of the buildup factor for ^{99m}Tc photons as a function of depth in a phantom which is radiologically equivalent to tissue.

Analytical

The buildup factor is clearly applicable also to radiation shielding design and many analytical methods of determining it have been derived in this field. Historically, the Boltzmann radiation transport equation has provided a means of calculating the buildup factor (Harima 1993). In terms of the photon energy fluence, the Boltzmann equation is,

$$\hat{\boldsymbol{\Omega}}\bullet\nabla\Psi\left(\mathbf{r},k,\hat{\boldsymbol{\Omega}}\right)+\mu(\mathbf{r},k)\Psi\left(\mathbf{r},k,\hat{\boldsymbol{\Omega}}\right)$$
$$=\int_0^\infty dk'\int d\hat{\boldsymbol{\Omega}}'\left(\frac{k}{k'}\right)\mu_{Sec}\left(\mathbf{r},k'\rightarrow k,\hat{\boldsymbol{\Omega}}'\rightarrow\boldsymbol{\Omega}\right) \quad (9.92)$$
$$\times\Psi\left(\mathbf{r},k',\hat{\boldsymbol{\Omega}}'\right)+S(\mathbf{r},k,\boldsymbol{\Omega})$$

The terms in this equation are the total photon interaction coefficient, $\mu(\mathbf{r},k)$, $\mu_{Sec}\left(\mathbf{r},k'\rightarrow k,\hat{\boldsymbol{\Omega}}'\rightarrow\boldsymbol{\Omega}\right)$ $dk'd\hat{\Omega}'$ is the probable number of secondary photons at $\mathbf{r}$ with an energy between k and k + dk and moving in a direction between $\hat{\boldsymbol{\Omega}}$ and $\hat{\boldsymbol{\Omega}} + d\hat{\boldsymbol{\Omega}}$ produced per unit length by an initial photon of energy k′ moving in the direction $\hat{\boldsymbol{\Omega}}'$ and $S(\mathbf{r},k,\boldsymbol{\Omega})$ is the production rate of photons at $\mathbf{r}$ of energy k moving in the direction $\boldsymbol{\Omega}$. Numerical solutions for the photon fluence at the point $\mathbf{r}$ (and integrated over photon energy and direction) are then used to provide a calculated buildup factor.

Monte Carlo

Calculation of the buildup factor is well-suited for the Monte Carlo simulation method described previously.

9.2.5.3 Analytical Representations of the Buildup Factor

Introduction

Tabulated values of buildup factors determined from measurement or previously calculated from numerical or Monte Carlo calculations exist and are readily available (see, e.g., the review article by Harima 1993). However, for practical dosimetry calculations, especially for repetitive computations, an analytical representation of the buildup factor is most useful. These analytical forms are fits to either these tabulated data or to bespoke measurements. Some analytical expressions for the buildup factor as functions of photon energy k and the thickness of the absorber in terms of the numbers of mean-free paths μr are summarized below.

[10] One will note the similarity between the ratio of (9.91) and the tissue-air ratio of radiotherapy.

Taylor Formula

This general expansion of the buildup factor can be given as the weighted sum of N exponential terms for a photon of energy k,

$$B(k, \mu r) = \sum_{i=1}^{N} A_i e^{-a_i(k)\mu r} \quad (9.93)$$

This expression is frequently limited the number of terms N to 2 to give the form (Harima 1993),

$$B(k, \mu r) = A e^{-a_1(k)\mu r} + (1 - A) e^{-a_2(k)\mu r} \quad (9.94)$$

Geometric Progression Formula

This analytical expression of the buildup factor was derived by Harima and Nishiwaki (1972),

$$B(k, \mu r) = 1 + (b(k) - 1)\frac{(K^{\mu r} - 1)}{(K - 1)} \quad K \neq 1 \quad (9.95)$$

$$B(k, \mu r) = 1 + (b(k) - 1)\mu r \quad K = 1 \quad (9.96)$$

where

$$K = c(k)(\mu r)^{a(k)} + d(k)\frac{\tanh\left(\frac{\mu r}{X(k)} - 2\right) - \tanh(-2)}{1 - \tanh(-2)} \quad (9.97)$$

where a(k), b(k), c(k), d(k), and X(k) are parameters to be fit (Yoshida 2006).

Meisberger Formula

Meisberger et al. (1968) presented a polynomial form of the buildup factor for use in brachytherapy,

$$B(k, \mu r) = e^{\mu r} \sum_{n=0}^{3} \kappa_n(k)(\mu r)^n \quad (9.98)$$

The calculation was, due to its required purpose, limited to a water medium as an approximation of soft tissue.

Leichner Formula

In his DPK calculations for internally-distributed β- and γ-emitting radionuclides, Leichner (1994) proposed an analytical form of $B_D(k, \mu r)$similar to that of Meisberger,

$$B_D(k, \mu r) = 1 + e^{-\kappa_0(k)\mu r} \sum_{n=1}^{3} \kappa_n(k)(\mu r)^n \quad (9.99)$$

The photon energy-dependent coefficients were also determined by fitting to Berger's data.

Kwok Formula

In their DPK calculations for internally-distributed β- and γ-emitting radionuclides, Kwok et al. (1985) proposed a nonlinear analytical fit to $B_D(k, \mu r)$ of the form,

$$B_D(k, \mu r) = \exp\left(\frac{\kappa_1(k) \times (\mu r)^{\kappa_2(k)}}{\mu r + \kappa_3(k)}\right) \quad (9.100)$$

where the $\kappa_i(k)$ are photon energy-dependent parameters. The values of these parameters for photon energies ranging between 15 and 400 keV are provided in their paper.

9.2.5.4 Effective Attenuation Coefficient

An alternative approach to accounting for the combined effects of attenuation and scatter due to an absorber is the effective attenuation coefficient which follows from the definition of the backscatter factor for an absorber of thickness t and a primary photon beam with linear attenuation coefficient μ,

$$\frac{\Psi_t}{\Psi_0} = B e^{-\mu t} \equiv e^{-\mu_{eff} t} \quad (9.101)$$

Solving for μ_{eff},

$$\mu_{eff} \equiv \mu - \frac{\ln B}{t} \quad (9.102)$$

9.2.6 Reciprocity Theorem

9.2.6.1 Introduction: Point Source and Target

The reciprocity theorem, which has been long known in radiation physics (e.g., King (1912)), summarizes the results following the exchange of the positions of a radiation detector and a radiation source within a medium or media. These are demonstrated in Fig. 9.18.

In Fig. 9.18a, the radiation source and detector are both contained within a homogeneous (uniform isotropic) medium of infinite extent. Reversing the positions of the source and detector does not alter the primary and scattered fluences at the radiation detector and the response remains unaffected. The response is reciprocal. In Fig. 9.18b, the homogeneity requirement is relaxed and there are now two media within this infinite volume. Thus, when the positions of the source and detector are reversed, the medium in which they reside is also reversed. Clearly, the primary radiation reaching the detector from the source is unaffected by an exchange of positions. Should the two media have different radiological properties, then the media will have an effect upon scattering and the scattered component of the radiation reaching the detector will differ depending upon the whether or not it is in medium 1 or 2. The degree of the effect of radiologically-dissimilar media upon the reciprocity theorem is dependent upon the magnitudes of the scattered component relative to the primary component. The reciprocity theorem may approximately hold in Fig. 9.18b if both media have low atomic numbers, the radiation is photons and Compton scatter dominates.

The main usefulness of the reciprocity theorem in nuclear medicine dosimetry calculations has been described by Cristy (1983) in the context of Monte Carlo calculations of the absorbed doses received by target small organs due to radioactivity contained within large source organs. This description was given in terms of the specific absorbed fraction, $\Phi(r_T \leftarrow r_S)$, which is the absorbed fraction $\phi(r_T \leftarrow r_S)$ normalized to the mass of the target region,

$$\Phi(r_T \leftarrow r_S) = \frac{\phi(r_T \leftarrow r_S)}{m_T} \tag{9.103}$$

$\Phi(r_T \leftarrow r_S)$ has units of inverse mass. If the target region is small (e.g., the ovaries), the Monte Carlo-derived value

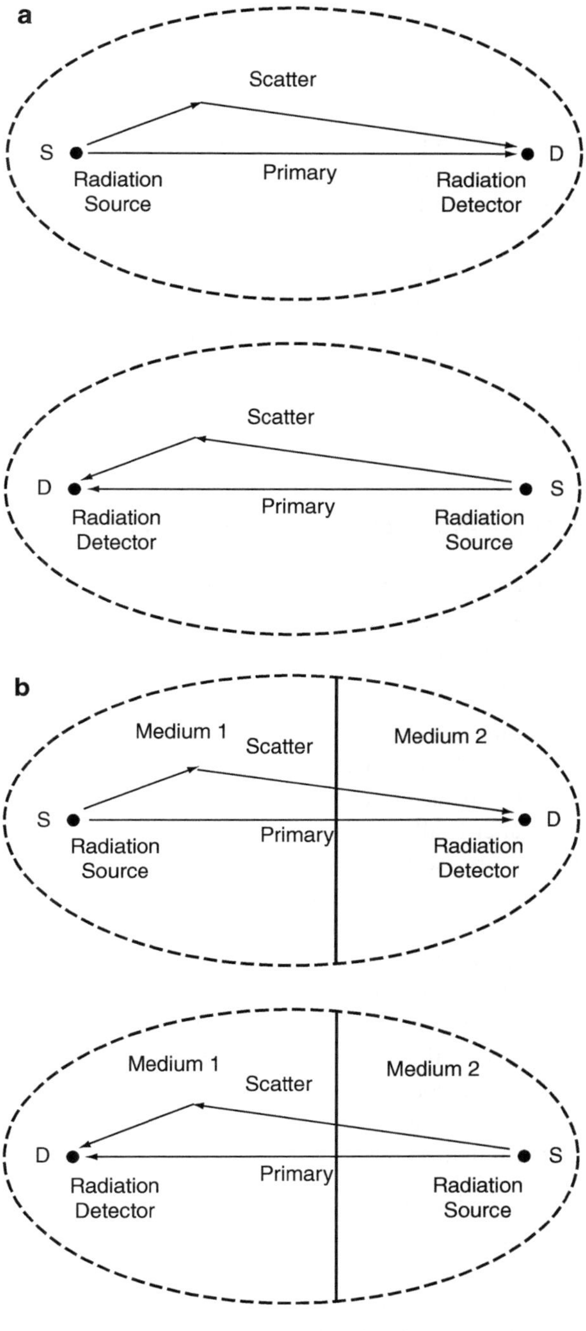

Fig. 9.18 The reciprocity theorem. In (**a**), both radiation source and detector are within the same homogeneous medium of infinite dimensions. As a result, exchanging their positions does not affect the radiation fluence detected. In (**b**), the source and detector are in different media. Exchanging their positions does not affect the primary radiation detected but can change the amount of scattered detected depending upon the differences between the radiological properties of the two media

of $\Phi(r_T \leftarrow r_S)$ may be statistically weak due to the limited number of interactions in the target region. From Poisson statistics, the relative standard deviation of the estimated absorbed dose to the organ is inversely proportional to the square-root of the number of energy deposition events which will be small due to the small organ size, thus leading to a high relative standard deviation. But the converse Monte Carlo-derived value of $\Phi(r_S \leftarrow r_T)$ may be more statistically reliable should the source region be a volume of larger size. A value of the specific absorbed fraction with greater statistical reliability can be calculated by a weighted sum of $\Phi(r_T \leftarrow r_S)$ and $\Phi(r_S \leftarrow r_T)$, with each weight proportional to $1/\sigma^2$, where σ^2 is the variance of the estimate.

9.2.6.2 Distributed Source and Target Regions

In his review of distributed radionuclide sources, Loevinger (1969) described those conditions under which the reciprocity theorem holds exactly or approximately:

- Uniform isotropic model : the source activity is uniformly distributed throughout an infinite homogeneous medium with a constant physical density. Scatter is allowed.
- Uniform scatterless model : the source activity is uniform in regions throughout a medium in which radiation is absorbed without scatter or photon buildup (physical density may vary but elemental composition remains constant).
- Uniform homogeneous model of finite size : the source activity is uniformly distributed within a homogeneous medium of finite size and constant physical density surrounded by a vacuum. Scatter is permitted.

The reciprocity theorem is not expected to vigorously hold for the third model, but is used in absorbed dose calculations. Loevinger stated that in the uniform isotropic and uniform scatterless models, the specific absorbed fraction is independent of what region is defined as a target and what region is designated as the source, or,

$$\Phi(r_T \leftarrow r_S) = \Phi(r_S \leftarrow r_T) = \Phi(r_T \leftrightarrow r_S) \quad (9.104)$$

where the $\leftrightarrow$ symbol signifies that either region can be the source or target.

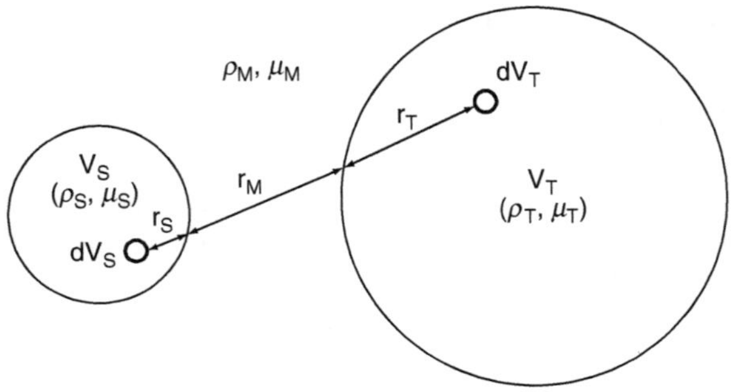

Fig. 9.19 The reciprocity theorem for distributed source and target

Mayneord (1945) considered source and target regions of extended dimensions within a medium, as shown in Fig. 9.19.

The source region has a volume V_S, physical density ρ_S, and attenuation coefficient μ_S and the corresponding quantities for the target region are V_T, ρ_T, and μ_T. Both of these regions are embedded within a third medium which is homogeneous with infinite extent and physical density ρ_M and attenuation coefficient μ_M. The source region contains a uniform distribution of a photon-emitting radionuclide of activity A_S and this radionuclide emits a single photon of energy k per disintegration. CPE is assumed to exist. A differential volume dV_S in the source thus contains the differential activity,

$$dA_S = (A_S/V_S)dV_S. \quad (9.105)$$

Scatter is neglected. The differential photon energy flux at the differential volume dV_T resulting from this differential amount of activity is,

$$d^2\Psi(dV_T \leftarrow dV_S) = k\frac{dA_S}{4\pi(r_S + r_M + r_T)^2}e^{-(\mu_S r_S + \mu_M r_M + \mu_T r_T)} \quad (9.106)$$

and the corresponding differential collision kerma rate,

$$\frac{d^2K_c(dV_T \leftarrow dV_S)}{dt} = \left(\frac{\mu_{en}}{\rho}\right)_T d^2\Psi(dV_T \leftarrow dV_S) \quad (9.107)$$

where $(\mu_{en}/\rho)_T$ is the mass energy-absorption coefficient for the target medium.

As CPE exists and the secondary electron range can be neglected, this differential collision kerma rate at dV_T will be equal to the differential absorbed dose rate at dV_T,

$$\frac{d^2D(dV_T \leftarrow dV_S)}{dt} = k\left(\frac{\mu_{en}}{\rho}\right)_T \frac{dA_S}{4\pi(r_S + r_M + r_T)^2} e^{-(\mu_S r_S + \mu_M r_M + \mu_T r_T)}. \tag{9.108}$$

The absorbed dose rate at dV_T due to all of the activity in the source region V_S is given by the integral of this result over the source region volume,

$$\frac{dD(dV_T \leftarrow V_S)}{dt} = \frac{k}{4\pi}\left(\frac{\mu_{en}}{\rho}\right)_T \frac{A_S}{V_S} \int_{V_S} \frac{dV_S}{(r_S + r_M + r_T)^2} e^{-(\mu_S r_S + \mu_M r_M + \mu_T r_T)}. \tag{9.109}$$

Finally, the absorbed dose rate to the entire target volume is given by averaging this absorbed dose rate over the entire target volume,

$$\frac{dD(V_T \leftarrow V_S)}{dt} = \frac{k}{4\pi}\left(\frac{\mu_{en}}{\rho}\right)_T \frac{A_S}{V_S V_T} \times \int\int_{V_S,V_T} \frac{dV_S\, dV_T}{(r_S + r_M + r_T)^2} e^{-(\mu_S r_S + \mu_M r_M + \mu_T r_T)} \tag{9.110}$$

If the source and target regions are interchanged, the target region is allowed to have the same activity as the original source region (i.e., $A_T = A_S \equiv A$) and the mass energy-absorption coefficients of both regions are equal, $(\mu_{en}/\rho)_T = (\mu_{en}/\rho)_S \equiv (\mu_{en}/\rho)$, then the absorbed dose to the original source region (but now without any contained activity) is,

$$\frac{dD}{dt}(V_T \rightarrow V_S) = \frac{k}{4\pi}\left(\frac{\mu_{en}}{\rho}\right) \frac{A}{V_S V_T} \int\int_{V_S,V_T} \frac{dV_S\, dV_T}{(r_S + r_M + r_T)^2} e^{-(\mu_S r_S + \mu_M r_M + \mu_T r_T)}. \tag{9.111}$$

Thus, under these conditions, which consider primary radiation only, the reciprocity theorem states that the absorbed dose rates in source and target regions are identical when they are interchanged.

9.2.6.3 Reciprocity Theorem Applied to Heterogeneous Media

The latter discussion of the reciprocity theorem was the closest in representing the conditions of soft tissue. But that calculation was of primary radiation only. The differing elemental compositions and physical densities and the vacuum-tissue interface used in Monte Carlo simulations affect scattered radiation and ensure that the reciprocity theorem will not generally be met exactly. Cristy (1983) stated that while under such conditions the reciprocity theorem does not hold, approximate reciprocity between the specific absorbed fractions can often hold in heterogeneous tissue and that such a feature could be referred to as a reciprocity principle.

The main applications of the reciprocity theorem of interest to the topic of this book are those in nuclear medicine radiation dosimetry. One of the first investigations of the reciprocity theorem/principle in heterogeneous media was by Ellett et al. (1965) who claimed that the reciprocity principle held to within about 3% in the homogeneous case of a 2.2 kg source region at the center of a 70 kg target for 40 and 662 keV photons. It was later reported by Snyder that, for soft tissue, the reciprocity principle was valid to within about a factor of 2. Such a significant difference in calculations based upon the assumptions of a reciprocity principle required further investigation. Cristy (1983) compared the specific absorbed fractions calculated using Monte Carlo simulations for a variety of organ pairs in his anthropomorphic pediatric phantoms. He concluded that the reciprocity principle was largely valid (to within 10%) for most soft-tissue organ pairs but that it was invalid at photon energies between 10 and 200 keV if one of the organ pairs was the skeleton.

9.2.7 Dose Point Kernels

The DPK concept as a means of calculating absorbed dose is applicable to both photons and electrons. In this calculation method, a source distributed throughout a volume is considered to be a spatial distribution of isotropic point sources. The DPK describes the radial distribution of absorbed dose or dose rate from a single point source and, through the superposition principle, these are combined to yield the total absorbed dose at a point. For example, for a β-emitting radionuclide, the absorbed dose rate at a distance r from a source of activity A is (Bardiès et al. 2003; Roeske et al. 2008),

$$\frac{dD(r)}{dt} = \kappa A \Delta \Phi(r) \quad (9.112)$$

where

$$\Delta = n\bar{T}. \quad (9.113)$$

n is the number of β particles emitted per decay and $\bar{T}$ is the mean energy of the β particles; hence, Δ is the mean energy released per radioactive transition. Φ(r) is the point kernel and κ is a conversion factor (equal to 1.6×10^{-8} g • cGy/MeV so as to give the absorbed dose rate in units of cGy/s if the units of activity, mean energy, and point kernel are Bq, MeV, and per gram, respectively). It will be noted that the point kernel, Φ(r), is the specific absorbed fraction where the source region is a point at r = 0 and the target region is a point at r, or $\Phi(r \leftarrow 0)$ in the medical internal radiation dose (MIRD) representation. In a more general form, the absorbed dose rate at a point a distance r away from a point source emitting monoenergetic electrons with energy T_0 is,

$$\frac{dD(r)}{dt} = \kappa A \Delta \Phi(r, T_0). \quad (9.114)$$

Here, $\Delta = nT_0$. Thus, in a medium of physical density ρ, $4\pi\rho r^2 \Phi(r, T_0)dr$ represents the fraction of energy emitted by the source which is absorbed in a spherical shell at radius r and thickness dr. Should the volume containing the source be larger than the electron range, all of the energy will be absorbed in the volume, and the normalization requirement is,

$$4\pi\rho \int_0^\infty dr\, r^2\, \Phi(r, T_0) = 1. \quad (9.115)$$

The DPK for a photon emitter can be similarly determined. Berger (1968) presented the DPK for a monoenergetic photon source emitting photons with energy k in a homogeneous medium of physical density ρ,

$$\frac{dD(r)}{dt} = \frac{\kappa}{4\pi r^2} k \frac{\mu_{en}}{\rho} e^{-\mu r} B_D(k, \mu r) \quad (9.116)$$

where the buildup factor, $B_D(k, \mu r)$, is used. From (9.114) and (9.116), replace the absorbed dose rate with the symbol ℵ to denote the DPK and use a subscript β or γ to denote if it is for electrons or photons when needed.

For a nonuniform distribution of activity through the volume described per unit volume by $\eta(\mathbf{r})$, the absorbed dose rate at a point is given by the convolution of the DPK and the activity distribution,

$$\frac{dD(\mathbf{r})}{dt} = \int d^3\mathbf{r}'\, \aleph(|\mathbf{r}' - \mathbf{r}|)\eta(\mathbf{r}) \quad (9.117)$$

where integration is over the source distribution. The solution of this convolution can often be difficult and is typically solved numerically or through other means, including the Fast Fourier Transform (Kwok et al. 1985; Akabani et al. 1997).

In practice, DPKs are typically determined through Monte Carlo simulation of point sources in a medium (e.g., Simpkin and Mackie 1990).

9.2.8 Cavity Theory

9.2.8.1 Introduction

The measurement of the absorbed dose to a medium requires, in most practical cases, the insertion of a probe or detector into the medium. This problem was first considered by Bragg (1910) and, later, by Gray (1929, 1936). Gray discussed the probe as being a gas-filled cavity, the ionization products generated within the gas as a result of irradiation being collected electrically and then used to estimate the absorbed dose to the medium at the point at which the probe is sited. This is a highly practical problem, although it should be noted that the detection medium in cavity theory need not be a gas as the dosimeter can be a semiconductor or a lithium fluoride crystal used in thermoluminscent dosimetry. In radiation dosimetry, a "cavity" is considered to be a "small" volume of a medium surrounded by a wall (which may or may not be of the same composition as the chamber medium) embedded within a larger volume of yet another medium. This is the typical configuration of an ionization chamber in which the cavity contains air (or another gas with radiological properties, other than physical density, equivalent to soft tissue), the walls are made of a material such as graphite or tissue-equivalent material and the cavity placed within water or another tissue-equivalent medium exposed to radiation. Generally, the chamber volume is sufficiently small so that there are negligible photon–air interactions within it and the chamber walls are sufficiently thick that most of the ionizations occurring within the air cavity are due to electrons produced by photon interactions in the wall.

This subsection presents the fundamentals of Bragg–Gray cavity theory to calculate the absorbed dose to a cavity of one type of material embedded in a medium of another. Detailed derivations, specific to clinical radiotherapy and including the refinements to the theory presented by the Spencer-Attix and Burlin theories, can be found in the texts by Johns and Cunningham (1983) and Attix (1986).

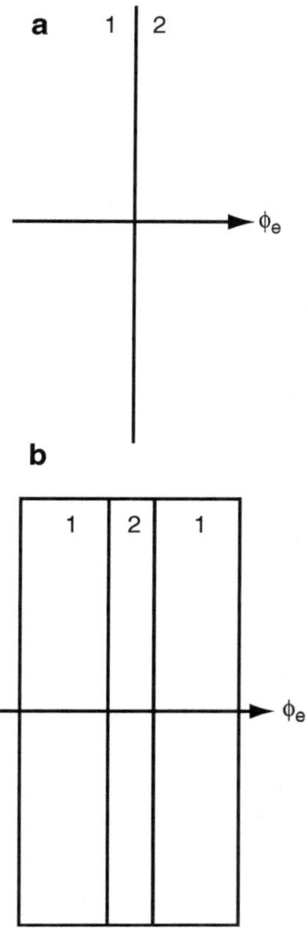

Fig. 9.20 In (**a**), an electron fluence crosses the interface between two different media (1) and (2); in (**b**) an electron fluence crosses through a thin layer of medium (2) embedded within medium (1)

9.2.8.2 Bragg–Gray Theory

Introduction

This is the original and simplest theory of deriving the absorbed dose to the medium within the cavity and relating it to the absorbed dose that would be at that point in the surrounding medium in the absence of the cavity.

Bragg–Gray Relation and Conditions

Recall the earlier discussion of the variation of electron fluence in the vicinity of an interface between two media exposed to photons and consider Fig. 9.20. A monoenergetic fluence of electrons is moving from left to right in the figure.

In Fig. 9.20a, the electron fluence crosses the interface between two different media, labeled as (1) and (2). The absorbed doses on either side of the media are the products of the fluence and the mass-collision stopping power appropriate for the medium and the electron energy,

$$D_1 = \Phi_e \left(\frac{dE}{\rho\, dx}\right)_{Col,1} \tag{9.118}$$

and

$$D_2 = \Phi_e \left(\frac{dE}{\rho\, dx}\right)_{Col,2}. \tag{9.119}$$

If both media have low atomic numbers, then backscatter can be neglected (recall the graph of electron fluences in the vicinity of the interface between two radiologically-dissimilar media of Fig. 9.8) and the electron fluence across the interface treated as being continuous. The ratio of the absorbed doses is then equal to the ratio of the mass-collision stopping powers,

$$\frac{D_1}{D_2} = \left(\frac{dE}{\rho\, dx}\right)_{Col}\Big|_2^1 \tag{9.120}$$

where an expression for the ratio of the mass-collision stopping powers has been defined,

$$\left(\frac{dE}{\rho\, dx}\right)_{Col}\Bigg|_2^1 \equiv \frac{\left(dE/\rho\, dx\right)_{Col,1}}{\left(dE/\rho\, dx\right)_{Col,2}} \qquad (9.121)$$

Now consider Fig. 9.20b in which a "thin" layer of medium (2) is now embedded within the medium (1). This layer is assumed to be sufficiently thin, relative to the electron range, that it does not perturb the electron fluence crossing it. This assumption of the lack of electron fluence perturbation is known as the first Bragg–Gray condition. The second Bragg–Gray condition is that the absorbed dose in medium (2) is due to only those electrons traversing it (i.e., if the electrons are primary products of photon interactions, the photons themselves do not interact with in the cavity). Having established these two conditions, we can now calculate the relationship between the absorbed doses in media (1) and (2). Allowing for generality, assume that the incident electron fluence exhibits an energy spectrum, $d\Phi_e/dE$, rather than being monoenergetic. This, of course, will be very much the case in practice. The energy spectrum-averaged mass-collision stopping powers in the two media are,

$$\overline{\left(\frac{dE}{\rho\, dx}\right)}_{Col,1} = \frac{\int_0^\infty dT \frac{d\Phi_e}{dT}\left(\frac{dE}{\rho\, dx}(T)\right)_{Col,1}}{\Phi_e} \qquad (9.122)$$

and

$$\overline{\left(\frac{dE}{\rho\, dx}\right)}_{Col,2} = \frac{\int_0^\infty dT \frac{d\Phi_e}{dT}\left(\frac{dE}{\rho\, dx}(T)\right)_{Col,2}}{\Phi_e}. \qquad (9.123)$$

The ratio of absorbed doses to the two media is equal to the ratio of mass-collision stopping powers of the two media,

$$\frac{D_1}{D_2} = \overline{\left(\frac{dE}{\rho\, dx}\right)}_{Col}\Bigg|_2^1 = \frac{\int_0^\infty dT \frac{d\Phi_e}{dT}\left(\frac{dE}{\rho\, dx}(T)\right)_{Col,1}}{\int_0^\infty dT \frac{d\Phi_e}{dT}\left(\frac{dE}{\rho\, dx}(T)\right)_{Col,2}}. \qquad (9.124)$$

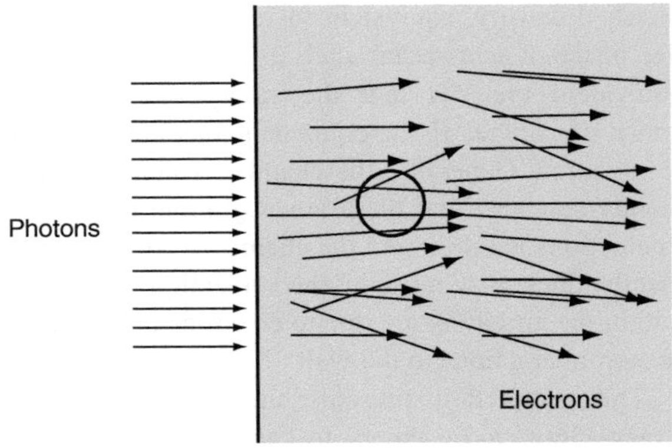

Fig. 9.21 An application of Bragg–Gray cavity theory: a cavity within a medium is crossed by the tracks of electrons set in motion by photon irradiation of the medium. The absorbed doses to the medium and to the medium contained within the cavity are related through the Bragg–Gray relation of (9.124), subject to the two Bragg–Gray conditions. These are that the cavity is sufficiently that it does not perturb the electron fluence and that there are no photon interactions within the cavity

This ratio of the absorbed doses to the surrounding medium to that in the cavity medium is known as the Bragg–Gray relation. Note that neither homogeneous irradiation of the cavity or the existence of CPE have been stipulated in this derivation. The only requirement is that the electron fluence Φ_e remains unperturbed in both media. Figure 9.21 presents an example of where the Bragg–Gray cavity theory can be applicable.

9.2.9 A Brief Overview of Microdosimetry

9.2.9.1 Introduction

Microdosimetry is a subspecialty of ionizing radiation dosimetry and is concerned with energy deposition in microscopic volumes. The dimensions of these volumes of interest range from those of mammalian cells (~20 μm) to that of the 2 nm diameter of the DNA double helix. At these dimensions, the inherently stochastic nature of energy deposition by ionizing radiation is dominant. Accurate accounting of the probability distributions of energy deposit is important as most fundamental biological responses to ionizing radiation are functions of the microscopic patterns of energy deposition (Rossi and Zaider 1991).

Most aspects of nuclear medicine dosimetry calculations are concerned with the transfer by, and deposition of energy from, ensembles of particles sufficiently large that they can be treated as being continuous and differentiable radiation fields. The volumes of interest in these cases are also large relative to those spatial dimensions associated with individual energy deposition/transfer processes. Hence, absorbed doses to individual organs and tissues of macroscopic dimensions are obtained from the summation of all energy deposits in a volume normalized to the mass of that volume. This calculation uses the mean (non-stochastic) values of radiometric and dosimetric quantities (e.g., stopping powers, photon interaction coefficients, particle fluences, etc.). However, the use of mean radiological quantities becomes invalid when the spatial dimensions of the volume of interest interacting with radiation become comparable to the separations between the individual ionization events. For example, consider an absorbed dose of 10 mGy due to electrons delivered to bulk soft tissue (which is assumed to be water equivalent). As the mass of a 10 μm-diameter spherical cell is of the order of 0.5 ng, this absorbed dose corresponds to a mean energy deposition of about 5 nJ per cell. Assuming that the electrons are minimally-ionizing (a linear stopping power of about 2 MeV/cm), the mean number of electrons traversing a cell is about 15, with a standard deviation of about 4. Figure 9.22 shows an example of individual energy deposition events in water for 10 keV and 1 MeV electrons. The 10 keV electron track could be typical of a high-energy Auger electron; the 1 MeV electron would be minimally ionizing. Note the subcellular spatial scale of these events as shown in the figure. The randomness of the energy depositions is readily apparent as is the inability to use the assumption of a continuous energy loss at such spatial dimensions in order to accurately describe the dosimetry.

In nuclear medicine, the applications of microdosimetry will arise primarily in the calculations of absorbed doses received by individual cells or their organelles as a result of the intracellular incorporation of radionuclides that emit α or β particles or Auger electrons. The DNA contained within the cell nucleus is the primary radiosensitive structure of the cell and the delivery of short-range and densely-ionizing charged particles to within the nucleus typically result in cell death. While this is desirable in radionuclide therapy, it has been of theoretical consideration with diagnostic radionuclides that emit Auger electrons (such as ^{99m}Tc and ^{111}In) and which are incorporated within the cell or nucleus. Here, the risk of interest is

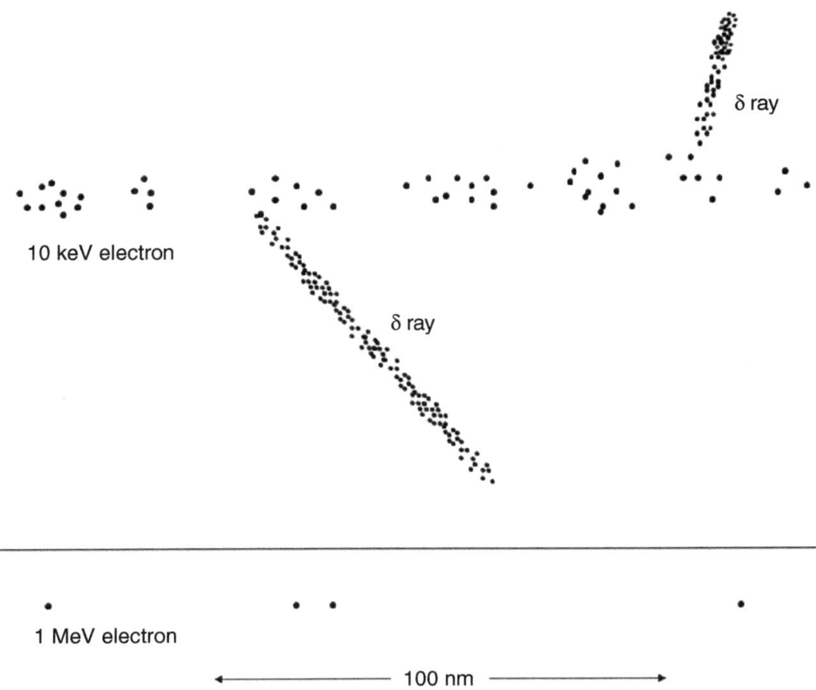

Fig. 9.22 Energy deposits in tissue along tracks of electrons with two different kinetic energies at subcellular dimensions. The *dots* represent individual ionization events

not that of cell death but that of radiation-induced mutation that can lead to carcinogenesis.

The above discussion implicitly highlights the need in ionizing radiation dosimetry to not only quantify the magnitude of the radiation yielding the absorbed dose but also its "quality." In microdosimetry, the quality of the radiation is characterized by the amount and spatial distribution of the energy deposition.

9.2.9.2 Linear Energy Transfer and Lineal Energy

Introduction

The simplest metric of the spatial distribution of energy deposition is that of the linear energy transfer (LET) which is the energy deposited at the point of interest per unit length of the charged particle track. This will be recognized as being the linear collision stopping power derived in Chap. 7. However, for the current application to small spatial dimensions, the quantity is typically expressed in units of keV/μm – a more natural unit to use to describe the rate of energy loss in microscopic dimensions. As with the collision stopping power, the LET can be considered in an unrestricted form, L_∞, which includes all energy transfers to electrons and a restricted form, L_Δ, which excludes those energy losses resulting in δ rays with kinetic energies exceeding an energy threshold Δ and carrying energy away from the interaction site. L_Δ is, hence, descriptive of local energy deposition. However, the L_Δ is a quantity defined by the cutoff in electron energy rather than by spatial dimension. The LET, as demonstrated in Chap. 7, is a non-stochastic quantity and is represented by the mean of the linear collision stopping power. This presents a deficiency in the use of LET in microdosimetry. Moreover, as to be shown in Chap. 10, biological responses to radiation are not linear with local energy deposition. This indicates that assessments of the biological responses to ionizing radiation require an understanding both of the statistical distribution of the energy deposition about its mean value.

LET Probability Distribution Functions

When a medium is irradiated by ionizing radiation, a flux of charged particles with a kinetic energy spectrum is generated within that medium. As L_∞ is a function of a charged particles' kinetic energy, L_∞ must also exhibit some nature of a spectrum which can be described by two distributions. The mean values of the non-stochastic L_∞ for these distributions are attempts to compress the considerable amount of information contained within the L_∞ spectrum to a single interpretable quantity. Neither of these simplifications are particularly satisfactory in practice.

The first descriptor is that of LET frequency distribution, for which the distribution function is given by the fraction of particle fluence that is associated with an LET with a value less than or equal to a specified value, L,

$$F(L) = \frac{\Phi_L}{\Phi}. \tag{9.125}$$

Φ is the total particle fluence and Φ_L is the fluence of particles with values of LET less than or equal to a specified value, L. If the charged particles are electrons, there arises some ambiguity about the role that δ rays play. The pdf of LET in fluence is,

$$f(L) = \frac{dF(L)}{dL}. \tag{9.126}$$

From these expressions, the mean LET is given by,

$$\begin{aligned} \overline{L}_F &= \int dL\, L\, f(L) \\ &= \int dL(1 - F(L)) \end{aligned} \tag{9.127}$$

where integration by parts has been used and the F subscript indicates that this is the mean defined from the frequency distribution.

A second means of specifying the LET probability distribution is through the absorbed dose from particles of a specified LET. The distribution function of absorbed doses from particles with an LET less than or equal to a specified value L is,

$$\Delta(L) = \frac{D_L}{D} \tag{9.128}$$

where D is the total absorbed dose and D_L is the absorbed dose due to particles with an LET less than or equal to L. The corresponding density function is,

$$d(L) = \frac{d\Delta(L)}{dL} \tag{9.129}$$

and the absorbed dose-averaged LET is,

$$\begin{aligned}\overline{L}_D &= \int dL\, L\, d(L) \\ &= \int dL(1 - \Delta(L)).\end{aligned} \quad (9.130)$$

From the above, the dose and frequency distributions can be shown to be related by,

$$d(L) = f(L)\frac{L}{\overline{L}_F} \quad (9.131)$$

Validity Conditions of LET

As ICRU Report 36 (1983) has noted, much experimental work comparing biological radiation effects has been based upon the use of the unrestricted LET, L_∞. This is of concerns as the use of the unrestricted LET in describing microscopic energy deposition is severely hampered by three major inherent limitations:

- *Energy straggling*: As the unrestricted LET L_∞ is the mean of the collision linear stopping power, it cannot quantify the variations in energy loss due to the stochastic energy straggling described in Sect. 7.4. As noted in that section, when charged-particle energy losses along smaller pathlengths are to be calculated for, the stochastic nature of these losses cannot be excluded, thus requiring the use Vavilov or Landau probability distribution functions of energy loss derived in Chap. 7.
- *δ rays:* As shown in Chap. 7, it is possible for the same value of L_∞ to be had for particles of differing charges and speeds. As an example, a 10 keV Auger electron and a 550 MeV α particle will both have the same value of L_∞ of 23 KeV/μm in liquid water. Despite this equality in L_∞, (2.24) and (2.27) demonstrate that the maximum δ-ray kinetic energies from these electron and α particle are 10 keV and 321 keV, respectively, with corresponding ranges of 2.5 μm and about 950 μm. Hence, the L_∞ presents a poor predictor of the lateral width of the ionization track which is defined by these δ rays.
- *Track length*: The length of the track relative to the dimension of the target of interest is not accounted for in L_∞. The charged particle can not only traverse the target volume, but can start or stop in it. If the volume is sufficiently large, then L_∞ can also vary as the charged particle traverses the volume.

The above brief discussion of the variation in track length indicates that the diameter of the site within which the charged particle traverses will also a variable in addition to the charged particle's kinetic energy to assessing energy deposition in that volume. A quantitative assessment of the suitability of the use of L_∞ in microdosimetry was performed by Kellerer and Chmelevsky (1975). They provided a graphical representation of the influences of energy straggling, δ-ray production and track length upon the use of L_∞ as a descriptor of energy deposition for spherical sites of diameter d ranging from 0.1 to 10 μm. Figure 9.23 shows such graphical representations for α particles

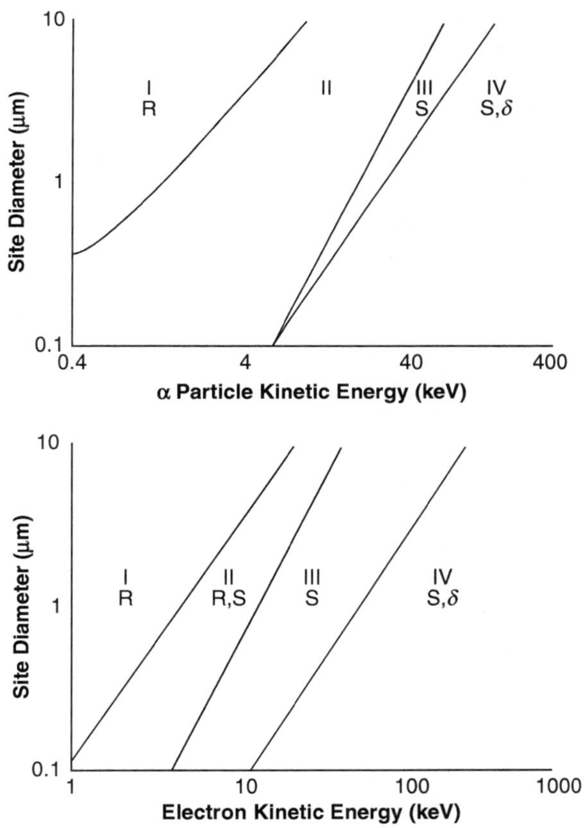

Fig. 9.23 Regions of charged-particle kinetic energy and site diameter in which the influences of particle range (R), energy straggling (S) and (δ) rays are strong for α particles and electrons upon dosimetry calculations based upon L_∞ (after Kellerer and Chmelevsky 1975)

and electrons, the charged particles which are of most interest in nuclear medicine. In each graph of particle kinetic energy vs. site diameter, four regions are identified.

First consider the plot for α particles (Fig. 9.23a):

- As noted above, the value of L_∞ can change along the trajectory passing through the site and the particle may even start and stop within that site. In Region I (corresponding to low α-particle kinetic energies and large site diameters), this charged-particle range effect leads to an error of more than 15% in the energy deposited in the site if calculated from L_∞ alone.
- The energy loss variations due to energy straggling can exceed those energy losses due to the distribution of track lengths across the site. Regions III and IV indicate those combinations of energy and site diameter for which the variance due to energy-loss straggling exceeds that of path-length variability.
- δ rays carry away energy and Region IV corresponds to the condition of that of over 10% of the energy being carried away from an ionization event within the site to be deposited at points beyond the site.

The remaining Region II contains the intermediate range of combinations of site diameter and α particle kinetic energy in which L_∞ is the dominant component in estimating the energy deposited within the site. In summary, for low α particle energies and large site dimensions (Region I), the limited particle range relative to the site dimension results in a reduction in the energy deposited from that estimated from L_∞. With increasing kinetic energy, energy loss straggling reduces the energy deposition from that expected from L_∞ in Regions III and IV. In Region IV, δ-ray energy transport reduces the energy deposition even further from that expected from L_∞. It is in Region II where energy deposition is dominated by L_∞ and the unrestricted LET can be a reasonable metric of energy deposition at the macroscopic scale for α particles.

The above descriptions of regions are also applicable to the graph for electrons. However, due to the low mass of the electron, the effects of energy straggling and particle range are greater than are those upon α particles. Regions I, III and IV, and the factors influencing them remain, although they are broader over both kinetic energy and site diameter. But the intermediate Region II is much more restricted for electrons

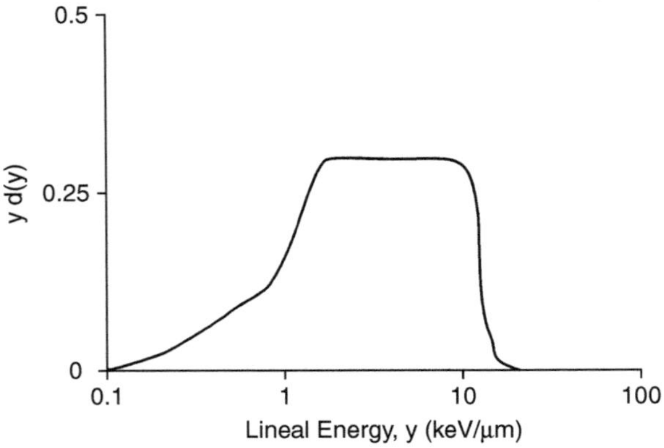

Fig. 9.24 Lineal energy transfer distribution for single events in a 1 μm diameter soft-tissue sphere exposed to 140 kVp X-rays. The ordinate is the weighted density of lineal energy (redrawn from Kellerer 1985)

than for α particles and is subject to the effects of both energy straggling and particle range. Hence, for electrons, there is no combination of site diameter below 10 μm and electron energy for which L_∞ dominates the description of energy deposition. Hence, the unrestricted LET is invalid for microdosimetric characterization of energy deposition in this combination of ranges of electron kinetic energy and site diameter.

Lineal Energy

As the LET is restricted in its accuracy in describing energy deposition by electrons in small-diameter sites, we can employ the lineal energy y, defined in (8.32) as the energy imparted in a single event within a volume normalized to the mean chord length resulting from the random interception of that sight by a straight line, $y = \varepsilon/\bar{l}$, which is the stochastic analog to the LET. For example, $\bar{l} = 4r/3$ for a sphere of radius r.

Figure 9.24 shows the probability distribution of the lineal energy received from single energy deposition events in a 1 μm diameter soft-tissue sphere exposed to 140kVp X-rays.

9.2.9.3 Specific Energy

The specific energy was defined by (8.31) and is applicable to any number of events that occur within the volume of interest. Assuming that the volume of interest is spherical with radius r and physical density ρ, the specific energy can be related to the lineal energy by,

$$z = \frac{4y}{4\pi r^2 \rho} \quad (9.132)$$

9.3 Radiation Dosimetry: Detection and Measurement

9.3.1 Introduction

Although the subject matter of this book is concerned with the theoretical principles behind nuclear medicine dosimetry, these principles cannot be discussed in the absence of an understanding of how the necessary physical and biological data required for such dosimetry calculations are obtained and of the associated precisions of these measured data.

In this section, the various radiation detectors used in measuring the absorbed doses of, and detecting the radiation emanating from, the nuclear medicine patient are presented along with their theories of operation. These fall into four main detector categories:

Gaseous ionization detectors: These are the ionization chamber, the proportional counter and the Geiger–Müller (GM) counter. The ionization chamber is ubiquitous in the radiopharmacy in the form of the dose calibrator and the proportional counter and GM counter are used for radioactive contamination monitoring in the nuclear medicine department. Historically, multiwire proportional chambers (MWPCs) have been used as position-sensitive detectors in PET. The theory of gas ionization and ionization product transport in gas is reviewed, along with each the theories of operation of these three subcategories of detectors.

Scintillation detectors: Scintillators dominate the means of nuclear medicine imaging. They are almost exclusively thallium-activated sodium iodide in conventional gamma cameras and the lanthanide compounds used in PET scanners. The theory of scintillation and the properties, advantages, and disadvantages of scintillator types in nuclear medicine applications are reviewed.

Semiconductor detectors: As discussed in the Introduction to this chapter, the use of these in nuclear medicine is growing, primarily through the development of zinc telluride (CZT) imaging platforms. However, of more immediate relevance to in vivo nuclear medicine radiation dosimetry are the use of MOSFET dosimeters, which were developed for in vivo dosimetry and have been used in nuclear medicine applications. An overview of the theory of operation of these devices is provided.

Thermoluminescent dosimeters: These are ubiquitous in personnel radiation exposure monitoring. They are also used in phantom measurements of radiation absorbed dose. An overview of the theory of thermoluminscence is provided.

Additional information regarding radiation detectors, including those applications specific to nuclear medicine, may be found in Tsoulfanidis (1983); Knoll (2000); Green (2000); Bailey et al. (2003); Hamilton (2004); and Wernick and Aarsvold (2004).

9.3.2 Gaseous Radiation Detectors

9.3.2.1 Introduction

Gaseous radiation detectors operate on the principle of collecting the electric charge produced by ionizations caused by radiation interactions in a gaseous medium through the application of an electric field across that medium. The fact that X-rays could ionize air was first noted by Thomson (1896a, b) shortly after their discovery by Röntgen in 1895. Air ionization detectors were subsequently used by Rutherford to discover α and β particles and by the Curies in their radioactivity work.

Figure 9.25 identifies the major regions of gaseous detector operations in terms of the logarithm of the output signal as a function of the electric field applied across the gas. The regions of detector operation are divided in terms of the multiplication factor, M, which is the ratio of the net collected charge to that produced in the initial ionization event.

Following an ionization in the gas, the released charges drift along the electric field lines, producing a current proportional to the rate of energy transfer in the gas (i.e., kerma rate). The drift speed of an ion or electron is set by the combination of the acceleration

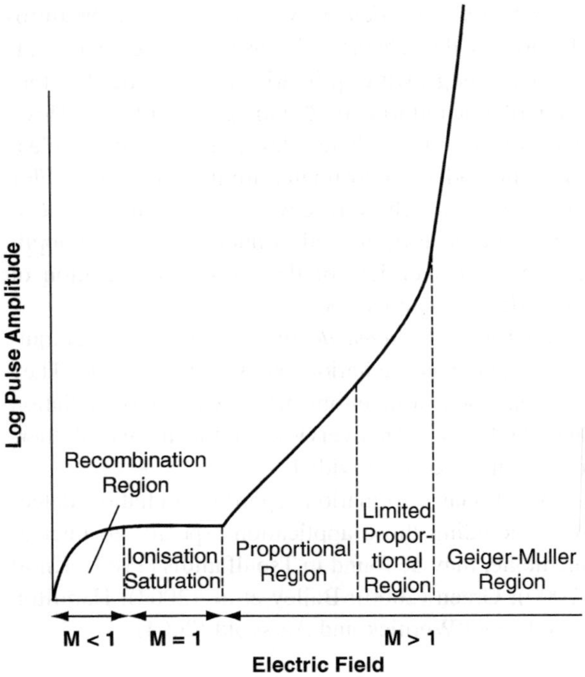

Fig. 9.25 The regions of operation of gaseous radiation detectors as a function of the electric field strength applied across the gas cavity, divided into three main regions defined by M, which is the "multiplication factor," equal to the ratio of the collected charge to that liberated by the initial ionization. Further subdivisions are discussed in the text. At electric field values exceeding those of the Geiger–Müller (GM) region, the detector enters a continuous discharge state

by the electric field, the random thermal motions the entity is subjected to and the mean-free path between collisions with neutral gas molecules whilst drifting towards the appropriate electrode. As a result, the drift speed is a characteristic of the particle or electron and is directly proportional to the applied electric field.

At low electric field values, the density of ionization products in the vicinity of the ionization site immediately following ionization is high as the products have not yet drifted sufficiently far away from each other. As a result, the recombination rate of the products is also high and the net collected charge will be less than that liberated ($M < 1$). With increasing electric field strength, the rate of recombination decreases and the collected charge increases to reach an ionization saturation plateau. At saturation, all of the liberated charge is collected ($M = 1$). This is the least sensitive operating region of a gaseous detector, which makes it useful in high fluence rate applications.

With increasing electric field strengths across the gas volume, the speeds of the electrons liberated by the initial ionizations become sufficiently high that collisions between these electrons and neutral gas molecules cause further ionizations and release additional electrons. These electrons, in turn, reach high drift speeds to ionize neutral gas molecules through further collisions leading to a Townsend avalanche of an exponentially-growing number of ionizations within the gas volume. The result is a collected electric charge much greater than that released in the original ionization and the multiplication factor is, hence, greater than unity. This $M > 1$ region can be subdivided in terms of detector operation. At the lower end of the field strength range, these avalanches are independent of each other. Consequently, the amplified signal is proportional to the magnitude of the original ionization; hence, this is known as the proportional region. A detector operated in the proportional region is far more sensitive than in the ionization saturation region. However, great care must be taken of the design of the chamber in order to yield the necessary high and stable electric fields and attention must paid to the nature of the fill gases in order that the multiplication factor M be independent of the size of the initial ionization event. The gas mixture is particularly important in that electron collision-induced excitations of gas molecules can result in the emission of ultraviolet (UV) photons which can ionize at a distance and lead to uncontrolled avalanches and a prolonged discharge. An additional gas is added to the fill gas to absorb these photons and to quench the discharge. Values of M in the proportional region are typically of the order of about 10^3–10^5. At greater electric field values, space-charge effects reduce the response proportionality. This is the limited proportionality region noted in Fig. 9.25.

At greater electric field strengths and, by allowing the emission of the UV from the excited gas molecules following collisions to create further avalanches, a significantly larger charge can be collected from a single ionization event. As the avalanches are no longer independent, the output signal is not proportional to the original amount of ionization caused in the gas. In this GM region, the magnitude of the output signal is the same regardless of the magnitude of the original ionization. This represents the most sensitive mode of operation for gaseous radiation detectors.

9.3.2.2 The Theory of Ionization in Gases

Magnitude of Initial Ionization Produced and the Fano Factor

As shown in Chap. 7, ionizing radiation interactions with gas molecules create ion-electron pairs which carry recoil kinetic energy, with the number of electron-ion pairs produced per unit energy (i.e., the reciprocal of W/e) being relatively constant with gas and radiation type. For a gas at standard conditions for temperature and pressure, charged particles with kinetic energies of a few MeV and below are thermalized in a period of time of about 10 ns or less. In the absence of an electric field, these liberated charges lose this recoil energy through multiple collisions with neutral gas molecules to eventually thermalize to a mean kinetic energy, (3/2)kT, where k is Boltzmann's constant and T is the gas temperature. For example, at a gas temperature of 293°K, the mean kinetic energy is $\bar{\varepsilon} \cong 0.03\,\text{eV}$. This will take about of the order of 1 μs for the liberated electrons. It will be recalled from the discussions in Chap. 7 that other nonionization energy channels exist so that the energy lost by a photon or charged particle traversing a gas will actually be greater than the ionization energy. The value for W/e of dry air is 33.85 eV per ion pair. Hence, a 100 keV electron stopped completely in a volume of air will produce about 3,000 ion-electron pairs in initial ionization events.

In the absence of an electric field, a group of ion-electron pairs produced by an ionization will simply diffuse randomly throughout the gas following a Gaussian distribution,

$$\frac{dN}{N_0} = \frac{e^{-x^2/4Dt}}{\sqrt{4\pi Dt}} dx \tag{9.133}$$

where dN/N_0 is the fraction of produced charges in spatial element dx at time t and D is the diffusion coefficient. For the example of argon gas (used in a dose calibrator), this diffusion coefficient is $D = 0.04\,\text{cm}^2/\text{s}$. Figure 9.26 shows the spatial distributions of charges in argon gas at different times postionization. These plots show that, shortly after ionization, the relative density of the ionization products around the position of the ionization event is high. As a result, there is a significant consequent probability of recombination of the ionization products, thus limiting the amount of collected charge. The amount of ionization resulting from the interaction is itself probabilistic in that the ratio W/e only defines the mean number of ion-electron pairs produced per energy lost. It would be expected that the number of ionization products produced per individual ionization event would follow Poisson statistics where the variance is equal to the mean number of ion-electron pairs created per event, N. However, it has been found empirically for gaseous detectors that the measured standard deviation of the number of electron-ion pairs produced is markedly less, by as much as a factor of 4, than that expected from the simple assumption of Poisson statistics. This empirical observation suggests that the processes producing an ion-electron pair are not statistically independent as assumed and that the Poisson model is not an appropriate one to use. This departure of the observed statistical variability from that expected on the basis of Poisson statistics is empirically quantified by the empirical Fano factor which is defined as the ratio of the measured signal variance to that expected from Poisson statistics,

$$F \equiv \frac{\sigma^2_{\text{Meas}}}{N}. \tag{9.134}$$

The magnitude of the Fano factor F is less than unity. If all of the energy of the stopped radiation in the medium yielded ion-electron pairs, then the number of these pairs would be consistent across these interactions and would have no variance. Consequently, the Fano factor would be equal to zero.

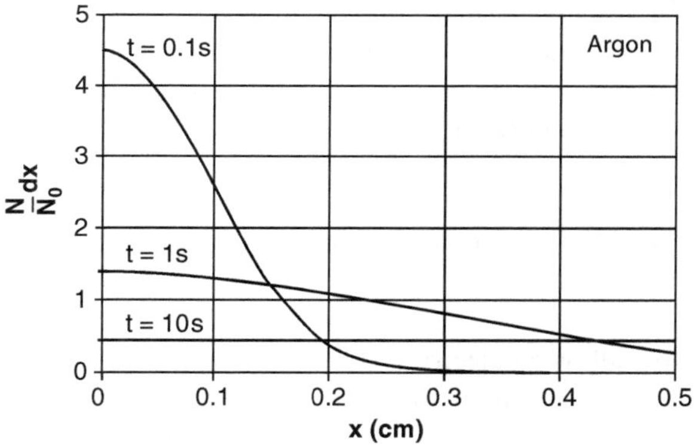

Fig. 9.26 Diffusion profiles at different times t for argon gas without the presence of an electric field calculated from (9.133)

But if only a small amount of energy lost by the stopped radiation was converted into producing ion-electron pairs, then Poisson statistics would be valid and the Fano factor would be equal to unity.

Ion and Electron Motions in a Gas With an Electric Field Applied

Now allow an electric field to be applied across a gas cavity exposed to ionizing radiation. Following an ionization in the gas, the different charged entities will drift away from the event site and across the cavity following the electric field lines: positively-charged ions to the cathode and electrons to the anode. A charged particle of charge e and mass m subject to an electric field E in vacuo will simply undergo constant acceleration equal to eE/m. In a gas, however, the net motion of the charged particle is a combination of that induced by the electrostatic force placed upon it, the random thermal motions it is subject to and the random collisions between it and the neutral gas atoms. These contributions to the net motion of a charged particle in a gas with an electric field applied are derived.

Consider the initial ionization events; as in the discussion of microdosimetry, it is possible to define a small volume of gas within which a number of ionizations occur. Immediately following the ionizations, the electrons and positively-charged ions are in close proximity with each other. The probability of recombination between these ionization products will increase with the densities of both species immediately following ionization. As high densities of the formed ion-electron pairs are associated with high ionizing radiation fluences, the recombination probability will increase with radiation fluence and, hence, absorbed dose. The rate of recombination, $\Re$ is proportional to the product of the concentrations of the positively- and negatively-charged moieties,

$$\Re = \alpha[n_+][n_-] \qquad (9.135)$$

where $[n_\pm]$ are the concentrations of the positively- and negatively-charged species and the constant of proportionality, α, is known as the recombination coefficient. Recombination can also occur between negatively- and positively-charged ions in which an electron is transferred between the two to yield two electrically-neutral entities; it is found that the recombination coefficient is several orders of magnitude greater between positively- and negatively-charged ions than between positively-charged ions and electrons. If, as expected at the time of initial ionization that, $[n_+] = [n_-] = [n]$, then, solving (9.135), yields the concentration as a function of time is,

$$[n(t)] = \frac{[n_0]}{1 + \alpha[n_0]t} \qquad (9.136)$$

where $[n_0]$ is the concentration of charged products at the time of initial ionization, $t = 0$.

While the charged products will carry some recoil energy from the energy transferred from the ionizing particle, they rapidly lose this energy to reach thermal speeds. Hence, in addition to the drift caused by the electric field, these ionization products will be subject to random thermal motion. The thermal speed of a particle of mass m at temperature T is given by,

$$v_{Th} = \sqrt{\frac{3kT}{m}} \qquad (9.137)$$

where k is Boltzmann's constant. For example, the thermal speed of a N_2 gas molecule is of the order of 500 m/s at STP whereas that of an electron would be of the order of 1.2×10^5 m/s.[11] In terms of velocity, the thermal velocity is randomly directed whereas the drift velocity has a specific direction along the electric field lines. In addition to the random thermal motion superimposed upon the drift velocity, random interruptions to the drift will occur due to collisions with surrounding neutral gas molecules. It is possible to derive an expression for the net drift speed of a gas ion in a gas medium by first considering the random thermal motion only and defining λ as the mean-free path between collisions. The mean time between collisions, τ, will be given by $\tau \sim \lambda/v_{Th}$. The mean-free path between collisions is proportional to the reciprocal of

[11]The thermal speeds of light gas species, such as helium, are about an order-of-magnitude greater than that of molecular nitrogen and comparable to the Earth's orbital escape speed. This is one reason why helium is not found in the Earth's atmosphere.

the interaction cross section, which at the low energies considered here, can be approximated by the geometrical cross section of the ion, $\sigma \approx \pi r_B^2$, where r_B is the ionic radius. Hence, the mean-free path for a typical gas molecule is of the order of 0.1–0.3 μm. As the magnitude of the acceleration of the charged moiety (of mass m and assumed here to be singly-charged) by an applied electric field strength is eE/m, the mean drift speed is,

$$\begin{aligned}\overline{v_d} &\approx \tau \frac{eE}{m} \\ &\approx \frac{\lambda}{v_{Th}} \frac{eE}{m}.\end{aligned} \tag{9.138}$$

Recall from Chap. 7 that the mean-free path between collisions is given by $\lambda = A/N_A\rho\sigma$. Combining this expression with (9.137) and (9.138), an expression for the mean drift speed of the gas ion can be written as,

$$\overline{v_d} \approx \frac{eEA}{N_A\rho\pi r_B^2\sqrt{3kTm}}. \tag{9.139}$$

Equation (9.139) shows that the drift speed will vary as $m^{-1/2}$. As the mass of a gas molecule is some 3×10^4 times greater than that of an electron and the interaction cross section smaller, the drift speed of an electron will be several hundred times to a thousand times greater than that of a gas molecule.

This relationship between the drift velocity and the electric field can be simplified dramatically. The electric field per unit density, E/ρ, with respect to STP is referred to as the reduced electric field,

$$\frac{E}{\rho} = E\frac{P_0}{P} \tag{9.140}$$

where $P_0 = 1$ atm at STP and P is the gas pressure. This enables (9.139) to be rewritten as,

$$\begin{aligned}\overline{v_d} &\approx \frac{eEA}{N_A\rho\pi r_B^2\sqrt{3kTm}} \\ &\approx \left(\frac{eAP_0}{N_A\pi r_B^2\sqrt{3kTm}}\right)\frac{E}{P} \\ &\approx \mu\frac{E}{P}\end{aligned} \tag{9.141}$$

where the constant of proportionality, μ, is known as the mobility of the drifting charged entity It can be seen from (9.141) that the mobility can be readily determined experimentally by measuring the drift velocity in a gas of pressure P with an electric field strength, E, applied across it. From our past immediate discussion, the mobility for a single electron will be much greater than those of an ionized gas molecule (by a factor of about 10^3) due to the differences both in mass and in interaction cross section. Experimentally, the mobility for an ion is found to be relatively constant over a range of gas types with, for example, values of 1.1 and 1.7 cm^2 Atm/V s for gaseous argon and CO_2, respectively. An argon ion drifting in argon gas at 1 atm pressure in a 100 V/cm electric field will attain a drift speed of the order of 2 μm/μs. The corresponding transit time over the dimensions typical of a gaseous detector is of the order of 1 ms. On the other hand, the corresponding electron drift speed would be of the order of 2 mm/μs and the transit time of an electron is much shorter at the order of 1 μs. This difference in drift speeds between the electron and positively-charged ion will dictate the time-dependence of the output pulse from the detector.

It is important to note that, from (9.137), the thermal speeds of a N_2 molecule and an electron are 500 μm/μs and 120 mm/μs, respectively. Hence, the electric-field induced drift speed is a small, but directionally constant, perturbation over the random thermal motions.

In addition to the effects described above upon charged particle motion, collisions with neutral gas molecules can lead to charge transfer if, in a mixture of gases, a gas molecule is present with an ionization energy lower than that of a positive ion. In a collision between the two, the latter can transfer its positive charge to the neutral molecule in a charge transfer collision by incorporating an electron from the neutral molecule. This is a process used in the internal quenching of proportional chambers. Negatively-charged ions can also be produced should a free electron be captured by a neutral gas atom. Oxygen, in particular, is electronegative, whereas other gas molecules in common use in gaseous detectors, such as the noble gases, hydrogen, or nitrogen, have low electron attachment coefficients. As a result, except for certain applications and in the ionization saturation region, electronegative gases are not used in gaseous radiation detectors.

9.3.2.3 $M \leq 1$ Regions: Ionization Chamber

As shown in Fig. 9.25, at low electric field strengths, the gaseous detector operates in what is known as the ionization region of operation. In this region, the resultant electron-ion pairs from an ionization drift along the electric field lines to the electrode of the appropriate polarity in the manner described in the previous subsection. The collected charge will be that liberated by the initial ionizations less the net charge of those products that recombine. As noted in the previous subsection, the degree of recombination is high for the low electric field strengths which are insufficient to separate the ionization products quickly enough to reduce the densities of these products and the subsequent recombination rate. Figure 9.25 demonstrates that, in this region, $M < 1$ due to recombination but that the collected charge increases with the electric field. Eventually, the collected charge saturates with electric field which reflects the status of all of the liberated electric charge being collected (i.e., $M = 1$). A detector operating in this region is referred to as being an ionization chamber or a unity-gain detector.

A simple parallel-plate detector operating in this region is shown in Fig. 9.27 in which two planar electrodes are enclosed within a gaseous medium with a constant bias voltage, V_B, applied across them. For parallel-plate electrodes, the magnitude of the electric field will be constant with a magnitude given by $E = V_B/d$, where d is the separation between the electrodes. The gas volume, within which the electrodes are placed, is exposed to constant ionizing radiation. This liberates electrons and positive ions at a constant rate. Assume that the rate of recombination of these ionization products is negligible and that the efficiency of collecting the ion-electron pairs by the electrodes is 100%. As a result, once equilibrium is reached, the ionization current flowing in the circuit is equal to the ionization rate produced within the gas chamber. For an incident photon beam, this saturated ionization current is proportional to the kerma rate which is related to the incident photon flux through the K/Φ ratio of (9.11).

At low bias voltages, the drift speed is small so that the distances traveled by the ion-electron pairs are correspondingly small and the densities of the ion-electron pairs near the ionization centers are high, thus allowing recombination. The net ionization current, I_C, is thus low. As V_B increases, the drift speed increases in linear proportion thus leading to a reduction in the densities in the electron-ion pair in the vicinity of the ionization event and a decrease in the rate of recombination. The result is that the ionization current increases. Eventually, at a sufficiently high V_B, all of the ion-electron pairs produced by the original ionization event are collected and I_C saturates and is independent of the bias voltage. It is in this saturation region that ionization chambers are operated. With increasing ionization rate and a fixed bias voltage, the ionization current will increase in proportion to the radiation rate when the chamber is operated in

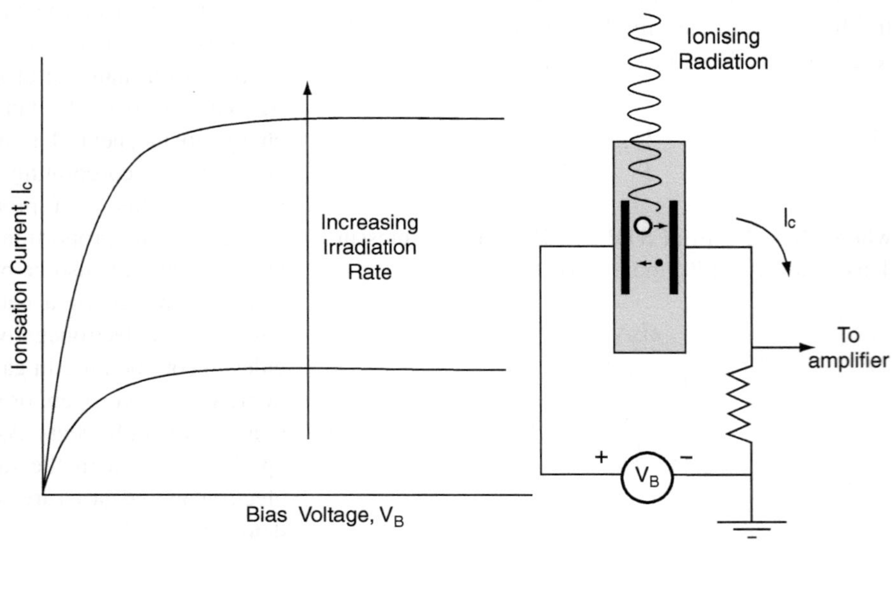

Fig. 9.27 Voltage-current characteristic curves for a simple parallel-plate ionization chamber as functions of irradiation rate

saturation mode. As, in this mode, all charges can be collected, the importance of the type of fill gas used in an ionization chamber is usually not significant. For example, air is frequently used as a fill gas. As the output of the ionization chamber is simply the total ionization released in the gas volume, its magnitude is low. The ionization chamber is thus invariably operated in direct current (DC) mode rather than pulse mode and the current integrated over time to yield a net collected charge. The magnitudes of the charges involved are small and a simple numerical example is telling. Consider an ionization chamber with an air medium and a volume of 50 mL. At STP, the mass of air contained within this volume is equal to 6.03×10^{-2} g and an in-air collision kerma rate of 0.1 mGy/h will result in an ionization current of only 0.05pA. Because of the low output of the ionization chamber, the ionization chamber is most often operated in DC mode where the ionization current is integrated over a set period of time to yield a net electrical charge as the output.

9.3.2.4 M > 1 Regions: Proportional Chamber and Geiger–Müller Counter

Introduction

In this subsection, the proportional and GM regions are grouped together as both can be defined by gas multiplication at high electric field strengths. As noted above, an ionization chamber has a low sensitivity due to its unity-gain characteristics. To increase the signal-to-noise ratio of a gaseous radiation detector, one takes advantage of the phenomenon of gas multiplication. Gas multiplication is a phenomenon which occurs in gases for electric field strengths extending beyond the ionization chamber plateau so as to exceed a few kilovolts per centimeter. The drifting electrons will gain sufficient kinetic energy in order to cause a variety of inelastic phenomena during their collisions with the uncharged gas molecules. The natures of these interactions between the electrons and the gas molecules are dependent upon the nature of the gas. For example, if the gas is made up of weakly-bound polyatomic molecules, such as a hydrocarbon, a number of inelastic energy transfer channels are presented through vibrational and rotational degrees of freedom. But, for all gas types, should the energy transfer from the electron to the molecule exceed the ionization potential of the gas molecule, the energy-transfer channel opens through which an additional ion-electron pair can be created with the primary electron continuing on to cause further ionizations. The secondary electron produced in the electron-gas molecule collision is then accelerated by the electric field and undergoes further collisions with gas molecules causing ionizations and leading to a chain reaction, or gas multiplication process, (Townsend avalanche). This avalanche eventually terminates when all of the electrons liberated by the initial ionization and by subsequent electron-gas molecule collisions are collected at the anode. While the net collected charge is proportional to the original charge released by the ionization, it will have been amplified through these subsequent Townsend avalanches.

A gaseous radiation detector operated in the proportional region is termed a "proportional chamber" or "proportional counter," as it is operated in the counting pulse mode (the amplitude of each signal is quantified). At the high electric fields and specific gas mixtures that lead to the loss of proportionality but high radiation detection sensitivity, the detectors are referred to as "GM counters."

Electric Field Requirements for Gas Multiplication

The creation of the required high electric fields (of the order of 10 kV/cm) to create Townsend avalanches must be considered in the design of a proportional chamber. The example of the simplistic parallel-plate geometry in the previous subsection is impractical for producing gas multiplication. In order to create the necessary electric field strength of 10 kV/cm to yield gas multiplication across a 1 cm gap a 10 kV bias voltage must be applied. The electrical design of the chamber would be complicated by the magnitude of this voltage in order to yield a safe practical instrument. However, by altering the chamber geometry, high electric field strengths can be obtained for small bias voltages. A simple, yet common, example is that of the cylindrical geometry consisting of two coaxial cylinders, as shown in Fig. 9.28. In proportional chamber design, the inner cylinder is the anode wire of the chamber. The inner and outer cylinders have radii r_a and r_b, respectively, and a bias voltage V_B is placed across

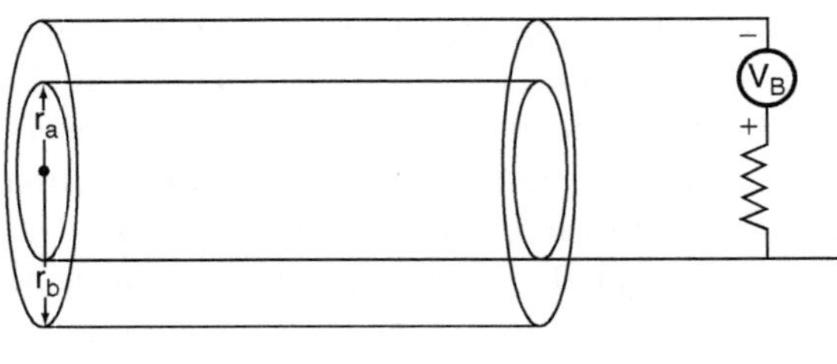

Fig. 9.28 Geometry for calculating the radial electric field of a proportional chamber composed of two coaxial cylinders. The *inner cylinder* represents the anode wire for a cylindrical proportional chamber

them. Because of symmetry, the electric field is radially directed and, from Gauss' law, will be,

$$E(r) = \frac{\rho_l}{2\pi\varepsilon_0 r} \tag{9.142}$$

where ρ_l is the surface charge per unit length (related to the capacitance per unit length). This expression for the electric field yields the potential difference between the two cylinders by,

$$\begin{aligned} V_B &= \int_{r_a}^{r_b} dr\, E(r) \\ &= \frac{\rho_l}{2\pi\varepsilon_0}\int_{r_a}^{r_b}\frac{dr}{r} \\ &= \frac{\rho_l}{2\pi\varepsilon_0}\ln\frac{r_b}{r_a} \end{aligned} \tag{9.143}$$

from which the surface charge per unit length is obtained,

$$\rho_l = \frac{2\pi\varepsilon_0 V_B}{\ln {}^{r_b}/_{r_a}}. \tag{9.144}$$

The ratio ρ_l/V_B is, in fact, the capacitance per unit length of the proportional chamber. Substituting this into (9.142) gives the radial electric field within the chamber,

$$E(r) = \frac{V_B}{r \ln {}^{r_b}/_{r_a}} \tag{9.145}$$

It will be noted that, unlike the parallel-plate geometry, the electric field is no longer constant with distance between the anode and cathode and varies as $1/r$.

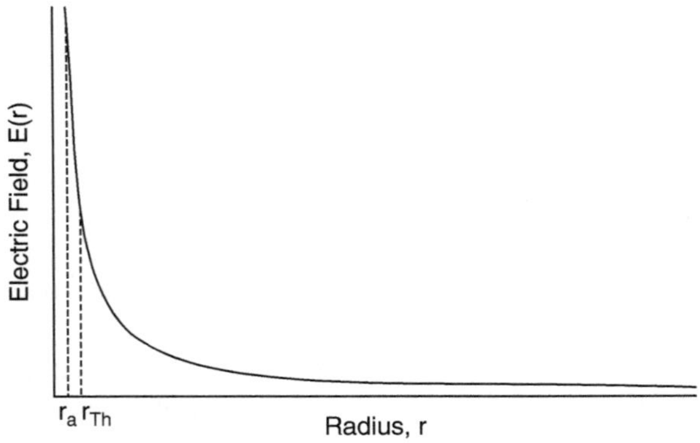

Fig. 9.29 Electric field in the cylindrical geometry of Fig. 9.28 as a function of radius. r_{Th} is the threshold radius at which the electric field below which the electric field is greater than the threshold value for gas multiplication Hence, the Townsend avalanches occur within the small cylindrical volume defined by r_a and r_{Th}

Thus the field increases dramatically with proximity to the central cylinder. For a practical geometry of a bias voltage of 1 kV, the anode (inner cylinder) being a wire of diameter of 0.15 mm and the cathode (outer cylinder) diameter of 2 cm, the electric field strength at the surface of the anode wire is 136 kV/cm. To achieve such an electric field strength for a parallel-plate geometry with the electrodes separated by 2 cm would require a bias voltage of 27.2 kV. The use of a cylindrical geometry and a thin anode wire can significantly increase the electric field in its vicinity for a low bias voltage.

In the cylindrical proportional design, Townsend avalanches will occur only very close to the anode wire. For example, the electric field threshold value for gas multiplication to occur is of the order of 10^6 V/m. For the above combination of anode wire and chamber diameters, this threshold value is reached at a radius of 0.204 mm (Fig. 9.29). Thus, in this example, gas multiplication occurs within about 0.005% of the total chamber volume.

Ionization and Space-Charge Effects

Figure 9.30 shows the spatial features of the combined drift, diffusion and multiplication processes in the vicinity of the central anode wire of a proportional chamber. After the initial ionization, the ions and electrons drift under the electric field towards the

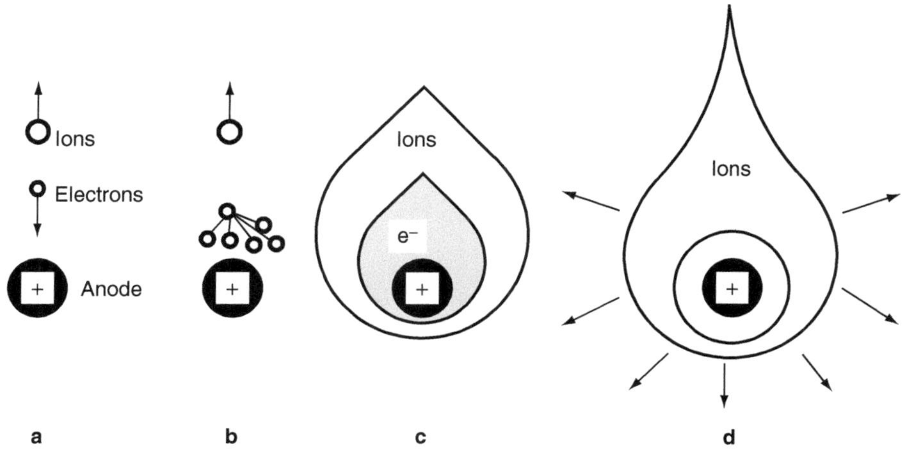

Fig. 9.30 Development of an avalanche around the anode wire of a cylindrical proportional chamber: (**a**) Initial ionization leading to ions drifting to the cathode and electrons drifting to the central anode; (**b**) as the electric field increases as r^{-1} in proximity to the anode, the electrons are accelerated and create further ion-electron pairs through collisions; (**c**) in addition to the electric-field induced diffusion, thermal diffusion is present and which broadens out the electron and ion clouds; and, (**d**) all of the electrons are collected by the anode and the slower ions continue to drift towards the cathode

cathode and anode, respectively. Due to their lower mobility, the ions move with a lower drift speed. As the electrons approach the anode, they experience increased acceleration due to the electric field which increases as $1/r$. These electrons reach speeds that create further ionizations through collisions with neutral gas molecules. As shown in Fig. 9.29, it is at points in very close proximity to the anode wire (not more than a few mean-free paths) that avalanches occur. Superimposed upon this drift behavior is the thermal diffusion which broadens out the distributions of electrons and ions. Because of their higher mobilities, all of the electrons quickly reach the anode and are collected leaving only the positively-charged ion cloud to drift slowly towards the cathode.

Once the electrons have been collected, the positively-charged ion cloud, which will be proximal to the anode (as this is where most of the gas multiplication has occurred) will reduce and distort the electric field in the vicinity of the anode. This has the potential to space-charge quench the net magnitude of electric field to a point below which gas multiplication cannot occur. This self-induced effect is dependent upon the chamber design (through the electric field) and the associated degree of gas multiplication.

Gas Multiplication

In the proportional region of Fig. 9.25, the electric charge collected at the electrodes in the proportional chamber is given by,

$$Q_{Coll} = Mn_0e \tag{9.146}$$

where n_0 is the original number of ion-electron pairs produced in the initial ionization event and M is the gas multiplication factor. This multiplication factor can be calculated through the application of a number of approximations. The first Townsend coefficient, α, is defined as the mean of the number of secondary electrons produced by a single free electron per centimeter pathlength. An expression for this coefficient can be had by first assuming that the probability of an electron interacting with a gas molecule is defined by Poisson statistics. That is, the probability that an electron would travel a distance x without a collision is given by $e^{-x/\lambda}$, where λ is the mean-free path between collisions. The energy gained by the electron traveling the distance x in an electric field E is given by the product, exE. In order for an ionization to occur as a result of a collision with a neutral gas molecule, this energy must exceed a mean ionization energy given by eV_{ion} (which is equal to 15.7 eV for argon). Hence, the probability that an ionization will occur is now given by $e^{-V_{ion}/E\lambda}$. As α is defined as the number of ionizations per electron per unit distance traveled, then the first Townsend coefficient is,

$$\alpha = \frac{e^{-V_{ion}/E\lambda}}{\lambda}. \quad (9.147)$$

Because λ will be inversely proportional to the gas pressure (at a constant gas temperature), the first Townsend coefficient can be written in the form,

$$\frac{\alpha}{P} = Ae^{-\frac{BP}{E}} \quad (9.148)$$

where A and B are constants. For argon, $A = 1.064\times 10^4$/cm • atm and $B = 1.37\times 10^5$ V/cm • atm; these values are valid for reduced electric fields of $E/P = 7.6 \times 10^4$ to 4.6×10^5 V/cm • atm.

Having defined the rate of ionization per unit pathlength, the logarithm of the gas multiplication factor is,

$$\ln M = \int_{r_a}^{r_{Th}} dr\,\alpha(r) \quad (9.149)$$

where r_{Th} is the radial distance from the anode beyond which the electric field strength is so weak that gas multiplication cannot occur (Fig. 9.29). This defines the critical radius below which the electric field is sufficiently strong to accelerate electrons to produce secondary ionizations. As α is the mean-free path of an electron between ionizations, the mean energy gained by an electron in an electric field E between collisions is E/α. Assuming that α is linearly proportional to the energy of the electrons, $\alpha = kN\varepsilon$, where N is the number of gas molecules per unit volume and k is a constant of proportionality, then an expression of α is,

$$\alpha(r) = \sqrt{\frac{kNCV_B}{2\pi\varepsilon_0}\frac{1}{r}} \quad (9.150)$$

where C is the capacitance per unit length of the cylindrical proportional chamber ($=\rho_l/V_B$). Substituting this into the integral of (9.149) gives,

$$\begin{aligned}\ln M &= \sqrt{\frac{kNCV_B}{2\pi\varepsilon_0}}\int_{r_a}^{r_{Th}}\frac{dr}{\sqrt{r}} \\ &= \sqrt{\frac{kNCV_B}{2\pi\varepsilon_0}}(\sqrt{r_{Th}} - \sqrt{r_a}) \\ &= \sqrt{\frac{kNCV_B r_a}{2\pi\varepsilon_0}}\left(\sqrt{\frac{r_{Th}}{r_a}} - 1\right).\end{aligned} \quad (9.151)$$

From this result, the gas multiplication factor can be written in the form,

$$M = \kappa_1 e^{\kappa_2 V_B} \quad (9.152)$$

if $V_B \gg V_{Th}$, where V_{Th} is the threshold voltage at which gas multiplication occurs. Thus, gas multiplication beyond the threshold voltage increases exponentially with bias voltage once that voltage exceeds the threshold voltage. This demonstrates the need for an extremely stable voltage supply for a proportional counter, a requirement which has contributed to the limited use of the proportional chamber in nuclear medicine applications. Another, but semiempirical, expression for the logarithm of the multiplication factor is presented by Knoll (2000) and is attributable to Diethorn,

$$\ln M = \frac{V_B}{\ln {r_b}/{r_a}}\frac{\ln 2}{\Delta V}\ln\left(\frac{V_B}{KPr_a \ln {r_b}/{r_a}}\right). \quad (9.153)$$

For the typical P-10 fill gas mixture of 90% argon and 10% methane used in proportional chambers, $K = 4.8 \times 10^4$ and $\Delta V = 23.6$ eV. As an example, Fig. 9.31 shows the multiplication factor calculated for the P-10 fill gas mixture at 1 atm pressure for a cylindrical ionization chamber as a function of bias voltage.

Fill Gas Requirements for the Proportional Region

The nature of the fill gas is important when the gaseous detector is operated in the $M > 1$ region, particularly so for the proportional region. In this region, it is essential to keep the output signal exactly proportional to the original ionization signal. As the basis of gas multiplication is the drifting of free electrons and the secondary ionizations they induce in collisions with

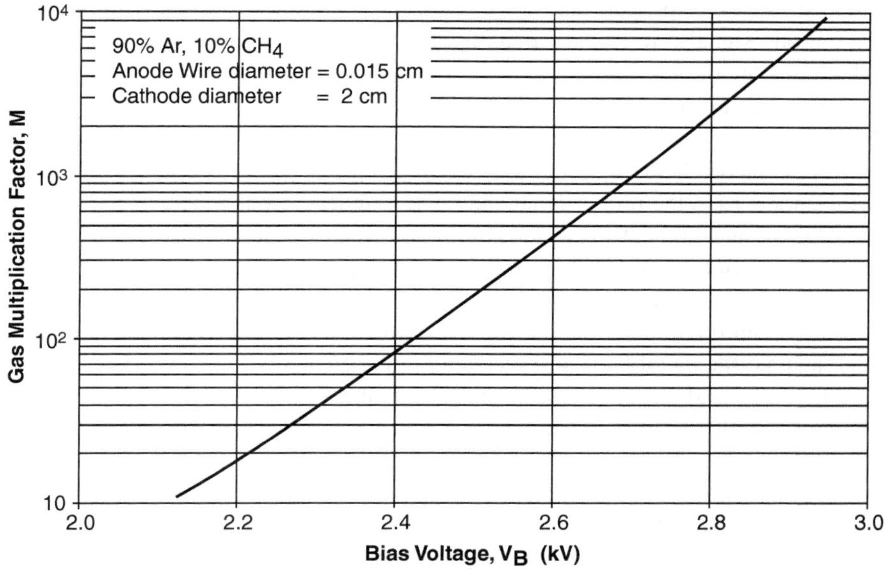

Fig. 9.31 Gas multiplication factor as a function of bias voltage for a cylindrical proportional chamber with a cathode diameter of 2 cm and an anode wire diameter of 0.015 cm and a fill gas mixture of 90% argon and 10% methane at 1 atm pressure calculated from the Diethorn empirical formula of (9.153)

neutral gas molecules, the fill gas must obviously have a low electron attachment coefficient. This contraindicates the use of electronegative gases such as oxygen and air. Common fill gases for a proportional chamber are thus the noble gases, most notably argon. For gas multiplication factors exceeding about 100, the user must also contend with nonionizing energy transfer channels present in a gas molecule. Following an electron collision with a monatomic gas, the atom is left in an excited state and does not contribute to the avalanche. This is immediately followed by de-excitation, a process including the possible emission of a UV photon which can cause an ionization within the gas. Even should this UV photon not be absorbed within the fill gas and instead impinge on the metal surface of the cathode, it can eject an electron from the surface which will then migrate towards the anode and thus extend the duration of the avalanches. In addition to these UV-induced creations of free electrons, when a cation is neutralized at the cathode by extracting an electron, there may be sufficient energy remaining in order to extract an additional electron from the cathode wall which again leads to further avalanches. It is necessary to quench these subsequent avalanches in order to attain stable operation of the proportional counter and to ensure proportionality of the output signal to the original ionization event. A means of internal quenching is to add a polyatomic gas, usually a hydrocarbon such as methane or ethane, with vibrational modes capable of absorbing the UV photons (a proportion of this absorbed energy may be reirradiated as infrared photons, but these are nonionizing). The quench gases can also contribute an electron to the cations, thus neutralizing them. The now positively-charged hydrocarbon ions drift to the cathode, but, due to their lower excitation energies, are less likely to extract an electron from the cathode. The most common proportional chamber gas fill is the P-10 mixture in which the hydrocarbon gas added is CH_4 at a 10% by volume to Ar.

Geiger–Müller Region

Recall that a proportional chamber is designed, in terms of electrical properties and fill gas composition, to ensure that the collected charge at the output of the chamber is proportional to the size of the initiating ionizing event. Polyatomic gas additives, such as the hydrocarbons, are used to quench the additional nonproportional ionization events due to ultraviolet emission from the primary fill gas molecules. In a GM counter, nonproportional ionization events are, on the contrary, sought in order to further increase the gas multiplication factor from typical values of about 10^4 for a proportional chamber to about 10^7. While this comes at a loss of proportionality between the input and output signals, it results in a highly sensitive radiation detector.

Following an initial ionization in the GM region, the electron migrates towards the anode wire under a very high electric field. Within a few mean-free paths of the anode, as in the proportional chamber, it undergoes collisions with neutral gas molecules and creates further ionizations. UV photons emitted from the de-excitation of postcollision states that do not result in ionization can travel elsewhere within the GM cavity (e.g., against the metal wall of the cathode, as shown in Fig. 9.32) and produce an ionization in almost synchrony with the first and inducing a Geiger discharge. There results a rapid chain reaction in the GM counter due both to the high electric field and the high electron speeds resulting and the ultraviolet photons. As a result of the lack of statistical independence between the individual avalanches, the output signal of a Geiger discharge is not proportional to the size of the initiating ionization. Indeed the signal is the same regardless of the magnitude of the initial ionization. After the electrons have been collected at the anode, the slow-moving ions are still drifting to the cathode and the magnitude of the net positive electrical charge of them creates a positive space-charge between the anode and cathode, thus reducing the effective electric field in the counter. This can quench any further gas multiplication. However, the arrival of these positive ions at the cathode can potentially liberate more electrons from the cathode and extend the Geiger discharge. This is avoided by the appropriate selection of fill gases and the selection of the time constant of the external detection electronics.

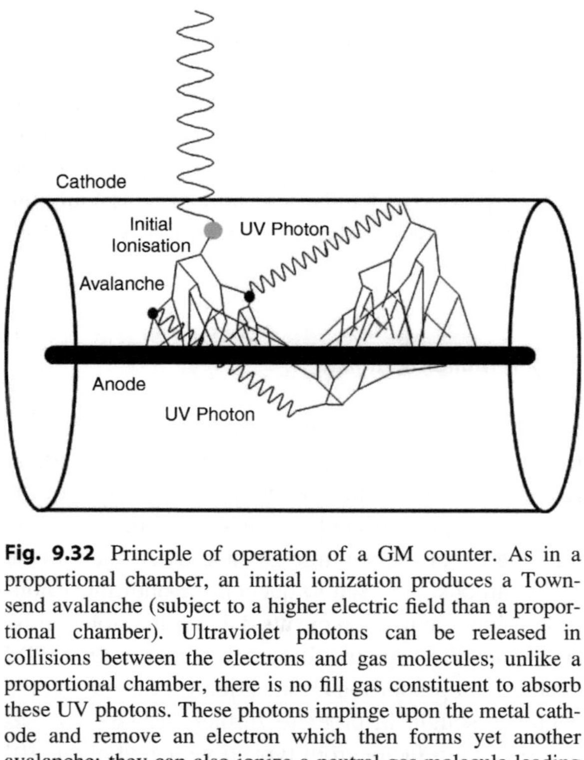

Fig. 9.32 Principle of operation of a GM counter. As in a proportional chamber, an initial ionization produces a Townsend avalanche (subject to a higher electric field than a proportional chamber). Ultraviolet photons can be released in collisions between the electrons and gas molecules; unlike a proportional chamber, there is no fill gas constituent to absorb these UV photons. These photons impinge upon the metal cathode and remove an electron which then forms yet another avalanche; they can also ionize a neutral gas molecule leading to another avalanche

Fill Gasses in the Geiger–Müller Region

The fill gas of a GM counter is primarily the same as that of a proportional chamber, i.e., usually a noble gas and no trace amounts of electronegative gases such as oxygen. UV-photon absorbing gases are, of course, not used. A degree of quenching is still sought as even following the neutralization of the cation at the cathode by the extraction of an electron, the energy difference between the ionization energy of the gas atom/molecule and the work function energy may exceed the work function energy, enabling the extraction of another electron from the cathode and which will migrate to the anode wire, causing a further Geiger discharge. Hence, the Geiger discharge could remain continuous.

As in the proportional mode, the internal quenching method uses a quenching gas added to the fill gas at a concentration of no more than a few percent. The quenching gas must have a lower ionizing potential than the primary fill gas and will frequently be organic, thus having a complex molecular structure. Following the initial ionization, the positive ions (predominantly those of the primary fill gas) drift towards the cathode and collide with other gas molecules, including those of the quenching gas. Because of the difference in ionization potentials between the primary and quenching gases, a collision between these will result in the transfer of an electron from the quench gas molecule, thus neutralizing the primary gas ion. The now-ionized quench gas molecule drifts to the anode where it is neutralized by extracting an electron from the cathode material. The excess energy is now provided a channel additional to the extraction of a second electron due to the complex form of the quench gas molecule which can absorb this energy and dissociate. This reduces significantly the likelihood of a second electron extraction. However, the dissociation of these quenching molecules means that the GM tube will have a finite lifetime, typically of the order of 10^9 pulses. Beyond this lifetime, repeat pulsing can no longer be quenched. In order to prolong the lifetime of the tube (indeed, make it indefinite) a halogen, such as chlorine or bromine, can be used as the quench gas. Like the hydrocarbon molecule, the halogen molecule can dissociate at the anode and thus prevent an additional extraction of an electron. But the dissociated halogen atoms can spontaneously recombine later.

A second (external) means of quenching the Geiger discharge is through the use of a relatively high RC-time constant at the input of the detector electronics (about 1 ms). This limits the time during which detection can occur.

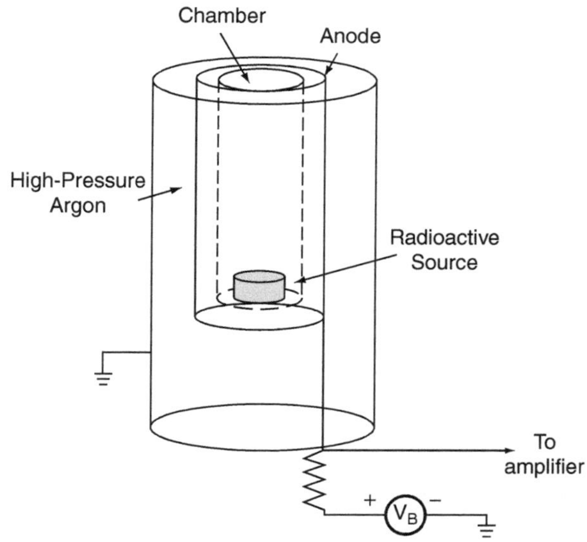

Fig. 9.33 Schematic diagram of the reentrant ionization chamber or "dose calibrator": a cylindrical ionization chamber used in the radiopharmacy to measure the activity of a radioactive sample placed within the chamber which is surrounded by argon gas at high pressure up to 20 atm. The design approximates a 4π detection geometry

9.3.2.5 Applications of Gaseous Radiation Detectors in Nuclear Medicine

Ionization Chambers

Ionization chambers are the least sensitive of the gaseous radiation detectors in that the collected charge is exactly equal to that liberated within the gas volume, excluding any recombination effects. The most common application of the ionization chamber in nuclear medicine is the reentrant ionization chamber, more commonly known as the "dose calibrator," used in the radiopharmacy to measure the amount of activity in a radiopharmaceutical prior to administration. A conceptual diagram of the dose calibrator is provided in Fig. 9.33. The sample to be assayed is placed within the chamber at a predefined position. The physical design is such that there is nearly a 4π collection geometry. Surrounding the sample holder is a collection volume containing argon ($Z = 18$) at high pressure (up to a few atmospheres). The high atomic number of argon and the high pressure increase the likelihood of photon interactions within the gas volume. A bias voltage of a few hundred volts is typically used. The charge collected within the volume is a function of the photon fluence incident to the gas volume which, in turn, is proportional to the activity of the radionuclide.

Because of the need to hold the argon at high pressure, the entrant walls of the chamber must be relatively thick, thus reducing the sensitivity of the device to the detection of low-energy photons. A typical photon energy range a commercial device is capable of measuring is from about 40 keV to about 2 MeV. These thick entrant walls restrict the dose calibrator to measuring photons above a certain energy; the energies of βs emitted by pure β-emitting radionuclides would be insufficient to cross the walls. In such cases (e.g., ^{90}Y), the *bremsstrahlung* generated as the β particles decelerate in the surrounding media is instead measured and used to infer the amount of ^{90}Y activity (Zimmerman et al. 2004). This is only practical for large activities for therapeutic uses: as gas multiplication is not used, the sensitivity of the dose calibrator will be limited; a typical range of activities that can be assayed by a commercial dose calibrator would be about 0.2 MBq to 70 GBq for ^{99m}Tc. Hence, a dose calibrator would be insufficient for, for example, measuring the small amounts of activity excreted by the nuclear medicine patient in the urine. In such a case, the more sensitive scintillator-based well counter is used.

The response of the reentrant ionization chamber is sensitive to a variety of physical and geometric factors. Recall that an ionization chamber measures the collected electric charge as a result of the kerma within its gas volume. Hence, the major sensitivity to photon energy is that of the kerma-to-fluence ratio, K/Φ, (Fig. 9.2) for argon. Indeed, if only photons were incident to the gas volume (and only photons), the K/Φ curve would by-and-large be the calibration curve of the dose calibrator. However, the walls across which the photons must traverse will preferentially attenuate low-energy photons, thus reducing the detector response for low-energy photons. Similarly, the nature of the container holding the radionuclide (e.g., syringe or glass ampoule) in terms of physical size and elemental composition will also affect the energy response of the calibrator. As a consequence, the energy response will be more complicated than that predicted simply by the K/Φ ratio for the fill gas. Monte Carlo estimates of the energy dependence have also been derived (Seneviratne et al. 2007). Indeed, if a calibration of a pure β-emitting radionuclide is to be based upon the *bremsstrahlung*, the radiological dependencies upon the media surrounding the sample integrated over the β-particle energy spectrum and the energy spectrum of the *bremsstrahlung* photons makes a calculable energy response highly impractical. The dose calibrator response is also sensitive to the geometric placement (axial and radial) of the sample within the chamber. In practice, these sensitivities to container volume, composition and placement and to photon energy are determined empirically and an appropriate calibration factor determined.

Another application of the ionization chamber in nuclear medicine is as a radiation survey instrument. The closed chamber contains a large volume of air, typically several hundred cc, in order to increase the device's sensitivity. For photon detection, the walls of the chamber are optimized to enable the detection of low-energy photons but also to maintain CPE. Shields over thin portions of the wall can be removed in order for the chamber to detect β particles. These devices are used in rate mode (i.e., the ionization current is measured rather than integrated to yield a net collected charge).

Proportional Chambers

Because of the need for a highly stable bias voltage source in order to maintain a constant gas multiplication factor, among other reasons, the proportional chamber has had limited applications in clinical nuclear medicine. It has had, however, a long history of use in experimental microdosimetry, allowing measurements of, for example, lineal energy spectra (Fig. 9.24) for various radiations. Figure 9.34 shows the conceptual design of what is known as the Rossi counter.

The Rossi proportional counter is a spherical device of about 15 mm diameter (Rossi 1968; Attix 1986). The walls are made of a plastic which is radiologically equivalent to tissue in terms of physical and electron density and effective atomic number. A tissue-equivalent gas flows through the chamber, the pressure of which is low (of the order of 10^{-2} atm and below) so that the energy deposited within the gas is similar to that deposited in very small volumes, such as that of the cell. The proportional counter is designed as a sphere so that there is no specificity in terms of sensitivity to radiation direction. To enable a uniform electric field within the spherical gas volume and a constant multiplication factor M along the length of the anode wire, the anode is surrounded by a helical grid which is biased to about $0.2V_B$. Electrons liberated by interactions with the wall drift to this grid by its inhomogeneous electric field; they pass through the grid and generate Townsend avalanches uniformly across the anode wire. As the Rossi proportional counter is operated in pulse mode, pulse-height analysis is used. Note the α-particle calibration source in Fig. 9.34; these α particles are directed along a known chord length, l_α, through the chamber volume. The mean absorbed dose to the gas due to a single α particle is,

$$\overline{D}_\alpha = \frac{(dE/\rho\, dx)_{Coll,gas}}{m_{gas}} \rho_{gas}\, l_\alpha \qquad (9.154)$$

where $(dE/\rho\, dx)_{Coll,gas}$ is the mass-collision stopping power for the α particle in the gas medium, m_{gas} is the mass of gas in the volume and ρ_{gas} is the physical density of the gas. A plot of the pulse-height spectrum could look like that of Fig. 9.35.

The total absorbed dose to the gas volume is obtained by integrating over the spectrum; as this is a discrete spectrum, then this absorbed dose is,

$$D_{gas} = \frac{\overline{D}_\alpha}{H_\alpha} \sum_{H=0}^{H_{max}} H \frac{dN}{dH}. \qquad (9.155)$$

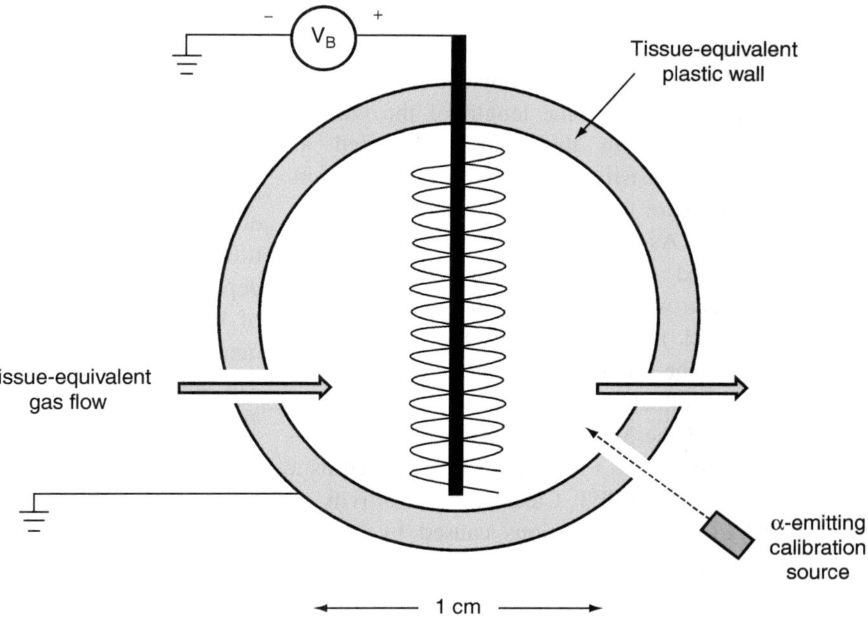

Fig. 9.34 Rossi proportional counter used to measure microdosimetric spectra. Refer to text for details

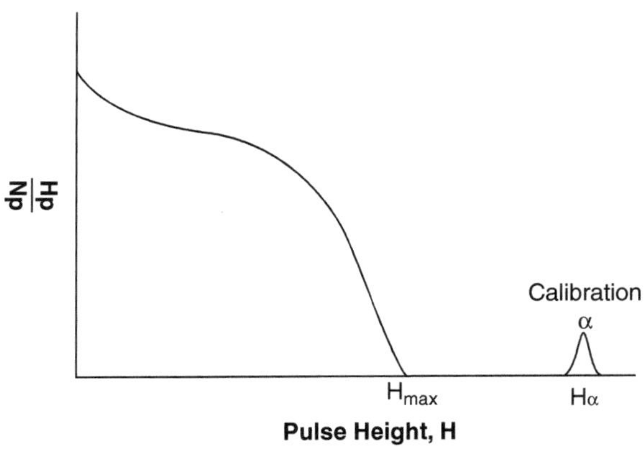

Fig. 9.35 The differential distribution of counts, dN/dh, measured in a Rossi proportional counter. The signal at H_{α} is that due to the calibration source; the peak is associated with the mean absorbed dose to the tissue-equivalent gas due to the α particle given by (9.154). The area of the spectrum is equal to that of absorbed dose for an assumed constant W/e value for all events occurring within the gas medium

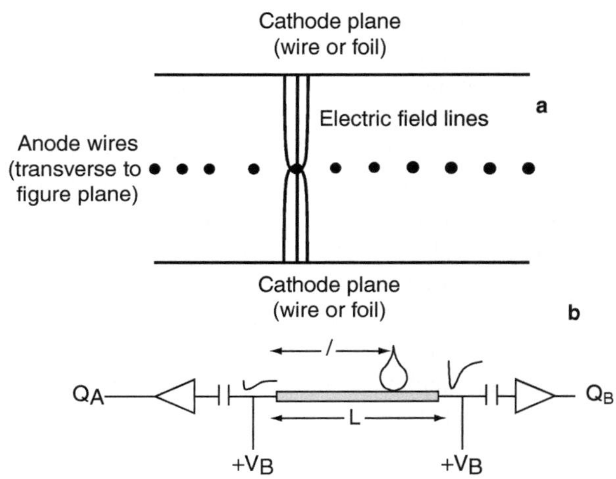

Fig. 9.36 Concept of MWPC operation. (**a**) A grid of anode wires is centered within a gas volume between a cathode plane; electric field lines for a single anode wire is shown. Identification of the ionization position along the position of the anode wire is (**b**) determined by measuring the differential between the collected charges from both ends of a resistive anode wire after amplification (after Attix 1986)

An historical application of proportional chambers has been those with multiple anode wires to allow spatial localization. These MWPCs were originally developed by Charpak et al. (1968), for which Charpak was awarded the1992 Nobel Prize in Physics. The original applications were for the detection of charged particles in high-energy physics experiments. The basic concept of the MWPC is shown in Fig. 9.36.

The position of the ionization along the plane of Fig. 9.36a is determined by which anode wire provides the greatest collected charged. This determines the ionization position to within a wire spacing; hence, spatial resolution is driven by the wire density. Spatial localization along the length of the wire is possible by measuring the charge collected at either end of the resistive anode wire, as shown in Fig. 9.36b. The wire must have a constant resistance per unit length. As a result of this resistance, the charges collected at either end are related to the distance between where the ionization was collected and the wire end. Hence, the ratio $Q_B/(Q_A + Q_B)$ is proportional to the length l along the anode that the ionization was collected.

The ability of the MWPC to encode spatial information led in the 1980s to its being investigated as a means of a detector in PET. Clearly, the sensitivity of an MWPC to detect ionizations caused by photon interactions within the gas volume is expected to be very low. Photon-detection sensitivity could be improved by using a high-Z fill gas and/or by increasing the gas pressure. One approach used (Marsden et al. 1986) to increase the efficiency of the MWPC in detecting the 511-keV γ rays resulting from electron–positron was to include a high-Z "converter," such as lead. Photon interactions with the lead resulted in electrons which were scattered along channels to be incident to the MWPC. This produced a spatially-linked electron from the incident photon, which was subsequently detected by the MWPC. Of course, in modern-day clinical PET technology, scintillation detectors are used exclusively.

Another application of the MWPC in nuclear medicine as a means of providing quantitative autoradiography in preclinical biodistribution studies (e.g., Petegnief et al. 1998)

Geiger–Müller Counters

The GM counter is used exclusively as a radiation survey instrument due to its high sensitivity. Such a survey meter provides a pulse count output which can be calibrated in terms of photon exposure, although it is quite possible to have errors of factors of 2 or more occurring due to the lack of proportionality between the original ionization and the final output signal. This is not of concern as it is an evaluated and consistent response.

9.3.3 Scintillation Detectors

9.3.3.1 Introduction

The scintillation detector is ubiquitous in nuclear medicine. It is present in the gamma camera, PET scanner, nonimaging thyroid probe and well-counter. The scintillator works on the principle of converting energy deposited in it through an ionization event into a pulse of low-energy photons (visible light) which is then detected and converted into an analog electrical signal. The amplitude of this signal is proportional to the amount of scintillation light and, by extension, the amount of energy deposited within the scintillating material.

There are six main requirements for a good scintillator:

- High scintillation efficiency in converting the energy deposited within it into visible light so as to yield high detection sensitivity.
- The scintillating material should be transparent to its emission photon wavelength to ensure good light collection.
- A rapid decay time of the scintillation so as to permit fast detection rates or, equivalently, operation in high radiation fluxes.
- A linear response between total light emitted and energy deposited.
- The output light spectrum must match the sensitivity of the photo-conversion device.
- The index of refraction must be suitable for allowing high photon collection.

There are two categories of scintillators: inorganic and organic. Inorganic scintillators dominate nuclear medicine applications due to their high photon detection efficiencies, linearity and high light output. The ubiquitous example is the alkali halide sodium iodide activated with thallium (NaI(Tl)), which is still present in nuclear medicine some six decades after its development by Hofstadter (1948). Organic scintillators have few applications in nuclear medicine, even though their temporal response tends to be superior to that of inorganic scintillators. These limitations are consequences of their low densities and low atomic numbers. However, these attributes which lend to tissue radiological equivalence, make organic scintillators ideal for dosimeters in radiotherapy. Liquid organic scintillators, on the other hand, do play a role in nuclear medicine in providing a nearly 100% efficient means of measuring the activity of a β-emitting sample which can be dissolved into the scintillating medium.

The scintillation process is made up of three components. The first is prompt fluorescence in which visible light is immediately emitted following an ionization within the scintillating material. Phosphorescence is the emission of light at a longer wavelength than that of prompt fluorescence with a slower decay time. Delayed fluorescence is the emission of light of the same wavelength as prompt fluorescence but after a time delay following the original excitation. The ideal scintillator will maximize the amount of energy released in prompt fluorescence and minimize that released in phosphorescence and delayed fluorescence.

This subsection on scintillation detectors begins with a review of the scintillation theories of inorganic and organic scintillators, with an overview of the attributes of inorganic scintillators used in gamma cameras and PET scanners. Liquid scintillation, using organic scintillators, is also briefly summarized. The transport of the scintillation light to a light detector is next considered. The means of converting scintillation light to an analog electrical signal in modern devices is dominated by the photomultiplier tube (PMT) and the operation of the PMT and the position-sensitive PMT (PSPMT) are derived.

9.3.3.2 Scintillation Theory

Inorganic Scintillators

The scintillation process for inorganic scintillators is dictated by the solid-state energy bands of the scintillating crystal, as shown in Fig. 9.37.

Small amounts of an impurity, called an activator, are added to the crystal during its formation in order to provide intermediate states within the gap between the valence and conduction bands of the crystal. As shown in Fig. 9.37, an ionization event elevates an electron from the valence band to the conduction band, leaving a remnant hole. The electron and a hole can migrate in their respective bands; the hole ionizes an activator site (i.e., an electron transits from the activator ground state to fill the hole in the valence band, leaving a hole in the ground state) and the electron in the conduction band can move into an excited state of the activator. If this resultant state provides an allowed transition, the recombination of the electron and hole occurs quickly between the excited and ground states of the activator and releases a quantum equal to the energy difference between the states. As this energy difference is less than that of the valence and conduction bands of the crystal, the released photon can be in the visible part of the spectrum, depending upon the activator chosen. Most importantly, the light is emitted at a frequency which is not absorbed by the bulk of the scintillator. The half-life of this decay of prompt fluorescence can be of the order of 0.1 μs. For example, sodium iodide on its own can scintillate at low temperatures only. However, about a 1 mM fraction of Tl added to high-purity NaI enables room-temperature scintillation. This is a result of the thallium ions randomly replacing the Na^+ ions in the cubic crystal lattice due to their similarity in size and charge (Esser 1998). These thallium ions interact strongly with the surrounding I^- in the crystal lattice, inducing a shift in the energy levels of the crystal lattice.

A competing process is one in which the electron is captured in an excited state of the activator for which the transition to the ground state is forbidden.

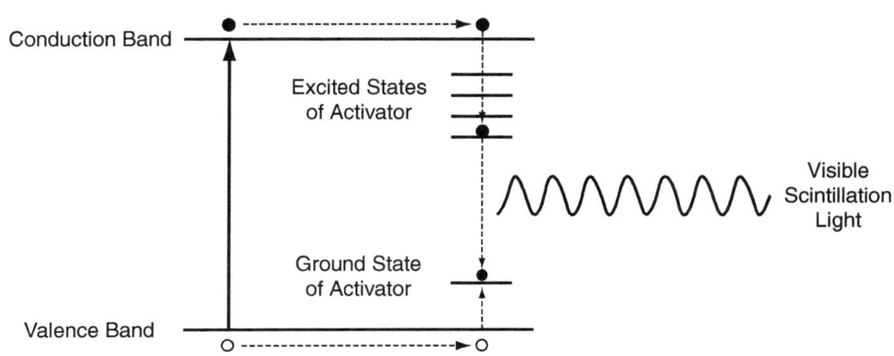

Fig. 9.37 Energy bands of an inorganic crystalline scintillator material containing an activator

Table 9.2 Physical properties of inorganic scintillator materials in common use in nuclear medicine (data are derived from Bailey et al. (2003); Particle Data Group (2004); and Lewellen (2008))

Material	Physical density (g/cm^3)	Effective atomic number	Scintillation efficiency (relative to NaI(Tl)) (%)	Intrinsic $\Delta E/E$ (%)	Decay constant (μs)
NaI(Tl)	3.67	50.6	100	6.6	0.23
$Bi_4Ge_3O_{12}$ (BGO)	7.13	74.2	15	10.2	0.30
Lu_2SiO_5:Ce (LSO)	7.4	65.5	75	9.1	0.04
Y_2SiO_5:Ce (YSO)	4.53	34.2	118	7.5	0.07
Gd_2SiO_5:Ce (GSO)	6.71	58.6	25	4.6	0.06

Thermal energy can elevate the electron into another activator excited state for which the transition to the ground state is allowed and, hence, light is emitted with a longer time delay (phosphorescence). Radiationless transitions between activator excited and ground states are also possible; these are referred to as quenching and represent an inefficiency by failing to convert the deposited energy into measurable light.

Table 9.2 presents a summary of the physical characteristics of inorganic scintillators used in nuclear medicine. As a medium for detecting photons, a scintillator should have a high density and effective atomic number (a combined measure of which is the attenuation length for the photon energy of interest).

NaI(Tl) has a very high light output (exceeded only by YSO in medical applications) and has a close to linear response to electrons with kinetic energies exceeding 20 keV. It has, however, an extended light decay constant which can limit its application in high-count rate studies: the prompt fluorescence of NaI(Tl) has a decay time of about 0.23 μs, which limits its application to particularly high counting rate environments. It has, in addition, it has a 150 ms phosphorescence component contributing about 10% of the total light yield.

Efficient photon detection using scintillators requires that the scintillator have a high density and atomic number and a thickness exceeding the attenuation length of the photon energy of interest. For imaging most single-photon emitting radionculides, NaI (Tl) is adequate. The thickness of a NaI(Tl) crystal in a typical gamma camera is typically 9.5 mm (3/8″); greater thicknesses are required if the gamma camera is used to detect the 511-kev γ rays resulting from electron–positron annihilations. In general, NaI(Tl) is not a suitable scintillator for such applications. The detection of the high-energy photons resulting from positron–electron annihilation requires a scintillator of high density and high atomic number in order to maximize the number of photon interactions in the scintillation medium. It is evident from Table 9.2 that, with the exception of YSO, the lanthanide oxyorthosilicates doped with caerium present advantages over NaI(Tl) in terms of γ ray absorption. In fact, the most common scintillators currently in use in PET are BGO and LSO. They also present the advantage of short light emission decay constants. On the other hand, they possess disadvantages in that their light outputs are considerably less than that of NaI(Tl). In terms of light output, the exception is YSO, but this advantage is coupled with the severe disadvantages of it having a physical density comparable to, and an atomic number less than, NaI(Tl). Other materials that have been considered for medical imaging applications are the caerium-doped lutetium garnet ($Lu_3Al_5O_{12}$:Ce) and lutetium perovskite ($LuAlO_3$:Ce) (Korzhik and Lecoq 2001). The latter is superior to LSO both in terms of physical density (8.34 g/cm^3) and decay constant (0.017 μs), but suffers from the disadvantage of having a light yield of about only 42% of that of LSO. Lead tungstate (PWO) is used as a scintillator in high-energy physics applications. It has advantages in terms of γ ray absorption with an attenuation length of 2 cm for 511-keV γ rays compared to 3.6 cm for LSO and 3 cm for BGO and a very short decay constant of 0.01 μs, making it ideal for high count-rate applications. Unfortunately, its light output is some two orders-of-magnitude less than LSO. This might be improved through appropriate doping and investigations are looking at this possibility.

Some scintillators in Table 9.2 have additional disadvantages. The most well known is that sodium iodide is hygroscopic and must be sealed to protect it from humidity. Lutetium-based scintillators are restricted from use in low-energy γ-ray detection due to the naturally-occurring isotope ^{176}Lu, which has a 2.6% abundance. ^{176}Lu has a half-life of 3.6×10^{10} years and decays through β^--decay and subsequent emission of photons and Auger electrons; in particular, it emits K-shell X-rays with energies of 88, 201, 307, and 401 keV with intensities of 14.5, 78, 94, and 0.4%, respectively. The activity concentration of ^{176}Lu in LSO is about 280 Bq/cm^3, resulting in a count rate of about 12 cps/g of LSO in a ±10% energy window set over the 511 140 keV photopeak (Bailey et al. 2003).

Organic Scintillators

Organic scintillators operate on a different principle than inorganics as the fluorescence is due to transitions between molecular energy levels in a single molecule and are not a result of an added activator. Moreover, organic scintillators do not require a regular crystal lattice structure as do the inorganics. Most organic scintillator molecules exhibit a π molecular bond, which is a covalent bond with a molecular energy structure shown in Fig. 9.38.

The energy spacing between the S_0 and S_1 levels is of the order of 3 eV and those between the vibrational modes is about 0.15 eV. When a charged particle deposits energy in the scintillator, the molecule is excited from the ground state into the higher singlet states which promptly decay to the S_1 state without the emission of photons. The transitions from this S_1 state to the S_0 state (and the associated vibrational states) are prompt with lifetimes of the order of 1 ns. A transition from the singlet S_1 state to the triplet T_1 state can occur with a phosphorescence emission occurring up to 1 ms after the initial excitation.

The organic scintillant is added at small concentrations to a bulk solvent. Upon exposure to ionizing radiation, energy absorbed by the solvent molecules is eventually transferred to a scintillant molecule with the result of light emission.

The almost exclusive application of organic scintillators to nuclear medicine is that of liquid scintillation counting. Here, the scintillator is a liquid in which the radioactive sample to be assayed is dissolved. This is especially useful for measuring the activity of low-energy β-emitting radionuclides such as ^{3}H and ^{14}C. Detection efficiencies approach 100%.

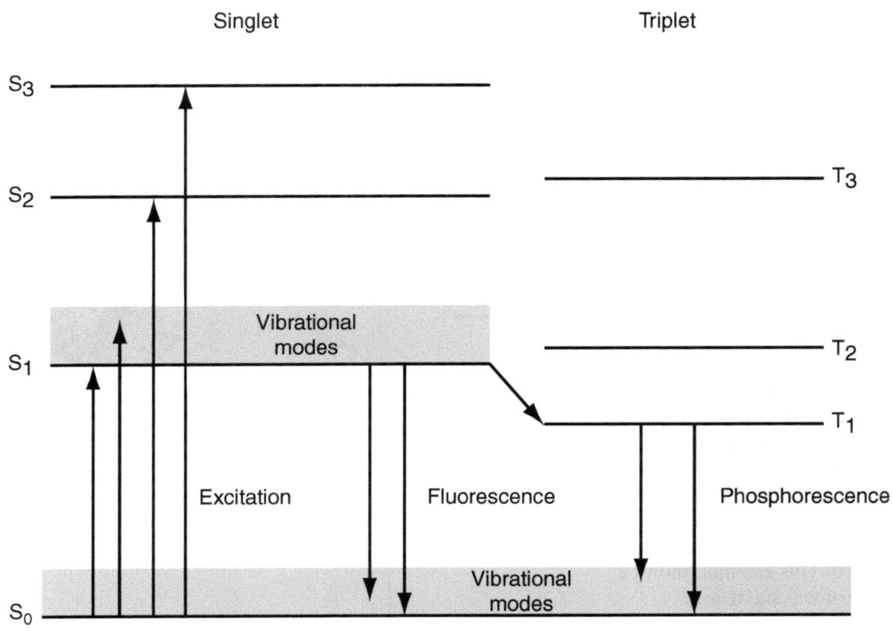

Fig. 9.38 Molecular energy levels for an organic scintillator molecule with the π molecular structure (after Birks 1964)

9.3.3.3 Light Collection

The fluorescence emitted by the scintillator must be transferred efficiently from the scintillator to the device converting the light to an analog electric signal. Scintillation light is emitted isotropically and may be lost either through self-absorption (which is highly limited in most scintillating media) or at interfaces with media with differing refractive index. As sodium iodide crystals are hygroscopic, they must be hermetically sealed to prevent contact with air. This requires the use of a quartz window to allow scintillation light to be detected by PMTs external to the crystal.

9.3.3.4 Light Conversion and Electron Multiplications

In nearly all applications in nuclear medicine, the light generated in the scintillator is converted to an analog signal using a PMT, a schematic diagram of which is shown in Fig. 9.39.

The PMT contains two major elements. The first is a semitransparent photosensitive layer, known as the photocathode, which absorbs the incident light photons and emits electrons through the photoelectric effect. The second is the dynode chain used to multiply the small number of electrons ejected from the photocathode by factors ranging from 10^5 to 10^7 in order to generate a measurable electrical signal.

Photocathode

The process of light conversion begins at the photocathode. There are three steps in this process:

- Absorption of the photon through the photoelectric effect and the ejection of an atomic electron into the photoemissive medium
- Migration of the photoelectron through the medium to the photocathode's back surface
- The overcoming of the work function between the photocathode surface and the vacuum by the photoelectron and its entrance into the dynode chain

The photocathode is typically either of the two bialkalis, Na_2KSb or K_2CsSb, doped with oxygen and caesium. The photocathode dictates the spectral response, quantum efficiency (QE) and overall sensitivity of the PMT. It also sets the level of "dark current," which is the signal received in the absence of any incident photons, due to thermionic electron emission. The spectral response of the photocathode is matched to the spectral output of the scintillator which is in the blue part of the visible spectrum (a

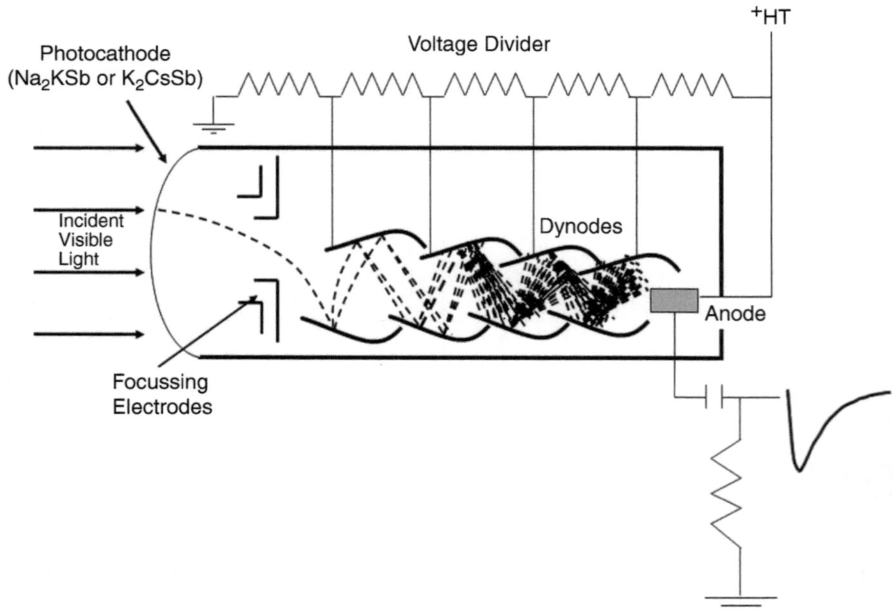

Fig. 9.39 Principle of operation of an 8-dynode photomultiplier tube (PMT). Visible light emitted by a scintillator is incident to a semitransparent photocathode; photoelectric absorption in the photocathode releases low-energy electrons which migrate through the photocathode. At the exit face into the vacuum, they are accelerated through a serial chain of dynodes, The *dashed lines* indicate the trajectories of electrons within the dynode chain. The signal at the anode is a negatively-going pulse as shown. This example shows a linear focusing dynode geometry. Refer to the text for details

wavelength of about 400 nm, corresponding to a photon energy of about 3 eV). At long photon wavelengths, the photon is insufficiently energetic to escape from the photocathode surface. The cut-off at short wavelengths is largely dictated by the transparent window the scintillation light passes through to impinge on the photocathode. For borosilicate glass, the most commonly-used window material, this cut-off is about 300 nm.[12] In those cases where the scintillator emits light in the ultraviolet, the entrance window is made from materials such as synthetic silica and MgF_2 in order to extend the transmission into the shorter wavelengths.

Care must be taken in the selection of the thickness of the photocathode: if it is too thin, photons are transmitted without absorption and if too thick, more photons are absorbed but too few electrons are ejected from its back surface.

The QE of the photocathode is the ratio,

$$\mathrm{QE} = \frac{\text{Number of photoelectrons emitted}}{\text{Number of photons incident}} \tag{9.156}$$

The QE is a function of at least four factors including the reflections of the incident light from the entrance window glass and from the photocathode, the photocathode material itself (which includes the likelihood for a photoelectric absorption to occur and the work function the photoelectron must overcome in order to escape) and the thickness of the photocathode and the resulting likelihood for the photoelectron to migrate to its rear face. As suggested in the previous discussion of the creation, migration, and emission of a photoelectron, the QE will also be a function of the incident light's wavelength. In practice, the QE can be as high as about 25%.

Dynode Chain

The number of electrons emitted by the photocathode is small and insufficient to generate an efficiently detectable electrical signal on their own. Hence, a chain of dynodes is used to amplify this signal. As shown in Fig. 9.39, each dynode is held at a progressively higher potential than the one prior; the voltage difference between each successive dynode is the same. The photoelectrons ejected from the photocathode are focused by the focusing electrodes and have a kinetic energy of about 1 eV only. These are attracted to, and bombard, the first dynode. By the appropriate choice of dynode material, a small bandgap between the valence and conduction bands can be selected. For example, if the bandgap were 2 eV, then for each electron incident to the dynode, 50 electrons are produced per 100 V potential applied to the dynode. The most commonly used material for dynodes is an alloy of antimony, potassium, and caesium (SbKCs).

The gain, or multiplication factor, per dynode is defined as the ratio,

$$\delta = \frac{\text{Number of electrons ejected}}{\text{Number of electrons incident}} \tag{9.157}$$

The overall gain for an N-stage dynode chain is

$$\Delta = \prod_{j=1}^{N} \delta_j \tag{9.158}$$

where δ_j is the multiplication factor for the jth dynode; should all of the dynodes have the same multiplication factor δ, then the overall gain is simply δ^N. For an 8-stage dynode chain shown in Fig. 9.39 where the multiplication factor per dynode is of the order of 5 (a low typical value), then the overall gain of the dynode chain is 5^8 or about 4×10^5.

The electron transit time is defined as the time between the arrival of a light photon at the photocathode (experimentally, this can be provided using a pulsed laser diode) and the peak of the output signal (Fig. 9.40).

Position-Sensitive Photomultiplier Tubes

PSPMTs are used in some PET scanners. A typical design of a PSPMT uses multiple linear dynode chains as shown in Fig. 9.41.

The multianode can be in either a matrix (2D) format or a linear (1D) format.

[12] Potassium-free borosilicate glass is used in low-count applications where it is necessary to remove background from the decay of primordial ^{40}K.

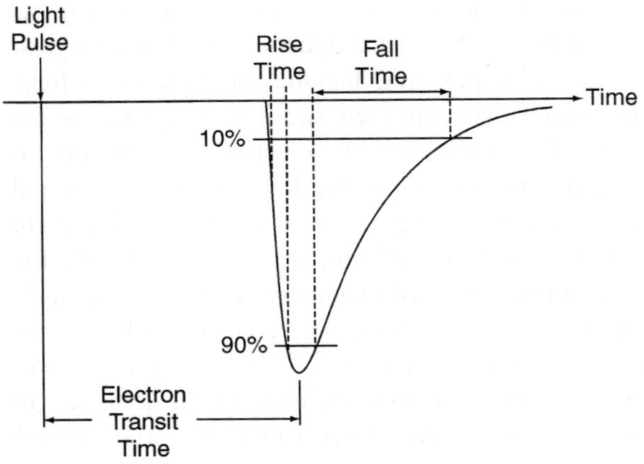

Fig. 9.40 Output waveform from a PMT showing the definitions of rise and fall times and the electron transit time

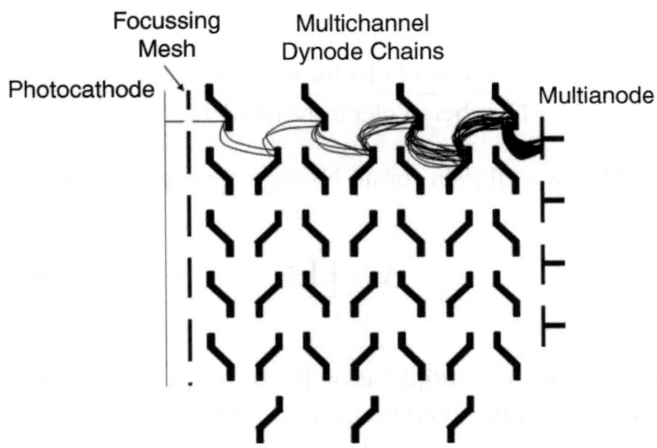

Fig. 9.41 Schematic structure of, and electron trajectories in, a PSPMT

9.3.3.5 Scintillation Spectroscopy

Scintillation spectroscopy is highly important in diagnostic nuclear medicine. For imaging, one wants to detect only those photons which have not been scattered within tissue and which are completely stopped in the scintillator. Such photons will have traveled directly from the point of emission (i.e., the in vivo site of the radionuclide) to the detector. Photons which have been scattered are to be excluded from imaging as the positional information they provide is not indicative of the spatial distribution of the radionuclide desired. This selection is achieved by measuring the

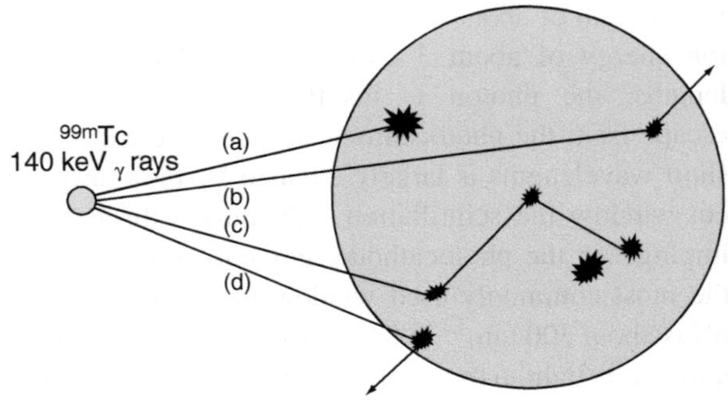

Fig. 9.42 Four different photon interactions in a NaI(Tl) crystal exposed to 140 keV γ rays from a ^{99m}Tc source. Refer to the text for details

energies of the photons incident to the scintillator and "counting" only those which satisfy the above condition.

As an example, consider the cases of a NaI(Tl) scintillator exposed to the 140 keV γ rays from a ^{99m}Tc source. The only energy-transfer processes that need be considered are photoelectric absorption and Compton scatter. Figure 9.42 shows the example in question in which four interactions are considered. The first (Interaction (a)) is the simplest: the photon is photoelectrically absorbed without prior interaction. The photoelectron is ejected with a kinetic energy equal to the incident photon energy (140 keV) minus the binding energy (which can be neglected). Any Auger electrons or characteristic X-rays resulting from the atomic relaxation following the photoelectric absorption are themselves absorbed within the crystal. The electron is stopped within the crystal and the resultant scintillations of such processes form a singular signal. This is known as the photopeak. In Interaction (b), the photon undergoes a Compton scatter within the crystal and escapes the crystal; the Compton-recoil electron is stopped within the crystal and the signal is proportional to the electron energy. The energy spectrum of these recoil Compton electrons are that of Fig. 6.15 up to a maximum electron energy corresponding to a backscattered photon (Fig. 6.10). For a 140 keV incident photon, this maximum recoil electron kinetic energy is equal to 49.6 keV. In Interaction (c), the incident 140 keV photon undergoes multiple Compton scatterings within the crystals until it is eventually photoelectrically absorbed. The net signal is the sum of the energy deposits of the

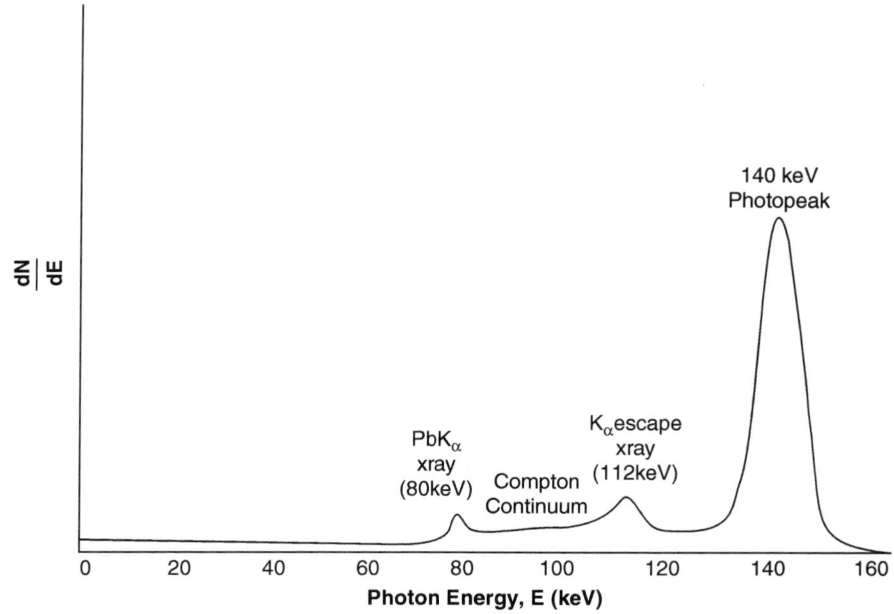

Fig. 9.43 Measured energy spectrum from a NaI(Tl) crystal exposed to the 140 keV photons from ^{99m}Tc. The photopeak corresponding to the absorption of the photons within the NaI(Tl) crystal dominates. The escape peak at 112 keV is also apparent (equal to the difference between the energies of the incident photon and the K_α X-ray of iodine). The slight peak at around 80 keV is due to photon absorption in the lead shielding around the NaI(Tl) scintillator with the subsequent characteristic X-ray emission of atomic relaxation

stopped Compton electrons and photoelectron and is proportional to the incident photon energy of 140 keV. Interaction (d) is the photoelectric absorption of the photon near the periphery of the crystal followed by the atomic relaxation in which the K_α X-ray is emitted. However, unlike Interaction (a), because the interaction site is near the edge of the detector, the K_α X-ray escapes from the crystal and the resulting energy signal is less than that expected for complete absorption, being equal to the difference between the 140 keV γ-ray energy and the 28 keV energy of the K_α X-ray. Should this process have happened at depth in the crystal and the K_α X-ray be stopped and absorbed within the crystal, the net signal would be proportional to 140 keV (minus the binding energy). The energy spectrum resulting from these interactions is shown conceptually in Fig. 9.43.

9.3.4 MOSFET

Solid-state detectors can present a considerable advantage over gaseous detectors due to their greater densities, of the order of three-orders-of-magnitude, which leads to greater detection efficiency. For example, a scintillator well-counter is used to measure low levels of activity which are below the detection limit of the reentrant ionization chamber. However, scintillators have relatively poor energy resolution, e.g., the full-width half-maximum (FWHM) of the 140 keV photopeak in NaI(Tl) is about 10%. This poor energy resolution is a consequence of it being a function of the number of information carriers which, in this case, are the photoelectrons. The output analog signal from the PMT is the result of the long serial chain beginning with the initial ionization and scintillation, the collection of the scintillation light, the conversion of this light to photoelectrons by the photocathode and their subsequent amplification through the dynode chain. As about 1 keV of energy must be absorbed in the scintillator in order to yield a single photoelectron, the total number of photoelectrons generated by a single ionization event is only a few thousand at best. Hence, the combined inefficiencies of the collection of the scintillation light and the conversion of this light by the photocathode leads to the reduced energy resolution.

An improvement in energy resolution is achievable if the information carriers were collected directly in a solid. This also takes advantage of the fact that solids have a lower $^W/_e$ value than gases. Semiconductors present a means of generating a high number of secondary information carriers per unit absorbed energy, thus improving energy resolution. Indeed, since about the 1960s, semiconductor detectors have been used in high-precision spectroscopy measurements. However, such detectors are operated cryogenically in order to reduce thermal noise, which is an impractical condition for clinical nuclear medicine. Certain room-temperature semiconductor devices have, though, proven useful in either in vivo measurement of radiation or emission imaging. In this subsection, one such device used in in vivo nuclear medicine dosimetry, the metal-oxide semiconductor field-effect transistor (MOSFET) is considered. Semiconductor detectors, such as CZT are used small field-of-view gamma cameras. Because of the small fields-of-view, such devices are currently limited to specific emission applications, such as scintimammography and cardiac imaging.

Figure 9.44 shows the basic operation of a p-channel MOSFET.[13] The device is constructed on an n-type silicon substrate; two of the three terminals (the source and drain) are placed over p-type silicon. The third terminal, the gate, is placed over an insulating SiO_2 layer which separates it from the n-type substrate. If a negative bias voltage, V_G, is placed at the gate as shown, minority p-type carriers are attracted to this region from both source and drain.

For a gate voltage above a threshold value, a source-to-drain current, I_{ds}, flows if a voltage is applied across the source and drain. During exposure to ionizing radiation, electron-hole pairs are generated in the silicon dioxide insulating layer. The junction potential leads to a net drift of electrons towards the gate and the holes towards the junction where they are trapped. These trapped positive charges lead to a shift in the threshold voltage for conduction; a greater negative gate potential is required to overcome the potential due to these trapped charges in order to allow conduction (Fig. 9.45). Hence, this shift in gate voltage is a measure of the amount of ionization within the SiO_2.

MOSFETs offer the advantages of a small radiosensitive area (dimensions of the order of 25 μm and less), which allow absorbed dose measurements at high spatial resolutions, stability, an absorbed dose reading which can be permanently stored and independence of the absorbed dose rate.

The in vivo applications of MOSFETs in nuclear medicine are discussed later.

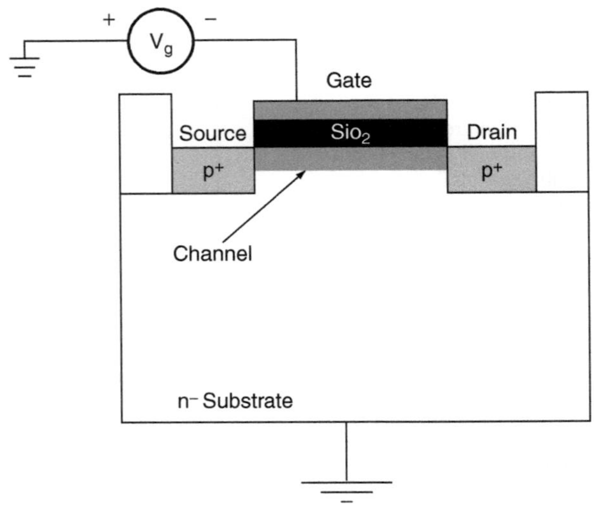

Fig. 9.44 Simplified structure of a p-channel MOSFET

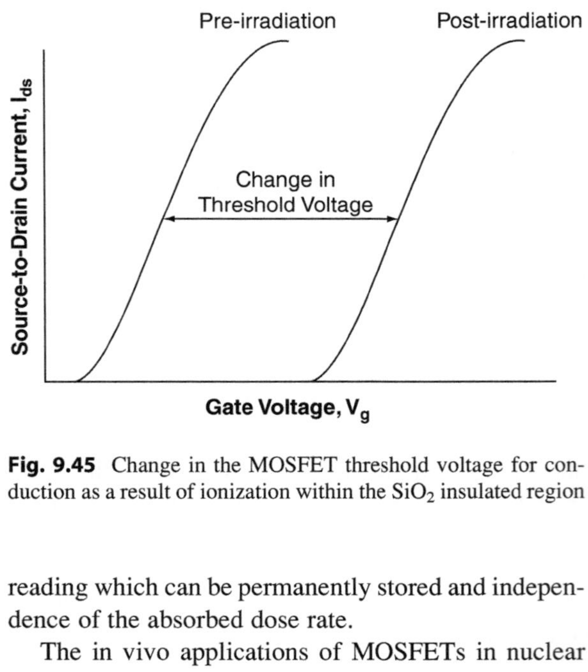

Fig. 9.45 Change in the MOSFET threshold voltage for conduction as a result of ionization within the SiO_2 insulated region

9.3.5 Thermoluminescent Dosimetry

9.3.5.1 Introduction

Thermoluminescent dosimetry provides a useful means of obtaining a record of integrated exposure to ionizing

[13] A p-type semiconductor is doped with a material to provide an excess of holes; an n-type semiconductor is doped with a material to provide an excess of electrons.

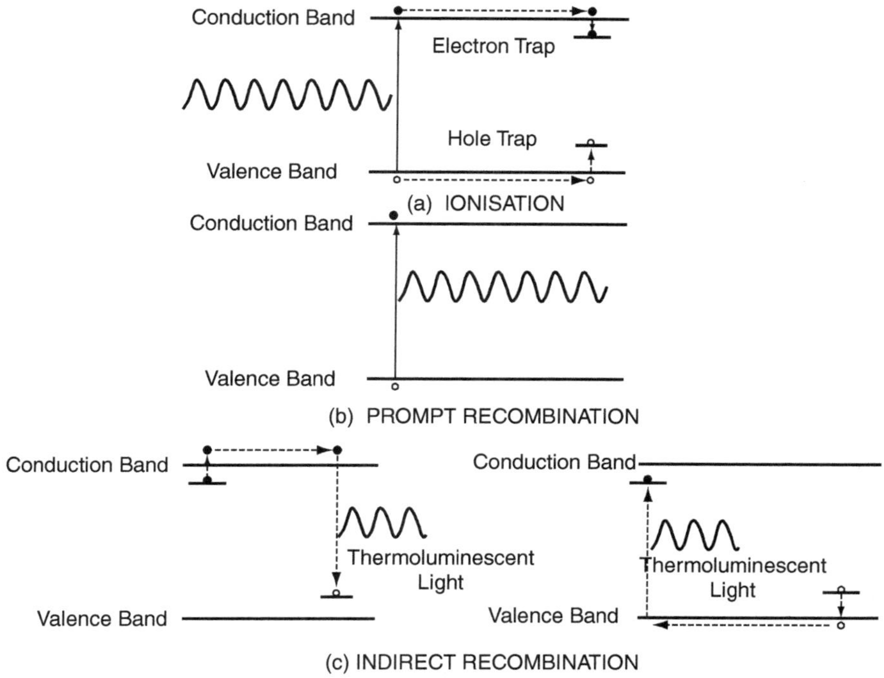

Fig. 9.46 Thermoluminescent dosimetry: an initial ionization in the crystal creates an electron-hole pair which can either be trapped in crystalline defects (**a**) or promptly recombine and releasing radiation with an energy equal to the difference between the conduction and valence bands (**b**). Following the application of heat the trapped electron or hole is released into the conduction or valence band (respectively) to recombine with the corresponding entity; the released photon has a longer wavelength in the optical range which is detected (**c**)

radiation. Medical applications of radiation detection using thermoluminescence have centered primarily on radiation protection (personnel exposure monitoring) and in situ measurements of patient absorbed doses in radiotherapy and radiographic imaging. In this subsection, the principles of thermoluminescent dosimetry are summarized; a detailed exposition of the thermoluminescence can be found in McKeever (1985).

9.3.5.2 Theory of Thermoluminescence

The principle of thermoluminescence differs markedly from that of the prompt fluorescence associated with scintillators. Scintillator detectors are doped so as to emit light upon exposure to ionizing radiation through prompt fluorescence. On the other hand, thermoluminscent dosimeters (TLDs), while also based upon inorganic crystals (including alkali-halides), use materials with high concentrations of "trapping" centers within the bandgap between the valence and conduction bands. Ionization produces the electron-hole pair, elevating the electron into the conduction band and leaving the hole in the valence band. This pair then migrates and both are captured in trapping centers, as shown in Fig. 9.46. Should the energy difference between the trapping center and the conduction band be sufficiently large that thermal excitation (at room temperature) is insufficient to elevate the electron back into the conduction band (or the hole into the valence band), the electron-hole pair is effectively trapped. Hence, exposure of such a crystal to ionizing radiation leads to a continuous increase in the population of trapped electrons and holes. The numbers of these trapped entities is proportional to the kerma within the crystal.

Following irradiation, these electrons can remain trapped indefinitely, depending upon temperature. A simple, two-level model of thermoluminescence was investigated by a number of authors in the 1940s (see McKeever (1985) for a review) and is shown in Fig. 9.47.

In addition to the energy levels defined in the legend to Fig. 9.47, the following quantities are defined:

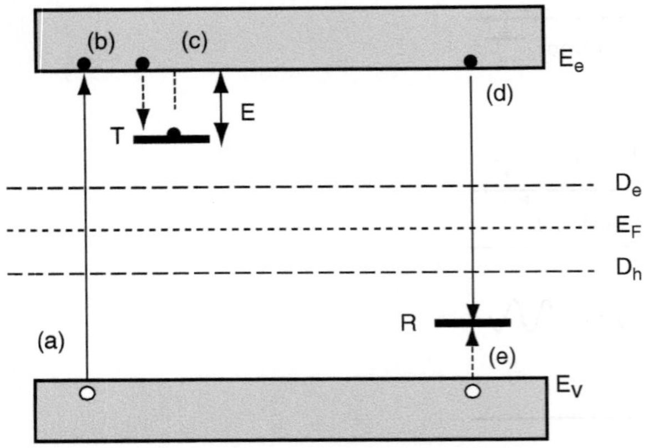

Fig. 9.47 Simple two-level for the thermoluminscence process. The transitions are: (*a*) initial ionization, (*b*) trapping of electron, (*c*) thermal release of electron, (*d*) radiative recombination of electron and hole with release of visible light and (*e*) trapping of hole. E_F us the equilibrium Fermi energy level, D_e and D_h are demarcation levels between electron and hole traps and E_e and E_v indicate the energy levels of the conduction and valence bands, respectively (after McKeever 1985)

f	Electron-hole production rate (volume per unit time) and corresponding to transition (a) in Fig. 9.47
$n_c(t)$	Number of electrons per unit volume in the conduction band
$n_v(t)$	Number of holes per unit volume in the valence band
N	Number of available electron traps per unit volume with energy E below the conduction band (T in Fig. 9.47)
n(t)	Number of trapped electrons per unit volume
$n_h(t)$	Number of holes per unit volume in recombination centers (R in Fig. 9.47)
N_h	number of available hole centers
A	transition coefficient (volume per unit time) for electrons being trapped from the conduction band (corresponding to transition (b) in Fig. 9.47)
A_h	Transition coefficient (volume per unit time) for holes in the valence band becoming trapped in the hole centers (corresponding to transition (e) in Fig. 9.47)
A_r	Recombination transition coefficient for electrons in the conduction band with trapped holes (corresponding to transition (d) in Fig. 9.47) and which result in the emission of visible light

The condition of charge neutrality leads to,

$$n_c + n = n_v + n_h \tag{9.159}$$

The rate equations of the electron and hole concentrations are,

$$\frac{dn_c(t)}{dt} = f - A_r n_c n_h - A n_c (N - n) \tag{9.160}$$

$$\frac{dn(t)}{dt} = A n_c (N - n) \tag{9.161}$$

$$\frac{dn_v(t)}{dt} = f - A_h n_v (N_h - n_h) \tag{9.162}$$

$$\frac{dn_h(t)}{dt} = A_h n_v (N_h - n_h) - n_c n_h A_r \tag{9.163}$$

Equation (9.160) gives the net production rate of electrons in the conduction band which is due to the initial ionization less the rates of recombination and trapping; this last rate is that of (9.161). Equation (9.162) describes the net production rate of holes in the valence band which is the initial ionization rate less the rate of trapping. Equation (9.163) provides the net production rate holes in recombination centers which is the difference between the rates of trapping and recombination.

A simplifying assumption to aid in solving the coupled first-order linear differential equations is that the irradiation has ceased before the electrons and holes have recombined is used – this is also the practical case in dosimetry measurements. In this case, transitions (a) and (e) in Fig. 9.47 can be neglected. In this case, the rate equations for the concentrations of electrons in the conduction band, electrons trapped and holes in recombination centers are,

$$\frac{dn_c(t)}{dt} = np - A_r n_c n_h - A n_c (N - n) \tag{9.164}$$

$$\frac{dn(t)}{dt} = A n_c (N - n) - np \tag{9.165}$$

$$\frac{dn_h(t)}{dt} = -n_c n_h A_r \tag{9.166}$$

where p is the probability per unit time that an electron is released from a trap. From thermodynamics, this probability can be expressed in the form of the Arrhenius equation,

$$p = \frac{e^{-E/kT}}{s} \tag{9.167}$$

where E is the energy difference between the trap and the conduction band (as shown in Fig. 9.47), k is Boltzmann's constant, T is the temperature and s is

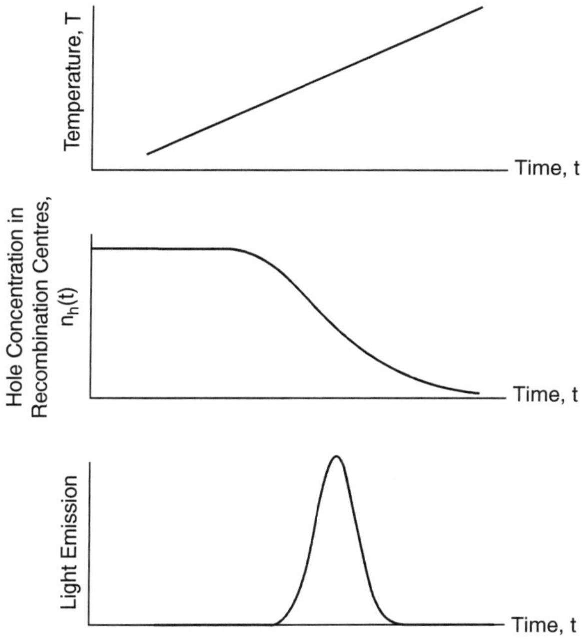

Fig. 9.48 The emission of thermoluminescence with temperature increasing linearly with time (after McKeever 1985)

commonly assumed to be a constant. Hence, the probability that an electron will be released, p, increases with T.

The intensity of the emitted thermoluminescence will be proportional to $-\,dn_h(t)/dt$, which is given by the negative of (9.166). Figure 9.48 shows the change in the concentration of holes in recombination centers as a function of time, along with the intensity of emitted thermoluminescence, for a linear increase in temperature of the crystal with time.

The above model is quite simple, but it does provide a basic insight into the thermoluminescence process. However, the measured light intensity ("glow curve") as a function of temperature during the heating of a postirradiated crystal is not given by a single peak; indeed, multiple peaks are ordinarily observed and this reflects the existence of multiple trapping centers with different energies. Figure 9.49 shows an example of the postirradiation glow curve from a commonly used thermoluminescent material, lithium fluoride, as a function of temperature.

In practice, a TLD crystal is annealed at high temperature prior to exposure in order to vacate all of the traps. After exposure to radiation, the crystal is often annealed at a lower temperature (e.g., 150°C) to vacate

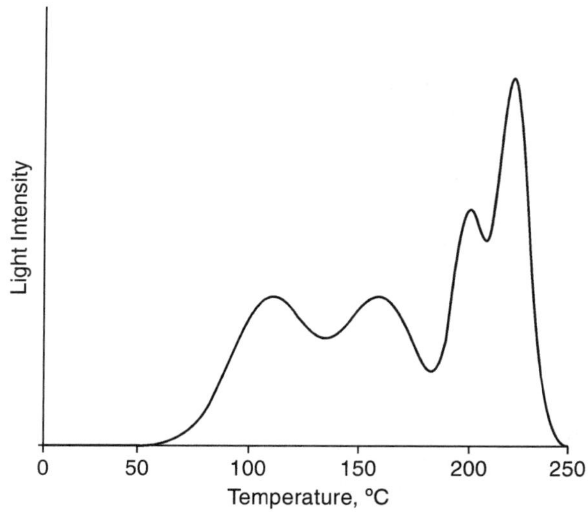

Fig. 9.49 Thermoluminescence output from LiF as a function of temperature

the unstable, lower-energy sites. During reading, the crystal is heated and the thermoluminescence detected by a PMT. The charge from the tube is integrated (for LiF, this is typically between crystal temperatures between about 220 and 275°C so as to measure the main peak of the glow curve). Hence, the recorded output of a TLD reader is frequently given in charge collected from the PMT.

9.3.5.3 Thermoluminescent Dosimetry Materials and Considerations

Here, a brief summary of common TLD crystals used for medical and radiological protection purposes and how they are used is provided.

The ideal requirements for a TLD material is that it have a density and effective atomic number as close as possible to that of tissue and have good light output and linearity with absorbed dose. Lithium fluoride is one of the most popular materials used in thermoluminescent dosimetry as it has an atomic number and density comparable to that of tissue. It does not require an activator to create traps or recombination centers as these are produced by naturally-occurring crystalline defects. LiF crystals are also useful for dosimetry measurements in mixed photon–neutron fields. Crystals with the lithium enriched in ^{6}Li, which has a high thermal neutron cross section via $^{6}Li(n,\alpha)^{3}H$, yield a signal proportional to both the neutron and photon fluences. Crystals with the lithium enriched in ^{7}Li have a much lower neutron sensitivity but a comparable photon sensitivity. Hence, a differential measurement using LiF crystals enriched in ^{6}Li and ^{7}Li can separate the photon and neutron contributions. Other popular thermoluminscent materials are $CaSo_4$:Mn (manganese being the activator to create the traps and recombination sites) and CaF_2:Mn.

References

Akabani G, Hawkins WG, Eckblader MB et al (1997) Patient-specific dosimetry using quantitative SPECT imaging and three-dimensional discrete Fourier transform convolution. J Nucl Med 38:308–314

Alm Carlsson G (1981) Absorbed dose equations. On the derivation of a general absorbed dose equation and equations valid for different kinds of radiation equilibrium. Radiat Res 85:219–237

Alm Carlsson G (1985) Theoretical basis for dosimetry. In: Kase KR, Bjärngard BE, Attix FH (eds) The dosimetry of ionizing radiation, vol I. Academic, Orlando

Attix FH (1986) Introduction to radiological physics and radiation dosimetry. Wiley, New York

Bailey DL, Karp JS, Surti S (2003) Physics and instrumentation in PET. In: Valk PE, Bailey DL, Townsend DW, Maisey MN (eds) Positron emission tomography: basic science and clinical practice. Springer, London

Bardiès M, Kwok C, Sgouros G (2003) Dose point-kernels for radionuclide dosimetry. In: Zaidi H, Sgouros G (eds) Therapeutic applications of Monte Carlo calculations in nuclear medicine. Institute of Physics, Bristol

Berger MJ (1968) Energy deposition in water by photons from point isotropic sources – MIRD Pamphlet No 2. J Nucl Med 9:15–25

Berger (1970) Beta-ray dose calculations with use of point kernels. In: Medical radionuclides: radiation dose and effects. US Atomic Energy Commission, Oak Ridge, pp 63–68

Berger MJ, Spencer LV (1959) Some radiological applications of gamma-ray transport theory. Rad Res 10:552–570

Bielajew A, Rogers DWO (1989) Variance reduction techniques. In: Jenkins T, Rindi A, Nahum A, Rogers DWO (eds) Monte Carlo transport of electrons and photons. Plenum, New York

Birks JB (1964) The theory and practice of scintillation counting. Pergamon, Oxford

Boston DA, Tattam D, McParland BJ (2001) Accurate determination of TG-43 dosimetry parameters of brachytherapy sources using Monte Carlo simulation. Proceedings of the Annual Meeting of the Institute of Physics and Engineering in Medicine, Belfast, Physica Medica XVII:168

Bragg WH (1910) Consequences of the corpuscular hypothesis of the gamma and x rays and the ranges of beta rays. Phil Mag 20:385–416

Brownell G, Ellet W, Reddy R (1968) MIRD Pamphlet No. 3: Absorbed fractions for photon dosimetry. J Nucl Med 9 (supp 1):27–39

Carlson B (1953) The Monte Carlo method applied to a problem in gamma-ray diffusion. Los Alamos Scientific Laboratory AECU Report 2857, Los Alamos

Cashwell ED, Everett CJ (1959) A practical manual on the Monte Carlo method for random walk problems. Pergammon, New York

Charpak G, Bouclier T, Bressani J, Favier J, Zupančič Č (1968) The use of multiwire proportional chambers to select and localize charged particles. Nuclear Instrum Meth 62: 202–226

Chauvie S, Lorenzo PM, Lechner A, Moscicki J, Pia MG (2007) Benchmark of medical dosimetry simulation using the Grid. IEEE Nucl Sci Symp Conf Rec 2:1100–1106

Clarkson JR (1941) A note on depth doses in fields of irregular shape. Br J Radiol 14:265–268

Cristy M (1983) Applying the reciprocal dose principle to heterogeneous phantoms: practical experience from Monte Carlo studies. Phys Med Biol 28:1289–1303

Cunningham JR (1972) Scatter-air ratios. Phys Med Biol 17: 42–51

Dutreix J, Bernard M (1966) Dosimetry at interfaces for high energy x and gamma rays. Br J Radiol 39:205–210

Ellett WH, Callahan AB, Brownell GL (1965) Gamma ray dosimetry of internal emitters. II. Monte Carlo calculations of absorbed doses from uniform sources. Br J Radiol 38:541–544

Ellett W, Humes R (1971) MIRD Pamphlet No. 8: absorbed fractions for small volumes containing photon-emitting radioactivity. J Nucl Med 12(supp 5):25–32

Emshoff JL, Sisson RL (1970) Design and use of computer models. MacMillan, New York

Esser PD (1998) An introduction to scintillation detector physics. In: Ljungberg M, Strand S-E, King MA (eds) Monte Carlo calculations in nuclear medicine – applications in diagnostic imaging. Institute of Physics, Bristol

Failla G (1921) Dosage in radium therapy. Am J Roentgenol 8:674–685

Fano U (1954) Note on the Bragg-Gray cavity principle for measuring energy dissipation. Radiat Res 1:237–240

Gray LH (1929) Absorption of penetrating radiation. Proc Roy Soc (Lond) A122:647–668

Gray LH (1936) Ionization method for the absolute measurement of gamma-ray energy. Proc R Soc (Lond) A156:578–596

Green D (2000) The physics of particle detectors. Cambridge University, Cambridge

Hamilton DR (2004) Diagnostic nuclear medicine – a physics perspective. Springer, Berlin

Harima Y (1993) An historical review and current status of buildup factor calculations and applications. Radiat Phys Chem 41:673–672

Harima Y, Nishiwaki Y (1972) An approximation of γ-ray buildup factors by transmission dose buildup geometrical progression. Nucl Eng Des 23:209–227

Hofstadter H (1948) Alkali halide scintillation counters. Phys Rev 74:100–101

ICRU (1980) Radiation quantities and units. Report 33. International Commission on Radiological Units and Measurements, Bethesda, MD

ICRU (1983) Microdosimetry – ICRU Report 36. International Commission on Radiation Units and Measurements, Bethesda

Jennings WA (2007) Evolution over the past century of quantities and units in radiation dosimetry. J Radiol Prot 27:5–16
Johns HE, Cunningham JR (1983) The physics of radiology, 4th edn. Charles C Thomas, Springfield
Kellerer AM (1985) Fundamentals of microdosimetry. In: Kase KR, Bjärngard BE, Attix AH (eds) The dosimetry of ionizing radiation, vol I. Academic, Orlando
Kellerer AM, Chmelevsky D (1975) Criteria for the applicability of LET. Radiat Res 63:226–234
King LV (1912) Absorption problems in radioactivity. Phil Mag 23:242–250
Knoll GF (2000) Radiation detection and measurement, 3rd edn. Wiley, New York
Korzhik M, Lecoq P (2001) Search for new scintillation materials for nuclear medicine applications. IEEE Trans Nucl Sci 48:628–631
Kwok CS, Prestwich WV, Wilson BC (1985) Calculation of radiation doses for nonuniformly distributed β and γ radionuclides in soft tissue. Med Phys 12:405–412
Leichner PK (1993) A unified approach to photon and beta particle dosimetry. J Nucl Med 35:1721–1729
Lewellen TK (2008) Recent developments in PET detector technology. Phys Med Biol 53:R287–R317
Ljungberg M, Strand S-E, King MA (eds) (1998) Monte Carlo calculations in nuclear medicine – applications in diagnostic imaging. Institute of Physics, Bristol
Loevinger R (1969) Distributed radionuclide sources. In: Attix FH, Tochlin E (eds) Radiation dosimetry, vol III, 2nd edn. Academic, New York
McKeever SWS (1985) Thermoluminescence of solids. Cambridge University, Cambridge
McParland BJ (1981) The effects of annular and conical inhomogeneities upon the central-axis scatter dose in a cobalt-60 photon beam. Thesis, University of British Columbia (unpublished)
Marsden PK, Bateman JE, Ott RJ, Leach MO (1986) The development of high-efficiency cathode converters for a multiwire proportional positron camera. Med Phys 13:703–706
Mayneord WV (1945) Energy absorption. IV: the mathematical theory of integral dose in radium therapy. Br J Radiol 18:12–19
Meisberger LL, Keller RJ, Shalek RJ (1968) The effective attenuation in water of the gamma rays of gold 198, iridium 192, cesium 137, radium 226 and cobalt 60. Radiology 90:953–957
Metropolis N, Ulam S (1949) The Monte Carlo method. J Am Stat Assoc 44:335–341
Morin RL, Raeside DE, Goin JE, Widman JC (1979) Monte Carlo advice. Med Phys 6:305–306
Petegnief Y, Petiet A, Peker MC, Bonnin F, Meulemans A, Le Guludec D (1998) Quantitative autoradiography using a radioimager based on a multiwire proportional chamber. Phys Med Biol 43:3269–3238
Raeside DE (1976) Monte Carlo principles and applications. Phys Med Biol 21:181–197
Particle Data Group (2004) Review of particle physics. Phys Lett B 592
Roeske JC, Aydogan B, Bardiès M, Humm JL (2008) Small-scale dosimetry: challenges and future directions. Semin Nucl Med 38:367–383
Rossi HH (1968) Microscopic energy distribution in irradiated matter. In: Attix FH, Roesch WG (eds) Radiation dosimetry, vol I. Academic, New York
Rossi HH, Zaider M (1991) Elements of microdosimetry. Med Phys 18:1085–1092
Seneviratne MDS, Reinhard MI, Baldock C (2007) The energy response of a T.P.A. Mk-II ionization chamber using GEANT4 Monte Carlo simulation. Phys Med Biol 52:3837–3846
Siegel JA (1985) The effect of source size on the buildup factor calculation of absolute volume. J Nucl Med 26:1319–1322
Siegel JA, Wu RK, Maurer AH (1985) The buildup factor: effect of scatter on absolute volume determination. J Nucl Med 26:390–394
Siegel JA, Stabin MG (1994) Absorbed fractions for electrons and beta particles in spheres of various sizes. J Nucl Med 35:152–156
Sievert RM (1930) Die gamma-strahlungsintensität an der oberfläche und in der nächsten umbgebung von radium nadeln. Acta Radiol 11:249–267
Simpkin DJ, Mackie TR (1990) EGS4 Monte Carlo determination of the beat dose kernel in water. Med Phys 17:179–186
Stabin MG, Konijnenberg M (2000) Re-evaluation of absorbed fractions for photons and electrons in small spheres. J Nucl Med 41:149–160
Stegun IA (1972) Miscellaneous functions. In: Abramowitz M, Stegun IA (eds) Handbook of mathematical functions. Dover, New York
Thomson JJ (1896a) On the discharge of electricity produced by the Roentgen rays, and the effects produced by these rays on dielectrics through which they pass. Proc R Soc Lond 59:274–276
Thomson JJ (1896b) The Roentgen rays. Nature 53:391–392
Tsoulfanidis N (1983) Measurement and detection of radiation, 2nd edn. Taylor and Francis, Washington, DC
Villard P (1908) The radiosclérometer. Arch Electr Méd Bordeaux 14:692
Wernick MN, Aarsvold JN (eds) (2004) Emission tomography: the fundamentals of PET and SPECT. Elsevier Academic, San Diego
Williamson JF, Morin RL, Khan FM (1983) Monte Carlo evaluation of the Sievert integral for brachytherapy dosimetry. Phys Med Biol 28:1021–1032
Wong JW, Henkelman RM, Fenster A, Johns HE (1981) Second scatter contribution to dose in a cobalt-60 beam. Med Phys 8:755–782
Worthley BW, Nicholls RL (1972) Computer evaluation of Sievert's integral by recursive formula. Phys Med Biol 17:854–857
Wu RK, Siegel JA (1984) Absolute quantitation of radioactivity using the buildup factor. Med Phys 11:189–192
Yoshida Y (2006) Development of fitting methods using geometric formulae of gamma-ray buildup factors. J Nucl Sci Technol 43:1446–1457
Zaidi H, Sgouros G (eds) (2003) Therapeutic applications of Monte Carlo calculations in nuclear medicine. Institute of Physics, Bristol
Zaidi H, Andreo P (2003) The Monte Carlo method: theory and computational issues. In: Zaidi H, Sgouros G (eds) Therapeutic applications of Monte Carlo calculations in nuclear medicine. Institute of Physics, Bristol
Zimmerman BE, Cessna JT, Millican MA (2004) Experimental determination of calibration settings for plastic syringes containing solutions of ^{90}Y using commercial radionuclide calibrators. Appl Radiat Isot 60:511–517

Biological Effects of Ionizing Radiation 10

Abstract In this chapter, the biological effects of ionizing radiation are presented, beginning with the radiobiology of the mammalian cell. This includes the chromosome and chromatid aberrations resulting from radiation-induced damage of DNA (primarily double-strand breaks, DSBs), the mechanisms, and categories of cell death and of germ-cell mutation. The models of various cell survival curves are described, culminating with the linear-quadratic (LQ) model. The effects of ambient oxygen and LET upon producing cell lethality are considered. Due to its clinical applications, including therapeutic nuclear medicine, the LQ model is studied in detail. The LQ model is derived from a model of DSB production and repair kinetics and the Lea–Catcheside dose-protraction factor and the biological equivalent dose are derived from this result. Human somatic effects of radiation, including epidemiological studies of irradiated populations which provide our estimates of radiation risk, are reviewed. The chapter concludes with consideration of the radiation protection of the nuclear medicine patient, which includes the derivation of the effective dose, and a brief introduction to radiobiology considerations in therapeutic nuclear medicine.

Contents

B.J. McParland, *Nuclear Medicine Radiation Dosimetry*,
DOI 10.1007/978-1-84882-126-2_10,

10.1 Introduction

Up to this point, only the physical effects resulting from exposure to ionizing radiation have been considered. The intents of these discussions were the ability to, first, calculate the radiation fluence in a medium and, second, calculate the resulting energy transfer and absorbed dose to that medium. The concept of a medium in the context of medical radiation dosimetry is more complex than has been considered as the medium is now a living organ or tissue which responds biologically to the energy deposited by radiation.

The division of nuclear medicine into diagnostic and therapeutic applications is a result of different considerations of the biological effects of ionizing radiation (BEIR). Diagnostic nuclear medicine seeks to avoid all biological effects by limiting the absorbed dose to a minimally-acceptable level, but whilst still maintaining the diagnostic efficacy of the study. As a result, the biological responses to these low absorbed doses are probabilistic and limited to oncogenesis and hereditary effects. Therapeutic nuclear medicine seeks neoplastic cell death subject to the optimization of minimizing the radiotoxicity that uninvolved healthy tissues can be subjected to. Despite the differing biological endpoints of interest to diagnostic and therapeutic nuclear medicine, a common understanding of cellular response to ionizing radiation is required for both. The challenges of such understanding are profound. In diagnostic nuclear medicine, one focuses on radiation-induced mutagenesis where ionizing radiation can, in the first case of a somatic cell, result in the production of a malignant cell. This transformation is the result of four or more genetic modifications:

- Overexpression of viral genes or proto-oncogenes
- Loss of apoptosis genes
- Mutations of tumor-suppressing genes (e.g., p53)
- Mutations of those genes necessary for DNA repair

In addition to these effects upon somatic cells are those inflicted upon germ cells. These effects will be, at low absorbed doses, the induction of mutations that can be expressed as inherited genetic effects manifested in the progeny. Despite popular folklore, the characteristics of these mutations are no different than those that occur naturally. Ionizing radiation can only increase the frequency of presentation of these otherwise naturally and spontaneously occurring mutations. At higher absorbed doses, the radiation-induced deaths of spermatogonia can lead to either temporary or permanent sterility in the male depending upon the magnitude of the absorbed dose conferred; in the female, temporary sterility is unachievable and death of the oocytes results in permanent sterility.

In therapeutic nuclear medicine, one seeks the optimization of the probabilities of tumor control and normal tissue complication (radiotoxicity). Tumor control is optimized through the selection of the appropriate vector to transport the radionuclide to the target cells (specificity), the physical half-life of the radioisotope so as to impede the ability for tumor cells to repair radiation damage and the appropriate isotope in terms of its radiation decay scheme. The large mean-free path of photons precludes photon emitters as radiotherapeutics and so interest is in those isotopes that emit charged particulate radiations: α or β particles or the very short-range Auger/Coster–Kronig electrons emitted following electron capture or internal conversion decays.

Consider the radiobiology of diagnostic nuclear medicine. One seeks the minimization, or perhaps even mitigation, of biological effects resulting from the exposure to low absorbed doses of radiation. The fundamental radiobiology of the mammalian cell is a foundation for estimating or modeling the subsequent risks presented to the patient. While knowledge of the response of the individual cell to radiation is required in order to mathematically model these risks, it is not sufficient. Epidemiological data derived from the consequences of populations exposed to ionizing radiation (deliberate or otherwise) are essential to deriving these risk estimates. In most cases, the absorbed doses received by such cohorts are much higher than those received by the medical imaging patient. Hence, it is necessary to extrapolate the observed risk, such as increased cancer incidence or mortality, at these higher absorbed doses to the lower doses more reflective of the diagnostic case. The extrapolation models used will rely greatly upon concepts derived from cellular radiobiology.

On the other hand, therapeutic nuclear medicine considerations have been historically much less reliant upon mathematical modeling and are often based purely upon clinical experience of radiotoxicity limitations. Administered activities of therapeutic radionuclides are still, generally, crudely applied in terms of a single value (MBq) or otherwise normalized to patient body surface area (MBq/m^2) or body weight (MBq/kg). Such approaches – which would not even be contemplated in the modern-day prescription of external beam radiotherapy absorbed doses – are being supplanted by more patient-specific prescriptions. Pretherapy imaging of the therapeutic moiety (at either a diagnostic level of activity if it should emit photons or of its replacement with a photon-emitting isotope, e.g., the replacement of ^{131}I with ^{123}I) allows an estimation of the uptake of the therapeutic agent in the region of interest. This can then be used to derive the amount of therapeutic activity required to obtain the absorbed dose and biological effect desired.

10.2 Radiobiology of the Mammalian Cell

10.2.1 Introduction

All biological processes in a unicellular organism such as a bacterium occur within that single cell. In multicellular organisms (metazoa), specific cell groups "specialize" in conducting distinct functions. As a result, the differential magnitudes of such specializations can frequently make it difficult to define a typical cell representative of all of those in the body. However, there is a significant degree of common internal structure (organelles) among cells that allow a general specification to be made.

Metazoans are defined as multicellular eukaryotic organisms. This chapter is concerned exclusively with mammalian cells. The first obvious reason for this distinction is that we are ultimately interested in the response of human tissues to ionizing radiation. The second, and subtle, reason is that nonmammalian cells exhibit markedly different responses to ionizing radiation than do mammalian cells due to the much higher deoxyribonucleic acid (DNA) content of the latter.

10.2.2 Structure of the Mammalian Cell

10.2.2.1 Cellular Structure

Despite their differentiation in terms of function, all mammalian cells have certain common attributes. With the exception of the erythrocyte, all contain subcellular structures known as organelles and are eukaryotic, i.e., have a nucleus containing DNA. The exceptional erythrocyte (red blood cell) contains a nucleus at its early stage of development, but extrudes it (along with other organelles) during maturation in order to increase the amount of hemoglobin content it may carry.

The intracellular matrix, or cytoplasm, contains the nucleus and other organelles. These organelles are:

Mitochondria: Ellipsoid-shaped structures responsible for energy production within the cell

Golgi apparatus: Responsible for the storage and transportation of secretory products to the external environment

Endoplasmic reticulum: A network of tubules and cisternae responsible for the folding and transportation of proteins to the cell membrane

Ribosomes: Spherical structures which are the site of protein synthesis and are either free in the cytoplasm or connected to the endoplasmic reticulum

Lysosomes: Contain enzymes responsible for the digestion of vacuoles formed by the phagocytosis of solid material (e.g., foreign bacteria) and for the eradication of worn-out organelles

Centrioles: Paired cylindrical structures involved in cell division (cytokinesis). During this stage, they move to opposite ends of the nucleus and form the ends of the mitotic spindle (bundles of microtubules of protein filaments)

10.2.2.2 Types of Mammalian Cells

Despite the commonality of a nucleus and organelles, mammalian cells differ markedly in terms of morphology, function and cell kinetic properties, and radiosensitivity. The four primary categories of tissues formed by mammalian cells are:

Connective: Fibrous tissue which holds organs in place and forms ligaments and tendons. It is subdivided further into categories of loose, dense, elastic, adipose, and reticular connective tissues.

Epithelial: Tissue consisting of cells lining cavities and lumen and organ surfaces. This tissue forms the skin and lines, for example, the gastrointestinal (GI) tract and glandular ducts.

Nervous: Tissue made up of neurons, which transmit electrical impulses, and the supportive neuroglia which are made up of microglia and macroglia cells.

Muscular: Contractile tissue made up of three categories – skeletal or voluntary muscle anchored by tendons to bone and which is under voluntary control; smooth or involuntary muscle found within the viscera such as the esophagus and the GI tract and the urinary bladder; and cardiac muscle which is similar to skeletal muscle although it is involuntary and is found in the heart alone.

10.2.2.3 DNA

DNA is a long polymer made up from repeating nitrogenous bases (adenine, guanine, thymine, and cytosine) and exists as an intertwined double helix with the bases binding the helices. The strand itself consists of alternating phosphate and sugar (deoxyribose) groups and has a diameter of the order of 2.5 nm. The nucleotides bind pair-wise specifically: adenine (purine) with thymine (pyrimidine) and cytosine (pyrimidine) with guanine (purine), as shown schematically in Fig. 10.1. The groupings of bases are thus complementary: adenine will bind with thymine and cytosine always binds with guanine.

10.2.2.4 Chromatin, Chromosomes, and Chromatids

As the length of the DNA contained within the eukaryotic cell is several thousand times the dimension of the cell, it must be compressed through sequential folding in order to fit within the nucleus and yet still be accessible. This folding of DNA forms the organized packed structure chromatin, which is a thread-like entity within the nucleus made up from DNA and protein. This is further integrated into the chromosome. The name chromosome is derived from the Greek *chroma* and *soma*, or "colored body," reflecting the chromosome's ability to take histologic stain. During mitosis, it condenses to form sister chromatids, along with the centromere, a site which does not take histologic stain. Its role in cell division, or mitosis, is to be explained below.

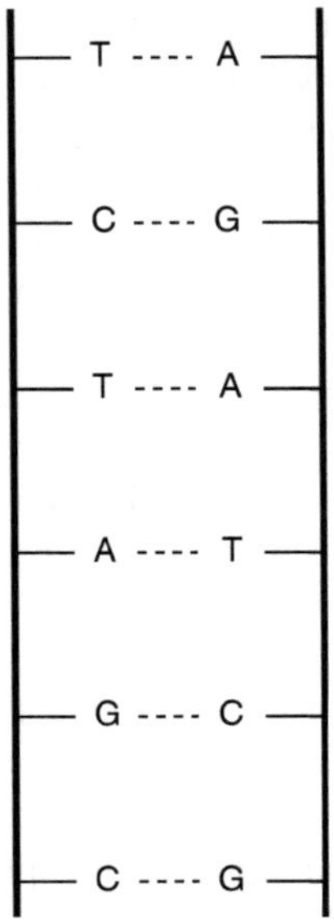

Fig. 10.1 Schematic representation of the binding between thymine (*T*)-adenine (*A*) bases and cytosine (*C*)-guanine (*G*) bases

10.2.2.5 Proliferation and Cell Cycle

A cell divides into two daughter cells following a cycle defined by sequential mitotic divisions. This cycle is partitioned into two phases of the short metaphase during which the cell undergoes mitosis and a longer interphase. Much activity goes on within the cell during the interphase so it is subdivided into three intermitotic phases: the two gap phases, G_1 and G_2, which follow and precede the mitotic phase, respectively, during which there is apparent cellular inactivity and the S DNA synthesis phase.

During the G_1 phase, the cell produces enzymes required for the S phase. The length of this is highly variable. Synthesis of DNA and replication of the chromosomes occurs during the S phase. In the G_2 phase, each chromosome is made up of two sister chromatids (Fig. 10.2).

During the mitotic (M) phase, the cell divides. This phase is made up of the prophase, prometaphase, metaphase, anaphase, and telophase (Hall and Giaccia

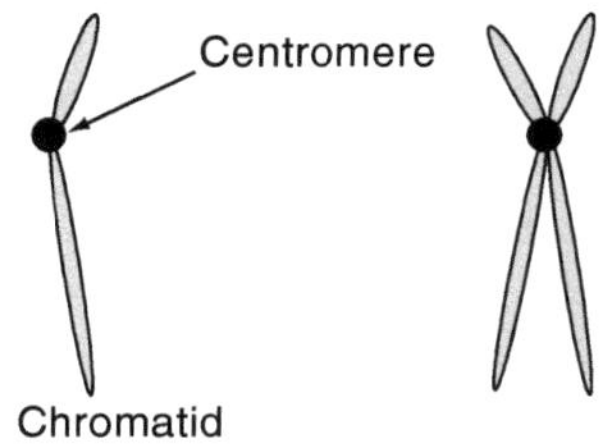

Fig. 10.2 Condensed chromosome morphology during (*left*) interphase and (*right*) the S phase of the cell cycle during which it has duplicated

2006). The subnuclear actions that occur during these mitotic phases are:

Prophase: During this phase, chromatin condenses to form the visible chromosome and ends with the maximal condensation and the disappearance of the nuclear membrane.

Prometaphase: The nuclear membrane and nucleoli having disappeared, the centrioles move to opposite sides of the nucleus and spindle fibers form between them.

Metaphase: Following the disappearance of the nuclear membrane, the cytoplasm and nuclear plasma intermix. The spindle fibers from the centrioles attach to the centromeres, the latter dividing to signal the end of metaphase.

Anaphase: The chromosomes are pulled to the opposing poles of the nucleus.

Telophase: The daughter chromosomes uncoil, the spindle fibers break off and nuclear membranes are formed around each ensemble of chromosomes and nucleoli regenerate. The chromosomes disperse into chromatin and the nucleus resumes its indistinct interphase appearance.

While the interphase of the cell cycle is that part when mitosis is not occurring, it accounts for other significant cell processes. The intermitotic phase which is experimentally most easily recognized is the S phase which, in most cases, is the only time during the cell cycle that DNA is synthesized.[1] On the other hand, RNA and protein synthesis can occur at any time throughout the interphase. Movement through the cell cycle is governed by proteins known as cyclins with "check points" at the G_1–S and G_2–M junctions. Transition through these junctions is enabled by cyclin-dependent kinases (Fig. 10.3).

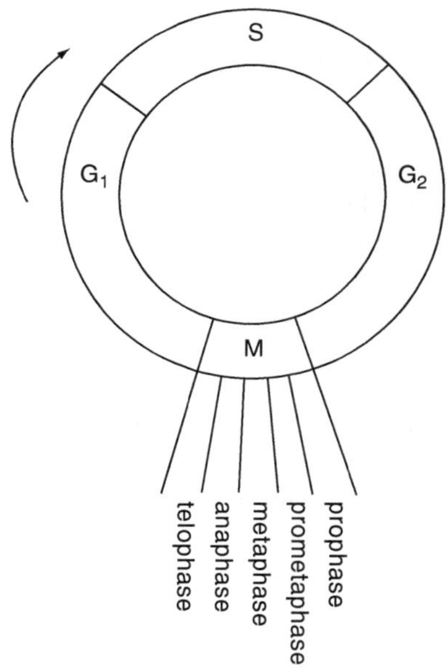

Fig. 10.3 The cell cycle where M is when mitosis occurs and the intermitotic phases (interphase) is made up of two gap phases, G_1 and G_2, and the S DNA synthesis phase. Some cells can enter a "resting" phase, G_0, following the M–G_1 junction before entering the G_1 phase. The length of the G_1 phase can be highly variable

10.2.3 Radiation-Induced Damage to the Cell

10.2.3.1 Introduction

Any discussion of cellular damage caused by ionizing radiation must begin with an understanding of what are the radiation-sensitive sites within the cell. Although mainly circumstantial, there is overwhelming evidence that the target is chromosomal DNA. That it is the cell nucleus that is sensitive to radiation and the cytoplasm insensitive has been demonstrated by two different categories of experiment. The first

[1]This statement is not strictly true as mitochondrial DNA synthesis can occur outside the S phase as can unscheduled nuclear DNA synthesis following radiation-induced damage.

uses the physical directing of ionizing radiation to the nucleus. This can be achieved by, for example, placing a needle tip, coated with an α-emitting isotope of polonium, adjacent to the nucleus (Munro 1961). Recall that such α particles have low energy and short ranges. Hence, with the appropriate selection of α-particle energy and design of the applicator, it is possible to ensure that the nucleus is the only organelle receiving an absorbed dose. Such experiments have demonstrated that the mean lethal absorbed dose to the nucleus is 1.5 Gy whereas an absorbed dose as high as 250 Gy delivered to the cytoplasm can have no effect upon cell proliferation. A similar, but more sophisticated, approach of subcellular irradiation is through a "microbeam" of charged ions bombarding subcellular targets in vitro (Folkard et al. 2007). The secondary type of experiment compares the effects upon the cell of tritiated water with those of tritiated thymidine (deoxythimidine). As thymidine is integral to DNA synthesis, labeling with tritiated thymidine will localize the short-range β particles emitted during the β decay of ^3H to ^3He directly to the DNA in the nucleus whereas the distribution of tritiated water within the cell is uniform and nonspecific throughout the cell. It is found that the radiation sensitivity of the cell to tritiated thymidine is several orders of magnitude greater than to tritiated water, implicitly suggesting that radiation damage to the nucleus is the center of cellular radiation damage.

Now that the nucleus has been established to be the most radiosensitive organelle, it is interesting to see if it is possible to delve deeper in terms of spatial resolution and identify a subnuclear structure that causes this radiosensitivity. This can be inferred by the spatial dimensions of the energy depositions of low-LET radiations, such as soft X-rays (energies below 1 keV). Recall from Chap. 7 that the energy deposition of the charged particles resulting from photon–matter interactions is stochastic, especially at small spatial dimensions, and is distributed nonuniformly along the trajectory of the particle. In the nomenclature promoted by Hall and Giaccia (2006), the energy deposition distributions are categorized here as either a "spur," which has a maximum energy deposition of 100 eV, is about 4 nm in diameter and typically contains three ion pairs, or a "blob" which has an energy deposition of between 100 and 500 eV, is about 7 nm in diameter and typically contains up to 12 ion pairs. For low-LET radiation, virtually all of the energy deposition is in the form of "spurs" and, as the spatial dimension of the spur is comparable to the 2 nm width of the DNA double helix, this provides further circumstantial evidence that nuclear DNA is the radiosensitive target.

10.2.3.2 Mechanisms of Radiation-Induced Damage

Indirect Effect

Photons and moving charged particles traversing a medium ionize atoms, leaving free electrons and ions in their wake. As the composition of the cell is about 70% water, the effects of the ionization of water by radiation (water radiolysis) will be of dominant interest,

$$H_2O \xrightarrow{\text{Radiation}} H_2O^{+\bullet} + e^-$$

where the superscripts + indicate a positive ion and • an unpaired electron. Another channel is the simple excitation of the water molecule,

$$H_2O \xrightarrow{\text{Radiation}} H_2O^*.$$

$H_2O^{+\bullet}$ is an ion radical[2] with a lifetime of the order of 0.1 ns and is highly reactive. The liberated electron can be subsequently hydrated, i.e., trapped by surrounding water molecules that it has polarized so as to form an aqueous electron,

$$e^- + H_2O \rightarrow (e^-)_{aq}.$$

As the electron is ejected in the ionization event with considerable energy so as to be displaced considerably from the point of ionization, the production of the $H_2O^{+\bullet}$ will not be in equilibrium and the free radical unable to recombine with the ejected electron.

[2]An atom or molecule is a free radical if it has an unpaired electron, even though it can also be electrically neutral. Free radicals are chemically reactive. An ion radical is both an ion and a free radical and, hence, is highly reactive.

The free radical can then either decompose to yield a free proton and a free neutral hydroxyl radical.

$$H_2O^{+\bullet} \rightarrow H^+ + OH^\bullet$$

Or interact with a water molecule to again form a free neutral hydroxyl radical,

$$H_2O^+ + H_2O \rightarrow H_3O^+ + OH^\bullet.$$

The aqueous electron can also interact with a water molecule to form,

$$(e^-)_{aq} + H_2O \rightarrow OH^\bullet + H^+$$

and it can interact with H^+ from other ionizations to form the hydrogen radical,

$$(e^-)_{aq} + H^+ \rightarrow H^\bullet.$$

While the above chain of reactions is, in fact, far more involved, the three most important reactive chemical species created and their relative yields per initial ionization are:

$$(e^-)_{aq}: \quad 45\%$$

$$OH^\bullet: \quad 45\%$$

$$H^\bullet: \quad 10\%.$$

Following their production, the reactive species (which have longer lifetimes of the order of 10 μs compared to the 0.1 ns lifetimes of ions) diffuse and, if the ionization is sufficiently close to the DNA double helix, can migrate to the helix and damage it. The $OH^\bullet$ is a particularly potent species in causing DNA damage as it can extract an electron from the DNA and leave behind a highly reactive site. This category of DNA damage by radiolytic products is known as the indirect effect.

It is also possible for these reactive radiolytic species to interact with each other, especially in the volume around the initial ionization event and prior to any diffusion, and neutralize. Two examples are:

$$OH^\bullet + H^\bullet \rightarrow H_2O$$

and

$$H^\bullet + H^\bullet \rightarrow H_2.$$

As a result, it is possible to modify the indirect effect, i.e., to sensitize or protect the cell from radiation effects by affecting these reactive species. Consider the dynamic equilibrium of an organic molecule ionized to form two free radicals,

$$RH \xrightarrow{\text{Radiation}} R^\bullet + H^\bullet.$$

Because of the proximity of $R^\bullet$ and $H^\bullet$ following the ionization event, there is a high probability that they will recombine immediately and, hence, cannot proceed to cause an indirect effect upon the DNA target. This interaction between $R^\bullet$ and $H^\bullet$ changes dramatically in the presence of oxygen. As oxygen is a free radical scavenger, it is possible in an oxygenated environment for the radicals to interact with the oxygen,

$$R^\bullet + H^\bullet + 2O_2 \rightarrow RO_2^\bullet + HO_2^\bullet.$$

Oxygen can thus "fix" the result of two free radicals which are then subsequently free to damage the DNA. Hence, hypoxia, which commonly occurs in tumors receiving insufficient vascularization, leads to radioresistance.

Direct Effect

Whereas the indirect effect of ionizing radiation upon DNA is through the intermediaries of radicals as a result of water radiolysis, it is also possible for DNA damage to be produced through the direct ionization of the DNA molecule by radiation.

Relative Contributions

The relative contributions of indirect and direct effects upon DNA damage are of practical importance in consideration of the above discussion regarding the use of radiation protectors or sensitizers to modify the indirect effect. It is believed that about two-thirds of the radiation-induced damage to the DNA is due to the indirect effect (Nais 1998).

10.2.3.3 Radiation-Induced DNA Lesions

Introduction

There are three categories of ionizing radiation-induced chemical changes to DNA that result in damage to the structure of the DNA.

Base Alterations

These are effects inflicted upon the purine and pyrimidine bases by ionizing radiation. Many of these insults are benign with no apparent effect postirradiation. Others result in miscoding during DNA replication leading to a mutation.

Single-Strand Breaks

A single-strand break (SSB) is the removal of one of a pair of bases through damage. Repair is possible during the DNA synthesis phase of the cell cycle as the remaining undamaged base will provide a complementary template for a base on opposing strands to form. Hence, cell lethality is not necessarily a consequence of the damage. Mutation, however, is possible if misrepair occurs or if repair is incomplete. SSBs on both strands can also be repairable if they are sufficiently separated, as shown in Fig. 10.4, as they may be considered independent breaks as such.

Double-Strand Breaks

A double-strand break (DSB) is one in which both DNA strands are broken at the same point or very close together, as shown in Fig. 10.4. DSBs are likely to be accompanied by extensive base damage and following such breaks, the chromatin splits into two segments. The production, repair, and misrepair of DSBs are of great importance to understanding radiation-induced cell lethality. The damaged chromosome may restitute (i.e., the damaged ends may reconnect) resulting in the repair of the physical integrity of the DNA chain but not reproducing the original nucleotide sequence. Binary misrepair of DSBs in two adjacent chromosomes can result in the illegitimate interaction of the two damaged chromosomes culminating in cell death. Of particular importance to the repair of cells damaged by low-LET radiation or low absorbed dose rates is that a DSB can result from two independent chromosomal lesions that occurred at different times but in close physical proximity. If the temporal separation between these two ionizations is sufficiently great, the cell is provided with the opportunity to repair the first lesion before the second, and potentially combinatorial fatal, lesion occurs.

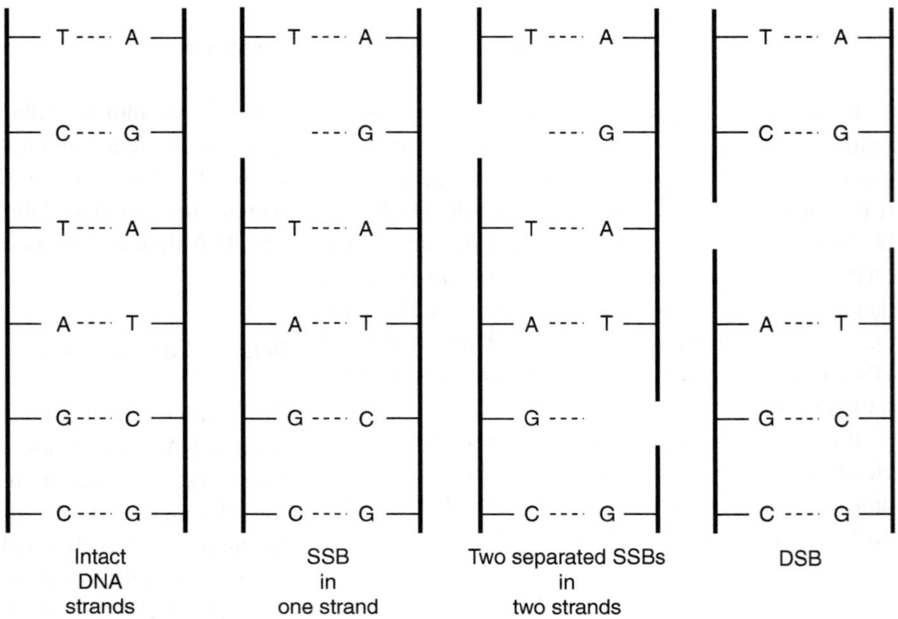

Fig. 10.4 SSB and DSB. An SSB or a pair of widely separated SSBs can be repaired using the undamaged remaining base as a template. A DSB leads the chromatin to separate into two

Summary

The probabilities of base alterations and SSBs occurring per unit dose are, as expected, greater than that of DSBs. Table 10.1 summarizes the incidences of damage per unit absorbed dose (for low-LET radiation) and the lethal consequences as the probability of cell death per lesion induction.

10.2.3.4 Chromosome and Chromatid Aberrations

Lethal

Three types of lethal chromosomal aberrations that can be induced by radiation through the interactions of a minimum of two strand breaks are considered. Further information on chromosomal aberrations can be found in Savage (1983) and in Hall and Giaccia (2006).

Table 10.1 Types of ionizing radiation-induced DNA lesions (low-LET radiation)

Lesion type	Incidence per unit absorbed dose (Gy^{-1})	Relative lethality (%)
Alteration of nucleotides	10^3	1
SSB	10^3	1
DSB	40	95

Dicentric

This type of aberration is the result of the replication of two chromosomes which were damaged in interphase with single breaks and which subsequently interacted, as shown in Fig. 10.5. Following replication during the S phase, the result is a chromosome with two centromeres and two fragments without centromeres. The latter discontinue at the subsequent mitosis as a centromere is required to move to a pole during anaphase in the M-phase of the cell cycle.

Centric Ring

A centric "ring" chromosome aberration can be the result of a DSB of a single chromosome which can recombine to form the cyclic structure shown in Fig. 10.6. The result following the S phase is a pair of overlapping ring chromosomes and a pair of acentric chromosome fragments which, as noted before, will be lost at mitosis.

Anaphase Bridge or Interarm Aberration

Whereas dicentric and ring aberrations are associated with changes to the chromosome (and are the result of presynthesis irradiation), the anaphase bridge is a chromatid aberration, as shown in Fig. 10.7, and is a result of postsynthesis irradiation during the G_2 phase. As separation of the replicated cells is impossible, this nature of aberration is fatal.

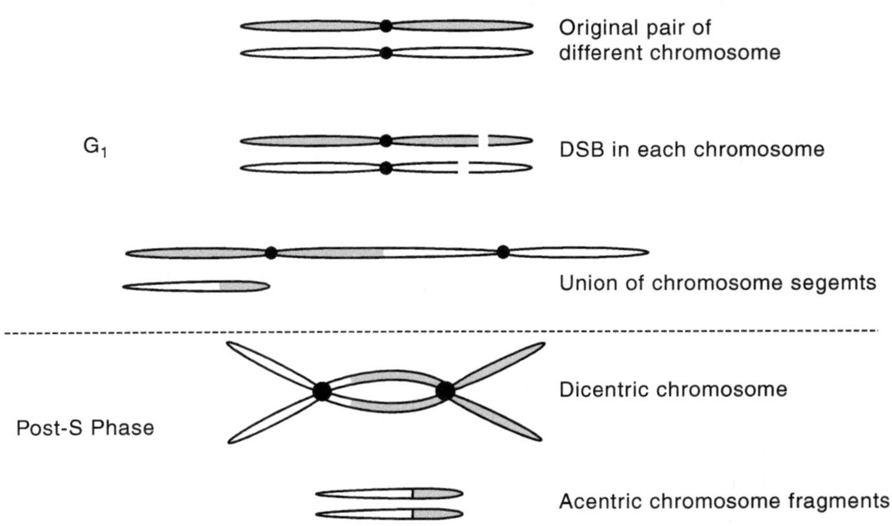

Fig. 10.5 Production of a dicentric chromosome aberration. Prior to replication, two chromosomes are "broken" by irradiation and then intercombine. Following replication, a dicentric chromosome (with two centromeres) is created in addition to two acentric chromosome fragments

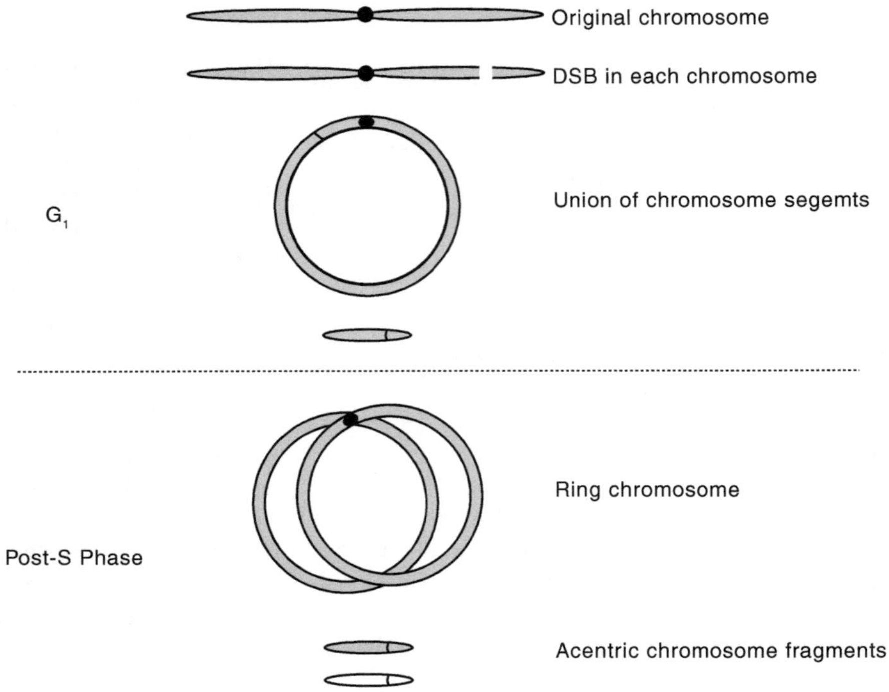

Fig. 10.6 Formation of a ring chromosome aberration. A DSB in a single chromosome is followed by a reconnection to form a ring chromosome and an acentric segment. Following replication, a pair of overlapping ring chromosomes is produced and a pair of acentric fragments which are lost at the subsequent mitosis

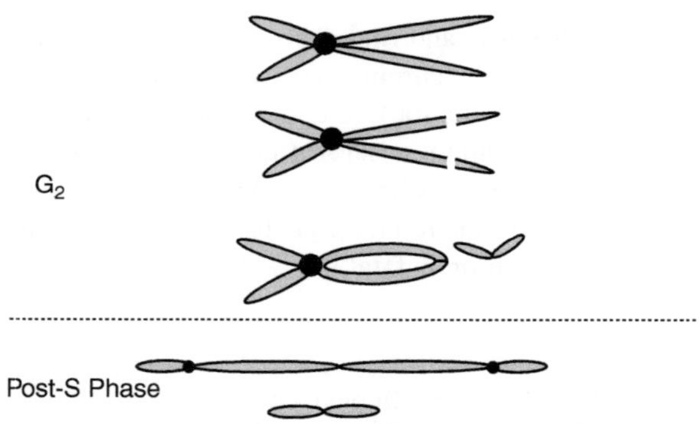

Fig. 10.7 Formation of an anaphase bridge. A chromosome in the G_2 phase is irradiated and a break occurs in both chromatids which recombine at the ends, forming an acentric chromatid fragment. Following anaphase, the centromeres will be attracted to each pole, thus stretching the chromatid between poles. The acentric chromatid segment will be lost

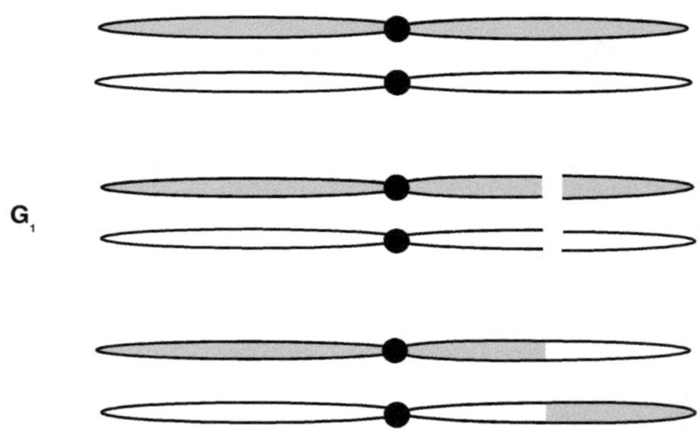

Fig. 10.8 Symmetric translocation. Following irradiation of two chromosomes in the G_1 phase, breaks are produced. The chromosome segments are exchanged between the chromosomes

Nonlethal

Symmetric Translocation

This is a not necessarily lethal chromosomal aberration resulting from the exchange of chromosome fragments formed during irradiation during the G_1 phase, as shown in Fig. 10.8. Such translocations can, however, activate an oncogene leading to a malignancy.

10.2.4 Radiation-Induced Cell Death

10.2.4.1 Introduction

The death of an irradiated cell is only one consequence of its exposure to ionizing radiation. In fact, the term "cell death" requires clarification and expansion. The most common mode of radiation-induced mammalian cell death is that in which the cell fails to undergo further mitosis and replicate, although it may remain metabolically active. Death can also occur during interphase due to apoptosis, frequently referred to as "programmed cell death" or "cell suicide." Radiation-induced cell death can be a consequence of the peculiar "bystander effect." In this effect, an unirradiated cell in proximity to an irradiated cell dies, presumably through the effects of some toxic agent released by the irradiated cell.

10.2.4.2 Mitotic Death

This is the most frequent means of radiation-induced cell death and is the result the of lethal chromosome asymmetric-type exchange aberrations previously discussed. Cell death results when the irradiated cell attempts mitosis; this can occur not only at the first attempt at mitosis postirradiation, but also during attempts at subsequent cell divisions. Mitotic death has been quantitatively demonstrated by Cornforth and Bedford (1987). They demonstrated that the logarithm of the surviving fraction (SF) of a population of irradiated cells was exactly equal to the mean number of lethal chromosomal aberrations per cell.

While mitotic processes can be halted as consequences of irradiation, it is usually possible for cell metabolism to continue. As a result, this procession of metabolism in the undivided cell leads to an exponential growth in cell size which eventually reaches a plateau. Even though such a "giant cell" will grow, it is considered to be "dead" as it has lost its proliferative capacity.

10.2.4.3 Interphase Death and Apoptosis

Cell death can also occur from irradiation during interphase, although this requires much higher absorbed doses than those which induced mitotic death. The mean absorbed dose required for mitotic death is of the order of 1–2 Gy in a single exposure. Should a population of cells be irradiated to an absorbed dose of up to 10^3 Gy in a single exposure, cellular metabolism and function ceases and necrotic death occurs. Such an extreme ionizing radiation insult is largely irrelevant to our considerations of diagnostic and therapeutic nuclear medicine. However, interphase death can occur at lower absorbed doses. Apoptosis is a naturally-occurring means of cell death in both tumor and normal tissue. It is also present in the developing embryo where obsolete tissues no longer required in its development are eradicated. It can be triggered by ionizing radiation in specific cell types, in particular, lymphatic and hematopoietic cells, through initiation by the p53 tumor-suppressor gene. Following the induction of radiation-induced DNA damage, the amount of p53 accumulates and leads to a delay in the progress to mitosis so as to allow time for the cell to repair this damage. If repair is not possible or is unsuccessful, the p53 gene can then initiate apoptosis in order to remove the nonfunctional cell from its environment. However, should the p53 gene be inactivated through a mutation, apoptosis cannot occur and cell immortality becomes possible.

A cell entering apoptosis begins this process by isolating itself from its neighboring cells. Chromatin condenses, the cellular nucleus fragments and the cytoplasm dehydrates with the result that the cellular volume shrinks. Phagocytosis follows to eradicate the fragmented apoptotic bodies.

10.2.4.4 Bystander Effect

This is an intriguing phenomenon in which a biological response is detected in cells that have not been irradiated but which are in close physical proximity to one that has (Sgouros et al. 2007). While evidence for the effect was first seen in the 1940s and 1950s (Hall 2003; Mothersill and Seymour 2001), current interest in was ignited by the work of Nagasawa and Little (1992) showing an elevated frequency of chromosomal damage relative to that expected for absorbed doses as low as 310 μGy following 3.3 MeV α-particle irradiation from a ^{238}Pu source. Thirty percent of the cells studied demonstrated an increase in sister chromatid exchanges, even though only 1% of the cells had actually been traversed by an α particle (see, also, Kahim et al. 1992). Since then, the bystander effect has been demonstrated for X-rays and protons. As unirradiated cells have been killed following the transfer of a medium which contained irradiated cells, it is presumed that the bystander effect is due to cytotoxins released by the irradiated cells.

10.2.4.5 Categories of Radiation-Induced Cell Damage

Introduction

The consequences following the irradiation of a cell are manifold and require specific definitions. This is necessary in order to differentiate between the results of the effects of ionizing radiation upon cells and the spectrum of normal cell variants. For example, mitotic cell death has already been described as the irradiated cell's loss of reproductive capacity caused by radiation damage. Yet, neurons do not undergo mitosis following full fetal development. Hence, the first effect of irradiation is the lack of the presentation of any damage whatsoever. While radiation-induced damage to the DNA may have occurred, it is possible that, even without repair, the DNA damage has had no impact upon the cell's ability to reproduce or upon the cell's functional ability.

When damage does occur and which can cause or lead to cell death, further differentiation of the damage is required. Clearly, the definition of lethal damage is immediately obvious. Nonlethal damage results in the retention of reproductive capacity, albeit perhaps with a reduced growth rate. The two categories of sublethal damage and potentially lethal damage are of great practical interest to radiobiology and are looked at in detail here.

Sublethal Damage

This is the category defined as the result of two independent radiation lesions. The first insult can undergo repair, but if a second radiation-induced insult is received prior to the first damage having been completely repaired, cell death results. For example, consider a DSB created in a single chromosome following irradiation. It is possible for the cell to be able to repair this damage by recombining the chromosomal fragments; the result is not necessarily lethal to the cell. However, if, before this DSB is repaired, another DSB is created in an adjacent chromosome, an illegitimate and likely lethal binary misrepair can result such as, for example, a dicentric. Thus, the first damage to have been incurred is referred to as being sublethal. The presence of sublethal damage can be demonstrated experimentally by irradiating cells in a split-dose regimen, as shown conceptually in Fig. 10.9. At low time separations between absorbed dose administrations, the probability of survival is reduced and, as the time separation between exposures increases, the SF increases to finally reach a plateau reflecting the complete repair of sublethal damage.

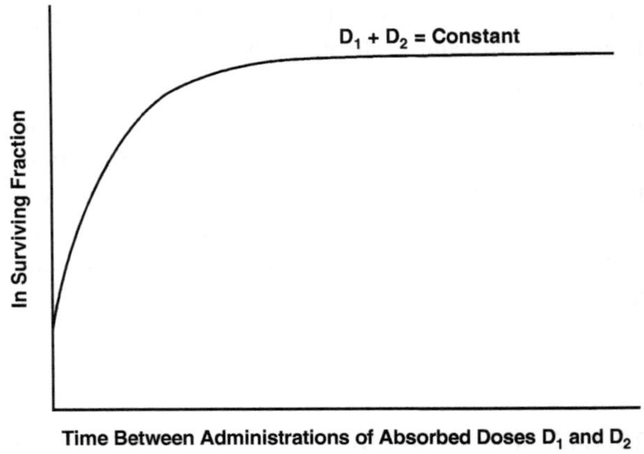

Fig. 10.9 The natural logarithm of the SF of cells irradiated to a constant absorbed dose separated into two separate administrations as a function of the time difference between the administrations of the two absorbed doses

Potentially Lethal Damage

This category of radiation-induced damage is somewhat more complicated than sublethal damage to define. Potentially lethal damage is that which would result in cell death was it not for some postirradiation modification of the cellular environment. In particular, if the cell cycle should be delayed prior to mitosis due to, for example, suboptimal growth conditions, the cell is provided with the opportunity to repair DNA damage. This mode of repair can be important for tumor cells which are quiescent due to a reduced nutritional support from neovasculature leading to an inherent radiation resistance.

10.2.5 Germ-Cell Damage

10.2.5.1 Introduction

Radiation-caused damage to a germ cell can be manifested in two ways. The first is through mutations which can be conferred to the irradiated individual's progeny following fertilization via an inherited genetic defect. These mutation-regulated defects caused by ionizing radiation are no different than those that appear naturally: only the frequency of mutations increases with exposure to ionizing radiation. As a result, it is only possible to detect radiation-induced hereditary effects by comparing the incidences of these effects in the progeny of an irradiated population against those in the progeny of a matched unirradiated population. Because of the low gonadal absorbed doses associated with medical imaging procedures, including diagnostic nuclear medicine, such hereditary effects have not been detected.

The second means of expressing radiation damage to the germ cell is that which leads to germ-cell lethality resulting in either permanent or temporary sterility. Temporary sterility in the male is a theoretically possible sequela of therapeutic nuclear medicine due to the high activity of radionuclide administered and should there be a sufficiently high testicular uptake of the radionuclide.

In this subsection, the latter effect only is considered. Hereditary effects attributable to ionizing radiation are discussed later in this chapter.

10.2.5.2 Oogenesis

In the adult human female, there are approximately 10^5 oocytes. These cells are nonproliferative and, consequently, permanent sterility through radiation-induced ovarian failure can be achieved through the deaths of all oocytes. The magnitude of the absorbed dose required to induce this is highly dependent upon age because, as the female ages, the number of oocytes decreases due to inherent degeneration and, to a lesser degree, ovulation. As a result, the absorbed dose required to cause permanent sterility will be lower in older women than in younger. For example, a typical sterilization absorbed dose to the ovaries in a prepubertal female is about 12 Gy whereas that in a premenopausal female is only about 2 Gy. It should also be noted that hormonal effects typical of the natural menopause accompany radiation-induced sterility.

10.2.5.3 Spermatogenesis

The adult human male gonadal kinetics is markedly different from those of the female, a result of which is the substantially different radiation response between male and female germ cells. Unlike oogenesis, production of sperm cells is a continuous process and spermatazoa are the end-product of several stages in which the spermatogonia (stem cells) lead through sequential differentiation to spermatocytes, spermatids and, finally, spermatozoa. This production process takes about 10 weeks in the human. As with all rapidly dividing cells, spermatogonia are more sensitive to radiation than the further-developed germ cells. This results in the effect of irradiation upon male reproductive capacity not necessarily being immediately evident. Following exposure to irradiation, the male may be only temporarily infertile as the mature sperm cells can remain unaffected. As these are depleted, azoospermia results and temporary sterility exists until the spermatogonia are able to repopulate. Azoospermia can occur for absorbed doses in excess of about 0.5 Gy and its duration is dependent upon the magnitude of the absorbed dose, ranging from about 1 year for absorbed doses of less than 1 Gy to about 3 years for absorbed doses exceeding 2 Gy. Permanent male sterility results for absorbed doses exceeding about 5 Gy given in a single exposure.

10.2.6 In Vitro Cell Survival Curves

10.2.6.1 Introduction

At the level of absorbed doses typical of medical imaging exposures where mitotic death is dominant, cell death is manifested by the lack of reproductive ability. Hence, the deaths of individual cells or colonies of cells due to ionizing radiation can be readily assessed through in vitro assays. The measurement of the fraction of cells that survive following in vitro exposure to ionizing radiation is fundamental to understanding cellular radiobiology and the environment that can affect the radiosensitivity of the cell.

In vitro measurement of a cell's response to radiation requires the excision of the tumor or tissue of interest, the fragmentation of the sample into individual cells which are then seeded into a culture dish with an appropriate growth medium and then incubated. Established cell lines can be formed in such a way and it is possible to extricate a number of cells using trypsin to cause the cells to detach from the dish. The cell number density (number of cells per unit volume of medium) can then be measured using, for example, a hemocytometer. As a result, a given number of cells can then be seeded in a growth medium, incubated and then viewed, following staining, after a period of about 10 days. Each individual cell has the potential to be clonogenic, i.e., to grow to form a colony which, when stained, is readily visible to the naked eye. The efficiency with which seeded cells eventually form colonies is defined as the plating efficiency,

$$\varepsilon_P = \frac{\text{Number of colonies}}{\text{Number of seeded cells}}. \tag{10.1}$$

ε_P is measured using an unirradiated culture and it is assumed that the plating efficiency is constant across the seeded cultures and is independent of radiation absorbed dose. In a radiobiology experiment, a number of cell cultures are formed and exposed to ionizing radiation. As each colony is the product of a single cell, the SF, accounting for the plating efficiency, represents the fraction of original cells that remain viable following irradiation,

$$SF = \frac{\text{Number of colonies}}{\text{Number of seeded cells} \times \varepsilon_P}. \tag{10.2}$$

The practical evaluations of (10.1) and (10.2) are shown in Fig. 10.10.

It is important to recognize that the SF is not only a function of the singular absorbed dose, but also of a wide variety of radiation and environmental factors which we shall explore shortly. Hence, the bridging from in vitro experimentation to the prediction of in vivo response must account for such factors which can evolve in the in vivo environment. In the following subsections, a variety of mathematical models of the probability of cell lethality following irradiation are developed.

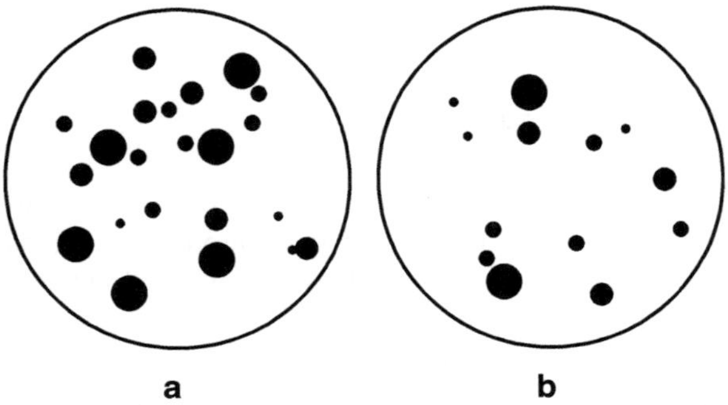

Fig. 10.10 Definitions of plating efficiency and survival fraction determined in vitro. (**a**) Control Petri dish not exposed to radiation; 25 single cells were seeded into it and 22 colonies have been produced resulting in a plating efficiency of 22/25 = 88%. (**b**) Petri dish in which 25 cells were also seeded followed by irradiation to yield 13 colonies; the resulting survival fraction is 13/(25 × 0.88) = 0.591

10.2.6.2 Single-Target Model

The first model of cell killing by ionizing radiation is the simplest in which a single deactivation (or *hit*) of some target within the cell is sufficient to result in cell death. To derive such a model, recall the Poisson discrete probability distribution function,

$$p(x;\mu) = \frac{\mu^x e^{-\mu}}{x!} \tag{10.3}$$

which gives the probability of x events occurring during a measurement interval for which the mean number of occurrences is μ. The probability of no events occurring (i.e., $x = 0$) during that interval is equal to $e^{-\mu}$. Hence, as the probability of no target deactivation must be equivalent to the probability of survival in this model, one can write the SF of cells following exposure to an absorbed dose D as,

$$SF(D) = e^{-(D/D_0)} \tag{10.4}$$

where $1/D_0$ is a constant of proportionality. It follows that D_0 is the absorbed dose at which the SF is equal to $e^{-1} \approx 0.37$. An example of a cell survival curve calculated for the single-target model with $D_0 = 0.25$ Gy is shown in Fig. 10.11. On the semilogarithmic plot

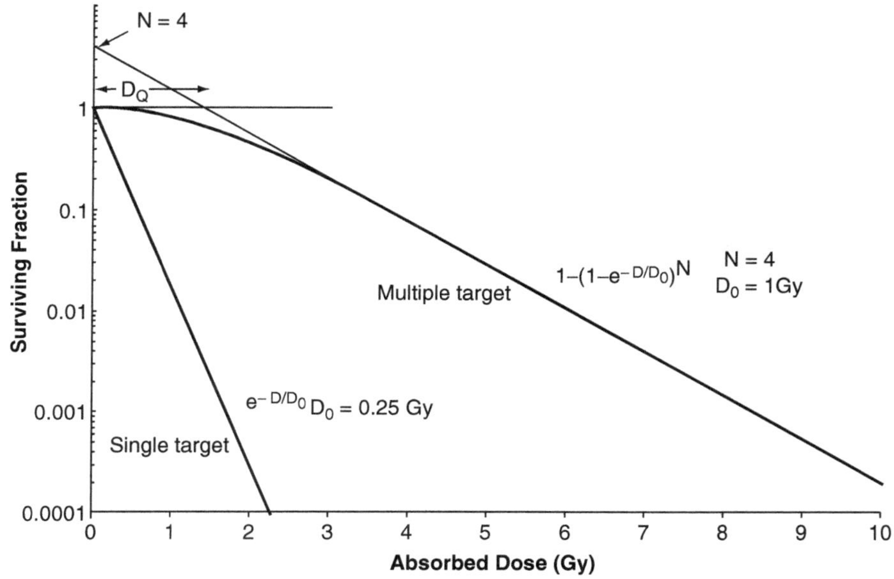

Fig. 10.11 Examples of cell survival curves calculated for the single-target model and a multiple-target model

(10.4), is a straight line reflecting a continuous killing of cells with absorbed dose. As to be shown, such a response reflects the lack of any cellular repair of radiation-induced damage. Such a survival curve is representative of the dose response to high-LET radiations (e.g., α particles) or high absorbed dose rates.

10.2.6.3 Multiple-Target Models

There is no immediate reason to assume that, under all conditions, a radiation-induced hit is sufficient to result in cell death. Table 10.1 would suggest that multiple lesions are required for cell death. Thus, consider the potential of cell death to be due to the inactivation of more than a single target. Equation (10.4) gives the probability that a single target is not hit; hence, the probability that the target is hit is equal to $1 - e^{-D/D_0}$. Assuming that N targets are required to be deactivated in order to cause cell death and that these inactivations are statistically independent, then the probability that cell death results from the N targets being hit is simply $\left(1 - e^{-D/D_0}\right)^N$. The probability of N targets not being deactivated is equal to the SF,

$$SF(D) = 1 - \left(1 - e^{-D/D_0}\right)^N \qquad (10.5)$$

It is clear that (10.5) reduces to the single-hit form of (10.4) for $N = 1$, as required. Figure 10.11 displays a plot of a hypothetical multiple-hit survival curve following (10.5) for an example of four inactivation targets ($N = 4$) and $D_0 = 4$ Gy. The curve has a shoulder at low absorbed doses and an initial slope of zero but which becomes exponential with increasing absorbed dose. At such high absorbed doses, the SF can be written as (using a binomial expansion),

$$SF(D) \approx Ne^{-D/D_0} \qquad (10.6)$$

As a result, the extrapolation of the curve at high absorbed doses back to $D = 0$ will give the number of assumed targets, N, required for cell inactivation. This is shown in Fig. 10.11. A measure of the width of the shoulder is provided by the so-called quasithreshold absorbed dose, D_Q, which is defined as the absorbed dose at which the extrapolation of (10.6) is equal to 1,

$$D_Q = D_0 \ln N \qquad (10.7)$$

This cannot be considered to be a real threshold absorbed dose as radiation damage will still occur at absorbed doses below D_Q. However, D_Q does provide a measure of the survival curve's shoulder at low absorbed doses that is observed in the experimental measurements of many types of mammalian cells.

10.2.6.4 Modified Multiple-Target Model

Measured mammalian cell survival curves indicate that there is in fact a nonzero slope at absorbed doses approaching zero. Contrary to this observation is that the slope of (10.5), expanding the exponentials to first order, is

$$\frac{dSF}{dD} = -\frac{N}{D_0}\left(\frac{D}{D_0}\right)^{N-1}\left(1-\frac{D}{D_0}\right) \quad (10.8)$$

which is equal to zero for $D = 0$. An improved fit to measured cell survival curves is obtained by incorporating the single-hit model,

$$SF(D) = e^{-D/D_0}\left(1-\left(1-e^{-D/D_1}\right)^N\right). \quad (10.9)$$

At low absorbed doses, the slope of this modified multiple-target is, again using the first-order expansions of the exponentials,

$$\frac{dSF}{dD} = -\frac{1}{D_0} \quad (10.10)$$

An example of a cell survival curve of this form is shown in Fig. 10.12 in comparison with the multiple-target model survival curve.

It will be noted that all three dose–response models presented so far asymptotically approach an exponential function of dose with a negative slope in agreement with radiobiological data.

10.2.6.5 Linear-Quadratic Model

Another approach to modeling the cellular response to irradiation is to recall from the discussion of chromosomal aberrations that radiation-induced mutations or cell lethality can result from a DSB produced by a single "hit," where a single ionization event leads to a lesion, or by two independent "hits," where two separate ionization events at different times but occurring in close physical proximity to each other interact to form a DSB. This latter means of lesion induction requires a second-order dose-dependent component to the probability of cell survival, or,

$$SF(D) = e^{-(\alpha D+\beta D^2)} \quad (10.11)$$

where α and β are constants. This is the linear-quadratic (LQ) model (Fowler and Stern 1960). An example of a hypothetical LQ cell survival model is shown in Fig. 10.13, along with the multiple-target example survival curve.

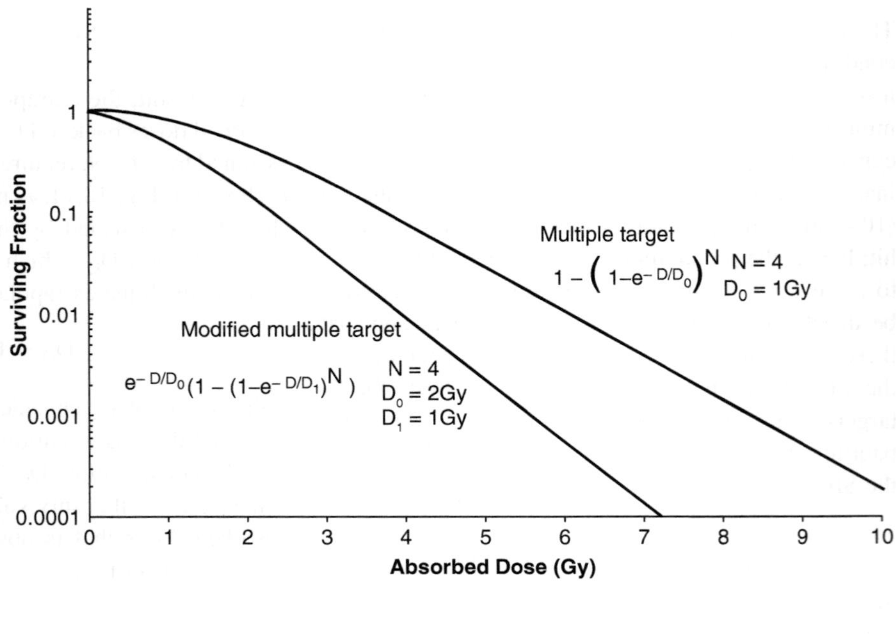

Fig. 10.12 Examples of multiple target and modified multiple-target survival curves. Note that the modified multiple-hit survival curve has a nonzero slope at $D = 0$

A significant problem that occurs with (10.11) at high absorbed doses is that the LQ model never approaches the pure e^{-D} dependence which is observed experimentally. This failure will tend to restrict the utility of the model to low absorbed doses where it does present an advantage over other models in that it requires only two parameters to enable a fit to measured radiobiological data. The ratio of the α and β parameters is equal to the absorbed dose at which the contributions of the linear and quadratic components are equal,

$$D_{Eq} = \frac{\alpha}{\beta}. \tag{10.12}$$

The coefficients of the LQ model can be determined from measured cell survival curves through (10.11),

$$-\frac{\ln SF}{D} = \alpha + \beta D. \tag{10.13}$$

A plot of D vs. $\ln SF/D$ will thus yield the coefficients of the LQ model where α is the intercept and β is the slope.

As the quadratic component represents cell lethality due to two DSB-inducing ionizations separated in time, one must allow for the possibility of the repair of the first sublethal DSB during the time before the second DSB occurs. This process will increase the SF over time. Should this second DSB occur before the repair of the first, the result is cell death. The inclusion of this sublethal damage repair is enabled by writing the LQ model of the survival fraction of (10.11) as,

$$SF(D) = e^{-\left(\alpha D + \beta G D^2\right)}. \tag{10.14}$$

G is the Lea–Catcheside factor where $G \leq 1$ and accounts for the reduction in cell lethality due to the repair of sublethal damage. This factor will obviously be a function of the difference in times between when the first radiation insult occurs and that of the second.

The LQ model has been extensively used in applications of clinical radiobiology to external beam radiotherapy, brachytherapy, and radionuclide therapy.

10.2.7 Radiation Sensitivity of Mammalian Cells

10.2.7.1 Introduction

The response of a given cell to radiation is a function of many biological, physical, and environmental variables such as cell cycle, the rate at which the radiation

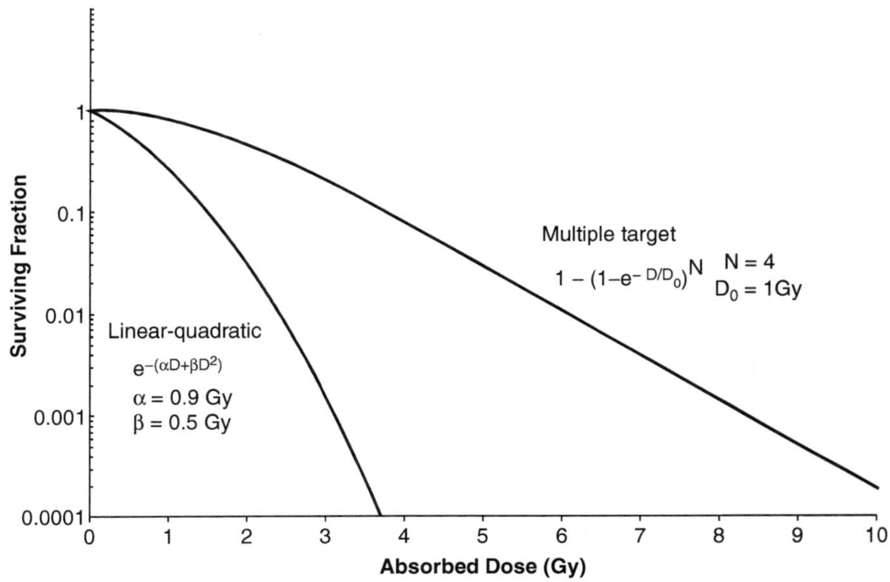

Fig. 10.13 Hypothetical cell survival curves for the multiple-target and linear-quadratic models. Note that the slope of the linear-quadratic curve in this semilogarithmic plot never approaches a straight line (i.e., it is not pure exponential), in contradiction to measured cell survival curves

absorbed dose is delivered and the ambient oxygen concentration. The last two factors are of particular importance in radionuclide therapy.

10.2.7.2 Cell Cycle and Age

The cell cycle, as defined by the period of time between subsequent mitoses, was introduced in Sect. 10.2.2.5. The sensitivity of the cell to radiation varies throughout the course of the cycle. The length of the G_1 phase, which precedes the DNA synthesis S phase, is indicative of the variation of the radiosensitivity of the cell throughout its cycle. Figure 10.14 shows two conceptual mammalian cell proliferative capacities and how they vary with the time of the cell cycle at which they receive the same absorbed dose. The cells are the most radiosensitive during the mitosis phase. For cells with long G_1 phases, the cell is relatively radioresistant during much of the course of G_1 but this sensitivity decreases as the G_1 phase enters the synthesis S phase; the radioresistance again increases to a maximum during the S phase and then decreases through the G_2 and M phases. On the other hand, for cells with short G_1 durations, radioresistance increases through the M and G_1 phases to reach a single maximum during the S phase to again decrease during the G_2 and M phases.

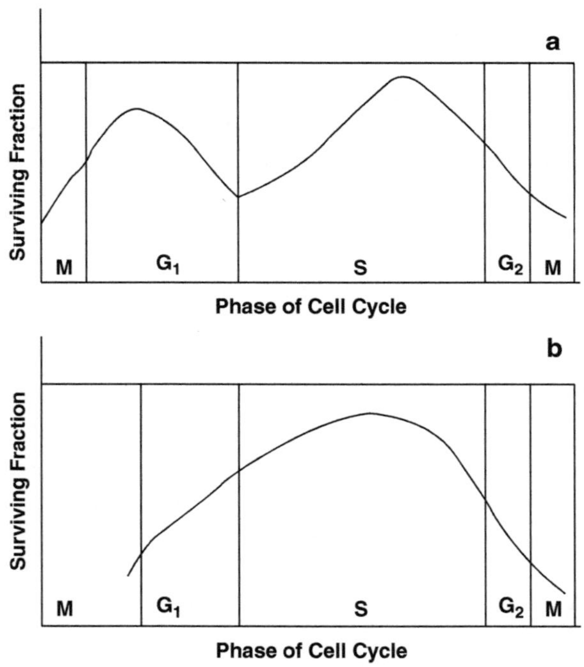

Fig. 10.14 Conceptual plots of mammalian cell survival, for two categories of (**a**) long and (**b**) short G_1 phases, for the same absorbed dose as a function of the phase of the cell cycle during which it was irradiated

This variability of radiosensitivity throughout the cell cycle enables the process of synchronization. Proliferative cells in situ or in culture will exist in different stages of the cell cycle without any synchronicity between them. Following irradiation, those cells which are in the radiosensitive phases will be preferentially killed whereas those in the radioresistant S phase will survive. Hence the culture of cells will, following irradiation, tend to be at the same phase of the cell cycle.

10.2.7.3 Relative Biological Effectiveness

The relative biological effectiveness (RBE) is a measure of the difference between the incidence or magnitude of biological effects incurred by exposure to a given radiation type or set of irradiation conditions relative to a low-LET reference radiation (commonly, the secondary electrons associated with 250 kVp X-rays) for a given biological endpoint, which is typically the survival fraction. This definition is shown in Fig. 10.15. An important point to note is that the RBE is a function of the endpoint selected, as can be seen by comparing the RBE values defined by 0.1 and 10% cell survivals.

10.2.7.4 Linear Energy Transfer

As described in Chap. 7, the linear energy transfer (LET) is a measure at which energy is deposited in a medium per unit length of travel by a charged particle. Increasing values of LET correspond to increasing ionization densities. The result is that, increasing LET (e.g., increasing from those of Compton electrons and photoelectrons through Auger/Coster–Kronig electrons to α particles), the probability of killing the irradiated cell will grow. That is, in terms of the cell survival curve, increasing LET will result in a reduction in the curve's shoulder and an increase in its slope, as shown in Fig. 10.16. These are the results

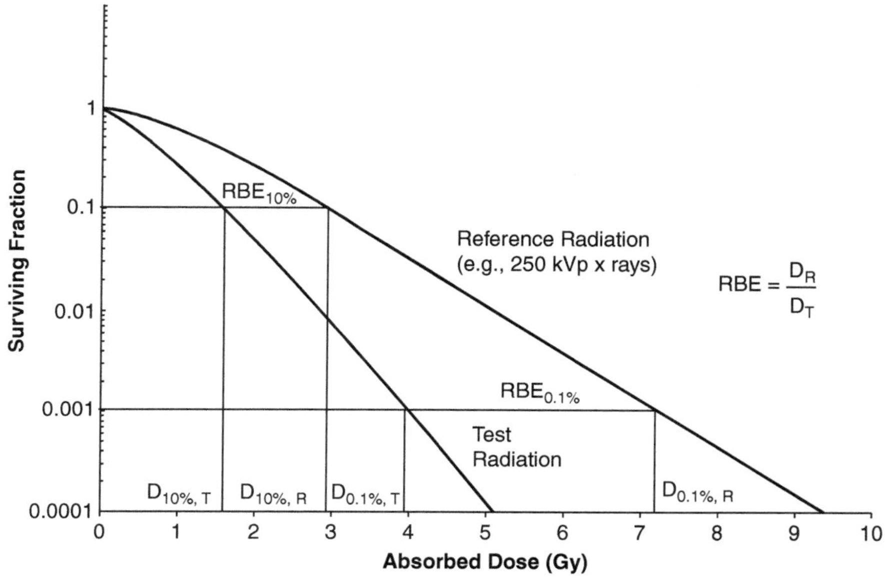

Fig. 10.15 Definition of the relative biological efficiency (RBE) as the relative reduction in absorbed dose from a specified low-LET reference radiation type for a test radiation/irradiation conditions to yield a given endpoint (here, cell SF). Note that the value of the RBE is dependent upon the endpoint selected (here, $RBE_{0.1\%} > RBE_{10\%}$)

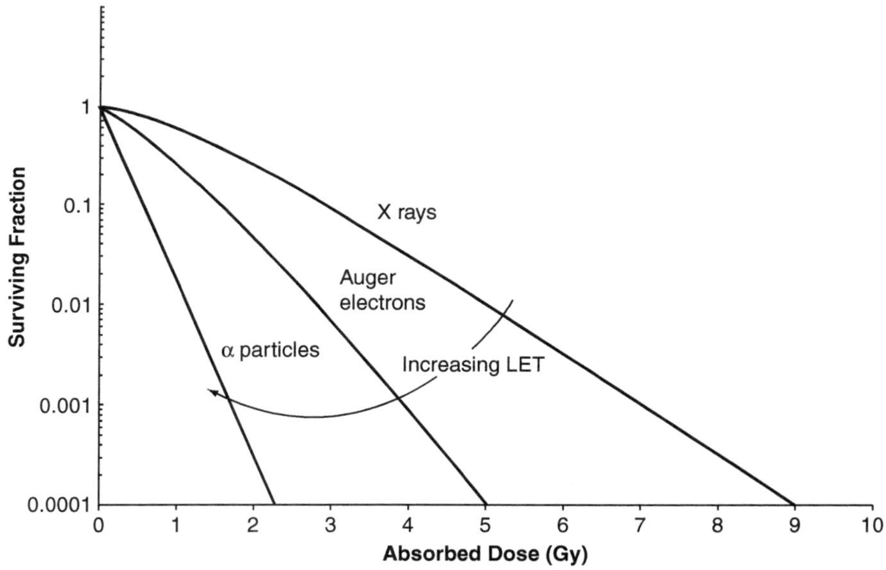

Fig. 10.16 Conceptual set of cell survival curves for radiations of varying LET values. Note that with increasing LET, the slope of the curve increases (becomes more negative) and the shoulder decreases

of a high incidence of cell killing and a reduced opportunity provided for sublethal damage repair.

10.2.7.5 Absorbed Dose Rate

Of importance to radionuclide therapy is the variation in biological response with absorbed dose rate. For a given fixed absorbed dose, the functional survival will decrease with an increase in the rate at which the dose is delivered. This is shown conceptually in Fig. 10.17. As a result of sublethal damage repair, the cell has a greater opportunity to restitute a DSB at low absorbed dose rates and the cell survival curve subsequently demonstrates a shoulder at low absorbed doses and a reduced slope. As the absorbed dose rate increases thus increasing the rate at which DSBs are produced, the ability of the cell to "keep up"

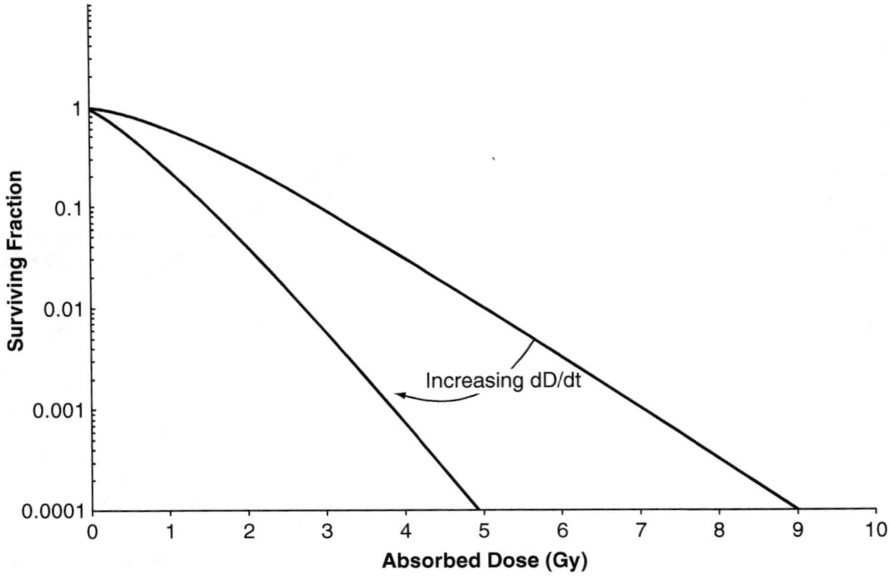

Fig. 10.17 Conceptual cell survival curves for identical low-LET radiation type but at different absorbed dose rates, dD/dt. For a fixed total administered absorbed dose, the fractional survival decreases with increasing absorbed dose rate due to the impairment of cellular repair of DNA damage

and repair DSBs as they are produced is diminished. The SF decreases as a result and the shoulder of the cell survival curve shortens and the slope increases.

Similar to the concept of varying the rate of a continuous exposure of ionizing radiation s the application of radiation administered in separated integral exposures, known as fractions. This permits irradiated normal tissue to repair between fractions. Figure 10.18 shows hypothetical dose–response curves for fractionated therapy with a time interval between each fraction sufficient to allow the repair of sublethal damage. The dashed line is the net dose response resulting from single exposures of 1 Gy each.

10.2.7.6 Hypoxia

In the discussion of indirect radiation-induced damage of the DNA, it was noted that the presence of oxygen can increase the radiosensitivity of the mammalian cell by impeding the immediate recombination of ion pairs following an ionization, thus "fixing" the presence of the free radicals produced by the ionization. Hypoxic conditions thus decrease the radiosensitivity of the cell, a feature which is of great practical concern in radiotherapy as a tumor frequently contains hypoxic regions.

A measure of this effect of oxygen concentration upon radiosensitivity is the *oxygen-enhancement ratio* (OER) defined as the ratio of the absorbed dose required to achieve a specified biological endpoint (e. g., survival fraction) under hypoxic conditions to that achieving the same endpoint under aerated conditions, as shown in Fig. 10.19.

10.2.7.7 RBE and OER as Functions of LET

Having now seen how fractional cell survival varies with OER and LET, it is of interest to see how both the RBE and OER vary with LET. Figure 10.20 shows hypothetical RBE and OER curves as functions of LET (although hypothetical, they demonstrate qualitatively those functions shown in Barendsen et al. (1966) and Barendsen (1968)).

The OER has a value of about 2.5–3 for low-LET radiations, but drops off precipitously for LET values in excess of about 30 KeV/μm to eventually "plateau" to reach unity at about 200 KeV/μm. On the other hand, the RBE has limited energy dependence, staying equal to unity with LET values less than about 10 KeV/μm, to reach a sharp maximum at about 100 KeV/μm before decreasing. At low values of LET, the number of ionizations is decreased and the presence of oxygen can fix the produced free

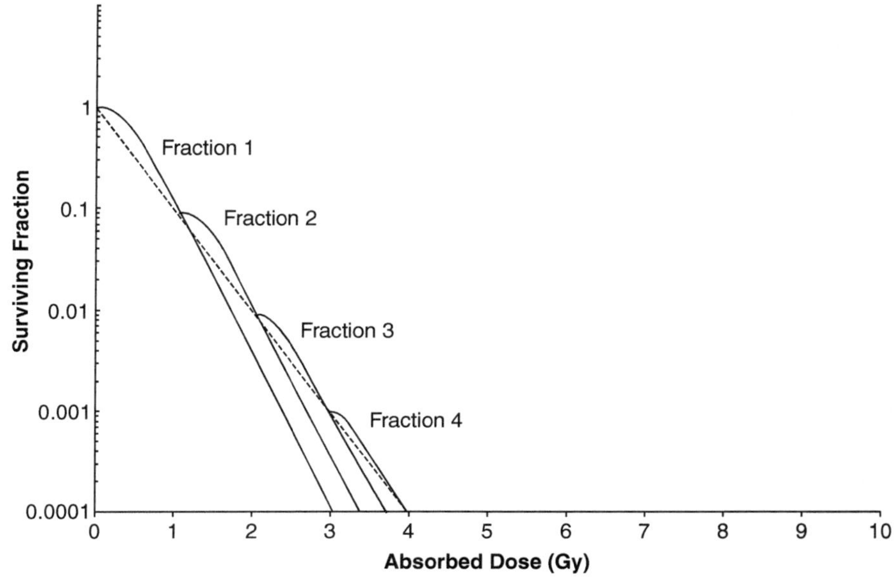

Fig. 10.18 Hypothetical dose–response curves for fractionated exposure. The cell survival is 10% for a single exposure of 1 Gy. The exposure is split into four fractions, each providing 1 Gy. The *solid lines* indicate the SFs resulting from an absorbed dose administered at a single time. The *dashed line* is the dose response obtained if the absorbed dose is administered in fractions of 1 Gy each. It can be seen that the SF for a given total absorbed dose is greater if the absorbed dose is given as smaller, fractionated exposures (e.g., 0.01% of the cells survive a 3 Gy absorbed dose given in a single exposure whereas 0.1% survive the same absorbed dose if given in three fractions)

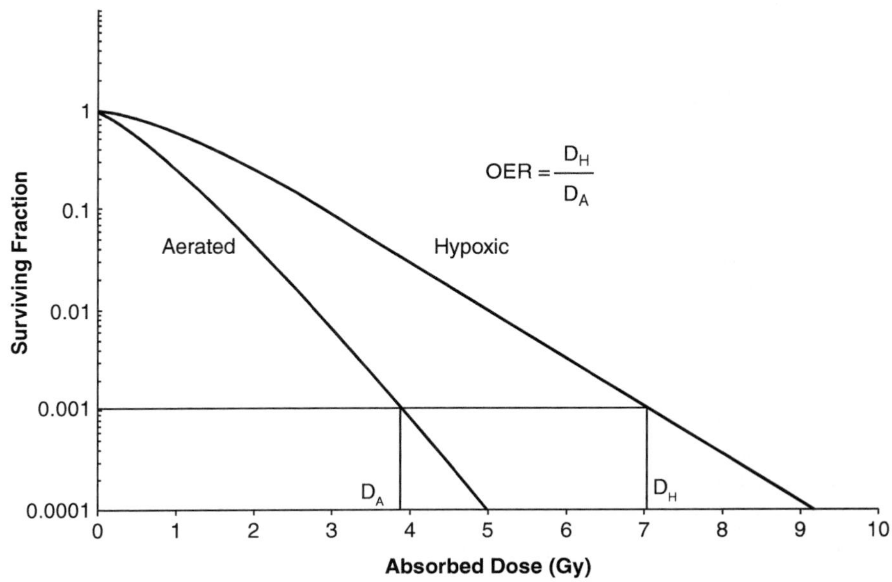

Fig. 10.19 Effect of hypoxia upon cell SF and the definition of the oxygen-enhancement ratio (OER) for two hypothetical cell survival curves; the endpoint used is arbitrary

radicals and, hence, the consequential OER is high. At high LET values, the ionization density within the intranuclear medium is sufficiently high that oxygen repair of indirect damage is insufficient and indirect damage is inhibited. The RBE remains relatively constant at low-LET values due to the large mean physical separation between ionization events. Because of this, the probability that an ionization will damage DNA through either direct or indirect mechanisms is small. At high LET, the separation between ionization events is small and the density of ionizations correspondingly higher so as to elevate the probability of an ionization near the DNA molecule leading to the increase in the damage incurrence

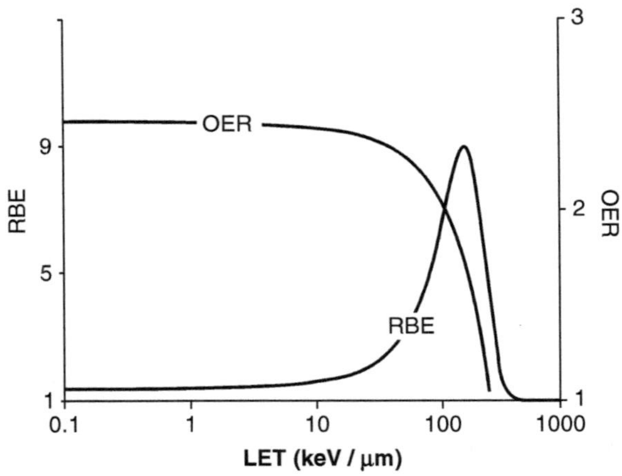

Fig. 10.20 Conceptual curves of OER and RBE as functions of LET. Note different ordinates for the quantities of interest

rate. The peaking of the RBE at about 100 KeV/μm can be explained by noting that the mean separation between energy deposition events at an LET value of 100 KeV/μm is about 2 nm, which, of course, is of the order of the DNA double helix's diameter. Hence, at this LET value, there is a high probability of a single charged particle inducing a DSB in the helix which can lead to cell death. At lower LET values, the mean separation between events is greater and the probability of a DSB occurring is correspondingly reduced. The decrease in RBE with LET increasing beyond 100 KeV/μm reflects "overkill" by densely-ionizing radiation. Multiples of DSBs are produced, but the increase in biological effect is limited whereas the absorbed dose increases with the net result that the RBE decreases. This RBE maximum at an LET value of 100 KeV/μm is of particular interest to therapeutic nuclear medicine of α-emitting radionuclides. From Table 10.2, it can be seen that the typical LET of the α particles emitted from a radionuclide is between 100 and 150 keV/μm and which is coincident with this RBE maximization and the minimization of the OER.

10.2.7.8 Cell Proliferation Kinetics and Radiosensitivity

The variation of cell sensitivity to ionizing radiation due to its proliferative capacity is of significant interest to both the induction of cellular effects and the minimization of radiation-induced sequelae. Tumor cells rapidly proliferate as do the blast cells of some normal tissues and, hence, would be expected to have comparable radiosensitivities.

Table 10.2 Typical values of LET and particle ranges

Particle	Typical LET (keV/μm)	Typical range (μm)
Secondary electrons resulting from Compton and photoelectric interactions with photons of energies typical of nuclear medicine	2–5	5,000
Electrons and positrons emitted in β decay	1	50–2,500
Auger/Coster–Kronig electrons	50	0.01
α particles emitted in α decay	100–150	50–100

It is possible to categorize normal cells in terms of their proliferation capacities:

No mitosis	Neurones
Low mitotic rate/limited cell renewal	Thyroid, liver, and connective tissue
High mitotic rate/ frequent cell renewal	Red bone marrow (erythroblasts), spermatogonia, and intestinal crypt cells

The sensitivity of a given cell to damage induced by ionizing radiation is commensurate with its proliferative capacity. Cells with a high mitotic index, which includes tumors, are particularly sensitive to ionizing radiation; those cells with a low mitotic index tend to be less radiosensitive.

10.2.8 Repair of Radiation-Induced Damage

10.2.8.1 Introduction

It has been estimated that the DNA of a single mammalian cell will experience normally some 10^5 lesions daily as the consequence of a variety of insults including replication errors, attacks by reactive chemical species or ionizing radiation. The fact that the natural mutation rate is so low demonstrates that intrinsic

DNA repair mechanisms exist. These mechanisms are applied at different phases of the cell cycle.

10.2.8.2 Repair of Sublethal Damage

Sublethal damage repair reflects the increase in cell survival if a given amount of absorbed dose is administered in multiple exposures rather than in a single one. Figure 10.21 shows a conceptual description of mammalian cell survival following two exposures (yielding the same total absorbed dose) separated by a time interval. The SF initially grows with increasing time between the two irradiations, reflecting the repair of sublethal damage caused by the first exposure (referred to as the conditioning dose). However, after about 2–4 h, the SF decreases to a minimum and then begins to increase again. This phenomenon is the result of radiation-induced synchronization, described in Sect. 10.2.7.2. As mammalian cells are most radioresistant during the S phase of the cell cycles so of a culture of cells irradiated in situ, the surviving cells will tend to be in the S phase.

After 6 h, the cells will be in the G_2/M phases which are radiosensitive. Irradiation during this time will decrease the SF; this is known as reassortment. At later times, the number of cells increases due to further cell divisions (repopulation).

10.2.8.3 The Four "Rs" of Radiobiology

The above discussions can be summarized by what are known as the four "Rs" of radiobiology and their applicability to fractionated or low-dose irradiation using low-LET radiation in radiotherapy.

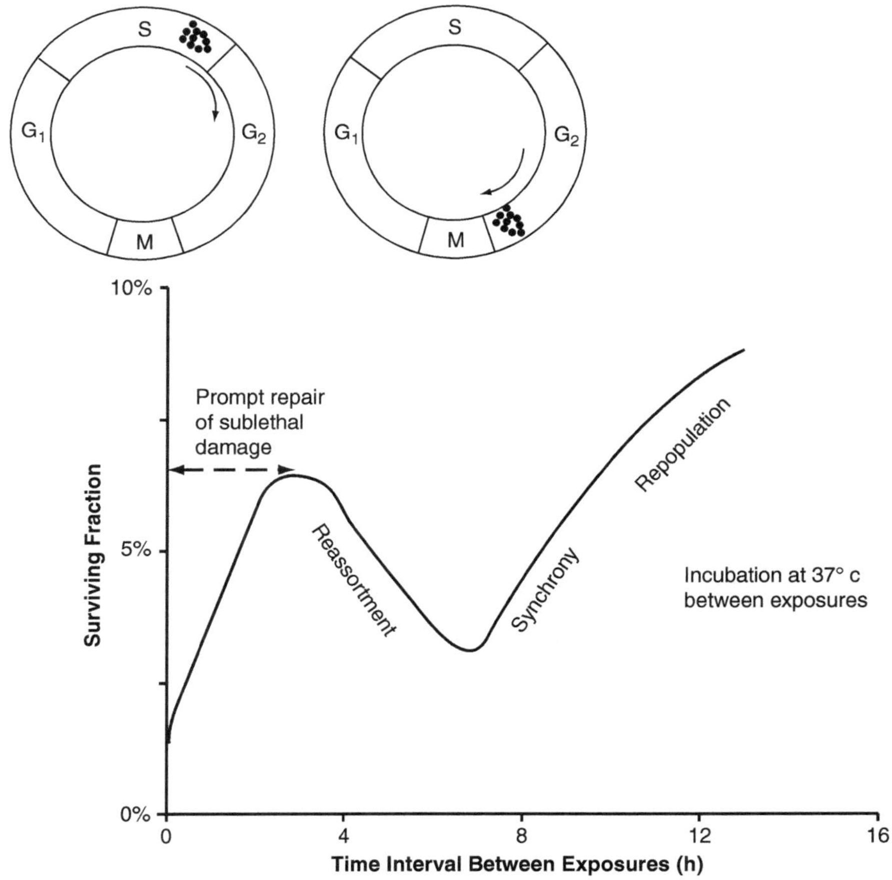

Fig. 10.21 SF of mammalian cells exposed to two fractions of radiation and incubated at 37°C for various lengths of time between the two exposures. Figure follows from Hall and Giaccia (2006). Refer to text for description

Repair of sublethal damage – The repair of sublethal damage occurs during the first few hours of lesion induction.

Reassortment – As shown in Fig. 10.20, subsequent irradiations of a cell population will preferentially lead to the dominance of radioresistant cells in the irradiated population.

Repopulation – This is a special concern in fractionated therapy. Tumor cells can repopulate between administration fractions, which is not desirable. On the other hand, repopulation of normal cells during these intervals is desirable so as to limit the normal tissue complication rate.

Reoxygenation – Reoxygenation is conceptually demonstrated in Fig. 10.22 in terms of the irradiation of a tumor cell.

In short, low-dose or fractional irradiation will spare normal tissue through the allowance of repair and repopulation; it will also allow the increased damage to tumor cells through reoxygenation and the reassortment of tumor cells into their radiosensitive phases. These results need to be reflected in the various considerations of radionuclide therapy. The extension of treatment time results in the reduction of immediate reactions of normal tissues to irradiation and, as demonstrated in Fig. 10.22, the reoxygenation of tumors and to increase their radiosensitivity.

10.2.9 Radiation-Induced Mutations

10.2.9.1 Introduction

The low absorbed doses associated with diagnostic nuclear medicine studies set the risks to the irradiated individual to be the probabilistic risks of radiocarcinogenesis and hereditary effects due to the induction of mutations. Tissue homeostasis is the outcome of the combination of controlled cell division and apoptosis. A tumor lacks these attributes and the progression from normal tissue to tumor is the result of mutations to three categories of genes (Hall and Giaccia 2006):

- The proto-oncogene which, in normal fashion, is a positive growth factor but, following mutation, produces an oncogene that ignores extracellular signals that would inhibit division.
- Tumor-suppressor genes (e.g., p53) which are negative growth factors.
- DNA stability genes which allow sensing of DNA damage and repair.

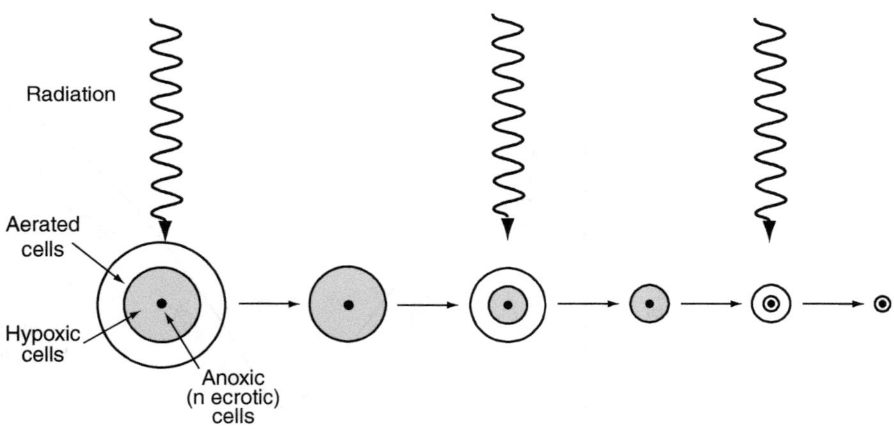

Fig. 10.22 The concept of reoxygenation in the fractionated/low absorbed dose rate therapy of a tumor. Due to inadequate vascularization, the tumor will have a necrotic core surrounded by a shell of hypoxic cells which is further surrounded by aerated cells in response to the degree of the radial dependence of vascularization and oxygen tension. Due to the oxygen-enhancement effect, radiation will preferentially kill the circumferential aerated cells leaving the hypoxic cells at the end of treatment. Following exposure, vascularization, and oxygenation of the outer rim of the tumor increases to produce aerated cells which are radiosensitive. Again, following irradiation, this outer shell is killed leaving a reduced hypoxic core. The process continues sequentially

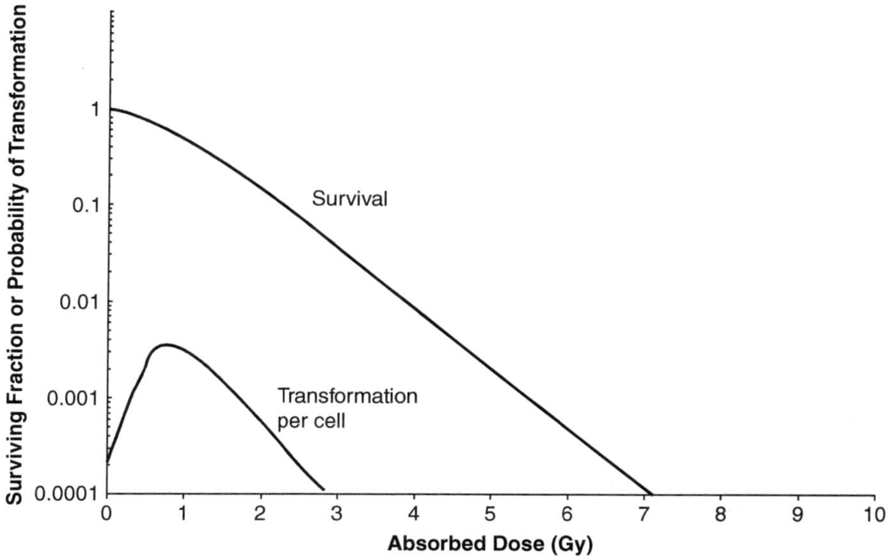

Fig. 10.23 A conceptual plot of the probability of a mutation per cell and the fractional survival of such cells in a population as a function of absorbed dose for low-LET radiation

Figure 10.23 compares the probabilities of survival and mutation as functions of absorbed dose for low-LET radiation. The likelihood of survival decreases exponentially whereas the rate of mutations, however, increases with absorbed dose to a maximum absorbed dose and then decreases with the same slope as that of the cell survival curve. This reflects the death of the mutated cells with increasing absorbed dose.

10.2.9.2 Oncogene Activation

The concept of the oncogene – a gene which can induce cancer – is based upon studies in the early twentieth century which demonstrated that tumors in mice, chickens, and rats were transmissible by the injection of the cell-free filtrate of the tumor into another animal. The transmitting factor has since been determined to be a retrovirus (a virus whose genome is composed of RNA). There are two models of the activation of oncogenes. The first is the viral oncogene model in which the retrovirus enters the cell and its RNA is then integrated into the host cell's genome and the genetic information leads to the synthesis of a protein which leads to a malignant transformation. The second model is that of the activation of proto-oncogenes through mutation. These are, in the first instance, normal genes but which, through mutation, lead to malignancy.

10.2.9.3 Inactivation of Tumor-Suppressor Genes

Uncontrolled growth of cells is inhibited through genes such as p53. A radiation-induced mutation of such a gene can lead to its inactivation followed by uncontrolled cell division leading to a tumor.

10.2.9.4 Germ-Cell Mutations

Mutation-induced activation of an oncogene or inactivation of a tumor-suppressor gene can only affect the individual in which these processes have occurred. The effect of a mutation of a germ cell, on the other hand, is not apparent in the exposed individual but is transmitted to its progeny and, as the mutation now enters the broader gene pool, can affect a multitude of individuals. Hence, radiation-induced hereditary effects have been of great interest as the affected population is much greater than the single exposed person.

Large absorbed doses to germ cells lead to permanent sterility. But at low absorbed doses received by germ cells, the viability of the cell is unaffected, but mutations can be induced. It is important to understand that mutations induced by ionizing radiation and which are expressed as hereditary effects are no different than those which occur naturally; ionizing radiation can only increase the frequency of such mutations and not influence their characteristics.

10.3 The Linear-Quadratic Dose–Response Model for Low-LET Radiation

10.3.1 Introduction

The LQ model was introduced in the discussion of cell survival curves within the qualitative context of a DSB produced by single and double ionizations. In the latter case, the temporal separation between the two separate ionizations allows the lesion to be repaired. Repair may be as simple as recombining the ends of the broken strands. This restitution restores the physical integrity of the helix, but not the original base pair sequence. Binary misrepair can also occur between damaged chromosome pairs. This illegitimate repair can result in the production of a dicentric chromosome and acentric chromosome fragments and centric rings, the latter being predominant.

In this section, the various applications of the LQ model in assessing and predicting tissue responses to radiation are investigated. The four main premises of the LQ formalism are (Brenner et al. 1998):

- The yield of radiation-induced DSBs is proportional to absorbed dose.
- The DSB repair rate is first-order.
- The binary misrepair of DSB pairs produced from two different radiation tracks competes with this first-order repair with a yield proportional to the square of the absorbed dose[3].
- Single radiation tracks can induce lethal lesions with a yield proportional to absorbed dose.

Discussion of the LQ model is limited to low-LET radiations (photons and electrons) where the scarcity of ionizations enables cells to repair radiation-induced damage. High-LET radiations (such as α particles or Auger electrons) have high ionization densities which fail to provide an opportunity for the cell to repair radiation-induced damage (e.g., Fig. 10.15).

10.3.2 DSB Repair Kinetics

10.3.2.1 First-Order Repair Kinetics

The basic kinetics of the induction of radiation-linked DSBs are first presented. Let U(t) be the mean number of DSBs present per cell at a time t following irradiation. If it is assumed that the number of repaired DSBs during a time interval δt is proportional to the total number of DSBs, given by the product $\mu U(t)\delta t$, then solving the first-order differential equation yields,

$$U(t) = U_0 e^{-\mu t} \tag{10.15}$$

where U_0 is the number of DSBs at the end of irradiation and μ is the first-order DSB repair rate constant.[4] Values of μ for most organs are of the order 0.2–1.4/h, corresponding to repair half-lives of between 0.5 and 3 h.

10.3.2.2 Binary DSB Misrepair

Whereas the majority of DSBs can be resolved through restitution in which the two ends of the DSB rejoin, some DSB repairs result in illegitimate binary unions. The rate at which binary misrepair occurs can be deduced by assuming that the probability distribution of the number of DSBs is Poisson. Let θ be the number of DSBs in a cell, where $U = \bar{\theta}$. The number of DSB pairs in the cell will thus be equal to $\theta(\theta - 1)/2$ (where, from the definition of U(t), a dicentric chromosome and the acentric fragments are counted as a single lesion). Each binary misrepair removes two DSBs. If κ is the rate of binary misrepair per DSB pair, the mean DSB-removal rate through binary misrepair is,

$$\begin{aligned} 2\kappa\left(\overline{\theta(\theta - 1)}/2\right) &= \kappa\bar{\theta}^2 \\ &= \kappa U^2 \end{aligned} \tag{10.16}$$

[3]The time difference between induction of the two lesions allows the first to be repaired before it would otherwise misrepair with the second.

[4]In some cases, a biexponential temporal behavior in the reduction of DSB number has been observed (Frankenberg-Schwager 1989). This is not considered here.

10.3.2.3 Kinetics of DSB Induction, Repair and Misrepair, and Cell Survival

In this subsection, the kinetics of DSB creation, repair, and misrepair are used to estimate the fraction of surviving cells in a population that has been irradiated. This leads naturally to the LQ expression for this fraction. However, to simplify the derivation, cell repopulation is considered to be negligible. From the premise that the induction of DSB is proportional to dose, the rate equation for the mean number of DSBs in a cell is,

$$\frac{dU}{dt} = \delta\frac{dD}{dt} - \mu U - \kappa U^2 \qquad (10.17)$$

where the first term is the DSB creation rate, the second term is the first-order repair rate, and the third term is the binary misrepair rate. δ is the mean number of produced DSBs per unit absorbed dose, the production rate of which is roughly 40 DSB/Gy (Sachs et al. 1997) and dD/dt is the absorbed dose rate.[5] The rate equation for the mean number of cells, N, in an irradiated population is,

$$\frac{dN}{dt} = -\left(\alpha\frac{dD}{dt} + \kappa U^2\right) N. \qquad (10.18)$$

The first term of (10.18) describes cell death due to a single ionizing radiation track and the second term describes the rate of cell death as a result of binary DSB misrepair. The κU^2 term is smaller than the other terms of (10.17) for absorbed doses of less than 5 Gy and, for the ease of solving this pair of coupled differential equations, is neglected. However, the corresponding term in (10.18) is not ignored. Neglecting κU^2 and integrating (10.17), the mean number of DSBs per cell is

$$U(t) = \delta\, e^{-\mu t} \int_{-\infty}^{t} dt' \frac{dD(t')}{dt'} e^{\mu t'}. \qquad (10.19)$$

Substituting this into (10.18) and integrating, one obtains, at time T, the logarithm of the SF,

$$\ln\left(\frac{N(T)}{N_0}\right) = -\alpha D - \frac{\delta^2\kappa}{2\mu} \times \left(\frac{2}{D^2}\int_{-\infty}^{T} dt\frac{dD(t)}{dt}\int_{-\infty}^{t} dt'\frac{dD(t')}{dt'}e^{-\mu(t-t')}\right) D^2$$
$$= -\alpha D - \beta G D^2 \qquad (10.20)$$

where $\beta \equiv \delta^2\kappa/2\mu$ and G is the Lea–Catcheside dose-protraction factor (Lea and Catcheside 1942) which describes the damage repair occurring between two separate ionizations. This factor is less than or equal to unity and is developed further in the following subsection. Equation (10.20) is the LQ model, modified so as to account for repair of sublethal damage.

Further examination of the β factor is of interest. It is proportional to the square of the mean number of DSBs induced per unit absorbed dose (δ), reflecting the effect of two independent ionizations. It is also proportional to the rate of binary DSB misrepair per DSB pair and inversely proportional to the first-order repair rate.

10.3.2.4 Lea–Catcheside Dose-Protraction Factor

The Lea–Catcheside factor is written in the more general form,

$$G = \frac{2}{D^2}\int_{-\infty}^{\infty} dt\frac{dD(t)}{dt}\int_{-\infty}^{t} dt'\frac{dD(t')}{dt'}e^{-\mu(t-t')} \qquad (10.21)$$

where

$$D = \int_{-\infty}^{\infty} dt\frac{dD(t)}{dt} \qquad (10.22)$$

The integrand of the second integral over t′ refers to the first of the two DSBs required to cause lethality; the exponential term describes the repair and subsequent

[5]"Saturable" repair mechanisms have been proposed as another means of describing the curvature of the LQ dose-response curve (e.g., Goodhead 1985). In saturable repair, μ has an absorbed-dose dependence, decreasing as dose increases and thus leading to a reduced repair efficiency with increased absorbed dose. Brenner et al. (1998) presented a modification of (10.17), $\frac{dU}{dt} = \delta\frac{dD}{dt} - U\sum_{i=1}^{2}\frac{\lambda_i}{1+\varepsilon_i U}$ where the first term in the summation corresponds to the creation of initial lesions and the second term corresponds to the production of lethal lesions from the initial lesions. These terms refer to the saturation of repair (essentially, this is the Michaelis–Menten equation applied to enzyme kinetics).

reduction of such a DSB. The integral over t refers to the second of the two DSBs that interacts with a remaining unrepaired first DSB to cause the inevitable lethal damage. For acute irradiation which does not enable the cell to repair radiation-induced damage, the Lea–Catcheside factor is $G = 1$. However, as the kernel $e^{-\mu(t-t')} \leq 1$ for prolonged irradiation, then $G < 1$ in such cases due to the ability for repair to occur.

The Lea–Catcheside factors for two types of absorbed dose rates are now derived for two cases of relevance.

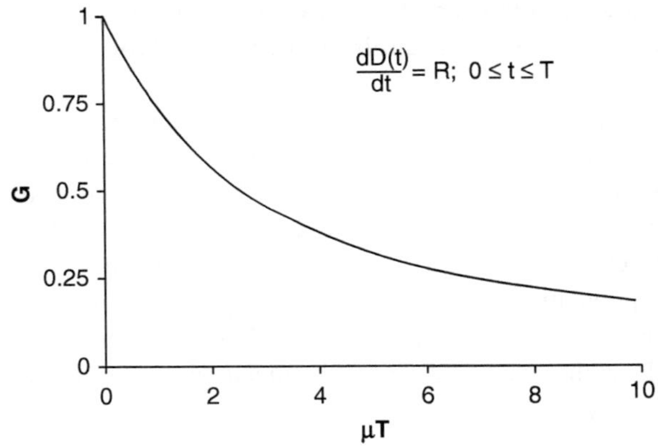

Fig. 10.24 Lea–Catcheside dose-prolongation factor G as a function of the product of μT for a constant absorbed dose rate for a finite time T and where μ is the first-order repair rate constant

Constant Absorbed Dose Rate for Finite Irradiation Time

In this case, the absorbed dose rate is given by,

$$\begin{aligned} \frac{dD(t)}{dt} &= R \quad 0 \leq t \leq T \\ &= 0 \quad \text{elsewhere} \end{aligned} \tag{10.23}$$

where R is a constant dose rate, T is the duration time of irradiation and the total absorbed dose is clearly the product of this dose rate and the irradiation time $D = RT$. Then, the Lea–Catcheside factor for this absorbed dose rate is, from its definition,

$$\begin{aligned} G &= \frac{2}{D^2} \int_{-\infty}^{\infty} dt \frac{dD(t)}{dt} \int_{-\infty}^{t} dt' \frac{dD(t')}{dt'} e^{-\mu(t-t')} \\ &= \frac{2}{(RT)^2} \int_0^T dtR \int_0^t dt' \, R\, e^{-\mu(t-t')} \\ &= \frac{2}{T^2} \int_0^T dt\, e^{-\mu t} \int_0^T dt' \, e^{\mu t'} \end{aligned} \tag{10.24}$$

Solving the integrals gives the dose-protraction factor for a constant absorbed dose rate delivered for a finite time T,

$$G = \frac{2}{(\mu T)^2} \left(\mu T + e^{-\mu T} - 1 \right) \tag{10.25}$$

It is straightforward to demonstrate that, by expanding the exponential to second-order, the Lea–Catcheside factor $G \to 1$ as the irradiation time $T \to 0$. In other words, the $\beta G D^2$ contribution to cell lethality increases to βD^2 as the absorbed dose is delivered more acutely. As $T \to \infty$, then $G \to 0$. A plot of G as a function of the product of the repair constant and irradiation time is shown in Fig. 10.24.

Exponentially-Decreasing Absorbed Dose Rate

For an example more representative of nuclear medicine, consider the Lea–Catcheside factor for an exponentially-decreasing absorbed dose rate,

$$\begin{aligned} \frac{dD(t)}{dt} &= Re^{-\lambda t} \quad 0 \leq t \leq T \\ &= 0 \quad \text{elsewhere.} \end{aligned} \tag{10.26}$$

For calculational purposes, a finite irradiation time has been allowed for, whereas in internally-administered absorbed dose, $T \to \infty$. The total absorbed dose is,

$$\begin{aligned} D &= \int_{-\infty}^{\infty} dt \frac{dD(t)}{dt} \\ &= R \int_0^T dt\, e^{-\lambda t} \\ &= \frac{R}{\lambda}\left(1 - e^{-\lambda T}\right). \end{aligned} \tag{10.27}$$

The Lea–Catcheside factor for this particular absorbed dose rate model is given by the evaluation of the double integrals,

$$
\begin{aligned}
G &= \frac{2}{D^2}\int_{-\infty}^{\infty} dt \frac{dD(t)}{dt} \int_{-\infty}^{t} dt' \frac{dD(t')}{dt'} e^{-\mu(t-t')} \\
&= 2\left(\frac{R}{D}\right)^2 \int_{0}^{T} dt\, e^{-\lambda t} \int_{-\infty}^{t} dt'\, e^{-\lambda t'}\, e^{-\mu(t-t')} \\
&= 2\left(\frac{\lambda}{(1-e^{-\lambda T})}\right)^2 \int_{0}^{T} dt\, e^{-(\lambda+\mu)t} \int_{-\infty}^{t} dt'\, e^{-(\lambda-\mu)t'} \\
&= \frac{2}{\lambda-\mu}\left(\frac{\lambda}{(1-e^{-\lambda T})}\right)^2 \int_{0}^{T} dt\, e^{-(\lambda+\mu)t}\left(1-e^{-(\lambda-\mu)t}\right) \\
&= \frac{2}{\lambda-\mu}\left(\frac{\lambda}{(1-e^{-\lambda T})}\right)^2 \left(\frac{1-e^{-(\lambda+\mu)t}}{\lambda+\mu} - \frac{1-e^{-2\lambda T}}{2\lambda}\right)
\end{aligned}
\tag{10.28}
$$

By again expanding the exponentials of (10.28) to second order, it is seen, after some algebraic manipulation, that the dose-protraction factor $G \to 1$ as $T \to 0$. For nuclear medicine dosimetry, one would want to calculate G_∞, which is the value for G as $T \to \infty$ and $t \to \infty$,

$$G_\infty = \frac{\lambda}{\lambda + \mu} \tag{10.29}$$

This G_∞ term is plotted in Fig. 10.25 as a function of the ratio of the decay to repair constants, λ/μ. From this figure, $G_\infty \to 1$ as $\lambda/\mu \to \infty$.

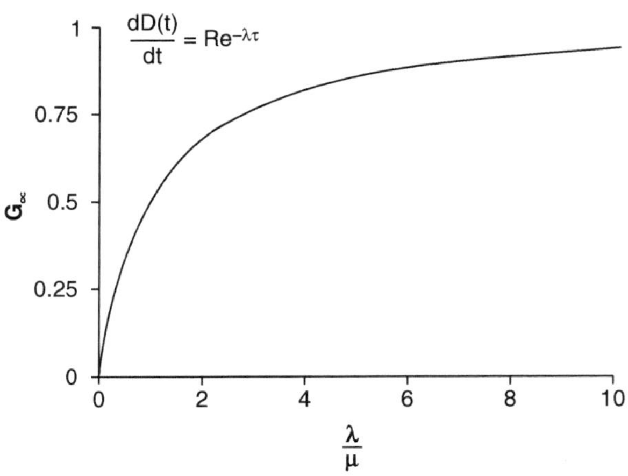

Fig. 10.25 Lea–Catcheside dose-protraction factor at infinite time for exponentially-decreasing absorbed dose rate as a function of the ratio of the effective decay constant (accounting for both physical decay and biological washout) to the repair time constant, λ/μ

The absorbed dose rate to a given tissue as a result of the administration of a radionuclide generally does not follow a simple monoexponential decrease It takes a finite amount of time for the tissue to uptake the radionuclide and the biokinetics are frequently such that washout follows a multiexponential temporal behavior. The Lea–Catcheside factor can be calculated by modeling the absorbed dose as a superposition of weighted exponential terms and calculating each separately and the final result obtained by the weighted summation of terms.

10.3.3 Biologically Equivalent Dose

From (10.14), the logarithm of the fraction of cells of a population that has been irradiated to an absorbed dose D is given as,

$$-\ln S = \alpha D + \beta G D^2 \tag{10.30}$$

The negative logarithm of the fractional survival is replaced by defining it as the biological effect of interest, $E \equiv -\ln S$. For a single acute absorbed dose given over a time duration much less than that time required for repair, we can then set the Lea–Catcheside factor $G = 1$ and the biological effect as,

$$E = \alpha D + \beta D^2. \tag{10.31}$$

Now, consider the case if the same net absorbed dose is administered but now expended over n fractions so as to enable damage repair between fractions. As the absorbed dose per fraction is $d = D/n$, the biological effect for fractionated radiotherapy is,

$$
\begin{aligned}
E &= n\left(\alpha d + \beta d^2\right) \\
&= \alpha n d\left(1 + \frac{d}{\alpha/\beta}\right) \\
&= \alpha D\left(1 + \frac{d}{\alpha/\beta}\right).
\end{aligned}
\tag{10.32}
$$

Dividing through by α gives,

$$\mathrm{BED} = D\left(1 + \frac{d}{\alpha/\beta}\right). \tag{10.33}$$

$BED \equiv E/\alpha$ is the biologically equivalent dose[6] and is equal to the product of the total absorbed dose and the factor $\left(1+\frac{d}{\alpha/\beta}\right)$. This latter factor is itself known as the relative effectiveness, RE, so that the definition of the BED can be written as,

$$BED = D \times RE. \tag{10.34}$$

The relative effectiveness for a constant absorbed dose rate R for time T is,

$$RE = 1 + \frac{R(\mu T + e^{-\mu T} - 1)}{\mu^2 T \, ^{\alpha}\!/_{\beta}} \tag{10.35}$$

For the case of the continuous and exponentially-decreasing absorbed dose rate of (10.26),

$$BED = D\left(1 + \frac{G_{\infty}R}{\lambda\left(^{\alpha}\!/_{\beta}\right)}\right). \tag{10.36}$$

Hence,

$$BED = \frac{R}{\lambda} \times RE \tag{10.37}$$

and the relative effectiveness is,

$$RE = 1 + \frac{R}{(\lambda + \mu)\left(^{\alpha}\!/_{\beta}\right)} \tag{10.38}$$

It would be useful at this point to employ an example to describe a practical use of these results. Consider the case of a tumor (with the ratio $\alpha/\beta = 10\,\text{Gy}$ and repair constant $\mu = 0.5\text{h}^{-1}$) to be treated with ^{131}I through internal administration. The effective half-life of the radionuclide in the tumor, accounting for both the biological washout and the physical decay of the isotope, is taken to be 5 days (i.e., $\lambda = 0.006\text{h}^{-1}$). One wishes to determine the initial absorbed dose rate required with the radionuclide therapy using ^{131}I so as to achieve the total absorbed dose required to achieve the same biological effect as if the tumor had been treated to a total absorbed dose of 50 Gy over a period of 5 days using a ^{137}Cs source. As the half-life of ^{137}Cs is far greater than the 5-day treatment time, we can treat it as a source of radiation providing a constant absorbed dose rate of 0.417 Gy/h to the tumor. Using (10.33), the biologically equivalent dose is 58.2 Gy. Substituting (10.34) into (10.37) and solving for the absorbed dose rate R from ^{131}I administered internally for this value of the BED, R = 0.324 Gy/h. The resulting total absorbed dose from the ^{131}I radionuclide therapy is 54 Gy, which is the absorbed dose required to yield the same biological effect upon the tumor as delivering 50 Gy to it continuously over 5 days by a ^{137}Cs source. Further discussion in relation to applications to the medical internal radiation dose (MIRD) schema can be found in Baechler et al. (2008).

10.3.4 *Effects of Repopulation*

The above derivations neglected cell population growth in order to simplify the various derivations. But, in reality, cells within a population (e.g., a tumor) can proliferate during and following irradiation which will result in a reduction of the BED (Dale 1996). This effect can be accounted for, admittedly crudely, by modifying the expression for the biologically effective dose to include a reductive factor in (10.37),

$$BED = \frac{R}{\lambda} \times RE - \frac{\ln 2}{\alpha}\frac{t}{T_{Pot}} \tag{10.39}$$

where it has been assumed that the clonogens increase exponentially over time t and where T_{Pot} is defined to be the potential doubling time of the cell population. Hence, the ratio t/T_{Pot} gives the number of cell doublings during t.

10.3.5 *Applications of the Linear-Quadratic Model to Internal Radiation Dosimetry*

10.3.5.1 Introduction

The applications of the LQ model to address problems in internal radiation dosimetry began with the seminal paper by Dale (1985).

[6] Also referred to as the extrapolated response dose (Wheldon and O'Donoghue 1990) or as the biologically effective dose.

10.3.5.2 α/β Ratios

Recall that the α/β ratio is the absorbed dose at which the linear and quadratic contributions to the biological effect are equal. Thus, in terms of the description offered by the cell survival curve, the curve will be shallower and more slowly bending for cells with high α/β ratios and correspondingly steeper and more curved for cells with low α/β ratios. In general, late reacting tissues are slowly dividing and have low α/β ratios of the order of 3 Gy or so. Acutely-reacting tissues, which are more rapidly dividing (such as neoplasia, skin, and intestinal epithelium), have higher α/β ratios, with values typically about 10 Gy.

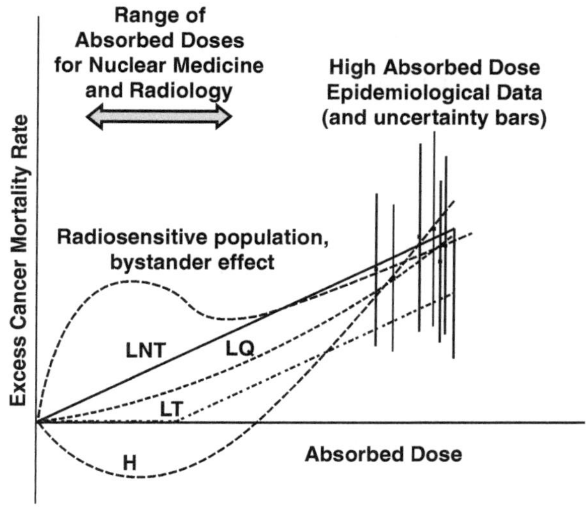

Fig. 10.26 Conceptual representation of the extrapolation of epidemiological dose–response data of ECMR from high absorbed doses to the low absorbed doses typical of medical imaging procedures. Five extrapolation models are shown: linear no threshold (LNT), linear threshold (LT), linear-quadratic (LQ), hormesis (H), and a bimodal model which predicts an elevated risk at low doses due to, for example, the bystander effect

10.4 Human Somatic Effects of Ionizing Radiation

10.4.1 Introduction

Cellular radiobiology enables both the estimation of the probability of cellular death or mutation as a consequence of exposure to ionizing radiation and the study of various influencing effects such as hypoxia and absorbed dose rate. While investigations of irradiated cells or colonies of cells are fundamental to understanding both the risks posed by ionizing radiation and how these risks can be modified or mitigated against, extrapolation of these cellular data to the metazoan level in order to predict unequivocally the effects of ionizing radiation upon the human is not possible. Consequently, the estimations of risk and effect resulting from medical exposures to ionizing radiation are reliant upon the experiences of exposures of human populations to ionizing radiation. In particular, because radiation-induced mutations and their consequences are no different than those that occur spontaneously or through other insults (e.g., chemical or viral), the effects of ionizing radiations at the low absorbed doses associated with medical imaging are assessed through epidemiological studies in which large cohorts of irradiated individuals are compared with large unirradiated populations in order to observe statistically the elevated incidence of a mutation or its effect resulting from ionizing radiation. These epidemiological sources are the subject of the following subsection.

A significant problem in estimating radiation risk is that many of these sources of human radiation exposure data are for absorbed doses that are much greater than those typical of medical diagnostic exposures. In order to use these data (e.g., elevated risk of cancer per unit absorbed dose) to estimate the risk to the patient receiving much lower doses, cellular radiobiology can be called upon to guide the development of dose–response models that can extrapolate from high absorbed dose epidemiological risks to those at the low absorbed doses associated with medical diagnostic procedures, including nuclear medicine. Figure 10.26 presents the difficulty of estimating the risk associated with the low absorbed doses due to medical imaging exposures (here, the risk is the excess cancer mortality rate (ECMR) beyond that observed in a population unexposed to radiation[7]) by using the risks determined through epidemiological studies at higher absorbed doses. Although these latter risks are subject to bias and confounding and are determined by

[7]Such a condition is hardly achievable as all populations are exposed to at least background radiation. Hence, the control population would be that that was not exposed to the test radiation.

comparing a population exposed to known absorbed doses against another matched unexposed population. Ideal matching must be in terms of age, sex, race, and exposure to other confounding carcinogens such as tobacco. Clearly, this isolation and matching of cohorts is difficult to achieve. Moreover, the uncertainty bars associated with the determined risk values are inherently large. Even assuming that perfect statistical matching of the populations is obtained, the observed risks must then be extrapolated to the low absorbed doses of interest accounting for the effects of absorbed dose rate, the LET of the radiation creating the absorbed dose and the above noted characteristics of the exposed and unexposed populations. Clearly, this is not a trivial problem to address and the interested reader is advised to consult, for example, the BEIR V and BEIR VII reports (National Research Council 1990, 2006) for a full exposition of the practicalities. What is of interest is, following from the earlier review of radiobiology, are the different types of radiobiological models used to extrapolate from the absorbed doses of the measured data to the range of absorbed doses of interest to us so as to provide an estimate of the risk associated with a diagnostic nuclear medicine study. Clearly, the first requirement of the extrapolation model is that it be able to reproduce the observed risks at higher absorbed doses and to then "sensibly" extrapolate to the low absorbed doses of interest to us. Radiobiology can guide the selection of an appropriate model. As a result, different estimates of risks at the absorbed dose range of clinical interest occur.

Five dose–response models for the ECMR are presented in Fig. 10.26.

Linear no threshold (LNT) response: This is the simplest dose-risk model in which the risk is considered to be proportional to absorbed dose,

$$\mathrm{ECMR} = \alpha_{\mathrm{LNT}}\,\mathrm{D} \tag{10.40}$$

where D is the absorbed dose and α_{LNT} is a constant of proportionality. This model predicts that there is *always* an excess risk of cancer induction and mortality as a function of dose and is only zero (i.e., equal to that which "naturally" occurs) at zero absorbed dose. This is perhaps the basic dose–response model in that it assumes that ionizing radiation produces irrepairable DNA lesions and that repair is neglected.

Linear threshold (LT) response: This dose–response model is a variant of the LT model, Here, the effect of sublethal cellular repair is accounted for in a simplistic fashion in that the model assumes that there is no risk of excess cancer mortality below some absorbed dose threshold, $\mathrm{D_T}$,

$$\begin{aligned}\mathrm{ECMR} &= 0 && \mathrm{D} \leq \mathrm{D_T} \\ &= \alpha_{\mathrm{LT}}\,\mathrm{D} && \mathrm{D} > \mathrm{D_T}.\end{aligned} \tag{10.41}$$

Our understanding of the repair of sublethal and potentially lethal damages does not account for the existence of a threshold absorbed dose, $\mathrm{D_T}$, below which stochastic radiation damage cannot occur. This requires more involved appreciations of the effects of the absorbed dose rate, times between fractionated radiation exposures and environmental conditions. Hence, this is a simplistic model which implies that, at absorbed doses less than $\mathrm{D_T}$, individual cells can repair radiation damage or, somatically, the body can eradicate any nascent tumors through, for example, an immunological response.

LQ response: From the earlier discussion of the LQ model, it is possible to hypothesize that the risk behaves in the form,

$$\mathrm{ECMR} = \alpha_{\mathrm{LQ}}\,\mathrm{D} + \beta_{\mathrm{LQ}}\,\mathrm{D}^2 \tag{10.42}$$

where the first term describes the effects of fatal DSBs and the quadratic term describes the effects of secondary DSBs. The Lea–Catcheside factor describing sublethal repair of initial DSBs is not included in this expression.

Hormetic (H) response: This is a controversial dose–response model in which it is assumed that the exposure to very low absorbed doses of radiation triggers an immunological response sufficient to eliminate any neoplasms. As a result, the cancer risk of the irradiated population is lower than that of a corresponding unirradiated population. This effect then reverses with increasing absorbed dose to present a risk that increases with absorbed dose (Luckey 1991).

Bystander response: It has been described how the bystander effect can lead to an elevated response at low absorbed doses. Should this model be valid, then there would be an increase in risk response at lower absorbed doses. However, such an effect would be difficult to discern in nuclear medicine due to the distribution of radioactivity within the body (Sgouros et al. 2007).

It is difficult to select from these models one that is appropriate for predicting effects at low absorbed doses and which is what would be regarded as a "reasonable" extrapolation from epidemiological data. For the purposes of the protection of the patient, the radiation worker or any other individual exposed to ionizing radiation, the most conservative model linking risk to absorbed dose is the LNT model. It is currently regarded that the scientific arguments and supporting epidemiological data for the use of the LT, LQ, bystander, and hormetic models in radiation protection are insufficient to cause abandonment of the LNT model. Hence, the use of the LNT model in radiation protection is still recommended in both International Commission on Radiological Protection (ICRP) Publication 103 (ICRP 2007) and the BEIR VII report of the National Academy of Sciences. This recommendation is, however, most certainly not without controversy. Whereas most epidemiological studies suggest that there is no evidence for cancer induction at effective doses (to be defined later) below about 150 mSv, which is 10–40 times greater than that received in most diagnostic nuclear medicine procedures, there has been indications that effective doses as low as even 10 mSv, which is of the order of magnitude of most diagnostic nuclear medicine studies, can be associated with the induction of solid tumors (Brenner and Hall 2007).

10.4.2 Epidemiological Sources of Human Data

Observations of the effects of ionizing radiation upon individual cells are insufficient to predict somatic effects. Inevitably, though, an understanding of the potential risk posed by exposure to ionizing radiation can only be obtained from analyses of populations of individuals exposed to ionizing radiation.

10.4.2.1 Nuclear Bombings of Hiroshima and Nagasaki

The largest cohort of individuals exposed to ionizing radiation is that of the survivors of the nuclear bombings of Hiroshima and Nagasaki at the end of the Second World War. The life span study (LSS) population consists of 120,321 individuals who were resident in those two cities in 1950, of whom 91,228 were present at the times of the detonations. This population has been followed up since that date. The population present at the detonations was made up of two subcohorts: one made up of survivors who were within 2.5 km of the detonations' hypocenters and a similar-sized control group who were between 3 and 10 km from the hypocenters and who received negligible absorbed doses.

This population presents three significant advantages in minimizing bias in estimating the risks resulting from exposure to ionizing radiation. First, it is a large heterogeneous population in terms of age and sex. Second, no bias is introduced in terms of disease (as in long-term survivors of radiotherapy) and occupation (as in radiation workers). Third, exposures were of the whole body enabling the assessment and comparison of cancer risks at different anatomical sites. However, there were a number of significant difficulties in estimating the absorbed doses received by individuals that were not resolved until 1980s (Radiation Effects Research 1987a, b; National Research Council 1990). The fissioning of both weapons produced neutrons and γ rays. Neutrons elastically scatter from protons in tissue and the secondary protons, due to their mass, present a high-LET radiation compared to the Compton electrons and photoelectrons set in motion by the γ rays. As Fig. 10.15 indicates, the LET-dependence of biological effects requires a clear separation of the contributions of neutrons and photons to the absorbed dose received in order to isolate their biological effects. The neutron fluences of both weapons differed because of their designs.[8] Earlier estimates of the neutron and photon absorbed doses grossly overestimated that due to neutrons due to, among other things, not allowing for attenuation of the neutron fluence by the water content (i.e., protons) of the humid air. The neutron and photon fluences have been estimated by Monte Carlo simulations, postdetonation measurements using neutron- and

[8]The "Little Boy" weapon used at Hiroshima consisted of two subcritical masses of ^{235}U, one of which was fired into the other to form a critical mass which subsequently fissioned. The "Fat Man" weapon used at Nagasaki consisted of a subcritical spherical mass of ^{239}Pu which was imploded to form the critical mass.

photon-emitting sources, and in situ measurements such as thermoluminescent dosimetry of building materials exposed to the blasts. Another factor of importance in being able to apply the survivor data (excess cancer incidence or mortality per unit absorbed dose) to provide risk estimates is that the exposure was essentially instantaneous. Recall that the incidence of a biological effect for a given fixed dose increases with dose rate. This, plus consideration of the extrapolation to low absorbed doses, requires that the application of the bomb survivor data incorporates a dose and dose-rate effectiveness factor (DDREF) to correct for the prompt exposure and arrive at a risk estimate more appropriate to protracted exposure to low absorbed doses. Figure 10.27 demonstrates how this factor was obtained in the BEIR VII report. The value of the DDREF is 2, but it is recognized that there is considerable uncertainty associated with this value (ICRP 2007).

There, however, remain two major complicating factors to be accounted for. First, during the conditions of war at the time, the Japanese population were malnourished and their intrinsic susceptibility to the effects of ionizing radiation could have been elevated thus increasing the estimates of risk per unit absorbed dose. The second, and countering effect, is that of the "healthy survivor." This effect predicts that the survivor had a predisposition to surviving prior to exposure

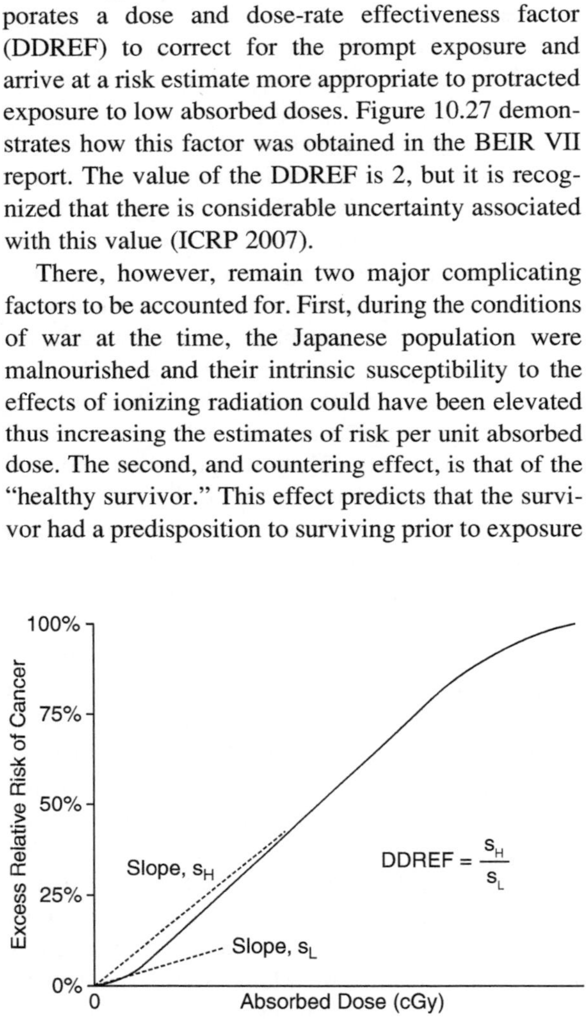

Fig. 10.27 The rationale for the dose and dose-rate effectiveness factor (DDREF) defined in the BEIR VII (2006) report for a hypothetical dose–response curve. A linear approximation at low absorbed doses (i.e., the tangent to the curve at zero absorbed dose) yields the slope s_L. At higher absorbed doses, where the response is linear, an extrapolated slope s_H is obtained. The DDREF is the ratio of these two slopes

to radiation and, as a result, would tend to decrease the estimate of risk per unit absorbed dose.

10.4.2.2 Medical Exposures: Examples

Retrospective analyses of cancer incidence in patient populations who have undergone medical exposures to ionizing radiation have, in general, the advantages of consistent and extended follow-up periods and the availability of accurate dosimetry data. Unfortunately, these populations can introduce a significant bias as a result of the disease for which they have been exposed to ionizing radiation.

There are two groups of medically-exposed subjects to be considered. The first is that made up of long-term surviving radiation oncology patients. Here, the exposed tissues and organs of interest are not those within the treatment portal, which receive a tumoricidal absorbed dose, but rather those extraneous to it and which receive only scattered radiation and much lower absorbed doses. Clearly, the requirement of the subjects to have survived for an extended period (beyond the latency period in order for a radiogenic cancer to be manifest) limits such cohorts to specific cancers and radiotherapy regimes with high survival probabilities. These include, for example, cancers of the cervix and breast and Hodgkin's disease (lymphoma). The second group is that of patients that have undergone diagnostic imaging procedures. While the magnitude of the absorbed doses received by this latter population are typical of those of interest to our application to nuclear medicine, it is generally far more difficult to retrospectively estimate the absorbed dose in these patients than for the therapeutic patients.

Secondary Neoplasia in Radiotherapy Patients

Patients suffering from cancers are frequently treated with ionizing radiation with curative intent; in fact, about half of all oncology patients receive radiotherapy during the course of treatment of their disease (Ron 1998). Radiotherapy is performed predominantly with X-rays or high-energy electrons produced by linear accelerators or γ rays from a radioisotope source (e.g., ^{60}Co, ^{137}Cs). Proton irradiation is growing as a means of therapy, particularly in the United States of

America, and, historically, pions and heavy ions have been used. Those patients with long survival times can express secondary tumors attributable to the therapeutic ionizing radiation they have received. These secondary malignancies can occur in regions outside the primary therapy portal (within which the radiation absorbed dose would have been intended to tumoricidal, with absorbed doses of the order of 50 Gy and greater) due to radiation scattered from the irradiated volume or from leakage radiation emitted by the treatment device. The magnitudes of the absorbed doses in these peripheral regions can be similar to those experienced in diagnostic imaging procedures and the detailed radiation prescriptions and treatment plans allow accurate retrospective estimation of doses at such sites. Even so, as the doses of interest for estimating the radiation risk associated with diagnostic levels of absorbed dose are of the order of tens of mGy compared to the therapeutic absorbed doses of tens Gy, accurate measurement, or calculation of such peripheral doses for radiotherapy patients receiving high absorbed doses within the treatment portal is challenging and frequently necessary (McParland and Fair 1992; Stovall et al. 1995).

Three populations of long-term surviving oncology patients treated with radiation have provided the most abundant data.

Cervical cancer: Long-term survivors treated for cervical cancer have provided a significant cohort to assess radiocarcinogenesis as a result of exposure to low absorbed doses. Radiotherapy is provided through intracavitary ^{137}Cs sources[9] within the cervix and/or uterus or external radiation beam therapy to give high absorbed doses to the cervix and uterus (several tens of Gy). Peripheral tissues will also receive a radiation absorbed dose due to the Compton scattering of photons from the treatment volume. Absorbed doses to the active bone marrow has been estimated to be as high as 7 Gy; absorbed doses to the breast and lung tissue are estimated to be of the order of 300 mGy and that to the thyroid is estimated to be 100 mGy (Kleinerman et al. 1995). Epidemiological studies by Boice et al. (1985) and extended by Kleinerman et al. were conducted for up to some 7,543 secondary cancers resulting from about 200,000 cervical cancer patients reviewed in eight countries. Specific details of the elevated risks of secondary cancers in these long-term survivors can be found in the original publications or in the BEIR V and VII reports.

Hodgkin's disease: Another cohort of long-term radiotherapy survivors are those treated by radiation for Hodgkin's disease. This is a lymphoma predominantly of younger populations and is frequently treated successfully with radiation, usually in conjunction with chemotherapy. As a result of its extended shape and location, the primary radiation portal is referred to as the "mantle field" due to its encompassing the axillary, mediastinal, and cervical lymph nodes. Additional radiation portals can be used to irradiate the spleen and Waldeyer's ring. This combination of a young age and a long survival time lends well to follow-up studies of secondary cancers attributable to scattered radiation received by tissues peripheral to the primary radiation field. In particular, the extended radiation field leads to the investigation of secondary cancers such as leukemia and solid cancers of the breast and lung.

Breast cancer: Long-term survivors of breast cancer treated with radiation provide yet another cohort of subjects at risk to secondary cancers attributable to the therapeutic radiation field. Lung cancer, leukemia, and contralateral breast cancer have been studied as sequelae due to radiotherapy of primary breast cancer with retrospective dose estimation derivable from measurement (McParland 1990).

Radiation therapy has also been used for benign conditions, of which three significant cohorts providing data of radiation-induced cancer risk are summarized.

Ankylosing spondylitis: This is a chronic arthritis affecting the spine and sacroiliac joints. Between 1935 and 1957, radiotherapy was used in the United Kingdom for the treatment of this disease. A cohort of 14,566 patients received spinal irradiation and was followed up into the 1990s (Weiss et al. 1994); the radiation dosimetry was estimated by Monte Carlo simulations of a sample of patients. The cancer mortality rate amongst the irradiated patients was significantly greater than those expected from rates in England and Wales, with significant increases seen in leukemia, non-Hodgkin's lymphoma, multiple myeloma, and solid tumors of the esophagus, pancreas, lung, urinary bladder wall, and kidney.

[9] In much earlier times during the twentieth century, ^{226}Ra sources were used. These were largely supplanted by safer ^{137}Cs sources in the 1970s and 1980s.

Tinea capitis: This is a fungal infection of the scalp and the use of X-rays to epilate the scalp was a frequent means of treating the infection. For example, between 1948 and 1960 in Israel, approximately 20,000 children were treated in such a manner (Ron et al. 1998). Epidemiological studies of children irradiated for tinea capitis revealed elevated incidences of brain tumors (glioblastoma and meningioma); a follow-up of 2,215 patients was performed for 25 years postirradiation (Shore et al. 1976).

Postpartum mastitis: Prior to the introduction of antibiotics and sulfonamides, irradiation was a frequent mode of treatment for inflammatory conditions. Some 601 women between 20 and 40 years of age were treated with radiation in the state of New York for postpartum mastitis during the 1940s and 1950s and received absorbed doses to breast tissue between 60 cGy and 14 Gy and exposures. This exposed cohort was compared against 1,239 women suffering from the same condition but who did not receive radiotherapy.

Cancers Arising from Diagnostic Imaging Procedures

Assessing the elevated risks of radiation-induced cancers from patient cohorts exposed to ionizing radiation in diagnostic imaging procedures can be difficult as accurate retrospective assessment of the absorbed dose and the anatomical sites being imaged is not as easy to achieve as with patients receiving therapeutic exposures. On the other hand, the sizes of populations undergoing diagnostic procedures are greater than those undergoing therapy and do not have the confounding factor of differentiating between "naturally-occurring" and radiation-induced secondary cancers arising in a patient having demonstrated a predisposition to malignancy. Moreover, they received absorbed doses with approximately the same magnitude for which one wishes to assess the risks associated with imaging. This latter point is a double-edged sword: although the low magnitudes of absorbed doses are comparable to the problem at hand, the concurrent reduction in cancer risk requires greater population sizes in order to derive a statistically-valid estimate of cancer risk per unit absorbed dose.

Fluoroscopy-aided artificial pneumothrax in treatment of pulmonary tuberculosis: A theoretical approach to treating patients suffering from pulmonary tuberculosis developed in the nineteenth century was to induce an artificial pneumothorax allowing drainage of the pleural space and cicatrization of pulmonary injuries. Beginning in the early part of the twentieth century, this therapeutic technique was coupled with fluoroscopic review which led to large numbers of patients being exposed to X-irradiation. Three epidemiological studies following up women having received multiple fluoroscopies in conjunction with artificial pneumothorax have provided significant data on radiation risk associated with the exposure of certain organs. The first consists of 31,710 women in Canada who received fluoroscopy-associated artificial pneumothorax therapy between 1930 and 1952, with many receiving multiple fluoroscopies with wide ranges of fractionations (e.g., BEIR VII reports a mean number of 92 fluoroscopy-aided artificial pneumothorax procedures per patient over periods of up to 2 years). Of this population, 8,380 received an absorbed dose to the breast exceeding 10 cGy (with a maximum exceeding 2 Gy). Of particular interest is the fact that patients in the Canadian province of Nova Scotia were generally imaged in the anterioposterior view (i.e., the X-rays incident to the patient's anterior) whereas the posterior–anterior view was prevalent elsewhere. The former patients would clearly receive a much higher absorbed dose to the breast.

Thorium-based vascular contrast medium: The radiation exposure of interest here is not that from the external X-ray beam, but rather that due to the radioactive thorium used in the vascular contrast. Thorotrast was a vascular contrast agent consisting of ThO_2 in a colloidal suspension used in the 1930s and 1940s. Thorium was selected for this application due to its high atomic number ($Z = 90$) and consequent high photoelectric absorption cross section allowing visualization of the vasculature in X-ray imaging. However, its dominant isotope is ^{232}Th which is an α emitter with a half-life of 14.1×10^9 years; the α particle has a kinetic energy of 4.1 MeV. There is a high uptake of the ^{232}Th by the liver, spleen, and bone. Liver washout is especially slow and in excess of 20 years. Follow-up studies in Europe, Japan, and the United States of America of those patients receiving Thorotrast have been conducted. In such studies, the absorbed dose to the liver from the α particles was estimated to be as high as 2 Gy and elevated incidences of angiosarcomas, biliary duct carcinomas, and hepatocellular carcinomas with a risk of the order of 3×10^{-2}/man-Gy for a latency period of 20 years.

In utero exposures: Of particular practical clinical nuclear medicine interest is the risk to the irradiated embryo or fetus. As proliferative cells are more radiosensitive than those that are not, the risk to the irradiated in utero being is expected to be elevated and, thus, of obvious concern. This has a specific clinical bearing as the leading risk of death during pregnancy is pulmonary embolism (PE) thus requiring imaging through either CT perfusion angiography or ventilation–perfusion scintigraphy, both of which contribute a radiation dose to the fetus. The Oxford Survey of Childhood Cancers (Stewart et al. 1956, 1958) and its reanalysis by Bithell and Stiller (1988) and a study conducted in the state of New England in the United States of America (MacMahon 1962) investigated the incidence of childhood cancer following in utero irradiation from diagnostic X-rays. The latter was extended to consider cancer mortality amongst 1,429,400 American children and demonstrated an excess in cancer incidence in those irradiated with X-rays in utero.

10.4.2.3 Occupational Exposures

Miners

One cohort exposed to radiation in the course of their occupation was miners. These individuals were chronically exposed to α particles resulting from the radioactive decays of ^{222}Rn and ^{226}Rn which are isotopes of radon which is an inert gas and a daughter product of ^{226}Ra.

Radiologists

The comparison of the cancer mortality rate of radiologists with those of other medical practitioners is based upon the expectation that radiologists practicing in the early part of the twentieth century were more likely to have received higher radiation absorbed doses due to then-primitive radiation protection practices than other physicians or radiologists who practiced in the latter part of the twentieth century.

Berrington et al. (2001) reported on an analysis of the causes of death of 2,690 radiologists registered in the United Kingdom and Ireland between 1897 and 1997 which found increased cancer mortality amongst radiologists registered up to 1954 compared to other physicians. After that year, there was no difference in the cancer mortality risks of radiologists and other physicians. The discussion, however, of radiologist risk has not been without debate and further examination (Cameron 2002; Doll et al. 2005).

Nuclear Workers

These cohorts are made up of individuals working in the civilian nuclear power industry and in shipyards building and servicing nuclear-propelled vessels such as submarines. A study of in excess of 95,000 workers in the United Kingdom, United States of America, and Canada demonstrated no excess risk in solid tumors, but a slight excess in leukemia in the United Kingdom cohorts was observed (Hall and Giaccia 2006).

Radium Dial Painters

This is perhaps the best-known cohort of individuals exposed to radiation through their occupation. These were workers who painted dials of watches and clocks with a radium-containing paint that fluoresced and was visible in the dark. They were predominantly women and who frequently shaped the paint brush with their lips, thus ingesting ^{226}Ra. Elevated incidences of osteosarcomas were observed (as there is high uptake of radium by bone) and of carcinomas in the paranasal sinuses and mastoid air cells, the latter presumably due to the radon gas product of the radium decay.

10.4.2.4 Chernobyl

The greatest recognized population accidentally exposed to ionizing radiation were those individuals living in the vicinity of the Chernobyl nuclear power reactor at the time of its explosion and of the firefighters sent to the site. Many of the latter suffered high radiation exposures and died, despite attempts at salvage through bone marrow transplantation, of symptoms reflective of the GI syndrome described in the following subsection. Of particular interests in the former population were the incidences of thyroid cancers due to the release of the fission product ^{131}I into the environment.

10.4.3 Radiation Pathologies

10.4.3.1 Introduction

Exposure of the whole body, or of specific sensitive organs or tissues, to high absorbed doses can culminate in acute effects leading to death. In diagnostic nuclear medicine, this acute radiation syndrome is not reached due to the very low amounts of administered activity and the consequent low organ absorbed doses. This is often not the case in therapeutic nuclear medicine where the radiotoxicities of red bone marrow (which leads to the hematopoietic syndrome) and kidney (leading to nephrotoxicity) can often limit the amount of administered activity.

Most data of the effects of total-body exposure to high absorbed doses of radiation are derived from animal experiments; human data are limited primarily to experiences of radiotherapy patients, the survivors of the Hiroshima and Nagasaki nuclear bombings and accidental exposures. Hall and Giaccia (2006) report that throughout the world some 120 fatalities resulted from radiation exposures arising from accidents involving radioactive material occurring between 1944 and 1999.

For a given population, the percentage of mortalities as a function of absorbed dose follows a sigmoidal curve, as shown in Fig. 10.28. The median lethal dose, labeled as LD_{50}, is about 400 cGy for humans (in the absence of therapy, such as bone marrow salvage).

Immediately following the exposure of a large part of the body to an absorbed dose exceeding the order of 50 cGy, early transitory symptoms appear. At absorbed doses comparable to the LD_{50} value given above, the resulting symptoms of irradiation include anorexia, nausea, and vomiting. At supralethal absorbed doses, the symptoms include diarrhea and hypotension, referred to as the prodromal radiation syndrome. This prodromal reaction is followed by a largely asymptomatic latent period, the duration of which is largely dictated by the kinetics of cell depletion in the irradiated tissues; an exception to this is the cerebrovascular syndrome for which the onset of death is rapid. Hematologic aplasia can be evident at absorbed doses as low as 50 cGy with a decrease in circulating lymphocyte counts.

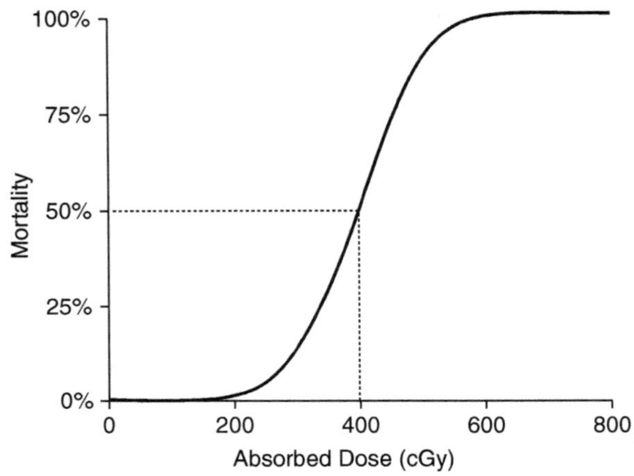

Fig. 10.28 A sigmoidal curve describing the mortality-absorbed dose relationship without salvage. The LD_{50} for the human is about 400 cGy

10.4.3.2 Cerebrovascular Syndrome

A whole-body dose of the order of 50–100 Gy will result in death within 48 h of exposure.[10] While all organ systems will be severely damaged as a result of such an irradiation, cerebrovascular damage brings death so rapidly that these other organ system effects cannot be manifested. The exact cause of death from the cerebrovascular syndrome is not clear. It has been typically attributed to cerebral edema, with extravasation of fluid, macrophages, and granulocytes into the brain and meninges. However, this description may not present the complete case as much higher absorbed doses are required to bring on this syndrome should only the brain be irradiated.

10.4.3.3 Gastrointestinal Syndrome

At whole-body absorbed doses of the order of 10 Gy, cells in the epithelial lining of the GI tract are

[10] In these discussions, we assume that the exposure is to photons (X or γ rays); due to the high-LET of the resulting recoil protons, the biological effects can be achieved at lower absorbed doses of neutrons.

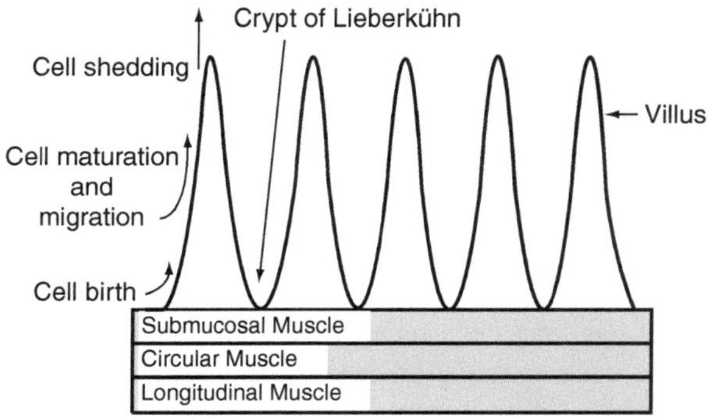

Fig. 10.29 Diagrammatic representation of the intestinal wall

depleted. The organization of this lining is that of self-renewing tissue. The mucosal walls of the intestine are covered by villi (as shown diagrammatically in Fig. 10.29) at a density of 20–40 villi/mm^2 and each villus projects about 0.5–1 mm and is covered by a single layer of columnar epithelium. The mucosa of the intestinal villi can be divided into four compartments. The stem cell compartment is at the base of the villus (the crypt of Lieberkühn) which is a region of great mitotic activity. Progressing up the villus, the cell differentiating compartment is found which also has a high mitotic index: this produces the functional cells which are found at the tip of the villus. Finally, at the villus' tip, the spent functional cells are sloughed off into the intestinal lumen.

Following irradiation to absorbed doses of about 10 Gy, the crypt and differentiating cells are killed off but the extrusion of the functional cells from the villi tips continues. As the villi surfaces are sloughed off due to the normal transit of material through the intestines, the villi begin to contract and, due to the non-replacement of the extruded cells, breaches of the intestinal lining occur in regions of denuded intestine leading to infection by intestinal flora. The presentation of denuded intestinal wall occurs a few days postirradiation in the human and the distal end of the irradiated small intestine is unable to resorb bile salts which enter the large intestine causing irritation and diarrhea. Because the magnitude of the absorbed dose will also lead to the individual being immunocompromised, any infections that occur within the intestinal wall can spread. Death in the human can subsequently occurs within a few days following irradiation.

10.4.3.4 Hematopoietic Syndrome

Whereas the cerebrovascular and GI syndromes are associated with high absorbed doses that are presented only in extreme circumstances, the hematopoietic syndrome can occur for absorbed doses comparable to those experienced in radiotherapy. Indeed, this syndrome presents the barrier of radiotoxicity to many approaches of radionuclide therapy. For absorbed doses of between 2.5 and 5 Gy, mitotically-active stem cells are killed off and the supply of erythrocytes and leukocytes is markedly diminished. The pathology resulting from the exposure to radiation does not occur until a number of weeks later when the circulating mature blood cells die off and the precursor cell supply is compromised.

10.4.4 Deterministic (Non-Stochastic) Effects

10.4.4.1 Introduction

This subsection reviews determinisitic effects associated with lower absorbed doses than those linked to the pathologies of the cerebrovascular, GI, or hematopoietic syndromes. The absorbed doses required to induce such pathologies are not typical of diagnostic medical procedures, although the radiotoxicity associated with the hematopoietic syndrome is a dose-limiting factor in therapeutic nuclear medicine. A deterministic effect is defined as a radiation effect in which there is an absorbed dose threshold of expression. Below this absorbed dose, there is no manifestation of an effect; but once this threshold is breached, the pathology becomes apparent and the degree of severity of the effect increases with absorbed dose, as shown in Fig. 10.30. Because of variations between individuals, the absorbed dose threshold is highly variably over a population. Table 10.3 summarizes absorbed dose levels at which such effects can become apparent.

10.4.4.2 Erythema and Epilation

These effects are observed in radiation therapy due to the high absorbed doses involved. They are, however,

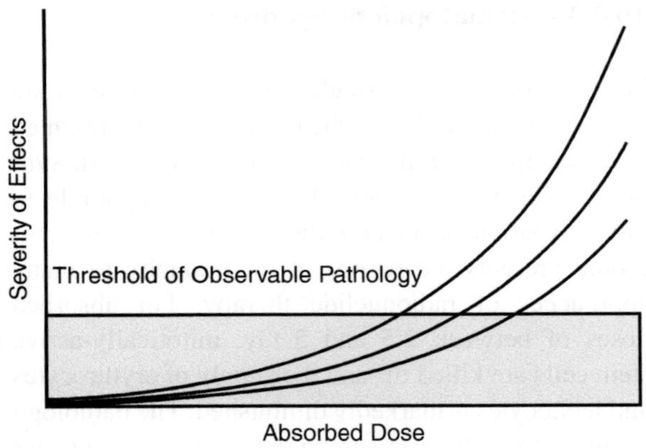

Fig. 10.30 Typical dose–response curves for a deterministic effect. Below a threshold, there is no observable pathology but, beyond which, the severity of the effect becomes manifest and increases with absorbed dose. The absorbed dose at which this effect is apparent varies among individuals as demonstrated by the three curves

Table 10.3 Absorbed dose thresholds for a selection of deterministic effects

Deterministic effect		Approximate typical threshold absorbed dose for acute exposure (Gy) (50% incidence in 5 years postirradiation)
Bone marrow: aplasia		0.5
Liver: hepatitis		40
Brain: infarction and necrosis		70
Lung: pneumonitis		35
Kidney: nephrosclerosis		25
Sterility	Female: permanent	60
	Male: temporary	0.15
	Male: permanent	50

not associated with either diagnostic or therapeutic nuclear medicine, but have been associated with some extreme forms of diagnostic exposures received in fluoroscopy examinations.

10.4.4.3 Sterilization

The effects of ionizing radiation upon individual germ cells have been described previously. Because of the difference between male and female gonadal kinetics, the induction of sterility differs between male and female in terms of degree and absorbed dose. The absorbed dose thresholds for temporary and permanent sterility are summarized in Table 10.3

10.4.4.4 Cataractogenesis

A cataract is any change in the transparency of the lens and it has long been recognized that exposure of the eye to ionizing radiation can result in the induction of a cataract. Cataractogenesis has been detected in both patient and medical practitioner as a result of diagnostic imaging irradiation. As protracted exposures require absorbed doses exceeding 5 Gy to the lens of the eye in order to induce a cataract, this effect is highly unlikely to result from a nuclear medicine procedure. However, it can be a potential risk associated with irradiation of the eye in diagnostic procedures such as cranial CT or fluoroscopy.

10.4.5 Stochastic Effects

10.4.5.1 Introduction

Stochastic effects are those for which the severity of outcome is independent of radiation dose but the probability of occurrence increases with absorbed dose. For example, consider a case of two women both exposed to ionizing radiation but from which one receives a higher absorbed dose to the breast than the other. The probability of breast cancer induction in the woman receiving the lower absorbed dose is expected to be less than that of the woman receiving the higher absorbed dose. However, the consequences of either morbidity or mortality are equal should both cancers be expressed in either woman, regardless of the probability of cancer induction. The stochastic effects resulting from exposure to ionizing radiation of interest are carcinogenesis and hereditary effects. Ionizing radiation does not lead to new stochastic effects attributable to radiation alone: all that can occur is the increase in the probability of the occurrence of deleterious effects which are manifested naturally.

There are two simple categories of epidemiological models which describe the elevation in risk resulting from exposure to ionizing radiation. The first is the absolute risk model in which exposure to ionizing radiation additively increases the incidence rate above the naturally-occurring rate. The second is the relative risk model in which the implicit assumption is that ionizing radiation merely increases the natural risk at all ages of exposure by a given factor. For the example of cancer, as the natural incidence of cancers increases with age, the relative risk model emphasizes a greater degree of radiocarcinogenesis in old age. A preferred modification of the relative risk model is to incorporate a time dependence. For example, let λ be the rate of cancer mortality of an irradiated population (e.g., the number of cancer deaths per unit population exposed to radiation) and λ_0 be that rate in an unirradiated population.

The epidemiological definition of the excess absolute risk (EAR) of cancer mortality is defined as

$$\mathrm{EAR} = \lambda - \lambda_0. \quad (10.43)$$

The relative risk of cancer mortality is,

$$\mathrm{RR} = \frac{\lambda}{\lambda_0} \quad (10.44)$$

and the excess relative risk (ERR) is, combining the above two results,

$$\begin{aligned} \mathrm{ERR} &= \frac{\lambda - \lambda_0}{\lambda_0} \\ &= \mathrm{RR} - 1. \end{aligned} \quad (10.45)$$

Note that both the relative and the ERRs are dimensionless quantities.

10.4.5.2 Radiation Carcinogenesis

The incidence of cancer following the exposure to low absorbed doses of ionizing radiation is of obvious concern to diagnostic nuclear medicine. This subsection considers the risks of three specific radiogenic cancers: leukemia, female breast, and thyroid. Models derived from epidemiological data and presented in the BEIR VII report are used in this subsection and considered in the context of low-LET radiation typical of diagnostic nuclear medicine. The risk is typically a function of the sex of the individual, the age at which exposure occurred, the age for which the risk is evaluated (i.e., time postexposure) and the absorbed dose received by the tissue of interest.

Leukemia

Radiation-induced leukemia was the first cancer to have been linked to the radiation exposures received by the survivors of the nuclear bombing of Hiroshima and Nagasaki (indeed, as from the deaths of many of the pioneers of the radiation sciences, such as Marie Curie). Moreover, it is recognized as being the cancer with the highest relative risk following exposure to ionizing radiation and is thus deserving of investigation. The BEIR VII parametric model of the ERR of radiation-induced leukemia is a function of absorbed dose, age of exposure, and of elapsed time following exposure. The sex-specific model of the ERR is,

$$\mathrm{ERR} = \beta D(1 + \theta D)e^{(\gamma\varepsilon^* + \delta \ln t/25 + \phi\varepsilon^* \ln t/25)} \quad (10.46)$$

where, for the consideration here of photon and electron radiation, D is the absorbed dose to the marrow (i.e., we need not consider the equivalent dose) and the mean estimated parameters are,

$$\begin{aligned} \beta &= 1.1\,\mathrm{Gy}^{-1} \quad \text{Male} \\ &= 1.2\,\mathrm{Gy}^{-1} \quad \text{Female;} \end{aligned}$$

$$\begin{aligned} \varepsilon^* &= \frac{\varepsilon - 30}{10} \quad \varepsilon \le 30 \\ &= 0 \quad \varepsilon > 30 \end{aligned}$$

where ε is the age at the time of exposure and the further parameters are,

$\gamma = -0.40$ per decade
$\delta = -0.48$
$\phi = 0.42$
$\theta = 0.87\ \mathrm{Gy}^{-1}$.

t is the time postexposure in units of years. Figure 10.31 shows a plot of the example of the ERR per Gy for a female 20 years of age at the time of exposure as a function of time following exposure,

The ERR per unit absorbed dose is elevated at early times postexposure, eventually subsiding to unity at about 45 years postexposure.

Breast Cancer

As the incidence of breast cancer incidence will vary with culture, diet, etc., most calculations of the ERR for radiogenic breast cancer are frequently referenced

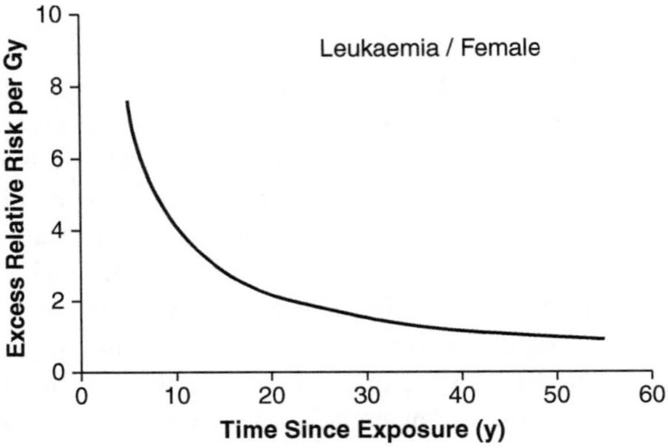

Fig. 10.31 Predicted ERR per Gy of leukemia for a 20-year-old female exposed to low-LET radiation as a function of time postexposure. Curve is calculated using (10.46) and mean values of parameters given in the BEIR VII report

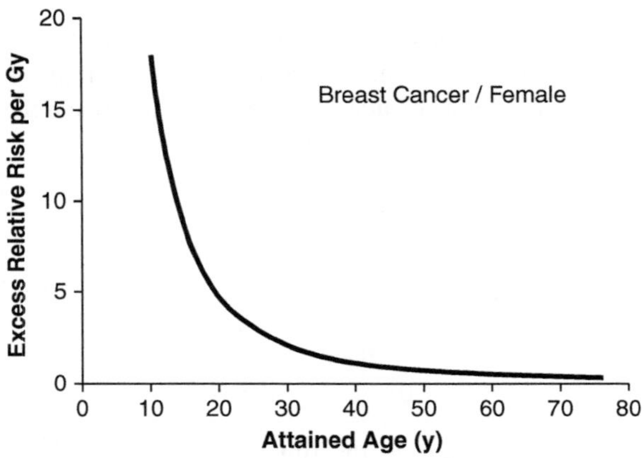

Fig. 10.32 Predicted ERR per Gy of breast cancer in the female as a function of attained age calculated from (10.47)

to North American women. Assuming, again, low-LET radiation, the ERR of breast cancer above the naturally-occurring rate, per Gy is modeled by the function in age,

$$\mathrm{ERR} = 0.51\left(\frac{60}{\mathrm{A}}\right)^2 \mathrm{Gy}^{-1} \qquad (10.47)$$

where A is the attained age in years. This ERR is shown in Fig. 10.32 as a function of attained age.

Thyroid Cancer

The induction of thyroid cancer as a result of the exposure of the thyroid gland to radiation due to, for

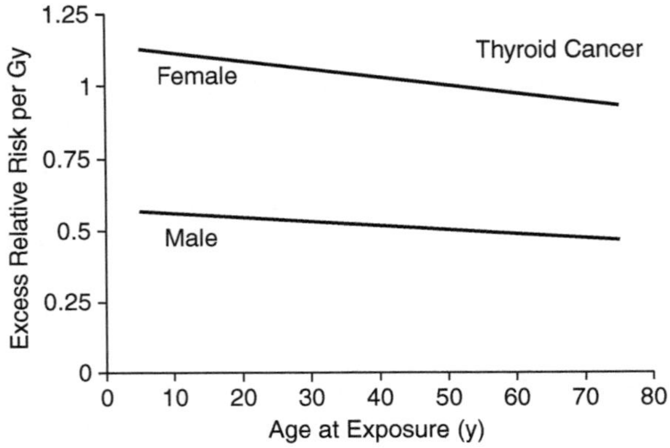

Fig. 10.33 Excess relative risk per unit absorbed dose of thyroid cancer incidence for males and females as a function of the age at exposure derived from the BEIR VII model

example, the high levels of radioactive iodine uptake by the population of the Marshall Islands who were subject to the radioactive fallout of the BRAVO thermonuclear bomb test in 1954 and those exposed to the radioactive fallout of the Chernobyl accident in 1986, is well quantified. For low-LET radiation, the BEIR VII model of thyroid cancer incidence as a function of age at the time of exposure is,

$$\mathrm{ERR} = \beta \mathrm{e}^{-0.083\left(\frac{\varepsilon-30}{30}\right)} \qquad (10.48)$$

where $\beta = 0.53$ for males, $\beta = 1.05$ for females and where ε is the age (in years) at the time of exposure. It should be recalled that ERR is multiplied by the naturally-occurring rate of thyroid cancer incidence to give the increased incidence due to ionizing radiation. This ERR as a function of the age at exposure is shown in Fig. 10.33.

It is immediately obvious that females are subject to a greater risk of radiogenic thyroid cancer than are males.

10.4.5.3 Hereditary Effects

Radiation-induced mutations in germ cells can lead to hereditary effects passed on to the progeny of the exposed individual. Such mutations do not lead to new or unique mutations that are specific to ionizing radiation. Hence, a convenient metric of quantifying the hereditary effects of ionizing radiation is through the "doubling dose," which is the absorbed dose required to increase the natural frequency of an hereditary effect by a factor of two. An estimate of the average doubling dose for low-LET radiation is 1.56 Gy (Hall and Giaccia 2006), whereas ICRP Publication 103 assumes the incidence of genetic risks up to the second generation following exposure to ionizing radiation as 0.2%/Gy over continuous low-dose rate exposure. Because of the naturally-occurring rate of hereditary effects, it is difficult to detect any radiation-induced mutations appearing in the progeny of irradiated individuals. In fact, to quote ICRP Publication 103: "There continues to be no direct evidence that exposure of parents to radiation leads to excess heritable disease in offspring" and that reports lend to the argument "...that the risk of heritable diseases tended to be overestimated in the past." Even so, prudence is necessary in assigning risks to hereditary effects occurring as a result of exposure to ionizing radiation. This is a consequence not necessarily of the risk itself but of its consequences. To explain this, one should consider the fact that a radiation-induced cancer can only affect the individual exposed. In the worst outcome, only a single individual dies. But in the case of a radiation-induced mutations passed onto the irradiated individual's offspring, a potentially deleterious mutation will have been inserted into the gene pool and more than just the irradiated individual can be affected. Hence, even though the risk of heritable disease as a consequence of exposure to ionizing radiation has not been demonstrated in the human, a risk coefficient for such effects is assigned.

10.5 Antenatal Effects

10.5.1 Introduction

Irradiation of the embryo and fetus is known to result in effects which may be manifest in either prenatal or neonatal states and which are dependent upon absorbed dose, gestational age, and absorbed dose rate. Recognizing that rapidly differentiating cells are more prone to radiation-induced damage, it is not unexpected that the embryo and fetus are acutely sensitive to the effects of ionizing radiation.

Medical irradiation of the pregnant patient is always of concern and presents the clinician of the starkest requirement of the optimization of risk and benefit of medical exposure to ionizing radiation. For example, most medical physicists practicing in nuclear medicine have been confronted with a case of a female patient in late pregnancy requiring a ventilation–perfusion scan to rule out PE, a condition which is the most important preventable cause of maternal death and with an incidence reaching 1.3% of all pregnancies (Chan and Ginsberg 1999). The need to estimate the absorbed dose to the fetus resulting from this procedure and contrasting the theoretical radiation-induced risk with the risks of failing to diagnose a present PE or the sequelae associated with the administration of an anticoagulant is almost always required in such situations. This subsection reviews the demonstrated risks of the result of maternal irradiation and the dose–response of these risks.

Prenatal sensitivity to ionizing radiation varies remarkably over the time of fertilization to implantation and gestation. During each cycle, within the ovarian cortex, a Graffian follicle containing a mature ovum moves to the surface of the ovary which ruptures to release the ovum into an awaiting end of the Fallopian tube which is likely brought into proximity of the ovary by chemotaxis. This process is known as ovulation. Once within the lumen of the Fallopian tube, the ovum is transported towards the uterus through ciliary action and peristalsis. Fertilization typically occurs shortly after intercourse (within 3 h) in the ampullary section of the Fallopian tube. The fertilized ovum is now a zygote and arrives at the uterus within about 5 days of fertilization. Whilst the zygote is still within the Fallopian tube, it begins to divide to form a mass of cells referred to as the morula, within which a cavity appears and the structure forms a blastocyst. Once the blastocyst has reached the uterus, it becomes embedded within the endometrium and extravasation of the maternal blood occurs around the blastocyst to provide its nutrition through maternofetal exchange. This description of the trail from ovulation to implantation and prenatal development is necessary for understanding the radiation sensitivities of the zygote, blastocyst, embryo, and fetus, and our characterizations of the risks. Following the use of Russell and Russell (1954), we simplify the discussion by dividing the prenatal phase into three components:

Preimplantation: That period of time (5–6 days) between fertilization in the Fallopian tube and the embedding of the blastocyst in the endometrium

Organogenesis: That period of time following implantation during which the major organs develop

Fetal period: That period of time during which the developed organs grow

10.5.2 Embryonic Death

Irradiation of the embryo can result in death. Absorbed doses of the order of 2–3 Gy delivered to the embryo in the first 20 days of gestation are likely to result in the resorption of the embryonic material or abortion of the embryo.

10.5.3 Microcephaly and Mental Retardation

Severe injuries to the developing human brain were documented in the survivors of the Hiroshima and Nagasaki bombings. The gestational age at which exposure occurred is critical to understanding the radiation-induced damage. Four categories of gestational age are aligned to development of the brain: 0–7 weeks, 8–15, 16–25, and greater than 26 weeks. The precursors of the neuroglia and neurons appear and are mitotically active during the first gestational stage which is followed by the second gestational stage which is marked by an increase in the number of neurons and their migration to their developmental sites subsequent to their cessation in mitotic capacity. During the third stage, synaptogenesis accelerates and the cytoarchitecture of the developing brain has been defined. Further advancements in cytoarchitecture differentiation and synaptogenesis occur in the final stage of gestational development.

The sensitivity of the developing brain is thus expected to be dependent upon the in utero absorbed dose and gestational age. Most of the epidemiological data to derive these results were obtained from children exposed in utero at the Hiroshima and Nagasaki bombings and the consequences of irradiation were microcephaly and mental retardation. Microcephaly is associated with in utero exposures at less than 15 weeks gestational age without an apparent threshold absorbed dose. The incidence of mental retardation is linearly related to the fetal absorbed dose during the critical gestational age of 8–15 weeks with an incidence risk coefficient of about 40%/Gy and which is commensurate with an absorbed dose threshold of about 15 cGy. Absorbed doses of such magnitudes are unachievable in diagnostic nuclear medicine.

10.5.4 Childhood Cancer

Childhood cancer as a consequence of in utero irradiation was recognized in studies of children irradiated in utero from radiological obstetric examinations and the Hiroshima and Nagasaki bombings. The sensitivity of childhood neoplasia due to prenatal irradiation peaks during the third trimester of pregnancy (Bithell and Stiller 1988). The EAR is estimated to be about 5%/Gy. In practical terms, ICRP Publication 103 (ICRP 2007) advises that the lifetime risk of cancer as a consequence of in utero exposure is no different from exposure during early childhood which is no more than three times that of the general population.

10.6 Radiation Risks Presented to the Diagnostic Nuclear Medicine Patient

10.6.1 Introduction

Consideration of the safety of the diagnostic nuclear medicine patient will clearly emphasize stochastic effects as deterministic effects of exposure to ionizing radiation are unachievable due to the low absorbed doses received by such a patient. Consequently, the foci of discussion will be on radiocarcinogenesis (of either the patient or the fetus should the patient be pregnant) and the potential for hereditary effects.

The fundamental problem in translating the risks determined from epidemiological studies to those of the diagnostic nuclear medicine patient is that both the subjects in the source data cohorts and the patient are, in general, irradiated nonuniformly. Different organs and different tissues have differing radiosensitivities, as demonstrated earlier in the ERRs of the breast, thyroid gland, and leukemia. As a result, the simple use of the physical quantity of absorbed dose is inadequate to derive the risk an individual is subject to as a result of nonuniform irradiation. Moreover, consideration must be taken of the sex and age of the individual and the time elapsed since exposure. It is immediately apparent that the mapping from the simple physical description of absorbed dose to an assessment of stochastic risk is highly complicated.

The ICRP has addressed these problems since the 1970s with simple, applicable and generic solutions which are periodically refined through the incorporation of new knowledge of the effects of ionizing radiation. These account for, first, the physical attributes of the radiation that the individual is exposed to (e.g., LET and absorbed dose rate) and, second, the inherent radiosensitivity of the individual organs and tissues irradiated. This latter consideration includes dependencies (which are averaged) upon sex, age, and absorbed dose rate. As a result, one must reflect upon the effective dose as a quantity which is rather unspecific and best suited as a means of comparing the relative risks between differing types of radiation exposure.

10.6.2 ICRP Recommendations

The ICRP publishes recommendations on radiation protection which is updated about every 5–15 years (ICRP 1959, 1964, 1966, 1977, 1991, 2007). The three publications of most modern-day relevance are those of Publication 26 (ICRP 1977), which was superseded by Publication 60 (ICRP 1991) which has been superseded, in turn, by Publication 103 (ICRP 2007).

Of particular importance to those imaging procedures incurring low doses of ionizing radiation, the ICRP has defined the quantity of the effective dose as a sex-, age-, and dose rate-independent measure of the stochastic risk presented to an individual as a result of a nonuniform irradiation. Note that this quantity is applicable to stochastic risk only. The ICRP Recommendations equate the stochastic risk of nonuniform radiation (of high- and/or low-LET radiation) to that resulting from uniform (i.e., total body) exposure: in other words, the effective dose due to nonuniform irradiation is numerically equal to the absorbed dose from a whole-body exposure to low-LET radiation that yields the same stochastic risk.

10.6.3 Equivalent (Radiation Weighted) Dose

10.6.3.1 Introduction

The ICRP Recommendations consider radiological protection in a wide variety of applications including medical (both practitioner and patient), occupational, and aviation exposures. As a result these must account for several different species of radiations: photons, electrons/positrons, neutrons, protons, α particles, and pions (the latter being a component of cosmic rays and consequently of interest to radiation exposure associated with high-altitude aviation). Following from the discussions of Chaps. 6 and 7, it will be recognized that these radiations will have different LETs resulting in, as shown earlier in this chapter, different RBEs. These are accounted for by defining the equivalent dose which is the absorbed dose weighted by a factor, w_R, to account approximately for the RBE relevant to the radiation of interest. For nuclear medicine dosimetry, these are photons and electrons/positrons resulting from β decay (α particle and Auger/Coster–Kronig electrons are relevant only to therapeutic applications associated with high absorbed doses).

10.6.3.2 Radiation Weighting Factor, w_R

The equivalent dose to a given tissue or organ is the weighted summation of the absorbed doses to that tissue or organ T from all of the radiations that it is exposed to,

$$H_T = \sum_R w_R D_{T,R} \tag{10.49}$$

where w_R is the radiation weighting factor for radiation species R and $D_{T,R}$ is the corresponding (physical) absorbed dose to the tissue from that radiation. The

Table 10.4 Radiation weighting factors used in ICRP Publications 26, 60, and 103

Radiation	Radiation weighting factor, wR		
	ICRP Publication 26 (1977)	ICRP Publication 60 (1991)	ICRP Publication 103 (ICRP, 2007)
γ, $e^{\pm}$, $\mu^{\pm}$ (excluding Auger electrons)	1	1	1
Protons	10	2	2
$\pi^{\pm}$	–	–	2
α particles, fission fragments, heavy nuclei	–	20	20
Neutrons	10	Function of neutron energy	Function of neutron energy

radiation weighting factor is an approximation of the LET of the radiation. The radiation weighting factors provided by ICRP Publications 26, 60, and 103 are provided in Table 10.4

The absorbed dose is averaged over the entire organ and the physical unit of the equivalent dose is that of the absorbed dose, Gy, but, because the quantity is committee-defined, is assigned the special name sievert (Sv).

As the weighting factor for photons and electrons/positrons[11] is unity, one can interpret the equivalent dose of a given combination of radiations to an organ or tissue as being equal to the (physical) absorbed dose in photons or electrons (with $w_R = 1$) that yields the same biological effect to the individual organ.

Auger/Coster–Kronig electrons are treated by the ICRP as a special case in which it is recognized that it is inappropriate to average the absorbed doses from these radiations over the total mass of DNA should such electron-emitting isotopes be incorporated within the DNA. If the Auger/Coster–Kronig electron-emitting isotope is external to the cell, the efficiency of producing any biological effect is negligible due to the very short ranges of Auger electrons. However, should the radionuclide be incorporated within the nucleus, the RBE increases. If Auger electrons are emitted from within the cell, but not from within the nucleus, the RBE can be as high as 8 (Kassis et al. 1988). For cases where the Auger electron-emitting radionuclide is incorporated within the nucleus and, subsequently, the DNA, RBE values in excess of 20 have been measured. The ICRP states that the biological effects of Auger electrons are omitted from the radiation weighting factors of Table 10.4 and must be dealt with using microdosimetry.

[11] Excluding Auger/Coster–Kronig electrons.

10.6.4 Effective Dose

10.6.4.1 Introduction

Medical radiation protection practice must contend with two important facts. First, almost all medical radiation exposures are nonuniform throughout the body and, second, tissues and organs have varying radiosensitivities. Thus, in order to estimate the stochastic risk associated with a given imaging study, one must know which organs/tissues have been irradiated, the equivalent doses received by them and their intrinsic radiosensitivities. Determining the absorbed doses received by organs and tissues in nuclear medicine is the fundamental topic of the following chapter. The intrinsic radiosensitivities are derived from the data described earlier in this chapter. The contribution of these to an overall estimate of radiation risk is modeled by the tissue weighting factor, w_T.

10.6.4.2 Tissue Weighting Factor, w_T

The effective dose is the sum of the equivalent doses over a defined ensemble of organs and tissues each weighted by a tissue weighting factor,

$$E = \sum_T w_T H_T \qquad (10.50)$$

In ICRP 26, this tissue weighting factor reflected only mortality risk whereas in ICRP 60 and ICRP 103 it reflected a quantity known as detriment which is an aggregate of four quantities:

- Probability of attributable fatal cancer
- Weighted probability of attributable nonfatal cancer

- Weighted probability of severe hereditary effects
- Relative amount of time of life lost

In ICRP 26, the weighted equivalent dose, which is a doubly-weighted absorbed dose, was assigned the name effective dose equivalent. The tissues and organs considered in the calculation of the effective dose have grown over the three publications reflecting the expanding knowledge of the radiosensitivities of these tissues. Table 10.5 summarizes the historical development of the weighting factors provided by ICRP Publications 26, 60, and 103.

The procedure recommended in ICRP Publication 103 of calculating the effective dose on the basis of the measured biodistribution of a radionuclide is shown in Fig. 10.34. The biodistribution is used to calculate the absorbed doses to specified tissues and organs of reference male and female anthropomorphic phantoms. The application of the radiation weighting factors converts these absorbed dose quantities to equivalent doses for each specified tissue and organ. Having calculated the equivalent doses, these are then averaged to yield an ensemble of organ and tissue equivalent doses for the male and female reference phantoms. These are then further averaged over both sexes to yield a single ensemble of tissue equivalent doses and the tissue weighting factors are then applied to yield the effective dose.

Table 10.5 Tissue weighting factors used in ICRP Publications 26, 60, and 103

Tissue or organ	Tissue weighting factor, w_T		
	ICRP Publication 26 (1977)	ICRP Publication 60 (1991)	ICRP Publication 103 (2007)
Brain	–	–	0.01
Salivary glands	–	–	0.01
Red bone marrow	0.12	0.12	0.12
Lung	0.12	0.12	0.12
Breast	0.15	0.05	0.12
Colon wall	–	0.12	0.12
Stomach wall	–	0.12	0.12
Esophagus	–	0.05	0.04
Bone surface	0.03	0.01	0.01
Skin	–	0.01	0.01
Thyroid gland	0.03	0.05	0.04
Gonads	0.25	0.20	0.08
Liver	–	0.05	0.04
Urinary bladder wall	–	0.05	0.04
Remaining tissues	0.30[a]	0.05[b]	0.12[c]

[a]The remaining tissues in ICRP 26 are the stomach, salivary glands, lower large intestine wall, and liver

[b]The remaining tissues in ICRP 60 are the adrenal glands, brain, upper large intestine wall, small intestine wall, kidney, muscle, pancreas, spleen, thymus, and uterus. Further modifications were recommended in later ICRP publications and are discussed in the text

[c]The remaining tissues in ICRP 103 are adipose tissue, adrenal glands, connective tissue, extrathoracic airways, gallbladder wall, kidneys, cardiac wall, lymphatic nodes, muscle, pancreas, prostate gland, small intestine wall, spleen, thymus, and uterus/cervix

Table 10.5 is an excellent example of displaying how the understanding of the effects of radiation dosimetry has evolved over the decades. For example, the most profound change has been the weighting factor assigned to the absorbed dose to the gonads which has decreased from 0.25 in ICRP Publication 26 to 0.20 in ICRP Publication 60 and, finally, to 0.08 in ICRP Publication 103. These reductions reflect the recognition over time that radiation-induced hereditary effects are much less likely than had been previously believed.

In diagnostic medicine, one is interested in a measure of the risks associated with the nonuniform exposure to low-dose ionizing radiation. The ICRP has provided estimates of the detriment per unit effective dose in both Publications 60 and 103. These are summarized in Table 10.6 (as the recommendations of these publications are applicable to those who are occupationally exposed to ionizing radiation, risk factors are calculated separately for adult workers and the general population, the latter including children).

The most marked change in stochastic risks between those specified by ICRP 60 and 103 is that assigned to severe hereditary effects. Any change in hereditary effects due to exposure ionizing radiation is measured by comparing the incidence of these effects in an irradiated population with a matched nonirradiated population. This reflects that the natural incidence of these effects in the natural (nonirradiated) population is not that markedly lower than that of the appearance of these effects in the matched irradiated population.

10.6.4.3 Additional Considerations

Following the release of ICRP Publication 60, the ICRP has published a number of refinements to the evaluation of the effective dose.

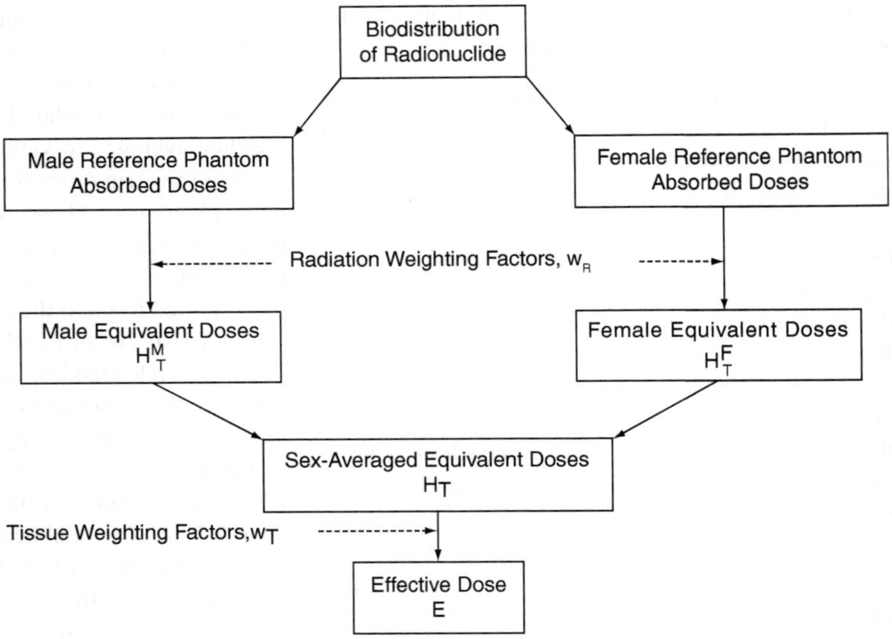

Fig. 10.34 Means of estimating the effective dose following the recommendations of ICRP Publication 103

Table 10.6 Nominal probability coefficients for stochastic effects from ICRP Publications 60 and 103

Exposed populations		Detriment (Sv^{-1})			
		Fatal cancer	Nonfatal cancer	Severe hereditary effects (%)	Total
Adult workers	ICRP 60	4.0%	0.8%	0.8	5.6
	ICRP 103	4.1% (combined)		0.1	4.2
Entire population	ICRP 60	5.0%	1.0%	1.3	7.3
	ICRP 103	5.5% (combined)		0.2	5.7

Gonadal Absorbed Dose

As the individual for whom the effective dose is often to be calculated for is an adult hermaphrodite, in such a case the absorbed doses to the gonads are to be averaged over the testicular and ovarian absorbed doses.

Esophagus/Thymus Absorbed Dose

The esophagus is a specified organ in the evaluation of the effective dose. However, the esophageal absorbed dose is not calculated for in the MIRD schema used in nuclear medicine (Chap. 11). To resolve this discrepancy, the absorbed dose to the thymus is used as a surrogate to that to the esophagus.

Colon Absorbed Dose

For radiation dosimetry purposes, the anatomy of the colon is considered to consist of:

- The ascending colon which is defined as being the extent of the bowel from the cecum to the hepatic flexure.
- The transverse colon which is defined as extending between the hepatic and splenic flexures.
- The descending colon which extends from the splenic flexure and includes the sigmoid colon and the rectum.

In ICRP Publication 60, the upper large intestine (ULI) wall is included within the remaining tissues category. However, the ICRP later recommended an alteration to the manner in which the absorbed dose to the walls of the colon and the corresponding tissue weighting factors contribution to E were to be evaluated. The absorbed dose to the colon is to be given by the mass-weighted absorbed doses to the walls of the ULI and lower large intestine (LLI),

$$D_{Colon} = 0.57\,D_{ULI} + 0.43\,D_{LLI}. \qquad (10.51)$$

The ULI is defined as that portion of the colon from the cecum, the ascending colon to the hepatic flexure; the transverse colon is that from the hepatic flexure to the splenic flexure; and, the LLI consists of the descending colon from the splenic flexure, the sigmoid colon, and the rectum.

10.6.4.4 Use of the Effective Dose in Nuclear Medicine

The use of the effective dose as a descriptor of radiation risk arising from medical exposures has long been contentious (e.g., Martin 2007; Brenner 2008; Dietze et al. 2009). This has been largely a consequence of the multiple acts of averaging over sex and age and the simplification of the biological effects of absorbed dose rate when using epidemiological data. Further, the effective dose does not account for the dependence of risk upon age at exposure and the tissue weighting factor represents a detriment, which is an amalgam of the endpoints of cancer mortality and incidence, hereditary risk, and reduction in life expectancy. Martin (2007) has suggested the means of how the effective dose should be used in assessing or comparing the risks associated with medical exposures. His key recommendation is that the use of the effective dose be simply as a generic indicator or risk to a reference hermaphroditic phantom. More specific risk estimates for a given patient should following the organ-, sex-, and age-specific risk models, examples of which were provided earlier.

10.7 Radiobiology Considerations for the Therapeutic Nuclear Medicine Patient

10.7.1 Introduction

Whereas the intent of the application of radiobiology to the diagnostic nuclear medicine patient is for estimating the potential for stochastic risk resulting from the exposure to low doses of low-LET radiation, the corresponding application to the therapeutic patient is markedly different. As noted several times before, biological effect in therapeutic nuclear medicine is sought in terms of killing tumor cells, whilst minimizing radiation-induced damage to normal tissue. This balance between tumor control probability (TCP) and radiotoxicity, or normal tissue complication probability (NTCP), is critical to the curative intent of therapeutic nuclear medicine.

10.7.2 Tumor Control Probability

One can consider that there are two endpoints for tumor response, those of sterilization and remission. Thus, tumor control is, as would be evident, related to the number of remnant viable tumor cells. Applying Poisson statistics, one can define the TCP as,

$$\mathrm{TCP} = e^{-N_t} \qquad (10.52)$$

where N_t is the number of surviving clonogenic tumor cells.

10.7.3 Normal Tissue Complication Probability

The NTCP is not only a function of the physical attributes of irradiation (LET and absorbed dose rate[12]), but also a function of the tissue's organizational structure. Normal tissue tolerance is the result of the ability of the clonogenic cells of the tissue to ensure an adequate number of mature cells required to enable normal organ or tissue function to continue. But this is also subject to the hierarchical structure of the tissue, hence the previous emphasis on mature cells. Rapidly proliferating stem cells are, as discussed earlier, sensitive to radiation. These stem cells differentiate to eventually produce functional cells which are less sensitive to radiation. Examples of this

[12] For external beam radiotherapy, the fractionation regime must also be considered.

hierarchical model are hematopoietic bone marrow and the intestinal epithelium. The resulting effects of radiation upon these tissues have a latent period: the pluripotential stem cells are killed off more readily by irradiation, but, due to the serial time development of the tissue, this is not apparent in terms of organ function until the mature cells have been depleted and there are no replacements from the descendants of the now-depleted stem cells.

Other tissues do not exhibit such a hierarchical structure, an example being the liver. Normally, the cells of such tissue rarely undergo mitosis but can be induced to do so following trauma or injury to the tissue. The tissue structure lacks a hierarchy and, as a result, all cells begin to divide following an insult.

Many organs or tissues may be considered to be made up of functional subunits (FSUs) which can be discrete and clearly defined structures contributing to the function of the organ and tissue. An example is the nephron in the kidney. The exchange of clonogenic cells between FSUs is not possible. In some tissues, the FSUs are not so clearly defined; examples include mucosa and the skin. Differentiation between the two types of FSU categorization is important as the responses to radiation differ between them. Due to each FSU being independent and small, depletion of the cells within it can be readily achieved through low absorbed doses. This, for example, explains the relatively low radiation tolerance of the kidney. On the other hand, migration of cells between structurally undefined FSUs is possible: an area of skin denuded due to radiation can be restored by the transfer of clonogenic cells from surrounding unaffected skin.

Tolerance doses are frequently quantified by $TD_{5/5}$ and $TD_{50/5}$ which are the tolerance doses yielding a 5% complication rate in 5 years and a 50% complication rate in 5 years, respectively. Values for these tolerance doses are, in general, derived from external beam radiotherapy and, hence, are complex functions of fractionation, absorbed dose rate, and fraction of the volume of organ/tissue irradiated. Meredith et al. (2008) provide a comparison of tolerance dose between external beam radiotherapy and radionuclide therapy using β-emitting radionuclides.

The tolerance doses of normal tissues will clearly vary between tissues and, indeed, between patients. For radionuclide therapy, the normal tissue which defines the radiotoxic limit is frequently the active bone marrow which would be recognized as having a tolerance dose of a few hundred cGy.

Meredith et al. (2008) have summarized the three main points associated with normal tissue tolerance to radionuclide therapy:

- The tolerance of normal tissue is affected by absorbed dose rate.
- The inhomogeneous distribution of absorbed dose affects the tolerance of the irradiated organ as a whole.
- The FSUs of different organs (e.g., kidney and spinal cord) have different tolerances.

Wessels et al. (2008) have recently shown how radiobiology concepts can be incorporated within the MIRD schema for estimating the risk of normal tissue complications to the kidney.

10.7.4 Selection of Isotopes for Radionuclide Therapy

The selection of a particular isotope for radionuclide therapy must consider several biological, chemical, and physical factors which can be distilled into three categories:

Radiation quality: For radiotherapy, particulate radiations (α particles, β particle, and Auger/Coster–Kronig electrons) are best suited due to the limited range and high LET of the particles. The limited ranges of these particles are advantageous in that this can limit radiotoxicity by reducing the absorbed doses to normal, healthy tissues provided that the vector carrying the radioisotope has a high specificity for the target. Kassis (2003) has noted the implication of the differences between the LET of α emitters (~100 keV/μm) with those of β particles (<1 keV/μm): approximately four traversals of the cellular nucleus by the former will result in cell death whereas several thousand traversals by the latter are required for the same result.

Specificity: The radionuclide must be delivered preferentially to the disease location with minimal uptake by normal tissues. This specificity is achieved through the choice of an appropriate vector. This can be simply the chemical form used. For example, radium chloride has a high uptake in bone and has been used (with various isotopes of radium) for the treatment of ankylosing spondylitis, bone tuberculosis, and osseous metastases. Monoclonal antibodies can be labeled with particulate-emitting radionuclides.

Physical half-life: From Chap. 5, the effective half-life is a function of the physical half-life of the radioisotope used and the biological half-life of the labeled moiety in the targeted organ or tissue. Here, we shall presume that the letter is a component of the specificity category above and consider the former. We have seen how the biological effects of irradiation are a function of the rate at which the absorbed dose is delivered and the α/β ratio of the tissue in question. For radionuclide therapy, this effect is demonstrated by the Lea–Catcheside dose-protraction factor. Howell et al. (1994) used the LQ model to incorporate the rate of repair, the rate at which the absorbed dose is delivered, the physical and biological half-lives of the radioactive moiety and the required absorbed dose to conclude that radionuclides with longer half-lives were more advantageous in radionuclide therapy. They concluded that the optimal physical half-life of the radionuclide should be two to three times greater than the biological clearance half-life. For example, consider the effective half-life given by (5.10) in which the physical half-life is $T_{1/2,\text{Phys}} = 3T_{1/2,\text{Biol}}$. Then,

$$T_{1/2,\text{eff}} = \frac{3}{2}T_{1/2,\text{Phys}} \tag{10.53}$$

The authors concluded that among the β-emitting isotopes, ^{32}P, with a 14.126 day half-life, would be optimal for radionuclide therapy. Rao and Howell (1993) demonstrated how time-fractionation from external beam radiotherapy can offer guidance on selection of the radionuclide half-life.

Currently, most radionuclide therapy is based upon the use of β-emitting isotopes such as ^{131}I and ^{90}Y with maximum kinetic energies of 610 keV and 2.28 MeV, respectively. Hence, the associated LET values will be low, although the range of these β particles are several millimeter which could be advantageous if there is not a specific uptake of the isotopes into the nucleus of the target tumor cell. α-emitting isotopes are progressing into the radionuclide therapy field, examples being ^{211}At and ^{226}Ra. α particles exhibit high LET and the typical range of them are several cell diameters. Hence, α-emitting isotopes can be more efficient at producing lethal effects.

Auger/Coster–Kronig electrons are particularly advantageous in delivering a high absorbed dose to a small volume. Even ^{99m}Tc has an Auger/Coster–Kronig electron component, releasing, on average, four such electrons per decay and depositing nearly 300 eV within a 5 nm sphere around the nuclide. ^{125}I emits an average of 20 Auger/Coster–Kronig electrons per decay which deposit an energy of about 1 keV within a 5 nm sphere. The high-energy depositions in small volumes associated with Auger/Coster–Kronig emissions demonstrates that the short ranges of these electrons requires that the radionuclides to be incorporated within the cellular nucleus in order to be in close proximity to the DNA target (or even incorporated within it) in order to yield tumor cell inactivation.

References

Baechler S, Hobbs RF, Prideaux AR et al (2008) Extension of the biological effective dose to the MIRD schema and possible implications in radionuclide therapy dosimetry. Med Phys 35:1123–1133

Barendsen GW, Koot CJ, van Kersen GR, Bewley DK, Field SB, Parnell CJ (1966) The effect of oxygen on impairment of the proliferative capacity of human cells in culture by ionizing radiations of different LET. Int J Radiat Biol Relat Stud Phys Chem Med 10:317–327

Barendsen GW (1968) Responses of cultured cells, tumors, and normal tissues to radiation of different linear energy transfer. Curr Top Radiat Res Q 4:293–356

Berrington A, Darbcy SC, Weiss HA, Doll R (2001) 100 years of observation on British radiologists: mortality from cancer and other causes 1897–1997. Br J Radiol 74:507–519

Bithell JF, Stiller CA (1988) A new calculation of the carcinogenic risk of obstetric x-raying. Stat Med 7:857–864

Boice JD Jr, Day NE, Andersen LA, Brinton R, Brown NW, Choi EA et al (1985) Second cancers following radiation treatment for cervical cancer: an international collaboration among cancer registries. J Natl Cancer Inst 74:955–975

Brenner DJ, Hlatky LR, Hahnfeldt PJ et al (1998) The linear-quadratic model and most other common radiobiological models result in similar predictions of time-dose relationships. Radiat Res 150:83–91

Brenner DJ, Hall EJ (2007) Computed tomography – an increasing source of radiation exposure. N Engl J Med 357:2277–2284

Brenner DJ (2008) Effective dose: a flawed concept that could and should be replaced. Br J Radiol 81:521–523
Cameron JR (2002) Radiation increased the longevity of British radiologists. Br J Radiol 75:637–639
Chan WS, Ginsberg J (1999) Management of venous thromboembolism in pregnancy. In: Oudkerk M, van Beek EJR, ten Cate JW (eds) Pulmonary embolism. Blackwell Science, Berlin
Cornforth MU, Bedford JS (1987) A quantitative comparison of potentially-lethal damage repair and the rejoining of interphase chromosome breaks in low passage normal human fibroblasts. Radiat Res 111:385–405
Dale RG (1985) The application of the linear quadratic dose-effect equation to fractionated and protracted radiotherapy. Br J Radiol 58:515–528
Dale RD (1996) Dose-rate effects in targeted radiotherapy. Phys Med Biol 41:1871–1884
Dietze G, Harrison JD, Menzel HG (2009) Letter to the editor. Br J Radiol 82:348–351
Doll R, Barrington A, Dary SV (2005) Low mortality of British radiologists. Br J Radiol 78:1057–1058
Folkard M, Prise KM, Vojnovic B (2007) Status of charged particle microbeams for radiation biology. J Phys 58: 62–67
Fowler JF, Stern BL (1960) Dose-rate effects: some theoretical and practical considerations. Br J Radiol 33:389–395
Frankenberg-Schwager M (1989) Review of repair kinetics for DNA damage induced in eukaryotic cells in vitro by ionizing radiation. Radiother Oncol 14:307–320
Goodhead DT (1985) Saturable repair models of radiation action in mammalian cells. Radiat Res Suppl 104: S58–S67
Hall EJ (2003) The bystander effect. Health Phys 85:31–35
Hall EJ, Giaccia AJ (2006) Radiobiology for the radiologist, 6th edn. Lippincott, Williams & Wilkins, Philadelphia
Howell RW, Goddu SM, Rao DV (1994) Application of the linear-quadratic model to radioimmunotherapy: further support for the advantage of longer-lived radionuclides. J Nucl Med 35:1861–1869
ICRP (1959) Recommendations of the International Commission on Radiological Protection. ICRP Publication 1. Pergamon, Oxford
ICRP (1964) Recommendations of the International Commission on Radiological Protection. ICRP Publication 6. Pergamon, Oxford
ICRP (1966) Recommendations of the International Commission on Radiological Protection. ICRP Publication 9. Pergamon, Oxford
ICRP (1977) Recommendations of the International Commission on Radiological Protection. ICRP Publication 26. Pergamon, Oxford
ICRP (1991) Recommendations of the International Commission on Radiological Protection. ICRP Publication 60. Pergamon, Oxford
ICRP (2007) The 2007 Recommendations of the International Commission on Radiological Protection. ICRP Publication 103. Pergamon, Oxford
Kahim MA, Macdonal DA, Goodhead DT, Lorimore SA, Marsden SJ, Wright EG (1992) Transmission of chromosomal instability after plutonium α-particle irradiation. Nature 355:738–740
Kassis AI, Havell RW, Sastry KSR, Adelstein SJ (1988) Positional effects of Auger decays in mammalian cells in culture. In: Baverstock KF, Charlton DE (eds) DNA: damage by Auger emitters. Taylor and Francis, London
Kassis AI (2003) Radiobiology aspects and radionuclide selection criteria in cancer therapy. In: Zaidi H, Sgouros G (eds) Therapeutic applications of Monte Carlo calculations in nuclear medicine. Institute of Physics, Bristol
Kleinerman RA, Boice JD, Storm HH, Sparen P, Andersen A, Pukkala E, Lynch CF, Hankey BF, Flannery JT (1995) Second primary cancer after treatment for cervical cancer: an international cancer registries study. Cancer 76:442–452
Lea DE, Catcheside DG (1942) The mechanism of the induction by radiation of chromosome aberrations in Tradescantia. J Genet 44:216–245
Luckey TD (1991) Radiation hormesis. CRC, Boca Raton
McParland BJ (1990) The effect of a dynamic wedge in the medial tangential field upon the contralateral breast dose. Int J Radiat Oncol Biol Phys 19:1515–1520
McParland BJ, Fair HI (1992) A method of calculating peripheral dose distributions of photon beams below 10 MV. Med Phys 19:283–293
MacMahon B (1962) Prenatal X-ray exposure and childhood cancer. J Natl Cancer Inst 28:1173–1191
Martin CJ (2007) Effective dose: how should it be applied to medical exposures? Br J Radiol 80:639–647
Meredith R, Wessels B, Knox S (2008) Risks to normal tissues from radionuclide therapy. Semin Nucl Med 38:347–357
Mothersill C, Seymour C (2001) Radiation-induced bystander effects: past history and future directions. Radiat Res 155:759–767
Munro TR (1961) Irradiation of selected parts of single cells. Ann NY Acad Sci 95:920–931
Nagasawa H, Little JB (1992) Induction of sister chromatid exchanges by extremely low doses of α-particles. Cancer Res 52:6394–6396
Nais AHW (1998) An introduction to radiobiology, 2nd edn. Wiley, Chichester
National Research Council (1990) Health effects of exposure to low levels of ionizing radiation BEIR V. National Academy Press, Washington, DC
National Research Council (2006) Health effects of exposure to low levels of ionizing radiation BEIR VII Phase 2. National Academy Press, Washington, DC
Radiation Effects Research Foundation (1987a) US-Japan joint reassessment of atomic bomb radiation dosimetry in Hiroshima and Nagasaki – final report, vol 1. Radiation Effects Research Foundation, Hiroshima
Radiation Effects Research Foundation (1987b) US-Japan joint reassessment of atomic bomb radiation dosimetry in Hiroshima and Nagasaki – final Report, vol 2. Radiation Effects Research Foundation, Hiroshima
Rao DV, Howell RW (1993) Time-dose-fractionation in radioimmunotherapy: implications for selecting radionuclides. J Nucl Med 34:1801–1810
Ron E, Modan B, Boice JD Jr (1988) Mortality after radiotherapy for ringworm of the scalp. Am J Epidemiol 127: 713–725
Ron E (1998) Ionizing radiation and cancer risk: evidence from epidemiology. Radiat Res 150:S30–S41

Russell LB, Russell WL (1954) An analysis of the changing radiation response of the developing mouse embryo. J Cell Physiol 43(S1):130–149

Sachs RK, Hahnfield P, Brenner DJ (1997) The link between low-LET dose-response relations and the underlying kinetics of damage production/repair/misrepair. Int J Radiat Biol 72:351–374

Savage JRK (1983) Some practical notes on chromosomal aberrations. Clin Cytogen Bull 1:64–76

Sgouros G, Knowx SJ, Joiner MC et al (2007) IRD continuing education: bystander and low-dose-rate effects: are these relevant to radionuclide therapy? J Nucl Med 48:1683–1691

Shore RE, Albert RE, Pasternack BS (1976) Follow-up study of patients treated by X-ray epilation for tinea capitis: re-survey of post-treatment illness and mortality experience. Arch Environ Health 31:17–24

Stewart A, Webb J, Giles D, Hewitt D (1956) Malignant disease in childhood and diagnostic irradiation in utero. Lancet 271 (6940):447–448

Stewart A, Webb J, Hewitt D (1958) A survey of childhood malignancies. Br Med J 1:1495–1508

Stovall M, Blackwell R, Cundiff J et al (1995) Fetal dose from radiotherapy with photon beams. Med Phys 22:63–82

Weiss HA, Darby SC, Doll R (1994) Cancer mortality following x-ray treatment for ankylosing spondylitis. Int J Cancer 59:327–338

Wessels BW, Konijnenberg MW, Dale RG, Breitz HB, Cremonesi M, Meredith RF et al (2008) MIRD Pamphlet No. 20: the effect of model assumptions on kidney dosimetry and response – implications for radionuclide therapy. J Nucl Med 49:1884–1899

Wheldon TE, O'Donoghue JA (1990) The radiobiology of targeted radiotherapy. Int J Radiat Biol 59:1–21

Nuclear Medicine Dosimetry

11

Abstract This chapter begins the introduction of the application of the radiation dosimetry theory developed so far to nuclear medicine. The phases through which the development of a radiopharmaceutical must pass in order to enter the market are summarized, with emphasis upon when radiation dosimetry evaluations are necessary. A brief overview of the history of internal radiation dosimetry is provided, looking at early attempts at measuring the biodistribution of an administered radionuclide, and reviewing the development of the Marinelli-Quimby-Hine dosimetry calculation method of the 1940s. The Medical Internal Radiation Dose (MIRD) schema is introduced and the fundamental MIRD equation derived from first principles. Similarly, the equivalent schema of the Internal Commission on Radiological Protection is introduced and parallels between it and the MIRD schema drawn. Finally, an overview of contemporary software codes used for nuclear medicine internal radiation dosimetry calculations is given.

Contents

11.1 Introduction

This chapter begins the development of the practical theory of nuclear medicine radiation dosimetry based upon the foundations derived in the previous ten. The adjective "practical" is used deliberately so as to ascribe to the theory the goal of determining absorbed doses to the organs and tissues of the nuclear medicine patient or of those subjects participating in the clinical trials of new radiopharmaceuticals. The applicability and degrees of accuracy required in these calculations are highly variable. In practice, there are three main applications within nuclear medicine for which radiation dosimetry calculations are required, depending upon the application of interest and the magnitudes of the radiation absorbed doses.

The first application is common to all radiopharmaceuticals and this is the regulatory requirement

B.J. McParland, *Nuclear Medicine Radiation Dosimetry*,
DOI 10.1007/978-1-84882-126-2_11,

to establish the radiation dosimetry profile of the radiopharmaceutical during the course of its preclinical and clinical development and before the product is licensed. This profile, in the form of tabulated organ absorbed doses and effective dose, is provided in the Summary of Product Characteristics (SPC) or Product Insert (PI) of the radiopharmaceutical which summarizes the properties of the product.

The second application is one which can arise within the diagnostic nuclear medicine clinic, often unexpectedly. This is the need to evaluate the radiation absorbed doses received by a specific patient as a result of, for examples, the postprocedural discovery that the patient was pregnant at the time of radiopharmaceutical administration or of a misadministration of a radiopharmaceutical. Such incidents are rare, but require estimates of the internal radiation dosimetry, typically using data provided in the SPC or PI.

The third application is of patient-specific radiation dosimetry in therapeutic nuclear medicine. Whereas the concern in the previous paragraph in diagnosis was stochastic risk, therapy requires a deterministic effect achieved through targeted cell kill. Sequellae resulting from the irradiation of uninvolved healthy tissue, such as bone marrow and kidney, resulting in radiation-induced myelotoxicity and nephropathy which can restrict the amount of administered activity and the consequent probability of tumor control. Hence, accurate patient-specific radiation dosimetry in therapeutic nuclear medicine is inevitably necessary. This application differs from that of the first. In the latter, a limited number of subjects are used, but all source organs and tissues are evaluated. In the third, all patients are involved, but the radiation dosimetry is evaluated for a limited number of source and target regions.

This chapter addresses the common radiation dosimetry calculation needs of the above and examines the theoretical basis of each. It is not intended to be a "how to do it" manual. The reader can readily obtain exquisite practical examples in Stabin's recent book (Stabin 2008) and in Pamphlet 16 published by the Medical Internal Radiation Dosimetry (MIRD) committee of the Society of Nuclear Medicine (SNM) in the United States (Siegel et al. 1999).

11.1.1 Development of Radiopharmaceuticals

11.1.1.1 Introduction

The development of a diagnostic or therapeutic radiopharmaceutical generally follows the typical linear process of Fig. 11.1 which is divided into four distinct phases. In each phase, the characteristics, or profiles,

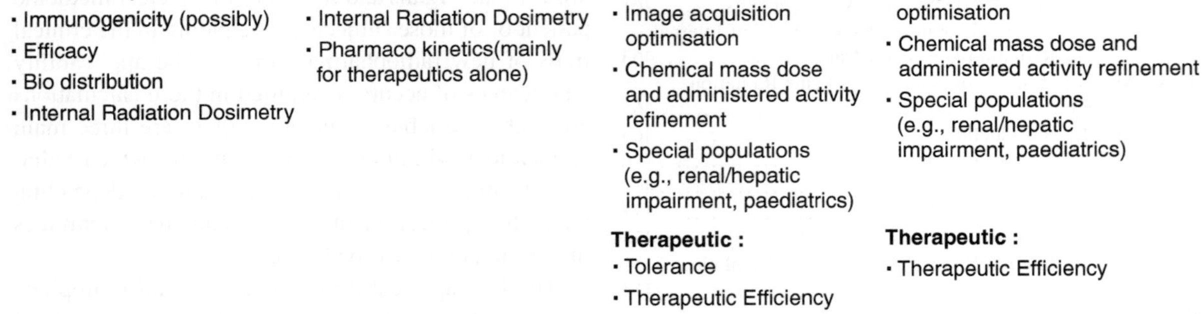

Fig. 11.1 The development phases of a radiopharmaceutical required to reach a new drug application (NDA). See text for description

of the radiopharmaceutical are determined with increasing detail as development progresses. Common to all phases is the expansion of the safety profile of the agent. In the human phases (I through III), this is done through the accrual over time of increasing numbers of subjects who have received the radiopharmaceutical.[1] This allows a growing base upon which to assess the rates of incidence of any physiological effects associated with the agent. In the early phases (I and II), more detailed safety assessments are typical, although their degrees of complexity and invasiveness usually diminish going into Phase III.

In each of the phases, pharmacological safety is paramount. But, especially with respect to diagnostic radiopharmaceuticals, one must recall the differences between the levels of diagnostic radiopharmaceutical administered and those of therapeutics. The latter are properly referred to as "drugs" as they carry intent to induce a physiological change. Diagnostic radiopharmaceuticals, on the other hand, must not have a physiological effect and, consequently, should not be referred to as "drugs." A common, but incorrectly applied as a generic, description of a diagnostic radiopharmaceutical is as a "tracer" to reflect this lack of physiological effect. Whereas the administered chemical dose per body mass of a therapeutic pharmaceutical is of the order of mg/kg, that of a diagnostic radiopharmaceutical is of the order of μg/kg. Indeed, if one looks further at the complexed ligand, i.e., that associated with the radioactive moiety used for imaging, the result is of the order of pg/kg.

At each juncture between phases, regulatory review and approval of the developmental program is required prior to proceeding to the next phase. In 2004, the American Food and Drug Administration (FDA) published guidelines advising on its interpretation of the means of how an imaging agent should be developed (FDA 2004a–c). Further information on the regulatory environment within which diagnostic radiopharmaceuticals are developed can be found in the FDA documents previously cited and which are available at the FDA website: http://www.fda.gov/cber/gdlns/medimagesaf.htm and through the guidelines provided by the European Medicines Agency: http://www.emea.europa.eu/pdfs/human/qwp/30697007enfin.pdf).

[1]A Phase IV study is one of an imaging agent which has already been granted marketing approval.

Additional contemporary descriptions of the European Union environment for the development of radiopharmaceuticals can be found in Verbruggen et al. (2008).

Stabin (2008) presents a particularly thorough review of the regulatory environment in the United States within which radiopharmaceuticals are developed.

11.1.1.2 Preclinical

The preclinical phase involves the use of nonhuman biological systems to characterize the safety and efficacy of the radiopharmaceutical prior to its entry into man. This is a mandatory requirement by all regulatory authorities in order to assure the maximization of confidence in the radiopharmaceutical's safety and efficacy prior to human trials. This characterizing can include the pharmacological profiling of the levels at which toxicity can become evident and the immunological responses associated with the agent. The internal radiation dosimetry profile of the radiopharmaceutical is an aspect of the agent's safety and, in the case of a therapeutic indication, its efficacy. Estimating the human biodistribution of the radiopharmaceutical and the associated radiation dosimetry is a critical activity of the preclinical phase.

The estimate of the human internal radiation dosimetry on the basis of preclinical studies is essential in the review of preclinical data prior to permissions being granted to proceed to Phase I.

11.1.1.3 Phase I

Phase I is the "first-time-into-man" stage of development. The subjects receiving the investigational medicinal product (IMP) may be healthy volunteers (which is the usual case for a diagnostic radiopharmaceutical) or subjects with the specific disease to be diagnosed or treated. Again, the investigation is focused on the safety profile of the agent and will generally include pharmacokinetic assessment, although this is unlikely for a diagnostic radiopharmaceutical due to the low chemical mass dose per administration. The human biodistribution of the pharmaceutical is measured and the radiation dosimetry evaluated. The effects of changes in chemical mass dose are always assessed for a therapeutic agent and also should be in the development of a diagnostic entity.[2]

Out of the radiation dosimetry profile generated for a diagnostic radiopharmaceutical, the stochastic risk associated with the IMP, as described by the effective dose, is evaluated. While this result is necessary for the safety profiling of the IMP, its applicability to clinical nuclear medicine requires caution in that the use of the effective dose (and its predecessor, the effective dose equivalent) in diagnostic nuclear medicine has not been without controversy. It has been noted that the genesis of the effective dose was the radiological protection of the radiation worker and, hence, would not be directly applicable to the individual diagnostic nuclear medicine patient (Poston 1993). More recently, its restricted applicability to diagnostic nuclear medicine risk assessment has been considered (Bolch et al. 2009). In any event, it must be recognized that the effective dose is an inappropriate metric by which to assess the radiation risk posed to the individual nuclear medicine patient. Nor, as the effective dose is a measure of stochastic risk, is it applicable to therapeutic nuclear medicine where deterministic effects are sought. In its current incarnation, the effective dose is the arithmetic average of the effective doses to International Commission on Radiological Protection (ICRP) Reference Man and Reference Woman both of which are modeled on 50th-percentile individuals. As discussed in Chap. 10, the tissue weighting factors used to calculate the effective dose are averaged over age and sex for which an assumed dose-rate factor is applied, thus leading to its inapplicability to a given individual. In the clinic, the effective dose is, at best, suitable only for comparing relative radiation risks between different diagnostic radiopharmaceuticals. In the development of diagnostic radiopharmaceuticals it is a required item within the SPC and PI and is a necessary quantity to provide when submitting clinical trial applications. But, to reiterate from Chap. 10, the user should be well aware of the limitations of what the effective dose attempts to represent.

On the other hand, deterministic effects are of concern in therapeutic applications of a radiopharmaceutical. This requires very accurate assessment of the dosimetry so as to maximize the absorbed dose to the lesion to be treated and to minimize that received by healthy radiosensitive tissues, such as the red bone marrow and the kidney.

11.1.1.4 Phase II

The Phase II stage of the clinical development of a radiopharmaceutical is usually that in which proof-of-concept /-mechanism is established and the efficacy of the IMP determined preliminarily. In diagnostic applications, the subjects in Phase II are not healthy volunteers but patients with the disease in question. The safety profile of the IMP is expanded and the imaging parameters (e.g., time postadministration of acquisition, duration of acquisition, administered activity) are investigated, refined and optimized. Dosimetry evaluations are usually not performed unless there are reasons to assume that the dosimetry can be altered as a consequence of the disease, although a simplified measurement of the biodistribution (e.g., through measurement of washout from blood or excretion in the urine) can be performed and compared with the measurements of the healthy volunteers in Phase I. However, it may be necessary to obtain imaging-based biodistributions in special patient populations. An example would be those of renally-impaired patients receiving a radiopharmaceutical which is excreted primarily in the urine. Here, the concern would be over whether a reduced rate of renal excretion could impede efficacy, through a reduced washout of background, or elevate organ absorbed doses. Again, as in Phase I, extensive safety evaluations must be performed.

11.1.1.5 Phase III

By this phase, the safety, dosimetry, and preliminary efficacy attributes of the IMP are now completely known. Larger populations of patients are recruited in these Phase III studies so that pivotal, statistically-significant data are obtained as required in the submission of the IMP for approval. For a diagnostic radiopharmaceutical, dosimetry calculations are generally not required in Phase III; this will not usually be the case for a therapeutic agent where patient-specific

[2]Recently, Ballinger et al. (2009) have emphasized the need to include information on the amount of chemical dose administered when a clinical trial of a diagnostic radiopharmaceutical is reported in the open literature.

dosimetry is required to plan for therapy and to assess the absorbed dose-responses of dose-limiting tissues (e.g., bone marrow).

11.1.2 Clinical Nuclear Medicine Applications

11.1.2.1 Diagnostic Radiopharmaceuticals

The main clinical applications of the internal radiation dosimetry profile for a diagnostic radiopharmaceutical are in the approximate assessments of the radiation absorbed doses of special cases (e.g., the pregnant nuclear medicine patient or mis-administrations), a nonspecific assessment of radiation risk or as a comparator with other imaging modalities. The level of accuracy in the determination of the radiation absorbed dose of a diagnostic radiopharmaceutical is relatively relaxed, thus allowing representative mathematical models of human anatomy to approximate the subject (Chap. 12). Such models are generally representative of the median of a given population (e.g., adult male Caucasians). For example, a typical calculated effective dose of a ^{99m}Tc-labeled diagnostic radiopharmaceutical at a typical administered activity will be of the order of 4 mSv and that of a ^{18}F-labeled radiopharmaceutical will be between 10 and 20 mSv. Recalling the discussion of Chap. 10, this range of effective doses would correspond to detriment risks of between 0.02 and 0.1%. Hence, uncertainties in the calculated effective dose of factors of 2–3 (due to inaccuracy in the dosimetry calculation or in the variability between patients) would not be regarded as significant in the overall estimate of individual risk.

11.1.2.2 Therapeutic Radiopharmaceuticals

The relaxation of accuracy requirements in the diagnostic nuclear medicine patient dosimetry is not, or should not be, applicable to the therapeutic patient. Ideally, as in contemporary external beam radiotherapy, the absorbed dose profile of the patient receiving radionuclide therapy should indicate the magnitude, and spatial distribution, of the absorbed doses in the region of interest (almost universally a tumor or metastasis) and that in healthy tissues which are acutely radiosensitive and can exhibit radiotoxicities which can inhibit potentially the administered activity and the tumor control probability. Unfortunately, this is currently not the status of common therapeutic nuclear medicine. While ^{131}I therapy of thyroid disease dominates the therapeutic field and posology is frequently based upon clinical experience, the development of new therapeutic radiopharmaceuticals is leading to patient-specific radiotherapy planning coming into vogue, at least within the research venue. This planning requires a pretherapy assessment of the expected biodistribution of the therapeutic moiety in order to allow the therapy planning physicist to predict the magnitude of the absorbed doses to the target regions and to the healthy tissues and for the clinician to then decide upon the amount of administered activity. The combination of clinical experiences of tissue responses arising from radionuclide therapy and external beam radiotherapy, coupled with accurate internal radiation dosimetry, will assist in reaching this decision.

11.2 History of Nuclear Medicine Radiation Dosimetry

11.2.1 Introduction

The two structures which have had the greatest influence upon the development of nuclear medicine radiation dosimetry have been the International Commission on Radiological Protection (ICRP) and the Medical Internal Radiation Dose (MIRD) Committee of the American SNM. The foci of both entities are, however, somewhat different. The work of the ICRP emphasizes the radiological protection of the worker who is occupationally exposed to ionizing radiation, although this is not exclusive, as commonly misbelieved. The ICRP has published numerous summaries of the radiation absorbed dose profiles of radiopharmaceuticals and recommendations on medical radiation exposure safety. On the other hand, the MIRD Committee's remit is exclusively that of the nuclear medicine patient.

The ICRP began life in 1928 as a Committee associated with the International Congress on Radiology

(Lindell et al. 2009), publishing its first report that year. The publications and recommendations of the ICRP have come to form the basis of most of the world's regulatory codes and regulations on radiological safety. The ICRP is composed of a Main Commission and four standing Committees. These are Committee 1 on radiation effects, Commission 2 on derived dose limits, Commission 3 on protection in medicine and Committee 4 on the applications of the ICRP recommendations. Task groups and working parties are created to assist on the planning and drafting of specific reports. The ICRP has published many recommendations which have become influential in nuclear medicine dosimetry. These would include ICRP Publications 26 (1977), 53 (1991a), 60 (1991b), and 103 (ICRP 2006), which defined the effective dose equivalent and its successor, the effective dose, and which provided data on the risks of deterministic and stochastic effects; ICRP Publication 30 (1979) which presented dosimetric models of the bone and bone marrow, the gastrointestinal (GI) tract, and the respiratory system and which was augmented or superseded by Publications 66 (1994) on the revised respiratory system, 70 on the revised skeletal model (1995) and 100 (2006) on the revised alimentary tract model; and those publications which provided a compilation of the absorbed dose profiles of a multitude of radiopharmaceuticals and radionuclides in medical use – Publication 53 and its addenda (ICRP 1987, 1998, 2008a). Another profound influence it has had on nuclear medicine radiation dosimetry has been through its introduction and refinement of Reference Man, Reference Woman and various reference pediatric phantoms (ICRP 1975).

The MIRD Committee was originally organized as an ad hoc entity and met for the first time in November 1964 and adopted its title at its second meeting of January 1965 (S-Stelson et al. 1995). Its first pamphlet was published in 1968 (Loevinger and Berman 1968) and described its schema of calculating internal radiation dosimetry. Since then, the MIRD committee has published numerous pamphlets, reports, and dose estimates, most of which are cited in this book. The MIRD pamphlet provides the scientific overview of a particular nuclear medicine application whereas the MIRD dose estimate provides dosimetric data specific for a given radiopharmaceutical.

The ICRP and the MIRD committee have developed parallel internal dose estimation algorithms, which will be considered in detail below. In fact, the algorithms are virtually identical with the main differences residing in the use of nomenclature and the target application. The MIRD committee has published a pamphlet to, in part, bridge this gap (Bolch et al. 2009).

The sophistication and accuracy of nuclear medicine radiation dosimetry calculations have grown in response to a variety of factors. These include the:

- Use of radionuclides moving away from naturally-occurring radionuclides to artificially-produced entities and their chelation to, or labeling of, pharmaceuticals.
- Increase in spatial and temporal precisions and the detection efficiencies of imaging devices.
- Development of more sophisticated mathematical models (phantoms) to represent the human anatomy so as to enable more accurate calculations of the absorbed dose distribution.
- Increase in computational power to allow detailed calculations of radiation transport and energy deposition in tissue.

Prior to the development of particle accelerators and nuclear reactors in the 1930s and 1940s, only naturally-occurring radioactive materials were available to be used in medicine. These were primordial radionuclides, such as ^{40}K, which has a half-life of 1.2×10^9 years, or the decay products (such as radium or radon) of ^{232}Th (half-life of 1.41×10^{10} years), ^{235}U (half-life of 7.04×10^8 years) and ^{238}U (half-life of 4.5×10^9 years). The major difficulties with the naturally-occurring radionuclides were the difficulty in isolating and concentrating them, their associated low specific activities, the nonpenetrating charged-particle emissions associated with the *extra corpus* detectable photons and, to a degree, the heavy metal nature of the isotopes leading to the associated imperfect biodistributions and potential toxicities.

The potential for artificially-created radioactive isotopes was heralded by the work of the Irish physicist, Walton, and his English colleague, Cockroft, in their construction of a proton accelerator with which, in 1932, they bombarded a lithium target to yield the reaction ^{7}Li(p,α)^{4}He. For this work, Walton and Cockroft won the Nobel Prize in Physics in 1951. This nuclear reaction using accelerated particles suggested immediately the possibility of artificially-induced radioactivity, which was achieved through the reaction

$^{27}Al(\alpha,n)^{30}P$ by the Joliot-Curies in 1934 which subsequently led to their being awarded the Nobel Prize in Chemistry in 1935. The α particles used were from an intense polonium source. Artificial radioactivity induced through neutron bombardment was shortly afterwards demonstrated by Fermi and his colleagues. But it was the work of Lawrence and others at the University of California through the invention of the cyclotron in the 1930s and its application to medical radioisotope production that accelerated the first use of artificially-produced radionuclides in medicine, including ^{24}Na, ^{32}P, ^{128}I, and ^{131}I.

In parallel with these developments were the design and construction of new radiation detectors which allow spatial information encoding. The Geiger-Müller detector, described in Chap. 9, was a commonly used device for early nonimaging quantification of in vivo uptake of activity (S-Stelson et al. 1995). Such a detector, or a scintillation probe, can provide a basic measure of the total body burden of activity which can provide a basic amount of data upon which a dosimetry calculation can be made. Implicit within this measurement is the assumption that the activity is uniformly distributed throughout the body (the probe measurement can only discern the body retention of a radionuclide or, in the specific case of iodinated radiopharmaceuticals, the uptake of radioactive iodine by the thyroid). Imaging-based quantification is essential for developing a biodistribution and evaluating a complete radiation dosimetry profile. Early attempts at imaging were inherently analog and involved the translation of a single radiation detector over the subject to be imaged (a rectilinear scanner) and encoding the change in intensity of photon detection as a function of spatial position in a Cartesian coordinate system (Hamilton 2004). Anger developed his eponymous camera and associated energy deposition and position-encoding logic in 1957 (Anger 1964). The camera, based upon the photoelectric absorption of collimated photons in a NaI (Tl) crystal and the conversion of the scintillation light into electrical signals by photomultiplier tubes, has remained in clinical use for over a half-century. The development of positron-emitting radionuclides, including ^{11}C and the ubiquitous ^{18}F, led to the development of the dedicated positron emission tomography (PET) scanner. Although coincidence detection of a positron-emitting radionuclide in a medical application can probably be dated to the early 1950s at Massachusetts General Hospital (Wrenn et al. 1951), the 1960s saw the proper birth of coincidence imaging techniques which have grown into the modern PET scanner (Jones 2003).

11.2.2 Early Biodistribution Measurements

11.2.2.1 Introduction

In order to evaluate the radiation absorbed doses to organs and tissues resulting from an administered radionuclide, the initial set of data to be obtained is that of how the radionuclide distributes within the body and is excreted over time. This data set is referred to as the biodistribution or biokinetics of the radiopharmaceutical. How these data are determined is a critical part of the internal dosimetry evaluation and modern means of obtaining these data are discussed later. In this subsection, early attempts made at measuring in vivo the biodistribution or biokinetics of an administered radioactive substance are briefly reviewed.

11.2.2.2 ADME

ADME is the acronym for administration, distribution, metabolism, and excretion used in the pharmacokinetic literature. Although it is emphasized here that measurements of the biodistribution are not pharmacokinetic, the above acronym is useful for partitioning the necessities of a biodistribution measurement.

Administration

A radiopharmaceutical (including radiolabeled antibodies) can be administered to the patient through a variety of routes largely depending upon whether or not the intent is diagnostic or therapeutic and upon the pathology or anatomy to be imaged. Administration can be intravenous, intraarterial, through direct infusion, intratumor, intracavitary, per os or by inhalation. The means of administration will dictate the type of biodistribution assessment.

Distribution

A radionuclide introduced to the vascular system will reach the right side of the heart, enter the pulmonary arteries and pass through the pulmonary capillary bed to reach the pulmonary veins and pass to the left-side of the heart where it then enters the arterial system for distribution throughout the body. Measurements of this biodistribution are made at many discrete times postadministration.

Metabolism

In pharmacokinetics, one is interested in how the body manages the pharmaceutical and how it is metabolized. This is not at all the case with a radiopharmaceutical biodistribution measurement. The metabolic outcome of an administered radiopharmaceutical cannot be determined through in vivo and in vitro measurements of the radionuclide linked to the radiopharmaceutical. These measurements can only indicate the amount of radionuclide present and cannot provide any information to answer questions on the chemical form of the radionuclide: i.e., is it still in the form of the unmetabolized original radiopharmaceutical moiety, is it part of a metabolite or is it free radionuclide? Such information can be determined, however, using high-performance liquid chromatography.

Excretion

The measurement of the amount of radionuclide excreted following administration of a radiopharmaceutical is essential to evaluating its associated radiation dosimetry. A radiopharmaceutical is designed to have as high specificity as possible for the pathology intended for it to detect or treat. For a diagnostic radiopharmaceutical, in particular, rapid and high excretion is frequently sought as a means of both reducing the nonspecific uptake and background which limits the diagnostic efficacy and of reducing the radiation absorbed dose. As renal excretion is much faster than that through the GI tract, the latter as a means of removing the radionuclide is sought if possible (although this is not possible for, for example, lipophilic radiopharmaceuticals where hepatic excretion into the gut is frequently predominant).

11.2.2.3 Excretion of Administered Radium

In the early twentieth century, radium was regarded popularly as a panacea for many diverse physical ailments and was widely used without control. Radium is a member of Group II (alkaline earth metals) of the periodic table of elements and has chemical properties similar to calcium. As a result, ingested radium is incorporated into bone and the resulting induction of osteosarcoma and other bone cancers are historically well-known consequences, as evidenced by the experiences of the cohort of radium dial-painters discussed in Chap. 10.

Radium and its radon daughter product were deliberately ingested or injected in the early 1900s for supposed health benefits. At this time, equally deliberate biodistribution measurements, in the form of experiments to assess the rate of excretion of injected or ingested radium were performed (Rowland 1994). For example, Evans (1933) reported a rate of excretion of 0.005% of administered radium per day in the feces. Seil et al. (1915) described the excretion of radium administered intravenously and per os (the latter to Seil himself). One subject received two intravenous administrations of radium, each of 3.7 MBq, 2 months apart. Seil self-administered radium per os in two instances of 1.85 MBq each, 7 days apart. Radium activity excreted in the urine and feces of both subjects was measured. Seil died many years later of diverticulitis and his body later exhumed. Radium burden measurements determined that he had had a systemic intake of 2.7 MBq of radium.

11.2.2.4 In Vivo Measurement of Blood Circulation Time

What is likely to have been the first in vivo measurement of the biodistribution of a radionuclide in the human was that performed by Blumgart and Yens (1926); Patton (2003). The study was the first to non-invasively measure the speed of blood flow, but was not, due to the limited radiation detection technology available in that era, an imaging-based study. Instead, a reciprocating Wilson cloud-chamber was used as a detector. The tracer employed was ^{214}Pb and ^{214}Bi, both β- and γ-emitters, adsorbed onto sodium chloride which was subsequently dissolved and injected into the antecubital vein of one arm of a subject.

The reciprocating cloud-chamber was placed at the contralateral arm with a lead shield placed between the chamber and the injection site. The time difference between that of injection and the detection of radioactivity at the contralateral arm was measured. This time difference was found to be in the range of 15–21 s in subjects with a normal cardiovascular system and in the range of 50–65 s in patients with cardiac decompensation.

11.2.2.5 Radionuclide Tracers: de Hevesy

The Hungarian chemist George de Hevesy is widely acknowledged as the pioneer of radioactive tracer science; moreover, the variety of apocryphal stories associated with his tracer work make him a most entertaining and fascinating individual. He received the Nobel Prize in Chemistry in 1943 for his work on the use of radionuclides as a means of noninvasively measuring metabolism in both plants and animals.

de Hevesy's interest in using radioisotopes famously began when he was working in Rutherford's Manchester laboratory in 1913. Rutherford had assigned him the task of separating one of the daughter products of radium decay, radium D, from lead. This was impossible to do as radium D, now known to be the isotope ^{210}Pb, cannot be chemically isolated from other lead isotopes. Inevitably, de Hevesy failed in this task but realized that this failure to separate radium D from lead allowed radium D to be used as a tracer of lead. Thus, the concept of the radiotracer was born. Chievitz and de Hevesy (1935) performed the first tracer experiment with an artificial radionuclide, ^{32}P, in the rat. The biodistribution of ^{32}P in the excreta, brain, spleen, kidneys, liver, skeleton, blood, muscle, and fat at 22 days post-administration was reported. This work provided the first demonstration that bone formation was a dynamic process: 30% of the ^{32}P atoms deposited within the skeleton were removed within 20 days.

11.2.2.6 Commentary

The above early in vivo measurements of the temporal distribution or excretion of radioactive material were not part of any attempts to estimate the consequent internal radiation dosimetry. At around the time of these measurements, the only algorithms for estimating the absorbed doses resulting from the placement of radioactive material in corpus were those associated with the introduction of encapsulated radium brachytherapy sources, the most widely used being derived from the Paterson-Parker rules (Paterson and Parker 1934). The first attempt at calculating the radiation dosimetry for an internally-distributed radionuclide is now described.

11.2.3 Marinelli–Quimby–Hine Method of Internal Radiation Dosimetry Calculations

Before the 1940s, the development of radiation dosimetry calculations was far more advanced in external beam radiotherapy and brachytherapy than in therapy using internal radiation sources. This however changed with three seminal papers (Marinelli 1942; Marinelli et al. 1948a, b). The first applications of administered radioactive substances were almost exclusively therapeutic; any diagnostic indications were restricted due to the limited amounts of photons released by most of the naturally-occurring radionuclides then in use. As a result, the calculation of the internal dosimetry was originally focused on therapy, although it was the Marinelli papers which first considered diagnostic applications and radiological safety. Further historical reviews of the development of internal radiation dosimetry can be found in the articles by S.-Stelson et al. (1995); Potter (2005); and Stabin (2006).

The first substantial treatise on internal radiation dosimetry calculations is that of Marinelli (1942) for β-emitting radionuclides, in particular ^{32}P in the treatment of leukemia. He presented a unique way of describing the energy deposited in tissue that was prescient of the modern concept of kerma. As the röntgen is defined only for the ionization caused by photon interactions in air, Marinelli defined a unit called the equivalent röntgen which was, essentially, the amount of energy associated with the kerma due to 1 röntgen exposure in 1 g of air deposited in 1 g of soft tissue. Knowing the amount of β-emitting radionuclide activity in a given mass of tissue and the mean β-particle kinetic energy, the amount of equivalent roentgens could thus be evaluated. This quantity would help enable the therapeutic clinician to prescribe the amount of activity to be administered in order to deliver a specified amount of kerma in terms of a known radiological quantity, the röntgen.

In this pioneering paper, Marinelli noted that knowledge of the biodistribution, in particular the rate of excretion of the radionuclide, was essential to accurate dosimetry. He even modeled the biological elimination of activity with a monoexponential function leading to the realization of the effective decay constant (or half-life), derived in Chap. 5. In addition, he preceded the MIRD formalism by calculating the cumulated activity (the time-integral of the radioactive decays in tissue to yield the total number of disintegrations) $\tilde{A}$ and noting that the largest absorbed dose in a tissue is given by a radionuclide for which the product $\tilde{A}\,\bar{T}_\beta$, where $\bar{T}_\beta$ is the mean kinetic energy of the β particles, is the greatest.

Marinelli continued this work, following the end of the Second World War, through two papers written in collaboration with Quimby and Hine (Marinelli et al. 1948a, b). The first paper (1948a) dealt with the physical and clinical aspects of both β- and γ-emitting isotopes. The second sentence of the abstract to this paper was predictive of medical internal radiation dosimetry by stating that the absorbed dose "... cannot, in general, be measured, but when the half-life, radiation energy, and biological uptake, and excretion are known, it can be calculated". Marinelli's previous work on β-emitting isotopes was reiterated with greater clarity. Noting that the result would only be approximate due to the lack of charged-particle equilibrium at the borders, Marinelli, Quimby, and Hine wrote that the "dose" (which we would recognize as being more closely related to kerma) to a sphere containing a β-emitting isotope is,

$$D_\beta = K_\beta C \tag{11.1}$$

where (and using the historical units of their paper),

$$K_\beta = 88\,\bar{T}_\beta\,T_{1/2}\ \text{e.r.} \tag{11.2}$$

where e.r. is equivalent röntgen, C is the initial number of mCi/g of activity and $T_{1/2}$ is the physical half-life in days. The above results are straightforward to obtain (refer to Chap. 9). They then extended this formalism to the internal radiation dosimetry of γ-emitting isotopes in a manner similar to dosimetry calculations for brachytherapy sources. Using, again, their historical units, the "dose" due to photons was,

$$D_\gamma = K_\gamma C g\ \text{e.r.} \tag{11.3}$$

where,

$$K_\gamma = \frac{T_{1/2}}{\ln 2}\frac{dX}{dt} \times 10^{-3} \tag{11.4}$$

where dX/dt is the in-air exposure rate in röntgens per hour measured at 1 cm from a 1 mCi point source of the radionuclide and C is the initial concentration of activity in tissue given in units of μCi/g. The factor g in (11.3) is a geometrical factor and which would be recognized as being related to a self-absorption factor. For the example of calculating the absorbed dose at a point P (Fig. 11.2) this factor is,

$$g = \rho\int d^3\mathbf{r}\,\frac{e^{-\mu r}}{r^2}. \tag{11.5}$$

where μ is the linear attenuation coefficient in the medium at the photon energy of interest. Two examples of calculating this geometrical factor are worthwhile to examine. A crude, but easy to calculate, approximation to the torso is a unit-density sphere of radius R. Then, for the calculation point at the center of the sphere,

$$\begin{aligned} g &= \rho\int d^3\mathbf{r}\,\frac{e^{-\mu r}}{r^2} \\ &= 4\pi\rho\int_0^R dr\,e^{-\mu r} \\ &= \frac{4\pi\rho}{\mu}\left(1 - e^{-\mu R}\right) \end{aligned} \tag{11.6}$$

and has units of gram per square centimeter.

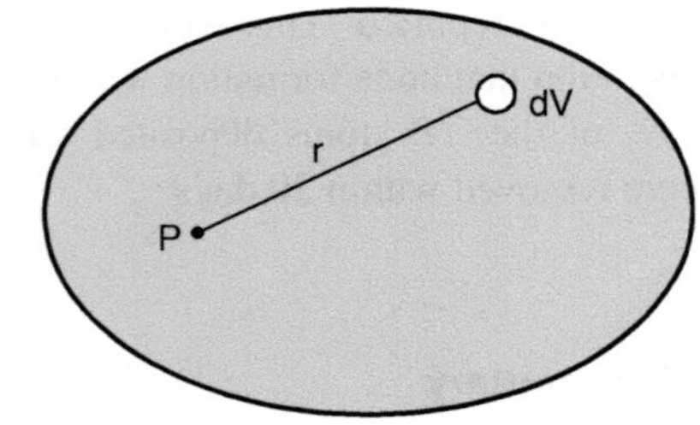

Fig. 11.2 Marinelli method for calculating the absorbed dose at the point P, in a volume V containing a uniformly-distributed γ-emitting isotope, due to the activity from the differential volume dV. This is determined by calculating the geometric factor, g, given by (11.5)

A model of the torso more realistic than that provided by a sphere is that of a right-cylinder of height H and radius R as shown in Fig. 11.3.

Applying cylindrical coordinates and assuming a uniformly-distributed radionuclide, the factor at point P in the center of the cylinder is

$$\begin{aligned} g &= \rho \int d^3\mathbf{r}\, \frac{e^{-\mu r}}{r^2} \\ &= 4\pi\rho \int_0^{H/2} dz \int_0^R dr\, r \frac{e^{-\mu\sqrt{r^2+z^2}}}{r^2+z^2}. \end{aligned} \tag{11.7}$$

The integral is not solvable analytically, but an approximate solution can be had by expanding the exponential to two terms,

$$e^{-\mu\sqrt{r^2+z^2}} \cong 1 - \mu\sqrt{r^2+z^2} + \frac{\mu^2}{2}\left(r^2+z^2\right). \tag{11.8}$$

Substituting this result into (11.7) gives, after a long and tedious calculation, an expression for g for a right cylinder of height H and diameter 2R,

$$\begin{aligned} g &= 4\pi\rho \int_0^{H/2} dz \int_0^R dr\, r\left(\frac{1}{r^2+z^2} - \frac{\mu}{\sqrt{r^2+z^2}} + \frac{\mu^2}{2}\right) \\ &= 4\pi \int_0^{H/2} dz \left(\int_0^R dr\, \frac{r}{r^2+z^2} \right. \\ &\quad \left. - \mu \int_0^R dr \frac{r}{\sqrt{r^2+z^2}} + \frac{\mu^2}{2}\int_0^R dr\, r\right) \\ &= \rho\left(4\pi R \tan^{-1}\frac{H}{2R} + \pi H \ln\left(1 + \left(\frac{2R}{H}\right)^2\right)\right. \\ &\quad - 2\pi\mu R^2\left(\frac{H}{2R}\sqrt{1+\left(\frac{H}{2R}\right)^2}\right. \\ &\quad \left. + \ln\left(\frac{H}{2R} + \sqrt{1+\left(\frac{H}{2R}\right)^2}\right)\right) \\ &\quad \left. - \frac{\pi\mu H^2}{2} + \frac{H}{2}\left(\frac{\mu R}{2}\right)^2\right). \end{aligned} \tag{11.9}$$

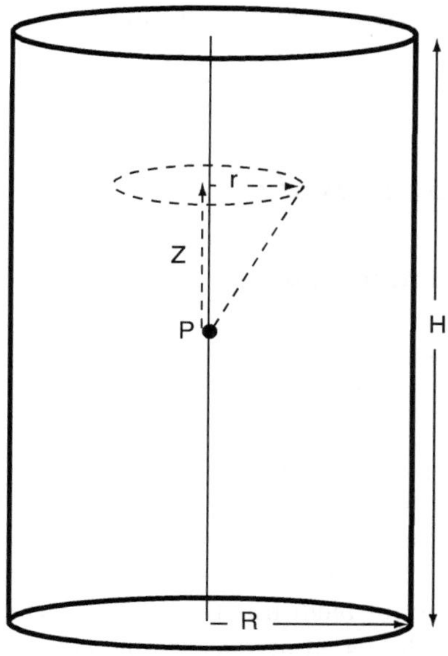

Fig. 11.3 Calculation of Marinelli–Quimby–Hine geometrical factor for a right-cylinder

The companion paper (1948b) discussed the practical applications of the results of the first and introduced the differential absorption ratio (DAR) to describe the differences in radionuclide uptake among tissues which lead to differences in absorbed dose. The DAR for a given tissue was defined as the ratio of the concentration of the isotope in that tissue to the average concentration in the whole body;[3] the DAR allowed one to understand the absorbed dose to an organ relative to the average absorbed dose. And it is in this paper that the effective half-life was defined.

In reading these three publications, one sees the genesis of many of the concepts that are routinely used in modern internal radiation dosimetry calculations.

11.3 Modern Methods of Nuclear Medicine Radiation Dosimetry

11.3.1 Introduction

This section provides an overview of modern means of calculating the internal radiation dosimetry associated

[3]One can see the similarity to the modern specific uptake value (SUV).

with a radiopharmaceutical using the MIRD and ICRP methods.

11.3.2 Sources of Data

11.3.2.1 Introduction

All methods of calculating internal radiation dosimetry require, as input, knowledge of the radiation characteristics of the administered radionuclide, the anatomy for which the dosimetry is being calculated for and, finally, the biokinetics of the radiopharmaceutical. The last requirement has already been discussed briefly in the review of the ADME concept and the early history of biodistribution measurements; it will shortly be discussed in considerable detail due to its importance in evaluating the dosimetry. The first two requirements are presented here.

11.3.2.2 Radiation and Nuclear Data

The radiation characteristics of the radionuclide are the first necessary data to be had. These data include:

- Nuclear decay scheme (decay type, emissions, yields, radiation energies, half-life)
- Radiation-matter interaction coefficients (attenuation coefficients, energy-transfer, and energy-absorption coefficients, unrestricted, and restricted collision stopping powers, radiative stopping powers)

These have been discussed in Chaps. 3, 4, 6, and 7.

Data for nuclear decay schemes are not static and refinements in these values are regularly published in the literature. Typically, the increased levels of precision and accuracy in these decay scheme updates have little direct effect upon the accuracy of an internal dosimetry calculation as these are dwarfed by uncertainties in other data, particularly those that are biological. Any refinement in values of interaction coefficients, on the other hand, is usually driven by improvements in theory. Again, any such changes largely have an insignificant impact upon a dosimetry calculation.

Nuclear, atomic and, radiological data are readily accessible at websites hosted by national, academic, or other centers. Examples include:

- National Nuclear Data Centre (Brookhaven National Laboratory): http://www.nndc.bnl.gov/
- Radiation Dose Assessment Resource (RADAR): http://www.doseinfo-radar.com/RADARHome.html
- International Atomic Energy Agency (IAEA): http://www.iaea.org/
- Ernest O Lawrence Berkeley National Laboratory Isotopes Project Nuclear Data Dissemination: http://ie.lbl.gov/toi.html
- National Institute of Standards and Technology (NIST): http://physics.nist.gov/PhysRefData
- Korea Atomic Energy Research Institute: http://atom.kaeri.re.kr/

Relevant nuclear decay data are also available in journal articles and reports. Examples include those of Stabin and da Luz (2002), Stabin and Siegel (2003) and ICRP Publication 107 (2008b). Data of the interactions between radiation and matter are available in a variety of ICRU publications which have been cited in Chaps. 6 and 7, some of which are available on the NIST website cited above.

11.3.2.3 Anatomical and Physiological Data

In addition to the requirements of physical data, nuclear medicine dosimetry calculations need artatomical and physiological data. If one is performing a patient-specific therapy calculation, then these data will be obtained from the individual patient through, for example, an acquired whole-body CT scan. However, such detailed calculations are relatively rare due to the inherent complexities and demands for high processing power in order to evaluate the radiation absorbed dose distribution to sufficient precision. For low-absorbed dose nuclear medicine applications, a representative set of anatomical data is usually sufficient. These are provided in the form of a Reference Man for a given ethnic population and which is described by an anthropomorphic phantom. These whole-body phantoms are described in Chap. 12.

11.3.3 MIRD Schema

11.3.3.1 Introduction

As described earlier, the MIRD Committee provided its first set of instructions on nuclear medicine dosimetry calculations in 1968. It is now ubiquitous in the field. As with all internal radiation dosimetry calculations, it provides a measure of the absorbed dose (i.e., the integral of the absorbed dose rate) to a target region due to a temporally-varying amount of activity in a source organ. The separate absorbed dose contributions to the target region are summed over the various source organs and the result provides the total absorbed dose to the target region.

11.3.3.2 The Fundamental MIRD Equation

Derivation of the Fundamental MIRD Equation

The MIRD approach to estimating the absorbed dose received by a tissue or organ due to activity contained within the organ itself or another source organ was presented originally in 1968 by Pamphlet 1 (Loevinger and Berman 1968). This pamphlet has been since been followed by the MIRD primer publications (e.g., Loevinger et al. 1988) and Watson et al. (1993) which have provided a succinct review of the formalism. In its purest form, the MIRD expression is remarkably simple and gives the absorbed dose to a target region r_T at time t from activity contained in a source region r_S as,

$$D_{r_T}(t) = \tilde{A}_{r_S}(t)S(r_T \leftarrow r_S; t) \qquad (11.10)$$

where $\tilde{A}_{r_S}$ is the total number of nuclear disintegrations in the source region that have occurred up to time t, and is known as the cumulated activity, and is further described shortly. $S(r_T \leftarrow r_S; t)$ is known simply as the "S-factor"[4] and relates the absorbed dose in the target region r_T to the cumulated activity in source region r_S at time t. In other words, it is the absorbed dose in the target region per nuclear disintegration in the source region. In practical nuclear medicine dosimetry calculations, the S-factor is a previously calculated entity for combinations of r_S and r_T and photon and electron energies. The time dependencies of absorbed dose, cumulated activity and S-factor are allowed in this general representation of the MIRD formula as both the volume and morphology of some source regions can vary with time. Two obvious examples are the urinary bladder, which fills and distends with inflowing urine and then empties and collapses with each void of urine, and tumors which can increase in size (in the case of the lack of therapy response) or decrease as a result of therapy. As a result, the irradiation geometry can change over time. Such a time-dependence in (11.10) is only a formalism as organ volumes and shapes are usually taken to be fixed in contemporary nuclear medicine dosimetry calculations. Empirically, one need only measure the cumulated activity $\tilde{A}_{r_S}$ of the source region and use the predetermined S-factor to derive its contribution to the absorbed dose received by the target region. However, as would be expected, the simplicity of (11.10) belies a considerable amount of physical detail. This detail will be presented through its derivation for the source region r_S and a target region r_T as shown in Fig. 11.4. The source region r_S, which can be an organ, tissue, or intracavitary contents, contains an amount of radioactive material and it is desired to calculate the total absorbed dose in the target region r_T resulting from the radiation emitted from r_S. It is possible for the source and target regions to be the same, as when calculating for self-dosing of an organ due to the radioactive material it contains. A fraction of the energies released by the radioactive decays in r_S, presented as either a photon energy or the kinetic energy of a charged particle, is deposited in r_T. Hence, the absorbed dose rate at r_T at time t due to the activity in r_S, $A_{r_S}(t)$ is,

$$\frac{dD_{r_T}(t)}{dt} = A_{r_S}(t)S(r_T \leftarrow r_S; t). \qquad (11.11)$$

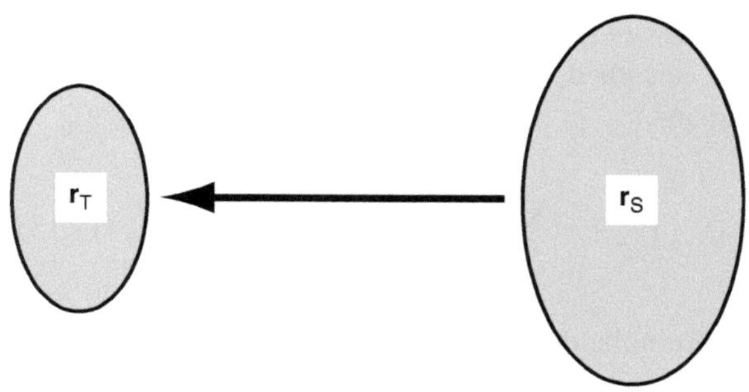

Fig. 11.4 Concept of the MIRD schema: the absorbed dose to target region r_T is calculated on the basis of the total number of radioactive decays occurring in source region r_S, $\tilde{A}_{r_S}$

[4]The S-factor is sometimes referred to as the dose factor DF or the dose-conversion factor in the literature.

To reiterate, the S-factor is the fraction of the energy released in a single radioactive decay occurring within r_S and which is deposited in r_T and is normalized to the mass of the target region r_T, m_{r_T}. Because of the dependence upon the distance separating r_S and r_T and the mass m_{r_T}, the S-factor is a function of the radionuclide and the sex- and age-dependent sizes of, and distances between, source and target regions. In the MIRD schema, these are defined as specific organs, tissues or contents of organs although there is no reason to restrict (11.11) to predefined source and target organs. For example, the schema can be applied to voxellated phantoms, where activity and absorbed dose distributions are evaluated for individual voxels derived from tomographic image data.

The total absorbed dose to r_T is the integral of the absorbed dose rate given by (11.11) over the exposure time, T_{Exp},

$$\begin{aligned} D_{r_T}(T_{Exp}) &= \int_0^{T_{Exp}} dt \frac{dD_{r_T}(t)}{dt} \\ &= \int_0^{T_{Exp}} dt\, A_{r_S}(t) S(r_T \leftarrow r_S; t). \end{aligned} \tag{11.12}$$

In nuclear medicine dosimetry calculations, the upper limit of the time integration T_{Exp} is set to ∞ whereas, in occupational radiation protection applications (those of the International Commission on Radiological Protection), T_{Exp} is set equal to 50 years for an adult (who is assumed to be aged 20 years at time of exposure) or else a variable time for pediatric exposures (Bolch et al. 2009). In practice, the target region will be irradiated by multiple source regions which requires (11.12) to be rewritten as the summation of the absorbed dose contributions from N_S source regions,

$$D_{r_T}(T_{Exp}) = \sum_{r_S=1}^{N_S} \int_0^{T_{Exp}} dt\, A_{r_S}(t) S(r_T \leftarrow r_S; t). \tag{11.13}$$

From the definition of the S-factor, one can write

$$S(r_T \leftarrow r_S; t) = \frac{\sum_i E_i Y_i \phi(r_T \leftarrow r_S; E_i; t)}{m_{r_T}(t)} \tag{11.14}$$

where the summation is over all nuclear decay channels. E_i is the energy of the ith decay (this is the released energy if a photon or the mean kinetic energy if the emission is of a β particle),[5] Y_i is the yield of that particular channel, $\phi(r_T \leftarrow r_S;E_i;t)$ is the absorbed fraction of the energy E_i emitted at time t in r_S which is absorbed by r_T (introduced in Chap. 9)[6] and $m_{r_T}(t)$ is the mass of the target region at time t. Neglecting the time dependence, the absorbed fraction is the fraction of energy E_i released by a radioactive decay at the source region r_S which is deposited in the target region, r_T. Stabin and Siegel (2003) have extended this expression of the S-factor (which they refer to as the dose factor, or DF) by including the radiation weighting factor from ICRP Publication 26 (1977) for the relevant product of the radioactive decay,

$$S(r_T \leftarrow r_S; t) = \frac{\sum_i w_{R,i} E_i Y_i \phi(r_T \leftarrow r_S; E_i; t)}{m_{r_T}(t)} \tag{11.15}$$

where $w_{R,i}$ is the radiation weighting factor for the ith emission. In nuclear medicine applications, $w_{R,i}$ is equal to unity except for Auger electrons and α particles.

The specific absorbed fraction is defined as the absorbed fraction per unit mass of the target region (thus having units of inverse mass),

$$\Phi(r_T \leftarrow r_S; t) = \frac{\phi(r_T \leftarrow r_S; E_i; t)}{m_{r_T}(t)} \tag{11.16}$$

allowing the expression for the S-factor given by (11.14) to be rewritten as,

$$S(r_T \leftarrow r_S; t) = \sum_i \Delta_i \Phi(r_T \leftarrow r_S; E_i; t) \tag{11.17}$$

where $\Delta_i = E_i Y_i$ is the mean energy emitted by each ith nuclear transition.[7] Equation (11.17) can also be written in continuous form,

[5]The symbols k and T were previously used for the photon energy and charged particle kinetic energy, respectively. Here, for simplicity, the symbol E will be used to describe the energies of both quantities.

[6]The absorbed fraction ϕ is sometimes referred to as AF in the literature.

[7]The specific absorbed fraction Φ is sometimes referred to as SAF in the literature.

$$S(r_T \leftarrow r_S; t) = \int_0^{E_{max}} dE\, E\, \Phi(r_T \leftarrow r_S; E; t) \quad (11.18)$$

which would be more appropriate for β-emissions in which the β-particle energy spectrum, described in Chap. 4, must be accounted for. Combining (11.13) and (11.17) and allowing the exposure time to go to infinity, the absorbed dose to the target region is,

$$\begin{aligned} D_{r_T} &\equiv D_{r_T}(\infty) \\ &= \sum_{r_S=1}^{N_S} \int_0^{\infty} dt\, A_{r_S}(t) S(r_T \leftarrow r_S; t) \\ &= \sum_{r_S=1}^{N_S} \int_0^{\infty} dt\, A_{r_S}(t) \int_0^{E_{Max}} dE\, E\, \Phi(r_T \leftarrow r_S; E; t). \end{aligned} \quad (11.19)$$

Equation (11.19) is a generic result in which both the morphology and size of the source organ are allowed to vary with time and the resulting effects of such variations appear in the specific absorbed fraction. In practice, the specific absorbed fraction is assumed to be constant with time, enabling (11.19) to be written as,

$$D_{r_T} = \sum_{r_S=1}^{N_S} \int_0^{E_{Max}} dE\, E\, \Phi(r_T \leftarrow r_S; E) \int_0^{\infty} dt\, A_{r_S}(t). \quad (11.20)$$

This time integral of the activity is the total number of nuclear decays that occur in the source region r_S and is known as the cumulated activity,

$$\tilde{A}_{r_S} = \int_0^{\infty} dt\, A_{r_S}(t). \quad (11.21)$$

Substituting (11.21) into (11.20) leads to the fundamental MIRD expression of (11.10), but without the time dependence. Note that the cumulated activity is dimensionless as it is the product of activity (per time) and time.

Evaluation of the S-Factor

The S-factor is the ratio of the absorbed dose in the target region r_T to the cumulated activity in the source region r_S. It is calculated for source region/target region combinations using Monte Carlo simulation of radiation transport and energy deposition to yield the absorbed fraction. As described in Chap. 9, it is possible in some circumstances to use the reciprocity theorem to simplify the calculation. S-factors have been evaluated typically for reference anthropomorphic phantoms. Patient-specific S-values can be derived, although this is currently in the research setting. Values of S-factors for combinations of source and target regions, radionuclides and radiations are found throughout the literature, especially the MIRD reports, the compilations by Stabin (2008) and the RADAR website (Eckerman 2002). In the conventional MIRD schema, the source and target regions are geometrical approximations of organs, tissues and organ intracavitary contents. The activity spatial distribution throughout the source region r_S is assumed to be constant and a Monte Carlo code is used to simulate the radiation transport from the radionuclide in r_S to the target region r_T. The individual energy depositions in the target region are then summed and normalized to the mass of the target region (effectively averaging the absorbed dose over the target region). While in the conventional MIRD schema, the target region is a specified organ or tissue or a suborgan entity (including cellular components), it is straightforward to have a voxel as a target region so that the absorbed dose spatial distribution within an organ or tissue can be determined. Such calculations are, as expected, immensely computationally intensive.

Absorbed Dose Contributions from β-Particle Bremsstrahlung

The above derivations are immediately applicable to the direct deposition of energy in the target region r_T due to photons and charged particles (electrons, positrons, and α particles) emitted by radionuclides in the source region r_S. However, as discussed in Chap. 7, charged particles decelerating in a medium will undergo radiative energy losses due to interactions with nuclear Coulomb fields (interactions with atomic electron Coulomb fields can be largely accounted for by replacing Z^2 in the *bremsstrahlung* yield calculation by $Z(Z + 1)$ where Z is the effective atomic number of the medium). Because the *bremsstrahlung* yield varies as m^{-}, where m is the charged particle's mass, the radiation emitted by a decelerating α particle at the energies typical of α decay (Sect. 4.2 in Chap. 4) in low-Z media such as tissue are negligible and are not considered here. Positrons emit *bremsstrahlung* but, as discussed in Chap. 7, their radiative yield becomes comparable to those of electrons only at kinetic energies exceeding about 75 keV. As β particles have a limited range, the absorbed fraction is very nearly equal to unity for most source regions (i.e., the absorbed fraction $\phi(r_S \leftarrow r_S) \approx 1$). However, depending upon the β-particle kinetic energy spectrum and the amount of activity of the β-emitting radionuclide, energy deposition at a distance due to *bremsstrahlung* may not be negligible. A common example of a β^--emitting radionuclide used in therapeutic nuclear medicine is ^{90}Y which has a physical half-life of 64.3 h. Therapeutic applications of this radionuclide have included radiation synovectomy (Nemec and Fridrich 1977), chelation with diethylenetriaminepentaacetic acid (DTPA) in intrathecal therapy of neuroleukaemia and central nervous system involvement in malignant lymphoma (Smith et al. 1976), the hepatic arterial infusion of ^{90}Y-labeled glass microspheres (Roberson et al. 1992) and, in more recent years, in monoclonal antibody therapy of non-Hodgkin's lymphoma (Wiseman et al. 2000). As the ^{90}Y β-particle energy spectrum has a maximum energy of 2.27 MeV, the range of these β particles in soft tissue is less than about 1 cm. Williams et al. (1989) first investigated a calculation of the *bremsstrahlung* contribution within the MIRD schema in relation to ^{90}Y. Their approach was to provide an approximate calculation of the S-factor by using the simple assumption that the *bremsstrahlung* spectrum could be adequately described by the Schiff triangular spectrum (Johns and Cunningham 1983). Folding this spectrum with the β-particle energy spectrum, they evaluated S-values for a uniform distribution of ^{90}Y within the contents of the urinary bladder. In addition, using an anthropomorphic phantom containing TLDs and a single ^{90}Y source placed within the urinary bladder section of the phantom, they compared measured absorbed dose values per unit administered activity with those calculated using the previously derived S-values. For example, they measured a mean absorbed dose within the bladder contents (not the urinary bladder wall) of 117 μGy/MBq vs. an estimated absorbed dose (excluding biological washout of activity) of 168 μGy/MBq. Although agreement between calculation and measurement was not exceptional, these results did demonstrate the need to account for the *bremsstrahlung* component in the internal radiation dosimetry calculations of β-emitting therapeutic radionuclides.

Stabin et al. (1994) modeled the *bremsstrahlung* absorbed dose by first writing the specific absorbed fraction at a point a distance r from a monoenergetic point source photon emitter of energy k,

$$\Phi(r,k) = \frac{\left(\frac{\mu_{en}}{\rho}(k)\right) e^{-\mu(k)r} B_{en}(\mu r, k)}{4\pi r^2} \qquad (11.22)$$

where $\mu_{en}(k)/\rho$ is the mass-energy absorption coefficient, $\mu(k)$ is the linear attenuation coefficient and $B_{en}(\mu x,k)$ is the energy absorption build up factor. When multiplied by the photon energy, this expression becomes the photon dose point kernel. For the poly-energetic beam of *bremsstrahlung*, the mean specific absorbed fraction is obtained by integrating over the photon spectrum,

$$\bar{\Phi}(r) = \int_0^{k_{Max}} dk\, \Phi(r,k) \frac{dN_\gamma(k)}{dk} \qquad (11.23)$$

where k_{Max} is the maximum photon energy. The *bremsstrahlung* spectrum is normalized so that,

$$\int_0^{k_{Max}} dk \frac{dN_\gamma(k)}{dk} = 1. \qquad (11.24)$$

Calculating the *bremsstrahlung* energy spectrum $dN_\gamma(k)/dk$ requires integrations over both the β-particle kinetic energy spectrum (from, e.g., the Fermi theory or measurement) and the photon production cross section (given by, e.g., the Bethe-Heitler form) for each point on the β-particle energy integration. Stabin et al. evaluated the specific absorbed fraction due to ^{90}Y *bremsstrahlung* and the absorbed doses to the MIRD-specified target regions (Table 11.1) for ^{90}Y uniformly distributed in the liver and spleen. They demonstrated that although the *bremsstrahlung*-related absorbed dose can be much smaller than that due to the β particles themselves (e.g., by a factor of 1,000 at a distance of 1 mm from the source in soft tissue), the absorbed doses of both mechanisms equate at distances of about 1 cm.

Table 11.1 Source and target regions used in the MIRD schema

Source regions		Target regions	
Adrenal glands		Adrenal glands	
Brain		Brain	
Breasts		Breasts	
Bone	Cortical	Gallbladder wall	
	Trabecular		
Gallbladder contents		Gastrointestinal tract walls	Stomach wall
Gastrointestinal tract contents	Stomach		Small intestine wall
	Small intestine		Upper large intestine wall
	Upper large intestine		Lower large intestine wall
	Lower large intestine	Heart wall	
Heart contents		Kidneys	
Heart wall		Liver	
Kidneys		Lungs	
Liver		Muscle	
Lungs		Osteogenic cells	
Muscle		Ovaries	
Ovaries		Pancreas	
Pancreas		Red marrow	
Red marrow		Skin	
Spleen		Spleen	
Testes		Testes	
Thymus gland		Thymus gland	
Thyroid gland		Thyroid gland	
Urinary bladder contents		Urinary bladder wall	
Uterus/uterine wall		Uterus	
Remaining tissues		Whole body	

The authors concluded, accordingly, that the *bremsstrahlung* absorbed dose contribution cannot always be considered negligible.

Normalized Cumulated Activity

For practical data tabulations, the physical quantities of absorbed dose and cumulated activity are normalized to the administered activity, A_0,

$$D_{r_T,\,Norm} = \sum_{r_S=1}^{N_S} \tilde{A}_{r_S,\,Norm} S(r_T \leftarrow r_S; t) \qquad (11.25)$$

where the normalized cumulated activity in the source region r_S is,

$$\tilde{A}_{r_S,\,Norm} = \frac{\tilde{A}_{r_S}}{A_0}. \qquad (11.26)$$

The dimension of the normalized cumulated activity is that of time which is why, historically, it was referred to as the residence time. As this latter term is potentially confusing and could be misinterpreted as biological half-life, effective half-life, or some other temporal quantity, the use of the term normalized cumulated activity to describe the quantity of (11.26) is now almost standard. The units of $D_{r_T,Norm}$ are of the absorbed dose per unit activity (e.g., μGy/MBq).

A feature of the normalized cumulated activity arises over the summation of the normalized cumulated activities in all regions containing activity. Consider the continuous case, rather than a discrete ensemble of organs, where $A(\mathbf{r_S}; t)$ is the activity at the position $\mathbf{r_S}$ and time t within a volume in which an activity A_0 has been administered. The total number of nuclear disintegrations over all time in that volume is given by the double integral, $\int_V d^3\mathbf{r_S} \int_0^\infty dt\, A(\mathbf{r_S}; t) = \int_0^\infty dt \int_V d^3\mathbf{r_S} A(\mathbf{r_S}; t)$. As the volume integral yields total activity within the volume, $\int_V d^3\mathbf{r_S} A(\mathbf{r_S}; t) = A_0 e^{-\lambda_P t}$, then

$$\int_V d^3\mathbf{r_S} \int_0^\infty dt\, A(\mathbf{r_S}; t) = \frac{A_0}{\lambda_P}. \qquad (11.27)$$

From (11.26), this result leads to,

$$\int_V d^3\mathbf{r_S} \tilde{A}_{r_S,Norm}(\mathbf{r_S}) = \frac{1}{\lambda_P}. \qquad (11.28)$$

In the discrete case of specified source regions, the integral is replaced by the summation of the normalized cumulated activities,

$$\sum_{r_S} \tilde{A}_{r_S,Norm} = \frac{1}{\lambda_P} \qquad (11.29)$$

where the summation is over all source regions containing activity, including the excreta.

Matrix-Vector Representation of the Fundamental MIRD Equation

Equation (11.25) can be written more compactly in matrix-vector form,

$$\mathbf{D}_{Norm} = \mathbf{S}\, \tilde{\mathbf{A}}_{Norm} \qquad (11.30)$$

where $\mathbf{D}_{\text{Norm}}$ is the row-vector of the target region absorbed doses of dimension N_T, which is the total number of target regions,

$$\mathbf{D}_{\text{Norm}} = \begin{pmatrix} D_{r_{T_1}} & D_{r_{T_2}} & \cdots & D_{r_{T_{N_T}}} \end{pmatrix}, \tag{11.31}$$

where $D_{r_{T_i}}$ is the absorbed dose to the ith target region. $\tilde{\mathbf{A}}_{\text{Norm}}$ is the column-vector of the source regions' normalized cumulated activities and is of dimension N_S, which is the number of source regions,

$$\tilde{\mathbf{A}}_{\text{Norm}} = \begin{pmatrix} \tilde{A}_{r_{S_1},\text{Norm}} \\ \tilde{A}_{r_{S_2},\text{Norm}} \\ \vdots \\ \tilde{A}_{r_{S_{N_S}},\text{Norm}} \end{pmatrix} \tag{11.32}$$

and $\mathbf{S}$ is the rectangular $N_T \times N_S$ matrix of S-factors between individual source and target regions,

$$\mathbf{S} = \begin{pmatrix} S(r_{T_1} \leftarrow r_{S_1}) & \cdots & S(r_{T_1} \leftarrow r_{S_{N_S}}) \\ \vdots & \ddots & \vdots \\ S(r_{T_{N_T}} \leftarrow r_{S_1}) & \cdots & S(r_{T_{N_T}} \leftarrow r_{S_{N_S}}) \end{pmatrix}. \tag{11.33}$$

In the current MIRD schema, there are 25 target and 26 source regions shown in Table 11.1, $\mathbf{S}$ is a rectangular 25×26 matrix (where the whole body is treated as being a target region on its own). In the event of the radionuclide emitting only charged particles, the S-factor matrix will be diagonal if *bremsstrahlung* and "cross-fire" between organs can be considered negligible (although not all of the diagonal elements will be non-zero). If the source and target regions are within an infinite medium and the radionuclide emits photons, then the reciprocity theorem suggests that $\mathbf{S}$ will be symmetric. This is valid only if all of the source regions are target regions also. As there is not such a perfect match between MIRD-specified source and target regions, this symmetry will be incomplete. Moreover, when the medium containing these regions is reduced in size to become the stylized and heterogeneous anthropomorphic phantoms to be discussed in Chap. 12, this symmetry becomes only approximate due to the loss of radiation equilibrium.

Variations in S-factor Values Due to Changes in Regions' Masses

The S-factors are calculated typically for predefined source and target regions in a known reference anthropomorphic environment. It is possible to modify a predefined S-factor in order to account for changes in source organ mass. For example, this can be necessary for estimating patient-specific dosimetry on the basis of that previously calculated for a reference phantom with known S-factor values or absorbed fractions (Yamaguchi 1978). Let m_{r_S} be the mass of the source region r_S and $S(r_T \leftarrow r_S)$ be the previously calculated S-factor and let m'_{r_S} and $S(r_T \leftarrow r'_S)$ be the corresponding quantities for the same source region but identified as r'_S as it now has a mass m'_{r_S}. Consider the case of source region self-irradiation due to charged particle emissions alone within it. Assuming that charged particle equilibrium exists, then the absorbed fractions $\phi(r_S \leftarrow r_S; E) = \phi(r'_S \leftarrow r'_S; E) = 1$. From (11.15), it follows that the S factor scales linearly with mass provided charged particle equilibrium exists,

$$S(r'_S \leftarrow r'_S) = \frac{m_{r_S}}{m'_{r_S}} S(r_S \leftarrow r_S) \quad \text{(Charged particles)}. \tag{11.34}$$

Consider now the case of a charged-particle emitting radionuclide uniformly distributed throughout the body and assume that charged particle equilibrium exists. The body has a mass m_{WB} and the energy released per single emission is E. The absorbed fraction to a target region r_T is $\phi(r_T \leftarrow WB; E) = 1$, so from (11.16), the specific absorbed fraction to the target region is,

$$\Phi(r_T \leftarrow WB) = \frac{1}{m_{WB}}. \tag{11.35}$$

The corresponding S-factor is, from (11.14),

$$S(r_T \leftarrow WB) = \frac{E}{m_{WB}} \tag{11.36}$$

Next consider the dependence of the self-irradiation S factor upon organ mass for a photon-emitting radionuclide uniformly distributed within the organ. Using the photon dose point kernel of (9.116) in Chap. 9, the specific absorbed fraction is given by the integral, assuming that the organ can be modeled by a sphere of radius r′,

$$\Phi(r_S' \leftarrow r_S'; k) = \frac{3}{4\pi r'^2 \rho} \int_0^{2\pi} d\varphi \int_{-1}^{1} d(\cos\theta) \times \int_0^{r'} dr\, r \frac{\mu_{en}(k)}{4\pi |r' - r|^2} e^{-\mu(k)|r'-r|} B_D(k, \mu |r' - r|) \tag{11.37}$$

where k is the photon energy and ρ is the physical density of the organ. In the condition of $\mu(k)r'$ being small, then the buildup factor $B_D(k, \mu|r' - r|) \approx 1$. This allows the integral to be solved in the fashion of (11.7) (Snyder et al. 1978),

$$\Phi(r_S' \leftarrow r_S'; k) = \frac{3}{4\pi r'^3 \rho} \frac{\mu_{en}(k)}{\mu(k)} \left(1 - e^{-\mu(k) r'}\right) \cong \frac{3\mu_{en}(k)}{4\pi r'^2 \rho} \tag{11.38}$$

where the last step is a consequence of the assumption of $\mu(k)r'$ being small. If (11.38) is adjusted for the absorbed fraction, then,

$$\phi(r_S' \leftarrow r_S') = \left(\frac{m_{r_S}'}{m_{r_S}}\right)^{1/3} \phi(r_S \leftarrow r_S). \tag{11.39}$$

For the case of the target and mass-scaled source regions differing and assuming that the physical distance between the source and target regions does not change significantly with the changes in the source and target region masses, then the specific absorbed fraction is independent of changes in the source and target region masses. However, for self-irradiation,

$$\Phi(r_S' \leftarrow r_S') = \left(\frac{m_{r_S}}{m_{r_S}'}\right)^{2/3} \Phi(r_S \leftarrow r_S). \tag{11.40}$$

Yamaguchi (1978) applied scalings of the photon absorbed fraction $\Phi(r_S \leftarrow r_T)$ in order to modify the S-factors previously calculated for the Caucasian ICRP Reference Man for an internal radiation dosimetry calculation particular to the Japanese population. Discussions of this application are defined to the subsection concerning non-Western population phantoms in Chap. 12.

MIRD Source and Target Regions

As shown in Table 11.1, the conventional MIRD schema currently specifies 26 source regions. Of these, 25 are specific tissues, organs or contents of organs. All other tissues are grouped into a "remaining tissues" category. In the current MIRD schema, 24 target regions are explicitly specified. The whole body is noted as being a target region: the S-factors for this target region yield the absorbed dose to the whole body (i.e., total energy absorbed by the body divided by the whole-body mass) and should not be confused with the effective dose, which is a stochastic risk-weighted sum of organ absorbed doses. Note also that the contents of the GI tract and the urinary bladder are treated as source regions and the walls of the GI tract and urinary bladder as target regions.

11.3.4 ICRP Method

The ICRP developed a dosimetry model almost identical to that of the MIRD committee (ICRP 1979). The main difference between the two methods is that, because of the focus of the ICRP is on the radiological protection of workers in the nuclear industry who are potentially exposed to radiations with a variety of LETs, the cumulative dose equivalent is instead calculated for rather than the absorbed dose. The dose equivalent, which has since been redefined and renamed as the equivalent dose in ICRP Publications 60 (1991b) and 103 (2007), was defined in Chap. 10 using the radiation weighting factors of ICRP Publication 30 (ICRP 1979). The cumulative dose equivalent is the total dose equivalent received by the target region r_T in a period of 50 years following uptake of a radionuclide. Using the units of ICRP Publication 30, the expression is,

$$H_{50,r_T} = k \sum_{r_S} U_{r_S} SEE(r_T \leftarrow r_S). \tag{11.41}$$

k is a constant for the matching of units (the cumulative dose equivalent has units of sieverts), U_{r_S} is the cumulated activity in the source region r_S and the specific effective energy, $SEE(r_T \leftarrow r_S)$, is equivalent to the MIRD S-factor but now modified by the radiation weighting factor w_R, as in the manner of Stabin and Siegel discussed earlier, but which was referred to in ICRP Publication 30 as the quality factor Q,

$$SEE(r_T \leftarrow r_S; t) = \frac{\sum_i Q_i E_i Y_i \phi(r_T \leftarrow r_S; E_i; t)}{m(r_T, t)}. \quad (11.42)$$

Q_i is the radiation weighting factor (labeled as w_R in Table 10.4) from ICRP Publication 26 (1977).

11.3.5 Suborgan Dimension Calculations

Interest in suborgan activity distributions is especially important in radionuclide therapy as "cold spots," or regions receiving inadequate absorbed dose, can lead to reduced tumor control probabilities and failure to cure or control disease. In such cases, the term "voxel dosimetry" is appropriate. In MIRD Pamphlet 17 (Bolch et al. 1999) "voxel dosimetry" is defined as the evaluation of the absorbed doses to tissue regions with submillimeter dimensions ranging up to dimensions of only a few centimeters. In particular, tomographic emission imaging enables the in vivo quantification at voxel dimensions of the order of 3–10 mm (dosimetry at spatial dimensions comparable to the cell or organelles are soon to be considered). Voxel dosimetry can be based upon three calculation methods. The first is founded on the use of the dose point kernels presented in Chap. 9. This calculation method uses the convolution of the kernel with the spatial distribution of activity which is the result of the superposition of individual point sources which are assigned to discrete voxels. The dose distribution is calculated usually through the means of the Fast Fourier Transform. For example, the absorbed dose at the point $\mathbf{r}'$ due to the activity volume distribution $\rho(\mathbf{r})$ is,

$$D(\mathbf{r}') = \int d^3\mathbf{r}\, \rho(\mathbf{r}) K(\mathbf{r} - \mathbf{r}') \quad (11.43)$$

Another means of calculating absorbed doses at the voxel dimension is that of Monte Carlo simulation of radiation transport and energy deposition. While clearly computationally intensive, this approach is a more accurate one in that it can account for the heterogeneous nature of tissue.

A third means, and one which is discussed in MIRD Pamphlet No. 17, applies the MIRD S-factor concept to voxel-to-voxel calculations. Here, (11.10) would represent the absorbed dose to the voxel at the location r_T due to radiation emitted from voxel at the location r_S. This approach, however, suffers from the inability to account for a heterogeneous medium due to the extremes of computation power required. Despite this limitation to homogeneous media, it presents computational ease through the use of precalculated S-factors.

11.3.6 Internal Radiation Dosimetry Calculation Software

11.3.6.1 Introduction

Table 11.1 demonstrated that the size of the ensemble of MIRD S-factors linking source and target regions can be significant (recognizing that the resulting S-factors are specific for a single radionuclide only). Manual calculations linking calculated cumulated activities in source regions to absorbed doses in target regions are daunting and time consuming. As a result, software presents a practical means of calculating target regions' absorbed doses. This become even more necessary when the organ-specific approach of the "classical" MIRD approach is extended to voxel-based calculations, as described earlier. In this subsection, various software platforms developed for use in internal dosimetry calculations are summarized.

11.3.6.2 MIRDOSE

This was the premiere software package for dosimetry calculations in nuclear medicine (Stabin 1996). In this package, the user provides normalized cumulated activities in MIRD source organs for a specific radionuclide and the absorbed doses per unit activity for MIRD-specified target organs are given as output.

The first version of the code (MIRDOSE 1) was developed in the 1980s and used internally only at the Radiation Internal Dose Information Center (RIDIC) based at the Oak Ridge National Laboratory. The next version MIRDOSE 2, was released for general use in the nuclear medicine community in 1987. This code was limited to internal radiation dosimetry calculations for 60 radionuclides and six anthropomorphic phantoms. The final version, MIRDOSE 3, was released in 1994 and allowed for calculations of more than 200 radionuclides and 10 anthropomorphic phantoms. The code was provided free-of-charge by RIDIC and gained widespread use. However, in 2000, a question arose of whether or not MIRDOSE 3 should be classified as a treatment planning device by the US FDA. As a result of this uncertainty, the RIDIC ceased distributing MIRDOSE 3.

11.3.6.3 OLINDA/EXM

A solution to the MIRDOSE 3 dilemma was provided by the development of the Organ Level Internal Dose Assessment/Exponential Modelling code and which is described by Stabin et al. (2005). OLINDA/EXM is currently unique in the field of nuclear medicine dosimetry in that it has a 510(k) exemption from the FDA. The code represents a considerable extension from the original MIRDOSE series in that it allows calculations for 814 radionuclides and a wide variety of adult, pediatric and pregnant female phantoms, tumors, and the prostate gland. It also contains a useful multi-exponential curve-fitting package allowing the user to obtain directly the cumulated activities from measured biodistribution data.

11.3.6.4 OEDIPE

OEDIPE is a graphical user interface (GUI) designed to create an input file for the MCNP Monte Carlo code on the basis of a voxellated phantom derived from either CT or MR image data sets (Chiavassa et al. 2006; Franck et al. 2007). The user provides the amounts of activities in voxels/organs as functions of time and the absorbed dose distribution is evaluated by the Monte Carlo code.

11.3.6.5 AIDE

This code (activity and internal dose estimates) (Bertelli et al. 2008) incorporates both in vivo biodistribution data and in vitro assay data to evaluate the internal radiation dosimetry. Its primary application is radiation safety and the estimates of absorbed doses associated with the occupational ingestion of radionuclides.

11.3.6.6 PLEIADES

This is a code (program for linear internal age-dependent doses) developed by the UK's Health Protection Agency (Fell 2007; Fell et al. 2007). It solves both the biokinetic and dosimetry problems and is intended primarily for radiological protection.

11.3.6.7 MABDOSE

While the acronym for this code evolves from monoclonal antibody dosimetry, the code is generic for solving the internal radiation dosimetry problem (Johnson et al. 1999a, b). It uses a GUI to allow definition of a tumor volume and a contained Monte Carlo package for evaluating S-factors.

11.3.6.8 MINERVA

This is a Monte Carlo-based treatment planning system (modality-inclusive environment for radiotherapeutic variable analysis), a component of which is used for radionuclide therapy (Descalle et al. 2003). It is based upon the Peregrine Monte Carlo code and allows for the absorbed dose distribution to be calculated for in a full three-dimensional fashion accounting for tissue heterogeneity and a spatially-variable distribution of activity.

11.3.6.9 CELLDOSE

This is a Monte Carlo-based code used to estimate the absorbed dose due to electrons at the cellular level (Champion et al. 2008; Hindié et al. 2009). The code is capable of following electrons to energies as low as 7.4 eV.

11.3.6.10 RADAR

While not necessarily an overt software package, the RADAR site[8] (Eckerman 2002) is a pearl for all those interested in internal radiation dosimetry. It is an easily accessible site of all forms of internal nuclear medicine dosimetry data.

References

Anger HO (1964) Scintillation camera with multichannel collimators. J Nucl Med 5:515–531

Ballinger JR, Mather SJ, O'Doherty MJ (2009) Radiopharmaceutical clinical trials – dose is more than sieverts and becquerels. Eur J Nucl Med Mol Imaging 36:1217–1218

Bertelli L, Melo DR, Lipsztein J, Cruz-Suarez R (2008) AIDE: internal dosimetry software. Rad Prot Dosim 130:358–367

Blumgart HL, Yens OC (1926) Studies on the velocity of blood flow: I. The method utilized. J Clin Invest 4:1–13

Bolch WE, Bouchet LG, Roberston JS, Siegel WBW, JA HRW, Erdi AV, Aydogan B, Costes S, Watson EE (1999) MIRD Pamphlet No. 17: the dosimetry of non-uniform activity distributions – radionuclide S values at the voxel level. J Nucl Med 40:11S–36S

Bolch WE, Eckerman KF, Sgouros G, Thomas SR (2009) MIRD Pamphlet Number 21: a generalized schema for radiopharmaceutical dosimetry – standardization of nomenclature. J Nucl Med 50:477–484

Champion C, Zanott-Fregonara P, Hindié E (2008) CELLDOSE: a Monte Carlo code to assess electron dose distribution – S-values for ^{131}I in spheres of various sizes. J Nucl Med 49:151–157

Chiavassa S, Aubineau-Lanièce BA, Lisbona A, Barbet J, Franck D, Jourdain JR, Bardiès M (2006) Validation of the a personalized dosimetric evaluation tool (Oedipe) for targeted radiotherapy based on the Monte Carlo MCNPX code. Phys Med Biol 51:601–616

Chievitz O, de Hevesy G (1935) Radioactive indicators in the study of phosphorus metabolism in rats. Nature 136: 754–755

Descalle M-A, Hartmann CL, Siantar H, Dauffy L, Nigg DW, Wemple CA, Yuan A, DeNardo GL (2003) Application of MINERVA Monte Carlo simulations to targeted radionuclide therapy. Cancer Biother Radiopharm 18:71–79

Eckerman K (2002) RADAR: the radiation dose assessment resource – an online source of dose information for nuclear medicine and occupational radiation safety (abstract). J Nucl Med 42(suppl):243P

Evans RD (1933) Radium poisoning: a review of present knowledge. m J. Public Health 23:1017–1023

FDA (2004a) Guidance for industry developing medical imaging drug and biological products. Part 1: conducting safety assessments. U.S. Food and Drug Administration, Washington, DC

FDA (2004b) Guidance for industry developing medical imaging drug and biological products. Part 2: clinical indications. U.S. Food and Drug Administration, Washington, DC

FDA (2004c) Guidance for industry developing medical imaging drug and biological products. Part 3: design, analysis and interpretation of clinical studies. U.S. Food and Drug Administration, Washington, DC

Fell TP (2007) The computation of ICRP dose coefficients for intakes of radionuclides with PLEIADES: biokinetic aspects. Rad Prot Dosim 127:220–222

Fell TP, Phipps AW, Smith TJ (2007) The internal dosimetry code PLEIADES. Rad Prot Dosim 124:327–338

Franck D, de Carlan L, Pierrat N, Broggio D, Lamart S (2007) OEDIPE: A new graphical user interface for fast construction of numerical phantoms and MCNP calculations. Radiat Prot Dosim 127:262–265

Hamilton D (2004) Diagnostic nuclear medicine – a physics perspective. Springer, Berlin

Hindié E, Champion C, Zanotti-Fregonara P, Rubello D, Colas-Linhart N, Ravasi L, Moretti J-L (2009) Calculation of electron dose to target cells in a complex environment by Monte Carlo code "CELLDOSE". Eur J Nucl Med Mol Imaging 36:130–136

ICRP (1975) Report on the task group on reference man – ICRP publication 23. International Commission on Radiological Protection, Pergamon, Oxford

ICRP (1977) Recommendations of the International Commission on Radiological Protection – ICRP publication 26. International Commission on Radiological Protection, Pergamon, Oxford

ICRP (1979) Limits for intakes of radionuclide workers – ICRP publication 30. International Commission on Radiological Protection, Pergamon, Oxford

ICRP (1987) Biokinetics and dosimetry: general considerations – ICRP publication 53. International Commission on Radiological Protection, Pergamon, Oxford

ICRP (1991a) Radiation dose to patients from radiopharmaceuticals: addendum 1 to ICRP publication 53. International Commission on Radiological Protection, Pergamon, Oxford

ICRP (1991b) Recommendations of the International Commission on Radiological Protection – ICRP publication 60. International Commission on Radiological Protection, Pergamon, Oxford

ICRP (1994) Human respiratory tract model for radiological protection – ICRP publication 66. International Commission on Radiological Protection, Pergamon, Oxford

ICRP (1998) Radiation dose to patients from radiopharmaceuticals: addendum 2 to ICRP publication 53. International Commission on Radiological Protection, Pergamon, Oxford

ICRP (1995) Basic anatomical and physiological data for use in radiation protection: the skeleton – ICRP publication 70. International Commission on Radiological Protection, Pergamon, Oxford

ICRP (2006) Human alimentary tract model for radiological protection – ICRP publication 100. International Commission on Radiological Protection, Pergamon, Oxford

ICRP (2007) Recommendations of the International Commission on Radiological Protection – ICRP publication 103.

[8]http://www.doseinfo-radar.com/RADARHome.html.

International Commission on Radiological Protection, Pergamon, Oxford

ICRP (2008a) Radiation dose to patients from radiopharmaceuticals: addendum 3 to ICRP publication 53. International Commission on Radiological Protection, Pergamon, Oxford

ICRP (2008b) Nuclear decay data for dosimetric calculations – ICRP publication 107. International Commission on Radiological Protection, Pergamon, Oxford

Johns HE, Cunningham JR (1983) The physics of radiology. Charles Thomas, Springfield

Johnson TK, McClure D, McCourt S (1999a) MABDOSE I: characterization of a general purpose dose estimation code. Med Phys 26:1389–1395

Johnson TK, McClure D, McCourt S (1999b) MABDOSE II: validation of a general purpose dose estimation code. Med Phys 26:1396–1403

Jones T (2003) Historical development of functional in vivo studies using positron-emitting tracers. In: Valk PE, Bailey DL, Townsend DW, Maisey MN (eds) Positron emission tomography. Springer, London

Lindell B, Dunster HJ, Valentin J (2009) International Commision on Radiological Protection: history, policies, procedures. http://www.icrp.org/docs/Histpol.pdf

Loevinger R, Berman M (1968) A schema for absorbed-dose calculations for biologically-distributed radionuclides – MIRD pamphlet number 1. J Nucl Med 9(suppl 1): 7–14

Loevinger R, Budinger T, Watson E (1988) MIRD primer for absorbed dose calculations. Society of Nuclear Medicine, New York

Marinelli LD (1942) Dosage determination with radioactive isotopes. Am J Roentgenol Radium Therapy 47:210–216

Marinelli LD, Quimby EH, Hine GJ (1948a) Dosage determination with radioactive isotopes I. Fundamental dosage formulae. Nucleonics 2:56–66

Marinelli LD, Quimby EH, Hine GJ (1948b) Dosage determination with radioactive isotopes II. Practical considerations in therapy and protection. Nucleonics 2:56–66

Nemec HW, Fridrich R (1977) Zur Frage der retention und der dosis bei der radiosynovierthese mit yttrium-90-silikatkolloid. Nuc Med 16:113–118

Paterson R, Parker HM (1934) A dosage system for gamma ray therapy. Br J Radiol 7:592–632

Patton DD (2003) The birth of nuclear medicine instrumentation: Blumgart and Yens, 1925. J Nucl Med 44:1362–1365

Poston JW (1993) Application of the effective dose equivalent in nuclear medicine patients. J Nucl Med 34:714–716

Potter CA (2005) Internal dosimetry: a review. Health Phys 88:565–578

Roberson PL, Ten Haken RK, McShan DL et al (1992) Three-dimensional tumor dosimetry for hepatic yttrium-90-microsphere therapy. Nucl Med 33:735–738

Rowland RE (1994) Radium in humans: a review of US studies. Report ANL/ER-3. Argonne National Laboratory, Argonne, IL

Seil HA, Voil CH, Ma G (1915) The elimination of souble radium salts taken intravenously and per os. NY Med J 101:896–997

Siegel J, Stubbs TS, Stabin M, Hays M, Koral K, Robertson J, Howell R, Wessels B, Fisher D, Weber D, Brill A (1999) MIRD Pamphlet No. 16: techniques for quantitative radiopharmaceutical biodistribution data acquisition and analysis for use in human radiation dose estimates. J Nucl Med 40:37S–61S

Smith S, Thomas RM, Steera HA, Beatty HE, Dawson KB, Pekham MJ (1976) Therapeutic irradiation of the central nervous system using intrathecal Y-90-DTPA. Br J Radiol 49:141–147

Snyder WS, Ford MR, Warner G (1978) Estimates of specific absorbed fractions for monoenergetic photon sources uniformly distributed in various organs of a heterogeneous phantom, Pamphlet No. 5. Revised. Society of Nuclear Medicine, New York

Stabin MG (1996) MIRDOSE: personal computer software for internal dose assessment in nuclear medicine. J Nucl Med 37:538–546

Stabin MG (2006) Nuclear medicine dosimetry. Phys Med Biol 51:R187–R202

Stabin MG (2008) Fundamentals of nuclear medicine dosimetry. Springer, New York

Stabin MG, Eckerman KF, Ryman JC, Williams LE (1994) Bremsstrahlung radiation dose in yttrium-90 therapy applications. J Nucl Med 35:1377–1380

Stabin MG, da Luz LCQP (2002) Decay data for internal and external dose assessment. Health Phys 83:471–475

Stabin MG, Siegel JA (2003) Physical models and dose factors for use in personal dose assessment. Health Phys 85: 294–310

Stabin MG, Sparks RB, Crow E (2005) OLINDA/EXM: the second-generation personal computer software for internal dose assessment in nuclear medicine. J Nucl Med 46: 1023–1027

S-Stelson AT, Watson WE, Cloutier RJ (1995) A history of medical internal dosimetry. Health Phys 69:766–782

Verbruggen A, Coenhen HH, Deverre JR, Guilloteau D, Langstrom B, Salvadori PA et al (2008) Guideline to regulations for radiopharmaceuticals in early phase clinical trials in the EU. Eur J Nucl Med Mol Imaging 35:2144–2151

Watson EE, Stabin MG, Siegel JA (1993) MIRD formulation. Med Phys 20:511–514

Williams L, Wong J, Findley D, Forell B (1989) Measurement and estimation of organ bremsstrahlung radiation dose. J Nucl Med 30:1373–1377

Wiseman A, White CA, Stabin M, Dunn WL, Erwin W, Dahlbom M, Raubitschek A, Karvelis K, Schultheiss T, Witzig TE, Belanger R, Spies S, Silverman DHS, Berlfein JR, Ding E, Grillo-Lopez AJ (2000) Phase I/II ^{90}Y-Zevalin (yttrium-90 ibritumomab tiuxetan, IDEC-Y2B8) radioimmunotherapy dosimetry results in relapsed or refractory non-Hodgkin's lymphoma. Eur J Nucl Med 27:766–777

Wrenn F, Good M, Handler P (1951) The use of positron emitting radioisotopes for the localization of brain tumors. Science 113:525–527

Yamaguchi H (1978) Estimation of internal radiation dose for various physiques using MIRD adult absorbed fractions. Acta Radiol Oncol 17:429–439

Anthropomorphic Phantoms and Models of Biological Systems

12

Abstract Except in some very unusual and rare cases, nuclear medicine radiation absorbed doses are calculated and not measured. Input measurements, in the form of the biodistribution of the radionuclide, are required but the dosimetry is evaluated through calculation. This evaluation is almost always currently performed for a representative anthropomorphic model, or phantom. This chapter describes the types of phantoms (stylized, voxellated/tomographic and hybrid) and surveys their development in increasing sophistication and complexity and ethnic diversities. Biological systems must also be modeled in the evaluation of the internal radiation dosimetry, especially those dynamic systems such as the urinary bladder and gastrointestinal tract where multi-compartmental transport of activity must be calculated for. Models of specific biological systems (respiratory system, gastrointestinal tract, heart and contents, kidney, urinary bladder, head and brain, bone and bone marrow, peritoneal cavity, rectum, prostate gland and spherical tumors) are reviewed and, where applicable, derivations of cumulated activity expressions provided.

Contents

12.1 Anthropomorphic Whole-Body Phantoms

12.1.1 Introduction

A phantom is a model of anatomy used in the calculation of the distribution of absorbed doses to tissues and organs as a result of exposure to radiation. The phantom can be physical or mathematical. Physical phantoms are used almost exclusively for external beam irradiation and are typically constructed from materials which are radiologically-equivalent to tissues. Radiation dosimeters (thermoluminescent dosimeters, radiographic film, or semiconductor detectors) are placed within the phantom for measuring the absorbed dose in situ. Because it is impractical to replicate the temporally-varying distribution of internal activities

B.J. McParland, *Nuclear Medicine Radiation Dosimetry*,
DOI 10.1007/978-1-84882-126-2_12,

due to an unconfined radioactive source, physical phantoms are rarely used for nuclear medicine dosimetry evaluations; mathematical phantoms are used instead. The sophistication of the morphology of the phantom and its radiological constituency can vary considerably and will be largely dictated by the complexity of the dosimetry problem at hand and by the computing power available for performing the calculation of interest. In general, modeling realism has increased over the decades from simple to more complex geometrical representations of the human anatomy to models derived from whole-body tomographic images. In modern times, the internal absorbed dose distribution is calculated using Monte Carlo simulations of radiation transport and energy deposition. The distribution of a radionuclide within the phantom is assigned to one or more source regions. Depending upon the spatial resolution of the phantom, the activities may be considered uniformly distributed within entire organs, allocated to individual suborgan structures or even to individual voxels so as to replicate nonuniform distributions. It is also common to use a source of monoenergetic photons or photons instead of a specified radionuclide; repeating the calculation for a range of photon or charged particle energies can then enable one to interpolate the resulting data to obtain results for a specific radionuclide. The transport of radiation and the deposition of energy to specified organs or tissues (target regions) are then simulated. The result is the distribution of absorbed doses which can be converted to absorbed fractions or specific absorbed fractions, leading to the calculation of S values for the source region / target region pair.

The earliest phantom of the human whole body used for dosimetry calculations was the physicist's common first-order bovine approximation, the homogeneous sphere, for which analytical dosimetry calculations are generally trivial. A right cylinder more closely approximates the whole body but, as demonstrated in Chap. 11 pertaining to the Marinelli–Quimby–Hine calculations, the complexity of the analytical calculation can demand necessary approximations. With increasing computer power and the availability of advanced Monte Carlo codes, more complex and heterogeneous phantoms were defined and more detailed dosimetry calculations performed.

As the phantom is intended to be a model of the individual for whom radiation dosimetry is to be evaluated, the patient or subject themselves can provide the anatomical and morphological data required for a bespoke dosimetry calculation. This has been a routine practice in external beam radiotherapy for many decades. The absorbed dose distribution can be estimated based upon a simple external contour taken from the patient and the assumption of homogeneous tissue. More complicated external beam treatment planning is based upon three-dimensional computed tomography (CT) image data from which the photon attenuation coefficients are and used in the treatment plan. This patient-specific approach has only come into research use in radionuclide therapy in recent years because the geometrical placements of the sources of irradiation in internal therapy are far more complex compared to external beam therapy with fixed radiation beams. Excellent historical reviews of phantoms and descriptions of contemporary mathematical anthropomorphic models can be found in the chapter by Poston et al. (2003) and the review article by Zaidi and Xu (2007).

There are three major classes of phantoms in contemporary use. The first is that described as being "stylized". These are phantoms containing a small ensemble of selected organs, the morphologies of which are modeled by simple geometrical structures such as cylinders, spheres, and ellipsoids. For example, an ellipsoid representing, say, a lung in Cartesian coordinates would be given by the parametric equation,

$$\left(\frac{x - x_0}{a}\right)^2 + \left(\frac{y - y_0}{b}\right)^2 + \left(\frac{z - z_0}{c}\right)^2 \leq 1 \quad (12.1)$$

where the parameters (x_0, y_0, z_0) and (a, b, c) would be specified through matching of the model to the general shape and dimensions of the organ.

The second class is categorized as tomographic, or voxellated, phantoms. These are derived from CT or magnetic resonance (MR) cross-sectional images of cadavers or living individuals, the latter being volunteers or patients from which medical tomographic examinations are used. The selection of the individual or cadaver to image is based upon its similarity to a particular reference individual specific to the ethnic population at question. Absorbed dose calculations can be performed using Monte Carlo simulations in a direct voxel-to-voxel approach or by the segmentation of macroscopic source and target regions followed by Monte Carlo calculations.

The third class of phantoms is the "hybrid" phantom which combines the features of the stylized and tomographic phantoms. The tomographic class is somewhat inflexible in its ability to adjust the dimensions and morphologies of organs or to account for cardiac and respiratory motions (which, however, are of limited importance to calculating the absorbed doses due to the internal distribution of a radionuclide). A hybrid phantom is mathematically based, but has a more realistic representation of an organ than would the stylized phantom be capable of. Organ sizes and shapes can be distorted in order to yield a desired result.

12.1.2 Reference Man

12.1.2.1 Introduction

Before embarking upon discussion of the various anthropomorphic phantoms used in internal radiation dosimetry, it is necessary to understand the concept of the very closely related example of Reference Man.[1] This is an individual with organs and tissues of sizes, masses and compositions, and a physiology representative of a given population. The still on-going development of Reference Man was begun by the International Commission on Radiological Protection (ICRP) after the end of the Second World War for use in radiological protection. Calculations of the absorbed doses to a single Reference Man allowed, for example, inter-comparison of different irradiation conditions, consistent radiation safety designs, and defined radiation protection standards. The ICRP construction was, and remains, representative of a Western individual and, as such, Reference Man cannot be considered representative of non-Western populations due to differences in physical stature, metabolisms, etc across differing ethnicities. This may need to be recognized in mapping dosimetry calculations for the ICRP Reference Man to different populations. The ICRP did not embark upon the task of defining additional Reference Men that were specific to other ethnicities. This was judged highly impractical because of the diversity among ethnic groups. Moreover, Reference Man is not a static model. Over time, changes in the means of anatomical and physiological measures will occur in a given population and this will be represented in Reference Man. Because of the effort involved in modifying Reference Man to represent changes over ethnicities and over time, the ICRP instead considered it to be the responsibility of the relevant local authorities to modify the ICRP Reference Man data to accommodate the specific population in question.

12.1.2.2 ICRP Reference Man

As described, the ICRP Reference Man is intended to be representative of the Western population. In particular, he is a Caucasian male of between 20 and 30 years of age, living in a temperate environ and has a diet and lifestyle typical of the North American and Western European populations. The first detailed description of the anatomy, composition, and physiology of ICRP Reference Man appeared in ICRP Publication 23 (ICRP 1975). In this instance, he had a whole-body mass of 70 kg and a height of 170 cm. Recently, the ICRP extended this description of Reference Man in its release of ICRP Publication 89 (ICRP 2002) which explicitly included a Reference Woman and five Reference Children and incorporated changes in anatomy and physiology considered typical as having occurred in the Western population during the time since the release of ICRP Publication 23. In ICRP Publication 89, Reference Man had a weight of 73 kg and a height of 176 cm.

12.1.2.3 Non-Western Reference Men

Recognizing that the ICRP Reference Man is not representative of all ethnicities and that the results of dosimetry calculations can vary with the physical sizes and masses of organs, a number of institutions have designed a Reference Man which is applicable to their local populations. Perhaps one of the most advanced and lengthy developments of a non-Western Reference Man is that of India, the development of which began at the Bhabha Atomic Research Centre in Mumbai by Venkataraman et al. (1963), with updates by Jain et al. (1995a, b) and new parameterizations by

[1] Here "man" is used in its generic sense and includes, in this discussion, the female and the child. Where sex or age specificity is required, the individual phantom being referred to will be noted explicitly.

Biju and Nagarajan (2000). In a more general development to cover the Asian continent, the International Atomic Energy Agency (IAEA) has led the development of a Reference Asian Man which is based upon reference individuals of Indian, Pakistani, Bangladeshi, Indonesian, and Vietnamese populations (IAEA 1998).

12.1.3 Stylized Whole-Body Phantoms

12.1.3.1 Introduction

The stylized phantom is equation-based and a composite of geometrical volumes that approximate the torso, head and neck, extremities, and internal organs. These volumes are typically made up from right circular cylinders, right elliptical cylinders, spheres, cones, and ellipsoids (and their intersections), all of which can be described by analytical functions that simplify Monte Carlo calculations of radiation transport. The first use of a stylized anthropomorphic phantom for internal radiation dosimetry calculations is traceable to the Manhattan Engineering District project of the Second World War and which subsequently evolved into the phantom of ICRP Publication 2 (ICRP 1959; Zaidi and Xu 2007). That phantom treated the human adult torso as a homogeneous sphere 30 cm in diameter with the internal organs considered as spheres of varying diameters. As computing capability increased so as to enable more realistic calculations of radiation transport, the complexities of the volumes used, in terms of both morphology and composition, grew accordingly.

12.1.3.2 Brownell–Ellet–Reddy Phantoms

These early phantoms were geometrically simplistic as necessitated by the limited computing power available during the 1950s and 1960s (Brownell et al. 1968). The dosimetry calculations performed were of the absorbed fractions at the centers of phantoms representing the adult human and made up of spherical, elliptical cylindrical, and right cylindrical shapes. The phantoms were assumed to be homogeneous with compositions radiologically-equivalent to tissue (hydrogen, carbon, oxygen, and nitrogen). Specific organs were not modeled.

12.1.3.3 Snyder and Snyder-Fisher Phantoms

Snyder began an effort at Oak Ridge National Laboratory in the late 1950s combining a simple anthropomorphic phantom with Monte Carlo calculations of the radiation transport within the phantom. The initial phantom was a right circular cylinder 60 cm in height and 30 cm in diameter so as to represent the adult human male torso (Poston et al. 2003). Further development by Fisher, Snyder, and others of anthropomorphic phantoms culminated in a heterogeneous hermaphroditic phantom. In this phantom, a right circular cylinder represented the combined torso, arms and pelvis; the head and neck were approximated by an elliptical cylinder; and, the legs and feet were modeled by a truncated elliptical cone.

By the late 1960s, Fisher and Snyder had developed a heterogeneous hermaphroditic model of the adult human (Snyder et al. 1969). The phantom was made up of three tissue types: skeletal, lung, and soft with physical densities of 1.5, 0.3, and 1 g/cm^3, respectively. Of particular interest is the fact that organ masses in this phantom were selected on the basis of those that were to eventually appear in the compilation of ICRP Reference Man (ICRP 1975). The phantom contained 22 internal organs modeled by geometrical shapes. Absorbed fractions were estimated using Monte Carlo simulation for monoenergetic photon sources (with energies between 10 keV and 4 MeV) considered to be uniformly distributed within source organs; only photon transport was considered and energy transferred in a photon–electron interaction (photoelectric absorption or Compton scatter) was assumed to be locally absorbed. These results were released in MIRD Pamphlet No. 5 (Snyder et al. 1978).

Improvements to this heterogeneous phantom were presented by Snyder and collaborators in the late 1970s (Snyder et al. 1978). The head and neck were represented by an elliptical cylinder; the upper extremities, torso and pelvis were represented by another elliptical cylinder; and, the lower extremities were modeled by a pair of truncated elliptical cones. Absorbed fractions calculated for this model appeared in MIRD Pamphlet No. 5, Revised.

Because of its adoption by the MIRD Committee, the Snyder phantom was frequently referred to as the MIRD phantom.

12.1.3.4 Oak Ridge National Laboratory Phantom Series

Introduction

As the genesis of many of the modern stylized phantoms were associated with the Oak Ridge National Laboratory, this series will be isolated as it reflects the beginning of the contemporary stylized models. Cristy (1980) reported on a new series of phantoms consisting of an adult male and five pediatric examples (newborn, 1-, 5-, 10-, and 15-years of age). Anatomical data from the ICRP Reference Man (ICRP 1975) were incorporated within this phantom.

Cristy–Eckerman Phantoms

The Cristy–Eckerman phantom series (Cristy and Eckerman 1987) is the series in most common use in current nuclear medicine dosimetry calculations and carried on from the work of Cristy noted above. It consists of six phantoms: the adult male and five pediatric examples (newborn, 1-, 5-, 10-, and 15-year old). The 15-year-old phantom was taken to be representative of the adult female. The adult male, which was based upon the ICRP Publication 23 Reference Man, is actually hermaphroditic. Specific absorbed fractions were calculated using the ETRAN Monte Carlo code for photon energies ranging from 100 keV to 4 MeV. The original Cristy and Eckerman technical reports describing these phantoms can be found at http://ordose.ornl.gov/documents/.

Stabin Phantoms of the Female Adult and the Pregnant Female

Stabin et al. (1995) extended the Cristy–Eckerman series by deriving a female phantom in nonpregnant and 3-, 6-, and 9-month pregnant forms. The nonpregnant female model was intended to supplant the use of the Cristy–Eckerman 15-year-old pediatric phantom as a model of the adult female. No attempt was made to model the fetus in the 3-month pregnant female in which the uterine contents were approximated simply by a homogeneous structure of soft tissue. For the 6- and 9-month old pregnant female, models of the fetus were explicitly included. The fetus was modeled as a cylinder capped with hemispheres; the morphology of the fetal skeleton was similarly modeled, although the material was radiologically comparable to bone. Specific absorbed fraction data were provided in that report.

12.1.3.5 Non-Western Populations

As discussed in the context of the development of Reference Man, its ICRP design was intended to be representative of the Western population and it was judged that the responsibility to modify this Reference Man to be more representative of a particular ethnic population was that of a local or national agency responsible for radiological protection in that area.

Most non-Western population phantoms presented in the literature are tomographic and based upon voxellated CT images acquired of non-Western subjects. However, efforts have been made in the past to scale the organ mass and subject physique from Caucasian phantoms to those appropriate for non-Western populations. Yamaguchi (1978) described the scaling of absorbed fractions obtained for a given phantom to those of another. This scaling of the absorbed fraction accounted for changes in the target region mass and the distances between the source and target regions. In his example, Yamaguchi reevaluated the absorbed fractions derived by Snyder to evaluate the dosimetry for a physique more representative of a Japanese male adult. Jain et al. (1995a) performed a similar scaling in order to evaluate the differences in absorbed doses and effective doses received by ICRP Reference Man, and Reference Men specific to China, Japan, and India as a result of the administration of three radiopharmaceuticals based on ^{99m}Tc and ^{131}I.

Park et al. (2006) describe a stylized phantom of an average Korean adult male that was based much upon using anthropometric and organ volume data derived from a national Korean survey (from body size measurements and MR assessments of internal organs) to modify the parametric equations defining the external body and internal organs of the ORNL adult male phantom. Specific absorbed fractions were calculated from Monte Carlo simulations of internal photon emissions with energies ranging from 0.01 to 4 MeV. The smaller organ volume and reduced inter-organ distances resulted in higher Φ values than in the corresponding ORNL phantom.

12.1.4 Voxellated (Tomographic) Whole-Body Phantoms

12.1.4.1 Introduction

The voxellated, or tomographic, whole-body phantom differs from the stylized phantom in that individual organs are not modeled by geometrical approximations. Instead, as in a conventional CT or MR image, the phantom is made up of individual voxels which can then be segmented and grouped to isolate specific organs or suborgan structures as source and target regions. The voxels can themselves be treated as individual source and target regions (although the computing power required linking the transport of radiation between the voxels would become phenomenal). The tomographic phantom provides a more detailed and realistic representation of the human body and this increase in realism makes the phantom suitable for patient-specific dosimetry calculations, especially if the phantom is derived from imaging data of the patient to be treated. This is one basis of patient-specific therapy. However, in diagnostic nuclear medicine applications, such specificity is not required as the amount of absorbed dose received by the patient undergoing a diagnostic procedure is so small that errors in absorbed dose calculation by factors of two to five are usually considered tolerable. Here, patient-specific calculations are not required and one need only have the phantom to mimic the patient as closely as possible. Should more accurate patient dosimetry calculation be desired, this can be achieved in the tomographic model by selecting the individual upon whom the phantom is to be based to be as close as possible to the desired reference in terms of size and weight. Organ deformation, as used in hybrid phantoms, is another option.

12.1.4.2 Construction Methods

Tomographic phantoms are created from CT and/or MR examinations of either living subjects or cadavers. Both sources of data have practical advantages and disadvantages. In the CT imaging of a cadaver, the radiation absorbed dose to the subject is not of concern and the entire body can thus be imaged to significant detail. However, a difficulty arises from the imaging of a cadaver postmortem in that the body degrades through pulmonary collapse and the body fluids will pool. In addition, X-ray contrast media cannot be introduced in order to enable vascular delineation in the cadaver. As a result of these disadvantages, the resulting image may not be sufficiently representative of the body in order to create the required phantom. These problems can be avoided through the CT examination of a living subject but this comes at the cost of the subject receiving a potentially high radiation dose and of the possibility of artifacts created through patient motion. MR imaging of a subject to create a tomographic phantom offers the advantage of the lack of ionizing radiation exposure, but suffers from its being time-consuming, its inability to accurately image bone and skeletal tissue and, in the case of imaging a frozen cadaver, the perturbations of the spin-lattice and spin–spin relaxation times (Poston et al. 2003). However, in either imaging case, once the image data have been acquired, organs and tissues must be consequently segmented and the tissue types and compositions assigned to the segmented organs. For large, macroscopically-visible tissues, this segmentation is relatively easy to do. However, for tissues with spatial dimensions comparable to or smaller than the image resolution (e.g., voxel dimension), special interventions are required in order to model the tissue. Examples would include the thickness of the skin or of the bone endosteum.

Because of its use in the development of a number of tomographic phantoms, the Virtual Human Project® (VHP) is briefly described here. This project is managed by the US National Library of Medicine[2] in which the cadavers of an adult male and adult female, both American, have been used to create three-dimensional representations of the human whole-body anatomy through CT and MR imaging, and through optical photographic cryosections. The CT data are provided in 512 × 512 pixel formats with 1-mm thick slices for the male and 0.33-mm thick slices for the female.

12.1.4.3 Zubal Phantoms

At Yale University, Zubal and colleagues developed voxellated phantoms of two live adult males (Zubal

[2] http://www.nlm.nih.gov/research/visible/visible_human.html.

et al. 1994) from CT image datasets of the head and torso and MR image sets of the head. Additional work was done to couple the arms from the Virtual Human Project® to the CT torso image set. The phantoms can be found at http://noodle.med.yale.edu/zubal/. Following postacquisition image processing, the CT data were based upon isotropic voxels with a side dimension of 4 mm.

Applications of the Zubal phantoms in nuclear medicine radiation dosimetry include the calculation of absorbed fractions (Yoriyaz et al. 2000) and specific absorbed fractions (Stabin and Yoriyaz 2001).

12.1.4.4 GSF Phantom Series

The German National Research Centre for Environment and Health Gesellschaft für Strahlenforschung (GSF) voxel phantom series is based upon CT and MR images of subjects (Zankl et al. 1988; Petoussi-Henss et al. 2002; Fill et al. 2004). The series consists of several tomographic phantoms of both sexes and a wide range of ages: BABY (female, 8-weeks old), CHILD (female, 7-years old), DONNA (female, 40-years old), FRANK (male, 48-years old), GOLEM (male, 38-years old), HELGA (female, 26-years old), IRENE (female, 32-years old), LAURA (female, 43-years old), and DONNA (female, 40-years old). The adult male from whom GOLEM was created was selected due to his similarity to ICRP Reference Man. GOLEM and LAURA have been selected by the ICRP to be the basis of new reference male and female models, respectively, to replace the MIRD-type stylistic phantom for radiation protection dosimetry (ICRP 2007). Anatomical data from ICRP Publications 89 have been used to adjust the dimensions of the phantoms.

12.1.4.5 University of Florida Pediatric Phantom Series

These are CT-derived pediatric phantoms in which the first group, referred to as UF Series A (Lee et al. 2005), consists of five phantoms, UF 9 month (male), UF 4 year (female), UF 8 year (female), UF 11 year (male), and UF 14 year (male). All were derived from torso or head CT images of live patients and voxel sizes range from $0.43 \times 0.43 \times 3$ mm^3 to $0.625 \times 0.625 \times 6$ mm^3. A second series, UF Series B (Lee et al. 2006a) was extended from Series A by the addition of scaled CT extremity image sets acquired from a Korean adult. Two additional early-age pediatric phantoms (Nipper et al. 2002) are based upon CT images of the cadavers of a healthy 6-day old female and a 6-month old male. The latter had died from cardiac failure arising from congenital heart disease and, due to the low patient weight at the time of death, the phantom was referred as a 2-month old.

12.1.4.6 NORMAN Whole-Body Phantom

NORMAN is a voxellated phantom of the adult human developed at the UK's National Radiological Protection Board (which is now part of the Health Protection Agency) developed from MR images of 2×2 mm^2 pixel size and 1 cm thick slices. It has been used for both nonionizing (Dimbylow 1997) and ionizing radiation dosimetry (Jones 1997).

12.1.4.7 MAX and FAX Whole-Body Phantoms

The male adult voxel (MAX) and female adult voxel (FAX) phantoms were developed by Kramer et al. (2003, 2004). MAX is based upon the Zubal segmented phantoms described above. The phantom's dimensions were adjusted to match ICRP Reference Man as closely as possible, with a mass of 74.65 kg and a height of 175.3 cm. New models for skin and skeletal dosimetry were incorporated.

FAX was derived from the combined CT scans (at a pixel resolution of 0.73×0.73 mm^2 and a slice separation of 2.5 mm) of the lower head, neck and torso of a 37-year-old female subject, with a mass of 63.4 kg and a height of 165 cm, and from CT scans of the legs and feet of a second woman. The arms and top of head of MAX were used after adjustment to fit. Adjustments were made to ensure FAX matched as closely as possible the ICRP Reference Woman defined in ICRP Publication 89 at a mass of 59.8 kg and height of 163 cm.

Kramer et al. (2006) revised MAX and FAX to form FAX06 and MAX06 to account for the new tissues specified in the new effective dose calculation of ICRP Publication 103 (2007).

12.1.4.8 VIP-MAN Whole-Body Phantom

The VIP-man (Virtual Photographic-Man) is a project initiated at the Rensselaear Polytechnic Institute in the United States and is based primarily upon the photographed cryosections of the VHP project discussed earlier. As it is based upon the cadaver of an individual who was not selected for the project on the basis of similarity to Reference Man, the VIP-Man is much heavier (104.3 kg) than the ICRP Reference Man (73 kg) but of comparable height (186 vs. 174 cm). Organ segmentation had been performed manually during the VHP project, with some further additions done by the research group (Xu et al. 2000).

12.1.4.9 Non-Western Populations

Introduction

As with our earlier discussion regarding Reference Man specific to a particular ethnic population, similar arguments arise for the development of anthropomorphic phantoms specific to different ethnicities. A literature review suggests that the non-Western anthropomorphic phantoms currently in existence are those representative of Japanese, Korean, and Chinese populations and are tomographic in nature.

Japanese Phantoms

A number of Japanese voxellated phantoms are described by various authors in the literature (Saito et al. 2001; Nagaoka et al. 2004; Sato et al. 2007). The male and female phantoms of Nagaoka et al. were based on MR images of healthy volunteers whereas those of Saito et al. and Sato et al. were based upon two males and two females, all healthy, who underwent whole-body CT. For example, the JM (Japanese Man) phantom of Sato et al. (2007) is based upon a 54-year-old Japanese adult male of a weight of 65 kg and height 171 cm, reported to be consistent with the corresponding population averages of 64 kg and 170 cm.

Korean Phantoms

KORMAN

This is a voxellated model of an adult Korean male (Lee et al. 2004). It is based upon whole-body MR images of a volunteer of a height (170 cm) and weight (68 kg) considered to approximate that of the reference Korean (170.9 ± 3.7 cm; 67.9 ± 4.8 kg). Organs were semiautomatically segmented.

KTMAN-1, KTMAN-2

These are additional whole-body tomographic phantoms developed from Mr (KTMAN-1) and CT (KTMAN) image sets of two healthy Korean adult males (Lee et al. 2006b).

HDRK-Man

This phantom ("High-Definition Reference Korean-Man") was developed not from CT or MR imagery, but from high-resolution color photographs of serial thin (0.22-mm thick) cryosections of the cadaver of a 33-year-old Korean male who was 164 cm in height and 55 kg in weight (Kim et al. 2008). The resulting imagery was of 8,590 images with a resolution of 0.1875×0.1875 mm^2. Organs were segmented by anatomists and the body weight and height, skeletal mass, and organ sizes were adjusted to fit those of Reference Korean-Man.

Chinese Phantom (CNMAN)

This phantom is of a Chinese adult male and was created, as was HDRK-Man, from color photography of slices from a cadaver (Zhang et al. 2007). A total of 1,759 1-mm thick slices were used to create the phantom. Imagery was color photography with a resolution of $3{,}072 \times 2{,}048$ pixels for each photograph. The reconstructed phantom has a minimum voxel size of $0.16 \times 0.16 \times 1$ mm^3. Physical properties of organs were assigned from published data.

12.1.5 Hybrid Phantoms

The stylized phantom and the voxellated/ tomographic phantom represent two extremes. Clearly, the stylized phantom is simplistic and representative of a defined Reference Man. Because of this, the stylized phantom is used primarily in diagnostic applications where the absorbed doses are limited so that the inherent inaccuracies in modeling a given patient with the stylized reference phantom are of no practical concern. On the other hand, it would be entirely inappropriate to use a reference phantom in, for example, radionuclide therapy planning. The resulting errors can lead to failure to deliver an adequate absorbed dose to the tumor or to excess irradiation of normal healthy tissue. While it is possible to scale the S-factors from the reference phantom to values more representative of the patient, such modifications would require validation.

The voxellated phantom is, clearly, far more realistic a representation of the anatomy and allows more comprehensive measures of the internal radiation dosimetry. As noted above, some authors have selected individual subjects with physical characteristics close to the reference individual of interest. In principle, although highly computationally extensive, a whole-body tomographic data set of a therapeutic patient can be used to provide a patient-specific dosimetry calculation. While the tomographic/voxellated phantom provides a far more spatially-detailed description of the absorbed dose distribution, it is also highly more specific and generally insufficiently generic to be applied to a given patient.

The recent developments of hybrid phantoms seek to address this problem through the combination of spatial detail and the ability to create a phantom applicable to any specified individual. The nonuniform rational B-spline (NURBS) model incorporates the advantages of the stylized and tomographic phantoms (Segars et al. 2004; Lee et al. 2007a, b). NURBSs are used to model three-dimensional surfaces and, through the use of mathematical transformations, can be used to model variations in anatomy and differences in organ size and morphology. Affine transformations can then be used to alter the surface (an affine transformation being one that preserves collinearity and the relative separations between points).

12.2 Models of Biological Systems

12.2.1 Introduction

In the static anthropomorphic model, the absorbed doses to target regions are calculated from known and static activities in source regions. An assumption of a temporally and spatially static and uniform distribution of activity is usually not valid for some organ systems, in particular the lungs, urinary bladder, and the gastrointestinal tract. The last two are of particular interest in diagnostic nuclear medicine in that high rates of excretion of a radionuclide are frequently sought in order to reduce the radiation dose burden and to reduce any background "signal" that can limit the efficacy of diagnosing the pathology of interest.

12.2.2 Respiratory System

12.2.2.1 Introduction

Activity can enter the respiratory system through either perfusion or via the inhalation of radioactive gases or aerosols. Perfusion is not considered in this subsection.

In recent times, the ICRP has presented two human respiratory system models. The first was described in Publication 30 (1979) and the second presented in Publication 66 (1994). The second model was an enhancement of the first and was designed to be coupled with the ICRP's new human alimentary tract model (HATM) to be described later. While the respiratory and GI tract models were developed for radiological protection purposes, they are applicable to nuclear medicine radiation dosimetry purposes. However, a major problem remaining in the evaluation the radiation dosimetry resulting from the inhalation of radioactivity is that of determining the cumulated activity. A model presented in the MIRD 18 report (Thomas 2001) considers this problem and is discussed below.

12.2.2.2 Anatomy

Only a brief overview only of the basic anatomy of the respiratory system relevant to the dosimetry discussion is provided here. The respiratory system consists of the nasopharynx extending to the larynx and the trachea which bifurcates into right and left principal bronchi which further segment into secondary and segmental (or tertiary) bronchi in the two lungs. This bronchial tree continues to branch into smaller bronchioles which terminate into the alveoli where the rapid exchange of gases between air and blood occurs. For radiation dosimetry purposes, interest in the respiratory systems focuses on the volume of the lung space and the exchange rates from the lung space into the vascular and lymphatic systems.

12.2.2.3 ICRP Models of the Respiratory System

ICRP Publication 30 Model

The model of the respiratory system derived in ICRP Publication 30 (1979) is specific to the inhalation of radioactive material and, hence, is primarily of radiation protection interest (although radioactive aerosols are used in pulmonary ventilation studies as described below). The respiratory system model is modeled by three compartments: the nasal passage, the trachea and bronchial tree and the pulmonary parenchyma, as shown in Fig. 12.1.

The absorbed dose received by the nasopharynx is neglected in the model as it is usually much smaller than those received by the other compartments. In Fig. 12.1, the parameters f_{NP}, f_{TB}, and f_P represent the fractions of inhaled material that are initially deposited in the nasal passage, the trachea and bronchial tree and the pulmonary parenchyma, respectively. The subcompartments A, C, and E are associated with absorption of radioactive material and transfer to body fluids whereas the subcompartments B, D, F, and G account for particle transport with the gastrointestinal tract. The pulmonary lymphatic system also aids in the removal of radioactive material: that contained in the parenchymal subcompartment H is transferred to subcompartment I which is subsequently transferred to the body fluids whereas that in subcompartment J is assumed to be retained within the lymphatics. ICRP Publication 30 provides the set of first-order linear differential equations describing the transport of radioactive material between compartments and subcompartments. The particulate size in the inhaled radioactive aerosol is considered as a variable.

ICRP Publication 66 Model

In 1994, the ICRP updated its respiratory model for use in radiological protection (ICRP 1994). Unlike the ICRP Publication 30 model, the radiation sensitivities of individual respiratory system tissues are considered and the absorbed fractions for charged particles (α particles, electrons and positrons) deposited within the lung are provided.

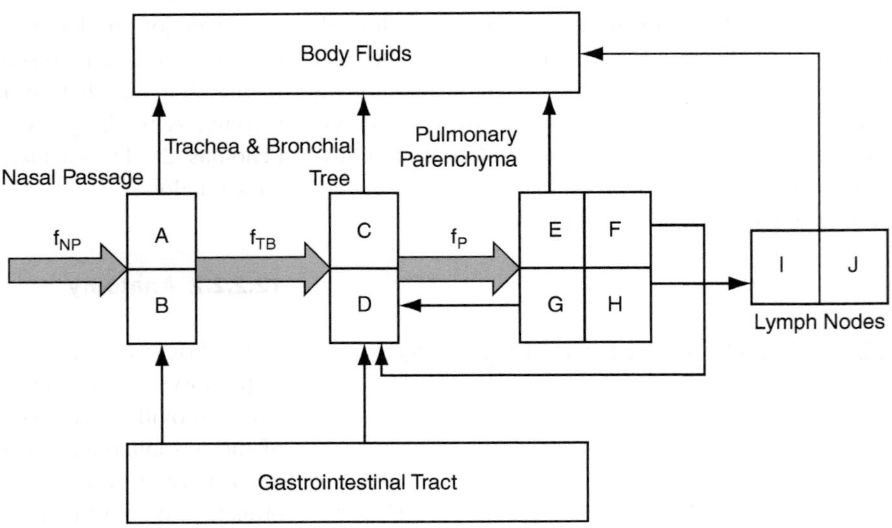

Fig. 12.1 Compartmental model of the respiratory system presented in ICRP Publication 30 (1979). A description of the model is provided in the text

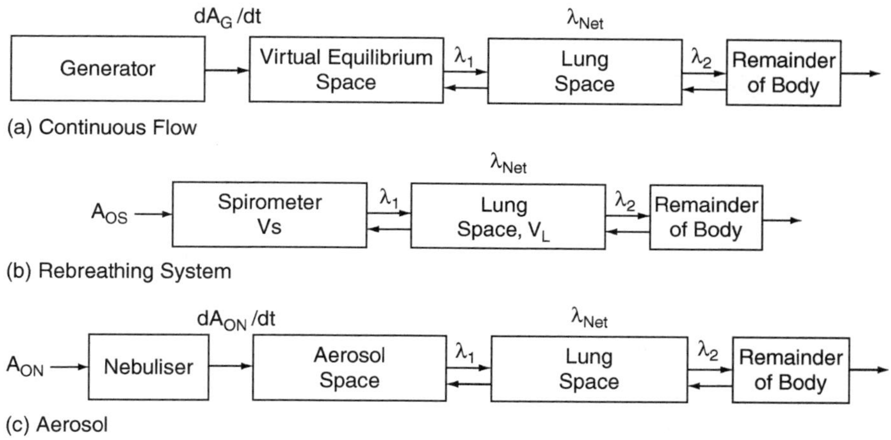

Fig. 12.2 MIRD Pamphlet No 18 compartmental models for ventilation studies based upon continuous flow, rebreathing and aerosol systems. Refer to text for details (after Thomas (2001))

12.2.2.4 MIRD Report 18 Ventilation Model

Introduction

Ventilation imaging following the inhalation of a radioactive gas or aerosol is used in conjunction with pulmonary perfusion imaging as part of the diagnosis of pulmonary embolism (Gray and Neilly 1999; Bajc et al. 2009). Radioactive gases used in such ventilation studies are ^{81m}Kr (with a half-life of only 13 s) or, more rarely, ^{133}Xe.[3] Radioactive aerosols include ^{99m}Tc-labeled DTPA[4,5] or an ultrafine dispersion of ^{99m}Tc-labeled graphite particles. Within laminar air flow, the aerosol behaves physically as a gas with the radioactive particles remaining airborne. However, once the air flow becomes turbulent, the particles are deposited (Hamilton 2004). Evaluating the dosimetry associated with inhaled radioactivity using the MIRD schema can be difficult as it is not necessarily a trivial problem to evaluate both the administered activity and the cumulated activity within the lung. This difficulty is due to both the dynamic natures of inhalation, expiration, and transfer across into the body and the means through which activity is introduced to the lung. MIRD Pamphlet No 18 (Thomas 2001) summarized the determination of the administered activity by considering the three ventilation procedures in common use: continuous-flow generator output, rebreathing, and aerosol delivery. For each procedure, a compartmental model was designed for the delivery system, the lung space receiving the radioactivity and the remainder of the body, as shown in Fig. 12.2.

In each compartmental model, λ_1 and λ_2 are the rate transfer constants into the lung space and the remainder of the body, respectively, and λ_{Net} is the net rate transfer constant incorporating the effects of λ_1 and λ_2. In the continuous flow system, the virtual equilibrium space consists of the facemask and associated tubing.

As shown in Fig. 12.3, the lung space activity for the continuous and rebreathing systems will increase to approach an asymptote. This is referred to as the wash-in or administration phase, and is represented by the equilibrium lung activity, $A_{Lung,Eq}$. Once administration of the radioactive gas has ceased at time t_{Admin}, the lung space activity decreases rapidly due to both exhalation of the radioactive gas and transepithelial transfer of the activity. The calculation model treats the lung activity as a single mean value and neglects the explicit calculation of the difference between inspiration and exhalation volumes and the mixing of radioactive and nonradioactive gases within the lung. On the other hand, in the aerosol method, administration is terminated before an equilibrium lung activity is achieved. The methodologies of evaluation the cumulated activities for all three systems are now derived.

[3]The radiation dosimetry associated with the radioactive xenon isotopes in pulmonary imaging was originally addressed in MIRD Dose Estimate Report 9 (Atkins et al. 1980).

[4]As ^{99m}Tc-DTPA is cleared from the alveolar region by transepithelial diffusion, the rate of its clearance can be used as an index of the alveolar epithelial membrane integrity where rapid clearance can be indicative of alveolar inflammation (Bondesson et al. 2007; Beadsmoore et al. 2007).

[5]The radiation dosimetry of ^{99m}Tc-DTPA aerosols is discussed in MIRD Dose Estimate Report No. 16 (Atkins et al. 1992).

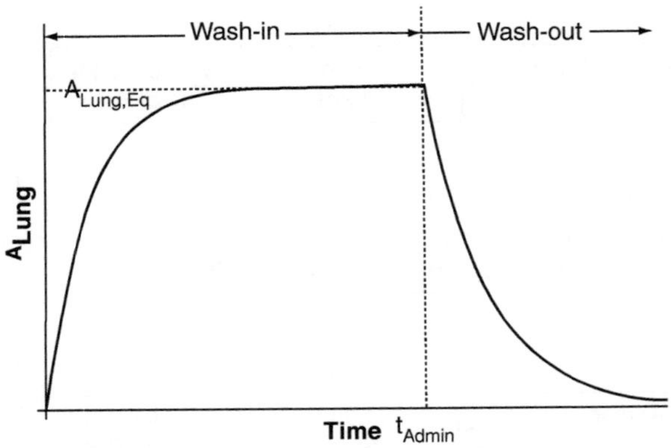

Fig. 12.3 Temporal dependence of the lung space activity. In the continuous flow and rebreathing systems, the equilibrium activity in the lung space, $A_{Lung,Eq}$, is achieved during the wash-in phase. At time t_{Admin}, administration of the gas is ceased and the activity washes out of the lung. For the aerosol delivery system, administration is ceased during the build-up portion of the curve, i.e., before equilibrium is achieved (after Thomas (2001))

Continuous Flow

The lung space activity is determined by assuming that the generator output activity rate, dA_G/dt, is constant. During the wash-in phase, the activity within the lung space for the continuous flow system is,

$$A_{Lung,CF}(t) = A_{Lung,Eq}\left(1 - e^{-(\lambda_{net}+\lambda_p)t}\right). \quad (12.2)$$

The pulmonary administered cumulated activity is that which occurs during the wash-in phase,

$$\tilde{A}_{Lung,CF,Admin} = \int_0^{t_{Admin}} dt\, A_{Lung,CF}(t) = \frac{dA_G/dt}{\lambda_P}\left(t_{Admin} - \frac{e^{-(\lambda_{net}+\lambda_p)t}}{\lambda_P}\right). \quad (12.3)$$

Rebreathing

Here, the equilibrium lung activity will be proportional to the activity contained within the spirometer,

$$A_{Lung,Eq} = \left(\frac{V_{Lung}}{V_{Lung}+V_S}\right)A_{OS} \quad (12.4)$$

where V_{Lung} and V_S are the volumes of the lung space and spirometer, respectively, and A_{OS} is the input activity to the spirometer. The lung space activity is averaged over the inspiration and exhalation volumes and physical decay is implicitly neglected. During the wash-in phase, the lung space activity is,

$$A_{Lung,S}(t) = A_{Lung,Eq}\left(1 - e^{-(\lambda_{net}+\lambda_p)t}\right). \quad (12.5)$$

The lung administered cumulated activity is determined by integrating (12.5) over the administration time, t_{Admin},

$$\tilde{A}_{Lung,S,Admin} = \int_0^{t_{Admin}} dt\, A_{Lung,S}(t) = \left(\frac{V_{Lung}}{V_{Lung}+V_S}\right)A_{0,S} \times \left(t_{Admin} - \left(\frac{1 - e^{-(\lambda_{Net}+\lambda_P)t_{Admin}}}{\lambda_{Net}+\lambda_P}\right)\right). \quad (12.6)$$

Aerosol

The lung administered cumulated activity is calculated in the same manner as was done for the continuous flow and rebreathing systems,

$$\tilde{A}_{Lung,Aerosol,Admin} = g\, A_{0,N}\left(t_{Admin} - \left(\frac{1 - e^{-(\lambda_{Net}+\lambda_P)t_{Admin}}}{\lambda_{Net}+\lambda_P}\right)\right). \quad (12.7)$$

where $A_{0,N}$ is the activity originally introduced to the nebulizer and g is that fraction of $A_{0,N}$ that is deposited within the lungs for infinitely-long administration times.

12.2.3 Gastrointestinal Tract

12.2.3.1 Introduction

The absorbed doses to the walls of the gastrointestinal tract (also referred to as the alimentary tract) are of great interest to nuclear medicine radiation dosimetry as these tissues have a notable stochastic radiosensitivity, as demonstrated by the allocation by ICRP Publications 60 and 103 of tissue weighting factors of 0.12 to the walls of the colon and stomach (Chap. 10). It bears reminding that, as the GI radiation syndrome has a threshold absorbed dose of the order of 10 Gy at which the depletion of the epithelial lining of the gut occurs, deterministic effects associated with GI tract wall are not of concern in any aspect of nuclear medicine.

The walls of the GI tract are target regions in the MIRD schema (Chap. 11) and, while they receive absorbed doses from other organs or tissues, of particular interest is the absorbed dose that they receive from radioactive contents transported through the GI tract. This special interest arises from the combination of the intimate contact between the GI tract lumenal contents and the walls and the protracted time it takes for the radioactive material to be transported through the gut.

There are three means through which activity can be introduced to the GI tract. The first is through the stomach, by either oral ingestion (e.g., per os administration of $Na^{131}I$) or the ingestion of a radioactive gas that has been expired from the lungs (e.g., following a ventilation scan). The second means, and the one which is of most general interest to nuclear medicine, is through introduction to the duodenum via the drainage of radioactive bile into the duodenum following parenteral administration of radiopharmaceutical. This is largely the case for radiopharmaceuticals which are lipophilic or have large molecular weights. A third is through resorption of activity present in body fluids through the GI tract walls. As activity in the GI tract contents is transported, dynamic models describing this transport through the different compartments making up the GI tract are required. Two models presented by the ICRP and used in nuclear medicine dosimetry calculations are discussed below.

12.2.3.2 Anatomy

Relevant aspects of the microanatomy of the GI tract have already been discussed in Chap. 10. Here, discussion is limited to a gross anatomical description of the gastrointestinal tract relevant to macroscopic internal dosimetry calculations.

The GI tract is a hollow tube of muscle lined with epithelium with varying diameters along its length and consisting of a number of compartments (Gray 1977; ICRP 2006):

- Oral cavity (including mouth, teeth, salivary glands, and the pharynx)
- Esophagus
- Stomach
- Small intestine (made of three components – duodenum, jejunum and ileum)
- Large intestine (described by three components – ascending, traverse, and descending colons)
- The sigmoid colon
- The rectum
- The anus

In the following discussion, interest will be limited to the source regions represented by the contents of the stomach, small and large intestines, the sigmoid colon, and the rectum.

The introduction, transport, and evacuation of material within the GI tract are now briefly described. Food is masticated in the mouth and, following deglutition, enters the stomach from the esophagus through the antrum cardiacum. Here, gastric digestion occurs. The stomach communicates the resulting semifluid product (chime) to the duodenum[6] through the pyloric sphincter. The duodenum is the first of the small intestine segments, the others being the jejunum and ileum. The small intestine is so-called due to its small intralumenal diameter compared to that of the large intestine, rather than from their comparative lengths. The fixation of the small intestine within the peritoneal cavity is through connective tissue, with the proximal aspect of the duodenum and the distal aspect of the ileum being relatively fixed leaving the remainder of the small intestine, including the whole jejunum, free within the cavity. The transition from the duodenum to the jejunum occurs at about 25 cm from the pylorus; of the remaining small intestine, 40% is considered to be jejunum and the distal 60% is taken to be the ileum (ICRP 2006). The total length of the small intestine from the pylorus to the ilea-caecal valve is about 6 m in the adult male.

The proximal segment of the duodenum receives both bile from the biliary tree through the common

[6] The word "duodenum" evolves from the Greek for "twelve fingers" which refers to its length being closely equal to twelve finger lengths.

bile duct and pancreatic secretions through the pancreatic duct. These two ducts unite to create a common entry into the duodenum, the ampulla of Vater (or hepatopancreatic ampulla) with the flow controlled by the sphincter of Oddi (or sphincter of papilla). Within the liver, bile produced by hepatocytes is secreted into bile canaliculi which further drain into bile ducts which, in turn, drain into hepatic ducts which transport the bile away from the liver to the duodenum through the common bile duct. The daily production volume of bile by the liver is between 250 and 1,500 mL and is continuous. The gallbladder is a sac-like organ which is attached to the inferior aspect of the liver, communicates with the common bile duct through the cystic duct and has a storage capacity of between 35 and 100 mL. When the duodenum is empty, the sphincter of Oddi closes and bile is refluxed through the cystic duct to be stored in the gallbladder. Subsequent gallbladder emptying is biphasic (Lawson et al. 1983) with an initial rate constant of 0.015/min followed by a later 0.005/min. The majority of bile salts released into the gut are resorbed through the wall in the distal ileum into the blood which then returns these to the liver to be resecreted into the duodenum (enterohepatic circulation).

The juncture between the small and large intestines occurs at the cecum, which is a pouch hanging below the ileo-caecal valve, which prevents reflux into the small intestine, to the right side of the body. The large intestine is structurally segmented into the cecum, colon, rectum, and anal canal; the colon itself is segmented into the ascending, transverse, descending, and sigmoid colons. The ascending colon extends superiorly from the cecum to the inferior hepatic surface where it bends to the left at the hepatic flexure to become the transverse colon which crosses towards the left side of the body and the splenic flexure where it then deflects downward to form the descending colon which deflects medially to form the sigmoid colon which leads to the rectum and anal canal. There are two other sets of nomenclature of the large intestine compartments that are used. In the first, which is used in the ICRP Publication 30 model of GI tract transit, the upper large intestine (ULI) is the combination of the ascending and transverse colons and the lower large intestine (LLI) is the combination of the descending and sigmoid colons and the rectum. In the other, which is used in the ICRP 100 model of the human alimentary tract, the right colon is defined to be the combined cecum, ascending colon, and proximal half of the transverse colon and the left colon is the combined distal half of the transverse colon, the descending colon, and the combination of the rectum and sigmoid, known as the rectosigmoid colon.

12.2.3.3 ICRP Publication 30 Model

ICRP Publication 30 presented a compartmental model of the GI tract within which the transfer of radioactive contents between compartments was described by first-order linear kinetics. This is shown in Fig. 12.4. Activity enters the GI tract through ingestion or hepatobiliary transport and exits through transfer to body fluids (via the walls of the small intestine only) or by defecation.

The ICRP Publication 30 model is based upon the seminal work by Dolphin and Eve (1966) and Eve (1966) in which the source regions are the compartmental contents. Whereas ICRP Publication 30 provides a generic set of solutions to sequential chains of first-order linear differential equations, the first-order kinetics of the GI tract compartmental model of Fig. 12.4 are derived explicitly here. Let $A_{St}(t)$, $A_{SI}(t)$, $A_{ULI}(t)$, and $A_{LLI}(t)$ be the activities, corrected

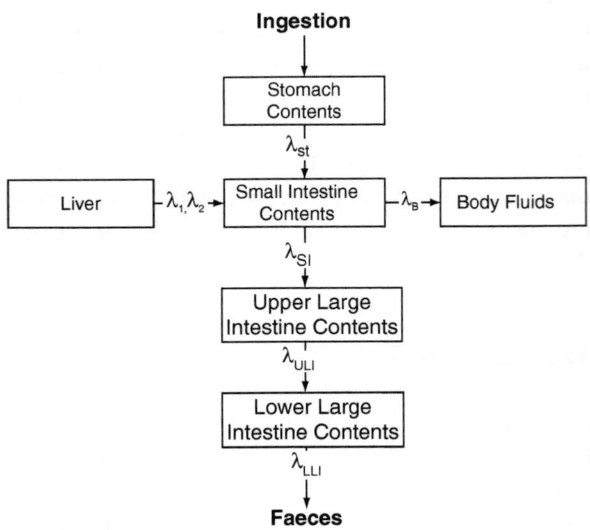

Fig. 12.4 ICRP Publication 30 compartmental model of the gastrointestinal tract. Parameters and kinetics are discussed in the text

for physical decay,[7] within the four GI tract compartments at time t and let $A_F(t)$ be the decay-corrected activity in the feces at time t. Activity enters the GI tract through ingestion (at a rate of $dI(t)/dt$) and through hepatobiliary transport which we assume to be biexponential,

$$H(t) = \sum_{i=1}^{2} k_i^{(HB)} e^{-\lambda_i t} \tag{12.8}$$

A set of first-order differential equations thus evolves from the model of Fig. 12.4,

$$\frac{dA_{St}(t)}{dt} = \frac{dI(t)}{dt} - \lambda_{St} A_{St}(t) \tag{12.9}$$

$$\frac{dA_{SI}(t)}{dt} = \frac{dH(t)}{dt} - (\lambda_{SI} + \lambda_B) A_{SI}(t) + \lambda_{St} A_{St}(t) \tag{12.10}$$

$$\frac{dA_{ULI}(t)}{dt} = \lambda_{SI} A_{SI}(t) - \lambda_{ULI} A_{ULI}(t) \tag{12.11}$$

$$\frac{dA_{LLI}(t)}{dt} = \lambda_{ULI} A_{ULI}(t) - \lambda_{LLI} A_{LLI}(t) \tag{12.12}$$

$$\frac{dA_F(t)}{dt} = \lambda_{LLI} A_{LLI}(t) \tag{12.13}$$

where the rate constants are defined in Fig. 12.4. For a radiopharmaceutical administered intravenously, gastric introduction of activity can be ignored and the only introduction of activity into the gut is through hepatobiliary transport into the duodenum. The rate constant for the start of enterohepatic circulation through the transfer of activity from the small intestine to body fluids, λ_B, can be estimated if the fraction of the stable element f_1 reaching the body fluids is known,

$$f_1 = \frac{\lambda_B}{\lambda_B + \lambda_{SI}}. \tag{12.14}$$

From this result,

$$\lambda_B = \left(\frac{f_1}{1 - f_1}\right) \lambda_{SI}. \tag{12.15}$$

In nuclear medicine dosimetry calculations, λ_B is typically set equal to zero. The rate transfer coefficients have been estimated from transit times of the GI tract contents. These times can be assessed in a variety of ways for aiding clinical diagnoses of unexplained gastrointestinal symptoms, including scintigraphic imaging following ingestion of ^{99m}Tc-labeled sulfur colloid, ^{111}In-DTPA or through a radiological study of a barium meal. Another means of estimating the oro-cecal transit time is to measure the rise in breath hydrogen following the ingestion of lactulose and resulting as a by-product of bacterial fermentation in the colon (Lin et al. 2005). Values of the rate constants λ_{SI}, λ_{ULI} and λ_{SSI} recommended for use in the ICRP Publication 30 model and taken from the work of Eve (1966) are presented in Table 12.1.

The above set of differential equations is elementary, but tedious, to solve for the activities of the individual compartments. Assuming that no activity is ingested into the stomach, these are,

$$A_{SI}(t) = \sum_{i=1}^{3} k_i^{(SI)} e^{-\lambda_i t} \tag{12.16}$$

$$A_{ULI}(t) = \sum_{i=1}^{4} k_i^{(ULI)} e^{-\lambda_i t} \tag{12.17}$$

$$A_{LLI}(t) = \sum_{i=1}^{5} k_i^{(LLI)} e^{-\lambda_i t} \tag{12.18}$$

$$A_F(t) = C_{GI} - \sum_{i=1}^{5} k_i^{(F)} e^{-\lambda_i t} \tag{12.19}$$

Table 12.1 Transfer rate constants of gastrointestinal tract contents used in the ICRP Publication 30 Model

Compartments	λ (d^{-1})
Stomach (St) → small intestine (SI), λ_{St}	24
SI → upper large intestine (ULI), λ_{SI}	6
ULI → lower large intestine (LLI), λ_{ULI}	1.8
LLI → feces, λ_{LLI}	1

[7] A correction for radionuclide physical decay is often used in nuclear medicine dosimetry so as to isolate the biokinetic features of the radiopharmaceutical. This is done simply by premultiplying the activity measured at time t by $\exp(\lambda_P t)$.

where C_{GI} is the total cumulative amount of activity that enters the GI tract and is excreted. For clarity in writing down these expressions, $\lambda_3 \equiv \lambda_{SI}$, $\lambda_4 \equiv \lambda_{ULI}$ and $\lambda_5 \equiv \lambda_{LLI}$. With the initial condition that there is no activity in the GI tract compartments at time $t = 0$, the coefficients and effective rate constants of the model are,

$$k_{i=1,2}^{(SI)} = k_i^{(HB)} \left(\frac{\lambda_i}{\lambda_3 - \lambda_i} \right) \tag{12.20}$$

$$k_3^{(SI)} = -\sum_{i=1}^{2} k_i^{(SI)} \tag{12.21}$$

$$k_{i=1,2}^{(ULI)} = k_i^{(HB)} \left(\frac{\lambda_i}{\lambda_3 - \lambda_i} \right) \left(\frac{\lambda_3}{\lambda_4 - \lambda_i} \right) \tag{12.22}$$

$$k_{i=3,4}^{(ULI)} = (-1)^i \left(\frac{\lambda_3}{\lambda_4 - \lambda_3} \right) \sum_{j=1}^{2} k_j^{(HB)} \left(\frac{\lambda_j}{\lambda_i - \lambda_j} \right) \tag{12.23}$$

$$k_{i=1,2}^{(LLI)} = k_i^{(HB)} \left(\frac{\lambda_i}{\lambda_3 - \lambda_i} \right) \left(\frac{\lambda_3}{\lambda_4 - \lambda_i} \right) \left(\frac{\lambda_4}{\lambda_5 - \lambda_i} \right) \tag{12.24}$$

$$k_{i=3,4}^{(LLI)} = (-1)^i \left(\frac{\lambda_3}{\lambda_4 - \lambda_3} \right) \left(\frac{\lambda_4}{\lambda_5 - \lambda_i} \right) \sum_{j=1}^{2} k_j^{(HB)} \left(\frac{\lambda_j}{\lambda_i - \lambda_j} \right) \tag{12.25}$$

$$k_5^{(LLI)} = \sum_{i=1}^{2} k_i^{(Input)} \left(\frac{\lambda_i}{\lambda_5 - \lambda_i} \right) \left(\frac{\lambda_3}{\lambda_5 - \lambda_3} \right) \left(\frac{\lambda_4}{\lambda_5 - \lambda_4} \right) \tag{12.26}$$

$$k_{i=1,2}^{(F)} = k_i^{(HB)} \left(\frac{\lambda_3}{\lambda_3 - \lambda_i} \right) \left(\frac{\lambda_4}{\lambda_4 - \lambda_i} \right) \left(\frac{\lambda_5}{\lambda_5 - \lambda_i} \right) \tag{12.27}$$

$$k_{i=3}^{(F)} = -\left(\frac{\lambda_4}{\lambda_4 - \lambda_3} \right) \left(\frac{\lambda_5}{\lambda_5 - \lambda_3} \right) \sum_{j=1}^{2} k_j^{(HB)} \left(\frac{\lambda_j}{\lambda_3 - \lambda_j} \right) \tag{12.28}$$

$$k_4^{(F)} = \left(\frac{\lambda_3}{\lambda_4 - \lambda_3} \right) \left(\frac{\lambda_5}{\lambda_5 - \lambda_4} \right) \sum_{j=1}^{2} k_j^{(HB)} \left(\frac{\lambda_j}{\lambda_4 - \lambda_j} \right) \tag{12.29}$$

$$k_5^{(F)} = -\sum_{i=1}^{2} k_i^{(HB)} \left(\frac{\lambda_i}{\lambda_5 - \lambda_i} \right) \left(\frac{\lambda_3}{\lambda_5 - \lambda_3} \right) \left(\frac{\lambda_4}{\lambda_5 - \lambda_4} \right) \tag{12.30}$$

$$C_{GI} = \sum_{i=1}^{2} k_i^{(HB)}. \tag{12.31}$$

The cumulative activities for each of the compartments are straightforward to determine through analytical integrations of the activity expressions,

$$\tilde{A}_{SI} = \sum_{i=1}^{3} \left(\frac{k_i^{(SI)}}{\lambda_i + \lambda_P} \right) \tag{12.32}$$

$$\tilde{A}_{ULI} = \sum_{i=1}^{4} \left(\frac{k_i^{(ULI)}}{\lambda_i + \lambda_P} \right) \tag{12.33}$$

$$\tilde{A}_{LLI} = \sum_{i=1}^{5} \left(\frac{k_i^{(LLI)}}{\lambda_i + \lambda_P} \right) \tag{12.34}$$

$$\tilde{A}_F = \frac{C_{GI}}{\lambda_P} - \sum_{i=1}^{5} \left(\frac{k_i^{(F)}}{\lambda_i + \lambda_P} \right). \tag{12.35}$$

In a biodistribution study of a radiopharmaceutical, it is frequently not possible to accurately assess in which GI tract compartment the measured in vivo activity is present. In practice, one instead measures the sum of the decay-corrected activities in the compartments determined from in vivo imaging of the GI tract contents and, if the radionuclide is sufficiently long-lived and sufficient activity is transported through the GI tract, in vitro measurements of activity present in the feces. In this case, the measurement is of the sum of (12.16)–(12.19). This sum is, in fact, equal to

$$A_{GI\,Total}(t) = C_{GI} - \sum_{i=1}^{2} k_i^{(HB)} e^{-\lambda_i t} \tag{12.36}$$

which asymptotically approaches the value C_{GI}, the total amount of activity that enters the GI tract and is eventually excreted in the feces. Hence, an analytical fit to (12.36) to the measured sum of the activity measured in vivo in the gut through imaging and, if present, in voided feces yields the values of C_{GI} and of $k_i^{(HB)}$ and λ_i for $i = 1, 2$. Using these, in conjunction

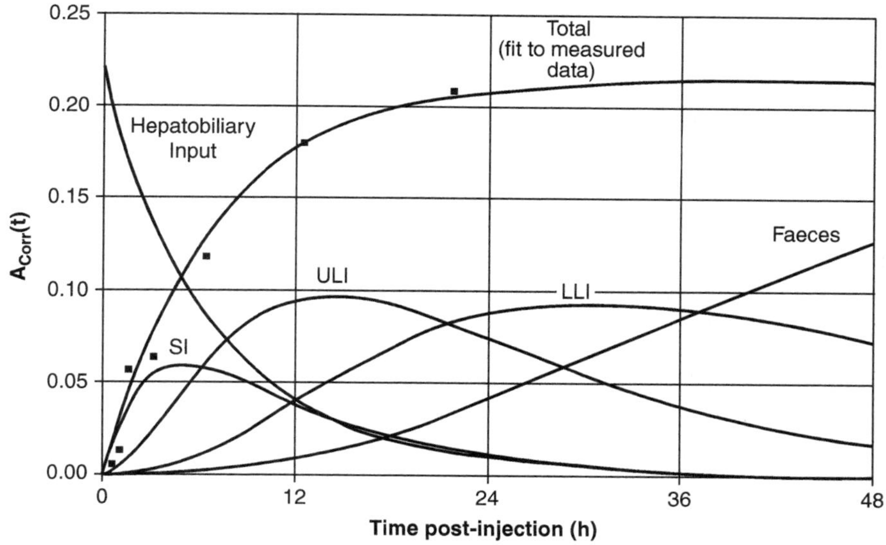

Fig. 12.5 ^{99m}Tc decay-corrected activities in the contents of the various GI tract compartments calculated using the ICRP 30 model from created data (*solid points*) reflecting the sum of the measured decay-corrected activities derived from in vivo imaging of GI tract contents and in vitro assay of voided feces. The asymptotic curve (marked "Total") is the fit to the measured data. The input of activity to the GI tract is assumed to be hepatobiliary only and described by a biexponential function

with the rate constants of Table 11.2 yields the cumulated activities of the GI tract content source regions.

Figure 12.5 shows an example of the temporal variations of the activities in the various GI tract content compartments as a function of time for ^{99m}Tc derived from artificial data. The sum of the decay-corrected activities determined from images of the GI tract contents and assays of voided feces were measured at 9 time points postinjection up to a time of 22 h and then fit by (12.36). The coefficients and effective rate constants were then determined for each of the GI tract content compartments and the corresponding cumulated activities calculated.

The MIRDOSE and OLINDA/EXM codes save the nuclear medicine physicist from the above tedious calculations and require the user to input only the decay-corrected fraction of the administered activity that enters the GI tract through the stomach or through hepatobiliary transport into the small intestine (i.e., the asymptote C_{GI} value of the analytical fit to the measured data). However, the user must approach this single-input method with some degree of caution as no temporal information regarding this transport of activity is incorporated.[8] In effect, it is implicitly assumed by the MIRDOSE and OLINDA/EXM codes that the activity enters the duodenum as a bolus. As shown in the previous discussion of the anatomy of the biliary tree, this is not the case. The temporal behavior of the introduction of activity from the liver to the duodenum will follow that of biliary transport: i.e., being generally continuous, unless the small intestine is devoid of food, at which point the sphincter of Oddi closes to force the bile into the reservoir presented by the gallbladder. When the sphincter relaxes, the gallbladder drains its contents into the duodenum with biphasic time dependence. Hence, the assumption of a bolus deliver of activity into the small intestine neglects the actual physiological time course during which the radionuclide undergoes physical decay. The consequence of this bolus method is that the bolus assumption will result in an overestimate of the activity entering the GI tract contents and a subsequent overestimation of the absorbed dose to the target regions (in particular, the walls of the small and large intestines). Thus, the user must reduce the input value of the amount of bolus activity assumed to enter the small intestine in order to compensate for physical decay during the actual protracted introduction. It is straightforward to determine this bolus activity by solving the set of differential equations for a bolus input to yield the same small intestine contents' cumulated activity as that derived above. This gives the result that instead of providing C_{GI} as the small intestine input in the MIRDOSE or OLINDA/EXM ICRP 30 input window, one should provide as the amount of

[8]The author is grateful to two colleagues, Drs RD Pickett and JF Burke, who brought this problem to his attention and led to this derivation. They achieved the same solution to the problem through a different and more intuitive means.

activity that enters the small intestine the value given by (for an assumed biexponential input from the liver),

$$A_{\text{Input}} = \sum_{i=1}^{2} k_i^{(HB)} \left(\frac{\lambda_i}{\lambda_i + \lambda_P} \right). \qquad (12.37)$$

As $C_{GI} = k_1^{(HB)} + k_2^{(HB)}$, this result demonstrates that each of the two input components is reduced by the ratio of the biological rate constant to the effective rate constant. Hence, this correction factor will be more significant for radionuclides with short physical half-lives, such as ^{11}C, ^{18}F and ^{99m}Tc, as the time scale for transport of activity from the liver to the duodenum will be of the order of the biphasic time constants of gallbladder emptying (0.005 and 0.015/min). The bolus input correction factors $(\lambda_i/\lambda_i) + \lambda_P$ of (12.37) are shown as a function of radionuclide half-life in Fig. 12.6. Two assumed rate constants are shown and the correction factors for ^{18}F, ^{99m}Tc, and ^{123}I highlighted. Clearly, failure to account for the bolus input correction will lead to significant overestimation of the activity within the GI tract contents.

12.2.3.4 ICRP Publication 100 Model

In 2006, the ICRP (2006) released a replacement of the Publication 30 model of the GI tract, which it then termed the HATM. The evolution of this new model was the result of a number of factors, including new gut transit times for adults and pediatric cases which had become available since the work of Eve and Dolphin in the 1960s. As with the earlier Publication 30 model, the intended focus of the new model was that of radiological protection (occupational exposure), but pediatric intakes were calculated for in the provision of age-dependent transfer rate constants. Additional changes that arise in the ICRP Publication 100 model include:

- Whereas the ICRP Publication 30 model accounted for transfer from the gut to body fluids only from the small intestine, the new model allowed for transfer from other compartments.
- Extensions of the modeling to include the oral cavity and esophagus in the alimentary tract and an explicit link to ingestion of expired radionuclides were made.
- The segmentation of the large intestine into upper and lower components was altered and the large intestine was instead segmented into right, left, and rectosigmoid colons.
- Liquid and solid GI tract contents were treated differently.

Another difference between the ICRP Publications 30 and 100 models is in terms of the dosimetry of the GI tract walls. Recall from Table 10.3 that at the time

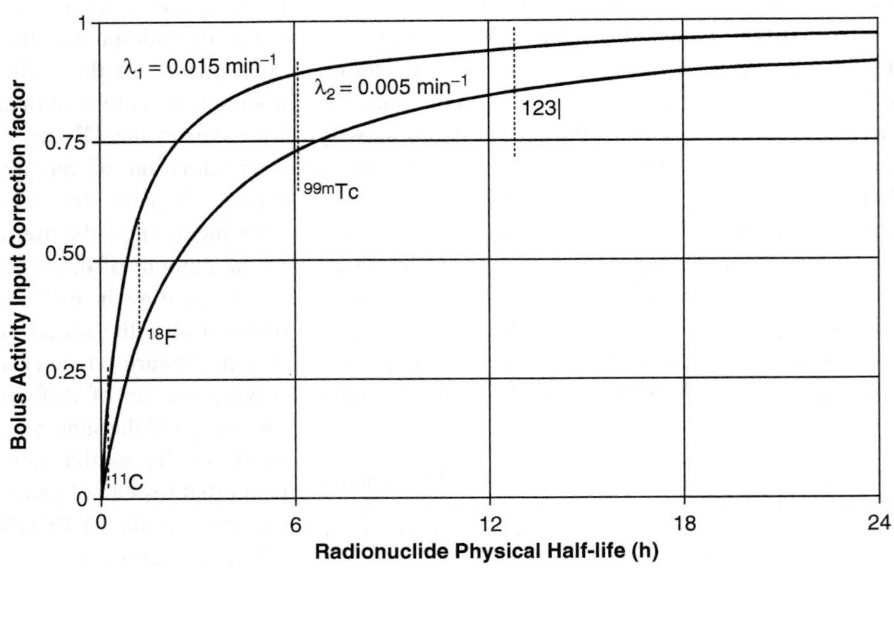

Fig. 12.6 The gastrointestinal tract bolus input correction factor of (12.37) as a function of radionuclide physical half-life for the two assumed rate constants of biliary drainage into the duodenum

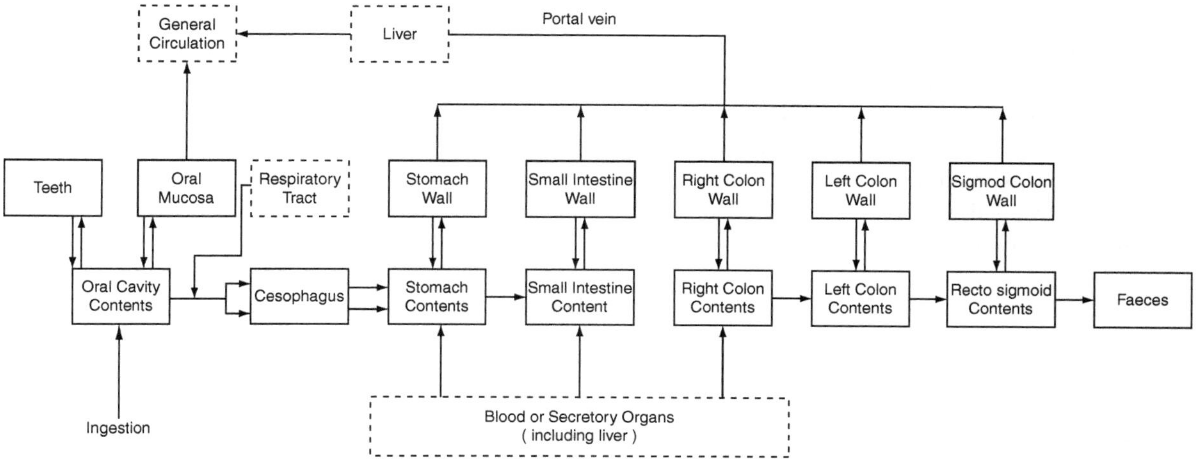

Fig. 12.7 ICRP Publication 100 compartmental model of the alimentary tract

the Publication 30 model was presented, there were no explicit tissue weighting factors assigned to the walls of the various GI tract compartments and only the walls of the stomach and LLI were grouped in the "remaining tissues" category. The Publication 100 model followed the ICRP Publication 60 recommendations (ICRP 1991) in which explicit weighting factors were specified for the stomach and colon walls, although the small intestine wall was placed in the "remaining tissues" category.[9]

The complexity of the model has, accordingly, increased considerably as shown in Fig. 12.7. Table 12.2 provides the new transfer rate constants shown in that figure. Of obvious interest is the inclusion of pediatric constants and the segmentation of the adult constants into male and female. A comparison of the adult values in Table 12.2 with those of Eve given in Table 12.1 will not reveal significant changes.

Table 12.2 Transfer rate coefficients for ICRP Publication 100 Model of the human alimentary tract

Compartment	Transfer rate coefficient (d^{-1})				
	3 Months	1 Year	5–15 Years	Adult Male	Adult Female
SI → right colon	6	6	6	6	6
Right colon → left colon	3	2.4	2.182	2	1.5
Left colon → rectosigmoid colon	3	2.4	2.182	2	1.5
Rectosigmoid colon → feces	2	2	2	2	1.5

12.2.3.5 Intestinal Wall as a Source Region

The current MIRD compilation of source regions does not include the GI tract walls. It is possible that the GI tract walls can contain activity and this may become apparent in an investigation of emission images of the gut. Typically, activity in the GI tract contents appears at an extended time after administration as the activity must first reach the liver and to then be transported into the small intestine through hepatobiliary drainage. Activity associated with the intestines seen at early time postadministration would likely reflect uptake in the soft tissue, rather than in the contents. Uptake in the walls may also appear highly diffuse and nonspecific. While it might be easier to consider this nonspecific uptake as part of the remaining tissues category, it may be possible to calculate explicitly the absorbed doses to the intestinal walls due to activity in the walls. This calculation may be performed directly or through an approximation.

Jönsson et al. (2002) derived S-factors for the small intestine wall, which was modeled as two

[9]This is in terms of stochastic effects. It is interesting to note that the wall of the small intestine is also the most important site of the gastrointestinal tract in terms of deterministic radiation injury (Nias 1998).

infinitely-long concentric cylinders. The S-factor values were not only of the self-dose due to activity uptake in the walls, but included the contributions resulting from irradiation by near-by surrounding loops of small intestine. Stabin (2008) describes a practical approximation using the output from OLINDA/EXM and the separation of electron and photon contributions.

12.2.4 Kidney

12.2.4.1 Introduction

As part of the urinary excretion route, the kidney is of considerable interest to the assessment of radiation absorbed dose due to the excretion of low-molecular weight radiopharmaceuticals through this pathway. Rapid and significant renal excretion of radioactivity is usually desirable of a diagnostic radiopharmaceutical in order to reduce nonspecific background accumulation of activity which can limit potentially the diagnostic efficacy of the radiopharmaceutical. Because of the low levels of administered activities associated with diagnostic applications, the potential for renal toxicity in such cases is nonexistent. This, however, may not be the case in therapeutic nuclear medicine. In fact, renal impairment as a consequence of radionuclide therapy has been described (e.g., Lambert et al. 2004). Thus, the kidney can be an absorbed-dose limiting organ through radiation nephropathy in therapeutic nuclear medicine. As a consequence, considerable effort has been expended on developing models of the response of the kidney to radiation. Typical conventional MIRD approaches have treated the kidney as a single compartmental model in which activity was assumed to be uniformly distributed within it. Such an assumption allows only a mean absorbed dose for the kidney to be calculated. In fact, due to its complex suborgan physiology, the activity distribution within the kidney is nonuniform, resulting in a nonuniform absorbed dose distribution (Bouchet et al. 2003a). As this nonuniformity can be imaged through emission tomography (and, to a much more limited degree, whole-body planar scintigraphy), kidney models of increasing complexity have been developed to account for this. In addition, as described below, radiation biology theory has been applied to these kidney models in an attempt to obtain more rigorous prediction of renal toxicity in therapeutic nuclear medicine.

12.2.4.2 Anatomy

The kidneys lie laterally to the vertebral column between the levels of the twelfth thoracic and third lumbar vertebrae. Due to its displacement by the liver, the right kidney is about 2 cm lower than the left. Each kidney is about 11 cm long, 6.5 cm wide and 3 cm thick; the hilum is the depression in the kidney along the medial border through which the renal vein and artery and the ureter penetrate. The kidney has two major sections, the renal cortex and renal medulla, and a cavity, the renal pelvis. The renal cortex contains the renal medulla which is made up of eight to fifteen conical renal pyramids, the apices (renal papillae) of which feed into the renal pelvis. In detail, these papillae project into minor calyces which unite to form major calyces which subsequently combine to form the hollow renal pelvis. The renal pelvis collects urine from the calyces and transports it to the ureter and into the urinary bladder. The nephron is the radiobiological functional unit that forms the urine and consists of urinary tubules and small blood vessels (there are about 10^6 nephrons per kidney). Blood enters the kidney through the renal artery, which subdivides into interlobar arteries which further subdivide into the arcuate arteries. Interlobular arteries radiate from the latter into the renal cortex, subdividing into microscopic afferent arterioles that feed into a capillary network known as the glomerulus (producing a filtrate that enters the renal papillae). The remaining blood feeds into the efferent arterioles which lead to a secondary capillary network known as the peritubular network. Blood is drained from this network into the venous system to leave the kidney through the renal vein into the inferior vena cava. The glomerular capillaries have large pores allowing water and dissolved solutes to pass from the blood plasma into the nephron tubules. This process of filtration has the effect of distributing activity within the renal cortex and then transferring it sequentially to the medulla, papillae, and renal pelvis. Variants of this simple transfer sequence can occur. For example, following water absorption in the medulla, activity can be concentrated in the renal pyramids. Regular peristaltic contractions of the urethral walls of between 1 and 5 times per minute transfer urine from the renal pelvis to the urinary bladder.

12.2.4.3 MIRD Renal Models

MIRD Pamphlet 5 and Revision

In this anthropomorphic model of the adult human (Snyder et al. 1969), the kidney was modeled by two symmetric and homogeneous ellipsoids cut by a plane. The kidney mass and volume were 144 g and 144 cm^3, as unit physical density was assumed, in reasonable agreement with the mass of 155 g of the kidney in ICRP Reference Man (ICRP 1975). Absorbed fractions were calculated for a photon-emitting radionuclide uniformly distributed within the kidney and for combinations of various target regions and photon energies. In the revision of MIRD Pamphlet 5 (Snyder et al. 1978), the physical density of the kidney tissue was reduced to 0.98 g/cm^3, but the volume retained so that the mass of each kidney was reduced to 142 g.

MIRD Pamphlet 19 Model

Bouchet et al. (2003a) recognized the nonuniform distribution of activity within the kidney and developed an age-dependent multiregional kidney model for six different ages. Absorbed fraction and S-factor values were calculated for these regions for photon and electron sources with energies ranging from 10 keV to 4 MeV. The outer region of the kidney was defined by an ellipsoid truncated by a plane, in agreement with the Cristy–Eckerman model to be described. The medullary pyramids and renal papillae were approximated by half-ellipsoids and the renal pelvis was modeled by an ellipsoid segment contained within the entire kidney. Source regions of the renal cortex, medullary pyramids, medullary papillae, and the renal pelvis were defined and the corresponding S-factors for 26 radionuclides were calculated.

12.2.4.4 Other Renal Models

McAfee Model

This was the first multiregional kidney model (McAfee 1969). The kidney was represented by three regions: the renal pelvis, modeled by a hollow wedge-shaped structure, with the renal cortex and medulla modeled by concentric ellipsoids. As in MIRD Pamphlet 5, unit physical density was assumed; the mass and volume of a single kidney being 151 g and 151 cm^3.

Blau Model

Blau et al. (1975) presented a multiregional kidney model used to calculate the kidney absorbed doses resulting from chlormerodrin labeled with the mercury isotopes, ^{197}Hg and ^{203}H.

Cristy–Eckerman Model

This model of the kidney is that of the Cristy–Eckerman series of phantoms discussed previously. The dual ellipsoid geometry of MIRD Pamphlet 5 was retained, but the renal physical density was increased to 1.04 g/cm^3 to result in a single kidney mass of 149.5 g.

12.2.4.5 Radiobiology Considerations

The developments of the multiregional models of the kidney by McAfee, Blau et al., and Bouchet et al. were instigated by the recognition that the complex internal anatomy of the kidney would result in a nonuniform distribution of activity within it which will lead to a nonuniform absorbed dose distribution within the kidney. Radiation-induced renal toxicity associated with small-ligand radiopharmaceuticals frequently leads to the kidney being the activity-limiting organ in therapeutic nuclear medicine. Because of the complexity of the kidney's internal absorbed dose distribution, it is difficult to obtain an absorbed dose response characteristic which can be predictive of radiation-induced nephropathy. Such predictive capability can be necessary in radionuclide therapy planning. To address this, MIRD Pamphlet 20 (Wessels et al. 2008) presented the application of radiation biology estimates of absorbed dose responses of the kidney. The MIRD Pamphlet 19 model was used in these calculations. The linear-quadratic (LQ) model of absorbed dose response (Chap. 10) was the basis in calculating the surviving nephron fractions in the renal cortex and medulla as functions of radionuclide concentration. The α/β ratio for the LQ model was estimated from published external beam radiotherapy data identifying renal complications. Temporal sensitivity to the physical half-life of the radionuclide was modeled using the biologically effective dose (BED) of Chap. 10. The authors concluded that the combination of the multiregional kidney model and the LQ/BED calculations were useful for predicting renal radiotoxicity as a sequela of radionuclide therapy.

12.2.5 Urinary Bladder

12.2.5.1 Introduction

In combination with the kidney, the urinary bladder is of particular interest to an internal radiation dosimetry calculation as a result of the renal excretion of activity. The radiosensitive tissue of interest, the urinary bladder wall, is in prolonged and intimate contact with the radioactive source, the contents of the urine. The urinary bladder is a dynamic organ as both its volume and morphology (and, hence, the amount of activity contained within it) vary over time due to filling and voiding. As a result, the S-factors which treat the urinary bladder contents as a source region r_S are temporally varying. This makes the calculation of the absorbed dose to the bladder wall more involved than that of a static organ and requires the assumption of a specific voiding interval pattern. Dynamic models of the urinary bladder have been developed with varying complexities to account for these variabilities in order to yield accurate S-factors.

12.2.5.2 Anatomy

The urinary bladder is a musculo-membranous reservoir for the urine with a shape and position dependent upon sex, age, and the degree of distension. It is placed posteriorly to the symphisis pubis and anteriorly to the rectum. In males, the prostate is positioned inferiorly to it and in females, the bladder is in contact with the vagina and uterus. The bladder wall, which is the tissue of dosimetric interest, is made up of four layers:

- Adventitia – a partial coating derived from the surrounding peritoneum.
- Muscularis – referred to as the *detrusor muscle* and consisting of three sublayers one of which is longitudinal followed by a middle layer of circularly-arranged muscular fibers and an internal layer which is also composed of longitudinal muscle fibers.
- Submucosa – a layer of areolar tissue which connects the muscular and mucous layers.
- Mucosa – a thin and smooth layer which is connected loosely to the muscular layer through the submucosal coat and, hence, tends to folds (called rugae) when the bladder is empty.

The urethra conveys urine from the bladder and is surrounded by two muscular sphincters. The *internal urethral sphincter*, which is formed from the *detrusor muscle*, is the upper and involuntary sphincter. The *external urethral sphincter* is lower and is made up of voluntary skeletal muscle.

The daily production of urine varies across different ethnic populations. For example, the Caucasian ICRP Reference Man is defined to void 1,600 mL of urine per day (ICRP 2002) whereas the Indian male adult is reported to void 2,200 mL/day (IAEA 1998). It is not known if this particular example is a reflection of ethnic physiological differences or of higher amounts of water intake in a tropical environment. A consequence of this difference is that the amount of retained radiopharmaceutical which has a high renal excretion fraction will be expected to be higher in the reference individual with the low urinary production rate.

12.2.5.3 Static Urinary Bladder Model

The MIRD Pamphlet 5 model of the urinary bladder is static, treating the organ as an ellipsoid with a constant volume (202.6 mL) and a wall volume of 45.73 mL. As the urinary bladder is a dynamic organ in terms of its physical size, morphology, and amount of activity contained within it due to filling, distension, and micturition, the full dosimetry description of the urinary bladder contents requires consideration of these effects.

12.2.5.4 Dynamic Urinary Models: The Cloutier Model

Derivation

Cloutier et al. (1973) proposed a model of the female urinary bladder in order to estimate the contribution of the activity contained within the urinary bladder contents to the absorbed dose that is received by the fetus in the pregnant patient. Their female phantom was that of the male phantom of Snyder et al. previously discussed, but scaled by the cube-root of the ratio of whole-body masses (equal to 0.94). The gravid uterus was modeled by the combination of a cone and a hemispherical cap and was fixed in place; the ellipsoidal urinary bladder was assumed to be always in contact with the uterus and to be displaced inferiorly as pregnancy progressed.

Of particular interest was the manner by which Cloutier et al. evaluated the variable cumulated activity of the urinary bladder as the result of the filling and regular voiding of urine. Assume that the rate at which activity (normalized to that administered) is transferred from the kidney to the urinary bladder can be described by a multiexponential series,

$$\frac{dA_{UB,Norm}(t)}{dt} = \sum_{i=1}^{N} k_i \lambda_i e^{-(\lambda_i+\lambda_P)t}. \qquad (12.38)$$

Further assume that the bladder is voided at a regular interval T_V, that there is no remnant activity following micturition and that the removal of activity from the bladder is instantaneous. The activity within the urinary bladder contents at a time t between the (M-1) and M voids is,

$$A^{(M)}_{UB,Norm}(t) = \int_{(M-1)T_V}^{t} d\tau \sum_{i=1}^{N} k_i \lambda_i e^{-(\lambda_i+\lambda_p)\tau} e^{-\lambda_p(t-\tau)}$$
$$= \sum_{i=1}^{N} k_i e^{-\lambda_p t}\left(e^{-\lambda_i(M-1)T_V} - e^{-\lambda_i(M-1)T_V}\right) \qquad (12.39)$$

where τ is the time at which activity enters the bladder and (t − τ) is the length of time between when the activity enters the urinary bladder and when it is voided. The normalized cumulated activity in the urinary bladder contents during the bladder filling during the time between the (M − 1)th and Mth voids is the time integral of (12.39),

$$\tilde{A}^{(M)}_{UB,Norm} = \sum_{i=1}^{N} k_i \int_{(M-1)}^{MT_V} dt e^{-\lambda_P t}\left(e^{-\lambda_i(M-1)T_V} - e^{-\lambda_i(M-1)T_V}\right)$$
$$= \sum_{i=1}^{N} k_i \left(\left(\frac{e^{(\lambda_i+\lambda_P)T_V} - e^{\lambda_i T_V}}{\lambda_P}\right) e^{-(\lambda_i+\lambda_P)MT_V} - \left(\frac{e^{\lambda_i T_V} - 1}{\lambda_i+\lambda_P}\right) e^{-\lambda_i MT_V}\right) \qquad (12.40)$$

The total normalized cumulated activity will be given by the infinite sum of voids,

$$\tilde{A}_{UB,Norm} = \sum_{M=1}^{\infty} \tilde{A}^{(M)}_{UB,Norm}. \qquad (12.41)$$

Applying this to (12.40) and recognizing that the geometric sum, $\sum_{M=1}^{\infty}(e^{-x})^M = \frac{e^{-x}}{1-e^{-x}}$, the total normalized cumulated activity in the urinary bladder contents is,

$$\tilde{A}_{UB,Norm} = \sum_{i=1}^{N} k_i \left(\frac{1-e^{-\lambda_P T_V}}{\lambda_P} - \frac{1-e^{-(\lambda_i+\lambda_P)T_V}}{\lambda_i+\lambda_P}\right) \bullet \frac{1}{1-e^{-(\lambda_i+\lambda_P)T_V}}. \qquad (12.42)$$

The Cloutier et al. model is widely used in nuclear medicine dosimetry. The parameters k_i and λ_i are determined by fitting the sum of the decay-corrected activities measured in vivo in the urinary bladder contents and the cumulative activity contained in the urine voided up to the time of the bladder contents measurement to the asymptotic form,

$$A_{UBVU,Corr,Norm}(t) = C_{VU} - \sum_{i=1}^{N} k_i e^{-\lambda_i t} \qquad (12.43)$$

where $A_{UBVU,Corr,Norm}(t)$ is the sum of the decay-corrected activities in the urinary bladder contents and voided urine normalized to that administered and

C_{VU} is the total amount of activity that is voided in the urine.[10] In practice, N = 2 is usually found to be sufficient to yield an accurate fit to the data.

The Cloutier et al. model has figured in the MIRDOSE and OLINDA/EXM codes. Simplifications inherent to the model were noted previously in its derivation: the urinary bladder is assumed to be morphologically static (200 mL), whereas its shape and volume will vary due to the degree of filling; retention of urine and activity within the bladder following voiding was assumed to be negligible. Such effects were examined in the models described in the following subsection.

Urinary Bladder Voiding Interval, T_V

The ICRP has recommended that internal radiation dosimetry calculations for adults be performed for a urinary voiding interval T_V of 3.5 h (ICRP 1998). It has also provided an age-dependent tabulation of urinary bladder voiding intervals, as summarized in Table 12.3.

Table 12.3 ICRP-recommended age-dependent urinary bladder voiding intervals (ICRP 1998)

Age (years):	Adult	15	10	5	1	Newborn
Urinary bladder voiding interval (h)	3.5	3.5	3.0	2.0	2.0	2.0

[10]It has been suggested (Stabin 2008) that image-based measurements of the urinary bladder contents should not be used in evaluating the normalized cumulated activity of the urinary bladder contents, due to a described difficulty in accurately extracting the value of the activity in the urinary bladder contents. Instead, it was advised that the ex vivo measurement of activity in the voided urine be the basis alone of determining the k_i and λ_i parameters. This method does indeed work. However, in the author's experience, the in vivo measurement of the activity within the bladder contents is readily possible through imaging and the method described here of fitting to the sum of the activities in the bladder contents and in the voided urine presents two advantages, particularly in a Phase I imaging study. Firstly, there are likely to be more imaging acquisitions than voids, thus providing a larger number of data points to fit (12.43) to. Secondly, if the subject is allowed to void immediately after the urinary bladder has been imaged, it is possible to compare the in vivo activity result with the ex vivo result and, allowing for any retained urine, test the quantitative agreement between the imaging- and nonimaging-based measurement techniques.

It is also a frequent practice (particularly in the United States) to calculate for voiding intervals of 2.4 and 4.8 h (corresponding to 10 and 5 micturitions per 24 h period, respectively). For a radiopharmaceutical with a high fraction of radionuclide excreted renally, it is useful to calculate for several voiding intervals in order to assess how radiation absorbed doses can be reduced by frequent micturition.

Figure 12.8 shows an example of the time-activity curve of the decay-corrected activity (normalized to that administered) in the urinary bladder (assuming constant voiding at 2-h intervals and zero postvoid retention of urine), that in the voided urine and the combined activity in the urinary bladder contents and voided urine. The activity in the urinary bladder contents is measured through in vivo imaging and that in the voided urine is determined through ex vivo assay as to be described. The sum of the two activities yields the asymptotic function shown and which is fit by (12.43) to yield the k_i and λ_i used in the dynamic bladder model estimate of the normalized cumulated activity of (12.42).

The effects of the combinations of voiding interval T_V, biological fill-up rate of the urinary bladder and the physical half-life of the radionuclide upon the total normalized cumulated activity of the urinary bladder contents are shown in Fig. 12.9. The ordinate is the total normalized cumulated activity of the urinary bladder contents for an assumed single exponential fill normalized to the fraction of administered activity that is excreted into the urine.[11] Curves are calculated for radionuclides over a wide range of physical half-lives: ^{18}F ($T_{1/2,Phys}$ = 109.7 min), ^{99m}Tc ($T_{1/2,Phys}$ = 6.02 h) and ^{111}In ($T_{1/2,Phys}$ = 2.83 days) and for urinary bladder filling half-lives of 0.25, 0.5, and 2.5 h. Voiding intervals of $T_V = 2.4$ h and $T_V = 4.8$ h (corresponding to 10 and 5 micturitions per 24-h period) and the ICRP-specified $T_V = 3.5$ h for adults are highlighted.

In general, the curves are very nearly linear in voiding interval, with the slope increasing with both radionuclide half-life and with bladder filling half-life. These features are important to minimizing the radiation absorbed dose to the urinary bladder wall and to

[11]The curves of Fig. 12.9 can be applied to each phase individually for a multiexponential function and then the results added as the normalized cumulated activity is a linear function of each of the individual components.

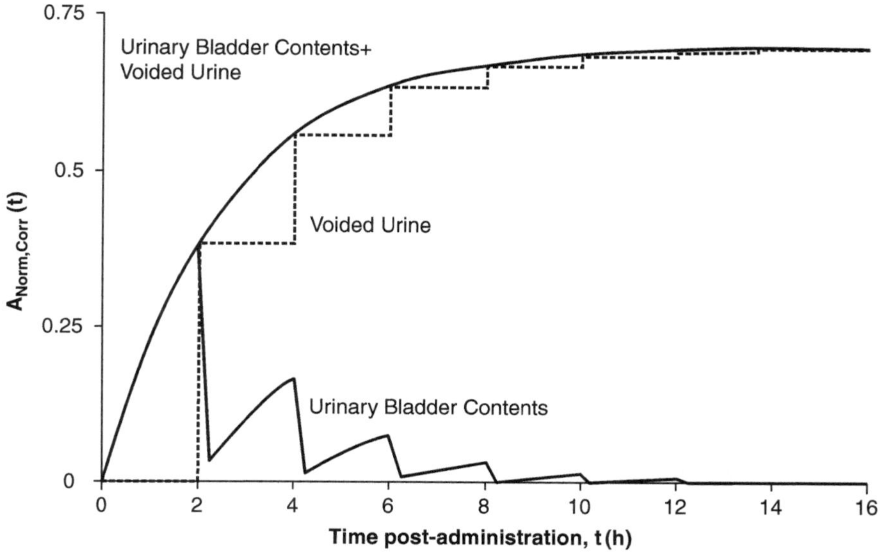

Fig. 12.8 The decay-corrected activity (normalized to that administered) in the contents of the urinary bladder, the voided urine and in the sum of the two. A constant voiding interval of 2 h is assumed

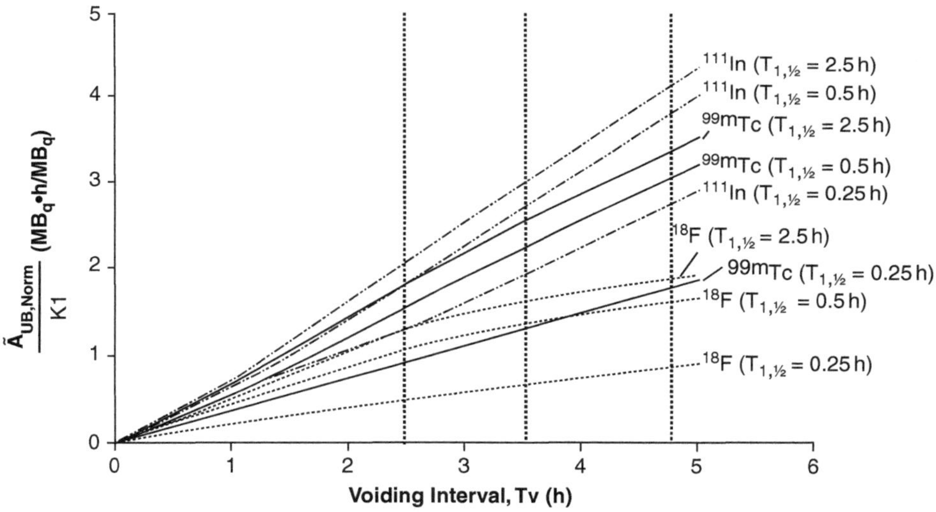

Fig. 12.9 The total urinary bladder contents normalized cumulated activity, normalized to the fraction of administered activity excreted in the urine for a single exponential clearance pattern, as a function of voiding interval. Curves are for ^{18}F (*dashed*; $T_{1/2,Phys} = 109.8$ min), ^{99m}Tc (*solid*; $T_{1/2,Phys} = 6.02$ h) and ^{111}In (*dash-dot*; $T_{1/2,Phys} = 2.83$ days) with bladder fill up half-lives of 0.25, 0.5, and 2.5 h. Voiding intervals of 2.4, 3.5, and 4.8 h are highlighted

surrounding tissues and organs in the lower abdomen, such as the gonads. Clearly, frequent micturition aids in reducing the cumulated activity within the urinary bladder contents, and this is especially critically for radionuclides with long physical half-lives. However, Fig. 12.9 does not tell the complete story as it only portrays the cumulated activity of the urinary bladder contents as a function of voiding interval. The nature of the radioactive decay of the radionuclide within the urinary bladder contents must be considered as, ultimately, the user is interested in the absorbed dose. This extension is considered in later in the discussion comparing the dosimetry of PET and single-photon-emitting radiopharmaceuticals.

12.2.5.5 Other Dynamic Models of the Urinary Bladder

Introduction

The Cloutier et al. model provided the first temporally-varying urinary bladder description. A number of simplifications, as noted above, were applied. In subsequent years, more comprehensive dynamic bladder models were developed. Two are described here; brief descriptions of others are provided in the summary of MIRD Pamphlet 14 (Revised) (Thomas et al. 1999).

Snyder and Ford Model

In the Snyder and Ford (1976) model of the urinary bladder, the organ is considered as an ellipsoid with a variable volume of between 0 and 500 mL. Continuous urinary transfer rates into the bladder (1, 1.4, and 2 L/day), residual volumes (i.e., urine retention following voiding) of between 0 and 30% were accounted for, as was a variable micturition frequency. The absorbed dose to the urinary bladder wall due to a charged particle-emitting radionuclide within the urine was estimated using a dose point kernel and that due to photons was calculated from a Monte Carlo code.

MIRD Pamphlet 14 and 14 (Revised) Models

In MIRD Pamphlet 14 and its revision (Thomas et al. 1992, 1999), the absorbed doses to the urinary bladder wall were estimated for a wide range of 19 different radiopharmaceuticals with the radionuclides of ^{18}F, ^{89}Sr, ^{99m}Tc, and ^{123,124,125,131}I. A more sophisticated dynamic model of the urinary bladder was developed than of those previous. The effects of urine retention, variable spherical volume, variable bladder wall thickness, variable urine transfer rates, initial bladder contents volume (i.e., the volume of urine within the bladder before activity enters it) and a nonregular voiding interval (i.e., to allow for delays during the night) were accounted for. A review of the results of this more sophisticated model concluded that it yielded a higher absorbed dose to the urinary bladder wall than did previous models. Most interestingly, particularly in the design of radiopharmaceutical clinical trials, was their observation that lower absorbed doses to the urinary bladder wall were achieved, as expected, through large initial urinary bladder volumes and higher transfer rates of urine into the bladder. Hence, absorbed dose reduction is expected for the subject who is well hydrated prior to administration of the radiopharmaceutical. Interestingly, it was also concluded that, depending upon voiding rates and hydration status, an early voiding interval would not necessarily be optimal for minimizing the urinary bladder wall dose. Depending upon the radiopharmaceutical, this optimum first voiding time could be anywhere from 40 min to 3 h postadministration.

12.2.6 *Head and Brain*

12.2.6.1 Introduction

The detail that could be desired in the internal radiation dosimetry calculation of the brain has grown through the development of radiopharmaceuticals which can lead to highly nonuniform absorbed dose distributions within the brain as a result of their high target specificities. Earlier models, such as that of the MIRD Pamphlet 5 (Revised), of the brain assumed only uniform distribution of radionuclide and homogeneity and were acceptable for estimating a mean absorbed dose to the brain.

12.2.6.2 MIRD Pamphlet 15 Model

This model (Bouchet et al. 1999a) is based upon the earlier brain and head model of Bouchet et al. (1996). The model segregated the head and brain into ten separate regions of the brain (caudate nuclei, cerebellum, cerebral cortex, cerebrospinal fluid contained within the cranium, lateral and third ventricles, lentiform nuclei, thalami, and white matter), three regions of the spine (spinal cord, cerebrospinal fluid within the spine and the spinal skeleton), the eyes, skin, teeth, thyroid, and upper face region (maxilla and zygomatic arch), mandible, and the cranium. The model was a stylized one and a heterogeneous nature of the tissue in terms of physical density and elemental composition was allowed.

Radiation transport was calculated using the EGS4 Monte Carlo code (Chap. 9), the results from which were derived absorbed fractions and S-factors were calculated for the above regions for 34 radionuclides.

12.2.7 Cardiac Wall and Contents

12.2.7.1 Models

Both the cardiac wall and the contents of the heart chambers are MIRD-specified source regions (Table 11.1). MIRD Pamphlet 13 (Coffey et al. 1981) presented a stylized model of the heart, the atria and ventricles and the contents. The atria and ventricles were represented by ellipsoid sections. The left ventricle was modeled by a semiellipsoid, the right ventricle and atrium by the shells between two quarter ellipsoids and the left atrium by the shell between two quarter ellipsoids.

Some modifications of the ICRP Pamphlet 5 (Revised) lung model were required in order to accept this new heart model. Specific absorbed fractions were calculated for monoenergetic photon sources with energies between 10 keV and 4 MeV within the cardiac wall and contents.

12.2.7.2 Isolating Activities in Cardiac Wall and Contents

Both the cardiac wall and the cardiac contents are treated as source regions in the MIRD schema and, hence, the cumulated activities of both are required for MIRD dosimetry calculations. While it is possible to isolate the activities in both regions in tomographic slices to define volumes-of-interest, such an approach is not practical in conjugate-view planar scintigraphy. In such a case, one may define a ROI over the entire cardiac volume which will include contributions from activities in the heart wall and the contents of the cardiac chambers. The activity in the blood contained within the cardiac chambers at a given time point can be estimated from the product of an assumed volume of blood within the chambers and the measured concentration of radionuclide activity in whole blood,

$$A_{\text{Heart Chambers}}(t) = V_{\text{Heart Chambers}}[A_{\text{Whole Blood}}(t)] \tag{12.44}$$

The volumes of blood in the heart chamber in the reference male and female are taken to be 9% of the corresponding total blood volumes, or 477 and 351 mL, respectively (ICRP 2002).

12.2.8 Bone and Red Bone Marrow

12.2.8.1 Introduction

The radiosensitivity of red bone marrow has been discussed in Chap. 10 in the context of the hematopoietic syndrome. As bone marrow is one of the major organs which exhibit radiotoxicity (a deterministic effect) and, hence, can limit the amount of administered activity in radionuclide therapy, the ability to accurately estimate the absorbed dose to marrow is paramount in therapeutic nuclear medicine. Evaluation of the marrow absorbed dose is also important in diagnostic nuclear medicine as it is a significant contributor to the effective dose and should be accurately known. Bone itself is sensitive to radiation, although this is of concern at low absorbed dose levels where the stochastic risk of radiation-induced osteosarcoma exists.

This section reviews dosimetric models of bone and bone marrow used historically and currently in nuclear medicine radiation dosimetry calculations. An additional source of a historical review of these models can be found in the chapter by Bouchet et al. (2003b).

12.2.8.2 Anatomy and Histology

Skeleton

The skeleton is composed of some 200 distinct bones (Gray 1977) which are divided into the four classes of long, short, flat, and irregular. The long bones make up the clavicle and those bones in the extremities (femur, tibia, fibula, humerus, radius, ulna, the metacarpal, and metatarsal bones and the phalanges). The short bones include the carpus and tarsus and the flat bones include the cranial bones and the scapulae. Those remaining bones that cannot be categorized into either of these three are grouped within the irregular class. These include the vertebrae, sacrum, and the coccyx. "Bone" is defined here as being the skeletal mineral material remaining following the removal of soft tissue components such as red and yellow marrow, cartilage, and periarticular tissue.

Further details of the skeleton, bone, bone marrow and their compositions, anatomy, and histology relevant to internal radiation dosimetry can be found in ICRP Publications 70 (1995) and 89 (2002).

Bone

At the microscopic scale, bone in the mature skeleton is classified by two distinct groups on the basis of porosity and microstructure. The first category is that of trabecular bone, which is also referred to as cancellous or spongy bone. Trabecular bone is present at the ends of the long bones (epiphysis), in the interior of the flat bones and makes up most of the vertebrae. It is so called due to its cavities (trabeculae) which contain bone marrow. The second bone category is that of cortical bone which, because it is hard and dense, is also referred to as compact bone. Cortical bone is found in the shafts of the long bones (diaphysis) and in the outer cortex of all bones. In the diaphysis, the cortical bone forms a cylinder surrounding the medullary cavity which is lined with a connective tissue, the endosteum. In the dosimetry literature, the endosteum is often referred to simply, and erroneously, as the bone surface, whereas it, in fact, has a finite thickness. In ICRP Publication 30, the endosteum is specified as being 10 μm thick; although recent studies suggest that the mean thickness may be greater at 50 μm (Gossner et al. 2000; Gossner 2003). The periosteum is dense connective tissue which surrounds the bone surface except at articulating surfaces. Trabecular bone is far more porous than cortical bone; the fraction of total volume taken up by bone tissue in trabecular bone can be as low as 5% and range up to a maximum of about 60%. It thus has a low density and strength, but a large overall surface area. The spongiosa is defined as the combination of the bone trabeculae and the marrow that they contain.

Differentiation between trabecular and cortical bone is not only important because of their differing anatomical and mechanical properties, but also in terms of their differences in retention of bone-seeking radionuclides, as will be discussed shortly, and the distribution of activity on and throughout them. In the neonate, virtually all of the bone is trabecular. The amount of cortical bone increases with age such that by the time of maturity, the ratio of cortical to trabecular bone, by mass, in the skeleton by mass is about 4:1 (and exactly so in the reference conditions specified by ICRP Publication 70). However, in terms of the ratio of the surface area to volume, that for trabecular bone is about four times greater than that of cortical bone with the result that the surface areas of trabecular and cortical bone in the adult have been often considered as being equal. However, the current reference values for the adult male skeleton provided in ICRP Publication 70, and which are reproduced in Table 12.4, show that the above simplifying assumptions underestimate the surface area of trabecular bone. For example, the ratio of surface area to volume is actually six times greater for trabecular bone than cortical bone. Further, 61.7% of the total bone area should be assigned to trabecular bone and 38.3% assigned to cortical bone, rather than the 50–50% assignation suggested by the simplistic argument above.

Both bone types have a hierarchical structure. The fundamental functional unit of cortical bone is the osteon (or Haversian system) which consists of lamellae of compact bone surrounding a central Haversian canal which is lined with a layer of endosteum. The osteons have a typical diameter of 200 μm (but which can vary considerably) and are each aligned parallel to the long shaft of the bone and each contains a nerve and blood vessels supplying the bone, osteoblasts, and undifferentiated cells. Interspersed between the lamellae are lacunae which are interconnected with canaliculi through which nutrients diffuse. The compact bone is penetrated by Volkmann's canals, orthogonal to the Haversian canals, which interconnect osteons with vasculature and nerves. Trabecular bone, as noted before, is made up of trabeculae, each with a typical diameter of about 50 μm. The trabeculae contain marrow, which is discussed in detail in the following section.

Osteogenic cells are found within the bone tissue in contact with the endosteum and periosteum; these

Table 12.4 Reference bone values for adult male (ICRP 1995)

Component	Volume (cm^3)	Surface area (m^2)	Surface area to volume ratio (cm^{-1})
Total bone	2,710	17	
Cortical bone	2,130	6.5	30
Trabecular bone	580	10.5	180

cells produce osteoblasts (bone-forming cells) and osteoclasts (bone-destroying cells). Osteocytes are mature osteoblasts which have synthesized organic material which subsequently undergoes mineralization and are present within the lacunae. The osteoclasts enzymatically break down bone and are essential for bone growth and remodeling.[12] The osteogenic cells are radiosensitive to the production of bone tumors.

Trabecular and cortical bone are both specified MIRD source organs (Table 11.1). Accordingly, one must specify measured bone activity to either of the two bone types. While this can be dealt by consideration of the metabolism of the radionuclide and its vector, the ICRP (1979) has provided two guidelines specific for alkaline earth metals, but which have been adopted for other radionuclides if specific bone uptake data are lacking. If the physical half-life of the radionuclide is 15 days or greater, then it should be assumed that the activity is distributed throughout the bone volume, otherwise the radionuclide is assumed to be distributed over the bone surfaces. Hence, using the data from Table 12.4, if the radionuclide physical half-life is 15 days or greater, then the activity would be assumed to be distributed within the bone volume with 21.4 and 78.6% assigned to trabecular and cortical bone, respectively. Similarly, if the radionuclide physical half-life is less than 15 days, then the activity would be assumed to be distributed over the bone surface with 38.3 and 61.7% assigned to cortical and trabecular bone, respectively.

Bone Marrow

Bone marrow is a connective tissue contained within the trabeculae and is one of the largest organs in the body with a mass comparable to that of the liver (1,500 g in total in ICRP Reference Man (ICRP 1975)). It can be considered to be made up grossly from four categories of tissues, the first of which is the hematopoietic component, or red (active) marrow. It is known as "red" due to the large numbers of erythrocytes it produces. This is the site of erythropoiesis which occurs within the trabeculae of all bones in children, but the extent of which diminishes with maturity. The red marrow contains pluripotent uncommitted stem cells and unipotent committed stem cells. The latter differentiate along the development paths shown in Fig. 12.10.

The physical placements of the various blood cell components within the marrow differ. For example, the granulocytic precursor cells (myeloblast and promyelocyte) are present along the endosteal surfaces of the trabeculae and approach the center of each trabeculum as they mature whereas the erythrocytic precursor cells tend to be at the center of the marrow space. The hierarchy of Fig. 12.10 indicates how, following the radiation-induced depletion of the stem cells, death can eventually result afterwards through immunocompromise, or the hematopoietic syndrome. On the other hand, at low absorbed doses, precursor cell depletion does not occur, but stochastic effects leading to mutagenesis are the risks of interest which are manifested through the leukemias.

The second category of intratrabecular tissue is comprised of an extracellular matrix and randomly-distributed stromal cells. These cells include adipocytes, thus leading to the description of this category as "yellow marrow". This is also referred to as "inactive marrow" as it does not lead to the production of blood cells, but instead provides structural support of the red marrow. The remaining categories comprise the venous sinuses and other intra-trabecular vasculature and additional supporting cells.

The bone marrow cellularity factor (CF) is defined as the fraction of the total bone marrow volume which is occupied by the active red bone marrow. This ratio is essentially equal to unity at birth and decreases with age. As pointed out by Bolch et al. (2002), this definition of the mallow CF ignores the volumes of the venous sinuses, the extracellular matrix, and marrow support cells. However, these volumes are comparable or smaller than the hematopoietic component and, clinically, difficult to segregate. The adipocytes (yellow marrow) are, however, easy to differentiate from the red marrow, thus allowing an approximation to the bone marrow CF as,

$$\mathrm{CF} \cong 1 - \text{Fraction of Volume Occupied by Adipocytes} \tag{12.45}$$

[12]The early tracer work of Chievitz and de Hevesy described in Chap. 11 using ^{32}P in the rat revealed the fact that bone was being constantly remodeled.

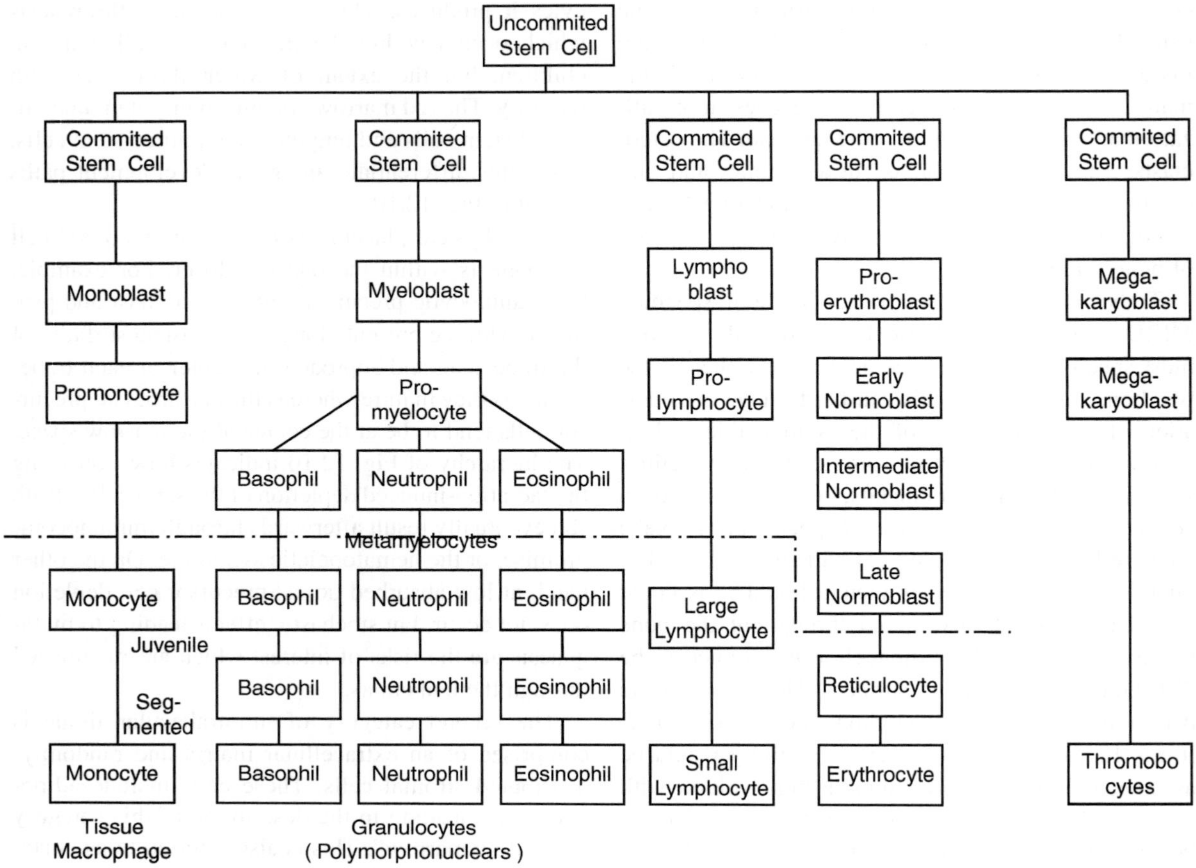

Fig. 12.10 Development and differentiation of the various blood cells from a bone marrow cell. Cells below the *dash-dot line* are also found in the peripheral blood (adapted from Ganong 1979)

The CF is also a function of bone type. For the adult, the marrow CF ranges from about 0.23 for the radius and ulna to 0.8 for the lumbar, thoracic and cervical spine (Eckerman and Stabin 2000). Details of the age- and skeletal site-distributions of hematopoietically-active marrow and CF are summarized in Tables 40 and 41 of ICRP Publication 70 (1995). Table 12.5 summarizes the skeletal distributions of red marrow and CF in a 40-year-old adult using data from these tables ICRP Publication 70 and which are actually attributable to Cristy (1981). Work at the University Florida is active in empirically characterizing the distributions and relative masses of hematopoietically-active marrow and adipocytic marrow (Bolch et al. 2002; Shah et al. 2003; Pichardo et al. 2007).

Table 12.5 Skeletal distribution of hematopoietically-active marrow and the cellularity factor in the 40-year-old human (ICRP 1995)

Skeletal site	Active marrow (% of total)	Cellularity factor, CF
Cranium	7.6	0.38
Mandible	0.8	–
Scapulae	2.8	0.38
Clavicles	0.8	0.33
Sternum	3.1	0.7
Ribs	16.1	0.7
Cervical vertebrae	3.9	0.7
Thoracic vertebrae	16.1	0.7
Lumbar vertebrae	12.3	0.7
Sacrum	9.9	0.7
Os coxae	17.5	0.48
Femora, proximal	6.7	0.25
Humeri, proximal	2.3	0.25

Estimation of the absorbed dose to red bone marrow is essential to radionuclide therapy, yet the correlation of absorbed dose alone to marrow response (e.g., myelotoxicity) has been incomplete. Methods of estimating bone marrow dose are discussed.

12.2.8.3 Bone and Bone Marrow Models

Spiers' Models

Between the 1940s and the 1970s, Spiers and his students at the University of Leeds established the foundations of the models of the radiation dosimetry of trabecular and cortical bone and of bone marrow (Spiers 1949; Whitwell and Spiers 1976; Beddoe et al. 1976; Beddoe 1977; Spiers et al. 1978). Avoiding the limitations set by the use of simple geometrical models of bone substructures upon which to base dosimetry evaluations, they instead used a bespoke optical scanner to measure the statistical distributions of pathlengths[13] in the internal cavities of trabecular and cortical bone using ex vivo thin slice sections. Monte Carlo modeling of the energy loss and electrons and positrons traversing these paths due to radionuclides contained in the bone (including ^{18}F, ^{32}P, ^{90}Sr, and ^{90}Y) was performed. There were, however, some limitations to this work. For example, Beddoe's calculation for cortical bone was for radionuclides within the bone and did not allow for any placement within the cavities, such as the Haversian canals. The studies by Spiers and his group established the first estimates of the electron absorbed fractions $\phi(r_T \leftarrow r_T)$ for bone.

Geometrical Models

Representations of bone and its structures by geometrical constructs were attempted by researchers in Canada in the 1960s (Charlton and Cormack 1963; Aspin and Johns 1963), particularly as a result of interest in absorbed doses to bone due to external beam radiotherapy.

MIRD 11 Model

The estimates of the self-dose absorbed fraction $\phi(r_T \leftarrow r_T)$ derived from Spiers' models were used in the S-factor calculations in MIRD Pamphlet 11. In these calculations, the target regions r_T were bone, red and yellow marrow and the endosteum and the source regions were trabecular and cortical bone and red and yellow marrow.

ICRP Publication 30 Model

In 1979, ICRP released its dosimetric model of bone (ICRP 1979). Although the focus of the model was on the radiological protection of individuals exposed to ionizing radiation as a result of their occupation, it has been used in nuclear medicine radiation dosimetry calculations. Hence, the dosimetric quantity of interest was largely the dose equivalent due to the risk of exposure to bone-seeking α-emitting radionuclides.[14] Regardless of the focus, the model is entirely relevant to nuclear medicine dosimetry. Much of the data used in the development of this model evolved from the work by Spiers and his group.

The cells recognized in the model as being at radiocarcinogenic risk are the hematopoietic stem cells, the endosteal osteogenic cells, and epithelial cells near to the surface of the bone. ICRP Publication 30 specified the location and spatial dimensions of what were to be regarded as the target regions. For those cells near the bone surface (i.e., endosteum), the target region was defined to be a 10 μm thick layer of tissue covering all endosteal surfaces and those surfaces of the bone lined with epithelial cells; the total mass of this target region was said to be 120 g. The other target region was the active red bone marrow within the trabeculae, with a

[13]Virtually all of the literature produced from outside the University of Leeds group that refers to these measurements describes them as being of a "chord length" rather than a "pathlength". One of Spiers' students, Darley, has written that the Spiers group consistently referred to these measurements being of pathlengths rather than of chord-lengths, as the latter implies the traversal of a continuous or closed structure (Darley 2006). To maintain that descriptive precision, these will be referred to as "pathlengths" in this discussion.

[14]It is entirely reasonable for α-emitting radionuclides to be included in the discussion as such radionuclides are used therapeutically (Lewington 2005).

mass of 1,500 g (ICRP 1975). The following radiations were considered:

- α emissions from radionuclides distributed uniformly within bone
- α emissions from radionuclides distributed uniformly on the bone surfaces
- β emissions from radionuclides distributed uniformly within bone
- β emissions from radionuclides distributed uniformly on the bone surfaces with
  - A mean kinetic energy of 200 keV or less
  - A mean kinetic energy exceeding 200 keV

Absorbed fractions for α-emitting radionuclides were obtained from the calculations by Thorne (1977); data for β-emitting radionuclides were derived, in much, from the work by Spiers and his research group, as referenced earlier. Table 12.6 provides a summary of the absorbed fractions given by ICRP Publication 30 for the combinations of particle emissions, anatomical location, and energy. It should be recognized that the assigned single-valued absorption factors with little or limited-energy dependencies are primarily for radiological protection calculation convenience and lack, for example, consideration of the marrow CF. These factors were, however, used in the MIRDOSE 2 code (Stabin 1996).

Eckerman and Stabin Model

Eckerman and Stabin (2000) reexamined the absorbed factors for β-emitting radionuclides for marrow and bone as target regions in a skeletal model that included age as a variable. Using the Cristy phantom described earlier and the pathlength method of Spiers, they evaluated the absorbed fractions for combinations of source regions of trabecular bone (volume and surface), red marrow and cortical bone (volume and surface) and of target regions of red marrow and bone surface for electron energies between 1 keV and 4 MeV for skeletal phantoms of the neonate, 1-, 5-, 10-, 15-year old and adult. The absorbed fraction within a region was calculated as the ratio of the amount of energy loss from an electron traversing the region along the pathlength as a fraction of the energy released in the β decay. As a result, the assignation of the absorbed fraction to red marrow could not be done explicitly. Instead, a marrow CF was assumed and the energy deposited within the contents of a trabecular volume scaled by it accordingly. As an example, Fig. 12.11 shows the absorbed fractions for the source region being a β-emitting radionuclide uniformly distributed over the trabecular bone surface or bone volume and the source regions being the bone surface and the red bone marrow. These are taken from the results

Table 12.6 Recommended values from ICRP Publication 30 (1979) for absorbed fractions for various bone source and target regions for α- and β-emitting radionuclides

Particle type	Energy	Source region (r_s)	Target region (r_T)	Recommended absorbed fraction $\phi(r_T \leftarrow r_S)$
α	All	Trabecular bone volume	Bone surface	0.025
			Red Bone marrow	0.05
		Cortical bone volume	Bone surface	0.01
			Red bone marrow	0
		Trabecular bone surface	Bone surface	0.25
			Red bone marrow	0.5
		Cortical bone surface	Bone surface	0.25
			Red bone marrow	0
β	All	Trabecular bone volume	Bone surface	0.025
			Red bone marrow	0.35
		Cortical bone volume	Bone Surface	0.015
			Red bone marrow	0
	$\bar{T} < 200\,\text{keV}$	Trabecular bone surface	Bone surface	0.25
	$200\,\text{keV} \leq \bar{T}$			0.025
	$\bar{T} < 200\,\text{keV}$	Cortical bone surface	Bone surface	0.25
	$200\,\text{keV} \leq \bar{T}$			0.015
	All	Trabecular bone surface	Red bone marrow	0.5
		Cortical bone surface	Red bone marrow	0

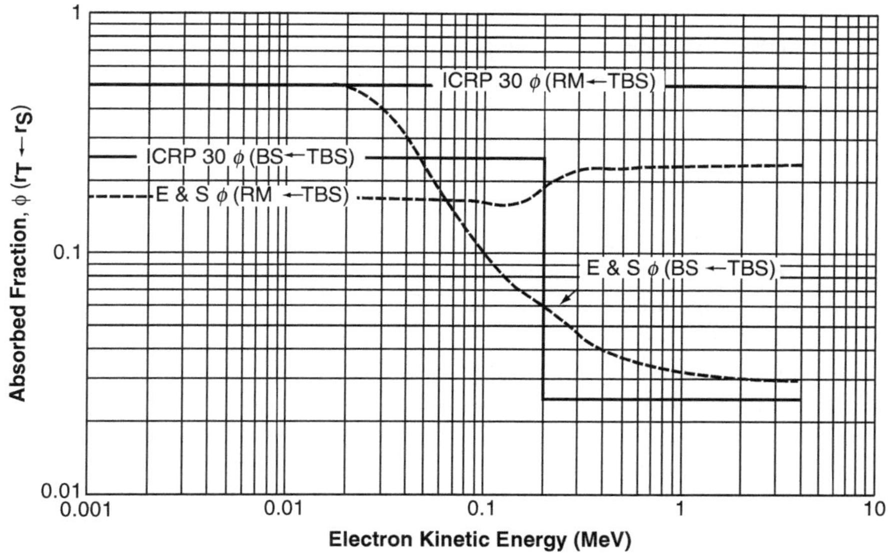

Fig. 12.11 Absorbed fractions calculated by Eckerman and Stabin (2000) (E&S), and compared with those of ICRP Publication 30 (1979), for the source region being a β-emitting radionuclide uniformly distributed over the trabecular bone surface (TBS) and the target regions being either the red marrow (RM) or the bone surface/endosteum (BS)

calculated by Eckerman and Stabin and from ICRP Publication 30.

Figure 12.11 demonstrates the simplicity of the absorbed fraction values provided by the ICRP Publication 30 for the source region being the trabecular bone surface. For example, the ICRP 30 approximation for the absorbed fraction for red marrow being the target region overestimates that determined by Eckerman and Stabin by about a factor of 3 for low electron energies, decreasing to about a factor of 2 for electron energies exceeding about 300 keV. The differences between the ICRP Publication 30 $\phi(\mathrm{RM} \leftarrow \mathrm{TBS})$ values and those of Eckerman and Stabin were attributed by the latter as being the result of differences in the assumptions of the marrow CF. The ICRP Publication 30 results were based upon the assumption that CF $= 1$, i.e., all of the marrow was active, whereas Eckerman and Stabin assumed that CF $= 0.6$.

The above results demonstrate that the ICRP Publication 30 approximations, whilst being appropriate for radiological protection applications, could likely be insufficiently accurate for bespoke radionuclide therapy calculations. But even the refinements of the Eckerman and Stabin results suffer. For example, no consideration is given of the lack of electronic equilibrium at the interface between marrow and bone. In Chap. 9, it was shown that radiations moving from a high-Z/high-density medium (e.g., bone) to a low-Z/low-density medium (e.g., marrow) will lead to an elevation of absorbed dose on the low-Z/low-density side of the interface. Kwok et al. (1991) demonstrated through both experimental measurement and calculation that absorbed dose enhancements of between 10 and 21% in the red marrow could occur at electron energies typically of many β-emitting radionuclides. Similar results were reported by Johnson et al. (1992) using Monte Carlo simulations of radiation transport arising from β-emissions from ^{153}Sm, ^{186}Re, and ^{166}Ho distributed on bone surfaces.

Bouchet Model

Bouchet et al. (1999b, c, 2000) developed a dosimetric model of bone and bone marrow using, again, the pathlength distributions measured by the Spiers group and Monte Carlo simulation of radiation transport in two- and three-dimensional models of bone and bone marrow. Their work culminated in the production of S-factors using masses of bone, marrow and endosteum derived from ICRP Publication 70 (1995). Further discussion of the model is provided below.

Comparison of Models of Eckerman and Stabin and Bouchet

The study of the radiation dosimetry of bone and bone marrow is a rich field to which many investigators have contributed over the past six decades. This section can provide only an overview of those developments that the author has judged as having had the most significant impact on internal radiation dosimetry. There are still many important facets to the field undergoing investigation, particularly in the application to radionuclide therapy. These two bone marrow models provide differing results when applied to dosimetry calculations. Recently, Stabin (2008) summarized the three differences between the models.

First, there is the difference in the use of the marrow CF. The use of this factor is necessary in a bone marrow dosimetry calculation as, while it is frequently the absorbed doses to the soft tissues within the trabeculae that are calculated, it is the red marrow which is the radiosensitive component within the trabecular marrow space. As both models were still largely based upon the pathlength concept which cannot differentiate between red and yellow marrow, explicit accounting of the differences between red and yellow marrow was either neglected or else done by the application of skeletal site-specific CF values. The Eckerman and Stabin model and the Bouchet et al. model differ in the calculation of the self-absorbed fraction due to a β-emitting radionuclide in the red marrow by the use of the CF,

$$\phi_i(\mathrm{RM} \leftarrow \mathrm{RM}) = \mathrm{CF}_i \phi_i(\mathrm{TM} \leftarrow \mathrm{TM}) \quad \text{Eckerman and Stabin} \tag{12.46}$$

$$\phi_i(\mathrm{RM} \leftarrow \mathrm{RM}) = \phi_i(\mathrm{TM} \leftarrow \mathrm{TM}) \quad \text{Bouchet et al.} \tag{12.47}$$

where RM refers to red marrow, TM refers to the total trabecular marrow space and the index i refers to the skeletal site. One can see that at low electron energies, where complete self-absorption will occur, then the absorbed fractions must be come equal, i.e., $\phi_i(\mathrm{RM} \leftarrow \mathrm{RM}) \rightarrow \phi_i(\mathrm{TM} \leftarrow \mathrm{TM})$. On the other hand, at higher electron energies, the electrons can traverse multiple trabeculae and the absorbed fraction will tend to the CF. These arguments would suggest that the Bouchet et al. model is applicable to low-energy electrons and the Eckerman and Stabin model to high-energy electrons. Bolch et al. (2002) summarized the expectations that, at intermediate electron energies, the Bouchet et al. model and the Eckerman and Stabin model would overestimate and underestimate, respectively, the self-absorbed fractions for red bone marrow. Using a three-dimensional trabecular bone model based upon high spatial resolution MR imaging of cadaveric sources of bone and simulation of adipocyte sites within the trabeculae, they used Monte Carlo simulations of radiation transport to evaluate the electron self-absorbed factor for red marrow. They demonstrated that, indeed, the Bouchet et al. model overestimates the $\phi_i(\mathrm{RM} \leftarrow \mathrm{RM})$ value by about 40% at an electron energy of 1 MeV whereas the Eckerman and Stabin model underestimates $\phi_i(\mathrm{RM} \leftarrow \mathrm{RM})$ by about 75% at very low electron energies (10 keV).

The second difference between the two models lay in the Monte Carlo description of the trajectories of electrons traversing the 10 μm thick endosteum. In the Eckerman and Stabin model, the endosteum is assumed to be penetrated by electrons with trajectories uniformly distributed in angle (between 0 and 180°). On the other hand, the Bouchet et al. model assumed a uniform distribution in the cosine of the angle (i.e., between −1 and +1). This affects the distribution of pathlengths and, subsequently, the energy deposition.

The third difference was in the assumption of the distribution of the electron source within the endosteum: Bouchet et al. assumed a uniform distribution within the endosteum whereas Eckerman and Stabin assumed a uniform distribution over the bone surface.

Stabin et al. (2002) used more refined anatomical data (Jokisch et al. 1998) in an attempt to reconcile the differences between the Eckerman and Stabin and the Bouchet et al. models and Bolch's group continued this study of the role of inactive yellow marrow in bone marrow dosimetry calculations (Shah et al. 2003).

Endosteum Thickness

ICRP Publication 30 specified a uniform endosteal thickness of 10 μm for dosimetry calculations and which was used in the Eckerman and Stabin and Bouchet et al. calculations. Work by Gossner and collaborators (Gossner et al. 2000; Gossner 2003) suggests that radiosensitive cells leading to osseous cancers can be as far as 50 μm from the surfaces of the trabeculae and the medullary cavities of cortical bone. Bolch et al. (2007) examined the effects of this increased endosteal thickness upon the specific absorbed fractions for β-emitting radionuclides.

Red Bone Marrow Dosimetry

Hematologic toxicity as a consequence of irradiation of active marrow is frequently a limiting factor in the amount of therapeutic activity administered. As a consequence, there is considerable interest in accurate evaluation of red bone marrow absorbed dose.

12.2.9 Peritoneal Cavity

The abdominopelvic cavity consists of the abdominal cavity (containing the small stomach, the small and most of the large intestines, liver, gallbladder, spleen, pancreas, and kidneys) and the pelvic cavity (containing the rectosigmoid colon, urinary bladder, uterus, ovaries, and prostate gland). Serous membranes (the peritoneal membranes) cover both cavities: the parietal peritoneum lines the abdominal wall and the visceral peritoneum covers the viscera. The space between these two membranes is the peritoneal cavity.[15] All of the abdominal organs are exterior to the peritoneal cavity. As ovarian metastases generally spread through the peritoneal cavity, radionuclide therapy has been used for intraperitoneal disease since the 1940s with the use of ^{198}Au and ^{32}P (Young 1985). Radioimmunotherapy has now been used for radionuclide therapy of the peritoneal cavity (AAPM 2001; McQuarrie et al. 2004).

Watson et al. (1989) developed a model for estimating the absorbed dose to the peritoneal space by modifying the Fisher-Snyder phantom. The peritoneal cavity was represented by two connected elliptical cylinders. Data for photon- and electron-emitting radionuclides were calculated.

12.2.10 Tumors (Spheres)

Localized tumors can be modeled by spheres, which permit a simple means of estimating the absorbed dose to the tumor resulting from its uptake of particulate-emitting radionuclides. The accuracy of this calculation is dependent upon the degree of charged particle equilibrium within the tumor which, in turn, is a function of the charged particles and the radius of the sphere. The absorbed fractions in a sphere of soft tissue, representing a tumor, as a function of the diameter for ^{90}Y and ^{131}I was presented Chap. 9. Bardiès et al. considered those cases of spheres, with radii ranging from 5 μm to 5 mm, which were labeled on the surface with β-emitting radionuclides (Bardiès et al. 1990) and spheres with a radii ranging between 5 and 200 μm, for α-emitting radionuclides (1990), in a simulation of radioimmunotherapy of small tumors and cells.

12.2.11 Prostate Gland

The prostate gland is approximately ellipsoid in shape with dimensions of 4 × 3 cm and with a mass of about 16 g. It is inferior to the urinary bladder and surrounds the proximal aspect of the urethra. As such, it is formed into lobules by the urethra and ejaculatory ducts extend throughout these lobules and communicate with the urethra.

As some radiopharmaceuticals have demonstrated prostatic uptake, Stabin (1994) developed a model of the prostate gland which was a sphere with a diameter of 3.08 cm and composed of soft tissue. Specific absorbed fractions were evaluated for photon energies ranging from 10 keV to 4 MeV. S-factors for a range of radionuclides were also derived and presented.

12.2.12 Rectum

A model of the lower colon and surrounding organs was presented by Mardirossian et al. (1999) as a modification of the rectosigmoid component of the Cristy–Eckerman Reference Man phantom. The rectum was not included within the LLI component of that phantom. In the new rectum model, the rectum was explicitly represented. Additional components added were the prostatic urethra and seminal ducts. The positions of the prostate gland and the urinary bladder were displaced, which resulted in changes in the absorbed doses to the testes and the LLI wall due to activity within the prostate gland.

[15] The kidneys and adrenal glands are posterior to the parietal peritoneum and are hence referred to as being retroperitoneal.

References

AAPM (2001) A primer for radioimmunotherapy and radionuclide therapy – AAPM Report No 71. American Association of Physicists in Medicine, College Park, MD

Aspin N, Johns HE (1963) The absorbed dose in cylindrical cavities within irradiated bone. Br J Radiol 36:350–362

Atkins HL, Robertson JS, Croft BY et al (1980) MIRD dose estimate report no. 9: radioxenons in lung imaging. J Nucl Med 21:459–465

Atkins HL, Weber DA, Susskind H, Thomas SR (1992) MIRD dose estimate report no 16: radiation absorbed dose from technetium-99m-diethylenetriaminepentaacetic acid aerosol. J Nucl Med 33:1717–1719

Bajc M, Neilly JB, Miniati M, Schuemichen C, Meignan M, Jonson B (2009) EANM guidelines for ventilation/perfusion scintigraphy. Eur J Nucl Med Mol Imaging 36:1356–3170

Bardiès M, Lame J, Myers MJ, Simoen JP (1990) A simplified approach to beta dosimetry for small spheres labelled on the surface. Phys Med Biol 35:1039–1050

Bardiès M, Myers MJ (1990) A simplified approach to alpha dosimetry for small spheres labelled on the surface. Phys Med Biol 35:1551–1561

Beadsmoore C, Cheow HK, Szezepura K, Ruparelia P, Peters AM (2007) Healthy passive cigarette smokers have increased pulmonary alveolar permeability. Nucl Med Commun 28:75–77

Beddoe AH (1977) Measurements of the microscopic structure of cortical bone. Phys Med Biol 23:298–308

Beddoe AH, Darley PJ, Spiers FW (1976) Measurements of trabecular bone structure in man. Phys Med boil 21: 589–607

Biju K, Nagarajan PS (2000) Computed normalised effective doses to an Indian adult in conventional x ray chest examinations. Radiat Prot Dosim 88:119–127

Blau M, McAfee JG, Rohrer RH, Snyder WS, Smith EM (1975) MIRD dose estimate report no. 6: Hg-197- and Hg-203-labeled chromerodrin. J Nucl Med 16:1095–1098

Bolch WE, Patton PW, Rajon DA, Shah AP, Jokisch DW, Inglis BA (2002) Considerations of marrow cellularity in 3-dimensional dosimetric models of the trabecular skeleton. J Nucl Med 43:97–108

Bolch WE, Shah AP, Watchman CJ, Jokisch DW, Patton PW, Rajon DA, Zankl M, Petoussi-Henss N, Eckerman KF (2007) Skeletal absorbed fractions for electrons in the adult male: considerations of a revised 50-μm definition of the bone endosteum. Radiat Prot Dosim 127:169–173

Bondesson E, Bengtsson T, Nilsson LE, Wollmer P (2007) Site of deposition and absorption of an inhaled hydrophilic solute. Br J Clin Pharmacol 63:722–731

Bouchet LG, Boch WE, Atkins WDA, HL PJW (1996) A revised dosimetric model of the adult head and brain. J Nucl Med 37:1226–1236

Bouchet LG, Bolch WE, Weber DA, Atkins HL, Poston JW (1999a) MIRD Pamphlet No. 15: radionuclide S values in a revised dosimetric model of the adult head and brain. J Nucl Med 40:62–101

Bouchet LG, Jokisch WJ, Bolch WE (1999b) A three-dimensional transport model for determining absorbed fractions of energy for electrons within trabecular bone. J Nucl Med 40:1947–1966

Bouchet LG, Jokisch WJ, Bolch WE (1999c) A three-dimensional transport model for determining absorbed fractions of energy for electrons within cortical bone. J Nucl Med 40:2115–2124

Bouchet LG, Bolch WE, Howell RW, Rao DV (2000) S values for radionuclides localized within the skeleton. J Nucl Med 41:189–212

Bouchet LG, Bolch WE, Blanco HP, Wessels BW, Siegel JA, Rajon DA, Clairand I, Sgouros G (2003a) MIRD Pamphlet No. 19: absorbed fractions and radionuclide S values for six age-dependent multiregion models of the kidney. J Nucl Med 44:1113–1147

Bouchet LG, Bolch WE, Stabin M, Eckerman KF, Poston JW, Brill AB (2003b) Monte Carlo methods and mathematical models for the dosimetry of skeleton and bone marrow. In: Zaida H, Sgouros G (eds) Therapeutic applications of Monte Carlo calculations in nuclear medicine. Institute of Physics Publishing, Bristol

Brownell GL, Ellet WH, Reddy AR (1968) MIRD Pamphlet No 3: absorbed fractions for photon dosimetry. J Nucl Med Suppl 29–39

De C, Cormack DV (1963) Energy dissipation in finite cavities. Rad Res 17:34–49

Cloutier RJ, Smith SA, Watson EA, Snyder WS, Warner GG (1973) Dose to the fetus from radionuclides in the bladder. Health Phys 25:147–161

Coffey JL, Cristy M, Warner GG (1981) Specific absorbed fractions for photon sources uniformly distributed in the heart chambers and heart wall of a heterogeneous phantom. J Nucl Med 22:65–71

Cristy M (1980) Mathematical phantoms representing children of various ages for use in estimates of internal dose. Report ORNL/NUREG/TM-367. Oak Ridge National Laboratory, Oak Ridge, TN

Cristy M (1981) Active bone marrow distribution as a function of age in humans. Phys Med Biol 26:389–400

Cristy M, Eckerman KF (1987) Specific absorbed fractions of energy at various ages from internal photon sources volumes I-VII. Report ORNL/TM-8381. Oak Ridge National Laboratory, Oak Ridge, TN

Darley PJ (2006) Bone structure studies at the University of Leeds (Letter to the Editor). Health Phys 90:176–177

Dimbylow PJ (1997) FTDT calculations of the whole-body averaged SAR in an anatomically realistic voxel model of the human body from 1 MHz to 1 GHz. Phys Med Biol 42:479–490

Dolphin GW, Eve IS (1966) Dosimetry of the gastrointestinal tract. Health Phys 12:163–172

Eckerman KF, Stabin MG (2000) Electron absorbed fractions and dose conversion factors for marrow and bone by skeletal regions. Health Phys 78:199–214

Eve IS (1966) A review of the physiology of the gastrointestinal tract in relation to radiation doses from radioactive materials. Health Phys 12:131–161

Fill UA, Zankl M, Petoussi-Henss N, Siebert M, Regulla D (2004) Adult female voxel models of different stature and photon conversion coefficients for radiation protection. Health Phys 86:253–272

Ganong WF (1979) Review of medical physiology, 9th edn. Lange Medical, Los Altos
Gossner W, Masse R, Stather JW (2000) Cells at risk for dosimetric modeling relevant to bone tumour induction. Radiat Prot Dosim 92:209–213
Gossner W (2003) Target cells in internal dosimetry. Radiat Prot Dosim 105:39–42
Gray H (1977) Anatomy, descriptive and surgical. Bounty, New York
Gray HW, Neilly JB (1999) The role and value of ventilation perfusion imaging in pulmonary embolism. In: Oudkerk M, van Beek EJR, ten Cate JW (eds) Pulmonary embolism. Blackwell Science, Berlin
Hamilton D (2004) Diagnostic nuclear medicine – a physics perspective. Springer, Berlin
IAEA (1998) Compilation of anatomical, physiological and metabolic characteristics for a Reference Asian Man. TEC-DOC 1005. International Atomic Energy Agency, Vienna
ICRP (1959) Report of Committee II on permissible dose for internal radiation – ICRP Publication 2. International Commission on Radiological Protection, Pergamon, NY
ICRP (1975) Report on the Task Group on Reference Man – ICRP Publication 23. International Commission on Radiological Protection, Pergamon, Oxford
ICRP (1979) Limits for intakes of radionuclide workers – ICRP Publication 30. International Commission on Radiological Protection, Pergamon, Oxford
ICRP (1991) Recommendations of the International Commission on Radiological Protection – ICRP Publication 60. International Commission on Radiological Protection, Pergamon, Oxford
ICRP (1994) Human respiratory tract model for radiological protection – ICRP Publication 66. International Commission on Radiological Protection, Pergamon, Oxford
ICRP (1995) Basic anatomical and physiological data for use in radiation protection: the skeleton – ICRP Publication 70. International Commission on Radiological Protection, Pergamon, Oxford
ICRP (1998) Radiation dose to patients from radiopharmaceuticals – ICRP Publication 80. International Commission on Radiological Protection, Pergamon, Oxford
ICRP (2002) Basic anatomical and physiological data for use in radiological protection: reference values – ICRP Publication 89. International Commission on Radiological Protection, Pergamon, Oxford
ICRP (2006) Human alimentary tract model for radiological protection – ICRP Publication 100. International Commission on Radiological Protection, Pergamon, Oxford
ICRP (2007) Recommendations of the International Commission on Radiological Protection – ICRP Publication 103. International Commission on Radiological Protection, Pergamon, Oxford
Jain SC, Reddy AR, Nagaratnam A (1995a) Radiation dose from radiopharmaceuticals in adults of different Asian countries. Bull Radiat Prot 18:96–99
Jain SC, Mehta SC, Kumar B, Reddy R, Nagaratnam A (1995b) Formulation of the Reference Indian Adult: anatomic and physiologic data. Health Phys 68:509–522
Jokisch DW, Patton PW, Inglis BA, Bouchet LG, Rajon DA, Rifkin J, Bolch WE (1998) NMR microscopy of trabecular bone and its role in skeletal dosimetry. Health Phys 75:584–596
Johnson JC, Langhorst SM, Loyalka SK, Volkert WA, Ketring AR (1992) Calculation of radiation dose at a bone-to-marrow interface using Monte Carlo modeling techniques (EGS4). J Nucl Med 33:623–628
Jones DG (1997) A realistic anthropomorphic phantom for calculating organ doses arising from external photon irradiation. Radiat Prot Dosim 72:21–29
Jönsson L, Liu X, Jönsson B-A, Ljungberg M, Strand S-E (2002) Dosimetry model for the small intestine incorporating intestinal wall activity and cross-doses. J Nucl Med 43:1657–1664
Kim CH, Choi SH, Jeong JH, Lee C, Chung MS (2008) HDRK-Man: a whole-body voxel model based on high-resolution color slice images of a Korean adult male cadaver. Phys Med Biol 53:4093–4106
Kramer R, Viera JW, Khoury HJ, Lima FRA, Fuelle D (2003) All about MAX: a male adult voxel phantom for Monte Carlo calculations in radiation protection dosimetry. Phys Med Biol 48:1239–1262
Kramer R, Khoury RJ, Vieira JW, Loureiro ECM, Lima VJM, Lima FRA, Hoff G (2004) All about FAX: a female adult voxel phantom for Monte Carlo calculation in radiation protection dosimetry. Phys Med Biol 49:5203–5216
Kramer R, Khoury HJ, Vieira JW, Lima VJM (2006) MAX06 and FAX06: update of two adult human phantoms for radiation protection dosimetry. Phys Med Biol 51:3331–3346
Kwok CS, Bialobzyski PJ, Yu SK (1991) Effects of tissue inhomogeneity on dose distribution of continuous activity of low-energy electrons in bone marrow cavities with different topologies. Med Phys 18:533–541
Lambert B, Cybulla M, Weiner SM et al (2004) Renal toxicity after radionuclide therapy. Radiat Res 161:607–611
Lawson M, Everson GT, Klingensmith W, Kern F (1983) Co-ordination of gastic and gallbladder emptying after ingestion of a regular meal. Gastroenterology 85:866–870
Lee C, Lee J, Lee C (2004) Korean adult male voxel model KORMAN segmented from magnetic resonance images. Med Phys 31:1017–1022
Lee C, Williams JL, Lee C, Bolch WE (2005) The UF series of tomographic computational phantoms. Med Phys 32:3537–3547
Lee C, Lee C, Williams JL, Bolch WE (2006a) Whole-body voxel phantoms of paediatric patients – UF Series B. Phys Med Biol 51:4649–4661
Lee C, Lee C, Park S-H, Lee J-K (2006b) Development of the two Korean adult tomographic computational phantoms for organ dosimetry. Med Phys 33:380–390
Lee C, Lodwick D, Hasenauer D, Williams JL, Lee C, Bolch WE (2007a) Hybrid computational phantoms of the male and female newborn patient: NURBS-based whole-body models. Phys Med Biol 52:3309–3333
Lee C, Lee C, Lodwick D, Bolch WE (2007b) NURBS-based 3-D anthropomorphic computational phantoms for radiation dosimetry applications. Radiat Prot Dosim 127:227–232
Lewington VJ (2005) Bone-seeking radionuclides for therapy. J Nucl Med 46:38S–47S
Lin HC, Prather C, Fisher RS, Meyer JH, Summers RW, Pimental M, McCallum RW, Akkermans LMA, Loening-Baucke V (2005) Measurement of gastrointestinal transit. Dig Dis Sci 50:989–1004

McAfee JG (1969) Problems in evaluating the radiation dose for radionuclides excreted by the kidneys. In: Cloutier RJ, Edwards CL, Snyder WS (eds). Medical radionuclides: radiation dose and effects, US Atomic Energy Commission, Oak Ridge, TN, pp 271–294

McQuarrie SA, Mercer JR, Syme A, Suresh MR, Miller GG (2004) Preliminary results of radiopharmaceuticals used in the radioimmunotherapy of ovarian cancer. J Pharm Pharmaceut Sci 7:29–34

Mardirossian G, Tagesson M, Blanco P, Bouchet LG, Stabin M, Yoriyaz H, Baza S, Ljungberg M, Strand S-E, Brill AB (1999) A new rectal model for dosimetry applications. J Nucl Med 40:1524–1531

Nagaoka T, Watanabe S, Sakurai K, Kunieda E, Taki M, Yamanaka Y (2004) Development of realistic high-resolution whole-body voxel models of Japanese adult males and females of average height and weight, and application of models to radio-frequency electromagnetic-field dosimetry. Phys Med Biol 49:1–15

Nias AHW (1998) An introduction to radiobiology, 2nd edn. Wiley, New York

Nipper JC, Williams JL, Bolch WE (2002) Creation of two tomographic voxel models of paediatric patients in the first year of life. Phys Med Biol 47:3143–3164

Park S, Lee JK, Lee C (2006) Development of a Korean adult male computational phantom for internal dosimetry calculation. Radiat Prot Dosim 121:257–264

Petoussi-Henss N, Zankl M, Fill U, Regulla D (2002) The GSF family of voxel phantoms. Phys Med Biol 47:89–106

Pichardo JC, Trindale AA, Brindle JM, Bolch WE (2007) Method for estimating skeletal spongiosa volume and active marrow mass in the adult male and adult female. J Nucl Med 48:1880–1888

Poston JW, Bolch WE, Bouchet LG (2003) Mathematical models of the human anatomy. In: Zaidi H, Sgouros G (eds) Therapeutic applications of Monte Carlo calculations in medicine. Institute of Physics, Bristol

Saito K, Wittmann A, Koga S, Ida Y, Kamei J, Funakibi J, Zankl M (2001) Construction of a computed tomographic phantom for a Japanese male adult and dose calculation system. Radiat Environ Biophys 40:69–76

Sato K, Noguchi H, Emoto Y, Koga S, Saito K (2007) Japanese adult male voxel phantom constructed on the basis of CT images. Radiat Prot Dosim 123:337–344

Segars WP, Tsui BMW, Frey EC, Johnson GA, Perr SS (2004) Development of a 4-D digital mouse phantom for molecular imaging research. Mol Imaging Res 6:149–159

Shah AP, Patton PW, Rajon DA, Bolch WE (2003) Adipocyte spatial distribution in bone marrow: implications for skeletal dosimetry models. J Nucl Med 44:774–783

Snyder WS, Ford MR, Warner GG, Fisher HL (1969) MIRD Pamphlet No. 5: estimates of absorbed fractions of monoenergetic photon sources uniformly distributed in various organs of a heterogeneous phantom. J Nucl Med 10(suppl 3):5–52

Snyder WS, Ford MR (1976) Estimation of dose to the urinary bladder and gonads. In: Cloutier RJ, Coffey JL, Snyder WS, Watson EE (eds). Radiopharmaceutical dosimetry symposium, HEW Publications (FDA) 76-8044, Oak Ridge, TN

Snyder WS, Ford MR, Warner G (1978) Estimates of specific absorbed fractions for monoenergetic photon sources uniformly distributed in various organs of a heterogeneous phantom, Pamphlet No. 5 Revised. Society of Nuclear Medicine, New York

Spiers FW (1949) The influence of energy absorption and electron range on dosage in irradiated bone. Br J Radiol 12:521–533

Spiers FW, Whitwell JR, Beddoe AH (1978) Calculated dose factors for radiosensitive tissues in bone irradiated by surface-deposited radionuclides. Phys Med Biol 23: 481–494

Stabin MG (1994) A model of the prostate gland for use in internal dosimetry. J Nucl Med 35:516–520

Stabin MG (1996) MIRDOSE: personal computer software for internal dose assessment in nuclear medicine. J Nucl Med 37:538–546

Stabin MG, Siegel JA, Sparks RB (2002) Sensitivity of model-based calculations of red-marrow dosimetry to changes in patient-specific parameters. Cancer Biol Radiopharm 17: 535–543

Stabin MG (2008) Fundamentals of nuclear medicine dosimetry. Springer, New York

Stabin MG, Watson EE, Cristy M, Ryman JC, Eckerman JF, Davis JL, Marshall D, Gehlen MK (1995) Mathematical models and specific absorbed fractions of photon energy in the nonpregnant adult female and at the end of each trimester of pregnancy. Report ORNL/TM-12907. Oak Ridge National Laboratory, Oak Ridge, TN

Stabin M, Yoriyaz H (2001) Photon specific absorbed fractions calculated in the trunk of an adult male voxel-based phantom. Health Phys 82:21–44

Thomas SR (2001) MIRD Pamphlet Number 18: administered cumulated activity for ventilation studies. J Nucl Med 42:520–526

Thomas SR, Stabin MG, Chen C-T, Samaratunga RC (1992) MIRD Pamphlet No 14 revised: a dynamic urinary bladder model for radiation dose calculations. J Nucl Med 33: 783–802

Thomas SR, Stabin MG, Chen C-T, Samaratunga RC (1999) MIRD Pamphlet No 14 revised: a dynamic urinary bladder model for radiation dose calculations. J Nucl Med 40:102S–123S

Thorne MC (1977) Aspects of the dosimetry of α-emitting radionuclides in bone with particular emphasis on ^{226}Ra and ^{239}Pu. Phys Med Biol 22:36–46

Venkataraman K, Somasundaram S, Soman SD (1963) An evaluation of radiation protection standards for Indian conditions. Health Phys 9:647–662

Watson EE, Stabin MG, Davis JL, Eckerman KF (1989) A model of the peritoneal cavity for use in internal dosimetry. J Nucl Med 30:2002–2011

Wessels BW, Konijnenberg MW, Dale RG, Breitz HB, Cremonesi M, Meredith RF, Green AJ, Bouchet LG, Bolch WE, Sgouros G, Thomas SR (2008) MIRD Pamplhet Number 20: the effect of model assumptions on kidney dosimetry and responses – implications for radionuclide therapy. J Nucl Med 49:1884–1899

Whitwell JR, Spiers FW (1976) Calculated beta-ray dose factors for trabecular bone. Phys Med Biol 21:16–38

Xu XG, Chao TC, Bozkurt A (2000) VIP-MAN: An image-based whole-body adult male model constructed from color

photographs of the visible human project for multi-particle Monte Carlo calculations. Health Phys 78:476–486
Yamaguchi H (1978) Estimation of internal radiation dose for various physiques using MIRD adult absorbed fractions. Acta Radiol Oncol 17:429–439
Young MEJ (1985) Distribution of radioactivity after intra-peritoneal administration of ^{32}P colloids. Br J Radiol 58: 100–101
Yoriyaz H, dos Santos A, Stabin M, Cabezas R (2000) Absorbed fractions in a voxel-based phantom calculated with the MCNP-4B code. Med Phys 27:1555–1562
Zaidi H, Xu XG (2007) Computational anthropomorphic models of the human anatomy: the path to realistic Monte Carlo modelling in radiological sciences. Annu Rev Biomed Eng 9:471–500
Zankl M, Veit R, Williams G, Schneider K, Petoussi N, Drexler G (1988) The construction of computer tomographic phantoms and their application in radiology and radiation protection. Radiat Environ Biophys 27:153–164
Zhang B, Ma J, Liu L, Cheng J (2007) CNMAN: a Chinese adult male voxel phantom constructed from color photographs of a visible anatomical data set. Radiat Prot Dosim 124:130–136
Zubal IG, Harrell CR, Smith EO, Rattner Z, Gindi G, Hoffer PB (1994) Computerized three-dimensional segmented human anatomy. Med Phys 21:299–302

The Biodistribution (I): Preclinical

13

Abstract The measurement of the biodistribution of a radiopharmaceutical must be performed in the preclinical setting in order to provide an initial estimate of the internal radiation dosimetry of the human prior to the first-into-man Phase I study. This chapter summarizes both the means of how the preclinical biodistribution data are derived and how these data are scaled from the animal to the human. Crucial to any preclinical research are the ethics of animal use. This is also reviewed.

Contents

13.1 Introduction

It is a regulatory requirement that the development of a radiopharmaceutical requires that a preclinical measure of the associated biodistribution be obtained prior to the agent's administration to a human subject in a Phase I clinical study. This preclinical result must then be used to predict the biodistribution in the human and, hence, the radiation dosimetry profile of the radiopharmaceutical in the human.[1] In this section, the methods of estimating the human data from the preclinical model and the associated legal, ethical and, regulatory requirements of animal studies are presented.

Whereas in vitro models can be used to assess some of the components of the safety profile of a radiopharmaceutical, they are clearly inadequate to allow adequate prediction of the biodistribution and radiation dosimetry of the radionuclide once the radiopharmaceutical is administered to the human. Preclinical models more advanced than in vitro measures are required as the use of higher-order species more closely approximates the human response. For the typical preclinical evaluation of the biodistribution of a radionuclide following the administration of a radiopharmaceutical, mammalian species are used. The model is typically a rat or mouse, although high-order species such as nonhuman primates can and have been used for biodistribution and radiation dosimetry

[1]Of course, preclinical studies of pharmacological safety are required in parallel.

B.J. McParland, *Nuclear Medicine Radiation Dosimetry*,
DOI 10.1007/978-1-84882-126-2_13, © Springer-Verlag London Limited 2011

studies using image-derived data. Indeed, in its report on the use of ionizing radiation in biomedical research, the Internal Commission on Radiological Protection (ICRP) advises that preclinical evaluation of new radiopharmaceuticals be performed on multiple species, one of which should be a nonhuman primate (ICRP 1991a, b).

The pharmacokinetics of the radiopharmaceutical will differ between the animal model and the human. This difference is a result of, among a multitude of others, the differences between species in body size (weight and surface area) and rates of metabolism. As a result, the biodistribution (i.e., the fraction of administered activity in a given organ or tissue as a function of time) will vary between the preclinical model and the human. Because of this manifestation of interspecies' differences, it can be advantageous that the biodistribution in the preclinical model be scaled to provide a more accurate prediction of that of the human. In some cases, it might be useful to measure the pharmacokinetics in different species with varying body masses in order to extrapolate to the expectation for the human. This scaling can be allometric, i.e., using a theoretical interspecies power-dependence of physiological or morphometric variables.

Finally, it is important to note that the discussion of the preclinical modeling of a given radiopharmaceutical is not restricted to the radiopharmaceutical alone, but must also include other radioactive entities such as excipients or radiolytic degradation products. In most cases, the magnitudes of the contributions of these entities to the radiation dosimetry profile can be neglected but still must be accounted for.

13.2 Ethical and Regulatory Requirements of Preclinical Research

The use of animals in medical research is recognized throughout the world to be an especially sensitive and emotive issue. It is fair to say that there are three parties with distinct views on the practice (Hepple et al. 2005). The first would be that of complete acceptance of animal studies and who would emphasize the benefits and contributions of such studies to the development and understanding of medicine. The second would be absolutist and unwilling to accept of any animal experimentation and of even the concept of transferring the results of animal studies to the human species. The third view could perhaps be considered to be intermediate, believing that sentient animals should not be subject to procedures that could potentially lead to suffering and pain, regardless of any new knowledge of biological processes that may be obtained.

The development of a new diagnostic radiopharmaceutical requires that estimated safety and radiation dosimetry profiles for the human species be established prior to the first administration into the human (EC 2003; FDA 2004). This clearly requires a need for a preclinical phase using nonhuman species in establishing these profiles. In any animal study, it is the ethical expectation that the investigator has thoroughly considered other options that would not require the use of animals (Festing et al. 2002). Indeed, within the European Union, this is a legal requirement in the use of vertebrates and other distinct species.[2] This leads to the requirement that the investigator intending to use animals in an experimental study consider what are known as the "3Rs" (Russell and Burch 1959). These are:

Replacement: Can the study be performed with a life form lower than that of a vertebrate? Can mathematical modeling of the physiological response to the administered agent be sufficiently accurate that it can be used as a substitute?

Refinement: If replacement of the animal model is not possible, then the study should be conducted in a manner so as to minimize the pain and suffering the animal is subjected to through, for example, the appropriate use of anesthesia or analgesia where required and the humane and painless killing of animals if and when needed.

Reduction: It is unquestionably necessary to minimize the number of animals used in a preclinical study whilst being consistent with achieving the desired outcome. In order to achieve this necessary goal, the investigator must understand both the objectives of the intended study and the sources of experimental variability and thus design the experiment accordingly.

In the United Kingdom, for example, the research use of animals is licensed through the Home Office.

[2]Such as *Octupus vulgaris* in the European Union.

The relevant current legislation is the Animals (Scientific Procedures) Act 1986 and may be found at: http://www.archive.official-documents.co.uk/document/hoc/321/321-xa.htm.

13.3 Means of Acquiring Preclinical Biodistribution Data

13.3.1 Introduction

Measurements of the biodistribution of activity following administration to an animal can be performed through two means, each with advantages and disadvantages. The first is that through postmortem dissection and the second is by in vivo imaging.

13.3.2 Dissection

In this approach, a cohort of animals are administered a radiopharmaceutical from which sub-cohorts are sacrificed at specified time points postadministration. In order to achieve sufficient statistical power in the measurement of the biodistribution, multiple animals are sacrificed at each time point. These sacrifices are followed by dissection, the harvesting of organs and tissues and subsequent assay of their radioactive content. As a result of this process and the need to obtain data as relevant as much as possible to the human, this approach is limited to mammalian species. Whilst being a nonlongitudinal process, it has the advantage of allowing clear excision of many organs and tissues for the assay of radioactive content. For dosimetry estimates, as many of the MIRD-specified source regions and excreta as possible should be collected and measured for this radioactive content. Additional organs or tissues which specifically have a high uptake of the radiopharmaceutical being studied should also be assayed if possible.

Consideration must be given as to how the activity content of an organ should be assayed. Depending upon the size of the organ, it may not be possible to either physically place the entire organ within the sampling volume of the activity counter. Further, the organ's size may be such that there is nonnegligible photon attenuation. In either case, an underestimate of the activity content is possible. One approach to resolving this problem is the excision of a sample of organ tissue and assaying its activity content. Assuming that the activity spatial distribution within the organ is uniform, the organ activity content can then be estimated by scaling the sample result by the ratio of the organ to sample masses. Another approach avoids the assumption of activity distribution uniformity by homogenizing the organ and repeating the above sampling procedure but by using a sample of the homogenate. However, unless there is further dissection and assaying of suborgan structures, these methods provide the activity content of the entire organ.

Some extended tissues and organs (e.g., muscle, skin, and bone) cannot be excised and assayed in toto. In such cases, a known sample mass of the tissue is assayed and the total activity content of the tissue is estimated on the basis of the measured activity per unit mass and an assumed whole mass of the given tissue based upon the total body mass of the animal (e.g., in the case of the rat, using the definitive data of Donaldson (1924)). For example, consider an excised tissue of mass m_{Sample} having been assayed to contain an activity A_{Sample}. One wishes to determine the total activity contained by the entirety of that tissue. Let $(m_{Tissue}/m_{WB})_{Ref}$ be the weight of the tissue in question relative to the whole-body mass for the reference animal. The total activity to be assigned to that tissue is then given by the scaling,

$$A_{Tissue} = \left(\frac{A_{Sample}}{m_{Sample}}\right)\left(\frac{m_{Tissue}}{m_{WB}}\right)_{Ref} m_{WB}. \qquad (13.1)$$

where m_{WB} is the whole-body mass of the actual animal. This approximation leads to a practical calculational problem as the assumptions of the relative weights of sampled tissues are unlikely to reflect the entireties of those in the individual animal in question. As a result, the estimate of total activity in the animal will not match the actual administered activity. This error, which is introduced by the above estimation, can be addressed by averaging this discrepancy over all sampled organs and tissues. For example, let $\sum_{i=1}^{N_S} A_i$

be the summation of measured activities over N_S organs and tissues, including the remnant carcass of the sacrificed animal and excreta, at time t postadministration and decay-corrected to the time of administration (i.e., premultiplied by $e^{+\lambda_P t}$), and let A_0 be the injected activity. If all of the activity has been completely accounted for, the ratio $\sum_{i=1}^{N_S} A_i/A_0$ should be equal to unity. In practice, however, this is unlikely to occur due to the assumptions of the fractions of total body mass represented by the sampled tissues. To force a normalization, the activity of each source region should then be divided by the ratio,$\sum_{i=1}^{N_S} A_i/A_0$ to ensure that the complete summation of decay-corrected and normalized activities is equal to unity.

13.3.3 Preclinical Imaging

Large animals, such as the pig (e.g., Jakobsen et al. 2006), dogs (e.g., Nimmagadda et al. 2005), and nonhuman primates (e.g., Acton et al. 2001), can be imaged on a conventional gamma camera or PET scanner, thus allowing longitudinal studies of the biodistribution (much as is done in a human trial) and not requiring sacrifice at each data collection time point. In recent years, with the development of purpose-built preclinical imaging devices such as micro-PET (e.g., Wang et al. 2006) and micro-SPECT scanners, some even with associated micro-CT scanners, these longitudinal studies have been extended to smaller species. This offers the potential advantage of using a smaller animal cohort size, in accordance with the ethical "reduction" requirement cited above. However, the in vivo measurement of activity in an animal is likely not to be as accurate as the ex vivo measurement of dissected tissues for a variety of reasons. These include partial volume effects, scatter and background corrections and the inability to completely isolate the activities amongst an ensemble of tissues and organs. But, at the same time, nonuniform distribution of activity within an organ has the potential to be more readily detected and measured in an imaging-based study than in one based upon dissection. Because preclinical imaging studies are inherently dealing with smaller spatial resolutions, stringent technical demands are placed upon the requirements of preclinical imaging devices.

13.3.4 Data Acquisition Times

The number and temporal distribution of times postadministration required to accurately assess a biodistribution in the preclinical model is dependent upon a variety of factors. There are two outcomes desired from a biodistribution study (preclinical or clinical). The first is an understanding of the biokinetics of the radionuclide and the second is the accurate measure of the cumulated activities in the source regions. Optimization of the number, and temporal distribution, of the measurement points is required in order to achieve these outcomes. In addition, if the dissection model is used to derive the radiopharmaceutical biodistribution, one must include within this optimization the need to minimize the number of animals sacrificed. The assumptions that the determination of the number and temporal distribution of times postadministration at which the biodistribution is measured in the preclinical model is essentially no different than that of the clinical model must be tempered. As to be shown, the difference in metabolic rates between the preclinical model and the human must be accounted for.

The time-dependence of the activity contained within an organ or tissue is frequently well-modeled by a least-square fit by a first- or second-order exponential function in time. Thus, a minimum of four biodistribution measurements will usually be required. Improvements to the fit are, of course, achievable with greater time points, although these could markedly increase the complexity of the analysis.

Having established the likely number of time points required, it is necessary to then assign their distribution over time. As the distribution of activity is most rapid following administration of the radiopharmaceutical, an improved fit is had by populating preferentially a greater number of measurement times early on, with a low density of measurement times following. The first measurement time point should be between 2 and 5 min postadministration in order to determine the distribution and uptake of the radionuclide. The time of the last biodistribution measurement is dictated primarily by the physical half-life of the radionuclide. A reasonable "rule-of-thumb" for short-lived radionuclides is that the time postadministration of the last biodistribution measurement is about four times the physical half-life, e.g., 24 h for ^{99m}Tc ($T_{1/2,Phys} = 6.02\,h$) or 8 h for ^{18}F ($T_{1/2,Phys} = 109.7\,min$). At these times, the biokinetics of the radionuclide will have largely stabilized. This type of selection of the last measurement time is not necessarily appropriate for longer-lived isotopes such as ^{111}In ($T_{1/2,Phys} = 2.83$ days) or ^{131}I ($T_{1/2,Phys} =$ 8.04 days) where the last measurement can be made at a shorter time postadministration (when the biodistribution has become largely static) and it can be assumed that the decrease in activity in a source region is due exclusively to physical decay with no contribution from biological washout. Siegel et al. (1999) and the Report 67 of the International Commission on Radiation Units and Measurements (ICRU 2002) have specified an initial distribution of sampling times on the basis of the whole-body effective half-life from Chap. 5. They recommended an ensemble of image acquisition starting times of 1/3, 2/3, 3/2, 3, and 5 multiples of the effective half life. Clearly in the preclinical setting, a prediction of $T_{1/2,Eff}$ is unavailable or at least difficult to estimate.

13.3.5 Sample Sizes

The reduction in the number of animals used in the preclinical evaluation of any entity, including radiopharmaceuticals, must always be sought for ethical and financial cost reasons. However, as sex-based differences in the biodistribution must be ascertained during the preclinical phase prior to entry into man, males, and females should be used in equal numbers. The sample size will be driven primarily by the number of time points at which the biodistribution is measured which, in turn, is defined by the biokinetics of the radiopharmaceutical.

13.4 Animal Phantoms

13.4.1 Introduction

Preclinical mathematical models (phantoms) can be used to estimate the absorbed doses in the animal model and which can be used to interpret the radiobiological responses of tissues to radionuclide therapy and to enable prediction of those that will occur in the human. These responses would include tumor regression and the therapy-limiting radiotoxicities of healthy tissues, such as kidney and bone marrow. However, the determination of the absorbed doses to tissues in species such as the rat or mouse due to administered radiopharmaceuticals is difficult. In vivo measurements using thermoluminescent dosimeters or MOSFETs are possible, although challenging. Discussions of these empirical measurements are deferred to the discussion of in vivo measurements in the human.

However, it is possible to apply the MIRD method to estimate the absorbed doses to organs, tissues, and tumors in species other than the human. Clearly, this requires the evaluation of species-specific S-factors. It would appear, to date, that nonhuman species phantoms and S-factor evaluations have been restricted to the mouse and rat.

13.4.2 Examples of Animal Phantoms

Athymic nude mice have been used to develop radioimmunotherapy to develop understanding of, in particular, radiotoxicity of bone marrow. As a result, this preclinical model was perhaps the first for which a phantom was developed. Hui et al. (1994) presented the first murine phantom and which was exclusively used for determining the internal absorbed doses due to antibodies labeled with the β^--emitting ^{90}Y. Ten athymic nude mice of about 25 g mass each were dissected and the spatial dimensions and masses of thirteen major organs measured. With the exception of the bone and marrow, the organs were modeled by ellipsoids; the skeleton was modeled by the femur. The bone and marrow were represented by concentric cylinders and a spherical tumor was included. The self-doses to organs were calculated using the dose point kernel method (Chap. 9) with the exception of lung, bone, and marrow which were calculated using Monte Carlo simulation due to their differing physical densities. Because the mean range of the ^{90}Y β particle (2.8 mm) is comparable to the murine organ dimensions, the absorbed fraction of a target region adjacent to a source region was taken to be proportional to the ratio of the overlapping area between the two regions to the total surface area of the source region. The mouse bone marrow model was extended by Muthuswamy et al. (1998) and applied to ^{131}I, ^{186}Re, and ^{90}Y. However, as the entire marrow of the mouse is reported to be active (Tavassoli and Yoffey 1983), a more complete accounting of the absorbed dose to marrow is necessary. Again, a cylindrical geometry was assumed for the bones of the upper and lower extremities. The flat bones (rib, clavicle, sternum, and pelvis) were modeled by slabs whereas the vertebrae and skull were represented by spheres. Absorbed doses were calculated using the kernel method.

An ellipsoid organ model of the Wistar rat[3] was developed by Konijnenberg et al. (2004) for the same purpose of preclinical radioimmunotherapy development. Again, measurements of organs harvested from dissected animals were used to define the dimensions of the ellipsoid models. Monte Carlo calculations were used to estimate the organ absorbed doses due to both β- and γ-emitting radionuclides. In this paper can be found Wistar rat S-values for ^{90}Y, ^{111}In, and ^{177}Lu.

More sophisticated and realistic preclinical models based upon high-spatial resolution MR and CT studies of the animal have been designed. For example, Johnson et al. (2002) have demonstrated that magnetic resonance microscopy (MRM) can enable visualization of the mouse at isotropic resolutions ranging from 25 to 110 μm. Segars et al (2004) have also used MRM of the mouse (at 110 μm resolution) to develop a high-spatial resolution murine phantom which also included respiratory and cardiac motions that had been derived from gated MR imaging of the mouse. Instead of modeling the organs as ellipsoids or other predefined geometrical shapes, they modeled these with nonuniform splines (recall the discussion of hybrid anthropomorphic phantoms of Chap. 12). Another approach, which is perhaps more immediately amenable to dosimetry calculations, was taken by Stabin et al. (2006) who described their voxellated murine models of a transgenic mouse and Sprague-Dawley rat. Images of the animals were acquired using a dedicated small-animal CT scanner with an isotropic voxel size of 200 μm for the mouse and an anisotropic voxel size of 300 × 300 × 500 μm for the rat. The kidneys, liver, lungs, spleen, heart, stomach, intestines, skeleton, urinary bladder, and testes were segmented from the images. Absorbed fractions for the segmented organs electrons with energies between 0.1 and 4.0 MeV and photons with energies between 0.01 and 4 MeV were derived for both models using Monte Carlo simulation. Importantly, they also provided comparisons of their measured mouse and rat organ masses with those of previous investigators. Another voxel-based approach was taken by Hindorf et al. (2004), but was instead based upon a mathematical model of the mouse in which organs were represented by geometrical shapes. MIRD S-factors for ^{90}Y, ^{99m}Tc, ^{111}In and ^{131}I were calculated in that work. The focus of that research was to determine the sensitivities of murine organ S-values to various parameters. In particular, they demonstrated that organ mass is the most significant parameter for self-absorbed S-factors. This result is an important consideration, as discussed below, when scaling from the preclinical model to the human: the weights of some organs, such as the thymus and testes, are of a higher percentage of total body weight in the mouse or rat than in the human. Finally, Kolbert et al. (2003) derived the MIRD S-factors and absorbed fractions for the athymic mouse based upon MR-derived segmented organs. Their approach was based upon the convolution of the image data and previously calculated dose point kernels. They provide photon and electron results for ^{32}P, ^{90}Y, ^{131}I, ^{153}Sm, and ^{188}Re.

13.5 Allometric Scaling of Animal Biodistribution Data to the Human

13.5.1 Introduction

Once having extracted the biodistribution of activity in the preclinical model, it is necessary to map these data over to the human in order to estimate the human biodistribution from which the human radiation dosimetry is evaluated. The extrapolation of preclinical animal data to the human is a complex field and one heavily researched by the pharmacological community (see, e.g., Riviere 1999). Clearly, one should not assume that the measured biodistribution in the preclinical animal model will exactly reflect that in the human, although it will be shown that such a simplistic assumption can provide a possible advantage in the planning of a Phase I study of a diagnostic radiopharmaceutical. The scaling of the biodistribution data from the preclinical model to the human can range markedly in terms of complexity. Methods have been

[3] A strain of albino rat which is one of the most frequent used in biomedical research.

investigated in the past for radiopharmaceuticals (e.g., Lathrop et al. 1989; Yorke et al. 1991; Onthank 2005).

Interspecies scaling of radionuclide biodistributions will be derived here from allometric scaling principles. Allometric scaling between species can be defined by the power-law (West et al. 1997),

$$P = km^b \tag{13.2}$$

where P is the physiological parameter to scale, m is the body or relevant organ mass and b is the allometric exponent. The constant of proportionality k is the allometric coefficient for the parameter. Linearizing this equation,

$$\ln P = \kappa + b \ln m \tag{13.3}$$

where $\kappa = \ln k$, provides a result which will be examined here for a number of parameter selections.

13.5.2 b = 0

By setting $b = 0$ in (13.2), any mass dependencies are neglected. In other words, all physiological parameters are the same in both the animal and the human. This is the simplest approach in which it is assumed that the fraction of the administered activity in corresponding organs and tissues is the same at all time points in the human and the preclinical animal model. However, this method, whilst easy to implement, neglects significant interspecies differences:

- The anatomies of the preclinical model and the human will not always match exactly. For example, the rat does not have a gall bladder which could complicate hepatobiliary mapping between the rat and the human.
- The mass of a given organ as a fraction of the whole-body mass typically differs between the preclinical model and the human. For example, Table 13.1 shows that, as a percentage of whole-body mass, the human brain is about twice as large as the rat's. Hence, excluding any brain uptake specificity, it is entirely reasonable to assume that the uptake of activity (as a percentage of that administered) will be twice as much in the human brain as in the rat's. More importantly, in terms of estimated radiation risk, is that Table 13.1 shows that the weight of the rat testes is about 23 times that of the ICRP Reference Man as a percentage of body mass. Assuming, again, nonspecificity, the assumption that the testicular uptake as a fraction of administered activity in both species is the same will lead to an overestimate of the actual uptake in the human testes by a factor of 23, with a subsequent overestimate of the human testicular absorbed dose. As the effective dose tissue weighting factor for the gonads is high, this will also result in an overestimate in the value of the effective dose and associated stochastic risk.
- The metabolism rates of the human and those species commonly used in biodistribution studies will typically differ.

While these differences are severe, there can be an advantage in neglecting allometric scaling in using preclinical biodistribution data in the planning of a Phase I study of a radiopharmaceutical in healthy volunteers. The goal of the optimization of the Phase I study planning is to minimize the organ absorbed doses and the effective dose. As noted above, the difference between the rat and human gonadal masses will, if it is assumed that the fraction of administered activity taken up by both is the same, result in an overestimation of the effective dose per unit administered activity to the human volunteer. This could be desirable as an effective dose constraint is imposed and, as a result, the administered activity is forced to be lower. As one cannot be entirely sure that the expected human biodistribution is valid, this could be considered by some as a conservative approach ensuring a minimized (perhaps excessively) effective dose to the healthy volunteer. However, as shown in Table 10.5 in Chap. 10, the tissue weighting factors for the gonads have been reduced from the value of 0.20 given in ICRP Publication 60 (1991a) to 0.08 in ICRP Publication 103 (2007). Using this reduced factor for the gonads largely will reduce the overestimation of the effective dose based upon the assumption of equivalent biodistributions.

13.5.3 $k = 1/m_{WB}$ and $b = 1$

In this simplification, it is assumed that the physiological parameter scales with the organ mass as a fraction

Table 13.1 Weights of organs and tissues for the reference human and a selected Wistar Norwegian Albino rat male/female pair (as percentages of whole-body weights) and the calculated activity scaling quantity, f_{Organ} (MIRD-specified source regions are in bold)

Organ		Reference human[a]		Rat[b]		f_{Organ}	
		Male[c] (%)	Female[d] (%)	Male[e] (%)	Female[f] (%)	Male	Female
Adrenal glands		0.02	0.02	0.02	0.03	1.13	0.81
Brain		1.99	2.17	0.93	1.06	2.14	2.05
Breasts		0.03	0.83				
Eyeball		0.02	0.03	0.13	0.15	0.15	0.17
Gallbladder contents		0.08	0.08	[g]			
Gallbladder wall		0.01	0.01				
GI tract contents	**Large intestine (right colon)**	0.21	0.27				
	Large intestine (left colon)	0.10	0.13				
	Large intestine (rectosigmoid colon)	0.10	0.13				
	Small intestine	0.48	0.47				
	Stomach	0.34	0.38				
GI tract walls	Small intestine wall	0.89	1.00				
	Stomach wall	0.21	0.23	0.57	0.61	0.36	0.38
	Large intestine (right colon)	0.21	0.24				
	Large intestine (left colon)	0.21	0.24				
	Large intestine (rectosigmoid colon)	0.10	0.12				
	Stomach, small and large intestines combined	0.71	0.83	4.56	4.77	0.16	0.17
Heart wall		0.45	0.42	0.40	0.41	1.13	1.00
Heart wall and contents		1.15	1.03				
Kidneys		0.42	0.46	0.85	0.87	0.50	0.53
Liver		2.47	2.33	5.01	5.22	0.49	0.45
Lungs (without blood)		0.68	0.70	0.59	0.60	1.16	1.16
Marrow	Total	5.00	4.50	3.02[h]		1.66	1.49
	Active (red)	1.60	1.50	0.97[i]	1.01	1.66	1.49
	Inactive (yellow)	3.40	3.00	2.05	2.01	1.66	1.49
Muscle (skeletal)		39.73	29.17	42.7[j]	43.3[k]	0.93	0.68
Esophagus		0.05	0.06				
Ovaries		0.02		0.03		0.65	
Pancreas		0.19	0.20	0.47	0.54	0.41	0.37
Pituitary gland		0.001	0.001	0.004	0.007	0.22	0.15
Prostate		0.02		0.08		0.29	
Salivary glands (submaxillary)		0.03	0.04	0.19	0.20	0.18	0.18
Skeleton		14.3	13.00	10.4[j]	13.2[k]	1.38	0.98
Skin (total)		4.52	3.83	19.2[j]	19.5[k]	0.24	0.20
Spleen		0.21	0.22	0.27	0.27	0.77	0.80
Testes and epididymus		0.0		1.41		0.038	
Testes only		0.05		1.09		0.044	
Thymus gland		0.03	0.03	0.12[l]	0.13[m]	0.28	0.25
Thyroid gland		0.03	0.03	0.02	0.02	1.70	1.70
Urinary bladder wall		0.07	0.07				
Uterine wall			0.23		0.23[n]		0.57

[a]Reference Man data from ICRP Publication 89 (2002)
[b]Rat anatomical data (with the exception of marrow) derived from data presented by Donaldson (1924)
[c]Reference man whole-body weight 73,000 g
[d]Reference woman whole-body weight 60,000 g
[e]Sample male rat weight of 200.8 g (except where noted)
[f]Sample female rat weight of 169.6 g (except where noted)
[g]Rats are absent a gallbladder
[h]Rat marrow percentage weight (whole) is from Fairman and Corner (1934) and is averaged over sexes
[i]Differentiation of active and inactive marrow over male and female rats is estimated using measured whole marrow and assuming the ratio of active to inactive marrow is the same in the rat as in the human
[j]Measured from a 218.7 g male rat
[k]Measured from a 183.5 g female rat
[l]Measured from a 194.1 g male rat
[m]Measured from a 175.7 g female rat
[n]Measured from a 185 g female rat

of the total body mass. In the context of activity uptake, Kirschner et al. (1975) proposed this method in terms of relative perfusion. This gives the normalized activity in a human organ as a fraction of that administered, $(A_{Organ}/A_0)_{Human}$, in terms of that measured in the corresponding animal organ, $(A_{Organ}/A_0)_{Animal}$. This scaling approach begins by assuming that there is a completely uniform distribution of activity throughout both the human and animal. This must be recognized to be a significant simplification, as will be addressed. The activity contained in the human organ is given by the product of the activity concentration (per unit mass) and the mass of the organ,

$$A_{Organ,Human} = [A]_{WB,Human}\, m_{Organ,Human} \tag{13.4}$$

where $[A]_{WB,Human}$ is the concentration given by the ratio of the total activity in the whole-body to the whole-body mass,

$$[A]_{WB,Human} = \frac{A_0}{m_{WB,Human}}. \tag{13.5}$$

Similarly, for the animal, the whole-body concentration is,

$$[A]_{WB,Animal} = \frac{A_0}{m_{WB,Animal}} \tag{13.6}$$

Solving for A_0 in (13.5), substituting the result into (13.6), rearranging and normalizing both sides to A_0 gives the relationship between the human and animal normalized activities in terms of the ratios of the organ mass to the whole-body mass,

$$\begin{aligned} A_{Norm,Organ,Human} &= A_{Norm,Organ,Animal}\left(\frac{m_{Organ,Human}/m_{WB,Human}}{m_{Organ,Animal}/m_{WB,Animal}}\right) \\ &= A_{Norm,Organ,Animal}\, f_{Organ} \end{aligned} \tag{13.7}$$

It then follows that the normalized cumulated activity scales linearly between species,

$$\tilde{A}_{Norm,Organ,Human} = \tilde{A}_{Norm,Organ,Animal}\, f_{Organ}. \tag{13.8}$$

Examples of the calculated organ relative mass scaling factors (f_{Organ}) from the Wistar rat to the ICRP Reference Adults are provided in Table 13.1. In deriving the data in this table, anatomical data for the male rat with a body weight as close as possible to 200 g and the female rat with a body weight as close as possible to 170 g were selected. Some data for the Reference Adults, which do not appear for the rat, are included for interest alone. A review of the entries in Table 13.1 will highlight a number of organs and tissues for which mass-scaling of the rat organ weight to that of the human is very obviously necessary. These include the brain, kidney, red bone marrow, ovaries, skin, testes, and thymus gland.

13.5.4 Metabolic Rate Scaling

Species of larger whole-body masses are known in general to have slower metabolic rates than those of lighter masses. Equation (13.1) can then be modified to account for this observation,

$$\begin{aligned} \lambda_{Human} &= \lambda_{Animal}\left(\frac{m_{WB,Animal}}{m_{WB,Human}}\right)^b \\ &= \lambda_{Animal}\, f_\lambda \end{aligned} \tag{13.9}$$

where λ_{Human} is the physiological rate in the human, etc. A simple consideration of surface area or volume would suggest that the allometric exponent of (13.9) should be an integral multiple of 1/3 or, b = 1/3, 2/3, or 1. However, empirical data suggest that b is, in fact, found in nature to be multiples of 1/4. For example, the rate of cell metabolism is reported to scale with $m^{-1/4}$ (West et al. 1997). Blood flow in the rat is about 15 s and about 50 s in the human (Lin 1998), a difference more closely reflected with b equal to 1/4 than 1/3. Further, there is no reason to necessarily restrict the mass ratio of (13.9) to that of whole-body mass (this is discussed in terms of permeability and blood flow below). This could be changed to those of the organs of specific interest, although the whole-body mass offers calculational simplicity. Using b = 1/4 and assuming 200 g and 73 kg for the whole-body masses of the rat and the human, then $f_\lambda = 0.23$. Hence, uptake and washout of the radiopharmaceutical is expected to be slower in the human than in the preclinical animal model and this difference must be accounted for when extrapolating the normalized cumulated activities across species.

While the fractional mass scaling method yielded a normalized cumulated activity that scaled linearly with f_{Organ}, this cannot be the case with the metabolic rate scaling method as only the physiological temporal variable is being altered. To demonstrate this, allow an *N*th-order multiexponential model of the decay-corrected activity normalized to that administered to the animal,

$$A_{Norm,Corr,Animal}(t) = C + \sum_{i=1}^{N} k_i e^{-\lambda_{i,Animal} t}. \quad (13.10)$$

As no interspecies changes in activity are accounted for, C and k_i do not vary. The decay-corrected normalized activity in the corresponding organ of the human is, from (13.9) and (13.10) is,

$$A_{Norm,Corr,Human}(t) = C + \sum_{i=1}^{N} k_i e^{-f_\lambda \lambda_{i,Animal} t}. \quad (13.11)$$

The normalized cumulated activity in the human is, thus,

$$\tilde{A}_{Norm,Organ,Human} = \left(\frac{C}{\lambda_P}\right) + \sum_{i=1}^{N} \left(\frac{k_i}{\lambda_P + f_\lambda \lambda_{i,Animal}}\right) \quad (13.12)$$

13.5.5 Combined Organ Mass and Metabolic Rate Scalings

The scaling of (13.8) and (13.12) can be combined (assuming an Nth-order multiexponential describing the activity) as,

$$\tilde{A}_{Norm,Organ,Human} = f_{Organ}\left(\left(\frac{C}{\lambda_P}\right) + \sum_{i=1}^{N} \left(\frac{k_i}{\lambda_P + f_\lambda \lambda_{i,Animal}}\right)\right) \quad (13.13)$$

13.5.6 Discussion

13.5.6.1 Permeability or Blood Flow Transfer

The above discussion was limited to two simple methods of allometrically scaling the normalized cumulated activity from the preclinical animal model to the human. These, as would be expected, are not the only approaches. As the parenterally-administered radiopharmaceutical transfers from the blood space into the various organ systems, classical biokinetic models can be used to guide the interspecies scaling. However, the transfer coefficients can be difficult to estimate a priori in order to extrapolate from the preclinical animal model to the human.

It can be reasonable to accept that in tissues which are not highly vascularized the blood-tissue exchange is dominated by permeability. The permeability surface area product (PSAP) is the product of the membrane permeability and the membrane's surface area and is in units of volume per unit time. Hence, it is linked to the rate constants of the multiexponential models proposed earlier. For permeability-dominant tissues, it can be assumed that the PSAP will scale between species by the surface area of the organ,

$$PSAP_{Organ,Human} = PSAP_{Organ,Animal}\left(\frac{SA_{Organ,Human}}{SA_{Organ,Animal}}\right) \quad (13.14)$$

If it is assumed that the surface area of the organ varies as the 2/3 power of the organ mass, (13.14) can be modified to,

$$PSAP_{Organ,Human} = PSAP_{Organ,Animal}\left(\frac{m_{Organ,Human}}{m_{Organ,Animal}}\right)^{2/3} \quad (13.15)$$

The 2/3 power reflects, generally, the case of where permeability dominates over blood flow. For tissues which are highly vascularized, such as the lung, spleen, and liver, and where blood flow would dominate, then the exponent could be reasonably set equal to unity.

13.5.6.2 Sum of Interspecies-Scaled Normalized Cumulated Activities

It must be recognized that (13.8) and (13.12) are approximations. In order to apply these in practical dosimetry calculations, corrections are required to compensate for the approximations inherent to the scaling models. These are considered in turn.

Relative Organ Mass Scaling

Recall that this method is based upon the assumption of uniform concentration of activity. As noted above, for the case of complete nonspecificity, if the human brain mass as a fraction of body mass were twice that of a rat's, then twice as much of the fraction of administered activity would be taken up by the human brain. At the other extreme, as exemplified by ^{99m}Tc-labeled macroaggregated albumin used in lung perfusion studies, virtually all of the administered activity will be eventually trapped in the lung capillaries regardless of the species. In such cases, the relative organ mass scaling method will be invalid. In addition, the requirement that the sum of all normalized cumulated activities be equal to the reciprocal of the physical decay constant (11.29) in both species leads to a dilemma as a result of the constant activity concentration assumption. That is, from (11.29),

$$\sum_{r_S} \tilde{A}_{r_S,\mathrm{Norm,Human}} = \sum_{r_S} \tilde{A}_{r_S,\mathrm{Norm,Animal}} = \frac{1}{\lambda_P}. \quad (13.16)$$

However, this in combination with (13.8) leads to the requirement,

$$\sum_{r_S} \tilde{A}_{r_S,\mathrm{Norm,Animal}} = \sum_{r_S} f_{r_S} \tilde{A}_{r_S,\mathrm{Norm,Animal}} \quad (13.17)$$

which is a condition unlikely to be met. It is possible for (13.16) and (13.17) to be simultaneously forced by splitting up the ensemble of source regions r_S into $N_S - 1$ explicitly-specified source regions (tissues or organs) and by defining the N_S^{th} source region be the remaining tissues category (i.e., the ensemble of tissues and organs, including the remnant carcass, that were not specified or measured for activity content individually). The scaling factors for the $N_S - 1$ explicitly-specified organs are calculated for from (13.8). The scaling factor for the N_S^{th} (remaining tissues) category is then solved for from the requirement that,

$$\sum_{r_S=1}^{N_S} f_{r_S} \tilde{A}_{r_S,\mathrm{Norm,Animal}} = f_{N_S} \tilde{A}_{N_S,\mathrm{Norm,Animal}} + \sum_{r_S=1}^{N_S-1} f_{r_S} \tilde{A}_{r_S,\mathrm{Norm,Animal}} = \frac{1}{\lambda_P}. \quad (13.18)$$

Solving for the weighting factor to apply to the remaining tissues category, f_{N_S},

$$f_{N_S} = \frac{1/\lambda_P - \sum_{r_S=1}^{N_S-1} f_{r_S} \tilde{A}_{r_S,\mathrm{Norm,Animal}}}{\tilde{A}_{N_S,\mathrm{Norm,Animal}}}. \quad (13.19)$$

Of course, this is an artificial result forced in order to achieve an expectation. However, allometric scaling from the animal to the human was always an approximation based upon equivalent and uniform activity concentrations. Whether or not such scaling is necessary or should be used and, if applied, the extent of its validity must be assessed for each application.

Metabolic Rate Scaling

This scaling will also suffer from the inability to satisfy the requirement of (11.29),

$$\sum_{r_S=1}^{N_S} \left(\left(\frac{C_{r_S}}{\lambda_P} \right) + \sum_{i=1}^{N_S} \left(\frac{k_{ir_S}}{\lambda_P + f_\lambda \lambda_{ir_S,\mathrm{Animal}}} \right) \right) = \frac{1}{\lambda_P} \quad (13.20)$$

This can be corrected, as before but with the additional assumption that the temporal-dependence of all source region activities can be adequately modeled by an Nth-order multiexponential function, by isolating the normalized cumulated activity of the remaining tissues category (N_S) and setting it equal to,

$$\tilde{A}_{N_S,Norm} = \frac{1}{\lambda_P} - \sum_{r_S=1}^{N_S-1}\left(\left(\frac{C_{r_S}}{\lambda_P}\right) + \sum_{i=1}^{N_S}\left(\frac{k_{ir_S}}{\lambda_P + f_\lambda \lambda_{ir_S,Animal}}\right)\right) \quad (13.21)$$

Combined Relative Organ Mass and Metabolic Rate Scaling

The above discussion regarding the metabolic rate scaling method is relevant here. The normalized cumulated activity of the remaining tissues category scaled to the human is given by,

$$\tilde{A}_{N_S,Norm} = \frac{1}{\lambda_P} - \sum_{r_S=1}^{N_S-1} f_{r_S}\left(\left(\frac{C_{r_S}}{\lambda_P}\right) + \sum_{i=1}^{N}\left(\frac{k_{ir_S}}{\lambda_P + f_\lambda \lambda_{ir_S,Animal}}\right)\right) \quad (13.22)$$

13.5.6.3 Validation

The applicabilities of these various allometric scaling methods have been assessed. In a meta-analysis of data presented 42 publications, Sparks and Aydogan (1999) compared the estimates of the human normalized cumulated activities for 115 organs estimated from preclinical studies of 11 different radionuclides and 33 radiopharmaceuticals using the methods of assuming that the organ's fractional uptake is the same in either animal or human, the relative organ mass scaling method, the metabolic rate scaling method and the combination of the last two methods. The types of animal species used in the various publications were not provided and the metric used was the ratio of the animal to human organ normalized cumulated activity. The authors determined that the ratio for no applied corrections tended to follow a log-normal distribution with a geometric mean of less than 1, with unity reflecting an exact scaling from the animal to the human. That the geometric mean was less than unity was not perhaps surprising owing to the expected rapid clearance in the animal. Of the combinations, the greatest agreement between animal and human normalized cumulated activities was observed when both organ-mass and measurement-time scaling were applied (geometric mean of 0.79 with a standard deviation of 3.5). The authors recognized that incomplete data were available from the literature sources and concluded that they were unable to identify a definite result as to which method was superior.

The above reflects the need for a more tightly-controlled comparison using a single or limited animal species and radionuclides, performed either prospectively or retrospectively. The endpoint would arguably be the dosimetry profile rather than the normalized cumulated activity.

References

Acton PD, Choi S-R, Hou C, Plössl K, Kung HF (2001) Quantification of serotonin transporters in nonhuman primates using [^{123}I]ADAM and SPECT. J Nucl Med 42:1556–1562

Donaldson HH (1924) The rat, 2nd edn, Philadephia

EC (2003) Directive 2003/83 of the European Parliament and of the Council 06 November 2001 on the community code relating to medicinal products for human use. http://www.emea.europa.eu/pdfs/human/pmf/2001-83-EC.pdf

FDA (2004) Guidance for industry developing medical imaging drug and biological products. Part 1: conducting safety assessments. U.S. Food and Drug Administration, Washington, DC

Fairman E, Corner GW (1934) The bone marrow volume of the albino rat. Anatom Record 60:1–4

Festing MFW, Overend P, Das PG, Borja MC, Berdoy M (2002) The design of animal experiments – reducing the use of animals in research through better experimental design. The Royal Society of Medicine Press, London

Hepple B et al (2005) The ethics of research involving animals. Nuffield Council on Bioethics, London

Hindorf C, Ljungberg M, Strand S-E (2004) Evaluation of parameters influencing S values in mouse dosimetry. J Nucl Med 45:1960–1965

Hui TE, Fisher DR, Kuhn JA, Williams LE, Nourigat C, Badger CC, Beatty BG, Beatty JD (1994) A mouse model for calculating cross-organ beta doses from yttrium-90 labeled immunoconjugates. Cancer 73(suppl):951–957

ICRP (1991a) Recommendations of the International Commission on Radiological Protection. ICRP Publication 60. Pergamon, Oxford

ICRP (1991b) Radiological protection in biomedical research – ICRP Publication 62. International Commission on Radiological Protection. Pergamon, Oxford

ICRP (2002) Basic anatomical and physiological data for use in radiological protection: Reference values – ICRP Publication 89. International Commission on Radiological Protection. Pergamon, Oxford

ICRP (2007) The 2007 recommendations of the International Commission on Radiological Protection. ICRP Publication 103. Pergamon, Oxford

ICRU (2002) Absorbed dose specification in nuclear medicine – report no 67. International Commission on Radiation Units and Measurements. Nuclear Technology, Ashford

Jakobsen S, Pedersen K, Smith DF, Jensen SB, Munk OL, Cumming P (2006) Detection of α_2-andrenergic receptors in brain of living pig with ^{11}C-yohimbine. J Nucl Med 47:2008–2015

Johnson GA, Cofer GP, Gewalt SL, Hedlund LW (2002) Morphologic phenotyping with MR microscopy: the visible mouse. Radiology 222:789–793

Kirschner A, Ice R, Beierwaltes W (1975) Radiation dosimetry of ^{131}I-19-iodocholesterol: the pitfalls of using tissue concentration data, the author's reply. J Nucl Med 16:248–249

Kolbert KS, Watson R, Matei C, Xu S, Koutcher JA, Sgouros G (2003) Murine S factors for liver, spleen and kidney. J Nucl Med 44:784–791

Konijnenberg MW, Bijster M, Krenning EP, de Jong M (2004) A stylized computational model of the rat for organ dosimetry in support of preclinical evaluations of peptide receptor radionuclide therapy with ^{90}Y, ^{111}In or ^{177}Lu. J Nucl Med 45:1260–1269

Lathrop KA, Tsui BM, Chen CT, Harper PV (1989) Multiparameter extrapolation of biodistribution data between species. Health Phys 57:121–126

Lin JH (1998) Applications and limitations of interspecies scaling and in vitro extrapolation in pharmacokinetics. Drug Metab Dispos 26:1202–1212

Muthuswamy MS, Roberson PL, Buchsbaum DJ (1998) A mouse bone marrow dosimetry model. J Nucl Med 39:1243–1247

Nimmagadda S, Mangner TJ, Sun H, Klecker RW, Muzik O, Lawhorn-Crews JM, Douglas KA, Collins JM, Shields AF (2005) Biodistribution and radiation dosimetry estimates of 1-(2′-deoxy-2′-^{18}F-fluoro-1-β-D-arabinofuranosyl)-5-bromouracil: PET imaging studies in dogs. J Nucl Med 46: 1916–1922

Onthank DC (2005) Prediction of "first dose in human" for radiopharmaceuticals/imaging agents based on allometric scaling of pharmacokinetics in pre-clinical animal models, Thesis. Worcester Polytechnic Institute, Worcester

Riviere JE (1999) Comparative pharmacokinetics: principles, techniques and applications. Iowa State Press, Ames

Russell WMS, Burch RL (1959) The principles of humane experimental technique. Universities Federation for Animal Welfare, Wheathampstead

Segars WP, Tsui BMW, Frey EC, Johnson GA, Perr SS (2004) Development of a 4-D digital mouse phantom for molecular imaging research. Mol Imaging Res 6:149–159

Siegel J, Stubbs TS, Stabin M, Hays M, Koral K, Robertson J, Howell R, Wessels B, Fisher D, Weber D, Brill A (1999) MIRD Pamphlet No. 16: techniques for quantitative radiopharmaceutical biodistribution data acquisition and analysis for use in human radiation dose estimates. J Nucl Med 40:37S–61S

Sparks RB, Aydogan B (1999) Comparison of the effectiveness of some common animal data scaling techniques in estimating human radiation dose. In: Stelson A, Stabin M, Sparks R eds. Sixth international radiopharmaceutical dosimetry symposium. Oak Ridage Associated Universities, Oak Ridge, TN, pp 705–716

Stabin MG, Peterson TE, Holbrun GE, Emmons MA (2006) Voxel-based mouse and rat models for internal dose calculations. J Nucl Med 47:655–659

Tavassoli M, Yoffey JM (1983) Bone marrow: structure and function. Alan R Liss, New York

Wang Y, Seidel J, Tsui BMW, Vaquero JJ, Pomper MG (2006) Performance evaluation of the GE Healthcare eXplore VISTA dual-ring small-animal PET scanner. J Nucl Med 47:1891–1900

West GB, Brown JH, Enquist BJ (1997) A general model for the origin of allometric scaling laws in biology. Science 276:122–126

Yorke ED, Beaumier PL, Wessels BW, Fritzberg AR, Morgan A (1991) Optimal antibody-radionuclide combinations for clinical radioimmunotherapy: a predictive model based on mouse pharmacokinetics. Nucl Med Biol 18:827–835

The Biodistribution (II): Human

14

Abstract The measurement of the distribution and excretion of the radionuclide following the administration of a radiopharmaceutical is necessary in order to evaluate the consequential internal radiation dosimetry. This chapter reviews various in vivo and in vitro means of acquiring these data. The conjugate-view method used commonly in planar scintigraphy is derived and the various compensatory techniques to account for attenuation and photon scatter are derived and examined. Quantitative single-photon emission computed tomography (SPECT) which, due to current limitations in technology, is limited to absorbed dose evaluations of small anatomical volumes is reviewed as are the corrections required for scatter and attenuation. Positron emission tomography (PET) is inherently quantitative and its greater sensitivity compared to SPECT permits whole-body biodistributions of positron-emitting radionuclides to be measured. The principles of PET data acquisition are reviewed in the context of the nuclear medicine physicist designing a protocol for a whole-body biodistribution measurement. Finally, quantitative *bremsstrahlung* imaging of β-emitting therapeutic radionuclides is a most challenging endeavor, and one which is not frequently performed. A review of the methodologies is provided.

Contents

14.1 Introduction

> The identity of the body is not dependent on the persistence of the same material particles. During life, by the process of eating and digesting, the body undergoes perpetual change
>
> St Thomas Aquinas

The calculation of the absorbed dose to a given target region first requires the determination of the

B.J. McParland, *Nuclear Medicine Radiation Dosimetry*,
DOI 10.1007/978-1-84882-126-2_14,

cumulated activities of the various source regions. As the cumulated activity is simply the total number of radioactive decays in the source region r_S, in practice one measures the activity in that region over time and integrates over time to yield the normalized cumulated activity, $\tilde{A}_{r_S,\mathrm{Norm}}$. In the previous chapter, preclinical biodistributions were largely determined through dissection and the harvesting of organs for activity assays, although the growth of preclinical imaging capability makes the future likelihood of longitudinal animal studies greater. On the other hand, human studies are almost exclusively imaging-based, with the limited exception of a single probe measurement which determines the whole-body activity burden without spatial information regarding the radionuclide. The methods of measuring these biodistributions in the human are presented here. Siegel et al. (1999) and Stabin (2008) both provide detailed discussions and examples of the practical implementations of these methods.

Red (active) bone marrow is a very special consideration in evaluating the human biodistribution. Due to its radiosensitivity, it can frequently be an administered activity-limiting tissue in radionuclide therapy. Moreover, because of the high tissue weighting factor applied to the tissue, accurate red bone marrow dosimetry is of interest even in diagnostic nuclear medicine. The cumulated activity of red bone marrow can be estimated from in vivo imaging or by other means. Consequently, the topic of red bone marrow activity acquisition and dosimetry is treated separately.

14.2 Data Collection

14.2.1 Introduction

The measurement of the biodistribution of a radionuclide in a human can be as limited to simply using a gamma scintillation probe so as to measure the retention of activity following the excretion phase. But for the complete measurements required of the biokinetics of an administered radionuclide for either a Phase I study of a new diagnostic radiopharmaceutical or the planning for radionuclide therapy, in vivo imaging and in vitro assays are required to quantify the activity and its distribution over time.

14.2.2 Source Regions: Definition and Segmentation

In any measurement of the biodistribution, source regions (organs, tissues, intracavitary contents, etc) containing activity which can be detected in an in vivo image or in vitro assays should be identified. Ideally, a compilation of source regions to examine should be prepared prior to actual image analysis on the basis of the preclinical biodistribution data. These regions would include the specified MIRD source regions. Additional organs or tissues expected to demonstrate uptake, on the basis of preclinical studies, should also be planned to be investigated even if, in the eventual dosimetry calculation, they are to be grouped within the remaining tissues category. These would include, for example, the salivary glands for which uptake could possibly indicate the presence of a radiochemical impurity in the case of a ^{99m}Tc-based radiopharmaceutical administration. The injection site should also be investigated as the presence of significant extravasation of the radiopharmaceutical may require correction for either the actual amount of radionuclide that has entered the vasculature or of the time course of delivery.

Regions-of-interest (RoIs), in planar images, or volumes-of-interest (VoIs) in tomographic data sets, can be defined manually by the user in order to segment organs or tissues. This manual approach has advantages and disadvantages. Perhaps the main advantage is the inclusion of the operator's skill and experience in the manual definition of RoIs. For a whole-body planar image set used for a biodistribution study, which may consist of images at no more than a dozen time points, this is likely to be considered reasonable. However, a disadvantage arises when a volume-of-interest (VoI) is to be defined from the concatenation of RoIs manually drawn on a tomographic data set consisting of, say, 128 or 256 coronal slices. Some degree of automated assistance may be required.

Automated segmentation techniques remove the reliance upon operator skill and experience in drawing RoIs. There are a number of techniques available (Fleming 1996; Lawson 1998). For example, a simple threshold defined by specifying a percentage of the maximum number of counts per pixel in an image and then including all pixels containing counts exceeding this threshold within the region is one approach. In practice, the threshold is set low and, in gamma camera planar scintigraphy, a highly-serrated contour will often result due to the contribution of surrounding background (although this noise tends to be reduced in tomography segmentation). Smoothing of the image data prior to thresholding is possible, although its inclusion is reported to require a further lowering of the threshold. As problems with overlap of organs in planar images (e.g., liver and right kidney) are expected, the approach will inevitably become semiautomatic and still require user correction. Spatial gradient-based segmentation approaches can also be employed), with an example shown by Fig. 14.1. The edge of the RoI can be defined by the points at which the count per pixel gradient is at a maximum (i.e., the first derivative of the variation of counts per pixel at the edge is equal to zero). Another modification is to define the edge where the second gradient is a maximum. This position occurs further out and will increase the number of pixels within the RoI and reduce the sensitivity of the activity calculation to the variations in RoI edge. However, there are difficulties associated with gradient searches in RoI/VoI segmentation. Firstly, they are sensitive to noise and some smoothing (noise averaging) is usually required prior to segmentation. Secondly, the example shown was for a simple one-dimensional case whereas gradients in planar and tomographic images are defined in two and three orthogonal directions, respectively.

Lawson (1998) has suggested approaches to allowing for multiple orthogonal gradients, including averaging the first-derivative gradients, or using the Laplacian,

$$\nabla^2 f(x,y,z) = \frac{\partial^2 f}{\partial x^2} + \frac{\partial^2 f}{\partial y^2} + \frac{\partial^2 f}{\partial z^2} \tag{14.1}$$

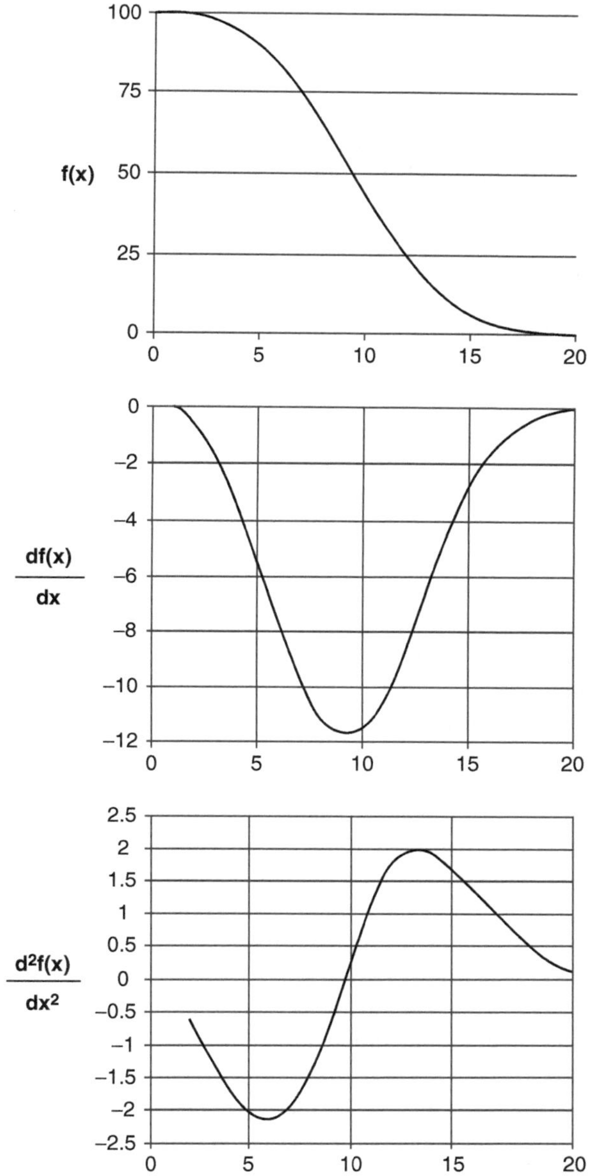

Fig. 14.1 Principle of edge gradient search in a simple one-dimensional model. The profile of the RoI is shown as f(x), which is the number of counts per pixel at x, with the first- and second-derivative of the profiles. The edge of the profile can be defined as that point where the gradient (first derivative) is equal to its maximum and where the second derivative is equal to zero. Another definition is that at which the second derivative is at a maximum

14.2.3 Data Acquisition Times

14.2.3.1 Image-Based Data

The problem of defining the optimum number and temporal distribution of data acquisition times in a Phase I trial is similar to that discussed in Chap. 13 in the context of the preclinical model. The effective half-life of the radionuclide, which encompasses both the physical decay of the radionuclide and the biokinetics, will be the main driving factor to consider. But, in a human study, there are additional factors to account for. As such a study is longitudinal and largely based upon imaging, the imaging device sensitivity will define image acquisition durations and, consequently, the number of image acquisition time points. As many biological processes can be considered as following first-order kinetics, a multiexponential function can be used to model the uptake and washout of activity from a compartment, as to be shown in Chap. 15. The order of this function and the number of image acquisition points will thus interplay. The acquisition duration will also be a function of subject compliance and tolerance. Long acquisition times risk the degradation of image quality as a result of subject motion. While dense sampling of the biodistribution is desirable at early time points, this will be restricted by the number of images that can be effectively acquired at these times. Hence, the nuclear medicine physicist planning, for example, the Phase I trial of a radiopharmaceutical must optimize the sensitivity of the imaging device (gamma camera or PET scanner) so as to both limit the amount of activity administered and to reduce the image acquisition times to tolerable levels. Typically, this would require a collimator with high sensitivity for gamma camera scintigraphy and often the use of the three-dimensional-acquisition mode in PET scanning. In both modalities, image acquisition durations should be increased to compensate for the reduced count rates at later times postadministration as a consequence of radioactive decay and excretion of activity. This will obviously reduce the temporal sampling rate at later times postadministration, although the biokinetics will by then have largely stabilized.

The determination of image acquisition times can be made on the basis of preclinical data or on human biokinetics of similar radiopharmaceuticals which may be predictive of those of the tested radiopharmaceutical. Using the previous arguments on sampling frequency on the basis of multiexponential fits, a minimum of four image acquisitions over a period of time equal to about four effective half-lives for a short-lived radionuclide (e.g., ^{18}F or ^{99m}Tc) would be required to generate a minimally-useful biodistribution. For longer-lived radionuclides (e.g., ^{111}In or ^{123}I), acquisitions can usually be terminated at much shorter times after the biodistribution has largely stabilized. Of course, the quality of the biodistribution and the information it can provide will improve with the number of image acquisitions, especially as it is necessary to increase the frequency of acquisitions at early times in order to accurately capture the temporal distribution of activity. It is important to note that the biokinetics are the most significant influence on image acquisitions number and temporal distribution, not the physical half-life alone. An example of the time ensemble postadministration to perform whole-body imaging for a ^{99m}Tc-labeled radiopharmaceutical would be: 5, 15, 30, 60, and 90 min; 3, 6, 12, and 24 h. Image acquisitions at early time points are critical to determining the early distribution of activity from the vascular space; otherwise the extrapolation of activity data back to the time of administration can lead to an overestimate of the cumulated activity. The acquisition durations of later imaging sessions should be extended in order to compensate for physical decay and excretion so as to ensure that they are adequate counting statistics in the image. As noted in Chap. 13, Siegel et al. (1999) and Report 67 of the International Commission on Radiation Units and Measurements (ICRU 2002) have specified an initial distribution of sampling times on the basis of the whole-body effective half-life. They recommended an ensemble of image acquisition starting times of 1/3, 2/3, 3/2, 3, and 5 multiples of the effective half life.

14.2.3.2 Assay-Based Data

Introduction

In addition to the activities measured in vitro by imaging, the complete biodistribution requires knowledge of how quickly activity transfers between the blood and other tissue compartments, if there is any binding of the radionuclide to whole blood and the rate at which activity is excreted from the body through the renal pathway and the gastrointestinal (GI) tract.[1] This is determined by measuring activities in vitro from samples acquired during the study.

[1]Excretion of activity through sweat and saliva is rarely measured and would only be done for a specific purpose such as estimating the radioactive contamination from a subject having received a high administered activity of a long-lived therapeutic radionuclide (e.g., ^{131}I).

Whole Blood and Plasma

The measurements of activity concentration in whole blood and plasma are useful for estimating the cumulated activities of the cardiac wall, cardiac contents, and the red bone marrow. They are also necessary for understanding the distribution of the radiopharmaceutical following injection and, for diagnostic radiopharmaceuticals, for allowing bridging studies between the biodistributions of the healthy volunteer cohorts of Phase I and the subjects, in Phase II, with the disease or pathology to be diagnosed.[2] The selection of time points at which blood should be sampled is largely driven by the limitation of the volume of blood that can be extracted. As the washout of activity from the blood/plasma can frequently be described by at least a biexponential, or frequently a tri-exponential, function in time, at least four to six blood sampling times are necessary. However, especially with small peptides administered intravenously, clearance from the vasculature is very rapid at early time points, so one should increase the number of measurements of blood/plasma activity concentrations at these early times. Blood sampling should thus preferably extend to about two to four physical half-lives of the radionuclide (i.e., 4–8 h for ^{18}F and 12–24 h for ^{99m}Tc)

Excreta

The measurement times of activity present in excreta is clearly set by the times at which they are voided from the body. As will be described, the inputs to the dynamic models of urinary and fecal excretion are the sums of activities in the respective excretory pathway and in the excreta. For dosimetry purposes, there is no point measuring the activity excreted following the last imaging time point during which the activity in the excretory pathways will have already been quantified.

[2]It is desired to know if there are significant differences between the biodistribution derived from the healthy volunteer cohort and those of patients that could affect the dosimetry estimate for the latter. As Phase II subjects are patients with disease, they are unlikely to be able to tolerate as complete an assessment of the biodistribution as performed in Phase I. Hence, comparison of the measurements of the rate at which activity is cleared from blood and/or is excreted can assist in assessing if the biodistribution is significantly changed in the diseased cohort.

14.2.4 Sample Size

A Phase I study of a radiopharmaceutical has two main objectives: measurement of the biodistribution of the radioactive moiety and an assessment of the pharmacological safety of the radionuclide. Fundamental to the design of the study is the determination of the sample size required. This has practical, statistical, and ethical issues associated with it. A study with too many subjects is unethical and, fortunately, rare. On the other hand, a study with an insufficient number of subjects will be unlikely to yield an answer to the hypotheses posed by the study and would also be regarded as unethical.[3] The sample size of a clinical study will be dictated by, among other factors, the power of the study, i.e., the probability that the study will lead to the rejection of the null hypothesis in favor of the alternative when the null hypothesis is, in fact, false. This argument cannot, however, be applied to a biodistribution study where no hypotheses are involved. However, Sparks (2005) used a Monte Carlo-type stochastic analysis to determine the minimum number of subjects required in a biodistribution study. He determined that, by assuming the organ absorbed doses and the effective dose equivalent to be normally-distributed, a sample size of six subjects would be adequate to yield a 95% confidence interval where the upper limit of this confidence interval is a factor of 2 greater than the estimated mean for an inter-subject variability of $\sigma/\mu = 1.25$. Indeed, a review of the literature reveals that the numbers of healthy volunteers recruited for Phase I biodistribution studies of diagnostic radiopharmaceuticals are almost always less than 10.

14.2.5 Nonimaging Quantification Methods

14.2.5.1 Introduction

Nonimaging in vivo measurements are uncommon in nuclear medicine dosimetry and any attempts at doing

[3]Everitt (1994) has challenged this last assertion by suggesting that the use of meta-analysis means that there are few clinical studies with insufficient numbers of subjects that cannot help to resolve important clinical questions.

so have been largely limited to therapeutic nuclear medicine. On the other hand, in vitro measurements of activity in whole blood, plasma, and excreta are essential to the determination of the biodistribution and internal radiation dosimetry of a radiopharmaceutical. In situ measurements in nuclear medicine of absorbed dose or activity are very unusual.

This section reviews the techniques used in these measurements.

14.2.5.2 In Vivo Nonimaging Techniques

Due to its nature, internal radiation dosimetry is very much based on imaging of the in vivo activity. There is relatively little scope for the physical in situ measurement of activity content or of absorbed dose. However, such measurements are required as a means of verifying absorbed dose calculation methodology, in particular for particulate-emitting radionuclides, or of determining spatially nonuniform absorbed dose distributions at very small distance scales.

Thermoluminescent Dosimetry

Thermoluminescent dosimeters (TLDs) are a means of measuring in vivo absorbed doses in both preclinical and clinical settings (Griffith et al. 1988; Strand et al. 1994a). Implanting the TLD crystal into tissue is a means of measuring the integrated absorbed dose only; as it is not a rate dosimeter, the TLD cannot yield the activity, which is proportional to absorbed dose rate. In general practice, $CaSO_4$:Dy has been the TLD material of choice for in vivo dosimetry measurement due to its high luminescence at low absorbed doses. However, the biological medium within which the TLD is placed will alter the response of the dosimeter as dissolution of the $CaSO_4$:Dy results when it is placed in a liquid medium. This TLD material, as a result, is sensitive to ambient pH and temperature, resulting in signal loss and supralinearity. A loss of sensitivity of as high as 70% after 9 days has been reported. Martin et al. (2000) described a micro-TLD design for in vivo dosimetry based upon a composite of LiF:Mg,Cu,P in the form of 400 μm diameter filaments which had a sensitivity 60% greater than $CaSO_4$:Dy and much greater stability when placed in liquid media with variable pH levels.

In vivo TLD measurements are additionally difficult due to the implantation and recovery processes, especially if the dosimeters are small. One interesting application of thermoluminescent dosimetry in nuclear medicine was that of Deloar et al. (1997) who used a method pioneered earlier by the group (Matsumoto et al. 1993) based on TLDs to estimate the absorbed doses to organs following inhalation of gaseous ^{15}O-labeled tracers. Because of the 122.2 s half-life of ^{15}O, image-based whole-body biodistribution measurements are not possible and, as a result, Deloar et al. placed TLDs on the skin over source organs of interest (brain, pharynx, thyroid, trachea, heart, lungs, liver, kidney, spleen, and urinary bladder). The absorbed doses recorded at the TLDs were proportional to the cumulated activities in the source organs multiplied by predetermined S-factors from the source organs to the positions of the TLDs.

MOSFET

Whereas a TLD is an integrating dosimeter, the MOSFET (Chap. 9) provides an absorbed dose rate measurement. Gladstone et al. (1994) described a MOSFET dosimeter of 1.6 mm diameter that could be inserted into tissue using a 16-gauge flexineedle. While all MOSFETs will suffer from nontissue radiological equivalence, a major shortcoming in the implantation of such devices is the necessity of leads to provide the gate voltage and to read out the source-to-drain current. A possible solution to this problem is described by Beddar et al. (2005) who described a glass-encapsulated MOSFET operated in the unbiased mode. Radiofrequency telemetry is used to power the MOSFET for absorbed dose measurements.

Scintillator Probe

This is the most common nonimaging means of measuring in vivo activity, especially that of thyroid gland uptake of ^{131}I. A scintillation probe is positioned at a fixed distance from the thyroid gland postadministration and the count rate measured using a multichannel analyzer and photon energy discrimination. The measurement is then repeated in a calibration setting in which a container of a known activity of ^{131}I is placed within a cylindrical plastic phantom and offset from the center. The count rate measurement is then repeated to yield a calibration factor from which the in vivo thyroid gland uptake is estimated.

Other Means

Other nonimaging methods of assessing absorbed dose include electron paramagnetic resonance measurements of mineralized tissue (Pass and Aldrich 1985; Schauer et al. 1994) and biological dosimeters such as the measurement of chromosome aberrations.

14.2.5.3 In Vitro Measurements

Introduction

The measurements of activity in fluids and substances obtained or excreted from a human subject having been administered a radiopharmaceutical are important components in obtaining the complete understanding of the biodistribution of the radionuclide. Measurements are typically made with a well scintillation counter, although quantitative imaging using a gamma camera can be used to quantify high-activity sources such as voided urine. The temporal variations of whole blood and plasma activity concentrations are used to understand the rapidity with which the radionuclide leaves the vascular compartment and can be used to estimate the activity within the contents of the cardiac chambers and, in some cases, the cumulated activity in red bone marrow. A comparison of the whole blood and plasma activity concentration will reveal any binding of the radionuclide to red blood cells.

Measurements of excreted activity are necessary in order to estimate the absorbed dose to the urinary bladder wall.[4] This is often of concern for diagnostic radiopharmaceuticals, in particular, as rapid excretion of activity is desirable both for reducing the absorbed dose and reducing any background tissue uptake which can impact on diagnostic efficacy. While the urinary pathway provides an ideal route for this excretion, it can result in a high absorbed dose to the urinary bladder wall. Measurements of excreted activity in the feces are usually of limited applicability as the time between when activity enters the GI tract (either through ingestion or hepatobiliary transport) and when it is excreted in the feces is much longer than the physical half-lives of many radioisotopes. However, such measurements may be necessary for longer-lived isotopes or if a particularly high amount of activity is introduced to the GI tract and fecal excretion is necessary to completely model the transport of activity within the tract contents.

Whole Blood

In a typical biodistribution study of a radiopharmaceutical, venous blood samples are acquired from the subject at multiple time points following administration.[5] Much of the sample volume will be dedicated to conventional safety analyses, however some blood must be assigned to radioactive content assay. Clearly, the times of administration, sample, and assay must be recorded so that the measured activity concentration can be decay-corrected to the time of administration. These times must be carefully judged by the nuclear medicine physicist planning the study to ensure that the activity being assayed cannot cause noncorrectable dead-time errors in the counting device. A conservative rule-of-thumb of estimating the maximum amount of activity in whole blood is to neglect physical decay and the transfer of activity from blood into tissues and assume a standard volume of blood. For example, in ICRP Reference Man (ICRP 2002), the volume of blood is 5,300 mL. Assuming that this is the amount of blood within the subject (and this value can be varied with body mass accordingly using data in ICRP Publication 89), then for a 100 MBq administered activity, the maximum activity concentration in whole blood is,

[4] This is not quite so true for short-lived radioisotopes, such as ^{11}C, excreted through the urinary pathway. If the physical half-life of the isotope is much less than the time between urinary bladder voids, the activity of the urinary bladder contents can be determined more directly through in vivo imaging.

[5] For activity measurements for biodistribution and dosimetry studies alone, there is no reason that blood samples cannot be arterial, should kinetic modeling using an arterial input function be concurrent as in a PET study.

$$[A_{BL}]_{max} = \frac{100,000\,kBq}{5,300\,mL} \approx 19\frac{kBq}{mL}$$

One need only use this value to obtain a rough estimate of the maximum activity concentration likely to be measured. Physical decay can also be applied to bring this estimate into a more reasonable range, although the above result is the most conservative.

The allocation of venous blood sample times can be estimated from the clearance of activity from blood in the preclinical animal. Although, in practice, this must often be considered in conjunction with required times for safety assessments and access to a venous catheter with the subject in the gamma camera or PET scanner. However, as the washout in a species smaller than the human is expected to be more rapid, one must be cognizant of the fact that the human blood sample will likely have a higher activity concentration than that estimated from the preclinical model.

It is good practice to allocate two aliquots from each venous blood sample to be assayed independently. Measurement errors are inevitable and an observed discrepancy between two aliquot measurements can be resolved by comparison with measurements at earlier and later time points. Because of the low activity values, the activity is usually measured in a high-sensitivity scintillator well-counter. Measurement corrections as described in Chaps. 5 and 9 are applied.

Plasma

Plasma activity concentrations can be obtained from the same venous blood samples; indeed, it is possible to obtain these from the same whole blood aliquots after these have been assayed. The whole blood is centrifuged and the supernatant removed, the volume measured, and the activity content assayed.

Urine

The activity in excreted urine should be measured as voided rather than accrued and measured as the total activity excreted renally. This excreted activity, as will be seen, is summed with that of the urinary bladder contents so that the cumulated activity of the sum of the urinary bladder contents and voided urine is yielded. The total activity in a urinary void at measurement time t, $A_{Urine,Void}(t)$, is estimated by measuring the activity concentration in an aliquot (or by measuring two and averaging, for good practice) of known volumes of urine and scaling this by the total volume of the void,

$$A_{Urine,Void}(t) = [A_{Urine,Aliq}(t)]V_{Void} = A_{Urine,Aliq}(t)\frac{V_{Void}}{V_{Aliq}}. \tag{14.2}$$

Another means of measuring the activity contained within a urine void is to quantify a conjugate-view image of the contents of the entire void within a container placed on a gamma camera.

Again, times of administration, excretion, and assay must be recorded in order to allow physical decay correction. Consideration must also be taken of the expected activity concentration so that the nuclear medicine physicist can plan the assay time postadministration. The activity eventually to be excreted in the urine can be estimated from the preclinical model (acknowledging the likely more rapid nature) and the volume of urine excreted per day can be estimated from data in ICRP Publication 89 and an assumed number of urinary bladder voids per day.

Feces

Measurements of the activity contained within the feces is relatively uncommon, but could be considered if the radioisotope has a physical half-life comparable to the GI tract content transit times or if there is a high amount of activity expected to enter the GI tract (e.g., for lipophilic radiopharmaceuticals expected to have increased liver uptake and excretion into the duodenum). For a given void, the mass of the voided stools is measured and a small sample taken and weighed. The sample is homogenized and the volume of the homogenate measured. A sample of known volume of the homogenate is then assayed. The activity in the assay volume is then scaled up to yield the activity in the total homogenate which is subsequently scaled up to yield the total activity voided in the feces. These tasks are best assigned to the unsuspecting undergraduate.

14.3 Imaging Quantification Methods

14.3.1 Introduction

The quantification of the in vivo activity at multiple times postadministration is the crux of determining the biodistribution of the radionuclide of interest. In almost all cases of the development of a diagnostic radiopharmaceutical, the biodistribution is initially measured in healthy volunteers. This is largely due to the impracticality of having patients undergo whole-body imaging at several acquisition time points. Imaging times can be long and patient compliance (e.g., in terms of lack of patient motion) may be difficult to attain if the patient is in physical discomfort.

As healthy volunteers are, by definition, without disease they can expect to receive no medical benefit from the study. Hence, unlike patients for whom clinical efficacy is to be assessed and who may receive possible medical benefits, healthy volunteers receive a much lower amount of administered activity than would a patient, the ethical issues of which are described in Chap. 16. This is necessary so as to minimize the radiation absorbed doses received by the healthy volunteer.[6] As a result, a Phase I biodistribution study using healthy volunteers must employ as sensitive a detection device as possible. For example, this would include the use of a high-sensitivity low-energy general purpose collimator on a gamma camera for ^{99m}Tc and ^{123}I studies and extended image acquisition times in order to maximize the number of detected photons for a given amount of activity. In this way, the amount of information acquired is maximized for an assumed minimized, and presumably acceptable to the subject, radiation risk.

A possible downside of the necessity to minimize the radiation exposure to the healthy volunteer is that the images generated will be noisy and nondiagnostic. This is not necessarily a problem as one is generally seeking to determine the biodistribution of activity in macroscopic organs rather than to identify changes in pathology. A second possible disadvantage is that it may not be possible to fit the entire body in the image for a combined low count-rate and limited acquisition time. For example, the lower extremities may need to be excluded from an image so as to make the image acquisition duration time tolerable for the subject or in order to be able to include a greater number of acquisitions. This requires the assumption that the activity distribution in the unimaged anatomy is both uniformly distributed and has a magnitude equal to the difference between the known administered activity and the sum of that measured in the imaged anatomy and that excreted (e.g., McParland et al. 2008).

Here, quantification methods suitable for biodistribution measurements using both single-photon-emitting radionuclides and positron-emitting radionuclides are described.

14.3.2 Single-Photon-Emitting Radionuclides

14.3.2.1 Conjugate-View Planar Scintigraphy

Method

The conjugate-view method of quantifying in vivo activity was first proposed by Thomas and collaborators in the 1970s (Thomas et al. 1972, 1976). Since then, a number of validation studies of the method using phantoms, animal measurements, and Monte Carlo simulations (e.g., Eary et al. 1988; Jonsson et al. 2005; He and Frey 2006) have appeared in the literature. In the conjugate-view approach, whole-body anterior and posterior emission scans of the subject are taken simultaneously with a multiple-head gamma camera. The camera heads are positioned so as to acquire anterior and posterior views of the subject who lies supine on the bed. Nearly all gamma cameras allow a continuous translation of the bed for a whole-body acquisition. If necessary, it is possible to generate a whole-body image suitable for a biodistribution measurement using contiguous static planar

[6] The patient would receive a higher administered activity in order to allow a diagnostic-quality image. Hence, the usual optimization of benefit versus exposure risk used in medical practice would apply.

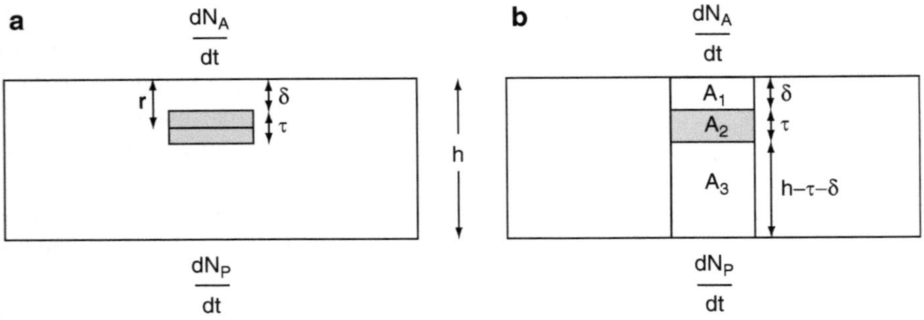

Fig. 14.2 Geometry describing the conjugate-view quantification method. In (**a**), a source region containing activity A is embedded in a medium of overall thickness h and which contains no activity. In (**b**), the surrounding medium now contains activity so the conjugate-view is now of three serial regions containing different amounts of activities

images axially distributed along the subject or, less conveniently, using diverging collimators.[7] Transmission scans with and without the subject present are acquired prior to administration of the radiopharmaceutical for attenuation measurements, as described below. The reader can find a detailed practical and comprehensive overview of the method and its practical implementation in MIRD Pamphlet No 16 by Siegel et al. (1999).

Consider the two-dimensional model of Fig. 14.2a which shows a region containing a uniformly-distributed activity A of a photon-emitting radionuclide within a medium. The overall medium has a thickness h and the activity-containing object has a thickness τ and lies at a depth δ. The first consideration is of the simplest case of where activity is present in the internal region only, the region and medium having the same radiological properties and scatter is neglected. Consider a differential element of the thickness of the activity-containing region, dr at a depth r as shown. This differential slice contains an amount of activity equal to dA where,

$$\mathrm{dA} = \mathrm{C_A dt} \tag{14.3}$$

where C_A is the amount of activity per unit thickness in the object,

$$\mathrm{C_A} = \frac{\mathrm{A}}{\tau} \tag{14.4}$$

The rate at which photons are detected from this differential thickness in what is identified as the "anterior" view is,

$$\frac{\mathrm{d^2N_A}}{\mathrm{dt\,dr}} = \kappa \mathrm{C_A e^{-\mu r}} \tag{14.5}$$

where κ is a calibration factor converting activity to detected counts per unit time and μ is the linear attenuation coefficient. κ can be conveniently obtained by imaging a source of the same radionuclide administered and of known activity at the same time as the subject. The count rate due to the activity in the entire region is given by integrating (14.5) over the depth of the object,

$$\begin{aligned}\frac{\mathrm{dN_A}}{\mathrm{dt}} &= \kappa \mathrm{C_A} \int_{\delta}^{\delta+\tau} \mathrm{dr\, e^{-\mu r}} \\ &= \frac{\kappa\, \mathrm{C_A}}{\mu} \mathrm{e}^{-\mu\delta}(1 - \mathrm{e}^{-\mu\tau}) \\ &= 2\frac{\kappa\, \mathrm{C_A}}{\mu} \mathrm{e}^{-\mu(\delta+\tau/2)} \sinh\frac{\mu\tau}{2}\end{aligned} \tag{14.6}$$

where $(1 - e^{-x}) = 2e^{-x/2}\sinh x/2$ has been used. Similarly, the rate at which photons are detected in the opposing "posterior" view is,

$$\begin{aligned}\frac{\mathrm{dN_P}}{\mathrm{dt}} &= \kappa \mathrm{C_A} \int_{\mathrm{h}-\tau-\delta}^{\mathrm{h}-\delta} \mathrm{dr\, e^{-\mu r}} \\ &= \frac{\kappa\, \mathrm{C_A}}{\mu} \mathrm{e}^{-\mu(\mathrm{h}-\delta)}(\mathrm{e}^{\mu\tau} - 1)\end{aligned} \tag{14.7}$$

[7] In principle, conjugate-view scintigraphy could be performed with a single-headed gamma camera: one view whole-body scan, followed by a rotation of the gamma camera head by 180° and a subsequent scan of the conjugate-view. However, error is introduced as the biodistributions of both views were not acquired at the same time. The error would be greatest at early times postinjection when still in the distribution phase.

The geometric mean of the two count rates is,

$$\begin{aligned}\left\langle\frac{dN}{dt}\right\rangle_{geom} &= \sqrt{\frac{dN_A}{dt}\frac{dN_P}{dt}} \\ &= \kappa A \frac{\sinh \mu\tau/2}{(\mu\tau/2)}\sqrt{\Im}\end{aligned} \tag{14.8}$$

where the attenuation of the emitted photons through the entire thickness of the medium is given by

$$\Im = e^{-\mu h} \tag{14.9}$$

and the total activity within the embedded region is found by rearranging (14.8),

$$\begin{aligned}A &= \frac{1}{\kappa\sqrt{\Im}}\left\langle\frac{dN}{dt}\right\rangle_{geom}\left(\frac{(\mu\tau/2)}{\sinh \mu\tau/2}\right) \\ &= \frac{1}{\kappa\sqrt{\Im}}\left\langle\frac{dN}{dt}\right\rangle_{geom} f_{Self-atten}\end{aligned} \tag{14.10}$$

where the self-attenuation factor, $f_{Self\text{-}atten}$, for the region containing activity has been defined.

If it is now assumed that the surrounding medium contains activity, one is presented with the more realistic case of Fig. 14.2b, in which there are now three regions with activities viewed in series. It is straightforward to repeat the above calculation to find that,

$$A_1 = \frac{1}{\kappa\sqrt{\Im}}\left\langle\frac{dN}{dt}\right\rangle_{geom}\left(\frac{(\mu_m\delta/2)}{\sinh \mu_m\delta/2}\right) \tag{14.11}$$

$$A_2 = \frac{1}{\kappa\sqrt{\Im}}\left\langle\frac{dN}{dt}\right\rangle_{geom}\left(\frac{(\mu_S\tau/2)}{\sinh \mu_S\tau/2}\right) \tag{14.12}$$

$$A_3 = \frac{1}{\kappa\sqrt{\Im}}\left\langle\frac{dN}{dt}\right\rangle_{geom}\left(\frac{(\mu_M(h-\tau-\delta)/2)}{\sinh \mu_M(h-\tau-\delta)/2}\right) \tag{14.13}$$

where the more general case in which the medium and source regions are associated with different linear attenuation coefficients, μ_M and μ_S, has been allowed for. In this case, the transmission factor is,

$$\Im = e^{-(\mu_M(h-\tau)+\mu_S\tau)} \tag{14.14}$$

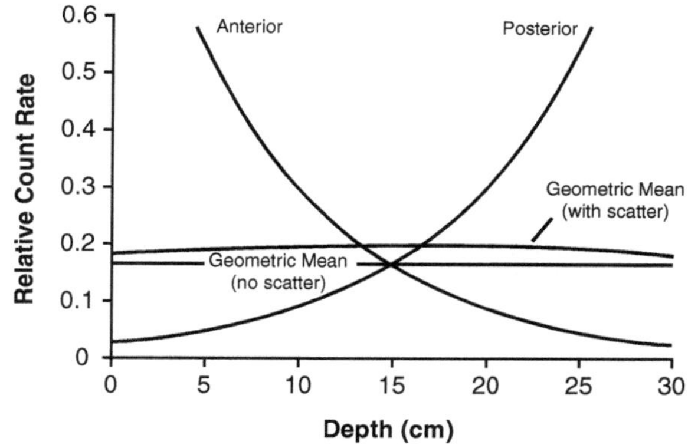

Fig. 14.3 Conceptualisation of the conjugate-view method

MIRD Pamphlet No 16 summarizes (14.11)–(14.13) for multiple overlying source regions in an imaged region-of-interest in the general form,

$$A_j = \frac{1}{\kappa\sqrt{\Im}}\left\langle\frac{dN}{dt}\right\rangle_{geom}\left(\frac{(\mu_j t_j/2)}{\sinh \mu_j\delta_j/2}\right) \tag{14.15}$$

for the jth source region.

A conceptualization of the conjugate-view method is shown in the simple representative diagram of Fig. 14.3 which summarizes calculations performed for a ^{99m}Tc point source in soft tissue. The plot shows the count rates measured on either side of a 30-cm thick homogeneous slab of soft tissue. The "anterior" and "posterior" count rates are exponentially dependent upon the thickness of material, whereas the geometric mean (without any scatter correction) is independent of thickness. As scatter has been neglected in these derivations, the approximate contribution of scattered photons is incorporated in the geometric mean estimate using results from the buildup factor calculation of Fig. 9.17.

Dead-Time Correction

The dead time of a radiation detector was discussed in Chap. 5 and was shown to increase with the actual count rate with the degree of increase being dependent on whether the dead time was paralysable or nonparalysable[8] and the length of the dead time following a single event. Measurement of the dead-time dependence of a gamma camera should be part of its normal quality control and methods of doing so are summarized in Reports 73 (IPEM 1996) and 86 (IPEM 2003) of the Institute of Physics and Engineering in Medicine (IPEM). Point or distributed sources are discussed, both of which involve measurement of the count rate as the source decays. The two-source method discussed in Chap. 5 can be used to assess the paralysable dead time of a gamma camera. In most cases of gamma camera quantification, especially in the low administered activity regime of diagnostic radiopharmaceutical Phase I trials, dead-time corrections can largely be ignored.

Transmission Factor

In planar scintigraphy, the transmission factor is typically measured by translating the subject over either a ^{99m}Tc flood source or a ^{57}Co source. Here, the transmission factor for a given RoI is,

$$\Im = \frac{dN/dt\big|_{\text{RoI,Subject}}}{dN/dt\big|_{\text{RoI, No Subject}}} \tag{14.16}$$

where the ratio is of count rates for the RoI with and without the subject present.

Equation (14.16) shows that the transmission is determined on a pixel-by-pixel basis which can then be averaged over macroscopic RoIs and compared with the scanning repeated without the subject present. Care must be taken if the radionuclide of the transmission source differs from the radionuclide being investigated. The transmission factor must be modified accordingly to allow for the difference between the linear attenuation coefficients of the two radionuclides. For example, let $\Im_{\text{Meas}}$ be the measured transmission factor for a radionuclide with linear attenuation coefficient, μ_{Meas}. The scaled transmission factor to use for the administered radionuclide is then,

$$\Im = e^{\frac{\mu}{\mu_{\text{Meas}}}\ln \Im_{\text{Meas}}} \tag{14.17}$$

[8] A gamma camera typically acts as a paralysable system.

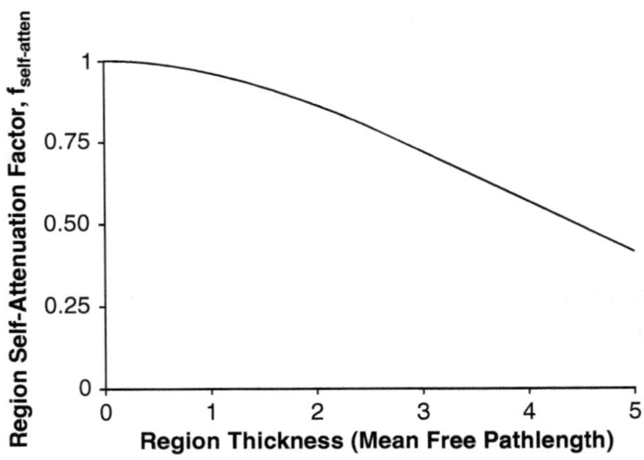

Fig. 14.4 The region self-attenuation factor of (14.10) as a function of the source region thickness

where μ is the linear attenuation coefficient in tissue for the administered radionuclide. Note that the broad-beam geometry values of μ should be used in (14.17) as scattered photons contribute to the signal.

Self-Attenuation Correction Factor

The self-attenuation correction factor, $f_{\text{Self-atten}}$, defined by (14.10), requires that the thickness of the region be known. Figure 14.4 shows $f_{\text{Self-atten}}$ plotted as a function of the source region thickness, in multiples of the mean-free pathlength (i.e., $\mu\tau$).

There are various ways of evaluating $f_{\text{Self-atten}}$. The first is to simply set it equal to unity (i.e., ignore it). Consider the case of 140 keV photons from ^{99m}Tc for which the mean-free path in soft tissue is approximately 7 cm. By neglecting self-attenuation and setting $f_{\text{Self-atten}} = 1$, errors in the determined activity of 2, 8, and 16% would occur for source region thickness of 5, 10, and 15 cm. These levels of uncertainties could be considered acceptable in some situations, especially when the effort required to determine the longitudinal thickness of the organ accurately enough is realized.

The second method is to determine $f_{\text{Self-atten}}$ directly by measuring the organ thickness in a transverse planar emission view or through SPECT or X-ray CT. The first two methods may not allow accurate measurement of the source region thickness, although the relative insensitivity of $f_{\text{Self-atten}}$ to the thickness for small dimensions (below about 1.5

mean-free pathlengths) can help to reduce the effect of this measurement inaccuracy. There are additional possible disadvantages to the use of emission imaging to determine transverse thicknesses. The planar transverse view may also have the complication of overlying regions of activity making the measurement difficult. SPECT suffers, in particular, if the subject is a healthy volunteer because in such a case, the amount of activity administered in a biodistribution study is minimal in order to reduce the radiation absorbed doses received by the subject. As a result, there will likely be an insufficient number of detected photons to yield a suitable SPECT image. On the other hand, the transverse thicknesses can be measured from X-ray CT. A diagnostic-quality CT data set may be available if the biodistribution is being measured in a patient and the CT is part of the subject's standard-of-care imaging. But it is highly unlikely that this would be the case for a healthy volunteer participating in the study. A nondiagnostic quality CT obtained from a SPECT–CT platform may yield images suitable not only for transverse thickness measurements but, also for providing crude anatomical localization and transmission data as noted previously.

A third method is to estimate $f_{Self\text{-}atten}$ from published anatomical data such as the Reference Man of ICRP Publications 23 (1975) and 89 (2002). Again, the relative insensitivity of $f_{Self\text{-}atten}$ to small changes in the thickness will help reduce the uncertainty in the activity estimation with this method.

It is perhaps worth noting that, in the specific case of the thorax, Macey and Marshall (1982) concluded that, due to its low physical density compared to soft tissue, quantification of the uptake in lung using the conjugate-view method did not require measurements of either $\Im$ and $f_{Self\text{-}atten}$. The latter conclusion is evident from Fig. 14.4, noting that the mean-free pathlength of photons in lung tissue is about three times greater than that in soft tissue.

Scatter Correction

Introduction

The above classical derivation of activity quantification in conjugate-view planar scintigraphy failed to account for the contribution of scattered photons. Narrow-beam geometry was implicitly assumed throughout the derivation. The mechanical collimation of gamma camera scintigraphy cannot exclude all of the scattered photons within the energy acceptance window. This is indeed a challenging, but absolutely essential, task in quantitative imaging (planar or tomographic). The detection of an unscattered photon provides the encoded spatial information of the source of activity, whereas a scattered photon generally comes from a position other than in the source and degrades the spatial resolution. In effect, the contribution of scatter will lead to an overestimate of the activity within the source region being imaged, image blurring and a reduced ability to quantify. In addition to collimation, other measures exist to exclude the contributions of Compton- and Rayleigh-scattered photons to the scintigraphic image formation process.

One method is by the analysis of the energies of the detected photons and the acceptance of only those photons with energies expected of unscattered photons. Compton-scattered photons have less energy following the scatter. Hence, setting a lower-energy limit below which detected photons are rejected from contributing to the image, serves to exclude a fraction of Compton-scattered photons. Due to the finite energy resolution of the detector, an excessively tight energy cut-off can exclude primary (i.e., unscattered) photons within the photopeak and thus reduce the detection efficiency. This is especially true for the NaI(Tl) scintillator, although new solid-state detectors, such as cadmium zinc telluride (CZT) with better energy resolutions are entering the nuclear medicine market. Energy discrimination becomes more complex if the radionuclide has multiple γ-ray emissions and imaging is performed using the dominant emission when there are photon emissions with higher energies. Two common examples of such radionuclides are ^{111}In and ^{131}I. Compton scatter and energy degradation of the higher-energy photon emissions increase the scattered photon component within the detector's photopeak.[9] A more sophisticated energy discrimination method is consequently required.

[9] While selecting the detector's energy discriminator to accept only the high-energy photon components will avoid this mechanism of scatter contribution, this comes at the impractical cost of thicker and heavier collimators.

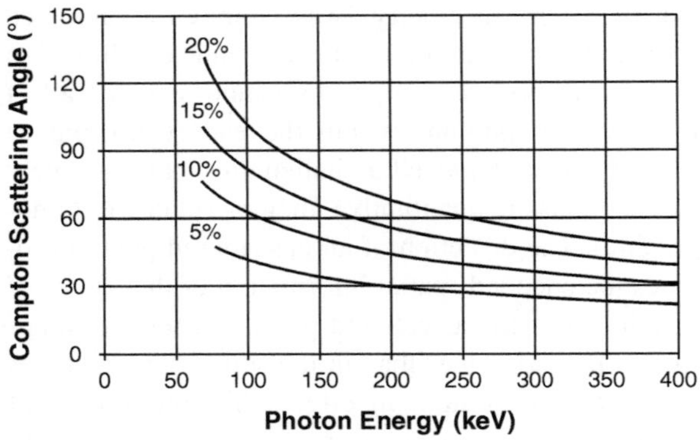

Fig. 14.5 Compton single-scattering angle as a function of photon energy for lower photon energy acceptance limits as a percentage decrease of the original photon energy. For example, for ^{99m}Tc photons, photons scattered by as much as 53.5° would have energies only 10% less than the original photon energy

The limitation of this type of energy discrimination analysis is inherent to the kinematics of Compton scatter. In particular, recall the Compton relationship between the scattered photon energy and the incident photon energy and the scattering angle. Figure 14.5 shows the Compton single-scatter angle as a function of photon energy through which scattered photons would still be accepted if the lower-energy window were 5, 10, 15, and 20% below the original photon energy. For example, if imaging 140 keV ^{99m}Tc photons and photons of energy below 126 keV were assumed to have been scattered at least once (i.e., the 10% curve of Fig. 14.5) and rejected, then photons that had been singly-scattered by as much as 53.5° would still be accepted and considered as unscattered photons.

Scatter compensation techniques can be conveniently placed into one of five categories:

- Photon energy discrimination or analysis (e.g., Buvat et al. 1994; Ichihara et al. 1993; Jaszczak et al. 1984; King et al. 1992; Koral et al. 1988)
- Factor analysis (Mas et al. 1990; Buvat et al. 1993)
- Convolution/subtraction methods (Axelsson et al. 1984; Willowson et al. 2008)
- The effective attenuation coefficient
- The buildup factor (Wu and Siegel 1984)

Many of these methods have been assessed and compared using Monte Carlo or experimental means of estimating the effects of scattered photons (e.g., Gilardi et al. 1988; Ljungberg et al. 1994; Buvat et al. 1995). The decision of which scatter compensation technique to use will be based largely upon the acquisition requirements (e.g., the full energy spectrum rather than selected energy windows) and the associated computational and implementation complexities which may or may not be manageable with the available hardware and software. Because of the great diversity of techniques available to scintigraphic imaging, only a selected sample will be presented here. The interested reader is referred to the original works cited above for discussion on the other techniques. In particular, the following discussion focuses on the energy window discrimination and buildup factor methods.

Photon Energy-Discrimination Methods

The simplest method of photon energy discrimination is, of course, to use a single energy channel centered on the photopeak corresponding to unscattered photons. For ^{99m}Tc, this is typically a ±10% energy window set over the 140 keV photopeak, i.e., only those photons with energies between 126 and 154 keV are accepted to form the image. The selection of the width of the energy acceptance window is critical. If it is set too "tight," then the sensitivity of the gamma camera diminishes rapidly; set too "broad," it accepts Compton-scattered photons which degrades both spatial resolution and the accuracy of quantification.

The dual-energy window method (Jaszczak et al. 1984, 1985) uses events detected within an energy window set below the photopeak window (e.g., between −35 and −10% of the ^{99m}Tc photopeak, or 95–126 keV), as shown in Fig. 14.6. It is assumed that there is a simple proportional relationship between the scatter contributions to the two energy windows. That is, for pixel with index i, the estimated scatter-free image is,

$$\hat{P}(i) = N_{PP}(i) - kN_{LEW}(i) \qquad (14.18)$$

where $N_{PP}(i)$ is the number of counts at pixel location i in the photopeak window and $N_{LEW}(i)$ is the number of counts at the same pixel location but in the lower-energy window. The parameter k represents the fraction of scattered events in the photopeak window (i.e., $0 \leq k \leq 1$). A commonly-accepted value is $k = 0.5$. Buvat et al. (1995) assessed this method using Monte Carlo simulations of NaI(Tl) scintigraphic imaging of

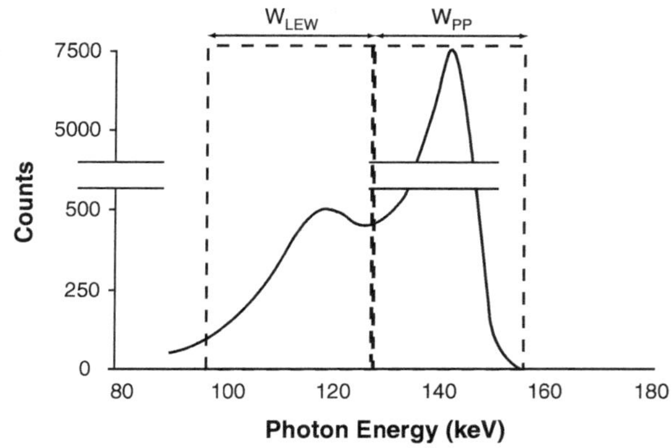

Fig. 14.6 Dual-energy window method for scatter compensation for the example of ^{99m}Tc. It is assumed that all of the photons detected in the lower-energy window (here 95–126 keV) are scattered and are in proportion to the total number of photons detected in the photopeak window (126–154 keV)

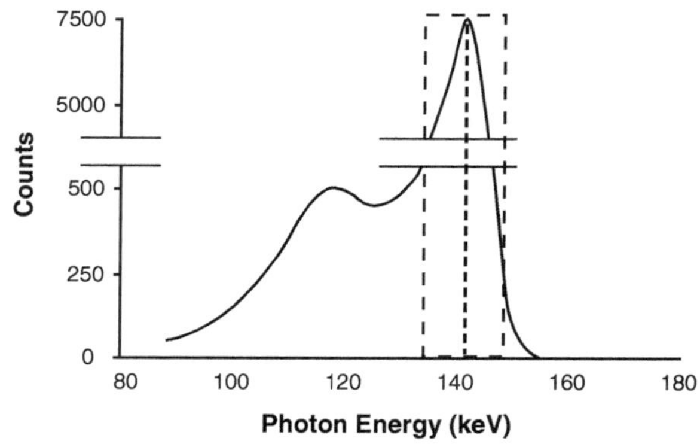

Fig. 14.7 Dual-photopeak window method. Two-equally wide and symmetric subwindows split the photopeak window

a phantom containing ^{99m}Tc. They concluded that the dual-energy window method was good for quantification provided that the number of unscattered photons per pixel exceeded 230.

A somewhat similar approach is the dual-photopeak window method as shown in Fig. 14.7. The photopeak window is split into two equal and symmetric nonoverlapping windows (King et al. 1992). The scatter component in the lower window will be greater than that in the upper window, as scatter results in energy degradation. The scatter fraction for the pixel with index i is parameterized by a polynomial function (Ljungberg et al. 1994) or by a power-function of the form,

$$\mathrm{SF(i)} = \alpha\left(\frac{\mathrm{N_{lw}(i)}}{\mathrm{N_{uw}(i)}}\right)^{\beta} + \gamma \tag{14.19}$$

where $N_{lw}(i)$ and $N_{uw}(i)$ are the total number of counts at pixel i in the upper and lower windows, respectively, and α, β and γ are parameters to fit. The number of scattered photons detected in the total photopeak window is simply,

$$\hat{\mathrm{S}}(\mathrm{i}) = (\mathrm{N_{lw}(i)} + \mathrm{N_{uw}(i)})\frac{\mathrm{SF(i)}}{1 + \mathrm{SF(i)}}. \tag{14.20}$$

The estimate of the counts in pixel i due to primary (unscattered) photons is,

$$\hat{\mathrm{P}}(\mathrm{i}) = (\mathrm{N_{lw}(i)} + \mathrm{N_{uw}(i)}) - \hat{\mathrm{S}}(\mathrm{i}). \tag{14.21}$$

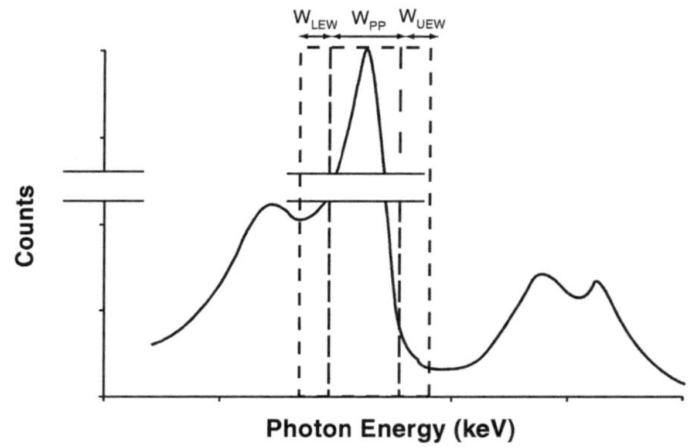

Fig. 14.8 Triple-energy window technique of photon scatter discrimination in which only primary photons arising within the energy window W_{PP} are to be used in forming the scintigraphic image and the down-scattered photons from higher-energy emissions are excluded

The coefficients can be estimated from Monte Carlo simulation or empirical measurement (Ljungberg et al. 1994; Buvat et al. 1995).

The triple-energy window (TEW) method is applicable to those cases where a radionuclide with multiple photon emissions of different energies occur and one wishes to discriminate against the higher-energy photons, e.g., ^{131}I (Koral et al. 1988; Ichihara et al. 1993). The configuration is demonstrated in Fig. 14.8.

The estimate of the counts in pixel i due to primary photons is,

$$\hat{\mathrm{P}}(\mathrm{i}) = \mathrm{P_{PP}(i)} - \mathrm{N_{LEW}(i)} - \mathrm{N_{UEW}(i)}. \tag{14.22}$$

where $N_{PP}(i)$ is the number of counts in the photopeak window and $N_{LEW}(i)$ and $N_{UEW}(i)$ are the counts in the lower and upper energy windows, respectively.

Buildup Factor Methods

The buildup factor provides another means of scatter compensation. Wu and Siegel (1984) presented a depth-dependent buildup factor method which corrected for both scatter and transmission. Returning to Fig. 14.2, the anterior and posterior count rates can be written as (neglecting, temporarily, the source region thickness so that it is assumed that $\tau = 0$),

$$\frac{dN_A}{dt} = \frac{dN_0}{dt}(SCD)B(\delta; SCD)e^{-\mu\delta} \tag{14.23}$$

$$\frac{dN_P}{dt} = \frac{dN_0}{dt}(SCD)B(h-\delta; SCD)e^{-\mu(h-\delta)} \tag{14.24}$$

where $dN_0/dt(SCD)$ is the count rate measured in air at the source-collimator distance (SCD), B(d;SCD) is the measured buildup factor for a SCD and for the source at a depth d in a scattering and attenuating medium and μ is the narrow-beam linear attenuation coefficient. Accounting for the thickness τ of the source region and following the derivations of (14.6) and (14.7), the anterior and posterior count rates are,

$$\frac{dN_A}{dt} = \frac{dN_0}{dt}(SCD)B(\delta; SCD)e^{-\mu\delta} \tag{14.25}$$

$$\frac{dN_P}{dt} = \frac{dN_0}{dt}(SCD)B(h-\delta; SCD)e^{-\mu(h-\delta)}e^{-\mu\tau}\frac{\sinh{}^{\mu\tau}/_2}{(^{\mu\tau}/_2)} \tag{14.26}$$

No transmission measurement of the subject need be made. The buildup factor is a premeasured quantity and is a function of photon energy, collimator type, energy acceptance window, and the depth, size, and overall thickness of the scattering medium. Using measured values of dN_A/dt and dN_P/dt and the predetermined buildup factor values, (14.25) and (14.26) are solved iteratively for both dN_0/dt and δ. The typical starting point is to use initial values of $B(h/2; SCD)$ and $\delta = h/2$. This method of using the buildup factor is referred to in MIRD Pamphlet No 16 as the "depth-dependent buildup factor."

Siegel et al. (1985a, b) extended this development of the buildup factor. Because of the contribution of scatter to the broad geometry inherent to scintigraphic imaging, photon attenuation is not purely exponential and a plot of the logarithm of the transmission against attenuator thickness will consequently not be a straight line but will display a shoulder. This is a common observation in the depth dose curves of external beam radiotherapy (Johns and Cunningham 1983). Siegel et al. (1985a) parameterized the transmission as a function of depth δ as,

$$\Im = 1 - \left(1 - e^{-\mu\delta}\right)^n \tag{14.27}$$

where μ is the narrow-beam linear attenuation coefficient. The form of (14.27) is the same as that of the multi-target fractional cell survival curve. As shown in (10.6), for large δ, a binomial expansion of this expression leads to the expression,

$$\Im \cong ne^{-\mu\delta} \quad (\text{large}\ \delta). \tag{14.28}$$

It is evident from (9.91) and (14.28) and Fig. 9.17 that the exponent n of (14.28) is the buildup factor at infinite depth, $B(\infty)$. Equation (14.28) is thus rewritten as,

$$\Im = 1 - \left(1 - e^{-\mu\delta}\right)^{B(\infty)} \tag{14.29}$$

and $\Im$ as a function of depth is demonstrated in Fig. 14.9.

This approach is the equivalent of the pseudo-extrapolation number method of Thomas et al. (1983) in which scatter compensation is accounted for by replacing $\Im$ in the denominator of (14.10) with $\sqrt{B(\infty)\Im}$. Again, allowing for a finite source thickness, the count rates are modified to,

$$\frac{dN_A}{dt} = \frac{dN_0}{dt}(SCD) \times \left(1 - \left(1 - e^{-\mu\delta}\right)^{B(\infty)}\right)e^{-\mu\tau}\frac{\sinh \mu\tau/2}{(^{\mu\tau}/_2)} \tag{14.30}$$

$$\frac{dN_P}{dt} = \frac{dN_0}{dt}(SCD) \times \left(1 - \left(1 - e^{-\mu(h-\delta)}\right)^{B(\infty)}\right)e^{-\mu\tau}\frac{\sinh \mu\tau/2}{(^{\mu\tau}/_2)}. \tag{14.31}$$

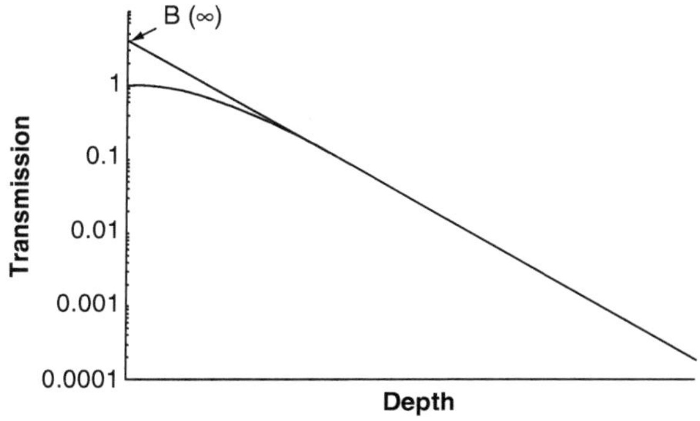

Fig. 14.9 Transmission curve for a broad-beam geometry allowing scatter. The extrapolation from the purely exponential portion of the curve to the ordinate yields the buildup factor at infinite depth

δ can be solved for numerically, subject to the constraint that $0 \leq \delta \leq h$, from the ratio of (14.30) and (14.31),

$$\frac{dN_A/dt}{dN_P/dt} = \frac{1-\left(1-e^{-\mu\delta}\right)^{B(\infty)}}{1-\left(1-e^{-\mu(h-\delta)}\right)^{B(\infty)}}. \tag{14.32}$$

Once δ has been determined, then dN_0/dt can be calculated from either (14.30) or (14.31).

Background Correction Methods

The background activity will contribute to the net number of detected photons in the anterior and posterior planar views and failure to account for this will lead to an overestimate of the activity within the organ. This subsection reviews a number of means of correcting for background.

Simple Subtraction

The simplest (and crudest) correction for this background source is to assume that the number of counts per pixel (count density) in an RoI set adjacent to the organ RoI is the same background count density in the organ RoI. The expression for the geometric mean is modified to,

$$\left\langle\frac{dN}{dt}\right\rangle_{geom} = \sqrt{\left(\frac{dN_A}{dt} - \left(\frac{dN_{A,Bkgd}}{dt}\left(\frac{W_{RoI}}{W_{Bkgd}}\right)\right)\right)} \times \sqrt{\left(\frac{dN_P}{dt} - \left(\frac{dN_{P,Bkgd}}{dt}\left(\frac{W_{RoI}}{W_{Bkgd}}\right)\right)\right)} \tag{14.33}$$

where $dN_{A,Bkgd}/dt$ and $dN_{P,Bkgd}/dt$ are the count rates measured from the background RoI in the anterior and posterior views, respectively, W_{RoI} is the number of pixels in the RoI of interest and W_{Bkgd} is the number of pixels in the background RoI. Neglecting the errors that will be introduced by the implicit assumption that this background signal remains constant between this RoI and the organ RoI, the method will overestimate the background contribution. This is a result of its inclusion of the contributions of photons emitted in the volume of surrounding tissue equal in thickness to that of the organ. If the background region is in close proximity to the source region, it will contain a contribution of scatter from the source region which will elevate the background counts. Similarly, there will be additional background counts in the source region RoI from surrounding tissues and the tendency will be for the background subtraction to partially compensate for this combination of scatter corrections (Fleming 1996).

The method also assumes that the background count rate is uniform over both the RoI of interest and the background RoI. The following two methods correct for this latter assumption.

Bilinear Interpolation

A bilinear interpolative approach to correct for the variability of the background count rate over the RoI of interest has been proposed by Goris et al. (1976) and can be described using Fig. 14.10.

A rectangular RoI is set about a target RoI. The counts at the two pixels at locations (k,m) and (l,n) (where k,l,m and n are all integers), which are outside the target RoI and, hence, are background are used to estimate the background contribution at the pixel at location (i, j) (where i and j are also integers). The total

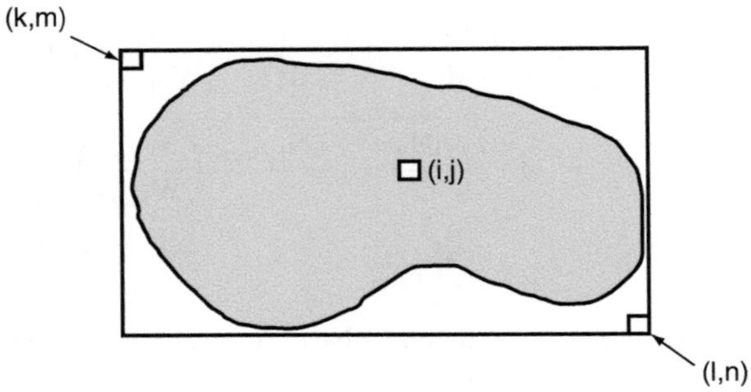

Fig. 14.10 Bilinear interpolative method for correcting for the background activity contribution. An organ region-of-interest is surrounded by a rectangular region-of-interest. See text for details

number of counts in the pixel at (i, j) is c_{ij} and the background contribution is,

$$b_{ij} = \frac{b_i + b_j}{2} \tag{14.34}$$

where b_i and b_j are given by the bilinear interpolations,

$$b_i = \frac{(j - m)c_{in} + (n - j)c_{im}}{n - m} \tag{14.35}$$

and

$$b_j = \frac{(i - k)c_{lj} + (l - i)c_{kj}}{l - k}. \tag{14.36}$$

The background-corrected counts in pixel (i, j) is, then,

$$\hat{c}_{ij} = c_{ij} - b_{ij} \tag{14.37}$$

Cauchy Integral Method

This approach uses the Cauchy integral theorem which states that, if the function f is analytic within and on a simple closed contour C, then the values of f within the contour C are determined by the values of f on the contour C. In other words, if z_0 is any point interior to C, then the value of the function at z_0 is given by the contour integral,

$$f(z_0) = \frac{1}{2\pi i} \oint_C dz \frac{f(z)}{z - z_0}. \tag{14.38}$$

Recasting this result, the background signal at a complex-valued point z_0 within the target RoI is given by,

$$b(z_0) = \frac{1}{2\pi i} \oint_C dz \frac{c(z)}{z - z_0} \tag{14.39}$$

where c(z) is the number of counts in a pixel at the complex-valued point z on an outside contour of the target RoI. Should the contour be circular with radius R (Nichols et al. 1987), then the background signal at the polar coordinate (r, θ) is given by the expansions,

$$b(r, \theta) = \frac{\alpha_0}{2} + \sum_{k=1}^{M} \left(\frac{r}{R}\right)^k (\alpha_k \cos k\theta + \beta_k \sin k\theta) \tag{14.40}$$

where the coefficients of the expansion are given by,

$$\alpha_k = \frac{2}{P} \sum_{j=1}^{P} b\left(R, \frac{2\pi j}{P}\right) \cos \frac{2\pi jk}{P} \tag{14.41}$$

$$\beta_k = \frac{2}{P} \sum_{j=1}^{P} b\left(R, \frac{2\pi j}{P}\right) \sin \frac{2\pi jk}{P} \tag{14.42}$$

and P is the number of discrete points (pixels) around the circular contour at which the integral is to be evaluated. Nichols et al. describe the practical implementations of the above equations, although the complexities of such implementations are evident.

Houston and Sampson (1989) have provided an experimental comparison of both the bilinear interpolative and Cauchy integral background methods using phantom and clinical data.

Correction for Over-Subtraction (I) Thomas Method

As noted above, there will be an overcorrection for background caused by the inclusion of the counts arising from the background ROI corresponding to the thickness of the organ. Three methods of varying complexity to correct for this over-subtraction are presented here. Thomas et al. (1976) extended the conjugate-view methodology to account for the activity "above" and "below" the object. In a straightforward, but tedious, manner the conjugate-view result is modified by a multiplicative factor accounting for background,

$$A = \frac{1}{\kappa\sqrt{\Im}} \left\langle \frac{dN}{dt} \right\rangle_{geom} f_{Self-atten} f_{Bkgd} \tag{14.43}$$

where

$$f_{Bkgd} = \sqrt{1 + \left(\frac{\gamma_1 f_2}{f_1}\right)^2 + \left(\frac{\gamma_3 f_2}{f_3}\right)^2 + 2\frac{\gamma_1\gamma_3 f_2^2}{f_1 f_3}} \times \sqrt{\cosh\left(\frac{\mu_1 t_1 + 2\mu_2 t_2 + \mu_3 t_3}{2}\right)} + \sqrt{2\frac{\gamma_1 f_2}{f_1}\cosh\left(\frac{\mu_1 t_1 + \mu_2 t_2}{2}\right)} + \sqrt{2\frac{\gamma_3 f_2}{f_3}\cosh\left(\frac{\mu_2 t_2 + \mu_3 t_3}{2}\right)}. \tag{14.44}$$

For conciseness, the subscripts 1 and 3 refer to the anterior and posterior background regions and $t_1 = \delta$, $t_2 = \tau$ and $t_3 = h - \delta - \tau$, different linear attenuation coefficients for the three regions are allowed,

$$f_i = \frac{\sinh \mu_i t_i/2}{(\mu_i t_i/2)} \quad i = 1, 2, 3 \tag{14.45}$$

and the ratios of the background activities in the anterior and posterior regions to that in the source region-of-interest are given by $\gamma_1 = A_1/A_2$ and $\gamma_3 = A_3/A_2$.

Correction for Over-Subtraction (II) Buijs Method

Buijs et al. (1992) proposed a modification which accounted for the thickness of the source organ,

$$\left\langle \frac{dN}{dt} \right\rangle_{geom} = \sqrt{\left(\frac{dN_A}{dt} - f\left(\frac{dN_{A,Bkgd}}{dt}\left(\frac{W_{RoI}}{W_{Bkgd}}\right)\right)\right)} \times \sqrt{\left(\frac{dN_P}{dt} - f\left(\frac{dN_{P,Bkgd}}{dt}\left(\frac{W_{RoI}}{W_{Bkgd}}\right)\right)\right)} \tag{14.46}$$

where f is the fraction of activity in the background RoI that is not from the thickness equivalent to the source RoI interest,

$$f = 1 - \frac{\tau}{h} \tag{14.47}$$

using the nomenclature of Fig. 4.2.

Correction for Over-Subtraction (III) Kojima Method

Kojima's method (Kojima et al. 1993) is similar in form to that of Buijs et al. in that it also modifies the background correction estimated from the background RoI by a multiplicative factor. However, a different multiplicative factor is used in the anterior and posterior views,

$$\left\langle \frac{dN}{dt} \right\rangle_{geom} = \sqrt{\left(\frac{dN_A}{dt} - f_A\left(\frac{dN_{A,Bkgd}}{dt}\left(\frac{W_{RoI}}{W_{Bkgd}}\right)\right)\right)} \times \sqrt{\left(\frac{dN_P}{dt} - f_P\left(\frac{dN_{P,Bkgd}}{dt}\left(\frac{W_{RoI}}{W_{Bkgd}}\right)\right)\right)} \tag{14.48}$$

where the f_A and f_P are calculated assuming that the concentrations of background activities leading to A_1 and A_2 are the same,

$$f_A = 1 - (1 - e^{-\mu\tau})\left(\frac{e^{-\mu(h-\delta-\tau)}}{1 - e^{-\mu h}}\right) \tag{14.49}$$

$$f_P = 1 - (1 - e^{-\mu\tau})\left(\frac{e^{-\mu\delta}}{1 - e^{-\mu h}}\right) \tag{14.50}$$

Buijs et al. (1998) have provided an experimental comparison of the simple background correction and the over-subtraction correction methods of Thomas, Buijs, and Kojima based upon a phantom study replicating a kidney within a uniform background activity distribution. They concluded, in this one example, that the Buijs and Kojima methods were superior and that the Kojima result was preferable at low source-to-background activity ratios.

Overlapping Regions of Activity

Despite its simplicity, the conjugate-view method has the disadvantage of the likelihood of overlapping regions each containing activity and the need to isolate the activities in each source region. Common examples are the overlap of the liver and the gallbladder, the spleen and the left kidney and the liver and right kidney. The second example can be further complicated by possible activity in the bowel overlapping the left kidney.

There are a number of possible approaches to approximately isolating the activity contents in the individual overlapping regions. The simplest is applicable to paired organs, such as the kidneys. One can measure the activity content in the organ with no overlap and assume that its partner has the same amount of activity.[10] Another method suggested by Stabin (2008) is to map the RoI drawn over source regions which are not overlapped in an image taken at a different time and apply it to the image when overlap occurs.

Another approach to isolating the overlapping activities is described here. Consider the general case of two organs overlapping a third (this could be the case of the right kidney overlapped by the liver and by bowel content activity). A simple example is shown in Fig. 14.11.

In this example, region 1 could correspond to liver, region 2 to the right kidney and region 3 to the bowel contents. A total of five RoIs would be drawn: three over the unoverlapped regions P_1, P_2, and P_3 and two over the overlaps P_{12} and P_{23}. The mean numbers of counts per pixel in the un-overlapped regions are $N_1/P_1, N_2/P_2,$ and N_3/P_3 and it is assumed that these mean numbers are the same in the overlapped regions. Hence, the actual numbers of counts in the three regions are simply,

$$N_1' = \frac{N_1}{P_1}(P_1 + P_{12}) \quad (14.51)$$

$$N_2' = \frac{N_2}{P_2}(P_2 + P_{12} + P_{23}) \quad (14.52)$$

$$N_3' = \frac{N_3}{P_3}(P_3 + P_{23}) \quad (14.53)$$

The above approach requires that the user be able to accurately draw the five RoIs or for the image analysis software to allow the user to draw only the three main RoIs and allow it to isolate the overlap and nonoverlap counts and pixels.

[10]The condition of an atrophic kidney with compensatory hypertrophy in the contralateral kidney, which is not unknown, would make this method unusable.

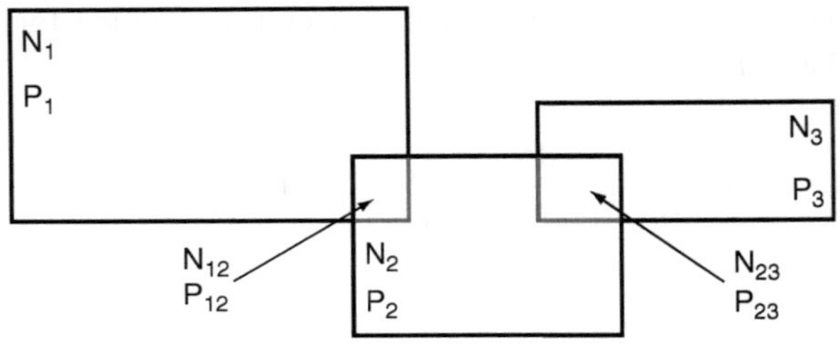

Fig. 14.11 An example of two regions (1 and 3) overlapping a third (2). The N_i and P_i ($i = 1, 2, 3$) are the numbers of counts and number of pixels in the nonoverlapping regions. The N_{ij} and P_{ij} ($i = 1, 2$; $j = 2, 3$) are those in the overlaps

Collimator Selection

The selection of the collimator to use in a Phase I study of a diagnostic radiopharmaceutical will be driven by the need to maximize detector sensitivity so as to minimize the amount of administered activity and the consequent absorbed dose to the (typically, healthy) volunteer. Hence, the collimator will be a high sensitivity or general purpose type with an energy selection based upon the radionuclide to be imaged. In nearly all cases, a parallel hole collimator is sufficient for extracting a biodistribution suitable for calculating the absorbed dose from. For a ^{99m}Tc- or ^{123}I-labeled radionuclide, a low-energy collimator is sufficient; a high-sensitivity collimator has a reduced spatial resolution but high sensitivity (about 400 cps/MBq), whereas a general purpose collimator will have a better spatial resolution but a reduced sensitivity (about 150 cps/MBq). In Phases II and III, the collimator selection will be defined by the clinical problem at hand to image.

Validation of the Conjugate-View Method

The conjugate-view method was tested empirically by Hammond et al. (1984) using a model of a 4 cm diameter lesion in a phantom for ^{131}I and errors of less than 10% were reported. An additional validation of the method was performed by Eary et al. (1989) for ^{131}I using two test models. The first was an anthropomorphic phantom with chambers representing organs containing known concentrations of activities. The second was through in vivo imaging of dogs injected with a ^{131}I-labeled antibody which were subsequently sacrificed and organs harvested and assayed for radioactive content. The ^{131}I activities in the organs were compared with those estimated through the application of (14.10). Higher levels of accuracy were stated.

An interesting validation of the conjugate-view method of quantification was performed by Jönsson et al. (2005) through Monte Carlo simulation of ^{99m}Tc photon transport in the Zubal voxel-based phantom described in Chap. 12 through to a NaI(Tl) crystal with a low-energy general purpose collimator. The biokinetics of the ^{99m}Tc in the phantom were modeled by measurements of the biodistribution of ^{99m}Tc-sestamibi in the Wistar rat. Two methods of scatter correction, the effective linear attenuation coefficient and the TEW method, were examined. The Buijs method of background correction was applied. Accuracy of the order of ±10% was obtained in nonoverlapping organs where scatter was corrected by the TEW method. However, overestimation of organ activities by more than 90% was possible in the case of overlapping organs.

14.3.2.2 Quantitative SPECT for Dosimetry Calculations

Introduction

Planar scintigraphy as a means of obtaining in vivo activity data has the two great advantages of relative ease of implementation and the lack of a need of high administered activities. This latter point is particularly important in diagnostic radiopharmaceutical research for which the Phase I component generally relies upon healthy volunteers, for whom the absorbed dose must be minimized. But planar scintigraphy suffers usually from the disadvantages of a lack of information on source organ thickness and of the overlaps between regions in the two conjugate views. While two correction factors for these two deficiencies have been discussed, tomographic imaging resolves these by providing three-dimensional image datasets.

Single-photon emission computed tomography (SPECT) suffers, itself, from distinct disadvantages. As SPECT is based upon the acquisition of multiple projections over an orbit (typically 64 or 128 projections) and over a clinically-practical acquisition duration, a higher amount of administered activity than that used in planar imaging is required in order to reduce the quantum noise in each image projection to acceptable levels. Depending upon the number of projections used and the number of heads on the gamma camera, the orbital period is typically about 30 min, but can range from about 15 min to about 45 min. Combining this protracted image acquisition duration with the fact that the axial length of the limited useful field-of-view in a typical conventional gamma camera, whole-body SPECT acquisitions are impractical and whole-body tomographic biodistributions cannot be obtained. These two limitations of SPECT thus currently restrict the use of quantitative SPECT in nuclear medicine dosimetry to therapeutic applications where the absorbed dose to a relatively small target volume need only be known. As with the planar scintigraphy methodology just discussed, quantitative SPECT requires careful consideration of image degradation due to attenuation and scattered photons. Spatial resolution is restricted by the partial-volume effect so that accurate quantitative data can be had only from structures about three times or greater than the total system spatial resolution (Hoffman et al. 1977). Even so, quantitative SPECT does play a role in nuclear medicine dosimetry, primarily in therapeutic applications. This subsection summarizes the principles of quantitative SPECT. An excellent detailed discussion can be found in the review by Rosenthal et al. (1995). Quantitative SPECT can be differentiated into relative and absolute quantification. In relative quantification, quantification is of the relative numbers of counts within RoIs within the same image or sets of images. Absolute quantification, however, yields an absolute measure of activity within the measure (e.g., Bq/voxel, Bq/mL, Bq/g, or Bq/organ). Examples of quantitative SPECT used in dosimetry calculations can be found in Strand et al. (1994b), Giap et al. (1995), and Gonzalez-Trotter et al. (2001).

The argument of whether or not SPECT can deliver sufficiently accurate quantification of in vivo activity remains active as this ability to deliver quantifiable data is dependent upon corrections for photon scatter and attenuation.

Scatter Compensation

Compensation for the effects of scattered photons is crucial in SPECT imaging as these lead to reductions in image contrast and spatial resolution and affect the relationship between the source region activity and the corresponding signal in the reconstructed image. The methods of scatter correction described in the above discussion of quantitative planar imaging are also applicable to SPECT; indeed many of the scatter correction methods noted were developed originally for SPECT imaging.

Attenuation Correction

The discussion of quantitative conjugate-view planar scintigraphy highlighted the need for the correction of photon attenuation in order to accurately quantify activity in vivo. Failure to account for attenuation will lead to an underestimation of the in vivo activity being measured and a degradation of image quality at depth.

Attenuation corrections in SPECT can be provided in a variety of ways. An analytical form is through the simple assumption of Chang (1978) that the imaged anatomy is homogeneous with a fixed linear attenuation coefficient applied. This correction is applied postreconstruction; for each reconstructed pixel, an average attenuation factor is calculated and the pixel value in the uncorrected tomographic image divided by this value. Clearly, this is a simplistic model and is not applicable to heterogeneous anatomy, such as the thorax.

Empirical attenuation correction methods are based upon attenuation measurements. These require a transmission source. This could be a radionuclide (e.g., ^{153}Gd which emits a range of X and γ rays, of which the 97.5 and 103 keV are used for transmission imaging) or polyenergetic X-rays. In modern SPECT–CT platforms, attenuation correction is provided by a CT scan acquired of the subject at very nearly the same time as the acquisition of the emission image. In all cases, the attenuation correction is obtained by first imaging without the subject present and then with. In CT-derived attenuation, the attenuation map is described by the matrix of CT numbers associated with each pixel in a tomographic slice. The CT number is a measure of the attenuation and is defined by,

$$\mathrm{CT\#} = \left(\frac{\mu}{\mu_{\mathrm{H_2O}}} - 1\right) \times 1000 \qquad (14.54)$$

where μ is the linear attenuation coefficient of the medium and $\mu_{\mathrm{H_2O}}$ is the linear attenuation coefficient for water. The unit of the CT number is the Hounsfield unit (HU); for air, the CT number is −1,000 HU and for water it is 0 HU. As the attenuation coefficient is required at the emission photon energy whereas that measured is for the polyenergetic X-ray beam from the CT scanner, conversion of the linear attenuation coefficients are required, as demonstrated by the calibration curve of Fig. 14.12.

Such a calibration curve is obtained empirically by imaging a phantom containing materials of different attenuation coefficients. In Fig. 14.12, the conversion for CT numbers less than zero is a straightforward proportionality,

$$\mu_{\mathrm{Tissue,Em}} = \frac{\mu_{\mathrm{H_2O,Em}}}{1000} \times \mathrm{CT\#} \qquad (14.55)$$

where the Em subscript refers to the emission photon energy. The proportionality is the consequence of the near-equivalence of atomic numbers between soft tissue and water. Bone yields CT numbers exceeding zero, but the slope changes. Hence, the linear mapping of the linear attenuation coefficient is

$$\mu_{\mathrm{Tissue,Em}} = \mu_{\mathrm{H_2O,Em}} + \mu_{\mathrm{H_2O,kVp}}\left(\frac{\mu_{\mathrm{Bone,Em}} - \mu_{\mathrm{H_2O,Em}}}{\mu_{\mathrm{Bone,kVp}} - \mu_{\mathrm{H_2O,kVp}}}\right) \times \frac{\mathrm{CT\#}}{1000} \qquad (14.56)$$

where the kVp subscript refers to the X-ray spectrum of the CT scanner.

Reconstruction

In order to use a SPECT image as a means of quantification, the SPECT data must reflect accurately the spatial distribution of activity (Rosenthal et al. 1996). This then places demands upon the reconstruction method to transfer the quantitative data from the projection image set to the tomographic image. SPECT reconstruction is performed using either backprojection or iterative methods (Bruyant 2002).

In basic backprojection, the measured projection data are reprojected into image space. This results in two different types of artifacts. The first is the streak

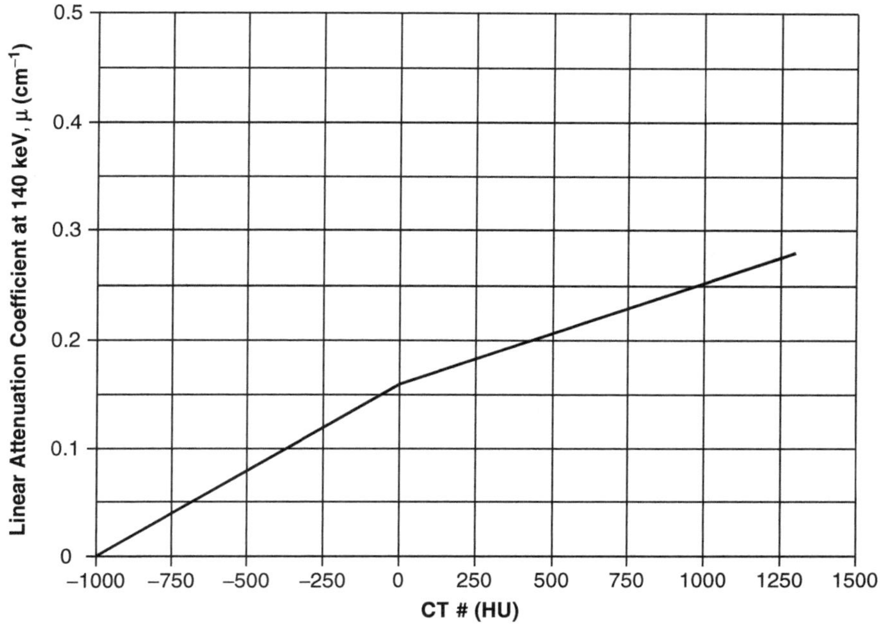

Fig. 14.12 Photon linear attenuation coefficient at 140 keV as a bilinear function of CT number

artifact which results in a "star-like" tomographic image. The other is due to quantum noise due to the random nature of radioactive decay. The former can result in counts appearing outside the subject's body; further, if the reconstruction method's normalization requires that the total number of counts from the projection data be preserved in the tomographic volume, then an underestimate of the activity within the body will be had. While the star artifact can be minimized by the use of higher numbers of projections, this will result in an increase in the amount of quantum noise in each projection if a constant orbit time is maintained. These effects can be reduced in image space by convolving the image data with appropriate filter, a procedure known as filtered backprojection (FBP). As convolution can be computationally intensive, the image data are Fourier transformed into the frequency domain. A ramp filter is applied to remove the streak artifacts in conjunction with a low-pass filter (e.g., Butterworth) to remove noise. The filtered data are then inverse Fourier transformed to yield the reconstructed volume, but this does come at a cost due to degraded spatial resolution as a result of the low-pass filtering. In addition, the reconstruction quality improves with the number of projections acquired.

Iterative reconstruction compares the projections from the reconstructed image with the acquired projections and iteratively modifies the reconstructed image in order to minimize the difference between the two projection sets. Although iterative reconstruction predated FBP in its applications to SPECT, it has become increasingly popular now due to the availability of increased computation power. A statistical approach to this type of iterative reconstruction is the maximum-likelihood-expectation maximization (MLEM) method (Dempster et al. 1977). This method is perhaps the most commonly-used iterative reconstruction method used in SPECT (Lange and Carson 1984). Newer approaches, such as ordered subsets expectation maximization (OSEM) method, partitions the data into subsets before modifying the image (Hudson and Larkin 1994). Iterative reconstruction offers the additional advantage of incorporating attenuation correction during the reconstruction process (e.g., Moore et al. 1982).

The reader can refer to the references provided in order to obtain a deeper knowledge of tomographic reconstruction. But what is of interest here is what reconstruction method to use when a tomographic data set is to be used for activity quantification. Clearly, the reconstruction should have as little noise as feasible so that organs can be identified and segmented for activity quantification. In addition, the reconstruction must be linear; in other words, the integrity of the activity determined in the projection data must be retained in the reconstructed image. For further practical advice in using SPECT quantification, the reader is referred to Rosenthal et al. (1996).

14.3.3 Positron-Emitting Radionuclides

14.3.3.1 Introduction

Positron emission tomography (PET) offers a number of significant advantages over SPECT, a key one being the use of electronic "collimation" rather than attenuative collimation. Because of the resulting high-acceptance solid angle, this yields a detection efficiency of up to two orders-of-magnitude greater than that of SPECT for comparable fields-of-view which, as a result, makes whole-body tomographic biodistribution measurements feasible for positron-emitting radionuclides, unlike the current technology of SPECT. On the other hand, PET is a more expensive technology than SPECT and demands an involved infrastructure for the production of positron-emitting radionuclides (Chap. 3) and the manufacturing of PET radiopharmaceuticals. One physical attribute of the majority of positron-emitting radionuclides that offers challenges in acquiring whole-body biodistributions is the relatively short half-life. For example, those of ^{11}C, ^{15}O, and ^{18}F are 20.38 min, 122.24 s and 109.77 min, respectively; that of ^{124}I is 4.18 days. Hence, accurate whole-body biodistributions necessary for internal radiation dosimetry calculations can be acquired for ^{18}F and ^{124}I, but would be challenging for ^{11}C. Clearly, ^{15}O would not be able to offer suitable data, although the ingenious method of external placement of TLDs and the use of precalculated S-factors by Deloar et al. (1997) provided organ absorbed dose estimates for a ^{15}O study as described earlier.

This subsection discusses the use of PET in biodistribution and radiation dosimetry evaluation and the dosimetric attributes of positron-emitting radionuclides. The discussion is, however, limited to providing an overview only of the attributes of PET that are relevant to the determination of a biodistribution of a positron-emitting radiopharmaceutical. It is not intended to be a detailed exposition of PET technology or data acquisition, reconstruction, and quantitative analysis; the reader is directed towards, for examples, the contemporary chapters by Bailey et al. (2003), Bailey (2003), Koeppe (2008), Meikle and Badawi (2003) and Turkington (2008), and the book by Saha (2005) for such comprehensive accounts. Rather, the limited intent of this subsection is to highlight those attributes of PET imaging that the nuclear medicine physicist need be aware of when measuring the biodistribution of a positron-emitting radionuclide for later evaluation of the associated internal radiation dosimetry. In particular, the impacts of scatter and random events and their corrections, two-dimensional and three-dimensional acquisition modes, corrections for attenuation, and that of image reconstruction upon quantitative whole-body biodistribution measurements are reviewed.

14.3.3.2 Photon Detection in PET

Event Types

The principle of PET is the detection within a coincidence time window of two collinear 511 keV photons resulting from an electron–positron annihilation event. Current-generation PET detectors are typically scintillators with high-atomic numbers so as to increase the photoelectric absorption of the 511-keV photon. In addition to the efficiency with which a photon is stopped, the ideal scintillator would have maximum light output in the prompt fluorescence response rather than through the phosphorescence and delayed fluorescence channels; i.e., have as rapid a rise time as possible and as small a decay constant as possible. This is necessary in order to maximize the temporal resolution of the detector. High conversion efficiency (i.e., high light output) is required; common scintillator types used in contemporary PET technology are BGO, LSO, and GSO (Table 9.2).

In modern-day PET scanners, the detectors are arranged contiguously in a ring around the field-of-view (FoV) to be imaged. Multiple rings are sited axially in order to provide an extended axial FoV. A typical radial (transaxial) FoV is about 60 cm and the typical axial FoV is about 15 cm. Tungsten or lead annular septa can be inserted between these rings so as to reduce the detection of scattered photons depending upon whether the acquisition is in two- or three-dimensional mode, as described below. All commercially-available PET scanners are now PET/CT hybrid platforms, where the CT component is used to provide transmission data for attenuation correction to the emission data and/or anatomical imaging for subsequent coregistration of functional and anatomical data. Attenuation correction using external radiation sources and the consequent radiation absorbed dose presented to the subject are discussed later.

Photon pairs are created as a result of an electron–positron annihilation, each with an energy equal to the positron/electron rest-mass, 511 keV. The annihilation site is unlikely to be coincident with the site of the positron emission (i.e., the position of the radionuclide), but will rather be at some distance away following the positron's moving through and slowing down in tissue. This sets a minimum uncertainty in the spatial localization of the radionuclide. The maximum distance between emission and annihilation sites is set by the range of the emitted positron. This, in turn, is a function of the positron kinetic energy spectrum which is of the form predicted by the Fermi theory and which is dictated by the transition matrix element, the kinematics of three-body final state and the atomic number of the postdecay nucleus (which presents a static repulsive Coulomb field to the positron). Superimposed upon this range are the effects of multiple scattering and range straggling. For example, the effective positron ranges for ^{18}F and ^{82}Rb are less than 1 mm and of the order of 10 mm, respectively. The two annihilation photons are exactly collinear only if the electron–positron annihilation occurs at rest; deviation from collinearity in dual-photon emission is a consequence of annihilation in flight (in which, collinearity exists only in the center-of-mass reference frame). As a result, there will be a spread in the photon emissions of approximately $\pm$ 0.25° around the 180° collinear trajectory. As described in Chap. 7, single photon emissions can result from positron annihilation on a bound atomic electron where the recoil nucleus takes up the remnant momentum and triple emission follows from the decay of the triplet state of positronium and the subsequent annihilation . These annihilations photons have energies differing from 511 keV and would not be recorded in the image-forming process through energy discrimination.

Consider the annihilation event A shown in Fig. 14.13 and which represents a "true" detected coincidence in which the directions traveled by the photons are 180° apart. A photon is detected first in a scintillator element and generates an electronic signal with a waveform of the shape of Fig. 9.40.[11] If this output exceeds a predefined amplitude threshold, a photon is defined as having been detected and a coincidence window of width T_{Window} triggered, as shown in Fig. 14.14. Should a second detection event occur in another detection element within the time T_{Window}, then a coincidence is defined to have occurred. In the case of the "true coincidence" of Event A, the line-of-response (LoR) joining the two detection elements S_{A1} and S_{A2} is coincident with the trajectories of the two annihilation photons. This would demonstrate correctly that the annihilation event occurred at some point along the LoR and that the annihilation photons did not undergo any nature of interaction between the annihilation site and the detectors.

It is also possible for one of the photons from an annihilation event to be detected in one element (S_{B1}), but for the second photon to undergo Compton scatter in vivo and be subsequently detected in another element (S_{B2}). This does not reflect a collinear detection (Event B) and is known as a scatter coincidence or

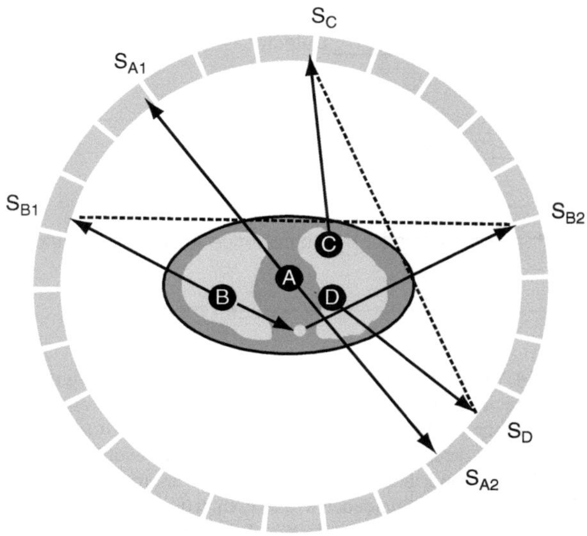

Fig. 14.13 True, scatter and random coincidences in PET. Refer to the text for a description. Although not shown, multiple coincidences result from a true coincidence and a random event occurring within the coincidence window T_{Window} of Fig. 14.14

[11]See, for example, Bailey (2003) for a description of the scintillation light photon-electron conversion detection element.

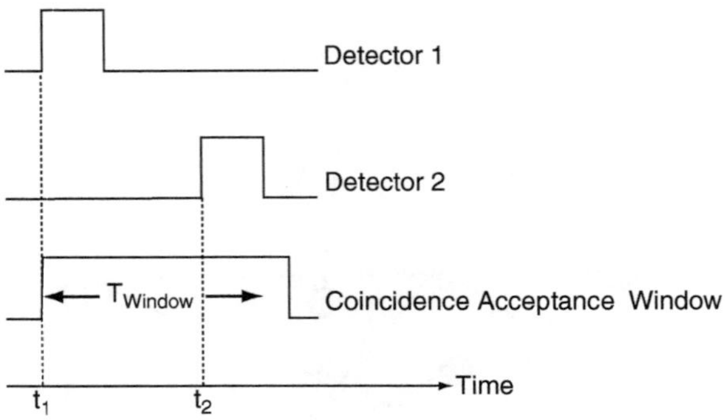

Fig. 14.14 PET coincidence timing. A scintillation signal in one detector, due to a detected photon, which exceeds a predefined threshold raises a start pulse at time t_1 which triggers a coincidence acceptance window of temporal width T_{Window}. Should a second detection event occur in another detector at time t_2 within the coincidence window, such that $t_2 - t_1 \leq T_{Window}$, then a coincidence between the two detector elements is defined to have occurred

event. The resulting LoR, shown by the dotted line in Fig. 14.13, does not include the correct position of the original annihilation event. Scattered coincidences add a background to the image created by true coincidences which changes relatively slowly over position and leads to reduced contrast and, importantly for biodistribution estimates, an overestimate of activity concentration.

A third possible coincidence scenario is demonstrated by two independent annihilation events occurring at points C and D and a photon from each annihilation being detected at elements S_c and S_D within the coincidence detection window. This is known as a random coincidence or event and, again, an erroneous LoR is generated and an overestimate of the activity concentration occurs.

A "prompt" event is defined as one in which any pair of photons, giving rise to true coincidences, scattered coincidences, or random coincidences are detected in two detectors within the coincidence time window.[12] The rate at which prompt coincidences occur is the sum of the rates of these three event types,

$$\dot{N}_{Prompt} = \dot{N}_{True} + \dot{N}_{Scat} + \dot{N}_{Rand} \tag{14.57}$$

In (14.57), $\dot{N}_{True}$ represents those events which are to be used in the tomographic reconstruction in order to yield a quantitative measurement of activity concentration. Hence, it is necessary to determine $\dot{N}_{Scat}$ and $\dot{N}_{Rand}$ and to remove these contributions from $\dot{N}_{Prompt}$.

Random Events

The setting of the length of T_{Window} is dictated by two main factors. The first is the variability set by the inherent timing resolution resulting from the light output rise time of the scintillator and the timing electronics. The second factor is a consequence of the combination of the photon time-of-flight and the radial FoV, the dimension which is limited by the scanner diameter. As the detector ring diameter is of the order of 100 cm for a whole-body scanner, the scanner geometry sets a minimum value of 3.3 ns for T_{Window} purely on the basis of photon transport. Typical values of T_{Window} are 12 ns for BGO scintillators decreasing to 8 and 6 ns for GSO and LSO scintillators, respectively.

The rate at which random coincidences are detected will increase with both T_{Window} and the amount of activity present. $\dot{N}_{Rand}$ is proportional to the product of the acceptance window length and, due to the statistically independent detection of photons by a detector pair, the square of the activity. These random events add uncorrelated data and reduce image contrast. On the other hand, true coincidence events are correlated and, hence, linearly proportional to the activity. As a result, the random coincidence event rate will exceed that of the true coincidences at high levels of activity.

The random coincidence rate can be determined by measuring the rate at which "single" events are detected. These are events detected without correlation to any of the paired coincidence detection. Consider two detectors identified by indices i and j and an event detected by detector i at time $t = 0$. A coincidence timing window of duration T_{Window} is generated. If there is an event in detector j during the time interval $-\frac{1}{2}T_{Window}$ and $+\frac{1}{2}T_{Window}$, then a coincidence between detectors i and j is concluded to have

[12] A scattered coincidence is, strictly speaking, a "true" coincidence as both detections arise from the same annihilation event, even though an incorrect LoR results. For the purpose of discussion here, a "true" coincidence is defined as that arising from the pair of unscattered photons created by the same electron–positron annihilation.

occurred and the line-of-response LoR_{ij} between the two detectors generated. Note that the total time during which a coincidence may occur is $2T_{Window}$ which is referred to as the resolving time. Let $\dot{N}_i$ be the detection rate of single events in detector i and $\dot{N}_j$ be that of detector j. It is assumed that no true coincidences occur so that $\dot{N}_i$ and $\dot{N}_j$ can be assumed to be statistically independent. Then, the rate of random coincidences is,

$$\dot{N}_{Rand} = 2T_{Window}\dot{N}_i\dot{N}_j \tag{14.58}$$

If the detector dead time is sufficiently small that it can be neglected, then (14.58) implies that the randoms event rate is proportional to the square of the activity whereas the true coincidences rate will be proportional to the activity. Equation (14.58) can be the basis for correcting for random events through the recording of individual events and performing a post hoc correction. Another means of correcting for random coincidences is the delayed coincidence method in which two coincidence circuits are used. One is the undelayed one shown in Fig. 14.14, whereas the other is delayed by several times the duration T_{Window}. As any correlations between events are removed by the time delay, the resulting coincidences must be due to random events. This rate $\dot{N}_{Rand}$ is then subtracted from the prompt event rate to yield a rate which is the sum of that due to true coincidences and that due to scattered events,

$$\dot{N}_{True} + \dot{N}_{Scat} = \dot{N}_{Prompt} - \dot{N}_{Rand}. \tag{14.59}$$

Much development in PET technology has focused on faster scintillator materials and faster electronics allowing T_{Window} to be reduced, thus lowering the random event rate without compromising the true coincidences rate.

Scatter Events

Introduction

The removal of the contributions from scattered photons, given by $\dot{N}_{Scat}$, is next required in order to yield the desired true coincidence rate. As in planar scintigraphy and SPECT, these scattered events degrade image contrast and lead to activity quantification overestimations. The contribution from scatter events can be reduced by inserting lead or tungsten annular septa between the detector rings in order to create the two-dimensional acquisition mode (to be discussed). This, however, reduces the sensitivity of the scanner, a result which may not be desirable if a whole-body biodistribution study were required to use minimal administered activity in order to reduce the radiation absorbed dose burden. Consequently, there has been considerable interest in the development of methods to remove the contributions of from photons that have been Compton-scattered in vivo when the scanner is operated in the three-dimensional acquisition mode (i.e., with the septa retracted). In such a mode, the signal-to-noise ratio is improved, but at a cost of an increase in the number of scatter events. Zaidi (2000) categorized three-dimensional PET scatter compensation techniques as:

- Photon energy discrimination
- Convolution/subtraction
- Mathematical estimation of the Compton scatter contribution
- Reconstruction based

Photon Energy Discrimination

The photon energy discrimination methods are based upon the multiple-energy window discrimination described earlier. The scatter-rejection ability of this method can be restricted due to the limited energy resolution of scintillators used and the need to avoid rejecting true coincidences. For example, Tarantola et al. (2003) report that a typical discrimination window setting for BGO is from between 300–350 and 650 keV. However, some investigators have investigated dual-energy (Grootonk et al. 1996) and triple-energy (Shao et al. 1994) discrimination in PET.

Convolution Method

This method (Bergstrom et al. 1983; Bailey and Meikle 1994; Bentourkia and Lecomte 1999) essentially treats the measured energy photopeak as being the convolution of the projection of the unscattered photons and a scatter function. For example (Meikle and Badawi 2003), the projected scatter contribution is, in the three-dimensional-acquisition mode,

$$P_{Sc} = \int_{-\infty}^{\infty} d\tau\, P_{Unsc}(\tau)h(s-\tau,\tau) \tag{14.60}$$

where $\int_{-\infty}^{\infty} d\tau\, P_{Unsc}(\tau)$ is the one-dimensional projection of the true activity distribution and $h(s, \tau)$ is the scatter contribution to the position s along the radial projection due to activity positioned at τ. This can be determined from measurements of a line source of activity within a scattering medium. However, $P_{Unsc}(\tau)$ cannot be measured and, in practice, is replaced by the observed projection $P_{Obs}(\tau)$ which preserves acceptable accuracy in activity quantification. Should the scatter contribution be spatially invariant, (14.60) is a convolution. The resulting scatter estimate is subtracted from the observed projection data.

In the three-dimensional-acquisition mode operated on a multi-detector ring platform with the septa retracted, this method is extended to an iterative approach where the unscattered projection data in the nth iteration is (Bailey and Meikle 1994),

$$P^{n}_{Unsc}(s, z) = P_{Obs}(s, z) - k\left(P^{n-1}_{Unsc}(s, z) \otimes h(s, z)\right) \tag{14.61}$$

where z is the axial variable and the convolution is performed in two dimensions.

Mathematical Estimation

The physics of Compton scatter and the underlying Klein-Nishina cross sections are well known. Consequently, the scatter contribution to the measured projection data can, in principle, be estimated and then subtracted (Ollinger 1996).

The scatter contribution can be calculated analytically using the Klein-Nishina cross sections. It is, however, limited in practice to estimating the contribution from single-order Compton scatter due to the extensive computation load and time required by calculation of higher-order multiple scatters. Monte Carlo simulation of photon transport can, in principle, yield a more accurate scatter estimate as it will account for multiple scatter contributions (Levin et al. 1995). Again, the computation load can be excessive in yielding a statistically powerful result.

Reconstruction Based

As in SPECT, the two tomographic reconstruction means used in PET are FBP and MLEM (OSEM). Scatter correction can be incorporated within the iterative reconstruction method. Zaidi (2000) discusses the hypotheses upon which the correction method is based. First, it is assumed that that spatial variability of the scatter component in the image is limited and, hence, represents a low-frequency component in Fourier space. The second is that the low-frequency components will converge faster than the high-frequency ones. A single iteration of the OSEM algorithm is then used to estimate the scatter component, which can then be removed from the total data set.

Performance Metrics

A useful metric of count performance is the noise-equivalent count rate, NECR, defined as (Strother et al. 1990),

$$\mathrm{NECR} = \frac{\dot{N}^2_{True}}{\dot{N}_{Prompt}} \tag{14.62}$$

where the prompt event count rate $\dot{N}_{Prompt}$ is given by (14.57). The NECR represents the count rate which would give the same signal-to-noise ratio in the detection data had the scattered and random coincidences not been present.

Figure 14.15 shows count rates due to the combination of true and scattered coincidences, random coincidences and the NECR. The "no counting loss" curve is the extrapolated count rate measured at low activity values where the contribution of random coincidences and detector dead-time are negligible. The best operating condition is for an activity concentration at which the NECR is maximized.

Another metric commonly used in characterizing PET scanner performance is the scatter fraction,

$$\mathrm{SF} = \frac{\dot{N}_{Scat}}{\dot{N}_{True} + \dot{N}_{Scat}} \tag{14.63}$$

The scatter fraction is measured using a low-activity source so that $\dot{N}_{Rand}$ can be considered negligible.

Dead-Time Correction

When a photon is absorbed in a scintillation crystal block, the light output is detected and Anger logic used to decode the spatial position of the interaction within the block. Energy discrimination is applied and the electronic processing of a coincidence event initiated if the photon energy is within the energy acceptance

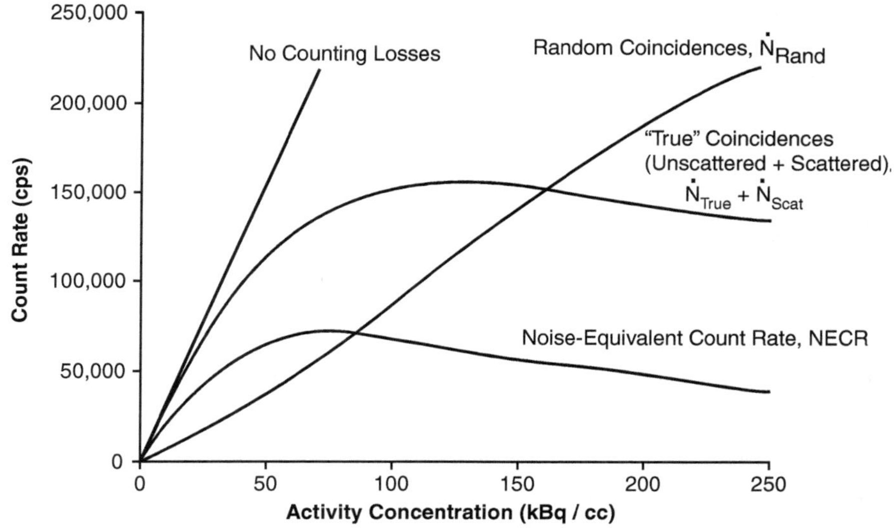

Fig. 14.15 Conceptual count-rate curves; see text for description (after Bailey 2003)

window. The system dead time is the combined times of these processes during which no further photon events can be processed. As described in Chap. 5, mathematical corrections can be applied to compensate for the losses due to dead time in the region where the measured count rate is linear, or very nearly linear, with the actual count rate. At high activities, where this linearity fails, this nature of correction is no longer robust.

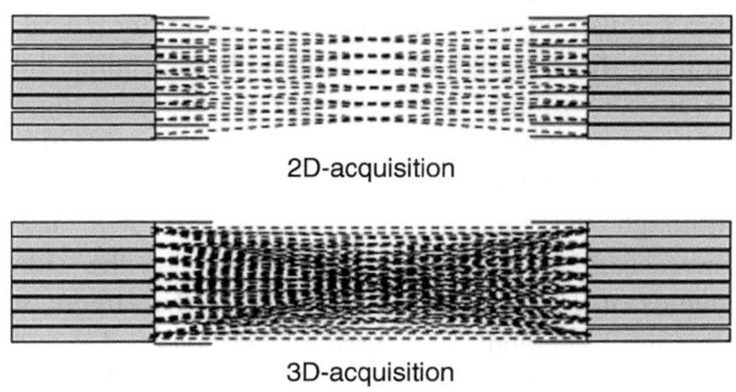

Fig. 14.16 Two- and three-dimensional acquisition configurations in axial cut-aways of a PET scanner (not drawn to scale). Septa slimit the number of cross-ring LoRs in the 2D-acquisition mode. The *dashed lines* indicate the LoRs

14.3.3.3 Data Acquisition and Corrections in PET

Two-Dimensional and Three-Dimensional Acquisition Modes

PET data can be acquired in either two-dimensional- or three-dimensional- acquisition modes, depending upon whether or not annular septa are inserted between the detection rings, as shown in Fig. 14.16. The significant difference between the two acquisition modes is the extent of the radial FoV which yields differences in sensitivities to true, random, and scattered events.

In the two-dimensional-acquisition mode, the annular septa restrict the number of LoRs between detector rings. Data from a few adjacent rings, the number depending upon the geometry, are grouped into discrete transaxial planes. Greater detection sensitivity, especially at the center of the axial FoV, can be achieved by removing these septa and operating in the three-dimensional-acquisition mode where LoRs between any two rings can be used. In this case, there is no grouping of data into individual planes. The increase in sensitivity between modes is by a factor of about 3–5. Reconstruction algorithms will also differ from those used in the two-dimensional-acquisition mode.

With the removal of the septa, the number of coincidences increases. The lack of septa increases the number of scatter coincidences, as shown in Fig. 14.17. With the septa present, the scattered photon is absorbed by a septum and a coincidence is not registered. With the septa removed, the same photon would have been detected and an erroneous LoR created as shown. The scatter correction methods

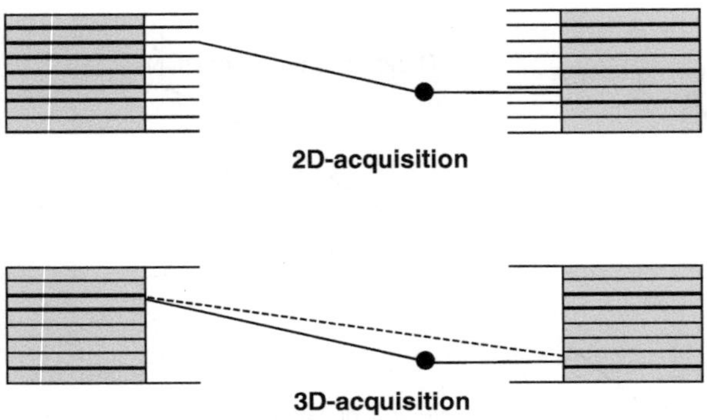

Fig. 14.17 Difference in scatter coincidences between 2D- and 3D-acquisition modes. The *dashed line* indicates the LoR

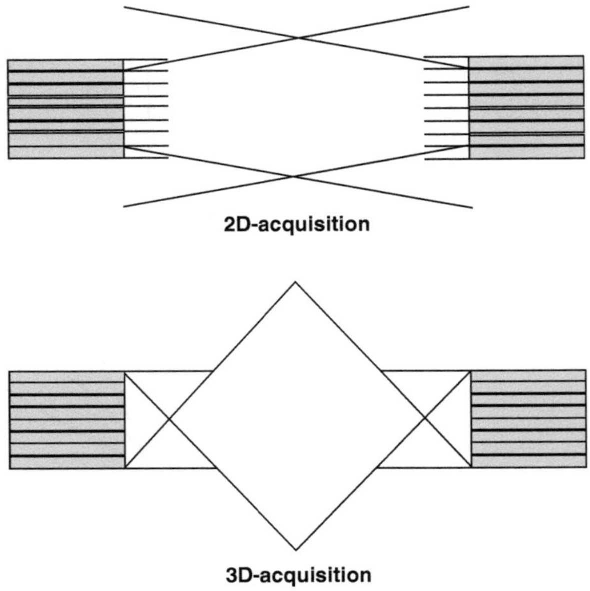

Fig. 14.18 The difference in the axial fields-of-view between 2D- and 3D-acquisition modes

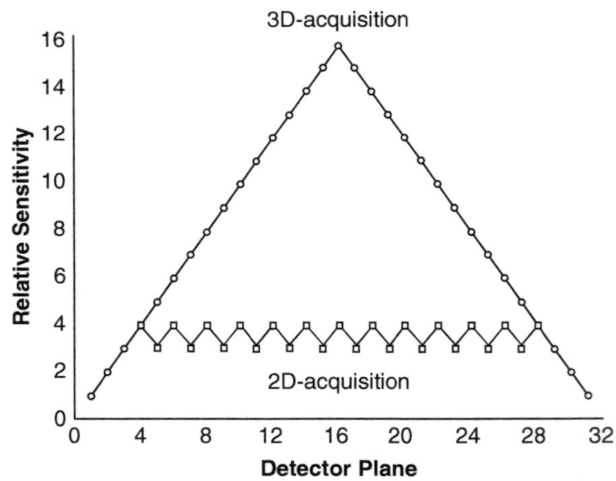

Fig. 14.19 Theoretical relative sensitivity along the radial planes for a PET scanner with 16 rings predicted from the number of LoRs. Here, coincidences in ±3 adjacent rings are summed in the 2D-acquisition mode

reviewed earlier are essential for both two-dimensional and three-dimensional PET acquisition but must deal with a much higher fraction in three-dimensional (35–40% as opposed to 15–20%).

Another important difference exists between the two acquisition modes as demonstrated in Fig. 14.18. In three-dimensional-acquisition mode, extending the axial FoV will increase the scanner's acceptance angle and consequently increase the scanner's sensitivities to random coincidences and scattered coincidences, although the rate of true coincidences will also increase. Extending the radial length of the side shields of the detector ring will reduce the effects of random coincidences occurring from events beyond the axial length of the detector rings, as suggested by Fig. 14.18 (Spinks et al. 1998).

From Figs. 14.16 and 14.18, it is clear that the detection sensitivity is greatest in three-dimensional-acquisition mode and that this sensitivity is axially-dependent. Figure 14.19 shows calculated sensitivity profiles along the axial FoV of a 16-ring PET scanner as a function of detector plane for both two-dimensional- and three-dimensional-acquisition modes. The peaked sensitivity of the three-dimensional-acquisition mode must be considered in whole-body PET acquisitions where whole-body images are generated by the concatenation of single axial FoVs. Significant overlaps of these FoVs are required in order to equalize the axial sensitivity.

Attenuation Correction

The correction of photon attenuation is, as in planar scintigraphy and SPECT, essential for both image quality and quantification, the latter being critical for biodistribution measurements. As two photons are detected in coincidence in order to generate the PET image, absorption, or scatter of either photon (i.e., attenuation) will result in the loss of a true coincidence. The result, in the absence of attenuation correction, would be reduced apparent radionuclide concentration from positron–electron annihilations at depth in tissue. The net probability of neither photon being absorbed is, for one photon having a pathlength d_1 in tissue and the other having a pathlength d_2 where $d_1 + d_2 = D$, where D is the total thickness along the projection,

$$P_{\mathrm{No\,Abs}} = e^{-\int_0^{d_1} dx\,\mu(x)}\, e^{-\int_0^{d_2} dx\,\mu(x)} = e^{-\int_0^{D} dx\,\mu(x)} \tag{14.64}$$

The line integrals of linear attenuation coefficients are determined in the same fashion as discussed previously in SPECT attenuation correction. The measured $\dot{N}_{\mathrm{True}}$ value along the pathlength is then scaled by $1/P_{\mathrm{No\,Abs}}$ to correct for photon number loss due to attenuation. The attenuation map is determined by measuring the transmission using either a positron-emitting radionuclide (e.g., ^{68}Ge) in which events are detected in coincidence, a single-photon-emitting radionuclide (e.g., ^{137}Cs) or, in modern PET/CT platforms, a polyenergetic X-ray beam. In the latter two cases, energy scaling of the attenuation coefficients to the 511 keV photon energy is required.

Normalization

The outcome of a PET biodistribution study is the mean fraction of administered activity present in organs (or suborgan components or even voxels). As the administered activity is a known quantity, this requires absolute quantification of activity in the PET image.

The sensitivities of LoRs will be variable and depend upon, for example, the efficiencies of the detectors and of the subtended solid angle. These sensitivities must be normalized and equated so that an absolute value of activity per voxel can be accurately obtained. Meikle and Badawi (2003) provide a succinct summary of normalization techniques to correct for the various factors that induce variable LoR sensitivities. A direct method is to irradiate all LoRs using a linear ^{68}Ge source; the individual LoR normalization coefficients are proportional to the reciprocal of the number of counts in each LoR. However this suffers from requiring extended irradiation times and a highly uniform linear source; further, the scattering conditions in the normalization scan will differ markedly from that in clinical imaging. A component-based model for normalization gives the activity along the LoR joining the ith detector in the jth ring with the kth detector in the *l*th ring as,

$$\begin{aligned} A^{(ijkl)} = \kappa\Big(&\Big(\dot{N}^{(ijkl)}_{\mathrm{Prompt}} - \dot{N}^{(ijkl)}_{\mathrm{Scat}} - \dot{N}^{(ijkl)}_{\mathrm{Rand}}\Big) \\ &\times \Big(1\big/\Im^{(ijkl)}\Big) T^{(ijkl)}_{\mathrm{Corr}}\, \eta^{(ijkl)}_{\mathrm{True}}\Big) \end{aligned} \tag{14.65}$$

where $\Im^{(ijkl)}$ is the measured transmission along the LoR (taken from (14.64)) and $T^{(ijkl)}_{\mathrm{Corr}}$ is the dead time correction factor (Chap. 5). $\eta^{(ijkl)}_{\mathrm{True}}$ is the normalization coefficient for true coincidences along the LoR and κ is a calibration factor to convert count rate to activity. As $\dot{N}^{(ijkl)}_{\mathrm{Scat}}$ cannot be measured but only estimated or calculated, special consideration is required. In two-dimensional-acquisition mode where scatter is a small component and slowly varying over separate the LoRs. This is not the case in the three-dimensional-acquisition mode and (14.65) is modified by the approximation,

$$\begin{aligned} A^{(ijkl)} = \kappa\Bigg(&\Bigg(\dot{N}^{(ijkl)}_{\mathrm{Prompt}} - \frac{\dot{N}^{(ijkl)}_{\mathrm{Scat,Calc}}}{\eta^{(ijkl)}_{\mathrm{Scat}}} - \dot{N}^{(ijkl)}_{\mathrm{Rand}}\Bigg) \\ &\times \Big(1\big/\Im^{(ijkl)}\Big) T^{(ijkl)}_{\mathrm{Corr}}\, \eta^{(ijkl)}_{\mathrm{True}}\Bigg) \end{aligned} \tag{14.66}$$

where $\dot{N}^{(ijkl)}_{\mathrm{Scat,Calc}}$ is the calculated scatter rate and $\eta^{(ijkl)}_{\mathrm{Scat}}$ is the normalization coefficient for the scatter coincidences, $\eta^{(ijkl)}_{\mathrm{Scat}} \neq \eta^{(ijkl)}_{\mathrm{True}}$ (Ollinger 1995). The true and scatter coincidences normalization coefficients are each the product of a variety of efficiency factors that can be determined empirically. The interested reader is referred to Meikle and Badawi (2003) for a detailed description of such measurements.

14.3.3.4 Biodistribution Acquisitions with PET

PET Scanner Characteristics Relevant to Biodistribution Acquisitions

As with any other imaging system, the main metrics of PET scanner performance are characterized by its spatial resolution and detection sensitivity. PET spatial resolution is defined through the quadrature sum of the effects due to the noncollinearity of the two γ rays created through positron annihilation, the finite positron range in tissue prior to annihilation and the intrinsic resolution set by the detector itself. This latter resolution is further defined by the separation between detectors, the transverse width of the detectors and the depth of interaction of the photon within the detector, the latter leading to parallax error in defining an LoR. These effects are all weighted by the nature of the tomographic reconstruction. In a dosimetric biodistribution study, the spatial resolution is usually not of significant consideration as one is evaluating typically the mean activities within organs in toto. Hence, spatial resolution is of limited concern except in unique cases where suborgan (voxel) absorbed dose evaluations are performed.

On the other hand, PET sensitivity is of much greater interest in a biodistribution study where, in many cases, one is seeking to maximize the sensitivity so as to minimize the administered activity, and associated radiation absorbed dose burden, provided to the imaged subject. In PET, sensitivity may be characterized as the ratio of the true coincidence detection rate to the actual amount of positron-emitting activity present. A simple and unit-less expression of this is,

$$S = \frac{A_{det}}{4\pi r^2} \varepsilon_{det}^2 \lambda_{atten} \qquad (14.67)$$

where A_{det} is the cross-sectional area of the detector, r is the detector ring radius, ε_{det} is the detector efficiency (which is squared due to the two detectors) and λ_{atten} accounts for attenuation between the source and detector. This expression displays, as expected, the increase in sensitivity from switching from two-dimensional-acquisition mode to three-dimensional-acquisition mode by the removal of septa and the increase in A_{det}.

The measured system sensitivity can be defined by the ratio of the measured $\dot{N}_{True}$ to the concentration of activity being imaged. In the current generation of PET scanners, the sensitivity is of the order of 5 cps/Bq/mL in the two-dimensional-acquisition mode and between 20 and 30 cps/Bq/mL in the three-dimensional-acquisition mode.

Whole-Body PET Acquisitions for Biodistribution and Dosimetry

Because of the increased sensitivity of the three-dimensional-acquisition mode to activity from outside the axial FoV, two-dimensional-acquisition is often used for imaging the torso so as to reduce the effects of scattered and random events. In whole-body imaging, this extraneous activity concentration can come from the torso and extremities. The decision of whether a two-dimensional- or three-dimensional-acquisition is used is based upon the type of scintillator and electronics used: e.g., GSO and LSO scintillators, with their fast response times, allow high coincidence count rates. While two-dimensional-acquisition reduces the confounding effects of scatter and randoms (e.g., a scatter fraction of about 15–20% for two-dimensional-acquisition mode vs. 35–40% in three-dimensional-acquisition mode), it also reduces the sensitivity to true coincidences which can be compensated for by the administration of higher amounts of activity.

For a whole-body biodistribution study typical of that of Phase I in which healthy volunteers are used, the amount of activity must be reduced in order to limit the radiation absorbed dose burden presented to the volunteer. Hence, the combination of high sensitivity (three-dimensional-acquisition mode) and reduced absorbed dose (low administered activity) is usually sought in whole-body biodistribution measurements. As a result, special considerations must be thought of in designing a PET whole-body biodistribution acquisition protocol. First is the combination of the relatively short axial FoV (typically 15–25 cm) and the three-dimension-acquisition mode axial sensitivity profile. In current PET scanner designs, data acquisition during continuous bed movement is not possible as in a gamma camera. Consequently, a whole-body PET image is generated by acquiring images at multiple contiguous bed positions and then retrospectively concatenating them. As the peaked axial profile in the three-dimensional-acquisition mode requires bed positions to be overlapped by up to 30–50% of the axial FoV (Tarantola et al. 2003), the effective axial FoV is reduced, thus requiring a larger number of bed positions to cover a specified total axial length.

PET has been used in radionuclide therapy planning. The therapeutic radionuclide is replaced with a positron-emitting isotope of the same element which is then imaged in order to quantify uptake in the disease to be treated or in healthy tissues for which radiotoxicity is to be minimized (Flux et al. 2006). Examples of such isotope exchanges are ^{124}I for ^{131}I in patient-specific planning of the treatment of thyroid carcinoma (Sgouros et al. 2004) and ^{86}Y for ^{90}Y in planning for therapy of somatostatin receptor-positive tumors (Rosch et al. 1999; Helisch et al. 2004). As ^{124}I has a very complicated decay scheme in which only about 23% of the decays result in positron emission, quantification is difficult but possible (Pentlow et al. 1996).

Biodistribution Evaluation from PET Images

The PET image set is intrinsically tomographic providing a three-dimensional representation of the biodistribution of the positron-emitting radionuclide. As PET is also inherently quantitative, the calibrated PET scanner will provide the user with the activity concentration within a specified VoI and the volume of that VoI. Corrections for the confounding effects of background subtraction and organ overlap required in quantitative planar scintigraphy are not needed in PET. However, the three-dimensional quantitative image data of PET can represent an embarrassment of riches. In evaluating the biodistribution (i.e., the amount of activity within an organ), the coronal view of the imaged subject is typically used as it is usually easier to identify the organs in this view. However, this coronal view will be of multiple tomographic slices, usually 128 or 256. This requires the operator to draw RoIs separately on multiple slices which are then coalesced to form a VoI. The combination of the large number of coronal slices, the multiple images acquired at time points postadministration and the number of subjects used to define the biodistribution makes such an approach of evaluating the biodistribution excruciatingly laborious. Moreover, the low event statistics for a single slice can make the drawing of an RoI subject to high variability. There are two possible solutions to this practical problem. One, for example, is to simply "collapse" the 128 or 256 coronal slices and add them to form a single slice (Sprague et al. 2007; McParland et al. 2008). In the second, Sprague et al. (2007) examined what they termed a "bisected model" in which the anterior and posterior halves of the slices were collapsed into two single slices and subsequently treated as anterior and posterior views as in conjugate-view scintigraphy. They demonstrated, for the ^{18}F-labeled radiopharmaceutical under investigation in their work, that the effective doses evaluated across all three methods (individual slices, bisected and collapsed) differed by less than 10% and concluded that the simpler and more manageable bisected and collapsed models were acceptable for pragmatic PET biodistribution evaluations for relatively broad biodistributions. In both approaches, the methods of quantitative planar scintigraphy developed in previously are applicable, except that the activities contained in the RoIs are provided directly. Background compensation and organ overlap corrections in the summed coronal slices are required and the user can employ the correction methods previously discussed or take advantage of the tomographic nature of the PET image and correct for these effects directly using, for example, sagittal images.

Radiation Dosimetry of Positron-Emitting Radionuclides

The limited range of positrons prior to annihilation can result in a high local absorbed dose due to their energy exchanges with the medium, as described in Chap. 7. Hence, one could initially presume that positron-emitting radionuclides necessarily deposit a higher absorbed dose per unit administered activity than do single photon-emitting radionuclides. This is not entirely true as the absorbed dose will, of course, be a function of the biokinetics of the radiopharmaceutical and the half-life of the radionuclide, among other variables. Moreover, most single photon-emitting radionuclides used in nuclear medicine also emit low-energy charged particles (e.g., Auger electrons) which can also result in elevated local absorbed doses. Of particular interest is the absorbed dose to the urinary bladder wall resulting from its intimate contact with the radioactive urine it contains. It is possible, as demonstrated in the earlier discussions of dynamic urinary bladder models, to be able to influence the normalized cumulated activity of the bladder contents and, hence, the absorbed dose to the bladder wall through modification of the micturition pattern. This can be achieved by the combination of the hydration of the patient prior to the administration of a radiopharmaceutical which has a high renal excretion fraction and the allowance of frequent voiding afterwards. Figure 14.20 shows the absorbed doses to the urinary bladder wall per unit administered activity calculated for three single photon-emitting radionuclides and five positron-emitting radionuclides as functions of the voiding interval. This figure should be considered in conjunction with Fig. 12.9 presented in the discussion of the Cloutier et al. dynamic urinary bladder model; that figure demonstrated only the normalized cumulated activity of the urinary bladder contents as a function of voiding interval. Hence, it reflected only the temporal functions of bladder filling and voiding and the physical half-lives of the radionuclides considered. Because of the emission of charged particles by positron-emitting radionuclides and, indeed, single-photon radionuclides, it is necessary to now expand the discussion to the absorbed dose which has the potential to make the urinary bladder wall the "critical tissue" (defined as that which receives the highest absorbed dose).

The normalized cumulated activities were calculated for each radionuclide using the Cloutier et al. model; in order to remove biological variability, it was assumed in each case that 25% of the administered activity was excreted renally and that the urinary bladder was filled in a single phase with a half-life of 15 min. These figures are not particularly untoward for a radiopharmaceutical with high renal excretion. The absorbed doses were then calculated using the OLINDA/EXM code (Stabin et al. 2005).

All of the absorbed dose curves exhibit a similar tendency to increase rapidly with voiding interval. This feature simply arises from the fact that reduced micturition frequency results in prolongation of exposure of the bladder wall to its radioactive contents. The absorbed dose due to ^{15}O is a factor of 10–100 times less than the other radionuclides, which is a consequence of its very short half-life of 122 s. With the exception of ^{11}C (again with a short half-life of 20.4 min), consider the remaining three single photon-emitting radionuclides (^{99m}Tc, ^{111}In, and ^{123}I) and three positron-emitting radionuclides (^{18}F, ^{64}Cu, and ^{124}I), separately. In Fig. 12.9, it was shown that the normalized cumulated activities of ^{111}In and ^{99m}Tc exceeded that of ^{18}F due to the shorter half-life of

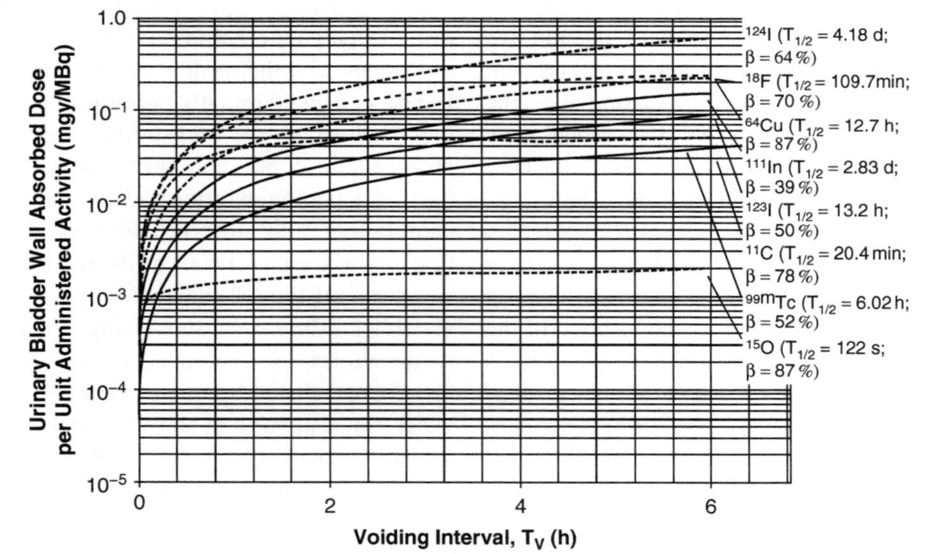

Fig. 14.20 Absorbed dose to the urinary bladder wall per unit administered activity as a function of voiding interval for eight radionuclides. *Solid-line* curves are for single-photon-emitting radionuclides (^{99m}Tc, ^{111}In and ^{123}I) and *dashed-line curves* are for positron-emitting radionuclides (^{11}C, ^{15}O, ^{18}F, ^{64}Cu, and ^{124}I). Data within the parentheses include the physical half-life of the radionuclide and the percentage of the total absorbed dose to the bladder wall which is due to charged particulate radiation (e.g., Auger electrons and electrons and positrons resulting from β decay). It is assumed in all cases that 25% of the administered activity is excreted renally and fills the urinary bladder in a single exponential phase with a 15-min half-life. This figure should be viewed in conjunction with Fig. 12.9

^{18}F. However, a different story appears when one considers the absorbed dose to the bladder wall. The single-photon-emitting radionuclides result in lower absorbed doses than the positron-emitters due to the fact that the absorbed dose contributions by β particles are higher for the latter (ranging from 64 to 87%) than for the former (ranging from 39 to 52%).

14.3.4 Imaging of Bremsstrahlung from β-Emitting Radionuclides and Activity Quantification

As charged particles have a short mean-free path in tissue and lose their kinetic energy in more compact volumes, they are the obvious particle of choice in therapeutic nuclear medicine. Radionuclide therapy planning requires knowledge of the fraction of administered activity that the target tumor uptakes and that which radiosensitive healthy tissues uptake. This is problematic as the charged particles released in vivo cannot be imaged directly. One possible solution is through the use of a radionuclide that emits both photons and β-particles, such as ^{131}I, thus enabling both therapy and image-based quantification. Another solution is to replace the therapeutic charged-particle emitting radiolabel with a photon-emitting analog (e.g., replacing ^{90}Y with ^{111}In) and image a diagnostic-level of the therapeutic vector that is labeled with the photon-emitting moiety. This permits a prediction of the amount of uptake in both the target and normal tissues and of the degree of excretion in order to plan the amount of administered activity of the therapeutic radionuclide required for the desired therapeutic effect.[13] However, it may be desired to verify directly the biodistribution of a β-emitting radionuclide.

Although not performed frequently, the measurement of the biodistribution and the quantification of the β-emitting therapeutic radionuclide activity is possible by imaging the *bremsstrahlung* emitted as the β particles decelerate in tissue (Williams et al. 1989; Siegel et al. 1992, 1995; Clarke et al. 1992; Shen et al. 1994; Minarik et al. 2008).[14] The difficulty is not to be underestimated due to the weak photon output and the continuous and broad energy spectrum resulting in limitations in both spatial resolution and quantification of *bremsstrahlung* imaging.

In vivo *bremsstrahlung* imaging has probably been performed mostly with ^{32}P (which emits a β particle with a mean energy of 695 keV) and ^{90}Y (which emits a β particle with a mean energy of 935 keV). The *bremsstrahlung* spectrum will be given by the folding of the *bremsstrahlung* differential cross section in photon energy with the β-particle energy spectrum. The resulting broad spectrum puts demands upon the selection of the appropriate collimator to use with the gamma camera, the energy acceptance window to use and the discrimination between scattered and source photons (e.g., Shen et al. 1994; Siegel et al. 1994). For example, in determining the combination of collimator and energy acceptance window to use, Shen et al. compared the ^{90}Y *bremsstrahlung* spectra measured by a gamma camera with a variety of collimators. In the absence of a collimator, the *bremsstrahlung* spectrum was, as expected, continuous and reaching a peak at about 100 keV. The inclusion of a collimator resulted in the spectrum displaying two peaks superimposed upon this reduced spectrum. The first was due to the 74.97 keV $K_{\alpha 1}$ fluorescent X-ray from the lead within the collimator and the second was attributed by the authors to be due to septal penetration. Shen et al. concluded that a combination of a medium-energy collimator with the widest energy acceptance window provided by the gamma camera's manufacturer yielded the best combination of resolution and sensitivity for ^{90}Y quantification.

14.4 Red Bone Marrow Activity

14.4.1 Introduction

In radionuclide therapy, active (or red) bone marrow often presents as an absorbed-dose limiting tissue. While the absorbed dose to the red bone marrow has,

[13] Although caution is required in the substitution approach as the vector conjugated with the therapeutic radionuclide may have differing biokinetics than when conjugated with the diagnostic radionuclide.

[14] Recall from Chap. 7 that *bremsstrahlung* from α particles with kinetic energies typically used in nuclear medicine is negligible.

itself, been reported to not be consistently predictive of hematologic toxicity, it is often necessary to evaluate accurately the cumulated activity in bone marrow as the marrow absorbed dose is likely to be one of a combination of factors, including pretherapy bone marrow reserve, that can predict myelosuppression (Siegel 2005). Moreover, because of the relatively high tissue weighting factor value of 0.12 assigned to the red bone marrow in the evaluation of the effective dose (Table 10.5), accurate estimation of the red bone marrow absorbed dose is necessary also in diagnostic nuclear medicine dosimetry. This section reviews means through which the cumulated activity to the haematopoietically active red bone marrow can be acquired.

14.4.2 In Vivo Imaging Estimation of Red Bone Marrow Activity

The amount of activity contained within the red marrow can be estimated using the image quantification techniques just described (Siegel et al. 1989). In this case, a single marrow-containing skeletal region is used to estimate the activity contained in all of the marrow. The lumbar vertebrae (L1–L5) are perhaps best for bone marrow imaging due to their relatively high percentage marrow content and their visibility due to separation, at least in planar scintigraphy, from organs which typically demonstrate radionuclide uptake, such as the liver and kidneys. Assuming that the imaged subject is an adult, one can then scale the activity measured in the lumbar vertebrae by the cellularity factor of 0.7 to determine the uptake by red marrow alone and then divide by the 12.3% of whole-body marrow content contained within the lumbar vertebrae in Reference Man (Table 12.5) to yield an estimate of the total activity within the red bone marrow. Clearly, this approach will yield, at best, an approximation as it is subject to the assumption that the imaged region is representative of all of the marrow and neglects the variabilities of marrow content throughout the skeleton (Sgouros et al. 1996) and of the distribution of the radionuclide within the marrow.

Moreover, for increased accuracy in the measurement of the marrow activity derived from planar scintigraphy, it may be necessary to account for any contributions resulting from activity within the blood contained in the overlapping aorta and inferior vena cava. Meredith et al. (2008) have described a method to correct for this as applied to patients receiving ^{111}In-based or ^{131}I-based radionuclide therapy in which the volumes of these overlapping vessels and their degrees of overlap with the lumbar vertebrae were determined from abdominal contrast-enhanced CT images. In vitro measurements of activity concentration in the blood were then used, in conjunction with the vessel volume measurements, to estimate the amount of vessel-contained activity overlapping the lumbar marrow activity. They estimated that a failure to account for the activities contained within overlapping large vessels for the radiopharmaceuticals under investigation could lead to an overestimate of the red bone marrow dose by as much as 76%.

Clearly, the tomographic information provided by quantitative SPECT is another means of isolating the vertebral column marrow activity from surrounding activity-containing anatomy.

14.4.3 In Vitro Estimation of Red Bone Marrow Activity

Sgouros (1993) has described the estimation of the cumulated activity in red bone marrow in radioimmunotherapy on the basis of measurements of activity concentration in whole blood or plasma and through measurement via core bone biopsies. An assumption of Sgouros' method is that there is rapid equilibration of the radionuclide-labeled vector within the marrow and that the vector does not bind to any constituent of blood, marrow, or bone. The radiolabeled antibody is assumed, due to its size, to equilibrate within the plasma and the extracellular fluid of the red marrow, liver, and spleen (Sgouros 2005). Hence, the concentrations of the antibody in the plasma and the extracellular fluid of the red marrow are taken to be the same. The concentration of activity in the red marrow normalized to the administered activity and corrected for physical decay is, allowing for an explicit time-dependence,

$$[A(t)]_{RM,Norm,Corr} = [A(t)]_{P,Norm,Corr} \frac{V_{EF}}{V_{RM}} \quad (14.68)$$

where RM refers to red bone marrow, P to plasma and EF to the extracellular fluid within the marrow. $[A(t)]_{P,Norm,Corr}$ is the activity concentration measured in plasma at time t postadministration normalized to the administered activity and decay-corrected to the time of administration, V_{EF} is the volume of the extracellular fluid within the marrow and V_{RM} is the total volume of red bone marrow. While an estimated value of the ratio V_{EF}/V_{RM} is 0.19, based upon measurements of the leporid femoral red bone marrow, this is not expected to be uniformly applicable in the human, especially in a patient who has a reduced bone marrow reserve as a result of therapy (Siegel 2005). $[A(t)]_{P,Norm,Corr}$ can be obtained directly from centrifuged blood or else estimated from the measured activity concentration in whole blood,

$$[A(t)]_{P,Norm,Corr} = \frac{[A(t)]_{BL,Norm,Corr}}{1 - h} \quad (14.69)$$

where BL refers to whole blood, $[A(t)]_{BL,Norm,Corr}$ is the measured activity in whole blood and h is the hematocrit, or the volume fraction of erythrocytes in whole blood. Standard values for the hematocrit are 47% in the male and 42% in the female, although this value can vary over the range of 0.2–0.6. Combining these two equations,

$$[A(t)]_{RM,Norm,Corr} = \frac{[A(t)]_{BL,Norm,Corr}}{1 - h} \frac{V_{EF}}{V_{RM}}. \quad (14.70)$$

Hence, the ratio of the activity concentrations in red bone marrow to blood is,

$$[A(t)]|_{Bl}^{RM} = \frac{V_{EF}/V_{RM}}{1 - h} = 0.36 \quad (14.71)$$

using the assumed ratio of the extracellular fluid and red marrow volumes noted above and a hematocrit value of 0.47. But it must be noted that this ratio of activity concentrations will vary with the values of the ratio of volumes and of hematocrit. Sgouros described the value given by (14.71) as being the baseline, and noted that it could range from 0.19 to 0.63. In comparison, a task group on radioimmunotherapy dosimetry of the American Association of Physicists in Medicine had recommended a first-order approximation of the red bone marrow specific activity to be 0.2–0.4 times that of whole blood. The task group recommended, in addition, that the red bone marrow activity washout be set equal to the same as that of blood and that the distribution of activity within the blood be assumed uniform. It should be reminded that the above derivation is applicable to radiolabeled antibodies which equilibrate within the extracellular fluid (Sgouros 2005). Should the radiopharmaceutical not be confined to the extracellular fluid of the red bone marrow but more widely distributed, then the activity concentration in whole blood is likely to be more representative of that in the red bone marrow.

Another means of measuring the activity contained within the red bone marrow is through core biopsy using a Jamshidi needle, which is a long, hollow needle tapered distally. The biopsy sample will contain red and yellow marrow, cortical and trabecular bone and blood. Hence, a direct measurement of the activity concentration (per unit mass) of the sample will result in an underestimation of $[A(t)]_{RM,Norm,Corr}$. A variety of correction factors must then be applied to the measured activity concentration of the biopsy sample to compensate for the contributions of the non-red marrow components. Sgouros gives simple expressions for these correction factors, but they do not account for multiple contributions (i.e., corrections are for biopsy samples consisting of red bone marrow and one other only contaminant).

The result of (14.71) implicitly assumes that the ratio of activity concentrations in red marrow to that in blood is constant. Neglecting this possibility of an additional time dependence, the red bone marrow normalized cumulated activity concentration is given by the time integral,

$$[\tilde{A}]_{RM,Norm} = \left(\frac{V_{EF}/V_{RM}}{1 - h}\right) \times \int_0^\infty dt\, e^{-\lambda_p t} [A(t)]_{BL,Norm,Corr} \quad (14.72)$$

The normalized cumulated activity in the red bone marrow is then obtained by scaling this result by the volume of red bone marrow,

$$\tilde{A}_{RM,Norm} = \left[\tilde{A}\right]_{RM,Norm} \frac{m_{RM}}{\rho_{RM}} \tag{14.73}$$

where m_{RM} is the total mass of red bone marrow in the body and ρ_{RM} is the physical density of red bone marrow.

14.4.4 Estimation of Red Bone Marrow Absorbed Dose

Although the subject of this chapter is the acquisition of the human biodistribution, it is natural to segue from the measurement of the red bone marrow cumulated activity to the estimation of the absorbed dose to the red bone marrow. The total absorbed dose to the marrow will be due to a self-irradiation component and to a "cross-fire" component due to irradiations from elsewhere in the body. Using the MIRD schema, the red bone marrow absorbed dose normalized to the administered activity is (Siegel et al. 2005; Siegel 2005),

$$\begin{aligned} D_{RM,Norm} &= D^{(Self)}_{RM,Norm} + D^{(Cross)}_{RM,Norm} \\ &= \tilde{A}_{RM,Norm} S(RM \leftarrow RM) \\ &\quad + \tilde{A}_{RB,Norm} S(RM \leftarrow RB) \\ &\cong \tilde{A}_{RM,Norm} S(RM \leftarrow RM) \\ &\quad + \left(\tilde{A}_{WB,Norm} - \tilde{A}_{RM,Norm}\right) S(RM \leftarrow RB) \end{aligned} \tag{14.74}$$

where WB and RB refer to "whole body" and "remainder of body" (i.e., the whole body excluding the red bone marrow). The S-factors are, typically, those calculated for a reference anthropomorphic phantom. For radionuclide therapy planning, where estimation of the absorbed dose to the red bone marrow is essential to avoiding radiation-induced myelotoxicity, patient-specific estimates of the absorbed dose to the red bone marrow are desired. Consequently, the absorbed dose calculation of (14.74) must be modified to represent that of the individual patient for whom the dosimetry is being recalculated. This result can be approximated by assuming that the blood and red bone marrow masses relative to the whole body mass are constants or,

$$\frac{m_{BL,Pat}}{m_{WB,Pat}} = \frac{m_{BL,Ref}}{m_{WB,Ref}} \tag{14.75}$$

where Pat refers to the patient in question and Ref refers to the reference phantom. Similarly for the red bone marrow,

$$\frac{m_{RM,Pat}}{m_{WB,Pat}} = \frac{m_{RM,Ref}}{m_{WB,Ref}} \tag{14.76}$$

However, it should be noted that Shen et al. (2002) found little correlation between m_{RM} and m_{WB}, thus suggesting the invalidity of (14.76). In contradiction, Woodard (1984) found a reasonable correlation between red bone marrow mass and whole body mass (1.37% in adult males and 1.16% in adult females). Hence, one must be aware of such questions when evaluating the red bone marrow radiation absorbed dose. Self-irradiation of the marrow by the marrow is assumed to be due to charged particles only as the mean-free pathlength of photons will greatly exceed the marrow cavity dimensions. Recall that, for charged-particle self-irradiation, the S-factors scale by mass,

$$S(RM \leftarrow RM)_{Pat} = S(RM \leftarrow RM)_{Ref} \frac{m_{RM,Ref}}{m_{RM,Pat}} \tag{14.77}$$

The S-factor for the "cross-fire" contribution from the remainder of the body can be approximated by (Siegel et al. 2005),

$$\begin{aligned} S(RM \leftarrow RB)_{Pat} &\approx S(RM \leftarrow WB)_{Ref} \frac{m_{WB,Pat}}{m_{WB,Pat} - m_{RM,Pat}} \\ &\quad - S(RM \leftarrow RM)_{Ref} \frac{m_{RM,Pat}}{m_{WB,Pat} - m_{RM,Pat}} \\ &= S(RM \leftarrow WB)_{Ref} \left(\frac{1}{1 - m_{RM,Pat}/m_{WB,Pat}} \right) \\ &\quad - S(RM \leftarrow RM)_{Ref} \left(\frac{1}{m_{WB,Pat}/m_{RM,Pat} - 1} \right) \\ &= S(RM \leftarrow WB)_{Ref} \left(\frac{1}{1 - m_{RM,Ref}/m_{WB,Ref}} \right) \\ &\quad - S(RM \leftarrow RM)_{Ref} \left(\frac{1}{m_{WB,Ref}/m_{RM,Ref} - 1} \right) \end{aligned} \tag{14.78}$$

An approximation to the patient-specific normalized absorbed dose to the red bone marrow can thus be obtained,

$$\begin{aligned} D_{\mathrm{RM,Norm,Pat}} &= \tilde{A}_{\mathrm{RM,Norm}} S(\mathrm{RM} \leftarrow \mathrm{RM})_{\mathrm{Ref}} \frac{m_{\mathrm{WB,Ref}}}{m_{\mathrm{WB,Pat}}} \\ &+ \left(\tilde{A}_{\mathrm{WB,Norm}} - \tilde{A}_{\mathrm{RM,Norm}}\right) \left(S(\mathrm{RM} \leftarrow \mathrm{WB})_{\mathrm{Ref}} \right. \\ &\times \left(\frac{1}{1 - m_{\mathrm{RM,Ref}}/m_{\mathrm{WB,Ref}}} \right) - S(\mathrm{RM} \leftarrow \mathrm{RM})_{\mathrm{Ref}} \\ &\times \left. \left(\frac{1}{m_{\mathrm{WB,Ref}}/m_{\mathrm{RM,Ref}} - 1} \right) \right). \end{aligned} \quad (14.79)$$

It should be noted that this calculation requires, in addition to that of the red bone marrow, measurement of the normalized cumulated activity of the whole body.

Another means of estimating the marrow absorbed dose is through an in vivo biological dosimeter. This would be the most patient-specific dosimeter and one that has clearly attracted investigators, although it would be a postadministration predictor of undesirable sequelae resulting from irradiation of the marrow. One could look at the survival of the hematological precursor cells (Fig. 12.10) following the onset of therapy (Ploemacher et al. 1989; Neben et al. 1993; Goddu et al. 1998), but this is necessarily an invasive procedure and, hence, limited in practice. Detection and quantification of radiation-induced changes in the easily-accessible peripheral blood cells would be preferable. Such changes would include the chromosomal aberrations discussed in Chap. 10. Lenarczyk et al. (2001) presented measurements of the frequency of micronuclei production in murine reticulocytes as a function of the administered activity of ^{32}P-labeled orthophosphate and ^{90}Y-labeled citrate. Such production had been previously calibrated against absorbed dose resulting from photon-irradiation by a ^{137}Cs source. They concluded that the micronuclear production frequency could estimate marrow absorbed dose.

References

Axelsson B, Msaki P, Israelsson A (1984) Subtraction of Compton-scattered photons in single photon emission computerized tomography. J Nucl Med 25:490–494

Bailey DL (2003) Data acquisition and performance characterisation in PET. In: Valk PE, Bailey DL, Townsend DW, Maisey MM (eds) Positron emission tomography – basic science and clinical practice. Springer, London

Bailey DL, Meikle SR (1994) A convolution-subtraction scatter correction method for 3D PET. Phys Med Biol 39:411–424

Bailey DL, Karp JS, Surti S (2003) Physics and instrumentation in PET. In: Valk PE, Bailey DL, Townsend DW, Maisey MM (eds) Positron emission tomography – basic science and clinical practice. Springer, London

Beddar AS, Salehpour M, Briere TM, Hamidian H, Gillin MT (2005) Preliminary evaluation of implantable MOSFET radiation dosimeters. Phys Med Biol 50:141–149

Bentourkia M, Lecomte R (1999) Energy dependence of nonstationary scatter subtraction-restoration in high resolution PET. IEEE Trans Med Imag 18:66–73

Bergstrom J, Eriksson L, Bohm C, Blomqvist G, Litton J (1983) Correction for scattered radiation in a ring detector positron camera by integral transformation of the projections. J Comput Assist Tomogr 7:42–50

Bruyant PB (2002) Analytic and iterative reconstruction algorithms in SPECT. J Nucl Med 43:1343–1358

Buijs WCAM, Massuger LFAG, Claessens RAMJ, Kenemans P, Corstens FHM (1992) Dosimetric evaluation of immunoscintigraphy of indium-111-labelled monoclonal fragments in patients with ovarian cancer. J Nucl Med 33:1113–1120

Buijs WCAM, Siegel JA, Boerman OC, Corstens FHM (1998) Absolute organ activity estimated by five different methods of background correction. J Nucl Med 39:2167–2172

Buvat I, Benali H, Frouin F, Bazin JP, Di Paola R (1993) Target apex-seeking in factor analysis of medical image sequences. Phys Med Biol 38:123–138

Buvat I, Benali H, Todd-Pokropek A, Di Paola R (1994) Scatter correction in scintigraphy: the state of the art. Eur J Nucl Med 21:675–694

Buvat I, Rodriguez-Vaillafuerte M, Todd-Pokropek A, Benali H, Di Paola R (1995) Comparative assessment of nine scatter correction methods based on spectral analysis using Monte Carlo simulations. J Nucl Med 36:1476–1488

Clarke LP, Cullom SJ, Shaw R et al (1992) Bremsstrahlung imaging using the gamma camera: factors affecting attenuation. J Nucl Med 33:161–166

Chang L-T (1978) A method for attenuation correction in radionuclide computed tomography. IEEE Trans Nucl Sci 25:638–643

Deloar HM, Watabe H, Nakamura T, Narita Y, Yamadera A, Fujiwara T, Itoh M (1997) Internal dose estimation including the nasal cavity and major airway for continuous inhalation of $C^{15}O_2$, $^{15}O_2$ and $C^{15}O$ using the thermoluminescent dosimeter method. J Nucl Med 38:1603–1613

Dempster A, Laird N, Rubin D (1977) Maximum likelihood from incomplete data via the EM algorithm. J Royal Stat Soc 39:1–38

Eary JF, Appelbaum FL, Durack L, Brown P (1988) Preliminary validation of the opposing view method for quantitative gamma camera imaging. Med Phys 16:382–387

Everitt BS (1994) Statistical methods for medical investigations, 2nd edn. Edward Arnold, London

Fleming JS (1996) Quantitative measurements for gamma camera images. In: Chandler ST, Thomson WH (eds) Mathematical techniques in nuclear medicine. Institute of Physics and Engineering in Medicine, York

Flux G, Bardiès M, Monsieurs M, Savolainen S, Strand S-E, Lassmann M (2006) The impact of PET and SPECT on dosimetry for targeted radionuclide therapy. Z Med Phys 16:47–59

Giap HB, Macey BJ, Podoloff DA (1995) Development of a SPECT-based 3-dimensional treatment planning system for radioimmunotherapy. J Nucl Med 36:1885–1894

Gilardi MC, Bettinardi V, Todd-Pokropek A, Milanesi L, Fazio F (1988) Assessment and comparison of three scatter correction ntechniques in single photon emission computed tomography. J Nucl Med 29:1971–1979

Gladstone DJ, Lu XQ, Humm JL, Bowman HF, Chin LM (1994) A miniature MOSFET radiation dosimeter probe. Med Phys 21:1721–1728

Goddu SM, Howell RW, Giuliani DC, Rao DV (1998) Biological dosimetry of bone marrow for incorporated ^{90}Y. J Nucl Med 39:547–551

Gonzalez-Trotter DE, Jaszczak RJ, Bowsher JE, Akabani G, Greer KKL (2001) High-resolution absolue SPECT quantification for I-131 distributions used in the treatment of lymphoma: a phantom study. IEEE Trans Nucl Sci 48:707–714

Goris ML, Daspit SG, McLaughlin P, Kriss JP (1976) Interpolative background subtraction. J Nucl Med 17:744–777

Griffith MH, Yorke ED, Wessels BW, deNardo GL, Neacy WP (1988) Direct dose confirmation of quantitative autoradiography with micro-TLD measurements for radioimmunotherapy. J Nucl Med 29:1795–1809

Grootonk S, Spinks TJ, Sashin D, Spyrou NM, Jones T (1996) Correction for scatter in 3D brain PET using a dual energy window method. Phys Med Biol 41:2757–2774

Hammond ND, Moldofsky PJ, Beardsley MR, Nulhern CB (1984) External imaging techniques for quantitation of distribution of I-131 $F(ab')_2$ fragments of monoclonal antibody in humans. Med Phys 11:778–783

He B, Frey EC (2006) Comparison of conventional, model-based quantitative planar, and quantitative SPECT image processing methods for organ activity estimation using In-111 agents. Phys Med Biol 51:3967–3981

Helisch A, Forster GJ, Reber H, Buchholz HG, Arnold R, Goke B, Weber MM, Wiedenmann B, Pauwels S, Haus U, Bouterfa H, Bartenstein P (2004) Pre-therapeutic dosimetry of biodistribution of ^{86}Y-DOTA-Phe1-Tyr3-octreotide versus ^{111}In-pentetreotide in patients with advanced neuroendocrine tumours. Eur J Nucl Med Mol Imaging 31:1386–1392

Hoffman EJ, Huang SC, Phelps ME (1977) Quantitation in positron emission computed tomography: 1. Effect of object size. J Comput Assist Tomogr 3:299–308

Houston AS, Sampson WFD (1989) Comparison of two interpolative background subtraction methods using phantom and clinical data. Nucl Med Comm 10:121–132

Hudson HH, Larkin RS (1994) Accelerated image reconstruction using ordered subsets of projection data. IEEE Trans Med Imag 13:601–609

Ichihara T, Ogawa K, Motomura N, Kubo A, Hashimoto S (1993) Compton scatter compensation using the triple-energy window method for single- and dual-isotope SPECT. J Nucl Med 34:2216–2221

ICRP (1975) Report on the Task Group on Reference Man – ICRP publication 23. International Commission on Radiological Protection, Pergamon, Oxford

ICRP (2002) Basic anatomical and physiological data for use in radiological protection: Reference values – ICRP publication 89. International Commission on Radiological Protection, Pergamon, Oxford

ICRU (2002) Absorbed dose specification in nuclear medicine – report no 67. International Commission on Radiation Units and Measurements. Nuclear Technology Publishing, Ashford

IPEM (1996) Mathematical techniques in nuclear medicine – report no 73. Chandler ST, Thomson WH (eds). Institute of Physics and Engineering in Medicine, York

IPEM (2003) Quality control of gamma camera systems – Report No 86. Bolster A (ed). Institute of Physics and Engineering in Medicine, York

Jaszczak RJ, Greer KL, Floyd CE, Harris CG, Coleman RE (1984) Improved SPECT quantification using compensation for scattered photons. Nucl Med 25:893–900

Jaszczak RJ, Floyd CE, Coleman RE (1985) Scatter compensation techniques for SPECT. IEEE Trans Nucl Sci 32: 786–793

Johns HE, Cunningham JR (1983) The physics of radiology, 4th edn. Charles C Thomas, Springfield

Jönsson L, Ljungberg M, Strand S-E (2005) Evaluation of accuracy in activity calculations for the conjugate view method from Monte Carlo simulated scintillation camera images using experimental data in an anthropomorphic phantom. J Nucl Med 46:1679–1686

King MA, Hademenos GJ, Glick SJ (1992) A dual-photopeak method for scatter correction. J Nucl Med 33:605–612

Koeppe RA (2008). Data analysis and image processing. In: Wahl RL (ed) Principles and practice of PET and PET/CT, 2nd edn. Lippincott Williams & Wilkins, Philadelphia

Kojima A, Takaki Y, Matsumoto M, Tomiguchi S, Hara M, Shimomura O, Koga Y, Takahashi M (1993) A preliminary phantom study on a proposed model for quantification of renal planar scintigraphy. Med Phys 20:33–37

Koral KF, Wang X, Rogers WL, Clinthorne NH, Wang X (1988) SPECT Compton-scattering correction by analysis of energy spectra. J Nucl Med 29:195–202

Lange K, Carson R (1984) EM reconstruction algorithms for emission and transmission tomography. J Comput Assit Tomogr 8:303–316

Lawson RS (1998) Application of mathematical models in dynamic nuclear medicine studies. Phys Med Biol 44: R57–R98

Lenarczyk M, Goddu SM, Rao DV, Howell RW (2001) Biologic dosimetry of bone marrow: induction of micronuclei in reticulocytes after exposure to ^{32}P and ^{90}Y. J Nucl Med 42:162–169

Levin CS, Dahlbom M, Hoffman EJ (1995) A Monte Carlo correction for the effect of Compton scattering in 3-D PET imaging. IEEE Trans Nucl Sci 42:1181–1188

Ljungberg M, King MA, Hademenos GJ, Strand S-E (1994) Comparison of four scatter correction methods using Monte Carlo simulated source distributions. J Nucl Med 35:143–151

McParland BJ, Miller MP, Spinks TJ, Kenny LM, Osman S, Khela MKK, Aboagye E, Coombes RC, Hui A-M, Cohen PS (2008) The biodistribution and radiation dosiemtry of the Arg-Gly-Asp (RGD) peptide [^{18}F]AH111585 in healthy volunteers. J Nucl Med 49:1664–1667

Macey DJ, Marshall R (1982) Absolute quantitation of radiotracer uptake in the lungs using a gamma camera. J Nucl Med 23:731–735

Martin S, Lisbona A, Richard J, Morteau S, Denizot B, Bardiès M (2000) Production of new thermoluminescent mini-dosimeters. Phys Med Biol 45:479–494

Mas J, Hannequin P, Younes RB, Bellaton B, Bidt R (1990) Scatter correction in planar imaging and SPECT by constrained factor analysis of dynamic structures (FADS). Phys Med Biol 35:1451–1465

Matsumoto M, Nakamura T, Watabe H, Itoh M, Hatazawa J (1993) Estimation of organ biodistribution and absorbed dose from experimental measurements with TLDs in PET studies. Med Biol Eng Comput 31:151–156

Meikle SR, Badawi RD (2003) Quantitative techniques in PET. In: Valk PE, Bailey DL, Townsend DW, Maisey MM (eds) Positron emission tomography – basic science and clinical practice. Springer, London

Meredith RF, Shen S, Forero A, Lobuglio A (2008) A method to correct for radioactivity in large vessels that overlap the spine in imaging-based marrow dosimetry of lumbar vertebrae. J Nucl Med 49:279–284

Minarik D, Skögreen Gleisner K, Ljungberg M (2008) Evaluation of quantitative ^{90}Y SPECT based on experimental phantom studies. Phys Med Biol 53:5689–5703

Moore SC, Brunelle JA, Kirsch A-M (1982) Quantitative multidetector emission computerized tomography using iterative attenuation compensation. J Nucl Med 23:706–714

Neben S, Anklesaria P, Greenberger J, Mauch P (1993) Quantitation of murine hematopoietic stem cells in vitro by limiting dilution analysis of cobblestone area formation on a clonal stromal cell line. Exp Hematol 21:438–443

Nichols K, Shrivastava PN, Powell OM, Adatepe MH, Isaacs GH (1987) Non-interventional background corrections for scintigrams. Phys Med Biol 32:605–613

Ollinger JM (1995) Detector efficiency and Compton scatter in fully 3D PET. IEEE Trans Nucl Sci 42:1168–1173

Ollinger JM (1996) Model-based scatter correction for fully 3D PET. Phys Med Biol 41:153–176

Pass B, Aldrich JE (1985) Dental enamel as an in vivo radiation dosimeter. Med Phys 12:305–307

Pentlow KS, Graham MC, Lambrecht RM, Daghigian F, Bacharach SL, Bendriem B, Finn RD, Jordan K, Kalaigian H, Karp JS, Robeson WR, Larson SM (1996) Quantitative imaging of iodine-124 with PET. J Nucl Med 37:1557–1562

Ploemacher RE, van der Sluijs JP, Voerman JSA, Brons NHC (1989) An in vitro limiting-dilution assay of long-term repopulating hematopoietic stem cells in the mouse. Blood 74:2755–2763

Rosch F, Herzog H, Stolz B, Brockmann J, Kohle M, Muhlensiepen H, Marbach P, Muller-Gartner HW (1999) Uptake kinetics of the somatostatin receptor ligand [^{90}Y]DOTA-Dphe(1)-Tyr(3)-octreotide ([^{86}Y]SMT487) using positron emission tomography in non-human primates and calculation of radiation doses of the Y-90-labelled analogue. Eur J Nucl Med 26:358–366

Rosenthal MS, Cullom J, Hawkins W, Moore SC, Tsui BMW, Yester M (1995) Quantitative SPECT imaging: a review and recommendations by the Focus Committee of the Society of Nuclear Medicine computer and Instrumentation Council. J Nucl Med 36:1489–1513

Saha GB (2005) Basics of PET imaging. Springer, New York

Schauer DA, Desrosiers MF, Le FG, Seltzer SM, Links JM (1994) EPR dosimetry of cortical bone and tooth enamel irradiated with x and gamma rays: study of energy dependence. Radiat Res 138:1–8

Sgouros G (1993) Bone marrow dosimetry for radioimmunotherapy: theoretical considerations. J Nucl Med 34: 689–694

Sgouros G (2005) Blood and bone marrow dosiemtry in radioiodine therapy of thyroid cancer (letter to the editor). J Nucl Med 46:899–900

Sgouros G, Jureidini IM, Scott AM, Graham MC, Larson SM, Scheinberg DA (1996) Bone marrow dosimetry: regional variability of marrow-localizing antibody. J Nucl Med 37:695–698

Sgouros G, Kolbert KS, Sheikh A, Pentlow KS, Mun EF, Barth A, Robbins RJ, Larson SM (2004) Patient-specific dosimetry for ^{131}I thyroid cancer therapy using ^{124}I PET and 3-dimensional-internal-dosimetry (3D-ID) software. J Nucl Med 45:1366–1372

Shao L, Freifelder R, Karp JS (1994) Triple energy window scatter correction technique in PET. IEEE Trans Med Imag 4:641–648

Shen S, DeNardo GL, Yuan A, DeNardo DA, DeNardo SJ (1994) Planar gamma camera imaging and quantitation of yttrium-90 bremsstrahlung. J Nucl Med 35:1381–1389

Shen S, Meredith RF, Duan J et al (2002) Improved prediction of myelotoxivity using a patient-specific imaging does estimate for non-marrow-targeting ^{90}Y-antibody therapy. J Nucl Med 43:1245–1253

Siegel JA, Wu RK, Maurer AH (1985) The buildup factor: effect of scatter on absolute volume determination. Nucl Med 26:390–394

Siegel JA (1985) The effect of source size on the buildup factor calculation of absolute volume. J Nucl Med 26:1319–1322

Siegel JA (2005) Establishing a clinically meaningful predictive model of hematologic toxicity in nonmyeloblative targeted radiotherapy: practical aspects and limitations of red marrow dosimetry. Cancer Biother Radiopharm 20:126–140

Siegel JA, Pawlyk LRE, DA HJA, Sharkey RM, Goldenberg DM (1989) Sacral scintigraphy for bone marrow dosimetry in radioimmunotherapy. Int J Rad Apll Instrum B16: 553–559

Siegel JA, Handy DM, Kopher KA, Zeiger LS, Order SE (1992) Therapeutic beta irradiating isotopes in bone metastasis: a technique for bremsstrahlung imaging and quantitation. Antibody Immunoconj Radiopharm 5:237–248

Siegel JA, Whyte-Ellis S, Zeiger LS, Order SE, Wallner PE (1994) Bremsstrahlung SPECT imaging and volume quantification with phosphorus-32. Antibody Immunoconj Radipharm 7:1–10

Siegel JA, Zeiger LS, Order SE, Wallner PE (1995) Quantitative bremsstrahlung SPECT imaging: Use for volume, activity, and absorbed dose calculations. Int J Radiat Oncol Biol Phys 31:953–958

Siegel J, Stubbs TS, Stabin M, Hays M, Koral K, Robertson J, Howell R, Wessels B, Fisher D, Weber D, Brill A (1999) MIRD Pamphlet No. 16: techniques for quantitative radiopharmaceutical biodistribution data acquisition and analysis for use in human radiation dose estimates. J Nucl Med 40:37S–61S

Siegel JA, Sparks RB, Sharkey RM (2005) Blood-based red marrow dosimetry: where's the beef? (letter to the editor). J Nucl Med 46:1404–1405

Sparks RB (2005) Data collection for dosimetry analysis in imaging studies. Presented at the Society of Nuclear Medicine Annual Meeting

Spinks TJ, Miller MP, Bailey DL, Bloomfield DM, Livieratos L, Jones T (1998) The effect of activity outside the direct field of view in a 3D-only whole-body positron tomography. Phys Med Biol 43:895–904

Sprague DR, Chin FT, Liow J-S, Fujita M, Burns HD, Hargreaves R, Stubbs JB, Pike VW, Innis RB, Mozley PD (2007) Human biodistribution and radiation dosimetry of the tachykinin NK_1 antagonist radioligand [^{18}F]SPA-RQ: comparison of thin-slice, bisected and 2-dimensional planar image analysis. J Nucl Med 48:100–107

Stabin MG (2008) Fundamentals of nuclear medicine dosimetry. Springer, New York

Stabin MG, Sparks RB, Crow E (2005) OLINDA/EXM: the second-generation personal computer software for internal dose assessment in nuclear medicine. J Nucl Med 46:1023–1027

Strand S-E, Strandh M, Ljungber M, Tagesson M (1994a) Mini-Tl-dosimeters for in vivo measurements. J Nucl Med 35:78P

Strand S-E, Ljungberg M, Tennvall J, Norrgren K, Garkavij M (1994b) Radio-immunotherapy dosimetry with special emphasis on SPECT quantification and extracorporeal immune-absorption. Med Biol Eng Comput 32:551–561

Strother SC, Casey ME, Hoffman EJ (1990) Measuring PET scanner sensitivity: relating countrates to image signal-to-noise ratios using noise equivalent counts. IEEE Trans Nucl Sci 37:783–788

Tarantola G, Zito F, Gerundini P (2003) PET instrumentation and reconstruction algorithms in whole-body applications. J Nucl Med 44:756–769

Thomas SR, Maxon HR, Keriakes JG, Saenger EL (1972) Quantitative external counting techniques enabling improved diagnostic and therapeutic decisions in patients with well-differentiated thyroid cancer. Radiology 122:731–737

Thomas SR, Maxon HR, Keriakes JG (1976) In vivo quantification of lesion radioactivity using external counting methods. Med Phys 3:253–255

Thomas SR, Gelfand MJ, Burns GS et al (1983) Radiation absorbed-dose estimates for the liver, spleen and metaphyseal growth complexes in children undergoing gallium-67 citrate scanning. Radiology 146:817–820

Turkington TG (2008) PET physics and PET instrumentation. In: Wahl RL (ed) Principles and practice of PET and PET/CT, 2nd edn. Lippincott Williams & Wilkins, Philadelphia

Williams LE, Wong JYC, Findley DO, Forell BW (1989) Measurement and estimation of organ bremsstrahlung radiation dose. J Nucl Med 30:1373–1377

Willowson K, Bailey DL, Baldock KC (2008) Quantitative SPECT reconstruction using CT-derived corrections. Phys Med Biol 53:3099–3112

Woodard HQ (1984) The relation of weight of haematopoietic marrow to body weight. Br J Radiol 57:903–907

Wu RK, Siegel JA (1984) Absolute quantitation of radioactivity using the buildup factor. Med Phys 11:189–192

Zaidi H (2000) Comparative evaluation of scatter correction techniques in 3D positron emission tomography. Eur J Nucl Med 27:1813–1826

The Biodistribution (III): Analysis

15

Abstract In this chapter, practical means of estimating the normalized cumulated activity from a measured discrete biodistribution set are reviewed. This involves the integration of these activity values from the time of administration to infinity. Least-squares fitting of the activity data by analytical functions, including the multiexponential form and the gamma variate, are reviewed. The analytical results of the integration of these functions are presented. In some cases, numerical integration is preferred and this is also reviewed in the context of the Trapezoidal Rule. As the end product of a biodistribution analysis is the absorbed doses to an ensemble of target regions, the various sources of uncertainty in the absorbed dose calculated for a target region using the MIRD schema, of which that of the biodistribution dominates, are analyzed.

Contents

15.1 Introduction

The two previous chapters have described how biodistributions of radiopharmaceuticals are measured in the preclinical model and in the human. Ultimately, the output of these acquisitions is the activities in source regions as functions of time. It is subsequently necessary to integrate these data in order to yield the cumulated activities i which, when coupled with the precalculated S-factors, yield the absorbed doses to the target regions. This chapter describes methods of yielding the cumulated activity and the uncertainties associated with the resulting absorbed dose.

15.2 Normalized Cumulated Activity

15.2.1 Introduction

In the MIRD schema, the absorbed dose per unit administered activity to a target region is calculated on the basis of measured normalized cumulated activities determined for the ensemble of source regions. The $\tilde{A}_{r_S,Norm}$ of the source region r_s, is the integral of the activity in the source organ, normalized to that administered, $A_{r_S,Norm}(t)$, from the time of administration to infinity. While $A_{r_S,Norm}(t)$ is a continuous

B.J. McParland, *Nuclear Medicine Radiation Dosimetry*,
DOI 10.1007/978-1-84882-126-2_15,

function, in practice it is determined as a series of discrete measurements at different times postadministration. In order to evaluate $\tilde{A}_{r_S,Norm}$, $A_{r_S,Norm}(t)$ must be integrated and the practical means of doing so are discussed here.

15.2.2 Analytical Fits to Activity Data

15.2.2.1 Introduction

The least-squares fitting of an analytical expression of $A_{r_S,Norm}(t)$ to measured data followed by the integration of this function is one possible means of evaluating the cumulated activity. An appropriate selection of the function to fit can also yield physiological information such as, for example, uptake and washout rates. However, it must always be noted that the measured activity data to which the function is to be fit reflect only the radionuclide and not the chemical entity to which it may or may not be bound. The in vivo or in vitro measurements cannot discern if the detected radionuclide is free, part of the administered radiopharmaceutical or a component of a metabolite fragment. Hence, pharmacokinetic interpretation of measured activity alone is inappropriate.

In order to isolate biological effects, the measured activity is corrected for physical decay by the premultiplication of each measured activity value at time t by $e^{+\lambda_{Phys}t}$, where λ_{Phys} is the physical decay constant. This physical decay-corrected result removes the effect of the physical decay of the radionuclide and isolates the temporal biological behavior of the radionuclide. The decay-corrected normalized activity for the source region r_S is,

$$A_{r_S,Norm,Corr}(t) = A_{r_S,Norm}(t)e^{+\lambda_{Phys}t}. \quad (15.1)$$

While this correction increases the magnitudes of the data points to fit at late time points, it also scales up the greater statistical uncertainties associated with these low-count data and this should not be forgotten. The result of the integration of the decay-corrected analytical fit must then be folded with the physical decay of the radionuclide, as shown below, in order to yield the corrected normalized cumulated activity.

15.2.2.2 Multiexponential Functions

Introduction

The simplest, and most common, means of describing the uptake and washout of activity with an analytical function amenable to the practical considerations of fitting to data is that of writing the decay-corrected and normalized activity as a series of exponential functions,

$$A_{r_s,Norm,Corr}(t) = C + \sum_{j=1}^{N} k_j e^{-\lambda_j t} \quad (15.2)$$

where C is a constant and k_j and λ_j are the amplitude and rate constant of the jth exponential term, respectively. Empirically, (15.2) is convenient to use and to fit to data and naturally results if the transfers of activity between compartments are described by first-order kinetics. In most practical cases, the number of exponential terms *N* does not exceed 1 or 2 in order to provide a good fit to an ensemble of activity data but in other cases, such as the washout from whole blood or plasma, $N = 3$ can sometimes be necessary.

The normalized cumulated activity for the source region r_s is determined by first applying a "decay-uncorrection" to the fit of (15.2) (i.e., premultiplying by $e^{-\lambda_{Phys}t}$) and then integrating from times $t = 0$ to ∞,

$$\tilde{A}_{r_s,Norm} = \int_0^\infty dt\, e^{-\lambda_{Phys}t} A_{r_s,Norm,Corr}(t) = \frac{C}{\lambda_{Phys}} + \sum_{j=1}^{N} \frac{k_j}{\lambda_j + \lambda_{Phys}}. \quad (15.3)$$

Fitting (15.2) to the measured normalized decay-corrected activity data can be accomplished in a large variety of ways.

Least-Squares Analysis

The least-squares approach (i.e., the minimization of the sum of the squared-differences between measured and fitted activity data) can be approached analytically or numerically/iteratively. A straight-forward analytical least-squares analytical fit of (15.2) can be had if a single exponential is appropriate for the data to be fit to. This fit is clearly accomplished by linearization through taking logarithms of both sides,

$$\ln A_{rs,Norm,Corr}(t) = \ln k_j - \lambda_j t \tag{15.4}$$

and performing linear regression to the data. Fits of cases where $N > 1$ and $C > 0$ can be achieved analytically using this linearization through the method of "exponential stripping," which is described below.

Numerical fits of (15.2) to measured data can be obtained by iteration. For the convenience of the following discussion, the variables of (15.2) will be written as,

$$y_i \equiv A_{rs,Norm,Corr}(t_i) \tag{15.5}$$

$$f(t_i, \boldsymbol{\beta}) \equiv C + \sum_{j=1}^{N} k_j e^{-\lambda_j t_i} \tag{15.6}$$

where the row vector $\boldsymbol{\beta}$ of dimension $2N + 1$ is defined as

$$\boldsymbol{\beta}^T \equiv (C; k_1, k_2, \ldots, k_N; \lambda_1, \lambda_2, \ldots, \lambda_N) \tag{15.7}$$

in order to make the derivation more concise. An iterative process is taken so as to minimize the sum of the squared-differences between measurement and fit,

$$\chi^2 = \sum_{i=1}^{M} (y_i - f(t_i, \boldsymbol{\beta}))^2 \tag{15.8}$$

where M is the number of times at which activity has been measured and is to be fit to. The iterations begin with the selection of an initial value of the coefficient vector $\boldsymbol{\beta}$. In many problems, this can be difficult but for fitting a multiexponential of the form of (15.2) in nuclear medicine biodistribution applications, an excellent initial guess can usually be had merely by inspection. If the activity does not monotonically grow to a constant or does not decrease to a constant value, then one would remove C from the vector of (15.7). Should the time-activity curve demonstrate a wash-in phase with activity increasing to a maximum and then decreasing as activity washes out of the source region, then it is reasonable to limit $N = 2$ and to fix $k_1 = k_2$. The magnitudes of the λ_j can also be approximated by inspection of the temporal behavior of the measured data. Following each iteration, the vector $\boldsymbol{\beta}$ is replaced by the estimate $\boldsymbol{\beta} + \boldsymbol{\delta}$, where the vector $\boldsymbol{\delta}$ represents the iteration in the coefficients and is to be solved for. In the gradient approach, the function $f(t_i, \boldsymbol{\beta})$ is linearized using a first-order Taylor series approximation,

$$f(t_i, \boldsymbol{\beta} + \boldsymbol{\delta}) \approx f(t_i, \boldsymbol{\beta}) + \frac{\partial f(t_i, \boldsymbol{\beta})}{\partial \boldsymbol{\beta}} \boldsymbol{\delta}. \tag{15.9}$$

Inserting (15.9) into (15.8), differentiating and setting the result equal to zero results in the matrix-vector form,

$$\boldsymbol{\delta} = \left(\mathbf{J}^T \mathbf{J}\right)^{-1} \mathbf{J}^T (\mathbf{y} - \mathbf{f}(\boldsymbol{\beta})). \tag{15.10}$$

$\mathbf{J}$ is the $(2N + 1) \times M$ Jacobian matrix,

$$\mathbf{J} = \begin{pmatrix} \frac{\partial f(t_1, \boldsymbol{\beta})}{\partial \beta_1} & \cdots & \frac{\partial f(t_M, \boldsymbol{\beta})}{\partial \beta_1} \\ \vdots & & \vdots \\ \frac{\partial f(t_1, \boldsymbol{\beta})}{\partial \beta_{2N+1}} & \cdots & \frac{\partial f(t_M, \boldsymbol{\beta})}{\partial \beta_{2N+1}} \end{pmatrix} \tag{15.11}$$

and the column vectors are,

$$\mathbf{y} = \begin{pmatrix} y_1 \\ \vdots \\ y_M \end{pmatrix} \tag{15.12}$$

and

$$\mathbf{f}(\boldsymbol{\beta}) = \begin{pmatrix} f(t_1; \boldsymbol{\beta}) \\ \vdots \\ f(t_M; \boldsymbol{\beta}) \end{pmatrix}. \tag{15.13}$$

Once $\boldsymbol{\delta}$ has been solved for, the new estimate of the coefficient vector is used in the next iteration which is continued until χ^2 has reached a minimum.

There are numerous additions and variants to the above iterative solution. It is likely that in most practical nuclear medicine dosimetry calculations the above iterative approach can be unnecessarily complicated due to the many matrix operations. Simplifications are possible which take advantage of the simplicity of linear regression. Two such methods are now described.

Exponential Stripping

The method of exponential stripping is a simple means of fitting the multiexponential form to measured data using linear regression and can easily be performed using the associated worksheet function in a spreadsheet (Lawson 1998). To demonstrate the process, consider the case of (15.2) for $C = 0$ and $N = 2$, which is a simple biexponential to fit,

$$A_{r_s,\mathrm{Norm,Corr}}(t) = k_1 e^{-\lambda_1 t} + k_2 e^{-\lambda_2 t}. \qquad (15.14)$$

In many cases, the decay constants are such that $\lambda_2 < \lambda_1$ so that at longer times postadministration, the second exponential term dominates over the first so that the approximation,

$$A_{r_s,\mathrm{Norm,Corr}}(t) \cong k_2 e^{-\lambda_2 t}(\text{large t}) \qquad (15.15)$$

may hold. An example of such a biexponential function is shown in Fig. 15.1. Equation (15.15) can then be fit to these data by taking logarithms of both sides and performing a simple linear regression of the logarithm of the measured $A_{r_s,\mathrm{Norm,Corr}}(t)$ at large t to yield estimates of $\ln k_2$ and λ_2. Let $\hat{k}_2$ and $\hat{\lambda}_2$ be the estimates of k_2 and λ_2 resulting from the linear regression. Returning these to (15.14) and rearranging,

$$A_{r_s,\mathrm{Norm,Corr}}(t) - \hat{k}_2 e^{-\hat{\lambda}_2 t} = k_1 e^{-\lambda_1 t}. \qquad (15.16)$$

Again, by taking logarithms of both sides and performing a linear regression on the remaining data points at short times, the estimates of $\hat{k}_1$ and $\hat{\lambda}_1$ are obtained.

For non-zero C, as in rise of activity to an asymptotic value, exponential stripping can also be used in practice (note that in this case, k_1 and k_2 will be negative). This is through fitting to the difference $\ln(C - A_{r_s,\mathrm{Norm,Corr}}(t))$. One selects a value of C, which must be greater than the maximum value of $A_{r_s,\mathrm{Norm,Corr}}(t)$, calculate the k_j and λ_j through exponential stripping and evaluate the reduced χ^2. In a spreadsheet, C can then be iterated and the process repeated until a minimum in χ^2 is found.

15.2.2.3 Nonmonotonic Biexponential/Gamma Variate

If in vivo measurements of activity in an organ are performed sufficiently early following administration, it is frequently possible to observe the uptake of activity by the organ followed by its washout. Figure 15.2 shows an example of such a time-activity curve. It would be desirable to avoid a least-squares fit to the

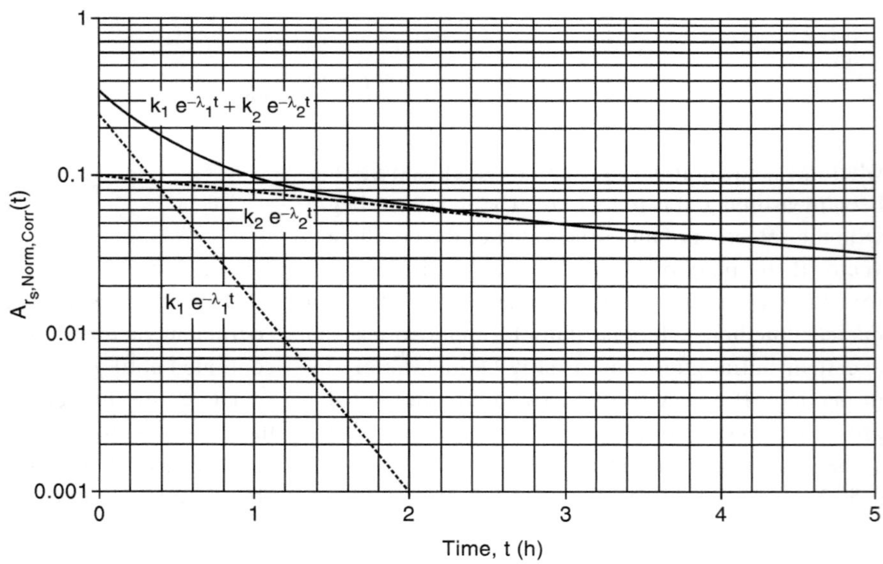

Fig. 15.1 Example showing principle of exponential stripping in fitting a biexponential function to the measured decay-corrected activity in source region r_S normalized to that administered. In this example, $k_1 = 0.25$ and $k_2 = 0.1$ and $\lambda_1 = 2.77\ h^{-1}$ and $\lambda_2 = 0.23\ h^{-1}$, corresponding to phase half-lives of 0.25 and 3 h, respectively

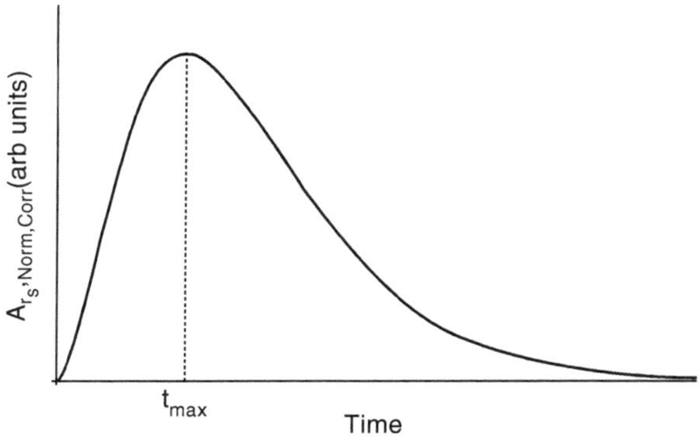

Fig. 15.2 Nonmonotonic uptake and washout of activity described by a gamma function

data using multiple regressions. There are two possible ways to achieve this. The first is to use the multi-exponential as a fit by fixing $C = 0$ and limiting $N = 2$ where $k_1 = -k_2 \equiv k$,

$$A_{r_S,\mathrm{Norm,Corr}}(t) = k\left(e^{-\lambda_1 t} - e^{-\lambda_2 t}\right). \qquad (15.17)$$

The activity at time $t = 0$ is equal to zero as required and the time at which the activity is at a maximum is given by,

$$t_{max} = \frac{\ln\left(\lambda_2/\lambda_1\right)}{\lambda_1 - \lambda_2}. \qquad (15.18)$$

In practice, exponential stripping can then be applied to these data in order to enable such a fit.

Another analytical function that can be used to fit to the above temporal variation is the gamma variate,

$$A_{r_S,\mathrm{Norm,Corr}}(t) = kt^{\alpha}e^{-t/\beta}. \qquad (15.19)$$

Madsen (1992) demonstrated a method through which this function can be reduced to a linear function and how the parameters k, α and β could be extracted through a linear regression. By setting the first derivative of (15.19) to zero, it is straight-forward to find that the product of α and β is equal to the time at which the function reaches a maximum,

$$\alpha\beta = t_{max}. \qquad (15.20)$$

Using this result and writing the constant k in terms of the maximum value of the activity,

$$k = A_{r_S,\mathrm{Norm,Corr}}(t_{max})t_{max}^{-\alpha}e^{\alpha}. \qquad (15.21)$$

Substituting this result into (15.19),

$$A_{r_S,\mathrm{Norm,Corr}}(t) = A_{r_S,\mathrm{Norm,Corr}}(t_{max})\left(\frac{t}{t_{max}}\right)^{\alpha}e^{\alpha(1-t/t_{max})}. \qquad (15.22)$$

By defining the reduced time, $\tau = (t/t_{max})$, this result is rewritten as,

$$A_{r_S,\mathrm{Norm,Corr}}(\tau) = A_{r_S,\mathrm{Norm,Corr}}(1)\tau^{\alpha}e^{\alpha(1-\tau)}. \qquad (15.23)$$

Taking the logarithms of both sides,

$$\ln\left(A_{r_S,\mathrm{Norm,Corr}}(\tau)\right) = \ln\left(A_{r_S,\mathrm{Norm,Corr}}(1)\right) + \alpha(1 + \ln\tau - \tau) \qquad (15.24)$$

where $A_{r_S,\mathrm{Norm,Corr}}(1)$, is the maximum value of the decay-corrected and normalized activity being fit to. Equation 15.24 is a linear function in the two variables of $\ln\left(A_{r_S,\mathrm{Norm,Corr}}(\tau)\right)$ and $(1 + \ln\tau - \tau)$. A linear regression of these two variables yields $\ln\left(A_{r_S,\mathrm{Norm,Corr}}(1)\right)$ and α. The parameter β is subsequently given by (15.20). This provides an analytical function representing the temporal behavior of the activity. The normalized cumulated activity is obtained by integrating the decay-uncorrected form of (15.19),

$$\tilde{A}_{r_S,\mathrm{Norm}} = k\int_0^{\infty} dt\, t^{\alpha}e^{-(\lambda_P+1/\beta)t}. \qquad (15.25)$$

This integral is be solved numerically or recast in the form of Euler's integral (Davis 1972),

$$\tilde{A}_{r_S,\mathrm{Norm}} = k\frac{\Gamma(1+\alpha)}{\left(\lambda_P + 1/\beta\right)^{(1+\alpha)}}. \qquad (15.26)$$

15.2.3 Numerical Methods

In some cases, analytical fits to the measured data are not sufficiently accurate or else are cumbersome to manipulate. Numerical integration methods can be used to calculate the cumulated activity using the discrete values of the measured activity as a function of time. In these cases, decay-correction must not be applied to the activity data to be integrated, as it cannot be folded out of the single-valued result afterwards. There are a wide variety of numerical integration formulae and a compendium can be found in Davis and Polonsky (1972). When deciding upon the type of numerical integration algorithm to employ, one must consider those which allow nonequidistant abscissa values.

The Newton-Cotes formulae are the most commonly-used types of numerical integration algorithms, often in the form of the Trapezoidal and Simpson's Rules. More involved methods, such as Gaussian Quadrature, are applicable but the simplicity of the Newton-Cotes approaches are desirable in clinical practice. Many numerical integration formulae are provided under the assumption that the sampling interval along the abscissa is constant, which is certainly not the likely case in a biodistribution case (and, indeed, undesirable). The simple Trapezoidal Rule can, in many cases, provide a suitable integration result.

An example of the application of the Trapezoidal Rule in numerically integrating an activity decreasing monotonically in time and sampled at five discrete time points is shown in Fig. 15.3. The shaded areas have an area of $\frac{1}{2}\sum_{i=2}^{5}(A(t_i)-A(t_{i-1}))(t_i-t_{i-1})$ which yields an approximation of the cumulated activity. One can immediately see from Fig. 15.3 two additional difficulties in the requirement to extrapolate the measured activity to those at time $t=0$ and as

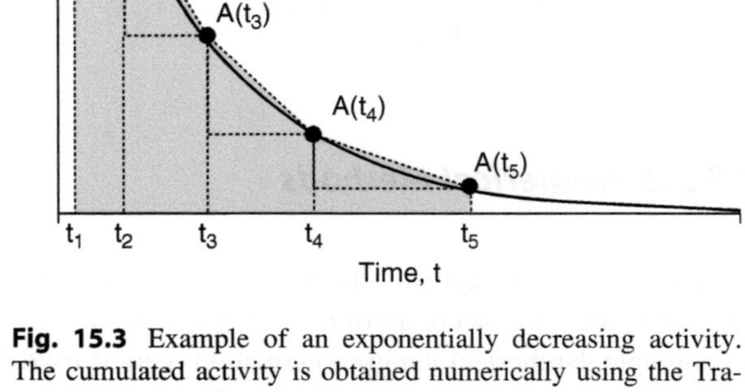

Fig. 15.3 Example of an exponentially decreasing activity. The cumulated activity is obtained numerically using the Trapezoidal Rule as described in the text

$t\rightarrow\infty$. In most practical biodistribution measurements, the first measurement time point is sufficiently soon after the time of administration that the extrapolation is simply modeled (e.g., in Fig. 15.3, a trapezoid could be formed by the combination of a rectangle and a triangle with its hypotenuse defined by the extrapolation of that formed between times t_1 and t_2). Extrapolation beyond the final timepoint can be more difficult and is discussed in the next subsection.

15.2.4 Extrapolation Beyond the Last Time Point

In any biodistribution measurement, the number of measurement points will necessarily be finite. However, as the calculation of the cumulated activity requires extrapolation to infinite time (e.g., after time t_5 in Fig. 15.3), some nature of extrapolation beyond the final time point is required. In the case of an analytical fit, this extrapolation is provided directly by the fit and will include the combined effects of the biological washout of activity from the source region and the physical decay of the radionuclide. If numerical integration has been used to estimate the cumulated activity up to the last measurement time, the means of obtaining the area under the curve to infinite time must be considered. Without analytical representation, it is not possible to include the biological washout component beyond the final measurement time. The simplest solution to this problem is to assume that the activity in the source region decays exponentially from the last measured time point with a rate constant equal to the physical decay constant alone. Then, the area under the curve from the final time point to infinite time can be solved simply through integration of that single exponential function and then added to the numerically-derived area to yield an estimate of the total cumulated activity. In this approach, as the biological washout is neglected beyond the final time point, it is implicitly assumed that all of the activity remains permanently bound within the source region and that the only decrease in activity within that region is due to physical decay. This assumption will lead to an overestimation of the cumulated activity in the source region and lead to an overestimate of the absorbed dose to it and the target regions. The magnitude of this error will depend upon the amount of activity remaining within the source region, the time postadministration at which extrapolation begins and the rate by which the radionuclide is biologically cleared. Should the biological decay constant be much less than the physical decay constant (i.e., $\lambda_{Biol} \ll \lambda_{Phys}$), then this approximation is entirely appropriate. Even if the two decay constants are approximately equal (i.e., $\lambda_{Biol} \approx \lambda_{Phys}$), the resulting overestimation may be judged acceptable as it would present a conservative result. Obviously, the individual performing this estimation must consider the requirements of the calculation at hand in order to determine the suitability of a method.

15.3 Uncertainty Analysis of the MIRD Formalism

15.3.1 Introduction

The net uncertainty in the calculated absorbed dose for target region r_T,

$$D_{r_T} = \sum_{r_S} \tilde{A}_{r_S} S(r_T \leftarrow r_S) \qquad (15.27)$$

will be the result of the combination of uncertainties in the physical and measured quantities and biological variability. A number of authors have attempted to estimate the net uncertainty in D_{r_T} in specific applications (e.g., Cloutier et al. 1975; Aydogan et al. 1999). Stabin (2008) has provided a more generalized and contemporary uncertainty analysis.

This subsection summarizes the magnitudes of the uncertainties and variabilities that factor into (15.27) to yield an estimate of the net uncertainty associated with D_{r_T}.

15.3.2 Sources of Uncertainty and Variability

15.3.2.1 Introduction

A calculation of the absorbed dose to a target region is subject to the variability in measurement of the in vivo and in vitro activities and the assumption of the appropriate anthropomorphic model to use. In order to understand the resulting variability in the target region's absorbed dose, (15.27) is expanded,

$$\begin{aligned} D_{r_T} &= \sum_{r_S} \tilde{A}_{r_S} S(r_T \leftarrow r_S) \\ &= \frac{1}{m_{r_T}} \sum_{r_S} \tilde{A}_{r_S} \sum_i Y_i E_i \phi_i(r_T \leftarrow r_S) \end{aligned} \qquad (15.28)$$

where the second summation is over the individual decays of the radionuclide of interest. Equation 15.28 isolates three categories of uncertainty/variability sources:

- Nuclear decay data uncertainties (Y_i, E_i)
- Biological variability, measurement error and modeling ($\tilde{A}_{r_S}$)
- Anthropomorphic phantom modeling (m_{r_T}, $\phi_i(r_T \leftarrow r_S)$)

Each of these uncertainties will be analyzed in turn.

The error analysis is simplified if a single pair of source and target regions and a single decay scheme with energy E, released per disintegration are considered. In this case, the absorbed dose to the target region is,

$$D_{r_T} = \frac{1}{m_{r_T}} \tilde{A}_{r_S} E \phi(r_S \leftarrow r_T). \qquad (15.29)$$

By assuming that the uncertainties of the four variables m_{r_T}, $\tilde{A}_{r_S}$, E, and $\phi(r_S \leftarrow r_T)$ are uncorrelated, the relative uncertainty in the absorbed dose to r_T is given by the quadrature sum,

$$\frac{\sigma_{D_{r_T}}}{D_{r_T}} = \sqrt{\frac{\sigma_E^2}{E^2} + \frac{\sigma_{\tilde{A}_{r_S}}^2}{\tilde{A}_{r_S}^2} + \left(\frac{\sigma_\phi^2}{\phi^2} + \frac{\sigma_{m_{r_T}}^2}{m_{r_T}}\right)} \qquad (15.30)$$

where the absorbed fraction has been written as ϕ for clarity. Each term in (15.30) is considered in sequence.

15.3.2.2 Uncertainty in Nuclear Decay Data

A large variety of academic and government institutions provide nuclear data. However, it is both the

National Nuclear Data Center (NNDC) based at the Brookhaven National Laboratory and the International Atomic Energy Agency based in Vienna which provide perhaps the most copious amounts of nuclear data and in a number of formats. These include access via an Internet browser or through Evaluated Nuclear Structure Data Files. Data have been evaluated and assessed from compiled measurements provided in the literature and the precision of the data, in many cases, is exquisite. For example, the NNDC reported value[1] of the major ^{99m}Tc γ ray is 140.5110 ($\pm$ 0.0010) keV, where the uncertainty contained within the parentheses is 1 standard deviation. Hence, in virtually all cases, excluding low-energy Auger electron emissions and low-yield emissions, any uncertainties associated with Y_i and E_i are negligible and can be ignored. As a result, the expression for the uncertainty in the calculated absorbed dose simplifies to,

$$\frac{\sigma_{D_{r_T}}}{D_{r_T}} = \sqrt{\frac{\sigma^2_{\tilde{A}_{r_S}}}{\tilde{A}^2_{r_S}} + \left(\frac{\sigma^2_\phi}{\phi^2} + \frac{\sigma^2_{m_{r_T}}}{m_{r_T}}\right)}. \quad (15.31)$$

This quadrature sum of uncertainties isolates the uncertainty in measurement and the uncertainty in anatomical modeling. Consider the case of a known (or single) phantom for which the dosimetry is to be calculated for. In this case, $\sigma^2_\phi = \sigma^2_{m_{r_T}} = 0$ and (15.31) reduces to,

$$\frac{\sigma_{D_{r_T}}}{D_{r_T}} = \frac{\sigma_{\tilde{A}_{r_S}}}{\tilde{A}_{r_S}}. \quad (15.32)$$

Thus the relative uncertainty in the target region's absorbed dose is equal to the relative uncertainty in the cumulated activity determined for the target region. As discussed next, $\sigma_{\tilde{A}_{r_S}}$ can be a significant uncertainty due to the compilation of measurement errors associated with it, as next discussed.

15.3.2.3 Uncertainty in the Derived Cumulated Activity Value

Recall that the cumulated activity is the total number of radioactive decays in a given source region. If the temporal biological behavior of the physical decay-corrected activity in that source region can be adequately modeled by an Nth-order exponential, $A_{r_S,Corr}(t) = \sum_{i=1}^{N} k_i e^{-\lambda_i t}$, then the cumulated activity can be expressed as,

$$\tilde{A}_{r_S} = \sum_{i=1}^{N} \frac{k_i}{\lambda_i + \lambda_{Phys}}. \quad (15.33)$$

As the uncertainty in the physical decay constant, λ_P, is negligible, the uncertainty in the cumulated activity calculation is driven by the fitting parameters, k_i and λ_i. These are ultimately set by the accuracies in the acquisitions of in vivo and in vitro activities. MIRD Pamphlet 16 (Siegel et al. 1999) presents an assessment of the various methodologies in acquiring these data. As the k_i and λ_i are obtained from fits to measured data, the uncertainty in $\tilde{A}_{r_S}$ is due to both measurement and fit or numerical integration. The measurement uncertainty arises from activity calibration inaccuracy and corrections for attenuation and scatter, among other factors, including reconstruction. Validation studies of the planar conjugate-view method have suggested an inaccuracy of the order of 10% for nonoverlapping organs increasing to as much as 90% for overlapping organs. Intersubject variability is another factor – differences as great as a factor of 2 in normalized cumulated activities for different subjects receiving the same radiopharmaceutical are common.

A detailed analysis of the uncertainties in the normalized cumulated activity values were presented by Kaplan et al. (1997) in the case of monoclonal antibody therapy planning by imaging ^{111}In. Three- and five-compartmental models were used in the analysis and a Monte Carlo approach used to estimate the variability in the cumulated activity resulting from the measured variances of the model's parameters. Perhaps not unexpectedly, the uncertainty in the cumulated activity was dependent upon the model used but was concluded by the authors to be generally less than 10%.

He and colleagues compared cumulated activity estimation methods, again for ^{111}In imaging, using both planar scintigraphy and SPECT through Monte Carlo (He et al. 2008, 2009a) and patient (He et al. 2009b) studies. They evaluated cumulated activity measurements obtained through quantitative SPECT

[1] http://www.nndc.bnl.gov/useroutput/AR_241919_.html

(using iterative reconstruction and correcting for scatter, attenuation, and partial-volume effects) and conjugate-view planar scintigraphy in the context of radioimmunotherapy planning. They reported that residence time errors of less than 3.8% were obtained using quantitative SPECT, but that these errors could extend to between 2 and 107% for planar scintigraphy alone. They concluded that conjugate-view planar scintigraphy as a means of dose quantification would be inadequate for radioimmunotherapy planning.

15.3.2.4 Uncertainty in Anatomic Modeling

Superimposed upon the methodology uncertainty briefly described in the previous subsection is the intersubject variation. This is well known (Siegel et al. 1999). Anyone experienced in evaluating the biodistributions of healthy volunteers in a Phase I study will be aware of intersubject differences of factors of 2 or greater. As an example, Aydogan et al. (1999) reported absorbed dose uncertainties evaluated for seven healthy volunteers who had been administered a ^{123}I-labeled radiopharmaceutical. They concluded that variations in individual biodistributions and organ masses could result in a factor of 2 increase in the 95%-confidence interval about a median organ absorbed dose estimate.

If the absorbed dose calculation is for a reference phantom, and it is recognized that the dosimetry is for that phantom, then the uncertainty arising from anatomic modeling is moot. However, one may be interested in how representative the dosimetry calculated is that of reality. This can be achieved through in situ dosimetry using micro-TLDs or MOSFET detectors, as discussed earlier. In most nuclear medicine applications, this is inconvenient and only the absorbed doses at discrete spatial sampling points are determined.

Another method of determining the uncertainty arising from the modeling of anatomy is to compare the absorbed doses calculated for the reference phantom of interest using the MIRD schema with those calculated for real subjects using Monte Carlo simulation of radiation transport and energy deposition. Divoli et al. (2009) have performed such an analysis by using the OEDIPE tool to calculate the S-factors for nine subjects receiving ^{131}I and comparing these with S-factors from the OLINDA/EXM code for the reference male phantom (Sect. 11.3.6.3 in Chap. 11). The subjects were male with ranges of whole-body masses of between 59 and 90 kg and heights of between 162 and 180 cm. Organ volumes were determined by CT volumetry and assumed tissue densities applied. They determined that the absorbed dose differences between the reference male and the subjects could be as significant as 140%. But, interestingly, the application of the mass-scaling factors of Chap. 11 between the reference phantom and the human subjects reduced these absorbed differences to less than 26%.

References

Aydogan B, Sparks RB, Stubbs JB, Miller LF (1999) Uncertainty analysis for absorbed dose from a brain receptor agent. In: Stelson A, Stabin M, Sparks R (eds) Sixth International Radiopharmaceutical Dosimetry Symposium. Oak Ridge Associated Universities, Oak Ridge, TN, pp 732–740

Cloutier RJ, Freeman LM, McAfee JG et al (1975) MIRD dose estimate report no. 4: summary of current radiation dose estimates with various liver conditions from ^{198}Au-colloidal gold. J Nucl Med 16:173–174

Davis PJ (1972) Gamma function and related functions. In: Abramowitz M, Stegun IA (eds) Handbook of mathematical functions. Dover, New York

Davis PJ, Polonsky I (1972) Numerical interpolation, differentiation and integration. In: Abramowitz M, Stegun IA (eds) Handbook of mathematical functions. Dover, New York

Divoli A, Chiavassa S, Ferrer L, Barbet J, Flux GD, Bardiès M (2009) Effect of patient morphology on dosimetric calculations for internal irradiation as assessed by comparisons of Monte Carlo versus conventional methodologies. J Nucl Med 50:316–323

He B, Wahl RL, Du Y, Sgouros G, Jacene H, Flinn I, Frey EC (2008) Comparison of residence time estimation methods for radioimmunotherapy dosimetry and treatment planning – Monte Carlo simulation studies. IEEE Trans Med Imag 27:521–530

He B, Du Y, Segars WP, Wahl RL, Sgouros G, Jacene H, Frey EC (2009a) Evaluation of quantitative imaging methods for organ activity and residence time estimation using a population of phantoms having realistic variations in anatomy and uptake. Med Phys 36:612–619

He B, Wahl RL, Sgouros G, Du Y, Jacene H, Kasecamp WR, Flinn I, Hammes RJ, Bianco J, Kahl B, Frey EC (2009b) Comparison of organ residence time estimation methods for radioimmunotherapy dosimetry and treatment planning – patient studies. Med Phys 36:1595–1601

Kaplan DD, Williams LE, Clarke KG, Odom-Maryon TL, Liu A, Lopatin G, Raubitschek AA, Wong JYC (1997) Estimating residence times and their associated errors in

patient absorbed-dose calculation. J Nucl Med Technol 25:264–268
Lawson RS (1998) Application of mathematical models in dynamic nuclear medicine studies. Phys Med Biol 44: R57–R98
Madsen MT (1992) A simplified formulation of the gamma variate function. Phys Med Biol 37:1597–1600
Siegel J, Stubbs TS, Stabin M, Hays M, Koral K, Robertson J, Howell R, Wessels B, Fisher D, Weber D, Brill A (1999) MIRD Pamphlet No. 16: techniques for quantitative radiopharmaceutical biodistribution data acquisition and analysis for use in human radiation dose estimates. J Nucl Med 40:37S–61S
Stabin MG (2008) Uncertainties in internal dose calculations for radiopharmaceuticals. J Nucl Med 49:853–860

16 The Ethics of Using Human Subjects in Clinical Trials Involving Radiopharmaceuticals

Abstract The participation of humans as subjects in any biomedical research is subject to ethical review of the appropriateness and value of the research, both to the individual and to society as a whole. When this research involves nuclear medicine, either as the primary focus of research or as an adjunct, considerations of the risks associated with the exposure to ionizing radiation must be taken. This chapter reviews the ethics of clinical studies involving humans, with an emphasis upon studies incorporating nuclear medicine.

Contents

16.1 Introduction

One of the intents of this book is to provide a deeper understanding of the calculation of the internal radiation dosimetry of subjects participating in clinical research studies in which a nuclear medicine procedure, diagnostic or therapeutic, is a component. Such studies could involve the development of new diagnostic or therapeutic radiopharmaceuticals or the use of a radiopharmaceutical with known dosimetry profile as a diagnostic imaging adjunct to other biomedical research. It is an ethical tenet in all such research using humans is that the benefit-to-risk ratio presented to the subject participating in such a study be maximized, or at least known, and made unambiguously apparent to the subject.[1] In nuclear medicine (neglecting the negligible possibility of chemical toxicity which is beyond the scope of this book), the risk is manifested by either stochastic risk at the low-absorbed doses of diagnostic radiopharmaceuticals or by deterministic risk at the higher-absorbed doses associated with therapeutic radiopharmaceuticals. Estimating the magnitudes of these risks requires that the internal radiation dosimetry be calculated as accurately as possible. Hence, the assurance of accurate radiation dosimetry is an ethical requirement in all nuclear medicine research.

The ethics of the use of animals in nuclear medicine research were outlined in Chap. 13. In this section, an overview of the ethical requirements of the use of humans in nuclear medicine research is provided.

[1]The benefit may not necessarily accrue to the participant in the study but can reflect a societal benefit.

B.J. McParland, *Nuclear Medicine Radiation Dosimetry*,
DOI 10.1007/978-1-84882-126-2_16, © Springer-Verlag London Limited 2011

16.2 Human Subjects in Biomedical Research: General Concepts

16.2.1 Introduction

The ethics of the use of human subjects in nuclear medicine research are no different than those associated with any other biomedical research. All biomedical research studies using humans are subject to both the Declaration of Helsinki and the principles of Good Clinical Practice (GCP). The latter are defined by the International Conference on Harmonisation[2] as "an international ethical and scientific quality standard for designing, conducting, recording, and reporting trials that involve the participation of human subjects" (ICH 1996). In the European Union, the implementation of GCP is set by the Directive 2001/20/EC of the European Parliament and the Council of the European Union[3] and member states are required to implement into law and regulation the requirements of this Direction.[4] In the United States of America, corresponding guidelines are provided in the US Code of Federal Regulations Title 21.

The nuclear medicine physicist participating in the design and execution of a clinical trial involving a radiopharmaceutical must be knowledgeable of these ethical and regulatory principles as the design considerations will be – and must be – very much driven by these principles.

16.2.2 Declaration of Helsinki

The Declaration of Helsinki is a policy statement of the World Medical Association (WMA) concerning the ethics of the involvement of human subjects in medical research, including identifiable human material and data.[5] It provides the genesis of regulatory and legislative requirements of the conduct of clinical trials involving humans. The first version was adopted in 1964 and has been amended some eight times since, the most recent being at the 59th WMA General Assembly held in Seoul in 2008. A primary expectation of the Declaration is that "(m)edical research involving human subjects must conform to generally accepted scientific principles, be based on a thorough knowledge of the scientific literature, other relevant sources of information, and adequate laboratory and, as appropriate, animal experimentation" (WMA 2008).

Central to all medical research as directed by the Declaration of Helsinki and manifested in all national regulatory and legislative requirements is the necessity of the freely given and informed consent of the subject participating in a study. This requires the subject to be made fully aware of, among other things, the risks and of the benefits associated with the clinical research he is invited to participate in. In the provision of this information in order to allow him to decide upon whether or not to consent to participate in the trial, the subject must not be influenced by the clinical investigator or trial staff.

The possible benefits of participating in the trial may accrue to the individual himself and/or to society in general. On the other hand, the subject will be the exclusive recipient of the risks. This requires that the trial be preceded by investigations of the predictable risks in order that knowledge of these can be conveyed to the subject. In a nuclear medicine trial, this requires accurate estimation of the expected internal radiation dosimetry. For a diagnostic radiopharmaceutical, this would be accompanied by an evaluation of the effective dose and the associated stochastic risk. Absorbed doses to radiosensitive tissues (such as red bone marrow and kidney) and estimates of radiation-induced sequellae would be part of the risk assessment of a trial of a therapeutic radiopharmaceutical.

---

[2]The full title is the International Conference on Harmonisation of Technical Requirements for Registration of Pharmaceuticals for Human Use (ICH) and which is a program which brings together regulatory authorities and representatives from the pharmaceutical industry from the European Union, Japan and the United States. It may be found at: http://www.ich.org.

[3]The Directive may be found at http://euopa.eu/eur-lex/pri/en/oj/dat/2001/l_121/l_12120010501en00340044.pdf.

[4]An example, in the UK, being The Medicines for Human Use (Clinical Trials) Regulations 2004.

[5]The current version of the Declaration of Helsinki may be found at http://www.wma.net/e/policy/b3.htm.

16.2.3 World Health Organization Recommendations

As in any biomedical research trial, the volunteer participating in a trial involving radiopharmaceuticals must be mentally competent to understand and appreciate the risks and benefits associated with the trial and to then make a reasoned judgment of whether or not to participate. The probabilities of stochastic risks or the potential magnitude of deterministic effects must be clearly communicated to the participating subject in written and verbal formats. Stochastic risks can be conveyed to the subject by, for example, comparison with other common radiological procedures or in terms of the effective dose received annually from normal background radiation. As, in a nuclear medicine trial, the subject is exposed to ionizing radiation, the volunteer should be an adult. In fact, the World Health Organization (WHO) proposed a guideline that subjects participating in studies involving exposure to ionizing radiation should be aged at least 40 years, where practicable, and preferably greater than 50 years (WHO 1977). The rationale for this is to reduce the risk of radiogenic cancer (if the latent period exceeds the life expectancy of the subject) and genetic effects. Because of the potential for fetal damage resulting from maternal irradiation, pregnant women should not participate in such studies. Of course, exceptions to these requirements may necessarily arise (ICRP 1991) leading to the need of the recruitment of members of disadvantaged or vulnerable populations. This can only occur if such a trial is responsive to the needs of that population and if there is a reasonable probability that the population will derive a benefit from the results of the research. If, for example, the trial has to be performed on mentally incompetent patients (an example being the development of a radiopharmaceutical for use in demented patients), then the informed consent for the subject to participate must come from the legally authorized representative of that subject; however, if the subject is capable of providing their own consent then they should. Similarly, there may be instances in pediatric medicine where the nuclear medicine procedure must be performed on a child. Clearly, informed consent from the parents or legally-appointed guardians of the child is required in such cases. In all situations, the subject is free to withdraw from the clinical trial at any time without reprisal.

16.2.4 Practical Considerations

The Declaration of Helsinki stipulates that any research trial involving humans must undergo review and approval by an independent Ethics Committee in addition by other competent authorities, prior to initiation. There are a number of practical considerations that must be incorporated in any clinical trial design involving radiopharmaceuticals. First, justification must be provided for the administration of the radiopharmaceutical. It is the responsibility of the Committee deciding upon whether or not to approve the application for a trial involving radionuclides to assess if substitution with an imaging modality that does not use ionizing radiation (e.g., ultrasound or magnetic resonance imaging) is possible. This is a realistic requirement only for those cases where the radiopharmaceutical is used as an imaging adjunct and is not itself the focus of the experiment. Should it be concluded that the administration of a radiopharmaceutical is necessary and unavoidable, the administration of radiation must be optimized through application of the ALARA (as low as reasonably achievable) in order to lower the amount of administered activity (and consequent absorbed dose burden) to a level at which the basic question the investigation seeks to answer can still be solved. The investigation should not be conducted at a lower administered activity as no possible benefit can be achieved, and the subject is confronted with a radiation absorbed dose and consequent theoretical risk.

The ALARA concept requires the implementation of appropriate quality assurance (QA) procedures both within the nuclear medicine clinic and the radiopharmacy. This is required to ensure that the imaging equipment is as sensitive as required (i.e., the count rate per unit activity imaged is maximized) and that the radiochemical purity of the administered substance is as high as achievable. ALARA extends not only to the individual absorbed dose burden, but also to that borne by a cohort of subjects. The number of subjects within a given cohort must be minimized through rigorous statistical justification, as discussed previously.

Finally, the imposition of an absorbed dose or effective dose limit is required. This limit is likely to be a function of the status of the volunteers (healthy or diseased, life expectancy, age), modality (due to the absorbed dose of the decelerating positrons, PET is a high-absorbed dose modality per unit activity than

Table 16.1 Levels of risks and benefits associated with the exposure to ionizing radiation in biomedical research (WHO 1977; ICRP 1991)

Level of risk	WHO risk category (order-of-magnitude total risk)	Corresponding effective dose (mSv)	Level of societal benefit
Trivial	Category I ($\sim 10^{-6}$ or less)	< 0.1	Minor
Minor to intermediate	Category IIa ($\sim 10^{-5}$)	0.1–1	Intermediate
	Category IIb ($\sim 10^{-4}$)	1–10	Moderate
Moderate	Category III ($\sim 10^{-3}$ or more)	≥ 10[a]	Substantial

[a]To be kept below deterministic effects' thresholds except for therapeutic studies

most single-photon imaging procedures) and culture (some societies are more tolerant of higher radiation exposures). It is highly unlikely that any one set limit can be specified to cover all conditions.[6]

16.3 Magnitudes of Risks Associated with Nuclear Medicine Research

In 1977, the WHO published a review and set of recommendations of the use of ionizing radiation and radionuclides on humans for biomedical research purposes (WHO 1977). Of particular interest were its descriptions of risk categories, which were subsequently refined by the International Commission on Radiological Protection (ICRP) in its Publication 62 (ICRP 1991). These are summarized in Table 16.1.

The WHO categorized risk in terms of the dose equivalent whereas ICRP Publication 62 used the more recent effective dose (Chap. 10). The latter defined the risk as the total detriment which is the sum of the probabilities of a radiation-induced fatal cancer, the weighted-probability of nonfatal cancers and the total probability over all succeeding generations of the occurrence of serious hereditary as a consequence of the irradiation.[7]

The risk categories of Table 16.1 correspond to stochastic risks of ionizing radiation exposure were derived by the WHO in consideration of the following:

- Category I: Within variations of normal background radiation
- Category II: Within "dose limits" for members of the general public
- Category III: Within "dose limits" for occupationally-exposed persons

The ICRP subsequently divided Category II into Categories IIa and IIb in which Category IIa was associated with a minor level of risk with the benefit of exposure expected to be related to increases in knowledge leading to health benefits. In Category IIb, the dose levels approach those of occupationally-exposed individuals with the benefit of exposure more directed at the cure or prevention of disease.

ICRP Publication 62 recommended how radiopharmaceuticals should be used in biomedical research:

- *Radiopharmaceutical with known internal radiation dosimetry profile*: The use of such a radiopharmaceutical must account for the radiation dosimetry which will be largely provided either by the agent's Summary of Product Characteristics / Package Insert or other published documentation.
- *New diagnostic radiopharmaceutical*: The research and development of a new diagnostic radiopharmaceutical must be supplemented by preclinical experiments prior to the introduction of the radiopharmaceutical to man. ICRP Publication 62 recommends that such preclinical evaluation be performed upon multiple species, of which one must be a non-human primate. A review of the contemporary literature would suggest that the last recommendation is not widely conformed to.
- *New therapeutic radiopharmaceutical*: Preclinical experiments with animal models should be performed and compared against human healthy

[6]Of interest is the proposal by Huda and Scrimger (1989) of a limit of 50 mSv to be applied to healthy volunteers participating in a nuclear medicine study. Such a value would not be tolerated in this era for healthy volunteers however it is possible to consider scenarios (e.g., studies involving multiple PET/CT examinations of patients with late-stage cancer) where such a limit might be considered acceptable in a clinical trial.

[7]ICRP Publication 62 characterized the total detriment due to ionizing radiation exposure to be 2–3 times greater in children than in adults and, for adults aged 50 years or more, to be 10–20% of that of younger adults. This reflects the difference between the latency period for a radiogenic cancer to express itself in the irradiated adult and the normal life expectancy of the same.

volunteer data for which the radiopharmaceutical has been administered at diagnostic levels.[8]

- *Other biomedical research:* These applications are those in which a nuclear medicine agent (or, colloquially, "tracer") is used as an associated diagnostic for the assessment of a biomedical procedure, including the administration of another agent. As, in the first case above, supplied data with the nuclear medicine agent should be used in order to estimate the absorbed radiation doses. If these are not available, or are insufficiently accurate, one should use, wherever possible, conservative estimates of the radiation dosimetry in order to lead to an overestimate of the effective dose per unit administered activity and force a drive to minimize the amount of administered activity.

16.4 Summary

Even though the intent of ICH is to minimize duplication of research in different geographies, biomedical research involving human beings is subject to many international and national regulations, recommendations, legislation, and expectations in order to assure the ethical conduct of such research. The use of radiopharmaceuticals requires the participating nuclear medicine physicist play an active and co-supervisory role in the design, execution, and analysis of the associated clinical trials. This participation must seek to minimize the ionizing radiation-associated risks and maximize the integrity of the data acquired through the use of the radiopharmaceutical. The nuclear medicine physicist engaged in these trials must be aware of the ethical requirements and the associate legislation and regulations.

References

Huda W, Scrimger JW (1989) Irradiation of volunteers in nuclear medicine. J Nucl Med 30:260–264

ICH (1996) ICH tripartite guideline for good clinical practice E6 (R1). International Conference on Harmonisation of Technical Requirements for Registration of Pharamecuticals for Human Use, Geneva

ICRP (1991) Radiological protection in biomedical research – ICRP Publication 62. International Commission on Radiological Protection. Pergamon, Oxford

WHO (1977) Use of ionizing radiation and radionuclides on human beings for medical research, training, and nonmedical purposes – WHO Technical Report Series 611. World Health Organization, Geneva

WMA (2008) World Medical Association Declaration of Helsinki – ethical principles for medical research involving human subjects. http://www.wma.net/e/policy/pdf/17c.pdf

[8]This recommendation by the ICRP clearly cannot be followed if the therapeutic radionuclide emits solely charged particles without any photon emissions that can aid external assessment of the biodistribution.

The Future of Nuclear Medicine Radiation Dosimetry

17

Abstract This chapter summarizes the directions along which the development of nuclear medicine radiation dosimetry progresses.

Contents

17.1 Introduction

The premise of this book is that, despite all of the scientific, medical, and technological advances in the past four decades, nuclear medicine radiation dosimetry is not yet a mature discipline. Many advances in the calculation and measurement of organ absorbed doses and in the prediction of biological effects remain to be achieved. This chapter summarizes areas considered likely to display advances in nuclear medicine dosimetry in the near future.

17.2 Single-Photon Radionuclide Imaging Technology

In the development of radiopharmaceuticals using single-photon emitting radionuclides, the planar conjugate-view method has been the usual method of estimating the associated internal radiation dosimetry. The disadvantages of the lack of knowledge of source-organ thickness and only an approximate ability to correct for organ overlap in the conjugate views have been noted in Chap. 14. While SPECT offers a solution to these dilemmas, it is limited by the detection sensitivities of contemporary gamma cameras which use NaI(Tl) scintillators to the acquisition of a single axial field-of-view in a clinically-realistic acquisition time. This disables the prospect of obtaining a three-dimensional whole-body biodistribution and, hence, would be limited to measuring the biodistribution within a limited anatomical volume.

Room temperature solid-state radiation detectors offer the promises of high-sensitivity and superior energy resolution and mechanical advantages such as reduced weight. The higher sensitivity evolves from the facts that the mean energy required to generate an ion pair is about an order of magnitude less in a solid-state medium than in a gaseous medium and that the number of channels that can reduce the detection efficiency is less than in scintillator. In many solid-state detectors, the thermal noise current can limit the sensitivity of the detector and necessitates the use of cryogenic cooling (e.g., liquid nitrogen) to reduce the random thermal noise. Some materials have a sufficiently low noise current at room temperature to make them practical for the nuclear medicine clinic. Materials such as cadmium zinc telluride (CdZnTe, or more commonly known as "CZT") and cadmium telluride (CdTe) have been used in nuclear medicine

B.J. McParland, *Nuclear Medicine Radiation Dosimetry*,
DOI 10.1007/978-1-84882-126-2_17,

applications. The size of the detectors are still small with, as a result, CZT imaging platforms currently limited to specific applications where such small fields-of-view are desired, such as in scintimammography, nuclear cardiology, and small-animal imaging. Rapid acquisition times have been reported with reductions in acquisition times from about 30 min for NaI(Tl) scintillator detectors to about 3–5 min for CZT detectors. If axial fields-of-view can be extended to 30–40 cm, it should be possible to imagine whole-torso (head to mid-thigh) tomographic imaging suitable for dosimetry evaluations requiring about 15–20 min for a single acquisition.

17.3 Reference Anthropomorphic Phantoms in Diagnostic Nuclear Medicine Radiation Dosimetry

Chapter 12 summarized past, contemporary, and future anthropomorphic phantoms and their roles in evaluating the internal radiation dosimetry of diagnostic nuclear medicine procedures. Patient-specific calculations are considered in the following subsection. The transition from reference phantoms based upon stylized phantoms (e.g., the Cristy-Eckerman series) to voxellated (e.g., the GSF series) phantoms is occurring in radiological protection, is noted below in the discussion pertaining to ICRP Publication 103 (2007). Such transitions can, no doubt, lead to more accurate estimates of the internal radiation dosimetry associated with diagnostic radiopharmaceuticals. However, at the same time, it must always be borne in mind that the dosimetry profile thus calculated is only for a reference phantom and is not specific to the individual patient. One could sensibly argue if such an improvement in accuracy will have an appropriate impact upon the comparative nature of the effective dose between radiopharmaceuticals in diagnostic nuclear medicine.

New phantom approaches, such as the use of non-uniform rational B-spline (NURBS), can allow a more diverse series of phantoms to represent the broad population of nuclear medicine patients in terms of sex, height, weight, and ethnicity. As a result, although the dosimetry calculation will be for a reference phantom alone, the calculation can become more representative of the patient in question.

17.4 Patient-Specific Radionuclide Therapy Planning

Siegel et al. (2002) advocated eloquently the need for radionuclide therapy planning to become increasingly patient-specific in order for it to benefit the therapeutic patient. This could require modifications of existing internal radiation dosimetry models, such as the MIRD method. However, with increasing computing power and the use of parallel processing, patient-specific Monte Carlo simulations of radiation transport and energy deposition using patient-derived image data (i.e., the radiological anatomy through whole-body CT and the biodistribution through, for example, a diagnostic-level administration of activity for imaging) should lead to increased accuracy in the evaluation of organ and tumor absorbed doses. Despite this expectation, Divoli et al. (2009) have compared patient absorbed dose values evaluated using the MCNPX code and patient CT image data sets through the OEDIPE system with those calculated using the MIRD schema (through OLINDA/EXM) but with the S-factors scaled by patient organ mass for ^{131}I. They concluded that the use of these mass-scaled S-factors could yield absorbed doses accurate to within 26%. As the uncertainty in the absorbed dose delivered in external beam radiotherapy is intended to be less than 5%, this would suggest that computationally-intensive patient-specific radionuclide therapy planning is a rich and open field for research.

17.5 New Radionuclide Delivery Vectors

In conventional nuclear medicine, the radionuclide administered to a patient through either its natural form for which it targets the organ or tissue of interest (e.g., sodium iodide for thyroid tissue), through chelation to a target-specific vector or as a replacement of the naturally-occurring nonradioactive moiety in a chemical agent. Some new delivery systems currently being researched for medical imaging are based upon nanostructures (Matsuura and Rowlands 2008). Such structures have physical dimensions of the order of a nanometer and are being investigated primarily as efficient means of transporting iodinated or gadolinium-bearing contrast for X-ray and MR applications.

17.6 New Means of In Vivo Radiation Dosimetry Measurement

Earlier, this chapter looked at conventional means, based upon small MOSFETs and small TLDs, of in vivo measurements of internal radiation dosimetry. The spatial resolution of such dosimeters are still relatively gross compared to the variation of internal radiation absorbed dose spatial variations. However, new possible means of obtaining such measurements through quantum dot (QD) dosimeters are being investigated (Stodilka et al. 2009). The QD is a semiconductor with physical dimensions of a few nm and can contain between 10^2 and 10^5 atoms only and can be made from materials such as cadmium selenide (CdSe), zinc sulfide (ZnS), indium arsenide (InAr), and indium phosphide (InP). A QD is defined by its spatial dimension being smaller than the Bohr exciton radius (DEFINE). Dai et al. (2004) have reviewed several metal chalcogenides (ZnS and CdSe/ZnS) as scintillators. The ability to fabricate small dosimeters from such devices is an intriguing means of measuring heterogeneous absorbed dose distributions.

17.7 New Estimates of Radiation Risk

In Chap. 10, the historical development of the effective dose equivalent/effective dose as a metric of stochastic risk was discussed through their presentation by ICRP Publications 26, 60, and 103 which were released in the 1970s, 1990s, and 2000s, respectively. The effective dose equivalent of ICRP 26 (1977) has at last been universally superseded by the effective dose with the tissue weighting factors of ICRP Publication 60 (ICRP 1991). Since then, ICRP Publication 103 (ICRP 2007) has presented updated tissue weighting factors based upon revised stochastic risk estimates resulting from continuing follow up of populations and individuals exposed to radiation. It could be expected that regulatory and legislative accounting of these updated tissue weighting factors will follow, much as they did in the 1990s following the release of ICRP Publication 60. As the expectation is that estimates of radiation-induced stochastic risk will be revised continuously as more epidemiological data are accrued, one should anticipate these changes and their impacts upon the estimate of this stochastic risk. But an important facet of ICRP 103 with highly practical implications to the nuclear medicine community is that the Commission has, for the first time, specified anthropomorphic phantoms to be used in the evaluation of the effective dose. As noted in that publication, the Snyder and Cristy-Eckerman phantoms have been used in the past in, for example, MIRD calculations, MIRDOSE, and OLINDA/EXM. However, ICRP Publication 103 uses the reference computational phantoms of the adult Reference Man and Reference Woman, with organ dimensions and weights as described in ICRP Publication 89 (2002). These phantoms are the voxellated GOLEM and LAURA from the GSF series (Chap. 12). It will be interesting to observe how or if, following the MIRD committee's recent summary of achieving congruence between MIRD and ICRP concepts (Bolch et al. 2009), the ICRP Publication 103 concepts will be adopted in nuclear medicine dosimetry calculations. This curiosity arises from the FDA's expectation that radiopharmaceutical dosimetry calculations should be derived from the OLINDA/EXM code for which a 510(k) exemption has been issued. As OLINDA/EXM currently uses the Cristy-Eckerman and Stabin et al. series of phantoms, a reconciliation of differences in the absorbed dose profiles evaluated using the GSF series of phantoms promoted by ICRP Publication 103 and the "conventional" phantom series of OLINDA/EXM would be of interest.

References

Bolch WE, Eckerman KF, Sgouros G, Thomas SR (2009) MIRD pamphlet number 21: a generalized schema for radiopharmaceutical dosimetry – standardization of nomenclature. J Nucl Med 50:477–484

Dai S, Saengkerdsub S, Im H-J, Stephan AC, Mahurin SM (2004) Nanocrystal-based scintillators for radiation protection. In: Trombka JL, Spears DP, Solomon PJ (eds) Unattended radiation sensor systems for remote applications. AIP Conf Proc 632:220–224

Divoli A, Chiavassa S, Ferrer L, Barbet J, Flux G, Bardiès M (2009) Effect of patient morphology on dosimetric calculations for internal irradiation as assessed by Monte Carlo versus conventional methodologies. J Nucl Med 50:316–323

ICRP (1977) Recommendations of the International Commission on Radiological Protection. ICRP Publication 26. Pergamon, Oxford

ICRP (1991) Recommendations of the International Commission on Radiological Protection. ICRP Publication 60. Pergamon, Oxford

ICRP (2002) Basic anatomical and physiological data for use in radiological protection: reference values – ICRP Publication 89. International Commission on Radiological Protection. Pergamon, Oxford

ICRP (2007) The 2007 recommendations of the International Commission on Radiological Protection. ICRP Publication 103. Pergamon, Oxford

Matsuura N, Rowlands JA (2008) Towards new functional nanostructures for medical imaging. Med Phys 35:4474–4487

Siegel JA, Stabin MG, Brill AB (2002) The importance of patient-specific radiation dose calculations for the administration of radionuclides in therapy. Cell Mol Biol 48: 451–459

Stodilka RZ, Carson JJL, Yu K, Zaman MB, Li C, Wilkinson D (2009) Optical degradation of CdSe/ZnS quantum dots upon gamma-ray irradiation. J Phys Chem C 113:2580–2585

Appendix

Appendix A Nuclear Form Factor for a Gaussian Charge Distribution

The Gaussian electrical charge density distribution is given by (3.113) where the normalization constant ρ_0 is determined from the normalization requirement,

$$4\pi\rho_0 \int_0^\infty dr\, r^2\, e^{-(r/R_N)^2} = 1 \tag{A.1}$$

where R_N is the nuclear radius. The integral is of the form, $\int_0^\infty dx\, x^2\, e^{-ax^2}$, which is solved by using the simpler integral $\int_0^\infty dx\, e^{-ax^2}$. This latter integral is solved by first squaring it,

$$\begin{aligned}\left(\int_0^\infty dx\, e^{-ax^2}\right)^2 &= \int_0^\infty dx\, e^{-ax^2} \int_0^\infty dy\, e^{-ay^2} \\ &= \int_0^\infty dx \int_0^\infty dy\, e^{-a\left(x^2+y^2\right)}\end{aligned} \tag{A.2}$$

Converting the double integral in Cartesian coordinates to one in polar coordinates,

$$\int_0^\infty dx \int_0^\infty dy\, e^{-a\left(x^2+y^2\right)} = \frac{1}{4}\left(2\pi \int_0^\infty dr\, r\, e^{-ar^2}\right) \tag{A.3}$$

and noting that $d(r^2) = 2r\, dr$,

$$\begin{aligned}2\pi \int_0^\infty dr\, r\, e^{-ar^2} &= \frac{\pi}{a} \int_0^\infty dt\, e^{-t} \\ &= \frac{\pi}{a}\end{aligned} \tag{A.4}$$

leading to $\int_0^\infty dx\, e^{-ax^2} = \frac{1}{2}\sqrt{\frac{\pi}{a}}$. Then,

$$\begin{aligned}\int_0^\infty dx\, x^2\, e^{-ax^2} &= \frac{d}{da} \int_0^\infty dx\, e^{-ax^2} \\ &= \frac{1}{4}\frac{\sqrt{\pi}}{a^{3/2}}\end{aligned} \tag{A.5}$$

From this, we obtain $\int_0^\infty dr\, r^2\, e^{-(r/R_N)^2} = \frac{\sqrt{\pi}}{4} R_N^3$ leading to, from (A.1),

$$\rho_0 = \frac{1}{\pi^{3/2} R_N^3} \tag{A.6}$$

and the normalized Gaussian electric charge spatial distribution,

$$\rho_e(r) = \frac{e^{-(r/R_N)^2}}{\pi^{3/2} R_N^3}. \tag{A.7}$$

The mean-square radius is,

$$\begin{aligned}\overline{r^2} &= \int d^3\mathbf{r}\, r^2\, \rho_e(r) \\ &= \frac{4}{\sqrt{\pi} R_N^3} \int_0^\infty dr\, r^4\, e^{-(r/R_N)^2}\end{aligned} \tag{A.8}$$

This integral is solved through the successive differentiations that led up to (A.5) to yield,

$$\int_0^\infty dr\, r^4\, e^{-(r/R_N)^2} = \frac{3\sqrt{\pi}}{8} R_N^5 \tag{A.9}$$

and,

$$\overline{r^2} = \frac{3}{2} R_N^2 \tag{A.10}$$

The nuclear form factor for the Gaussian electrical charge spatial distribution derived above is,

$$F(q) = \left(\frac{\hbar c}{q}\right) \frac{4}{\sqrt{\pi} R_N^3} \times \int_0^\infty dr\, r \sin\left(\frac{qr}{\hbar c}\right) e^{-(r/R_N)^2}. \tag{A.11}$$

The integral is rewritten as a complex integral of the form,

$$\int_0^\infty dx\, x \sin bx\, e^{-ax^2} = \frac{1}{2} \mathrm{Im} \int_{-\infty}^\infty dx\, x\, e^{-\left(ax^2 - ibx\right)} \tag{A.12}$$

and the exponent in the integrand is written as a square,

$$ax^2 - ibx = \left(\sqrt{a}x - i\frac{b}{2\sqrt{a}}\right)^2 + \frac{b^2}{4a}. \tag{A.13}$$

The integral of interest is,

$$\int_{-\infty}^\infty dx\, x\, e^{-\left(ax^2 - ibx\right)} = e^{-b^2/4a} \int_{-\infty}^\infty dx\, x\, e^{-\left(\sqrt{a}x - ib/2\sqrt{a}\right)^2}. \tag{A.14}$$

Using the substitution of the variable, $w = \sqrt{a}x - ib/2\sqrt{a}$, the integral can be split into two simpler integrals,

$$\int_{-\infty}^\infty dx\, x\, e^{-\left(ax^2 - ibx\right)} = \frac{e^{-b^2/4a}}{a} \left(\int_{-\infty - ib/2\sqrt{a}}^{\infty - ib/2\sqrt{a}} dw\, w\, e^{-w^2} + i\frac{b}{2\sqrt{a}} \int_{-\infty - ib/2\sqrt{a}}^{\infty - ib/2\sqrt{a}} dw\, e^{-w^2} \right) \tag{A.15}$$

Both integrals are solved for using the Cauchy integral theorem, $\oint_\Gamma dz\, f(z) = 0$, providing that $f(z)$ is analytic within a region containing the closed contour, Γ. The simpler of the two integrals is solved for first,

$$\int_{-\infty - ib/2\sqrt{a}}^{\infty - ib/2\sqrt{a}} dw\, e^{-w^2} = \lim_{R \to \infty} \int_{-R - ib/2\sqrt{a}}^{R - ib/2\sqrt{a}} dw\, e^{-w^2}. \tag{A.16}$$

The contour for both integrals is $\Gamma = \Gamma_1 \cup \Gamma_2 \cup \Gamma_3 \cup \Gamma_4$, as shown in Fig. A.1 where $w = u + iv$. Equation (A.16) is,

$$\int_{-R - ib/2\sqrt{a}}^{R - ib/2\sqrt{a}} dw\, e^{-w^2} = \int_{\Gamma_1} dw\, e^{-w^2} = -\int_{\Gamma_2} dw\, e^{-w^2} - \int_{\Gamma_3} dw\, e^{-w^2} - \int_{\Gamma_4} dw\, e^{-w^2} \tag{A.17}$$

The integral over the contour element Γ_2 is,

$$\begin{aligned} \int_{\Gamma_2} dw\, e^{-w^2} &= \int_{R - ib/2\sqrt{a}}^{R} dw\, e^{-w^2} \\ &= i \int_{-b/2\sqrt{a}}^{0} dv\, e^{-(R+iv)^2} \\ &= 0 \quad \text{for} \quad R \to \infty \end{aligned} \tag{A.18}$$

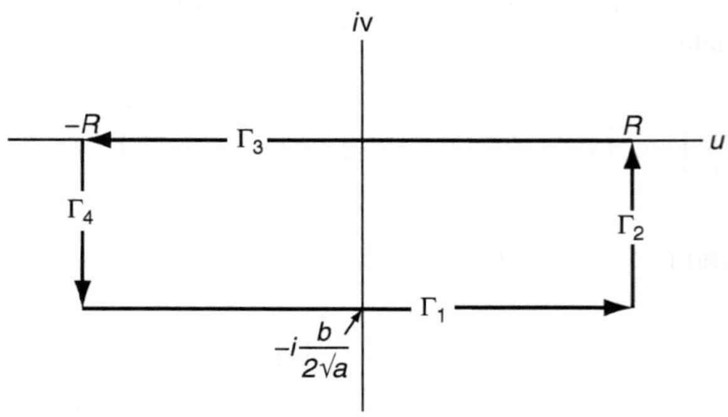

Fig. A.1 Contour for evaluating the integrals of (A.15)

That over Γ_3 is,

$$\int_{\Gamma_3} dw\, e^{-w^2} = \int_{R}^{-R} du\, e^{-u^2} = -\sqrt{\pi} \tag{A.19}$$

and that over Γ_4 is,

$$\int_{\Gamma_4} dw\, e^{-w^2} = \int_{-R}^{-R-ib/2\sqrt{a}} dw\, e^{-w^2} = 0 \tag{A.20}$$

via a similar manipulation to lead to (A.18). Hence, the integral is,

$$\int_{\infty-ib/2\sqrt{a}}^{\infty-ib/2\sqrt{a}} dw\, e^{-w^2} = \sqrt{\pi}. \tag{A.21}$$

The other integral is solved in a similar fashion,

$$\begin{aligned}\int_{-R-ib/2\sqrt{a}}^{R-ib/2\sqrt{a}} dw\, w\, e^{-w^2} &= \int_{\Gamma_1} dw\, w\, e^{-w^2} \\ &= -\int_{\Gamma_2} dw\, w\, e^{-w^2} - \int_{\Gamma_3} dw\, w\, e^{-w^2} \\ &\quad - \int_{\Gamma_4} dw\, w\, e^{-w^2}.\end{aligned} \tag{A.22}$$

As before, the integrals over Γ_2 and Γ_4 go to zero as $R \to \infty$. The integral over Γ_3 is, using the substitution $y = w^2$ which leads to $dy = 2w\, dw$,

$$\begin{aligned}\int_{\Gamma_3} dw\, w\, e^{-w^2} &= \int_{R}^{-R} dw\, w\, e^{-w^2} \\ &= \frac{1}{2}\int_{R}^{-R} dy\, e^{-y} \\ &= 0\end{aligned} \tag{A.23}$$

Thus,

$$\int_{-\infty-ib/2\sqrt{a}}^{\infty-ib/2\sqrt{a}} dw\, w\, e^{-w^2} = 0 \tag{A.24}$$

Combining all of the above results in,

$$\int_{-\infty}^{\infty} dx\, x\, e^{-\left(ax^2-ibx\right)} = i\frac{\sqrt{\pi}}{2}\frac{b\, e^{-\left(b^2/4a\right)}}{a^{3/2}}. \tag{A.25}$$

From (A.12),

$$\int_{0}^{\infty} dx\, \sin(bx)\, e^{-(ax)^2} = \frac{\sqrt{\pi}}{4}\frac{b\, e^{-\left(b^2/4a\right)}}{a^{3/2}} \tag{A.26}$$

which leads to,

$$\begin{aligned}&\int_{0}^{\infty} dr\, r\, \sin\left(\frac{qr}{\hbar c}\right) e^{-\left(r/R_N\right)^2} \\ &= \frac{\sqrt{\pi}}{4}\frac{qR_N^3}{\hbar c} e^{-\left(qR_N/2\hbar c\right)^2}\end{aligned} \tag{A.27}$$

and an expression for the form factor associated with a Gaussian electric charge distribution,

$$F(q) = e^{-\left(qR_N/2\hbar c\right)^2} \tag{A.28}$$

Appendix B Nuclear Form Factor for the Woods–Saxon Charge Distribution

The normalization constant of the Woods–Saxon electrical charge density distribution is determined from the requirement,

$$\rho_0 = \frac{1}{4\pi \int_0^{\infty} dr \left(\frac{r^2}{1+e^{\left(\frac{r-R_{50}}{a}\right)}}\right)}. \tag{B.1}$$

The integral is not particularly easy to solve and few solutions are provided in the literature. The solution is,

$$\int_0^\infty dr \frac{r^2}{1+e^{\left(\frac{r-R_{50}}{a}\right)}} = 2a^3 \mathrm{Li}_3\left(-e^{\left(\frac{r-R_{50}}{a}\right)}\right) - 2ra^2 \mathrm{Li}_2\left(-e^{\left(\frac{r-R_{50}}{a}\right)}\right) + \frac{r^2}{3}\left(r - 3a\ln\left(1+e^{\frac{r-R_{50}}{a}}\right)\right)\Bigg|_{r=0}^{r=\infty} \tag{B.2}$$

where $\mathrm{Li}_n(z)$ is the polylogarithm (also known as Jonquière's function) of order n defined by,

$$\mathrm{Li}_n(z) = \sum_{k=1}^{\infty} \frac{z^k}{k^n} \quad |z| < 1 \tag{B.3}$$

where $z = x + iy$. By analytical continuation for $|z| > 1$, the dilogarithm and trilogarithm are,

$$\mathrm{Li}_2(z) = -\mathrm{Li}_2\left(1/z\right) - \frac{(\ln(-z))^2}{2} - \frac{\pi^2}{6} \quad |z| > 1 \tag{B.4}$$

and

$$\mathrm{Li}_3(z) = \mathrm{Li}_3\left(1/z\right) - \frac{(\ln(-z))^3}{6} - \frac{\pi^2}{6}\ln(-z) \quad |z| > 1. \tag{B.5}$$

From (B.4), for $r \to \infty$,

$$\mathrm{Li}_2\left(-e^{\left(\frac{r-R_{50}}{a}\right)}\right) = -\mathrm{Li}_2\left(-e^{-\left(\frac{r-R_{50}}{a}\right)}\right) - \frac{\left(\frac{r-R_{50}}{a}\right)^2}{2} - \frac{\pi^2}{6} \to \frac{r^2}{2a^2} - \frac{\pi^2}{6} \quad \text{as } r \to \infty \tag{B.6}$$

and, from (B.5) for $r \to \infty$,

$$\mathrm{Li}_3\left(-e^{\left(\frac{r-R_{50}}{a}\right)}\right) = \mathrm{Li}_3\left(-e^{-\left(\frac{r-R_{50}}{a}\right)}\right) - \frac{\left(\frac{r-R_{50}}{a}\right)^3}{6} - \frac{\pi^2}{6}\left(\frac{r-R_{50}}{a}\right) \to -\frac{r^3}{6a^3} - \frac{\pi^2 r}{6a} \quad \text{as } r \to \infty \tag{B.7}$$

Thus, for $r \to \infty$,

$$2a^3 \mathrm{Li}_3\left(-e^{\left(\frac{r-R_{50}}{a}\right)}\right) - 2ra^2 \mathrm{Li}_2\left(-e^{\left(\frac{r-R_{50}}{a}\right)}\right) + \frac{r^2}{3}\left(r - 3a\ln\left(1+e^{-\left(\frac{r-R_{50}}{a}\right)}\right)\right) \to 2a^3\left(-\frac{r^3}{6a^3} - \frac{\pi^2 r}{6a}\right) - 2ra^2\left(\frac{r^2}{2a^2} - \frac{\pi^2}{6}\right) + \frac{r^2}{3}\left(r + 3a\left(\frac{r}{a}\right)\right) = 0 \tag{B.8}$$

and, for $r \to 0$

$$2a^3 \mathrm{Li}_3\left(-e^{\left(\frac{r-R_{50}}{a}\right)}\right) - 2ra^2 \mathrm{Li}_2\left(-e^{\left(\frac{r-R_{50}}{a}\right)}\right) + r^2\left(r - 3a\ln\left(1+e^{\left(\frac{r-R_{50}}{a}\right)}\right)\right) \to 2a^3 \mathrm{Li}_3\left(-e^{-\left(\frac{R_{50}}{a}\right)}\right) \tag{B.9}$$

These two results lead to,

$$\int_0^\infty dr \frac{r^2}{1+e^{\left(\frac{r-R_{50}}{a}\right)}} = -2a^3 \mathrm{Li}_3\left(-e^{-\frac{R_{50}}{a}}\right) \tag{B.10}$$

and the normalization,

$$\rho_0 = -\frac{1}{8\pi a^3 \mathrm{Li}_3\left(-e^{-\frac{R_{50}}{a}}\right)}. \tag{B.11}$$

A feature of the trilogarithm is, as $a \to 0$ (i.e., the diffuse edge of the nucleus becomes more defined),

$$a^3 \mathrm{Li}_3\left(-e^{-\frac{R_{50}}{a}}\right) \to -\frac{R_{50}^3}{6} \tag{B.12}$$

leading to,

$$\rho_0 \rightarrow \frac{3}{4\pi R_{50}^3} \quad \text{(B.13)}$$

which is equivalent to the normalization for the hard-edge sphere model of the nucleus.

In summary, the normalized Woods–Saxon charge distribution is,

$$\rho_e(r) = -\frac{1}{8\pi a^3\left(1+e^{\left(\frac{r-R_{50}}{a}\right)}\right)\mathrm{Li}_3\left(-e^{-\left(\frac{R_{50}}{a}\right)}\right)}. \quad \text{(B.14)}$$

The Woods–Saxon mean-square radius is determined from the general formula for the moments of the Woods–Saxon distribution,

$$\frac{1}{n!}\int_0^\infty dr \frac{r^n}{1+e^{\left(\frac{r-R_{50}}{a}\right)}} = -a^{n+1}\mathrm{Li}_{n+1}e^{-\left(\frac{R_{50}}{a}\right)}. \quad \text{(B.15)}$$

The mean-square radius is,

$$\begin{aligned}\overline{r^2} &= 4\pi\int_0^\infty dr\, r^4\, \rho_e(r)\\ &= -\frac{1}{2a^3\mathrm{Li}_3\left(-e^{-\frac{R_{50}}{a}}\right)}\int_0^\infty dr \frac{r^4}{1+e^{\left(\frac{R_{50}}{a}\right)}} \quad \text{(B.16)}\\ &= 12a^2\frac{\mathrm{Li}_5\left(-e^{-\left(\frac{R_{50}}{a}\right)}\right)}{\mathrm{Li}_3\left(-e^{-\left(\frac{R_{50}}{a}\right)}\right)}\end{aligned}$$

The nuclear form factor for the Woods–Saxon electrical charge spatial distribution is,

$$\begin{aligned}F(q) = &-\left(\frac{1}{2a^3\mathrm{Li}_3\left(-e^{-\left(\frac{R_{50}}{a}\right)}\right)}\right)\left(\frac{\hbar c}{q}\right)\\ &\times\int_0^\infty dr \frac{r\,\sin\left(\frac{qr}{\hbar c}\right)}{\left(1+e^{\left(\frac{r-R_{50}}{a}\right)}\right)} \quad \text{(B.17)}\end{aligned}$$

As no closed analytical solution exists, the integral must be solved numerically. On the other hand, another approach to solving for the Woods–Saxon form factor is to write the Woods–Saxon charge distribution as the finite sum of weighted Gaussians,

$$\begin{aligned}\rho_e(r) &= \frac{\rho_0}{1+e^{\left(\frac{r-R_{50}}{a}\right)}}\\ &\cong \sum_{i=1}^{Z} w_i e^{-k_i(r-R_i)^2}\end{aligned} \quad \text{(B.18)}$$

Appendix C Pair Production Energy Threshold

In this Appendix, it is demonstrated that electron–positron pair production in the nuclear Coulomb field, while a nonnegligible process in the dosimetry of external photon beam radiotherapy, can be largely neglected in diagnostic and therapeutic nuclear medicine where photon energies of interest are much less than 1 MeV. This is due to the fact that the photon energy required to produce the lepton pair exceeds $2m_e$ of 1.022 MeV. The initial state consists of the photon and the nucleus and the final state consists of the recoil nucleus and the electron–positron pair considered to be grouped (Fig. C.1). In the centre-of-mass frame (labeled by the * superscript), the threshold covariant four-vector momenta in the initial and final states are,

$$\boldsymbol{p}^*_{\mu,\gamma} = (k^*, -\mathbf{k}^*) \quad \text{(C.1)}$$

$$\boldsymbol{p}^*_{\mu,N} = (E^*_N, \mathbf{k}^*) \quad \text{(C.2)}$$

$$\boldsymbol{p}^{*'}_{\mu,e} = (2m_e, 0) \quad \text{(C.3)}$$

$$\boldsymbol{p}^{*'}_{\mu,N} = \left(E^{*'}_N, 0\right) \quad \text{(C.4)}$$

and the corresponding covariant four-vector momenta in the laboratory frame are,

$$\boldsymbol{p}_{\mu,\gamma} = (k - \mathbf{k}) \quad \text{(C.5)}$$

$$\boldsymbol{p}_{\mu,N} = (m_N, \mathbf{0}) \quad \text{(C.6)}$$

$$\boldsymbol{p}'_{\mu,e} = \left(E'_e, \mathbf{p}_e\right) \quad \text{(C.7)}$$

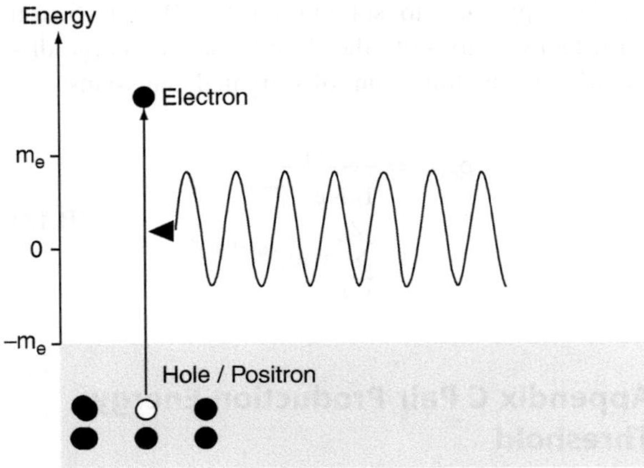

Fig. C.1 Electron–positron pair production described by Dirac's "hole theory"

$$p'_{\mu,N} = \left(E'_N, \mathbf{p}'_N\right). \tag{C.8}$$

The Mandelstam s variable is defined as,

$$s \equiv (p_\gamma + p_N)^2 = \left(p^*_\gamma + p^*_N\right)^2 \tag{C.9}$$

where the equality follows as this is a relativistic invariant. We have,

$$\begin{aligned}(p_\gamma + p_N)^2 &= p_{\mu,\gamma}p^\mu_\gamma + p_{\mu,N}p^\mu_N + p_{\mu,N}p^\mu_\gamma + p_{\mu,\gamma}p^\mu_N \\ &= m_N^2 + 2km_N.\end{aligned}$$

As in the centre-of-mass system, $p^*_\gamma + p^*_N = p^{*'}_e + p^{*'}_N$, then, and recalling that the kinematic quantities with subscript e refer to the electron–positron pair which has a rest mass of $2m_e$,

$$\begin{aligned}\left(p^*_\gamma + p^*_N\right)^2 &= \left(p^{*'}_e + p^{*'}_N\right)^2 \\ &= p^{*'}_{\mu,e}p^{\mu*'}_e + p^{*'}_{\mu,N}p^{\mu'}_N + p^{*'}_{\mu,N}p^{\mu'}_e + p^{*'}_{\mu,e}p^{\mu*'}_N \\ &= m_N^2 + 4m_e^2 + 4m_N m_e.\end{aligned} \tag{C.10}$$

Equating (C.9) and (C.10) and solving for the threshold photon energy for electron–positron pair production,

$$k_{Thr} = 2m_e + 2\left(\frac{m_e^2}{m_N}\right). \tag{C.11}$$

This result shows that, in order to simultaneously conserve momentum and energy, the recoil nucleus takes away momentum which fixes the photon threshold energy to be greater than twice the electron rest mass by an amount $2\left(m_e^2/m_N\right)$. This excess energy amount is equal to a few tens of eV and is negligible. Hence, in practice, it is possible to specify that the threshold energy for pair production in the nuclear Coulomb field is 1.022 MeV.

Index

B

MIX
Papier aus verantwortungsvollen Quellen
Paper from responsible sources
FSC® C105338

If you have any concerns about our products,
you can contact us on
ProductSafety@springernature.com

In case Publisher is established outside the EU,
the EU authorized representative is:
Springer Nature Customer Service Center GmbH
Europaplatz 3, 69115 Heidelberg, Germany

Printed by Libri Plureos GmbH
in Hamburg, Germany